T0141847

Advances in Intelligent Systems and Computing

Volume 672

Series editor

Janusz Kacprzyk, Polish Academy of Sciences, Warsaw, Poland
e-mail: kacprzyk@ibspan.waw.pl

The series "Advances in Intelligent Systems and Computing" contains publications on theory, applications, and design methods of Intelligent Systems and Intelligent Computing. Virtually all disciplines such as engineering, natural sciences, computer and information science, ICT, economics, business, e-commerce, environment, healthcare, life science are covered. The list of topics spans all the areas of modern intelligent systems and computing.

The publications within "Advances in Intelligent Systems and Computing" are primarily textbooks and proceedings of important conferences, symposia and congresses. They cover significant recent developments in the field, both of a foundational and applicable character. An important characteristic feature of the series is the short publication time and world-wide distribution. This permits a rapid and broad dissemination of research results.

More information about this series at http://www.springer.com/series/11156

Vikrant Bhateja · Bao Le Nguyen
Nhu Gia Nguyen · Suresh Chandra Satapathy
Dac-Nhuong Le
Editors

Information Systems Design and Intelligent Applications

Proceedings of Fourth International Conference INDIA 2017

Editors
Vikrant Bhateja
Department of Electronics and Communication Engineering
Shri Ramswaroop Memorial Group of Professional Colleges
Lucknow, Uttar Pradesh
India

Suresh Chandra Satapathy
Department of CSE
PVP Siddhartha Institute of Technology
Vijayawada, Andhra Pradesh
India

Bao Le Nguyen
Technology and Engineering Division
Duy Tan University
Da Nang
Vietnam

Dac-Nhuong Le
Faculty of Information Technology
Hai Phong University
Hai Phong
Vietnam

Nhu Gia Nguyen
Duy Tan University
Da Nang
Vietnam

ISSN 2194-5357 ISSN 2194-5365 (electronic)
Advances in Intelligent Systems and Computing
ISBN 978-981-10-7511-7 ISBN 978-981-10-7512-4 (eBook)
https://doi.org/10.1007/978-981-10-7512-4

Library of Congress Control Number: 2017960784

Printed on acid-free paper

This Springer imprint is published by Springer Nature
The registered company is Springer Nature Singapore Pte Ltd.
The registered company address is: 152 Beach Road, #21-01/04 Gateway East, Singapore 189721, Singapore

Preface

This book is a collection of high-quality peer-reviewed research papers presented at the Fourth International Conference on Information Systems Design and Intelligent Applications (INDIA 2017) held at Duy Tan University, Da Nang, Vietnam, during June 15–17, 2017.

INDIA 2012 was organized by Computer Society of India (CSI), Vizag Chapter, and had been well supported by Vizag Steel, RINL, and Government of India. Its sequel INDIA 2015 has been organized by Kalyani University, sponsored by Microsoft and technically cosponsored by IEEE Kolkata Section and IEEE Computational Intelligence Society, Kolkata Chapter. INDIA 2016 was again held in Visakhapatnam, India, during January 2016. Various past editions of this conference were enriched by numerous special sessions floated in specific domains of image processing, pattern recognition, machine learning, multi-criteria decision analysis, and wireless sensor networks. All papers of INDIA 2012, INDIA 2015, and INDIA 2016 were published by Springer AISC Series. Presently, INDIA 2017 is the fourth edition of this conference series held at Duy Tan University, Da Nang, Vietnam.

INDIA 2017 had received a good number of submissions focusing on intelligent applications and various system design issues. The published papers cover a wide range of topics of computer science and information technology discipline ranging from image processing, database application, data mining, grid and cloud computing, bioinformatics, and many others. These papers are shortlisted after a rigorous peer-review process with the help of our program committee members and external reviewers (from the country as well as abroad). The review process has been very crucial with minimum 2 reviews each, and in many a cases 3–5 reviews along with due checks on similarity and overlaps as well. In addition to the main track, the conference featured eight special sessions in various cutting-edge technologies of specialized focus including computer vision, advanced fuzzy theory, autonomic sustainable computing, nature-inspired computing, Internet of Things, and natural language processing. These theme-based sessions were organized and chaired by eminent professors from various countries like India, Sri Lanka, Vietnam, Iran, Bangkok, Norway, USA, Taiwan, Malaysia, Indonesia, and South Korea.

The conference featured many distinguished keynote addresses by eminent speakers like Dr. Suresh Chandra Satapathy, PVPSIT, Vijayawada, India, who delivered a talk on Social Group Optimization (SGO): A New Population Evolutionary Optimization Technique. Dr. Le Hoang Son, Vietnam National University, Hanoi, Vietnam, discussed on Advanced Fuzzy-based Optimization Techniques. Last but not the least, the session on "Publishing Ethics and Author Services?" by Mr. Aninda Bose, Senior Publishing Editor, Springer India, embraced the huge toll of audience of students, budding researchers as well as delegates. We are thankful to Springer India for sponsoring the eight Best Paper Awards being awarded to delegates from each of the eight oral presentation tracks of the conference.

We thank the Honorary and General Chairs: Dr. Le Cong Co and Bao Le Nguyen from Duy Tan University, Vietnam, for their continuous support to overcome various difficulties in the process of organizing this conference. We extend our heartfelt thanks to the Organizing and Program Chairs of this conference for being with us from the beginning to the end of this conference; without their support, this conference could never have been successful. We would also like to thank the entire organizing team of Duy Tan University, Da Nang, Vietnam, for coming forward to support us to organize the fourth edition of this conference series. Involvements of faculty coordinators and student volunteers under the leadership of Dr. Nhu Gia Nguyen have been praiseworthy in every respect. We are confident that in future too we would like to organize many more international-level conferences in this beautiful campus.

We take this opportunity to thank authors of all submitted papers from 17 different countries across the globe for their hard work, adherence to the deadlines, and patience with the review process. We congratulate the winners of the Best Paper Awards for their motivation, scholar quality work, and dynamic presentation. The quality of a referred volume depends mainly on the expertise and dedication of the reviewers. We are indebted to the program committee members and external reviewers who not only produced excellent reviews but also did these in short time frames. All the efforts are worth and would please us all, if the readers of this proceedings and participants of this conference found the papers and conference inspiring and enjoyable. Our sincere thanks to all press print and electronic media for their excellent coverage of this conference.

We take this opportunity to thank all keynote speakers, track and special session chairs, and delegates for their excellent support to make INDIA 2017 a grand success in Da Nang, Vietnam.

Lucknow, India	Dr. Vikrant Bhateja
Da Nang, Vietnam	Dr. Bao Le Nguyen
Da Nang, Vietnam	Dr. Nhu Gia Nguyen
Vijayawada, India	Dr. Suresh Chandra Satapathy
Hai Phong, Vietnam	Dr. Dac-Nhuong Le

Organization

Honorary Chairs

Le Cong Co, Duy Tan University, Vietnam

General Chairs

Bao Le Nguyen, Duy Tan University, Vietnam
Suresh Chandra Satapathy, PVPSIT, Vijayawada, India

Organizing Chair

Nhu Gia Nguyen, Duy Tan University, Vietnam

Publication Chairs

Vikrant Bhateja, SRMGPC, Lucknow, UP, India
Viet Hung Dang, Duy Tan University, Vietnam

Program Chairs

Chakchai So-In, Khon Kaen University, Thailand
Binh Nguyen, Duy Tan University, Vietnam; IIASA, Austria
Nilanjan Dey, Techno India, Kolkata, India

Do Nang Toan, VNU, Vietnam
V. Santhi, VIT University, Vellore, India
Le Hoang Son, VNU, Vietnam

Publicity Chairs

Son Van Phan, Duy Tan University, Vietnam
Duong Nguyen Thanh, Duy Tan University, Vietnam
Man Nguyen Duc, Duy Tan University, Vietnam
Trung Dang Ngoc, Duy Tan University, Vietnam
Dac-Nhuong Le, Hai Phong University, Vietnam

Secretariat

Chung Le, Duy Tan University, Vietnam
Nin Ho Le Viet, Duy Tan University, Vietnam
Giang Nguyen Hong, Duy Tan University, Vietnam

Web Administrators

Huy Nguyen Dang Quang, Duy Tan University, Vietnam
Thanh Nguyen Trong, Duy Tan University, Vietnam

Steering Committee

Nguyen Quang Thanh, Da Nang DIC

Track Chairs

Machine Learning Applications: Steven L. Fernandez, SCEM, Mangalore, India.
Image Processing and Pattern Recognition: V. N. Manjunath Aradhya, SJCE, Mysore, India.
Big Data, Web Mining & IoT: Sireesha Rodda, Gitam University, Visakhapatnam, India.
MANETs and Wireless Sensor Networks: Pritee Parwekar, ANITS, Visakhapatnam, India

Special Session Chairs

Computer Vision and Pattern Recognition (CVPR): V. N. Manjunath Aradhya, B. S. Harish, SJCE, Mysuru, and S. Manjunath, Samsung Electro-Mechanic Co. Ltd, Bengaluru
Knowledge-driven Computing (KDC): Sireesha Rodda, GITAM University, Visakhapatnam, India, and Pritee Parwekar, ANITS, Visakhapatnam, India
Advanced Fuzzy Theory and Computational Intelligence (AFT): Le Hoang Son, Vietnam National University, Vietnam, and Bui Cong Cuong, Vietnam Academy of Science and Technology
Computational Intelligence to Autonomic Sustainable Computing through Data Sciences (CIASCD): Tanupriya Choudhury and Praveen Kumar, Amity University, UP, India
Software Engineering and its Applications (SEA): Suma V, Dayananda Sagar College of Engineering, Bengaluru, India
Machine Learning in Internet of Things (MLIoT): Chintan Bhatt, Charotar University of Science and Technology, India, and Dac-Nhuong Le, Haiphong University, Vietnam
Natural language Processing (NLP): Ali Reza Afshari, Islamic Azad University, Shirvan, Iran
Advanced Machine Learning Algorithms (AMLA): Dr. Amit Joshi, CSI, Udaipur, India

Technical Program Committee/International Reviewer Board

A. Govardhan, India
Aarti Singh, India
Abhishek Kumar Mishra, India
Ali Reza Afshari, Iran
Amartya Mukherjee, India
Almoataz Youssef Abdelaziz, Egypt
Amira A. Ashour, Egypt
Amulya Ratna Swain, India
Anand Nayyar, India
Ankur Singh Bist, India
Anuradha Bhatia, India
Arvind Kumar Pandey, India
Aparna Tripathi, India
Athanasios V. Vasilakos, Athens
Banani Saha, India
Bhabani Shankar Prasad Mishra, India
Bharat Gaikawad, India
Bharath Bhushan, India

Binh Nguyen, Austria
B. Tirumala Rao, India
Carlos A. Coello, Mexico
Chakchai So-In, Thailand
Charan S. G., India
Chirag Arora, India
Chilukuri K. Mohan, USA
Chung Le, Vietnam
Dac-Nhuong Le, Vietnam
Dac Binh Ha,Vietnam
Do Nang Toan, Vietnam
Delin Luo, China
Dhananjay Singh, India
Dhruba Ghosh, India
Hai Bin Duan, China
Hai V. Pham, Vietnam
Heitor Silvério Lopes, Brazil
Igor Belykh, Russia
J. V. R. Murthy, India
K. Parsopoulos, Greece
Kamble Vaibhav Venkatrao, India
Kailash C. Patidar, South Africa
Koushik Majumder, India
Krishnendu Guha, India
Lalitha Bhaskari, India
Jeng-Shyang Pan, Taiwan
Juan Luis Fernández Martínez, California
Le Hoang Son, Vietnam
Leandro Dos Santos Coelho, Brazil
L. Perkin, USA
Lingfeng Wang, China
M. A. Abido, Saudi Arabia
Maurice Clerc, France
Meftah Boudjelal, Algeria
Monideepa Roy, India
Mukul Misra, India
Naeem Hanoon, Malaysia
Nhu Gia Nguyen, Vietnam
Nikhil Bhargava, India
Oscar Castillo, Mexico
P. S. Avadhani, India
Rafael Stubs Parpinelli, Brazil
Ramgopal Kashyap, India
Ravi Subban, India
Roderich Gross, England

Saeid Nahavandi, Australia
Sahaj Saxena, India
Sankhadeep Chatterjee, India
Sandip Das, India
Sandeep Singh Sengar, India
Sanjay Sengupta, India
Santosh Kumar Swain, India
Saman Halgamuge, India
Saparkhojayev Nurbek Pazharbekovich, Kazakhstan
Sayan Chakraborty, India
Shabana Urooj, India
S. G. Ponnambalam, Malaysia
Sk Md Obaiduallh, India
Srinivas Kota, Nebraska
Srinivas Sethi, India
Sriparada Ramshree, India
Sumanth Yenduri, USA
Suberna Kumar, India
T. R. Dash, Cambodia
Tri Gia Nguyen, Vietnam
Vasavi S.
Vipin Tyagi, India
V. Santhi, India
Viet Hung Dang, Vietnam
Vikram Puri, India
Vimal Mishra, India
Vinh Le Trong, Vietnam
Walid Barhoumi, Tunisia
X. Z. Gao, Finland
Ying Tan, China
Zong Woo Geem, USA
M. Ramakrishna Murthy, India
Tushar Mishra, India
And many more.

Contents

About the Editors

Vikrant Bhateja is Associate Professor, Department of Electronics and Communication Engineering, Shri Ramswaroop Memorial Group of Professional Colleges (SRMGPC), Lucknow, India. Presently, he is also the Head (Academics & Quality Control) in the same college. He completed his doctorate in Biomedical Imaging and Signal Processing and has a total academic teaching experience of 14 years with more than 120 publications in reputed international conferences, journals, and online book chapter contributions. His areas of research include: digital image processing, computer vision, and medical imaging. However, the core publications are focused on design of image denoising filters, enhancement algorithms, multi-resolution image processing, image fusion, and image fidelity analysis. He has chaired the editorial teams and also organized special sessions from the above domain in international conferences of IEEE, ACM and Springer. He has published 8 books currently with Springer under AISC, LNEE, LNNS, SIST and SCI Series. He is associate editor in the International Journals of Rough Sets and Data Analysis (IJRSDA), Synthetic Emotions (IJSE), Ambient Computing and Intelligence (IJACI) under IGI Global Publications. He is in the editorial board of International Journal of Image Mining (IJIM) and International Journal of Convergence Computing under Inderscience Publishers. He has been guest editor for a special issue with Springer: Arabian Journal for Science and Engineering (AJSE). Recently, he has received the "Certificate of Recognition" for outstanding contribution to Editorial Board for being Active Reviewer for the year 2016 in International Arab Journal of Information Technology (A SCIE-indexed journal with IF = 0.8). His recent publication on Multi-Spectral Image Fusion published in Review of Scientific Instruments (A SCIE-indexed journal with IF = 1.6 published by American Institute of Physics) has been selected as the "Editors' Pick Article" of the August 2016 issue.

Bao Le Nguyen is the Vice Provost of Duy Tan University (DTU), Vietnam, where he is in charge of the Technology and Engineering Division. His research domain includes data warehousing, 3D animation, and online marketing. Under his design and supervision, software teams at DTU have completed the construction of various academic, HR, and financial information systems at DTU over the past 10 years. He also brought about the adoption of conceive–design–implement–operate (CDIO) and problem-based learning (PBL) models at DTU since 2011, and has helped sustain the university-wide CDIO and PBL effort of DTU until now.

Nhu Gia Nguyen received Ph.D. degree in Computer Science from Hanoi University of Science, Vietnam. He serves as Vice Dean of the Graduate School at Duy Tan University. His experience includes over 17 years of teaching, and he has more than 40 publications in various conference proceedings and international journals of repute. His research interests include algorithm theory, medical imaging, network optimization, and wireless security. He has been organizing chair of (Springer LNAI) IUKM-2016 held at Vietnam. He is associate editor in International Journal of Synthetic Emotions (IJSE) under IGI Global.

Suresh Chandra Satapathy received his Ph.D. in Computer Science and is currently working as Professor and Head, Department of CSE, PVPSIT, Vijayawada, Andhra Pradesh, India. He held the position of the National Chairman Div-V (Educational and Research) of Computer Society of India for 2015–17, which is the largest professional society in India having more than 2 lakh members and having 51 years of history of IT and Computing. He also held the position of Secretary and Treasurer of Computational Intelligence Society under IEEE Hyderabad Chapter. He is a senior member of IEEE. He has been instrumental in organizing more than 18 international conferences in India as Organizing Chair and edited more than 30 book volumes from Springer LNCS, AISC, LNEE, and SIST Series as Corresponding Editor. He is quite active in research in the areas of swarm intelligence, machine learning, and data mining. He has developed a new optimization algorithm known as Social Group Optimization (SGO) published in Springer journal. He has delivered number of keynote address and tutorials in his areas of expertise in various events in India. He has more than 100 publications in reputed journals and conference proceedings. He is Editor-in-Chief of Neutorosphic Computing and Machine Learning—An International Journal of Information Science published by New Mexico University, USA. He is in editorial board of IGI Global, Inderscience, Growing Science journals, and also Guest Editor for Arabian Journal for Science and Engineering published by Springer.

Dac-Nhuong Le received his Ph.D. in Computer Science and Engineering and is Deputy-Head of Faculty of Information Technology, Haiphong University, Vietnam. Presently, he is also serving as the Vice Director of Information Technology Apply Center in the same university. His areas of research include soft computing, network communication, network optimization, network security, cloud computing, and image processing in biomedical. He has around 50 quality publications in reputed international conferences, journals, and online book chapter contributions. He has been on TPC and chaired some sessions from the above domain in international conferences of Springer. He has been the track chair and served in the core technical teams for international conferences: FICTA 2014, CSI 2014, IC4SD 2015, ICICT 2015, INDIA 2015, IC3T 2015, INDIA 2016, FICTA 2016, IC3T 2016, ICDECT 2016, IUKM 2016 under Springer ASIC/LNAI Series. Presently, he is serving in the editorial board of international journals under AIRCC Publishing Corporation, Wireilla Scientific Publications.

Pharmacophore Modeling and Docking Studies of SNCA Receptor with Some Active Phytocompounds from Selected Ayurvedic Medicinal Plants Known for their CNS Activity

Preenon Bagchi[1,2(✉)], M. Anuradha[1], and Ajit Kar[2,3]

[1] Padmashree Institute of Management and Sciences, Bengaluru, India
{prithish.bagchi,pimsprincipal}@gmail.com
[2] Sarvasumana Association, Bengaluru, India
{ajitkarsatsang}@gmail.com
[3] Satsang Herbal Research Laboratory, Satsang, Deoghar, India

Abstract. Neurodegeneration is an alarming problem all over the globe. The significant role of SNCA receptor has been well established in several neurodegenerative disorders including various types of progressive dementia—most commonly in Parkinson's disorder (PD) and Lewy body dementia. Phytocompounds selected from ayurvedic medicinal plants are thoroughly screened against the SNCA receptor (*in silico*). The gene receptor responsible for Parkinson's disorder (PD), SNCA, was taken for this work. Mutated mammalian SNCA implicated as one of the factors responsible for PD was taken from NCBI; templates as retrieved from BLAST were downloaded from PDB. The 3D structure of SNCA was modeled; the 3D structures of phytocompounds (unknown ligands) were taken from various online databases; phytocompounds were virtually screened against the SNCA receptor, and the best ligand was selected.

Keywords: Parkinson's disorder (PD) · SNCA · Pharmacophore · Modeling Ramachandran plot · Docking

1 Background

From very ancient days, different medicinal plants dealing with health care are part and parcel of major populations of India, and other Asian countries.

Parkinson's disorder (PD) is a movement disorder having well-expressed symptoms like slowing of movement, tremor, rigidity or stiffness, and balancing problems. As per Dr. Dickson, the typical patient with PD has Lewy bodies (aggregates of protein, alpha synuclein or SNCA) in the brain neurons. Research suggests that the above central nervous system (CNS) disorder, PD, is a combination of environmental factors and multi-gene mutation [1–6].

In the present study for the preliminary screening, the following medicinal plants have been selected to study the scope and activity of different compounds on CNS using the above-mentioned bioinformatic parameter. These are *Hydrocotyle asiatica/Centella*

V. Bhateja et al. (eds.), *Information Systems Design and Intelligent Applications*, Advances in Intelligent Systems and Computing 672,
https://doi.org/10.1007/978-981-10-7512-4_1

asiatica, Bacopa monnieri, Convolvulus pluricaulis, Mucuna pruriens, Ocimum sanctum, Tinospora cordifolia, Curcuma longa, Nardostachys jatamansi, Gloriosa superba, Colchicum autumnale, etc., especially considered in this work.

Mutations in the SNCA (alpha synuclein) receptor and is noted as causal factors for many CNS disorders is used in this work [7–9].

SNCA

Alpha synuclein is expressed in the brain and integrates presynaptic signals and membrane traffics, mutations and defects of are implicated in the pathogenesis of PD [10, 11]. Unconventionally spliced transcripts encoding different isoforms have been identified for this gene [12]. Researchers are still much unclear about the function of SNCA, though it is seen it generally maintains a supply of synaptic vesicles in presynaptic terminals by clustering synaptic vesicles and controls the release of dopamine which is an important neurotransmitter which critically controls the voluntary and involuntary movements [13].

In this work, the 3D structure of the SNCA receptor is modeled by homology modeling, using high-throughput screening (HTS), and its ligand is selected from Indian Ayurvedic Herbs.

2 Methodology

The 3D structure of the SNCA receptors was modeled using modeler software [14]. The SNCA receptor's amino acid sequence is downloaded from NCBI; its homologous templates were selected by BLAST (Table 1). The receptor and their corresponding templates were submitted to modeler software to model their 3D structure. Using Rampage Ramachandran plot server [15], the generated five models generated by modeler were evaluated and the selection of the best model was done (Table 2). The 3D structures of the unknown ligands (phytocompounds as in Table 3) were downloaded from various other online databases, and a phase database is generated [16]. Structure-based pharmacophore model is a unique procedure for generating energy-optimized e-pharmacophores. This was done by selecting the regular features of the 3D structure of SNCA receptor interacting with the known ligands; thus, phores were generated in the 3D structure of the SNCA receptor at the interaction sites with the above known ligands. Hence, these e-pharmacophores were used as queries for virtual screening [17, 18].

Docking was performed by PATCHDOCK server by selecting the best model (model 4) with the ligand selected by pharmacophore modeling, colchicine, to get the docked structure [19, 20].

3 Results and Discussions

3.1 Homology Modeling

The amino acid sequences of SNCA receptor were downloaded from NCBI (Table 1). Its homologous templates were selected by BLAST (Table 1).

Table 1. PD receptors with their GenBank accession number and homologous templates

Receptor	Accession number	Homologous templates
SNCA	P06241	2H8HA, 1Y57A, 1FMKA

The amino acid sequences of the receptors along with their homologous templates were submitted to modeler software for the generation of the 3D structures of the receptor using the principle of homology modeling [14]. Modeler generated five models for each receptor. The 3D models generated by modeler of SNCA (Table 3) are submitted to Rampage Ramachandran Plot server for model verification [15], and the best (Figs. 1 and 2) model was selected.

Table 2. Ramachandran plot analysis of SNCA receptor's modeler generated models

	Number of residues in favored region	Number of residues in allowed region	Number of residues in outlier region	
Model 1	493 (92.1%)	28 (5.2%)	14 (2.6%)	
Model 2	499 (93.3%)	27 (5.0%)	9 (1.7%)	
Model 3	497 (92.9%)	26 (4.9%)	12 (2.2%)	
Model 4	500 (93.5%)	26 (4.9%)	9 (1.7%)	Selected
Model 5	496 (92.7%)	28 (5.2%)	11 (2.1%)	

3.2 Structure-Based Pharmacophore

Pharmacophore sites were created in the SNCA receptor (model 4) using the known ligands, viz. phenothiazine, N-acylaminophenothiazine, N-alkylphenothiazine, stimovul, etorphine, propoxyphene, and pentazdine. The above ligands are established ligands for SNCA receptor [21, 22].

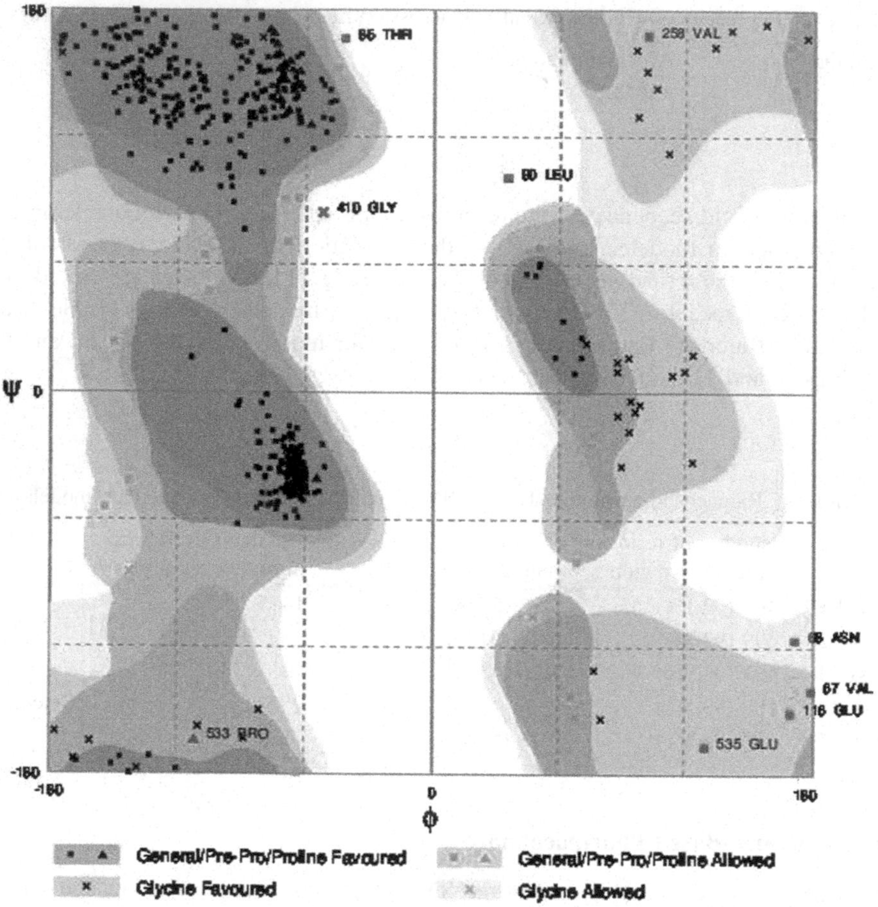

Fig. 1. Ramachandran plot analysis of SNCA receptor model 4

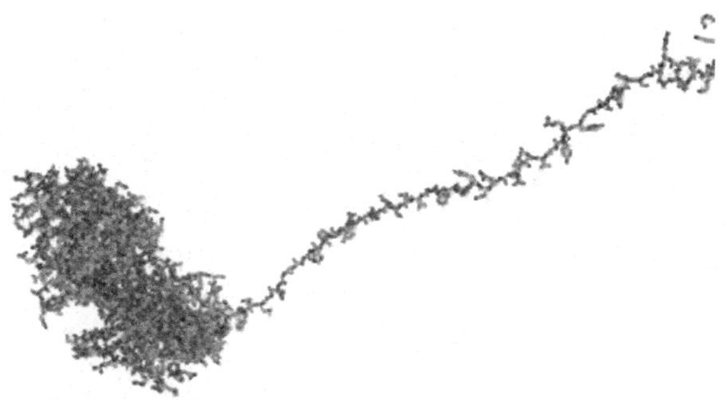

Fig. 2. 3D structure of SNCA receptor model 4

Based on preliminary screening, special emphasis is given on the following herbs having known phytocompounds as given in Table 3 were screened.

Table 3. Ayurvedic herbs (the source of phytocompounds) used in this work.

Source Natural remedies, Bengaluru, India, and Bhesaj Uddyan, Satsang, India
Plant Name/(with known phytocompounds)
Acorus calamus (acoradin)
Asparagus racemosus (racemosol)
Rauwolfia serpentina (rescinnamine)
Withania somnifera (withanone)
Andrographis paniculata (andrographolide)
Gloriosa superba (colchicine)
Colchicum autumnale (colchicine)
Centella asiatica (asiaticoside)
Convolvulus pluricaulis (shankhapushpine)
Ocimum sanctum (eugenol)

As per structure-based pharmacophore results, phytocompound having the fitness score 1.089598 was selected as the best fitted ligand (Fig. 3, Table 4) and the further docking studies were done using this phytocompound.

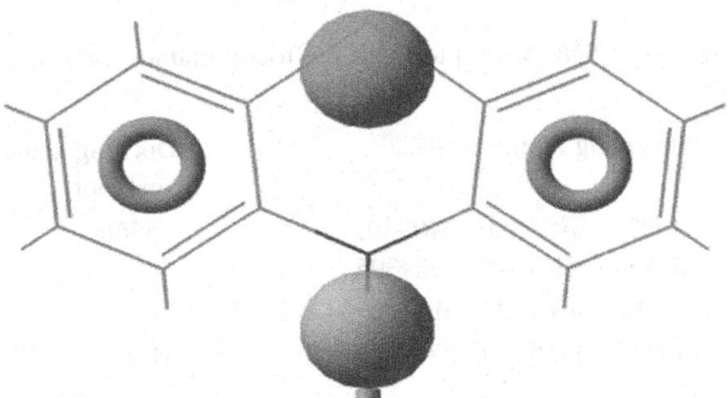

Fig. 3. Pharmacophore features of phenothiazine

Table 4. Results of pharmacophore modeling

Known ligand	Fitness score	Phytocompound	Plant name
Phenothiazine	1.089598	Colchicine	*Gloriosa superba and Colchicum autumnale*

3.3 Molecular Docking

SNCA receptor (model 4) was docked with the phytocompound using software PATCHDOCK [19, 20]. It was seen that SNCA receptor docks with the phytocompound, colchicine, with a docking score of −5566 kcal/mol (Fig. 4, Table 5).

From the above result is compared with the docking interactions of SNCA receptor

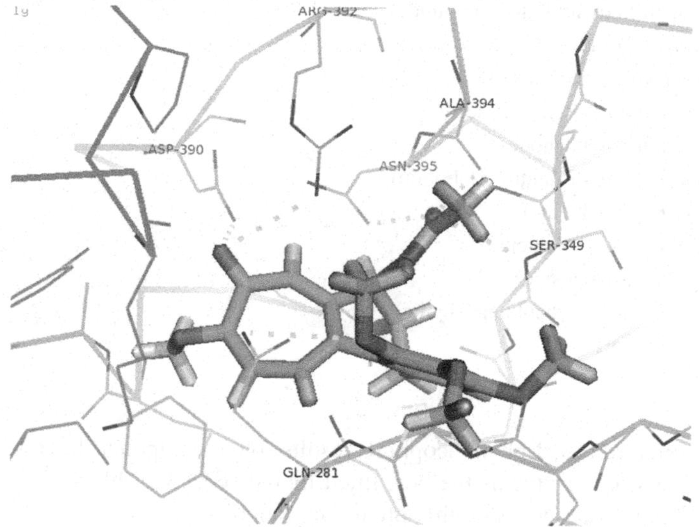

Fig. 4. Docked structure of SNCA receptor with phytocompound colchicine

Table 5. Docking results of SNCA receptor with phytocompound colchicine, jatamanin11, and bacopaside I

Phytocompound	Interacting amino acids	Docking score (in kcal/mol)	Docking
Colchicine	GLN281, ASP390, ARG392, ALA394, SER349, ASN395	−5566	Yes
Jatamanin11	ASP121, ASN122, GLU126, TYR125	−7.33	Yes
Bacopaside I	LEU350, LYS299, TYR420	5106	Yes

with known phytocompounds Jatamanin11 from *Valeriana jatamansi* (Fig. 5) [23, 24] and Bacopaside I from *Bacopa monnieri* (Fig. 6) [25] which were established as ligands for PD [23, 25] in our previous works.

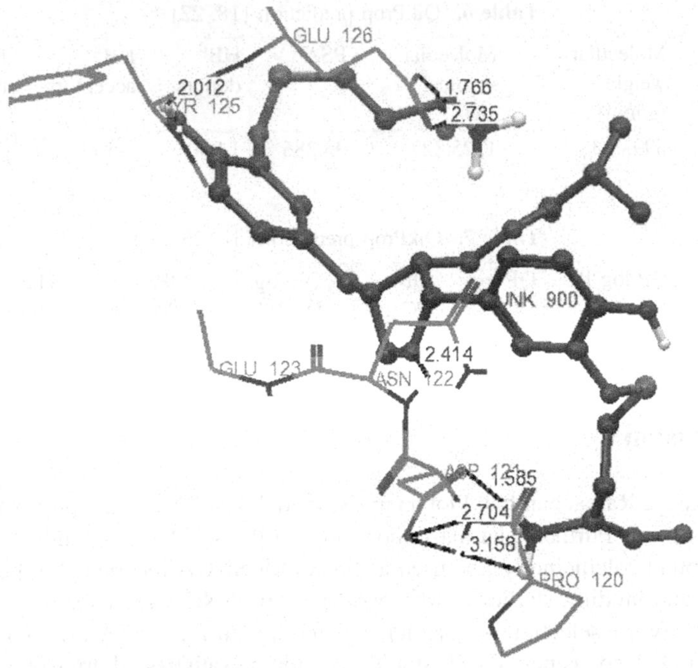

Fig. 5. Docked structure of SNCA receptor with phytocompound jatamanin11 [23, 24]

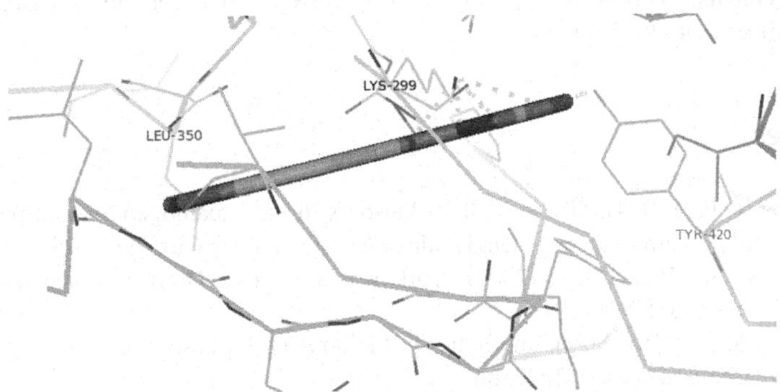

Fig. 6. Docked structure of SNCA receptor with phytocompound bacopaside I [25]

3.4 ADME Screening

ADME is an abbreviation for absorption, distribution, metabolism, and excretion. It is used to calculate the drug-like properties of the molecules [16].

QikProp generated the following output (Tables 6 and 7) [18, 23] for the phytocompound colchicine.

Table 6. QikProp prediction [18, 23]

Lead molecules	Molecular weight[a] (g/mol)	Molecular volume[b] (Å)	PSA[c]	HB[d] donors	HB[e] acceptors	Rotatable bonds[f]
Colchicine	399.443	1225.827	93.286	1.000	7.500	5.000

Table 7. QikProp prediction [18, 23]

Lead molecule	QP log P (o/w)[a]	QP log S[b]	QP PCaco[c]	QP log HERG[d]	QP PMDCK[e]	Human oral absorption[f]
Colchicine	2.545	−3.809	550	−3.180	483	91

4 Conclusion

As per Rampage Ramachandran Plot analysis, model 4 of SNCA receptor was selected as the best model. Further, pharmacophore and molecular docking studies prove that phytocompound colchicine can be used as ligand for SNCA receptor. Further, in vitro receptor–ligand binding studies can be performed on SNCA receptor with the above ligand to justify the selection of colchicine as the ligand for SNCA receptors and as a remedy for PD. Also, as per ADME studies, colchicine satisfies all drug-like properties which correlate with the results of Bagchi et al. [26].

Acknowledgements. This work was supported by SERB-NPDF for Preenon Bagchi, author reference number PDF/2015/000047.

References

1. Alexander, G.E., (2004), "Biology of Parkinson's disease: pathogenesis and pathophysiology of a multisystem neurodegenerative disorder", Dialogues Clin Neurosci., 6(3):259–280.
2. Thomas, B. and Beal, M.F., (2007) "Parkinson's disease, Human Molecular Genetics", 2007, 16(2): R183–R194.
3. Saha, A., (2013), "Psychiatric morbidity in Parkinson's Disease: a case report", American Journal of Life Sciences, 1(2):27–30.
4. Djamshidian, A. and Lees, A., (2012), "Impulsive Compulsive Behaviours In Patients With Parkinson's Disease Treated With Dopamine Agonists", Focus on Parkinson's Disease, 23 (1):16–21.
5. Stacy, M., (2009), "Impulse control disorders in Parkinson's disease", F1000 Med Rep. 1: 29.
6. Stacy, M., Galpern, W., Samuel, M., Lang, A., (2008), "Impulse control disorders in Parkinson's disease abstract supplement", Movement Disorders, 23(9):1332–1351.
7. Klein, C. and Westenberger, A., (2012), "Genetics of Parkinson's Disease", Cold Spring Harb Perspect Med., 2(1):1–15.
8. Dauer, W. and Przedborski, S., (2003), "Parkinson's Disease: Mechanisms and Models", Neuron, 2003, 39:889–909.

9. Rachakonda, V., Pan, T.H. and Le, W.D., (2004), "Biomarkers of neurodegenerative disorders: How good are they?", Cell Research, 14:347–358.
10. George JM (2002). "The synucleins". Genome Biol. 3 (1): REVIEWS3002.
11. Polymeropoulos MH, Lavedan C, Leroy E, Ide SE, Dehejia A, Dutra A, et al. (1997). "Mutation in the alpha-synuclein gene identified in families with Parkinson's disease". Science. 276 (5321): 2045–7.
12. Goedert M (July 2001). "Alpha-synuclein and neurodegenerative diseases". Nat. Rev. Neurosci. 2 (7): 492–501.
13. Shah, M., Doctoral Thesis, 2013, The cytoskeletal linker protein, Ezrin, inhibits α-synuclein fibrillization and toxicity by a novel mechanism, Freien Universität Berlin.
14. Sali, A. and Blundell, T.L., (1993), "Comparative protein modelling by satisfaction of spatial restraints" J. Mol. Biol., 234:779–815.
15. Laskoswki, R.A., MacArthur, M.W., Moss, D.S. & Thorton, J.M., (1993), "Procheck: a program to check the stereochemical quality of protein structures", J. Appl. Cryst. 26:283–291.
16. Schrödinger Suite (2010) Protein Preparation Guide, SiteMap 2.4; Glide version 5.6, LigPrep 2.4, QikProp 3.3, Schrödinger, LLC, New York, NY.
17. Taha MO, Dahabiyeh LA, Bustanji Y, Zalloum H, Saleh S, (2008)Combining Ligand-Based Pharmacophore Modeling, Quantitative Structure-Activity Relationship Analysis and in Silico Screening for the Discovery of New Potent Hormone Sensitive Lipase Inhibitors. J Med Chem 51:6478–6494.
18. Singh KhD, Kirubakaran P, Nagarajan S, Sakkiah S, Muthusamy K, Velmurgan D, Jeyakanthan J (2012) Homology modeling, molecular dynamics, e-pharmacophore mapping and docking study of Chikungunya virus nsP2 protease. J Mol Model 18: 39–51.
19. Duhovny D, Nussinov R, Wolfson HJ, (2002), "Efficient Unbound Docking of Rigid Molecules", In Gusfield et al., Ed. Proceedings of the 2'nd Workshop on Algorithms in Bioinformatics (WABI) Rome, Italy, Lecture Notes in Computer Science 2452, pp. 185–200, Springer Verlag.
20. Schneidman-Duhovny D, Inbar Y, Nussinov R, Wolfson HJ, (2005), "PatchDock and SymmDock: servers for rigid and symmetric docking", Nucl. Acids. Res. 33: W363–367.
21. Yu L, Cui J, Padakanti PK, Engel L, Bagchi DP, Kotzbauer PT, and Tua Z, (2012), "Synthesis and in vitro evaluation of α-synuclein ligands", Bioorg Med Chem., 20(15): 4625–4634.
22. Jayaraj RL, Ranjani V, Manigandan K And Elangovan N, (2013), "Insilico Docking Studies To Identify Potent Inhibitors Of Alpha-Synuclein Aggregation In Parkinson Disease", Asian Journal of Pharmaceutical and Clinical Research, 6(4): 127–131.
23. Preenon Bagchi, Waheeta Hopper 2011 "Virtual Screening of compounds from *Valeriana jatamansi* with α-Synuclein" IPCBEE, 5:11–14.
24. Preenon Bagchi, Somashekhar. R and Ajit Kar, 2015, "Scope of some Indian medicinal plants in the management of a few neuro-degenerative disorders *in-silico*: a review", International Journal of Public Mental Health & Neurosciences, 2(1):41–57.
25. Preenon Bagchi, Anuradha M, Ajit Kar, 2017, "Ayur-Informatics: Establishing an Ayurvedic Medication for Parkinson's Disorder" Int'l Journal of Advances in Chemical Engg., & Biological Sciences, 4(1): 21–25.
26. Bagchi P, Venkatramana DK, Mahesh. M, Somashekhar. R and Kar A, 2014, "Identification of Novel Drug Leads for Receptors Implicated in Migraine from Traditional Ayurvedic Herbs Using in silico and in vitro Methods", Journal of Neurological Disorders, 2(6):2–6 ISSN: 2329-6895.

Sociopsycho-Economic Knowledge-Based System in E-Commerce Using Soft Set Theory Technique

P. Vijayaragavan[1(✉)], R. Ponnusamy[2], and M. Arramuthan[3]

[1] Department of Computer Science and Engineering, Jawaharlal Nehru Technological University, Kakinada 533003, AP, India
vijayaragavanpm@yahoo.co.in

[2] Department of Computer Science and Engineering, CVR College of Engineering, Vastu Nagar, Mangalpalli (V), Ibrahimpatan (M), Hyderabad 501510, Telungana, India
r_ponnusamy@hotmail.com

[3] Department of Information Technology, Perunthalaivar Kamarajar Institute of Engineering and Technology (PKIET), Nedungudi, Karaikudi 609603, Tamil Nadu, India
aramudhan1973@yahoo.com

Abstract. Fast creating of the present territory, a few of the wonders are broadly seen to flourishing the significant fulfillment to the behavior of the information to the punter. The prerequisite from client satisfaction is made as a design to achieve sociopsycho-economic knowledge. The substance structure approach gathers the informational collection to accomplish the objective of consumer necessity. The system contains the client fulfillment to online market conduct to watch the customer request on every single tick of client, based on every single arrangement of snap and hunt information, thereby constructing the model. The representation supports the sociopsycho behavior to watch the request limits of the customers with few ways to achieve the state of a client. Earlier frameworks do not thoroughly investigate the human sociopsycho based Web-based promoting. The designed structure provides investigating the human sociopsycho completely with the assistance of soft set theoretic hypothesis technique. The strategy assesses every single substance of the client request and investigates it.

Keywords: Multi-agent · Punter · Psycho-Economics · Psychology
Negotiation system

1 Introduction

The snare is the best medium of correspondence in upcoming businesses. Various concerns are rethinking their business models to enhance the business yields. It is intriguing to note that apparent convenience is an inconsequential impact Internet

V. Bhateja et al. (eds.), *Information Systems Design and Intelligent Applications*, Advances in Intelligent Systems and Computing 672,
https://doi.org/10.1007/978-981-10-7512-4_2

shopping conduct. Finding uncovered buy expectation massive emphatically changes the Internet shopping behavior drastically. Future researches, tested from various working adults thereby identifying different factors with online shopping that incorporates the limits from examining predisposition [1]. Businesses over the Web give a chance to clients and partners where their items and particular businesses are deployed. These days online business breaks the obstruction of time and space when contrasted with the physical office. Immense organizations around the globe understand that online business is not merely acquirement and offering over the Internet, and to some degree, it enhances the effectiveness to contend with new monsters in the market [2]. Progress in innovation has opened new doors for online shopping, where clients can interface through a Web-based business site. The procedure for online shopping follows the use of User Behavior Analysis strategy [3]. Multi parameter investigation approach identifies numerous clients who have been proposed and executed for data recovery, selling, facilitating, arranging, and so on by the customer record set at the site components [4].

2 The Negotiation Process

An essential part of any E-business framework is the proficient, precise hunt and determination of items a client wishes to buy in the best possible approach. Creating the E-trade MAN* (Fig. 1) will give proficient pursuit strategy to the scan calculation for determining the client necessity.

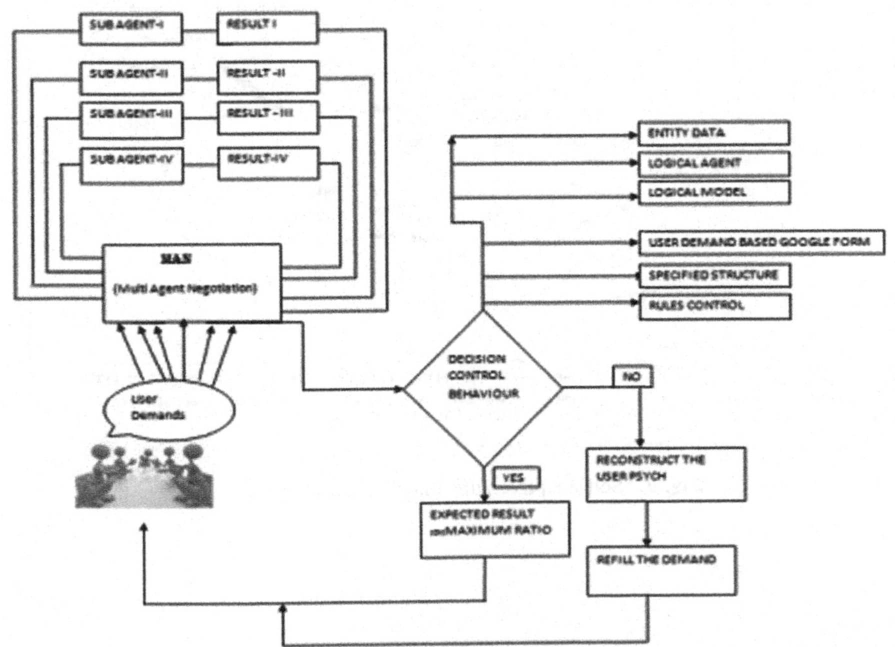

Fig. 1. MAN* architecture model

Fathoming the statical inquiry approach by the outline of the structure is to improve online business and fast response. By the regular monitoring of the customer's needs to reformulate their queries results in gainful viability data. Consumers and known data are used in the three stage approach to deal with building trust and making them to buy [5]. This paper shows the design considering proficient item data recovery and better determination of the E-trade framework and administrations [6]. It will distinguish the necessities and prerequisite of clients. With the help of the Google Form, the customer requirement is assessed.

2.1 Agent-Based E-Commerce Modeling

The apparent benefit of the master vendors, and non-master vendors vary in their impression of the estimation of online sales [7]. The MAN* Architecture (Fig. 1) operator is preparing two wise capacity one is it's passing the client requests to various specialists and accepting the distinctive sort of reports may correct match with the question or entirely filled by the proportion [8]. The MAN* (Fig. 2) chooses the last arrangement from the different kinds of reports accumulated from the multi-specialist in view of some practical conduct utilizing gentlest hypothesis method. That particular outcome may coordinate with the correct request or greatest full ll rate of the right client requirement. The second procedure assumes the nobody report matches with the exact request. The framework won't give the negative report or answer to the client. Rather than that it will trade o the client to buy the other asset most final match with

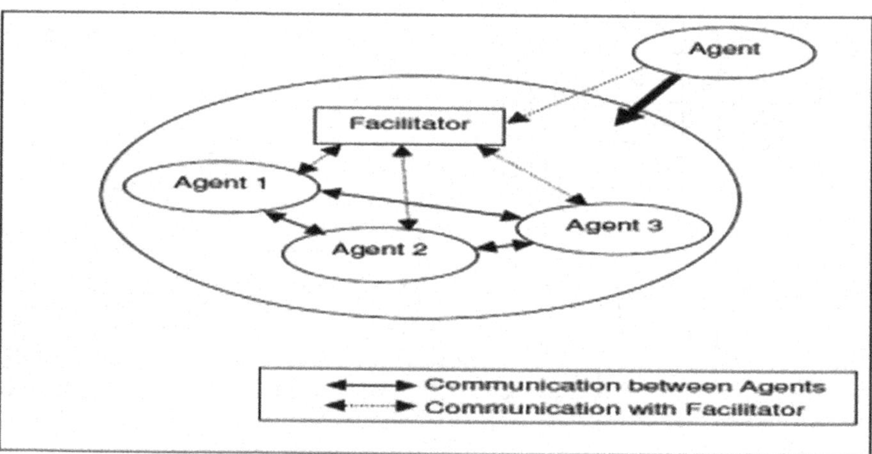

Fig. 2. Self-adaptive auto match protocol process

real prerequisite or refreshing the learning for comment the customer psycho like financial based, innovation-based, and so forth. This methodology conceivable to actualize. All things considered, the comprehensive research did in this eld created different hypotheses. The transaction comprises an arrangement of principles that administer the cooperation among specialists for advantages of E-Commerce, System based item was exchanging. sociopsycho-financial aspects are learning based framework in online business; Electronic commerce is a space where operator advancements are appropriate [9]. The inquiry and recovery of item data are vital in each E-business framework. While a few frames give the client only a settled choice of items, others offer a wide assortment of issues with various qualities [10]. The direct approach to recognize and eyewitness client requests through all customer contribution on each snap and uncheck handle. It will see by the base points of interest of client which incorporate the fundamental of the client utilizing shape portion prepare.

3 Implementation of Approach

Through that framework will get the points of interest of the client to satisfy the procedure necessity [11]. While getting the passage of client, the design and plan will examination the useful behavior of client.

```
Definition 1.
   Action    P SYECOAP P (root; Succ; para);
   Again   (rootcheck ! node)
   f
      While(root ! checknodeallthreads)
      f
         S = intsucc(root)
         P = logicanalysis(threadallmethod)
      g
      Free(root)
   g
   If  rootnode
```

```
Definition 2.
    check = process1
    P roduct = Not select
    While(check <= max round)&(check = nextitem)
        IfNewproduct > NextpossibleItem
                increase_rate = userdataformcheck
                Datavalidate = checkthemandatoryentity (process1+increase datavalidation)
                Budgetprice > Betterspecification
                A=Ctoformrequirement = Gooditems
            ElseIf      check = true
                Itempurchase > buyersatisfied
                Ratio  ofproduct = Purchaseratio
            Else
                gotonextproduct
            EndIf
        Else
            product = true
        EndIf
        check = check + process1
        If      customer  parches < parches offer
                arches_offer = seller offer (process1 customer concession rate) If
                parches offer < customer offer
                    P arches offer = customer satisfied P
                    roduct = true
        Else if parches_offer < newparches_added New
            parchesentry = customer_satisfaction
        Else
            NewCartNotFound
        End If
Else
    check = true
End If
    check = check +
process1 End While
```

```
Definition 3.
    Start :
        Value : checkproblem; itemfromrealworld
    Else
        Localparameter : checkcurrentuserdemand; facilator
        Facilator : accessdatafromalloftheirrunningagents
        CurrentstatementAccessBehaviour(facilator[parameter(problem)]
            Loop
                    Doconditioncheckvalidateformdate(userinput)
                    Ifuserdata(conditionstatementsatisfied)
                    Returncheck(condition)
                    Currentstatement[AccessBehaviour]
            End;
```

Definition 4.
F : AP (U)
W here : U ! universal set
E ! parameter
P (U)DenotesetandAE A; F ()
Approximate
Elementof(F; A)
(i)AB
(ii)A; F ()

In this given MAN* model, the user demand is predicted. The different approaches in the MAN* model reflect the total strategies. On the realm of chance that broker learns excellent choices in a convenient and very much considered way (Fig. 3).

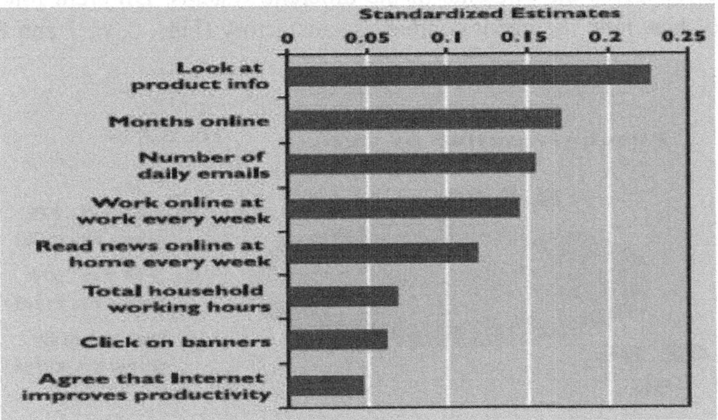

Fig. 3. Prediction of consumer behavior

The customer is searching for the particular item, and he/she is getting to the excessive number of search results, regardless of not knowing the need of the customer (Fig. 4). Here, the MAN* engineering will occur, and it will help the customer to locate their particular item. It gives the all conduct way to deal with the client, though they can pick the particular item [12]. It takes the specific approach by the Google client data, however (Fig. 1).

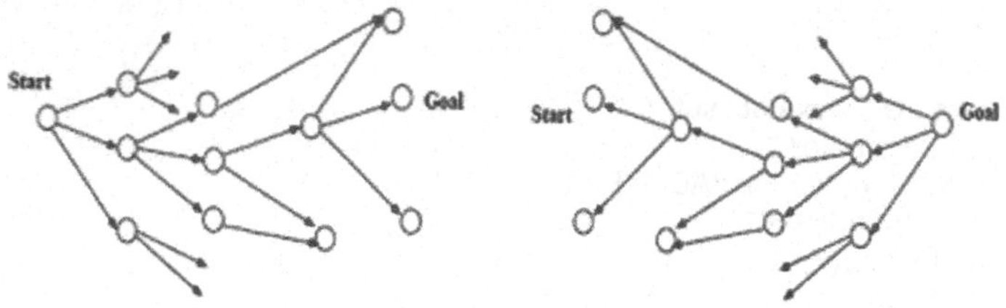

Fig. 4. Customer helping agents

3.1 Sociopsycho-Economic Behavior Analysis

Sociopsycho-economic behavior analysis that checks the real data through the client. It makes the sensible conduct to maintain the particular information [13]. Setting the online overview gives the outcome of the different systems. Different information is collected to show how the client conduct is satisfactory (Figs. 5, 6, 7 and 8).

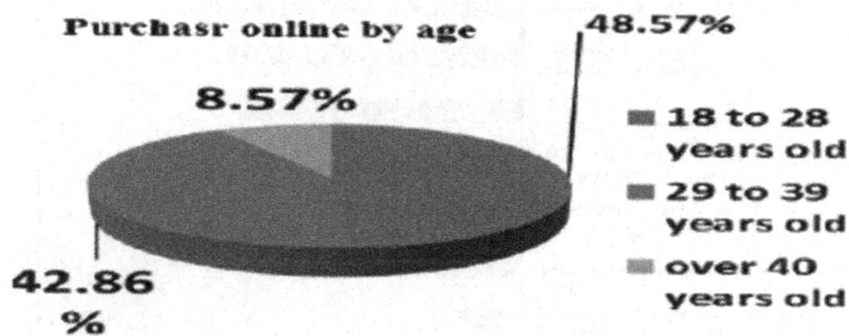

Fig. 5. Age-wise online purchase of customer behavior

Fig. 6. Recommended online shopping ratio

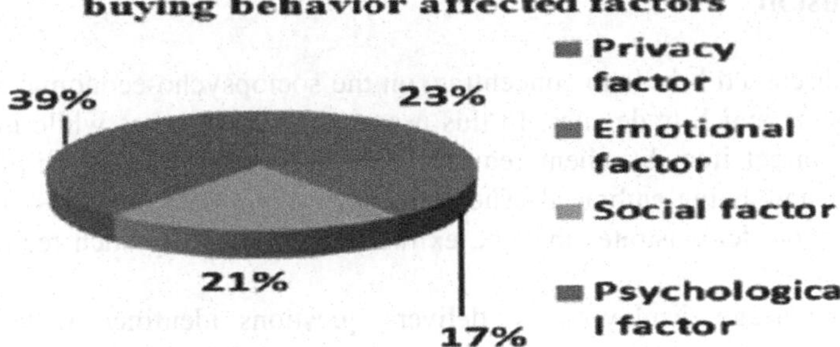

Fig. 7. Customer buying behavior of affected factor

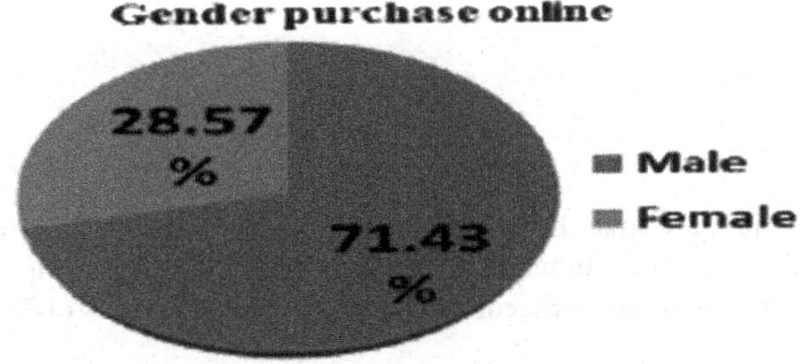

Fig. 8. Gender-wise online purchase ratio

3.2 Experimental Analysis

In MAN* model, the technique is applied to distinguish the huge number of the information input. While accepting the different of combinational information to get to the particular item. In the approving the informational index of client, it investigates the customer behavior with the assistance of certain Google frame (Fig. 9).

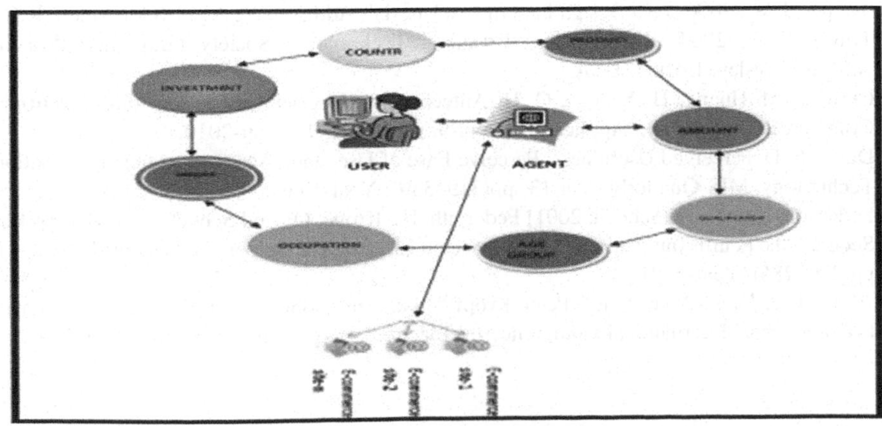

Fig. 9. Model of users and agents

4 Conclusion

The model discussed here is to concentrate on the sociopsycho-economic analysis of online shopping and E-trade sites. In this usage, the case focuses while in transit to purchase a correct item by client request. This approach to being helpful for the further work now being embraced. The element of client data demands. The execution examination demonstrates that the extra costs initiated by such techniques are adequate.

While the usage displayed just delivers questions identified with item data recovery and introduction, this further going review will incorporate the execution of the considerable number of operators and parts of the engineering applying different specialist stage to recover the data through client contribution to analyze as the physiology.

References

1. Yi Jin Lima, Abdullah Osman B, Shahrul Nizam Salah Uddin*, Abdul Rahim Rom-led, Sazal Abdullah,—"Factors Influencing Online Shopping Behavior: The Mediating Role of Purchase Intention", Procedia Economics and Finance Vol-35, pp (401–410), Year-2006.
2. Putro and Budhi Haryanto Factors Affecting Purchase Intention of Online Shopping in Zalora Indonesia Haryo Bismo, British Journal of Economics, Management & Trade, Article no. BJEMT.18704 ISSN: 2278-098X, pp(1–12), Year-2015.
3. P. Vijayaragavan and R. Ponnusamy—"Socio-Economic Psycho Knowledge-Based In-telligent Agents for Automated E-Commerce Negotiation," Springer, HCI2011, LNCS, Vol-6764, pp(274–284), Year-2011.
4. Amalia Pirvanscu, Costin Badica, Marcin Paprzycki, "Developing A JADE-Based Multi-Agent E-Commerce Environment", IADIS, pp(1–8), Year-2005.
5. Aghdaie, S. F., Piraman, A., Fathi, S. An Analysis of Factors Affecting the Consumer's Attitude of Trust and their Impact on Internet Purchasing Behaviour. International Journal of Business and Social Science, Vol-2, pp(147–158), Year-2011.
6. Cooper, J. W., 2000. Java Design Patterns. A Tutorial. Addison-Wesley, USA Laudon, K.C., Traver, C.G., 2004. E-Commerce. Business, Technology, Society (2nd ed.). Pearson Addison-Wesley, Boston, USA.
7. Chiu, C. M. Huang, H. Y. Yen, C. H. Antecedents of trust in online auctions. Electronic Commerce Research and Applications, Vol-9, pp(148–159), Year-2010.
8. Davis, F. D. Perceived Usefulness, Perceive Ease of Use, and User Accep-tance Information Technology. MIS Quarterly, Vol-13, pp(319–339), Year-1989.
9. Federrath, Krone, and Schoch, 2001] Federrath, H., Krone, O., and Schoch, T., "Making Jini Secure", the Fourth International Conference on Electronic Commerce Research (ICECR-4), pp(276–286), Year-2001.
10. Abbas Kia, Esma Ameur, and Peter Kropf-"Akia Automobile: a Multi-Agent Sys-tem in E-Commerce," International Conference on Electronic Commerce Research, Year-2002.

11. N.C. Agman, S. Enginoglu, and F. C. Itak—"FUZZY SOFT SET THE-ORY AND ITS APPLICATIONS", Iranian Journal of Fuzzy Systems Vol. 8, No. 3, pp. 137–147,2011.
12. Howard, J., J. Sheth. The Theory of Buyer Behavior, Wiley. JADE. Java Agent Development Framework. See http://jade.cselt.it, Vol-1, pp(32–47), Year-1969.
13. H. Chen, R.H.L. Chiang, and V.C. Storey, Business intelligence and analytics: from big data to big impact, MIS Quarterly, Vol-36 No.4, pp(1165–1188), Year-2012.

RSSI-Based ZIGBEE Independent Monitoring System in Prison for Prisoners

P. Vijayaragavan[1(✉)], R. Ponnusamy[2], M. Arramuthan[3], and T. Vijila[4]

[1] Department of Computer Science and Engineering, Jawaharlal Nehru Technological University, Kakinada 533003, AP, India
vijayaragavanpm@yahoo.co.in

[2] Department of Computer Science and Engineering, Sri Lakshmi Ammal Engineering College, Thiruvanchery, Tambaram East, Chennai 600126, Tamil Nadu, India
rponnusamy@hotmail.com

[3] Department of Information Technology, Perunthalaivar Kamarajar Institute of Engineering and Technology (PKIET), Nedungudi, Karaikudi 609603, Tamil Nadu, India
aramudhan1973@yahoo.com

[4] Department of Computer Science and Engineering, Aarupadai Veedu Institute of Technology, Vinayaka Missions University Paiyanoor, Chennai, Tamil Nadu, India
omvijila@yahoo.co.in

Abstract. Law-breakers around the world are put behind the bars, whereby various chances of escaping from the prison are available. Several secure technologies are available in and around the world, but none of them helps to reduce the chances of escaping from the prison. So a system was created to protect the prisoners from escaping the prison by using the method of received signal strength indicator (RSSI) which monitors the distance between a pair of nodes in a network. Another method called Electron-ICS enables the development of microsensor which manages the wireless communication network.

Keywords: Microcontroller · Wireless sensor network · RSSI · Localization Nerve stimulator · ZIGBEE · Electron-ICS

1 Introduction

The localization in wireless sensor networks is based on received signal strength indicator (RSSI) which uses ZIGBEE to estimate the distance between the pairs of beacon nodes with the help of Electron-ICS for deploying and managing microsensors across the wireless communication network [1]. There exists a nerve simulator which is used to simulate the vibration of a particular person, and the information gets transformed to the control room. For security purposes, the vibrating sensors are used for measuring, displaying, and analyzing their linear velocity, displacement, and acceleration. Vibration however. The vibrating sensor and nerve stimulator are connected to the ARDUINO microcontroller, and the power supply is given to the controller. The code is written

V. Bhateja et al. (eds.), *Information Systems Design and Intelligent Applications*, Advances in Intelligent Systems and Computing 672,
https://doi.org/10.1007/978-981-10-7512-4_3

in the ARDUINO language (EMBEDDED C), and the whole database and record of the prisoner are also connected to the control room. Then, the ARDUINO microcontroller is connected to the RSSI ZIGBEE which estimates the distance between a pair of nodes.

Environments having low signal strengths in the wireless Personal Area Net-work Standard, the RSSI signal can be used to estimate the distances be-tween beacon nodes accurately. ZIGBEE is very low-power radio based on IEEE 802.15.4 where all the database of the prisoner are stored on the PC (server room) [2].

2 Related Work

2.1 Received Signal Strength Indicator (RSSI)

In telecommunications, received signal strength indicator (RSSI) is a measurement of the power present in a received radio signal. However, because signal strength can vary greatly and affect functionality in wireless networking. RSSI is often done in the intermediate frequency (IF) stage before the IF amplier. RSSI is usually expressed in decibels from 0 (Zero) to −120 db and closer to zero, which indicates a stronger signal. The RSSI consists of strong strength within the range of 30–100 m using ZIGBEE [3]; proposed system consists of four antennas to keep the signal strong without affecting the signal at varying weather environments.

The proposed RSSI ZIGBEE system will transmit the information to the controlling section which can able to identify the distance between the two nodes. During the distance estimate step [4], the RSSI method has advantages like low cost, low power, and accessibility. So they are commonly used in diverse systems for increasing the signal strength.

2.2 RSSI ZIGBEE

ZIGBEE is a wireless technology developed as an open standard to address the unique needs of low-power, low-cost M2M networks. The ZIGBEE standard operates on the IEEE 802.15.4 physical radio specification and operates in unlicensed bands including 2.4 GHz, 900 MHz, and 868 MHz, respectively. The protocol allows devices to communicate in a variety of network topologies and can have battery life lasting for several years. The ZIGBEE protocol is mainly used to communicate data through RF environment that are common in commercial and industrial applications [5].

2.3 ARDUINO Microcontroller

The ARDUINO microcontroller consists of 14 digital I/O pins and 6 analog input pins [6]. The ARDUINO programming language is a simplified version of C/C++. We are creating a program for vibrating sensor and nerve simulator. Suppose the prisoner tries to escape from the prison or crosses the limit node, the sensor indicates to the control room through alarm and the person gets sudden nerve vibration. For the above process, we used ARDUINO programming language (embedded C). An important feature of ARDUINO is that we can create a control program on host PC, download it to the ARDUINO, and it will run automatically [7]. After the installation of the program to the sensor the

ARDUINO microcontroller connected to the RSSI ZIGBEE, if any prisoner crossed limit, then the signal or buzzer indicates to the control room (Figs. 1, 2, and 3).

The ARDUINO microcontroller is mainly connected to the vibration sensor and nerve stimulator. ARDUINO can be used to develop a variety of sensors.

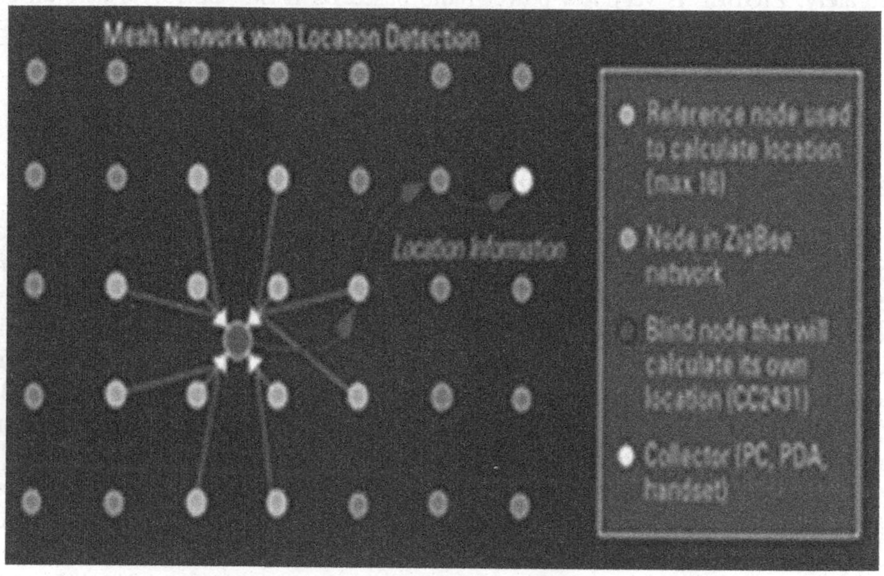

Fig. 1. RSSI location detection

Fig. 2. RSSI ZIGBEE

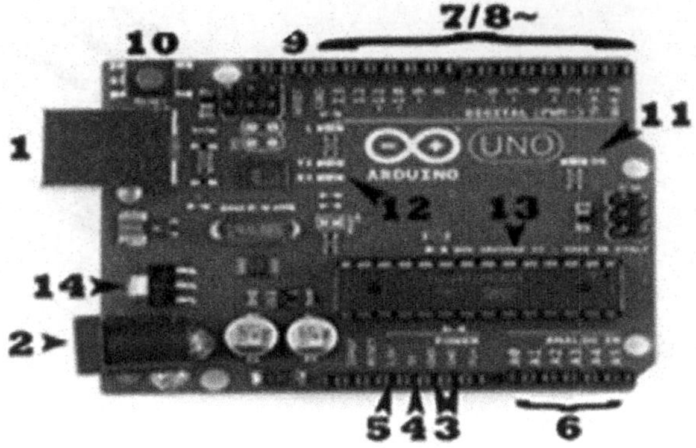

Fig. 3. ARDUINO microcontroller

2.4 Vibration Sensors

The mini sense 100 from measurement specialties is a low cost cantilever type vibration sensor loaded by a mass to o er high sensitivity at low frequencies. Useful for detecting vibration and tap inputs from a user. A small AC and large voltages is created when the lm moves back an forth. Can also be used for impact sensing or a flexible switch. Comes with solder pins that allows for vertical mounting. The features of vibrate sensor are wide range, breadboard-friendly leads.

2.5 Nerve Stimulator

Nerve stimulators are used to determine the level of neuromuscular blocking agent might have in their system during anesthesia. We are here to use nerve stimulator giving shock to the person who is trying to escape. The shock given to the prisoner is not high voltage; the TOF (train of four) ratio or percentages like the STIMPOD NMS 410 are provided for locating nerves for regional anesthesia. STIMPOD 410 is safe (Fig. 4).

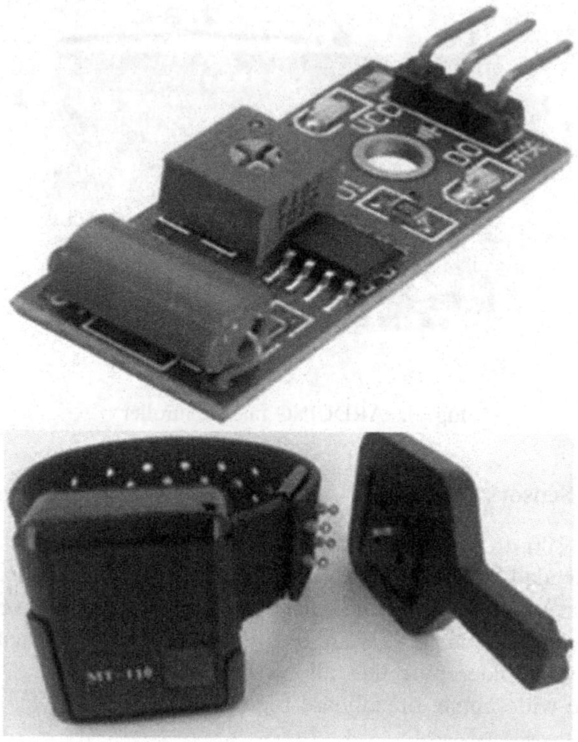

Fig. 4. Vibration sensors

3 Goals and Principles

Our goal is to provide security to the prisoner without escaping from the prison, human kit, control section and server rooms are used. For example, if any one person from jail is trying to escape, the nerve stimulator is mainly used to stimulate the vibration to the particular person. Then, the information is transformed to the control room. The vision is mainly based on security for which vibrating sensors are connected programmed with all the prisoners database. To create the program, we are connecting this vibrating sensor and nerve stimulator to the ARDUINO microcontroller. The power supply is given to the controller.

The program is written in the ARDUINO language (EMBEDDED C), and the whole database and record of the prisoner are also connected to the control room. Then, the ARDUINO microcontroller is connected to the RSSI ZIGBEE. The RSSI estimates the distance between a pair of nodes which actually sense and indicate the control room. A sturdy RSSI signal can be achieved even if the surrounding environment is poor [4]. The RSSI signal is at its best for particular distances where the wireless personal area network standard is low. ZIGBEE is very low-power radio based on IEEE 802.15.4 for all the databases of the prisoner that are stored on the PC (server room).

3.1 Block Diagram of Human Kit

The power supply is applied to the ARDUINO microcontroller; the power supply is converted into digital signal, and then, controller is connected to the ZIGBEE. ZIGBEE is mainly used to transfer data. Vibrate sensor consists of 3 V, and power supply consists of 5 V. TOSCI is mainly used for clock frequency that is mainly for delay purpose. Suppose if we want to transfer data of a prisoner within first 10 s and next 10 s delay and then continue simultancously (Fig. 5).

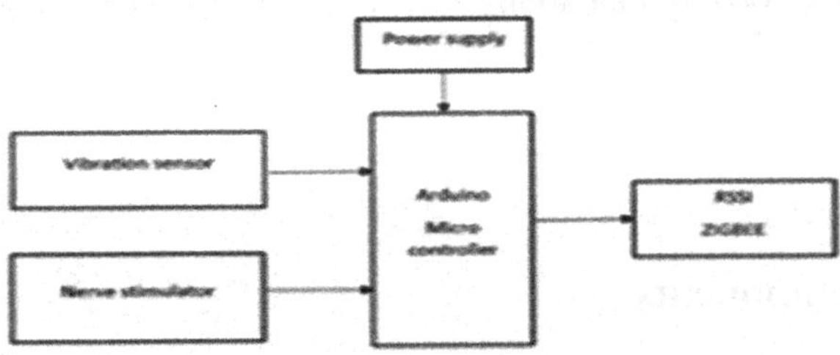

Fig. 5. Human kit

3.2 Block Diagram of Control Section

All the databases of the prisoner are stored in personal computer and connected to the RSSI ZIGBEE. The ZIGBEE is mainly used to transfer data within particular place or the surrounding environment [15] (Fig. 6).

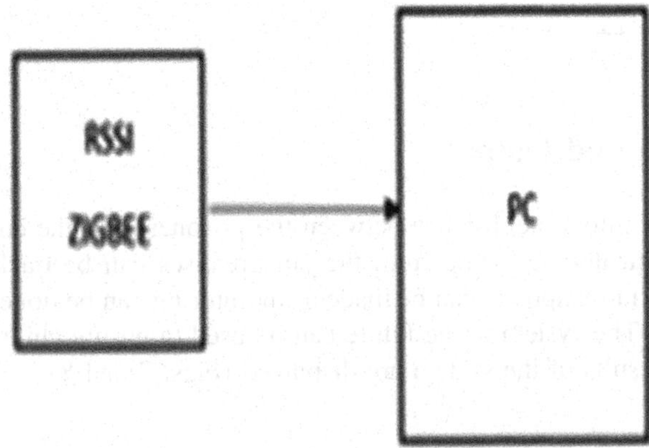

Fig. 6. .

Figure 3.2 Control Section Crystal oscillator is mainly used for clock frequency. The clock frequency is 19.2 MHz [8].

4 Design and Work

The sensors are designed as a band and the quartz could transform mechanical energy into electrical energy as output. The ability of certain crystals to exhibit electrical charge under mechanical loading has no practical application. The development of the application is mainly based on the piezoelectric quartz crystal oscillator. The ARDUINO software work is described as below:

1. Open ARDUINO IDE.
2. Select the serial port from the tool.
3. Select the ARDUINO board from tools.
4. Write the ARDUINO IDE.
5. Compile the code.

5 System Components

The RSSI ZIGBEE transmits the information to the controlling section which identifies the distance between the two beacon nodes. The nerve stimulator is used to stimulate the vibration of a particular person escaping from the prison. The information is then transferred to the control division. The hardware and software components include:

1. ARDUINO microcontroller
2. RSSI ZIGBEE
3. Vibration sensor
4. Nerve stimulator
5. PC
6. Embedded C/C++.

6 Conclusion and Output

RSSI ZIGBEE monitors the distance between the prisoners and the control room and sends a status indicator. Escaping from the jail premises can be tracked to a higher extent. Reduction in manpower can be made as monitoring can be done easily with the proposed system. The system in the future can be used to ensure child safety, women safety, etc. The results of the system are displayed (Figs. 7 and 8).

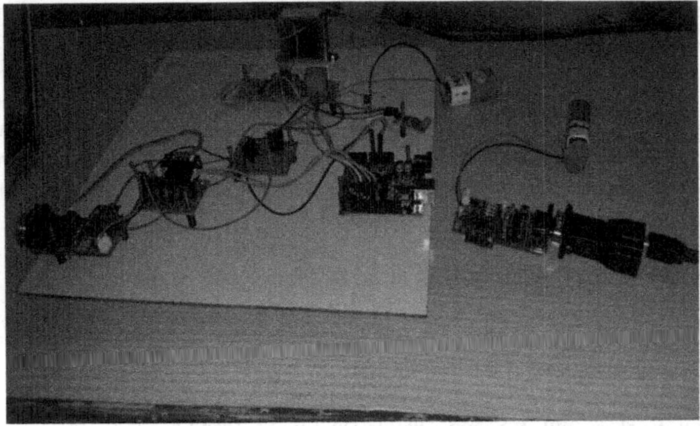

Fig. 7. Monitoring system

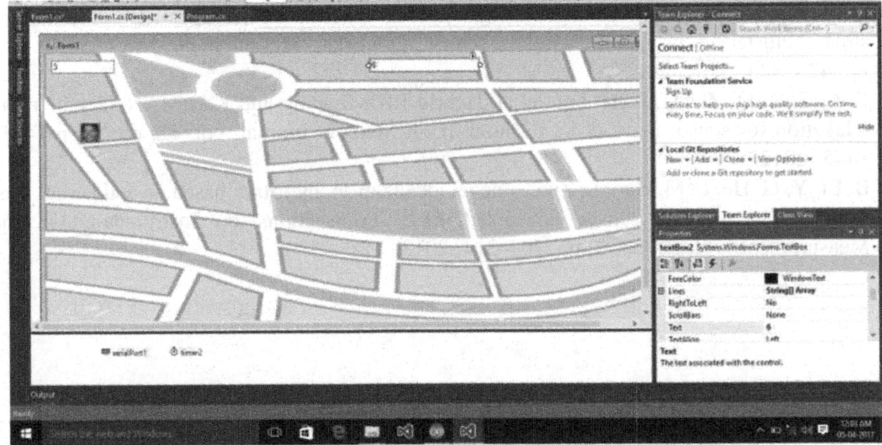

Fig. 8. RSSI distance node

References

1. A. VEMPATY, O. OZDEMIR, K. AGARWAK, et al, Localization in Wireless Sensor Networks: Byzantines and Mitigation Techniques, IEEE Trans. on Signal Process., vol. 61, no. 6, pp. 1495–1508, Jun., 2013.
2. X.F. Zhang, Q.M. Cui, Y.L. Shi, X.F. Tao, Robust Localization algorithm for solving neighbor position ambiguity, Electronics Letter, Vol. 49, no. 17, pg.1106–1107, Aug 2013.
3. J. A. Jiang, X. Y. ZHENG, Y. F. Chen, et al, A Distributed RSS-Based Localiza-tion Using a Dynamic Circle Expanding Mechanism, IEEE Sensors Journal, vol. 13, no. 10, pp. 3754–3766, October, 2013.
4. Q. H. Luo, Y. Peng, X. Y. Peng Ei Saddik, Uncertain Data Clustering-based Distance Estimation in Wireless Sensor Networks, Sensors, vol. 14, no. 4, pp. 6584–6605, April., 2014.

5. M. F. SNOUSSI, H. Richard, Interval-Based Localization using RSSI Comparison in MANETs, IEEE Trans. Aerospace and Electronic Systems., vol. 47, no. 4, pp. 2897–2910, Apr., 2011.
6. P. Bahl, V. N. Padmanabhan, RADAR: An In-Building RF-based User Loca-tion and Tracking System, in IEEE INFOCOM 2000, Tel-Aviv, Israel, March, 2000, pp. 775–784.
7. W. Kim, J. Park, J. Yoo, H.J. Kim, C.J. Park, Target Localization using Ensemble Sup-port Vector Regression in Wireless Sensor Networks, IEEE Trans. Cybernetics, Vol. 43, no. 4, pg. 1189–1198, Apr. 2013.
8. Y. Ji, S. Bias, S. Pandey, P. Agrawal, ARIADNE: A dynamic indoor signal map construction and localization system, In ACM MobiSys06, Uppsala, Sweden, June, 2006, pp. 151–164.
9. W. L. Zhang, Q. Y. Yin, H. Y. Chen, et al, Distributed Angle Estimation for local-ization in wireless sensor networks, IEEE Transactions on Wireless communications, vol. 12, no. 2, pp. 527–5367, Feb., 2013.
10. P. K. Sahu, E. H. K. Wu, J. Sahoo, DURT: Dual RSSI Trend based localization for wireless sensor networks, IEEE Sensors Journal, vol. 14, no. 3, pp. 3115–3123, Mar. 2013.
11. Y. M. Ji, C. B. Yu, J. M. Wei, et al, Localization bias reduction in wireless sensor networks, IEEE Transactions on Industrial Electronics, vol. 62, no. 5, pp. 3004–3016, May, 2015.
12. Q. H. Luo, X. Z. Yan, J. B. Li, Y. Peng, dynamic distance estimation using uncertain data stream clustering in mobile wireless sensor networks, Measurement, vol. 55, no. 40, pp. 423–433, Sept., 2014.
13. B. C. Seet, Q. Zhang, C. H. Foh, et al, Hybrid RF Mapping and KALMAN altered spring relaxation for sensor network localization, IEEE Sensors Journal, vol. 12, no. 5, pp 1427–1445. 2012.
14. B. Li, Y. G. He, F. M. Guo, L. Zuo, A novel localization algorithm based on and partial least squares for wireless sensor networks, IEEE Transactions on Instrumen-tation and Measurement, vol. 62, no. 2, pp. 304–314, 2013.

Automatic Target Acquisition and Discreet Close Surveillance Using Quad Copter with Raspberry Pi Support

P. Vijayaragavan[1(✉)], R. Ponnusamy[2], and M. Arramuthan[3]

[1] Department of Computer Science and Engineering, Jawaharlal Nehru Technological University, Kakinada 533003, AP, India
vijayaragavanpm@yahoo.co.in

[2] Department of Computer Science and Engineering, CVR College of Engineering, Vastu Nagar, Mangalpalli (V), Ibrahimpatan (M), Hyderabad 501510, Telungana, India
r_ponnusamy@hotmail.com

[3] Department of Information Technology, Perunthalaivar Kamarajar Institute of Engineering and Technology (PKIET), Nedungudi, Karaikudi 609603, Tamil Nadu, India
aramudhan1973@yahoo.com

Abstract. The paper is about making a micro modular drone (MMD) which can be used by armed forces to automatically acquire targets in a wide environment and can maintain discreet close surveillance. A simple quad copter is used to maintain contact with the hostiles. To acquire targets, we use a PI sensor camera which is programmed by Raspberry Pi to automatically acquire targets. For real-time tracking and surveillance, we use a cloud to store coordinates which are continuously sent by the quad copter. A simple application (RM7) is used to acquire data from cloud which can be used by the quick reaction force (QRF) to effectively neutralize threats without taking too much risks and damages. The application is made in a way so that it is compatible with any handheld device and can be used anywhere. Thus, we make a drone which is modular, does not require air eld and much knowledge to operate, and can be used in border with fewer resources.

Keywords: Micro modular drone (MMD) · PI sensor · RM7 · Modular Resources

1 Introduction

Most drones are heavy and large in size, which require resources such as solar cells, maintenance, manpower, and often costly to deploy in border. Most of the time, the borders are not close to air eld which reduces the drones endurance time. To provide a solution to that, we employ a micro modular drone (MMD). The MMD has four basic structural modules. These four modules are aerodynamic and stability augmentation system, target acquisition, cloud data link, and QRF application management, i.e., RM7. The MMD has a lot of advantages over conventional drone due to certain

V. Bhateja et al. (eds.), *Information Systems Design and Intelligent Applications*, Advances in Intelligent Systems and Computing 672,
https://doi.org/10.1007/978-981-10-7512-4_4

characteristics. For combating insurgency in the border, close surveillance has to be maintained 24 $*$ 7. Deploying border staff on regular patrol is inefficient as it is waste of manpower. Other factors would include human error, endurance limits, and of course, great risks of losing life in the border. Instead of placing patrol, a drone can be used to maintain surveillance. A central monitoring team from the control room can control the area from a remote place. If a target does infiltrate through the border, it would be unwise for the QRF to directly confront the intruder, as the intruder maybe armed and may even be hidden in the terrain. The modules of MMD are made in a way to suit the needs of the QRF and central monitoring team in the control room. The systems are made redundant so that even if there is malfunction, the drone may either transfer its function to other MMDs present in the area or can continue to operate with damages. The modules are doubly redundant so that during damages, data and functions can be recovered without any error present in the cloud data link layer [1]. The application RM7 is made versatile. It can be used in any handheld devices such as phone or tablet through which the coordinates of the target can be acquired from the cloud with ease. The data link between the drone and the cloud is secured using an 8-bit encryption system. Ample security is given to the cloud so that, it would be difficult to be hacked and the data inside it cannot be manipulated by the infiltrators. The application can also be combined with auto artillery programs that lie far away to deploy missiles on the targets. The application is extremely useful for the QRF to make effective decisions such as either sending a squad to neutralize the threat or to land smart munitions such as Joint Defense Attack Munitions (JDAM) directly on the hostiles using the coordinates acquired by the MMD [2]. In the following section, the four modules are briefed well in detail. In the final section, the MMD is compared with the conventional drone surveillance to highlight the advantages of MMD.

2 The Basic Modules

The four basic modules are aerodynamic and stability augmentation system, target acquisition, cloud data link, and QRF application management, i.e., RM7. Each of these is designed based on the environment in which the MMD would be functioning. The modules are integral part of the MMD.

A. Aerodynamics and stability augmentation systems

A quad copter has four rotors. To make it airborne, all the four rotors rotate to produce lift in a horizontal direction [3]. The environment it would be facing would be heavy rains, dust, snow, sleet, and of course, enemy fire if detected by the hostiles. For heavy rains and winds, we increased rotor speed and the thrust is applied to counter to the wind direction. Another way for protection against wind is applying translation force by tilting the axis thus maintaining attitude equilibrium [4]. Snow does not possess much problem as the rotor produce torque force to withstand enough snow and air. The mainframe is made of carbon materials which are lightweight and can withstand heavy amount of damage. The drone is RF-controlled [5]. It can work autonomously and can be overridden using RF control if the command center feels the need to. The MMD is made fault-tolerant [6]. Even if one or two rotors completely fail due to enemy fire or due environment fault, an automatic augmentation system tilts the quad copter to maintain attitude equilibrium [5]. The augmentation systems programmed automatically and do not need to be manually engaged during failure (Fig. 1).

Fig. 1. Quad copter in controlled fight despite losing one rotor. Stability augmentation system is based on this model

The payload of the copter consists of:

1. Rashberry pi—300 g
2. Solar panels—50 g
3. 5000 mAh battery—1000 g
4. PI camera—100 g.

Total weight = 1450 g.

The thrust-to-weight ratio is ample enough to make it maneuverable.

The payload is enclosed in a carbon box to protect it against snow, rain, sleet, and enemy fire. However, the MMD cannot protect itself against heavy explosives such as SAMs, anti-aircraft artillery (AAA) guns, and 40-mm bullets.

B. Target acquisition

To acquire targets, we use PI sensor in the Raspberry Pi. The PI sensor captures an area and converts it into digital image. The digital image is processed by the Raspberry Pi. Raspberry Pi has its own processor and RAM to quickly scan for targets [7]. The targets are acquired using color differentiation method. Once a target is acquired, the PI sensor changes from search to tracking mode automatically. While in track mode, preset settings such as distance, altitude are programmed so as discretion is maintained in the hostile area [8]. If the target is lost due to LOS or terrain or signal interference, then the PI sensor goes back to search mode. It searches back in the same last location. While automatically acquiring target, an alert message is already given to the central command. For overriding, the basic controls are replaced by RF commands given by central command. The drone has a built-in GPS system which simultaneously calculates the coordinates [9]. The coordinates and digital image data are multiplexed and sent to the cloud.

C. Cloud data link

The data link from the cloud and drone is made using 3 g or a 4 g connection. If a tower is nearby then 3 g connection is enough, if it is not there then 4 g is available for data transmission [10]. The digital image and coordinates are sent to the cloud. The central command is able to maintain close surveillance to the target by acquiring data from the cloud [11]. The data link is protected by 8-bit encryption. The cloud is configured using cloud tools and thus does not require much maintenance. A central command has systems to which the cloud is connected through a pipe [12]. Real-time surveillance is maintained by the control panel via the cloud (Fig. 2).

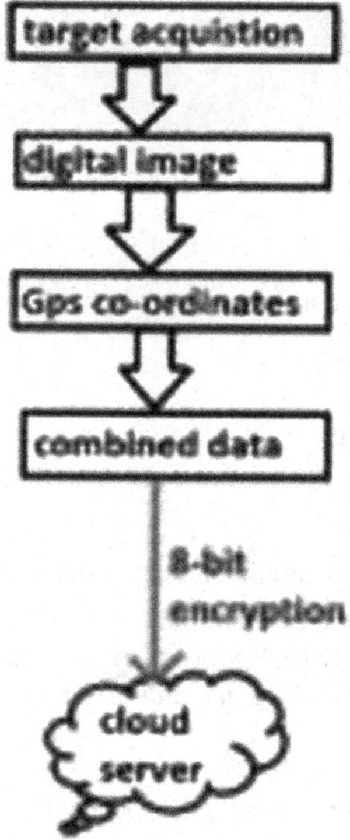

Fig. 2. Data link from drone to the cloud for maintaining real-time surveillance

D. QRF application management RM7

The QRF application, i.e., RM7 is an application that gives real-time coordinates of targets. It is a simple smartphone app that downloads the coordinates from the cloud. The app works in a very simple way. Coordinates from the cloud are plotted on a map application, and thus the location of the targets is known. The app focuses on acquiring target location only. RM7 is designed to work with many different systems. It is extremely versatile as the application is based on Java (Fig. 3).

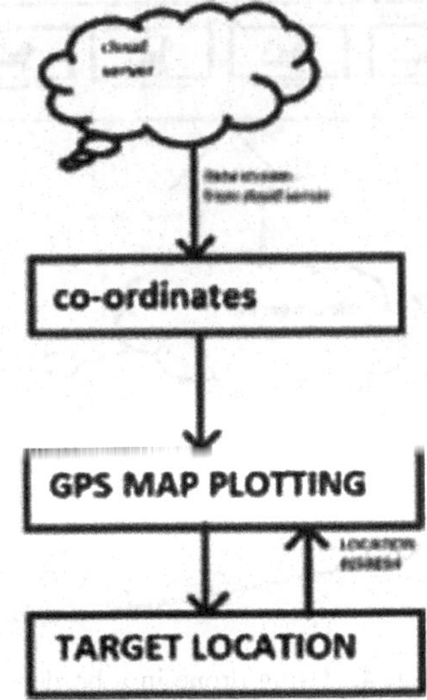

Fig. 3. RM7 application functioning

3 Drone Operation and Functioning

The drone is launched by the border force and sent on a routine patrol mission. A predefined path is already specified toward which the drone scans for target using its PI sensor. The target-found message is given to central command once the intruder has been acquired by the drone [13]. The central command then overrides the control function to RF control to gain control of the drone [14]. The target is tracked, and live stream video of the intruder is acquired via the cloud. The coordinates are obtained in the phone app using the RM7 application.

Some of the characteristics of MMD drone are as follows:

A. Equipment

The equipment required for MMD is very less. They cost less and can be used anywhere. Only cost would involve buying cloud and optimizing it for application purpose.

B. Modular

Every part of the drone is modular. Even if any part or modules get damaged, they can immediately be replaced. The parts cost very less. Turnaround time is very less, and the MMDs can be made combat-ready in a very short time.

C. Accuracy

The data obtained from the drones are extremely accurate as they are taken immediately after coverage. The data link speed depends on the type of connection maintained from cloud to the MMDs (Fig. 4).

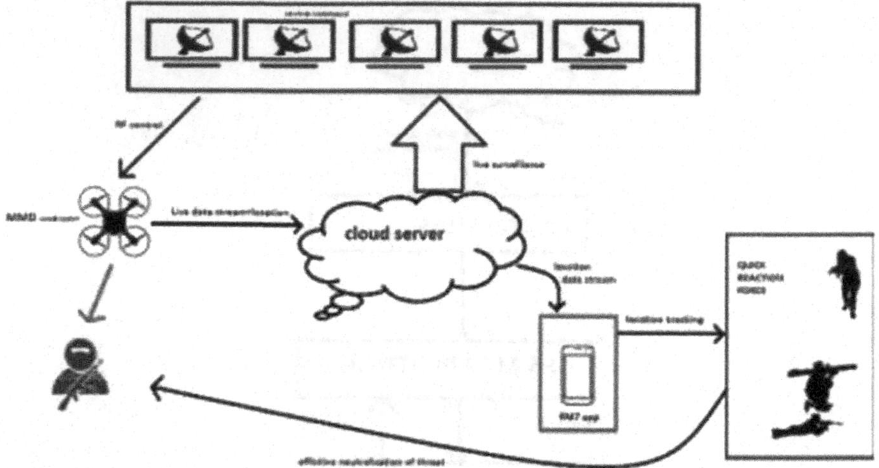

Fig. 4. Using drone in a border

D. Resources

MMD is a drone that operates without fuel. It does not require huge air force bases for taking off and landing. The MMD is solar-powered. It uses only 5000 mAh battery to perform all functions. The running costs of MMD are extremely low. Many MMDs can be inducted to maintain a fleet of drones for surveillance. All of them can be tracked and controlled by a specific set of drone pilots from a remote central command.

4 Comparison of MMDs to Conventional Drones

Drone versus MMDs

5 Conclusion

The paper discusses making a micro modular drone that can be used in many places. Here, we compare with conventional drone for making MMDs application border security area. We see that MMDs have a lot of advantages over conventional drone. The only drawback in a MMD is that it lacks armament.

In future, we hope wide number of MMDs would be deployed in the border which would reduce huge amount of defense spending and would lower the risks of border security force. We hope the RM7 app would help commander to make effective tactical decisions that would lower risks of QRF fighting around the borders (Table 1).

Table 1 Comparision between Micro Modular Drone and Conventional Drone

Characteristics	Micro Modular Drone	Conventional drones
Size	Very small in size	It is very big wings
Airfield	Does not require airfield	Requires airstrip to take off and land
Shootdown chances	The radar cross section is lesser for SAM's system to fire missiles	The RCS is enough for SAMs to acquire it in their radars to send a missile
Cost	Costs around $5000 approx	Costs around 1.8 million
Fuel	It runs on solar power maintenance	It runs on solar power and the panels need maintenance
Turnaround time	Turnaround time is less	Turnaround time is high
Tracking	Close surveillance is possible	Close surveillance is not possible
Pilot requirement	Does not require much experience	Flying experience is required
Altitude limits	Can fly at any altitude to track	It has to fly high to avoid detection. It has to fly in circle over the target area
Application	Application support is available	Application support is available
Camera and sensor	It uses PI sensor and can be replaced easily	It uses close CCD which is very large
Weight	The weight of the MMD is around 2–5 kg	The weight of conventional drone is very high, i.e., in tonnes
Armament	It provides no support for armament	The payload mostly consists of A–G missiles
Discretion	Discretion can be maintained as rotors do not produce much sound	Very high discretion can be maintained. However at lower altitude, detection is very much possible
Airspace requirements	Requires very less airspace	Requires lots of airspaces

Acknowledgements. The MMDs' creation is possible with the module drawn idea. I thank the Aurora project team for formulating the ideas and deploying them.

I thank Mr. Mark Weller (IEEE) for providing research paper on stability and control of quad copter despite the loss of one or two rotors.

References

1. R. Mahony, V. Kumar, and P. Corke, Aerial vehicles: Modeling, estimation, and control of quadrotor, IEEE robotics & automation magazine, vol. 19, no. 3, pp. 20–32, 2012.
2. P. Martin and E. Salaun, The true role of accelerometer feedback in quadrotor control, in IEEE International Conference on Robotics and Automation, 2010, pp. 1623–1629
3. P. Pounds, R. Mahony, P. Hynes, and J. Roberts, Design of a four-rotor aerial robot, in Australasian Conference on Robotics and Automation, vol. 27, 2002, p. 29.
4. A. Chamseddine, Y. Zhang, C. A. Rabbath, C. Join, and D. Theilliol, Flatness-based trajectory planning/replanning for a quadrotor unmanned aerial vehicle, IEEE Transactions on Aerospace and Electronic Systems, vol. 48, no. 4, pp. 2832–2848, 2012.

5. http://yingmachinearena.org/wp-content/publications/2014/mueIEEE14.pdf.
6. D. P. Bertsekas, Dynamic Programming and Optimal Control, Athena Scientific 2007, Vo-II, 2007.
7. Y. Zhang, A. Chamseddine, C. Rabbath, B. Gordon, C.-Y. Su, S. Rakheja, C. Fulford, J. Apkarian, and P. Gosselin, Development of advanced FDD and FTC, techniques with application to an unmanned quadrotor helicopter testbed, Journal of the Franklin Institute, 2013.
8. D. P. Bertsekas, Dynamic Programming and Optimal Control, Vol. II. Athena Scientific, 2007.
9. N. A. Chaturvedi, A. K. Sanyal, and N. H. McClamroch, Rigid-body attitude control, Control Systems, IEEE, vol. 31, no. 3, pp. 3051, 2011.
10. J. Fink, N. Michael, S. Kim, and V. Kumar, Planning and control for cooperative manipulation and transportation with aerial robots, The International Journal of Robotics Research, vol. 30, no. 3, pp. 324–334, 2011.
11. M. Ranjbaran and K. Khorasani, Fault recovery of an under-actuated quadrotor aerial vehicle, in IEEE Conference on Decision and Control. IEEE, pp. 4385–4392, Year-2010.
12. D. Scaramuzza, M. Achtelik, L. Doitsidis, F. Fraundorfer, E. Kosmatopoulos, A. Martinelli, M. Achtelik, M. Chli, S. Chatzichristo, L. Kneip et al. Vision-controlled micro ying robots: from system design to autonomous navigation and mapping in gpsdenied environments. [online] http://robotics.ethz.ch/scaramuzza/DavideScaramuzzales/publications/pdf/IEEERAMsubmitted.pdf, Year-2013.
13. Ravello Darea, Rosela d cruz, Stability of quadcopters using RF, IEEE, 2014.
14. J F. Fraundorfer, L. Heng, D. Honegger, G. H. Lee, L. Meier, P. Tanskanen, and M. Pollefeys, Vision-based autonomous mapping and exploration using a quadrotor MAV, in IEEE/RSJ International Conference on Intelligent Robots and Systems. IEEE, pp. 4557–4564, Year-2012.

Mining Frequent Fuzzy Itemsets Using Node-List

Trinh T. T. Tran[1,2](✉), Giang L. Nguyen[3], Chau N. Truong[4], and Thuan T. Nguyen[1,2]

[1] Graduate University of Science and Technology—VAST, Hanoi, Vietnam
thuytrinh85@gmail.com, nguyentanthuan2008@yahoo.com

[2] Duy Tan University, Da Nang, Vietnam

[3] Institute of Information Technology—VAST, Hanoi, Vietnam
nlgiang@ioit.ac.vn

[4] Danang University of Technology, Da Nang, Vietnam
truongngocchau@yahoo.com

Abstract. Data mining plays an important in knowledge discovery in databases; many types of knowledge and technology have been proposed for data mining. Among them, association rule mining is the problem important not only in data mining task but also in many practical applications in different areas of life. These previous studies mostly focused on showing the transaction data with binary values. However, in real-world applications, transactions also contain uncertain and imprecise data. To solve the above-mentioned problem, fuzzy association rule mining algorithms are developed to handle quantitative data using fuzzy set. In this paper, we present proposed algorithm NFFP, an improved fuzzy version of PPV algorithm for discovering frequent fuzzy itemsets using Node-List structure.

Keywords: Fuzzy frequent itemsets · Fuzzy sets · Node-List · Quantitative databases

1 Introduction

Owing to the speedy development of data, in many enterprises, enormous amount of data have been collected and stored in the database. In the fact, many important business information is hidden in data, so efficient tools are necessary for acquiring useful information. For this reason, data mining which is used to find information and knowledge from the incomplete, unclear data has developed with a variety of techniques. Since being introduced in [1], the technique of association rule mining (ARM) has received interest by the data mining community and a lot of researchers in the world. ARM detects interesting associations, correlations, frequent patterns, and relationships of data values in the transaction databases. One of them, frequent itemset

V. Bhateja et al. (eds.), *Information Systems Design and Intelligent Applications*, Advances in Intelligent Systems and Computing 672,
https://doi.org/10.1007/978-981-10-7512-4_5

mining (FIM), is an elemental task of this research field. The previous studies in [2–4] mostly focused on representing the transaction data of binary value that is only concerned about the presence or absence of these items. When dealing with quantitative data, it is difficult to discover the frequent itemsets with crisp values. To resolve this issue, the approach of fuzzy set has been applied to handle the quantitative databaseQuery.

The earlier algorithms used in mining fuzzy association rules are mostly fuzzy versions of the Apriori algorithm. With this approach, it has to scan database repeatedly to generate a large set of candidates and then check the support of them, so it is usually slow and ineffective in the case of huge database. The algorithm uses the principles of memory dependences as FFP_Growth [5–7] and MFFP_Tree [8] gain benefits over the Apriori approach finding out the frequent fuzzy itemsets without generating candidates. However, these strategies extract recursively frequent fuzzy itemsets from the tree structure throughout the entire process of algorithm; therefore, it tends to require big memories to store the temporal trees. In addition, with this approach, it must access database frequently; this will cause damage to the performance.

In this paper, the proposed algorithm NFFP uses the data structure called Node-List to extract the frequent fuzzy itemsets. In this approach, we only scan database two times: The first time is to convert the crisp value in quantitative database into fuzzy sets. The second time is to calculate the support of each fuzzy region in updated database and build the FPPC_tree. The tree is used to generate PP_code for each node by traversing the tree in pre- and post-order. After finishing its task, FPPC_tree will be deleted and the process of mining the fuzzy itemsets will be executed based on PP_-code, so it can reduce the memory usage requirements.

The remaining part of the paper includes the following: Part 2 reviews the basic concepts and related works, Part 3 presents the proposed algorithm NFFP, and Part 4 states the conclusion.

2 Basic Concepts and Related Work

2.1 Frequent Fuzzy Itemset Mining Problem

Let be a set of items, is the quantitative database of transactions. Each T_q in QD is where TID is a transaction identifier and X is an itemset that contains several items with their quantities and is a minimum support threshold (minsup) and is membership function defined by user.

Definition 1: [9] Fuzzy set

Assume that where each (fuzzy term represented in natural language) is the element in the fuzzy set of *i* and are, respectively, their fuzzy values (defined by the membership function). The fuzzy set is presented as follows:

Definition 2: [9] The support of fuzzy item

The support of the fuzzy item denoted is defined as follows:

Where is the updated fuzzy database after converting quantitative values into fuzzy values.

Definition 3: [9] In this paper we uses τ—norm in fuzzy set such as *a* τ *b* are *min* (*a*, *b*)for finding the fuzzy support value. The support of a fuzzy k-itemset, or, is the minimum of the fuzzy values of the fuzzy items, is:

Lemma 1 [8] If there are n transactions, then we have.

Problem statement

Frequent fuzzy itemset mining (FFIM) is the problem that extracts all frequent fuzzy itemsets as:

Example 1: Table 1 presents the quantitative database (QD) which will be used for illustration in this paper. QD has six transactions with five items indicated; the minsup is set to 30%. Assume that we use the triangular membership function for all items shown in Fig. 1, and quantities are set into one of fuzzy terms: Low, Middle, and High.

Table 1. Quantitative database

TID	Items
1	(I_1:5) (I_3:10) (I_4:2) (I_5:9)
2	(I_1:8) (I_2:2) (I_3:3)
3	(I_2:3) (I_3:9)
4	(I_1:5) (I_2:3) (I_3:10) (I_5:3)
5	(I_1:7) (I_3:9) (I_4:3)
6	(I_2:2) (I_3:8) (I_4:3)

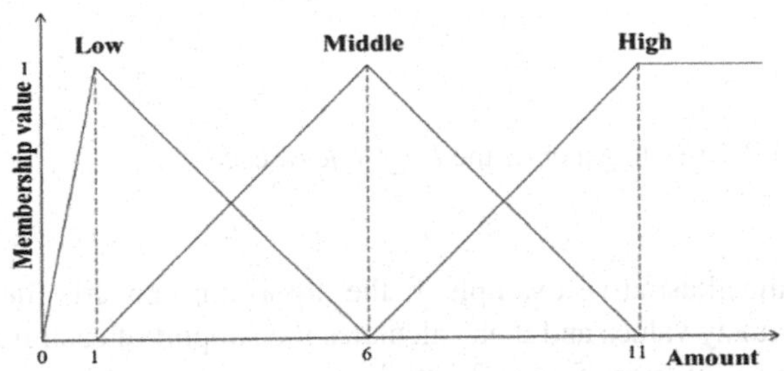

Fig. 1. Triangular membership function for all items

2.2 FPPC_Tree

The FPPC_tree is built on integrating of fuzzy concepts and PPC_tree [2] like approach.

Definition 4: FPPC_tree is a tree with a structure contains one root and a set of item prefix subtree; each node in a subtree comprises five elements: *f_fuzzy_term, f_support, children_list, fpre_code,* and *fpost_code,* which are the fuzzy items, the support of *f_fuzzy_term* in this node, list of child nodes, the pre- and post-order codes of the node, respectively.

The FPPC_tree_building algorithm is presented in Algorithm 1.

Algorithm 1 (FPPC_tree_building)

Input: A quantitative database QD, membership function, and minsup .

Output: A FPPC-tree (*FTr*), the set of frequent fuzzy 1-itemset ().

1. Convert the value of each in transaction into a fuzzy set as in (1).

2. Scan updated database containing fuzzy values to compute the support of each fuzzy item in the transaction as in (2).

3. Check if , put the in . That is .

4. Sort the frequent fuzzy items in in support decreasing order.

5. If , delete from all

6. Generate the root of the FPPC_tree and label it as "null"

7. **for** each in {

8. Sort the remaining fuzzy items in support decreasing order.

9. Insert the fuzzy items into FFPC_tree (this process is similar to MFFP_tree [14])

10. }

11. Traverses FPPC-tree to generate the *PP_Code* of each node.

Example 2: In the illustrative example 1, the algorithm converts the value () of all items in QD into fuzzy values and then calculates the support of each fuzzy items in the updated database as given in Tables 2 and 3.

Then, the algorithm discards all fuzzy items, whose supports are less than the minsup, and sorts the rest of fuzzy items in support decreasing order (Table 4). The updated transactions with frequent fuzzy1-itemsets are presented in Table 5.

Next, the algorithm inserts the fuzzy items in each updated transaction into FPPC_tree. Lastly, the algorithm walks through the FPPC_tree (Fig. 2) to generate PP_code of each node.

Table 2. Fuzzy database after converting quantitative values into fuzzy values

TID	Fuzzy items
1	$\left(\frac{0.2}{I_2,Low}+\frac{0.8}{I_1,Middle}\right),\left(\frac{0.2}{I_3,Middle}+\frac{0.8}{I_3,High}\right),\left(\frac{0.8}{I_4,Low}+\frac{0.2}{I_4,Middle}\right),\left(\frac{0.4}{I_5,Middle}+\frac{0.6}{I_5,High}\right)$
2	$\left(\frac{0.6}{I_2,Middle}+\frac{0.4}{I_1,High}\right),\left(\frac{0.8}{I_2,Low}+\frac{0.2}{I_2,Middle}\right),\left(\frac{0.6}{I_3,Low}+\frac{0.4}{I_3,Middle}\right)$
3	$\left(\frac{0.6}{I_2,Low}+\frac{0.4}{I_2,Middle}\right),\left(\frac{0.4}{I_3,Middle}+\frac{0.6}{I_3,High}\right)$
4	$\left(\frac{0.2}{I_2,Low}+\frac{0.8}{I_1,Middle}\right),\left(\frac{0.6}{I_2,Low}+\frac{0.4}{I_2,Middle}\right),\left(\frac{0.2}{I_3,Middle}+\frac{0.8}{I_3,High}\right),\left(\frac{0.6}{I_5,Low}+\frac{0.4}{I_5,Middle}\right)$
5	$\left(\frac{0.8}{I_2,Middle}+\frac{0.2}{I_1,High}\right),\left(\frac{0.4}{I_3,Middle}+\frac{0.6}{I_3,High}\right),\left(\frac{0.6}{I_4,Low}+\frac{0.4}{I_4,Middle}\right)$
6	$\left(\frac{0.8}{I_2,Low}+\frac{0.2}{I_2,Middle}\right),\left(\frac{0.6}{I_3,Middle}+\frac{0.4}{I_3,High}\right),\left(\frac{0.6}{I_4,Low}+\frac{0.4}{I_4,Middle}\right)$

Table 3. Support of fuzzy items

Fuzzy item	Support	Fuzzy item	Support	Fuzzy item	Support
I_1.Low	0.4	I_3.Low	0.6	I_5.Low	0.6
I_1.Middle	3.0	I_3.Middle	2.2	I_5.Middle	0.8
I_1.High	0.6	I_3.High	3.2	I_5.High	0.6
I_2.Low	2.8	I_4.Low	2.0		
I_2.Middle	1.2	I_4.Middle	1.0		
I_2.High	0	I_4.High	0		

Table 4. Resulted frequent fuzzy 1-itemsets based on support calculation

Frequent fuzzy items	Support
I_3.High	3.2
I_1.Middle	3.0
I_2.Low	2.8
I_3.Middle	2.2
I_4.Low	2.0

2.3 Node-List

Zhihong Deng and Zhonghui Wang [4] presented some important definitions of Node-List and some properties associated with it. We summarize and apply these concepts to our fuzzy values as shown below.

Table 5. Updated transactions with frequent fuzzy1-itemsets

TID	Fuzzy items
1	,
2	,
3	,,
4	,
5	,
6	,,

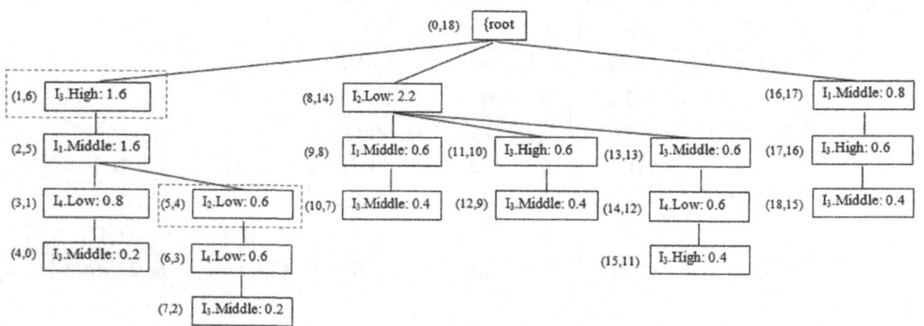

Fig. 2. FPPC_tree created from QD' with δ = 30%

Definition 5: The PP_code of each node in a FPPC_tree is C_i = .

Example 3: The highlighted nodes N_1 and N_2 in Fig. 2 have the PP_codes: C_1 = < 1, 6, 1.6 > and C_2 = < 5, 4, 0.6>

Property 1: Node N_1 is called an ancestry of node N_2, if and only if and

Definition 6: Let C_1 be and C_2 be, C_1 is an ancestry of C_2 if and only if $pr_1 < pr_2$ and $po_1 > po_2$.

Definition 7: The Node-List of a fuzzy item is a chain of PP_codes of nodes representing the fuzzy items in the FPPC_tree. The PP_codes are sorted in *fpre_code* increasing order. Each PP_code is indicated by.

Example 4: The Node-List of I_1.*Middle* includes three nodes: Figure 3 shows the Node-List of fuzzy frequent 1-itemsets in the above example.

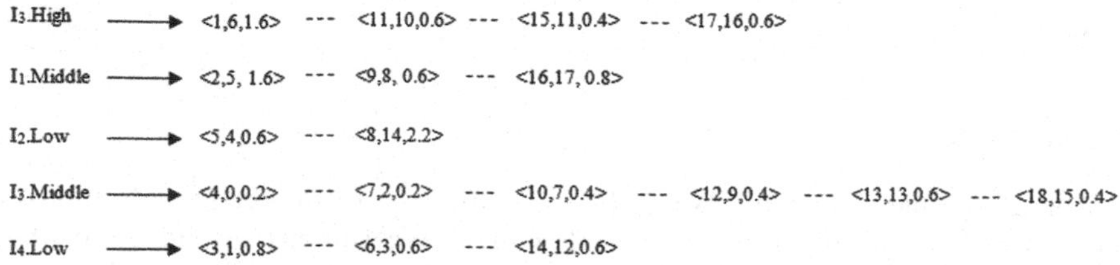

Fig. 3. Node-List of frequent fuzzy items

After building the FPPC_tree, we traverse the FPPC_tree in *fpre_code* order. For each node N_i, we insert into the Node-List of the item specified by N_i and gain the Node-List of each fuzzy frequent 1-itemset. The construction of the Node-List of all frequent fuzzy 1-itemsets is shown in Algorithm 2.

Algorithm 2: FNodelist_Generation

Input: FPPC-tree (FTr) and F_1 (the list of frequent fuzzy 1-itemsets)

Output: NL_1 (the set of Node list of frequent fuzzy 1-itemsets)

1. Assume that $NL_1[k]$ is the node list of an kth element in F_1.

2. Traverse all nodes N_i in the FPPC_tree by

3. **if** (N_i.*f_fuzzy_term* = F_1[k].*f_fuzzy_term*)

4. insert into NL_1[k];

5. **return** ;

Property 2: Given and, if N_1. *fpre_code* < N_2. *fpre_code*, then N_1. *fpost_code* < N_2. *fpost_code*.

Proof: If, by traversing across the tree, N_1 is traversed before N_2. But N_1 cannot be the ancestry of N_2 because they have the same *f_fuzzy_term*, so N_1 must locate the left side of N_2 on the tree. In the contrary traversal, we also traverse the tree from left to right, so.

Property 3: If the Node-List of fuzzy item is, then the support of a fuzzy item is.

Example 5: The Node-List of I_1. *Middle* includes, and the support of I_1. *Middle* is *3.0.* All fuzzy items are true to this property.

Definition 8: (relation) Given two frequent fuzzy items and (. when if and only if is before in.

3 Proposed NFFP Algorithm for Mining Frequent Fuzzy Itemsets

In this section, we introduce the NFFP algorithm and present two main contributions in this paper: (1) improve Node-List intersection of frequent fuzzy k-itemsets from [4] and (2) propose the new approach using Node-List structure to find the frequent fuzzy itemsets; this approach helps lessen the memory utilization requirements, since storing FPPC_tree during frequent fuzzy itemset mining process is not needed.

3.1 Node-List Intersection

In paper [4], the author proposed the Code_intersection method for determining the Node-List of k-itemsets which is only consistent with crisp values. In this case, PPC_tree is constructed according to itemsets arranged in their descending order of frequency. If $i_1$is before $i_2$in L_1then all nodes of i_1 are always ancestry of nodes of i_2. So author only check whether Node-List of is ancestry of Node-List of i_2. This work proposes an improved method that offers the Node-List intersection of frequent fuzzy k-itemsets to obtain the Node-List of frequent fuzzy (k + 1) itemsets.

Definition 9: (Node-List of frequent fuzzy 2-itemsets).
Suppose that two different fuzzy frequent items in which (is before in), and their Node-List respectively are: and. For any PP_code and. If and has the ancestry—descendant relationship of PP_Code, then insert the descending PP_Code into the Node-List of.

Example 6: The Node-List of two itemsets I_1.Middle and I_2.Low is shown in Fig. 4

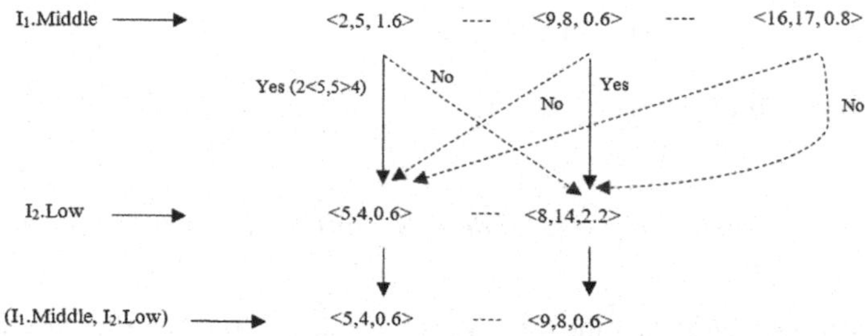

Fig. 4. Node-List of two items (I_1.Middle, I_2.Low)

Based on Definition 9, we generalize to the concept of the Node-List of a frequent fuzzy k-itemsets ($k \geq 3$) in below.

Definition 10: (Node-List of frequent fuzzy k-itemsets): Let be an frequent fuzzy itemsets (), the Node-List of be, the Node-List of be. For any PP_code and. If and has the ancestry—descendant relationship of PP_Code, then insert the descendant PP_Code into the Node-List of.

Example 7: We have known that the Node-List of (I_1.Middle, I_2.Low) and (I_1.Middle, I_4.Low) is < 5, 4, 0.6 >, < 9, 8, 0.6 > and < 3, 1, 0.8 >, < 6, 3, 0.6 >. The Node-List of (I_1.Middle, I_2.Low, I_4.Low) is shown in Fig. 5.

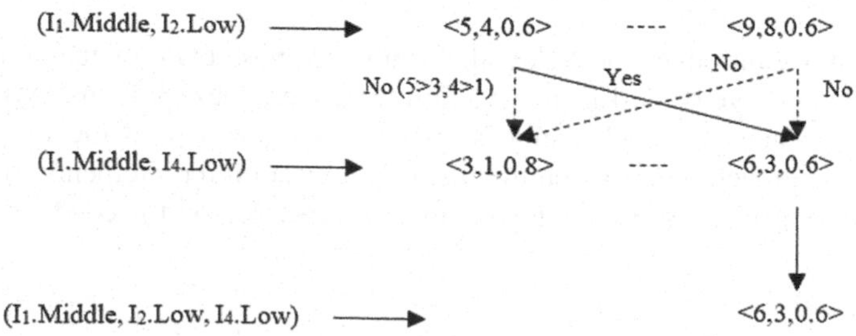

Fig. 5. Node-List of two items (I_1.Middle, I_2.Low, I_4.Low)

Property 4: For any Node-List of fuzzy k-itemsets, represented by, the support of P is.

Proof: With k = 1: In accordance with Property 3, the conclusion is true.
With: The designed algorithm sorts the fuzzy items in a in their supports decreasing order in each transaction for building the FPPC_tree structure and uses - norm in fuzzy set such as (as the intersection operator so we can obtain the minimum values between fuzzy regions from support of the child nodes. Given any fuzzy k-itemsets have the Node-List then according to Definitions 9 and 10, is a child node. So the support of P following the Definition 4 is.

The improved Node-List intersection method is presented in Algorithm 3.

Algorithm 3: FNodelist_Intersection

Input: and where , are the node list of two frequent fuzzy k-itemsets.

Output: , the node list of frequent fuzzy (k+1) itemsets.

```
for
        for
            if {
                if
                    insert into ;
            }
            else {
                        if
                            insert into ;
                }
        }
}
return
```

3.2 Mining Frequent Fuzzy Itemsets Using Node-List

According to the main idea of process sequence, the proposed NFFP algorithm includes four main steps: (1) construct FPPC tree and identify all fuzzy frequent 1-itemsets, (2) construct Node-List of fuzzy frequent 1-itemsets, (3) perform Node-List intersection of k-1-itemsets to build Node-List of k-itemsets, and (4) mining all frequent fuzzy itemsets.

Algorithm 4: NFFP

Input: A quantitative database QD, membership function, and minsup .

Output: The complete set of frequent fuzzy itemsets (FFIs)

1. Call *FPPC_tree_building (D,)* to generate FPPC – tree (R), F_1
2. Call *FNodelist_Generation (R, F_1)*;
3. Call *Find _FFI (, F_1, NL_1)*;
4. Return FFIs

Procedure Find_FFI (, F_1, NL_1)

```
1. For   do begin
2.       For all , where , {
3.
4.             If all k-1 subsets of P are in   {
5.                  = FNodelist_Intersection (, );
6.                 Calculate ;                 // Use Property 4
7.                           If   {
8.                  ;
9.                  ;
10.              }
11.           }
12.     }
13.        Delete
14.  }
15.
```

Example 8: Table 6 shows the final results of the frequent fuzzy itemsets in illustrative example.

Table 6. Final results of frequent fuzzy itemsets

Fuzzy frequent 1-itemsets	Support	Fuzzy frequent 2-itemsets	Support
I_3.High	3.2	{I_3.High, I_1.Middle}	2.2
I_1.Middle	3.0	{I_3.High, I_4.Low}	1.8
I_2.Low	2.8		
I_3.Middle	2.2		
I_4.Low	2.0		

4 Conclusion

In this paper, the proposed algorithm NFFP uses a FPPC_tree to store the quantitative database with descending order membership values. Based on the FPPC_tree, we detect the Node-List of each frequent fuzzy item. Then, NFFP algorithm obtains Node-Lists of the frequent fuzzy (k + 1) itemsets by intersecting the Node-Lists of frequent fuzzy k-itemsets and then extracts the fuzzy frequent (k + 1)-itemsets. The advantage of this algorithm is that the FPPC_tree is used to generate pre–post code for each node to get the Node-List of each frequent fuzzy item, and after that, it will be deleted so it can reduce the memory usage requirements. However, suppose that the data used in this paper is static, in the real applications, data may be changed with time and automatically inserted into database. In the future works, we will attempt to solve this problem of fuzzy data mining.

References

1. Agrawal R., Srikant R.: Fast algorithms for mining association rules. In proceedings of 20th International Conference on Very Large Databases, Santiago, Chile (1994).
2. Deng Zhihong, Wang ZhongHui and Jiang JiaJian: A new algorithm for fast mining frequent itemsets using N-lists. Science China Information Sciences, Vol. 55 No.9, (2012) 2008–2030.
3. Farah Hanna AL-Zawaidah, Yosef Hasan Jbara: An improved algorithm for mining association rules in large database. World of Computer Science and Information Technology Journal (WCSIT), Vol. 1, No. 7, (2011) 311–316.
4. Zhihong Deng, Zhonghui Wang: A new fast method for mining frequent patterns. International Journal of Computational Intelligence Systems, Vol.3, No. 6, (2010) 733–744.
5. De Cock, M., Cornelis, C., Kerre, E.E.: Fuzzy Association Rules A Two-Sided Approach. In: FIP, (2003) 385–390.
6. Reza Sheibani, Amir Ebrahimzadeh, Member, IAUM: An Algorithm For Mining Fuzzy Association Rules. Proceedings of the International MultiConference of Engineers and Computer Scientists 2008 Vol I, (2008) 486–490.
7. Yan, P., Chen, G., Cornelis, C., De Cock, M., Kerre, E.E.: Mining Positive and Negative Fuzzy Association Rules. In: KES, Springer, (2004) 270–276.

8. Tzung-Pei Hong, Chun-Wei Lin and Tsung-Ching Lin: The MFFP-tree fuzzy mining algorithm to discover complete linguistic frequent itemsets. Computational Intelligence, Vol.0, No.0, (2012).
9. Chun-Wei Lin, Philippe Fournier-Viger, Tzung-Pei Hong: A fast algorithm for mining fuzzy frequent itemsets. Journal of Intelligent and Fuzzy Systems (2015).

A New Approach for Query Processing and Optimization in Fuzzy Object-Oriented Database

Thuan T. Nguyen[1,2](✉), Ban V. Doan[3], Chau N. Truong[4], and Trinh T. T. Tran[1,2]

[1] Graduate University of Science and Technology—Vietnam Academy of Science and Technology (VAST), Da Nang, Vietnam
[illegible]@yahoo.com, thuytrinh85@gmail.com

[2] Duy Tan University, Da Nang, Vietnam

[3] Institute of Information Technology—VAST, Da Nang, Vietnam
doanban@gmail.com

[4] Da Nang University of Technology, Da Nang, Vietnam
truongngocchau@yahoo.com

Abstract. For enhancing the efficiency of processing users' queries, all database management systems (DBMSs) must conduct query preprocessing, or query optimizing. This paper proposes a new model for the Fuzzy Object-Oriented DBMS (FOO-DBMS), which optimizes the query statements and processes the data before returning back to users based on fuzzy object algebra and equivalent transformation rules. Discussions on this model are also presented with computation and analysis.

1 Introduction

The fuzzy object-oriented database (FOO-DB) model contains unclear, uncertain, and inaccurate information. Hence, in order to deal with the ever-increasing and complex data, it is always time- and resource-consuming. Typically, a database management system works well if its components perform efficient processing (less processing time, less resources). This paper proposes a new approach for optimizing and processing query based on the fuzzy object algebra (FOA) expression and the rules of equivalent transformation. This approach is the main idea of our proposed model which is analyzed and discussed later in the paper.

A few propositions have been introduced on query translation and optimization [1]. Similar research in OODB has also been presented in [2, 3] and is still very useful for the development of the concerning part of our study which is used to translate user queries from FOQL to fuzzy object algebra.

V. Bhateja et al. (eds.), *Information Systems Design and Intelligent Applications*, Advances in Intelligent Systems and Computing 672,
https://doi.org/10.1007/978-981-10-7512-4_6

The remaining part of the paper includes: part 2 that introduces about fuzzy algebraic operations in FOODB, part 3 proposes a new model for the fuzzy object-oriented DBMS (FOO-DBMS), which optimizes the query statements and processes the data before returning back to users based on fuzzy object algebra and equivalent transformation rules, the analysis on several experiments using the proposed algorithm shows better performance of query processing, which proves the efficiency enhancement of our method are presented in part 4, and the conclusion is stated in part 5.

2 Algebraic Operations in FOODB

2.1 Association Operators

A new fuzzy class $\left(\widetilde{FC}\right)$ can be created by combining two existing fuzzy classes. Depending on the relationships between the attributes of the combination classes, there are six types of binary combination and two types of unary operations which are: (1) fuzzy union $(\widetilde{\cup})$, (2) fuzzy intersection $(\widetilde{\cap})$, (3) fuzzy join $(\widetilde{\bowtie})$, (4) fuzzy cross-product/Cartesian $(\widetilde{\times})$, (5) fuzzy difference $(\simeq)$, (6) fuzzy division $(\widetilde{\div})$, (7) fuzzy projection $\left(\widetilde{\Pi}\right)$, and (8) fuzzy selection $(\widetilde{\sigma})$ and are denoted in [4]. Let $\widetilde{FC_1}$ and $\widetilde{FC_2}$ be any two fuzzy class with sets of attributes $FA(fA_1)$and$FA(fA_2)$, respectively. Thus, a new fuzzy class $\widetilde{FC}$ is created from the combination of two classes $\widetilde{FC_1}\&\widetilde{FC_2}$ and is defined as follows:

1. $\widetilde{FC} = \widetilde{FC_1} \,\widetilde{\cup}\, \widetilde{FC_2}$, *if* $FA'\left(\widetilde{FC_1}\right) = FA'\left(\widetilde{FC_2}\right)$
2. $\widetilde{FC} = \widetilde{FC_1} \,\widetilde{\cap}\, \widetilde{FC_2}$, *if* $FA'\left(\widetilde{FC_1}\right) = FA'\left(\widetilde{FC_2}\right)$
3. $\widetilde{FC} = \widetilde{FC_1} \,\widetilde{\bowtie}\, \widetilde{FC_2}$, *if* $FA'\left(\widetilde{FC_1}\right) \cap FA'\left(\widetilde{FC_2}\right)$
4. $\widetilde{FC} = \widetilde{FC_1} \,\widetilde{\times}\, \widetilde{FC_2}$, *if* $FA'\left(\widetilde{FC_1}\right) \cap FA'\left(\widetilde{FC_2}\right) = \emptyset$ *and* $FA'\left(\widetilde{FC_1}\right) \neq FA'\left(\widetilde{FC_2}\right)$
5. $\widetilde{FC} = \widetilde{FC_1} \simeq \widetilde{FC_2}$, *if* $FA'\left(\widetilde{FC_1}\right) = FA'\left(\widetilde{FC_2}\right)$
6. $\widetilde{FC} = \widetilde{FC_1} \,\widetilde{\div}\, \widetilde{FC_2}$, *if* $FA'\left(\widetilde{FC_1}\right) = FA'\left(\widetilde{FC_2}\right)$
7. $\widetilde{FC''} = \widetilde{\Pi}_{sub}\left(\widetilde{FC'}\right)$
8. $\widetilde{FC''} = \widetilde{\sigma}_{sub}\left(\widetilde{FC'}\right)$

In which, $FA'\left(\widetilde{FC}_1\right)\&FAtr'\left(\widetilde{FC}_2\right)$ are achieved from $FA\left(\widetilde{FC}_1\right)\&FA\left(\widetilde{FC}_2\right)$, by removing the degree of their respective fuzzy attributes $FA\left(\widetilde{FC}_1\right)\&FA\left(\widetilde{FC}_2\right)$. The $\mu_{\widetilde{FC}}$ represents the membership degree of attributes which belong to class $\widetilde{FC}$. Assume we have an object $\mu_{\widetilde{FC}}(o)$ of $\widetilde{FC}$, and $\mu_{\widetilde{FC}}(o)$ is used to represent the value of o on $\mu_{\widetilde{FC}}$. Let fA_i be the attribute of class $\widetilde{FC}$ · $\mu_{\widetilde{FC}}(o)(A_i)$ is the fuzzy value of object $\mu_{\widetilde{FC}}(o)$ on FA_i. If $FA = \{fA_i, fA_i, \ldots, fA_m\}$ is a set of fuzzy attributes, then $\mu_{\widetilde{FC}}(o)(FA)$ stands for all values of $\mu_{\widetilde{FC}}(o)$ object on attributes in FA. More generally, $\mu_{\widetilde{FC}}(o)\left(FC'\right)$ represents all values of o on attributes of $\mu_{\widetilde{FC}}(o)$ on attributes of $\widetilde{FC}$. The formal conditions of fuzzy association operations are shown as follows:

1. Fuzzy union $(\widetilde{\cup})$: Through fuzzy union, new class $\widetilde{FC}$ is made up of two fuzzy classes $\widetilde{FC}_1\&\widetilde{FC}_2$, requires $FA'(fA_1) = FA'(fA_2)$, it means that all related attributes of $\widetilde{FC}_1\&\widetilde{FC}_2$ have the identical weights. Suppose, the new $\widetilde{FC}$ is the fuzzy union of $\widetilde{FC}_1\&\widetilde{FC}_2$. Then the $\mu_{\widetilde{FC}}(o)$ of $\widetilde{FC}$ are created from three types of $\mu_{\widetilde{FC}}(o)$. First, two types of $\mu_{\widetilde{FC}}(o)$ are derived from the component class (for example, class $\widetilde{FC}_1$) and do not coincide with any $\mu_{\widetilde{FC}}(o)$ in the rest of component class (e.g., $\widetilde{FC}_2$) that satisfies the given threshold. Let δ be the threshold.

$$\widetilde{FC} = \widetilde{FC}_1 \,\widetilde{\cup}\, \widetilde{FC}_2$$
$$= \left\{ \mu_{\widetilde{FC}}(o) \;\middle|\; \begin{array}{c} \left(\forall \mu_{\widetilde{FC}}(o)''\right)\begin{pmatrix} \mu_{\widetilde{FC}}(o)'' \in \widetilde{FC}_2 \wedge \mu_{\widetilde{FC}}(o) \in \widetilde{FC}_1 \wedge \\ SE\left(\mu_{\widetilde{FC}}(o)\left(\widetilde{FC}_1\right), \mu_{\widetilde{FC}}(o)''\left(\widetilde{FC}_2\right)\right) < \delta \end{pmatrix} \vee \\ \left(\forall o \mu_{\widetilde{FC}}(o)'\right)\begin{pmatrix} \mu_{\widetilde{FC}}(o)' \in \widetilde{FC}_1 \wedge \mu_{\widetilde{FC}}(o) \in \widetilde{FC}_2 \wedge \\ SE\left(\mu_{\widetilde{FC}}(o)\left(\widetilde{FC}_2\right), \mu_{\widetilde{FC}}(o)'\left(\widetilde{FC}_1\right)\right) < \delta \end{pmatrix} \vee \\ \begin{pmatrix} \left(\exists \mu_{\widetilde{FC}}(o)'\right)\left(\exists \mu_{\widetilde{FC}}(o)''\right)\left(\mu_{\widetilde{FC}}(o)' \in \widetilde{FC}_1 \wedge \mu_{\widetilde{FC}}(o)'' \in \widetilde{FC}_2\right) \wedge \\ SE\left(\mu_{\widetilde{FC}}(o)'\left(\widetilde{FC}_1\right), o\mu_{\widetilde{FC}}(o)''\left(\widetilde{FC}_2\right)\right) \geq \\ \delta \wedge \mu_{\widetilde{FC}}(o) = merge\left(\mu_{\widetilde{FC}}(o)', \mu_{\widetilde{FC}}(o)''\right) \end{pmatrix} \end{array} \right\} \tag{1}$$

In which, $\mu_{\widetilde{FC}}(o)', \mu_{\widetilde{FC}}(o)''$ are two $\mu_{\widetilde{FC}}(o)$ of $\widetilde{FC}$, and $\mu_{\widetilde{FC}}(o) = merge\left(\mu_{\widetilde{FC}}(o)', \mu_{\widetilde{FC}}(o)''\right)$. Then $\mu_{\widetilde{FC}}(o)\left(\widetilde{FC}\right) = \mu_{\widetilde{FC}}(o)'\left(\widetilde{FC}\right)$ or $\mu_{\widetilde{FC}}(o)\left(\widetilde{FC}\right) = \mu_{\widetilde{FC}}(o)''\left(\widetilde{FC}\right)$ and $\mu_{\widetilde{FC}}(o) = max\left(\mu_{\widetilde{FC}_1}(o)', \mu_{\widetilde{FC}_2}(o)''\right)$.

2. Fuzzy intersection $\left(\widetilde{\cap}\right)$: Through fuzzy intersection, new class $\widetilde{FC}$ is made up of two fuzzy classes $\widetilde{FC}_1$ & $\widetilde{FC}_2$, and requests $FA'(fA_1) = FA'(fA_2)$. Let δ be the threshold.

$$\widetilde{FC} = \widetilde{FC}_1 \widetilde{\cap} \widetilde{FC}_2 = \left\{ \mu_{\widetilde{FC}}(o) \middle| \begin{array}{c} \left(\forall \mu_{\widetilde{FC}}(o)''\right)\left(\begin{array}{c} \mu_{\widetilde{FC}}(o)'' \in \widetilde{FC}_2 \vee \mu_{\widetilde{FC}}(o) \in \widetilde{FC}_1 \wedge \\ SE\left(\mu_{\widetilde{FC}}(o)\left(\widetilde{FC}_1\right), \mu_{\widetilde{FC}}(o)''\left(\widetilde{FC}_2\right)\right) < \delta \end{array}\right) \wedge \\ \left(\forall \mu_{\widetilde{FC}}(o)'\right)\left(\begin{array}{c} \mu_{\widetilde{FC}}(o)' \in \widetilde{FC}_1 \vee \mu_{\widetilde{FC}}(o) \in \widetilde{FC}_2 \vee \\ SE\left(\mu_{\widetilde{FC}}(o)\left(\widetilde{FC}_2\right), \mu_{\widetilde{FC}}(o)'\left(\widetilde{FC}_1\right)\right) < \delta \end{array}\right) \wedge \\ \left(\begin{array}{c} \left(\exists \mu_{\widetilde{FC}}(o)'\right)\left(\exists \mu_{\widetilde{FC}}(o)''\right)\left(\begin{array}{c} \mu_{\widetilde{FC}}(o)' \in \widetilde{FC}_1 \vee \mu_{\widetilde{FC}}(o)'' \in \widetilde{FC}_2 \vee \\ SE\left(\mu_{\widetilde{FC}}(o)'\left(\widetilde{FC}_1\right), \mu_{\widetilde{FC}}(o)''\left(\widetilde{FC}_2\right)\right) \end{array}\right) \geq \\ \delta \wedge \mu_{\widetilde{FC}}(o) = minimize\left(\mu_{\widetilde{FC}}(o)', \mu_{\widetilde{FC}}(o)''\right) \end{array}\right) \end{array} \right\} \tag{2}$$

In which, $\mu_{\widetilde{FC}}(o)', \mu_{\widetilde{FC}}(o)''$ are two $\mu_{\widetilde{FC}}(o)$ of $\widetilde{FC}$ and $\mu_{\widetilde{FC}}(o) = merge\left(\mu_{\widetilde{FC}}(o)', \mu_{\widetilde{FC}}(o)''\right)$. We have $\mu_{\widetilde{FC}}(o)\left(\widetilde{FC}\right) = \mu_{\widetilde{FC}}(o)'\left(\widetilde{FC}\right)$ or $\mu_{\widetilde{FC}}(o)\left(\widetilde{FC}\right) = \mu_{\widetilde{FC}}(o)''\left(\widetilde{FC}\right)$ and $\mu_{\widetilde{FC}}(o) = minimize\left(\mu_{\widetilde{FC}_1}(o)', \mu_{\widetilde{FC}_2}(o)''\right)$.

3. Fuzzy join ($\widetilde{\bowtie}$): Through fuzzy join, new class $\widetilde{FC}$ is made up of two fuzzy classes $\widetilde{FC}_1$&$\widetilde{FC}_2$, in which $FA'\left(\widetilde{FC}_1\right) \cap FA'\left(\widetilde{FC}_2\right) \neq \emptyset$ *and* $FA'\left(\widetilde{FC}_1\right) \neq FA'\left(\widetilde{FC}_2\right)$. $FA'\left(\widetilde{FC}_1\right) \cup \left(FA'\left(\widetilde{FC}_2\right) - \left(FA'\left(\widetilde{FC}_1\right) \cap FA'\left(\widetilde{FC}_2\right)\right)\right)$ are attributes for $\widetilde{FC}$ and the membership degree of the attributes. $\mu_{\widetilde{FC}}(o)$ of $\widetilde{FC}$ are made up of a combination of $\mu_{\widetilde{FC}}(o)$ in $\widetilde{FC}_1$&$\widetilde{FC}_2$ that are equivalent in semantics on $FA'\left(\widetilde{FC}_1\right) \cap FA'\left(\widetilde{FC}_2\right)$ under a given threshold. Let δ be the threshold.

$$\begin{aligned} &\widetilde{FC} = \widetilde{FC}_1 \widetilde{\bowtie} \widetilde{FC}_2 \\ &= \left\{ \mu_{\widetilde{FC}}(o) \middle| \begin{array}{c} \left(\exists \mu_{\widetilde{FC}}(o)'\right)\left(\exists \mu_{\widetilde{FC}}(o)''\right)\mu_{\widetilde{FC}}(o)' \in \widetilde{FC}_1 \wedge \mu_{\widetilde{FC}}(o)''\widetilde{FC}_2 \wedge \\ \left(\begin{array}{c} SE\left(\begin{array}{c} \mu_{\widetilde{FC}}(o)'\left(FA'\left(\widetilde{FC}_1\right) \cap FA'\left(\widetilde{FC}_2\right)\right), \\ \mu_{\widetilde{FC}}(o)''\left(FA'\left(\widetilde{FC}_1\right) \cap FA'\left(\widetilde{FC}_2\right)\right) \end{array}\right) \geq \delta \wedge \mu_{\widetilde{FC}}(o)\left(FA'\left(\widetilde{FC}_1\right)\right) = \\ \mu_{\widetilde{FC}}(o)'\left(\widetilde{FC}_1\right) \wedge \mu_{\widetilde{FC}}(o)\left(FA'\left(\widetilde{FC}_2\right) - \left(FA'\left(\widetilde{FC}_1\right) \cap FA'\left(\widetilde{FC}_2\right)\right)\right) = \\ \mu_{\widetilde{FC}}(o)''\left(FA'\left(\widetilde{FC}_2\right) - \left(FA'\left(\widetilde{FC}_1\right) \cap FA'\left(\widetilde{FC}_2\right)\right)\right) \wedge \mu_{\widetilde{FC}}(o) = \\ op\left(\left(\mu_{\widetilde{FC}_1}(o)', \mu_{\widetilde{FC}_2}(o)''\right)\right) \end{array}\right) \end{array} \right\} \end{aligned} \tag{3}$$

In which, operations of op are not defined. In most cases, $op\left(\mu_{\widetilde{FC}_1}(o)', \mu_{\widetilde{FC}_2}(o)''\right)$ either $min\left(\mu_{\widetilde{FC}_1}(o)', \mu_{\widetilde{FC}_2}(o)''\right)$ or $\mu_{\widetilde{FC}_1}(o)' \times \mu_{\widetilde{FC}_2}(o)'$.

4. Fuzzy cross-product/Cartesian ($\widetilde{\times}$): Through fuzzy product, new class $\widetilde{FC}$ is made up of two fuzzy classes $\widetilde{FC}_1 \& \widetilde{FC}_2$ and membership degree attribute.

$$\widetilde{FC} = \widetilde{FC}_1 \,\widetilde{\times}\, \widetilde{FC}_2 = \left\{ \mu_{\widetilde{FC}}(o) \middle| \begin{array}{c} \left(\forall \mu_{\widetilde{FC}}(o)'\right)\left(\forall \mu_{\widetilde{FC}}(o)''\right) \mu_{\widetilde{FC}}(o)' \in \widetilde{FC}_1 \wedge \mu_{\widetilde{FC}}(o)\left(FA'\left(\widetilde{FC}_1\right)\right) = \\ \left(\begin{array}{c} \mu_{\widetilde{FC}}(o)'\left(\widetilde{FC}_1\right) \wedge \mu_{\widetilde{FC}}(o)\left(FA'\left(\widetilde{FC}_2\right)\right) = \\ \mu_{\widetilde{FC}}(o)''\left(\widetilde{FC}_1\right) \wedge \mu_{\widetilde{FC}}(o) = op\left(\mu_{\widetilde{FC}_1}(o)', \mu_{\widetilde{FC}_2}(o)''\right) \end{array} \right) \end{array} \right\} \tag{4}$$

In which, operations of op are not defined. In most cases, $op\left(\mu_{\widetilde{FC}_1}(o)', \mu_{\widetilde{FC}_2}(o)''\right)$ either $min\left(\mu_{\widetilde{FC}_1}(o)', \mu_{\widetilde{FC}_2}(o)''\right)$ or $\mu_{\widetilde{FC}_1}(o)' \times \mu_{\widetilde{FC}_2}(o)'$.

5. Fuzzy difference ($\simeq$): The fuzzy difference of $\widetilde{FC}_1 \& \widetilde{FC}_2$ requests $FA'(fA_1) = FA'(fA_2)$, which pulls all the properties in $\widetilde{FC}_1 \& \widetilde{FC}_2$ with the same weight numbers. For new class, $\widetilde{FC}$ is the fuzzy subtraction of $\widetilde{FC}_1 \& \widetilde{FC}_2$. Then, the objects of $\widetilde{FC}$ are made up of two types of objects. The first type of object is derived from class $\widetilde{FC}_1$ and does not appear in $\widetilde{FC}_2$, satisfying the given threshold. The second type of object obtained by moving duplicate objects in $\widetilde{FC}_1$ satisfies the given threshold. Let δ be the threshold.

$$\widetilde{FC} = \widetilde{FC_1} \simeq \widetilde{FC_2} = \left\{ \mu_{\widetilde{FC}}(o) \middle| \begin{array}{c} \left(\forall \mu_{\widetilde{FC}}(o)''\right) \left(\begin{array}{c} \mu_{\widetilde{FC}}(o)'' \in \widetilde{FC}_2 \wedge \mu_{\widetilde{FC}}(o) \subset \widetilde{FC}_1 \wedge \\ SE\left(\mu_{\widetilde{FC}}(o)\left(\widetilde{FC}_1\right), \mu_{\widetilde{FC}}(o)''\left(\widetilde{FC}_2\right)\right) < \delta \end{array} \right) \vee \\ \left(\begin{array}{c} \left(\exists \mu_{\widetilde{FC}}(o)'\right)\left(\exists \mu_{\widetilde{FC}}(o)''\right)\left(\mu_{\widetilde{FC}}(o)' \in \widetilde{FC}_1 \wedge \mu_{\widetilde{FC}}(o)'' \in \widetilde{FC}_2\right) \wedge \\ SE\left(\mu_{\widetilde{FC}}(o)'\left(\widetilde{FC}_1\right), \mu_{\widetilde{FC}}(o)''\left(\widetilde{FC}_2\right)\right) \geq \delta \wedge \mu_{\widetilde{FC}}(o) = \\ remove\left(\mu_{\widetilde{FC}}(o)', \mu_{\widetilde{FC}}(o)''\right) \end{array} \right) \end{array} \right\} \tag{5}$$

6. Fuzzy division ($\widetilde{\div}$): The fuzzy division is written as ($\widetilde{\div}$), to extract data tuples of a relational relation with all remaining relations.

$$\widetilde{FC} = \widetilde{FC}_1 \tilde{\div} \widetilde{FC}_2 = \left\{\mu_{\widetilde{FC}}(o)[fA_1, \ldots, fA_n] : \mu_{\widetilde{FC}}(o) \in \widetilde{FC}_1 \wedge \forall \widetilde{FC}_2 \in \widetilde{FC}_2\left(\left(\mu_{\widetilde{FC}}(o)\left[fA_1, \ldots, fA_n \cup \widetilde{FC}_2\right) \in \widetilde{FC}_1\right)\right\} \quad (6)$$

In which, the $\tilde{\div}$ of $\widetilde{FC}_1$ *and* $\widetilde{FC}_2$ requires $FA'(fA_1) = FA'(fA_2)$, and $\widetilde{FC}$ is composed with attribute $[fA_1, \ldots, fA_n]$ and $\mu_{\widetilde{FC}}(o)[fA_1, \ldots, fA_n]$ is the limitation of $\mu_{\widetilde{FC}}(o)$.

7. Fuzzy projection $\left(\tilde{\Pi}\right)$: The new $\widetilde{FC}''$ composed with membership degree attribute is created from the fuzzy projection on the attribute subset sub.

$$\begin{aligned}\widetilde{FC}'' &= \tilde{\Pi}_{sub}\left(\widetilde{FC}'\right) \\ &= \left\{\mu_{\widetilde{FC}}(o) | \left(\forall \mu_{\widetilde{FC}}(o)'\right)\left(\mu_{\widetilde{FC}}(o)' \in \widetilde{FC}' \wedge \mu_{\widetilde{FC}}(o) = \cup_{\Im} \mu_{\widetilde{FC}}(o)'\right)\right\}\end{aligned} \quad (7)$$

Here, the removal of redundant $\mu_{\widetilde{FC}}(o)$ in the fuzzy set of $\mu_{\widetilde{FC}}(o)'$ is used by the $\cup_{\Im} \mu_{\widetilde{FC}}(o)'$ operation.

8. Fuzzy selection $(\tilde{\sigma})$: The new $\widetilde{FC}''$ composed with membership degree attribute is created from the fuzzy selection on the attribute subset sub.

$$\widetilde{FC}'' = \tilde{\sigma}_{sub}\left(\widetilde{FC}'\right) = \left\{\mu_{\widetilde{FC}}(o) | \mu_{\widetilde{FC}}(o) \in \widetilde{FC}', \varphi\left(\mu_{\widetilde{FC}}(o)\right)\right\} \quad (8)$$

Here, the removal of redundant $\mu_{\widetilde{FC}}(o)$ in the fuzzy set of $\mu_{\widetilde{FC}}(o)'$ is used by the $\varphi o'$ operation.

2.2 Structured Fuzzy Object Query Language

Structure of a fuzzy object-oriented query consisting of three clauses:

SELECT < attribute list > FROM < Class WITH threshold Class WITH threshold>

WHERE < query condition WITH threshold > .

Example for a $\widetilde{FC}$ OldSalesPersons as follows:

CLASS OldSalesPersons WITH DEGREE OF 1.0 INHERITS SalesPersons WITH DEGREE OF 1.0 ATTRIBUTES ID: TYPE OF string WITH DEGREE OF 1.0

Name: TYPE OF string WITH DEGREE OF 1.0 Age: FUZZY DOMAIN {very young, young, old, very old}:

TYPE OF integer WITH DEGREE OF 1.0 Sex: FUZZY DOMAIN {male, female}: TYPE OF character WITH DEGREE OF 1.0 DOB: FUZZY DOMAIN {day, month, year}:

TYPE OF integer WITH DEGREE OF 1.0 Membership_Attribute name

WEIGHT w (ID) = 0.1w (Name) = 0.1w (Age) = 0.9w (Sex) = 0.1w (DOB) = 0.6

METHODS END

A query based on the class is issued by using:

SELECT Name

FROM OldSalesPerson as O, SalesPersons as S WITH 0.6

WHERE O.foid O. foid AND O. Age = 'very old' WITH 0.7.

3 Proposed Fuzzy Query Processing Architectures and Optimization

To unify the execution steps of a query in a certain process, we propose a query processing architecture for FOODB, represented in Fig. 1.

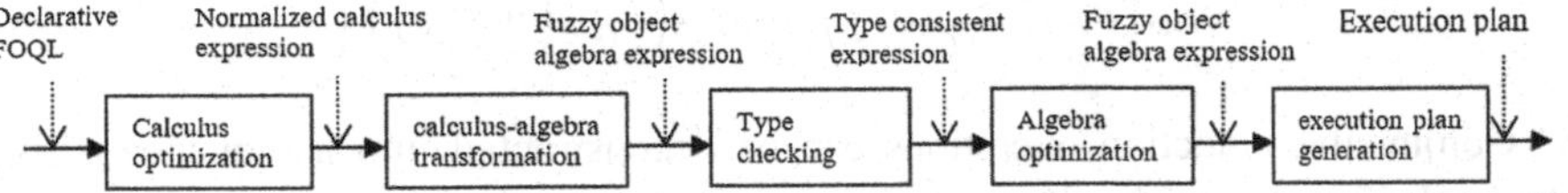

Fig. 1. Fuzzy query processing architecture

When the user submits a FOQL, it is first parsed by the FOODBSs, which verify the syntax and type correctness of the FOQL. Being a declarative language, FOQL does not suggest concrete ways to evaluate its queries. Therefore, a parsed FOQL has to be converted into a fuzzy object algebra expression, which can be evaluated directly using the algorithms. A typical FOQL query such as

SELECT Name FROM OldSalesPerson as O, SalesPersons as S WITH 0.6 WHERE O.foid = S.foid AND O. Age = 'very old' WITH 0.7.

Is normally translated into the following fuzzy object algebraic expression:

$$\tilde{\pi}_{Name}\left(\begin{array}{c}\tilde{\sigma}_{\text{OldSalesPersons}.foid=SalesPersons.foid \wedge \text{OldSalesPersons}.Age='very\ old'}\\ (\text{OldSalesPersons} \,\tilde{\bowtie}\, SalesPersons)\end{array}\right)$$

The above architecture is described through the steps as follows. First, users submit a fuzzy query that they do not need knowledge of fuzzy objects. Next, the system receives the fuzzy query, performs the removal of duplicate predicates, and applies the identities and rewriting. Next, the system performs the conversion of the query to the fuzzy algebraic expression. Next, the system expresses this fuzzy algebraic expression a nested expression in the form of a fuzzy algebraic tree. Next in the fuzzy query processing process is to apply the same rewrite rules equivalent to preserving the equivalent of algebraic expressions for fuzzy–fuzzy objects. Finally, an implementation plan that takes into account the implementation of fuzzy objects is generated from the optimized fuzzy algebraic expression.

3.1 Equivalent Transformation Rules

Assume that $\mu_{\widetilde{FC}}(o), \mu_{\widetilde{FC}_1}(o), \mu_{\widetilde{FC}_2}(o), \mu_{\widetilde{FC}_3}(o)$ are the set for fuzzy object: e, f, g, h are algebraic expressions, operations $op \in \{union, diff\}$. These rules apply only on the fuzzy object operations, math sets, set operations, and multiset operations (bag). On signs, we only use math notations in a form [4] operations can be setup with changes in a number of different models.

R1. Selection operations are commutative:

$$\sigma_{\lambda t.g}\left(\sigma_{\lambda s.f}\left(\mu_{\widetilde{FC}}(o)\right)\right) = \sigma_{\lambda s.f}\left(\sigma_{\lambda s.g}\left(\mu_{\widetilde{FC}}(o)\right)\right)$$

R2. Conjunctive selection operations can be deconstructed into a sequence of individual selections; cascade of σ:

$$\sigma_{\lambda s.(f \wedge g \wedge \ldots h)}\left(\mu_{\widetilde{FC}}(o)\right) = \sigma_{\lambda s.f}\left(\sigma_{\lambda t.g}\left(\cdots\left(\sigma_{\lambda u.h}\left(\mu_{\widetilde{FC}}(o)\right)\right)\cdots\right)\right)$$

R3. Only the final operations in a sequence of projection operations is needed, the others can be omitted; cascade of Π:

$$\Pi_{(a_1,\,\ldots,\,a_n)}\left(\Pi_{(b_1,\,\ldots,\,b_n)}\left(\mu_{\widetilde{FC}}(o)\right)\right) = \Pi(a_1,\,\ldots,\,a_n)\left(\mu_{\widetilde{FC}}(o)\right)|\{a_1,\ldots,a_n\} \subset \{b_1,\ldots,b_n\}$$

R4. Permutation selection and projection:

$$\sigma_{\lambda s.e}\left(\Pi_{(a_1,\ldots,a_n)}\left(\mu_{\widetilde{FC}}(o)\right)\right) = \Pi_{(a_1,\ldots,a_n)}\left(\sigma_{\lambda s.e}\left(\mu_{\widetilde{FC}}(o)\right)\right)$$

R5. Permutation and a projection over union, on a set/multiset:

$$\Pi_{(a_1,\ldots,a_n)}\left(\mu_{\widetilde{FC}_1}(o) op\, \mu_{\widetilde{FC}_2}(o)\right) = \Pi_{(a_1,\ldots,a_n)}\mu_{\widetilde{FC}_1}(o)\; op\, \Pi_{(a_1,\ldots,a_n)}\mu_{\widetilde{FC}_2}(o)$$

R6. The selection operation distributes over the union, and difference, on a set/multiset

$$\sigma_{\lambda s \cdot f}\left(\mu_{\widetilde{FC}_1}(o) op\, \mu_{\widetilde{FC}_2}(o)\right)$$
$$= \sigma_{\lambda s \cdot f}\left(\mu_{\widetilde{FC}_1}(o)\right) op\, \mu_{\widetilde{FC}_2}(o),\ \textit{if f is related to}\ \mu_{\widetilde{FC}_1}(o)$$

Generality:

$$\sigma_{\lambda s \cdot (f \wedge g \wedge h}\left(\mu_{\widetilde{FC}_1}(o) op\, \mu_{\widetilde{FC}_2}(o)\right)$$
$$= \sigma_{\lambda u \cdot h}\left(\sigma_{\lambda s \cdot f}\left(\mu_{\widetilde{FC}_1}(o)\right) op\, \sigma_{\lambda t \cdot g}\left(\mu_{\widetilde{FC}_2}(o)\right)\right),\ \textit{if f is related to}\ \mu_{\widetilde{FC}_1}(o),$$
$$\textit{g is related to}\ \mu_{\widetilde{FC}_2}(o)\ \textit{and}\ \text{h related to both}\ \mu_{\widetilde{FC}_1}(o)\ \text{and}\ \mu_{\widetilde{FC}_2}(o)$$

R7. Permutation between selection operation and apply operation: if conditions only contain attributes selected by the operation returns apply:

$$apply_{\lambda_{\mu s \cdot e}}\left(\sigma_{\lambda t \cdot f}\left(\mu_{\widetilde{FC}}(o)\right)\right) = \sigma_{\lambda t \cdot f}\left(apply_{\lambda_{\mu s \cdot e}}\left(\mu_{\widetilde{FC}}(o)\right)\right)$$

R8. Permutation between flat and apply on set and multiset: suppose that $\mu_{\widetilde{FC}}(o)$ is an instance of a class and x is a complex set of attributes of the class:

$$flat\left(apply_{\lambda s \cdot \left(apply_{\lambda t \cdot e}\left(\Pi_{(X)}\left(\Pi_V\left(\mu_{\widetilde{FC}}(o)\right)\right)\right)\right)}\left(\mu_{\widetilde{FC}}(o)\right)\right)$$
$$= apply_{\lambda t \cdot e}\left(flat\left(apply_{\lambda s \cdot \Pi(X)}\left(\Pi V\left(\mu_{\widetilde{FC}}(o)\right)\right)\left(\mu_{\widetilde{FC}}(o)\right)\right)\right)$$

R9. Set union is associative:

$$\left(\mu_{\widetilde{FC}_1}(o)\ union\ \mu_{\widetilde{FC}_2}(o)\right) union \mu_{\widetilde{FC}_3}(o) \mu_{\widetilde{FC}_1}(o)\ union\left(\mu_{\widetilde{FC}_2}(o)\ union \mu_{\widetilde{FC}_3}(o)\right)$$

R10. The inheritance laws to allow the selection and apply: if $\widetilde{FC}_2$ is a subclass of $\widetilde{FC}_1$, instance of $\mu_{\widetilde{FC}_2}(o)$ is a subset of instance of $\mu_{\widetilde{FC}_1}(o)$.

$$\sigma_{\lambda s \cdot f}\left(\mu_{\widetilde{FC}_1}(o)\right) union \sigma_{\lambda s \cdot f}\left(\mu_{\widetilde{FC}_1}(o)\right) = \sigma_{\lambda s \cdot f}\left(\mu_{\widetilde{FC}_1}(o)\right) apply_{\lambda s \cdot e}\left(\mu_{\widetilde{FC}_1}(o)\right)$$
$$union\ apply_{\lambda s \cdot e}\left(\mu_{\widetilde{FC}_2}(o)\right) = apply_{\lambda s \cdot e}\left(\mu_{\widetilde{FC}_1}(o)\right)$$

3.2 FOQL to Fuzzy Object Algebra Translation

The transformation equivalence between FOQL queries and fuzzy object algebra.

Definition. If E is a fuzzy object algebraic expression and Q fuzzy query object is FOQL together define sets of fuzzy object, we say Q represent E and the opposite; we call E equivalent to Q. Symbol $E \approx Q$.

Equal representation between the query language and algebra FOQL fuzzy object is expressed through two theorems 1 and 2 as follows:

Theorems 1. Every algebraic expression is fuzzy object represented by the object query in FOQL.

Theorems 2. Every fuzzy object in FOQL queries are represented by algebraic expressions fuzzy object

Thus, rewrite a given query into algebraic expressions with algebraic set objects are equivalent. The algebraic expressions can be estimated with different abatement costs. So theoretically, we wanted to find an algebraic expression equivalent to a query so that it can achieve a plan for more effective enforcement. However, in the solution installed, because the number of queries equivalent too large, we only need a subset of this query. Therefore, in order to find other similar queries, we will need a set of rules to transform the equivalent algebraic expressions. So, we wanted to prove that the transformation preserved on a basis equivalent algebraic fuzzy objects that may be acceptable. Some transformation rules is presented [4].

3.3 Heuristic Optimization Based on Algebraic Equivalences

1. Search space and transformation rules.

The most important advantage of processing and optimizing fuzzy algebra is that through the algebraic expression of the object, we can use algebraic properties such as transformation, distribution. Therefore, each fuzzy query has the number of different equivalent expressions that depend on the input of the query from the user's request. These expressions are corresponding to the results they created, but different from their costs. However, fuzzy query optimizers modify fuzzy query expressions by using algebraic transformation rules to achieve the same results at the possible lowest cost. The transformation rules depend a lot on specific objects, as they are determined properly for each object algebra and their combinations.

2. Search algorithm

Heuristic

1. The parser of a high-level fuzzy query creates an initial internal representation;
2. Apply heuristics rules to optimize the internal representation.

3. A fuzzy query execution plan is generated to execute groups of operations based on the access paths available on the files involved in the fuzzy query.
   The main heuristic is to apply first the operations that reduce the size of intermediate results.
   For example, apply fuzzy SELECT and fuzzy PROJECT operations before applying the fuzzy JOIN, or other binary operations.

Outline of a heuristic fuzzy object algebraic optimization algorithm:

1. Using rule R2, break up any select operations with conjunctive conditions into a cascade of select operations.
2. Using inheritance laws for projection (R3), the selection and allows apply (R10) combination of projection, select a projection and a selection.
3. For each selection, use the law (R4, R6, R7, R10) "pushed" to allow the selection components to classes or "through" connection nodes and allows creation group.
4. For each projection (objects, sets, sets), use legislation (R3, R4, R5) to projection move down as far as possible. If the projected attributes include all the attributes of the expression, we remove that projection.
5. Using the law (R8, R9, R10) on the object class, to remove duplicate elements in the object class; move allows flattened (flat), lets remove duplicates in multiple files (bagtoset) ahead of the group or connection operations.
6. Creating a sequence of steps for estimating change in an order every star team for no group is evaluated; its subgroups.

3.4 Fuzzy Query Execution Plans

After translating the fuzzy query into a fuzzy algebra expression, the query processor passes the expression to the query optimizer, generates different execution plans, or a combination of operators.

There are some of algebraic transformations that are performed in the query optimizer to create equivalent (rational) query plans. By removing ($\widetilde{\times}$) and Push ($\widetilde{\sigma}$).

1. Execution plan

The query processor converts the query to an equivalent fuzzy algebra expression for the input query and forward it to the query optimizer. The first fuzzy algebraic object created by the query processor involves in the fuzzy Cartesian product called execution plan execution.

Example 1: returns the list of names of old salespersons whose Age are 'very old.'
In FOQL, it can be represented as:
SELECT Name FROM OldSalesPerson as O, SalesPersons as S WITH 0.6
WHERE O.foid = S.foid AND O. Age = 'very old' WITH 0.7.

In Fuzzy object Algebra above FOQL statement is represented as:

$$\tilde{\pi}_{Name}\left(\tilde{\sigma}_{\text{OldSalesPersons}.\boldsymbol{foid}=salesPersons.\boldsymbol{foid}\wedge\text{OldSalesPersons}.\boldsymbol{Age}='\boldsymbol{very\ old}'}\left(SalesPersons\ \tilde{\times}\ \text{OldSalesPersons}\right)\right)$$

The above expression is represented by the algebraic tree as Fig. 2:

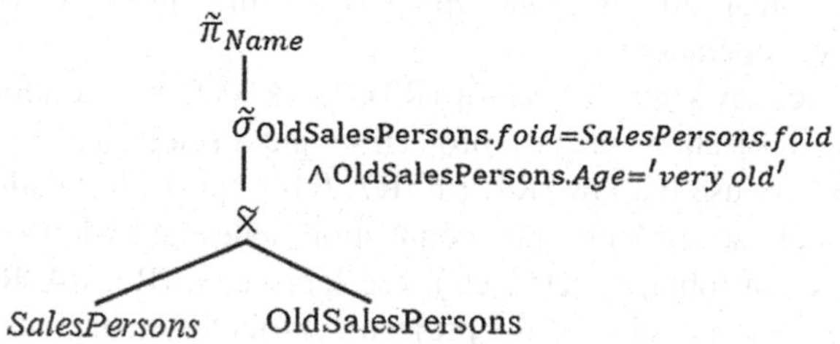

Fig. 2. Implementation plan concern in $(\tilde{\times})$

2. Elimination of $(\tilde{\times})$

The $(\tilde{\times})$ operations can be combined with $(\tilde{\boldsymbol{\sigma}})$ operations (and sometimes, with fuzzy projection operations) which use data from both relations to form joins. After replacing $(\tilde{\times})$ with $(\tilde{\bowtie})$, fuzzy object algebra for the query given in example 1 can be presented as:

$$\tilde{\pi}_{Name}\left(\tilde{\sigma}_{\text{OldSalesPersons}.\boldsymbol{Age}='\boldsymbol{very\ old}'}\left(SalesPersons\ \tilde{\bowtie}_{\text{OldSalesPersons}.\boldsymbol{foid}=SalesPersons.\text{foid}}\left(\text{OldSalesPersons}\right)\right)\right)$$

Figure 3 shows gives the operator tree of the above algebraic expression:

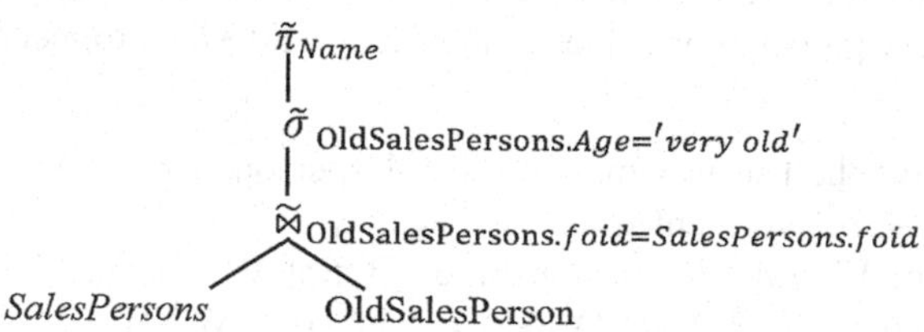

Fig. 3. Implementation plan using $(\tilde{\bowtie})$

3. Push $(\tilde{\sigma})$

By pushing $(\tilde{\sigma})$ operation down the expression tree, we actually reduce the size of relations we need to do before. The fuzzy object algebra for the fuzzy query written in example under push fuzzy selection strategy can be presented as:

$$\tilde{\pi}_{Name}\left(\begin{array}{c} SalesPersons \ \tilde{\bowtie}_{\text{OldSalesPersons}.foid=SalesPersons.\text{foid}} \\ \left(\begin{array}{c}\tilde{\sigma}_{\text{OldSalesPersons}.Age='very\ old'} \\ (\text{OldSalesPersons})\end{array}\right)\end{array}\right)$$

Figure 4 is a description of their above fuzzy object algebra expression:

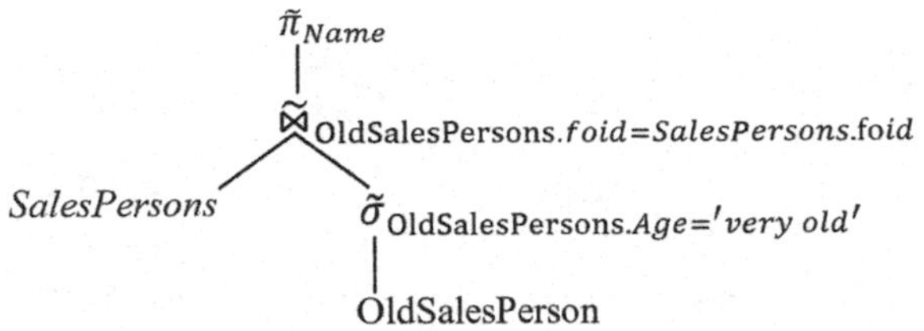

Fig. 4. Pushing $(\tilde{\sigma})$ down the tree

4 Performance Evaluation

To provide preliminary performance evaluation on implementation of query, processing has been proposed based on fuzzy object algebra [4]. We defined the three queries processed condition extract filter data for two cases of single conditions, most conditions and implement them on the same dataset.

The fuzzy query processor first extract filter data for single condition processing cases. Request query processing engine returns all employees age is very old. Such queries are written as follows:

SELECT * FROM OldSalesPersons, SalesPersons WITH 0.6

WHERE AND OldSalesPersons.Age = 'very old' WITH 0.7.

The second query processing extracts filter data for single-case conditions and enables a natural join. Request query processing engine return all employees age is very old. Such queries are written as follows.

SELECT Name FROM OldSalesPerson as O, SalesPersons as S WITH 0.6 WHERE O.FIOD = S.FOID AND O.Age = 'very old' WITH 0.7.

The third query processing the extract filter data for single-case conditions and enable a natural join. After performing the optimization algebra objects. Request query processing engine return all employees age is very old. Such queries are written as follows.

SELECT Name FROM SalesPersons as S inner join OldSalesPersons as O on O. FIOD = S.FOID WITH 0.6 WHERE O.Age = 'very old' WITH 0.7.

From the above experiments, results achieved confirm that the performance of this method is effective. As an example, we evaluate the query according to this approach from the chart the way the query results shown in Fig. 5.

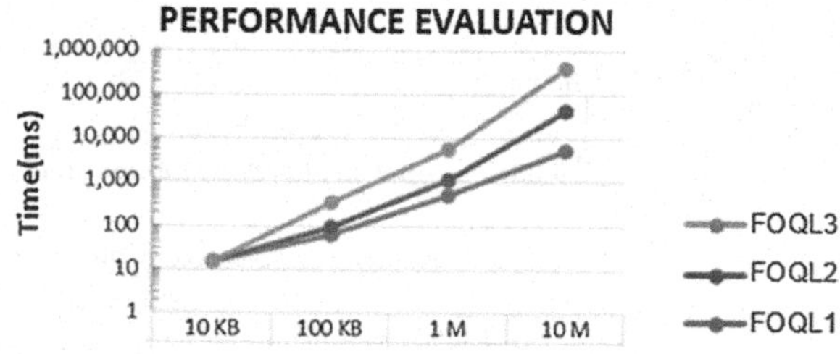

Fig. 5. Query performance

5 Conclusion

This paper presents a new model for optimizing the efficiency of query processing by semantic analyzing and FO algebra transforming. Specifically, we develop a heuristic fuzzy object algebraic optimization algorithm relied on equivalent transformation rules and fuzzy object algebra transformation. Analysis on several experiments using the proposed algorithm shows better performance of query processing, which proves the efficiency enhancement of our method.

References

1. Selee Na.: A Process of Fuzzy Query on New Fuzzy Object Oriented Data Model, In IEEE Tranon Knowledge and Data Engineering. 1(2010), 500–509.
2. Stefano Ceri, Georg Gottlob.: Translating SQL Into Relational Algebra: Optimization, Semantics, and Equivalence of SQL Queries, Software Engineering, IEEE Transactions, vol. SE-11, issue 4, 4(1985). 324–345.

3. XU Silao, HONG Mei.: Translating SQL Into Relational Algebra Tree-Using Object-Oriented Thinking to Obtain Expression Of Relational Algebra, IJEM, vol.2, no.3, (2012). 53–62.
4. Truong Ngoc Chau, Nguyen Tan Thuan.: A Approach New In The Algebra Fuzzy Object, Proceedings of the @ Conference, Viet Nam. 11(2013), 204–209.

A Hybrid Threshold Group Signature Scheme with Distinguished Signing Authority

Dao Tuan Hung[1(✉)], Nguyen Hieu Minh[2], and Nguyen Nam Hai[2]

[1] National Laboratory of Information Security, 34A Tran Phu, Ba Đinh, Hanoi, Vietnam
daotuanhung@gmail.com
[2] Academy of Cryptography Technique, Hanoi, Vietnam
{hieuminhmta, hainamnguyen}@gmail.com

Abstract. This paper proposes a new hybrid threshold group signature scheme with distinguished signing authority to provide all proof of member signing processes in case of dispute internally and internal integrity of multisignature generation process. In practical, the proposed scheme has more controls to an organization by using the threshold mechanism and allowing a limited number of members who can authorize transactions while allowing the group to grow. Moreover, the risk of losing group secret either by an APT attack or by any subset of corrupt members can be eliminated. The proposed scheme is secure based on the hardness of elliptic curve discrete logarithm problem (ECDLP).

1 Introduction

Digital signatures are widely used in many aspects of electronic life. They are designed to be part of security services such as authentication, data integrity and non-repudiation. To date, many schemes such as multisignature [1], group signature [2], traditional signature [3, 4] have been proposed. A multisignature scheme is designed so that a group of users can sign a single document [1]. Multisignatures can be categorized into such as without signing authority or with distinguished signing authority [5, 6].

Group signatures, first introduced by Chaum and van Heyst in [2]. In a group signature scheme, any group member of a given group can sign an electronic document on behalf of the group in an anonymous and unlinkable way. On the other side, anyone only needs the group public key to verify the validity of a group signature. In case of a dispute, only group manager can reveal a member who signed, while other group members neither can identify the identity of the signer nor determine whether multiple signatures are produced by the same group member. To prevent a single corrupt member illegally authorizing a transaction, the threshold signature scheme can be used. A large number of studies were published on (t, n) threshold signature schemes. Schemes at [7–11] based on various hard problems such as RSA system, discrete logarithm (DLP), Chinese Remainder Theorem (CRT), ECDLP. However, schemes at [7, 8, 10] are not secure ones [12–14]. Signature of scheme at [11] cannot be verified by just one verifier and therefore is not practical. At [15] presents an idea of masking group's private key to prevent group members who can collaborate to recover it but

V. Bhateja et al. (eds.), *Information Systems Design and Intelligent Applications*, Advances in Intelligent Systems and Computing 672,
https://doi.org/10.1007/978-981-10-7512-4_7

need a trusted party that use all member private keys to construct group signature and so one can argue that doesn't meet requirement for non-repudiation.

Moreover, previous schemes often assume the number of users being controlled by an adversary less than threshold number [9, 10, 14, 16] in order to keep group's private key safe. However, if the number of members grow, secret shared group keys will be delivered to more and more people. Therefore, there are more chances for group signature scheme to be unsecure. Previous group signature schemes lack mechanisms to maintain a balance between security and scaling of group. Especially, when considering the situation, a company might suffer Advanced Persistent Threat (APT) attacks. This leads to a valid security concern that group secret key might be lost by either corrupt members who can collaborate and recover the key. Another bad situation is many personal computers were targeted and compromised under an APT attack by hackers or state-sponsored APT campaigns that cause the group secret key being steal undetected.

Research at [17] proposed a group signature schemes that have distinguished signing authorities based on the multisignature protocols. Scheme at [17] requires a group manager to collect and issue signature.

This paper proposes a new threshold group signature protocol based on ECDLP that is highly secure, constant length and short signature, distinguished signing authority. The proposed scheme can protect group's private key from being revealed by any set of corrupt signers or hacker's threat. The proposed scheme allows group secret key shares to be kept on limited privilege signers only while allowing new people to join the group without recalculating group public key and easy revocation.

2 Proposed Group Signature Protocol

Currently, cryptographic protocols based on elliptic curves (EC) over finite field have been applied. In the proposed scheme, we use the EC, which order contains a sufficiently large prime divisor q (more than 256 bits) and a point G having order equal to q.

System initialization: Assume that a large group has n privilege signers who can keep company's secret key shares (for example: directorate board) and any number of normal staffs. Only privilege signers have shared company's secret key shares. Group's policy requires that at least t $(t < n)$ privilege signers must join signing process to make a valid group signature. Here are four roles in the proposed scheme:

Group Manager (*GM*): Group manager is a trusted party of the group signature scheme. He creates the secret parameters for the group, calculates and distributes secret key shares to privilege members; add, removes group members, and reveals the identity of the group member in a special case.

Distributed Center (*DC*): special hardened servers of the group that communicate with all signers during signing process. DC calculates some secret parameters needed by signers to create signatures for each transaction. Moreover, all signer's shared signatures are safely stored on DC. Only GM can open DC when needed.

Normal signers: digitally sign on their own work inside large group document.

Privilege signers: digitally sign on their own work inside large group document. With enough t signatures of them, a signature of the group can be generated.

An example of this could be: A complex CAD design files of a construction company need to be internally signed by different people including signatures of t important people such as head of financial office, planning office, directorate board to form a valid group signature. The company wants to hide its internal structure. Head of financial office, planning office, and member of directorate board are privilege signers. In the case design defects are found, the company can traceback and see who is responsible for defect parts of the design.

System preparation phase:
Group manager (GM) chooses two random integers A_0, SE $(1 < A_0 < q,\ 1 < SE < q)$. A_0 is group's private key which is unchanged. SE is another secret number but can be changed to another value when GM decides to redistribute secret key shares. GM calculates secret key shares for n privilege signers following the cryptographic technique of Shamir's perfect secret sharing scheme [18].

$$\begin{aligned} &f(x) = (SE * A_0 + s_1 x + s_2 x^2 + \cdots + s_{t-1} x^{t-1}) \bmod q \\ &(v_i, y_i),\ i = 0,\ 1,\ \ldots,\ n; y_i \equiv f(v_i) \bmod q \end{aligned} \tag{1}$$

Values $s_1, s_2, \ldots, s_{t-1}$ are random integers with $(1 < s_i < q)$. These values are known only by GM. n values $(v_i,\ y_i)$ are secretly sent to n privileges, where y_i is secret shared value of signer i, and v_i is i-th signer's identity. All v_i from (1) are published inside privilege group. Value $A_0 * SE$ can be recovered by any t privilege people or devices who hold secret shares [18], while any number of privileges less than t can reveal nothing about a value $A_0 * SE$:

$$A_0 * SE \equiv \sum_{k=1}^{t} y_k \left[\prod_{i=1, i \neq k}^{t} \frac{-v_i}{v_k - v_i} \right] \bmod q \tag{2}$$

Each privilege who sign will use this equation to calculate of his share during signing process:

$$f_j \equiv y_j \Big(\prod_{i=1, i \neq j}^{t} \frac{-v_i}{v_j - v_i} \Big),\ (j = 1,\ 2,\ \ldots,\ n) \tag{3}$$

GM calculates public key of the group as an EC point: $P_{gm} = A_0 * G$ and another EC point $P_{DC} = (SE - 1) * A_0 * G$. Point P_{gm} is group public key, which can be used by anyone to verify group signatures. GM keeps values $h_{SE} = h(SE)$ and P_{DC} on DC.

Key generation phase:

Each member i-th in the group generates their private key as a random number $k_i (1 < k_i < q)$, and then public key computed as the point $P_i = k_i G$, $i = (1, 2, \ldots, N)$.

Group signature generation phase:

1. Assume N people including t privilege signers together sign the document set $M = m_1 || m_2 || \ldots || m_N$. M is sent to DC prior to the signing process, the DC

calculates values $h_i = H(m_i)$, $z_i = H(h(M)||h_i||P_i||h_{SE})$. Then values (z_i, h_i) are sent to corresponding signer i-th.

2. The DC calculates an EC point as follow:

$$U = h_1 z_1 P_1 + h_2 z_2 P_2 + \cdots + h_N z_N P_N \tag{4}$$

U is the first element of group signature.

3. Each signer i-th chooses a random integer t_i $(1 < t_i < q)$, and calculates $R_i = t_i G$, then sends R_i to DC.
4. DC calculates an EC point:

$$R = R_1 + R_2 + \cdots + R_N \tag{5}$$

and the second element of group signature:

$$e = H(M||x_R||x_U) \tag{6}$$

where x_R and x_U are x-coordinates of EC points R and U, respectively. DC sends the value e to the group members who initiated the protocol.

5. Each signer (privilege or normal) computes their signature share s_i on his assigned part (m_i) of the document differently as follow:
If i-th signer is normal signer he computes:

$$s_i = t_i + h_i k_i z_i e \bmod q \tag{7}$$

If i-th signer is privilege signer he computes an EC point $V_i = f_i e G$ and then s_i:

$$s_i = f_i e + t_i + h_i k_i z_i e \bmod q \tag{8}$$

Normal signer sends (s_i) to DC, privilege sends two values $(s_i,\ V_i)$ to DC.

6. DC verifies s_i of a normal signer (s_i is sent by U_i) if DC received s_i only by checking following equation:

$$R_i = s_i G - z_i h_i e P_i \tag{9}$$

DC verifies s_i of a privilege signer (if DC received two values $(s_i,\ V_i)$) by checking following equation:

$$R_i = s_i G - V_i - z_i h_i e P_i \tag{10}$$

7. If the equation holds for all s_i, DC computes the third, fourth elements of group signature $P_V = eP_{DC}$ and:

$$s = s_1 + s_2 + \cdots + s_N \bmod q \tag{11}$$

Group signature of M is a tuple $(U,\ P_V,\ e,\ s)$, which consists of two EC points and two integer values.

Group Signature verification:

1. Verifier computes the hash of the document $M = m_1||m_2||\cdots||m_N$ as $h = H(M)$.
2. Verifier uses the group public key P_{gm} and the signature $(U,\ P_V,\ e,\ s)$ to compute an EC point $\tilde{R} = sG - P_V - e(U + P_{gm})$, and value $\tilde{e} = H(M||x_{\tilde{R}}||x_U)$. Accept the signature only if $\tilde{e} \equiv e$.

3 Analysis of the Proposed Group Signature Scheme

3.1 Proof of Correctness

1. Share signature verification equation (for privilege signer i-th):

$$\begin{aligned} R_i &= s_iG - V_i - eh_iz_iP = Gf_ie + t_iG - V_i + k_ih_iz_ieG - k_ih_iz_ieG \\ &= V_i + t_iG - V_i + k_ih_iz_ieG - ez_ih_ik_iG = t_iG \equiv R_i \end{aligned}$$

2. Share signature verification equation (for normal signer i-th):

$$\begin{aligned} R_i &= s_iG - eh_iz_iP = t_iG + k_ih_iz_ieG - k_ih_iz_ieG \\ &= t_iG + k_ih_iz_ieG - k_ih_iz_ieG = t_iG \equiv R_i \end{aligned}$$

3. Signature verification equation:

With total N signers including t privilege signers, and equations at (2), (3), (7), (8) we have:

$$\begin{aligned} \tilde{R} &= sG - P_V - e(U + P_{gm}) = sG - e(U + P_{DC} + P_{gm}) \\ &= \left(\sum_{i=1}^{N} s_i\right)G - e\left(P_{gm} + P_{DC} + \sum_{i=1}^{N} h_iz_iP_i\right) \\ &= \left(\sum_{i=1}^{N}(t_i + k_ih_iz_ie) + \sum_{i}^{t} f_ie\right)G - e\left(A_0G + (SE-1)A_0G + \sum_{i=1}^{N} k_ih_iz_iG\right) \\ &= \left(\sum_{i=1}^{t} f_ie + \sum_{i=1}^{N} t_i + \sum_{i=1}^{N} k_ih_iz_ie - eA_0 + e(SE-1)A_0 - \sum_{i=1}^{N} k_ih_iz_ie\right)G \\ &= \left(\sum_{i=1}^{t} f_ie + \sum_{i=1}^{N} t_i - eA_0 + e(SE-1)A_0\right)G = \left(SEeA_0 + \sum_{i=1}^{N} t_i - SEeA_0\right)G \\ &= \sum_{i=1}^{N} t_iG = R \Rightarrow \tilde{e} = H(M||x_{\tilde{R}}||x_U) = H(M||x_R||x_U) = e. \end{aligned}$$

If number of privilege signers who participated less than t or simply absent, above equation does not hold and signers cannot create a valid group signature.

Signature length: Signature of a document is a tuple of two integers and two EC point (U, P_V, e, s), in the case of 128-bit security, q can be chosen with size around 256 bits and signature length will approximately 1536 bits. Compared with group schemes in [19], the proposed scheme has shorter signature length. If choose 80-bit security signature length will approximately 960 bits, with $|q| = 160$ bits).

3.2 Security Analysis

Theorem 1: *Protection of private keys and secret key shares.*

Proof. Normal and privilege signer use private key k_i to sign on a partial message m_i follow (7) and (8) respectively. In both cases two secret random values are used t_i and e. Using adaptive message attack is invalid with the scheme. *Therefore, private keys and secret key shares are protected from other members.*

Theorem 2: *Any subset of t privileged signers out of n to generate a valid signature of the group, but they cannot recover private key of group* A_0.

Proof. If all privileged signers are curious, they can get a value $A_0 * SE$ by following (2). In order to find A_0 from $A_0 * SE$, they have to try each possible guess value of SE' to get A'_0 and check if $A'_0 * G = P_{gm}$, with assumption of Elliptic curve problem is hard, this task is computational infeasible. Compared with previous works [9, 15, 20], the proposed scheme can protect group secret with any number corrupt members. Therefore, the proposed scheme is secure against conspiracy attack [12, 21].

Theorem 3: *Signers cannot bypass DC to create signature.*

Proof. An element of signature $P_V = e * P_{DC}$,which P_{DC} is a private EC point kept on DC only and e (6) is a value related to the document and signer public keys. Without P_{DC}, a signature cannot pass verification process. Often, a group wants to keep records of all transactions. If signers in a group signature scheme can collaborate without a system to keep track of all activities, this situation might cause issues for large group. At DC, a company can place more security protections than it can do with individual personal devices.

Theorem 4: *Suffering an APT attack, company group secret remains safe.*

Proof. During signing and verification process, group secret A_0 is not reconstructed at any step. So, if suffering an APT attack many computers might be compromised, but hackers cannot use memory forensic technique or network sniffer to find A_0. Assume hacker that can get all shares secrets of n privilege signers and following (2), they can recover $A_0 * SE$. They cannot get A_0 directly from $A_0 * SE$ and P_{DC} because of ECDLP problem. Values $h_{SE} = h(SE)$ and P_{DC} are stored on DC, but they are produced of safe hash function and multiplication on elliptic curve, respectively. *Group secret is protected with APT attack.*

Traceability
In the case of dispute, group manager needs to convince that specific signers signed sessions of document. In order to identify signer, GM can show values that related to only signer i-th (*privilege or normal signer*): h_i, R_i, $z_i = H(h(M)||h_i||P_i||h_{SE})$, $s_i = t_i + h_i k_i z_i e \bmod q$ $s_i = f_i e + t_i + h_i k_i z_i e \bmod q$ and an EC point $V_i = f_i eG$. These values satisfy check equations for normal signers (9) or privilege signers (10) so only signer i-th is responsible for document session with $h_i = h(M_i)$. Thus, the scheme provides distinguished signing authority feature internally. Disclosure of R_i, V_i is safe because they are produce by the multiplication on Elliptic curve.

Unforgeability
Signer i-th needs approval from DC to get z_i to calculate his share signatures that pass a verification equation at (9) or (10) for normal or privilege signer, respectively. Generating group signature needs cooperation of DC with t privilege signer members and only DC can produce a valid group manager with valid member's shared signatures.

Unlinkability
Identifying the two different signatures generated by one member (or group of members) is impossible, except for the group manager.

Exculpability
In the proposed scheme, no member (or many corrupt members work together) can forge signatures of other. This is because signature of member is calculated not only by private key but also on R_i, z_i, e which are calculated for specific signer. Therefore, to forge the signature of a group member, they need to pass the signature check equation for each member of the group manager. That means they must break the ECDLP.

4 Conclusion

In this paper, a new threshold group signature scheme based on usage of Elliptic curve is proposed. The new scheme has these new practical benefits:

1. Scaling group without worrying about group secret loss; enables only a limited number people can hold secret key shared while allows number normal members to grow; Practical revocation and joining group.
2. Compared with previous threshold group signature schemes, no chance for an adversary or dishonest group of signers can steal group secret.
3. Reduce the risk of unexpected transaction of threshold group signature scheme by using a trusted DC.
4. Provides distinguished signing authority feature of multisignature internally.

The size of the output signature is comparable with known schemes. In practically, the proposed protocol provides more control to an organization by threshold mechanism and allowing a limited number of members who can authorize transactions. The scheme possesses many security advantages compared with previous works.

References

1. Harn, L.: Digital multisignature with distinguished signing authorities. Electronics Letters. Vol.35 (1999) 294–295.
2. Chaum, D., Van Heyst: Group signatures. Advances in Cryptology—EUROCRYPT 1991. LNCS. Springer Heidelberg (1991) 257–265.
3. Rivest, Ronald L., Adi Shamir, Leonard Adleman.: A method for obtaining digital signatures and public-key cryptosystems. Communications of the ACM. Vol.21 (1978): 120–126.
4. ElGamal, Taher.: A public key cryptosystem and a signature scheme based on discrete logarithms. IEEE transactions on information theory. Vol.31 (1985): 469–472.
5. Harn, L.: New digital signature scheme based on discrete logarithm. Electronics Letter, Vol. 30. (1994) 396–398.
6. Huang, Hui-Feng, Chin-Chen Chang.: Multisignatures with distinguished signing authorities for sequential and broadcasting architectures." Computer Standards & Interfaces. Vol.27. (2005) 169–176.
7. Wang, Ching-Te, Chu-Hsing Lin, Chin-Chen Chang.: Threshold signature schemes with traceable signers in group communications. Computer Communications. Vol.21. (1998) 771–776.
8. Harn, Lein.: Group-oriented (t, n) threshold digital signature scheme and digital multisignature. IEE Proceedings-Computers and Digital Techniques 141.5 (1994): 307–313.
9. Harn, Lein, and Feng Wang.: Threshold Signature Scheme without Using Polynomial Interpolation. IJ Network Security. Vol.18. (2016): 710–717.
10. Yu, Yuan-Lung, and Tzer-Shyong Chen.: An efficient threshold group signature scheme. Applied Mathematics and Computation Vol.167. (2005) 362–371.
11. Mante, Ganesh, and S. D. Joshi.: Discrete logarithm based (t, n) threshold group signature scheme. International Journal of Computer Applications 21.2. (2011) 23–27.
12. Michels, Markus, and Patrick Horster.: On the risk of disruption in several multiparty signature schemes. International Conference on the Theory and Application of Cryptology and Information Security. Springer Berlin Heidelberg, (1996).
13. Tseng, Yuh-Min, and Jinn-Ke Jan.: Attacks on threshold signature schemes with traceable signers. Information Processing Letters Vol.71 (1999) 1–4.
14. Shao, Z.: Repairing Efficient Threshold Group Signature Scheme. International Journal of Network Security, Vol.7, No.2, (2008) 218–222.
15. Zhao, Lin-Sen; LIU, Jing-Mei. (t, n) Threshold Digital Signature Scheme with Traceable Signers against Conspiracy Attacks. In: Intelligent Networking and Collaborative Systems (INCoS), 2013 5th International Conference on. IEEE, (2013) 649–651.
16. Bozkurt, Ilker Nadi, Kamer Kaya, Ali Aydın Selçuk.: Practical threshold signatures with linear secret sharing schemes. International Conference on Cryptology in Africa. Springer Berlin Heidelberg. (2009) 167–178.
17. Tuan, H.D., Nguyen, H.M., Tran, C.M., Nguyen, H.N., Adreevich, M.N.: Integrating Multisignature Scheme into the Group Signature Protocol. Advances in Information Communication Technology: Proceedings of the International Conference, ICTA 2016. Springer International Publishing. (2017) 294–301.
18. Shamir, A.: How to Share a Secret. Communications of ACM, 22. (1979) 612–613.
19. Laguillaumie, Fabien, et al.: Lattice-based group signatures with logarithmic signature size. International Conference on the Theory and Application of Cryptology and Information Security. Springer Berlin Heidelberg. (2013) 41–61.

20. Bozkurt, Ilker Nadi; Kaya, Kamer; Selçuk, Ali Aydın.: Practical threshold signatures with linear secret sharing schemes. In: International Conference on Cryptology in Africa. Springer Berlin Heidelberg. (2009) 167–178.
21. Boldyreva, Alexandra.: Threshold signatures, multisignatures and blind signatures based on the gap-Diffie-Hellman-group signature scheme. In: International Workshop on Public Key Cryptography. Springer Berlin Heidelberg. (2003) 31–46.

Implementation and Analysis of BBO Algorithm for Better Damping of Rotor Oscillations of a Synchronous Machine

Gowrishankar Kasilingam[1,2](✉), Jagadeesh Pasupuleti[2], and Nithiyananthan Kannan[3]

[1] Department of ECE, Rajiv Gandhi College of Engineering and Technology, Pondicherry, India
gowri200@yahoo.com

[2] Department of Electrical Power, Universiti Tenaga Nasional (UNITEN), Kajang, Malaysia

[3] Department of Electrical Engineering, Karpagam College of Engineering, Othakkal Mandapam, TamilNadu, India

Abstract. In this paper, a new evolutionary algorithm, namely the biogeography-based optimization (BBO) algorithm, is proposed to tune the Proportional Integral Derivative (PID) controller and power system stabilizer (PSS) parameters. The efficiency of BBO algorithm is verified on the single-machine infinite-bus (SMIB) system for various operating conditions. The proposed method suppresses the (0.1–2.5 Hz) low-frequency electromechanical oscillations and increases the power system stability via minimizing the objective function such as integral square error (ISE). Simulations of BBO-based PID, BBO-based PSS, BBO-based PID-PSS and conventional PSS are performed by using MATLAB/Simulink. The results show that BBO-based PID-PSS method improves the system performance and provides better dynamic performance.

1 Introduction

A stable power system is able to preserve the frequency and voltage levels at the desired level after it is subjected to external disturbances such as abrupt increase of load, severe faults, voltage collapse [1]. In general, power system is a nonlinear and complex system. Sometimes, it exhibits electromechanical oscillations (0.1–2.5 Hz) due to deficient damping torque caused by the poor operating conditions. Insufficient damping may cause system separation due to the amplifications of these oscillations. The synchronous generators are generally equipped with PSS so as to promote damping and to provide feedback signals to stabilize the excitation system [2]. Controller such as PID is extensively used in many control applications. It is easy to implement and robust. Therefore, additional PID damping controller is combined with PSS to suppress low-frequency electromechanical oscillations in the SMIB system under different vulnerable conditions [3]. Many efforts had been devoted to tune the PID controller and PSS by using advanced control systems such as intelligent/adaptive

V. Bhateja et al. (eds.), *Information Systems Design and Intelligent Applications*, Advances in Intelligent Systems and Computing 672,
https://doi.org/10.1007/978-981-10-7512-4_8

control to enhance system stability and oscillation damping. Though there are merits associated with these advanced approaches, power system utilities still opt for the simple PID controller and lead–lag PSS structures. Modern control methods such as artificial intelligence techniques such as artificial neural network (ANN) and fuzzy logic (FL) offer smart results for power system stabilization. Zhang et al. [4] reported the application of ANN-based PSS. The ANN-based PSS provides excellent damping and improves the performance of the system. Later, Segal [5] designed a self-tuning PSS using ANN. Here, the PSS parameters are tuned in real time. It had been shown that self-tuning ANN-based PSS was quite robust under various loading conditions. Gandhi and Joshi [6] proposed two intelligent techniques while designing PSS. Here, ANN technique is used to design the PSS, while the genetic algorithm was adopted to adjust the network. To improve the dynamic power system stability, Kamalesh et al. [7] designed a FL power system stabilizer. After that, El-Metwally [8] presented a hybrid FL-based PSS. It is well accepted that FL controller has capability to provide automatic adjustment of PID gains. Related works on designing PSS using AI tools had been reported as well. Even though there are many advantages associated with this method, it is unable to deal with uncertain models. The population-based algorithms such as genetic algorithm (GA), evolutionary programming and differential evolution (DE) had been used methods to tune the values of PSS such as parametric uncertainty and nonlinearity. By using these methods, the LFO can be damped out and improves the stability [9–12] over a wide range of operating conditions. These methods have been used to find solution for high-dimensional problems in power system. However, the drawbacks of these algorithms are poor convergence, long response time and complexity in computation. More recently, swarm intelligence (SI)-based optimization algorithms had been adopted to design more sophisticated PSS. These algorithms rely on collective intelligence to achieve a global goal. Theja et al. [13] reported their ABC algorithm to optimize the parameters of PID controller equipped with PSS. They found that this approach provided better damping in rotor oscillations because of mild discrepancies in loads and generation. ACO algorithm had been suggested as well to improve the dynamic performances of single-machine power systems [14]. Also, PSO algorithm had been used to tune the values of PSS in order to design a robust power system [15]. From the survey reports, seemingly, there are advantages and drawbacks associated with each technique used to tune the PID and PSS parameters. Moreover, it appears that more emphasis had been given to optimizing the PSS parameters or PID gains. In the current work, a new method biogeographical-based optimization (BBO) algorithm is suggested to elevate the gain of PID controller and the PSS parameters to suppress the oscillations in the power system.BBO algorithm has been applied by many researchers to tune the PID gains for several applications. Mohammed Salem [16] used BBO to optimize the PID controller for nonlinear systems, and the method was tested over an inverted pendulum and mass-spring-damper system. Then, the BBO algorithm was implemented by Wang et al. [17] to solve the premature convergence problem. Later, Karthikeyan [18] designed a robust PID controller using BBO algorithm to enhance the rotor angle stability of the synchronous machine. This technique was experienced on the SMIB system for a various operating conditions. Furthermore, Gowrishankar and Jagadeesh [19] found that BBO is performing better than those of PSO and adaptation law while optimizing the PID gains to damp out LFO of synchronous machine. Therefore, this paper focuses on designing the values of PID and the PSS damping controller of a synchronous machine via MATLAB. The aim is to enhance the power system stability when system subjected to different operating conditions. The PID controller and PSS parameters will be optimized using a novel BBO algorithm. Also, the effectiveness of the proposed BBO-based PID-PSS system will be evaluated and compared with those of BBO-based PID, BBO-based PSS and conventional PSS. Meanwhile, system parameters such as speed deviation for different case studies are analysed by using Simulink. It is interesting to note that the proposed BBO algorithm is able to improve the system damping and increase the power system stability.

2 Related Work

2.1 Power System Stabilizer (PSS)

Power system stabilizer (PSS) produces a feedback stabilizing control signals to the excitation system to suppress unwanted oscillations. On the other hand, conventional power system stabilizer (CPSS) is designed based on stabilizer gain (K_{stab}), washout time (T_w) and lead–lag compensator (T_1 and T_2). In CPSS, the parameter values are optimized and prescribed for certain operating conditions to provide better damping over different operating conditions. The speed deviation signal ($\Delta\omega$) is given as input signal to PSS, and the output is stabilizing signal (ΔV_{PSS}). PSS is connected to the excitation system by employing the automatic voltage regulator (AVR). The stabilizing signal from PSS is then combined to the reference voltage of the excitation system [20]. The block diagram of CPSS is given in Fig. 1.

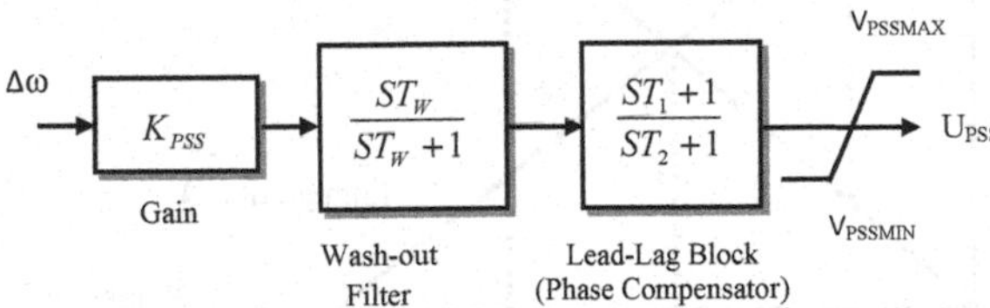

Fig. 1. Power system stabilizer (PSS)

2.2 Proportional Integral Derivative (PID) Controller

PID design involves three important parameters: proportional (K_P) gain used to define the steady-state error, integral (K_I) gain used to improve the rate of change of error and derivative (K_D) gain used to determine the reactions based on the rate of the change in error. Tuning of PID controller is performed so as to meet the specifications of the closed-loop system performance. The mathematical expression of PID controller consists of control signal u(t) and control error e(t) which can be expressed as follows:

$$u(t) = K_P e(t) + \frac{1}{T_i} \int_0^t e(\tau) d\tau + T_d \frac{de(t)}{dt} \quad (1)$$

Here, T_i—integral time and T_d—derivative time. Also, $K_D = K_P T_d$ and $K_I = K_P / T_i$.

3 Biogeography-Based Optimization (BBO) Algorithm

Biogeography-based optimization (BBO) was firstly introduced by Dan Simon in 2008 [21]. It was designed based on a new population-based technique called evolutionary algorithm (EA). Biogeography is a part of the biological study, where it relies on the data collected from the earth sciences, population biology, systematic and ecology [22]. The model of BBO algorithm explains the movement of species, formation of new species and their extinction. The term Habitat Suitability Index (HSI) defines the suitability of a place for a species to reside. High HSI indicates good performance of an optimization problem. The features in each habitat are called the Suitability Index Variable (SIV), which is corresponding to the problem dimension. SIV and HSI are generally treated as independent and dependent variables, respectively. An island with a good HSI has a high emigration rate as shown in Fig. 2.

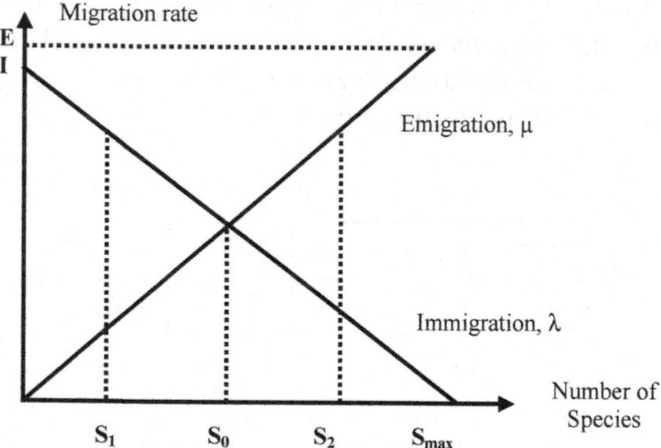

Fig. 2. Model of immigration rate and emigration rate

Here, S is the total number of species at equilibrium; S_{max} is the maximum number of species, λ is immigration rate, and μ is emigration rate. The values of emigration rate and immigration rate can be evaluated from Fig. 2:

$$\lambda = I \cdot \left(1 - \frac{S}{S_{\max}}\right) \quad (2)$$

$$\mu = \frac{E \cdot S}{S_{max}} \tag{3}$$

The general procedure of BBO involves:

Step 1: Parameters such as mutation probability, island modification probability and elitism parameter have to be designed.
Step 2: Initialize the number of population (island).
Step 3: Compute the value of λ and μ.
Step4: Choose the emigration and immigration island, respectively, based on the migration rate.
Step 5: Choose the values of SIVs randomly based on the selected islands.
Step 6: Mutation was performed based on mutation probability.
Step 7: Estimate the objective function for each island.
Step 8: Check the optimization solution. Otherwise, move to step 3.

The BBO algorithm consists of two important subalgorithms which are migration and mutation. Models of migration and mutation algorithms are developed to optimize the parameters of the PID controller. Figure 2 shows a simplified model of biota of an island used to provide general relationships of immigration and emigration. To model the concepts of BBO in detail, the case of habitat containing only S species is considered. Here, changes from time t to as below:

$$P_S(t+\Delta t) = P_S(t)(1 - \lambda_S\Delta t - \mu_S\Delta t + P_{S-1}\lambda_{S-1}\Delta t + P_{S+1}\mu_{S+1}\Delta t) \tag{4}$$

where λ and μ are the rates of immigration and emigration when S species reside in the habitat.

3.1 Migration

The parameter values such as K_P, K_I, K_D, K_{PSS}, T_1 and T_2 appeared in the solution vector are measured as SIV. To check the solution quality, Habitat Suitability Index (HSI) is computed. Here, to optimize the PID and PSS values, HSI is considered as the objective function which may be defined as ITAE, IAE, ITSE or ISE. In this work, the integral square error (ISE) of the speed deviation (Δω) is considered as the objective function. ISE tends to produce smaller overshoots and oscillations than IAE (Integral of the Absolute Error) or ITAE (Integral Time Absolute Error). Therefore, the PID parameters are tuned by using ISE.

The fitness function is defined as follows:

$$ISE : J = \int_0^{\infty} \Delta\omega^2(t)dt, \; \infty = t_{sim} \tag{5}$$

where t_{sim} is the simulation time.

The speed deviation (Δω) is chosen to assess the performance of the design system. As the random values of K_P, K_I, K_D K_{PSS}, T_1 and T_2 are generated during the initialization of problem space, these values are fed into the PID controller and the PSS. Then, the speed deviation is determined by computing the performance index, J. Finally, those K_P, K_I, K_D K_{PSS}, T_1 and T_2 values that give the minimum J value are treated as the optimal values of PID and PSS. Thus, the main problem in tuning PID and PSS parameters is to choose the best habitat (solution) in order to minimize the performance index, J. As mentioned, habitat with high HSI has more species and vice versa. In other words, HSI indicates the immigration rate and the emigration rate of each habitat.

3.2 Mutation

Mutation in BBO is known as SIV mutation. The species count probability is used to determine the mutation rate. Mutation is likely to occur if a habitat has medium HIS. Elitism is used to save the features of the habitat that has the optimal K_P, K_I, K_D K_{PSS}, T_1 and T_2 values in the BBO process. Therefore, even if the mutation ruins the HSI, process reversal is possible based on the saved features.

$$m = m_{\max}\left(1 - \frac{P_S}{P_{\max}}\right) \tag{6}$$

where P_S is the probability of each island containing the S species, $P_{\max}$ is the maximum value of P_S, $m_{\max}$ is the user-defined maximum mutation rate, and m is the mutation rate.

The ranges of the optimized PID parameters are as follows:

$$\begin{aligned}
&K_P^{\min} \le K_P \le K_P^{\max} = 0.5 \le K_P \le 80\\
&K_I^{\min} \le K_I \le K_I^{\max} = 0.2 \le K_I \le 30\\
&K_D^{\min} \le K_D \le K_D^{\max} = 0.1 \le K_D \le 15\\
&K_{PSS}^{\min} \le K_{PSS} \le K_{PSS}^{\max} = 1 \le K_{PSS} \le 60\\
&T_1^{\min} \le T_1 \le T_1^{\max} = 0.2 \le T_1 \le 2\\
&T_2^{\min} \le T_2 \le T_2^{\max} = 0.2 \le T_2 \le 2
\end{aligned}$$

Figure 3 shows the flow chart of BBO algorithm used to optimize the PID parameters [19]. The parameters of BBO used to optimize the PID controller and the PSS are given in Table 1.

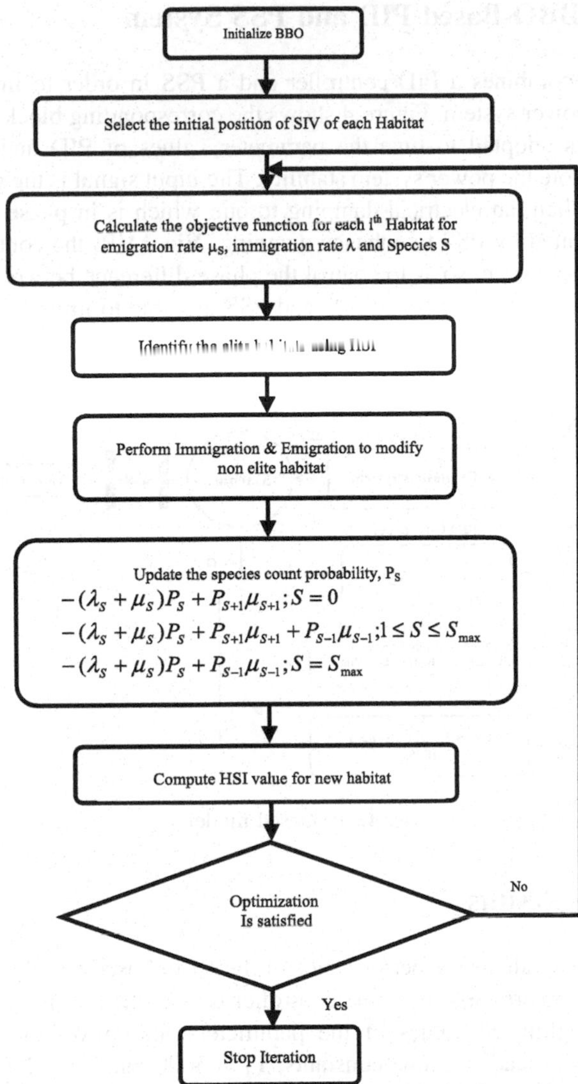

Fig. 3. Flow chart of BBO algorithm

Table 1. Parameters for tuning PID gains and PSS using BBO

Parameters	Size
Habitat modification probability	1
Population number	50
Mutation rate	0.05
Iteration count	50
Number of elite habitat	6
Max. emigration and immigration rate	1

4 Proposed BBO-Based PID and PSS System

The new system combines a PID controller and a PSS in order to improve the performance of the power system. Figure 4 shows the corresponding block diagram. Here, BBO algorithm is adopted to tune the parameter values of PID and PSS damping controller to promote the power system stability. The input signal is the generator speed deviation (Δω). Then, an electrical damping torque which is in phase with the speed deviation is generated by PSS to enhance damping. Via AVR, the controller output is set to the excitation system so as to control the phase difference between the generator and the load. In general, BBO-based PID and PSS are able to improve the stability of the power system.

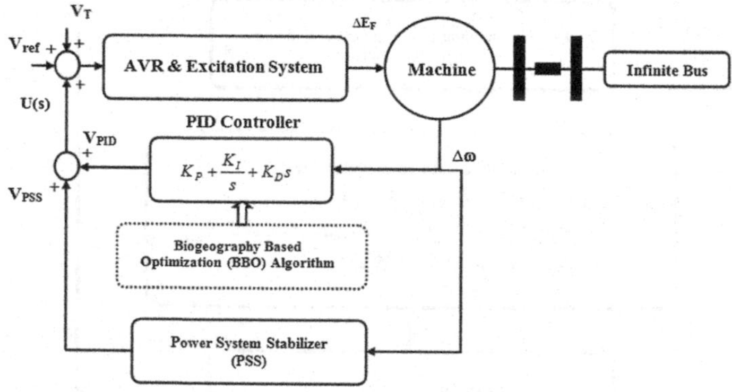

Fig. 4. Proposed model

5 Simulation Results

Simulink is used to evaluate the performance of the BBO-based coordinated controller. The robustness of the proposed damping controller is examined for numerous operating conditions. The optimized values of the parameters for conventional PSS are [3] $K_{PSS} = 125$; $T_w = 2$; lead–lag time constants, $T_1 = 5000$ and $T_2 = 2000$. The parameter values of the PID and the PSS damping controllers obtained by using the BBO algorithm are presented in Table 2.

The optimization results are computed, and the convergence characteristics of BBO and PSO methods are shown in Fig. 5. From the convergence plots, BBO algorithm

Table 2. Parameters of PID controller and PSS using BBO algorithm

Tuning—method	PID gains			PSS parameters		
	K_P	K_I	K_D	K_{PSS}	T_1	T_2
BBO algorithm	52.6	20.2	9.73	31.7	0.80	0.38

exhibits better convergence than PSO algorithm. Also, the dynamic behaviours and the convergence characteristics of the algorithms can be analysed by using the mean (M) and the standard deviation (Δω):

$$M = \frac{\sum_{i=1}^{n} f(K_i)}{n} \tag{7}$$

$$\sigma = \sqrt{\frac{1}{n}\sum_{i=1}^{n} (f(K_i) - M)^2} \tag{8}$$

where $f(K_i)$ is the fitness value of an individual K_i and n is the population size.

The BBO-based PID-PSS results give better fitness value as compared to other controllers as shown in Table 3. As observed, BBO-based PID-PSS shows better convergence characteristic than BBO-PID and BBO-PSS. Therefore, the parameters

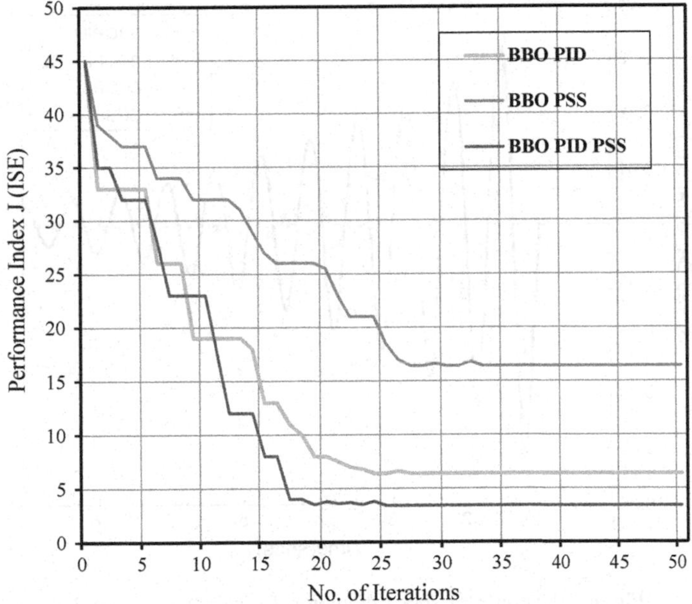

Fig. 5. Comparison of fitness function

can be optimized efficiently by using the proposed controller. The performances of BBO PID-PSS, BBO-PID, BBO-PSS and CPSS are simulated and analysed in the MATLAB/Simulink environment for different operating conditions.

Table 3. Comparison of computational efficiency

Optimization Methods	Max.	Min.	Range	Mean (M)	Standard. deviation (σ)
BBO PSS	45	16.4	28.6	23.90	8.360
BBO PID	45	6.22	38.78	12.363	9.8884
BBO PID-PSS	45	3.4	41.6	10.376	11.356

The following test cases are simulated.

Case (i): For base load (200 MVA) with 3-ϕ Fault:

Here, the synchronous machine is exposed to base load (active power P = 200 MVA; inductive reactive power Q_L = 70 MVAr; capacitive reactive power Q_C = 20 MVAr). A 3-ϕ fault is then introduced to the transmission line, where the fault switching of phase A, phase B and phase C is activated. The initial status of the fault breaker is usually 0 (open). Here, the fault is applied at t = 0.6/60 s and terminated at t = 6/60 s. Figure 6 shows the response of speed deviation for case 1.

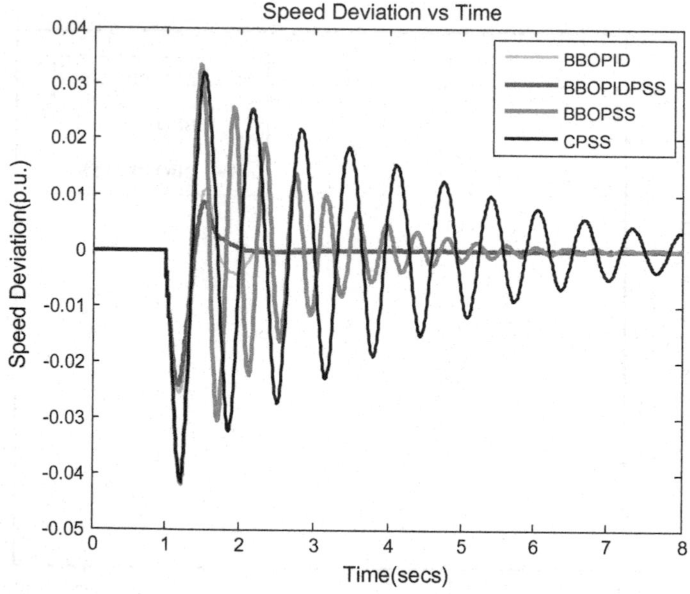

Fig. 6. Response of speed deviation for case 1 and case 2

Case (ii): For heavy load (600 MVA) with ground fault:

In this case, a heavy load (i.e. 3 times × normal load: active power P = 600 MVA; inductive reactive power Q_L = 210 MVAr; capacitive reactive power Q_C = 60 MVAr) is introduced to the synchronous machine with ground fault condition in the transmission line. At each transition time, the selected fault breakers are opened and closed depending on the initial state. The ground fault is applied at t = 0.6/60 s and closed at t = 6/60 s in the transmission line. The response of speed deviation for case 2 is shown in Fig. 7, respectively.

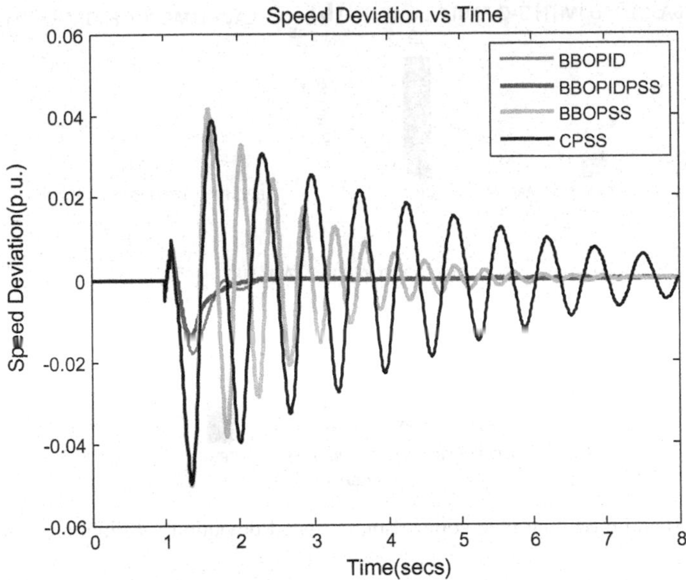

Fig. 7. Response of speed deviation for case 1 and case 2

Case (iii): For heavy load (600 MVA) with 3-ϕ fault:
The response of speed deviation for case 3 is shown in Fig. 8, respectively. The trends are similar to those of cases 1 and 2.

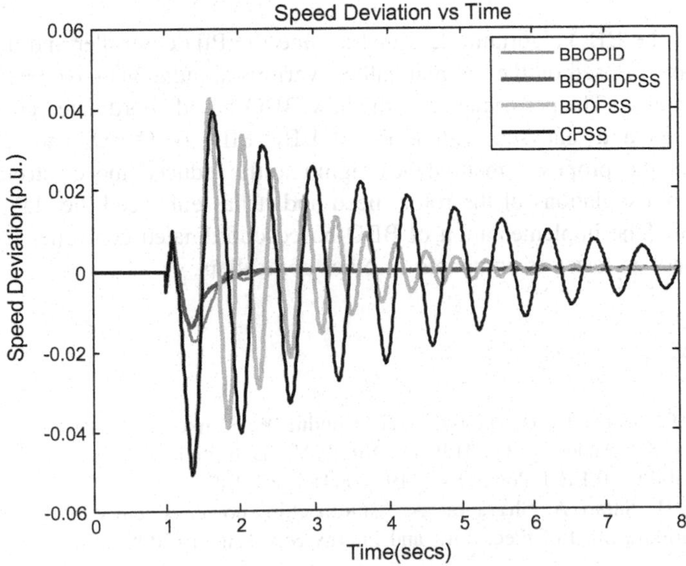

Fig. 8. Response of speed deviation

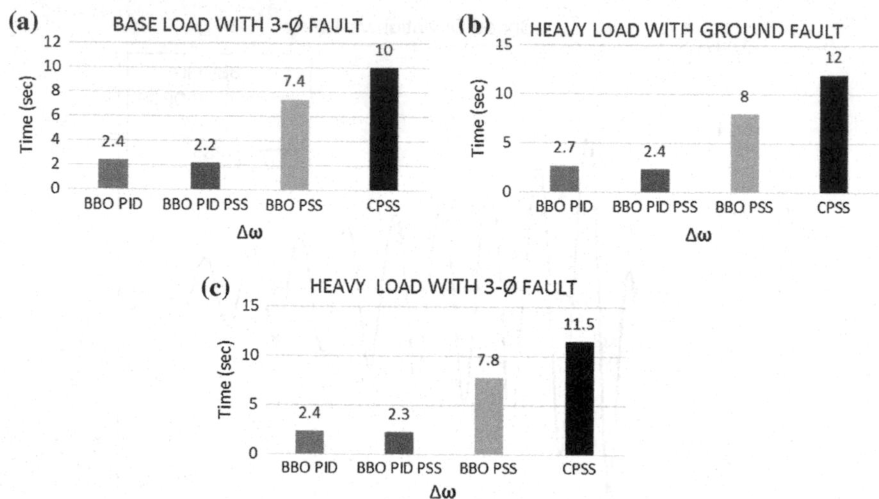

Fig. 9. **a**, **b** and **c** Settling time comparison of speed deviation for different case studies

The settling time comparison of speed oscillations of different controllers under various case studies is graphed in Fig. 9a–c. Also, in this case, the BBO PID-PSS attains best settling times over other damping controllers. By observing these settling time comparison plots, it is concluded that the BBO PID-PSS overtakes the effectiveness of other controllers.

6 Conclusion

In this paper, the BBO algorithm designed to tune the PID controller and the PSS has been developed. Based on the optimal values, various simulations have been executed by using Simulink. The performance of the new BBO-based coordinated controller has been compared and analysed with those of BBO-PID, BBO-PSS and CPSS. It is observed that the proposed method has significantly reduced the electromechanical low-frequency oscillations of the rotor speed and it has enhanced the stability of the power system. The implementation of BBO-based coordinated controller is easy, and its convergence is faster as compared to other controllers.

References

1. Bose, A., Canizares, C., Hatziargyriou, N., Kundur, P., Stankovic, A., Paserba, J., Ajjarapu, V., Taylor, C., Anderson, G., Hill, D., Vittal, V.: Definition and classification of power system stability. IEEE T Power Syst. 19 (2004) 1387–1401.
2. Shayeghi, H., Safari, A., Shayanfar, A.: Multimachine power system stabilizer design using PSO algorithm. Int. J of Elec. Pow. and Energy Sys. Eng. 1 (2008) 226–233.

3. Gowrishankar, K., Ragavendiran, R., Gnanadass, R.: Modelling and performance analysis of proportional integral derivative controller with PSS using simulink. 4th IEEE Int. Conf. Comp. Appl. Electr Eng. Recent Adv. (2010) 5–8.
4. Zhang, Y., Chen, G.P., Malik, O.P., Hope, G.S.: An artificial neural network based adaptive power system stabilizer. IEEE Trans. on Energy Conv. 8 (1993) 71–77.
5. Segal, R., Sharma, A., Kothari, M.L.: A self-tuning power system stabilizer based on artificial neural network. Int. J of Elec. Pow. and Energy Sys. 26 (2004) 423–430.
6. Gandhi, P.R., Joshi, S.K.: Design of power system stabilizer using genetic algorithm based neural network. J of Elec. Eng. (2012) 1–12.
7. Kamalesh Chandra Rout, Panda P.C.: Power system dynamic stability enhancement of SMIB using fuzzy logic based power system stabilizer. Pow. Electr. and Inst. Engineering, Springer 102 (2010) 10–14.
8. El-Metwally, K.A.: A Fuzzy Logic? Based PID for Power System Stabilizer. Electr. Pow. Comp. and Systems 29 (2010) 659–669.
9. Ahmed I. Baaleh, Ahmed F. Sakr.: Application of genetic algorithms in power system stabilizer design. Multiple approaches to Intelligent Sys., Springer 1611 (1999) 165–174.
10. Serhat Duman, Ali Ozturk.: Robust design of PID controller for power system stabilization by real coded genetic algorithm. Int. Rev. of Elec. Eng. 5 (2010) 2159–2170.
11. Abido, M.A, Abdel-Magid, Y,L.: Optimal design of power system stabilizers using evolutionary programming. IEEE Trans. on Energy Conv. 17 (2002) 429–436.
12. Zhe Sun, Ning Wang, Dipti Srinivasan, Yunrui Bi.: Optimal tuning of type -2 fuzzy logic power system stabilizer based on differential evolution algorithm. Int. J of Elec. Pow. & Energy Sys. 62 (2014) 19–28.
13. Theja, B.S., Rajasekhar, A., Kothari, D.P., Das, S.: Design of PID controller based power system stabilizer using Modified Philip—Heffron's model: An artificial bee colony approach. IEEE Sym. on Swarm Intelligence (SIS) (2013) 228–234.
14. Sheeba, R., Jayaraju. M., Kinattingal.: Performance enhancement of power system stabilizer through colony of foraging ants. Electr Pow. Comp. and Sys. 42 (2014) 1016–1028.
15. Al Habri, W., Azzam, M., Chaklab, M., Al Dhaheri, S.: Design of PID controller for power system stabilization using particle swarm optimization. IEEE Int. Conf. on Electr. Pow. and Energy Conv. Sys. (2009) 1–6.
16. Khelfi, M.H., Salem, M.: Application of biogeography based optimization in tuning of PID controller for non-linear systems. IEEE Int. Conf. on Complex Sys. (ICCS) (2012) 1–6.
17. Xiang, L.C., Jiangtao, C., Fuli, W.: Local search strategy biogeography based optimization algorithm for self-tuning of PID parameters. 32nd Chinese Control Conf. (2013) 4306–4310.
18. Lakshmi, P., Karthikeyan, K.: Optimal design of PID controller for improving rotor angle stability using BBO. Int. Conf. Modelling Optimization & Comp, Elsevier (2012) 889–902.
19. Gowrishankar, K., Jagadeesh, P.: BBO algorithm-based tuning of PID controller for speed control of synchronous machine. Turk. J of Elec. Eng. & Comp. Sci. 24 (2016) 3274–3285.
20. Gowrishankar, K., Jagadeesh, P.: A Comparative Study of the Z-N, Adaptation Law and PSO Methods of Tuning the PID Controller of a Synchronous Machine. Int. Rev. of Modeling and Simulations (2014) 919–926.
21. Simon, D. J.: Biogeography based optimization. IEEE T on Evolut. Comput. 12 (2008) 702–713.
22. Brown, J. H., Riddle, B. R., Lomolino, M. V.: Bio-geography. 3rd ed. Sunderland, Massachusetts, UK: Sinauer Associates Inc. (2009).

Eye Blink Artefact Cancellation in EEG Signal Using Sign-Based Nonlinear Adaptive Filtering Techniques

Sruthi Sudha Nallamothu[1(✉)], Rama Kotireddy Dodda[2], and Kanthi Sudha Dasara[3]

[1] Department of ECE, JNTUK, Kakinada, AP, India
nallamothusruthi@gmail.com
[2] Department of Instrument Technology, Visakhapatnam, AP, India
rkreddy_67@yahoo.co.in
[3] Department of ECE, VNR VJIET, Hyderabad, India
kanthisudha@gmail.com

Abstract. This paper presents the filtering of the noise from the Electroencephalogram (EEG) using an adaptive filtering approach known as Nonlinear LMS algorithm. The noise presence in the signals makes it difficult to analyze the EEG as well as to get the correct representation of the signal. It, therefore, becomes important to design the signal filters to minimize the noise in such EEG signals. Nonlinear adaptive filter is implemented to minimize the ocular artefact/eye blink from the EEG in which the parameters can be adjusted to maintain the input–output relationship. Further, within Nonlinear adaptive filter approach, different sign-based versions such as Nonlinear sign Least Mean Square (NSSLMS), Nonlinear sign Least Mean Square (NSLMS), Nonlinear sign regressor Least Mean Square (NSRLMS) are developed to minimize the eye blink noise along with reducing the computational complexity. After this, a comparison of the filtered output signal is done with the output from the conventional LMS filter. Based on the parameters such as signal-to-noise ratio (SNR), misadjustment ratio (Madj) and excess mean square error (EMSE), we noticed that the filtered output from the Nonlinear adaptive filter is considerably more efficient. The main benefit of preferring NLMS in our analysis is its superior computational speed with multiplier free weight update loops. Also, it eliminates the need to make assumptions regarding data distribution and its size. Finally, to test the performance of the proposed algorithm, the same is applied on EEG signals extracted from CHB-MIT database. On comparing the results of the proposed algorithms with the conventional LMS, we noticed that NSRLMS outperforms the current realizations in reducing the noise.

Keywords: EEG · Nonlinear adaptive algorithms · Computational complexity Adaptive noise canceller · SNR · EMSE

V. Bhateja et al. (eds.), *Information Systems Design and Intelligent Applications*, Advances in Intelligent Systems and Computing 672,
https://doi.org/10.1007/978-981-10-7512-4_9

1 Introduction

Any brain activity produces electrical signals (measured in volts) because of ionic current transmitting between the neurons. Any deviations in these electrical signals help diagnose the disorders in the brain like sleep disorders, epilepsy, coma. The commonly used electrode placement is International 10–20 system. However, analyzing these brain activity electrical recordings using an Electroencephalogram is a chief concern in the area of neuroscience. One of the key reasons is the electrical signals from the brain originate from different sources and it becomes difficult to identify the desired signal. Hence, it becomes critical to remove the noise or unwanted signals from the electrical recordings for a couple of reasons: 1: for easier interpretation of data signal and 2: to retrieve the signal that flawlessly represents the functioning of the brain. Ocular artefacts which are triggered by blinking of eyes are one of the most common and frequently occurred during the recordings of the EEG signals. A number of authors have proposed various techniques for removing EEG artefacts. For instance, Fortgens [1] developed a technique to remove eye movement artefacts from the non-cephalic reference EEG signal. Croft [2, 3] had proposed a few approaches for removing ocular artefacts. Using algorithms based on independent component analysis (ICA), Romo-Vazquez et al. [4] have compared different combinations of wavelet denoising (WD) for ocular artefacts removal in EEG and proposed that source separation by second-order blind identification (SOBI-RO) as the highly effective method, followed by wavelet denoising by SURE thresholding. Romero [5] developed an adaptive filtering method called regression and blind source separation for ocular noise reduction in EEG signals. Carlos et al. [6] used a combination of ICA and an adaptive filtering technique recursive least squares (RLS) to remove eye movement artefacts. Kirkove et al. [7] developed 16 new methods to identify and correct ocular-based artefacts. Recently, Sruthi Sudha et al. [8] presented a technique for detection and removal of artefacts from EEG signal using sign-based LMS adaptive filters. An adaptive filtering technique NLMS-based approach using double density wavelet transform and ICA was proposed by Vandana Roy et al. [9] to eliminate noise from EEG signals. In [10], NLMS algorithm with declining step size was proposed by Ching-An Lai. Nallamothu Sruthi et al. [11] introduced sign-based normalized adaptive filtering techniques for denoising PLI and respiration artefacts from EEG signals. Many Nonlinear LMS algorithms were developed by Douglas [12, 13] for reducing the noise. These are specially classified for biomedical applications.

In this paper, we have developed a Nonlinear LMS algorithm with numerous sign-based modifications, for example, NSSLMS, NSLMS, NSRLMS. Adaptive noise cancellers are designed based on the Nonlinear LMS algorithm which basically limits the mean-squared error between the noisy EEG (primary input) and the reference input. The reference input can either be a noise signal that is associated with the primary signal's noise or an input signal that is only associated with EEG in the primary input. Finally, to evaluate the performance of these filter structures, we carried out simulations on CHB-MIT database [14]. SNR, EMSE and M_{adj} are the different metrics used for judging the performance of the different algorithm that we developed. The flow is outlined in the following sections. The basics of proposed NLMS algorithm, its sign variant algorithms for eliminating the noise due to eye blink are discussed in Sect. 2. Using MATLAB, the simulation results for LMS, NLMS, NSRLMS, NSLMS and NSSLMS algorithms are projected in Sect. 3. Section 4 captures the conclusions from our findings.

2 Proposed Implementation

The basic structure of adaptive filter is shown in Fig. 1. The primary input to the adaptive filter is s1 signal which is an EEG, combined with n1 noise. Another input called reference input n2 is fed to the adaptive filter which is just noise. This noise is generated from another system and is highly associated with error n_1 in the primary input.

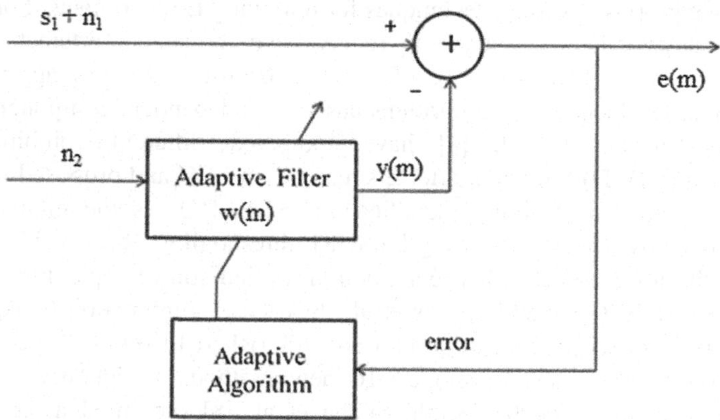

Fig. 1. Structure of adaptive filter

2.1 Conventional LMS Algorithm and its Sign Variant Algorithms

The weighted equation of conventional Least Mean Square (LMS) algorithm is

$$w(m+1) = w(m) + v\,a(m)e(m) \tag{1}$$

Signed Regressor LMS Algorithm (SRLMS): In the conventional LMS recursion, if the tap-input vector a(m) is replaced with the vector sgn{a(m)}, it is referred to as the signed regressor algorithm. In this case, the weighted equation is mentioned below:

$$w(m+1) = w(m) + v\,\mathrm{sgn}\{a(m)\}e(m) \tag{2}$$

Sign LMS Algorithm (SLMS): In the conventional LMS recursion, if we replace e (m) with sgn{e(m)}, it is referred to as the signed LMS algorithm. In this case, the weighted equation is mentioned below:

$$w(m+1) = w(m) + \nu a(m)\mathrm{sgn}\{e(m)\} \tag{3}$$

Sign–Sign LMS Algorithm (SSLMS): A combination of sign recursions and signed regressor, results in sign–sign LMS algorithm. In this case, the weighted equation is mentioned below:

$$w(m+1) = w(m) + \nu\,\mathrm{sgn}\{a(m)\}\mathrm{sgn}\{e(m)\} \tag{4}$$

2.2 Proposed Nonlinear LMS (NLMS) Algorithm

A key problem with the conventional LMS adaptive filter is its design followed by implementation mainly due to (ν) which is a step-size selection. In the stationary period, LMS algorithm comes close to the mean if the (ν) is between 0 and 2/λ_max and comes close to the mean square if the (ν) is between 0 and <2/(tr(R_x)).

However, it becomes necessary to estimate either Rx or λ_max as both of them are unknown. The step-size range for the mean square algorithm is

$$0 < \nu < \frac{2}{\boldsymbol{a}^T(m)\boldsymbol{a}(m)}$$

Moreover, the upper bound is given as

$$\nu(m) = \frac{\nu}{\boldsymbol{a}^T(m)\boldsymbol{a}(m)} = \frac{\nu}{||\boldsymbol{a}(m)||2}$$

In order to overcome the issue of small tap-input vector a(m), we change the above recursion by adding a small positive constant ε. The parameter ε is basically added to ensure that the denominator is not too small which otherwise results in a bigger step-size parameter. Now the step-size parameter is written as,

$$\nu\ (m) = \frac{\nu}{\varepsilon + ||\boldsymbol{a}(m)||2} \tag{5}$$

ν (m) is a Nonlinear step size where ν value lies between 0 and 2. Substituting ν in the weighted Eq. (1) of LMS with ν(m) results in Nonlinear LMS, this is mentioned below as:

$$\boldsymbol{w}(m+1) = \boldsymbol{w}(m) + \frac{\nu}{||\boldsymbol{a}(m)||2} e(m)\boldsymbol{a}(m) \tag{6}$$

The problem related to amplification of the noise is eliminated in the Nonlinear LMS by normalizing the step size of the LMS by $||a(m)||^2$. Even though the Nonlinear LMS algorithm evades the problem related to amplification of the noise, another problem emerges once ||a(m)|| becomes minute. To avoid this, an alternate approach is arrived where following change is done to Nonlinear LMS

$$w(m+1) = w(m) + \frac{\nu}{\varepsilon + ||a(m)||2} e(m)a(m) \tag{7}$$

2.3 Extension to Sign-Based Realizations of Nonlinear LMS Algorithm

Less computational complexity is attained by trimming either the input data or error estimate. The LMS algorithms based on clipping of error or data are SRLMS, SLMS and SSLMS. The combination of these three simplified algorithms with Nonlinear algorithms provides fast convergence and reduced computational complexity. The benefit of Nonlinear LMS over the conventional LMS is that both tap weight count and the power of the input signal does not influence the step size (ν). Therefore, this results in better steady-state error and convergence rate. Moreover, the conventional LMS execution involves lot of added computations to derive ν (n) which in turn increase complexity. Hence, we developed techniques such as Nonlinear sign sign Least Mean Square (NSSLMS), Nonlinear sign Least Mean Square (NSLMS), Nonlinear sign regressor Least Mean Square (NSRLMS) to minimize the eye blink noise along with reducing the computational complexity.

2.3.1 Nonlinear Sign Regressor LMS (NSRLMS) Algorithm

Nonlinear signed regressor Least Mean Square algorithm is an extension of NLMS and is derived on the lines of SRLMS. Here no multiplication operation is required to compute the normalization factor. NSRLMS enjoys the benefits of both NLMS and SRLMS algorithms. The normalizing factor presence makes the steady state error independent of the power output of the reference signal. By combining (2) and (7) the weight update recursion for NSRLMS is

$$w(m+1) = w(m) + \mu(m)e(m)Sign\{a(m)\} \tag{8}$$

2.3.2 Nonlinear Sign LMS (NSLMS) Algorithm

In this algorithm, signum is considered for the error signal. This algorithm guarantees a certain convergence as well as it is easy to implement. This algorithm being the combination of SLMS and NLMS enjoys the benefits of both less complexity and fast convergence. Using (3) and (7), the weighted equation for NSLMS can be written as,

$$w(m+1) = w(m) + \mu(m)Sign\{e(m)\}a(m) \tag{9}$$

2.3.3 Nonlinear Sign LMS (NSSLMS) Algorithm

In this algorithm, signum is considered for both error signal and reference input. Similar to NSRLMS and NSLMS, NSSLSM can be obtained by combining SSLMS and NLMS algorithm. Using (4) and (7), the recursion for NSSLMS is:

$$w(m+1) = w(m) + \mu(m)Sign\{e(m)\}Sign\{a(m)\} \tag{10}$$

2.4 Computational Complexity of Nonlinear LMS Algorithms

Along with removing noise in the EEG signal efficiently, Nonlinear LMS algorithms which are sign based are exempt from multiplier–accumulator computations. The number of computations executed for filtering process for different methods/algorithms proposed in this paper is listed out in the below table. On comparing these number of computations, it is being observed that of all proposed algorithms, Nonlinear LMS algorithm is slightly more complicated with more MAC (2K + 1) operations as well as a division computation executed in its weighted equation. Among sign-based methods, both NSLMS and NSRLMS require one division computation; however, the former also requires K number of MAC computations while the latter is independent from filter length with only 1 MAC computation. NSSLMS needs K additions with sign check (ASC) computations and a division computation to execute the weighted equation. Since ASC requires far less complex circuit design in comparison to MAC operations, among all the sign-based Nonlinear algorithms discussed in this paper, NSLMS is most complex Table 1.

Table 1. Computational complexity comparison for normalized adaptive algorithms

S.No.	Algorithm	MACs	ASC	Divisions
1.	NLMS	2K + 1	Nil	1
2.	NSRLMS	1	Nil	1
3.	NSLMS	K	Nil	1
4.	NSSLMS	Nil	K	1

3 Simulation Results

To demonstrate the real efficacy of sign-based NLMS algorithms in analytical conditions, the EEG data with extensive range of wave morphology is corroborated with these algorithms. The 10–20 system of EEG was followed while collecting data. Of all trials, a set of five EEG signals (chb01, chb02, chb03, chb04 and chb05) were cumulated to cinch the pliability of results. As a first step in simulation, 600 samples of EEG signal are collected and then corrupted with eye blink noise. The corrupted signal is given as initial input signal to the adaptive filter shown in Fig. 1. The ref signal (n2) is an eye blink noise while output being the discretized signal. The experiment is carried out with the data set and average SNR is evaluated to compare the performance of the algorithms. These results for chb01 are shown in Fig. 2. In this simulation, the filter length is selected as 5 and step size for all the filters is chosen as 0.001. Table 2 shows the SNR values for the data set. From the SNR values obtained, it is established that NLMS outperforms conventional LMS algorithm with an average SNR of

14.3630 dB. And NSRLMS gives high SNRI 13.7784 dB among all the sign variants of NLMS. Figure 3 depicts the EMSE behaviour of Nonlinear adaptive filters in the eye blink noise cancellation. Figure 4 shows the convergence curves of NLMS and its sign-based realizations. The NSRLMS results can be seen concurring with that of NLMS.

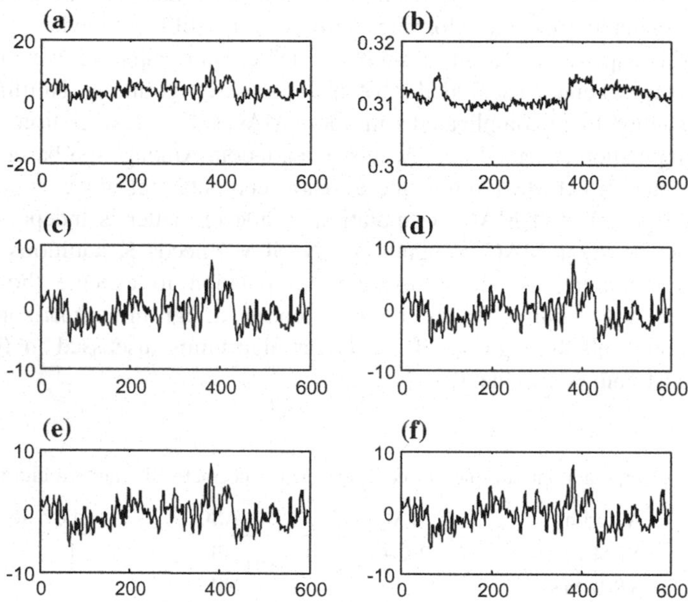

Fig. 2. Typical filtering results for eye blink cancellation using data Nonlinear adaptive filtering techniques: **a** EEG signal (chb01) with eye blink, **b** real eye blink noise, **c** retrieved signal with NLMS algorithm, **d** retrieved signal with NSRLMS algorithm, **e** retrieved signal with NSLMS algorithm, **f** retrieved signal with NSSLMS algorithm

Table 2. Comparison of SNRI using Nonlinear algorithms for the removal of eye blink artefact (in dB)

Rec. no.	LMS	NLMS	NSRLMS	NSLMS	NSSLMS
Chb01	6.7289	13.7293	12.5643	11.3206	10.9295
Chb02	7.3538	14.8472	14.1936	12.7382	11.4826
Chb03	7.5329	14.7483	14.3792	12.9348	10.7813
Chb04	6.4368	13.9638	13.4748	11.4395	10.5384
Chb05	7.8539	14.5267	14.2803	12.4524	10.8312
Avg. SNR	**7.1812**	**14.3630**	**13.7784**	**12.1771**	**10.9126**

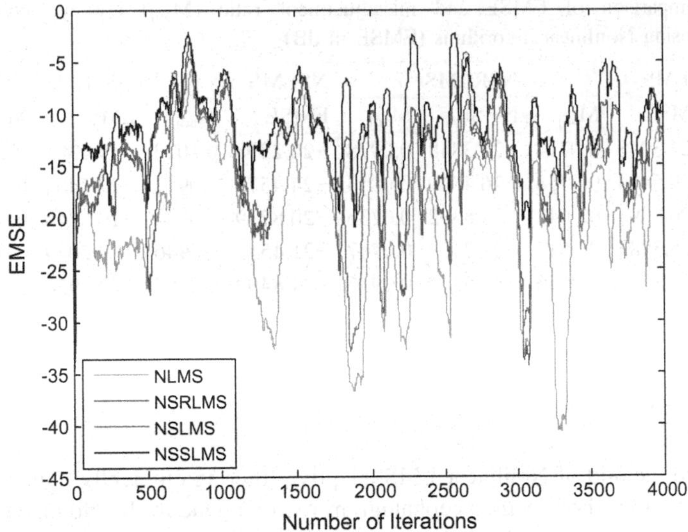

Fig. 3. EMSE behaviour in the eye blink noise cancellation using Nonlinear adaptive filtering techniques

Table 3 shows the comparison of EMSE and misadjustment (M_{adj}) using Nonlinear algorithms. In the reduction of eye blink artefacts, NLMS performs better in terms of SNRI; however, NSRLMS with single multiplication achieves SNRI slightly less than that of NLMS. Similarly, the EMSE and M_{ad} of NSRLMS are slightly inferior to that of NLMS. However, we consider NSRLMS as the best adaptive noise canceller for eye blink artefacts with reduced computational complexity.

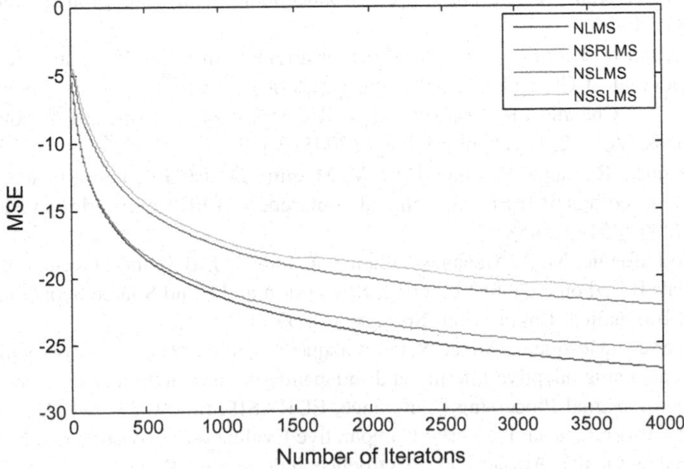

Fig. 4. Convergence curves of NLMS and its sign-based realizations

Table 3. Comparison of EMSE and misadjustment ratio (M_{adj}) for eye blink artefact cancellation using Nonlinear algorithms (EMSE in dB)

Rec. no.	NLMS		NSRLMS		NSLMS		NSSLMS	
	EMSE	M_{adj}	EMSE	M_{adj}	EMSE	M_{adj}	EMSE	M_{adj}
Chb01	−24.3048	0.0714	−22.3848	0.0978	−20.2398	0.1023	−19.4301	0.1147
Chb02	−28.5494	0.0758	−26.4849	0.0853	−24.4572	0.0957	−20.7484	0.1054
Chb03	−26.8894	0.0648	−24.8392	0.0736	−20.8729	0.0934	−19.5748	0.0945
Chb04	−25.3983	0.0547	−23.7429	0.0747	−21.4541	0.0946	−19.2049	0.0843
Chb05	−26.8298	0.0656	−25.5942	0.0936	−24.8493	0.0985	−22.5938	0.1037
Average	**−26.394**	**0.0664**	**−24.6090**	**0.0851**	**−22.3750**	**0.0969**	**−20.3102**	**0.1005**

4 Conclusion

Our proposed method of Nonlinear LMS adaptive filtering efficiently removes the eye blink artefacts from the Electroencephalogram records. Due to the Nonlinear filtering, the μ (step size) value is not fixed, so faster convergence rate can be attained along with good filtering capability. The newly developed algorithm greatly increases the convergence rate. The performance of the filtering process can be evaluated with the help of 3-key metrics; SNR, M_{adj}, EMSE and the NLMS adaptive filter perform better across all the three metrics. Among the proposed sign-based algorithms, NSRLMS method is the top choice for removing eye blink artefact as it is less complex compared to NLMS but at the same time does not compromise much on the convergence rate.

References

1. C. Fortgens and M. D. Bruin: Removal of eye movement and ECG artifacts from the non-cephalic reference EEG, Vol. 56. Electroencephalography and Clinical Neurophysiology, (1983) 90–96.
2. R. J. Croft and R. J. Barry: Removal of ocular artefacts from the EEG: a review, Vol. 30. Elsevier, Journal of Clinical Neurophysiology, (2000) 5–19.
3. R. J Croft, J. S Chandler, R. J Barry, and N. R Cooper: EOG correction: A comparison of four methods, Vol. 42. Psychophysiology, (2005) 16–24.
4. Romo-Vazquez R, Ranta R, Louis-Dorr V, Maquin D: EEG ocular artefacts and noise removal, Proceedings of IEEE International conference- IEEE Engineering in Medicine and Biology, (2007) 5445–5458.
5. S. Romero Lafuente, M. A. Mananas Villanueva, and M. J. Barbanoj: Occular Reduction in EEG Signals Based on Adaptive Filtering, Regression and Blind Source Separation, Vol. 37. Annals of Biomedical Engineering, Springer, (2009) 176–191.
6. Carlos Guerrero-Mosquera, Angel Navia Vazquez: Automatic removal of ocular artifacts from eeg data using adaptive filtering and independent component analysis, Proceedings of 17th European Signal Processing Conference, EURASIP, (2009) 2317–1321.
7. M. Kirkove, C. Franois, and J. Verly: Comparative Evaluation of Existing and New Methods for Correcting Ocular Artifacts in Electroencephalographic Recordings, Vol. 98. Signal Processing, (2014) 102–120.

8. N. SruthiSudha, D. V. RamaKoti Reddy: Detection and Removal of artefacts from EEG signal using sign based LMS Adaptive Filters, International Journal of Scientific & Engineering Research, vol. 8, no. 2, (2017) 950–954.
9. Vandana Roy, Shailja Shukla: A NLMS Based Approach for Artifacts Removal in Multichannel EEG Signals with ICA and Double Density Wavelet Transform, Proceedings of IEEE International conference on Communication Systems and Network Technologies (CSNT), (2015).
10. Ching-An Lai: NLMS algorithm with decreasing step size for adaptive IIR filters, Vol. 82 IEEE Transactions on Signal Processing, (2002) 1305–1316.
11. Nallamothu Sruthi Sudha, Rama Koti Reddy Dodda: Electroencephalogram Enhancement using Sign based Normalized Adaptive Filtering Techniques, International Journal of Engineering Sciences & Research Technology, vol. 6, no. 3, (2017) 101–107.
12. S. C. Douglas and Teresa H. -Y. Meng: Normalized Data Nonlinearities for LMS Adaptation, Vol. 42. IEEE Transactions on Signal Processing, (1994) 1352–1365.
13. S. C. Douglas: A Family of Normalized LMS Algorithms, Vol. 1. IEEE Signal Processing Letters, (1994) 1352–1365.
14. PhysioNet, The Massachusetts Institute of Technology—Children's Hospital Boston (CHB-MIT) Scalp EEG Database, Available: https://physionet.org/pn6/chbmit/.

No-Key Protocol for Deniable Encryption

Nam Hai Nguyen[1], Nikolay Andreevich Moldovyan[2], Alexei Victorovich Shcherbacov[3], Hieu Minh Nguyen[1(✉)], and Duc Tam Nguyen[1]

[1] Academy of Cryptography Techniques, Hanoi, Vietnam
{hainnvn61, hieuminhmta, nguyenductamkma}@gmail.com

[2] St.Petersburg Institute for Informatics and Automation of Russian Academy of Sciences, 199178 St.Petersburg, Russia
nmold@mail.ru

[3] Institute of Mathematics and Computer Science of Academy of Sciences of Moldova, Academiei str. 5, 2028 Chishinau, Moldova
scerb@math.md

Abstract. There is proposed a new method for deniable encryption based on commutative transformations. The method has been used to design the deniable encryption protocol resistant to the passive coercive attacks, which uses no pre-shared secret keys and no pre-exchanged public keys. The protocol begins with the stage at which the sender and receiver exchange their single-use public keys and compute the single-use shared secret key. Then, it is performed pseudo-probabilistic three-pass protocol with simultaneous commutative encryption of the fake and secret messages. Resistance of the proposed protocol to coercive attacks is provided by its computational indistinguishability from probabilistic no-key three-pass protocol used to send securely the fake message. To perform commutative encryption, it used exponentiation cipher. To provide security against active coercer, the protocol is to be complemented with procedure for authenticating the sent messages.

1 Introduction

Protocols and algorithms for deniable encryption (DE) allow one to solve a number of specific practical problems for providing information security of the information technologies [1,2]. The DE cryptoschemes are characterized in their resistance to the coercive attacks in the model of which it is supposed that the coercive attacker (coercer) intercepts ciphertext transmitted via an open communication channel and has possibility to force sender and receiver of the message to disclose the plaintext and the used secret keys, including private keys in the case of using the public key encryption algorithms. In the literature, the public key DE protocols [3] and the shared-key DE ones [4] are described. Recently, it has been proposed DE protocol based on using commutative encryption algorithm and shared secret key [5]. Commutative encryption procedures (commutative ciphers) are very interesting for practical application in the case

V. Bhateja et al. (eds.), *Information Systems Design and Intelligent Applications*, Advances in Intelligent Systems and Computing 672,
https://doi.org/10.1007/978-981-10-7512-4_10

of passive potential attacks, since they can be put into the base of no-key encryption protocols that permit one to transmit a secret message via a public channel without using public and secret keys shared by the receiver and the sender.

To provide resistance to passive attacks, the no-key encryption protocol should be based on some commutative cipher that is secure to the known-plaintext attack. The Pohlig–Hellman exponentiation cipher [6] represents a commutative encryption algorithm that satisfies the indicated requirement. Usually, when considering the DE protocols, it estimated their resistance to potential coercive attacks implemented by passive attacker. For the first time, the problem of providing security of the DE protocols to active coercive attacks had been discussed in papers [7,8]. To provide security to attacks of active coercer, it had been proposed to include in the DE protocols procedure of mutual authentication of the sender and receiver. Remaining in the framework of the model of the passive coercive attacks, it represents theoretical and practical interest to construct no-key DE protocol.

This paper first provides a method for no-key DE and describes a protocol that implements this method. The designed protocol begins with the execution of the public key agreement that provides the sender and receiver the secret message with shared single-use secret value Z. Then, it is implemented simultaneous encryption of the secret T and fake M messages, the encryption being computationally indistinguishable by the ciphertexts from the probabilistic no-key encryption of the message M. The value Z is used as a parameter of the probabilistic no-key encryption protocol on which the probabilistic no-key encryption depends. In the case when both the sender and the receiver are coerced, they disclose the fake message M, the value Z, and local keys used for commutative encryption of the message M and intermediate ciphertexts relating to M. They also declare and insist that during the communication session, they used the probabilistic no-key encryption protocol in order to send securely the massage M. Using the disclosed values, it is computationally impossible for the coercer to insist reasonably that the sender and the receiver used the no-key DE method.

In this paper, Sect. 2 presents an overview of the used cryptosystems. Section 3 presents the no-key deniable encryption method. Section 4 describes the no-key probabilistic encryption protocol. Section 5 presents the no-key deniable encryption protocol that implements secure transmission of the secret message T via a public channel, which is resistant to passive coercive attacks. Section 6 summarizes the results of the paper.

2 Used Cryptosystems

The proposed method for no-key DE includes as its three basic components the following cryptographic primitives: the Diffie–Hellman public key agreement protocol, Pohlig–Hellman commutative encryption algorithm, and no-key encryption protocol.

In the Diffie–Hellman protocol [9] it used a sufficiently large prime number p (having size, e.g., not less than 2,464 bits, which provides the 128-bit security), such that number (p - 1) contains a large prime divisor r (e.g., having size at least 256 bits), and the number α that is a primitive element modulo p. Each user chooses a private key as a random number x ($0 < x < p - 1$) having size more than 256 bits, and computes his public key y in accordance with the formula: $y = \alpha^x \bmod p$.

Then, the owner of the public key registers his public key in a specially created certification center, called certificate authority (CA). All public keys are placed in a public directory which is signed by a CA in order to avoid possible attacks with substituting public keys or imposing of the false public keys. If two users A and B want to establish secret communication, they proceed as follows. User A takes public key of user B out a public key directory and computes the shared secret Z_{AB}:

$$Z_{AB} \equiv y_B^{x_A} \equiv (\alpha^{x_B})^{x_A} \equiv \alpha^{x_B x_A} \bmod p \quad (1)$$

where y_B is user B's public key and x_A is user A's private key. The users have no need to transmit the shared secret key Z_{AB} via a communication channel, since calculates the value Z_{AB} by a similar formula:

$$Z_{AB} \equiv y_A^{x_B} \equiv (\alpha^{x_A})^{x_B} \equiv \alpha^{x_B x_A} \bmod p \quad (2)$$

where y_A is user A's public key; x_B is user B's private key. It is assumed that a potential attacker knows the values of $y_B = \alpha^{x_B} \bmod p$ and $y_A = \alpha^{x_A} \bmod p$, available in a public directory or digital certificates (signed by a CA) that are exchanged between users A and B via a public channel. However, in order to calculate the value Z_{AB}, the attacker has to solve computationally difficult problem of the discrete logarithm. Shared secret Z_{AB} can be used by users to encrypt the session secret key with which secret message can be encrypted using some symmetric encryption algorithm.

The Pohlig–Hellman commutative encryption algorithm [6,10] represents performing operation of raising the plaintext to a secret degree e (minimum length of the value e is equal to 256 bits) modulo a large prime p (requirements to the prime p coincide with the requirements to the modulus p in the previous cryptoscheme). Encryption and decryption are performed as raising to different degrees e and d, respectively. Encryption of the message $M < p$ is described as computation of the ciphertext C using the formula $C = M^e \bmod p$. Decryption is represented by the formula $M = C^d \bmod p = M^{ed} \bmod p$. The correctness of the decryption is provided with the condition $ed = 1 \bmod (p - 1)$ which should be implemented while generating the secret key (e, d). To be able to fulfill the last condition, it should be selected the number e that is relatively prime to the number $(p - 1)$. Then using the extended Euclidean algorithm, it is easy to compute the respective inverse value $d = e^{-1} \bmod (p - 1)$.

Thus, the Pohlig–Hellman exponential cipher represents the commutative encryption function, the encryption procedure $E_K(M)$ of which is described by the formula:

$$C = E_K(M) = M^e \bmod p \quad (3)$$

The corresponding decryption procedure D_K is as follows:

$$M = D_K(C) = C^d \bmod p \tag{4}$$

where $D_K = E_K^{-1}$ and the encryption key $K = (e, d)$. Security of the Pohlig–Hellman algorithm to known-plaintext attack is as high as computational difficulty of the discrete logarithm problem.

No-key encryption protocol uses some persistent commutative encryption function $E_K(M)$, where M is the input message and K is the encryption key, i.e., the function for which the following equality holds:

$$E_{K_A}(E_{K_B}(M)) = E_{K_B}(E_{K_A}(M)) \tag{5}$$

where K_A and $K_B (K_B \neq K_A)$ are different encryption keys. The property of commutativity of the encryption function $E_K(M)$ is exploited in Shamir's no-key protocol (also called Shamir's three-pass protocol) that includes the following three steps [10]:

1. Sender of the message M generates a random key K_A and calculates the ciphertext $C_1 = E_{K_A}(M)$. Then, he sends C_1 to the receiver via an open channel.
2. Receiver generates a random key K_B, encrypts the ciphertext C_1 with the key K_B as follows $C_2 = E_{K_B}(C_1) = E_{K_B}(E_{K_A}(M))$, and sends C_2 to the sender.
3. Sender, using decryption procedure $D = E^{-1}$, calculates the ciphertext $C_3 = D_{K_A}(C_2) = D_{K_A}(E_{K_B}(E_{K_A}(M))) = D_{K_A}(E_{K_A}(E_{K_B}(M))) = E_{K_B}(M)$ and sends C_3 to the receiver of the message M.

Using the received ciphertext C_3, the receiver recovers message M according to the formula $M = D_{K_B}(C_3) = D_{K_B}(E_{K_B}(M)) = M$.

In this protocol used encryption keys K_A and K_B are local parameters of commutative transformations; therefore, one can call them local keys. Since the sender and the receiver do not use any shared key, the protocol is called the no-key protocol.

3 No-Key Deniable Encryption Method

Suppose some remote user A wishes to send a secret message $T < p$ to a remote user B, using the no-key encryption protocol, so that they can securely open the local keys K_A and K_B, if passive coercer intercepting ciphertexts C_1, C_2, and C_3 will attack them. In this case, conserving secrecy of the message T is possible, if the ciphertexts C_1, C_2, and C_3 are produced as simultaneous ciphering two different messages, the message T and some fake message M. Besides, the encryption process should look like probabilistic ciphering of the fake message M in the frame of no-key protocol that uses probabilistic commutative encryption function. It is proposed that the encryption method can include the following steps:

1. In accordance with the Diffie–Hellman protocol, the sender and the receiver generate session (single-use) public keys and exchange with them. Then they compute the single-use (session) shared secret key Z that is actual only in the current communication session.
2. User A generates a fake message $M < p$.
3. Users A and B perform the no-key encryption protocol using a commutative encryption function allowing to perform simultaneous ciphering the messages M and T. During the commutative ciphering, each of the users applies two different local keys for encrypting messages T and M.

At step three, it is to be used the encryption function that is computationally indistinguishable from the probabilistic commutative encryption function applied to the fake message. In other words, the ciphertexts C_1, C_2, and C_3 produced with the commutative encryption function for ciphering simultaneously the messages M and T could be potentially produced with the probabilistic commutative encryption function applied to the fake message, the probabilistic encryption process being dependent on the single-use shared key. Since the key Z is produced during the communication session without using any pre-agreed keys (secret or public), the communication protocol can be attributed to the class of no-key protocols. Possibility to connect the ciphertexts C_1, C_2, and C_3 with the probabilistic no-key protocol allows users, in case of the passive coercive attacks, to disclose only local keys used for transformation of the fake message M. To catch the users that they are cheating should be computationally impossible for the passive coercer while using the disclosed local keys, the single-use private keys, the single-use shared secret Z, and fake message M. Section IV introduces appropriate probabilistic no-key protocol, and in Section V, it proposed no-key DE protocol satisfying the last requirement.

4 No-Key Probabilistic Encryption Protocol

Two approaches can be used to provide no-key encryption protocols resistance to attacks based on chosen source message. The first approach is to select a prime modulus p, such that the number of $(p-1)/2$ is prime. The second approach is to embed probabilistic mechanisms in the original protocol. Thus, the users have reasonable motivation to use the probabilistic no-key protocol and this is significant for assigning probabilistic protocols with the DE protocols. Let us consider mechanism for providing security to chosen plaintext attacks.

When ciphering procedure depends on randomly selected values and the single-use shared key Z, a potential attacker performing the chosen plaintext attack is not able to eliminate the influence of random parameter on the produced ciphertext; therefore, his attack is inefficient, when encryption procedure is properly composed. If the key Z is produced during the communication, then a common assumption that prior to the execution of the no-key encryption protocol the sender and the receive do not share any secret values (keys) and have no public keys registered in the CA. (Otherwise, there is no need to use a no-key encryption protocol to send a secret message via a public channel, since to solve the problem one can use symmetric or public encryption.) Thus, applying the single-use shared keys does not contradict the notion of the no-key encryption.

Using exchange of the single-use public keys and computation of the single-use shared secret value, it has been designed the following protocol implementing the probabilistic no-key encryption of the message $M < p$:

1. User A generates a random value $k_A < (p-1)$, which plays the role of his private single-use key, computes his public single-use key $R_A = \alpha^{k_A} \bmod p$, and sends the value R_A to the user B.
2. User B generates a random value $k_B < (p-1)$ as his private single-use key, computes his public single-use key $R_B = \alpha^{k_B} \bmod p$, and sends the value R_B to the user A.
3. User A generates his local key $K_A = (e_A, d_A)$, where $d_A = e_A^{-1} \bmod (p-1)$ calculates the single-use shared secret $Z = R_B^{k_A} \bmod p$, generates a random value ρ, and computes the ciphertext $C_1 = (C'_1, C''_1)$ as the solution of the following system of linear equations with the unknown C'_1 and C''_1:

$$\begin{cases} C'_1 + C''_1 = \rho \bmod p \\ C'_1 + ZC''_1 = M^{e_A} \bmod p \end{cases} \tag{6}$$

Then, user A sends the ciphertext C_1 to the user B.

1. User B generates his local key $K_B = (e_B, d_B)$, where $d_B = e_B^{-1} \bmod (p-1)$ calculates the single-use shared secret $Z = R_A^{k_B} \bmod p$ and the value $S_1 = M^{e_A} \bmod p = (C'_1, ZC''_1) \bmod p$, generates a random value ρ', and calculates the ciphertext $C_2 = (C'_2, C''_2)$ as the solution of the following system of linear equations with the unknowns C'_2 and C''_2:

$$\begin{cases} C'_2 + C''_2 = \rho' \bmod p \\ C'_2 + ZC''_2 = S_1^{e_B} \bmod p \end{cases} \tag{7}$$

Then, user B sends the ciphertext C_2 to the user A.
2. User A generates a random value ρ'' and calculates value $S_2 \equiv S_1^{e_B} \equiv (C'_2 + ZC''_2) \bmod p$ and ciphertext $C_3 = (C'_3, C''_3)$ as solution of the following system of equations with the unknowns C'_3 and C''_3:

$$\begin{cases} C'_3 + C''_3 = \rho'' \bmod p \\ C'_3 + ZC''_3 = S_2^{e_A} \bmod p \end{cases} \tag{8}$$

Then, user A sends the ciphertext C_3 to the user B.
Having received the value C_3, user B computes the message M as follows: $M = (C'_3 + ZC''_3)^{d_B} \bmod p$.

5 No-Key Deniable Encryption

Using general construction scheme of the no-key deniable encryption protocol described in Sect. 3 and probabilistic no-key encryption protocol presented in Sect. 4, it is easy to write down the following protocol that implements secure transmission of the secret message $T < p$ via a public channel, which is resistant to passive coercive attacks:

1. Sender of the message T generates randomly his single-use private key k_A, calculates his single-use public key $R_A = \alpha^{k_A} \bmod p$, and sends R_A to the receiver.
2. Receiver generates randomly his single-use private key k_B, calculates his single-use public key $R_B = \alpha^{k_B} \bmod p$, and sends R_B to the user A.
3. Sender generates his local keys $K_A = (e_A, d_A)$, where $d_A = e_A^{-1} \bmod (p-1)$, and $Q_A = (\varepsilon_A, \delta_A)$, where $\delta_A = \varepsilon_A^{-1} \bmod (p-1)$, calculates the single-use shared secret $Z = R_B^{k_A} \bmod p$, forms a fake message $M < p$, and calculates the ciphertext $C_1 = (C_1', C_1'')$ as a solution of the following system of equations with the unknowns C_1' and C_1'':

$$\begin{cases} C_1' + Z^2 C_1'' = T^{\varepsilon_A} \bmod p \\ C_1' + Z C_1'' = M^{e_A} \bmod p \end{cases} \tag{9}$$

   Then, the sender sends the ciphertext C_1 to the receiver.
4. The receiver generates his local keys $K_B = (e_B, d_B)$, where $d_B = e_B^{-1} \bmod (p-1)$, and $Q_B = (\varepsilon_B, \delta_B)$, where $\delta_B = \varepsilon_B^{-1} \bmod (p-1)$, calculates the single-use shared secret $Z = R_A^{k_B} \bmod p$, and calculates the values $S_1 \equiv M^{e_A} \equiv (C_1' + ZC_1'') \bmod p$ and $U_1 \equiv T^{e_A} \equiv (C_1' + Z^2 C_1'') \bmod p$ and the ciphertext $C_2 = (C_2', C_2'')$ as solution of the following system of equations with the unknowns C_2' and C_2'':

$$\begin{cases} C_2' + Z^2 C_2'' = U_1^{e_B} \bmod p \\ C_2' + Z C_2'' = S_1^{e_B} \bmod p \end{cases} \tag{10}$$

   Then, the receiver sends the ciphertext C_2 to the sender.
5. The sender calculates the values $S_2 \equiv S_1^{e_B} \equiv (C_2' + ZC_2'') \bmod p$ and $U_2 \equiv U_1^{e_B} \equiv (C_2' + Z^2 C_2'') \bmod p$ and ciphertext $C_3 = (C_3', C_3'')$ as solution of the following system of equations with the unknowns C_3' and C_3'':

$$\begin{cases} C_3' + Z^2 C_3'' = U_2^{\delta_A} \bmod p \\ C_3' + Z C_3'' = S_2^{e_A} \bmod p \end{cases} \tag{11}$$

   Then, the sender sends the value C_3 to the receiver.

After receiving the ciphertext C_3, the receiver computes the message T:

$$T = (C_3' + Z^2 C_3'')^{\delta_B} \bmod p \tag{12}$$

If necessary (in the case of coercive attack), the receiver can also be calculated fake message M as follows:

$$M = (C_3' + ZC_3'')^{d_B} \bmod p \tag{13}$$

The proof of the correctness of the protocol:
Recovery of the secret message:

$$(C_3' + Z^2 C_3'')^{\delta_B} \equiv \left(U_2^{\delta_A}\right)^{\delta_B} \equiv (U_1^{\varepsilon_B})^{\delta_A \delta_B} \equiv (T^{\varepsilon_A})^{\varepsilon_B \delta_A \delta_B} \equiv T \bmod p \tag{14}$$

Recovery of the fake message:

$$(C_3' + ZC_3'')^{d_B} \equiv \left(S_2^{d_A}\right)^{d_B} \equiv (S_1^{e_B})^{d_A d_B} \equiv (M^{e_A})^{e_B e_A d_B} \equiv M \bmod p \quad (15)$$

When being coerced by a passive attacker, the sender and receiver of the message disclose the fake message M and the keys k_A, R_A, k_B, R_B, Z, (e_A, d_A), and (e_B, d_B). They also say that for securely sending the message M, they used a probabilistic no-key encryption protocol. Since the intercepted by the attacker by procedures specified by the probabilistic no-key encryption protocol associated with no-key deniable encryption protocol, the attacker has the following two possibilities: (i) to agree with the users and (ii) to prove the ciphertexts were produced with the no-key DE protocol. However, the second possibility is computationally infeasible, since to show the difference between the values $\rho_i = (C_i' + C_i'') \bmod p (i = 1, 2, 3)$ and random values the coercer has to compute one of the local keys Q_A or Q_B and to recover the message T. Computing one of the local keys Q_A or Q_B is connected with solving the problem of finding the discrete logarithm modulo p. The last is selected so that computing discrete logarithm is computationally impracticable (see Sect. 2).

In comparison with the known public key DE protocols [11,13] in which the message is encrypted consecutively bit by bit (each bit is sent in the form of 1024-bit pseudorandom number) in the proposed protocols, the message is transformed as a single data block that provides significantly higher performance.

6 Conclusion

Applying commutative encryption algorithm, it has been proposed a method for no-key DE implemented as simultaneous ciphering two messages, secret and fake ones, which is based on public agreement of the single-use shared key with exchange of the single-use public keys of the participants of communication protocol. An important point of the method is fulfillment of the requirement of computationally indistinguishable from probabilistic no-key encryption protocol. To implement the method as an practical no-key DE protocol, it designed a probabilistic no-key protocol in the encryption process of which it used the single-use shared key. Then, the no-key DE protocol has been constructed as pseudo-probabilistic no-key encryption protocol.

The proposed method and protocol for no-key DE provide resistance to passive coercive attacks. In cases when it is required to provide resistance to potential attacks performed by active coercer that impersonates the sender or receiver of secret message, one should imbed in the proposed protocol mechanism for verifying the authenticity of the data sent via communication channel. For example, one can imbed steps for authentication of the single-use public keys. The authentication mechanism can be implemented with using short (having size 16 to 56 bits) pre-shared secret keys, like in the protocol described in [5]; however, in this case, one will have a shared-key DE protocol.

The proposed no-key DE protocol provides sub-exponential deniability. To get the exponential deniability to passive coercive attacks, one can implement the proposed method using computations on elliptic curves; however, detailed consideration of this item represents a topic of an individual work.

References

1. Canetti, R., Dwork, C., Naor, M., Ostrovsky, R.: Deniable Encryption. Proceedings Advances in Cryptology-CRYPTO 1997. Lectute Notes in Computer Science. Springer-Verlag. Berlin, Heidelberg, New York, (1997), vol. 1294, 90–104
2. Meng, B.: A Secure Internet Voting Protocol Based on Non-interactive Deniable Authentication Protocol and Proof Protocol that Two Ciphertexts are Encryption of the Same Plaintext. Journal of Networks. (2009), vol. 4, no. 5, 370–377
3. Ishai, Y., Kushilevits, E., Ostrovsky, R.: Efficient Non-interactive Secure Computation. Advances in Cryptology - EUROCRYPT 2011. Lecture Notes in Computer Science. Springer-Verlag. Berlin, Heidelberg, New York. (2011), vol. 6632, 406–425
4. Wang C., Wang, J.A.: Shared-key and Receiver-deniable Encryption Scheme over Lattice. Journal of Computational Information Systems. (2012), vol. 8, no. 2, 747–753
5. Moldovyan, N.A., Moldovyan, A.A., Shcherbacov, A.V.: Deniable-encryption protocol using commutative transformation. Workshop on Foundations of Informatics. (2016) 285–298
6. Hellman M.E., Pohlig, S.C.: Exponentiation Cryptographic Apparatus and Method. U.S. Patent No 4, 424, 414. (1984)
7. Moldovyan, N.A., Berezin, A.N., Kornienko, A.A., Moldovyan, A.A.: Bi-deniable Public-Encryption Protocols Based on Standard PKI. Proceedings of the 18th FRUCT and ISPIT Conference, Technopark of ITMO University, Saint-Petersburg, Russia. FRUCT Oy, Finland. (2016) 212–219
8. Moldovyan, A.A., Moldovyan, N.A., Shcherbakov, V.A.: Bi-Deniable Public-Key Encryption Protocol Secure Against Active Coercive Adversary. Buletinul Academiei de Stiinte a Republicii Moldova. Mathematica. (2014), no. 3, 23–29
9. Diffie W., Hellman, M.E.: New Directions in Cryptography. IEEE Transactions on Information Theory. (1976), vol. IT-22, 644–654
10. Menezes, A.J., Oorschot, P.C., Vanstone, S.A.: Applied cryptography. CRC Press, New York, London, (1996)
11. Ibrahim, M.H.: A method for obtaining deniable Public-Key Encryption. International Journal of Network Security. (2009), vol. 8, no. 1, 1–9
12. Barakat, M.T.: A New Sender-Side Public-Key Deniable Encryption Scheme with Fast Decryption. KSII Transactions on Internet and Information Systems. (2014), vol. 8, no. 9, 3231–3249
13. Dachman-Soled, D.: On minimal assumptions for sender-deniable public key encryption. Public-Key CryptographyPKC 2014: 17th International Conference on Practice and Theory in Public-Key Cryptography. Lecture Notes in Computer Science. SpringerVerlag. Berlin, Heidelberg, New York. (2014), vol. 8383, 574–591

JCIA: A Tool for Change Impact Analysis of Java EE Applications

Le Ba Cuong[1,2](✉), Van Son Nguyen[2], Duc Anh Nguyen[2], Pham Ngoc Hung[2], and Dinh Hieu Vo[2]

[1] Academy of Cryptography Techniques, 141 Chien Thang, Hanoi, Vietnam
[2] VNU University of Engineering and Technology, 144 Xuan Thuy, Hanoi, Vietnam
cuonglb@hocvienact.edu.vn, sonnguyen@vnu.edu.vn, nguyenducanh@vnu.edu.vn, hungpn@vnu.edu.vn, hieuvd@vnu.edu.vn

Abstract. This paper presents a novel tool for change impact analysis of Java EE applications named *JCIA*. Analyzing the source code of the Java EE applications is a big challenge because of the complexity and large scale of the applications. Moreover, components in Java EE applications are not only in Java language but also in different languages such as XHTML, XML, JSP. This tool analyzes source code of Java EE applications for building the dependency graphs (called JDG). The main idea for generating JDG is based on developing the source code analyzers for the typical technologies of Java EE such as JavaServer Faces, Context and Dependency Injection, Web services. Based on the obtained JDG and the given change sets, *JCIA* calculates the corresponding impact sets by applying the change impact analysis (CIA) based on change types and Wave-CIA method. The calculated impact sets help managers in planning and estimating changes, developers in implementing changes, and testers with regression testing.

Keywords: Change impact analysis · Static analysis · source code analysis · Java EE applications

1 Introduction

In the software evolution, change requests are daily and unavoidable activities because of various reasons including adding new features, changing businesses, improving quality. Unfortunately, a small change to the software may cause several unpredictable effects and even potential errors on the evolving software. As a result, the change requests propose several challenges for software maintainers. The managers are faced with the cost and time estimation, management of the changes. Moreover, the developers have many difficulties in identifying which modules need to be modified in order to accomplish a change. And the testers are suffered from minimizing the number of selected test cases for regression testing. Thus, one of the main issues in the software evolution is how to identify

V. Bhateja et al. (eds.), *Information Systems Design and Intelligent Applications*, Advances in Intelligent Systems and Computing 672,
https://doi.org/10.1007/978-981-10-7512-4_11

source code components affected by the changes. The change impact analysis (CIA) methods have been known as the potential solutions to deal with this problem [10]. CIA plays a crucial role in software evolution and especially in enterprise applications because of the complexity in architecture and large scale of these applications.

The Java Platform Enterprise Edition (Java EE) is recently very important and popular in developing the enterprise applications. Besides that, the maintenance of these applications is a complex process, high cost, and time. However, the current CIA methods do not support Java EE applications because they handle only pure Java applications [10]. The CIA methods can be performed by analyzing the static and dynamic information about the software systems [9]. The static analysis methods are often performed by analyzing the syntax and semantic, or evolutionary dependence of the applications. JRipples [3] and Impact-Miner [4], which are famous CIA tools plugged in Eclipse, have been developed by using the static source code analysis and the co-change analysis from source code repositories, respectively. The dynamic analysis is performed on the information collected during application execution. Chianti [13] is an Eclipse plug-in that uses dynamic information of the applications for predicting the impacts of the changes. However, there have been no CIA tools for the Java EE applications. Unlike pure Java applications, the logic of Java EE applications has been controlled by not only interactions between components in Java language but also combination of source code written in different programming languages such as XHTML, XML, JSP. via the configuration files and annotations [7]. Moreover, several emerging technologies have been used in developing these applications such as JSF, CDI, Web services [7]. As a consequence, a change to the software may affect the components written not only in the same programming language but also in other languages. In addition, CIA for the applications in general is hard to deal with, even more difficult problem for Java EE applications in particular because of the complexity and daily business changes of the applications. In software industry, software companies are difficult to understand the architectures of the applications before the changes have been implemented because of lacking or incomplete software documentations. Furthermore, the change impact prediction, cost estimation, regression testing, etc., have been encountered many challenges because of the complexity and large scale of the applications. Thus, developing a CIA tool for Java EE applications under analyzing is still an open and interesting problem.

This paper introduces a novel tool for CIA of the Java EE applications called *JCIA*. This tool is developed based on the static analysis of the dependencies in Java EE technologies such as J2SE, JSF, CDI, Web services in order to construct the dependency graphs of the Java EE applications. The obtained graph and the given change sets are used to calculate the potential impact sets based on the change types [1] and Wave-CIA method [11].

The rest of this paper is organized as follows. At first, we show an overview of the method in Sect. 2. In Sect. 3, an implementation and experiments of the tool for CIA of the Java EE applications are presented. Finally, we conclude the paper in Sect. 4.

2 CIA Method for Java EE Applications

This section gives an overview of the change impact analysis methodology of Java EE applications. Given source code project of a Java EE application, the proposed method firstly generates a dependency graph which represents dependencies among components in the Java EE application at class member level. In second step, given a change set, the method computes a potential impact set corresponding to the change set based on the dependency graph.

Definition 1. (*Java EE Dependency Graph*). Given a Java EE project, a Java EE dependency graph, denoted JDG, is defined as a pair $\langle V, E\rangle$, where $V = \{v_0, v_1, \ldots, v_n\}$ is a list of nodes representing components such as packages, files, classes, methods, attributes, and $E = \{(v_i, v_j) \mid v_i, v_j \in V\} \subseteq V \times V$, is a list of directed edges. Each edge $(v_i,\ v_j)$ represents a dependency between v_i and v_j that means v_i depends on v_j.

2.1 Generating Dependency Graphs

The given source code project is preprocessed in order to construct a directory tree at file level. The corresponding Java EE dependency graph defined in Definition 1 is constructed from the tree. The vertexes and edges of JDG had been obtained from those of the tree. Additionally, the vertexes representing the Java source files are analyzed for identifying their child nodes such as classes, methods, attributes, and the dependencies between them in order to add to the JDG. Moreover, the dependencies among components related to the Java EE technologies such as JSF, CDI, and Web services are identified for adding to the edges of JDG. As a result, the JDG corresponding to the given project is generated.

With each Java source file, the corresponding abstract syntax tree (AST) has been built by using the Java development tool (JDT[1]). Based on the obtained AST, the components at lower level such as classes, methods, attributes, and the dependencies among them are discovered and added to JDG.

JavaServer Faces (JSF) technology is a framework for developing and building user interface components at server side for Web applications. JSF technology [7] is based on the model-view-controller (MVC) architecture. In the model layer, the JavaBean components contain business functions of the application and object data. The view layer including XHTML, JSP files, etc. deals with the user interfaces. The controller layer handles the user interactions and communication between the view and the model layer. The components at different layers interact with each other not only in the same language but also in different languages. The interactions between the managed beans implemented in Java files are identified by the annotations (e.g., @ManagedBean). Moreover, the components in html/jsp files interact with each other or with managed beans in Java files via configuration files (e.g., Web.xml, faces-config.xml). Thus, the dependencies among these components are determined by analyzing the XML

[1] http://www.eclipse.org/jdt/.

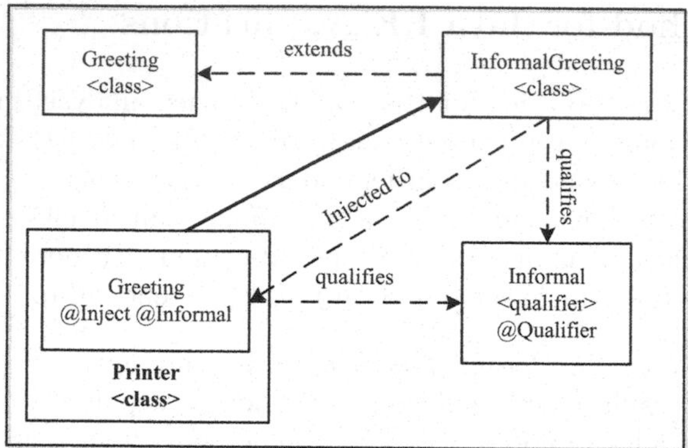

Fig. 1. The *simplegreeting* CDI example

tags such as <navigation-rule>, <managed-bean> in the configuration files or the annotations in the managed beans.

Context and dependency injection (CDI) is an integral part of Java EE that fills the gaps between JSF in the view layer and enterprise beans in the business layer. CDI technology [7] allows Java EE components such as EJB session beans and JSF-managed beans to be injected and to interact in a loosely coupled mechanism. In this paper, the injected beans are defined as injectees and the others are injectors. A change in the injectees may cause several effects on the injectors. Thus, the dependencies between the injectors and the injectees are identified. Figure 1 shows a *simplegreeting*[2] project as a CDI example. In this project, the *InformalGreeting* class extends the *Greeting* class. The *Printer* class, which is a JSF managed bean, declares a field named *greeting*. By default, *greeting* is an instance of *Greeting*. However, *greeting* is injected by using two annotations such as @Inject and @Informal, where @Informal is defined as Qualifier in *InformalGreeting*. As a result, *greeting* is an instance of *InformalGreeting* class. Therefore, *Printer* class depends on *InformalGreeting* class.

The dependencies between the injectors and the injectees in CDI are identified based on injection points which are defined by the annotations such as @Inject, @Produces. The injectors are the components such as classes, methods, and attributes containing the injection points. With each injector, the corresponding injectee is identified by analyzing the semantic annotations in CDI specification. For example, an instance of Greeting class (named *greeting*) is replaced by the instance of *InformalGreeting* in case *greeting* is injected by two annotations such as @Inject and @Informal. The dependencies from the injectors to the injectees are added to JDG.

Web services technologies include Java API for XML Web Services (JAX-WS) and Java API for RESTful Web Services [7] (JAX-RS). In this paper, we only consider the first one which is more popular than the other. In JAX-WS technology, developers specify the Web service operations by defining methods

[2] https://docs.oracle.com/cd/E19798-01/821-1841/gjbju/index.html.

in an interface and implement these methods in a corresponding Java class. On the client side, a proxy, which is a local object representing the service, is automatically generated. Then, the client application invokes service methods on the proxy. A statement on the client application calls a service method mean that a service operation on the provider side is invoked. As a result, the methods invoking the service operations depend on the corresponding operations on the provider side. The client methods using services are identified by finding an annotation named @WebServiceClient. Based on the invoked methods at client side, the corresponding service providers are determined.

2.2 Analyzing Change Impacts

The change impact analysis approaches can be performed based on the static information analysis or the dynamic information analysis of under analyzing applications. The static analysis includes three major techniques such as structural static analysis [1,2,8,14], textual analysis [12], and historical analysis [15]. Most of the current researches focused on the structural analysis which usually consists of two phases [10] such as extracting intermediate representations as dependency graphs of applications and calculating potential change impact sets based on the graphs. There are several algorithms to calculate the impact set such as graph traverse [2], Wave-CIA [11], CIA based on the change types [1]. Combining CIA based on the change types and Wave-CIA has been known as particularly suitable solution for multiple changes at method/attribute level which occurs regularly in software industry.

The change impact analyzer has been implemented based on the impact mechanism of the change types [1] at class and method/attribute levels and Wave-CIA [11] method. Given a change set, before applying Wave-CIA method, the change impact analyzer calculates an initial impact set from the change set according to the impact mechanism of the change types. After that, the obtained initial impact set is used as an input of Wave-CIA method for calculating the potential impact set. The Wave-CIA method is based on the idea of water wave propagation. The ripples on water surface will meet each other and generate several spreading center. Like this natural process, the method calculates a core set corresponding to the spreading center based on JDG. The core set is constructed from the initial impact set and elements impacted by more than one element in the initial impact set. Then, the potential impact set is extended from the core set by adding components generated from the propagation analysis of the core.

3 JCIA Implementation and Experiments

3.1 JCIA Implementation

JCIA has been implemented using Java technology. This tool consists of four major modules named preprocessor, dependency graph generator, change impact analyzer, and visualization (Fig. 2). The first module analyzes the source code project of Java EE applications for building the directory trees. The second module includes four dependency analyzers corresponding to the typical Java EE

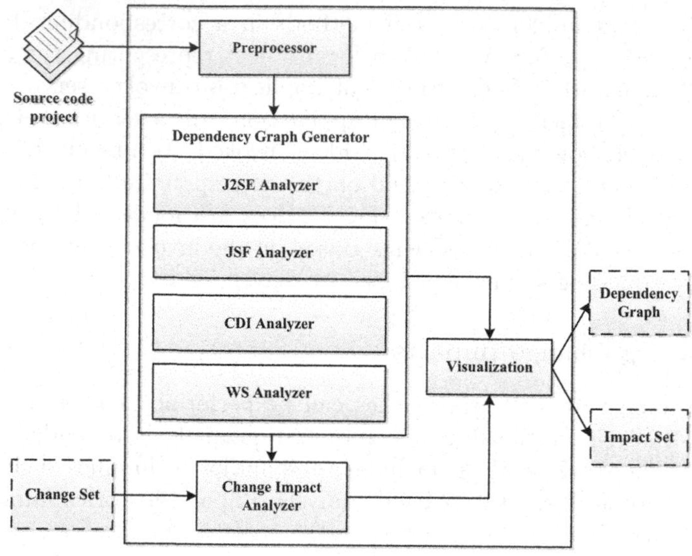

Fig. 2. JCIA architecture

technologies including J2SE, JSF, CDI, and Web services. This module generates the Java EE dependency graphs from the directory trees. The change impact analyzer has realized the CIA method shown in Sect. 2.2. The analyzer is used for calculating the potential impact set corresponding to the given change set. The last module has been implemented by extending *D*3 Javascript library [6]. It is responsible for visualizing the change impact informations such as the Java EE dependency graphs, the potential impact sets.

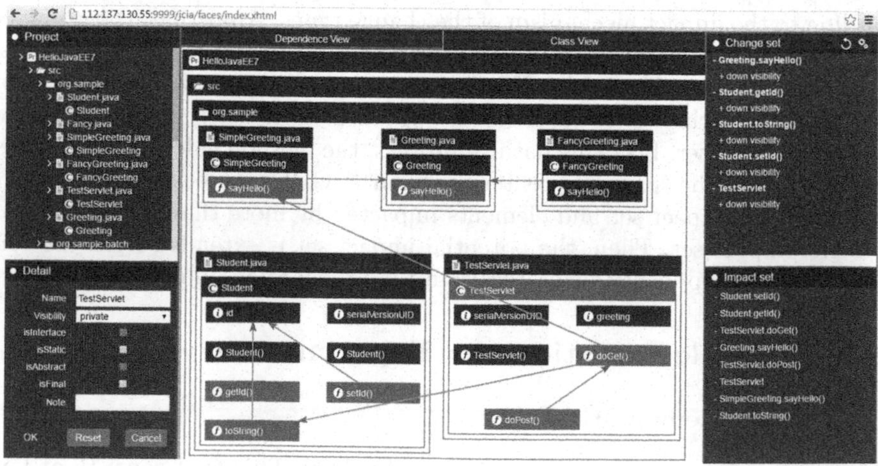

Fig. 3. JCIA with the sample application

Figure 3 shows a snapshot of a JCIA session. In this view, the *Dependency* tab displays the generated JDG where users can choose the changed components.

Table 1. Results of the sample experiments

No	Changes	Predicted	Correctly Predicted	Really Impacted	P (%)	R (%)
1	3	6	6	7	100	86
2	5	8	8	10	100	80
3	7	10	10	11	100	91

The *Class* tab shows the output of the tool at class level. The tool is only supported by the changes at class and method/attribute levels. The users can choose these changed components by their change types from the *Detail* tab. As a result, these changes are listed on the *Change Set* tab. After applying our change impact analysis on the change set, the potential impact set is calculated and viewed on the *Impact Set* tab. Additionally, the directory structure of the Java EE application is represented on the *Project* tab in order to provide another view of the application.

3.2 Experiments

In order to evaluate the practical usefulness and correctness of *JCIA*, we used two well-known measures including precision and recall [5]. The precision (denoted P) and recall (denoted R) shows the percentage of the number of components correctly predicted corresponding to the number of predicted components and the number of really impacted components, respectively.

In this research, we tested this tool by applying several Java EE applications in order to show the effectiveness of the tool. For this purpose, a sample Java EE application named *HelloJavaEE7*[3] is selected for evaluating the accuracy of the tool. The application had been analyzed by *JCIA* in order to generate the corresponding dependency graph shown in Fig. 3. Based on the obtained graph, we tested the change impact calculation with three change sets in different sizes. The first change set has three changes included change return type of *Student.id*, down visibility of *Greeting.sayHello*() and *FancyGreeting.sayHello*(). The second change set includes all changes in the first set and two other changes included up visibility of *TestServlet.greeting* and down visibility of *Student* class. And the last one contains five changes in the second set and two other changes included down visibility of *SimpleGreeting.sayHello*() and *Student.toString*(). On the one hand, each change set would be implemented by our group of students in order to identify the really modified components. On the other hand, each change set is used as an input of *JCIA* for calculating the potential impact set. Table 1 shows the experimental results of this tool. In this experiment, we evaluate the size of the change sets (denoted *Changes*), the number of the tests (i.e., *No*), the size of the predicted set (i.e., *Predicted*), the size of the correctly predicted set (i.e., *Correctly Predicted*), the number of really impacted components (i.e., *Really Impacted*), the precision (i.e., *P*), and the recall (i.e., *R*). As

[3] https://github.com/javaee-samples/javaee7-eclipse.

it can be seen from Table 1, the precision of this tool is always equal to 100% and the recall is also high accuracy with more than 80% in all cases. That means, all predicted impacts are correct and the tool also identifies almost real impacts. Moreover, the running time of this tool is an approximate real time.

We also selected several practical Java EE projects included Java EE documents[4] for testing *JCIA*. In which, the project named *Duke'sforest* is the typical case study for Java EE with approximately 15 KLOC. With this project, on the one hand, by manually, we showed that there are six sub-projects such as *dukes-store*, *dukes-payment*, *dukes-shipment*, *dukes-resources*, *entities*, and *events* along with all their packages (112 packages), files (164 source code files), and 64 Java classes. Moreover, the dependencies among these components have been found by manual analyzing. On the other hand, by applied JCIA, this tool represented exactly the expected above results. The running time of the tool for generating the dependency graph and calculating the potential impact set is acceptable. However, checking the correctness of the calculating potential impact set from the given change set is a hard problem because of the large scale and complexity of the application and lack of the expert knowledge about the application.

The experimental results show clearly that the proposed method is sound, efficient, and easy to be applied because of fully automated and supporting most of the technologies in Java EE applications. The dependency graphs can help understanding of the applications in case of lacking or incomplete software documentations which are popular in software industry. Moreover, the potential impact set is used not only for maintenance task such as planning changes, assessing the cost of changes, implementing changes but also for the regression testing to minimize the selection of test cases.

3.3 Discussion

JCIA is deployed as a Web application that helps users easy-to-use anytime and anywhere with Internet connection. This is the first CIA tool supported for almost all practical Java EE applications. The output of this tool can help managers to decide among various change solutions, estimate the costs, time, and resources, etc. Moreover, developers can use this output in order to understand the applications and implement the changes. Additionally, the tool also supports testers to minimize test cases selection for regression testing. One another advantage of the tool is that the dependency graphs of Java EE applications (JDGs) is generated by applying the static analysis techniques. Therefore, the JDGs contain almost all possible dependencies among components of the Java EE applications. Furthermore, the running time of JCIA on the current server with normal configurations is acceptable. In the future, we strongly believe that the tool will be applied widely in software companies to support change management and maintenance process in order to improve software quality.

---

[4] http://docs.oracle.com/javaee/6/tutorial/doc/gkgjw.html.

JCIA has been restricted by some disadvantages. In complex Java EE applications, any dependencies may not to be discovered by the tool such as the dependencies between source code components and database components, etc. Therefore, the tool cannot predict the effects of changes in these cases. Furthermore, for several practical Java EE applications which are large scale and complex, the calculated impact sets may contain a lot of components. However, all dependencies among the components in the applications are considered with similar weight. So, the tool has not yet ranked these impacted components by impact levels. In practical, the dependencies among the components have different weights because these components have the different impact levels. Therefore, the potential impact sets can be ranked in order to help the maintainers considering the impacted components which are high priority. As a result, the cost and time for the maintenance process are reduced.

4 Conclusion

This paper has presented a tool for change impact analysis of Java EE applications based on the static analysis. The analyzers of this tool including J2SE, JSF, CDI, and Web services analyzer are applied on the source code projects in order to generate the Java EE dependency graphs at method/attribute level, which is used for calculating the potential impact sets corresponding to the given change sets. In this tool, the change impact analyzer calculates the impact sets from the change sets selected by the users. Every changes at class or class member level cause the certain impact on several components based on the change types. So, the analyzer calculates the initial impact sets including the certain impact components, and then Wave-CIA method has been applied on the dependency graphs for identifying the potential impact sets corresponding to the initial impact sets. Moreover, this tool is also integrated a visualization module which is extended from *D*3 Javascript library [6]. The visualization provides the ability to visualize the results of the method and to interact with the users such as selection of the change sets.

In the future, we plan to upgrade the tool in order to fully support Java EE applications. We are also investigating other CIA methods for Java EE applications in order to identify the best solution or propose a novel CIA method. Moreover, we would like to consider exploring a coupling measurement in Java EE applications for ranking the impact sets. Additionally, we also intend to research on how to apply the resulted impact set for software maintenance activities such as change schemes, selecting regression testing cases. Finally, we are exploring the wide application of this tool in software companies for supporting the maintenance tasks.

Acknowledgements. This work is supported by the project no. QG.16.31 granted by Vietnam National University, Hanoi (VNU).

References

1. L. Badri, M. Badri, and N. Joly. Towards a change impact analysis model for java programs: An empirical evaluation. *Journal of Software*, 10(4):441–453, 2015.
2. L. Badri, M. Badri, and D. St-Yves. Supporting predictive change impact analysis: a control call graph based technique. In *APSEC'05*, pages 167–175. IEEE, 2005.
3. J. Buckner, J. Buchta, M. Petrenko, and V. Rajlich. Jripples: A tool for program comprehension during incremental change. In *Proceedings of the 13th International Workshop on Program Comprehension*, IWPC '05, pages 149–152, Washington, DC, USA, 2005. IEEE Computer Society.
4. B. Dit, M. Wagner, S. Wen, W. Wang, M. Linares-Vásquez, D. Poshyvanyk, and H. Kagdi. Impactminer: A tool for change impact analysis. In *Companion Proc. of the 36th Int. Conf. on Soft. Engineering*, ICSE Companion 2014, pages 540–543, New York, NY, USA, 2014. ACM.
5. L. Hattori, D. Guerrero, J. Figueiredo, J. Brunet, and J. Damasio. On the precision and accuracy of impact analysis techniques. In *Proc. of the 7th IEEE/ACIS Int. Conf. on Computer and Information Science*, pages 513–518. IEEE, 2008.
6. A. Jain. Data visualization with the d3. js javascript library. *Journal of Computing Sciences in Colleges*, 30(2):139–141, 2014.
7. E. Jendrock, R. Cervera-Navarro, I. Evans, D. Gollapudi, K. Haase, W. Markito, and C. Srivathsa. *The Java EE6 Tutorial: Advanced Topics.* Addison-Wesley, 2013.
8. M. Lee, A. J. Offutt, and R. T. Alexander. Algorithmic analysis of the impacts of changes to object-oriented software. In *Tech. of OO Languages and Systems, 2000. TOOLS 34. Proc. 34th Int. Con. on*, pages 61–70, 2000.
9. S. Lehnert. A taxonomy for software change impact analysis. In *Proc. of the 12th Int. Workshop on Principles of Soft. Evol. and the 7th Annual ERCIM Workshop on Soft. Evolution*, pages 41–50, New York, NY, USA, 2011. ACM.
10. B. Li, X. Sun, H. Leung, and S. Zhang. A survey of code-based change impact analysis techniques. *Soft. Testing, Verification and Reliability*, 23(8):613–646, 2013.
11. B. Li, Q. Zhang, X. Sun, and H. Leung. Wave-cia: a novel cia approach based on call graph mining. In *Proc. of the 28th Annual ACM Sym. on Applied Computing*, pages 1000–1005. ACM, 2013.
12. D. Poshyvanyk, A. Marcus, R. Ferenc, and T. Gyimóthy. Using information retrieval based coupling measures for impact analysis. *Empirical software engineering*, 14(1):5–32, 2009.
13. X. Ren, F. Shah, F. Tip, B. G. Ryder, and O. Chesley. Chianti: A tool for change impact analysis of java programs. *SIGPLAN Not.*, 39(10):432–448, Oct. 2004.
14. X. Sun, B. Li, C. Tao, W. Wen, and S. Zhang. Change impact analysis based on a taxonomy of change types. In *2010 IEEE 34th Annual Computer Software and Applications Conference*, pages 373–382. IEEE, 2010.
15. T. Zimmermann, A. Zeller, P. Weissgerber, and S. Diehl. Mining version histories to guide software changes. *IEEE Trans. on Soft. Engineering*, 31(6):429–445, 2005.

On the Robustness of Cry Detection Methods in Real Neonatal Intensive Care Units

Manh Chinh Dang[1(✉)], Antoni Martínez-Ballesté[2], Ngoc Minh Pham[1], and Thanh Trung Dang[3]

[1] Institute of Information Technology, 18 Hoang Quoc Viet Road, Hanoi, Vietnam
{dangmanhchinhbkhn, pnminh2001}@gmail.com

[2] Universitat Rovira i Virgili, Av. Paisos Catalans 26, Tarragona, Catalonia, Spain
antoni.martinez@urv.cat

[3] Electric Power University, 235 Hoang Quoc Viet Road, Hanoi, Vietnam
thanhtrungepu@gmail.com

Abstract. The detection of cry is crucial in intelligent computerized systems that aim at assessing the well-being of neonates during their hospitalization periods. Moreover, a precise characterization of cry allows its classification (e.g., hunger, pain, tiredness…). Although several cry detection and characterization techniques can be found in the literature, there is no testing of such techniques in real-life environments such as hospital intensive care units. In this article, we first summarize the problem of background noise in intensive care units that may prevent the operation of cry detection algorithms from succeeding. Second, we implement a specific cry detection technique that is based on some of the relevant cry detection proposals that have been found in the literature. Finally, we test this method using audio samples recorded in a real neonatal intensive care unit.

1 Introduction

Some newborns have to spend a period of time in neonatal intensive care units (NICUs). Most of them are preterm newborns who suffer from some immaturity or disease. During hospitalization, newborns might suffer from pain (for instance, due to invasive procedures) and if this pain is not treated correctly, the child could suffer from some neurological disorders in the future. In order to detect whether neonatal patients are suffering from pain or discomfort, health professionals proceed by directly observing physiological parameters (i.e., heartbeat rate, temperature) and behavioral parameters (frowning, crying…) and using some assessment scales to figure out if some pain must be relieved. These procedures are done in an hourly basis, by a reduced number of people that also take care of the other typical duties in the NICUs.

Moreover, current trends in nursing for the neonates focus on personalized observations of the newborn (for instance, the progressively adopted NIDCAP, Newborn Individualized Developmental Care and Assessment Program [1]).

V. Bhateja et al. (eds.), *Information Systems Design and Intelligent Applications*, Advances in Intelligent Systems and Computing 672,
https://doi.org/10.1007/978-981-10-7512-4_12

In order to improve the quality in this aspect of healthcare, some attempts to automatize the detection of neonate's cry have been published. Accurately characterizing the cry in real time could be used in both automatic pain assessment [2] and NIDCAP procedures. In fact, if some neonate in large NICUs is crying, it could be useful to trigger alarms that request the attention of the nurses.

Regarding methods and techniques to analyze baby's cries, the literature deals with "neat environment tests": Only cry sounds are analyzed, and to the sake of our knowledge, proposals do not consider the background noise. This topic attracts the attention of some researches, merely on measuring typical sounds and noises in NICUs and how they affect the well-being of the neonates. In our work, we took a look at papers to get the brief knowledge of background noise in NICUs environments.

1.1 Background Noise in Neonatal Intensive Care Units

For cry analysis in NICUs, background noise is an important aspect which is taken into account. Background noise can come from several sources, namely equipment or machines in hospital rooms, ventilator systems, sounds generated when interacting with incubators, staff talking, phone ring, alarm noise. There are some standards about noise level, for example, the standard of American Academy of Pediatrics. If the noise level inside NICUs exceeds these standards for a long period, it will cause some problems on newborns; it can affect neurodevelopment or other important aspects.

In [3], authors describe these environments in terms of background noise, in a recent contribution. Measures were conducted in both an old and a new hospital. Their sound environment was surveyed for 24 h periods in Melbourne and compared to Australian recommendation, which states maximum and minimum sound levels. For instance, the maximum sound level should be less than 65 dB and no more than 6 min in an hour. They also consider sound peaks (defined as jumps of more than 12 dB). The granularity for noise measuring is 5 s, for which averages of decibels are computed. The results show that in general that recommendation is not achieved in a half of observation periods, and the results are not different from the old building to the new one.

Moreover, in [4], authors assess background noise according to American Academy of Pediatrics, which recommends that sound levels should not exceed the maximum acceptable level of 45 dB. Authors concluded that the sound environment in the NICUs is louder than most home or office environments and contains disturbing noises of short duration and at irregular intervals. Elevated levels of speech are needed to overcome the noisy environment in the NICU, thereby increasing the negative impacts on staff, newborns, and their families. High noise levels are associated with an increased rate of errors and accidents, leading to decreased performance among staff. The aim of interventions included in this review is to reduce sound levels to 45 dB or less. This can be achieved by lowering the sound levels in an entire unit, treating the infant in a section of a NICU (i.e., in a "private" room), or in incubators in which the sound levels are controlled, or reducing the sound levels that reach the individual infant by using earmuffs or earplugs.

Finally, in [5], authors describe background noise both quantitatively (measuring sound) and qualitatively (by means of interviews). They compare data collected during day and night periods and conclude that there is no statistically significant difference.

1.2 Contribution and Plan of This Paper

In this paper, we analyze the robustness of a cry detection method against real-life sounds that occur in a neonatal intensive care unit. To that end, we implement a method, described in Sect. 2, that is representative of the most outstanding ones found in the literature. In Sect. 3, we test this implementation using audio samples from a cry database, mixed with real audio samples of NICU's background noises. Finally, Sect. 4 concludes the paper.

2 A Cry Detection Technique

In this section, the method implemented for cry detection is described. The goal of the method is to detect cry sounds in the waves obtained in the real neonatal intensive care unit (NICU). For the sake of simplicity, we have implemented software that analyzes the recorded waves. However, in a real deployment of the system, it would be able to analyze the sound over real-time acquired audio.

2.1 Key Concepts

Cry analysis methods make use of several concepts or techniques that play a key role in the process. Sampling is the first step in analysis of audio from real environments. Sounds to be analyzed are in the form of samples of b-bit resolution (usually, $b = 16$ for generic audio and $b = 8$ for voice-targeted applications). Samples are taken at a specific sampling rate (being 44.100 Hz a typical value for generic audio and 8.000 Hz the value for voice-targeted applications). Hence, sampling consists of the reduction of a continuous-time signal to a discrete-time signal.

Most audio analysis applications rely on frequency estimation. In this step, a window is applied to select a certain interval of data to analyze. Normally, a cry duration has the minimum length of 200 ms, so that the window size normally is 50 ms. After determining the window size, a set of data is collected consecutively from the sampling data set which is obtained from the sampling step. Then, a time to frequency transformation technique, such as fast Fourier transform, is applied.

Besides sampling and frequency estimation, the following concepts play a key role in the reviewed cry analysis techniques [6–8]:

- **Fundamental frequency**: It is the most significant frequency contribution of the analyzed audio sample. In terms of frequency analysis of periodic waves, it is the lowest frequency in the sum.
- **Short time energy (STE)**: It is the average value of the samples, considering its absolute value, for a generally short number of samples. This granularity is related to the precision of the implementation.

- **Melodic shape**: It is the result of the observation of successive values of fundamental frequencies. Hence, the raising and falling of the cry sound can be characterized by analyzing the features of this melodic shape.
- **Glide**: It is a very rapid change of fundamental frequency.

2.2 Description of the Method

The method works with a WAVE audio file. We assume that it has been obtained using a high-sensitivity microphone placed next to the incubator. The audio file is recorded at a sampling rate of r samples per second, b = 16 bit. The file is divided into a succession of N windows $\{w_1, w_2, \ldots, w_N\}$ of length S, being the latter a parameter of the system. Thus, w_i is a succession $\{s_1^i, s_2^i, \ldots, s_S^i\}$ of samples.

The first step consists of detecting a succession of non-silent windows. Note that several successions might be found in a file, but each succession will be analyzed individually. In order to classify w_i into silent or non-silent, the STE is utilized. Then, if STE (w_i) > T, i.e., a specific threshold, w_i is classified as non-silent; otherwise, it is considered a silent window and therefore is discarder for further analysis.

The problem in the first step is related to detecting as non-silent window sound samples that belong to a near source of background noise. For instance, if a breathing machine is working next to the incubator, the microphone will record this noise but with a high STE. In order to avoid this problem and assuming that cry frequencies are between 150 and 900 Hz [6], we might proceed with a preprocessing. Preprocessing consists of eliminating from w_i all the frequencies that according to [6] do not belong to cry. Naturally, higher frequencies pass the filtering, and as a result, this sound could be confused with cries. To that end, a second step analysis, that considers the entire succession of non-silent windows, is mandatory.

In the second step we analyze the entire succession of non-silent windows. This succession's length is L, and is variable in length. The goal of the second step is to compare the wave of the non-silent windows succession with the waves in a so-called Cry Dictionary. The Cry Dictionary contains the wave features of several newborn cries, each with different durations, and in terms of fundamental frequencies.

Hence, in this second step, the fundamental frequencies of each window are obtained. As a result, a succession F of L fundamental frequencies $\{f_1, f_2, \ldots, f_L\}$ is obtained. In this step, we apply a filter algorithm to eliminate the fundamental frequency less than 150 Hz. Because the main purpose of our method is detecting cry unit in background noise environment, we only care about the fundamental frequency related to cry fundamental frequency. The fundamental frequency of background noise after filtering step can have values in the range of cry fundamental frequency, but it never has the shape of changing in frequency like cry.

Finally, the wave in the Cry Dictionary at a nearest distance to F is found. If this distance is below a specific threshold, the succession of L non-silent windows is considered a cry. For the sake of simplicity, our detection technique is based on distances.

3 Experimental Results

The aim of experimental results is to assess if the cry detection algorithm implemented detects sounds not related to crying (e.g., machinery, conversations…) as false cries. Hence, we have tested our implementation with neonatal cries in the scenario of a real NICU.

3.1 Sound Samples for Testing

Sound samples used for testing have been generated using separate samples for background noise and newborn cries. On the one hand, it is not straightforward to obtain a representative sample of newborn cries. On the other, there is a database of neonate: The Baby Chillanto Data Base is a property of the Instituto Nacional de Astrofisica Optica y Electronica—CONACYT, Mexico, and has been used in the literature to assess proposals on cry detection and characterization [9]. We have selected four samples from this database, namely "Normal_Cry_1," "Normal_Cry_2," "Pain_Cry," and "New_Cry." The three first samples have been used to build the Cry Dictionary.

Background noises have been recorded in a real NICU scenario: the neonatal unit in Hospital Universitari Joan XXIII. Samples, with a duration of 10 s each, were recorded during November 2016 using a high-sensitivity cardioid condenser microphone and an audio interface attached to a Mac Book Pro. Samples have been recorded by Audacity software.

We have selected the following background noise samples:

- "Plain NICU noise," background noise that includes humming of the machinery in the unit.
- "Alarm NICU noise," background noise that includes an alarm sound.
- "Conversation NICU noise," background noise that includes a conversation between adults.
- "Children NICU noise," background noise that includes some little girl speaking.
- "Telephone NICU noise," background noise that includes a telephone ringing sound and the voice of a nurse attending the call.

In order to generate the final samples that are used in the tests in this work, we have mixed cry sounds with the background noise at different proportions specifically 80% cry plus 20% background and 50% cry plus 50% background. As a result, we test our algorithm with $5 \times 3 \times 2 = 30$ samples with a duration of 10 s.

For each sample, we know the exact time in which cry starts, and hence, a false positive (i.e., false cry) will consist of detecting a cry outside the boundaries of the parts of the sample where we know the real cry is.

3.2 Robustness Against NICU Noise

Tests have been conducted using a laptop equipped with a core i5-5200U processor and 4 GB DDR3 RAM. Each sample took around 3–5 s to be evaluated. In this result, the accuracy (in time) is calculated by total time of cry in detection/total time of cry in one benchmark. Of course, we also consider these cry units are at the correct position. Tables 1, 2, and 3 show some of the results, for the sake of brevity. The list of the complete tests can be obtained from the document downloadable from the link http://smarthealthresearch.com/docs/cryanalysisreport.pdf.

For the "Normal_Cry_1" (Table 1) tests, no false positives were found, and thus, our method is shown to be valid since it does not consider noises as cries. For the "Normal_Cry_2" (Table 2), the result of detecting in first situation of mixing with proportion 20–80% is quite good; there is no false positive in this situation. However, for the second case of mixing with proportion of 50–50%, there exists false positive. The minimum distance of frequency when false positive occurs is 150, and this value could contribute to fine tune our detection algorithm using the Cry Dictionary and our current simple detection method.

Finally, in the case of "Pain_Cry" (Table 3), the results are optimal since no false positives have been detected.

3.3 Detection of a Cry Not in the Dictionary

Until now, we have described the success of our method in terms of not detecting false positives. In addition, we consider assessing the validity and robustness of our method about detecting new cries that are not in the dictionary. To that end, we have tested our method using the fourth sample from the database and creating mixes with background noise from the real NICU. Table 4 shows that our given method detects cry in new cry sample correctly and also does not detect false positives.

3.4 Discussion

In most of the cases, our method can correctly detect cry regardless the background noise, even in case of mixing 50% of noise. False positive only exists in some tests of second cases with 50–50% mixing noise.

In case of "Normal_Cry_1" and "Pain_Cry," which we already know are the good cry recording, the quality of the method is correct. It gives very high accuracy in case of 20–80% of mixing noise. In "Normal_Cry_1" case, it always gives the accuracy more than 80%. In case of "Pain_Cry," the result is even better, around 100% of accuracy.

In case of "Normal_Cry_2", we knew that this file—i.e., the sample from the cry database— also contains a noticeable amount of background noise. Thus, the detecting result is not good as the others. In situation of 20–80% mixing background noise, it only gives the result with accuracy around 60%, but in situation of 50–50% mixing background noise, the result is even worse and the false positive appears. It is not good enough to become the standard for detecting cry. However, the value of minimum distance of frequency obtained is very valuable for estimating the standard fundamental frequency threshold in the future.

The result in the situation of mixing 20–80% mixing background noise is always better than the result in the situation of mixing 50–50%. In case of "Normal_Cry_1" and "Pain_Cry," situation of 20–80% gives result of over 80% and nearly 100%,

Table 1. Some representative results in the case of "Normal_Cry_1"

Sample	Cry interval	Noise interval	Non-silent interval	Accuracy	False positive
20_80_Alarm_Noise_Normal_Cry_1	4.0–4.4 4.8–5.8 6.0–6.4 7.6–7.8	1.0–1.2; 2.2–2.4 4.4–4.6; 5.8–6.0 6.6–7.2; 7.4–7.6 7.8–8.0; 8.4–8.8	1.0–1.2; 2.2–2.4 4.0–4.6; 4.8–6.4 6.6–7.2; 7.4–8.0 8.4–8.8	91	No
20_80_Conversation_Noise_Normal_Cry_1	4.0–4.4 5.0–5.8 6.0–6.4 7.6–7.8	1.2–1.4; 2.8–3.0 4.4–5.0; 5.8–6.0 6.6–7.2; 7.4–7.6 7.8–8.0; 8.4–8.8	1.2–1.4; 2.8–3.0 4.0–6.4; 6.6–7.2 7.4–8.0; 8.4–8.8	82	No
50_50_Alarm_Noise_Normal_Cry_1	4.0–4.4 4.8–5.8 6.2–6.4	1.0–1.4 2.2–2.6 7.0–7.8 9.4–9.6	1.0–1.4; 2.2–2.6 4.0–4.4; 4.8–5.8 6.2–6.4; 7.0–7.8 9.4–9.6	73	No
50_50_Telephone _Noise_Normal_Cry_1	4.2–4.4 5.0–5.8 6.2–6.4	0.0–0.2; 0.8–1.0 2.4–2.6; 4.0–4.2 4.8–5.0; 6.4–6.6 6.8–8.6; 9.2–9.4	0.0–0.2; 0.8–1.0 2.4–2.6; 4.0–4.4 4.8–5.8; 6.2–6.6 6.8–8.6; 9.2–9.4	54	No

Table 2. Some representative results in the case of "Normal_Cry_2"

Sample	Cry interval	Noise interval	Non-silent interval	Accuracy	False positive
20_80_Alarm_Noise_Normal_Cry_2	4.6–4.8 5.2–5.6 5.8–6.2 6.4–6.8 7.0–7.2	4.2–4.6 4.8–5.0 5.6–5.8 6.2–6.4	4.2–5.0 5.2–6.8 7.0–7.2	62	No
20_80_Conversation_Noise_Normal_Cry_2	4.2–4.4 5.8–6.8 7.0–7.2	0.8–1.0; 1.2–1.4 2.8–3.0; 4.0–4.2 4.4–5.8; 6.8–7.0 7.2–7.4; 8.6–8.8	0.8–1.0 1.2–1.4 2.8–3.0 4.0–7.4 8.6–8.8	54	No
50_50_Alarm_Noise_Normal_Cry_2	5.8–6.2 7.0–7.2	0.0–1.6 1.8–5.8 6.2–7.0 7.2–8.2 8.4–10.0	0.0–1.6 1.8–8.2 8.4–10.0	25	No
50_50_Telephone_Noise_Normal_Cry_2	5.4–5.6 5.8–6.0 6.2–6.6 7.0–7.2 8.2–8.4 8.6–8.8	0.2–0.6; 1.8–2.0 2.4–3.2; 3.4–4.0 4.2–5.0; 5.2–5.4 5.6–5.8; 6.0–6.2 6.6–7.0; 7.4–7.6 8.0–8.2; 8.8–9.2	0.2–0.6; 1.8–2.0 2.4–3.2; 3.4–4.0 4.2–5.0; 5.2–7.2 7.4–7.6; 8.0–8.4 8.6–9.2	35	Yes

Table 3. Some representative results in the case of "Pain_Cry"

Sample	Cry interval	Noise interval	Voiced interval	Accuracy	False positive
20_80_Alarm_Noise_Normal_Cry_1	4.0–5.2 5.8–6.6 7.0–7.8	1.0–1.4; 2.2–2.6 5.2–5.8; 6.6–7.0 7.8–8.0	1.0–1.4 2.2–2.6 4.0–8.0	100	No
20_80_Conversation_Noise_Normal_Cry_1	4.0–5.2 5.8–6.6 7.0–7.8	0.8–1.0; 1.2–1.4 2.8–3.0; 5.2–5.8 6.6–7.0; 7.8–8.0 8.6–8.8	0.8–1.0 1.2–1.4 2.8–3.0 4.0–8.0 8.6–8.8	100	No
50_50_Alarm_Noise_Normal_Cry_2	4.2–5.2 5.8–6.6 7.0–7.6	1.0–1.4; 2.2–2.6 4.0–4.2; 5.4–5.6 6.6–7.0; 7.6–7.8	1.0–1.4; 2.2–2.6 4.0–5.2; 5.4–5.6 5.8–7.8	86	No
50_50_Conversation_Noise_Normal_Cry_2	4.0–4.6 5.8–6.6 7.0–7.6	2.8–3.0 4.6–5.2 5.4–5.6 7.6–7.8	2.8–3.0 4.0–5.2 5.4–5.6 5.8–6.6 7.0–7.8	72	No

Table 4. Results in the case of "New_Cry"

Sample	Cry interval	Noise interval	Non-silent interval	False positive
NewCry_Plain _Noise	4.2–6.0 6.4–7.2	4.0–4.2; 6.0– 6.4 7.2–7.4; 7.6–7.8 8.0–8.2	4.0–7.4 7.6–7.8 8.0–8.2	No
NewCry_Telephone_Noise	4.0–4.6 4.8–5.8 6.2–7.2	2.8–3.0; 4.6– 4.8 6.0–6.2; 7.2–7.4 7.6–7.8; 8.0–8.2 8.8–9.0	2.8–3.0; 4.0–5.8 6.0–7.4; 7.6–7.8 8.0–8.2; 8.8–9.0	No

respectively. But the situation of 50–50% only gives result of average 60% and 80%, respectively.

In case of "New Cry," our method gives a good result of detecting cry. There is no false positive in this case even the amplitude of noise is quite high. Although it still cannot detect all the cry duration in the audio file, it can detect the most important cry signals correctly.

4 Conclusions and Future Work

In this paper, we have addresses the robustness of cry detection techniques against real-life neonatal intensive care units. Our method has correctly detected cry regardless background noise. In our work, we determined that fundamental frequency is the most important feature of newborn infant cry. Thus, we focused on characterizing the feature of the changing and the shape in fundamental frequency. Comparing to the other method, it could reduce the computational cost but still gives us the good result.

Although we get good result in detecting cry unit in most of the case, but there still exists some false positives. The reason is insufficient quality of Cry Dictionary. The cry sample with the name "Normal Cry 2" containing noise inside, when creating standard cry unit from it, reduces the quality of the Cry Dictionary. It lets to some false positives in detecting cry in other cry samples. In future, we will investigate more about building the good and standard Cry Dictionary and make it to be a good benchmark for other researches.

In our method, the value of some importance parameters, which are power threshold, minimum distance in fundamental frequency, are still not fixed. Because we do not have the standard scenario for recording cry, the cry samples are mixed by other audio software. In near future, we will design a standard scenario, and the device for recording cry of the newborn infant will be put near the incubator to warranty that the background should have the lower energy than the cry.

In this work, we just implemented the basic method to classifying cry unit based on calculating the similarity in the characteristic of fundamental frequency. It can be still improved because fundamental frequency is not the unique characteristic of cry unit. In the future, we will implement artificial intelligent method like neural network, supported vector machine to classify cry unit based on the set of input attributes including fundamental frequency, some resonance frequencies, and average power frequency ratio. The new method will take into account all the features of cry signal; thus, the quality of cry detection should be improved.

Acknowledgements. This paper was completed by financial supporting from VAST project (Vietnam Academy of Science and Technology): "**Design and development of a remote visual monitoring system applied for security applications**" which project's code is: *VAST01.10/17-18*. Authors also thank Hospital Universitari Joan XXIII and the personnel in its neonatal intensive care unit for allowing them the recording of audio samples in their facilities. Finally, we like to thank Dr. Carlos A. Reyes-Garcia, Dr. Emilio Arch-Tirado and his INR-Mexico group, and Dr. Edgar M. Garcia-Tamayo for their dedication to the collection of the infant cry data base.

References

1. Nidcap.org, A.: What is NIDCAP?—NIDCAP. http://nidcap.org/en/families/what-is-nidcap/ (Retrieved 13 January 2017).
2. Martinez-Ballesté, A., Casanovas-Marsal, J. O., Solanas, A., Casino, F., & Garcia-Martinez, M., A.: An autonomous system to assess, display and communicate the pain level in newborns. In Medical Measurements and Applications (MeMeA), 2014 IEEE International Symposium on. p. 1–5. IEEE, 2014.
3. Shoemark, H., Harcourt, E., Arnup, S. J., & Hunt, R. W., A.: Characterising the ambient sound environment for infants in intensive care wards. Journal of paediatrics and child health, 52.4, 436–440 (2016).
4. Almadhoob, A., & Ohlsson, A., A.: Sound reduction management in the neonatal intensive care unit for preterm or very low birth weight infants. The Cochrane Library (2015).
5. Darcy, A. E., Hancock, L. E., & Ware, E. J, A.: A descriptive study of noise in the neonatal intensive care unit ambient levels and perceptions of contributing factors. Advances in Neonatal Care, 8(3), 165–175 (2008).
6. Manfredi, C., Bocchi, L., Orlandi, S., Spaccaterra, L., & Donzelli, G. P., A.: High-resolution cry analysis in preterm newborn infants. Medical Engineering and Physics, 31.5, 528–532 (2009).
7. Cohen, R., & Lavner, Y, A.: Infant cry analysis and detection. In Electrical & Electronics Engineers in Israel (IEEEI), 2012 IEEE 27th Convention of. p. 1–5, IEEE (2012).
8. Díaz, M. A. R., García, C. A. R., Robles, L. C. A., Altamirano, J. E. X., & Mendoza, A. V., A.: Automatic infant cry analysis for the identification of qualitative features to help opportune diagnosis. Biomedical Signal Processing and Control, 7.1, 43–49 (2012).
9. Reyes-Galaviz, O. F., Cano-Ortiz, S. D., & Reyes-García, C. A., A.: Evolutionary-neural system to classify infant cry units for pathologies identification in recently born babies. In Artificial Intelligence, 2008. MICAI'08. Seventh Mexican International Conference on. p. 330–335. IEEE (2008).

Energy-Efficient Resource Allocation for Virtual Service in Cloud Computing Environment

Nguyen Minh Nhut Pham[1(✉)], Van Son Le[2], and Ha Huy Cuong Nguyen[3]

[1] Vietnam Korea Friendship Information Technology College, Danang, Vietnam
nhutpnm@viethanit.edu.vn
[2] Danang University of Education, Danang University, Danang, Vietnam
levansupham@yahoo.com
[3] Quangnam University, Quangnam, Vietnam
nguyenhahuycuong@gmail.com

Abstract. For the past several years, using cloud computing technology has become popular. With the cloud computing service providers, reducing the physical machine number providing resources for virtual service in cloud computing is one of the efficient ways to decrease the energy consumption amount which in turn enhance the performance of data centers. In this study, we propose the resource allocation problem to reduce the energy consumption. $ECRA-SA$ algorithm was designed to solve and evaluate through CloudSim simulation tool compared with an FFD algorithm. The experimental results indicate that the proposed $ECRA-SA$ algorithm yields a higher performance in comparison with an FFD algorithm.

Keywords: Resource allocation · Simulated annealing · Virtual Service · Cloud computing · Energy consumption

1 Introduction

The development of virtual technology and the applicability of cloud computing have led to the rapid growth of the need for physical machines in data centers. The amount of energy consumed by data centers was between 1, 1 and 1, 5% of the total worldwide electric power consumption in 2010 [16]. This is consistents with the typical yearly electricity consumption of 120 million households, producing negative greenhouse effects and CO_2 footprints. As a result, energy consumption increases more and more rapidly which can threat the environment and cause serious problems. A way to help with this issue is by finding energy-efficient techniques and algorithms to manage computing resources [3–5,8,11,14].

Eugen Feller et al. [6] showed a resource allocation model grounded on a homogeneous platform to provide the number of physical machines (PMs) for virtual machines (VMs) and proved that the system's energy consumption is

V. Bhateja et al. (eds.), *Information Systems Design and Intelligent Applications*, Advances in Intelligent Systems and Computing 672,
https://doi.org/10.1007/978-981-10-7512-4_13

reduced if the number of used PMs is diminished. We will explore the heterogeneous platforms such as resources of the PMs are not the same [12,13]. Inside, [12] established the resource allocation problem and used an *FFD* algorithm to solve. But, Thomas Setzer [11] demonstrated that this algorithm tends to wasting capabilities. According to some reviews [3–5,8,11,14] pointed out some practices of system′s resources allocation with the minimal energy consumption and just focused on power utilization on the CPU of PMs. We believed that this burning is not only upon CPU but also over other appliances such as hard disk, bandwidth, RAM.

Therefore, this study aims to solve the energy efficiency resource allocation problem (physical resources) for virtual services on the heterogeneous server platform which involve in minimizing the system's energy consumption. The main results are as follows:

(a) Stating the problem of resource allocation for virtual services as an optimal problem: minimizing the energy consumption of the system;
(b) Developing the $ECRA - SA$ algorithm, which are based on the Simulated Annealing algorithm [13], using CloudSim tool [3] to solve;
(c) Comparing the energy consumption and the execution time between an $ECRA - SA$ and an FFD algorithm.

The following sections are proceeded as follows: Sect. 2 reviews the related works. Section 3 presents the energy consumption resource allocation problem. Section 4 proposes an $ECRA - SA$ algorithm for problem solving. Section 5 provides experimental results. Section 6 points out some conclusions and suggests the further works.

2 Related Work

Numerous researchers have paid attention to address the problem of energy efficiency in which focused on the Dynamic Voltage and Frequency Scaling mechanism [2]. This technique is performed in modern processors that include the combined change of the provided voltage and the frequency of the CPUs. The key point is to decrease the voltage and the clock frequency by scaling it down when it is not fully utilized and doing the opposite actions when it is being entirely utilized. Although the hardware's energy-efficient scheduling has been improved, overall energy consumption keeps growing owing to the rising requirements toward computing resources.

Armbrust et al. examined the request scheduling problem for Web applications in virtualized heterogeneous platform so as to keep the energy consumption to a minimum while still qualifying the performance requirements [14]. They suggested a heuristic algorithm for solving the multidimensional packing problem as an approach of workload integration.

In previous studies [7,8], we proposed two algorithms using to detect a deadlock mechanism for allocating resources in heterogeneous distributed environment. We designed the deadlock detection algorithms in order to resolve the resource optimization problems built on the recovery of allocated resources. The core of these articles was set to explore the effectiveness of scheduling method in allocating resources.

Sotomayor et al. [15] introduced a lease-based model and proposed the First-Come-First-Serve scheduling algorithm and a greedy-based algorithm for VMs in order to map leases that consist of some VMs with/without start and duration time when setting homogeneous PMs. This research is mostly conducted on homogeneous PMs.

Along with hardware design, energy efficiency is influenced on how the software manages the resources. Energy-efficient resource management techniques were first introduced on mobile devices, where it has a direct impact on battery lifetime [2]. These techniques can be adapted for servers and data centers.

Ranjana et al. [1] studied the strategies of power-aware VMs placement techniques in a survey. They have discussed the used optimization algorithms to save power. They have classified the energy-saving techniques in a data center into static and dynamic methods.

Feller et al. [6] have presented a power-aware optimization technique for VMs placement. They applied an ant colony optimization in dynamic workload placement to save energy. Nevertheless, the cost of computation time is high as a result of optimum placement searching.

Therefore, when proposing energy-efficient resource allocation, one needs to be aware of the SLA to avoid performance degradation of the consumer applications, including increased response times, timeouts, or even failures. Consequently, cloud providers have to establish an QoS requirements to avoid SLA violations and meet the QoS requirements while minimizing energy consumption.

3 Energy Consumption Resource Allocation Problem

As stated in our previous work [12], each type of resource provisioned from a physical machine (PM) may compose of one or more resources of a single element (i.e., single real CPU, single real memory) and aggregate resources. Accordingly, the PM′s resources reached as an ordered pair of multidimensional resource vectors (C^e, C^a). Inside, $C^e = \{c^e_{jk} \mid c^e_{jk} \in \mathbb{Q}^+; j, k, M, D \in \mathbb{N}^+; j = 1, \ldots, M; k = 1, \ldots, D\}$, and $C^a = \{c^a_{jk} \mid c^a_{jk} \in \mathbb{Q}^+; j, k, M, D \in \mathbb{N}^+; j = 1, \ldots, M; k = 1, \ldots, D\}$ are vectors which represent the elementary resource and aggregate resource of a PM j for a resource type k, respectively. Hence, M is the number of PMs and D is the number of resource types.

Similarly, the virtual service′s resource needs are performed by involving the elementary and aggregate one. Indeed, the resource needs of each virtual service are categorized into two groups: *rigid needs* and *fluid needs* [12]. Hence, the rigid needs of a virtual service i for a resource type k are represented by a first ordered vector pair (R^e, R^a) which determines the resource demands in running the virtual service at the acceptable level. Inside, $R^e = \{r^e_{ik} \mid r^e_{ik} \in \mathbb{Q}^+; i, k, N, D \in \mathbb{N}^+; i = 1, \ldots, N; k = 1, \ldots, D\}$ and $R^a = \{r^a_{ik} \mid r^a_{ik} \in \mathbb{Q}^+; i, k, N, D \in \mathbb{N}^+; i =$

$1, \ldots, N; k = 1, \ldots, D\}$ are to denote the elementary and aggregate rigid needs of a resource type k of a virtual service i, respectively. Hence, N is the number of virtual services. The fluid needs of a virtual service i for a resource type k are represented by a second ordered vector pair (F^e, F^a), which displays the additional resource demand when running the virtual service at the maximum performance level. Inside, $F^e = \{f^e_{ik} \mid f^e_{ik} \in \mathbb{Q}^+; i, k, N, D \in \mathbb{N}^+; i = 1, \ldots, N; k = 1, \ldots, D\}$ and $F^a = \{f^a_{ik} \mid f^a_{ik} \in \mathbb{Q}^+; i, k, N, D \in \mathbb{N}^+; i = 1, \ldots, N; k = 1, \ldots, D\}$ are to denote the elementary and aggregate fluid needs of a resource type k of a virtual service i, respectively. Considering that $Q = \{q_{ij} \mid q_{ij} \in \mathbb{Q}^+; i, j, N, M \in \mathbb{N}^+; i = 1, \ldots, N; j = 1, \ldots, M\}$ is a vector of additional factor of virtual services when the users requests. Correspondingly, the resource needs of a type k of a virtual service i are indicated by an ordered vector pair$(R^e + Q \times F^e, R^a + Q \times F^a)$.

Assuming that each virtual service constitutes a single virtual machine having a rigid resource need. An energy consumption resource allocation (ECRA) problem for virtual services is formulated as detailed below:

Let N be virtual services, $N = \{i \mid i = 1, \ldots, N; N > 0\}$, and M be PMs whose resources are configured differently, $M = \{j \mid j = 1, \ldots, M; M > 0\}$. Each PM provisions resource type D, $D = \{k \mid k = 1, \ldots, D\}$. And r^e_{ik} and r^a_{ik} are used to denote the elementary and aggregate rigid needs of a virtual service i for a resource type k, respectively. f^e_{ik} and f^a_{ik} are to denote the elementary and aggregate fluid needs of a virtual service i for a resource type k, respectively. c^e_{jk} and c^a_{jk} represent the elementary resource and aggregate resource of a PM j for resource type k, respectively. And q_{ij} stands for the additional factor of a virtual service i on a PM j. A variable $x_{ij} \in \{0, 1\}$ is set to be equal to 1 in case a PM j allocates the resource for a virtual service i and 0 otherwise. Lastly, $P^{busy}_j \in \mathbb{Q}^+$ and $P^{idle}_j \in \mathbb{Q}^+$ are the power amount of PM j in the maximum utility state and idle state, respectively. Given these notations, an ECRA problem is described as follows:

$$x_{ij} \in \{0, 1\}; q_{ij}, r^e_{ik}, r^a_{ik}, f^e_{ik}, f^a_{ik}, c^e_{ik}, c^a_{ik} \in \mathbb{Q}^+, \quad \forall i, j, k \tag{1}$$

$$\sum_{j=1}^{M} x_{ij} = 1, \quad \forall i \tag{2}$$

$$(r^e_{ik} + q_{ij} \times f^e_{ik}) \times x_{ij} \le c^e_{jk}, \quad \forall i, j, k \tag{3}$$

$$\sum_{i=1}^{N} (r^a_{ik} + q_{ij} \times f^a_{ik}) \times x_{ij} \le c^a_{jk}, \quad \forall j, k \tag{4}$$

$$\left(\sum_{j=1}^{M} \left(\sum_{k=1}^{D} \left(\left(P^{busy}_j - P^{idle}_j \right) \times \frac{\sum_{i=1}^{N} (r^a_{ik} + q_{ij} \times f^a_{ik}) \times x_{ij}}{c^a_{jk}} + P^{idle}_j \right) \right) \right) \times \Delta t \rightarrow \min \tag{5}$$

Constraint (1) indicates the domain of the variables. Constraint (2) presents the state in which resources for a virtual service i are supplied exactly by a PM

j. Constraint (3) and Constraint (4) imply that the elementary resource and aggregate resource, respectively, of a PM j are not overcome. Lastly, formulation (5) shows the optimization objective of minimizing the energy consumption in a time period Δt.

4 Simulated Annealing Algorithm-Based Solution

The Simulated Annealing (SA) algorithm [13] is a well-studied meta-heuristic-based approach. The main idea of this algorithm is that it allows a mechanism to avoid and escape local optima with the belief of identifying global optimum for the optimization problems.

Algorithm 1 *ECRA-SA*

Input:
- Number of virtual services, N; type of resources, D and additional factor, Q;
- Number of PMs, M.

Output: list of used PMs, S_{best}.

```
double T ← T0; Tmin;
S_best ← initSolution(M, N); //Execute initialization solution by an FFD.
E_best ← according to Eq.(5);
while (T > Tmin) do
    for l:= 1 to L do //L is the number of iterative
        S_current ← S_best;
        E_current ← according to Eq.(5);
        S_neighbor ← currNeigSolution(M, N);//Execute the current
                                             //neighboring solution
        E_neighbor ← according to Eq.(5);
        if(E_neighbor < E_current) then
            S_current ← S_neighbor;
            E_current ←according to Eq.(5);
        end if
        if (exp((E_neighbor − E_current)/T) > random(0, 1)) then
            S_current ← S_neighbor;
            E_current ←according to Eq.(5);
        end if
        if(E_current < E_best) then
            S_best ← S_current;
            E_current ←according to Eq.(5);
        end if
    end for l := 0 to L
    T ← T × (1 − CR); // CR is a Cooling Rate
end While
return S_best;
```

In order to demonstrate the SA algorithm′s specific features for discrete optimization problems, numerous explanations are needed to be figured out. Let Y be the solution space. Let a mapping $f : X \to \mathbb{R}$ be an objective function identified in the solution space. The problem aims at finding a global minimum Y^* (i.e., $Y^* \in X$ such that $f(Y^*) \leq f(Y)$ for all $Y^* \in X$). The objective function must be delimited to make certain that Y^* exists. Let $E(Y)$ be the neighborhood function for $Y \in X$. Thus, taken up with every solution, $Y \in X$, are the neighboring solutions, $E(Y)$, that can be attained in a single iteration of a local search algorithm.

The SA algorithm begins with an initial solution $Y \in X$. Next, a neighboring solution $Y' \in E(Y)$ is generated. This algorithm is depended on the Metropolis acceptance criterion, which models the way a thermodynamic system moves from the current solution $Y \in X$ to a candidate solution $Y' \in E(Y)$, in which minimizing the energy content. The candidate solution, Y', is regarded as the current solution relied on the acceptance probability.

Therefore, the algorithm is used a temperature global variable T. Initially, T is a very high value and which is decreased to the lower value after each of iteration. In the specific loop, the cost function is considered for two solutions: the current solution and the neighboring solution. If the neighboring solution performs better than the current one, the current solution will be selected. In contrast, the neighboring one can still be admitted in hopes that the local extreme for searching global extreme is escaped with a probability depending on the cost function value of the current one, so far best explanation and parameter value T. The algorithm will stop after looking for a global optimal value or a temperature as the value T_{min}. In summary, the pseudo code of an *ECRA-SA* algorithm to solve an ECRA problem is shown as an Algorithm 1.

To begin with, we initialize the temperature T_0, T_{min}. We achieve the initial solution S_{best} by an *FFD* algorithm and calculate the objective function E_{best}. Next, an initial solution S_{best} is performed and a current solution $S_{current}$ is showed. We use the random resource allocation approach to generate an appropriate neighboring solution $S_{neighbor}$, calculate the objective function $E_{neighbor}$, and determine the Metropolis condition (i.e., If $(E_{neighbor} < E_{current})$, the new solution $S_{neighbor}$ is accepted and regarded as the initial solution for next loop. Otherwise, if $(exp\left(\frac{E_{neighbor} - E_{current}}{T}\right) > random\,(0,1))$, then the new solution $S_{neighbor}$ can be accepted). In case the iteration value L is not enough, we continue. Finally, reduce the temperature T if the stopping temperature T_{min} is higher than the new temperature, and stop and output the optimal solution S_{best}.

Mitra et al. [18] have shown that a SA algorithm limitedly converges to a globally optimal solution with probability 1. So, the *ECRA-SA* algorithm will converge after the finite iteration steps (i.e., $T = T_{min}$). Let N be the number of virtual services, M be the number of PMs, D be the type of resources, T be an initialized temperature, and L be the number of loops. The complexity of algorithms is calculated as follows: the initializing solution: $O(D \times N \times M)$, the neighboring solution: $O(D \times N \times M)$, the estimating of cost function: $O(D \times M \times$

N). So, the complexity of algorithms: $O(D \times N \times M) + O(T \times L \times (O(D \times N \times M) + O(D \times M \times N)))$. If $D, T, andL$ are regarded as constants, this algorithm complexity is $O(N \times M)$.

5 Numerical Result and Evaluation

5.1 Simulation

So as to assess the *ECRA-SA* algorithm, we compared this algorithm to the *FFD* [12] by using the CloudSim tool [3]. CloudSim is used to help to model and simulate the cloud computing systems as well as the application provision environments. This tool supports modeling the components of cloud systems, including virtual machines (VMs), hosts, data centers, and resource provision policies (Fig. 1).

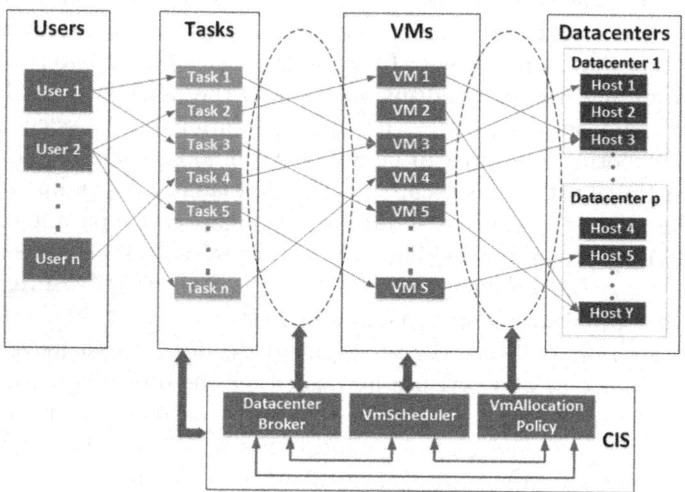

Fig. 1. CloudSim components [3]

As presented in Sect. 3, each virtual service is one separate virtual machine (VM). So, we inherited the *Vm* layer to expand the resource need features of the VM and the *Host* layer to expand the PMs resource features. Meanwhile, we inherited the *VMAlloctonPolicy* layer to perform the resource allocation policy for virtual machine grounded on the *ECRA-SA* algorithm. The experimental data are extracted from the real data as indicated in [1,2]. Table 1 presents the resource characteristics and the power of PMs. The resource characteristics of VMs, which are similar to the VMs of Amazon EC2 cloud, are adjusted to suit the problem, shown in Tables 2 and 3.

The parameters of the *ECRA-SA* algorithms are used: $T_0 = 1000$; $CR = 5$; $T_{min} = 0$; $L = 100$. The number of VMs (represents the set of virtual services) in the experiment is 100; 200; 300; 400; 500, respectively, and the energy consumption in a time period of $\Delta t = 24\,\text{h}$. Measure unit of energy consumption is kWh, and the runtime of algorithm is calculated by second (s). The performed time is measured by a PC having an Intel(R) Core(TM) i5-3235M 2.60 GHz, RAM 4Gb microprocessor.

Table 1. The resource characteristics and power consumption of PM

Type of physical machine	CPU (MHz)	RAM (GB)	BW (GB/s)	Disk (GB)	P_{idle}	P_{busy} (kW)
HP proliant G4	2 core x 1860	4	1	20	86	117
HP proliant G5	2 core x 2660	4	1	40	93.7	135
IBM Server x 3250	4 core x 2933	8	1	600	46.1	113
IBM Server x 3550	6 core x 3067	16	1	800	58.4	222

Table 2. The CPU resource and RAM resource characteristics of VM

Type of virtual machine	CPU(MHz)				RAM(GB)			
	f^e_{cpu}	r^e_{cpu}	Num. of element	Total	f^e_{ram}	r^e_{ram}	Num. of element	Total
VM 1	1000	1500	1	2500	0.4	0.45	1	0.85
VM 2	1000	1000	1	2000	1.0	2.75	1	3.75
VM 3	500	500	1	1000	0.7	1.0	1	1.7
VM 4	250	250	1	500	0.113	0.5	1	0.613

Table 3. The BW resource and DISK resource characteristics of VM.

Type of virtual machine	BW(MHz/S)				DISK(GB)			
	r^e_{bw}	f^e_{bw}	Num. of element	Total	f^e_{disk}	r^e_{disk}	Num. of element	Total
VM 1	0.20	0.25	1	0.45	0.5	2.00	2	5.00
VM 2	0.10	0.25	1	0.35	2.0	3.00	2	10.0
VM 3	0.10	0.15	1	0.25	2.5	5.00	2	15.0
VM 4	0.05	0.10	1	0.15	5.0	5.00	2	20.0

5.2 The Experimental Results

The results in Table 4 and Fig. 2 showed that the *ECRA-SA* algorithm reduced consumed energy better than an *FFD*. This is due to the fact that while an *FFD* algorithm tends to use a large set of PMs, the *ECRA-SA* algorithm uses a more efficient global search solution. Therefore, the energy consumption when using the *ECRA-SA* algorithm is lower than *FFD*. However, the runtime of the *ECRA-SA* algorithms is larger than *FFD*. This is explained by the effects of parameters of the iteration number in *ECRA-SA* algorithms. This restriction is a general trend of global optimization algorithms.

Table 4. The used physical machines and consumed energy

Num. of VM	Algorithm	Times (s)	Energy (kWh)
100	*FFD*	0,032	201.284
	ECRA-SA	0,038	193.000
200	*FFD*	0,078	396.706
	ECRA-SA	0,088	392.490
300	*FFD*	0,116	392.490
	ECRA-SA	0,125	597.989
400	*FFD*	0,144	793.411
	ECRA-SA	0,160	772.185
500	*FFD*	0,200	994.694
	ECRA-SA	0,218	972.439

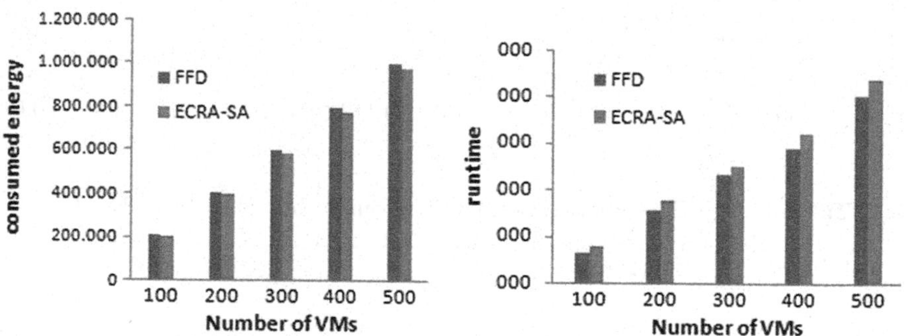

Fig. 2. The graph of runtime and consumed energy

6 Conclusion and Future Works

This study proposed the resource allocation problem for virtual services with minimum energy consumption. We also set up, assessed, and compared with an *FFD* algorithm through two parameters: consumed energy and runtime. The experimental outstanding shows that an *ECRA-SA* algorithm reduced the energy consumption better than an *FFD* algorithm. Future research should extend the proposed approach for multi-objective resource allocation.

References

1. Arianyan, E., Taheri, H., Sharifian, S.: Novel energy and SLA efficient resource management heuristics for consolidation of virtual machines in cloud data centers. Computers Electrical Engineering 47, pp. 222–240. Elsever (2015).
2. Beloglazov, A., Buyya, R.: Optimal online deterministic algorithms and adaptive heuristics for energy and performance efficient dynamic consolidation of virtual machines in cloud data centers. Concurr. Comput. : Pract. EXper. 24(13), pp. 1397–1420. John Wiley and Sons Ltd (2012).
3. Calheiros, R.N. and et al.: Cloudsim: A toolkit for modeling and simulation of cloud computing environments and evaluation of resource provisioning algorithms. Softw. Pract. EXper. 41(1), pp. 23–50. John Wiley and Sons Ltd (2011).
4. Cao, Z., Dong, S.: Dynamic VM consolidation for energy-aware and SLA violation reduction in cloud computing. In: Parallel and Distributed Computing, Applications and Technologies (PDCAT), 2012 13th International Conference on, pp. 363–369. IEEE(2012).
5. Farahnakian, F. and et al.: Energy-aware dynamic VM consolidation in cloud data centers using ant colony system. In: Cloud Computing (CLOUD), 2014 IEEE 7th International Conference on, pp. 104–111. IEEE (2014).
6. Feller, E., Rilling, L., Morin, C.: Energy-aware ant colony based workload placement in clouds. In: Grid Computing (GRID), 2011 12th IEEE/ACM International Conference on, pp. 26–33. IEEE (2011).
7. Ha Huy Cuong Nguyen, et al.,: A New Technical Solution for Resource Allocation in Heterogeneous Distributed Platforms, in Advances in Digital Technologies, 2015, IOS Press, The Netherlands: The University of Macau, Macau. pp. 184–194 (2015).
8. Ha Huy Cuong Nguyen, et al.,: Deadlock Detection for Resource Allocation in Heterogeneous Distributed Platforms, in Recent Advances in Information and Communication Technology 2015, pp. 285–295. Springer (2015).
9. Luo, L. and et al.: A resource scheduling algorithm of cloud computing based on energy efficient optimization methods. In: Green Computing Conference (IGCC), 2012 International, pp. 16. IEEE (2012).
10. Quan, D.M. and et al.: Energy Efficient Resource Allocation Strategy for Cloud Data Centres, Computer and Information Sciences II: 26th International Symposium on Computer and Information Sciences, chap., pp. 133–141. Springer (2012).
11. Setzer, T., Stage, A.: Decision support for virtual machine reassignments in enterprise data centers. In: Network Operations and Management Symposium Workshops (NOMS Wksps), 2010 IEEE/IFIP, pp. 88–94. IEEE (2010).
12. Nguyen Minh Nhut Pham, Thu Huong Nguyen, Van Son Le: Resource Allocation for Virtual Service Based on Heterogeneous Shared Hosting Platforms. In: 8th Asian Conference ACIIDS 2016, Da Nang, Vietnam 2016, pp. 51–60. Springer (2016).
13. Kirkpatrick, S.: Optimization by simulated annealing: Quantitative studies. Journal of Statistical Physics 34(5), 975–986. Kluwer Academic Publishers-Plenum (1984).
14. Armbrust, M., et al.,: Above the Clouds: A Berkeley View of Cloud Computing.Technical Report No. UCB EECS-2009-28, the University of California at Berkley, USA, (2009).
15. Sotomayor, B.: Provisioning Computational resources using virtual machines and leases. Ph.D. Thesis submitted to The University of Chicago, USA, (2010).

16. Koomey, J.: Growth in data center electricity use 2005 to 2010. A report by Analytical Press, completed at the request of The New York Times, Vol. 9, (2011).
17. Ranjana, R., Raja, J.: A survey on power aware virtual machine placement strategies in a cloud data center. In 2013 International Conference on GreenComputing, Communication and Conservation of Energy (ICGCE), pp. 747–752, (2013).
18. Mitra, D., Romeo, F., and Vincentelli, A.S.: Convergence and Finite-Time Behavior of Simulated Annealing. Advances in Applied Probability, Vol. 18(3), pp. 747–771, (1986).

Novel Adaptive Neural Sliding Mode Control for Uncertain Nonlinear System with Disturbance Estimation

Thiem V. Pham[1,2] and Du Dao Huy[2(✉)]

[1] CReSTIC, University de Reims Champagne Ardenne, Reims, France
phuthiem@tnut.edu.vn, van-thiem.pham@etudiant.univ-reims.fr
[2] Electronics Faculty, TNUT, Thai Nguyen, Vietnam
daohuydu@tnut.edu.vn

Abstract. The paper deals with the problem of tracking control for a class of nonlinear systems in presence of the disturbances. The developed formation for the tracking control is taken into account as an adaptive neural sliding mode. A chattering phenomenon will be eliminated by reducing a norm of disturbance based on disturbance estimation and feed-forward correction. The set of controller's parameter, which is a satisfy Hurwitz polynomial, is then updated by adaptive laws via a model reference system. In addition, the unknown nonlinear functions are estimated by radial basis functions neural network. The adaptive updated law based on radial basis functions neural network and a feed-forward correction is proposed to estimate both estimation errors of nonlinear functions and external disturbances, which is called lumped disturbances. An asymptotic stability of a closed loop system is illustrated by Lyapunov theory. And lastly, to demonstrate an efficiency of our approach, an illustrative example, a coupled-tank liquid system, is shown.

Keywords: Neural network · Disturbance estimation · Sliding mode control · Adaptive control · Coupled-tank liquid system

1 Introduction

In an uncertain nonlinear system in presence of exogenous disturbances, the tracking control problem is a challenging issue and it has been received many considerations. To enhance a performance of the tracking under uncertainties and disturbances, several methodologies are combined with together in the control design approach such as robust design, adaptive update method, feed-forward, sliding mode control. In the recent decades, many control methodologies [1] are

V. Bhateja et al. (eds.), *Information Systems Design and Intelligent Applications,* Advances in Intelligent Systems and Computing 672,
https://doi.org/10.1007/978-981-10-7512-4_14

derived from a feedback linear controller as gain scheduling. In [2], based on the significant properties of Lie algebra of a vector field, Krener, and D. Cheng [3] showed that there is always existence a coordinate transformation with the aim that the nonlinear system transform into a controllable linear. From which, K. Khalil proposed the exact feedback linearization methodology to figure out a control strategy under several assumptions such as a model of a nonlinear system and all states are known clearly and measured, respectively. In following, there are many adaptive update control schemes for nonlinear with known or the unknown parameters is investigated [4]. However, there are several inherent difficulties associated with these approaches. First of all, the plant dynamic structure may not be known exactly, which is common in practice, this is cause the linearization method faced with serious obstacles in applying the adaptive update algorithms in practice. Second, it was shown that some of the these designed methods may lack robustness against uncertainties [5]. Another trend, algorithms for online function approximation with piecewise based on fuzzy logic system or/and an artificial neural network is interested in [10,14,18]. Based on the analysis of radial basis function neural network's advantage in [7,14, 15], in which the output of radial basis function (RBF) neural network only depends linearly on the matrix weight, which describes a relationship between the output's layer and the neural network's hidden layer, and by modifying the radial basis function's centers and widths, the structure of neural network transfers linearly into nonlinearly parameterized. In this work, the RBF neural network is thus taken into account in designing an adaptive update law.

Sliding mode control (SMC) has received extensive attention in the past. Based on the advantages of SMC and Indirect Adaptive Fuzzy Control (IAFC), a number of papers had combined the two achieving more superior performances such as overcoming some limitations of the traditional SMC in [6,12,13,16]. To ensure the robustness against the disturbances and modeling uncertain, the sliding mode control is used as a powerful tool and efficiency. In addition, the chattering phenomena due to high gain and high-speed switching control are considered as a sufficient condition of existence of sliding mode control. In which the constraint $\lambda \leq \rho$ (*where* λ *is a coefficient of sliding mode control and a norm of disturbance* ρ) must be satisfied. The undesirable chattering problem may effect previously unmodeled dynamics of system and damage actuators, which results in unpredictable stability of a closed loop system. To overcome this problem, [11] employ the proportional-integral (PI) control term was proposed such as $u_p = \phi_{p1}s + \phi_{p2}\dot{s}$. After that, $\phi_p = (\phi_{p1}\ \phi_{p2})^T$ is adjustable parameter vector. And [18] using the sliding mode techniques and approximated a *sign* function by consider functions such as *sat* function. In other trends, [7–9] sliding mode control component is approximated by neural network control and fuzzy logic control method and so on. However, to reduce the chattering phenomena, then a norm of disturbance need to reduce. To do this, adding an estimation component $\bar{d}(x,t) \approx d(x,t)$ such as $\sup_{0\leq\tau\leq t} |\bar{d}(\boldsymbol{x},\tau) - d(\boldsymbol{x},\tau)| \leq \rho$. Following, the constant sliding surface ρ is still selected satisfying $\sup_{0\le\tau\le t}|\bar{d}(\boldsymbol{x},\tau)-d(\boldsymbol{x},\tau)|\le\lambda\le\rho$.

Motivated by the above work, the problem of tracking control for the uncertain nonlinear system with disturbances is considered. The nonlinear function is estimated by adaptive RBF neural network. Therefore, the main contributions of this paper are summarized as follows: First, we develop a sliding mode control combined with adaptive RBF neural network based on state feedback controller and ensure the stability of the closed loop system. Second, to eliminate the chattering phenomena, the radial basis function neural network and feed-forward correction $\boldsymbol{\Gamma}$ is proposed to estimate the lumped disturbances, which contains unknown input disturbances and the error of structure. Third, the set of controller's parameter, which is a satisfy Hurwitz polynomial, is then updated by adaptive laws via a model reference system. Consequently, our proposed approach will be to satisfy simultaneously some targets such as a closed loop function is stable; anti-disturbance and uncertain components; and improve the performance of tracking error.

The paper is organized as follows. In Sects. 2 and 3, the adaptive neural sliding mode control with disturbance estimation based on adaptive RBF neural network (DEVRS) and a novel adaptive RBF neural via sliding mode control and disturbance estimation (NASC-DE) are presented. Simulation results are given in Sect. 4, and Sect. 5 concludes this work.

2 Adaptive Neural Sliding Mode Control with Disturbance Estimation

The class of nonlinear system in presence of disturbance $d(\boldsymbol{x},t)$ is consider as

$$\begin{cases} y^{(n)} = F(\boldsymbol{x}) + H(\boldsymbol{x})u + d(\boldsymbol{x},t); \;\; F(\boldsymbol{x}); H(\boldsymbol{x}) \in \mathcal{C}^{\infty} \\ \boldsymbol{x} = (x_1 \;\ldots\; x_n)^T \in \mathcal{R}^n; \; u \in \mathcal{R}^{n\times m}; \|d(\boldsymbol{x},t)\| \le \rho \end{cases} \tag{1}$$

Assumption 1. $d(\boldsymbol{x},t)\in\mathcal{C}^{\infty}$; $\|d(\boldsymbol{x},t)\|_{\infty}=\sup_{\boldsymbol{x},t}|d(\boldsymbol{x},t)|=\rho<\infty$

In [10], the radial basis function (RBF) neural network could be taken into account as a three layers network with one hidden layer. The relationship between output layer and input layer is expressed as follows $a(\boldsymbol{x}) = \sum_{j=1}^{m}\psi_j(\boldsymbol{x})\Phi_j = \boldsymbol{\Phi}^T\boldsymbol{\psi}(\boldsymbol{x})$ in which $\boldsymbol{\psi}=(\psi_1\;\psi_2\;\ldots\;\psi_m)^T$; $\boldsymbol{\Phi}=(\Phi_1\;\Phi_2\ldots\;\Phi_m)^T$, $\psi_j(x)=\phi(\|\boldsymbol{x}-c_j\|_2)$, $\boldsymbol{x}$ is the input vector, $\boldsymbol{\phi}$ is a radial basis function, Φ_j are connection weights which are defined the link between the hidden layer and the output layer, c_j are centers of the basis function, m is the number of basis function. The common basis function is chosen as the Gaussian function

$$\phi(\boldsymbol{x}) = \exp\left(\frac{-r^2}{2\sigma^2}\right) \tag{2}$$

where $r=\|x-c_j\|_2$, σ represents the width of Gaussian function.

The RBF neural network will use to approximate the nonlinear function $F(\boldsymbol{x}), H(\boldsymbol{x})$, in which the vector $\boldsymbol{\Phi}^T$ is the defined parameters by adaptive law. The nonlinear system (1) can be rewritten as follows

$$\begin{aligned} y^{(n)} &= \boldsymbol{\Phi}_F^T \psi_F(\boldsymbol{x}) + \boldsymbol{\Phi}_H^T \psi_H(\boldsymbol{x}) u_c + w(\boldsymbol{x}, t) \\ w(x, t) &= d(\boldsymbol{x}, t) + \Delta_F(\boldsymbol{x}) + \Delta_H(\boldsymbol{x}) u_c \end{aligned} \tag{3}$$

where $w(\boldsymbol{x}, t)$ is called a lumped disturbance, $\hat{F}(\boldsymbol{x}) = \boldsymbol{\Phi}_F^T \psi_F(\boldsymbol{x}), \hat{H}(\boldsymbol{x}) = \boldsymbol{\Phi}_F^T \psi_F(\boldsymbol{x})$, and $\Delta_F(\boldsymbol{x}), \Delta_H(\boldsymbol{x})$ are the structure error which are satisfy the under assumption.

Assumption 2.

$$\|\Delta_F(\boldsymbol{x})\| \le \rho_F; \|\Delta_H(\boldsymbol{x})\| \le \rho_H; \tag{4}$$

A feedback control law is constructed to ensure the error dynamic of system approach asymptotically to the zero. We assume that all states of the nonlinear system (1) available for feedback. In case of $d(\boldsymbol{x}, t) = 0$, the mentioned controller based on the state feedback controller:

$$u_c = H^{-1}(\boldsymbol{x})[-F(\boldsymbol{x}) + y_m^{(n)} + \sum_{i=1}^{n-1} k_i e^{(i)}] \; ; \; e(t) = y_m - y \tag{5}$$

where k_i is the coefficients of Hurwitz polynomial

$$h(s) = k_1 + k_2 s + \cdots + k_{n-1} s^{n-2} + s^{n-1} \tag{6}$$

In this following, we take into account $w(\boldsymbol{x}, t) \neq 0$ satisfying the assumption 1. The control algorithm (5) can only drive the output system (1) to the neighborhood of the desired trajectory. However, if a norm of disturbance and error approximator are large. The closed loop system should be unstable. Therefore, in this work, another controller combined disturbance estimation and sliding mode control is proposed. In which, $w(\boldsymbol{x}, t)$ is estimated by the adaptive RBF neural network $\hat{w}(\boldsymbol{x}, t) = \boldsymbol{\Phi}_w^T \psi_w(\boldsymbol{x})$.

To anti-uncertain components, the sliding mode control will be aided into the controller (5), in which a sliding surface is selected as

$$s(e) = k_1 e + k_2 e + \cdots + k_{n-1} e^{n-2} + e^{n-1} = \boldsymbol{\Phi}^T \boldsymbol{e} \tag{7}$$

in which, $\boldsymbol{\Phi}^T = (\boldsymbol{\Phi_k} \; 1); \boldsymbol{\Phi_k} = (k_1, k_2, \ldots, k_{n-1})$. The state feedback controller approach is proposed as follows

$$u(t) = u_c(t) + u_s(t) + u_d(t) \tag{8}$$

in which,

$$\begin{aligned} u_s(t) &= \hat{H}^{-1}(\boldsymbol{x}) \lambda sgn(s); u_d(t) = -\hat{H}^{-1}(\boldsymbol{x}) \hat{w}(x, t) \\ u_c(t) &= \hat{H}^{-1}(\boldsymbol{x})(-\hat{F}(\boldsymbol{x}) + y_m^{(n)} + \sum_{i=1}^{n-1} k_i e^{(i)}) \end{aligned}$$

Assumption 3. The nonlinear function $H(\boldsymbol{x}), \hat{H}(\boldsymbol{x})$ are bounded

$$0 < H_{low} \leq H(\boldsymbol{x}), \hat{H}(\boldsymbol{x}) \leq H_{up} < \infty \tag{9}$$

Theorem 1. *Consider the nonlinear system (1) satisfying the assumptions 1–3 with the controller is constructed as (8) in which, $\boldsymbol{\Phi}_H^T, \boldsymbol{\Phi}_F^T$ are defined by the adaptive updated law*

$$\dot{\boldsymbol{\Phi}}_F = -\boldsymbol{Q}_F^{-1}\boldsymbol{\psi}_F(\boldsymbol{x})s(e); \ \dot{\boldsymbol{\Phi}}_H = -\boldsymbol{Q}_H^{-1}\boldsymbol{\psi}_H(\boldsymbol{x})s(e)u_c$$

and the disturbance estimation

$$\hat{w}(\boldsymbol{x}, t) = -\int_0^t \boldsymbol{\psi}_w^T(x)(\boldsymbol{Q}_w^{-1})^T s(e)dt\boldsymbol{\psi}_w(x) - \boldsymbol{\Gamma}\boldsymbol{e}(t) \tag{10}$$

where $\boldsymbol{\Gamma}$ is root of

$$\boldsymbol{\Gamma}\boldsymbol{A}_d + \boldsymbol{\Gamma}\boldsymbol{B}_d\boldsymbol{\Gamma} + \boldsymbol{\psi}_w^T(\boldsymbol{x})\boldsymbol{Q}_w^{-1}\boldsymbol{\psi}_w(\boldsymbol{x})\boldsymbol{a}_d = 0 \tag{11}$$

then, for all $t \geq 0$, $\lim\limits_{t\to\infty} e(t) \to 0$, $\lim\limits_{t\to\infty} (\hat{w}(\boldsymbol{x}, t) - w(\boldsymbol{x}, t)) \to 0$ and the closed loop system is stable with

$$\sup_{0\leq\tau\leq t} |w(\boldsymbol{x}, \tau)| \geq \lambda \geq \sup_{0\leq\tau\leq t} |\hat{w}(\boldsymbol{x}, \tau) - w(\boldsymbol{x}, \tau)| \tag{12}$$

Proof. There are two parts in our proof. First, the stability of closed loop system will be given by Lyapunov function. After that, an explaining to a feed-forward correction $\boldsymbol{\Gamma}$ is obtained by (10) and a proof of convergence of estimated disturbance error. First of all, a dynamic of error will be expressed

$$\begin{aligned} y^{(n)} &= -(\hat{F}(\boldsymbol{x}) - F(\boldsymbol{x})) - (\hat{H}(\boldsymbol{x}) - H(\boldsymbol{x}))u_c + \\ &+ y_m^{(n)} + \sum_{i=1}^{n-1} k_i e^{(i)} + H(\boldsymbol{x})(u_d + u_s) + d(\boldsymbol{x}, t) \end{aligned}$$

From (7), a derivative of sliding surface can be presented

$$\dot{s} = \tilde{\boldsymbol{\Phi}}_F^T\boldsymbol{\psi}_F(x) + \tilde{\boldsymbol{\Phi}}_H^T\boldsymbol{\psi}_H(\boldsymbol{x})u_c + \tilde{\boldsymbol{\Phi}}_w^T\boldsymbol{\psi}_w(\boldsymbol{x}) - \hat{w}(\boldsymbol{x}, t) - H(\boldsymbol{x})u_d - H(\boldsymbol{x})u_s$$

where $\tilde{\boldsymbol{\Phi}}_F^T = \left(\boldsymbol{\Phi}_F^T - \boldsymbol{\Phi}_F^{*T}\right); \tilde{\boldsymbol{\Phi}}_H^T = \left(\boldsymbol{\Phi}_H^T - \Phi_H^{*T}\right); \tilde{\boldsymbol{\Phi}}_w^T = \left(\boldsymbol{\Phi}_w^T - \boldsymbol{\Phi}_w^{*T}\right)$. The Lyapunov function will be chosen as follows

$$V = \tfrac{1}{2}s^2 + \tfrac{1}{2}\tilde{\boldsymbol{\Phi}}_F^T\boldsymbol{Q}_F\tilde{\boldsymbol{\Phi}}_F^T + \tfrac{1}{2}\tilde{\boldsymbol{\Phi}}_H^T\boldsymbol{Q}_H\tilde{\boldsymbol{\Phi}}_H^T + \tfrac{1}{2}\tilde{\boldsymbol{\Phi}}_w^T\boldsymbol{Q}_w\tilde{\boldsymbol{\Phi}}_w^T$$

in which, $\boldsymbol{Q}_{F,H} = \boldsymbol{Q}_{F,H}^T > 0$; $\boldsymbol{Q}_w^T = \boldsymbol{Q}_w > 0$. Using the derivative of V and notice that $\dot{\tilde{\boldsymbol{\Phi}}}_i = \dot{\boldsymbol{\Phi}}_i$,

$$\begin{aligned} \dot{V} &= \tilde{\boldsymbol{\Phi}}_F^T(s\boldsymbol{\psi}_F(\boldsymbol{x}) + \boldsymbol{Q}_F\dot{\boldsymbol{\Phi}}_F) + \tilde{\boldsymbol{\Phi}}_H^T(s\boldsymbol{\psi}_H(\boldsymbol{x})u_c + \boldsymbol{Q}_H\dot{\boldsymbol{\Phi}}_H) - s\hat{w}(\boldsymbol{x}, t) + \\ &+ \tilde{\boldsymbol{\Phi}}_n^T(s\boldsymbol{\psi}_n(\boldsymbol{x}) + \boldsymbol{Q}_n\dot{\boldsymbol{\Phi}}_n) - sH(\boldsymbol{x})(u_d + u_s) \end{aligned} \tag{13}$$

To guarantee the stability of the closed loop system, a vector of parameter will be chosen as follows

$$\begin{aligned} s\boldsymbol{\psi}_F(\boldsymbol{x}) + \boldsymbol{Q}_F\dot{\boldsymbol{\Phi}}_F = 0 \Rightarrow \dot{\boldsymbol{\Phi}}_F = -\boldsymbol{Q}_F^{-1}\boldsymbol{\psi}_F(\boldsymbol{x})s(e) \\ s\boldsymbol{\psi}_H(\boldsymbol{x})u_c + \boldsymbol{Q}_H\dot{\boldsymbol{\Phi}}_H = 0 \Rightarrow \dot{\boldsymbol{\Phi}}_H = -\boldsymbol{Q}_H^{-1}\boldsymbol{\psi}_H(\boldsymbol{x})s(e)u_c \end{aligned} \tag{14}$$

$$s\boldsymbol{\psi}_w(\boldsymbol{x}) + \boldsymbol{Q}_w\dot{\boldsymbol{\Phi}}_w = 0 \Rightarrow \dot{\boldsymbol{\Phi}}_w = -\boldsymbol{Q}_w^{-1}\boldsymbol{\psi}_w(\boldsymbol{x})s(e) \tag{15}$$

From which,

$$\dot{V} = -s\hat{w}(\boldsymbol{x},t) - sH(\boldsymbol{x})u_d - sH(\boldsymbol{x})u_s \tag{16}$$

Substitute $u_d(t)$; $u_s(t)$ and (4) into (16)

$$\dot{V} \le |s|\left(\frac{H(\boldsymbol{x})}{H_{up}} - 1\right)\|\hat{w}(\boldsymbol{x},t)\| - |s|\,(\lambda - \sup_{0\le\tau\le t}|\Delta_w(\boldsymbol{x},\tau)|)\frac{H(\boldsymbol{x})}{H_{up}} \le 0$$

where $\lambda \ge \sup\limits_{0\le\tau\le t}|\Delta_w(\boldsymbol{x},\tau)|$.

Since V is quadratic function and $\dot{V} \le 0$, the control system is proven to be stable. It is clear that $V \in L^\infty$, which implies $s \in L^\infty$ and $\tilde{\boldsymbol{\Phi}}_i \in L^\infty$. We have $\dot{s} \in L^\infty$. Consequently, from the definition of and (7), we have $e^{(j)} \in L^\infty, j = 1,\ldots,r-1$. Also, since $\int_0^t \dot{V}dt = V(t) - V(0) < \infty$, we have $e \in L^2$. Since $e \in L^2$, we have $\dot{e} \in L^\infty$. From the facts that $\dot{e} \in L^\infty$ and $e \in L^\infty \cap L^2$ according to Barbalats Lemma [17], we have $\lim\limits_{t\to\infty} e(t) \to 0$. Because of limit space, the next step is similar in [10].

Remark 1. With the constraint (12), it is obvious that, the chattering phenomena will be reduce dramatically by choosing $\lambda \le \sup\limits_{0\le\tau\le t}|w(\boldsymbol{x},\tau)|$. In ideal case, the disturbance component is estimated exactly and then the chattering phenomena is removed completely with $\lambda = 0$.

3 Novel Adaptive RBF Neural via Sliding Mode Control and Disturbance Estimation (NASC-DE)

With the estimated nonlinear function as Sect. 2, the closed loop system including state feedback controller (8) and uncertain nonlinear system with disturbance (1) can be expressed

$$e^n = -k_1e - k_2\dot{e} - k_3\ddot{e} - \cdots - k_{n-1}e^{n-1} - \lambda sgn(s) - \Delta_w(\boldsymbol{x},t) \tag{17}$$

in which, the parameters $k_1; k_2; \ldots; k_{n-1}$ are chosen in (6). They only ensure the stability of closed loop system but, a tracking quality of closed loop is not considered. *For example*, the roots of the linear ordinary differential equation with $n = 2$

$$\begin{aligned} e_0(t) = k_1\exp(0.5t\sqrt{-k_2^2 + 4k_1} + k_2) + k_2\exp(0.5t\sqrt{k_2^2 + 4k_1} \\ e(t) = e_0(t) - k_1^{-1}\sup_{\boldsymbol{x},t}|\Delta_w(\boldsymbol{x},t)| - \lambda sgn(s)k_1^{-1} \end{aligned} \tag{18}$$

For all t, $e_0(t) \geq 0$, at the time $t = t_d$,$k_1^{-1} \sup_{\boldsymbol{x},t=t_d} |\Delta_w(\boldsymbol{x}, t_d)| = k_1^{-1}\rho > e_0$, from that $e(t = t_d) < 0$ and $sgn(s) < 0$, the $-k_1^{-1}\lambda sgn(s) = k_1^{-1}\lambda$ will be compensated and attracted its error into an original and otherwise.

On the other hand, it is clear that the performance of closed loop system (the settling time, overshoot, and static error) depends on the parameters k_1 and k_2. Therefore, to enhance the tracking performance of system, we need to find the optimal parameters or the adaptive parameters. In this paper, we propose an adaptive law for the parameters $k_1; k_2; \ldots; k_{n-1}$. Specifically, the equation (17) can be rewritten

$$\begin{aligned} e^n &= -[k_{10} + k_1(t)]e - [k_{20} + k_2(t)]\dot{e} - [k_{30} + k_3(t)]\ddot{e} - [k_{(n-1)0} + k_{n-1}(t)]e^{n-1} \\ &= k_{10}e + k_{20}\dot{e} + \cdots + k_{(n-1)0}e^{(n-1)} + \boldsymbol{\Phi}_k^T \boldsymbol{e} \end{aligned} \tag{19}$$

where $k_{10}, k_{30}, k_{30}, \ldots, k_{(n-1)0}$ are defined in (6), and $k_1(t); k_2(t); \ldots; k_{n-1}(t)$ can be defined by adaptive updated law.

The u_c controller component in (8) is rewritten as follows

$$u = \hat{H}^{-1}(\boldsymbol{x})(-\hat{F}(\boldsymbol{x}) + y_m^{(n)} + \boldsymbol{\Phi}_k^T \boldsymbol{e}) \tag{20}$$

where $\boldsymbol{\Phi}_k^T = (k_1, k_2, \ldots, k_{n-1})$; $\boldsymbol{e} = (e \ \dot{e}, \ldots, e^{(n-1)})^T$. The $\boldsymbol{\Phi}_k^T$ is calculated via the adaptive law in Theorem 2.

Theorem 2. *Consider the nonlinear system (1) is satisfy the assumptions 1–3. The controller is defined as (8) where $\boldsymbol{\Phi}_H^T, \boldsymbol{\Phi}_F^T$ are found by the adaptive updated law*

$$\dot{\boldsymbol{\Phi}}_F = -\boldsymbol{Q}_F^{-1}\boldsymbol{\psi}_F(\boldsymbol{x})\varepsilon; \ \dot{\boldsymbol{\Phi}}_H = -\boldsymbol{Q}_H^{-1}\boldsymbol{\psi}_H(\boldsymbol{x})\varepsilon u_c$$

the parameters of controller are updated by

$$\begin{aligned} &k_1(t) = k_{10} + \tfrac{1}{q_{k11}} \textstyle\int_0^t (y_m - y)\varepsilon dt \\ &\vdots \\ &k_{n-1}(t) = k_{(n-1)0} + a_{n-1} \textstyle\int_0^t (y_m^{(n-1)} - y^{(n-1)})\varepsilon dt \end{aligned} \tag{21}$$

where $a_{n-1} = \frac{1}{q_{k(n-1)(n-1)}}$, and the disturbance estimation and the feed-forward correction in (10), (11), respectively. Then, the closed loop system is sable and $\lim_{t\to\infty} (\hat{w}(\boldsymbol{x}, t) - w(\boldsymbol{x}, t)) \to 0$ for all $t \geq 0$.

Proof. Similar as the proof of Theorem 1.

4 A Coupled-Tank Liquid Level

A coupled-tank liquid level system and parameters in [13] shown as follows

$$\frac{d}{dt}\begin{pmatrix} x_1 \\ x_2 \end{pmatrix} = \begin{pmatrix} \frac{-b_{12}a_{12}\sqrt{2g|x_1-x_2|}}{A_1} \\ \frac{1}{A_2}\left(b_{12}a_{12}\sqrt{2g\,|x_1 - x_2|} - b_2a_2\sqrt{2gx_2}\right) \end{pmatrix} + \begin{pmatrix} \frac{k}{A_1} \\ 0 \end{pmatrix} u(t) + d(\boldsymbol{x}, t)$$
$$y = x_2; \boldsymbol{x} = (h_1 \ h_2)^T$$

The control target is that tracking control the liquid level of Tank 2 is at the set point. In which the input control is opened value ratio of pump 1. There are two uncertain parameters (*the cross-sectional area interaction pipe* a_{12}*; and the flow rate of liquid into tank 2* a_2) as in Fig. 4.

The nonlinear function $\hat{H}(\boldsymbol{x}); \hat{F}(\boldsymbol{x}); \hat{w}(\boldsymbol{x}, t)$ are through the adaptive RBF neural network with one hidden layer, which has 9 neurons. $\boldsymbol{x} = (x_1\, x_2)^T$ is the input vector of RBF neural network, the Gaussian functions are considered as radial basis functions $\phi_{bi}(\boldsymbol{x}) = \exp\left(-\frac{(x_1 - c_{1i})^2 + (x_2 - c_{2i})^2}{2\sigma^2}\right)$ where the width for each function is $\sigma = 0.6$ and the centers of the basis function are evenly distributed in state space as Fig. 1 and $\boldsymbol{\Phi}^T$, connection weight matrix, is updated by adaptive law in Theorem 1 for DEVRS and Theorem 2 for NASC-DE. Chosing the fixed coefficients $k_p = 1$; $k_d = 2$ of DEVRS and initial parameters $k_p(0) = 1$; $k_d(0) = 2$ of NASC-DE.

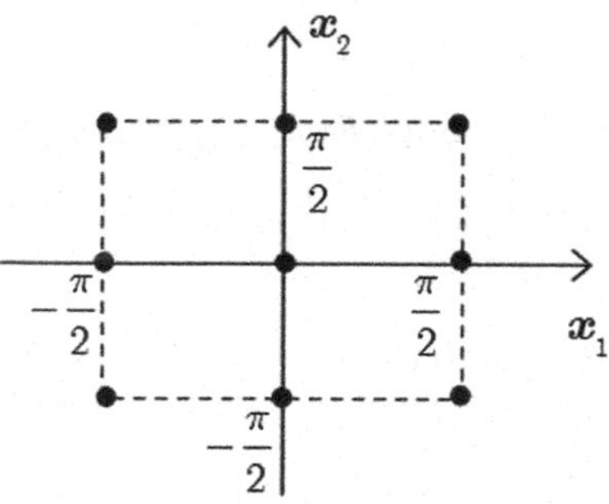

Fig. 1. Centers of basis function

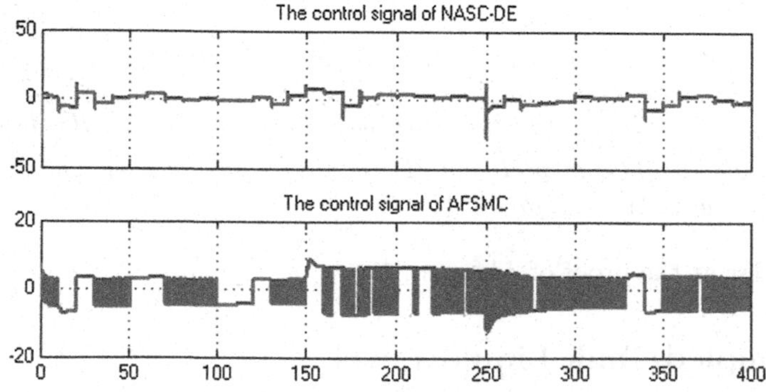

Fig. 2. Control signal of NASC-DE and AFSMC

In first case, our approach, NASC-DE, is compared to Adaptive Fuzzy Sliding Mode Control (AFSMC) in [16]. The disturbance estimation and feed-forward

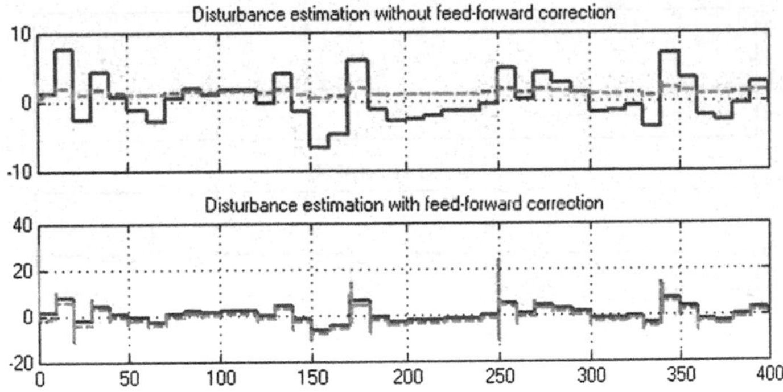

Fig. 3. Disturbance estimation without Γ and with Γ

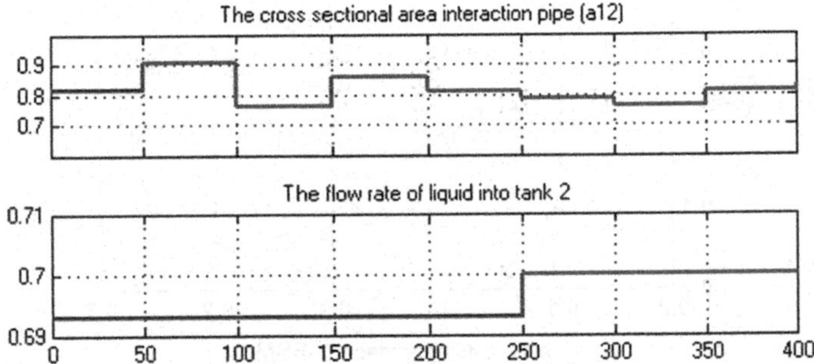

Fig. 4. Two uncertain parameters of system

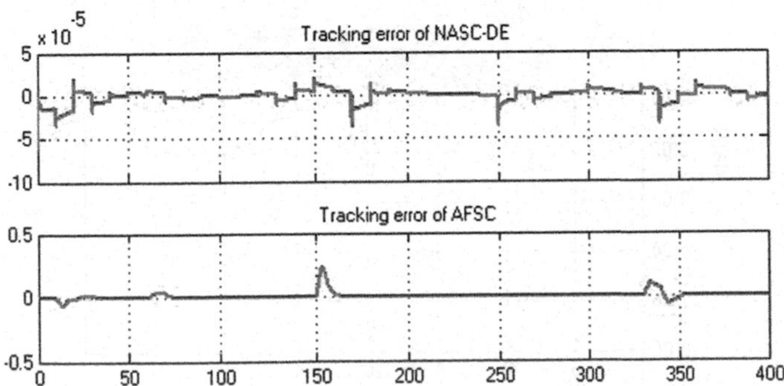

Fig. 5. Tracking error of NASC-DE and AFSMC

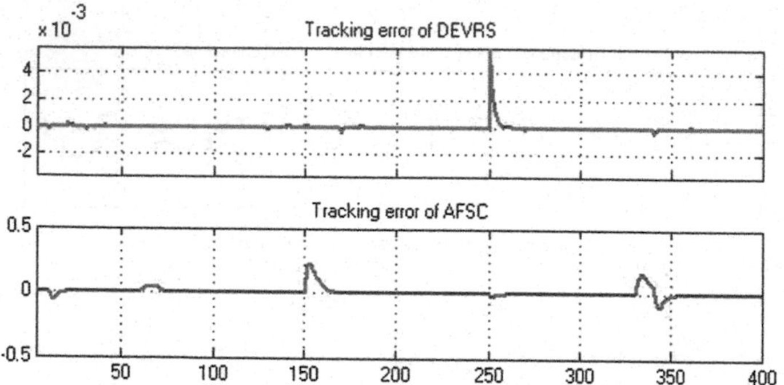

Fig. 6. Tracking control of DEVRS and AFSMC

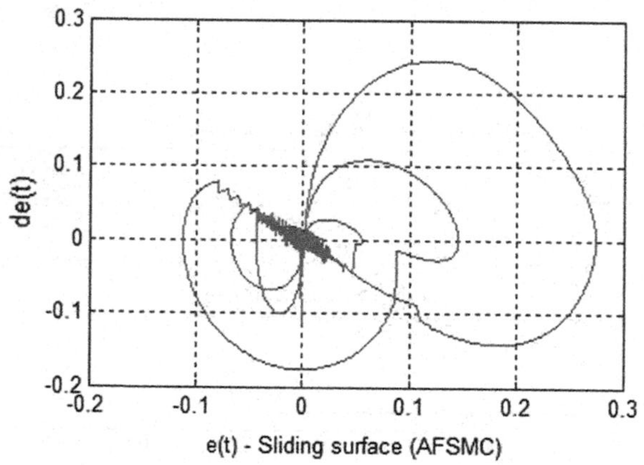

Fig. 7. Sliding surface of AFSMC

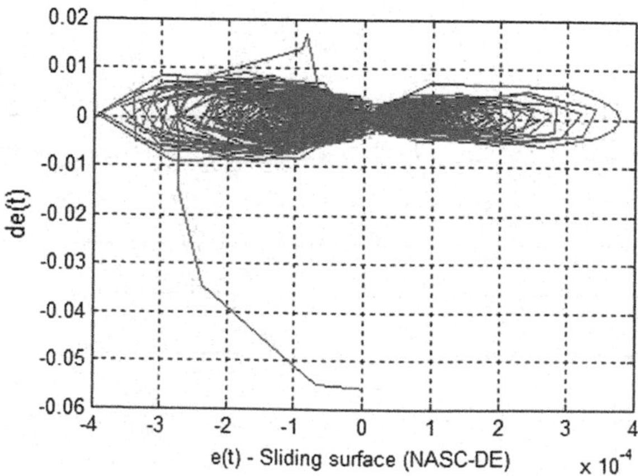

Fig. 8. Sliding surface of NASC-DE

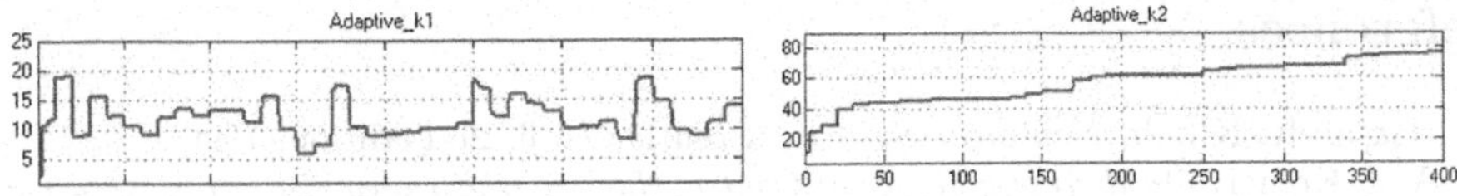

Fig. 9. Adaptive parameters of the NASC-DE

correction component $\boldsymbol{\Gamma}$ are computed following Theorem 2

$$\begin{aligned}&\boldsymbol{A}_d = \begin{pmatrix} 0 & 1 \\ -k_1 & -k_2 \end{pmatrix}; \boldsymbol{a}_d = (\,-k_1 \; -k_2\,); \boldsymbol{\Gamma} = (\,\Gamma_1 \; \Gamma_2\,) \\ &\boldsymbol{B}_d = (\,0\;1\,)^T; \boldsymbol{Q}_w = 0.01\boldsymbol{I}_{(9\times 9)}; \boldsymbol{\psi}_w(x) = (1\;\mathbf{0}_{(1\times 8)}) \\ &k_1(t) = k_1(0) + \tfrac{1}{q_{k11}} \int (y_m - y)\varepsilon dt; k_2(t) = k_2(0) + \tfrac{1}{q_{k22}} \int (\dot{y}_m - \dot{y})\varepsilon dt\end{aligned} \tag{22}$$

Based on Fig. 3, we easy see that the disturbance is almost exactly and the adaptive component $\boldsymbol{\Phi}_k^T = (k_1 \; k_2)$ of the NASC-DE is changed if the nonlinear system is influenced by the disturbance (see Fig. 9). The result of liquid level control uses NASC-DE and AFSMC is shown. In Fig. 5, the tracking error of the NASC-DE is better than that of AFSMC. In addition, AFSMC always exits the chattering phenomena in which $\lambda = 0,4 \geq \rho$ in Fig. 2 with the sliding surface as Fig. 7, and this phenomena almost eliminate by choosing $\sup\limits_{0\leq\tau\leq t} |\Delta w(\boldsymbol{x}, \tau)| \leq \lambda = 0.01 \leq \rho$ with sliding surface as Fig. 8.

In second case, DEVRS is shown. The disturbance estimation is computed in Theorem 1 in which the term $\boldsymbol{\Gamma}$ is chosen as

$$\begin{aligned}&\boldsymbol{A}_d = \begin{pmatrix} 0 & 1 \\ -1 & -2 \end{pmatrix}; \boldsymbol{a}_d = (\,-1 \; -2\,); \boldsymbol{\Gamma} = (\,\Gamma_1 \; \Gamma_2\,) \\ &\boldsymbol{B}_d = (\,0\;1\,)^T; \boldsymbol{Q}_w = 0.01\boldsymbol{I}_{(9\times 9)}; \boldsymbol{\psi}_w(x) = (1\;\mathbf{0}_{(1\times 8)})\end{aligned} \tag{23}$$

and tracking error of DEVRS is less than AFSMC (see Fig. 6) but larger than NASC-DE.

5 Conclusion

The paper showed the method which can eliminate the chattering phenomena by disturbance estimation associated with the feed-forward correction. And the performance of nonlinear system is improved by giving the adaptive parameter vector of controller. In addition, a robust of system is guaranteed by sliding mode technique. Our approaches are developed for a class of nonlinear affine systems under uncertainties and disturbances.

References

1. Hassan K. Khalli, *Nonlinear Systems*, Prentice Hall, 2nd edition, 1996
2. A. J. Krener. *On the Equivalence of Control System and Linearization of Nonlinear Sys-tems.* In: SIAM J. Control Optim. 11. pp 670–676, 1973
3. D. Cheng, A. Isidori, W. Respondek, and T. J. Tarn. *Exact Linearization of Nonlinear with Ouput.* Math, System Theory 21. pp 63–68 ,1998
4. Astolfi. A and Marconi. L, *Analysis and Design of Nonlinear Control Systems*, Springer Verlag, 2008
5. Chia-Shang Liu, Huei Peng, *Disturbance Observer Based Tracking Control*,Transactions of the ASME, Vol. 122, June 2000.
6. Z. Bouchamaa, N. Essounboulib, M.N. Harmasc, A. Hamzaouib, K. Saoudi, Reaching phase free adaptive fuzzy synergetic power system stabilizer. *International Journal of Electrical Power and Energy Systems*, 77, 43–49, 2016
7. Yong Tao, Jiaqi Zheng, Yuanchang Lin, A Sliding Mode Control-based on a RBF Neural Network for Deburring *Industry Robotic Systems. International Journal of Advanced Robotic Systems*, 2016
8. R M Nagarale; B. M. Patre, Decoupled neural fuzzy sliding mode control of nonlinear systems. *Conference on Fuzzy Systems*,pp 1–8, 2013
9. Hur Abbas, Sajjad and Shahid Qamar, Sliding Mode Control of coupled tank liquid level control system. *IEEE 10th International Conference on Frontires of Information Technology*, 2012
10. Thiem V Pham, Lai K Lai, Quynh T.T Nguyen, Minh T. Nguyen, Disturbance Estimation Combined with New Adaptive RBF Neural Network for Uncertain System with Disturbance. *International Conference on Advanced Technologies for Communication (IEEE-ATC2016)*, pp 247–252, 2016
11. H.F. Hoa, Y.K. Wonga, A.B. Rad Adaptive fuzzy sliding mode control with chattering elimination for nonlinear SISO systems. *Simulation Modelling Practice and Theory*, 17(7), pp 1199–1210, 2009
12. Roopaei, M, Mansoor Zolghadrib, Sina Meshksarc, Enhanced adaptive fuzzy sliding mode control for uncertain nonlinear systems. *Communications in Nonlinear Science and Numerical Simulation*, 14(910), pp 3670–3681, 2009
13. Thiem P.V., Lai L.K., Quynh N.T.T. Improved Adaptive Fuzzy Sliding Mode Control for Second Order Nonlinear System. *Advances in Information and Communication Technology. ICTA 2016. Advances in Intelligent Systems and Computing, vol 538. Springer, Cham*, 2017
14. Mahamed Bahita, Khaled Belarbi, Neural Network Linearization Adaptive Control for Affine Nonlinear Systems Based on Neural Network Estimator. *Serbian journal of electrical engineering*, vol 8, No, 3, November, pp 307–323, 2011
15. D. R. Hush, B. Horne, Efficient Algorithms for Function Approximation with Piece wise, *IEEE Transactions on Neural Networks*, Vol. 9, No. 6, Nov. 1998, pp. 1129–1141, 1998
16. Shaojiang Wang, Li Hou, Lu Dong, Huajun Xiao Adaptive fuzzy sliding mode control of uncertain nonlinear SISO systems. *Procedia Engineering*, 24, pp 33–37, 2011
17. Popov. V. M, Hyperstability of Control Systems, *Springer-Verlag, NewYork*, 1973
18. Ouriagli, M. Neural Network Sliding Mode Controller for a Class of Nonlinear Uncertain Systems, *Multimedia Computing and Systems (ICMCS), 2012 International Conference*, pp 0–3, 2012

A Modification of Solution Optimization in Support Vector Machine Simplification for Classification

Pham Quoc Thang[1(✉)], Nguyen Thanh Thuy[2], and Hoang Thi Lam[1]

[1] Tay Bac University, Sonla, Vietnam
{thangpq, lamht}@utb.edu.vn
[2] VNU University of Engineering and Technology, Hanoi, Vietnam

Abstract. The efficient classification ability of support vector machine (SVM) has been shown in many practical applications, but currently it is significantly slower in testing phase than other classification approaches due to large number of support vectors included in the solution. Among different approaches, simplification of support vector machine (SimpSVM) accelerates the test phase by replacing original SVM with a simplified SVM that uses significantly fewer support vectors. Nevertheless, the final aim of the simplification is to try keeping the simplified solution as similar as possible to the original one. To ameliorate this similarity, in this paper, we present a modification of solution optimization in SimpSVM. The proposed approach is based on stochastic optimization. Experiments on benchmark and sign language datasets show improved results by our modification.

Keywords: Support vector machines · Simplification of support vector machine · Stochastic optimization · Solution optimization

1 Introduction

Over the years, the classification problem has been widely studied. Many factors can affect the results of classification such as incomplete data, selecting parameter values for a particular model. There are many proposed methods to solve this problem more efficiently: the back propagation neural network, the decision tree and the support vector machine [1].

SVM (Vapnik, 1995) is an efficient machine learning method to solve classification and regression problems. SVM has been shown to be successful in many pattern recognition problems such as speech recognition, digits recognition, handwriting recognition... By combining with the kernel function method, SVM provides an effective model for classification and nonlinear regression problems in practice. However, SVM is significantly slower in testing phase than other learning methods with similar generalization performance such as decision trees, neural networks [2–5].

V. Bhateja et al. (eds.), *Information Systems Design and Intelligent Applications*, Advances in Intelligent Systems and Computing 672,
https://doi.org/10.1007/978-981-10-7512-4_15

A SVM has the solution consisting of a set of training vectors called support vectors, and a set of corresponding weights. To classify a new test example, SVM compares it with these support vectors through kernel calculations. If the number of support vectors is large, this number of comparison becomes very expensive because it is proportional to the number of support vectors. While SVM is a robust classification technique, one of the obstacles to its practical application is the large size and slow testing time. Therefore, reducing this comparison will increase the speed of the testing phase.

There have been some proposed algorithms to reduce this computational complexity, such as creating a smaller set of new vectors, called reduced vector set, so that the loss in generalized performance is acceptable. Burges [2] has developed the first constructive reduced set methods, by approximately the original SVM with a new one includes a reduced vector set. This approach is also described in [5] and further developed in [6]. Refs. [4,6] builds step by step the reduced set by finding vectors which minimise the difference between the reduced set expansion and the original vector one. The authors in [7] extended this method by using the same stratagem to create reduced vector set, then share it between multiple binary-class SVMs in a multi-class SVM by retraining them. Nguyen and Ho [8] proposed a reduction process that would iteratively replace two support vectors of the same class with a newly constructed reduced vector for simplifying binary-class SVM. Nguyen et al. [9] extend this method to multi-class case by calculating the optimal combination of two multi-weighted support vectors.

However, it would be better if the simplification keeps the simplified solution as similar as possible to the original one. To ameliorate this similarity, [9] proposed adjust the whole reduced vectors globally in respect of minimising the difference between norms of solution's hyperplanes and then the authors use a gradient descent for minimising this difference but the speed is quite slow. In this paper, we will introduce our modified version of SimpSVM algorithm in [9]. It has idea of stochastic gradient descent method to solve the above solution optimization problem. Experimental results on different benchmark and sign language data sets show that the modified SimpSVM can reduce the time for simplifying SVM while preserving as much as possible of its prediction performance.

The remainder of this paper has been organised as follows. Sections 2 and 3 present Support Vector Machine, Simplification of Support Vector Machine. Section 4 describes our modified SimpSVM algorithm. Section 5 presents experimental results and conclusion is presented in the last section.

2 Support Vector Machine

SVM works in feature space F by using a kernel function $K(x,y) = \Phi(x) \cdot \Phi(y)$ with $\Phi : R^n \rightarrow F$ as a mapping function from a input space to a feature space with higher dimensional [1]. $K(x,y) = \Phi(x) \cdot \Phi(y)$ is a kernel function that returns the dot product of two vectors $\Phi(x)$ and $\Phi(y)$ in feature space.

For a binary-class classification problem, the decision function via kernel function K:

$$f(x) = sign\left(\sum_{i=1}^{Ns} \alpha_i K(x_i, x) + b\right) \tag{1}$$

where $x_i, i = 1, \ldots, N_S$ is a subset of the training set, they are called *support vectors*. N_S is the number of support vectors, α_i is weight of support vector x_i, b is the bias and x is the input vector needed to classify.

SVM classification is basically a binary-class classification technique and it has to be modified to solve the multi-class missions. Let M be the number of classes, two primary multi-class techniques are one-vs-all and one-vs-one. The one-vs-all approach builds M binary one-vs-all SVMs, mth SVM separates class mth (positive class) out of the other classes (negative class). The one-vs-one approach builds $M(M-1)/2$ one-vs-one SVMs, each SVM splits a pair of classes. The last decision is then based on M functions

$$f_m(x) = \sum_{i=1}^{Ns} \alpha_{mi} K(x, x_i) + b_m, m = 1, \ldots, M \tag{2}$$

with α_{mi} is the weight of the support vector x_i for the mth SVM [9].

3 Simplification of Support Vector Machine

For SVM, when classifying a new data point x, the most time-consuming process is to compare it with all support vectors through kernel function K. This number of comparison usually linearly scales with the number of support vectors N_S. To decrease this computation cost, as well as to accelerate the test phase, SVM simplification methods try to replace N_S original support vectors by N_Z ($N_Z < N_S$) new vectors, they are called *reduced vector* set. Then the decision functions come to be

$$f'_m(x) = \sum_{i=1}^{Nz} \beta_{mi} K(z_i, x) + b_m, m = 1, \ldots, M \tag{3}$$

There have been some studies done for constructing reduced vectors in which [8] has suggested iteratively choosing a pair of support vectors (x_i, x_j) then replace it by a combined vector z that is found by solving

$$min \leftarrow \|(\alpha_i \Phi(x_i) + \alpha_j \Phi(x_j)) - \beta \Phi(z)\|^2 \tag{4}$$

For a multi-class classification problem, if we need to replace a pair of multi-weighted support vectors $((x_i, \alpha_{mi}), (x_j, \alpha_{mj}))$ by one new single-vector $(z, \beta_m), m = 1, \ldots, M$, we have to solve the solution optimization problem for all single SVMs

$$min \leftarrow \sum_{m=1}^{M} \|(\alpha_{mi} \Phi(x_i) + \alpha_{mj} \Phi(x_j)) - \beta_m \Phi(z)\|^2 \tag{5}$$

The simplification procedure iteratively selects a pair of support vectors (x_i, x_j) and replaces it with one new vector z by using the method as described in [9].

The combination criterion (5) of the procedure described above is individually optimized for the pair of selected support vectors. Nevertheless, it would be better if the simplification keeps the simplified solution as similar as possible to the original one. To ameliorate this similarity, [9] proposed adjust the whole reduced vectors globally in respect of minimising the differences between norms of solution's hyperplanes as follows:

$$min \leftarrow F = \sum_{m=1}^{M} \left\| \sum_{i=1}^{Ns} \alpha_{mi} \Phi(x_i) - \sum_{i=1}^{Nz} \beta_{mi} \Phi(z_i) \right\|^2 \tag{6}$$

or can be rewritten as

$$min \leftarrow F = \sum_{m=1}^{M} F_m \tag{7}$$

with

$$F_m = \left\| \sum_{i=1}^{Ns} \alpha_{mi} \Phi(x_i) - \sum_{i=1}^{Nz} \beta_{mi} \Phi(z_i) \right\|^2 \tag{8}$$

In [9], the authors applied the gradient descent to minimise F in respect of the whole reduced vectors z_i, $i = 1, \ldots, N_Z$ with the directions of search are as follows:

$$\frac{\partial F}{\partial z_i} = \sum_{m=1}^{M} \frac{\partial F_m}{\partial z_i} \tag{9}$$

4 Modified SimpSVM Algorithm

Consider the problem of minimising an objective function that is written as a sum of differentiable functions:

$$F(w) = \frac{1}{m} \sum_{i=1}^{m} F_i(w) \tag{10}$$

where the parameter w which minimises $F(w)$ is to be estimated, each function F_i is typically associated with the ith observation.

In optimization theory, there are many different methods to solve above problem. Of which, the gradient descent method performs updating the weight w on the basis of the gradient of $F(w)$ in each iteration:

$$w^{(t+1)} = w^{(t)} - \eta \nabla_w F(w^{(t)}) = w^{(t)} - \eta \sum_{i=1}^{m} \nabla_w F_i(w^{(t)}) \tag{11}$$

where η is the step size.

Stochastic gradient descent is a drastic simplification algorithm in machine learning. The true gradient of $F(w)$ in stochastic gradient descent is approximated by a gradient at a single observation:

$$w^{(t+1)} = w^{(t)} - \eta \nabla_w F_i(w^{(t)}), i \in [1, m] \tag{12}$$

Solving the problem (7) is equivalent to solve the following problem:

$$min \leftarrow F^* = \frac{1}{M} \sum_{m=1}^{M} F_m \tag{13}$$

The authors in [9] proposed using the gradient descent method, optimized by updating z_i at each iteration as follows:

$$z_i^{(t+1)} = z_i^{(t)} - \eta \sum_{m=1}^{M} \frac{\partial F_m}{\partial z_i^{(t)}}, i = 1, \ldots, N_z \tag{14}$$

Our proposed method is inspired by stochastic gradient descent. At each iteration update z_i, instead of having to calculate the true gradient of F^* using all single SVMs, we estimate this gradient on the basis of a single randomly picked mth SVM

$$z_i^{(t+1)} = z_i^{(t)} - \eta^{(t)} \frac{\partial F_m}{\partial z_i^{(t)}}, i = 1, \ldots, N_z \tag{15}$$

where $\eta^{(t)}$ is a step size at iteration t and adapted according to the training process. The update process will be performed until the objective function converges to the smallest value. The proposed method is summarized in Algorithm 1.

Algorithm 1. Modified SimpSVM Algorithm

Input: set of N_S support vectors x_i, $i = 1, ..., N_S$
Output: set of N_Z reduced vectors z_k, $k = 1, ..., N_Z$ with $N_Z < N_S$
1. **while** *Stopping Condition* is not satisfied **do**
2. Select a pair (x_i, x_j)
3. Replace (x_i, x_j) with a new constructed vector z using the method as described in [9]
4. **end while**
5. Re-compute coefficients for all reduced vectors
6. Optimize the whole reduced vector set as (15)

This approach has computational complexity which scales linearly with the number of SVMs (or the number of classes) of the problem, allow to effectively classify data with large classes.

5 Experiments

In this section, we evaluated the effectiveness of our modified SimpSVM algorithm. We have implemented it by using programming languages C/C++. We compared the performance of the modified SimpSVM with original one on eight benchmark and a sign language data sets. All the programs were run on a PC which has CPU Intel Core i3 3.3 GHz and 2 GB of RAM.

5.1 Benchmark Datasets

The performance of the modified SimpSVM has been studied for eight benchmark data sets, namely the "*DNA*", "*Letter Recognition*", "*Shuttle*", "*Satimage*", "*Vowel*", "*Pendigits*", "*Usps*" and "*Mnist*". All data sets are publicly available from the UCI Machine Learning Repository [10], and their details are given in Table 1. The selection of parameters C, γ for SVM, SimpSVM models is very important. We used grid search to select training parameters so that standard SVM classifiers achieve the highest predictive accuracy on testing data sets. We also listed the parameter values used in our experiments in Table 1.

Table 1. Characteristics of datasets and parameter setting

Name	# Attributes	# Class	# Training	# Testing	Parameters
DNA	180	3	1400	1186	$C = 10, \gamma = 0.01$
Letter	16	26	10500	5000	$C = 10, \gamma = 2$
Shuttle	9	7	30450	14500	$C = 10, \gamma = 0.1$
Satimage	36	6	3104	2000	$C = 10, \gamma = 0.1$
Vowel	10	11	528	462	$C = 2, \gamma = 2$
Pendigits	16	10	7494	3498	$C = 4, \gamma = 0.25$
Usps	256	10	7291	2007	$C = 10, \gamma = 0.0078$
Mnist	780	10	60000	10000	$C = 10, \gamma = 0.0128$

Table 2. Comparing the results of modified SimpSVM versus original one

Dataset	Original		Modified	
	Time (s)	Accuracy (%)	Time (s)	Accuracy (%)
DNA	63	94.52	32.19	94.44
Letter	2013	97.00	119	96.68
Shuttle	334	98.98	76	98.98
Satimage	76	89.25	7.08	89.25
Vowel	33.63	60.82	2.89	61.04
Pendigits	89	98.48	26.44	98.48
Usps	315	95.42	169	95.42
Mnist	17896	98.34	9196	98.28

In the first experiment, we compared the accuracy and reducing time of modified SimpSVM and original one on the prepared test data sets. This comparison is shown in the Table 2. As can be seen in the Table 2, the modified SimpSVM can reduce the time for simplifying SVM while preserving as much as possible of its prediction performance. Special, on the "*letter*", "*vowel*" and "*satimage*" data sets, modified SimpSVM can run faster than original one to 10 times while the prediction accuracy has not been changed much.

In our next experiment, we compared performance of modified SimpSVM, original SimpSVM and the SVM simplification methods as described in [2,7] on the "*usps*" handwritten digit data set, the results are as shown in Fig. 1. In this experiment, standard SVM classifier gives 1459 support vectors and the accuracy is 95.32%. As can be seen in Fig. 1, modified SimpSVM gives competitive performance in terms of accelerating rate while preserve its prediction accuracy.

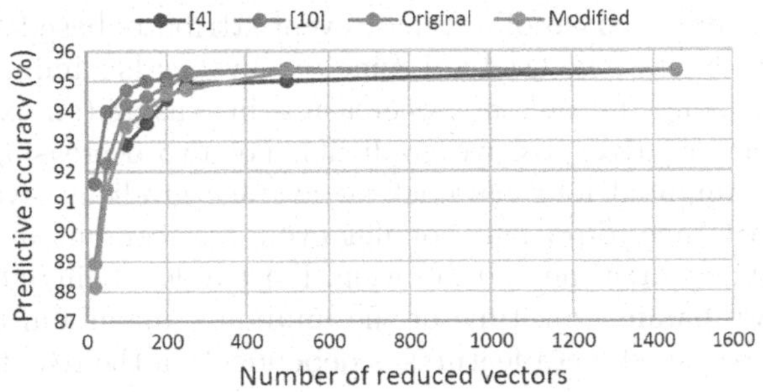

Fig. 1. Comparison of number of reduced vectors and predictive accuracy between simplified SVMs

5.2 Sign Language Recognition

There are many sign languages in the world, in which Auslan is the sign language used by the Non-vocal and Deaf community in Australia. In Auslan, to express a meaning, each sign is expressed as a series of gestural patterns. There are multiple signs in Auslan language so this is a multi-class classification problem (Fig. 2).

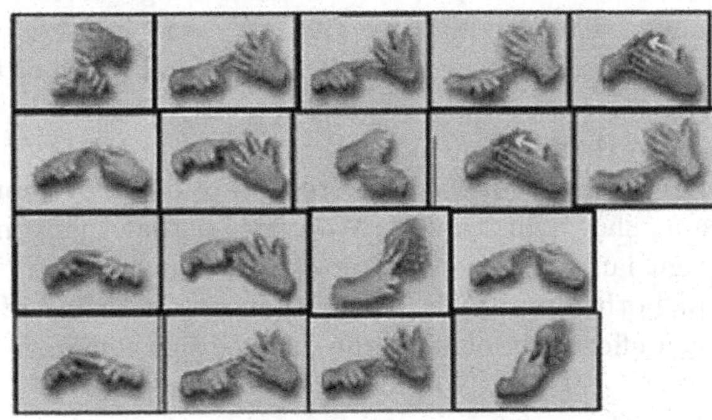

Fig. 2. Some sample images of signs in Auslan sign language

Data. We took the data set from the UCI Machine Learning Repository [10]. In this data set, each data sample contains a series of 22-dimensional value vectors: x, y, z, yaw, pitch, roll, Little finger bend, Ring finger bend, Middle finger bend, Forefinger bend, Thumb bend... for both hands and their details are described in [11]. The data set consists of a total of 2565 samples belonging to 95 distinctive signs with an average of 27 samples per sign.

Feature Extraction. There have been many suggested methods to extract various features from the 22 channels of information. In this experiment, we used the method as described in [11] to extract global features and meta features.

First, we computed global features by extracting various stream features including: minima/maxima/mean (of each channel). From 22 channels of information, we computed above three information to achieve 66 global features [12].

Next, we computed meta features by extracting original events and then clustering them to generate a basis of synthetic event attributes based on a confidence metric. For each channel, we used following five events: localmin, localmax, flat (no perceptible change in gradient), decreasing/increasing (an extended period with considerable negative/positive gradient). For two decreasing and increasing events, we computed four parameters: average gradient, average value of the channel, start time, duration. For flat event, we computed three parameters: average value, start time and duration. For two localmin/localmax events, we computed two parameters: time of the minimum/maximum and the value. The number of extracted meta features is dependent on the results of clustering events [12].

Finally, the data from the global features and meta features are recombined in a form suitable for classification.

Sign Classification. For ease of comparison with the results of other studies that were previously published, we used 5-fold cross validation procedure for experimentation.

In Fig. 3, we compared the predictive performance and reducing time of modified SimpSVM and original one on the three types of features: global features, meta features and both. The results show that both methods increase the classification accuracy when the number of features increases. If using a combination of both feature types, modified SimpSVM obtains the highest classification accuracy (98.32%) compared to original one (98.28%) and it needs only 356 s for reducing time while original one needs to 4011 s. On all types of features, modified SimpSVM can run faster than original one 10 times while the prediction accuracy has not been changed much.

We also compared this result with the results of other previously published authors, they are as shown in Table 3. We can see that many machine learning methods on sign language recognition can obtain over 90% accuracy, and higher accuracy of both SimpSVMs. From then on, both modified SimpSVM and original one can effectively classify the data of sign language recognition.

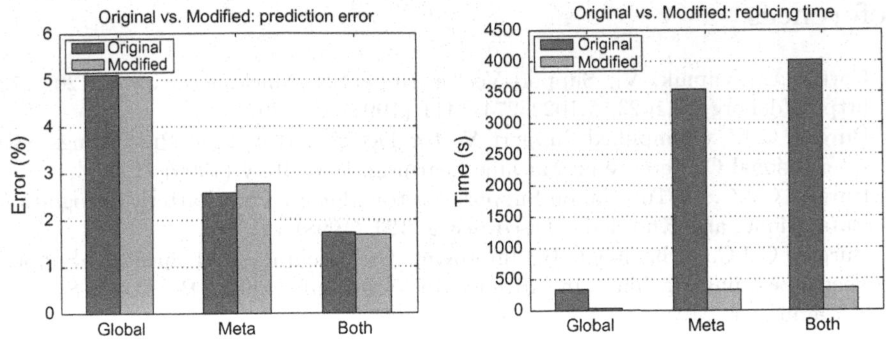

Fig. 3. The error rates and reducing times of models learned by modified SimpSVM and original one on Auslan data sets

With these above results, we believe that the modified SimpSVM can reduce the time for simplifying SVM, since allow to improve overall training time of SimpSVM.

Table 3. Comparing the results of both SimpSVM with other machine learning methods on the Auslan data set

Methods	Accuracy (%)
J48 [11]	85.5
Bag [13]	90.6
AdaBoost [14]	93.6
Nave segmentation [14]	94.5
Hidden Markov model [14,15]	87.1
Stochastic ensemble [13]	97.9
Decision lists [16]	85.0
RBF networks (by boosting) [17]	97.5
Original SimpSVM	98.28
Modified SimpSVM	93.32

6 Conclusion

In this paper, we have presented a modified SimpSVM algorithm based on stochastic optimization for classification problem. The capability of this algorithm was investigated through the performance of several experiments on the standard benchmark and sign language datasets. Experimental results show the effectiveness of our approach that it can reduce the time for simplifying SVM since allow to improve overall training time of SimpSVM while preserving as much as possible of its prediction performance.

The future work will be to certify our algorithm on other data sets and real life problems.

References

1. Cortes, C., Vapnik, V.: Support-Vector Networks. Machine Learning, 20: 273, https://doi.org/10.1023/A:1022627411411 (1995) 273–297
2. Burges, C.J.C.: Simplified Support Vector Decision Rules. In: Proceedings 13th International Conference on Machine Learning. Bari, Italy (1996) 71–77
3. Burges, C.J.C.: A Tutorial on Support Vector Machines for Pattern Recognition. Data Mining and Knowledge Discovery, 2: 121 (1998) 121–167
4. Burges, C.J.C., Schoelkopf, B.: Improving the Accuracy and Speed of Support Vector Learning Machines. In: Advances in Neural Information Processing Systems 9. Cambridge, MA: MIT Press (1997) 375–381
5. Liu, C., Nakashima, K., Sako, H.: Handwritten Digit Recognition: Benchmarking of State-of-the-art Techniques. Pattern Recognition, 36 (2003) 2271–2285
6. Schlkopf, B., Mika, S., Burges, C.J.C., Knirsch, P., Mller, K.R., Rtsch, G., Smola, A.: Input Space vs. Feature Space in Kernel-Based Methods. IEEE Trans. Neural Networks 10 (1999) 1000–1017
7. Tang, B., Mazzoni, D.: Multiclass Reduced-set Support Vector Machines. In: Proc. Int'l Conf. Machine Learning. ICML'06, ACM, New York, USA (2006) 921–928
8. Nguyen, D.D., Ho, T.B.: An Efficient Method for Simplifying Support Vector Machines. In: Proceedings of the 22nd International Conference on Machine Learning, ICML 2005, Bonn, Germany, Vol. 119. ACM, New York, USA (2005) 617–624
9. Nguyen, D.D, Kazunori, M., Kazuo, H., Yasuhiro, T., Daichi, T., Masahiro, T.: Multi-class Support Vector Machine Simplification. In: Proceedings of the 10th Pacific Rim International Conference on Artificial Intelligence. Trends in Artificial Intelligence, PRICAI 2008, Hanoi, Vietnam (2008) 799–808
10. https://archive.ics.uci.edu/ml/datasets.html
11. Kadous, M.W.: Temporal Classification: Extending the Classification Paradigm to Multivariate Time Series. In: PhD Thesis. The UNSW (2002)
12. Pham, Q.T., Nguyen, D.D., Nguyen, T.T.: A Comparison of SimpSVM and RVM for Sign Language Recognition. In: Proceedings of the 2017 International Conference on Machine Learning and Soft Computing, ICMLSC'17, Ho Chi Minh City, Vietnam, January 13–16, 2017. ACM, New York, USA (2017) 98–104
13. Kadous, M.W, Claude, S.: Constructive Induction for Classifying Time Series. In: European Conference on Machine Learning (ECML) (2004)
14. Kadous, M.W, Claude, S.: Classification of Multivariate Time Series and Structured Data using Constructive Induction. Machine Learning Journal, 58 (2005)
15. Yale, S., Ying, Y.: Sign Language Recognition. In: 6.867 Machine Learning Term Paper (2008)
16. Juan, J.R., Carlos, J.A., Henrik, B.: Learning First Order Logic Time Series Classifiers: Rules and Boosting. Principles of Data Mining and Knowledge Discovery. Lecture Notes in Computer Science, Volume 1910 (2000) 299–308
17. Juan, J.R., Carlos, J.A.: Building RBF Networks for Time Series Classification by Boosting. Pattern Recognition and String Matching Combinatorial Optimization, Volume 13 (2002) 135–153

A Two-Stage Detection Approach for Car Counting in Day and Nighttime

Van-Huy Pham[1(✉)] and Duc-Hau Le[2,3]

[1] Faculty of Information Technology, Ton Duc Thang University, No. 19 Nguyen Huu Tho Street, Tan Phong Ward, District 7, Ho Chi Minh City, Vietnam
phamvanhuy@tdt.edu.vn

[2] Vinmec Research Institute of Stem Cell and Gene Technology, Hanoi, Vietnam
hauldhut@gmail.com

[3] School of Computer Science and Engineering, Thuyloi University, 175 Tay Son, Dong Da, Hanoi, Vietnam

Abstract. We developed a car counting system using car detection methods for both daytime and nighttime traffic scenes. The detection methods comprise two stages: car hypothesis generation and hypothesis verification. For daytime traffic scenes, we proposed a new car hypothesis generation by rapidly locating car windshield regions, which are used to estimate car positions in occlusion situations. For car hypothesis at nighttime, we proposed an approach using k-means clustering-based segmentation to find headlight candidates to facilitate the later pairing process. Counting decision is made from Kalman filter-based tracking, followed by rule-based verification. The results evaluated on real-world traffic videos show that our system can work well in different conditions of lighting and occlusion.

1 Introduction

In applications of traffic monitoring, vehicle detection is usually the first and key step for the success of the systems. To be more efficient, the detection is usually comprised of two stages: hypothesis generation and hypothesis verification [1]. The hypothesis stage, where possible car bounding boxes should be located, rapidly searches for potential car candidates and filters out unnecessary regions to facilitate the next stage where the presence of vehicles is verified by a more complicated algorithm. In the literature, there are various hypothesis generation methods for cars. Symmetry-based method has successfully applied to car detection [2, 3]; contour and color are good appearance features used in [4, 5], while some other authors also used shadow [6] and texture [7] for the detection. The mentioned approaches show best results on normal circumstances of car appearance but they did not pay much attention to occlusion situations.

In vision-based car counting systems, the camera is usually fixed on a suitable position to capture vehicles passing the field of view. The most difficulty is that captured images may have a variety of appearance of shapes, colors, illumination

V. Bhateja et al. (eds.), *Information Systems Design and Intelligent Applications*, Advances in Intelligent Systems and Computing 672,
https://doi.org/10.1007/978-981-10-7512-4_16

conditions, and occlusions, etc. Unsupervised methods were successfully applied in real-time systems on highway such as edge detection [8], background subtraction [9, 10], frame differencing, optical flow [11], etc. In case of dense traffic flow, the performance of these detection methods is decreased because of car occlusion or slow movement [12]. For example, car detection using background modeling methods [13, 14] may fail to segment car parking or move slowly, and nearby cars could be clustered as one object.

Supervised techniques have been widely used to deal with the challenges of classical methods. These techniques make use a feature extraction process to train a classifier from samples. In the recent object detection applications, a variety of feature descriptions have been used. In our application, we choose histograms of oriented gradient (HOG) proposed in [15] to represent car image features for the training process because HOG is good for rigid object like cars and robust with clutter backgrounds and lighting conditions.

Assuming a static monocular monitoring camera, we aim to develop a system that is robust in dense traffic situations, where occlusion is a challenge, and in different illumination conditions, e.g., daytime and nighttime [16–18]. For that purpose, we proposed two different approaches for car counting in daytime [19] and nighttime.

The remaining sections describe our methods in detail. In Sect. 2, car detection method in daytime traffic scenes is introduced, including hypothesis generation using car windshield regions and verification of car hypothesis using HOG-based detector. In Sect. 3, headlight-based car hypothesis generation for nighttime traffic scenes is described followed by headlight pairing process to verify car candidates satisfying the defined constrains. Section 4 describes in detail our reliable counting system using the detection results and Kalman filter. The experiments on videos of the two mentioned illumination conditions are shown in Sect. 5. Section 6 gives some conclusions from our system.

2 Car Detection in Daytime Traffic Scenes

2.1 Car Hypothesis Generation

Shadow-based [6] or symmetry-based [2, 3] methods of car hypothesis generation work well in the situation of non-occluded cars that most of car parts are exposed. As our mentioned assumption, we used a fixed camera looking down the streets to capture car images from a distance. As a result, in dense traffic flow, occlusion usually happens that shadow-based or symmetry-based methods could miss potential car candidates. In such situation, using car windshield as a cue to locate car in the traffic flow is a promising solution.

Through the observation that windshield region of cars has a boundary of trapezoid, we developed a method for car hypothesis generation by just rapidly locating trapezoid regions in a traffic scene. Such trapezoid regions are almost homogeneous regions and may appear varying in size and at arbitrary positions in the region of interest. The method of using multiple scales commonly used in object detection is very slow for the hypothesis generation stage which should be fast enough for the verification stage.

Here, we propose an algorithm that avoids using scales and sliding window by simply based on edge detection and morphological operations and Hough transform.

2.1.1 Car Windshield Region Localization

Grayscale traffic image is used as input for the localization process as shown in Fig. 1. Firstly, an edge detection method, e.g., Canny method [20], is applied to find the edge map from the input. In most situations, windshield regions give strong edge response that ensures all windshield candidates could be found. The trapezoid boundary found by the edge detector may not be perfect in terms of a closed boundary because of noise and the surrounding conditions. To suppress edge gaps, the edge map is dilated to connect nearby edge segments. The dilated edges are not directly used to find boundary of candidate windshields but it could help to isolate the inner regions limited by the dilated edges from surroundings. By simply inversion of the dilated edge maps, region maps can be obtained rapidly. Candidate regions are obtained after filtering out noisy ones by region size. As an illustration, an experimental result is shown in Fig. 2, where most of windshield regions were detected, but roofs of cars could also be wrongly detected as windshield regions.

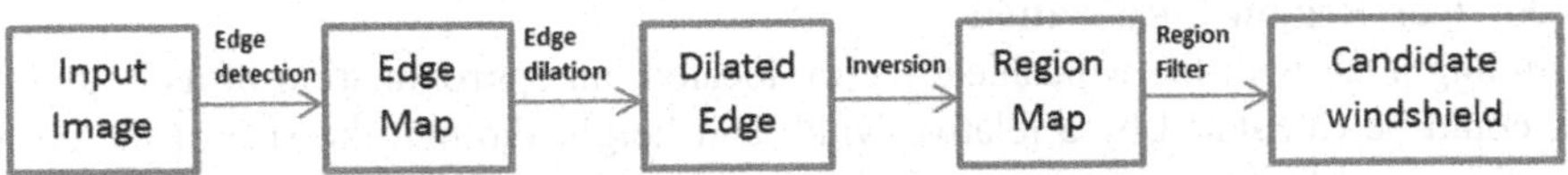

Fig. 1. Car windshield region localization process

Fig. 2. Windshield region detection: **a** a region map and **b** detected candidate windshields

2.1.2 Car Windshield Verification

In the previous step, rectangle regions of candidate windshields can be located in the edge map thanks to some morphological operations. Each rectangle region location is then used to extract edges from the edge map with some loose-fitting margins for the next step to verify if it has a shape of trapezoid. To overcome the imperfect shape of candidate trapezoids, we make use of Hough transform-based line detection, which considers lines in parameter space facilitating the verification process. The verification is mainly based on the angle and distance between lines as shown in Fig. 3.

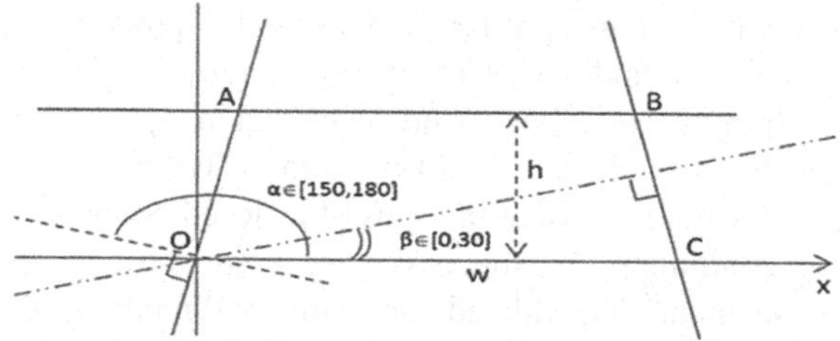

Fig. 3. Parameter ranges for trapezoid shape detection in Hough space

Figure 4a shows a trapezoid shape successfully verified by using our proposed method. Some noisy edges exist but thanks to a suitable parameters range, the region could be identified as a trapezoid shape.

2.1.3 Car Region Localization

Assuming a car with the windshield region located, an approximation of its bounding box could be calculated by a relative width and height through the size of the corresponding windshield region. In our application, we choose the size of $3/2W_{ws}$ x $3H_{ws}$, where $W_{ws}xH_{ws}$ is the size of the car windshield, as illustrated in Fig. 4b.

The approximated bounding boxes could have some extra margins, and all possible car regions should be found; e.g., we allow false positives, but minimize false negatives, because car regions will be verified by a classifier using HOG feature in the next stage.

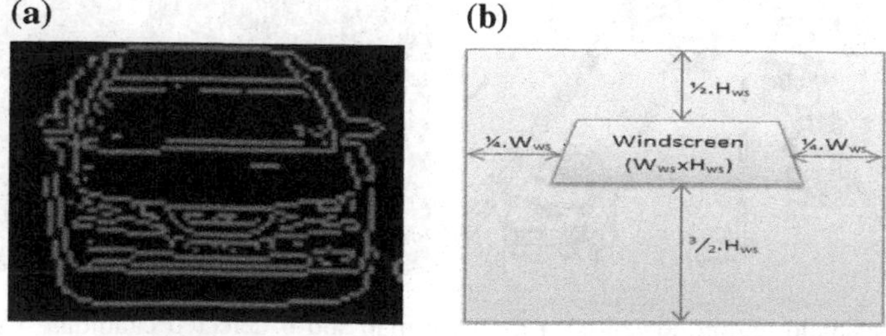

Fig. 4. **a** A detected car windshield and **b** estimation for car position

2.2 HOG-Based Car Verification

To verify car hypotheses generated by the previous state, we trained a classifier from a set of collected car images. The classifier makes use of HOG descriptors proposed by Dalal [15], which is widely applied in the field of object detection [21, 22]. We use the size of 64 × 64 pixels for all samples, and the sliding window with spatial binning of each cell is 8 × 8 pixels and blocks of 2 × 2 cells and 9 bins for orientation.

Figure 5 shows a visualization of a HOG feature from a positive sample. As the dimension of HOG feature vectors is large (1764 in our application), support vector machines (SVMs) are properly used to train the final classifier.

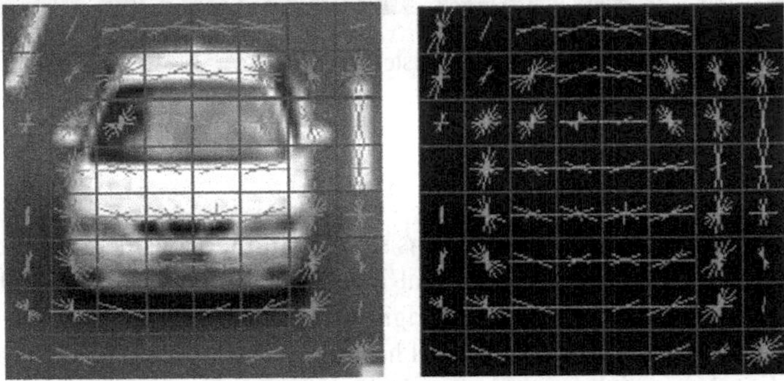

Fig. 5. HOG visualization of a car image (the gradients are rotated 90°)

3 Car Detection in Traffic Scenes at Nighttime

A supervised method is not useful for car detection at nighttime. The only dominant cue for the detection is car headlights which is used to generate car hypothesis followed by a pairing process in our method.

3.1 Car Headlights Segmentation

Traffic backgrounds at different hours at nighttime appear to be from darker to dark and from bright to brighter, and the only consistent visible things are headlights of some car which appear to be the brightest regions. Similar regions are the regions of light reflection from the nearby headlights, which should be considered in the detection.

Our idea is to divide the scenes into four layers from the highest to the lowest of brightness: car headlights, headlight reflection on background, light background, and dark background. A naïve approach for this segmentation is using four fixed thresholds but it is not robust with the time-by-time illumination changes. Instead, we use k-means clustering to adaptively threshold the gray levels of the traffic images. In this approach, k = 4 and the initial seeds are easily selected by four points equally distributed in the range [0, 255]. The values of 0 and 255 should be selected initially for grouping the headlight and dark background pixels. Our experiments showed that the method works well for the goal of brightness segmentation. Figure 6 shows some segmentation results by k-means method with different values of k for better comparison.

Fig. 6. Brightness layers from k-means clustering with k = 3, 4, 5 (from left to right, respectively)

3.2 Car Detection Based on Headlight Pairing

The segmented headlights from the previous step are further considered to verify if they are belonging to some car or not. Spatial constraints are proposed for the relative positions between headlights as they are rigid of objects.

Our method is to pair nearby individual headlights that hold for specific pre-defined constraints. For the spatial constraints to be more robust at different distances, perspective transformation is applied to minimize distance errors, as illustrated in Fig. 7.

Fig. 7. A nighttime traffic scene with a perspective transformed region

We considered the following constraints:

Size Constraint

Noisy regions should be filtered out by the size thresholds. The headlight region size should be limited by the bounding parameters as indicated by (1):

$$A_{min} < A_{Blob} < A_{max} \tag{1}$$

where A_{min} and A_{max} are acceptable minimum and maximum for the area of a headlight regions; A_{Blob} is the area of the detected blob corresponding to the headlight.

Distance Constraint

Headlights that belong to a car must be at an appropriate distance from each other in both horizontal and vertical terms, which are indicated by (2):

$$\left|P_X^i - P_X^j - \theta\right| < \varepsilon_X, \left|P_Y^i - P_Y^j\right| < \varepsilon_Y \tag{2}$$

where P_X^i and P_Y^i are horizontal and vertical positions of the ith headlight and θ is the pre-defined distance template between headlights. ε_X and ε_y are used as tolerance (in pixels) in horizontal and vertical directions, respectively.

4 4 A Car Counting System

We developed a counting system using the proposed car detection methods. Car detection in daytime has an advantage of using a supervised method that slowly moving or parking cars or even occluded cars could be detected at each frame independently. The counting system accuracy depends closely on the detection outputs which are hard to avoid a detection failure. Kalman filter is used to estimate a car position whenever a failure in detection is found to aid the counting system. Figure 8 shows the flowchart of our counting system.

When a car detection failure happens, an estimated car position is used for the continuously tracking process. The estimated bounding box is not only to support the detection failure but also to increase the tracking accuracy by its estimated centroids for distance evaluation used in the tracking process. The velocity component is also included in the Kalman filter configuration to tackle the case of cars parking or moving in backward direction.

For the counting purpose, a virtual trigger line is set across the tracking region as illustrated in Fig. 9. Moving cars are tracked continuously until exiting the trigger line. In our application, a count is recorded whenever the centroid of a car's bounding box has passed the trigger line in terms of its horizontal coordinate.

Some failures from the detection stage such as false positives (cars without bounding box), false negatives (bounding boxes without car), false fusions (redundant bounding boxes) could increase counting error and degrade the overall performance. We take the current frame counts of the cars being tracked into consideration to improve the counting system accuracy. An example of a detection failure is that a car bounding box suddenly disappears after several continuous frames being tracked, then we applied the following rule:

Car detection
HOG + SVM
New car?
True
False
Car matching
KF Initialization
Matched?
True
KF Predict & correct
False
Predicted position as car position
Car disappeared
False
Car out?
True
Car counting

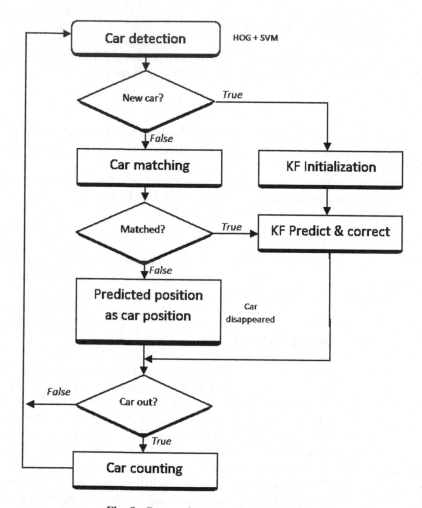

Fig. 8. Proposed car counting system flow

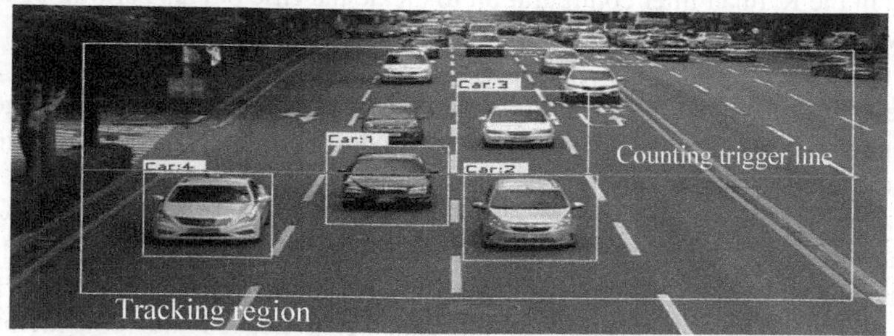

Fig. 9. A tracking region and trigger line used in the counting system

$$IF\left(\sum DetectedFrameCount > a\,threshold\right)\ and$$
$$\left(\sum DetectedFrameCount >= \sum MissedFrameCount\right)\ THEN\ count.$$

In nighttime traffic scenes, car counts are estimated by their headlights which usually appear in pairs. To facilitate the pairing process, each headlight is associated with a Kalman filter during its existence. Pairs of headlights resulting from the pairing process are also tracked based on the track of each of the headlight composition. With the same idea from daytime counting system, the counts from the system are finally checked by some pre-defined rules to filter out unexpected counts. Figure 10 shows our method flow for the nighttime counting system.

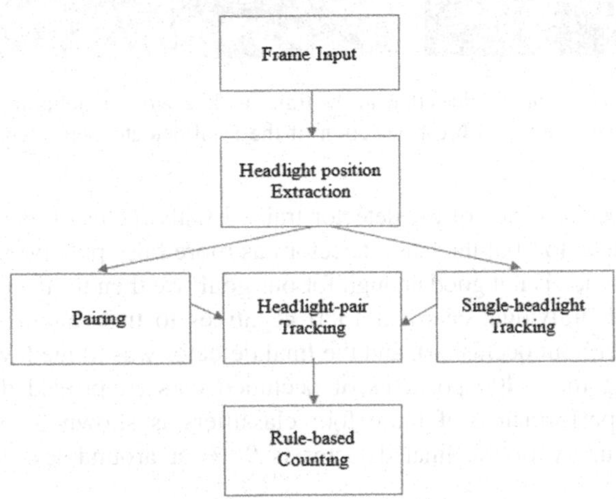

Fig. 10. Nighttime car counting system flow

5 Experiments and Discussion

In our system setup, a CCD camera was fixed at 6 meters above and set to look down the streets to capture car images at a fixed angle. The dataset used to train a classifier for car detection was collected from real-life traffic videos at the resolution of 640 × 480 pixels (30fps) by the camera, so that they can work well with HOG. We used a dataset of 6400 positives (including 400 occlusions) and more than 1300 negatives to train a classifier for car detection in daylight. The system was tested on some minutes of real traffic videos of the same resolution on various illumination conditions. In hypothesis generation stage, with a suitable setting parameter set for edge detection, morphological operations, and Hough transform, cars are localized at high accuracy rate. One important note is that, the parameter set should be chosen so that false negatives must be minimized while allowing higher false positive rate. Figure 11 shows detection results from our system.

For evaluation, we trained an initial detector using a dataset of 500 positives and 300 negatives (initial detector without occlusion). For the detector to cope with occlusion, we then added 50 positives of occluded car images (initial detector with occlusion). The performance of the detector trained with occlusion is almost the same with the former detector, but the latter detector has more false positives. The two initial detectors performance is not good enough for our goal; we then used a bigger dataset of 6000 positives of individual cars and 1380 negatives to train another classifier (improved detector without occlusion), and the final detector was trained with an extended dataset by adding more 400 positives of occluded cars (improved detector without occlusion). The performance of these four classifiers is shown by ROC curves in Fig. 12. The accuracy of the final detector is 98% at around less than 20 missed

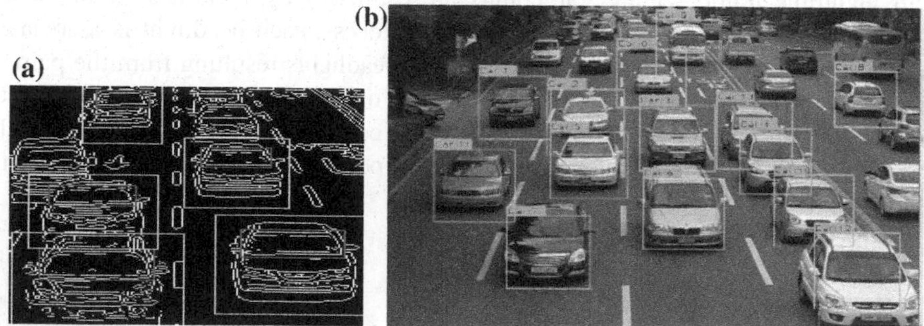

Fig. 11. An experiment of car detection in daytime: **a** detected car bounding boxes based on trapezoid shape localization and **b** car detection of the final detector with occlusion

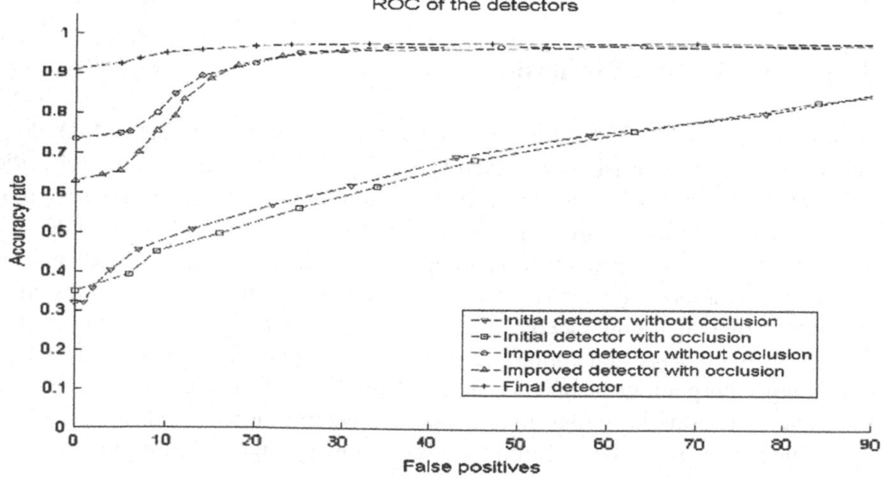

Fig. 12. Performance evaluation of the five detectors

detections among thousands of positives. An experimental result from our system is shown in Table 1, where counts recorded on a testing video are listed. The tracking processing contributes to the success of the counting thanks to Kalman filter prediction. The proposed system is reliable that 99% of cars were captured.

Table 1. Counts recorded on a 20-min testing traffic video

Situations	Counts
Human-aided counts	672
System counts on car exit	653
Extra counts by reasoning	12
False positive counts detected	120

A result from our proposed nighttime car detection is shown in Fig. 13 where segmented and paired headlights are connected by lines and then show up in the original traffic image for car existence confirmation. That the accuracy of the proposed counting system at nighttime on videos at 640 × 480 resolution (15fps) of real-life traffic experimentally reaches 98%, has proved the efficiency of our methods.

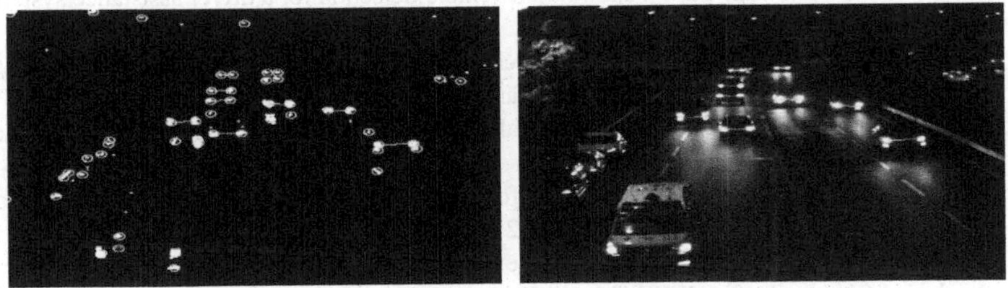

Fig. 13. A result of car headlight detection and pairing

6 Conclusion

We introduced methods of car detection in both daytime and nighttime traffic scenes. For car detection, we employed a two-stage approach: car hypothesis generation and verification. In daytime, car windshields are located rapidly in the image based on trapezoid regions. The idea of the proposed windshield-based hypothesis generation is that car windshield is the dominant feature in most situations including occluded cars. Car candidates are estimated from windshield locations and then fed into a trained classifier for verification. In nighttime traffic scenes, car headlights are extracted by using k-means method to segment image brightness into four illumination levels. A highly reliable counting system is designed by utilizing Kalman filter to keep track of moving cars continuously by prediction and rule-based reasoning. The results evaluated on real world traffic videos show that our system can work well in different conditions and is robust with various situations of illumination lighting and occlusion. More enhanced machine-learning techniques such as deep learning will be considered as our future work.

References

1. Bin, T., Qingming, Y., Yuan, G., Kunfeng, W., Ye, L.: Video processing techniques for traffic flow monitoring: A survey. In: Intelligent Transportation Systems (ITSC), 14th International IEEE Conference on, pp. 1103–1108. (2011).
2. Zielke, T., Brauckmann, M., Vonseelen, W.: Intensity and Edge-Based Symmetry Detection with an Application to Car-Following. Cvgip-Imag Understan 58, 177–190 (1993).
3. Teoh, S.S., Braunl, T.: Symmetry-based monocular vehicle detection system. Machine Vision and Applications 23, 831–842 (2012).
4. Luo-Wei, T., Hsieh, J.-W., Kao-Chin, F.: Vehicle detection using normalized color and edge map. In: IEEE International Conference on Image Processing, pp. II-598–601. (2005).
5. Betke, M., Haritaoglu, E., Davis, L.S.: Real-time multiple vehicle detection and tracking from a moving vehicle. Machine Vision and Applications 12, 69–83 (2000).
6. Tzomakas, C., Seelen, W.v.: Vehicle Detection in Traffic Scenes Using Shadows. (1998).
7. Bucher, T., Curio, C., Edelbrunner, J., Igel, C., Kastrup, D., Leefken, I., Lorenz, G., Steinhage, A., von Seelen, W.: Image processing and behavior planning for intelligent vehicles. IEEE T Ind Electron 50, 62–75 (2003).
8. Kuo, Y.C., Pai, N.S., Li, Y.F.: Vision-based vehicle detection for a driver assistance system. Comput Math Appl 61, 2096–2100 (2011).
9. Stauffer, C., Grimson, W.E.L.: Adaptive background mixture models for real-time tracking. In: IEEE Conference on Computer Vision and Pattern Recognition, pp. 252. (2000).
10. Boninsegna, M., Bozzoli, A.: A tunable algorithm to update a reference image. Signal Process-Image 16, 353–365 (2000).
11. Tyrer, J.R., Lobo, L.M.: An optical method for automated roadside detection and counting of vehicle occupants. P I Mech Eng D-J Aut 222, 765–774 (2008).
12. Buch, N., Velastin, S.A., Orwell, J.: A Review of Computer Vision Techniques for the Analysis of Urban Traffic. IEEE T Intell Transp 12, 920–939 (2011).
13. Thou-Ho, C., Yu-Feng, L., Tsong-Yi, C.: Intelligent Vehicle Counting Method Based on Blob Analysis in Traffic Surveillance. In: International Conference on Innovative Computing, Information and Control, pp. 238–238. (2007).
14. Ying-Li, T., Lu, M., Hampapur, A.: Robust and efficient foreground analysis for real-time video surveillance. In: IEEE Conference on Computer Vision and Pattern Recognition, pp. 1182–1187 vol. 1181. (2005).
15. Dalal, N., Triggs, B.: Histograms of oriented gradients for human detection. In: IEEE Conference on Computer Vision and Pattern Recognition, pp. 886–893. (2005).
16. Taktak, R., Dufaut, M., Husson, R.: Vehicle detection at night using image processing and pattern recognition. In: Image Processing.Proceedings. ICIP-94., IEEE International Conference, pp. 296–300 vol.292. (1994).
17. Wang, J., Sun, X., Guo, J.: A region tracking-based vehicle detection algorithm in nighttime traffic scenes. Sensors 13, 16474–16493 (2013).
18. Cucchiara, R., Piccardi, M., Mello, P.: Image analysis and rule-based reasoning for a traffic monitoring system. Intelligent Transportation Systems, IEEE Transactions, 119–130 (2000).
19. Van Pham, H., Lee, B.-R.: Front-view car detection and counting with occlusion in dense traffic flow. Int. J. Control Autom. Syst. 13, 1150–1160 (2015).

20. Canny, J.: A computational approach to edge detection. IEEE transactions on pattern analysis and machine intelligence 8, 679–698 (1986).
21. Felzenszwalb, P.F., Girshick, R.B., McAllester, D., Ramanan, D.: Object detection with discriminatively trained part-based models. IEEE transactions on pattern analysis and machine intelligence 32, 1627–1645 (2010).
22. Hoang, V.D., Le, M.H., Jo, K.H.: Hybrid cascade boosting machine using variant scale blocks based HOG features for pedestrian detection. Neurocomputing 135, 357–366 (2014).

An Effective FP-Tree-Based Movie Recommender System

Sam Quoc Tuan, Nguyen Thi Thanh Sang(✉), and Dao Tran Hoang Chau

School of Computer Science and Engineering, International University – Vietnam National University, Ho Chi Minh City, Vietnam
itiu09030@gmail.com, {nttsang, dthchau}@hcmiu.edu.vn

Abstract. Movie recommender systems play an important role in introducing users to the most interesting movies efficiently. It is useful for users to find what they want in a large numerous of various movies on the Web quickly. The performance of movie recommendation is influenced by many factors, such as user behavior, user ratings. Therefore, the aim of this study is to mine datasets of user ratings and user behaviors in order to recommend the most suitable movies to active users. User behaviors are sequences of users' movie viewing activities which can be discovered by a frequent-pattern tree (FP-Tree). The FP-tree is then modified with rating data and an effective recommendation strategy can improve the recommendation performance of the FP-tree. A MovieLens dataset which is public and popular for evaluating movie recommender systems is observed and examined for assessing the proposed method.

1 Introduction

Due to the information explosion on the Internet, users are facing with an enormous number of choices. A pool of spam data and inaccurate information may require a lot of time for searching relevant information. Recommender systems (RSs) have become indispensable tools to filter available data and provide the user with the most relevant information. Most RSs use a hybrid approach, which is a combination of content-based and collaborative approaches. Collaborative filtering algorithms [1] assume that users with similar tastes will rate items similarly. A content-based recommendation system [2] uses the user's history to recommend new items.

In this study, the data of user's viewing history and ratings are considered. Datasets are collected from MovieLens site (https://www.movielens.org). We use FP-tree [3] to present a novel clustering method which generates frequent patterns of movies and recommend appropriate movies. The FP-tree was built based on users' movies watching activities. The rating data is discovered and then modified into the FP-tree in order to remove low rated movies and improve recommendation performance. The system will be evaluated by various experiments.

The remaining of this article is organized as follows. Related works are considered in Sect. 2, and a new movie recommendation framework is introduced in Sect. 3. Section 3 presents the methodology of using FP-Tree, and Sect. 4 shows its experimental results and discussions. Finally, Sect. 5 concludes and discusses future work.

V. Bhateja et al. (eds.), *Information Systems Design and Intelligent Applications*, Advances in Intelligent Systems and Computing 672,
https://doi.org/10.1007/978-981-10-7512-4_17

2 Related Work

In the field of recommender systems, many approaches have been developed and some recent ones are listed below.

Hybrid Multigroup Co-Clustering (HMCoC) [4]
A Hybrid Multigroup Co-Clustering recommendation framework has been proposed to achieve meaningful user-item groups by extracting user-item rating records, user social networks, and item features from DBpedia knowledge base. This framework is composed of three main modules: information fusion, hybrid multigroup co-clustering, and the top-n recommendation module. Information fusion integrates data from the rating matrix, user social networks, and item's topic. After that, it utilizes some publicly available knowledge base. Then, it uses a uniform graph model to represent the integrated information. HMCoC co-clusters users and items into multiple groups simultaneously. It combines the one-sided and two-sided clustering techniques and presents a fuzzy c-means-based clustering method to discover user-item clusters with different information sources. As a result, it merges the predictions from each cluster, and then makes top-n recommendations to the target users.

User-specific Feature-based Similarity Models (UFSMs) [5]
User-specific Feature-based Similarity Models take historical user preferences into account for building a personalized user model. To build this model, global similarity functions are learned by combining linearly user-independent similarity functions. As a result, we will have item similarity functions learned. Furthermore, these global similarity functions are combined linearly and personalized for users. It is proved to outperform both regression-based latent factor modeling (RLFM) and attribute-to-feature mapping (AFM) methods in cold-start top-n item recommendations.

Recommender Systems using Category Correlations based on WordNet Similarity [6]
In these recommender systems, genres of movies in the database are considered to draw genre correlations among movies. Each movie will be then assigned a new score by computing its average rating and genre correlations. Through the users' inputs, user's preferred genres are gained, and from the ratings of each movie, average movie ratings are estimated. From that, the scores are sorted in descending order and the high position movies can be recommended to active users.

Empirical Study of User Preferences Based on Rating Data of Movies [7]
This study represents a hyper-network of users and movies. In this network, a node of a user connects to many movies, and a node of a movie connects to many users. Rating data was used to calculate distances between movies. If a user rated movie *a*, then it is predicted that the user will rate movie *b*, when movies *a* and *b* are first-order h-neighbors. The idea was that if movie *a* was the first-order h-neighbor of movie *b*, the opinion of a user about movies *a* and *b* are almost the same. A user preference model with two tunable parameters has been introduced after many analysis results.

Grouping Like-Minded Users for Ratings' Prediction (GLER) [8]
Principal components analysis (PCA) and K-Means were used to group like-minded users. A recommender system was trained using those user groups, and for each group we build a specific model. GLER algorithm will give ratings predictions of user-movie based on the previous ratings of that user and others in the same group using this group's model. The model of closest group will be used for a new user. MovieLens-100K data set and SVD++ (Singular Value Decomposition) were used to evaluate this algorithm. Root mean squared error (RMSE) and mean absolute error (MAE) were also used as the two evaluation metrics.

3 Methodology

3.1 Framework

Figure 1 illustrates the framework of the proposed recommender system including three main process units: (1) preprocessing, (2) building FP-tree, and (3) recommendation engine.

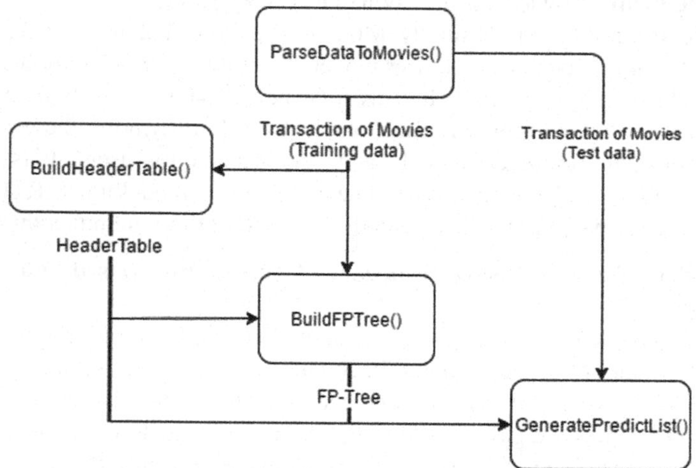

Fig. 1. Recommender system framework

(1) Preprocessing (ParseDataToMovies() in Fig. 1)
There are two ways to achieve movie transactions from MovieLens dataset in this phase: (1) Low rating movies are retained and (2) low rating movies are removed. We are going to have an experimental comparison about accuracy and runtime between retaining low rating movies and removing them.

The following presents the two algorithms of processing data.

Parameters:

- Timestamp represents seconds since midnight Coordinated Universal Time (UTC) of January 1, 1970. It is the time when a movie was viewed.

- Constant period represents time interval between two transactions, that is, movies are viewed continuously.

Method 1: With Low Rating Movie

```
Input: users' movie viewing data
Output: Processed data
Data processing:
FOR each line of data
    FOR each movie of a user
        IF (timestamp of current movie) - (timestamp of
        previous movie) < period THEN
            Add current movie into current transaction.
        ELSE
            Add current movie into a new transaction.
```

Method 2: Without Low Rating Movie

```
Input: users' movie viewing data, and rating data
Output: Processed data
Data processing:
FOR each line of data
    FOR each movie of a user
        IF (timestamp of current movie) - (timestamp of
        previous movie) < period THEN
            Add current movie into current transaction.
        ELSE
            Add current movie into a new transaction.
REMOVE all movies whose rating is less than 2.5
```

(2) FP-Tree

FP-tree is a compact structure compressing a large database of event sequences into a tree for complete frequent pattern mining. It also avoids scanning data repeatedly, that is very costly. Generally, an FP-tree structure is defined as follows: "one root labeled as "null," a set of item prefix sub-trees as children of the root, and a frequent-item header table." Each node in the item prefix sub-trees is constituted by three fields: (1) item-name registering which item this node represents, (2) count being the number of transactions, i.e., the portion of the path reaching this node, and (3) node-link linking to the next node carrying the same item-name. Each entry in the frequent-item header table is constituted by two fields: item-name and head of node-link that points to the first node carrying the item-name.

The algorithm of FP-tree construction is described as follows:

a. Scan the set of transactions *S* once; Retrieve *F*, which is the set of frequent items, and compute the support of each item; Sort *F* in support-descending order.
b. Create the root node of an FP-tree, tree, and label it as "null"; For each transaction *T* in *S* do the following:

- Find frequent items in *T*. Let the selected frequent-item list in *T* be [*e* | *E*], where e is the first element and E is the remaining list. Call *insert-tree*([*e* | *E*], *Tree*).
- *insert-tree*([*e* | *E*], *Tree*): If tree has a child *C*: *N.item-name* = *e.item-name*, then increase C's count by 1; else create a new node C, and set its count to 1, link its parent-link to tree, and link its node-link to the nodes carrying the same item-name. If *E* is nonempty, call *insert-tree*(*E*, *C*) recursively.

(3) Recommendation Engine (GeneratePredictList() in Fig. 1)
The data sets are firstly preprocessed and clean. Firstly, all movies that each user watches were added into a transaction. Then we divide this transaction into smaller transactions using *timestamp*. The reason is that in a period of time, a user may like a set of movies but in another period this user may be interested in another set of movies. After having these transactions, we are able to build FP-tree.

A movie recommendation algorithm is proposed as follows: When a testing transaction is input, the algorithm would apply top-n recommendations based on the built FP-tree. That means subsequences of movies in the transaction are matched with patterns in the FP-tree, the longest matching patterns will be considered. Movies in the matching patterns and movies in children nodes will be candidates for recommendation. They are sorted in descending weights (counts) and top-n movies are recommended.

3.2 Evaluation Methods

This study applies two evaluation metrics: precision and satisfaction for assessing the performance of the movie recommendation system. Precision is the proportion of number of correct predictions in the next step to the number of matching times. While, satisfaction is the proportion of the number of correct predictions in next m-steps to the number of matching times. The following describes evaluation algorithms for performance measure.

Performance evaluation:
Testing data is a subset of users' movie viewing sequences, in which, a transaction is a sequence of viewing movies continuously. In other words, in a transaction, time interval between two viewed movies should not be longer than a predefined period.

a. Calculate precision:

```
set positive = 0; and set negative = 0;
FOR each transaction in the testing data.
    FOR i = 1 to length of current transaction.
        Get the list of recommended movies for the active
        sequence of watched movies from 1 to index = i.
        IF movie at index = i+1 in current transaction
        CONTAINS at least one movie in the recommendation list
        THEN INCREMENT positive
        ELSE INCREMENT negative
Precision = positive / (positive + negative) *100%
```

b. Calculate satisfaction with m = 5

```
FOR each transaction in the testing data.
    FOR i = 1 to length of current transaction.
        Get the list of recommendation movies for the active
        sequence of watched movies from 1 to index = i.
        IF movies at index = i+1 to i+5 in current transaction
        CONTAINS at least one movie in the recommendation list
        THEN INCREMENT positive
        ELSE INCREMENT negative
Satisfaction = positive / (positive + negative) *100%
```

Return Precision and Satisfaction

4 Experiments

In order to evaluate the performance, fivefold cross-validation is applied. Besides, a number of values are set for the min_support and the timestamp to examine several different points of view. The min_support is used to filter less interesting movies. Removing low rating movies can return good experimental results. Applying the precision and statistical metrics is very effective to evaluate the system's performance.

4.1 Dataset

A popular MovieLens dataset[1] containing 20 million of movies is used. In this dataset, we focus on ratings.csv (rating data) and tags.csv (users' movie viewing behaviors). Each line after the header row in file ratings.csv represents one movie's rating by each user following this format: userId, movieId, rating, timestamp. The lines within this file are ordered first by userId, then, within a user, by movieId. Ratings are based on a five-star scale, with half-star increments (0.5 stars–5.0 stars). Timestamps represent seconds since midnight Coordinated Universal Time (UTC) of January 1, 1970.

The datasets are firstly preprocessed and cleaned before building FP-Tree. We add all movies within a user into a transaction. Then timestamp is used to divide a transaction into smaller transactions. The period of each smaller transaction is predefined by observing the datasets. After having these transactions, we can build the FP-tree.

4.2 Experimental Results

In the following, we conduct six experimental cases.

a. Keep low rating movies in the dataset, and the period of each examined transaction is set to 1000000 s. The satisfaction and precision are measured at different min_support values: 7, 8, and 9%. Table 1 shows the results of satisfactory measure with m = 5.

Table 1. Satisfactions examined for three min_supports 7%, 8%, and 9%

Min_support (%)	Satisfaction (m = 5) (%)	Runtime (h)
7	100	40
8	99	12
9	99	3

Table 2. Precisions examined for three min_supports 7, 8, and 9%

Min_support (%)	Precision (%)	Runtime (h)
7	99	40
8	99	12
9	99	3

Table 1 shows that choosing min_support is not easy. We can achieve 100% satisfaction for min_support of 7%, but it costs too much time. For min_support of 9%, the satisfaction decreases 1%, but it takes less time. It is similar when measuring precisions, but we can just obtain 99% of accuracy at min_support = 7% (Table 2).

[1] https://www.movielens.org, https://www.imdb.com.

b. Keep low rating movies in the dataset, and the min_support is set to 8%. The performance is evaluated in two cases: the chosen periods of each examined transaction are 1000000 and 500000s (Tables 3 and 4).

Table 3. Satisfaction and runtime for the two different chosen periods

Timestamp (s)	Satisfaction (m = 5) (%)	Runtime (h)
1000000	99	12
500000	97	2

Table 4. Precision and runtime for the two different chosen periods

Timestamp (s)	Precision (%)	Runtime (h)
1000000	100	12
500000	97	2

The chosen period affects the accuracy (satisfaction and precision) and runtime, as in Tables 3 and 4. Period 100000s gives higher accuracy (99%) but it takes 12 h to complete all testing data. Period 500000s gives a lightly lower accuracy (97%) but it takes only 2 h.

c. Low rating movies can be removed or not from the dataset. It is supposed that low rating movies may be not interested by users, so they should not be taken into account. The chosen period of each examined transaction is 1000000s, the min_support is 8%, as shown in Tables 5 and 6.

Table 5. Satisfaction and runtime for two cases: with and without low rating movies

	Satisfaction (m = 5) (%)	Runtime (h)
Without rating movies	99	12
With low rating movies	99	4

Table 6. Satisfaction and runtime for two cases: with and without low rating movies

	Precision (%)	Runtime (h)
Without rating movies	99	12
With low rating movies	99	4

The results in both Tables 5 and 6 give the same accuracies but the runtimes are significantly different between the two cases with and without low rating movies. Thus, removing low rating movies offers much more benefit.

d. Removing low rating movies and the chosen period of each examined transaction is 500000s. The min_support is set to 8%. Table 7 shows that we receive the same precision and satisfaction (m = 5), i.e., accuracy, with the same runtime. Thus, it proves that the proposed method can achieve significantly high performance.

Table 7. Accuracy and runtime when applying the precision and satisfaction (m = 5) metrics

Metric	Accuracy (%)	Runtime (h)
Precision	96	1
Satisfactory with m = 5	96	1

e. Removing low rating movies and the chosen period of each examined transaction is 500000s. The min_support is set to 8%. In this experiment, we apply fivefold cross-validation to accuracy estimate (Table 8).

Table 8. Accuracy (precision and satisfactory) and runtime when applying fivefold cross-validation

Metric	Accuracy (%)	Runtime (h)
Precision	98	7
Satisfactory with m = 5	98	7

In fivefold cross-validation, the experimental data is divided into fivefolds, onefold is used for testing while the remaining fourfolds are used for training. Each fold has once it becomes testing data and four times it is a part of the training data. Applying fivefold cross-validation in Table 8 will make the accuracy much more reliable. As a result, the performance achieves higher accuracy (98%).

f. Comparing with the experimental results in [5] using the same dataset, our results are remarkable.

The same data set MovieLens-1M was used for our recommendation engine. We also applied the same Recall at n metric (Rec@n) to compare the performance of our method with UFSM. User profile was the input for UFSM while we used user history to recommend movies. Given top-n recommendation movies for a user, Rec@n of the user is computed as: *Rec@n = number of matched movies / n*. Rec@n is computed for each user and then averaged over all test users.

Figure 2 shows that the performance of top-n recommendation of USFM and our method for values of n = 5, 10, 20. The accuracy of our method are much higher than USFM. About the training time, if we run on the MovieLens-20M dataset, it will take 68.5 minutes. While, USFM takes about eight times of training the MovieLens-1M dataset, which is 20 times smaller than the one used in our experiments. Table 9 shows how efficient our method is.

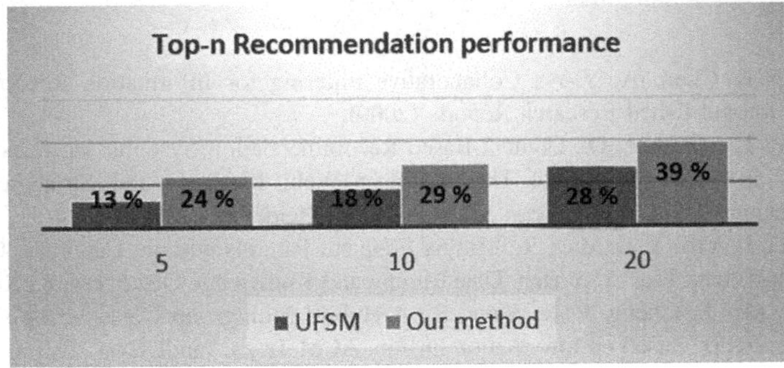

Fig. 2. Performance of top-n recommendation

Table 9. Training time in minutes

Method	Training time in minutes
USFM	566.31
Our method	68.5

5 Conclusions

In this study, we have proposed a recommendation system with an advanced sequence mining method using FP-tree and the effective recommendation strategy. This methodology compresses a large database into a compact FP-tree. Based on that, movies are recommended efficiently. In particular, the largest dataset (20M) in MovieLens is used to test its performance.

The experimental results have showed that if we remove low rating movies from the dataset, we can save a lot of time with a slightly decreased accuracy rate (about 1%). The proposed methodology is proved to achieve high performance, and the precision is almost acceptable.

In the future, we will consider the levels of rating movies which helps rank the movies and recognize which movie a user likes most. The count of a node in FP-tree will be replaced by the weight. Ratings will be used to calculate the weight. Since ratings are based on five-star scale, we choose the average of rating 2.5 to be equal to 1 unit of weight. The function *insert-tree*([*e* | *E*], *Tree*) will have some change according to the weight.

Acknowledgements. The authors would like to thank the anonymous reviewers for their valuable comments and suggestions to improve the quality of this article.

References

1. Mcleod, D. & Chen, A. Y.-A., Collaborative Filtering for Information Recommendation Systems. Non-published Research Reports (2009).
2. Pazzani, M. J. & Billsus, D., Content-Based Recommendation Systems. In: Brusilovsky, p., Kobsa, a. & Nejdl, W. (eds.) The Adaptive Web: Methods and Strategies of Web Personalization. Berlin, Heidelberg: Springer Berlin Heidelberg (2007).
3. Han, J., Pei, J., Yin, Y. & Mao, R., Mining Frequent Patterns without Candidate Generation: A Frequent-Pattern Tree Approach. Data Mining and Knowledge Discovery, 8 (2004) 53–87.
4. Huang, S., Ma, J., Cheng, P. & Wang, S., A Hybrid Multigroup Coclustering Recommendation Framework Based on Information Fusion. ACM Trans. Intell. Syst. Technol., 6 (2015) 1–22.
5. Asmaa Elbadrawy and George Karypis, User-Specific Feature-Based Similarity Models for Top-n Recommendation of New Items, ACM Transactions on Intelligent Systems and Technology, Vol. 6, No. 3 (2015).
6. Choi, S.-M., Cho, D.-J., Han, Y.-S., Man, K. L. & Sun, Y. Recommender Systems Using Category Correlations Based on WordNet Similarity, International Conference on Platform Technology and Service (PlatCon), 26–28 Jan. (2015) 5-6.
7. YingSi Zhao, Bo Shen, Empirical Study of User Preferences Based on Rating Data of Movies, PLoS ONE 11(1): e0146541. https://doi.org/10.1371/journal.pone.0146541, January 6 (2016).
8. Jaffali S., Jamoussi S., Hamadou A.B., Smaili K., Grouping Like-Minded Users for Ratings' Prediction. In: Czarnowski I., Caballero A., Howlett R., Jain L. (eds) Intelligent Decision Technologies 2016. Smart Innovation, Systems and Technologies, vol 56. Springer, Cham (2016).

Metadata-Based Semantic Query in Relational Databases

Ch. V. S. Satyamurty[1(✉)], J. V. R. Murthy[2], and M. Raghava[1]

[1] CVR College of Engineering, Hyderabad, India
satyamurty@cvr.ac.in, satyamurtycvs@yahoo.co.in, raghava.m@cvr.ac.in
[2] JNTUK, Kakinada, India
mjonnalagedda@gmail.com

Abstract. The retrieval of data from various semantically equivalent databases having different schemas is long been an important issue. In this context, the proposed WordNet-based model demonstrates the semantic data retrieval capabilities from different databases using metadata available with them and publishes the results.

Keywords: Semantic query · WordNet · RDBMS · Metadata

1 Introduction

Relational databases are playing more important role in industry. The data collected are stored in a well-designed schema of the database. Different organizations expected to follow different schema's despite the data presented in these different schemas of these databases are likely to be semantically equivalent [1–4]. Consolidation of the information pertaining to different organizations and generating a comprehensive report from the consolidated data is a challenging task in the absence of proper metadata processing. Clear indexing and preparation of mapping tables us as a collaborative activity of domain experts of the respective organizations, and it is typically one of the classical ways to retrieve valid information from the databases. However, these approaches will have subjective influence and more error prone when original designer or design document is not available. Hence, it is imperative to develop adoptive methods for addressing the semantic gaps in the query and the data present in the databases.

This paper proposes a framework which prompts the query from the user in a raw form, explores the metadata of respective databases using Semantic WordNet, and builds the query automatically. The results of the automatically constructed query prove to compensate the naive knowledge of the user on the data model.

V. Bhateja et al. (eds.), *Information Systems Design and Intelligent Applications*, Advances in Intelligent Systems and Computing 672,
https://doi.org/10.1007/978-981-10-7512-4_18

1.1 Related Work

Jarunsree Salee and Veera [4] used query-graph-based approach to extract data from the relational databases. The graph consists of relational lists, attribute lists, joining conditioning, and selection condition. Using this, ranking condition applies to tuples in the database. Jyotimor et al. [5] concerned with semantic query optimization using inductive learning approach, and they concentrated more on join order and parallelization of query. The inductive learning approach is implemented in SQL using SQL hints. This keyword-based query processing in relational databases returns tuples as a connected components based on the way they are associated. DBXplorer [10] and DISCOVER [11] implemented the Candidate Networks approach and BANKS [12] applied the Steiner Tree approach. These approaches have some drawbacks. Mariana Soller Ramada, Joao Carlos da Silva, Plínio de Sa Leitao-Junior [6] implemented query-based approach by semantically analyzed before applying to databases for this they computed intrinsic weight computation, using synonyms as keywords, Weight normalization for sub-matrix and proximity between the keywords. Lipyeow Lim, Haixun Wang, Min Wang [9] implemented query as a graph structure using ontology-constructed graph structure. The graph consists of nodes as concepts which are interlinked. Using this extract, the features and from these features they are learning semantic queries using SVM classification technique.

2 Proposed Model

WordNet [7] is a thesaurus for the English language based on psycholinguistics studies and was developed by Princeton university. It consists of a set of interconnected nodes as concepts and the links connecting the nodes as various types of relations between the concepts, such as synonymy, homonymy, holonymy, and hypernym. It contains lexical semantics relation of the words called synsets. The synsets are very helpful for obtaining lexical items with similar meaning. For example, a word father and a word begetter are grouped in the father, begetter and father also grouped in don as the godfather.

2.1 Frame Work

In this section, we will present the overall view of the model as shown in Fig. 1 and subsequently details of the model. The user interacts with graphical user interface, which is realized by the first layer of the model. The users' keyword is converted into SQL queries using the WordNet as this is the second layer of the model. Then in turn, the access to various databases is implemented through the corresponding data model interface, retrieves the information, and presents it to the end user. This is the third layer of the model. In our model, the first layer prompts the user to enter the keyword, and using that keyword, it interacts with the WordNet API of English language. WordNet is responsible for the retrieval of the synonyms of the keyword. The WordNet will be acting as a mediator to convert the raw SQL query into semantically equivalent multiple queries based on the synonyms and is evaluated sequentially until one query instance is successful. The synonyms extracted from the WordNet are considered to generate SQL query of the keyword, for each word extracted from the WordNet. Then it checks the database metadata for the availability of a table with the synonym. If the check is successful, then it constructs the query based on the synonyms also in a sequential manner and retrieves the data. It repeats the same exercise for every synonym until it either succeeds or ends up with an error.

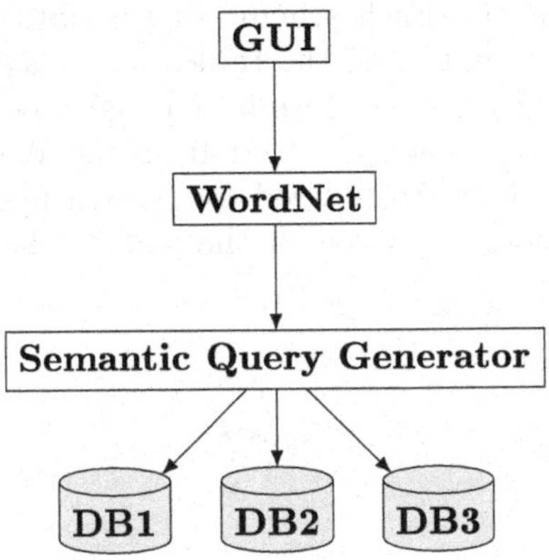

Fig. 1. Proposed framework

2.2 Graphical User Interface

It allows user to enter the keyword and displays the returned results.

2.3 WordNet

This contains lexical data relationships of English. Using this, we can extract synonyms of the given keyword.

2.4 Query Generator

This takes the keyword or synonym of the keyword and generates the query corresponding to the word.

2.5 Database(s)

The databases of different institutions contain semantically equivalent data of the keyword at different formats.

3 Experimental Results

We considered educational domain as our interest, and the institution we consider has many departments. Each schema of the database maintains data in different formats, and the names of the tables also may look different but are semantically equivalent. The keyword we have considered for experimentation is 'student' and the extracted synonyms to it from the WordNet such as scholar, book man, and Educated are considered for generating the find related SQL query. The keyword student is given in the graphical user interface which is shown in Fig. 2.

Fig. 2. Keyword specification by the user through the GUI

For each of the keyword, the model which searched the database metadata and the table corresponding to the synonym is accessed. The sample results of a query related to the keyword student is shown in Fig. 3.

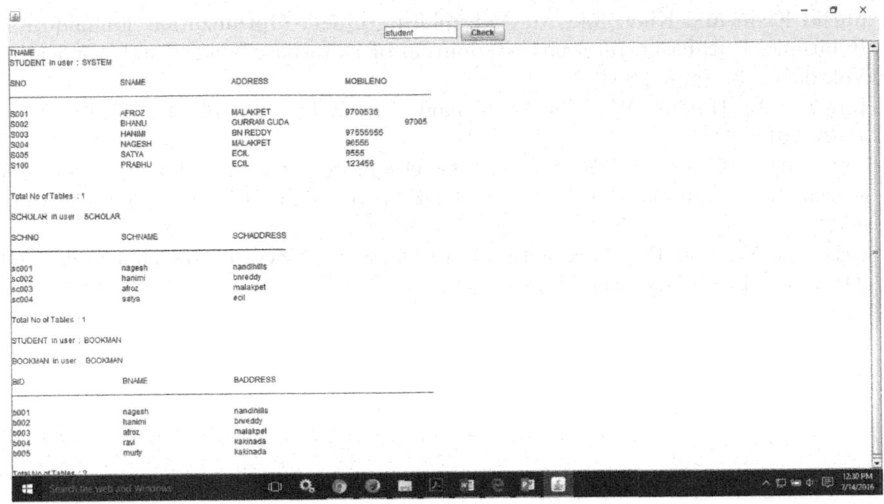

Fig. 3. Experimental results showing various semantically equivalent information retrieved from the database

4 Summary

This paper describes the architectural framework and each component of the framework. It describes middle layer about WordNet or dictionary to know about synonyms. Using this layer, a semantically equivalent query is generated for the input query and helps to retrieve the data from various databases. The results of the generated query processing are presented at a centralized location.

References

1. Kumar, P., Mohan.,Vaideeswaran, J. Semantic based Efficient Cache Mechanism for Database Query Optimization, International Journal of Computer Applications, Volume 43 No. 23, page(s): 14–18, April 2012.
2. Saini, M., Sharma D., Gupta, P., K. Enhancing Information Retrieval Efficiency Using Semantic-based-Combined-Similarity-Measure, International Conference on Image Information Processing (ICIIP 2011), IEEE Computer Society, 2011.
3. Hsu, C., Knoblock, A. Semantic query optimization for query plans of heterogeneous multidatabase systems, Knowledge and Data Engineering, 12(6):959978, 2000.
4. Jarunee, S., Veera, B., A Metadata Search Approach to Keyword Query in Relational Databases, IJCA vol. 69 may 2013
5. Jyoti, Mor., Indu Kashyap ., RK pathy. Implementing Semantic Query Optimization in Relational Databases, IJCA vol. 52 No 9 Aug 2012.
6. Mariana, S, R., Joao, C, da, Silva. Data Extraction from Structured Databases using Keyword-based Queries, 29th SBBD, ISSN 2316-5170 , October 2014
7. Miller, G. Nouns in WordNet: A Lexical Inheritance System, International Journal of Lexicography, vol. 3, no. 4, 1990.

8. Majid Khan and Khan, M., N., A. Exploring Query Optimization Techniques in Relational Databases, International Journal of Data-base Theory and Application Vol. 6, No. 3, June, 2013
9. Lipyeow, L., Haixun, W., Min, W. Semantic query by Example, EDBT 13, March 18-22 2013.
10. Agrawal, S., Chaudhuri, S., and Das, G. Dbxplorer: A system for keyword-based search over relational databases. In ICDE, pages 516. IEEE Computer Society, 2002.
11. Hristidis, V., and Papakonstantinou, Y. Discover: Keyword search in relational databases. In VLDB, pages 670681, 2002.
12. Aditya, B., Bhalotia, G., Chakrabarti, S., Hulgeri, A., Nakhe, C., Parag, and Sudarshan, S. Banks: Browsing and keyword searching in relational databases. In VLDB, pages 10831086, 2002.

Performance Analysis of Information Transmission Systems Over Indoor LED Lighting Based Visible Light Communication Channels

Trung Ha Duyen(✉) and Tuan Do Trong

School of Electronics and Telecommunications, Hanoi University of Science and Technology, 405/C9, No. 1, Dai Co Viet St., Hanoi, Vietnam
{trung.haduyen, tuan.dotrong}@hust.edu.vn

Abstract. This paper presents performance analysis of full-duplex indoor Visible Light Communication (VLC) system for text and image information transmission over Light Emitting Diode (LED) lighting. The system and channel models are firstly studied. The designed system consists of transmitters and receivers then investigated that can achieve information data rate of 161.2 Kbps with error free at the distance of 92 cm between for real-time text and image transmission over indoor environment. Experiment results have shown that the transmission delays for text and image transmission at different distances of 0.8 and 2.4 m are almost constant. In addition, symbol error rate (SER) do not vary data rate at some transmission distances. The bit error rate (BER) confirms that the longest transmission distance is achieved at distance of 107.5 cm with bit rate of 57.6 kbps for the forward channel and 38.4 kbps for the reverse channel without any symbol error.

1 Introduction

Light Emitting Diodes (LEDs) has recently been expecting to replace tradition illumination sources due to more advantages compared to the existing incandescent in terms of low power consumption long life expectancy and high tolerance to humidity. Their applications consists of numeric displays, flash-lights, vehicle brake lights, traffic signals and the ubiquitous power-on indicator light [1–4].

Besides the above mentioned unique advantages of the while LED as an illumination light source, requirements for indoor LED lighting based information communication systems have been increasingly growing because there are many devices using lightings in daily life such as offices buildings, home appliances. Typical LEDs have characteristics to light on/off very fast at ultra-high speed. When applying these visible lights for the data information transmission, related problems in the field of wireless communications need to be resolved such as transmitting signal processing at ultra-high speed whereas harmless for human eye [5, 6]. An indoor LED lighting based visible light communication systems is investigated as a convergence telecommunication technology which is integrated device used for both lighting and information transmission [7].

V. Bhateja et al. (eds.), *Information Systems Design and Intelligent Applications*, Advances in Intelligent Systems and Computing 672,
https://doi.org/10.1007/978-981-10-7512-4_19

There are a number of VLC research projects have been implemented [8–11]. In [8], the VLC Consortium (VLCC) established in 2003 in Japan with industrial companies with goal of standardizing the VLC technology. VLCC has been performed and evaluated in the various industry fields to provide the communication capability through LED lightings in homes and offices such as LED commercial displays, LED lamps and LED traffic signals. Then, the European OMEGA project [9], the Wireless World Research Forum (WWRF) [10] and a forthcoming IEEE standard for VLC [11] have dedicated to this research area. In addition, research works on high speed data transmission, channel characteristics and modulation schemes have been proposed for the capability of designing different indoor VLC transmission models that based on the current infrastructure [12–16]. In [12] Vucic et al. reported a wireless visible-light link operating for the first time at 125 Mb/s over 5 m indoor distance using on-off-keying (OOK) modulation at the BER less than 2.10^{-3}. Hoa et al. described in [13] a high-speed VLC data rate of 10 Mb/s using a white-light LED and OOK nonreturn-to-zero (NRZ) modulation. In [14] authors reviewed the research directions of increasing transmission speed and bandwidth. In [15] Mesleh et al. proposed an efficient pulsed modulation technique base on power and bandwidth. Moreover, in [16], the data rate of 20 Mb/s is achieved at the distance of 1 m free space transmission using the quaternary amplitude shift keying (4-ASK) modulation to improve the direct modulation speed of white LED.

Most recently, there are a number of research interests in VLC transceiver system design and demonstration depending on the indoor application scenarios [17–21]. In [17] and [18], authors focused on wireless optical transceiver with high data rate in the short range visible communication channel. It expected to be used as a peripheral interface of hand-held devices such as mobile phones, notebooks, digital cameras. In [19] Png et al. presented VLC audio transmitter and receiver circuits that can be integrated in LED lights on board for airlines' passengers. Png also described in [20] a circuit construction of VLC mass-storage prototypes that allow file transfers between the SD (Secure Digital) card and the PC through LED lights. The designed system consists of a pair of identical white-LED transceivers that is connected to the PC and the SD-card sub-circuit, simultaneously. File transfer operates at 19200 bps. In [21] Rajbhandari et al. presented divide constraints and design considerations of high-speed integrated VLC system to demonstrate results of a multiple-input multiple-output (MIMO) configuration system transmitting at 1 Gb/s data rate. In addition, VLC transceiver designs were studied on a field programmable gate array (FPGA) [22–24], in which the digital basebands for the transceivers are implemented using two separate FPGA kits. These existing works mainly focused on basic functionality demonstration or signal processing techniques to increase the data rate. However, the detail fundamental consideration of VLC information transmission including system model, channel model, circuit transceiver modules, experimental scenarios of indoor full-duplex transceiver, text and image demonstration using the self Graphic User Interface (GUI) software have not been clarified.

In this paper, we therefore analyze performance of a real time full-duplex information transmission system using LED between two computers that provides 2 Mbps data rate for indoor visible light application scenarios. Self-written software is developed to transmit and receive the real-time information transmission by text and image.

Design and analysis of an integrated VLC receiver with an USB 2.0 interface for PC/Note Book has been studied in [25] and [26]. USB interface is a universal standard of external bus to specify the connection and communication between PC and electronic devices.

The rest of the paper is organized as follows: Sect. 2 introduces the LED lighting based VLC transmission system including system model, channel model. Section 3 provides VLC information transmission system. Performance analysis is introduced in Sect. 4. Finally, conclusions are given in Sect. 5.

2 LED Lighting Based VLC Transmission System

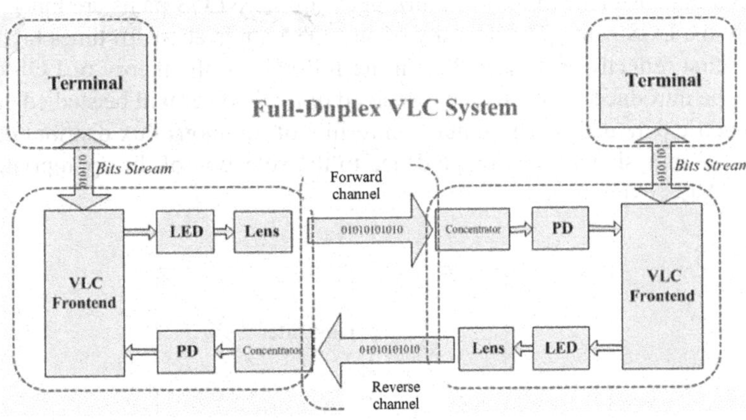

Fig. 1. A fundamental VLC system architecture

2.1 VLC System Model

A fundamental VLC system architecture is shown in Fig. 1. The system consists of VLC front-ends which connected to communication terminals at one side and to LED and Photodiode (PD) at the other side. LED plays an important role of a light source which emits lights and data transmission, simultaneously. Data is transmitted between two or more terminals. Each front-end device consists of a transmitter and a receiver. The transmitter transmits data into free-space channel and that is received by the receiver at the other terminal.

Communication terminal permits users to transmit or receive data in types of text or images. Terminals could be a Laptop/PC, embedded equipments or even a smart phone/tablet. Necessary software will be installed in these terminals. Software is used to generate a bit stream pushing to physical layer of the VLC transmitter in transmitting direction and to receive the bit stream from physical layer of VLC receiver in the reverse direction. Software also is used to analyze the communication delay or throughput over VLC channels. VLC Front—End is a hardware part which designed to convert data in term of bit streams at the input to suitable signal for controlling light intensity of light source by turn on and off the LED. The third part at transmitting side is LED and lens to generate and direct the light source for high luminous flux efficiency. At the receiving side, a PD is used convert optical signal into electrical signal. Concentrators could be also used to reduce interference from other light sources or to limit the received light outside of field of view of the PD.

2.2 VLC Channel Model

Indoor VLC channels are classified into two main types of Line-of Sight (LOS) and Non-LOS (NLOS) as shown in Fig. 2. However, for indoor VLC links in a typical room with a dimension of (5 × 5 × 3 m), lights in the NLOS paths are much weaker than that in the LOS paths. The intensity of an LOS path is about 10 times higher than that of the first reflective path [19, 20]. In the followings, the theory of LOS channel model will be introduced. Path loss and received optical power will be studied based on the photometric parameters. The simulation results of luminous flux distribution based on the theory analysis will provide guidance to the selection of the appropriate LEDs used in the VLC experiments.

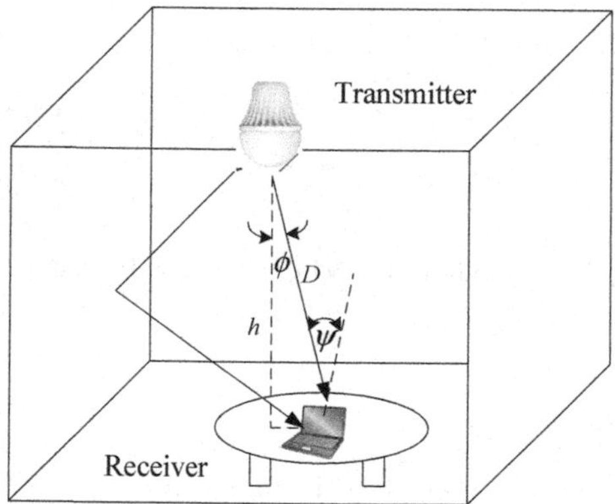

Fig. 2. VLC channel model

The relationship between receive power and transmit power for LOS channel can be expressed as

$$P_r = P_t L_L, \tag{1}$$

where L_L is path loss of the communication channel, which can be given by [5]

$$L_L = \frac{(m+1)A_r}{2\pi D^2} \cos\psi \ \cos^m\phi \qquad (2)$$

where $m = -\ln 2/\ln\left(\cos\phi_{1/2}\right)$, A_r is the physical area of the PD's detector, D is the distance between transmitter and receiver, ϕ denotes the angle of irradiance at transmitter, ψ is the angle of incidence at receiver, m is the order of Lumberton emission, $\phi_{1/2}$ is the transmitter semi-angle defined at half power, and it physically determines the illuminating beam width of the a single LED.

In case of multiple LOS channels, the receive power is summed up of all receive power for each LOS that is given as

$$P_r = \sum_i^n P_{ri} \qquad (3)$$

The electrical power distribution at receiver depends on the illuminance distribution of light source. It expresses the brightness of the illuminated surface. According to the Lambert's Cosine Law, the luminous intensity with the angle ϕ is given by [5]

$$I(\phi) = I(0)\ \cos^m(\phi). \qquad (4)$$

Almost every commercial LED is produced following Lambert's Cosine Law. The luminous intensity reduces when the angle ϕ increases. In Eq. (4), $I(0)$ is the LED's centre luminous intensity, ϕ is the angle of irradiance and m is the order of Lumberton emission. The horizontal illuminance, E_{hor}, at the point (x, y) is defined as

$$E_{\text{hor}} = \frac{I(0)\ \cos^m(\phi)\ \cos(\psi)}{D^2}. \qquad (5)$$

The illuminance intensity distribution is investigated through simulation in Matlab™ software. A LED Lambert Luxeon 1 W is used to illuminate light source in the space with dimension of 0.5 × 0.5 × 0.7 m. This is corresponding to the dimension of VLC testbed frame shown in Fig. 3a. A light source (transmitter) is located at position (x = 0.25, y = 0.25, z = 0.7). The Photodiode (receiver) is place on different position on the plane of z = 0.

The relative position of transmitter and receiver in simulation scenario is illustrated in Fig. 1. In the case of $\phi = \psi$, $\cos(\phi)$ and D can be, respectively, calculated as

$$\cos\phi = \cos\psi = \frac{h}{\sqrt{h^2 + (x-0.25)^2 + (y-0.25)^2}}, \qquad (6)$$

$$D = \sqrt{h^2 + (x - 0.25^2 + (y-0.25)^2}, \qquad (7)$$

where h is the vertical distance between transmitter and receiver. Replace Eqs. (6) (7) into Eq. (5), the horizontal illuminance is given as

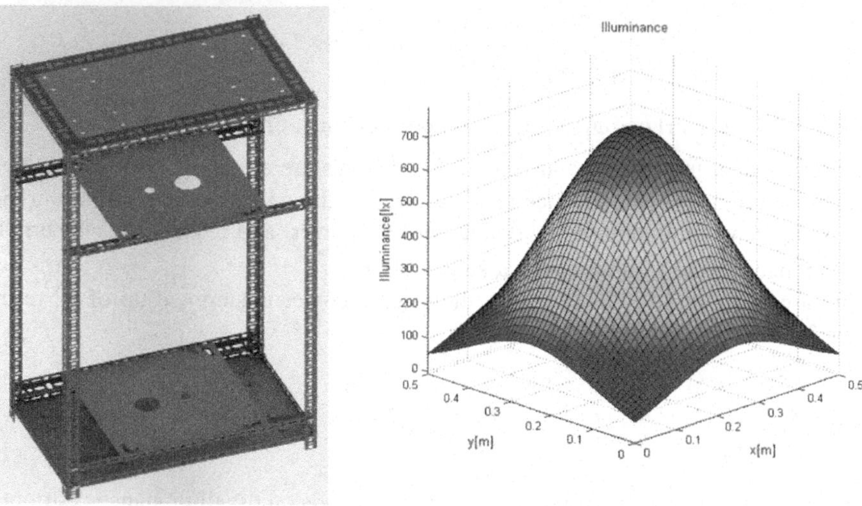

(a) VLC experimental structure. (b) The illuminance intensity distribution.

Fig. 3. VLC testbed structure and illuminance intensity distribution

$$E_{\text{hor}}(x, y, 0) = \frac{358h^{m+1}}{\left[h^2 + (x - 0.25)^2 + (y - 0.25)^2\right]^{\frac{m+3}{2}}}. \quad (8)$$

.

Base on the theoretical analysis above, the simulation of the illuminance intensity distribution was presented in Fig. 3b. It can be seen that the illuminance concentrated with luminous intensity at the center of the frame and reduces gradually to the edge. The luminous intensity varies a small amount when receiver located nearby center of the ground plane. The luminous flux at center of 700 lx is suitable for both lighting and information transmission for the small space of the target VLC experimental frame.

3 VLC Information Transmission System

For practical experiment purposes, the system architecture consists of all elements of VLC overall system architecture as shown in Fig. 4, including the communications realized by Laptops, Smartphone and Embedded Computers. In the testbed presented in Fig. 4, LED and PD are placed in the vertical direction. In this model, the interference from other light sources of environment is limited because LED and PD are place in vertical axis. In addition, PD is located in the bottom of a PVC pipe with the length of 8 cm. In this testbed, the experiments can be done at any time without effecting by surrounded light sources. Practical experiments have shown that this feature leads to many advantages compared to our previous testbed model. The data in type of text and image is transferred in real-time between two virtual terminals corresponding to two software modules which installed in the Laptop.

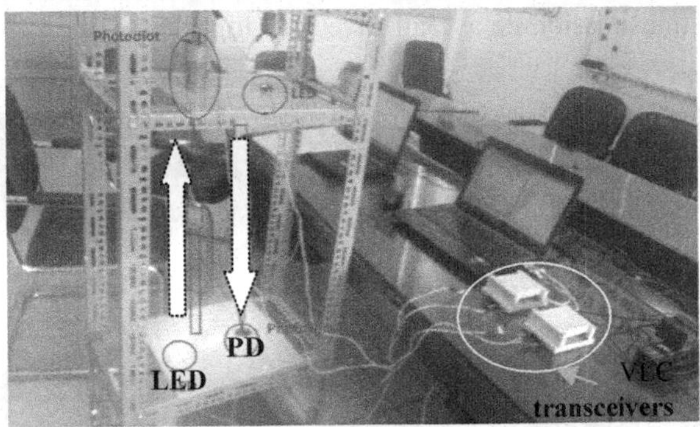

Fig. 4. VLC experimental system of indoor full-duplex: 1LED × 1PD and transmission distance of 0.8 m

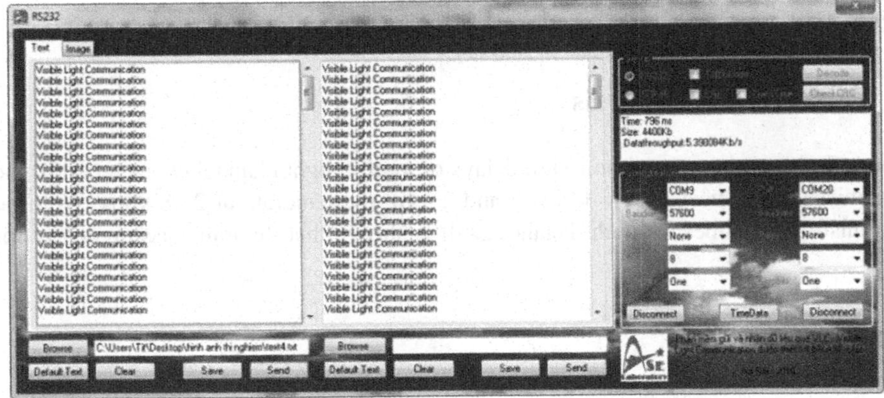

Fig. 5. GUI experimental software: text-information transmission mode at distance of 1.7 m and baudrate of 28.8 kbps

The experimental framework (Fig. 4) helps us measuring exactly value of communication transmission delay due to elimination of clock offset in case of using two Laptop or two different physical devices. The character string "visible light communication" is used for text transmission, where as the file "foreman" of 7.7 kB (248 × 203 pixels) is used for image transmission. The software is developed by ourselves for experiments with VLC testbed (Figs. 5 and 6), which is designed for experiments of real-time data communications with following main functions: (1) measure transmission delay using customized "Ping" command, wherein the timestamp of short "Ping" packets at moment of sending at transmitter and receiving moment at receiver are recorded. The delay value the is calculated based these timestamp and save to the log files; (2) transfer text in both way with different bit rates, measure transmission delay and save into log files; (3) transfer images in both forward and backward links, with different format, resolution and bit rates, display received images, measure transmission delay and save into log files.

Fig. 6. GUI experimental software: image-information transmission mode at distance of 2.4 m, file size of 17.133 kb, and baudrate of 28.8 kbps

4 Performance Analysis

Figure 7 shows the text transmission delays of visible light channel characteristics, the transmission distances of D = 0.8, 1.7 and 2.4 m, the baudrate of 28.8 kbps. We used 10 transmission times for each distance D. It is shown that the transmission delays for

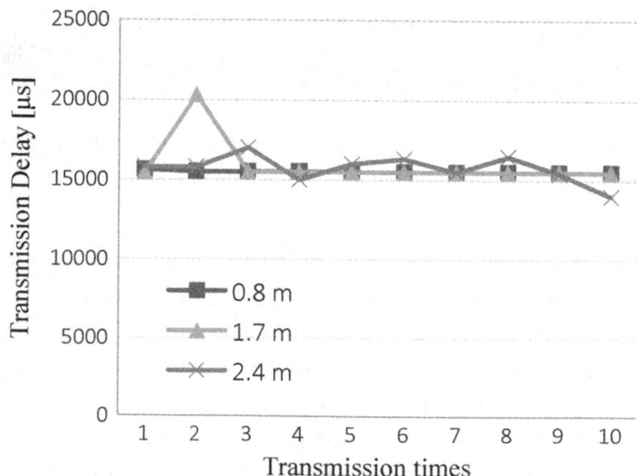

Fig. 7. Comparison of text transmission delay at the distances of 0.8, 1.7 and 2.4 m, baudrate of 28.8 kbps

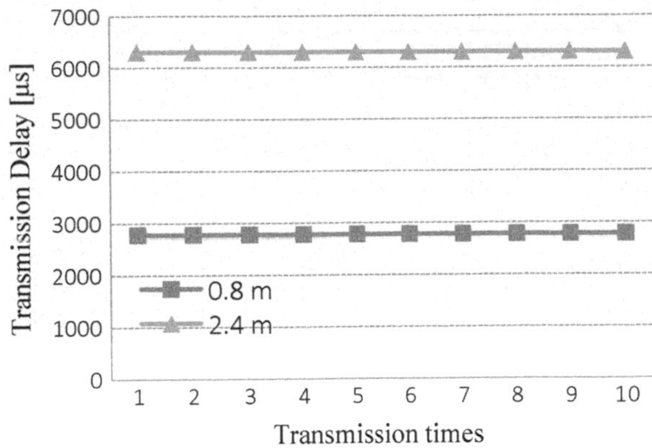

Fig. 8. Comparison of image transmission delay at the distances of 0.8 m and 2.4 m, and baudrate of 28.8 kbps

both distances were around 15 ms. Error transmission with delay of 20 ms at D = 1.7 m was observed. Similarly, Fig. 8 illustrates the image transmission delay versus 10 transmission times. It can be seen that the constant values of 2780 and 6300 µs are achieved at the distances of 0.8 m and 2.4 m, respectively.

Figure 9 shows the SER for forward channel and reverse channel at different transmission bit rate. The transmission distances were varied from 46.5 to 107.5 cm for different bit rates of error free (Fig. 10). It can be observed that at the longest transmission distance of 107.5 cm, bit rate of 57.6 kbps for forward channel and 38.4 kbps for reverse channel without any symbol error were achieved. The highest bit rate of 161.2 and 134 kbps was archived at transmission of 92 cm for forward channel and 38.4 kbps for reverse channel when SER equal to zero.

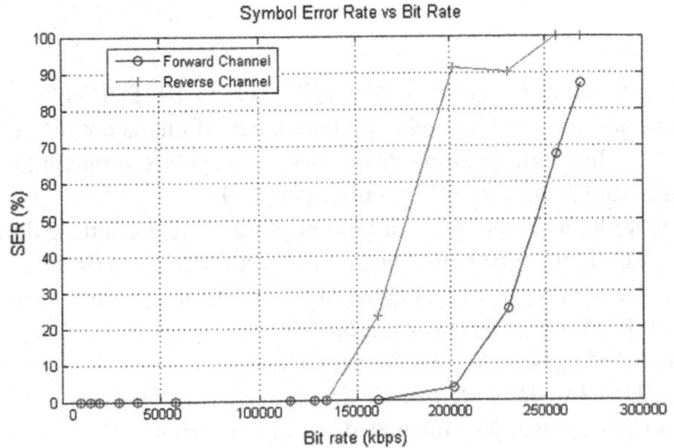

Fig. 9. Symbol error rate versus bit rate

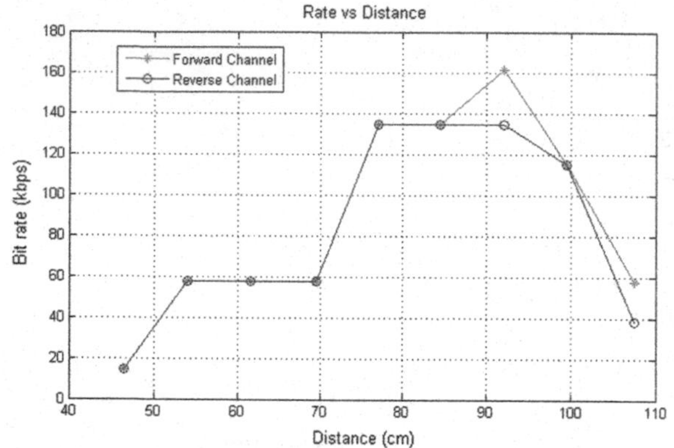

Fig. 10. Bit rate versus transmission distance

5 Conclusions

The practical transceiver performance analysis of indoor VLC systems, including transmitter and receiver devices was presented and discussed. Some experimental scenarios were demonstrated and analyzed in terms of transmission range, symbol error rate and transmission delay to confirm that information data (text and images) transmission for indoor applications can be implemented using white LEDs. In future works, we will improve effectiveness of designed VLC transceivers for real-time data transmission capability in various scenarios with considerations to light interferences from other light sources as well as effects of receiver concentrators.

Acknowledgements. This research is funded by the Hanoi University of Science and Technology (HUST) under project number T2016-LN-12.

References

1. Mukai T., and Nakamura, S., White and W LEDs. OYO BUTURI 68(1999), 152–155.
2. Tamura, T., Setomoto, T., and Taguchi, T., Fundamental characteristics of the illuminating light source using white light-emitting diodes based on InGaN semiconductors, *IEEJ Trans. on Fundamentals and Materials*, 120 (2000), 244–249.
3. Taguchi, T., Technological innovation of high-brightness light emitting diodes (LEDs) and a view of white LED lighting system, *Optronics*, 19(2000), 113–119.
4. Nakamura, T., Development of ZnSe-based white Light emitting diodes with longer lifetimes of over 10,000 hr, *Electrical Engineering in Japan*, 154(2006), 42–48.
5. Komine, T., and Nakagawa, M., Fundamental analysis for visible light communication system using LED lights, *IEEE Trans. on Consumer Elec.*, 50(2004), 100–107.
6. Komine, T., and Nakagawa, M., Integrated system of white LED visible-light communication and power-line communication, *IEEE Trans. on Consumer Electronics* 49 (2003) 71–79.
7. Haruyama, S., Visible light communication, *IEICE Trans.* on J86-A (2003) 1284–1291.

8. VLCC, Visible Light Communications Consortium, 2008.
9. Home Gigabit Access project, funded by European Framework 7." http://www.ict-omega.eu/.
10. Wireless World Research Forum, http://www.wireless-world-research.org/.
11. IEEE, IEEE P802.15 Working Group for Wireless Personal Area Networks (WPANs), 2008.
12. J. Vucic, C. Kottke, S. Nerreter, and K. Habel, A., Buttner, K. D. Langer and J. W. Waleski, 125 Mbit/s over 5 m Wireless Distance by Use of OOK-Modulated Phosphorescent White LEDs. *Proc. of 35th European Conf. of Opt. Commun.* (2009) 1–2.
13. H. Le-Minh, D. O'Brien, G. Faulkner, L. Zeng, K. Lee, D. Jung, Y. Oh and E. T. Won., 100-Mb/s NRZ Visible Light Communications Using a Postequalized White LED, *IEEE Photonics Tech. Lett.* 21 (2009) 1063–1065.
14. Y. Zheng and M. Zhang, Visible Light Communications Recent Progresses and Future Outlooks, *Proc. of Photonics and Optoelectronics Conf.* (2011) 1–6.
15. R. Mesleh, H. Elgala and H. Hass, Optical Spatial Modulation, *Journal of Opt. Commun. and Netw.* 3 (2011) 234–244.
16. C. H. Yeh, Y. F. Liu, C. W. Chow, Y. Liu, P. Y. Huang and H. K. Tsang, Investigation of 4-ASK Modulation with Digital Filtering to Increase 20 Times of Direct Modulation Speed of White-Light LED Visible Light Communication System, *Optics Express*, 20 (2012) 16218–16223.
17. A. Burton, C. Amiot, H. L. Minh and Z. Ghassemlooy, Design of an integrated Optical Receiver for Mobile Visible Light Communications, *Proc. PGNet*, (2011).
18. H. Shin, S.-B. Park, D.K. Jung, Y.M. Lee, S. Song, J. Park, VLC Transceiver Design for Short-Range Wireless Data Interfaces, *Proc. of International Conf. on Information and Commun. Tech. Convergence, ICTC* (2011) 689–690.
19. L. C. Png, S. X. Lim, A. R., B.-W. Chan, F. A. Hazman, Designs of VLC Transceiver Circuits for Reading Light Transmission of High-Quality Audio Signals on Commercial Airliners, *Proc. International Conf. on Consumer Electronics, ICCE* (2014) 97–98.
20. L. C. Png, N. L. Minh, L. Chen, K. S. Yeo, Designs of a Free-Space White-LED Mass-Storage Transceiver for SD-Card File Transfer, *Proc. 3rd IEEE Workshop on Opt. Wire. Commun.* (2012) 1260–1263.
21. S. Rajbhandari, et. al., High-Speed Integrated Visible Light Communication System: Device Constraints and Design Considerations, *IEEE J. on Slec. Areas in Commun.* 33 (2015) 1750–1757.
22. D. Terra, et. al., Design, Development and Performance Analysis of DSSS-based Transceiver for VLC. *Proc. EUROCON* (2011) 1–4.
23. F. Che, B. Hussain, L. Wu, C. P. Yue, Design and Implementation of IEEE 802.15.7 VLC PHY-I Transceiver, *Proc. International Conf. on Solid-State and Integrated Circuit Tech., ICSICT* (2014) 1–4.
24. B. Hussain, et. al., Visible Light Communication System Design and Link Budget Analysis, *J. of Ligh. Tech.* 33 (2015) 5201–5209.
25. L. Ding, F. Liu, Y. He, H. Zhu, Y. Wang, Design of Wireless Optical Access System using LED, *Optics and Photonics Journal*, 3 (2013) 148–152.
26. I. E. Lee, et. al., Design and Development of A Portable Visible-Light Communication Transceiver for Indoor Wireless Multimedia Broadcasting, *Proc. 2nd International Conf. on Electronic Design, ICED* (2014) 20–24.

Collective Signature Protocols for Signing Groups

N. K. Tuan[1(✉)], V. L. Van[2], D. N. Moldovyan[3], H. N. Duy[4], and A. A. Moldovyan[5]

[1] Duy Tan University, Da Nang, Vietnam
nkimtuan@gmail.com
[2] Pedagogical College of Dalat, Da Lat, Vietnam
vlvandalat@gmail.com
[3] St. Petersburg State Electrotechnical University LETI, Professora Popova street 5, St. Petersburg 197376, Russia
namoldovyan@gmail.com
[4] Department of Information Technology, Hanoi, Vietnam
hoduy027@gmail.com
[5] ITMO University, Lomonosova street 9, St. Petersburg 191002, Russia
aamoldovyan@gmail.com

Abstract. To extend practical applicability of the group signature protocols, there are introduced signature schemes of two novel types: (i) collective signature shared by a set of signing groups and (ii) combined collective signature shared by several signing groups and several individual signers. The protocol of the first type is constructed and described in detail. It is also shown a possible modification of the described protocol which allows transforming the protocol of the first type into the protocol of the second type. The proposed collective signature protocols have significant merits, one of which is connected with possibility of their practical using on the base of the existing public key infrastructures.

1 Introduction

Digital signature (DS) protocols are widely used in the information technologies to process electronic legal messages and documents. The DS protocols are based on DS schemes that represent a mathematical technique applied in public key cryptography to validate the authenticity of digital messages or documents. Such validation is connected with the fact that DS as some redundant information can be computed only by using the private key that is known only to one person, i.e., to the signer. Verification of the signature validity is performed with signer's public key that is known publicly. To solve a variety of different practical tasks, different types of signature are proposed: usual (individual) signature [1]; blind signature [2–5]; collective signature [6]; group signature [7]. The group signature protocols can be used in applications where the recipient only needs to know that the signature came from a signing group. List of applications that benefit from group signatures includes, for example, vehicle safety communications, electronic voting, electronic sales, electronic mails, system to

V. Bhateja et al. (eds.), *Information Systems Design and Intelligent Applications*, Advances in Intelligent Systems and Computing 672,
https://doi.org/10.1007/978-981-10-7512-4_20

preserve the privacy of its users, anonymous attestation, bidding electronic cash transactions [8, 9].

The notions of *collective signature* and *group signature* are essentially different. The collective signature refers to a signature generated with participation of each of individual signers included in some declared set of signers. Validity of collective signature to some electronic document *M* means that *M* is signed by each of them. To generate a collective DS, it is needed each of the mentioned individual signers uses this private key. The procedure of the verification of collective DS is performed using public keys of each signer. The collective DS protocols can be practically used on the base of the public key infrastructure (PKI) existing in practice to support the widely used individual DS protocols. Another merit of the collective DS protocols relates to possibility to implement them using many official DS standards [10], for example, the Russian standard GOST R 34.10–2012 [11].

The group signature refers to a signature formed on behalf of group of signers (signing group) headed by a person called group manager or dealer. The group digital signature (GDS) to an electronic message is generated by a group member. To verify the group signature, group public key needs to be used. Except the group manager, nobody can disclose which particular group member signed the document. The group signature has the following important properties. Firstly, only group members can sign a document. Secondly, group manager, who has both document and valid group signature, can reveal the group members signed the document. Finally, non-group members could not reveal the original signers, who generate the group signature. The group manager is a trusted party of the group signature protocol. He creates the secret parameters used by the signers to generate signature. If the group signature scheme allows the group members to dynamically join the group and revoke the group, then it is called dynamic group signature [12]. In usual group signature schemes, the group signature is generated by one group member. In threshold group signature schemes, the group signature can be formed only by a set of group members, including not less than a given number of individual signers [6]. GDS protocols proposed in papers [7, 13] represent special interest since they can be practically used on the base of existing PKI created to support applications that use the individual signature protocols. Besides, those protocols need no procedure in order for a group member to join the group and to revoke from the group. But only one person is fixed in the signing group, the group manager who can arbitrary distribute works for preparing and signing electronic documents among arbitrary subsets of the persons having public keys registered in certification authority.

Combining the main properties of the collective and digital schemes in frame of some single DS protocol is very actual in the case when a document is to be processed and signed by several different signing groups. In such case, it is reasonable to create a single signature that notifies that each of the given set of the signing groups has signed the document. In this paper, we propose the collective signature protocol providing such possibility. In another practical scenario, an electronic document is to be processed and signed by several different signing groups and by several different individual signers. We also propose a combined collective DS protocol that provides possibility to generate a single DS notifying that some given document is signed by several signing groups and several individual signers.

Thus in the present paper, we introduce protocols of GDS of two novel types, which ensure the expansion of the functionality of the group signature protocols, which consist possibility of signing an electronic document by (i) an arbitrary number of group signers and (ii) arbitrary number of group and individual signers with a single signature, the size of which does not depend on the number of signers (from one to many hundreds). When constructing new protocols of GDS, we used the idea of combining in the single cryptographic scheme the mechanisms for generating both the collective DS and the group DS based on masking of public keys of signers. The results of the paper represent further development of the ideas used in the DS protocols described earlier in papers [5, 7, 13, 14].

2 The Protocol of Collective Digital Signature for Group Signers

For the purpose of developing a protocol of collective DS for group signers based on the computational difficulty of the problem of finding discrete logarithm modulo a prime, it has been used in the protocol described in [7] as prototype. A basic scheme of GDS was developed, which is described as follows.

The following parameters are used in the protocol: (1) sufficiently large prime p (with length more than 2464 bits), such that number $p - 1$ contains a large prime divisor q (with length $\geq$ 256 bits); (2) number α which order is equal to q modulo p.

The GDS protocol presupposes the formation of a digital signature to some electronic document on behalf of some collegial body (group of signers, i.e., signing group), which is headed by a group manager. Each representative of a group of signers generates his private key x (with length $|x| \geq 256$ bits) and his public key $y = \alpha^x \bmod p$. The public key Y of the group manager is a public key of the group and is calculated as follows $Y = \alpha^X \bmod p$, where X is manager's private key. The value Y is also the public key of the group, i.e., the value Y is used to verify authenticity of the GDS.

Let m group members (having public keys $y_i = \alpha^{x_i} \bmod p$ and corresponding private keys x_i, $i = 1, 2, \ldots, m$) wish to sign the document M. The group signature protocol is described as follows.

1. The group manager computes hash value from document $H = F_H(M)$, where F_H is some specified hash function, and calculates masking coefficients

$$\lambda_i = F_H(H||y_i||F_H(H||y_i||X)), \tag{1}$$

and sends each value λ_i to the corresponding ith group member, for $i = 1, 2, \ldots, m$. Then the group manager computes the first element of the group signature

$$U = \prod_{i=1}^{m} y_i^{\lambda_i} \bmod p. \tag{2}$$

2. Each ith group member (i = 1, 2, …, m) generates a random number $k_i < q$, computes the value $R_i = \alpha^{k_i} \bmod p$, and sends R_i to the group manager.
3. The group manager generates the random number K < q and computes the values

$$R' = \alpha^K \bmod p,$$
$$R = R' \prod_{i=1}^{m} R_i \bmod p = \alpha^{K+\sum_{i=1}^{m} k_i}, \tag{3}$$

and

$$E = F_H(M||R||U), \tag{4}$$

where E is the second element of the group signature. Then he sends value E to all group members who have initiated the protocol.

4. Each ith group member (i = 1, 2, …, m) computes his signature share

$$S_i = k_i - x_i \lambda_i E \bmod q \tag{5}$$

and sends it to the group manager.

5. The group manager verifies the correctness of each share S_i by checking equality

$$R_i = y_i^{\lambda_i E} \alpha^{S_i} \bmod p. \tag{6}$$

If all signature shares S_i satisfy the last verification equation, then he computes his share

$$S' = K - XE \bmod q \tag{7}$$

and the third element of the group signature

$$S = S' + \sum_{i=1}^{m} S_i \bmod q. \tag{8}$$

The verification procedure includes the following steps:

1. The verifier computes the hash function value from the document M: $H = F_H(M)$. Using the group public key Y and signature (U, E, S), he computes value

$$\tilde{R} = (UY)^E \alpha^S \bmod p. \tag{9}$$

2. He computes value

$$\tilde{E} = F_H(M||\tilde{R}||U). \tag{10}$$

3. He compares the values E and $\tilde{E}$. If $\tilde{E} = E$, then the verifier concludes that the group signature is valid. Otherwise, he rejects the signature.

It is easy to see that the value

$$D = \sum_{i=1}^{m} S_i \bmod q,$$

can be considered as a "group pre-signature" approving of which is performed by the group manager with adding (modulo q) his signature share S' (see formula (8)). The value is actually calculated analogously for computation of the collective signature in the protocols [5, 14]. The main difference between the described GDS protocol and collective DS protocols [5, 14] is using the masking coefficients λ_i at time of generating the collective public key U in accordance with formula (2), which is used as the first element of the GDS. The value U conserves the information about all group members participated in the process of generating the GDS. It is easy to see that only the group manager can open the GDS, using the value U, since only he can compute the masking values λ_i.

The signature randomization parameter R in formula (3) represents the single-use collective signature computed as product of the single-use public keys R_i of each individual signer and the single-use public keys R' of the group manager. The last uses the values R_i and formula (6) to verify validity of the signature share S_i of each ith individual signer, which is computed using the formula (5) and private key of the ith signer. The group manager computes his share as value S' in formula (7) and then he calculates the third element of the GDS as value S in formula (7). The GDS verification procedure includes computations in accordance with formulas (9) and (10).

In the protocol developed in this paper, it also used the mechanism of the formation of the collective DS. Namely, this mechanism is used in the following two ways: (i) to form a pre-signature and (ii) to form a collective signature shared by several signing groups. Let g signing groups with public keys $Y_j = \alpha^{X_j} \bmod p$, where $j = 1, 2, \ldots, g$; X_j is the secret key of the jth group manager, have intention to sign the document M. Suppose also the jth signing group includes m_j active individual signers (persons appointed to act on behalf of the jth signing group). The protocol of collective signature for group signers is described as follows.

The signature generation procedure relating to the proposed collective DS protocol for signing groups

1. Within the framework of the GDS protocol described above, the manager of each j group of signers (j = 1, 2, …, g) generates masking parameters λ_{ji} (see formula (1)) for the signers of his group and computes the value

$$U_j = \prod_{i=1}^{m_j} y_{ji}^{\lambda_{ji}} \bmod p \tag{11}$$

(where i = 1, 2, …, m_j) as the jth share in the first element of the collective group signature and the randomizing parameter

$$R_j = R'_j \prod_{i=1}^{m_j} R_{ji} \bmod p. \tag{12}$$

Then he sends values U_j and R_j to all other managers.

2. Each jth group manager (j = 1, 2, …, g) computes values

$$U = \prod_{j=1}^{g} U_j \bmod p,$$

$$R = \prod_{j=1}^{g} R_j \bmod p = \alpha^{\sum_{j=1}^{g} K_j} \bmod p,$$

and

$$E = F_H(M||R||U),$$

where E and U are the first and second elements of the group signature.

3. Each jth group manager (j = 1, 2, …, g) computes signature share of his group

$$S_j = S'_j + \sum_{i=1}^{m_j} S_{ji} \bmod q, \tag{13}$$

where S_{ji} is the signature share of the ith individual signer in the ith signing group, and sends it to other group managers.

4. Each jth group manager can verify the correctness of each share S_j by checking equality

$$R_j = (U_j Y_j)^E \alpha^{S_j} \bmod p.$$

If all shares S_j satisfy the last verification equation, then the third element S of the collective signature is computed:

$$S = \sum_{j=1}^{g} S_j \bmod q.$$

The tuple (U, E, S) generated by the above procedure presents the collective signature (to the document M) shared by g signing groups. The verification procedure is described as follows:

The signature verification procedure relating to the proposed collective DS protocol for signing groups

1. Compute the collective public key shared by all signing groups:

$$Y_{\text{col}} = \prod_{j=1}^{g} Y_j \bmod p = \alpha^{\sum_{j=1}^{g} X_j} \bmod p. \quad (14)$$

2. Compute the value

$$\tilde{R} = (UY_{\text{col}})^E \alpha^S \bmod p. \quad (15)$$

3. Compute the value

$$\tilde{E} = F_H\left(M||\tilde{R}||U\right).$$

4. Compare the values E and $\tilde{E}$. If $\tilde{E} = E$, then one concludes that the group signature is valid. Otherwise, the signature is rejected.

The first element U of the collective signature contains information about all the group members of each signing group who signed the document M. The identification procedure (the disclosure of the group signature) is carried out by analogy with the procedure for disclosing the group signature described in [7]. It should be noted that the procedure for identifying individual signer requires the participation of the group managers of each group who share the collective signature. At the same time, the computational complexity of this procedure is relatively high and rapidly increases with the growth of number of the signing groups that share collective signature.

In the proposed collective DS scheme, the signature verification procedure includes the steps of the verification procedure in the group signature scheme and additional initial step for computing the collective public key in accordance with formula (14). In the signature verification Eq. (15), it used the collective public key Y_{col} instead of the group public key.

3 Protocol of Collective Digital Signature for Group and Individual Signers

Another important practical scenario relates to the processing document M by several individual signers and by several group signers. Construction of the collective signature protocol for such case can be implemented in full correspondence with the collective signature protocol for group signers described in Sect. 2, if it is accepted an agreement that for individual signers the value U_j is equal to 1. Actually in the protocol of the collective signature for group signers, it should be taken the value $U_j = 1$, if the jth signer is an individual one, instead of performing computations defined by formula (11). Besides, computation of the value R_j is performed using formula $R_j = \alpha^{k_j} \bmod p$ instead of (12). Respectively computation of the value is performed using formula $S_j = k_j - x_j E \bmod q$, where x_i is private key of the jth individual signer, instead of (13).

It is evident that only all group managers act in the procedure of disclosing the collective group signature (identification of the individual signers acted in frame of each group signer).

The proposed protocols are based on computational difficulty of the discrete logarithm problem in the ground finite field $GF(p)$. It seems possible to design collective DS protocol for signing groups and combined collective DS protocol on the base of computational difficulty of finding roots modulo p, degree of which is equal to a sufficiently large prime number π such that π^2 divides the number $p - 1$ and $|\pi|$ 160 bits [15]. In paper [15], it is proposed collective DS protocol can be used as initial point for construction of the group signature scheme and then for construction of the collective DS protocols of the two proposed types.

4 Conclusion

In this paper, it is proposed an extension of the notion of collective signature to the case when a single DS is shared by arbitrary number of different signing group. Such type of DS has fixed size independent of the number of signing groups and notifies that each of the lasts has signed the document to which such collective signature relates. Other proposed extension of the mentioned notion relates to combined collective DS that is shared by arbitrary number of signing groups and by arbitrary number of individual signers. The protocol implementing the first extension has been described. Like collective signature protocols [5, 14] and group signature protocols [7, 13], the proposed protocol can be practically used on the base of the already existing PKI. Construction of the protocol implementing the notion of the combined collective signature represents a natural modification of the described protocol of collective signature for signing groups, which consists in accepting the agreement to assign to the share in the collective signature element U the value equal to 1, if the share relates to individual signer.

The proposed protocol is based on computational difficulty of the problem of finding discrete logarithm in the ground finite field $GF(p)$. The proposed design can be implemented using computations on an elliptic curve. In such case, one can significantly reduce the signature length and increase security of the protocol and its performance. Detailed consideration of constructing the collective DS protocol for signing groups and combined collective DS protocol represents topic for independent work covering the implementations based on existing DS standards.

We also hope in our future work to develop the collective DS schemes of the proposed types, in which the signature contains only two elements E and S, like in DS standards [1, 10, 11] based on the computational problem of finding discrete logarithms.

References

1. National Institute of Standards and Technology. Digital Signature Standard. FIPS Publication 186-3, 2009.
2. *Chaum D.* Blind Signatures for Untraceable Payments. Advances in Cryptology: Proc. of CRYPTO'82. Plenum Press, 1983, p. 199–203.
3. Camenisch J.L., Piveteau J.-M., Stadler M.A. Blind Signatures Based on the Discrete Logarithm Problem. In: Advances in Crypology – EUROCRYPT'94 Proc. Lecture Notes in Computer Science, Vol. 950. Springer-Verlag, Berlin Heidelberg New York (1995) 428–432.
4. Minh N. H., Binh D. V., Giang N. T., Moldovyan N. A. Blind Signature Protocol Based on Difficulty of Simultaneous Solving Two Difficult Problems. Applied Mathematical Sciences. 6 (2012) 6903–6910.
5. Moldovyan N.A.Blind Signature Protocols from Digital Signature Standards. Int. Journal of Network Security. 13 (2011) 22–30.
6. Pieprzyk J., HardjonoTh., Seberry J.: Fundamentals of Computer Security. Springer-verlag. Berlin (2003).
7. Moldovyan A.A., Moldovyan N.A. Group signature protocol based on masking public keys. Quasigroups and related systems. 22 (2014) 133–140.
8. Seetha R., Saravanan R. Digital Signature Schemes for group communication: A Survey. International Journal of Applied Engineering Research. 11 (2016) 4416–4422.
9. Enache A.-C. About Group Digital Signatures. Journal of Mobile, Embedded and Distributed Systems. IV (2012) 193–202.
10. International Standard ISO/IEC 14888-3:2006(E). Information technology –Security techniques – Digital Signatures with appendix – Part 3: Discrete logarithm based mechanisms.
11. GOST R 34.10-2001. Russian Federation Standard. Information Technology. Cryptographic data Security. Produce and check procedures of Electronic Digital Signature. Government Committee of the Russia for Standards, 2012 (in Russian).
12. Rajasree R.S. Generation of Dynamic Group Digital Signature. International Journal of Computer Applications. 98 (2014) 1–5.
13. Moldovyan N.A., Nguyen Hieu Minh, Dao Tuan Hung, Tran Xuan Kien. Group Signature Protocol Based on Collective Signature Protocol and Masking Public Keys Mechanism. International Journal of Emerging Technology and Advanced Engineering. 6 (2016) 1–5.
14. Moldovyan N.A. Blind Collective Signature Protocol. Computer Science Journal of Moldova. 19 (2011) 80–91.
15. Moldovyan N.A. Digital Signature Scheme Based on a New Hard Problem. Computer Science Journal of Moldova. 16 (2008) 163–182.

Deniability of Symmetric Encryption Based on Computational Indistinguishability from Probabilistic Ciphering

Nikolay Andreevich Moldovyan[1], Ahmed Al-Majmar Nashwan[2], Duc Tam Nguyen[3], Nam Hai Nguyen[3], and Hieu Minh Nguyen[3(✉)]

[1] St. Petersburg Institute for Informatics and Automation of Russian Academy of Sciences, St. Petersburg 199178, Russia
nmold@mail.ru

[2] Ibb University, Ibb, Yemen
almojammer2015@gmail.com

[3] Academy of Cryptography Techniques, Hanoi, Vietnam
{nguyenductamkma, hainnvn61, hieuminhmta}@gmail.com

Abstract. It is proposed as a novel interpretation of the notion of the shared-key deniable encryption, extended model of the coercive adversary, set of the design criteria, and a new practical approach to designing the shared-key deniable encryption algorithms, which is characterized using computational indistinguishability from probabilistic ciphering. The approach is implemented in several described algorithms relating to the plan-ahead shared-key deniable encryption schemes. The algorithms encrypt simultaneously secret and fake messages and produce the single cryptogram that is computationally indistinguishable from the ciphertext produced by some probabilistic cipher, while encrypting the fake message. The proposed algorithms are based on block conversion functions (hash-functions and block ciphers) and satisfy criterion of complete coincidence of the algorithms for recovering the fake and secret messages. Due to possibility to perform the inverse transformation the block ciphers used as the base block conversion function provide higher speed of the deniable encryption. It is also proposed as a general design of fast block deniable encryption algorithms satisfying the proposed design criteria.

Keywords: Symmetric encryption · Computational indistinguishability · Probabilistic ciphering · Shared-key deniable encryption

1 Introduction

The notions of public-key deniable encryption and of shared-key deniable encryption were introduced by Canetti et al. in 1997 [1]. These important cryptographic primitives are applied in cryptographic protocols to resist coercive attacks. In the

V. Bhateja et al. (eds.), *Information Systems Design and Intelligent Applications*, Advances in Intelligent Systems and Computing 672,
https://doi.org/10.1007/978-981-10-7512-4_21

concept of deniable encryption, there are considered sender-deniable, receiver-deniable, and sender- and- receiver-deniable (bi-deniable) schemes in which coercive adversary attacks the party sending message, the party receiving message, and the both parties, correspondingly. In the model of the coercive attack, it is supposed that coercive adversary has power to force a party or the both parties simultaneously to open the cryptogram (ciphertext) after it has been sent.

Paper [1] initiated a lot of investigations on developing secure and efficient methods for public-key deniable encryption [2] in which no pre-shared information is used. Some of papers propose public-key deniable encryption combined with sharing secret key (the sender and the receiver initially share a common secret key) and plan-ahead encryption (the fake message is selected at the stage of encryption) [3]. Detailed attention of the researchers to this direction in the area of deniable encryption is explained by the applicability of the public-key deniable encryption to prevent vote buying in the Internet-voting systems [4] and to provide secure multiparty computations [5].

Practical applications of the plan-ahead shared-key deniable encryption can be attributed to the case of the information protection against unauthorized access in computer and communication systems in the case of coercive attacks. As it is noted in [1], for some models of such attacks the plan-ahead shared-key deniable encryption schemes can be trivially constructed as using l different keys and composing the ciphertext as concatenation of l cryptograms that are results of encrypting l different messages. Such naive shared-key deniable encryption is applicable to create hidden volumes in computer security systems, for example, like BestCrypt [6,7] and FreeOTFE [8,9]. In that systems plausible deniability of the encryption is based on assumption that the coercer will be not able to recognize the hidden volume in a huge-size pseudorandom data array. Actually such application of the naive deniable encryption represents some combination of the cryptographic and steganographic mechanisms. Paper [10] presents results of the analysis of the possibility to detect the hidden volume in the 2008 versions of the systems, TrueCrypt, BestCrypt, and FreeOTFE, using forensic analysis tools. It had been shown the possibility of data leakage from the hidden volume during normal operation of the software. Such data leakage indirectly indicates presence of a hidden volume; however, no useful information can be gained from the volumes critical data block if the last is fully encrypted.

Unfortunately, the naive interpretation of the notion of the shared-key deniable encryption is not suitable for many other practical applications, for example, for protecting messages sent via telecommunication channels. Indeed, the coercive attacker will ask about the parts of cryptogram that are not used for disclosing the fake message. Presence of such parts of the cryptogram indicates the cryptogram contains some additional message that is encrypted with a key different from the fake key, i.e., the deniability is not satisfactory. Thus, to extend the application areas one should introduce an advanced interpretation of the notion of the plan-ahead shared-key deniable encryption which should be oriented to providing resistance in the case when coercer can follow the use of all bits of the full cryptogram during the decryption process. Besides, it is reasonable to suppose that the potential coercer can have possibility to measure the execution time of the decryption procedure in the case of disclosing the secret message or he can know the decryption algorithm.

Present paper proposes an advanced understanding of the plan-ahead shared-key deniable encryption notion and considers the coercive-attack model against which the known trivial and naive construction of the shared-key deniable encryption does not provide sufficient deniability. To resist the considered potential coercive attack, it is proposed as a novel method for the plan-ahead shared-key deniable encryption methods producing cryptogram that is computationally indistinguishable from the ciphertext produced by some probabilistic cipher.

2 Model of Adversary and Design Criteria

It is assumed that after ciphertext has been sent the adversary has possibility to force both the sender and the receiver to open the following:

- the plaintext corresponding to the ciphertext;
- encryption and decryption algorithms;
- the source software code used for performing decryption of the ciphertext;
- the encryption key with which encryption of the opened message yields all bits of the ciphertext.

Thus, in the considered extended model of the coercive attack the sender and the receiver are coerced to open parameters and algorithm of the ciphering procedure with which each bit of the sent ciphertext has been produced depending on the opened message (plaintext).

Security against the described attack can be provided using the symmetric deniable encryption algorithm that produces the ciphertext like cryptogram produced as result of probabilistic encryption of the fake message with fake key. The ciphers satisfying the last criterion are called pseudo-probabilistic ciphers (PPC). Construction of the symmetric PPC can be implemented using the following design criteria:

- Symmetric deniable encryption should be performed as simultaneous encryption of two messages, secret one and fake one, using secret and fake keys (which are shared by sender and receiver);
- a probabilistic encryption algorithm should be associated with the SDE algorithm;
- the associated probabilistic encryption algorithm should transform the fake message with the fake key into the same ciphertext that is produced by the SDE algorithm;
- the algorithms for recovering the fake and secret messages should completely coincide;
- high security of the deniable encryption of the messages having arbitrary length should be provided while using the fixed-size shared keys.

3 Implementation Using Hash-Functions

Suppose F_H be a secure hash-function (e.g., one of the candidates for standard SHA-3 [11] or one of the hash-functions using an n-bit block cipher algorithm [12]) and T be a secret message represented as sequence of u-bit symbols (data blocks) $t_i : T = \{t_1, t_2, ..., t_i, ..., t_z\}$, for example u = 1 to 16.

The following algorithm performs probabilistic encryption with using F_H and the 128-bit secret key K.

Algorithm 1.

1. Set counter $i = 1$ and random 128-bit initialization vector V. (The value V is not secret and is to be sent by sender to receiver of the secret message T.)
2. Set counter $j = 0$.
3. If $j < 2^{k+1}$, then generate a random k-bit value $r(k > u)$. Otherwise output the message "The ith data block is not encrypted," increment the counter $i = i + 1$ and go to step 2.
4. Compute the value $t = F_H(K||V||i||r) \bmod 2u$, where $||$ is the concatenation operation. If $t\#t_i$, then increment $j = j + 1$ and go to step 3.
5. Set $ri = r$. If $i\#z$, then increment $i = i + 1$ and go to step 2. Otherwise STOP.

The described algorithm outputs the ciphertext $R = \{r_1, r_2, ..., r_i, ..., r_z\}$ represented as sequence of k-bit data blocks r_i.

Algorithm 1 represents the associated probabilistic encryption algorithm corresponding to the following SDE algorithm that performs simultaneous encryption of the secret message T and fake message $M = \{m_1, m_2, ..., m_i, ..., m_z\}$, where data blocks m_i have size u bits (for example $u = 1$ to 8), using the 128-bit secrete key K and the 128-bit fake key Q.

Algorithm 2.

1. Set counter $i = 1$ and random 128-bit initialization vector V.
2. Set counter $j = 0$.
3. If $j < 2^{k+1}$, then generate a random k-bit value $r(k > 2u)$. Otherwise output the message "The ith data block is not encrypted," then increment $i = i + 1$ and go to step 2.
4. Compute the values $t = F_H(K||V||i||r) \bmod 2^u$ *and* $m = F_H(Q||V||i||r)$ *mod* 2^u.
5. If $t\#t_i$ or $m\#m_i$, then increment $j = j + 1$ and go to step 3.
6. Set $r_i = r$. If $i\#z$, then increment $i = i + 1$ and go to step 2. Otherwise STOP.

Algorithm 2 outputs the ciphertext $R = \{r_1, r_2, ..., r_i, ..., r_z\}$ decryption of which is performed with Decryption Procedure 1 that is used also for decrypting the ciphertext produced by Algorithm 1:

Decryption Procedure 1.

1. Set counter $i = 1$, decryption key $X(X \leftarrow K$ or $X \leftarrow Q)$, and random 128-bit initialization vector V.
2. Compute the value $w_i = F_H(X||V||i||r_i) \bmod 2^u$.

4. If $i \# z$, then increment $i = i + 1$ and go to step 2. Otherwise STOP.

The decryption algorithm outputs the sequence of u-bit data blocks w_i : $W = \{w_1, w_2, ..., w_i, ..., w_z\}$. The correctness of the decryption procedure is evident and $W = M$, if $X = Q$, or $W = T$, if $X = K$.

On the average, to encrypt a data block there required 2^u and 2^{2u} trials for selecting appropriate random value r at step 3 of Algorithm 1 and Algorithm 2, correspondingly. To define sufficiently small probability that at step 3 the encryption algorithms output the message "The ith data block is not encrypted" one should use sufficiently large values k. For example, in the case of Algorithm 2 $k = 32$, if $u = 8$, or $k = 6$, if $u = 1$.

4 Implementation Using Block Ciphers

In frame of the method presented in Sect. 3, one can replace using hash-function F_H by using some n-bit block encryption algorithm E. In this case, the probabilistic encryption algorithm is described by Algorithm 1 in which step 4 takes, for example, the form as follows:

4. Compute the value $t = E_{K \oplus V}(V^{'}||i||r) \bmod 2^u$, where E is a secure 128-bit block cipher (e.g., AES [13] or Eagle-128 [14]); r is the 32-bit ciphertext block; $\oplus$ is the XOR operation; i is the 56-bit counter; $V^{'} = V \bmod 2^{40}$. If $t \# t_i$, then increment $j = j + 1$ and go to step 3.

The deniable encryption algorithm looks like Algorithm 2 in which steps 4 and 5 are replaced, for example, by the following two steps:

4. Compute the values $t = E_{K \oplus V}(i||r) \bmod 2^u$ and $m = E_{Q \oplus V}(i||r)$, where E is a secure 64-bit block cipher (e.g., IDEA [15] or one of fast data-driven 64-bit block ciphers [16,17]); r is the 24-bit ciphertext block; $\oplus$ is the XOR operation; i is the 40-bit counter.

5. If $t \# t_i$ or $m \# m_i$, then increment $j = j + 1$ and go to step 3.

Such straightforward use of the block ciphers in the proposed method for SDE gives only small increase in the performance. To obtain more significant encryption speed of the SDE, one can use the procedure of simultaneous encryption both the block encryption function E and the respective decryption function $D = E^{-1}$. This idea is implemented in the following SDE algorithm for simultaneous encryption of the secret message T and the fake message M:

Algorithm 3.
1. Set counter $i = 1$ and random initialization vector V.
2. Set counter $j = 1$.
3. Generate a random k-bit value ρ and set the counter $r \leftarrow \rho$.
4. Compute the n-bit value $c_i = E_{Q \oplus V}(m_i||r)$.
5. Compute the value $D_{K \oplus V}(c_i) = (t||r')$, where the n-bit value of the decryption function is interpreted as concatenation of the u-bit value t and k-bit value r' (in this algorithm, the values u and k are used such that $n = u + k$, where, e.g., $u = 4$ to 16; $n = 32$ to 256).
6. Compare the values t and $t - i$. If $t = t - i$, then set the ciphertext block $c - i = c$ and go to step 7, otherwise go to step 3.

7. If $j < 2^{2u}$, then increment $j \leftarrow j+1$, increment the counter $r \leftarrow r+1$, and go to step 4. Otherwise output the message "Encryption of the pair of input data blocks t_i and m_i has not been fulfilled".

8. If $i < z$, then increment $i \leftarrow i+1$ and go to step 2, otherwise STOP.

Algorithm 3 outputs the ciphertext $C = \{c_1, c_2, ..., c_i, ..., cz\}$ decryption of which is performed as follows:

Decryption Procedure 2.

1. Set counter $i = 1$, decryption key $X(X \leftarrow K$ or $X \leftarrow Q)$, and initialization vector V.
2. Compute the value $w_i = E_{Q \oplus V}(c_i)$ *div* 2^u.
3. If $i \# z$, then increment $i = i+1$ and go to step 2. Otherwise STOP.

This decryption algorithm outputs the sequence of u-bit data blocks $w_i : W = \{w_1, w_2, ..., w_i, ..., w_z\}$. The correctness of the decryption procedure is evident and $W = M$, if $X = Q$, or $W = T$, if $X = K$.

On the average, to encrypt a data block there required 2^u trials for selecting appropriate random value r at step 3 of Algorithm 3, i.e., the average number of trials is 2^u times lower than the average number of trials at step 3 of Algorithm 2 (and in its modification with using a block cipher E instead of the hash-function F_H).

Probabilistic encryption of the fake message associated with the SDE defined by Algorithm 3 is as follows:

Associated Probabilistic Encryption Algorithm.

1. Set counter $i = 1$ and random initialization vector V.
2. Generate a random k-bit value r.
3. Compute the n -bit value $c_i = E_{Q \oplus V}(m_i || r)$.
4. If $i < z$, then increment $i \leftarrow i+1$ and go to step 2, otherwise STOP.

5 Discussion

The SDE algorithms described in Sects. 3 and 4 represent the implementation of a new approach characterized in using computational indistinguishability from probabilistic ciphering to justify deniability. Bi-deniability of the described algorithms in Sects. 3 and 4 is based on existing respective probabilistic algorithms that are associated with the first ones. Security of the SDE algorithms is based on using secure hash-functions and secure block ciphers. Indeed, insecurity of Algorithm 2 (or insecurity of the probabilistic encryption algorithm associated with Algorithm 2) to the known or to the chosen plaintext (or ciphertext) attack means that output of the hash-function is not uniformly pseudorandom, i.e., insecurity assumption about the proposed SDE algorithm leads to the same assumption about the hash-function put into the base of the SDE method. Respectively, insecurity assumption about Algorithm 3 (or about the probabilistic encryption algorithm associated with Algorithm 3) to the mentioned attacks leads to the insecurity assumption about the block cipher put into the base of the SDE method.

Performance of the decryption procedure is significantly larger than performance of the encryption procedure defined by Algorithms 1, 2, and 3, since the decryption does not require performing many trials to select appropriate random value r used in the encryption process. The decryption speed $S_H d$ of Decryption procedure 1 can be estimated as $S_{Hd} = S_{H/\rho}$, where ρ is the ration of the size of the value $X||V||i||r_i$ to value u and S_H is the performance of the used hash-function F_H. Performance of Algorithm 1 and Algorithm 2 can be estimated with formulas, correspondingly:

$S_{SDE-1} = 2^{-u} S_{H/\rho}$ and

$$S_{SDE-2} = 2^{-2u} S_{H/\rho} \tag{1}$$

The encryption speed S'_{SDE-2} of the modified version of Algorithm 2 (in which instead of the hash-function F_H there is used an n-bit block cipher E) can be estimated approximately with the following formula

$$S'_{SDE-2} = (2^{-2u-1} u/n) S_{H/\rho} \tag{2}$$

where S_E is the performance of the n-bit block cipher E.

The decryption speed $S_E d$ of Decryption procedure 2 can be estimated as $SEd \approx uSE/n$. The encryption speed S_e of Algorithm 3 can be estimated approximately with the following formula

$$S_{SDE-3} \approx (2^{-2u-1} u/n) S_E \tag{3}$$

Thus, comparing formula (3) with (1) and (2) one can come to conclusion that in the case $S_E \approx S_H$ the deniable encryption performed with using Algorithm 3 is about 2^u times faster than with using Algorithm 2 as well as with using the mentioned modification of Algorithm 2.

To obtain faster SDE algorithms it represents interest to use hardware efficient algorithms based on data-driven operations [14,16,17] and software-suitable block ciphers described in [18,19]. This provides the following potential values of the performance of Algorithm 3:

$S_{SDE-3} = 10^5 - 2.10^7$ bit/s (for hardware implementations) and $S_{SDE-3} = 10^4 - 2.10^6$ bit/s (for software implementations).

The proposed SDE methods represent practical interest for some particular applications, problem of designing SDE algorithms providing higher performance is actual though. We estimate that the proposed approach can be used to construct significantly faster SDE algorithms and some new design ideas are to be applied. For such purpose, one can use permuting ciphertext blocks of the secret and fake messages encrypted with a block cipher, the permutation being performed with a permutation network, like in [17,20].

A general approach to design fast SDE algorithms appears to be as the following one. Suppose E' and E'' be two secure block ciphers with n-bit and $(2n)$-bit input data block, correspondingly; M' and T' be n-bit data blocks of the fake and secret message. Then perform the following encryption steps:

1. Using the fake key Q' and the cipher E' transform the block M': $C_M = E'_{Q'}(C_M)$.

2. Using the secret key K and the cipher E' transform the block T': $C_T = E'_{K'}(C_T)$.

3. Using the fake key Q'' and the cipher E'' transform the $(2n)$-bit block $C_M||C_T$ representing concatenation of the n-bit blocks of intermediate ciphertexts C_M and $C_T : C = E''_{Q''}(C_M||C_T)$.

The respective associated probabilistic encryption algorithm is as follows:

1. Using the key Q' and the cipher E' transform the n-bit block $M' : C_M = E'_{Q'}(C_M)$.
2. Generate a uniformly random n-bit string R.
3. Using the key Q'' and the block cipher E'' transform the $(2n)$-bit block $C_M||R : C = E''_{Q''}(C_M||C_T)$.

The decryption algorithm includes the next steps:

1. Using the key Q'' and the $(2n)$-bit block decryption function $D'' = E'' - 1$ decrypt the ciphertext block $C : C_M||C_T = D''_{Q''}(C)$.
2. Using the fake key Q' (or secret key K) and the n-bit block decryption function $D' = E''^{-1}$ decrypt the left (right) n-bit part P of the intermediate ciphertext block $C_M||C_T : W = D'_X(P)$, where $W = M'$ or $W = T'$; $X = Q'$ or $X = K$; $P = C_M$ or $P = C_T$, correspondingly.

The described general design of the SDE algorithms of block type potentially provides high performance and satisfies the criterion of computational indistinguishability from probabilistic block ciphering the fake message block M' with the fake key including subkeys Q' and Q''. However, the decryption algorithms for recovering the fake and secret messages do not completely coincide, i.e., at step 3 of the decryption algorithm it should be taken the left or right half of the $(2n)$-bit intermediate ciphertext block $C_M||C_T$. Because of such difference the design criterion of complete coincidence of the algorithms for recovering the fake and secret messages is not satisfied. To implement this criterion one can propose to use a key-dependent map function F_X mapping the pairs of the n-bit intermediate ciphertext blocks C_M and C_T into the single $(2n)$-bit output ciphertext block C. Besides, the inverse map is to be performed with function $F_X^{-1}(C)$ mapping the $(2n)$-bit argument into n-bit output value: $F_X^{-1}(C) = C_M$, if $X = Q'$, and $F_K^{-1}(C) = C_T$, if $X = K$ (it is supposed to use no additional subkey Q''). In our future research, we plan to propose concrete implementations of the map function F_X.

Currently, the main focus of researchers is on the development of the area of the public-key deniable encryption algorithms and protocols [21–23] We hope that the results of this article will initiate more active research in the field of the plan-ahead shared-key deniable encryption schemes.

6 Conclusion

It has been proposed as an advanced understanding of the shared-key deniable encryption as the process that produces the ciphertext computationally indistinguishable from the ciphertext produced by some probabilistic encryption. It has been proposed to construct the SDE procedure as process of simultaneous encryption of secret and fake messages using secret and fake keys, the SDE procedure being associated with some probabilistic encryption algorithm. To satisfy the design criterion of computational indistinguishability from the associated probabilistic encryption, the constructed SDE algorithms are associated with evidently defined probabilistic ciphers that decrypt the ciphertext C produced by the SDE into a fake message. The associated cipher potentially encrypts the fake message into the ciphertext C, while using fake key and respective values of probabilistic parameter r (see step 3 in Algorithm 1, 2, and 3).

Such approach allows one to construct the SDE algorithms with fixed-size key, which provide possibility to encrypt messages of arbitrary length. The approach has been illustrated with concrete SDE that has been constructed as transformation of the hash-functions and block ciphers into the probabilistic deniable encryption algorithm.

The designed SDE algorithms can be used for encrypting messages sent via public channels and individual files. For using in the computer security systems to perform on the fly transparent disk encryption, for example, like in the system FreeOTFE, there are required faster SDE algorithms. In future research on the SDE, we plan to develop fast deterministic and fast probabilistic SDE algorithms satisfying the proposed design criteria and having performance up to several Gbit/s. The future research is to be focused on constructing computationally efficient map functions F_X. Another interesting problem for future research relates to design SDE algorithms of the block type, with which it will be possible to encrypt simultaneously data blocks of the fake and secret message having different bit length (preferably having arbitrary ratio of their sizes).

References

1. Canetti, R., Dwork, C., Naor, M., Ostrovsky, R.: Deniable Encryption. Proceedings Advances in Cryptology-CRYPTO 1997. Lectute Notes in Computer Science. Springer-Verlag. Berlin, Heidelberg, New York, (1997), vol. 1294, 90–104
2. O'Neil, A., Peikert, C., Waters, B.: Bi-Deniable Public-Key Encryption. Advances in Cryptology-CRYPTO 2011. Lectute Notes in Computer Science. Springer-Verlag. Berlin, Heidelberg, New York, (2011), vol. 6841, 525–542
3. Klonowski, M., Kubiak, P., Kutylowsk, M.: Practical Deniable Encryption. SOFSEM 2008: Theory and Practice of Computer Science, 34th Conference on Current Trends in Theory and Practice of Computer Science, Nov Smokovec, Slovakia, January 19–25, (2008) 599–609
4. Meng, B.: A Secure Internet Voting Protocol Based on Non-interactive Deniable Authentication Protocol and Proof Protocol that Two Ciphertexts are Encryption of the Same Plaintext. Journal of Networks. (2009), vol. 4, no. 5, 370–377
5. Ishai, Y., Kushilevits, E., Ostrovsky, R.: Efficient Non-interactive Secure Computation. Advances in Cryptology – EUROCRYPT 2011. Lectute Notes in Computer Science. Springer-Verlag. Berlin, Heidelberg, New York, (2011), vol. 6632, 406–425
6. https://www.jetico.com/products/personal-privacy/bestcrypt-container-encryption

7. https://www.jetico.com/web_help/bc8/index.php?info=html/03_new_features/01_new_features.htm
8. https://www.download.cnet.com/FreeOTFE/3000-2092_4-10656559.html
9. https://www.softpedia.com/get/Security/Encrypting/FreeOTFE.shtml
10. Irvin, A., Hunt, R.: Forensic Methods and Techniques for the Detection of Deniable Encryption, https://www.cosc.canterbury.ac.nz/ray.hunt/deniable_encryption_tool_a_survey, (2003) 657–667
11. Andreeva, E., Bogdanov, A., Mennink, B., Preneel, B., Rechberger, C.: On security arguments of the second round SHA-3 candidates. International Journal of Network Security. (2012), vol. 11, Issue 2, 103–120
12. ISO 10118-2:2010. Information technology-Security techniques-Hash-functions-Part 2: Hash-functions using an n-bit block cipher algorithm, https://www.iso.org/standard/44737.html
13. Announcing Approval of Federal Information Processing Standard (FIPS) 197, Advanced Encryption Standard (AES), https://www.federalregister.gov/documents/2001/12/06/01-30232/announcing-approval-of-federal-information-processing-standard-fips-197-advanced-encryption-standard
14. Lai, X., Massey, J.L: A proposal for a new block encryption standard. Advances in Cryptology – EUROCRYPT 1990. Lectute Notes in Computer Science. Springer-Verlag. Berlin, Heidelberg, New York, (1991), vol. 473, 389–404
15. Moldovyan, N.A., Moldovyan, A.A., Eremeev, M.A., Sklavos, N.: New class of Cryptographic Primitives and Cipher Design for Network Security. International Journal of Network Security. (2006), vol. 2, no. 2, 114–125
16. Moldovyan, N.A.: On Cipher Design Based on Switchable Controlled Operations. International Journal of Network Security. (2008), vol. 7, no. 3, 404–415
17. Moldovyan, N.A., Moldovyan, A.A.: Data-driven block ciphers for fast telecommunication systems. Auerbach Publications. Talor & Francis Group. New York, London. (2008)
18. Moldovyan, N.A., Moldovyanu, P.A., Summerville, D.H.: On Software Implementation of Fast DDP-Based Ciphers. International Journal of Network Security. (2007), vol. 4, no. 1, 81–89
19. Moldovyan, N.A., Moldovyan, A.A.: A method for encrypting a message. Russian patent #2459275. (2012) (in Russian).
20. Moldovyan, A.A., Moldovyan, N.A., Moldovyanu, P.A.: Architecture Types of the Bit Permutation Instruction for General Purpose Processors. Springer LNGC. 3d Int. Workshop IF & GIS'07 Proc. St.Petersburg, (2007), vol. 14, 147–159
21. Barakat, M.T.: A New Sender-Side Public-Key Deniable Encryption Scheme with Fast Decryption. KSII Transactions on Internet and Information Systems. (2014) 3231–3249
22. Moldovyan N.A., Moldovyan A.A., Shcherbacov V.A.: Generating Cubic Equations as a Method for Public Encryption. Buletinul Academiei de Stiinte a Republicii Moldova. Matematica. (2015) 60–71
23. Dachman-Soled, D.: On minimal assumptions for sender-deniable public key encryption. Public-Key Cryptography-PKC 2014: 17th International Conference on Practice and Theory in Public-Key Cryptography. Lecture Notes in Computer Science. Springer-Verlag. Berlin, Heidelberg, New York. (2014), vol. 8383, 574–591

Opinion Extraction from Quora Using User-Biased Sentiment Analysis

Akshi Kumar, Satyarth Praveen(✉), Nalin Goel(✉), and Karan Sanwal(✉)

Department of Computer Science and Engineering, Delhi Technological University, Delhi, India
akshikumar@dce.ac.in, {satyarth934,nalingl994, karansanwal}@gmail.com

Abstract. Opinion extraction is a field of computer science which deals with understanding the context of textual data and further forming an opinion on behalf of the user. In this paper, we present an opinion extraction model based on user's profile. The opinion formulation algorithm is governed by factors that vary among users. A user-biased sentimental analysis technique is introduced, which mines the answers written on various topics on the popular Web site Quora and provides an opinion based on user's preferences. The generics work independently. For implementation, a personal assistant to assist students in selecting a university for graduate studies based on their preferences, of course, the return of investment expectations, importance to university ranks, etc., was created. The algorithm achieved optimal performance and hence can be used as a reliable method to form opinions on behalf of the user.

1 Introduction

The amount of data available to the public has increased exponentially since the advent of World Wide Web. The large amount data has made it possible to theoretically make informed decisions based on the vast information available on the Internet today. It may, however, not be feasible for a user to scan through the entire sheer volume of data before forming an opinion. Artificial intelligence methods which scan the data on behalf of the user and then form an opinion have hence been gaining immense popularity.

Most opinion extraction algorithms [1–5] rely on natural language processing methods to understand the context along with a sentiment analysis model that assigns the context a polarity or a generally predicted likeliness. In this paper, however, we acknowledge the fact that opinion formulation depends on the factors that differ amidst users. For example, it is possible for two users to form mutually exclusive opinions on the same textual information. Understanding the user profile is hence a crucial step prior to forming an opinion on behalf of the user and is hence gaining popularity in the literature [6, 7].

V. Bhateja et al. (eds.), *Information Systems Design and Intelligent Applications*, Advances in Intelligent Systems and Computing 672,
https://doi.org/10.1007/978-981-10-7512-4_22

The proposed algorithm expounds a user's opinion formulation mechanism by uniquely defining a model to form an opinion for the user. After an exhaustive study of the pertinent literature, we realized that it is much easier for a machine to model the opinion formulation process of a user by monitoring how the opinion of the user changes dynamically related to the text in concern. We thus subject the user to certain predefined constraints and then closely study how the opinion of the user varies with the introduction of virtual facts and use this variation to define an opinion formulation model that closely resembles that of the user. It is important to note that sentiment analysis is performed independently with respect to the user, but the way the results of the sentiment analysis is interpreted changes from user to user. This makes sense, because though "Apples are fresh" is a sentence that should always be taken in the context that the apples are indeed fresh irrespective of the user in question, but whether that is a good, neutral, or bad piece of information would depend on whether the user really wants fresh apples or if the user was indeed looking for stale apples. Other domains explored in this paper are word similarity mappings. The content of the World Wide Web is a contribution by a large base of users who have different paraphrasing, writing styles, and vocabulary. While a conversion of the entire content to a similar, machine-understandable format is one of the major areas of research concerning the development of a Semantic Web, we have solved the problem in hand by using similarity indexes between words for mapping similar words under the same cluster of concern.

As an implementation of the model, we have developed a personal assistant aimed at helping students to find the right university for graduate studies by modeling their requirements and opinion formulation process and further searching the exhaustive yet reliable content in the popular question and answer Web site Quora and forming an opinion on behalf of the user regarding a university of interest.

The rest of the paper is organized as follows: Sect. 2 mentions about previous works in literature, Sect. 3 discusses the dataset Quora, Sect. 4 discloses the proposed approach in detail, and Sect. 5 discusses the results and inferences drawn from the results of the work done and suggests some future work directions.

2 Related Work

Gojali [8] emphasized the relevance of aspect-based sentiment analysis as compared to the general opinion of the whole document. They proposed a system which extracts the potential aspect and its sentiment. It further finds its polarity and classifies it accordingly. Tirath et al. [9] demonstrated a system to determine the polarity of movie reviews and applied semantic methods for preprocessing of data. They further used feature impact analysis to calculate the importance and to reduce the features. The proposed algorithm was evaluated on an IMDB dataset, and it achieved a result of 88.95% accuracy. Gokulakrishnan et al. [10] presented a model for opinion mining and sentiment analysis. They first classified data as subjective–objective or irrelevant and then applied the positive/negative sentiment analysis, which according to their result produced increased accuracy. Sun et al. [11] presented a series of opinion mining techniques suitable for different situations like document-level opinion mining, sentence-level opinion mining, cross-domain opinion mining. They later on also listed the comparative and deep learning approaches of the same. Piryani et al. [12] presented a detailed mapping of the research works done on opinion mining with sentiment analysis during the period of 2000–2015. The team conducted a scientometric analysis of the work done during the period. The analysis included metrics such as publication patterns, publication growth, collaboration patterns, global contribution patterns. It also identifies the approaches that grew popular in the field over time.

Though a lot of previous work on opinion mining has been proposed with respect to a general model of a user for whom the opinion is being formed, not enough study on how the general opinion varies amidst different users has been done. This paper works on a model that considers the profile of the user on whose behalf the opinion is being formed and hence is more useful in the real world where different users consider different factors with varying priorities before arriving at an opinion.

3 Dataset

Quora is a popular question and answer Web site available on World Wide Web. Its large community of users interacts with each other through asking and answering questions as well as upvoting other's answers. The numbers of upvotes/downvotes act as a metric of agreement between the users. As of now, Quora is considered as a reliable source of information. For our model, we have scrapped answers from Quora which belong to questions corresponding to the university that the user is interested in.

4 Proposed Approach

The proposed algorithm aims to mimic its user's opinion formulation mechanism and hence forms an opinion on a large dataset on behalf of its user. The domain of prerequisite knowledge depends on the type of data on which the opinion has to be formed on. For our implementation, we have considered questions and answers relating to universities in the popular Web site Quora as our reliable dataset. Our algorithm hence reads through the numerous questions and answers on Quora about a particular university and then forms an opinion regarding that university on behalf of the user. This will hence help students shortlist universities for their higher education. The algorithm consists of four steps as shown in Fig. 1.

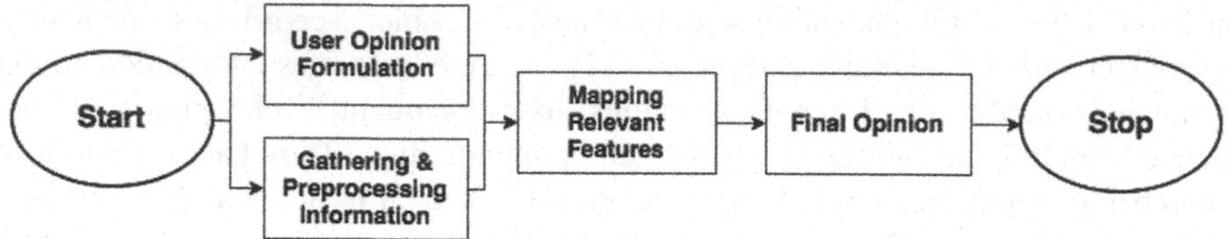

Fig. 1. Flowchart

4.1 Modeling User Opinion Formulation Mechanism

This step is the heart of the algorithm. It aims to model the mechanism by which a user forms an opinion regarding a particular university (Fig. 2). We had shortlisted the following seven major factors to act as a universal umbrella set covering most of the factors that describe a university:

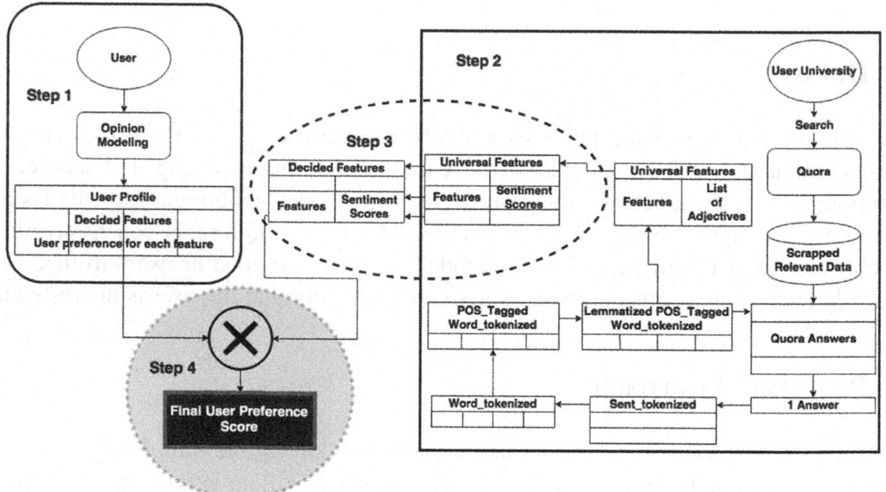

Fig. 2. Architecture of the proposed approach

- University ranking.
- Research.
- Placement.
- Fee.
- Scholarship options.
- Faculty.
- Campus.

To model the opinion formulation mechanism, the importance of these individual factors is to be known for the user. Since "importance" of a factor is a vague term, quantizing it is a trivial task even through direct interaction with the user. We hence try to obtain the relative importance of these features with respect to each other as that would be sufficient to understand the preference and priorities of the user. We scrutinize the variance of the user opinion and interpret the reasons behind opinion change. Hence changes in user opinion are used to model the user's opinion formulation mechanism. Each factor is finally assigned a weight that is representative of the importance of that factor. Initially, all factors are assigned the weight 0. The user is then introduced to a non-existing university XYZ that helps us to model the user's opinion formulation mechanism. Every iteration of this step introduces a previously unknown fact regarding the university to the user, and the user is then asked to rate the university with all of the known knowledge up to that point on a scale of 1–10. The scale is seen as consisting of two sets: a set of ratings (1–5) is termed as the rejection zone, while the set [5–10] is termed as the acceptance zone. By default, initially, the university is rated as 4. Every iteration aims to present the user with a fact that aims to push the user's rating to the zone opposite of what the current zone is. For example, if the current university rating by the user is 9, then a negative fact about the university, such as high fee, will be presented to the user and his change in rating will be recorded and observed. Consider another example where the user has currently rated the university as 4, then a positive fact about the university such as stellar research in the field of interest of the user will be mentioned and again the rise in user ratings will be recorded and observed. In our implementation, facts based on factors were presented in the same order as the mentioned factors and in the mood depending on the current university ratings. Whenever there is a change of zone (rejected to accepted or vice versa), then all the facts presented to the user since the last zone change are re-examined and the weights relating to the factors that have contributed to this change of zone are increased/decreased based on the rise/fall of the ratings they were able to achieve.

This approach achieves three important characteristics of user's opinion formulation mechanism:

(1) *It helps the machine understand the relative importance of the factors as more important factors will cause larger changes in the university ratings.*
(2) *It covers the case where an opinion toward a university might be a contribution from a collection offactors that individually do not affect the opinion. For example, the user might be okay with an expensive university and also be okay with a university with a poor world rank, but the user might not like a university that is both expensive and also has a poor world rank.*
(3) *It covers the case when an individual factor is sufficient for the user's opinion toward the university. For example, if it was mentioned to the user that the university has an excellent faculty, and that is all that matters to the user, then the*

university ratings will ideally never change to an extent that it changes zone and all of the subsequent factors will hence maintain their default weight—importance.

The algorithm 1 in Sect. 4.5 presents the approach in the form of a pseudocode.

4.2 Obtaining and Preprocessing Information from the Dataset

Here, information is scrapped from the question–answer Web site, Quora, and the scrapped information is then processed. Each question is seen as a collection of answers, each answer is seen as a collection of the sentence, and each sentence is seen as a collection of words. The sentences are then parts of speech (POS)-tagged where a learned model identifies the parts of the input sentence. This step helps in identifying the nouns in the sentence and the adjectives that describe the nouns that are in their proximity, and they are scrutinized later; the lemmatized form of those words is then added to a set describing the noun. Semantic analysis is then carried out on these adjectives to assign a score to them. Each of these nouns hence becomes a factor with a score indicating how good the presence of the particular factor is for the particular university.

4.3 Mapping Relevant Features

The content of the World Wide Web is a contribution by a large base of users who have different paraphrasing, writing styles, and vocabulary. While a conversion of the entire content to a similar, machine-understandable format is one of the major areas of research concerning the development of a Semantic Web, we have solved the problem in hand easily by using similarity indexes between words for mapping similar words under the same cluster of concern. Step 1 assigned weights to a set of predefined factors, and step 2 consisted of quantizing the semantics of a wide number of factors as described in our dataset.

4.4 Final Opinion Formulation

The importance of every factor is multiplied by its score, and the addition of these products is representative of how good the university is for the particular user. Higher the score, higher will be the chances that the user will like a particular university. Note that the predilection of the university may vary from user to user because of variations in the user opinion formulation process that was modeled in step 1. Hence, our algorithm is able to scan through a large dataset and form an opinion that closely represents the opinion of the user had he/she read the entire dataset.

The algorithm 2 in Sect. 4.5 represents the working of steps *4.2*, *4.3*, and *4.4*.

4.5 Program Code

Algorithm 1

Input:

- *newUnivRate*: user input is required to formulate their personal preference
- *FACTORS*: the opinion of the user revolves around these factors.

Output:

- *rateChange*: a hashmap that contains the weights formulated for each factor based on the inputs of the user.

```
begin
  univRate := 4
  flag := 2
  for each factor in FACTORS:
    listFactors.add(factor)
    if(univRate < 5):
      present factor in +ve mood
      newUnivRate := userInput()
      rateChange(factor) := |univRate - newUnivRate|
      univRate := newUnivRate
      if(flag == 1):
        for each factor in listFactors:
          factor.importance := rateChange(factor)
        listFactors.empty()
      flag := 2
    if(univRate >= 5):
      present factor in -ve mood
      newUnivRate := userInput()
      rateChange(factor) := |univRate - newUnivRate|
      univRate := newUnivRate
      if (flag == 2):
        for each factor in listFactors:
          factor.importance := rateChange(factor)
        listFactors.empty()
      flag := 1
end.
```

Algorithm 2

Input:

- *userUniv*: the university that the user wishes to check his preferences with
- *ansVec*: the vector containing all the scrapped answers from Quora.

Output:

- *userBiasedScore*: the user-biased score that denotes how suitable the university is based on the preferences of the user.

```
begin
  userUniv := userInput()
  ansVec := scapeDataFromQuora()

  // intFeatSent: contains extracted features from Quora
  intFeatSent := blank initialization
  for each ans in ansVec:
    sentVec := sentenceTokenize(ans)
    for sent in sentVec:
      wordVec := wordTokenize(sent)
      wordVecPOS←posTagging(wordVec)
      wordVecLem := [Lemmatize(w[0]) for w in wordVecPOS]

      // populating intermediate feature-sentiment pair
   vector
      intFeatSent←popSentFeat(wordVecLem)
  featSentScore←computeScore(intFeatSent)
  for each df in decidedFeatureVec:
    dfCluster←getCluster(df, featSentScore)

    // sentiment aggregate of dfCluster for current
decidedFeature(df)
    dfScores[df] ←sentimentAggregate(dfCluster)
  userBiasedScore←finalScore(userProfile, dfScores)
end.
```

5 Evaluation

Figure 3 shows the user's interests corresponding to the seven factors for the universities, namely, "University Ranking" (Ra), "Research" (Re), "Placement" (Pl), "Fee" (Fe), "Scholarship" (Sc), "Faculty" (Fa), and "Campus" (Ca). Table 1 shows the exact weight values for each factor obtained after the opinion formulation.

The solid line depicts the general preference of the public as extracted from Quora, and the dotted lines depict the interest of two different users, "User 1" and "User 2." From the graph, it is noticeable that the overall shape of the users' plot is quite similar to the general preference; only the intensity or weights of factors vary for a different

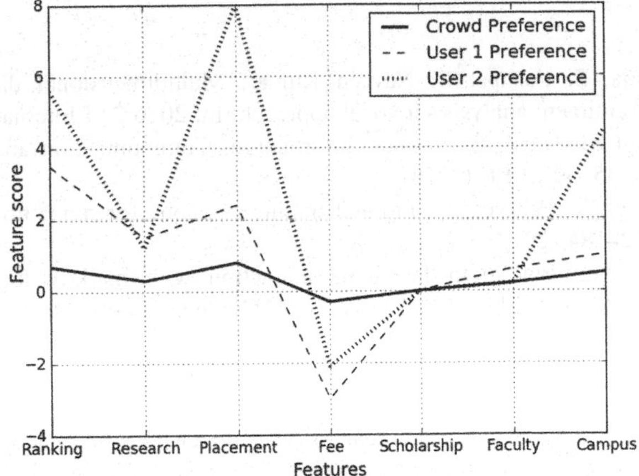

Fig. 3. User-biased preference profile

Table 1. Factor preference weights

Feature / Source	Ra	Re	Pl	Fe	Sc	Fa	Ca
Quora	0.7	0.3	0.8	−0.3	0.0	0.2	0.5
User 1	3.5	1.5	2.4	−3.0	0.0	0.6	1.0
User 2	5.6	1.2	8.0	−2.1	0.0	0.2	4.5

user. But this does not guarantee the shape of the plot of every user as the general preference is a kind of aggregation of each user's preference.

6 Conclusion

The algorithm proposed in this paper was adept to understand and capture the preferences and priorities of the user and use them to form an opinion regarding various universities on behalf of the user. Two different users obtained diverse opinions on the same university which depicts that the algorithm is indeed casting opinions on behalf of its user. Such an algorithm is useful in cases where the dataset is too large (for example, Quora) for the user to go through manually. The algorithm depends on the quality of modeling user opinion formulation mechanism for its optimal performance. A possible direction of future work can be to consider the number of upvotes in an answer to determine its validity. Also, an AND–OR tree for determining relationships between factors can be created for better representation.

References

1. D. Kumar Raja, S. Pushpa, B. Naveen Kumar: Multidimensional distributed opinion extraction for sentiment analysis - a novel approach. In: 2016 2nd International Conference on Advances in Electrical, Electronics, Information, Communication and Bio-Informatics (AEEICB). pp. 35–39. IEEE (2016).
2. Z. Yan, X. Jing, W. Pedrycz: Fusing and mining opinions for reputation generation. Inf. Fusion. 36, 172–184 (2017).
3. L. Mojica, V. Ng: Fine-grained opinion extraction with markov logic networks. In: Proceedings - 2015 IEEE 14th International Conference on Machine Learning and Applications, ICMLA 2015. pp. 271–276. IEEE (2016).
4. Akshi Kumar, T. M. Sebastian: Sentiment Analysis: A Perspective on its Past, Present and Future. Int. J. Intell. Syst. Appl. 4, 1–14 (2012).
5. Akshi Kumar, T. M. Sebastian: Sentiment Analysis on Twitter. Int. J. Comput. Sci. 9, 372–378 (2012).
6. A. Rashid, I. Albert, D. Cosley, D., S. K. Lam, S. M. McNee, J. Konstan, J. Riedl: Getting to Know You: Learning New User Preferences in Recommender Systems. Int. Conf. Intell. User Interfaces, IUI 2002. 127–134 (2002).
7. H. Sakagami, T. Kamba: Learning personal preferences on online newspaper articles from user behaviors. Comput. Networks ISDN Syst. 29, 1447–1455 (1997).
8. S. Gojali: Aspect Based Sentiment Analysis for Review Rating Prediction. (2016).
9. T. P. Sahu, S. Ahuja: Sentiment analysis of movie reviews: A study on feature selection & classification algorithms. 2016 Int. Conf. Microelectron. Comput. Commun. 1–6 (2016).
10. B. Gokulakrishnan, P. Priyanthan, T. Ragavan, N. Prasath, A. Perera: Opinion Mining and Sentiment Analysis on a Twitter Data Stream. Int. Conf. Adv. ICT Emerg. Reg. (ICTer 2012). 182–188 (2012).
11. S. Sun, C. Luo, J. Chen: A review of natural language processing techniques for opinion mining systems. Inf. Fusion. 36, 10–25 (2017).
12. R. Piryani, D. Madhavi, V. K. Singh: Analytical mapping of opinion mining and sentiment analysis research during 2000–2015. Inf. Process. Manag. 53, 122–150 (2017).

Smart Surveillance Robot for Real-Time Monitoring and Control System in Environment and Industrial Applications

Anand Nayyar[1], Vikram Puri[2], Nhu Gia Nguyen[3(✉)], and Dac Nhuong Le[4]

[1] Department of CA&IT, KCL IM&T Jalandhar, Jalandhar, India
anand_nayyar@yahoo.co.in
[2] Department of Electronics and Communications Engineering, GNDU, Jalandhar, India
vikrampuri@acm.org
[3] Graduate School, Duy Tan University, Da Nang, Vietnam
nguyengianhu@duytan.edu.vn
[4] Faculty of Information Technology, Haiphong University, Haiphong, Vietnam
nhuongld@hus.edu.vn

Abstract. The current ongoing revolution of Internet of Things (IoT), is now integrated with Robotics in various diverse fields of everyday life is making up new era i.e. Internet of Robotics (IoR). Internet of Robotics is on the mature stage of development and is currently surrounded by various challenges to be solved for more implementations, i.e., design, security, sensors, and long-range communication systems. The main objective of this paper is to propose an Internet-of-Things-based Internet of Robot, i.e., InterBot 1.0. InterBot 1.0 is efficient in terms of real-time environmental monitoring in terms of temperature, humidity, and gas sensing and is equipped with long-range communication system via 2.4 GHz 6-channel remote and also short range via HC-05 (Bluetooth module). InterBot 1.0 is IoT-based via ESP8266, and all the data can be viewed in live graphs via ThingSpeak.com. The Results state the efficiency of Interbot 1.0 in monitoring real-time environments.

Keywords: Robotics · Internet of Things · Internet of Robotics · Environmental monitoring · Sensors

1 Introduction

The next wave dominating the computing era will be completely different from the present stage of computer and mobile computing. The next era wave transforming computing forever will be regarded as Internet of Things. Seeing the scenario from the past few months, a new term is always reflected in almost every component of research dominating the computing "*Internet of Things*" (IoT). Today, around five million people around the world use Internet for doing tons of tasks like browsing Web, e-mail, chatting, online gaming, multimedia streaming, social media. More and more people

V. Bhateja et al. (eds.), *Information Systems Design and Intelligent Applications*, Advances in Intelligent Systems and Computing 672,
https://doi.org/10.1007/978-981-10-7512-4_23

have access to the global information and communication infrastructure, i.e., Internet which is also becoming a strong backbone to communicate, compute, and network among each other for sharing of information.

IoT [1, 2], a novel paradigm, is gaining rapid percentage. As per the recent predictions by Gartner [3] by 2025, 40 billion IoT objects will be connected and installed everywhere. Generally, IoT [4] refers to the networked interconnection of everyday objects, which are often equipped with ubiquitous intelligence. IoT will increase the ubiquity of Internet by integrating every object for interaction via embedded systems, which will lead to the development of highly distributed network of devices communicating and coordinating with human beings as well as other devices [5, 6]. IoT is regarded as very complex platform facilitating the connection of things based on objects being tagged for their identification, but also includes sensors, actuating elements, and other technologies. Connectivity of things on Internet is enabled by various open wireless technology standards like Bluetooth, *radio-frequency identification* (RFID), Wi-Fi, and smart wireless microcontroller board technologies like Arduino, Raspberry Pi, ESP8266. IoT technology is playing a crucial role in advancements of various commercial sectors like business management, manufacturing, smart intelligent transport system, agriculture, automobiles, and even robotics.

Seeing the current advancement and benefits of IoT, lots of IT organizations and research centers are investing lots of money in automation systems enabled via IoT. CISCO has projected about 30 billion devices to be connected to IoT by 2020 [3], and Morgan Stanley estimated more than 80 billion devices by 2020, and lots of IT giants like Microsoft, Facebook, Google, and other robotics companies like Kuka are spending billions of dollars in research in IoT and are upcoming with various products like Microsoft HoloLens, Facebook, Oculus Rift, Google self-driving autonomous cars, and even various companies are working on IoT-based robots and drones. One of the main areas of prime focus of researchers and IT organizations throughout the planet is development of smart real-time robots based on IoT, facilitating real-time monitoring and doing day-to-day activities autonomously using smart sensors and even integrating the concept of "*Cloud Computing*" in their overall data management.

Research trends nowadays [7], in the area of Robotics, are shifting toward more advanced and efficient advancement, i.e., Internet of Robots [8] and IoT-aided robotics applications. The main aim of this research paper is to propose an IoT-based real-time environmental monitoring robot: InterBot 1.0—Smart Surveillance Robot for Monitoring and Control for environment and industrial applications. In terms of Internet of Robots and IoT-aided robots [7], applications will make up a digital ecosystem comprising humans, robots, and IoT nodes. But the area of IoT-based robots is surrounded by various issues and challenges in terms of security, energy efficiency, and reliable wireless communication. So, lots of research is required in order to develop efficient IoT robots for several applications to build a smart, pervasive, and secure environment.

The rest of the paper is organized as follows: Sect. 2 will cover comprehensive description of Internet of Robots; Sect. 3 will give detailed description regarding InterBot 1.0: Smart Surveillance Robot for Monitoring Environment and Industrial Applications—Components (*hardware, sensors, modules, mechanical components, software*), architecture, and other overall description of features of robot developed. Section 4 will highlight live demonstration of InterBot 1.0: working, data gathering via sensors, and online data analysis using cloud computing. Section 5 includes conclusion and future scope.

2 Internet of Robot

With the tight integration of IoT, robotics applications are successfully implemented in diverse areas. Robotics, in the modern world, are tightly integrated with sensing, computing, and communication hardware which enables the robots to do all types of complex and coordinated operations. With the integration of IoT in robotics, several units complement the robot works like smart objects, sensors in the field areas, servers, and all sorts of network communication hardware. IoT-based robots these days are implemented in health care, industrial plants, military applications, research centers, and even automobile-based production units. In the real world, lots of researchers around the world are working on real-time automation IoT robotics product development for as such implementation in diverse areas.

3 InterBot 1.0: Smart Surveillance Robot for Monitoring Environment and Industrial Applications

In this section, InterBot 1.0: A real-time environmental and industrial monitoring Internet of Robot (IoR) is highlighted. This section will highlight all the components in terms of hardware used, architecture-cum-circuit working of the robot.

3.1 Components

In this section, hardware, modules-cum-sensors being used for the development of InterBot 1.0: IoR is highlighted. The hardware Arduino Nano [9–11] is regarded as compact, complete, and breadboard-friendly development board based on ATmega328.

3.2 Sensors

DHT11 [12] is low-cost digital temperature and humidity sensor. It consists of a capacitive humidity sensor and a thermistor to measure air and gives output of digital signal on the data pin. LPG Sensor MQ-6 [13, 14], a suitable gas sensor for sensing butane and propane gases. The sensor has high sensitivity and fast response time. MQ-3 sensor [15] is a gas sensor which detects alcohol, gasoline, and vapors.

3.3 Modules

Three-axis MEMS accelerometer [16] is a triple hub MEMS accelerometer with less commotion and power dissemination at 320 μA. It is appropriate to measure the static speeding up the gravity in tilt-sensing applications, and additionally dynamic increasing speed coming about because of movement or stun. Its high determination (4 mg/LSB) empowers estimation of slant changes under 1.0°.

ESP8266 [17] Wi-Fi module is SOC with TCP/IP protocol stack integrated which facilitates any microcontroller to access Wi-Fi network. ESP8266 module is cost-effective module and supports APSD for VOIP applications and Bluetooth coexistence interfaces.

DC motor is an apparatus get together appended to the engine. The speed of motor is counted in terms of rotations of the shaft per minute and is termed as RPM. The apparatus gathering helps in expanding the torque and diminishing the speed.

T6A 2.4 GHz system [18] is a passage-level transmitter offering the dependability of 2.4 GHz flag innovation and a recipient with six channels. This transmitter requires a PC to alter any of the channel variables factor including blending and servo switching.

Specialized specifications: 6-channel 2.4 GHz transmitter with servo turning around, easy to utilize control for fundamental models, includes 6-channel beneficiary and trainer framework choice.

A relay [19] is used to control a circuit by a different low-control flag, or where a few circuits are to be connected by one flag. Specialized specifications: high-affectability (250 mW) and high-limit (16 A) rendition. Designed for cooking and HVAC controls: blower engine, damper, dynamics air refinement, pipe stream support fans, and so forth: Conforms to VDE (EN61810-1). UL recognized/CSA guaranteed; Meets EN60335-1 requirements for family unit items Clearance and crawl age separate: 10 mm/10 mm; Tracking resistance: CTI > 250; Coil Insulation system: Class F; RoHS Compliant.

Solar Panel: High-execution sun-oriented board uses very proficient crystalline sun-oriented cells to build light assimilation and enhance productivity. Specialized specification: 0.53 mA: voltage: 11.2 V.

A 16 × 2 LCD display is very basic module and is very commonly used in various devices and circuits. These modules are favored more than seven segments and other multi-segment LEDs.

3.4 Software Components

Arduino IDE [20–22] is stage free base for Arduino equipment and can keep running on various working framework stages. It is essentially a cross-stage application in light of Java innovation and has establishment of preparing programming dialect and wiring ventures. Arduino IDE is a strong platform for all researchers, programmers, and other industry project development professionals to develop projects on Arduino controllers and other sensors.

The latest software which is available till date is Arduino 1.6.5. Arduino IDE is bundled with software library called "wiring" to facilitate easy I/O operations. The entire program structure can be written in main functions:

- setup(): This function is used for initialization of settings and executes at least once at execution of program.
- loop(): This function is executed iteratively till powering off the main board.

IoT-based Web Services: ThingSpeak.com [20, 23] is regarded as IoT-based open-source application platform and programming interface to store and retrieve data from things via HTTP protocol over Internet. Thingspeak.com allows users to analyze and plot graphs using data coming from sensors over Internet. It provides integrated support for MATLAB software. It works accurately with Arduino, BeagleBone [24], Raspberry Pi [25], biosensors, and many more.

3.5 InterBot 1.0—Block Diagram and Circuit Diagram

In this section, block diagram and circuit diagram of InterBot 1.0 will be covered.

3.5.1 Block Diagram

See Fig. 1.

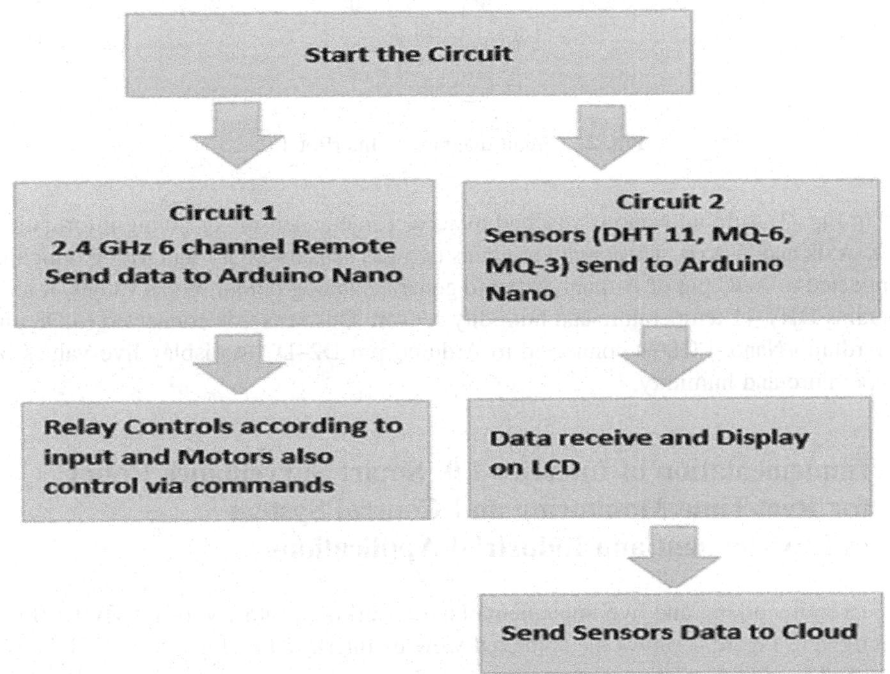

Fig. 1. Block diagram of InterBot 1.0

3.5.2 Circuit Diagram of InterBot 1.0

In Fig. 2, circuit diagram of InterBot 1.0 relays are connected to Arduino Nano pin 2–9 to control motors of InterBot and robotics ARM. There is a use of ULN2803 (transistor array) between Arduino Nano and relay to provide sufficient current between them. The whole circuit is controlled via 2.4 GHz remote. 2.4 GHz receiver is connected to pin 10, 11, 12, and analog pin A0.

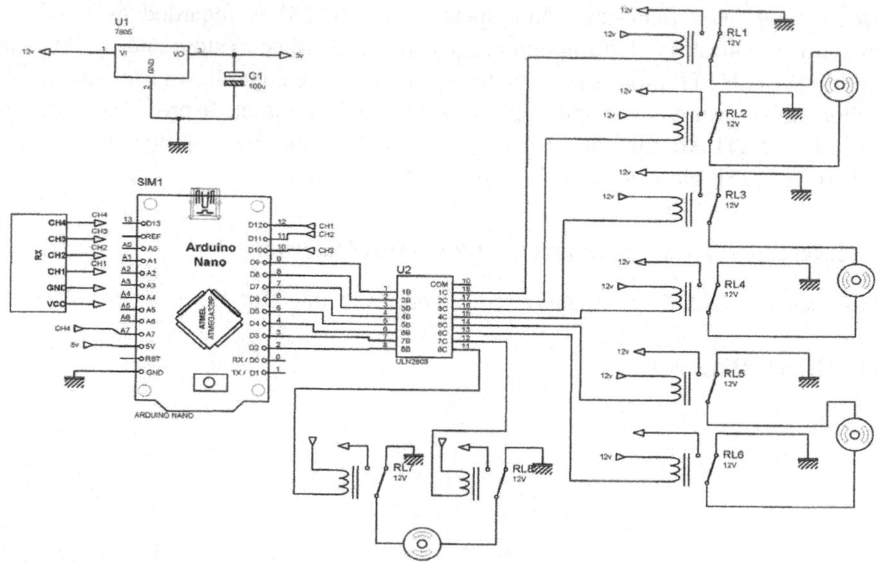

Fig. 2. Circuit diagram of InterBot 1.0

In Fig. 3, Arduino Nano is attached to accelerometer sensor for giving information of X-Axis and Y-Axis. This circuit contains two gas sensors MQ-3 and MQ-6 which is connected to ADC pin of Arduino Nano to generate analog output of gas values. It also contains DHT-11 temperature and humidity sensor. This sensor is connected to D8 pin of Arduino Nano. LCD is connected to Arduino pin D2–D7 to display live values of temperature and humidity.

4 Implementation of InterBot 1.0: Smart Surveillance Robot for Real-Time Monitoring and Control System in Environment and Industrial Applications

The overall working and live implementation of working prototype of InterBot 1.0 are highlighted. Figure 4 shows the animated view of InterBot 1.0 (Figs. 5, 6, 7, 8, 9, 10, 11, 12, 13, and 14).

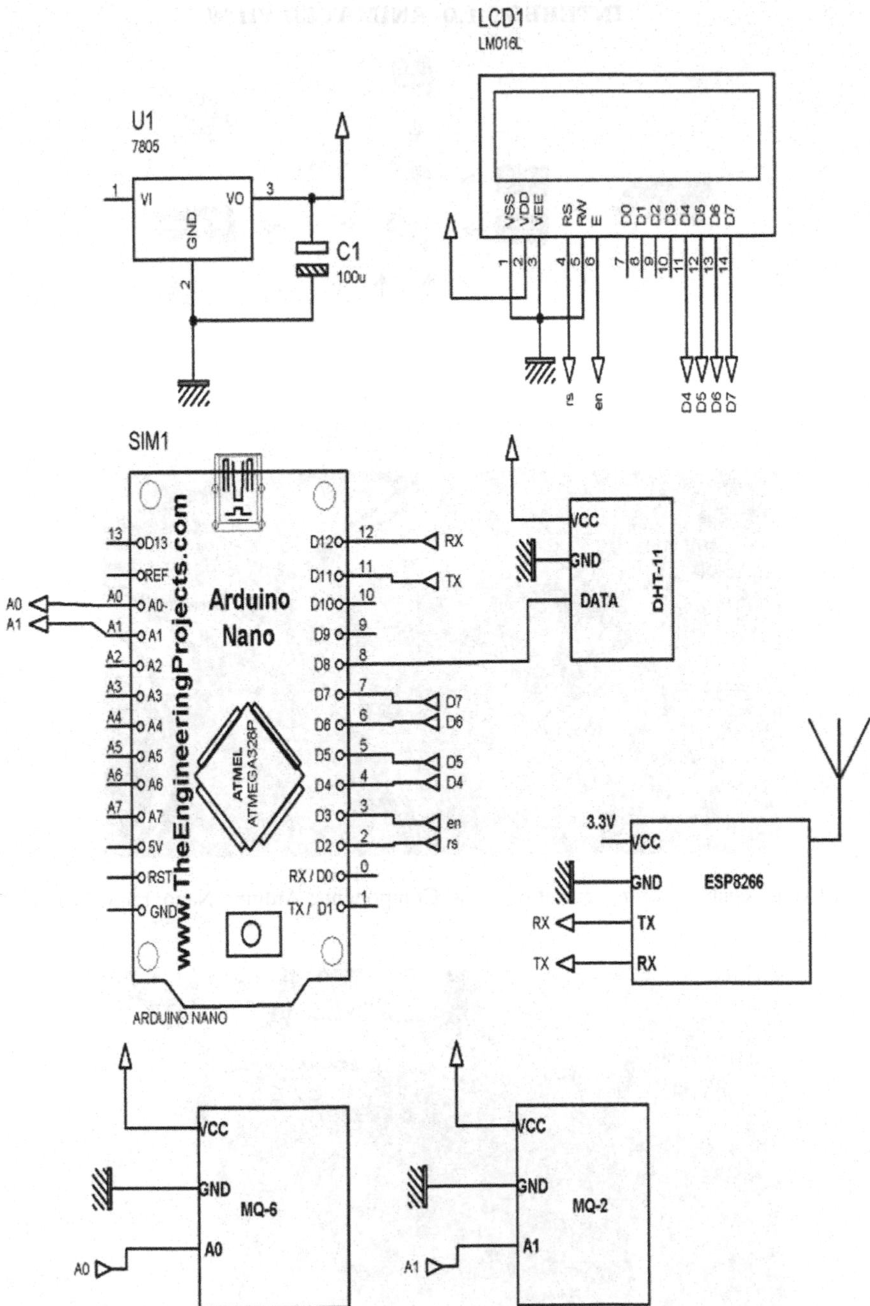

Fig. 3. Circuit diagram of ESP8266 and sensors

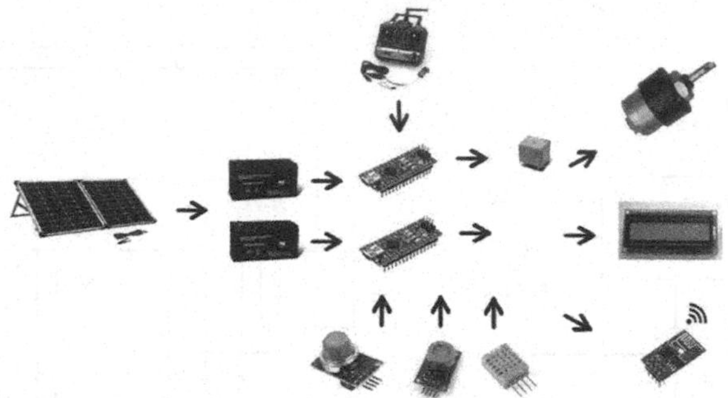

Fig. 4. Animated view of InterBot 1.0

Fig. 5. Circuit enabling of movement of robot. Components: Arduino Nano, relays, ULN2803

Fig. 6. Circuit of IoT consists of Arduino Nano, 3-axis accelerometer, ESP8266, DHT-11, MQ-3, MQ-6, and LCD

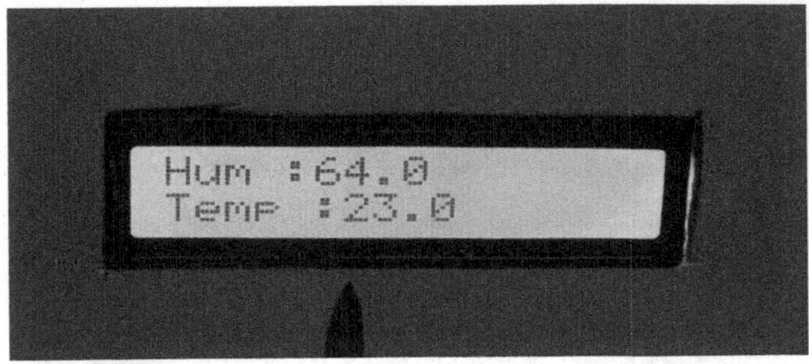

Fig. 7. Temperature and humidity display on LCD

Fig. 8. DHT-11 sensor on InterBot 1.0

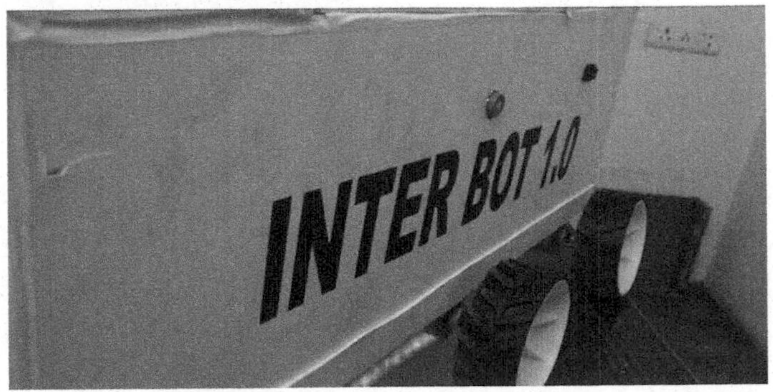

Fig. 9. MQ-3 sensor on InterBot 1.0

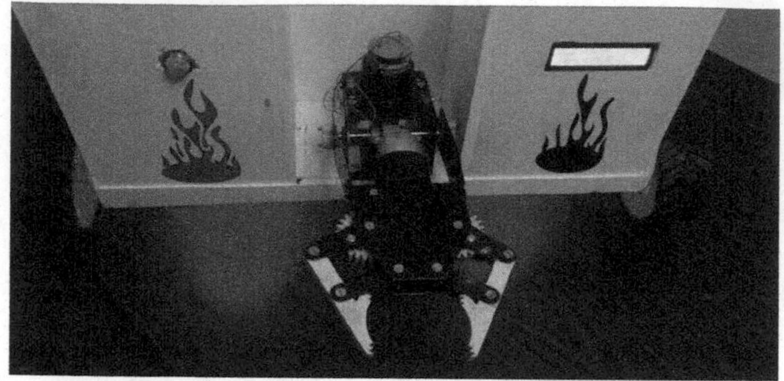

Fig. 10. MQ-6, LCD screen, and robotic ARM on InterBot 1.0

Fig. 11. Solar panel placed on InterBot 1.0

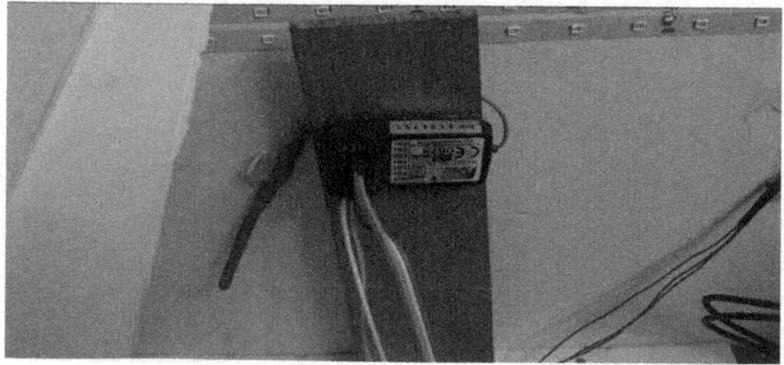

Fig. 12. 2.4 GHz 6-channel receiver

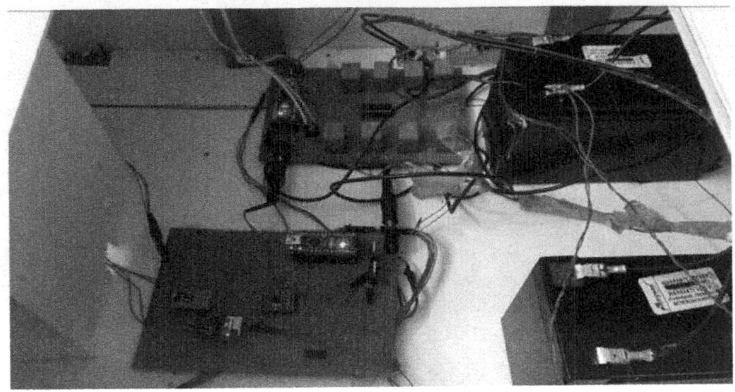

Fig. 13. Internal circuit setup with batteries

Fig. 14. InterBot 1.0 robot with 2.4 GHz 6-channel transmitter

Sensor based Live Monitoring of Environmental Data: Temperature, Humidity, LPG-Gas and Robot Movements (X and Y Axis) results are given by online website: https://www.thingspeak.com (Figs. 15, 16, 17, 18 and 19).

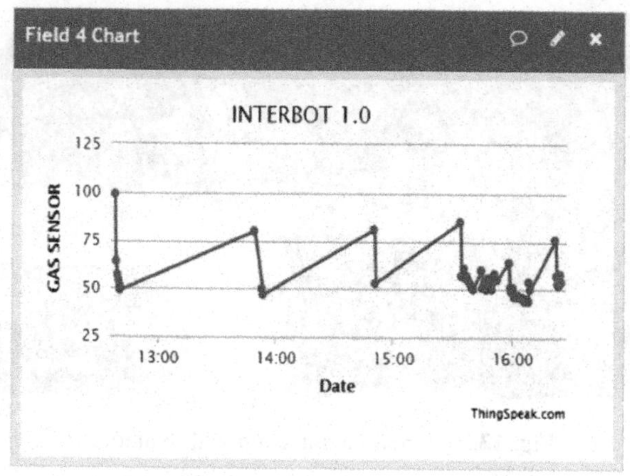

Fig. 15. Results of gas sensor (MQ-3)

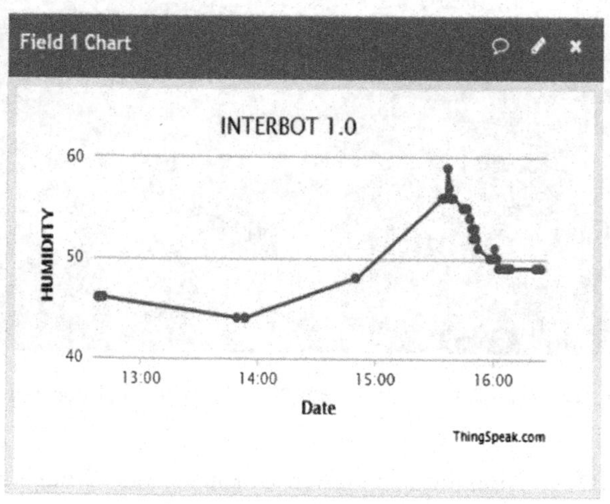

Fig. 16. Result of humidity sensor

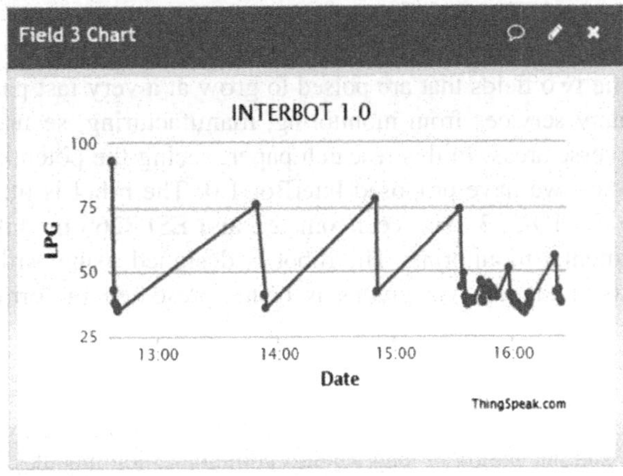

Fig. 17. Results of LPG sensor

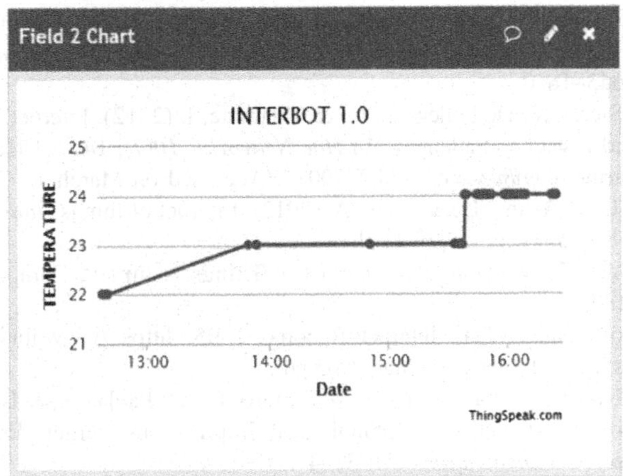

Fig. 18. Results of temperature

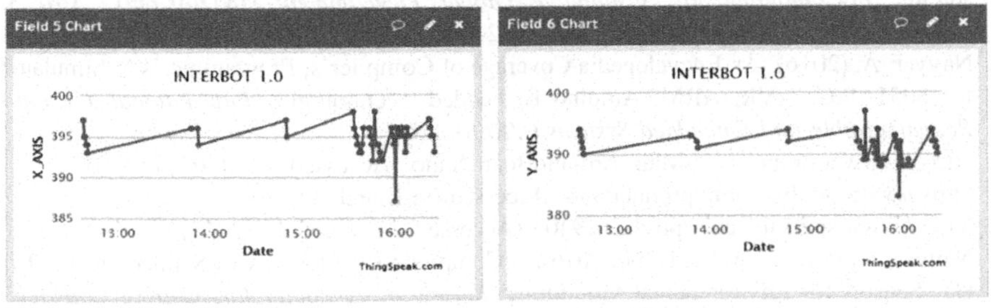

Fig. 19. Output of accelerometer X-Axis and Y-Axis

5 Conclusion and Future Scope

IoT and IoR are the two fields that are poised to grow at a very fast pace and have the ability to offer many services from monitoring, manufacturing, security surveillance, etc., in various diverse areas. In this research paper, seeing the potential of Internet of Things and Robotics, we have proposed InterBot 1.0. The robot is fitted with sensors like DHT11, MQ-6, MQ-3, 3-axis accelerometer, and ESP8266 module for providing real-time environmental monitoring. The robot is designed using Arduino microcontroller, and live feed of data via graphs is being presented in form of graphs via ThingSpeak.com.

Future Scope. To fully exploit the potential of advanced technology, i.e., IoT in the coming years, more research would be done in both protocol development, energy efficiency, more sensors integration, and integration of long wireless communication modules.

References

1. Gubbi, J., Buyya, R., Marusic, S., & Palaniswami, M. (2013). Internet of Things (IoT): A vision, architectural elements, and future directions. *Future Generation Computer Systems, 29*(7), 1645–1660.
2. Miorandi, D., Sicari, S., De Pellegrini, F., & Chlamtac, I. (2012). Internet of things: Vision, applications and research challenges. *Ad Hoc Networks, 10*(7), 1497–1516.
3. https://www.gartner.com/newsroom/id/2970017 (Accessed on March 15, 2015).
4. Xia, F., Yang, L. T., Wang, L., & Vinel, A. (2012). Internet of things. *International Journal of Communication Systems, 25*(9), 1101.
5. L. Atzori, A. Iera, G. Morabito, The Internet of Things: a survey, Comput. Netw. 54 (15) (2010) 2787–2805.
6. The Internet of Things, ITU Internet Reports, 2005. https://www.itu.int/net/wsis/tunis/newsroom/stats/The-Internet-of-Things-2005.pdf.
7. Grieco, L. A., Rizzo, A., Colucci, S., Sicari, S., Piro, G., Di Paola, D., & Boggia, G. (2014). IoT-aided robotics applications: Technological implications, target domains and open issues. *Computer Communications, 54*, 32–47.
8. M. Waibel, M. Beetz, J. Civera, R. D'Andrea, J. Elfring, D. GalvezLopez, K. Haussermann, R. Janssen, J. Montiel, A. Perzylo, B. Schiele, M. Tenorth, O. Zweigle, R. D. Molengraft, Roboearth, IEEE Robotics & Automation Magazine 18 (2) (2011) 69–82.
9. Nayyar, A., & Puri, V. (2016, March). A review of Arduino board's, Lilypad's & Arduino shields. In *Computing for Sustainable Global Development (INDIACom), 2016 3rd International Conference on* (pp. 1485–1492). IEEE.
10. Nayyar, A. (2016). An Encyclopedia Coverage of Compiler's, Programmer's & Simulator's for 8051, PIC, AVR, ARM, Arduino Embedded Technologies. *International Journal of Reconfigurable and Embedded Systems (IJRES), 5*(1).
11. https://www.arduino.cc/en/Main/ArduinoBoardNano (Accessed on March 15, 2017).
12. https://www.adafruit.com/product/386 (Accessed on March 15, 2017).
13. https://www.sparkfun.com/products/9405 (Accessed on March 15, 2017).
14. Nayyar, A., Puri, V., & Le, D. N. (2016). A Comprehensive Review of Semiconductor-Type Gas Sensors for Environmental Monitoring. *Review of Computer Engineering Research, 3* (3), 55–64.

15. https://www.sparkfun.com/products/8880 (Accessed on March 15, 2017).
16. https://www.sparkfun.com/products/9269 (Accessed on March 15, 2017).
17. https://www.sparkfun.com/products/13678 (Accessed on March 15, 2016).
18. http://www.etechpk.net/shop/multicopter-accesories/flysky-fs-ct6b-2-4ghz-6-channel-transmitter-receiver-radio-system/ (Accessed on March 15, 2017).
19. https://www.parallax.com/sites/default/files/downloads/400-00052-Omron-12V-Relay-Datasheet.pdf (Accessed on March 15, 2017).
20. Nayyar, A., & Puri, V. (2016). Smart farming. In *Communication and Computing Systems* (pp. 673–680). CRC Press.
21. Nayyar, A., & Puri, V. (2016). Data Glove: Internet of Things (IoT) Based Smart Wearable Gadget. *British Journal of Mathematics & Computer Science, 15*(5).
22. https://www.arduino.cc/en/Main/Software (Accessed on March 15, 2017).
23. https://thingspeak.com/ (Accessed on March 15, 2017).
24. Nayyar, A., & Puri, V. (2015). A review of Beaglebone Smart Board's-A Linux/Android powered low cost development platform based on ARM technology. In *Future Generation Communication and Networking (FGCN), 2015 9th International Conference on* (pp. 55–63). IEEE.
25. Nayyar, A., & Puri, V. (2015). Raspberry Pi- A Small, Powerful, Cost Effective and Efficient Form Factor Computer: A Review. *International Journal of Advanced Research in Computer Science and Software Engineering, 5*(12), 720–737.

Agile Team Assembling Supporting High Cooperative Performance

Sylvia Encheva(✉)

Western Norway University of Applied Sciences, Bjørnsonsg. 45, 5528 Haugesund, Norway
sbe@hvl.no

Abstract. Modern organizations are repeatedly immersed in rapidly changing situations requiring a large variety of skills and expertise which more often than not implies bringing together new teams. In this paper, we present an approach that can significantly speed up team assembling processes where each required skill is possessed by more than one team member. Another important feature is providing automated information about members of other teams who possess specific currently requested skills when there is a need for it.

Keywords: Teams · Cooperation · Bipartite graphs

1 Introduction

Managers in service-providing organizations often face situations requiring urgent reassignment of employees to new tasks. Finding the right people at the right time can be a lot on the manager's plate while being in charge of a large number of employees or, e.g., being new in the job. Thus, the need for a quick screening procedure of up-to-date stuff skills and expertises is obvious.

Bipartite graphs were previously employed for solving assignment and transportation problems [2,16]. In most of the previous work, the main focus was on looking for an answer to problems where no changes were assumed. Thus, the well-known personal assignment problem referred to also as the Hungarian method is used to find an assignment of jobs to workers at a minimum cost [12]. Here, we are concerned with providing an approach that can handle dynamic changes while reassigning employees to new projects. Visualization of staff members grouped according to skills they possess is further illustrated based on n-dimensional hypercubes. Bipartite graphs can be factored as a product of two graphs one of which is a hypercube and the other is a complete graph [1].

V. Bhateja et al. (eds.), *Information Systems Design and Intelligent Applications*, Advances in Intelligent Systems and Computing 672,
https://doi.org/10.1007/978-981-10-7512-4_24

2 Related Work

Effective skills management is of vital importance to any organization. It is pointed in [13] that 'one of the big disasters in any company would be placing people in wrong roles and making failures of successful people.' The significance of assigning the right employees to the right projects as well as the need of up-to-date information about workers skills is discussed in [6,14].

The n-dimensional cube Q_n is the graph for which $V(Q_n)$ is the set of all 0–1 sequences of length n, with two vertices being adjacent if and only if they differ in exactly one position [1]. The authors investigate the problem of factoring Q_n as $Q_n = H \times K_2$. According to [5,10], Q_n can be presented as a direct product $Q_n = H \times K_2$ where H is a hypercube of dimension $n-1$ and K_2 is a complete graph. A bipartite graph is a graph whose vertices can be divided into two disjoint sets where every edge connects a vertex in one set to a vertex in the other set, [7]. For further reading on bipartite graphs, we refer to, e.g., [4,8,9].

Bipartite graphs have special value for running of iterative enhancement processes [3]. They have been employed in algorithm design [11,15] and used in social network analysis [17].

3 Team Assembling

Assume an organization where employees' skills or areas of expertise are used to gather teams contributing to high cooperative performance. The question we address here is how to organize the available and eventually non-available but possibly needed skills in way that supports efficient resource management.

In this scenario skills in an organization, or in each of its sections, are summarized in eight sets denoted by 'a, b, c, d, e, f, g, h.' A set can contain a single skill, several skills or being currently empty. These eight sets are further arranged in sixteen quadruples where:

- the first position is occupied by either a or e
- the second position is occupied by either b or f
- the third position is occupied by either c or g
- the forth position is occupied by either d or h

and any two quadruples differ in at least one position. The quadruples are $\{e,b,c,h\}$, $\{a,b,c,h\}$, $\{a,f,c,h\}$, $\{e,f,c,h\}$, $\{e,b,c,d\}$, $\{a,b,c,d\}$, $\{a,f,c,d\}$, $\{e,f,c,d\}$, $\{e,b,g,d\}$, $\{a,b,g,d\}$, $\{a,f,g,d\}$, $\{e,f,g,d\}$, $\{a,b,g,h\}$, $\{a,f,g,h\}$, $\{e,f,g,h\}$, $\{e,b,g,h\}$.

Each of these quadruples is, further on, associated with a vertex of the graph shown in Fig. 1 originally presented in [1] as graph H. Keep in mind that to the rest of this paper we are only interested in efficient representation of people and their skills and therefore not restricted by all requirements in graph theory.

A vertex belonging to any of the four squares

- $\{e,b,c,h\}$, $\{a,b,c,h\}$, $\{a,f,c,h\}$, $\{e,f,c,h\}$

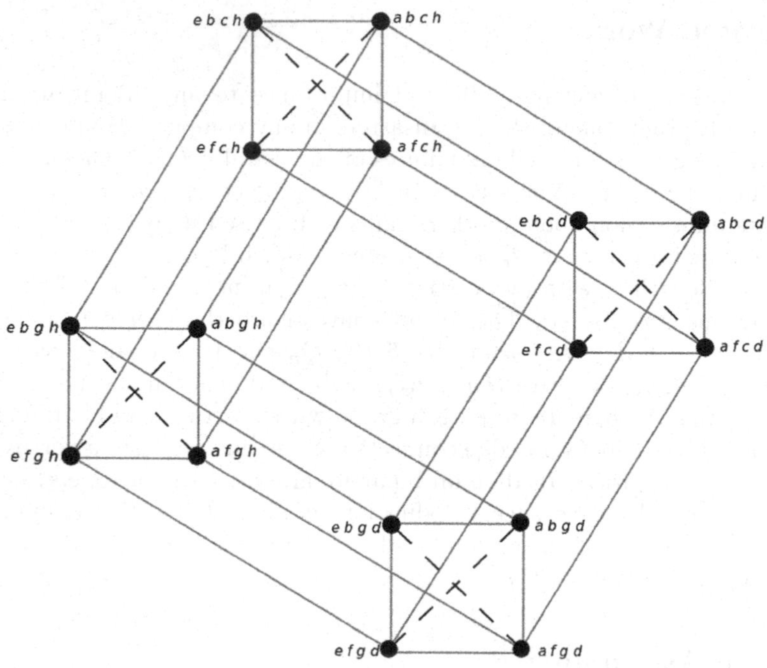

Fig. 1. Skills representation

- $\{e, b, c, d\}$, $\{a, b, c, d\}$, $\{a, f, c, d\}$, $\{e, f, c, d\}$
- $\{e, b, g, d\}$, $\{a, b, g, d\}$, $\{a, f, g, d\}$, $\{e, f, g, d\}$
- $\{e, b, g, h\}$, $\{a, b, g, h\}$, $\{a, f, g, h\}$, $\{e, f, g, h\}$,

is connected to any of the other three vertices in the same square and to two vertices belonging to two other squares. Thus, two vertices connected by an edge which is not a diagonal in a square differ in one position only, while two vertices connected by an edge which is a diagonal in a square differ in two positions. Interpretation: Two persons placed in two vertices which differ in one position only can assist or substitute each other in case any of the three skills they share are required. If however they are placed in two vertices which differ in two positions, they can assist or substitute each other in case any of the two skills they share are required.

There are two diagonals in each of the above-mentioned four squares in Fig. 1. This indicates that smaller teams consisting of two groups associated with each square can be assembled when two sets of skills are required only. A dashed line is used to emphasize connections between vertices which differ in exactly two positions.

The representation in Fig. 1 allows assembling of different teams. We are now looking at the cases where all members of a team share exactly two of the eight skills 'a, b, c, d, e, f, g, h.' This is visually illustrated by parallelograms with vertices associated with quadruples sharing two skills, e.g., quadruples $\{a, b, c, h\}$,

$\{a, b, c, d\}$, $\{a, b, g, d\}$, $\{a, b, g, h\}$ share skills $\{a, b\}$ (highlighted in blue) while quadruples $\{e, b, c, d\}$, $\{e, f, c, d\}$, $\{e, f, g, d\}$, $\{e, b, g, d\}$ share skills $\{e, d\}$ (highlighted in read), Fig. 2. These are only two out the 28 parallelograms in Fig. 1. This demonstrates excellent potentials for interactivity, i.e., whenever a new team is needed in a way that its members share specific skills a manager can straight away call on the corresponding people, placed in the vertices of above-described parallelograms.

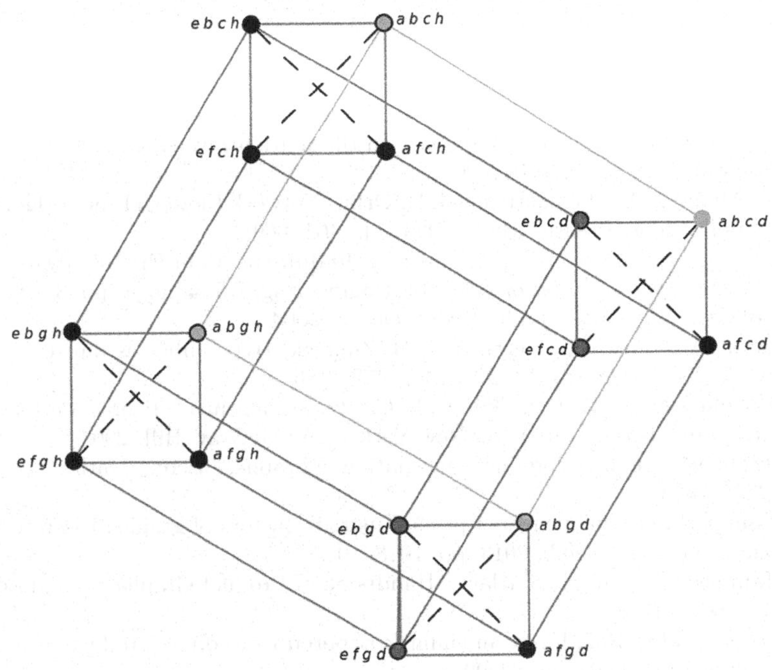

Fig. 2. Quadruples sharing skills a, b and e, d, respectively

Such representation of available skills and areas of expertise can be used to monitor market demands of a particular work force and subsequently extend the competence level of current staff members and or employ new appropriate personel. Visibility of available and non-available skills can be emphasized by different coloring of vertices and edges as, e.g., in Fig. 2.

Remark: If we think about a quadruple as a vector with four coordinates with values 0 or 1, then the binary sum of two such vectors will vary from 1 to 4. Such types of presentations turn out to be very handy in the process of programming an app using the above-described approach for reassignment of employees.

4 Conclusion

The main idea in this work is to provide assistance to managers and project leaders in cases when they quickly have to assemble a team with necessary skills and competences for completing a new task or being able to find a substitute for a member who is currently involved in other projects or for some other reason is unavailable. In addition, the presented approach can be used for developing a mobile application facilitating efficient restructuring of a workforce.

References

1. G. Abay-Asmerom, R. H. Hammack, C. E. Larson, and D. T. Taylor, Direct Product Factorization of Bipartite Graphs with Bipartition-reversing Involutions, SIAM J. Discrete Math., 23(4), 2042–2052.
2. R. K. Ahuja, T. L. Magnanti, and J. B. Orlin, Network Flows: Theory, Algorithms, and Applications, Prentice Hall, (1993) pp. 461–509.
3. T. S. Asgar and T. M. King, Formalizing Requirements in ERP Software Implementations, Lecture Notes on Software Engineering, vol. 4, no. 1, pp. 34–40, 2016.
4. B. Bollobas: Extremal Graph Theory, Dover, 2004.
5. B. Brear, W. Imrich, S. Klavar, and B. Zmazek, Hypercubes as direct products, SIAM J. Discrete Math., 18 (2005), pp. 778–786.
6. L. R. Gomez-Mejia, D. B. Balkin, R. L. Cardy: Management: People, Performance, Change, 3rd edition. New York, New York USA: McGraw-Hill, 2008.
7. R. Hammack, On direct product cancellation of graphs, Discrete Math., 309 (2009), pp. 2538–2543.
8. R. Hammack, On uniqueness of prime bipartite factors of graphs, Discrete Mathematics, Vol. 313, Issue 9, 2013, pp. 1018–1027.
9. R. Hammack, W. Imrich, S. Klavar, Handbook of Product Graphs, Second Edition, CRC Press, 2011.
10. W. Imrich and D. Rall, Finite and infinite hypercubes as direct products, Australas. J. Combin., 36 (2006), pp. 83–90.
11. J. Kleinberg, and E. Tardos, Algorithm Design, Addison Wesley, (2006), pp. 94–97.
12. H. W. Kuhn, Variants of the Hungarian method for assignment problems, Naval Research Logistics Quarterly, 3, 1956, pp. 253–258.
13. J. Ramu: The Changing Talentscape, (2014).
14. G. Ritchie, T. Collins, and A. Andrews: Skills Management: Lessons Learned from the Real World, (2010).
15. R. Sedgewick, Algorithms in Java, Part 5: Graph Algorithms (3rd ed.), Addison Wesley, (2004), pp. 109–111.
16. J. Xu, Theory and Application of Graphs, Springer, 2003.
17. S. Wasserman and K. Faust, Social Network Analysis: Methods and Applications, Structural Analysis in the Social Sciences, 8, Cambridge University Press, 1994, pp. 299–302.

Accelerating Establishment of a Balanced Structure in New Organizations

Sylvia Encheva(✉)

Western Norway University of Applied Sciences, Bjørnsonsg. 45, 5528 Haugesund, Norway
sbe@hvl.no

Abstract. When several organizations agree upon establishment of a new larger one, they have to solve a number of organizational problems. Several of them are related to assembling administrative and academic units across campuses. Some difficulties, arising while aiming at a flexible structure ready to support tomorrow's needs, can be avoided by providing visual representation of employees' skills and expertise.

Keywords: Merger · Distribution of skills · Organizational structure

1 Introduction

Merging educational organizations struggles quite a lot when it comes to creation of a new institute structure. New policies are to be followed, former educational and research groups have to be considered, geographical distances cannot be ignored regardless current digitalization, and somewhat balanced distribution of skills and expertise ought to be obtained. In this paper, we focus on how to provide visual representation of employees' skills and expertise. This type of services is really helpful when new subjects, study lines and programmes on both undergraduate and graduate levels are to developed. It is definitely not sufficient to know what single individuals can; it is necessary to know which skills and expertise they share. This way several of them can join forces completing a larger task or they can replace each other when necessary. In order to facilitate processes within human resource management which most of the time rely on incomplete data, we suggest involvement of Grey theory [7–10].

V. Bhateja et al. (eds.), *Information Systems Design and Intelligent Applications,* Advances in Intelligent Systems and Computing 672,
https://doi.org/10.1007/978-981-10-7512-4_25

2 Background

Definition 1. Let X be the universal set. Then a grey set G of X is defined by its two mappings $\overline{\mu}_G(x)$ and $\underline{\mu}_G(x)$.

$$\begin{cases} \overline{\mu}_G(x) : x \to [0,1] \\ \underline{\mu}_G(x) : x \to [0,1] \end{cases}$$

$\overline{\mu}_G(x) \geq \underline{\mu}_G(x), x \in X, X = R$, $\overline{\mu}_G(x)$ and $\underline{\mu}_G(x)$ are the upper and lower membership functions in G, respectively.

When $\overline{\mu}_G(x) = \underline{\mu}_G(x)$, the grey set G becomes a fuzzy set. It shows that grey theory considers the condition of the fuzziness and can deal flexibly with the fuzziness situation.

Priority ranking and consensus formation are considered in [5], while aggregation of numerical readings can be found in [6,11]. If factors like importances and frequencies are to be involved in the decision processes, one may use the method of incorporating quantitative weights into aggregation [4].

Graph theory is very well presented in [1–3].

3 Human Resources

Administrative and academic stuffs possess a variety of skills that may not be completely known in a new organization. In this section, we attempt to summarize such information employing graphs. People are affiliated with graph's vertices while an edge represents a skill. Furthermore, it is possible to assign a grey number to each employee reflecting previous experience and current performance. This can be used when assembling new teams.

For the sake of clarity, we assume that there is one person associated with a vertex in any of the graphs considered in this article. In real life, however, there is no limit on the number of people associated with a vertex.

The Brinkmann graph in Fig. 1 illustrates how to organize 21 experts in seven (not disjoint) groups:

- (E1, E8, E9, E17, E21),
- (E2, E9, E10, E15, E18),
- (E3, E10, E11, E16, E19),
- (E4, E11, E12, E17, E20),
- (E5, E12, E13, E18, E21),
- (E6, E13, E14, E15, E19),
- (E7, E8, E14, E16, E21)

with three levels of expertise. Each group has one member placed on level 1 (the most outer circle composed of experts E1, E2, E3, E4, E5, E6, E7). Thus, level 1 can represent group leaders. A person on level 1 shares two skills with two other persons on level 1 (both belong to two different groups) and two skills with two other persons from its own group. Each group has two members placed on level 2 and two members placed on level 3. All members appearing on level 2 and level 3 belong to two groups.

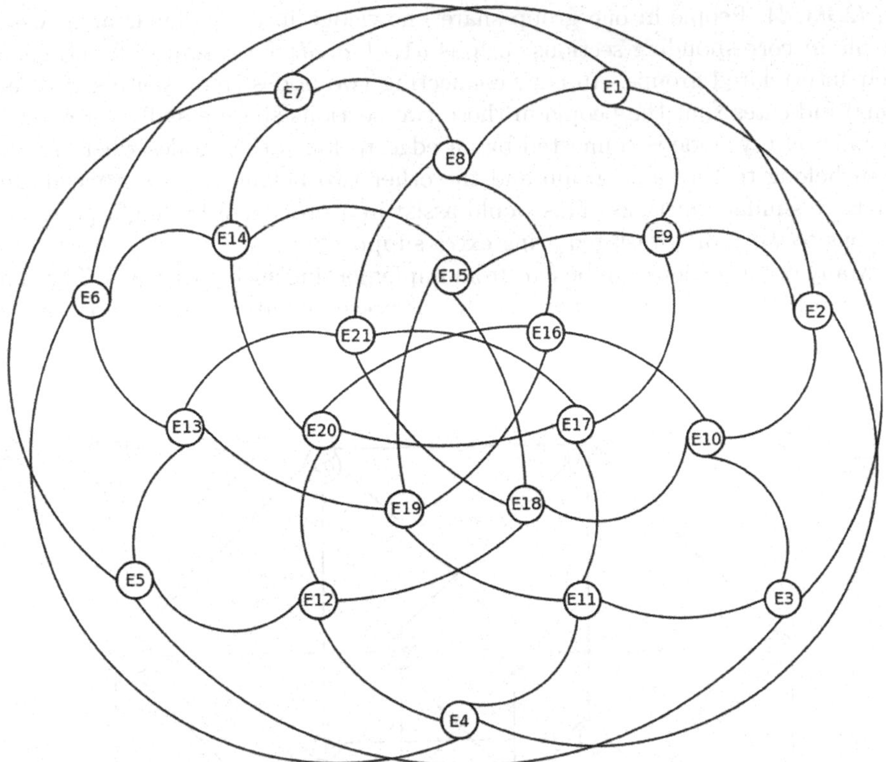

Fig. 1. Brinkmann graph

- An expert on level 2 shares two skills with experts on level 1 and another two skills with experts on level 3.
- An expert on level 3 shares two skills with experts on level 2 and two skills with experts on level 3.

Experts on level 3 belonging to one group share one skill. Experts on level 2 are the only ones that do not share skills among themselves, i.e. they cannot replace each other but they possess a combination of skills characteristic of both level 1 and level 2. Different groups can be within one organization or with several organizations.

Remark: Grey correlation analysis can be applied for predicting which option is the most adequate for solving current problems. Grey numbers can be used to facilitate decision processes related to choosing an option.

Below we focus on integration of four existing groups of employees into one unit, Fig. 2. The latter is known as a Hadamard graph [1]. Each of these groups is further on presented by four sections. Thus, the first group consists of sections $a1, a2, a3, a4$, the second group consists of sections $b1, b2, b3, b4$, the third group consists of sections $c1, c2, c3, c4$, and the fourth group consists of sections

$d1, d2, d3, d4$. People in one group share knowledge in a particular area, while people in corresponding sections such as $a1, b1, c1, d1$ have similar functions in their interrelated groups. An edge connecting two nodes (representing two sections) indicates that the people in those two sections share a skill or a number of skills. Every node is connected by an edge to four other nodes where two of them belong to the same group and the other two belong to two other groups but have similar functions. This would assist in a process of finding a person or persons to work on a task requiring extra support.

Many organizations can benefit from implementing such structures. The similarity in forming sections or departments as presented above allows more or less equal distribution of man power and easy visualization of current status and opens for quick restructuring if a situation demands it.

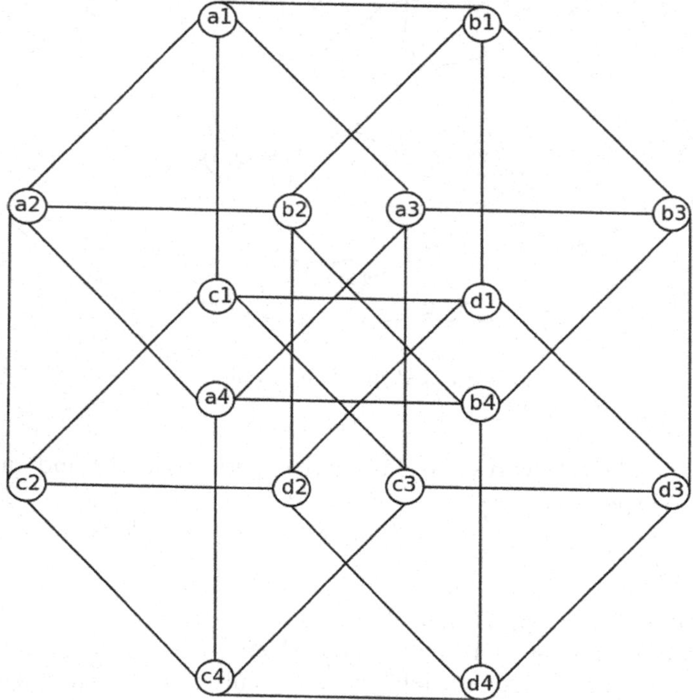

Fig. 2. Sixteen sections

4 Conclusion

Digitalization of services in educational sector has a long history when it comes to teaching and learning. In this work, we address the problem of organizing effective and flexible stuff member units that can support current developments as well as new undertakings. Graph structures have been employed to assure balanced distribution of workforce and provide visual representation of available expertise.

References

1. Bollobas, B., Extremal Graph Theory, Dover, 2004.
2. Brinkmann G., Generating Cubic Graphs Faster Than Isomorphism Checking, Preprint 92-047 SFB 343. Bielefeld, Germany: University of Bielefeld, 1992.
3. Brinkmann G. and Meringer M., The Smallest 4-Regular 4-Chromatic Graphs with Girth 5, Graph Theory Notes of New York 32, 40–41, 1997.
4. Calvo, T., Mesiar, R., Yager, R.: Quantitative weights and aggregation. IEEE T. Fuzzy Systems 12(1), (2004), 62–69.
5. Cook, W. D. and Seiford M. L.: Priority Ranking and Consensus Formation, Management Science, 24(16), (1978), 1721–1732.
6. Dubois, D. and Prade, H., A review of fuzzy sets aggregation connectives, Information Science, 36, (1985) 85–121.
7. J.L Deng, *Control problems of grey systems*, System and control letters, 5, 1982, 288–294.
8. J.L. Deng, *Introduction to grey system theory*, Journal of grey systems, 1, 1989, 1–24.
9. Y.C. Hu, *Grey relational analysis and radical basis function network for determining costs in learning sequences*, Applied mathematics and computation, 184, 2007, 291–299.
10. S. Liu, Y. Lin, *Grey information*, Springer, 2006.
11. Yager, R. R. and Rybalov, A.: Noncommutative self-identity aggregation. Fuzzy Sets and Systems, 85, (1997), 73–82.

Interpretations of Relationships Among Knowledge Assessment Tests Outcomes

Sylvia Encheva(✉)

Western Norway University of Applied Sciences, Bjørnsonsg. 45, 5528 Haugesund, Norway
sbe@hvl.no

Abstract. Recently developed knowledge assessment tests provide students and educators with information about degrees to which certain skills are mastered, terms and concepts are understood, and to which extend abilities to solve predefined problems efficiently and in a timely fashion without difficulty are demonstrated. In case of partially correct or incorrect answers, students are suggested appropriate theory, examples, and possibilities to take new tests addressing those specific issues. Provided help is primarily related to a problem a student has not solved, a question not being answered correctly, or an answer has been omitted. In this work, we attempt to unveil hidden correlations among correct and wrong answers from students in a test.

Keywords: Tests · Knowledge · Dependencies

1 Introduction

Digitalization of knowledge assessment is an ongoing process. Earlier works were dedicated to electronic delivering av tests for both formal and self-assessment purposes as well as offering automated and to some extend personalized help in case of failure [3,7,16].

After being taken, most tests provide students and educators with information about degrees to which certain skills are learnt, terms and concepts are understood, and abilities to solve predefined problems are exhibited. In case of partially correct or incorrect answers, they are suggested appropriate theory, examples, and eventually are offered to take a new test with similar questions. Such help is primarily related to the problems a student has not solved or respectively to the questions she has not answered correctly or simply has omitted to answer. This work is an attempt to use a somewhat different approach to finding explanations for students inabilities to provide correct responses to test questions.

V. Bhateja et al. (eds.), *Information Systems Design and Intelligent Applications*, Advances in Intelligent Systems and Computing 672,
https://doi.org/10.1007/978-981-10-7512-4_26

2 Related Work

The n-dimensional cube Q_n is the graph for which $V(Q_n)$ is the set of all 0–1 sequences of length n, with two vertices being adjacent if and only if they differ in exactly one position [1]. The authors investigate the problem of factoring Q_n as $Q_n = H \times K_2$. According to [6,14], Q_n can be presented as a direct product $Q_n = H \times K_2$ where H is a hypercube of dimension $n-1$ and K_2 is a complete graph. A bipartite graph is a graph whose vertices can be divided into two disjoint sets where every edge connects a vertex in one set to a vertex in the other set [11]. Bipartite graphs can be factored as a product of two graphs one of which is a hypercube and the other is a complete graph [1]. For further reading on bipartite graphs, we refer to f. ex. [5,12,13]. Bipartite graphs have special value for running of iterative enhancement processes [4]. They have been employed in algorithm design [15,18] and used in social network analysis [22]. Bipartite graphs were previously employed for solving assignment and transportation problems [2,20].

Let P be a non-empty ordered set. If $sup\{x, y\}$ and $inf\{x, y\}$ exist for all $x, y \in P$, then P is called a *lattice* [10]. In a lattice illustrating partial ordering of knowledge values, the logical conjunction is identified with the meet operation and the logical disjunction with the join operation.

Association rules are used to analyze data for frequent if/then patterns and to identify the most important relationships [19].

Knowledge assessment has been of interest to many authors [8,17,21]. Learning preferences describe person's predispositions in receiving, processing, and delivering information. Learning styles are considered to be one of or a mixture of sensory, intuitive, visual, verbal, active, reflective, sequential, and global type [9].

3 Interpretation of Tests Outcomes

One of the foremost tasks of knowledge assessment is discovering what is causing students inability to provide correct answers to test questions. Thus, if all or the majority of wrong answers indicate misunderstanding, misconception, and misinterpretation of terms, then a revision of corresponding learning materials seems to be an appropriate action. When wrong answers are submitted in a following test on the same topic, it can be an idea to consider students learning preferences and provide them with assistance in an appropriate form. Amount of wrong answers might also be reduced by reformulation of answer alternatives, since if the latter are unclear or ambiguous, this can mislead students.

We are now looking at a test with four questions where the correct answers are denoted with 'e, f, g, l' and wrong answers are denoted with 'p, q, r, s'. The latter can be due to misconception, inability to apply a combination of skills, insufficient factual or explicit knowledge, or both, etc. Note that all concerns about protected exam environment, permutation of questions, number of correct answers following a question, partially correct answers, and so on are not in the scope of this work, and students have not been directly involved in this study.

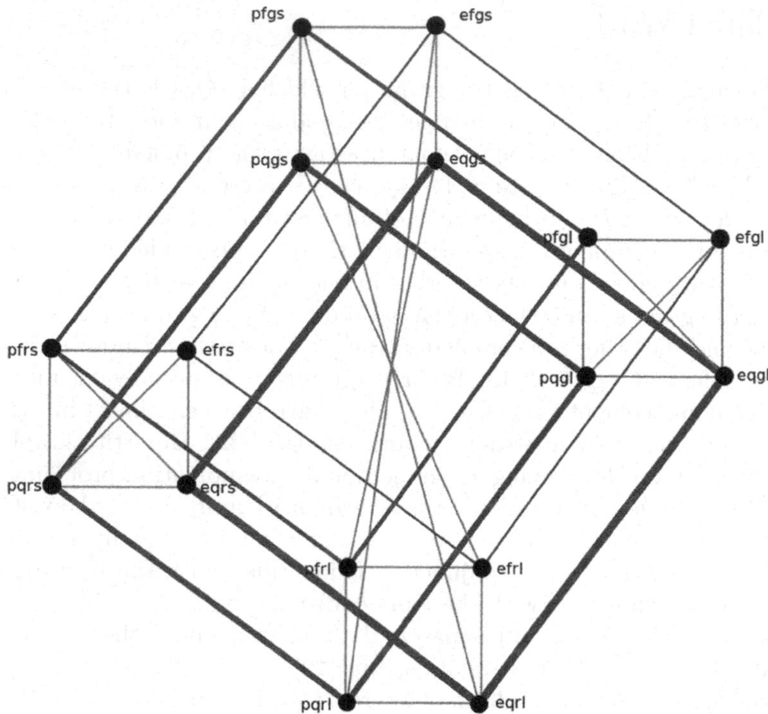

Fig. 1. Answers concerning the first and second questions

Students' responses to test questions are ordered in sixteen quadruples where the first position is reserved for an answer to the first question, the second position is reserved for an answer to the second question, and so on. This way the first position is occupied by either e or p, the second position is occupied by either f or q, the third position is occupied by either g or r, the forth position is occupied by either l or s, and any two quadruples differ in at least one position. This particular property allows use of n-dimensional cubes. A quadruple is assigned to a vertex of the graph shown in Fig. 2 originally presented in [1] as graph H.

A number of possibilities can be considered while searching for dependencies among students answers to questions in such a test.

Any of the four squares in Fig. 1 (edges are highlighted in blue with different thickness to simplify visual recognition) relates responses that coincide in the first two positions in a quadruple:

- $\{p, f, g, s\}$, $\{p, f, g, l\}$, $\{p, f, r, l\}$, $\{p, f, r, s\}$ the answers to the first question are wrong and to the second question are correct
- $\{e, f, g, s\}$, $\{e, f, g, l\}$, $\{e, f, r, l\}$, $\{e, f, r, s\}$ the answers to the first and second questions are both correct

- $\{p, q, g, s\}$, $\{p, q, g, l\}$, $\{p, q, r, l\}$, $\{p, q, r, s\}$ the answers to the first and second questions are both wrong
- $\{e, q, g, s\}$, $\{e, q, g, l\}$, $\{e, q, r, l\}$, $\{e, q, r, s\}$ the answers to the first question are correct and to the second question are wrong

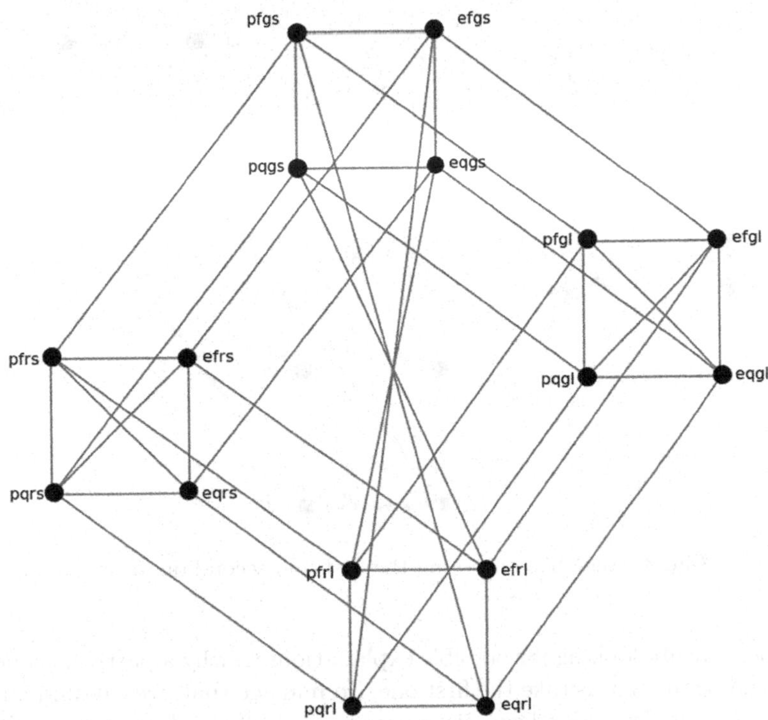

Fig. 2. Answers concerning the third and fourth questions

Any of the four squares in Fig. 2 (vertices indicated with black circles) relates responses that coincide in the last two positions in a quadruple:

- $\{p, f, g, s\}$, $\{e, f, g, s\}$, $\{e, q, g, s\}$, $\{p, q, g, s\}$ the answer to the third question is correct and to the fourth is wrong
- $\{p, f, g, l\}$, $\{e, f, g, l\}$, $\{e, q, g, l\}$, $\{p, q, g, l\}$ the answers to the third and fourth questions are both correct
- $\{p, f, r, l\}$, $\{e, f, r, l\}$, $\{e, q, r, l\}$, $\{p, q, r, l\}$ the answer to the third question is wrong and to the fourth is correct
- $\{p, f, r, s\}$, $\{e, f, r, s\}$, $\{e, q, r, s\}$, $\{p, q, r, s\}$ the answers to the third and fourth questions are both wrong.

Similar parallelograms relate responses to any other couples of questions, e.g., $\{p, f, g, s\}$, $\{p, q, g, s\}$, $\{p, q, r, s\}$, $\{p, f, r, s\}$ the answers to the first and fourth questions are both wrong, Fig. 3.

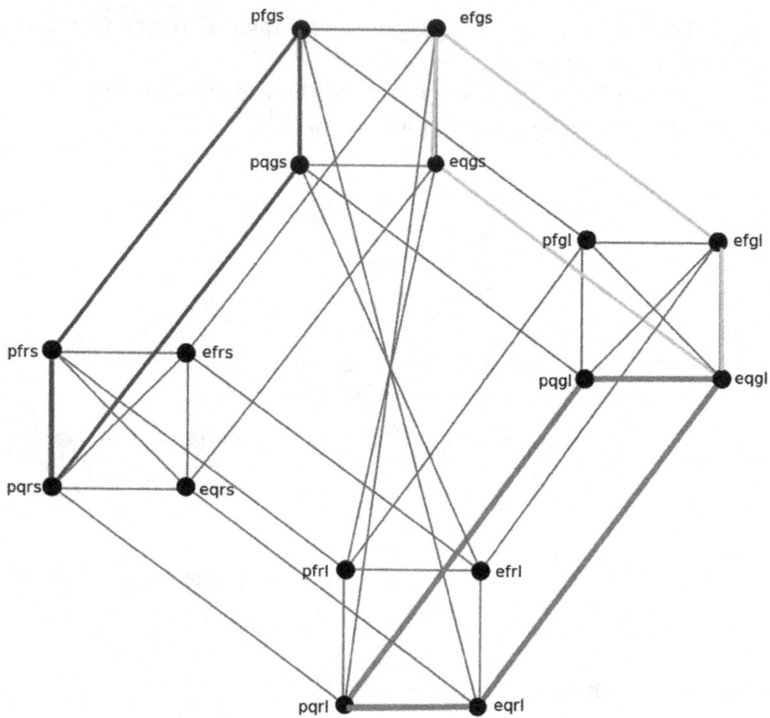

Fig. 3. Answers concerning the first and second questions

Suppose we are looking for possible explanations to why a particular question is answered wrong, let us take the first one. To manage that, we can find all cases with the highest similarity level of responses, i.e., all quadruples that differ in one position only and contain the same response to the first question, (highlited in green) in Fig. 4. In order to manage that, we should consider the three vertices connected by a single edge to the vertex we have in mind: For example, $\{p, f, g, s\}$ is connected to $\{p, f, r, s\}$, $\{p, q, g, s\}$, and $\{p, f, g, l\}$. Focusing on the groups of students placed in these four vertices may provide useful information about which types of difficulties they experience and what can be done to help them overcome these difficulties. Similar inquiries can be executed beginning at any vertex in Fig. 4.

Another gripping question is related to responses placed in quadruples that differ in exactly two positions. This will be illustrated by the following example. Let us begin with quadruple $\{e, f, r, s\}$. The quadruples that differ in exactly two positions from $\{e, f, r, s\}$ can be divided into two sets:

(1) $\{e, f, r, s\}$, $\{e, q, r, l\}$, $\{p, f, r, l\}$
(2) $\{e, f, r, s\}$, $\{e, q, g, s\}$, $\{p, f, g, s\}$.

Note that there is one more quadruple $\{e, f, g, l\}$ that differs in exactly two positions from $\{e, f, r, s\}$. This is a trivial case where all provided answers are correct and no helpful conclusions can be made.

In set (1), all students provided wrong answer to the third question and answered correctly to two of the remaining questions. In set (2), all students provided wrong answer to the fourth question and answered correctly to two of the remaining questions. Such ways of looking at students responses might throw different light on where the learning problems are. It is important to follow changes in the number of students placed in different vertices since this can give indications about what can be done to improve students knowledge and understanding of new concepts.

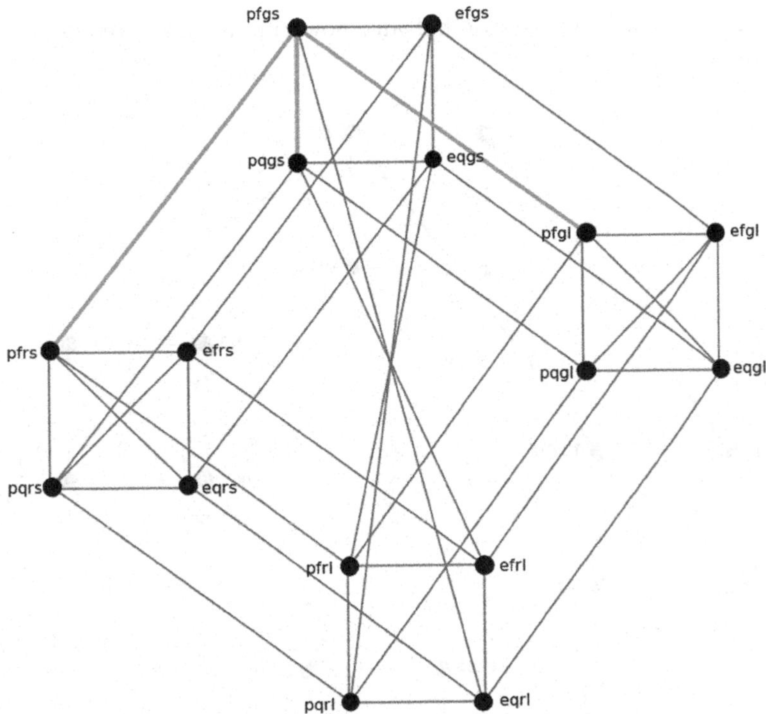

Fig. 4. Answers concerning the first and second questions

Remark: In terms of graph theory, corresponding vertices will be at distance two. Calculations modulo 2 will simplify digitalization of this approach; see Fig. 5. Correct answers are denoted by '1' and wrong answers are denoted by '0'. In these notations, the two above discussed sets are

(1) {1100}, {1001}, {0101}
(2) {1100}, {1010}, {0110}.

The paths from quadruple {1100} to any of the remaining four quadruples are highlighted in different colors, in addition to one edge with double line illustrating two paths. The n-dimensional cube Q_n structure can be used to reflect on tests outcomes with larger number of questions while any number of students can be placed in a vertex. Responses to any four of a test's questions can also be placed in quadruples and thus take advantage of the above discussed reasoning possibilities. Conclusions can be further on used to make predictions of students performance and provide automated personalized help when there is a need for it.

Students background should also have influence on the process of forming patterns. This includes level of previous education, number of years since it has been completed, relevant work experience, results from other exams and or tests, in order to provide individual help. It is important to know what they have been working with when they pass or fail and how this is connected to the other answers in a test.

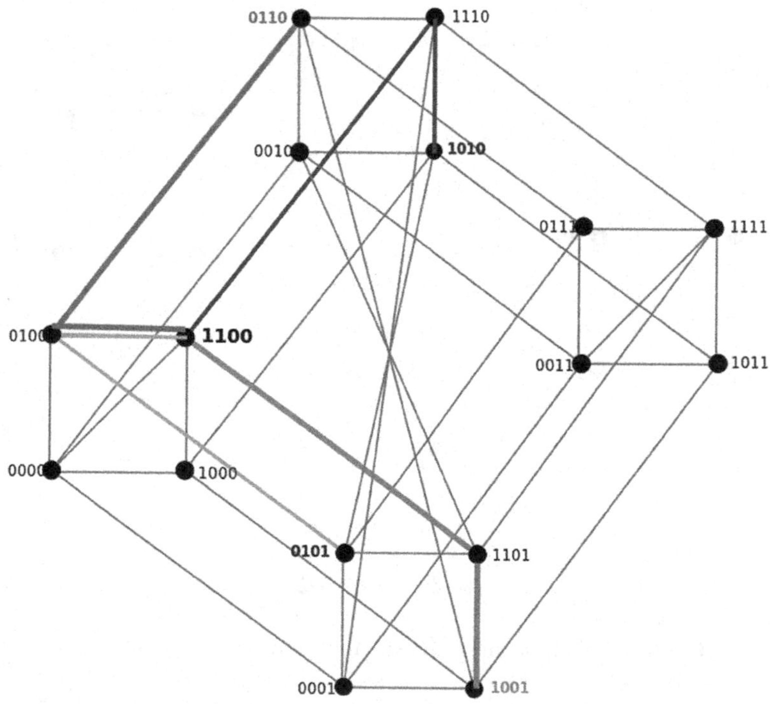

Fig. 5. Examples of quadruples at distance two

Formal concept analysis can be employed to discover more dependencies among answers and possible misconceptions of students placed in a single vertex, in a set of vertices that differ in exactly one position, and in a set of vertices that differ in exactly two positions. This way educators can consider reasonings based on smaller amount of more specific attributes as well as reasonings based on a larger amount of attributes that can point to connections that are otherwise difficult to notice. Association rules applied to the same dataset may indicate important tendencies.

4 Conclusion

Evaluation of students knowledge is an ongoing research that requires further work. Our belief is that involvement of well-established mathematical theories and structures contributes for employing a systematic approach that can be used in different areas. Visualization of tests outcomes may improve understanding of main reasons for insufficient learning since both students and educators can benefit from discovering new patterns in students responses.

References

1. G. Abay-Asmerom, R. H. Hammack, C. E. Larson, and D. T. Taylor, Direct Product Factorization of Bipartite Graphs with Bipartition-reversing Involutions, SIAM J. Discrete Math., 23(4), 2042–2052.
2. R. K. Ahuja, T. L. Magnanti, and J. B. Orlin, Network Flows: Theory, Algorithms, and Applications, Prentice Hall, (1993) pp. 461–509.
3. H. L. Andrade, and G. L. Cizek, eds., Handbook of Formative Assessment. New York, NY: Routledge, 2010.
4. T. S. Asgar and T. M. King, Formalizing Requirements in ERP Software Implementations, Lecture Notes on Software Engineering, vol. 4, no. 1, pp. 34–40, 2016.
5. B. Bollobas: Extremal Graph Theory, Dover, 2004.
6. B. Bresar, W. Imrich, S. Klavar, and B. Zmazek, Hypercubes as direct products, SIAM J. Discrete Math., 18 (2005), pp. 778–786.
7. D. C. Briggs, A. C. Alonzo, C., Schwab, and M. Wilson, Diagnostic assessment with ordered multiple choice items, Educational Assessment, vol. 11, no. 1, 2006.
8. C. Chang, and K. Tseng, Use and performances of Web-based portfolio assessment, British Journal of Educational Technology, vol. 40, No. 2, 2009.
9. Felder, R.M., and Spurlin, J.: Applications, reliability and validity of the index of learning styles. Internation Journal of Engineering Education, 21(1), 2005, pp. 103–112.
10. B. Ganter and R. Wille, *Formal Concept Analysis*, Springer, 1999.
11. R. Hammack, On direct product cancellation of graphs, Discrete Math., 309 (2009), pp. 2538–2543.
12. R. Hammack, W. Imrich, S. Klavar, Handbook of Product Graphs, Second Edition, CRC Press, 2011.
13. R. Hammack, On uniqueness of prime bipartite factors of graphs, Discrete Mathematics, Vol. 313, Issue 9, 2013, pp. 1018–1027.
14. W. Imrich and D. Rall, Finite and infinite hypercubes as direct products, Australas. J. Combin., 36 (2006), pp. 83–90.
15. J. Kleinberg, and E. Tardos, Algorithm Design, Addison Wesley, (2006), pp. 94–97.
16. J. H. McMillan, Classroom Assessment: Principles and Practices for Effective Standards-Based Instruction. New York, NY: Pearson, 2011.

17. A. Rupp, J. Templin, and R. Henson, Diagnostic Measurement: Theory, Methods, and Applications. New York, NY: Guilford Press, 2010.
18. R. Sedgewick, Algorithms in Java, Part 5: Graph Algorithms (3rd ed.), Addison Wesley, (2004), pp. 109–111.
19. P.-N. Tan, S. Michael, and V. Kumar: Introduction to Data Mining. Addison-Wesley, (2005).
20. J. Xu, Theory and Application of Graphs, Springer, 2003.
21. B. Zimmerman, and D. Schunk, eds., Handbook of Self-regulation of Learning and Performances. New York, NY: Routledge, 2011.
22. S. Wasserman and K. Faust, Social Network Analysis: Methods and Applications, Structural Analysis in the Social Sciences, 8, Cambridge University Press, (1994) pp. 299–302.

Implementing Complex Radio System in Short Time Using Cognitive Radio

Madhuri Gummineni(✉) and P. Trinatha Rao

GITAM University, Hyderabad 502329, Telangana, India
madhuri.vijay2003@gmail.com, trinath@gitam.in

Abstract. Usually, the frequency use of wireless system is characterized by static spectrum allocation. In order to cope with high data rate and high quality service, new wireless communication cognitive radio is emerged. The major issue in communication is declared by FCC is that the licensed band remains unused. These problems are addressed by the Mitola (Software radio, wireless architecture for twenty-first century [1]), and as a result, the cognitive radio technology has been proposed to improve the efficiency in spectrum utilization, by exploiting frequencies that are not been used by licensed users at a given time and location. Modulation determines the performance of communication system. The primary purpose of this paper is to experiment various modulation techniques of the wireless channel using software-defined radio. The paper emphasizes the cognitive radio and its significance compared with conventional wireless communication and also provides a comprehensive literature review of modulation techniques and its implementation using the universal software radio peripheral (USRP) device which acts as a transmitter, a receiver, and a relay.

1 Introduction

The emphasis on software-defined radio can be better understood with the application in current research. A complex polyphase structure is implemented by GNU Radio Companion (GRC) a graphical simulation tool [2], one to many communication using UDP Block which is a low-cost communication networking [3] using GNU Radio is demonstrated in real time using GRC and USRP [4], rather than theoretical concepts, implementing reconstruction filter can be analyzed using software-defined radio (SDR) [5]. All the digital and analog experiments can be performed using USRP [6]. Frequency modulation continuous wave radar is implemented using software-defined radio (SDR) for target tracking [7]. And the various applications of USRP are an APC025 compatible transmitter/receiver and decoder, RFID reader, testing equipment, a cellular GSM base station, a GPS receiver, an FM radio transmitter, and a digital television (ATSC) decoder, etc. Several applications motivate us to implement wireless communication system with new methods and new techniques and can be evaluated in short time using SDR. The research in the field of cognitive radio network (CRN) helps to demonstrate and characterize the performance of number of techniques given in the literature in the real-world scenario through USRP.

V. Bhateja et al. (eds.), *Information Systems Design and Intelligent Applications*, Advances in Intelligent Systems and Computing 672,
https://doi.org/10.1007/978-981-10-7512-4_27

1.1 Software-Defined Radio (SDR)

Instead of implementing on hardware [8], all the signal manipulations and processing works in radio communication are done in SDR. Figure 1 shows the concept of software radio (SR). SDR is equipped with computer, instead of general hardware the software environment GNU Radio upgrades the required modulation techniques, filters, coders etc, therefore an analog-to-digital converter (ADC) is required preceded by some form of RF front end. The electromagnetic wave captured by the antenna will be digitized by ADC and pass it to the baseband processor for further process, i.e., demodulation, source coding, and channel coding.

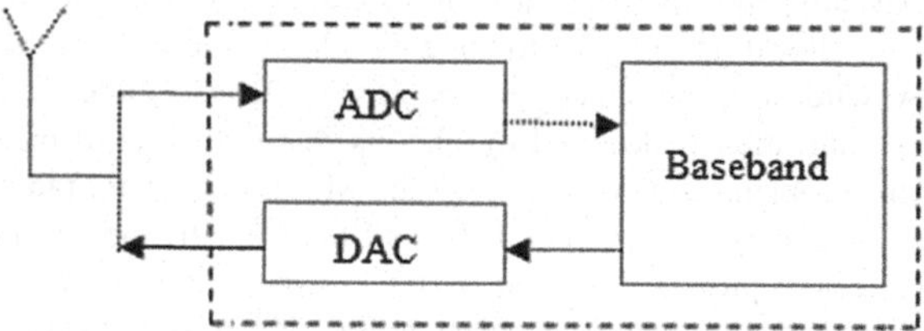

Fig. 1. Software radio (SR) block diagram

1.2 Cognitive Radio (CR)

A cognitive radio (CR) is one sort of integrated framework that incorporates the computational and physical resources for application-driven tasks. Parameter adjustments in CR mostly depend on closed-loop control between transmitter and receiver with an objective to maintain low interference from the primary user (PU). After CR devices gather their needed information from the radio environment, they can dynamically change their transmission parameters according to the sensed environment variations and achieve optimal performance. In order to boost the channel efficiency with low interference from the PU, the secondary user (SU) needs to sense the spectrum and decide the optimum transmission power value. Information transfer between controllers such as sensed information, measured QoS, and channel gain is presented in Fig. 2 using dashed line.

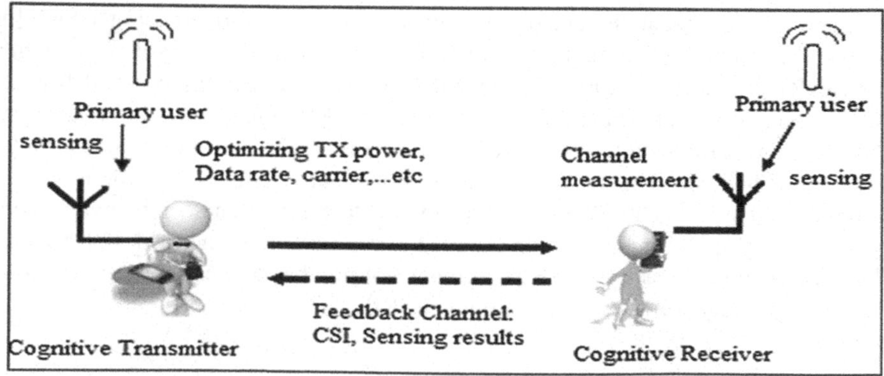

Fig. 2. Open- and closed-loop control applications in cognitive radio (CR)

The remainder of this paper is organized as follows: Section 2 presents the comparison between CR and conventional wireless communication system (CWCS). Section 3 discusses the USRPN210 and GNU Radio. The modulation technique implementation and experimental results have also been discussed in Sect. 4.

2 Comparison Between CR and CWCS

In communication system for signal transmission and reception in order to compare various modulation techniques, we usually rely on individual hardware modules for conducting the test in the laboratory. There is a need [9] of low-cost, high-speed wireless communication system that offers greater flexibility and can adapt the parameters to implement various modulation techniques as shown in Table 1.

Table 1. Comparisons between cognitive radio (CR) and conventional wireless communication system (CWCS)

Issues	Conventional wireless communication system (CWCS)	Cognitive radio (CR)
Handling different radio signals	Need to build extra circuitry per signal	No need for the extra circuitry
Time to market and development cost	Time and cost based on model	Supplements many needs
Upgradable	Not facilitated	Instantaneously parameters can be upgraded
Fading, shadowing effects on communication system	Can be reduced by handoff, multipath propagation, but leads to delay spread	Dynamic spectrum allocation that addresses these issues
Application	Intended design limits various applicabilities	Multiple applications

3 USRP and GNU Radio

A CR is a specially designed radio that is capable to modify certain operations such as frequency, bandwidth, and code. CR consists of USRP hardware platform works with the software GNU Radio [1]. A personal computer (PC) is connected through gigabit cable to the USRP. The information is passed between the signal processing blocks once the flow graph blocks are connected and initialized within the python thread. USRP is compatible with GNU Radio and is best suitable for scholarly research.

The universal software radio peripheral (USRP) products are commonly used with GNU radio software. GNU Radio is a tool kit where digital signal processing blocks are written in C++ and connected to each other with python. This makes it easy to develop more sophisticated signal processing systems, because many blocks already written and can quickly put them together to create a complete system. The functions offered by a software platform GNU radio include the following:

- FFT/IFFT blocks.
- Encoder/decoder.
- Interpolation/decimation.
- Data type conversions (float, byte, etc.).
- Filters (high pass, low pass, FIR, IIR, Hilbert, etc.).
- Mathematical operations (add, subtract, multiply, etc.).
- Modulation techniques (AM, FM, PSK, QAM, GMSK, etc.).
- Signal sources and sinks (noise, random, audio, message, file, etc.).
- Graphical sinks (constellation sink, histogram sink, waterfall sink, FFT sink, etc.), control blocks (threshold block, automatic gain control block, detect peak block).

The primary purpose of this experiment is to observe modulation techniques of the wireless channel using SDR. For this basic experiment, the following system parameters have been employed (Fig. 3 and Table 2).

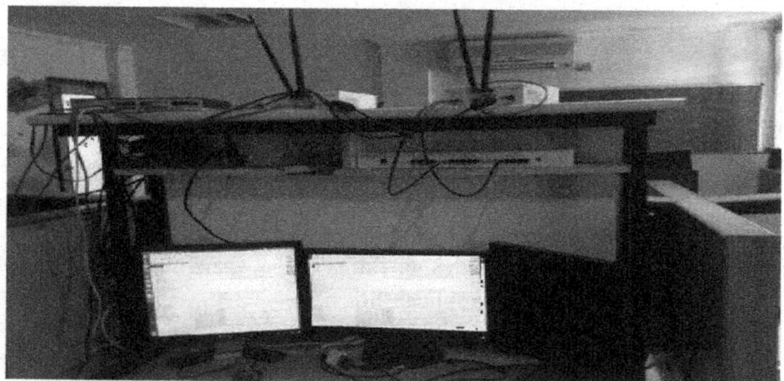

Fig. 3. Setup of transmitting and receiving USRPN210

Table 2. Simulation parameters

Spectrum	68.75–2200 MHz
Hardware	USRP Model-USRPN210
Interface	Gigabit Ethernet
Number of PU	1
Number of SU	1
Modulation schemes	QAM-PSK, ODFM, GMSK
FFT size	512
Number of transmitting antenna	1
Number of receiving antenna	1
Software	GNU RADIO (GRC)

In this work, the various parameters and modulation techniques and its dependence and influence on the cognitive radio are evaluated by using the hardware structure USRPN210 and GNU Radio as programming stages for the improvement of SDR (Figs. 4, 5, 6, 7, 8, and 9).

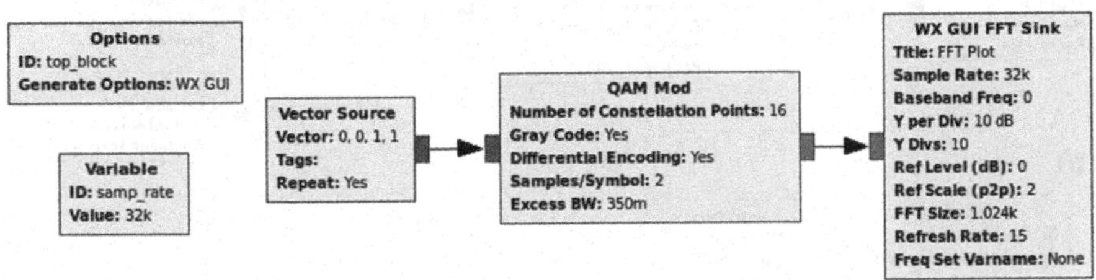

Fig. 4. GRC implementation of QAM modulation

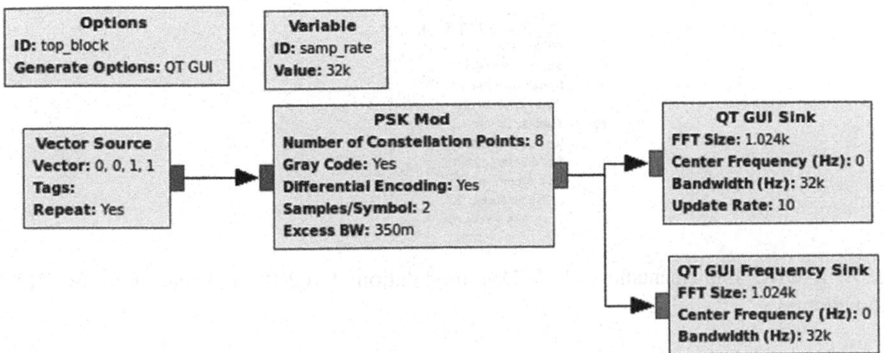

Fig. 5. GRC implementation of PSK modulation

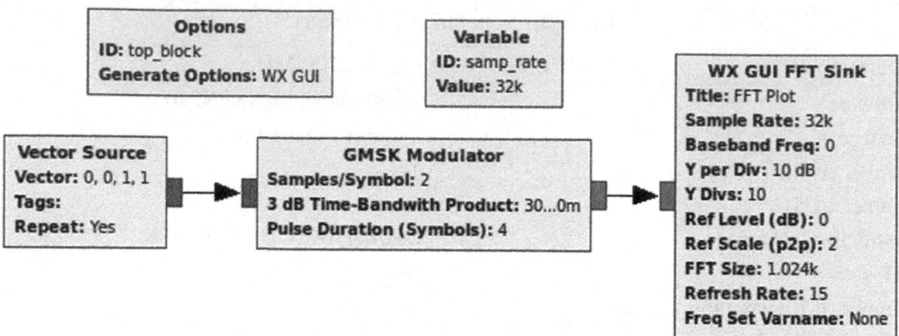

Fig. 6. GRC implementation of GMSK modulation

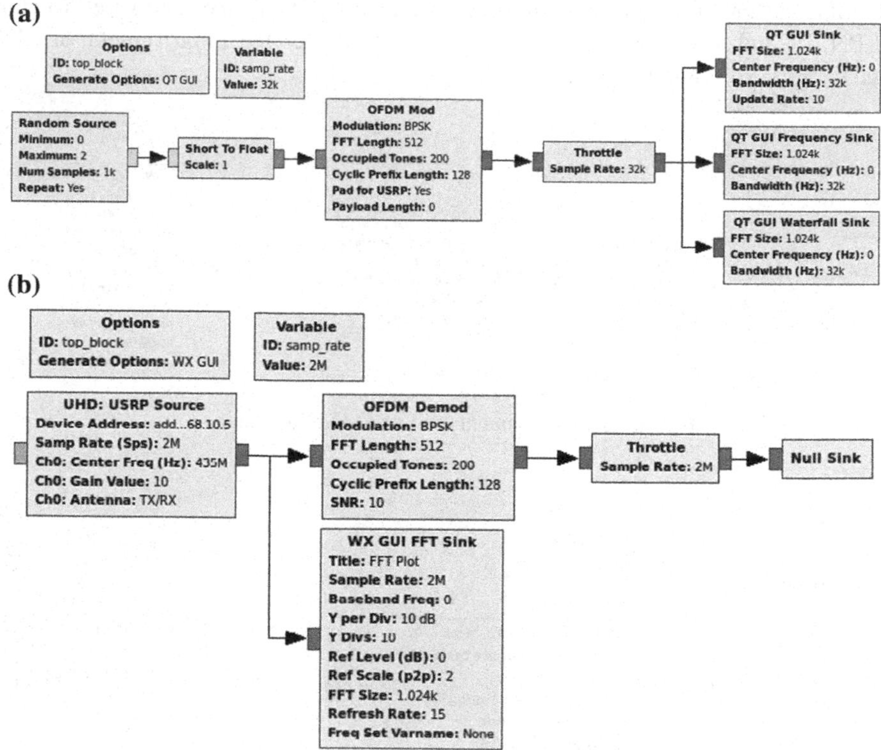

Fig. 7. **a** GRC implementation of OFDM modulation. **b** GRC implementation of OFDM demodulation

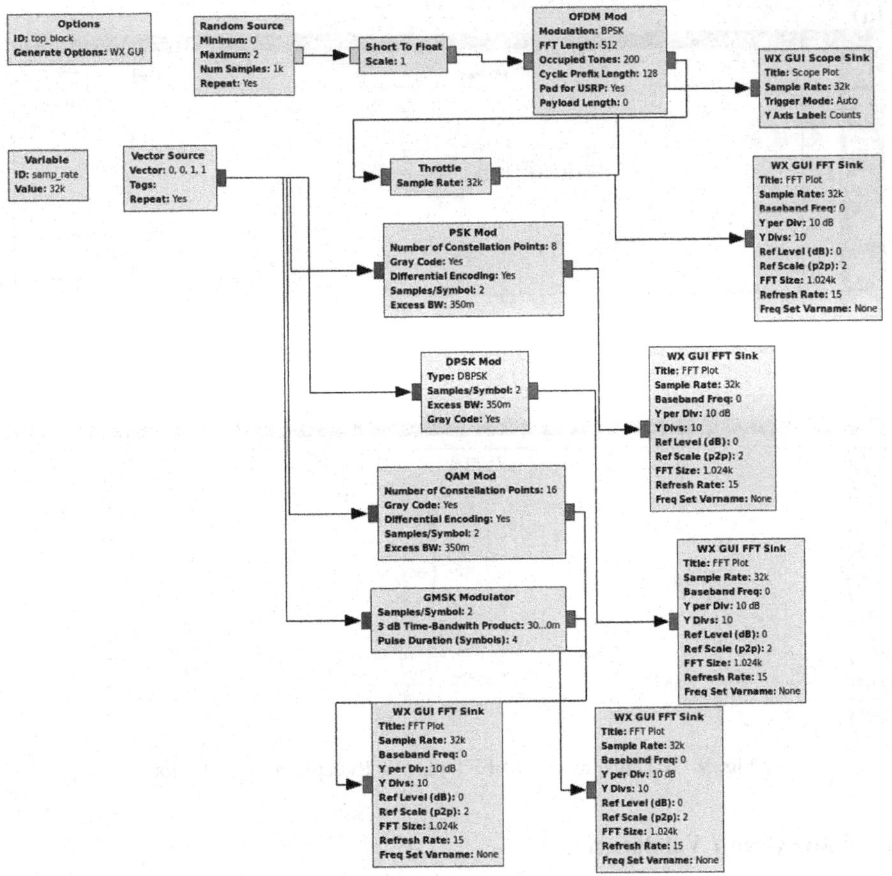

Fig. 8. GRC implementation of modulation techniques

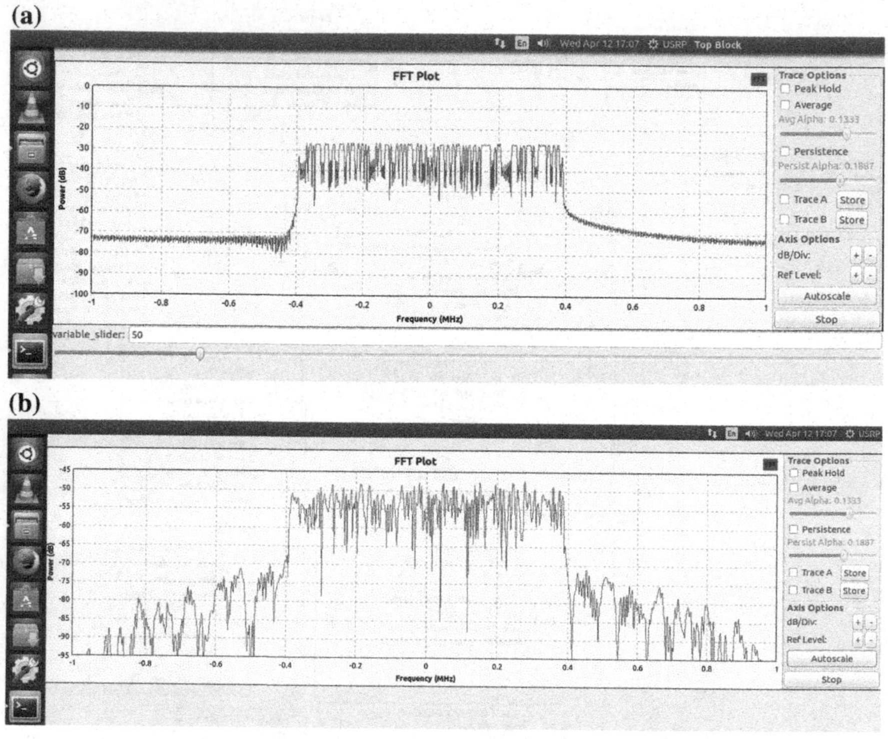

Fig. 9. **a** Transmission of FFT plot. **b** Reception of FFT plot

3.1 Flow Graph Validation

1. Connections are given between input and output sockets of the same data type. Invalid connections are highlighted in red.
2. Disconnected sockets cause their signal block to have a red label.
3. All signal block parameters must be valid. Invalid parameters have colored red labels and cause their signal block to have a red label.

4 Conclusion and Future Scope

In our paper, we deploy USRPN210 to develop some fundamental modulation techniques. We also transmitted and received by using two USRPN210, by giving device address and successfully received at 435 MHz, and its FFT signal has been plotted using GRC. This research project led us to learn the concepts of communication systems more easily with short time, to improve spectrum efficiency and signal quality in real-world noisy environment, and to characterize the performance.

Finding the interrelation between the parameters of cognitive radio adapts the system for improving the spectrum efficiency and also minimizes the interference.

Research in CRN provides a context for the evaluation of spectrum sensing algorithms, which act as the feedback for avoiding spectrum congestion.

References

1. Mitola j software radio, wireless architecture for 21st century[J], Mitola's satisfaction ISBN 0-9671233-0-5.
2. K. R. Jijeesh, V. Anupama, Chakravarthula Raghavachari, R. Gandhiraj. Study of polyphase structure made easy using GNU Radio 2014 International Conference on Green Computing Communication and Electrical Engineering (ICGCCEE) 2014, https://doi.org/10.1109/ICGCCEE.2014.6922232.
3. Sruthi M B, Abirami M, Akhil Manikkoth, Gandhiraj R, Soman K P Low cost digital transceiver design for software defined radio using RTL-SDR https://doi.org/10.1109/iMac4s.2013.
4. M. Abirami, V. Hariharan, M B Sruthi, R. Gandhiraj, K P Soman Exploiting GNU radio and USRP: An economical test bed for real time communication systems 2013 Fourth International Conference on Computing, Communications and Networking Technologies (ICCCNT) https://doi.org/10.1109/ICCCNT.2013.6726630.
5. Lekshmi Kiran. S, Parvathy. G. Mol, Hisham. P. M, Gandhiraj. R, Soman. K. P. Study of multirate systems and filter banks using GNU Radio 2015 International Conference on Communications and Signal Processing (ICCSP) https://doi.org/10.1109/ICCSP.2015.7322816.
6. Song, wenmiao ConFig Cognitive Radio using GNU Radio and USRP Microwave Antenna Propagation and EMC Technologies for Wireless Communication, 3rd IEEE International Symposium, on 27–29 Oct. 2009 IEEE DOI:978-1-4244-4076-4/09/$25.00©2009.
7. Sundaresan S, Anjana C, Tessy Zacharia, Gandhiraj R. Real time implementation of FMCW radar for target detection using GNU radio and USRP 2015 International Conference on Communications and Signal Processing (ICCSP) https://doi.org/10.1109/ICCSP.2015.7322772.
8. Mohd Adib Sarijari, Arief Marwanto, Norsheila Fisal, Sharifah Kamilah Syed Yusof, Rozeha A. Rashid, Muhammad Haikal Satria Energy Detection Sensing based on GNU Radio and USRP: An Analysis Study https://doi.org/10.1109/MICC.2009.5431525 2009 IEEE.
9. Zhe Huang, Weidong Wang, Yinghai Zhang Design and implementation of cognitive radio hardware platform based on USRP Proceedings of ICCTA 2011.

Communication in Internet of Things

Vivek Hareshbhai Puar[1], Chintan M. Bhatt[1], Duong Minh Hoang[2], and Dac-Nhuong Le[3](✉)

[1] Chandubhai S. Patel Institute of Technology, Gujarat, India
chintanbhatt.ce@charusat.ac.in
[2] Da Nang Radio and Television, Da Nang, Vietnam
[3] Haiphong University, Haiphong, Vietnam
nhuongld@hus.edu.vn

Abstract. Internet of Things (IoT) is merging all things together and make this world better and make it easy to live, actually with the help of IoT we can achieve those goals which do not till now. IoT will create a very big network of huge numbers of "*Things*" that will communicate with each other. Use of IoT tools is the simple way to make things smart. Contiki OS is one of the simulator on which one can run programs and simulate them. One of the examples given here which is border router using InstantContiki simulator. In this paper, we look forward to research how we can connect the border router to internet and how it is connected to other motes.

Keywords: Internet of Things · Communication · Simulation tools Contiki OS · Border router

1 Introduction

The Internet of Things (IoT) is a transformative development [1–3]. From its starting, the Internet was a network of networks, associating different government and academic PCs together to share information. What has changed progressively in the course of recent decades is the capacity to interface remote and versatile "*things*" or "*machines*" or "*resources*" to the Internet or corporate Intranets using remote interchanges and low-cost sensors or computers or storage. The confluence of productive remote protocols, enhanced sensors, less costly processors, and varied new businesses and built-up organizations build up the very important administration and application programming has at long last created the concept of the IoT standard [4–6]. IoT is an open and intensive

V. Bhateja et al. (eds.), *Information Systems Design and Intelligent Applications*, Advances in Intelligent Systems and Computing 672,
https://doi.org/10.1007/978-981-10-7512-4_28

system of clever things that have the capability to auto-sort out, offer data, information, and assets responding and acting in face of circumstances and changes in the earth [7,8].

We can allude IoT as developing the present Internet and giving association, communication, and between systems administration amongst gadgets and physical items, or Things [9]. The IoT worldview remains for all intents and purposes interconnected items that are identifiable and furnished with sensing, computing, and correspondence capacities. The IoT paradigm indicates the pervasive and omnipresent interconnection of billions of installed gadgets that can be interestingly recognized, confined, and conveyed [10]. The IoT is a recent correspondence worldview that imagines a not thus distant future, in every the ventures of normal life are going to be supplied with microcontrollers, handsets for computerized correspondence, and acceptable convention stack that may create them able to speak with one another and with the clients, turning into a basic part of the Web [11]. The new idea of the IoT brings an open door for the production of creative applications that coordinate the very well-known conventional computerized advances. IoT is about interfacing these self-governing gadgets to communicate without human mediation and produce incorporated information. The IoT is the thing that happens when regular normal items have interconnected microchips inside them. These microchips help monitor different items, as well as a hefty portion of these gadgets sense their encompassing and report it to different machines and in addition to the people [12].

The IoT is regards as a technology and economic wave of information industry after internet. The IoT is an astute system which interfaces all things to the Web with the end goal of trading data and conveying through the data-detecting gadgets in accordance with agreed protocols. It is associated with extension and enlargement of internet-based network, that expands the communication from human to human (H2H), human to machine (H2M) and machine to human (M2H) [13,14]. The IoT is made up of hardware and software technologies. The equipment comprises of the associated gadgets which range from basic sensors to cell phones and wearable gadgets and the networks that connects them, for example, 4G Long-Term Evolution, WiFi, and Bluetooth. Software part incorporate information storage platforms and examination programs that present data to users [15–18].

2 Tools in IoT

IoT covers a colossal range of industries nowadays and use cases that scale from a solitary compelled gadget up to gigantic cross-stage deployments of inserted innovations and cloud frameworks interfacing continuously. IoT contains various protocols at various levels like infrastructure layer contains protocols like 6LowPAN (*IPv6 over Low power Wireless Personal Area Networks*), RPL (*Routing Protocol for Low-Power and Lossy Networks*), IPv6, identification layer contains

EPC, uCODE, URIs, Transport layer contains Wifi, Bluetooth, LPWAN, Data Protocols contains MQTT, CoAP, AMQP, Node [2,19,20]. IoT contains various operating systems like Contiki OS, Tiny OS, Mantis OS, RIOT [21].

Contiki [22] is an open source, exceedingly compact, multi-entrusting working system for memory-productive networked implanted systems and remote sensor networks. Contiki has been utilized is an assortment of projects, for example, road tunnel fire monitoring, interruption discovery, wildlife monitoring, and in reconnaissance networks. Contiki is designed for microcontrollers with less amount of memory. A typical Contiki configuration is 2 kilobytes of RAM and 40 kilobytes of ROM. TinyOS (open source) is BSD-licensed working system intended for low-control remote gadgets those are utilized as a part of sensor systems, ubiquitious registering, individual zone systems, shrewd structures, and brilliant metres. Mantis OS kernel provides services like thread management, communication device management, and input–output device management. The I/O device management service provides a UNIX like uniform interface to the underlying hardware. RIOT is a well-disposed working system organized, memory-compelled systems with an attention on low-control remote Internet of Things. RIOT is a developer friendly. In RIOT OS, we can write programs in C or C++ as we are used to in other OS. It is planned especially to meet prerequisites of IoT which contains low memory impression, high-vitality productivity, ongoing abilities, a secluded and configurable correspondence stack, and backing for an extensive variety of low-power gadgets. RIOT provides a microkernel which comprises thread management, a priority-based scheduler, a powerful API for *inter-process communication* (IPC), a system timer, and mutexes. RIOT gives features like adaptable memory administration, high determination, IPv6, 6LoWPAN, UDP, RPL, CoAP, the local port permits to run RIOT as-is on Linux, BSD, and MacOS [19]. Various occurrences of RIOT running on a solitary machine can likewise be interconnected by means of a straightforward virtual Ethernet bridge.

3 Contiki OS

Contiki [18] is an open source sensor node operating system having event driven kernel. It gives dynamic loading and emptying of projects and services. Contiki is an OS similar to Microsoft Windows and Linux, except for a awfully purpose and primarily centred on things within IoT. The goal of Contiki is to satisfy requirements of the smallest devices such as smart dust. Things ought to be ready to communicate a few bits of data to one another. Contiki is an open source programming venture with the vision to make a moderate and working OS for wide deployment. Contiki gives intense low-control Internet communication. Contiki underpins completely standard IPv6 and IPv4, alongside the late low-control remote norms: 6lowpan, RPL, CoAP. With Contiki's ContikiMAC and sleepy routers, even wireless routers will be battery operated. Contiki applications are written in standard C, with the Cooja simulator. Contiki can be unreservedly utilized both as a part of business and non-business systems. Contiki is intended to keep running in little measures of memory. A regular system with full IPv6, organizing with sleepy routers and RPL routing needs under 10 kilobytes of RAM and 30 kilobytes of ROM. Contiki OS offers numerous features. It is intended for small systems, having just a couple of kilobytes of memory accessible. Contiki is along these lines exceedingly memory proficient and gives a set of mechanisms for memory distribution. It gives a full IP network stack, with standard IP protocols, for example, UDP, TCP, and HTTP, notwithstanding the extraordinary failure power benchmarks like 6lowpan, RPL, and CoAP. Contiki underpins dynamic loading and connecting of modules at run-time.

4 Border Router in Contiki OS

Here, we will see how Border router gets connected to the internet. The border router is used to interface IP network with an RPL 6LoWPAN network. Web-servers tries the border router to connect to the Internet sky motes. Before that we will discuss about mote. The node of IOT where wireless transceivers are combined with sensors is known as mote. Every mote consists of an address in order to uniquely identify it and also to find it. System that manages the address ability of the motes is called Identity of Things.

Here, we will use InstantContiki 2.7 simulator.[1] First open cooja in Instant-Contiki simulator and create new simulation as shown in Fig. 1 and name it.

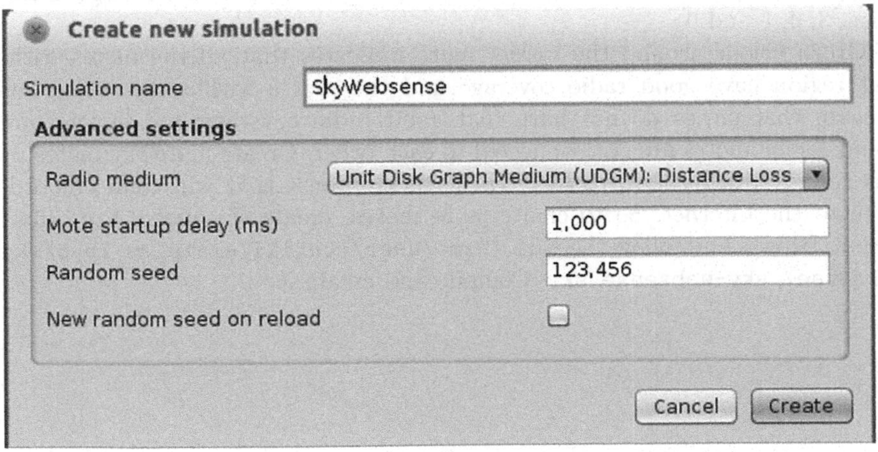

Fig. 1. Create new simulation interface

[1] www.contiki-os.org.

In second step, create a sky mote as shown in Fig. 2.

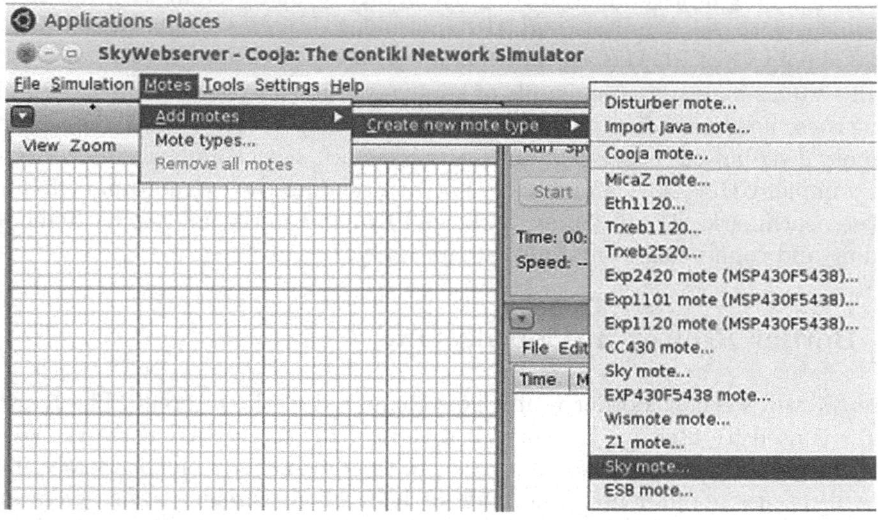

Fig. 2. Creation of sky mote

In third step, browse and follow the path `/home/user/contiki/examples/ipv6/rpl-border-router/border-router.c`. Now compile it and after it gets compiled, click on create. It will contain only one border router, so create 1 mote (Figs. 3, 4, 5 and 6).

Green colour around the border mote indicates that all the motes within that region have good radio coverage and the motes within the grey colour indicate that motes do not have that good radio coverage and beyond grey colour region motes are not connected to each other. Create more sky motes but this time it will be motes which will have Websense that will connect border mote to the Internet. So to create more motes, create sky motes from motes option. Browse and follow the path `/home/user/contiki/examples/ipv6/sky-websense/ sky-websense.sky`. Compile and create it.

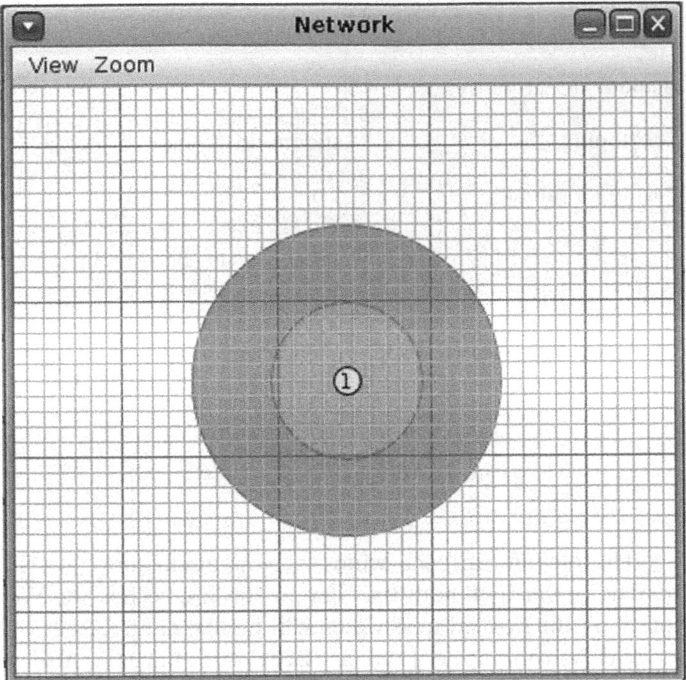

Fig. 3. Network window

Now we will place 3 motes in green region and 2 motes in grey region. Select mote type from view option to differentiate the motes. Open radio messages from tools. It will show what happens to radio messages. Right click on border mote and select serial socket from mote tools for Sky 1. It will monitor serial socket for border router. Now in order to start the simulation follow the steps as mentioned further. Open new terminal and type the following commands.

1. `cd`
2. `cd contiki-2.7/examples/ipv6/rpl-border-router`
3. `Make connect-router-cooja`
4. Press enter and it will ask for password which is *user* itself
5. Now what you will see is tun slip. It is one type of protocol which is used to capture the routing packets and it will aslo show the ip address and all in detail in terminal. Now move to the created simulation and start the simulation from the simulation control window.

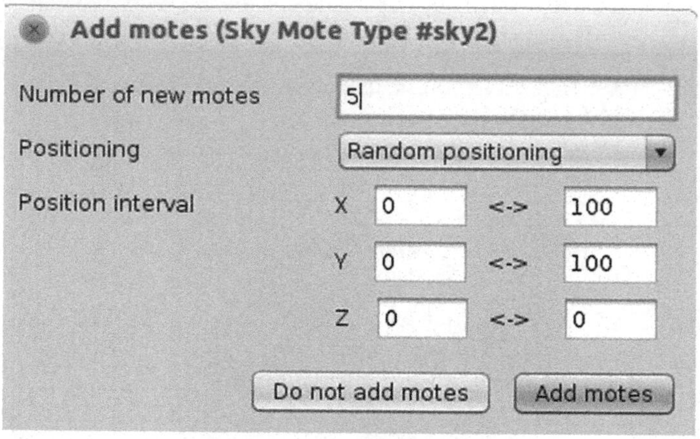

Add motes (Sky Mote Type #sky2)

Number of new motes 5

Positioning Random positioning

Position interval X 0 <-> 100

Y 0 <-> 100

Z 0 <-> 0

Do not add motes Add motes

Fig. 4. Addition of sky mote

We will see the outputs in each window. Now again move to the terminal and we will see the server ipv6 address. Now to check whether the Web server is working or not, copy the first address which will be like `aaaa::212:7401:1:101`. Open browser and write `http://[ aaaa::212:7401:1:101]`. And we will get output as the ipv6 address of neighbour motes and routes of motes.

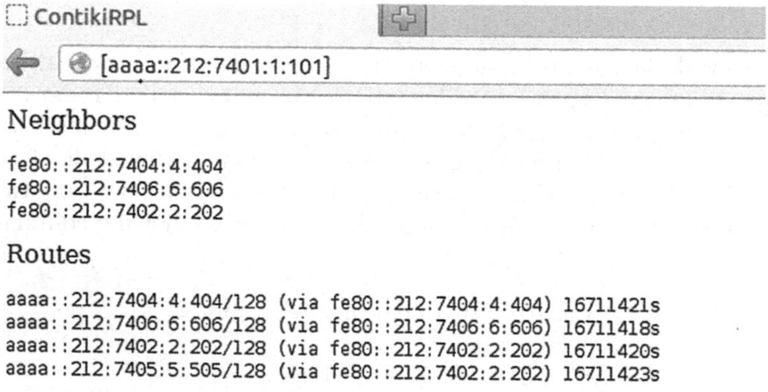

ContikiRPL

[aaaa::212:7401:1:101]

Neighbors

```
fe80::212:7404:4:404
fe80::212:7406:6:606
fe80::212:7402:2:202
```

Routes

```
aaaa::212:7404:4:404/128 (via fe80::212:7404:4:404) 16711421s
aaaa::212:7406:6:606/128 (via fe80::212:7406:6:606) 16711418s
aaaa::212:7402:2:202/128 (via fe80::212:7402:2:202) 16711420s
aaaa::212:7405:5:505/128 (via fe80::212:7402:2:202) 16711423s
```

Fig. 5. Route status, as visible to the border router

Now when we will write address of other motes, i.e. `[aaaa::212:7402:2:202]`, we will get output as shown in figure. It shows light and temperature at that mote.

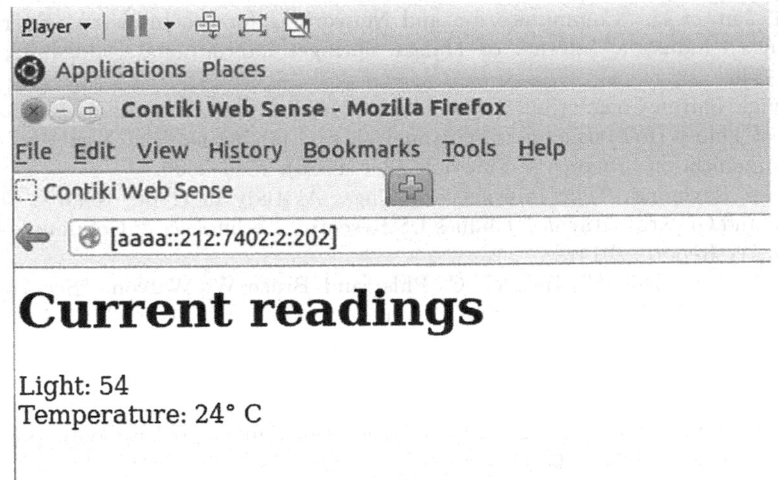

Fig. 6. Mote readings

Similarly, we can check the light and temperature of every mote in internet. Thus in simple terms, we can call this example as gateway, i.e. gateway connected to the internet.

5 Conclusions

In this paper, we discussed about IOT and its impact in our life in brief, all the tools and OS that are available in IOT like Contiki OS, Tiny OS. IOT contains several technologies like communication technology and low-power electronics. The development in IoT brought many new challenges including unclear architecture, immature standards, and lack of fundamental theory. We saw about Contiki OS and how we can use cooja simulator. Contiki gives a full IP network stack, with standard IP protocols, for example, UDP, TCP, and HTTP. We saw example of border router which saws how we can connect all the border mote and other sky mote with each other through internet. An interesting future add-on to the simulator may be a additional advanced radio medium.

Acknowledgements. The authors would like to thank Professor Chintan Bhatt and Dr Dac-Nhuong Le for the support and guidance in contributing the paper on IOT for further reference.

References

1. Walport, Mark. "The Internet of Things: making the most of the Second Digital Revolution." London: UK Government Office for Science (2014).

2. Rth, Jan, et al. "Communication and Networking for the Industrial Internet of Things." Industrial Internet of Things. Springer International Publishing, 2017. 317–346.
3. Luong, Nguyen Cong, et al. "Data collection and wireless communication in Internet of Things (IoT) using economic analysis and pricing models: A survey." IEEE Communications Surveys & Tutorials 18.4 (2016): 2546–2590.
4. James, Raymond. "The Internet of Things: A study in Hype, Reality, Disruption, and Growth." Raymond James US Research, Technology & Communications, Industry Report (2014).
5. Staudemeyer, Ralf C., Henrich C. Phls, and Bruce W. Watson. "Security and Privacy for the Internet of Things Communication in the SmartCity." Designing, Developing, and Facilitating Smart Cities. Springer International Publishing, 2017. 109–137.
6. Dhillon, Harpreet S., Howard Huang, and Harish Viswanathan. "Wide-area wireless communication challenges for the Internet of Things." IEEE Communications Magazine 55.2 (2017): 168–174.
7. Madakam, Somayya, R. Ramaswamy, and Siddharth Tripathi. "Internet of Things (IoT): A literature review." Journal of Computer and Communications 3.05 (2015): 164.
8. Bello, Oladayo, and Sherali Zeadally. "Intelligent device-to-device communication in the internet of things." IEEE Systems Journal 10.3 (2016): 1172–1182.
9. Bizer, Christian, Tom Heath, and Tim Berners-Lee. "Linked data-the story so far." Semantic services, interoperability and web applications: emerging concepts (2009): 205–227.
10. Kantarci, Burak, and Hussein T. Mouftah. "Trustworthy sensing for public safety in cloud-centric internet of things." IEEE Internet of Things Journal 1.4 (2014): 360–368.
11. Zanella, Andrea, et al. "Internet of things for smart cities." IEEE Internet of Things journal 1.1 (2014): 22–32.
12. Dlodlo, Nomusa, et al. "The state of affairs in Internet of Things research." Academic Conferences International Ltd, 2012.
13. Chen, Shanzhi, et al. "A vision of IoT: Applications, challenges, and opportunities with china perspective." IEEE Internet of Things journal 1.4 (2014): 349–359.
14. Botta, Alessio, et al. "Integration of cloud computing and internet of things: a survey." Future Generation Computer Systems 56 (2016): 684–700.
15. Ritz, John, and Zane Knaack. "internet of things." Technology & Engineering Teacher 76.6 (2017).
16. Bhayani, Malay, Mehul Patel, and Chintan Bhatt. "Internet of Things (IoT): in a way of smart world." Proceedings of the International Congress on Information and Communication Technology. Springer Singapore, 2016.
17. Bhatt, Chintan, Nilanjan Dey, and Amira S. Ashour. "Internet of Things and Big Data Technologies for Next Generation Healthcare." (2017).
18. Bhatt, Yesha, and Chintan Bhatt. "Internet of Things in HealthCare." Internet of Things and Big Data Technologies for Next Generation Healthcare. Springer International Publishing, 2017. 13–33.
19. Salman, Tara, and Raj Jain. "NETWORKING PROTOCOLS AND STANDARDS FOR INTERNET OF THINGS." Internet of Things and Data Analytics Handbook (2017): 215–238.
20. Atzori, Luigi, Antonio Iera, and Giacomo Morabito. "The internet of things: A survey." Computer networks 54.15 (2010): 2787–2805.

21. Miorandi, Daniele, et al. “Internet of things: Vision, applications and research challenges.” Ad Hoc Networks 10.7 (2012): 1497–1516.
22. Dunkels, Adam, et al. “The Contiki OS: The Operating System for the Internet of Things.” Online], at http://www.contikios.org (2011).

A Proposed Model to Integrate Business Intelligence System in Cloud Environment to Improve Business Function

Manas Kumar Sanyal, Biswajit Biswas(✉), Subhranshu Roy, and Sajal Bhadra

Department of Business Administration, University of Kalyani, West Bengal, India
{manassanyal123, biswajit.biswas0012, sajal.bhadra}@gmail.com, subhranshu_81@yahoo.com

Abstract. Business intelligence system (BIS) has gained a high momentum and played a significant role in enhancing the business environment especially with the help of cloud. The BIS with cloud computing is the solution to overcome complex problems where large amounts of data (market analysis, customer's view, feedback and organization response time) are required to be processed. The objective of this paper is to show the realistic perspective of the possibilities for maximizing benefits and minimizing risks in Indian business. The authors have highlighted the importance of cloud computing in business and tried to develop a framework for integrating BIS with cloud computing, and an attempt has also been made to analyse the efficiency of the proposed model which improves performance over the cloud. Some of the challenging issues as well as the future research directions have also been discussed.

Keywords: Business Intelligence System (BIS) · Cloud computing · Data warehouse · Risk · Profit

1 Introduction

In any traditional business, sometimes it is very difficult for an organization to take some right decisions at the right time causing failure to achieve their goals. In this paper, authors have proposed an integrated cloud BIS model that can be used in an organization to store data, customers' views and feedbacks in data warehouse and analyse the same with the existing data. As we all know that business intelligence always offers solutions for complex business operations. It is sometimes very expensive for an organization to install and maintain on-premise infrastructure considering both hardware and software. To reduce this, "cloud computing" solutions can be developed that help the smooth execution of the business processes with reduced cost. The cloud computing environment is a successful way for information sharing. This system helps an organization to utilize the complex database management system (DBMS) in order to execute the solutions that they needed. Cloud providers take care of their clients, both in hardware and software services. It helps to increase the profit and improves

V. Bhateja et al. (eds.), *Information Systems Design and Intelligent Applications*, Advances in Intelligent Systems and Computing 672,
https://doi.org/10.1007/978-981-10-7512-4_29

return on investment (ROI). Cloud computing technologies are helping to transform the way information technology (IT) is deployed and managed. It reduces costs and complexity of implementation and maintenance of applications, along with the ability of scaling high-performance applications and infrastructures on demand. In almost every aspect, using a cloud environment is always a gain for an organization.

1.1 Definitions of Cloud Computing

According to Gartner [1], cloud computing is a platform where clients are able to use frequently the online facilities with the help of Internet technologies. The National Institute of Standards and Technology (NIST) [2] defined cloud computing as on-demand network access to a shared set of configurable computing servers, storage areas, applications and services that can be smartly provisioned and released with nominal management of efforts and services from its provider. As per Furht et al. [3], cloud computing is a present trend of computing in which vigorously scalable and frequently virtualized effects are provided as a facility over the Internet. Yuvraj and Vijay defined [4] cloud computing as easy, usable and accessible virtualized resources (such as hardware, development platforms and/or services) to retrieve any real-time information, enquired by client, in a fraction of a second. Thus, it implies that "cloud computing" refers to the on-demand delivery of information technology resources and applications via the Internet with pay-as-you-go pricing. Organizations can use cloud computing environment to avoid to make large investments as well as to spend a lot of efforts to maintain on-premise hardware and software. In cloud computing system, one can access as many resources as it is needed, but pay for what is used [5]. Cloud computing technology will be emerging as the next-generation business model, considering its utility as "every things into one" and "pay use manner to user".

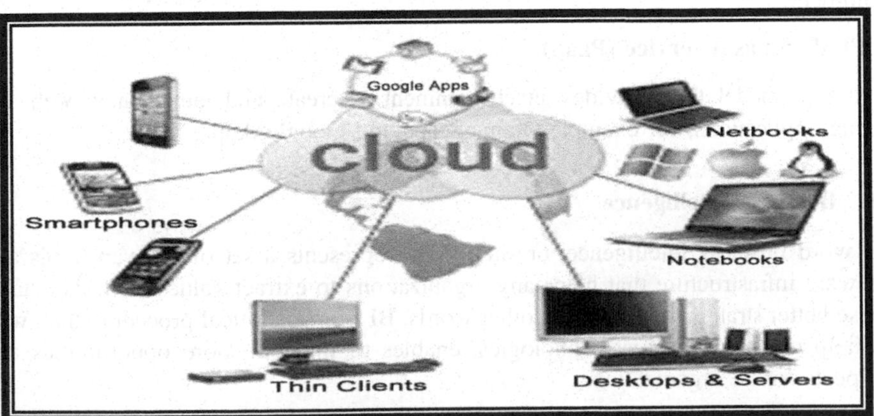

Fig. 1. Cloud users

1.2 Cloud Computing Deployment Model

- **Public Cloud**

It is open for public use. Generally, we can use this cloud without paying any cost. In this type of cloud, the main concern is privacy because providers do not assure it (Fig. 1).

- **Private Cloud**

This type of infrastructure is used by only single user or organization. It has similar benefits like public cloud, but the security is higher than public cloud. Private cloud is devoted to a single organization.

- **Hybrid Cloud**

This is an arrangement of private and public cloud. By this cloud, users can identify which data reside on public cloud and which on private cloud. By using hybrid cloud, users can reduce their expenses.

1.3 Cloud Service Model

There are three service models available for cloud computing services.

- **Software as a service (SaaS)**:

In this model, users can be able to use different software from different servers through the Internet.

- **Infrastructure as a Service (IaaS)**:

It is a model that fulfils the prime goal of an organization to reduce time and money required. It allows users to rent computing resources, servers and storage network equipment.

- **Platform as a service (PaaS)**:

It is a model that provides an environment to create and launch new web and mobile applications, for example OrangeScape and Google Apps.

1.4 Business Intelligence

The word business intelligence, or simply BI, represents a set of software tools and hardware infrastructure that helps any organizations to extract values from its data to devise better strategic planning. In other words, BI is an analytical procedure that, with the help of data-intensive technologies, enables to find out more opportunities and prospects for an organization.

1.5 Integrated BIS in Cloud Environment

Authors propose a model that integrates the business intelligence system on cloud environment where the organization has the ability to extract business values from the applications placed on cloud environment which provides reduced cost and improved return on investment. It will help the organization to take better decisions which in turn creates new opportunities.

2 Related Works

Microsoft [6] has termed business intelligence as a tool that simplifies value of information finding and analysis, making it possible for decision-makers at top levels of an organization with easy access and support to recognize, analyse and act on information, on anytime and anywhere basis. Gartner Inc. [7] suggested that BI platforms for an organization have mainly three types of roles: "analysis (online analytical processing (OLAP)), information transfer (reports and dashboards) and raised area integration (BI metadata supervision and an expansion environment)". The main issue in cloud computing is its data security. Along with it, the cloud users may face security threats. Mai Kasem [8] said cloud business intelligence is a novel concept of business process that is built on cloud architecture with a lower cost. Hans Peter Luhn [9] highlighted that a business intelligence system may act as the controller of management information system (MIS) work on BIS (Fig. 2).

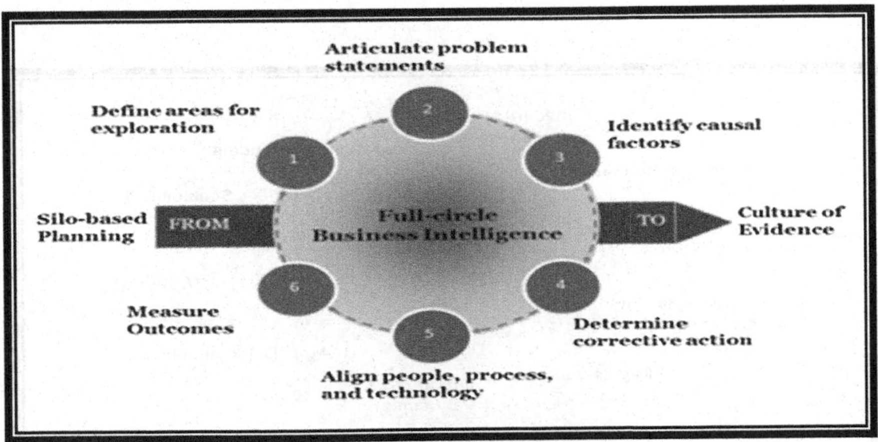

Fig. 2. BIS cycle

In the research paper, Martens and Teuteberg [10] scrutinized cloud service with a mathematical decision model from various sources. They considered that cost and risk were the key to take decision. The managerial proposal of their model was related to sustainable valuation support and inclusive evaluation approach. It is very crucial for an organization to trade-off the benefits and risk of the BIS on cloud.

3 Risk in Cloud Environment Business

In cloud environment business, it is necessary to categorize the risk according to their importance in the system of cloud infrastructure provider (CIP) and cloud service provider (CSP). This risk may be divided into three stages:

(i) Identification: During this stage, an organization can establish and identify its potential risk.
(ii) Description: During this stage, an organization completes a comprehensive risk evaluation using various techniques.
(iii) Estimation: An organization estimates the possibility.

4 Benefits of Cloud Computing

Cloud computing technology provides multifaceted benefits for any organizations which are represented in Fig. 3.

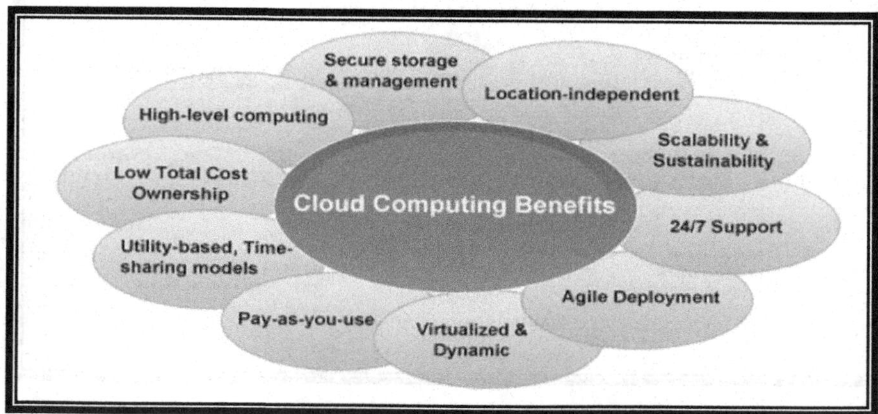

Fig. 3. Benefits of cloud computing in business

5 Adoption Rate for Various Cloud Deployment Model

Adoption rate for cloud deployment is rising day by day. Any organizations, irrespective of their size (low, medium or large scale), are moving their IT systems to cloud environment to avail the above-mentioned benefits. Here, the authors have presented some statistics of adoption rate for various cloud computing models for last three years (Fig. 4). The data were captured by Right scale [2015, 2016 and 2017] by survey method. They interviewed various individuals like IT managers, implementation experts, system user and top managements from various organizations across the globe. Tables 1 and 2 show the distribution of respondents and adoption of various deployment models, respectively.

Table 1. Comparisons of year-wise respondents

Year	All respondents	Enterprises respondents	Small medium business respondents (SMB)[a]
2015	930	306	624
2016	1060	433	627
2017	1002	480	517

[a]Here, less than 1000 employees in an organization are considered as SMB and more than 1000 are considered as Enterprise

Table 2. Number of different cloud users

Year	Total users	Public cloud users	Private cloud users	Hybrid cloud users	Any cloud users
2015	930	818	586	539	439
2016	1060	943	816	753	1007
2017	1002	892	721	671	952

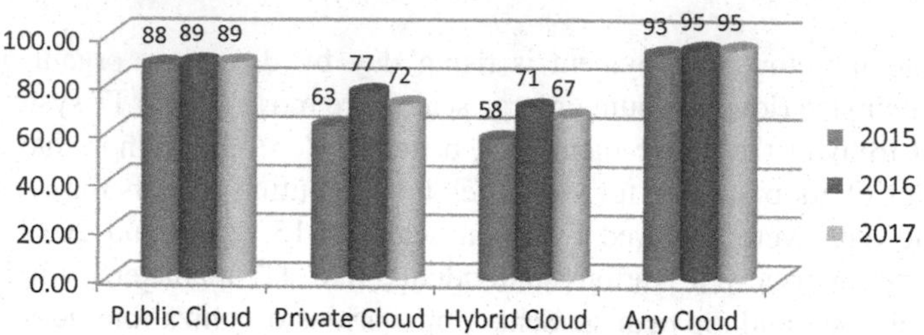

Fig. 4. Respondents are shown in percentage (%)

6 Proposed Work

The authors divide their proposed work into two parts that are mentioned below:

6.1 Proposed Model

The authors propose the below model (Fig. 5) for BIS on cloud computing environment which will help organizations to minimize the installation and operational costs of the system with the smooth execution of their intended business processes. This will indeed help them to maximize the profit along with minimizing the risks.

BIS system on cloud gathers data from the cloud system, and depending on the data, it tries to extract value or decision based on some objective parameters like revenue increase, profit margin or return on investment (ROI), etc.

In Fig. 5, by says for exam, we can set profit margin as an objective parameter. As per Fig. 5, initially, the profit margins are set as the standard point that the organization has achieved last year. On the basis of the cloud data, it updates the velocity and profit margin on time-to-time basis. If the velocity decreases or stays at the same point, business needs to identify and analyse the root cause to mitigate the risk. Thus, BIS on cloud can act as a risk monitoring system.

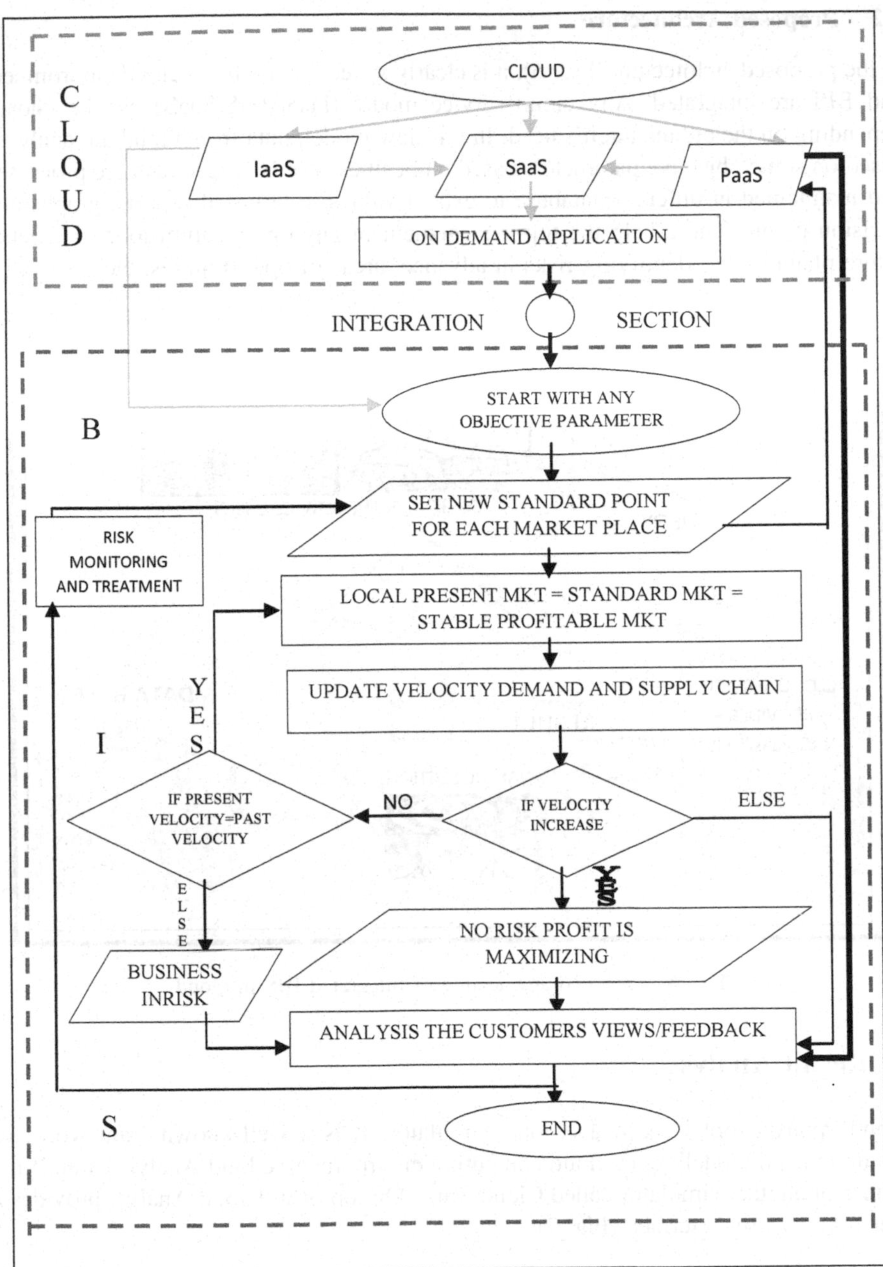

Fig. 5. Proposed model of integrated BIS in cloud environment

6.2 Proposed Architecture

In the proposed architecture (Fig. 6), it is clearly stated that both the cloud environment and BIS are integrated. Any cloud service model (IaaS/PaaS/SaaS) can be chosen depending on the organization's need. In this new model, data from cloud are analysed based on some objective parameters based on business needs. These data are processed and maintained in different database to extract values in terms of reports, graphs and decision points. These help the top management of any organization to devise better future planning by identifying risks in advance, areas of opportunities, etc.

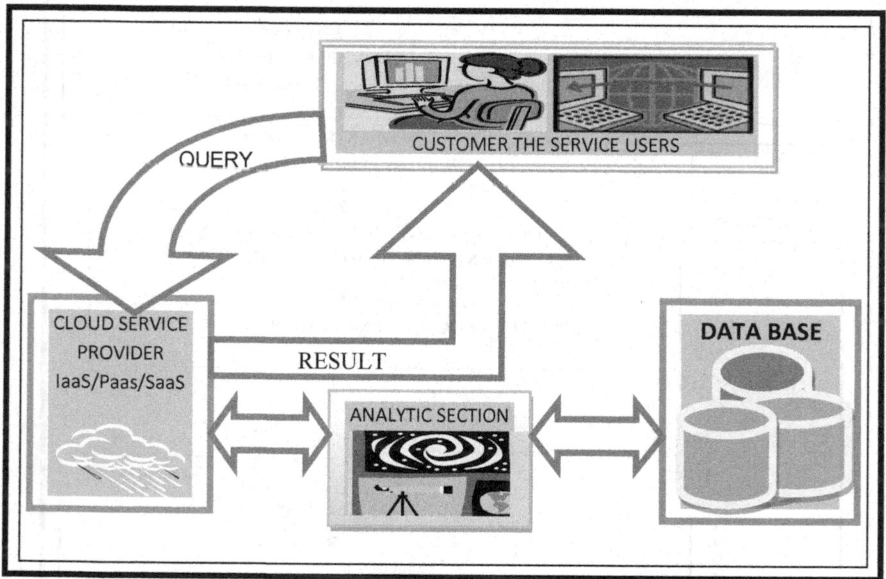

Fig. 6. Proposed architecture of integrated BIS in cloud

7 Result Analysis

Cloud Analyst tool is used as Cloud Simulator. It is a well-known framework for simulation and modelling of cloud computing environment. Cloud Analyst extends the feature of another simulator called Cloud Sims. On top of it, Cloud Analyst provides a user-friendly user interface (Fig. 7).

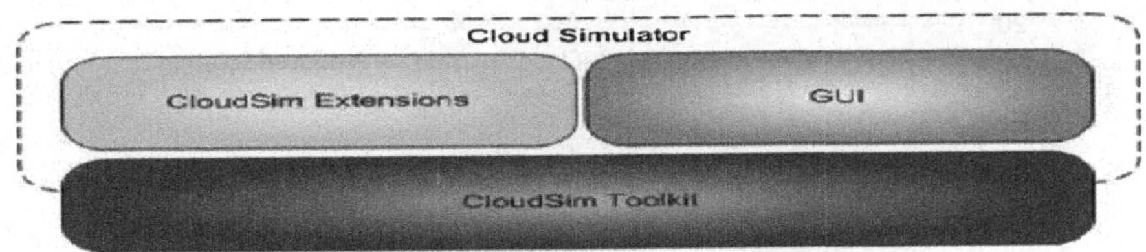

Fig. 7. Cloud Analyst block diagram

Authors have used Cloud Analyst to simulate some of the data points like risk factor and response time for three super markets having three types of Cloud Solutions, i.e. IaaS, PaaS and SaaS, respectively. Data points are also simulated for the proposed model using Cloud Analyst (Table 3).

Table 3. Data received as output from Cloud Analyst

	Cloud solution		Proposed solution	
Cloud systems	Average response time in ms	Profit margin increase in terms of %	Average response time in ms	Profit margin increase in terms of %
Super market-1 with SaaS model	641	17.5	559	18.21
Super market-2 with PaaS model	249	9.35	199	8.74
Super market-3 with IaaS model	213	8.45	178	7.91

The simulated data using Cloud Analyst (Figs. 8 and 9) clearly showed the improvement of using business intelligence system on cloud. The proposed model is not only going to increase the response time and profit margin, but also helps any organizations to identify the risks in advance so that top management can take some remediation plan to mitigate the inherent risks.

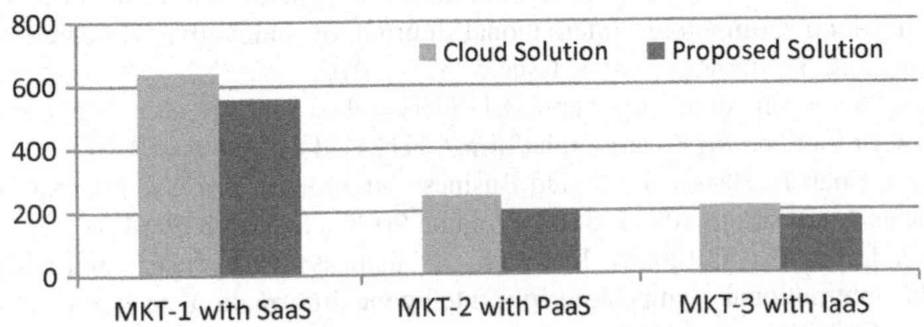

Fig. 8. Comparison of response time in ms

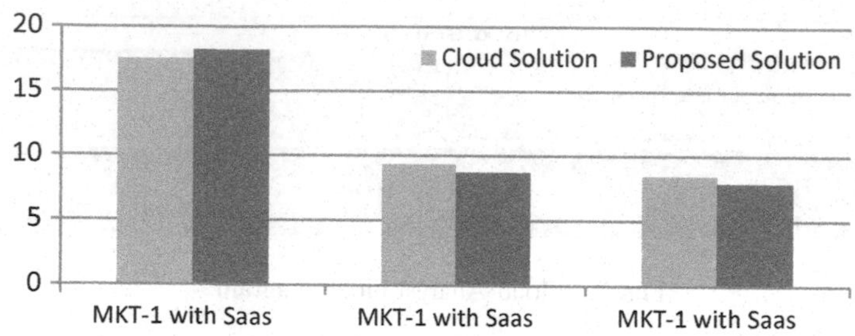

Fig. 9. Comparison of profit in %

8 Conclusion and Future Scope

From our studies, it is observed that BIS and cloud computing system are siblings in the business environment. It also established the importance of data warehouse in real-time decision-making system. The proposed model also supports to minimize the decision-taking time of any organization. The customer's feedback and behaviour are also taken into account to update the exiting supply chain along with future plan of business. It is also shown in our work how we monitor risk and treatment with analysis of customers' feedback. Our studies increase the effectiveness and productivity in an organization to reduce cost.

In future, the authors can include more and more parameters to improve the performance of this model by simulating the same using Cloud Analyst.

References

1. Baars, H. Kemper H-G: "Management support with structured and unstructured data, an integrated business intelligence framework," ISM, 2008, Vol.(25), 132–148. http://www.gartner.com/itglossary/bi-platforms/, accessed on January 2014.
2. Sabherwal, R., Becerra-Fernandez, I. Business Intelligence: Practices, Technologies, and Management. John Wiley & Sons, Inc.: Hoboken, NJ. 2011.
3. Frolick, M.N., Ariyachandra, T. "Business Performance Management: One Truth." Information Systems Management, 2006, 23(1), 41–48.
4. Sanjay P. Ahuja A View of Cloud Computing: http://mars.ing.unimo.it/didattica/ingss/Lec_SaS/CloudView.pdf.
5. Hari Pandey et al, Research on Profit Maximization Scheme with Guaranteed Quality of Service in Cloud Computing; International Journal of Innovative Research in Science, Engineering and Technology; vol.5, Issue 5, May 2016.
6. http://technet.microsoft.com/en-us/libery/cc811595(v=office.12).aspx, accessed on January 2014.
7. http://www.gartner.com/itglossary/biplatforms/, accessed on January 2014.
8. Mai Kasem, Ehab E. Hassanein: Cloud Business Intelligence Survey: International Journal of Computer Applications (0975–8887) Volume 90-No 1, March 2014.
9. Hans Peter Luhn, Fu, T. (2008). Research on business intelligence pattern based on the BiaaS. In International symposium on intelligent information technology application workshops, China.
10. Martens and Teuteberg: The Journal of Strategic Information System 2012, Third paper, science direct. Published by Elsevier Ltd.

Combined Center-Symmetric Local Patterns for Image Recognition

Bhargav Parsi(✉), Kunal Tyagi, and Shweta R. Malwe

Indian Institute of Technology (Indian School of Mines), Dhanbad 826004, India
bhargav265@gmail.com, kunaltyagi12@gmail.com, shweta.malwe26@gmail.com

Abstract. Local feature description is gaining a lot of attention in the fields of texture classification, image recognition, and face recognition. In this paper, we propose Center-Symmetric Local Derivative Mapped Patterns (CS-LDMP) and eXtended Center-Symmetric Local Mapped Patterns (XCS-LMP) for local description of images. Strengths from Center-Symmetric Local Derivative Pattern (CS-LDP) which is gaining more texture information and Center-Symmetric Local Mapped Pattern (CS-LMP) which is capturing nuances between images were combined to make the CS-LDMP, and similarly, we combined CS-LMP and eXtended Center-Symmetric Local Binary Pattern (XCS-LBP), which is tolerant to illumination changes and noise were combined to form XCS-LMP. The experiments were conducted on the CIFAR10 dataset and hence proved that CS-LDMP and XCS-LMP perform better than its direct competitors.

Keywords: Local binary pattern · Local mapped pattern · Local derivative pattern · Background subtraction · Image recognition

1 Introduction

One of the most important characteristics for image analysis is its texture. Background subtraction (BS) Wu et al. [1] is one of the main steps in computer vision applications. Many challenging situations were faced by BS such as illumination changes, dynamic environments, inclement weather, camera jitter, noise, and shadows. For effective analysis, a descriptor histogram should be made that allows efficient matching of images. A good descriptor should have the following features: (i) tolerance to illumination changes, (ii) invariance to scale and rotation, (iii) robust to noise, and (iv) ability to capture nuances between images. Numerous local texture descriptors recently have attracted considerable attention, especially the local binary pattern (LBP) because it is simple to implement. However, the LBP operator produces a long feature set since it only adopts the first-order gradient information between the center pixel and its neighbors.

V. Bhateja et al. (eds.), *Information Systems Design and Intelligent Applications*, Advances in Intelligent Systems and Computing 672,
https://doi.org/10.1007/978-981-10-7512-4_30

In this paper, we propose to modify G. Xue et al. [2], by combining it with the technique suggested by C. T. Ferraz et al. [3] so that the resulting method can capture nuances between images while preserving the advantages of G. Xue et al. [2]. In the same way, we also propose to modify C. Silva et al. [4] with the help of C. T. Ferraz et al. [3] technique.

The rest of the paper is organized as follows. Section 2 gives a brief literature review of the presently followed local description methods. Section 3 gives the basic idea and definitions of a few existing descriptors. Our proposed methods would be present in Sect. 4. We have considered a few existing local descriptors and compared their performances with our proposed methods in Sect. 5. Finally, we conclude our paper in Sect. 6.

2 Related Work

Various methods for texture classification are proposed in the literature. Firstly, in T. Ojala et al. [5] LBP was introduced as an approach for texture classification. After witnessing the success of LBP in texture classification, LBP finds its way into various classification problems such as facial recognition as proposed in T. Ahonen et al. [6]. The original LBP was rotation variant which sometimes can be seen as a disadvantage. To tackle this issue, in T. Ojala et al. [7] extended version of LBP was introduced which was rotation invariant. Later it was found out that LBP was sensitive to global intensity variations and also to local intensity along edge components. To handle these shortcomings in Bongjin Jun and Daijin Kim [8] a new method was introduced called local gradient pattern (LGP). LGP was shown to have higher discriminant power than LBP in the case of facial recognition. To improve recognition rate using LBP, in T. Ahonen et al. [6] an efficient method of facial representation was used. This new method of facial representation divides the image into small regions, and then on each region, LBP is applied and eventually the descriptor of each region is combined to form feature vector of the image. In M. Heikkilä et al. [9] a new method was introduced which combines the strengths of both LBP and SIFT operator called Center-Symmetric Local Binary Pattern (CS-LBP). It has many advantages such as low-size feature vector, tolerance to illumination changes, and robustness on flat image areas. Later, in C. Silva et al. [4] a new method called eXtended Center-Symmetric Local Binary Pattern (XCS-LBP) was introduced which combines both the strengths of CS-LBP and CS-LDP operator. Also in C. T. Ferraz et al. [3] a method called Center-Symmetric Local Mapped Pattern (CS-LMP) is presented which outperforms CS-LBP. Now to make this classification computationally efficient in Y. Zheng et al. [10] two approaches were proposed, i.e., dense Center-Symmetric Local Binary Patterns (CS-LBP) and pyramid Center-Symmetric Local Binary/Ternary Patterns (CS-LBP/LTP). These methods are computationally efficient than LBP and CS-LBP. Next, in A. Abdesselam [11] two LBP histograms were constructed, one for edge pixels and one for non-edge pixels. The final feature vector is the weighted combination of the two histograms.

3 Local Descriptors

In this section, we define the CS-LBP, CS-LDP, CS-LMP, and XCS-LBP and give brief explanation about them.

3.1 Center-Symmetric LBP Operator

The CS-LBP operator is a simple extension to the LBP operator which reduces the number of bins in the histogram drastically. For example, if $P = 8$ and $R = 1$ in LBP we get 256 bins. Whereas in CS-LBP only 16 bins would be needed. As mentioned in M. Heikkilä et al. [9], the CS-LBP operator produces features which are tolerant to illumination changes and robust in flat image areas and are straightforward to compute. T is a threshold value which is user defined. In our experiments, we always took it as 0. Equation 1 gives the mathematical model of CS-LBP.

$$CS - LBP_{(P,R,T)}(x,y) = \sum_{i=0}^{\frac{P}{2}-1} s(n_i - n_{i+\frac{P}{2}})2^i \tag{1}$$

where

$$s(x) = \begin{cases} 1 & \text{if} \, x > T \\ 0 & \text{otherwise} \end{cases} \tag{2}$$

3.2 Center-Symmetric LDP Operator

The CS-LDP operator was introduced by G. Xue et al. [2], and it claims that CS-LBP extracts only the first derivative and cannot extract multilevel derivative features. Hence, CS-LDP was made so that the problem could be alleviated and therefore captures more detailed texture information. This issue is because in CS-LBP the value of the center pixel was not considered whereas in CS-LDP it was considered. The following equation gives us the mathematical model of CS-LDP. Equation 3 gives the mathematical model of CS-LDP.

$$CS - LDP_{(P,R)}(x,y) = \sum_{i=0}^{\frac{P}{2}-1} t((n_i - n_c).(n_c - n_{i+\frac{P}{2}}))2^i \tag{3}$$

where

$$t(x_1.x_2) = \begin{cases} 0 & \text{if} \, x_1.x_2 > 0 \\ 1 & \text{otherwise} \end{cases} \tag{4}$$

3.3 Center-Symmetric LMP Operator

Another issue of CS-LBP as pointed out by C. T. Ferraz et al. [3] is the following: The symmetric difference was thresholded by 1 or 0 and hence cannot capture the minute differences between images. This problem was mitigated by introducing a sigmoid function instead of a Heaviside step function. In our experiments, the value of β was always considered to be 1 for simplicity. It represents the slope of the sigmoid function. The value b is one less than the number of histogram bins. Equation 5 gives the mathematical model of CS-LMP.

$$CS-LMP_{(P,R,f)}(x,y) = round(\frac{\sum_{i=0}^{\frac{P}{2}-1} f(n_i - n_{i+\frac{P}{2}})2^i}{\sum_{i=0}^{\frac{P}{2}-1} 2^i})b \tag{5}$$

where

$$f(x) = \frac{1}{1+e^{\frac{-x}{\beta}}} \tag{6}$$

3.4 Extended Center-Symmetric LBP Operator

This operator produces a shorter histogram compared with LBP, and the length of the histogram is equal to that of CS LBP. However, since the CS LBP method does not make use of the center pixel C. Silva et al. [4], this approach was modified so that the center pixel value is considered and therefore extracting more information than CS-LBP. Equation 7 gives the mathematical model of XCS-LBP.

$$XCS-LBP_{(P,R)}(x,y) = \sum_{i=0}^{\frac{P}{2}-1} s(g_1(i,c) + g_2(i,c))2^i \tag{7}$$

where

$$s(x_1+x_2) = \begin{cases} 1 & \text{if}\, x_1+x_2 \geq 0 \\ 0 & \text{otherwise} \end{cases} \tag{8}$$

$$g_1(i,c) = n_i - n_{i+\frac{P}{2}} + n_c, g_2(i,c) = (n_i - n_c)(n_{i+\frac{P}{2}} - n_c) \tag{9}$$

The advantages of both CS-LMP, CS-LDP, and XCS-LBP motivated us to make methods which use the features of CS-LMP and CS-LDP, CS-LMP, and XCS-LBP, respectively. For example, the CS-LMP method does not use the center pixel value, and it does not give more detailed texture information. But we know that CS-LDP uses the center pixel value for the histogram calculation which gives more texture information. Therefore, using the features of CS-LMP and CS-LDP, we have discovered a new method which gives more texture information and captures nuances between images.

4 Proposed Local Descriptor Methods

In this section, we propose methods which use the advantages of multiple methods and finally prove it experimentally that it is indeed the case. Firstly, we offer Center-Symmetric Local Derivative Mapped Patterns (CS-LDMP), and next, we recommend eXtended Center-Symmetric Local Mapped Patterns (XCS-LMP).

4.1 Center-Symmetric Local Derivative Mapped Patterns (CS-LDMP)

Here, we propose to modify the threshold function of the CS-LDP operator to an amended version of the sigmoid function. The reason behind this is, as, in CS-LBP, CS-LMP did not consider center pixel values and hence useful texture information was not captured. Moreover, CS-LDP was not able to capture nuances between images. Hence, the proposed CS-LDMP overcame the difficulty and hence performed better than both CS-LDP and CS-LMP. The above results were tested on the CIFAR10 dataset as proposed in A. Krizhevsky and G. Hinton [12].

It is interesting to observe that the original definition of sigmoid function was not used but a modified version was used. In Eq. 3 we can see that the threshold function is a bit different from others like CS-LBP or XCS-LBP. We have changed the sigmoid function so that it looks like a continuous representation of the CS-LDP threshold function. The value b is one less than the number of histogram bins. Equation 10 gives the mathematical model of CS-LDMP.

$$CS-LDMP_{(P,R,f)}(x,y) = round(\frac{\sum_{i=0}^{\frac{P}{2}-1} f((n_i - n_c).(n_c - n_{i+\frac{P}{2}}))2^i}{\sum_{i=0}^{\frac{P}{2}-1} 2^i})b \quad (10)$$

where

$$f(x_1.x_2) = \frac{1}{1+e^{\frac{x_1 \cdot x_2}{\beta}}} \quad (11)$$

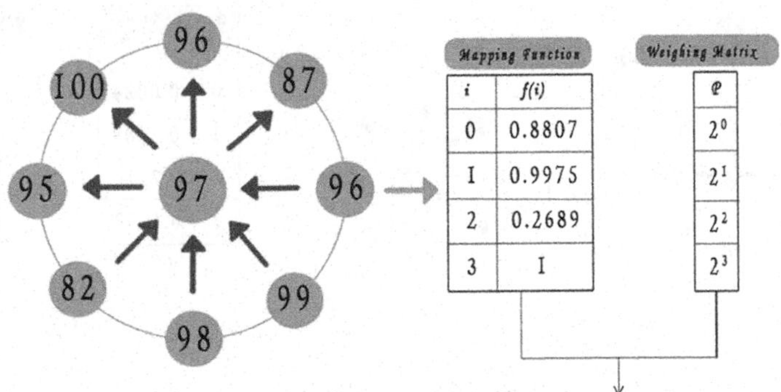

Fig. 1. Working of CS-LDMP

In Fig. 1 the calculation proceeds as follows,

$$CS-LDMP_{8,1,f} = round((\frac{0.8807x2^0 + 0.9975x2^1 + 0.2689x2^2 + 1x2^3}{15})x15) = 12$$

4.2 eXtended Center-Symmetric Local Mapped Patterns (XCS-LMP)

In this section, we propose to modify the threshold function of XCS-LBP operator which gives us a new operator XCS-LMP. Similar to the above method, we use the sigmoid function instead of a regular Heaviside function. It would appear that this method can capture minute differences between images, which the earlier XCS-LBP method could not accomplish.

The method of calculation is as follows. First, we calculate the g_1 and g_2 function values and then send the result to the sigmoid function. The results from the sigmoid function from different pairs of pixels are then multiplied with the weighing factors to produce a floating-point number. The result is divided by sum of weighing factors, and then it is rounded off to the nearest integer value. The value b is one less than the number of histogram bins. Equation 12 gives the mathematical model of XCS-LMP.

$$XCS-LMP_{(P,R,f)}(x,y) = round(\frac{\sum_{i=0}^{\frac{P}{2}-1} f(g_1(i,c)+g_2(i,c))2^i}{\sum_{i=0}^{\frac{P}{2}-1} 2^i})b \tag{12}$$

where

$$f(x_1+x_2) = \frac{1}{1+e^{\frac{-(x_1+x_2)}{\beta}}} \tag{13}$$

$$g_1(i,c) = n_i - n_{i+\frac{P}{2}} + n_c, g_2(i,c) = (n_i - n_c)(n_{i+\frac{P}{2}} - n_c) \tag{14}$$

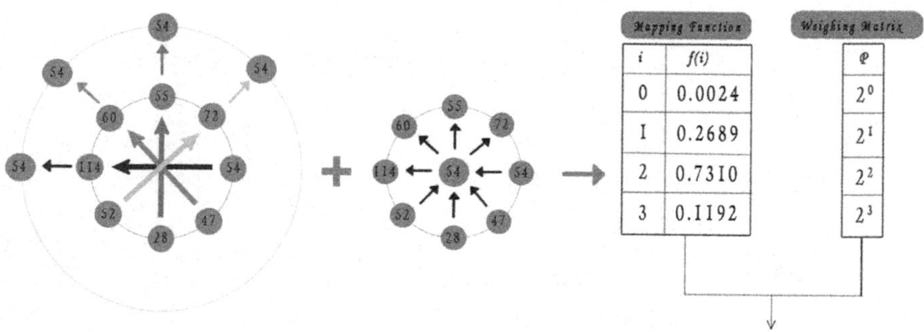

Fig. 2. XCS-LMP

In Fig. 2 the calculation proceeds as follows,

$$XCS-LMP_{8,1,f} = round((\frac{0.0024x2^0+0.2689x2^1+0.7310x2^2+0.1192x2^3}{15})x15) = 3$$

Table 1. Accuracy table

Est	CS-LBP	CS-LDP	CS-LMP	Proposed CS-LDMP	XCS-LBP	Proposed XCS-LMP
10	0.2878	0.309	0.2614	**0.2996**	0.3578	**0.3604**
50	0.3061	0.3304	0.2878	**0.3329**	0.3906	**0.3949**
100	0.3135	0.3415	0.3001	**0.35**	0.4086	**0.4058**
200	0.3178	0.3457	0.3057	**0.3614**	0.4126	**0.4121**
400	0.32	0.3515	0.3125	**0.3673**	0.4154	**0.4162**
800	0.3191	0.3518	0.3117	**0.3714**	0.4181	**0.4179**
1000	0.3174	0.3507	0.3121	**0.37**	0.4168	**0.4173**
1500	0.3149	0.3477	0.3154	**0.3658**	0.4155	**0.4154**
2000	0.3156	0.3453	0.3122	**0.3649**	0.4129	**0.413**
2500	0.3141	0.3455	0.3123	**0.3636**	0.4108	**0.4108**

5 Experimental Results

Numerous experiments were conducted to illustrate the performances of both of the proposed operators. For coding purposes, we use Python 3.6 with the xgboost library on a Windows 10 OS equipped with Intel(R) Core(TM) i7-6700HQ CPU @ 2.60Ghz 2.60 GHz and 1600MHz DDR3L RAM. The dataset used is CIFAR10 as in A. Krizhevsky and G. Hinton [12]. They contain images of objects such as an airplane, bird, dog. We give the comparison of performances of the following operators: (i) CS-LBP (ii) CS-LDP (iii) CS-LMP (iv) CS-LDMP (v) XCS-LBP (vi) XCS-LMP

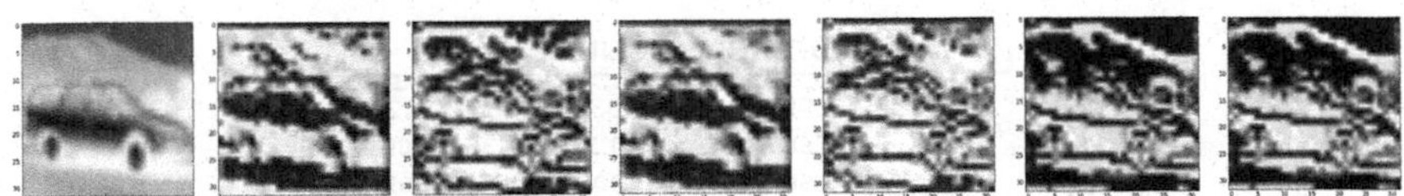

Fig. 3. Comparison of transformed images of a car starting with original image, CS-LBP, CS-LDP, CS-LMP, Proposed CS-LDMP, XCS-LBP, Proposed XCS-LMP from left to right

Color images of the CIFAR10 dataset are first converted to grayscale, for which luminance was used. The following formula for luminance given in C. Kanan and G. W. Cottrell [13] was used, that is, given in Eq. 15.

$$G_{luminance} = 0.3R + 0.59G + 0.11B \tag{15}$$

where $R, G, and\, B$ are the intensity values of red, blue, and green, respectively. Secondly, we convert the original image to a different image using the local

descriptor. Pixels are always compared in a clockwise fashion starting with the top left pixel. After that, we calculate the histogram of the transformed image and store it separately. Finally, we use the Xgboost algorithm as in T. Chen and C. Guestrin [14] for training the algorithm on the extracted features from the algorithm from the extracted features from the corresponding operator.
For analyzing the performances, three evaluation metrics were used: (i) accuracy, (ii) precision vs recall curve, and (iii) receiver operator characteristics curve (ROC Curve). ROC curve is simply the plot between true positive rate and false positive rate. The definitions of the terms are given in Eqs. 16, 17, 18, and 19.

$$Accuracy = \frac{TP + TN}{P + F} \tag{16}$$

$$Precision = \frac{TP}{TP + FP} \tag{17}$$

$$Recall = TruePositiveRate = \frac{TP}{TP + FN} \tag{18}$$

$$FalsePositiveRate = \frac{FP}{TN + FP} \tag{19}$$

where TP = No. of True Positives, TN = No. of True Negatives, P = No. of Positive examples, F = No. of Negative examples, FP = No. of False Positives, FN = No. of False Negatives. The above definitions would be valid for a binary classification problem. Since we would be doing a multi-class classification problem, we should be following a different strategy, i.e., micro-averaged evaluation metric as given in V. Van Asch [15]. Firstly, the accuracy table is given, followed by precision vs. recall curve (PR) and receiver operating characteristics curve (ROC) with the area under each curve (AUC). The PR curve and ROC curve are

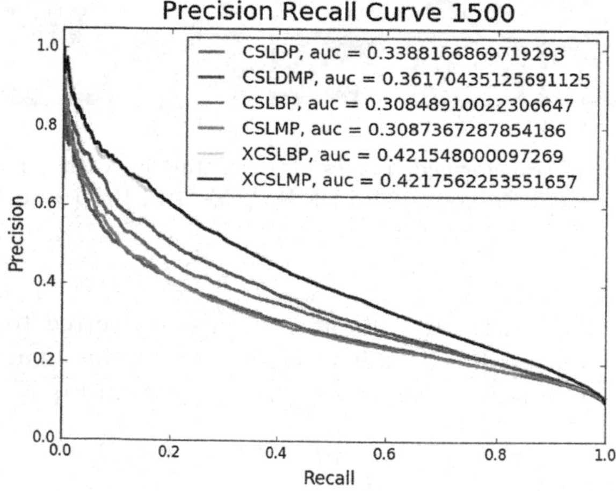

Fig. 4. Precision-recall curve for 1500 estimators

provided for a specific number of estimators of the xgboost library due to space constraints. Here, Est. indicates the number of estimators used in the xgboost classifier. In Fig. 3, transformed images of a car are given, by different operators.

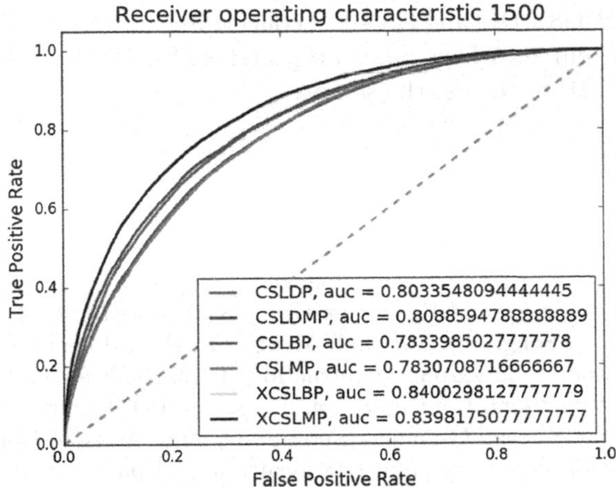

Fig. 5. Receiver operating characteristics for 1500 estimators

Now, performances are analyzed in the following order: accuracy, PR curve, ROC curve. The proposed CS-LDMP operator outperforms the CS-LBP, CS-LDP, and CS-LMP operators concerning accuracy as given in Table 1. When $est = 800$, CS-LDMP's accuracy is better than CS-LMP by 20%. Moreover, we can also see that the proposed XCS-LMP operator is giving slightly better accuracy than XCS-LBP operator, and hence better accuracies when compared with the other operators. Secondly, the Precision-Recall Curve in Fig. 4 is observed. It can be for plainly said that the proposed CS-LDMP outperforms CS-LBP, CS-LDP, and CS-LMP operators as the area under the curve (AUC) is greater by 20% with CS-LBP and CS-LMP and 5% higher than CS-LDP. Moreover, at any fixed recall, the precision of the proposed CS-LDMP is greater than CS-LBP, CS-LDP, and CS-LMP. Additionally, XCS-LMP is slightly better than XCS-LBP, which is 0.04 percent regarding AUC, and we can even say that when the recall value is small, precision of XCS-LMP appears to be greater than XCS-LBP. Finally, in the ROC curve as shown in Fig. 5, proposed CS-LDMP has AUC 2.5% more than CS-LBP and CS-LMP and 0.6% more than CS-LDP. However, the AUC of proposed XCS-LMP is very slightly less than XCS-LBP.

6 Conclusion

In this paper, two new local descriptors were presented, proposed CS-LDMP and XCS-LMP. CS-LDMP uses the advantages of CS-LDP and CS-LMP and therefore performs better than them. Therefore, trivially it performs better than

CS-LBP. It captures more texture information by using the center pixel information, which is the strength of CS-LDP and captures minute differences between images as in CS-LMP. Proposed XCS-LMP uses the powers of XCS-LBP, robustness to noise, tolerance to illumination changes, and CS-LMP capturing nuances between images. Experimentally, we have shown that proposed CS-LDMP outperforms CS-LBP, CS-LDP, and CS-LMP concerning accuracy, AUC of PR curve and AUC of ROC curve. Additionally, proposed XCS-LMP has a slight improvement over XCS-LBP regarding the same.

References

1. Hefeng Wu, Ning Liu, Xiaonan Luo, Jiawei Su, and Liangshi Chen. Real-time background subtraction-based video surveillance of people by integrating local texture patterns. *Signal, Image and Video Processing*, 8(4):665–676, 2014.
2. Gengjian Xue, Li Song, Jun Sun, and Meng Wu. Hybrid center-symmetric local pattern for dynamic background subtraction. In *Multimedia and Expo (ICME), 2011 IEEE International Conference on*, pages 1–6. IEEE, 2011.
3. Carolina Toledo Ferraz, Osmando Pereira Jr, and Adilson Gonzaga. Feature description based on center-symmetric local mapped patterns. In *Proceedings of the 29th Annual ACM Symposium on Applied Computing*, pages 39–44. ACM, 2014.
4. Caroline Silva, Thierry Bouwmans, and Carl Frélicot. An extended center-symmetric local binary pattern for background modeling and subtraction in videos. In *International Joint Conference on Computer Vision, Imaging and Computer Graphics Theory and Applications, VISAPP 2015*, 2015.
5. Timo Ojala, Matti Pietikäinen, and David Harwood. A comparative study of texture measures with classification based on featured distributions. *Pattern Recognition*, 29(1):51 – 59, 1996.
6. T. Ahonen, A. Hadid, and M. Pietikainen. Face description with local binary patterns: Application to face recognition. *IEEE Transactions on Pattern Analysis and Machine Intelligence*, 28(12):2037–2041, 2006.
7. T. Ojala, M. Pietikainen, and T. Maenpaa. Multiresolution gray-scale and rotation invariant texture classification with local binary patterns. *IEEE Transactions on Pattern Analysis and Machine Intelligence*, 24(7):971–987, 2002.
8. Bongjin Jun and Daijin Kim. Robust face detection using local gradient patterns and evidence accumulation. *Pattern Recognition*, 45(9):3304 – 3316, 2012. Best Papers of Iberian Conference on Pattern Recognition and Image Analysis (IbPRIA'2011).
9. Marko Heikkilä, Matti Pietikäinen, and Cordelia Schmid. Description of interest regions with center-symmetric local binary patterns. In *Computer vision, graphics and image processing*, pages 58–69. Springer, 2006.
10. Yongbin Zheng, Chunhua Shen, Richard Hartley, and Xinsheng Huang. Pyramid center-symmetric local binary/trinary patterns for effective pedestrian detection. In *Asian Conference on Computer Vision*, pages 281–292. Springer, 2010.
11. Abdelhamid Abdesselam. Improving local binary patterns techniques by using edge information. *Lecture Notes on Software Engineering*, 1(4):360, 2013.
12. Alex Krizhevsky and Geoffrey Hinton. Learning multiple layers of features from tiny images. 2009.

13. Christopher Kanan and Garrison W Cottrell. Color-to-grayscale: does the method matter in image recognition? *PloS one*, 7(1):e29740, 2012.
14. Tianqi Chen and Carlos Guestrin. Xgboost: A scalable tree boosting system. In *Proceedings of the 22Nd ACM SIGKDD International Conference on Knowledge Discovery and Data Mining*, pages 785–794. ACM, 2016.
15. Vincent Van Asch. Macro-and micro-averaged evaluation measures [[basic draft]], 2013.

A Laguerre Model-Based Model Predictive Control Law for Permanent Magnet Linear Synchronous Motor

Nguyen Trung Ty[1], Nguyen Manh Hung[1], Dao Phuong Nam[1(✉)], and Nguyen Hong Quang[2]

[1] Hanoi University of Science and Technology, Hanoi, Vietnam
trungty.bk@gmail.com, manhhung050795@gmail.com, nam.daophuong@hust.edu.vn
[2] Thai Nguyen University of Technology, Thai Nguyen, Vietnam
nhquang.tnut@gmail.com

Abstract. Linear motors have many advantages in comparison with rotary motors due to directly creating linear motion without gears or belts. The difficulties of designing the controller are that we need not only the tracking of position and velocity but guarantee that the voltage control and its variation are small enough as well. Model predictive control (MPC) is an advanced method of control that needs the corresponding predictive model. In this work, we propose the model predictive control based on Laguerre function with the constraints of voltage control and its variation. The numerical simulation generated by MATLAB–Simulink validates the performance of the proposed controller.

1 Introduction

The dynamic model of permanent magnet linear synchronous motor was presented in [1]. In this paper, authors use the model based on cascade structure and the current loop (inner loop) is designed on the dynamic coordinate. The position loop (outer loop) is simply deal with the controller designed based on Hurwitz polynomial although results in [1,2] can get the tracking of position and speed of motor. However, the constraints of the voltage input and its variation have not been considered in [1,2]. Thereby, the voltage input and its variation maybe out of the allowable values. In fact, it results in several problems, such as there does not exist any actuator that can pursue the signal from the designed controller. Authors in [3] proposed a MPC method that can deal with periodic disturbances, the repetitive tracking error is treated, but authors neglect the constraints of the control signal and state variables. The results of MPC in [4] can guarantee the tracking of position and speed of motor as well as keep the fluxes and currents within permissible values. However, the case of infinite prediction horizon is not considered.

V. Bhateja et al. (eds.), *Information Systems Design and Intelligent Applications*, Advances in Intelligent Systems and Computing 672,
https://doi.org/10.1007/978-981-10-7512-4_31

In this study, the model predictive control based on Laguerre function is designed. We also take into account the constraints of voltage input and it variation aim at dealing with above difficulties. Section 2 (main content) is divided into three subsections. In Sect. 2.1, we present the model predictive control based on Laguerre function. Section 2.2 applys the control law in Sect. 2.1 to the current loop of linear motor. Position and velocity of motor are controlled by controller supposed in Sect. 2.3.

2 Main Content

2.1 Model Predictive Control Based on Laguerre Function

In this section, we consider the discrete-time linear system described by:

$$x_m(k+1) = A_m x_m(k) + B_m u(k) \tag{1}$$

where x_m is vector of state variable and u_k is the input at the time k. Convert Eq. (1) as follows:

$$\begin{aligned} & x_m(k+1) - x_m(k) = A_m x_m(k) - A_m x_m(k-1) + B_m u(k) - B_m u(k-1) \\ & \Leftrightarrow \Delta x_m(k+1) = A_m \Delta x_m(k) + B_m \Delta u(k) \end{aligned} \tag{2}$$

We denote that $\Delta x_m(k+1) = x_m(k+1) - x_m(k)$ and $\Delta u(k) = u(k) - u(k-1)$ From (1), (2) and by letting $x(k+1) = [\Delta x_m(k+1) \quad x_m(k+1)]^T$, we arrive at:

$$x(k+1) = Ax(k) + B\,\Delta u(k) \tag{3}$$

where $A = \begin{bmatrix} A_m & \Theta_{n\times n} \\ A_m & I_{n\times n} \end{bmatrix} andB = \begin{bmatrix} B_m \\ B_m \end{bmatrix}$.

Predictive model of (3) at k_i is as follows:

$$x(k_i + l + 1|k_i) = Ax(k_i + l|k_i) + B\Delta u(k_i + l) \tag{4}$$

with $l = 0, \ldots N_p - 1$ and N_p is the prediction horizon. From (4), the sake of designing the MPC controller is finding the sequence of input signal $u(k)$ minimizing this under cost function:

$$J = \sum_{j=1}^{N_p} x(k_i + j|k_i)^T Q x(k_i + j|k_i) + \sum_{j=0}^{N_p} \Delta u(k_i + j)^T R \Delta u(k_i + j) \tag{5}$$

with $Q \geq 0, \quad R > 0$.

2.1.1 Finite Prediction Horizon

Consider (5), let:

$$\Delta u(k_i + k) = \sum_{j=1}^{N} c_j(k_i) l_j(k) \tag{6}$$

where c_j are the coefficients which depend only on the initial of the prediction horizon. $l_j(k)$ are the orthogonal Laguerre functions presented in [5]. By letting:

$$\eta = [c_1(k_i) \quad c_2(k_i) \ldots \; c_N(k_i)]^T \tag{7}$$

$$L(k) = [l_1(k) \quad l_2(k) \ldots \; l_N(k)]^T \tag{8}$$

We rewrite (6) to: $\Delta u(k_i + k) = L(k)^T \eta$

From model (4), we can obtain:

$$\begin{aligned} x(k_i + j|k_i) &= A^j x(k_i) + \sum_{i=0}^{j-1} A^{j-i-1} B \Delta u(k_i + i) \\ &= A^j x(k_i) + \sum_{i=0}^{j-1} A^{j-i-1} B L(i)^T \eta \\ &= A^j x(k_i) + \phi(j)^T \eta \end{aligned} \tag{9}$$

where $\phi(j)^T = \sum_{i=0}^{j-1} A^{j-i-1} B L(i)^T$. Substituting into the cost function, we obtain:

$$J = \sum_{j=1}^{N_p} \left(A^j x(k_i) + \phi(j)^T \eta \right)^T Q \left(A^j x(k_i) + \phi(j)^T \eta \right) + \sum_{j=0}^{N_p} (L(j)^T \eta)^T R (L(j)^T \eta)$$

$$\begin{aligned} J = &\sum_{j=1}^{N_p} \left(A^j x(k_i) + \phi(j)^T \eta \right)^T Q \left(A^j x(k_i) + \phi(j)^T \eta \right) \\ &+ \sum_{j=0}^{N_p} (L(j)^T \eta)^T R (L(j)^T \eta) \\ = &\sum_{j=1}^{N_p} \left(A^j x(k_i) + \phi(j)^T \eta \right)^T Q \left(A^j x(k_i) + \phi(j)^T \eta \right) \\ &+ \sum_{j=0}^{N_p} (L(j)^T \eta)^T R (L(j)^T \eta) \\ = &\eta^T \left(\sum_{j=1}^{N_p} \phi(j) Q \phi(j)^T + \sum_{j=0}^{N_p} L(j) R L(j)^T \right) \eta \\ &+ 2\eta \left(\sum_{j=1}^{N_p} \phi(j) Q A^j \right) x(k_i) + \sum_{j=1}^{N_p} x(k_i)^T (A^T)^m Q A^m x(k_i). \end{aligned} \tag{10}$$

Let $\Omega = \sum_{j=1}^{N_p} \phi(j)Q\phi(j)^T + \sum_{j=0}^{N_p} L(j)RL(j)^T$ and $\psi = \sum_{j=1}^{N_p} \phi(j)QA^j$ we call:

$$\eta^* = \arg\min J = \arg\min \left(\eta^T \Omega \eta + 2\eta^T \psi x(k_i)\right) \tag{11}$$

• Unconstrained MPC controller

In this case, (3) was considered without any constraints on inputs and state variables, we have:

$$\frac{\partial J}{\partial \eta} = 2\eta^T \Omega + 2\psi x(k_i) \tag{12}$$

Lead to:

$$\eta^* = -\Omega^{-1}\psi x(k_i) = -\left(\sum_{j=1}^{N_p} \phi(j)Q\phi(j)^T + \sum_{j=0}^{N_p} L(j)RL(j)^T\right)^{-1} \sum_{j=1}^{N_p} \phi(j)QA^j.x(k_i) \tag{13}$$

So that, the optimal input signal at time k_i:

$$\Delta u(k_i) = L(0)^T \eta^* \tag{14}$$

• Constrained MPC controller

We suppose that (1) has constraints as follows:

$$u_{\min} \leq u(k) \leq u_{\max} \tag{15}$$

We consider prediction model (4) and constraints (15) at time k_i:

$$u_{\min} - u(k_i - 1) \leq \Delta u(k_i|k_i) \leq u_{\max} - u(k_i - 1) \tag{16}$$

Furthermore, $\Delta u(k_i) = L(0)^T \eta$. So we rewrite (16) as:

$$M\eta \leq \gamma \tag{17}$$

Thus, we need to optimize:

$$J = \eta^T \Omega \eta + 2\eta^T \psi x(k_i) \tag{18}$$

Subject to: $M\eta \leq \gamma$.

Since J is the quadratic function, optimal issue in (12) can be solved by quadratic programming (QP) to obtain the solution η^*, the control signal is calculated by:

$$\Delta u(k_i) = L(0)^T \eta^* \tag{19}$$

2.1.2 Infinite Prediction Horizon

In this case, the under cost function is taken into account as:

$$J_\infty = \sum_{j=1}^{\infty} x(k_i + j|k_i)^T Q x(k_i + j|k_i) + \sum_{j=0}^{\infty} \Delta u(k_i + j)^T R \Delta u(k_i + j) \tag{20}$$

Theorem 1. *Consider system (1) without constraints, [6] demonstrated that this controller* $\Delta u(k) = Kx(k)$ *optimizes above cost function.*
with $K = (R + B^T P_\infty B)^{-1} B^T P_\infty A$ *and* P_∞ *satisfies:*

$$A^T \left[P_\infty - P_\infty B(R + B^T P_\infty B)^{-1} B^T P_\infty\right] A + Q - P_\infty = 0 \tag{21}$$

Remark 1. Without constraints, it is not difficult to find the solution of above algebraic Riccati equation. However, it is the hard problem when the constraints are considered.

Remark 2. In the presence of constraints, finding the solution in case of infinite prediction horizon is more difficult than in finite prediction horizon.

Theorem 2. *([6]) There exist* $\alpha > 1$ *to both of cost functions as described in (3), which have the same optimal solution:*

$$\begin{aligned}
J_\infty &= \sum_{j=1}^{\infty} x(k_i + j|k_i)^T Q x(k_i + j|k_i) + \sum_{j=0}^{\infty} \Delta u(k_i + j)^T R \Delta u(k_i + j) \\
\hat{J}_\infty &= \sum_{j=1}^{\infty} \alpha^{-2j}\, \hat{x}(k_i + j|k_i)^T Q_\alpha \hat{x}(k_i + j|k_i) \\
&+ \sum_{j=0}^{\infty} \alpha^{-2j}\, \Delta u(k_i + j)^T R_\alpha \Delta u(k_i + j)
\end{aligned} \tag{22}$$

where x *is state variable along the trajectory of this system:*

$$\begin{cases} \hat{x}(k_i + l + 1|k_i) = \frac{A}{\alpha}\hat{x}(k_i + l|k_i) + \frac{B}{\alpha}\, \Delta u(k_i + l|k_i) \\ \hat{x}(k_i) = x(k_i) \\ l = 0, 1, \ldots N_p - 1 \end{cases} \tag{23}$$

with $Q_\alpha = \alpha^{-2}Q + (1 - \alpha^{-2})P_\infty$, $R_\alpha = \alpha^{-2}R$.
By resorting Theorem 2, we perhaps find the solution of $\hat{J}_\infty$ *for (13) in place of finding the solution of* J_∞. *Due to suitable* α, *we have:*

$$\alpha^{-2N_p}\, \hat{x}(k_i + N_p|k_i)^T Q_\alpha \hat{x}(k_i + N_p|k_i) + \alpha^{-2N_p}\, \Delta u(k_i + N_p)^T R_\alpha \Delta u(k_i + N_p) \approx 0 \tag{24}$$

Thereby, we can get:

$$\hat{J}_{N_p} \approx \sum_{j=1}^{N_p} \alpha^{-2j}\, \hat{x}(k_i + j|k_i)^T Q_\alpha \hat{x}(k_i + j|k_i) + \sum_{j=0}^{N_p} \alpha^{-2j}\, \Delta u(k_i + j)^T R_\alpha \Delta u(k_i + j) \tag{25}$$

Thus, we would find the solution of $\hat{J}_{N_p}$ *with finite prediction horizon instead of* $\hat{J}_\infty$. *In the finite prediction horizon, let:* $\bar{x}(k_i + j|k_1) = \alpha^{-j}\, \hat{x}(k_i + j|k_1)$ *and* $\Delta\bar{u}(k_i + j|k_1) = \alpha^{-j}\, \Delta u(k_i + j|k_1)$.
We arrive at:

$$\begin{aligned}
\bar{x}(k_i + j + 1|k_1) &= \alpha^{-j-1}\left(\hat{A}\hat{x}(k_i + j|k_1) + \hat{B}\Delta u(k_i + j)\right) \\
&= \alpha^{-1}\hat{A}\bar{x}(k_i + j|k_1) + \alpha^{-1}\hat{B}\Delta\bar{u}(k_i + j)
\end{aligned} \tag{26}$$

$$\hat{J}_{Np} = \sum_{j=1}^{N_p} \bar{x}(k_i + j|k_i)^T Q_\alpha \bar{x}(k_i + j|k_i) + \sum_{j=0}^{N_p} \Delta\bar{u}(k_i + j)^T R_\alpha \Delta\bar{u}(k_i + j) \quad (27)$$

2.2 MPC Controller for Current Subsystem

As mentioned in [2], current loop model is described by:

$$\begin{cases} \frac{di_{sd}}{dt} = -\frac{R_s}{L_{sd}} i_{sd} + \left(\frac{2\pi}{\tau} v\right) \frac{L_{sq}}{L_{sd}} i_{sq} + \frac{U_{sd}}{L_{sd}} \\ \frac{di_{sq}}{dt} = -\frac{R_s}{L_{sq}} i_{sq} - \left(\frac{2\pi}{\tau} v\right) \frac{L_{sd}}{L_{sq}} i_{sd} - \left(\frac{2\pi}{\tau} v\right) \frac{\psi_p}{L_{sq}} + \frac{U_{sq}}{L_{sq}} \end{cases} \quad (28)$$

where $L_{sd} and L_{sq}$ are the d-axis, q-axis inductance, v is the velocity of motor, $i_{sd}, i_{sq}, U_{sd}, U_{sq}$ denote the dq components of the primary. ψ_p is the value of pole flux. Denote that i_{sd}^d and i_{sq}^d are desired outputs and $e_{sd} = i_{sd} - i_{sd}^d$, $e_{sq} = i_{sq} - i_{sq}^d$. Substituting into (28):

$$\begin{cases} \frac{de_{sd}}{dt} = -\frac{R_s}{L_{sd}} e_{sd} + \left(\frac{2\pi}{\tau} v\right) \frac{L_{sq}}{L_{sd}} e_{sq} + \frac{U_{sd}}{L_{sd}} \\ \qquad - \left(-\frac{R_s}{L_{sd}} i^d{}_{sd} + \left(\frac{2\pi}{\tau} v\right) \frac{L_{sq}}{L_{sd}} i^d{}_{sq}\right) \\ \frac{de_{sq}}{dt} = -\frac{R_s}{L_{sq}} e_{sq} - \left(\frac{2\pi}{\tau} v\right) \frac{L_{sd}}{L_{sq}} e_{sd} + \frac{U_{sq}}{L_{sq}} \\ \qquad - \left(\frac{R_s}{L_{sq}} i^d{}_{sq} - \left(\frac{2\pi}{\tau} v\right) \frac{L_{sd} i^d{}_{sd} + \psi_p}{L_{sq}}\right) \end{cases} \quad (29)$$

Define:

$$A_m(t) = \begin{bmatrix} -\frac{R_s}{L_{sd}} & \left(\frac{2\pi}{\tau} v(t)\right) \frac{L_{sq}}{L_{sd}} \\ -\frac{R_s}{L_{sq}} & -\left(\frac{2\pi}{\tau} v(t)\right) \frac{L_{sd}}{L_{sq}} \end{bmatrix}, B_m(t) = \begin{bmatrix} 1 & 0 \\ 0 & 1 \end{bmatrix}$$

$$u(t) = \begin{bmatrix} \frac{U_{sd}}{L_{sd}} - i_{sd}^d - \left(-\frac{R_s}{L_{sd}} i^d{}_{sd} + \left(\frac{2\pi}{\tau} v(t)\right) \frac{L_{sq}}{L_{sd}} i^d{}_{sq}\right) \\ \frac{U_{sq}}{L_{sq}} - i_{sq}^d - \left(\frac{R_s}{L_{sq}} i^d{}_{sq} - \left(\frac{2\pi}{\tau} v(t)\right) \frac{L_{sd} i^d{}_{sd} + \psi_p}{L_{sq}}\right) \end{bmatrix}$$

(29) is rewritten as:

$$\frac{dx}{dt} = A_m(t)x(t) + B_m u(t) \quad (30)$$

Discretize system (30) we obtain:

$$x(k+1) = \widehat{A}_m(k)x(k) + \widehat{B}_m u(k) \quad (31)$$

where

$$\widehat{A}_m(k) = e^{AT_s}$$

$$\widehat{B}_m = \int_0^{T_s} B_m dt = T_s B_m$$

Then we can apply MPC controller designed in Sect. 2.1 to current loop.

2.3 Design Outer Controller for Position Loop

In [1], the model of position loop is given by:

$$\begin{cases} \frac{dv}{dt} = \frac{p}{m}(F - F_c) \\ F = \frac{3\pi}{p\tau}\left[\psi_p i_{sq} + (L_{sd} - L_{sq})\, i_{sd} i_{sq}\right] \\ \frac{dx}{dt} = v \end{cases} \tag{32}$$

where F is the magnetic force impacted on the primary part of motor, F_c is the resisting force, m is the mass of the primary part, p is the number of pole. The control law for this subsystem is simply chosen:

$$\begin{cases} v_c = \dot{x}_r - k_1(x - x_r) \\ i_{sq}^r = \frac{\hat{F}_c + \frac{m}{p}\dot{v}_c - k_2(v - v_c)}{\frac{3\pi}{p\tau}\left[\psi_p + (L_{sd} - L_{sq}) i_{sd}\right]} \end{cases} \tag{33}$$

Using above controller, [1] demonstrated that the position and velocity asymptotically converge to its desired values.

3 Simulation Results

Based on the above conclusions, the simulation model of PMLSM and controller are constructed in MATLAB environment. The parameters of PMLSM are given:

Number of poles	2
Pole step	72 mm
Rotor mass	3.5 kg
Phase coil resistance	8.5 Ω
d-axis inductance	4.1 mH
q-axis inductance	4.1 mH
Flux	0.8 Wb

Considering the desired trajectory of motors is expressed by: $x^d(t) = t$, with $T_s = 0.01$, we obtain the following coefficients:

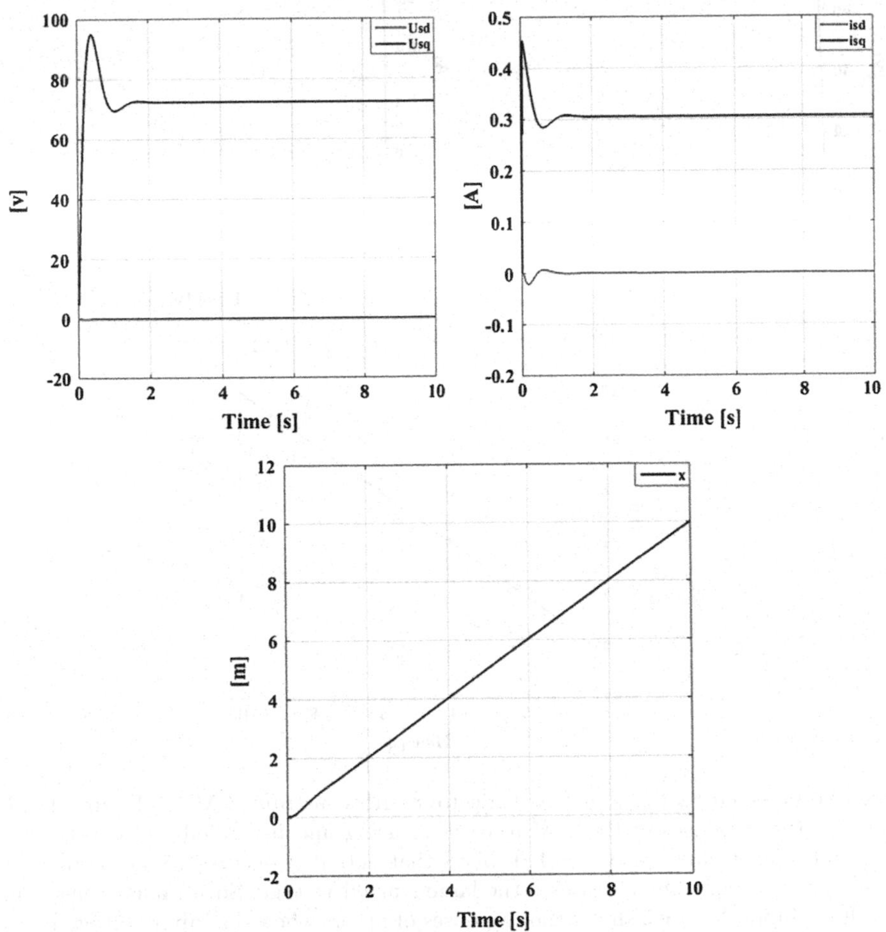

Fig. 1. Voltages, current, and actual trajectory with unconstrained MPC

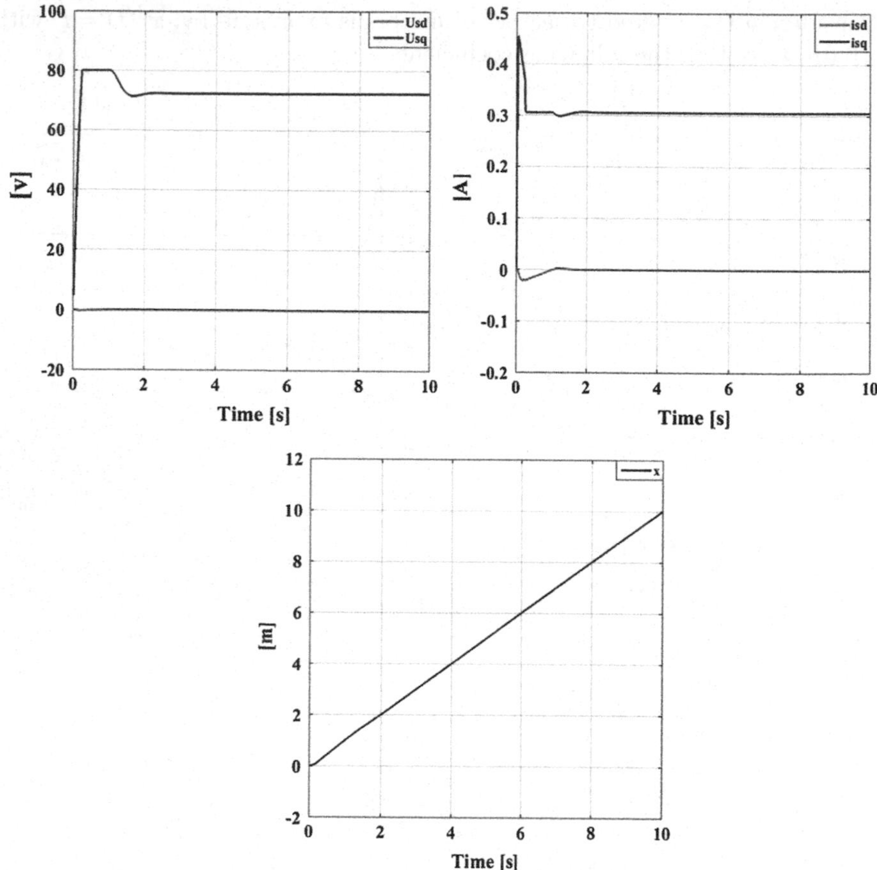

Fig. 2. Voltages, current, and actual trajectory with constrained MPC. Figures 1 and 2 describe the responses of PMLSM in cases of using unconstrained and constrained MPC controller. In first case, Fig. 1 displays that actual trajectory of motor pursues the desired trajectory but it requires the large started voltage. So we must constraint the voltage input. Figure 2 shows the responses of motor when the input voltage u_{sq} is limited by $-80V \leq u_{sq} \leq 80V$ and $-20V \leq \Delta u_{sq} \leq 20V$

4 Conclusions

In this work, MPC controller is designed for linear motor with constraints on voltage inputs and state variables. Laguerre model is used to find the control law for optimizing cost function. Multiplier coefficient α is applied to approximate the optimal problem with infinite prediction horizon by finite prediction horizon which can be applied the results of Laguerre model to overcome several difficulties. Finding the multiplier coefficient α meet some problems that can be dealt by LMIs.

References

1. Quang H. Nguyen, Nam P. Dao, Hung M. Nguyen, Ty T. Nguyen (2017): Design an Exact Linearization Controller for Permanent Stimulation Synchronous Linear Motor Polysolenoid.
2. Quang H. Nguyen, Nam P. Dao, Ty T. Nguyen, Hung M. Nguyen (2016): Flatness Based Control Structure for Polysolenoid Permanent Stimulation Linear Motors.
3. Runzi Cao, Kay-Soon Low (2007): Repetitive Model Predictive Control of a Precision Linear Motor Drive, The 33rd Annual Conference of the IEEE Industrial Electronics Society (IECON) Nov. 5–8, 2007, Taipei, Taiwan.
4. A.A. Hassan, J. Thomas (2008): Model Predictive Control of Linear Induction Motor Drive, The International Federation of Automatic Control Seoul, Korea, July 6–11, 2008.
5. A. Bakan, T. Craven, G. Csordas, Interpolation and the Lagurre polya-class.
6. Liuping Wang (2009): Model Predictive Control System Design and Implementation Using MATLAB.
7. F.L. Lewis, D. Vrabie, V.L. Syrmos (2012) optimal control-3rd. edition.

Hybrid Model of Self-Organized Map and Integrated Fuzzy Rules with Support Vector Machine: Application to Stock Price Analysis

Duc-Hien Nguyen[1(✉)] and Van-Minh Le[2]

[1] Hue University of Sciences, Hue University, Hue, Vietnam
ndhien@cit.udn.vn
[2] College of Information Technology, University of Danang, Danang, Vietnam
vanminh.le246@gmail.com

Abstract. Prediction of stock price is always an interesting task. However, it is not easy to make this prediction with high accuracy. Recently, plenty of combinations of statistical methods have been proposed. The main direction of these methods is that combination of regression learner (e.g., SVM) and a clustering of data (e.g., SOM). While these methods make relative success, their extensibility is still under discussion. In this paper, we propose an hybrid model of self-organized map and integrated fuzzy rules with support vector machine. The proposition method is evaluated to be a good approach to apply to stock price analysis. Moreover, this method provides interpretable rules which can be understood, calibrated, and modified by experts in order to direct the learning phase.

Keywords: Support Vector Machine · Fuzzy model · Self-Organized Map · Machine learning · Stock prediction

1 Introduction

As far as we know that prediction of stock price is always an interesting but difficult task. Plenty of combinations of statistical methods have been recently proposed. These methods make relative success but their extensibility is still under discussion.

Support Vector Machine was introduced by Vapnik in 1995 [1]. This machine learning model basing on statistical learning theory is proposed to solve plenty of classification problems [1–3]. Many other studies then propose to apply SVM to solve optimization problem by using regression, which is called Support Vector Machine Regression (SVR) [4–7]. Moreover, SVM is also used to extract fuzzy models which are useful for solving regression problems [7–10] as well as classification ones [3,8,9,11].

The most important difference between fuzzy SVM and Original SVM is the "interpretability" provided by fuzzy SVM. This ability makes the fuzzy SVM

V. Bhateja et al. (eds.), *Information Systems Design and Intelligent Applications,* Advances in Intelligent Systems and Computing 672,
https://doi.org/10.1007/978-981-10-7512-4_32

model more understandable than Original SVM [11,12]. However, the accuracy of all SVM models depends on the number of Support Vectors (generated by training phase), which means that more Support Vectors are expected to bring a more accurate result but cause a problem of complexity. In the case of fuzzy SVM is used, the number of fuzzy rules must be increased [8,9,11,13]. Thus, in order to increase the accuracy of the model and to maintain the complexity, we should reduce "interpretability" of the fuzzy model. Therefore, we propose an approach of Hybrid model of self-organized map and integrated fuzzy rules with support vector machine.

This paper is organized as follows: in the Sect. 2, we present the related work to our study; Sect. 3 is the proposition; in Sect. 4, our proposition approach is implemented and evaluated. The last section is the conclusion.

2 Related Work

First, our work is about prediction using machine learning algorithm. Then, we begin with the well-known methods for prediction. The method Support Vector Machine (SVM) introduced by Vapnik in 1995 [1] is one of the most famous methods. In order to improve the performance of prediction phase, most studies focus on combination SVM with other approaches.

The most direct combination is the (Self-Organized Map) SOM and SVM (proposed [14–16]). These works received relative success because they take advantage of SOM to cluster the relative data, which reduces the space and also the noise. In another case of SOM and SVM, a proposed approach is to directly use fuzzy SVM to improve the performance of prediction [13].

Another direction which provides promising outcome is the combination SOM with a neural network learner. In some cases, the artificial neural network fuzzy inference system (ANFIS) works well, which leads to the combination SVM-ANFIS [17,18].

From all the ideas of related work, we think about a different approach which integrates of rules into the learning phase. We all know that machine learning needs an expert factor to train because the statistical training might lead the prediction to the bad result. Once we extract the rules from the data, the expert can calibrate them to direction prediction.

3 Proposition

In this section, we bound our proposition to the problem of prediction of stock price. It is also true that our method is extensible for other problems but we focus on this problem in order to clarify the explanation.

In this case, our approach to solve the problem is the hybrid of self-organized map and integrated fuzzy rules with support vector machine. Then, the model is separated into three parts: SOM clustering part, fuzzy rules extraction part, and Support Vector Machine learning part. Figure 1 shows the structure of this model.

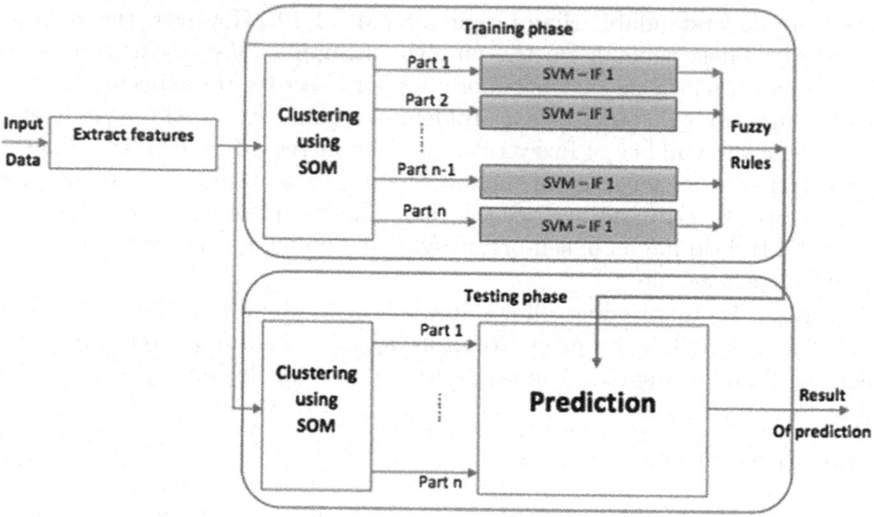

Fig. 1. Proposed model

In our model, the SOM [19,20] is used to cluster the data which are similar to each other. Then, each cluster is fed to the SVM-IF to extract the fuzzy rules. Thus, all the fuzzy rules are used to train the model. The result of training model is used in prediction phase. This model works in three following steps.

3.1 Selection of Input Attributes

According to the analysis Cao and Tay in [6], it is useful for the prediction if we extract the data basing on the 5-day relative difference in percentage of price RDP. This extraction is promised to bring a significant accuracy to the prediction, which motivates us to use this approach for extracting the attributes of our data. Another reason for us to use this extraction is that we want to compare our result with that of the studies proposed in [6,21].

Table in Fig. 2 shows the attributes that we extracted and the formula corresponding each attributes to compute the quantitative values. In these formulas, $P(i)$ is the index of closing session of the day i, and $EMA_m(i)$ is m-day exponential moving average of the price in closing session of that day.

3.2 Clustering Using SOM

It is true that we want to reduce the number of attributes by selecting the useful ones instead of using all of attributes. It also reduces the complexity if we group the similar attributes. We decided to make clustering on the data after the first phase.

Variable	Attributes	Formula
x_1	EMA100	$P_i - \overline{EMA_{100}(i)}$
x_2	RDP-5	$(P(i) - P(i-5))/P(i-5) * 100$
x_3	RDP-10	$(P(i) - P(i-10))/P(i-10) * 100$
x_4	RDP-15	$(P(i) - P(i-15))/P(i-15) * 100$
x_5	RDP-20	$(P(i) - P(i-20))/P(i-20) * 100$
y	RDP+5	$(\overline{P(i+5)} - \overline{P(i)})/\overline{P(i)} * 100$ $\overline{P(i)} = \overline{EMA_3(i)}$

Fig. 2. Extract attributes from input data

Recently, plenty of studies propose to use SOM to cluster the data, especially the data of stock market [21, 22]. In this work, we also use SOM to cluster the data which brings us two advantages:

1. Reducing size of the data, which accelerates the computation
2. Grouping the data basing on statistical values of these data, which reduces the noise.

3.3 Extract Fuzzy Rules by SVM-IF Algorithm

At this phase, each cluster of data, which comes from SOM, is fed the Support Vector Machine - Interpretable Fuzzy Rules (SVM-IF) in order to extract the fuzzy rules. These fuzzy rules are considered the result of training phase, which is used in the next predicting phase. The most important contribution of these fuzzy rules is the interpretable meaning which can be understood and modified by experts in order to adapt the model into real application. Moreover, the interpretability provides experts a method to re-evaluate the rules which might not be suitable in some certain cases of data. Figure 3 shows the SVM-IF Algorithm which makes this training phase.

3.4 Prediction Phase

At this phase, all fuzzy rules from SVM-IF training are used to predict. In our study, we use this model to predict the stock prices. The detail of this implementation is described in the next section.

Procedure ModelExtraction($\mathcal{H}$, k, tol)
Initialize: $C, \varepsilon, \sigma, step$
while error>tol **do**
$f(x) = \sum_{i=1}^{l} (\alpha_i - \alpha_i^*) K(x_i, x) + b$
$SV = \{(\alpha_i - \alpha_i^*) : (\alpha_i - \alpha_i^*) \neq 0, i \in \{0, \ldots, l\}\}$
InterpretabilityTest(SV, n, σ, k)
Calibrate kernel: $H' = \begin{bmatrix} D' & -D' \\ -D' & D' \end{bmatrix}$, with $D'_{ij} = \frac{\langle \varphi(x_i), \varphi(x_j) \rangle}{\sum_j \langle \varphi(x_i), \varphi(x_j) \rangle}$
$error = E[\|f(x) - \mathcal{H}\|^2]$
$\varepsilon = \varepsilon + step$
end while
$\sigma_i(t+1) = \sigma_i(t) + \delta \varepsilon_{1,i} \left[\frac{(x-c)^2}{\sigma^3} \exp\left(-\frac{(x-c)^2}{2\sigma^2}\right) \right]$
$c_i(t+1) = c_i(t) + \delta \varepsilon_{1,i} \left[\frac{-(x-c)}{\sigma^2} \exp\left(-\frac{(x-c)^2}{2\sigma^2}\right) \right]$
return $f(x) = \frac{\sum_{i=1}^{l} (\alpha_i - \alpha_i^*) K(x_i, x)}{\sum_{i=1}^{l} (\alpha_i - \alpha_i^*)}$
end procedure
Procedure InterpretabilityTest(SV, n, σ, k)
repeat
Compute similarity between Fuzzy sets: $S^G(A_i, A_j) = \frac{\sigma e^{-\frac{d^2}{\sigma^2}}}{2\sigma - \sigma e^{-\frac{d^2}{\sigma^2}}}$
Select set A_i^* and A_j^* such that $S^G(A_i^*, A_j^*) = max_{i,j}\{S^G(A_i, A_j)\}$
if $S^G(A_i^*, A_j^*) > k$ **then**
Combine A_i^* and A_j^* into new Fuzzy Set A_k
end if
until no Fuzzy Sets with similarity $S^G(A_i, A_j) > k$
end procedure

Fig. 3. SVM-IF Algorithm

4 Implementation and Evaluation

4.1 Description of Case Study

In this section, we present the implementation and evaluation of our proposition model. The model is implemented with MATLAB [23] by using some library of this software.

First, the input data are clustered by using SOM toolbox developed by Alhoniemi et al. [19]. Then, the data are passed to LIBSVM developed by Chang et al. [2] to do the SVM phase. The library LIBSVM helps us to generate Support Vectors from the training data. These Support Vectors are the base for the SVM-IF Algorithm to extract the fuzzy rules. Finally, we use function EVALFIS in the fuzzy logic library developed by Ahmed [24].

Name	Time	Size of training set	Size of testing set
IBM Corporation stock (IBM)	03/01/2000 - 30/06/2010	2409	200
Apple inc. stock (APPL)	03/01/2000 - 30/06/2010	2409	200
Standard & Poor's stock index (S&P500)	03/01/2000 - 23/12/2008	2028	200
Down Jones Industrial Average index (DJI)	02/01/1991 - 28/03/2002	2352	200

Fig. 4. Description of data set

#	Rules
r1	IF x1=Gaussmf(0.10,-0.02) and x2=Gaussmf(0.10,-0.08) and x3=Gaussmf(0.10,0.02) and x4=Gaussmf(0.10,0.04) and x5=Gaussmf(0.10,0.02) THEN z=-0.02
r2	IF x1=Gaussmf(0.10,0.02) and x2=Gaussmf(0.09,-0.00) and x3=Gaussmf(0.10,0.06) and x4=Gaussmf(0.10,0.05) and x5=Gaussmf(0.09,0.00) THEN z=0.04
r3	IF x1=Gaussmf(0.09,-0.04) and x2=Gaussmf(0.10,0.07) and x3=Gaussmf(0.09,-0.16) and x4=Gaussmf(0.09,-0.14) and x5=Gaussmf(0.11,-0.05) THEN z=0.16
r4	IF x1=Gaussmf(0.09,0.01) and x2=Gaussmf(0.10,0.08) and x3=Gaussmf(0.09,-0.06) and x4=Gaussmf(0.09,-0.09) and x5=Gaussmf(0.09,-0.04) THEN z=0.01
r5	IF x1=Gaussmf(0.09,-0.05) and x2=Gaussmf(0.09,0.04) and x3=Gaussmf(0.10,-0.13) and x4=Gaussmf(0.10,-0.08) and x5=Gaussmf(0.08,-0.04) THEN z=-0.18

Fig. 5. Fuzzy Rules extracted

The case study of our work is the prediction of stock price of some corporation in USA such as IBM Corporation stock (IBM), the Apple inc. stock (APPL), the Standard & Poors stock index (S&P500), and the Down Jones Industrial Average index (DJI). The data are acquired from Yahoo Finance http://finance.yahoo.com/. Figure 4 presents the table of data which are used for this work.

4.2 Extraction of Fuzzy Rules

As we mentioned in the previous section, we integrate the fuzzy rule extraction in the training phase. Then, it is necessary to know how these rules look like.

In this part, we present a short demonstration of the fuzzy rules extracted from the data. Figure 5 shows the fuzzy rules which are extracted by training the data set of stock index S&P500. In the table, there are five rules which are extracted from only one cluster of data. The more clusters are generated by SOM, the more rules would be considered in the next phase. However, it is much better than f-SVM approach (proposed in [13]) which outcomes more than 450 rules for each cluster. This result of experiment is acceptable for what we expected because less rules reduce complexity.

Index	# clusters	SOM + ANFIS			SOM + SVM		
		NMSE	MAE	DS	NMSE	MAE	DS
IBM	6	1.2203	0.0617	47.74	1.1054	0.0564	48.05
APPL	55	2.8274	0.0650	49.75	1.0877	0.0474	52.27
SP500	6	1.7836	0.1421	48.24	1.1100	0.1200	51.25
DJI	35	1.7602	0.1614	49.75	1.0660	0.1104	50.86

Fig. 6. First quantitative result table

Index	# cluster	SOM + f-SVM				SOM + SVM-IF		
		NMSE	MAE	DS	#Rules	NMSE	MAE	DS
IBM	6	1.1028	0.0577	44.22	5*6	**1.0324**	**0.0554**	**50.75**
APPL	55	1.1100	0.0445	52.76	5*55	**1.0467**	**0.0435**	**53.27**
SP500	6	1.1081	0.1217	52.76	5*6	**1.0836**	**0.1207**	**53.27**
DJI	35	1.0676	0.1186	50.25	5*35	**1.0459**	**0.1181**	**51.76**

Fig. 7. Second quantitative result table

4.3 Quantitative Evaluation

It is true that a proposed approach needs quantitative evaluation. In this case, we make compassion between our approach SVM-IF and SOM+AFIS, SOM+SVM (in [21,22]) and model of SOM+f-SVM propose in [13].

Tables in Figs. 6 and 7 show the comparison of our approach and other works. In this case, we use normalised mean square error (NMSE) [25], mean absolute error [26] and the distances as measurement. The implementations run on 200 entries of data.

According to the tables, our proposition prevails significantly the SOM+ANFIS. It is also better than original SOM+SVM and SOM+f-SVM. More important contribution of this approach is that it reduces many rules which accelerate the executions.

4.4 Quantitative Evaluation

There is another contribution of our approach that is not yet evaluated quantitatively. This is the interpretability of the fuzzy rules. As we can see in Fig. 5, the extracted rules are readable by human, which is useful for experts to inject their own rules in order to direct the learning of the machine. This contribution opens another perspective of this research.

5 Conclusion

In this paper, we propose a Hybrid model of self-organized map and integrated fuzzy rules with support vector machine. Inside our proposition is the SVM-IF Algorithm which allows us to extract the priori knowledge from data. This knowledge is formalized in the form of fuzzy rules (extracted the relations of Support Vectors which are clustered from SOM). These rules are used for prediction in the problem of classification as well as regression problems.

Our proposition method is implemented and evaluated by a case study of prediction of stock price. In this case, our proposition proves to be better than other methods.

The extracted fuzzy rules are also another contribution of our word which opens another perspective. These rules are interpretable, which can be calibrated, modified by the experts and then injected into the training phase in order to direct the learning. This idea becomes interesting for ones who want to hybrid the human and machine in machine learning domain.

References

1. C. Cortes and V. Vapnik, "Support-vector networks," *Machine learning*, vol. 20, no. 3, pp. 273–297, 1995.
2. C.-W. Hsu, C.-C. Chang, C.-J. Lin *et al.*, "A practical guide to support vector classification," 2003.
3. J.-H. Chiang and P.-Y. Hao, "Support vector learning mechanism for fuzzy rule-based modeling: a new approach," *IEEE Transactions on Fuzzy systems*, vol. 12, no. 1, pp. 1–12, 2004.
4. A. J. Smola and B. Schölkopf, "A tutorial on support vector regression," *Statistics and computing*, vol. 14, no. 3, pp. 199–222, 2004.
5. B. Schölkopf, A. J. Smola, R. C. Williamson, and P. L. Bartlett, "New support vector algorithms," *Neural computation*, vol. 12, no. 5, pp. 1207–1245, 2000.
6. L.-J. Cao and F. E. H. Tay, "Support vector machine with adaptive parameters in financial time series forecasting," *IEEE Transactions on neural networks*, vol. 14, no. 6, pp. 1506–1518, 2003.
7. V. Uslan and H. Seker, "Support vector-based takagi-sugeno fuzzy system for the prediction of binding affinity of peptides," in *Engineering in Medicine and Biology Society (EMBC), 2013 35th Annual International Conference of the IEEE*. IEEE, 2013, pp. 4062–4065.
8. D. Martens, J. Huysmans, R. Setiono, J. Vanthienen, and B. Baesens, "Rule extraction from support vector machines: an overview of issues and application in credit scoring," in *Rule extraction from support vector machines*. Springer, 2008, pp. 33–63.
9. N. Barakat and A. P. Bradley, "Rule extraction from support vector machines: a review," *Neurocomputing*, vol. 74, no. 1, pp. 178–190, 2010.

10. W.-H. Hsu, Y.-Y. Chiang, W.-Y. Lin, W.-C. Tai, and J.-S. Wu, "Svm-based fuzzy inference system (svm-fis) for frequency calibration in wireless networks," in *Proceedings of the 3rd International Conference on Communications and information technology*. Citeseer, 2009, pp. 207–212.
11. J. L. Castro, L. Flores-Hidalgo, C. J. Mantas, and J. M. Puche, "Extraction of fuzzy rules from support vector machines," *Fuzzy Sets and Systems*, vol. 158, no. 18, pp. 2057–2077, 2007.
12. J. Yen and R. Langari, *Fuzzy logic: intelligence, control, and information*. Prentice-Hall, Inc., 1998.
13. D.-H. Nguyen and M.-T. Le, "A two-stage architecture for stock price forecasting by combining som and fuzzy-svm," arXiv preprint arXiv:1408.5241, 2014.
14. T.-S. Li and C.-L. Huang, "Defect spatial pattern recognition using a hybrid som–svm approach in semiconductor manufacturing," *Expert systems with Applications*, vol. 36, no. 1, pp. 374–385, 2009.
15. S. Patra and L. Bruzzone, "A novel som-svm-based active learning technique for remote sensing image classification," *IEEE Transactions on Geoscience and Remote Sensing*, vol. 52, no. 11, pp. 6899–6910, 2014.
16. W. Wei, L. Xin, X. Min, P. Jinrong, and R. Setiono, "A hybrid som-svm method for analyzing zebra fish gene expression," in *Pattern Recognition, 2004. ICPR 2004. Proceedings of the 17th International Conference on*, vol. 2. IEEE, 2004, pp. 323–326.
17. Z. Avdagic and A. Midzic, "The effects of combined application of som, anfis and subtractive clustering in detecting intrusions in computer networks," in *Information and Communication Technology, Electronics and Microelectronics (MIPRO), 2014 37th International Convention on*. IEEE, 2014, pp. 1435–1440.
18. V. Nourani, M. T. Alami, and F. D. Vousoughi, "Hybrid of som-clustering method and wavelet-anfis approach to model and infill missing groundwater level data," *Journal of Hydrologic Engineering*, vol. 21, no. 9, p. 05016018, 2016.
19. J. Vesanto, J. Himberg, E. Alhoniemi, and J. Parhankangas, "Som toolbox for matlab 5," *Helsinki University of Technology, Finland*, 2000.
20. T. Kohonen, "The self-organizing map," *Proceedings of the IEEE*, vol. 78, no. 9, pp. 1464–1480, 1990.
21. F. E. H. Tay and L. J. Cao, "Improved financial time series forecasting by combining support vector machines with self-organizing feature map," *Intelligent Data Analysis*, vol. 5, no. 4, pp. 339–354, 2001.
22. S.-H. Hsu, J. P.-A. Hsieh, T.-C. Chih, and K.-C. Hsu, "A two-stage architecture for stock price forecasting by integrating self-organizing map and support vector regression," *Expert Systems with Applications*, vol. 36, no. 4, pp. 7947–7951, 2009.
23. W. L. Martinez, A. R. Martinez, A. Martinez, and J. Solka, *Exploratory data analysis with MATLAB*. CRC Press, 2010.
24. N. Ahmed, "Fuzzy logic control using matlab part ii," 2003.
25. Y. Ephraim and D. Malah, "Speech enhancement using a minimum-mean square error short-time spectral amplitude estimator," *IEEE Transactions on Acoustics, Speech, and Signal Processing*, vol. 32, no. 6, pp. 1109–1121, 1984.
26. C. J. Willmott and K. Matsuura, "Advantages of the mean absolute error (mae) over the root mean square error (rmse) in assessing average model performance," *Climate research*, vol. 30, no. 1, pp. 79–82, 2005.

Translation of UNL Expression into Vietnamese Compound Sentence Based on DeConverter Tool

Thuyen Thi Le Phan(✉) and Hung Trung Vo

The University of Danang, Danang, Vietnam
thuyenptl@gmail.com, vthung@dut.udn.vn

Abstract. DeConverter is one of the core software of the Universal Networking Language (UNL) translation system. Automatic translation system based on UNL consists of two components: EnConverter and DeConverter. EnConverter is used to translate a sentence from natural language into an equivalent UNL expression, and DeConverter is used to create a sentence on natural language from an UNL expression. The UNL system has supported over 48 different languages, but not much as certain research has been done for Vietnamese. In this paper, we describe the process of the translation from UNL expressions with scope into Vietnamese sentences by analyzing Vietnamese grammar and the semantic relationships in UNL expressions with scope node and how rules of translation from UNL expressions to equivalent Vietnamese sentences are built. We tested the UNL–Vietnamese–English translation on 500 different expressions alternatively by the DeCoVie tool and the English server, which was then compared to the manual translations. The results showed that the two machine translations are synonymous and 90% are identical in terms of grammar.

1 Introduction

Along with the development of information technology, people have created an enormous amount of information that can be found on the Internet. Despite the abundance of information on the Internet, people still cannot fully exploit it due to a variety of reasons; one of the most important reasons is language barriers. It promotes the need to build a tool that can translate from a natural language into another natural language; this process is called machine translation. Machine translation through an intermediate language has recently appealed to several experts of auto-translators as it reduces the cost of building the system for translated pairs from n*(n−1)/2 down to 2*n. In addition, compared to other methods, the method of translation through an intermediary language solves the case of lacking language resources [1].

One of the projects following this approach was called UNL, which was proposed and implemented by United Nations University, Tokyo, Japan [2]. The mission of UNL is to allow people from all over the world to access information on the Internet in their own language. UNL is an artificial language developed independently based on the mediated method for machine translation. It can express all the knowledge of natural language without any ambiguity. UNL includes elements of the natural language: Universal Word

V. Bhateja et al. (eds.), *Information Systems Design and Intelligent Applications*, Advances in Intelligent Systems and Computing 672,
https://doi.org/10.1007/978-981-10-7512-4_33

(*UW*), relation, attributes, and UNL Knowledge Base (*UNLKB*). UNL links UWs based on semantic and attribute relationships to create sentences. These links are called relations; the attributes are used to describe the speaker's perspectives, and the linguistic knowledge is used to define possible relations between concepts in language.

Currently, researchers in many countries have been participated in the project and are developing a UNL system for their languages. The purpose is to reduce translations of document archives between language pairs when they are integrated into the system. For example, Vietnamese documents are converted to UNL and stored in this format. When needed, the UNL documents will be used to convert to any other languages if the server is integrated into the system. Each language server has two components: EnConverter and DeConverter. EnConverter is used to convert a natural language sentence to an equivalent UNL expression, and DeConverter is used to create a natural language sentence from an input UNL expression.

DeConverter operates independently in using the grammar rules of the target language and leaning on bilingual dictionaries to convert UNL expressions to natural language [3]. Bilingual dictionaries contain all word forms defined with the corresponding UNL concepts. Lexical items in a dictionary containing morphological information at the shallow level of the analysis of a language are used as references during the conversion by the DeConverter tool. The grammar rules are based on the semantic and grammatical relationships of the target language to create and order the output words.

In this paper, we present the study result for the method of converting a UNL expression with the scope node into equivalent Vietnamese sentences. To perform the conversion, we built a DeConverter tool for Vietnamese describing the conversion process including parsing UNL expressions, linking entries from the UNL–Vietnamese dictionary, morphological analysis of Vietnamese sentences, inserting words, and arranging words of the output sentences.

The paper is organized into the following sections: Following the introduction is the description of the DeConverter tool; the third section presents some related research; the fourth section is our suggestion to convert a UNL expression with a scope node to a Vietnamese sentence; the fifth section presents the test results including testing and evaluation; finally, the conclusion presents the achievements and the development trends.

2 DeConverter

A DeConverter is software that provides morphologies and syntaxes to create any natural language from UNL expressions based on dictionaries and conversion rules of the respective language. The dictionaries contain information of the words of the natural language with the corresponding Universal Words of the UNL expression and the grammatical attributes. This information is used to describe the characteristics of the word types. The grammar rule set will describe how to convert a UNL expression into a complete natural language sentence. The structure of DeConverter is described in Fig. 1 [4, 5].

The operation of the tool is as follows: First, it handles the binary relation of the UNL expression to make a directed graphical structure called Node-net. The root node of the Node-net is called the entry node. The conversion process is performed on Node-list nodes; the process of inserting nodes from the Node-net into the Node-list is

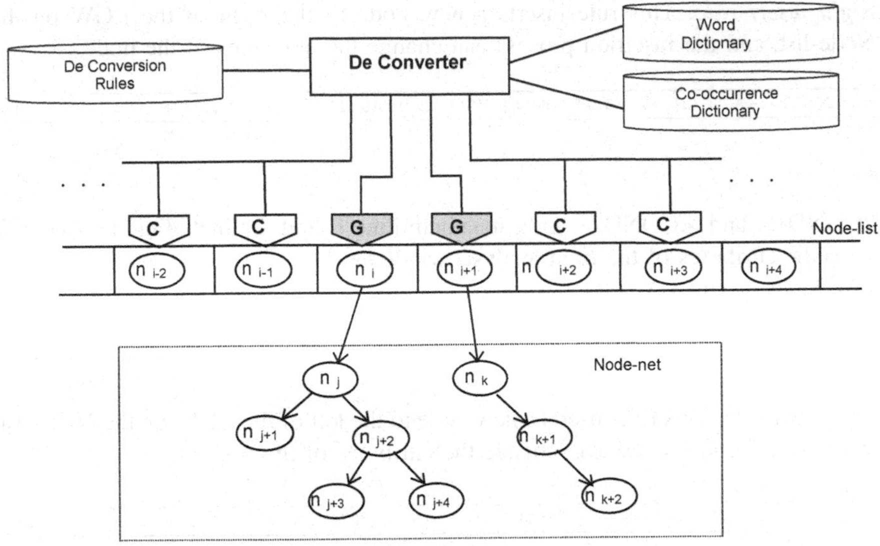

Fig. 1. Structure of DeConverter

done through windows. There are two types of windows: the condition window (*CW*) and the generated window (*GW*). CW checks the adjacent nodes of the two GWs to determine whether neighboring nodes satisfy conditions to apply decoding rules. This process constantly continues until all nodes of Node-net are inserted into the Node-list. Node-list nodes are the target sentences.

During the conversion, UWs are replaced by the corresponding headwords. If it is impossible to clearly identify the correct headword of a UW, the co-occurrence dictionary will be used. The co-occurrence dictionary contains much more semantic information for suitable selections, avoiding ambiguity.

3 Related Works

3.1 UNL Expression

A UNL expression is a binary relationship list described generally as follows:

```
<relation> ( <uw1>, <uw2> )
```

where <relation> is one of the relations, and <uw1> and <uw2> are two UWs to have relation through <relation>.

3.2 Decoding Rules Format

We built DeCoVie tool with encode rules format which is designed based on UNL EnConverter Specifications [3]. To convert a UNL expression to one or more UNL expressions, we use the following four basic rules:

1. *Right insert rule*. This rule inserts a new node to the right of the LGW on the Node-list, and the insertion process can change the attributes of the node.

```
:{<COND1>:<ACTION1>:<RELATION1>}"<COND2>:<ACTION2>: <RELATION2>";
```

where

- <COND1> and <COND2> indicate conditions 1 and 2, including lexeme and semantic attributes of the right analysis windows.
- <ACTION1> and <ACTION2> indicate actions to be taken incase of proper corresponding to conditions.
- <REL1> and <REL2> indicate possible relations between two analysis windows.

2. *Left insert rule*. This rule inserts a new node to the left of the LGW on the Node-list, and the insertion process can change the attributes of the node.

```
:"<COND1>:<ACTION1>:<RELATION1>"{<COND2>:<ACTION2>: <RELATION2>};
```

3. *Rule delete left node (DL)*. This is to delete the nodes of LGWs from the Node-list. After applying the rule, the left node of the root node is cleared to hold the LGW position and the right node of the root node is cleared to hold the RGW position.

```
DL{<COND1>:<ACTION1>:<RELATION1>}"<COND2>:<ACTION2>: <RELATION2>";
```

4. *Rule delete right node* (*DL*). This rule allows to delete the nodes of RGWs from the Node-list. After applying the rule, the left node of the root node is cleared to hold the LGW position and the right node of the root node is cleared to hold the LGW position.

```
DR"<COND1>:<ACTION1>:<RELATION1>"{<COND2>:<ACTION2>: <RELATION2>};
```

3.3 UNL–Vietnamese Dictionary Structure

A entry of dictionary consists of three basic components based on EnConverter Specifications [3]: a headword, a vocabulary, and a set of grammatical attributes. The entry format for UNL–Vietnamese dictionary is as follows [6]:

```
[HW]"UW"(ATTR,ATTR,…)<FLG,FRE,PRI>;
```

where HW (*headword*) is the entry of the natural language, UW is the Universal Word of UNL, ATTR is the grammatical and semantic attributes, FLG is the flag of natural language, FRE is the frequency rule, and PRI is the priority of the rule.

4 Proposed Solution

To convert UNL expressions into Vietnamese, we describe using the diagram shown in Fig. 2.

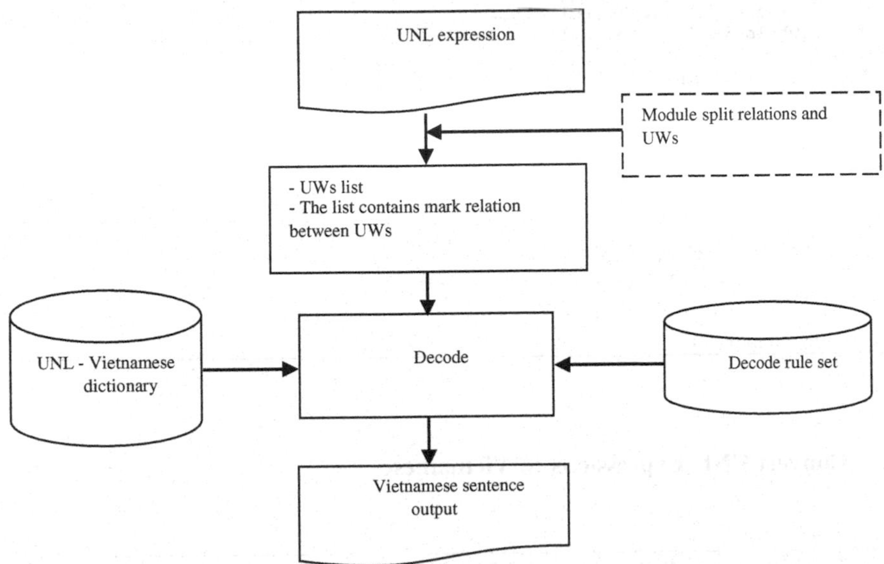

Fig. 2. Conversion of UNL expressions into Vietnamese sentences

The conversion process of the DeCoverter tool is described as follows: On the input UNL expression which is a set of binary relations, the DeCoverter tool breaks those binary relations by alternatively splitting each relation between the UWs and the semantic relations. Each UW will be stored as a list of nodes (n_1, n_2, … n_n). The semantic relations stored on the list are called Rel (r_1, r_2,… r_n) and are associated with two nodes of the Node-list. The list of Node-lists used contains nodes, which are the output of the Vietnamese sentence. On the Node-list, there are two active windows for moving and inserting nodes: The left window is called LGW (*Left Generation Window*) and the right one is called RGW (*Left Generation Window*).

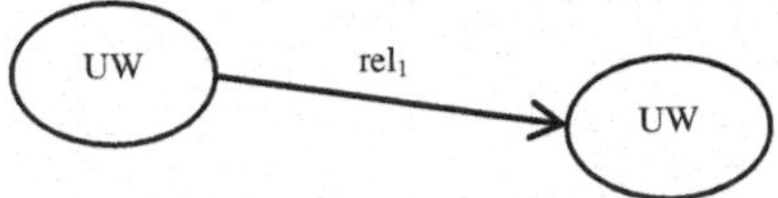

LGW and RGW operate on the condition of two adjacent nodes, and if two windows satisfy a condition rule, then one rule is executed. This rule may be node insertion rule into the Node-list and changing node attribute, or deleting a node in Node-list.

4.1 Solution for Parsing UNL Expressions

```
Algorithm 4.1. Parse the UNL expression
Input: UNL expression
Output: The UWs and relations
{
    create node list; rel list;
    While (UNLexp not null)
  {
   Read (UNLexp);
   i=1;
   rel_i = relation(UW_1;UW_2);
   If (UW_1 not in node)
     {
       add UW_1 into node_i;
       ++i;
     }
   If (UW_2 not in node)
      add UW_1 into node_i;
   ++i;
   }
return(output);}
```

4.2 Convert UNL Expressions to Vietnamese

```
Algorithm 4.2. Convert UNL expressions to Vietnamese
Input: node{n_1,n_2,…,n_n}and rel{r_1,r_2,...,r_n};
Output: Vietnamese sentences
{
link(node,dict);
Create node-list contain three node ({<<}, {nroot},{>>});
LGW =node-list_1; RGW=node-list_2;
While (LGW = ">>") do
   If ({find rule})
    {
      If (insert left rule)
       { Insert node into left RGW;
          Add node attribute;
          Delete relation in rel;
          LGW → left; RGW → left;
          Delete node in node; }
    Else
       { Insert node into left RGW;
          Add node attribute;
          Delete relation in rel;
          LGW → left; RGW → left;
          Delete node in node; }
  Else
    {
     RGW → right; LGW→ right;
    }
 Delete attribute in node-list;
 Output (node-list);}
```

4.3 Solution for the UNL Expression Has a Scope Node

```
Algorithm 4.3. Solution for the UNL expression has a scope node
Input: node{n_1,n_2,…,n_n} and rel{r_1,r_2,...,r_n};
       scope =(:01;:02;...:256);
Output: Vietnamese compound sentences
{
node-list =n_root;
While (node not null)
    if (ni.link=scope)
      Insert scope node into node-list;
While (node not null)
  {
   n_i=nroot;
   While (n_i.link not null)
    {
      Visit(n_i.link);
      if (n_i.link<>scope)
       {
         If ({n_i.link not null})
           {
             Insert n_i vào node-list;
              Delete n_i;
             n_i=n_i.link;
            }
           Else
            {
             Insert n_i into node-list;
              Delete n_i;
            }
         }
       else
         {
         Insert node contain @entry into node-list;
         Delete scope node in node and node-list;
         }
     }
 }
Output (node-list);}
```

5 Testing and Evaluation

5.1 Testing

1. Test data

Example for UNL expressions containing a scope node with relations "*obj*(*v, vs;p, pp*)," "*cob*(*v, vs;n, nt*)," "*agt*(*v, vt;p, pp*)," and "*tim*(*v, vt;a, ap*)" is analyzed to built rules as follows.

In the semantic relations "*obj*(*v, vs;p, pp*)," the pronoun (p) is a word class for replacement and pointing; it replaces things that have been named, said, and known before. Pronouns can undertake the syntactic functions of replaceable words. The personal pronoun (pp) used to replace speaker is divided into distinct hierarchies: the first (*tôi, tao, ta, tớ*), the second indicating the second partner (*mày, cậu, mi*), the third indicating the absent person or thing that is spoken (*nó, hắn, y, ả, ta, chúng ta, chúng tôi*), second person plural (*chúng mày, chúng bay*), third plural (*họ, chúng, chúng nó*). The stative verb (vs) is used to indicate the state, way of existence of things; these verbs act as main factors. The rule breaks the relations between two UWs and inserts a new node with the attribute "p, pp" to the left node that has the attribute "v, vs" as follows:

```
: "v,vs:null:obj"{p,pp:null:null};
```

In the semantic relations "*cob(v, vs;n, nt)*," nouns (n) can be main elements in noun phrases, which are often the subject of the sentence. Countable noun (nt) usually refers to relatives, positions, occupations. The rule breaks the relations between the two UW and inserts a new node with the "n, nt" attribute to the right of the node that has the "v, vs" attribute as follows:

```
:{v,vs:null:null}"n,nt:null:cob";
```

In the semantic relations "*cnt (v,vs;scope),*" the rule breaks the relations between the two UWs and inserts a new node whose "scope, coma" attribute to the right of the node has the "v, vs" attribute as follows:

```
:{v,vs:null:null}":01:+scope,+comma:cnt";
```

In the semantic relations "*agt(v, vt;p, pp)*," transitive verb (vt) is a verb that only acts bridging the things outside of it. Though these verbs have role as the predicate main of the sentence but it requires a direct/indirect complement to explain clearly the meaning of the sentence. The rule breaks the relations between two UW and inserts a new node with the "p, pp" attribute to the left of the node with the "v, vt" attribute as follows:

```
:"v,vt:null:agt"{p,pp:null:null};
```

In the semantic relation "*tim(v, vt;a, ap)*," property adjective has the meaning of qualities, density, length, weight, shape, color, taste, sound. The rule breaks the relations between two UW and inserts a new node with the "a, ap" attribute to the right of the node that has the "v, vt" attribute as follows:

```
:{v,vt:null:null}"a,ap:null:tim";
```

6 Decode Tool

The DeCoVie tool consists of three windows: an input window, the rule search window, and a result window (Fig. 3).

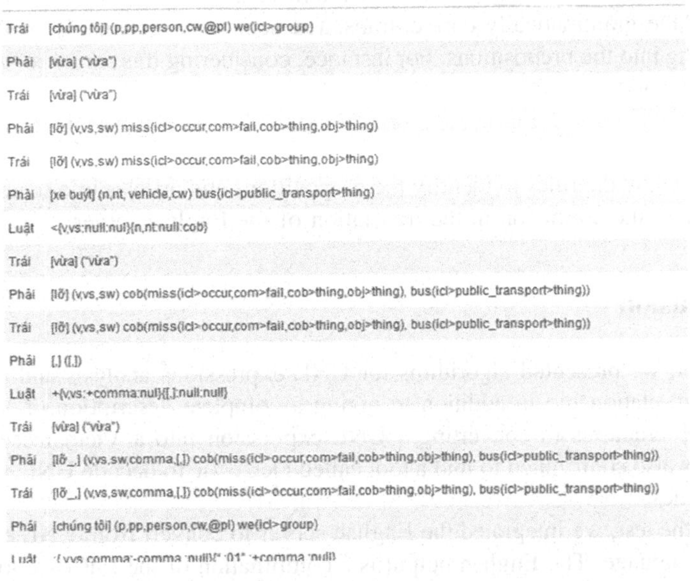

```
Trái   [chúng tôi] (p,pp,person,cw,@pl) we(icl>group)
Phải   [vừa] ("vừa")
Trái   [vừa] ("vừa")
Phải   [lỡ] (v,vs,sw) miss(icl>occur,com>fail,cob>thing,obj>thing)
Trái   [lỡ] (v,vs,sw) miss(icl>occur,com>fail,cob>thing,obj>thing)
Phải   [xe buýt] (n,nt, vehicle,cw) bus(icl>public_transport>thing)
Luật   <{v,vs:null:nul}{n,nt:null:cob}
Trái   [vừa] ("vừa")
Phải   [lỡ] (v,vs,sw) cob(miss(icl>occur,com>fail,cob>thing,obj>thing), bus(icl>public_transport>thing))
Trái   [lỡ] (v,vs,sw) cob(miss(icl>occur,com>fail,cob>thing,obj>thing), bus(icl>public_transport>thing))
Phải   [,] ([,])
Luật   +{v,vs:+comma:nul}{[,]:null:null}
Trái   [vừa] ("vừa")
Phải   [lỡ_,] (v,vs,sw,comma,[,]) cob(miss(icl>occur,com>fail,cob>thing,obj>thing), bus(icl>public_transport>thing))
Trái   [lỡ_,] (v,vs,sw,comma,[,]) cob(miss(icl>occur,com>fail,cob>thing,obj>thing), bus(icl>public_transport>thing))
Phải   [chúng tôi] (p,pp,person,cw,@pl) we(icl>group)
Luật   :{,v,vs,comma:-comma:null}{" :01" :+comma:null}
```

Fig. 3. Window finds the rules

7 Results

We conducted a test on the DeCoVie tool with two UNL expression cases for 500 expressions. In order to evaluate the output quality, the following process was done: We had the DeCoVie tool to translate UNL expressions into Vietnamese, and we use the English server (http://www.unl.ru/deco.html) to translate into English sentences. In comparison with the original Vietnamese and English versions, the output is considered 100% semantically accurate and 90% grammatically correct (Table 1).

Table 1. Change rate of sentences on two translators

Translation ways	Rate of sentences	
	No change	Change
English server: UNL → tiếng Anh	450 (90%)	50 (10%)
DeCoVie: UNL → tiếng Việt	490 (98%)	10 (2%)

7.1 Evaluation

We have studied the semantic relationships and the attributes of Vietnamese words and specified the equivalence between word classes, the attributes in Vietnamese sentences with the attributes of the UW, and the relations between UWs in UNL expressions. According to the analysis results, we have built a set of rules for decoding some UNL expressions with associated nodes into Vietnamese sentences.

90% of the grammatically correct-translated sentences are mainly in the English server, falling into the prepositions. For instance, considering this sentence "*If he learns studiously, he will pass the semester*", instead of using the article "*the*" to identify the examination, the translator used the preposition "*in*" for the target sentence which is "*If he learns studiously he will pass in the semester.*" The concepts which are not completely defined by the rules to identify the propositions and articles in a sentence then it will be created the confusion in the translation of the English server.

8 Conclusion

In this paper, we presented algorithms for UNL expressions analysis and determined the semantic relationship in addition to giving a complete description of Vietnamese grammatical structures to converting a UNL expression into a Vietnamese sentence. Our findings also contributed to find a tool called DeCoVie to decode UNL expressions into Vietnamese.

During the test, we integrated the English server to convert from UNL expressions to English language. The English output is a combination of the EnCovie tool [7] and the English server, which is the first step marking the integration of the Vietnamese server into a multilingual UNL-based translation system.

In the future, we will study the remaining semantic relations of UNL to translate into Vietnamese compound sentence. In addition, we will develop DeCoVie tools for many other cases.

References

1. P. T. L. Thuyen and V. T. Hung: Results comparison od machine translation by direct translation and by through intermediate language. International Journal of Advance Research in Computer Science and Management Studies, Volume 3, Issue 4, (2015).
2. P. T. L. Thuyen and V. T. Hung, "Automatic translation for Vietnamese based on UNL language", International Conference on Electronics Information and Communication (ICEIC 2016), page 628– 632.
3. UNL centre: DeCoverter Specifications. Version 2.7, http://www.undl.org (2002).
4. P. Kumar and R. K. Sharma: Punjabi DeConverter for generating Punjabi from Universal Networking Language. Journal of Zhejiang University-SCIENCEC ISSN 1869-1951 (Print), ISSN 1869-196X (Online), papes 179–196 (2013).
5. A. K. Saha, M. F. Mridha and J. K. Das: Semantic Analysis of Bangla Language for Developing A UNL Deconverter. International Journal of Advanced Research in Computer Science and Software Engineering, papers 273–278 (2012).

6. P. T. L. Thuyen and V. T. Hung, "Expand data on UNL – Vietnamese dictionary of UNL Explorer", Journal of Science and Technology, University of Danang, No 56, 2014.
7. P. T. L. Thuyen and V. T. Hung, "Automatic translation of Vietnamese simple sentences based on UNL", 3rd National Foundation for Science and Technology Development Conference on Information and Computer Science, pages 218 – 222 (2016).

Filling Hole on the Surface of 3D Point Clouds Based on Reverse Computation of Bezier Curves

Van Sinh Nguyen(✉), Khai Minh Tran, and Manh Ha Tran

School of Computer Science and Engineering, International University of HCMC, Ho Chi Minh City, Vietnam
{nvsinh, tkminh, tmha}@hcmiu.edu.vn

Abstract. Reconstructing the surface of a 3D object is an important step in geometric modeling. This paper presents a proposed method for filling the holes on a surface of 3D point clouds based on reverse computation of Bezier curves. The novelty of the method is processed directly on the 3D point clouds consisting of three steps. In the first step, we extract the exterior boundary of the surface. In the second step, we detect the hole boundary and its extended boundary. In the third step, we fill the holes based on the reverse computation of Bezier curves and surface patch to find and insert missing points into the holes. Our method could process very fast comparing to the existing methods, fill all holes on the surface, and obtain a reconstructed surface that is watertight and close to the initial shape of the input surface.

1 Introduction

The development of 3D data gathering techniques can help us obtain an object that approximates its real model. However, in some cases, the acquired data points are sparse or distributed irregularly on the surface of the object. It leads to what we call "hole," and we need to develop a method for such surface reconstructing. Filling the holes in a triangular mesh has been developed and widely used in the reconstruction of a graphical model in recent years. However, it is still a challenge to fill the holes in a surface of 3D point clouds. The existing methods for filling the hole on a triangular mesh can be processed based on the triangular information (i.e., vertices, edges, faces). To work on a surface of 3D point clouds, the neighborhood information for each point is very important. Our problem is to process a big data coming from a seismic data acquisition technique. After processing these data by the method of Verney [1], they are organized and stored in the 3D volume; the surfaces have contained some holes. We thus decided to start our processes directly on these holes of the surfaces. In order to fill the holes, we first identified the hole boundary in the previous work [2]. This process is performed after extracting the exterior boundary of the surface [3]. The continuous work in this research is that we propose a method to fill the holes on a surface of 3D point clouds based on reverse computation of Bezier curves for building a surface patch. This computation starts with a rough Bezier curve that is formed by the hole boundary points and their extended neighboring points. Some missing points on

V. Bhateja et al. (eds.), *Information Systems Design and Intelligent Applications*, Advances in Intelligent Systems and Computing 672,
https://doi.org/10.1007/978-981-10-7512-4_34

the rough Bezier curve are then computed and inserted to obtain a smooth Bezier curve. We apply the same process to all Bezier curves following both directions of *xoz* and *yoz* planes for each surface patch to fill hole. The hole is then refined to adapt the local curvature of the surface.

The rest of paper is organized as follows: We first review the methods to fill the hole in a triangular mesh and a surface of 3D point clouds in Sect. 2. Our proposed method is presented in Sect. 3. Section 4 presents the implementation and results. Discussion and evaluation are showed in Sect. 5. The last Sect. 6 is our conclusion.

2 Related Works

In this section, we study the methods in the state of the art for filling the holes on the 3D surface. These methods are classified into two main kinds: filling the holes on a triangular mesh and filling the holes on a 3D point cloud. However, in practice, most of these methods are performed after triangulating the surface of 3D point clouds. We have also revised the Bezier curves, surfaces, and their application in the field of surface reconstruction [4–9]. Triangulating the 3D surface has been proved that is an efficient solution to reconstruct such an approximating surface [10, 11]. Filling the holes is also an important step to obtain an object that is close to its initial shape.

Van et al. [12] and Jima [11] presented their work for filling the holes of a triangular mesh based on the Delaunay triangulation. The holes are then refined to obtained surfaces that are approximated and very close to the initial model. In the previous work [13], we described a completed method for reconstructing an elevation surface of 3D point clouds. At first, we identified the hole based on characteristic of triangular faces on the hole boundary (i.e., the edge of triangle is only shared with one face). The hole is then filled and refined by computing inserted points which are formed with the boundary edges of the hole to create new triangles. Marta [7] introduced his method to fill the holes based on the B-spline surface. He first used a B-spline surface of degree (3, 2) that fits boundary curve segments of the hole. Then, the hole is refined with a fairing condition, according to prescribed normal curvature values at the closing point to adapt the local curvature of the surface. Syed et al. [6] presented a method for computing the inverse points of a Bezier curve. It shows how to compute control points of a Bezier curve based on a set of given points that formed a Bezier curve. Yann et al. [14] presented a method for filling hole in a digitized point cloud. The method is based on a mesh deformation to fill the digitized holes after identifying these holes. This method can be applied in a complex shape (torsion, bending, stretching) and preserved the curvature of the surface. Hongwei et al. [15] presented a method for fitting a triangular model smoothly stitching bi-quintic Bezier patches. The input mesh is first segmented into a set of quadrilateral patches with whose boundary forming a quadrangle mesh. The boundary of each quadrilateral patch is then computed based on its normal curve and a boundary-fitting curve which fit the normal and position of its boundary vertices, respectively. The last step is computing the interpolation of Bezier patch to obtain an initial smoothly stitching with G1 continuity. Amitesh et al. [16] developed an algorithm for patching hole on the surface. The method is based on NURBS surface to generate smooth triangulated surface patches. This method can process some simple holes on discrete surface models to obtain a watertight 3D model from input surfaces. However, the Delaunay criterion (i.e., edge swap) is used to compute inserted points of the hole that proved an expensive computation. Xiao et al. [17] described a method to fill the hole of a triangular mesh. The method is based on computation of the marked interpolation points by using RBF (radial basis function). However, this method is difficult to process with the folds of the surface. To overcome this problem, Genli et al. [18] proposed a method to fill complex hole by subdividing the hole into many simple parts and processing on each part in the whole hole. Pavel et al. [19] proposed a method to fill the hole of a surface of 3D point clouds. After determining all boundary points of a hole, an algebraic surface patch is used to fit the hole based on the neighboring points of the boundary points. The hole is then refined to fix the local surface curvature.

The above methods could work well on a triangular mesh but were time consuming. Our proposed method works directly on a 3D point cloud to obtain faster computation.

3 Proposed Method

3.1 Overview

We first process the surface by extracting the exterior boundary [3]. The holes on the surface are then determined based on their boundaries [2]. For filling the hole, we apply the inverse computation of a Bezier curve [6]. After determining the points on the hole boundary p_b, we compute their extended boundary points p_{eb} toward the outside of the hole. These points (p_b, p_{eb}) are considered as control points and decided the shape of a rough Bezier curve (see Fig. 1). For each hole, we compute Bezier curves following both horizontal and vertical directions (i.e., the curves that cover the hole on both directions of *xoz* and *yoz* plane; see Fig. 4). We start from four original points p_{1e}, p_1, p_2, p_{2e} (two boundary points (p_1, p_2) and their extended boundary points (p_{1e}, p_{2e})); see Fig. 2. The shape of a rough Bezier curve (C) is then formed passing through four points. In the next step, we compute two control points (q_1, q_2) based on C. In the last

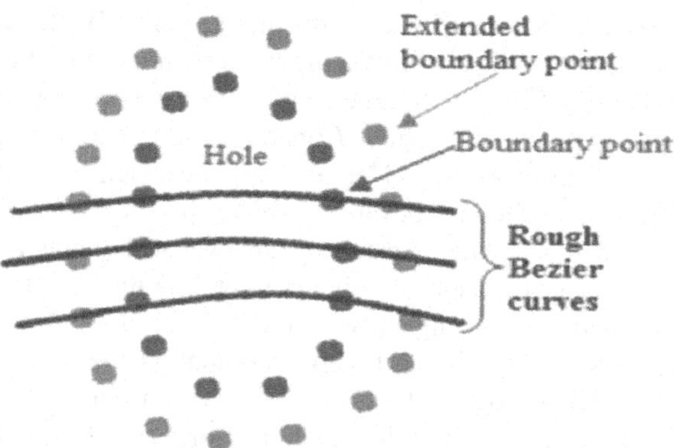

Fig. 1. Boundaries of the hole

step, we compute inserted points p_i based on q_1, q_2 such that p_i lies on C (see Fig. 3). This process is repeated on each curve covering the hole. We refine the curves by adjusting the z-coordinate of the inserted points to obtain a surface patch adapted to the local curvature of the surface.

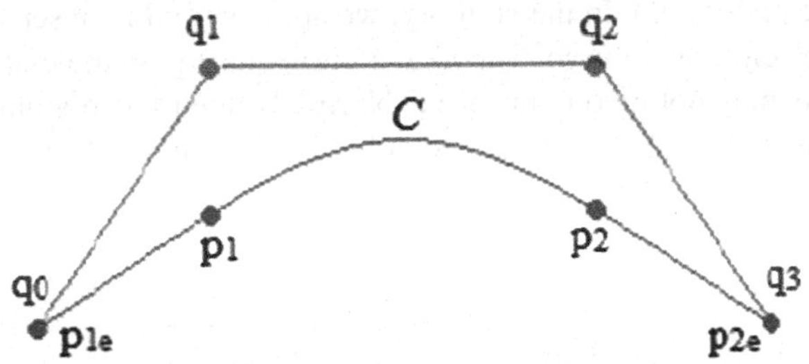

Fig. 2. The curve C is formed by four points: p_{1e}, p_1, p_2, p_{2e}

3.2 Filling the Hole

In this section, we present in detail our method following the below steps:

(a) *Identifying an extended boundary of hole*: After the hole boundary has been found, we loop on the ring of boundary points. For each boundary point, we base on its four-connectivity to determine extended boundary points (i.e., the points locate on the next ring toward the outside of the hole boundary ring); see Fig. 1.

(b) *Overview of Bezier curve*: Bezier curve is first proposed by the mathematician Paul De Casteljau in 1959. It is then widely used and became famous when a French engineer Pierre Bezier [4] published his research in the late 1960s. Following the definition of Bezier curve in [5, 20], it is a special case of a non-uniform rational basis spline. It is determined by a control polygon following the properties: The shape of curve generally follows the shape of control polygon; the first and the last points on the curve are coincident with the first and the last points of the control polygon; a planar Bezier curve is constructed from a set of points $P_i \in R^2$ (i run from 0 to n) and depicted as the below formula:

$$C(t) = \sum_{i=0}^{n} P_i B_{n,i}(t) \quad \text{where } 0 \leq t \leq 1 \tag{1}$$

where P_i are control points; $B_{n,i}(t)$ is the Bezier (or called: Bernstein polynomial, basis, blending function). C(t) is computed as follows:

$$C(t) = \sum_{i=0}^{n} \frac{n!}{i!(n-i)!} t^i (1-t)^{n-i} P_i \tag{2}$$

We realize an important characteristic of the Bezier curve is that the approximation of curve depends on the value of t (i.e., if t is small (between 0 and 1), the number of points that lies on the curve is increasing, and the curve is therefore more smooth).

(c) *Reverse Bezier curve*: We study the inverse case of building a Bezier curve [6]. In order to create a Bezier curve, we need a set of control points (at least three, including the endpoints). In the contrary, we are now finding a set of control points based on a given Bezier curve. The Bezier curve must pass the endpoints but some control points may not lie on that curve. Normally, the most popular case of Bezier curve is used in the third order (called cubic Bezier curve). They are fully defined in terms of four control points that can be computed from (2) as follows:

$$B(t) = (1-t)^3 q_0 + 3(1-t)^2 t q_1 + 3(1-t)t^2 q_2 + t^3 q_3 \tag{3}$$

where q_0, q_1, q_2, q_3 are control points; q_0 and q_3 are coincident with p_{1e}, p_{2e} (see Fig. 3).

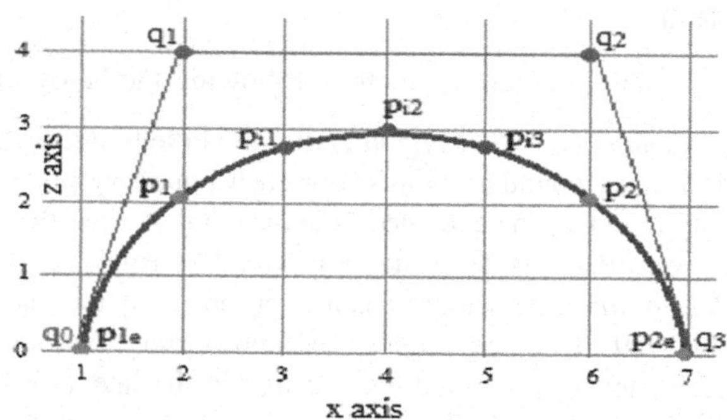

Fig. 3. Computing the inverse Bezier curve based on *xoz* plane

From Eq. (3), we can see that if t = 0, the curve will pass through q_0; and if t = 1, the curve will pass through q_3. In addition, q_0 and q_3 are coincident with p_{1e} and p_{2e}, respectively (see Fig. 3). Hence, we can rewrite (3) as follows:

$$B(t) = (1-t)^3 p_{1e} + 3(1-t)^2 t q_1 + 3(1-t)t^2 q_2 + t^3 p_{2e} \tag{4}$$

To compute q_1 and q_2 (they are two control points), we will mathematically reason as follows: Suppose the curve pass through p_1 at t = u; the curve passes through p_2 at t = v, where 0 < u < v < 1. We first apply (4) to compute $p_1(u)$ and $p_2(v)$ as follows:

$$\begin{aligned} p_1(u) &= (1-u)^3 p_{1e} + 3(1-u)^2 u q_1 + 3(1-u)u^2 q_2 + u^3 p_{2e} \\ p_2(v) &= (1-v)^3 p_{1e} + 3(1-v)^2 v q_1 + 3(1-v)v^2 q_2 + v^3 p_{2e} \end{aligned} \tag{5}$$

Eq. (5) is then simplified into:

$$\begin{aligned} 3(1-u)^2 u q_1 + 3(1-u)u^2 q_2 = c_1 \\ 3(1-v)^2 v q_1 + 3(1-v)v^2 q_2 = c_2 \end{aligned} \tag{6}$$

where

$$\begin{aligned} c_1 = p_1 - (1-u)^3 p_{1e} - u^3 p_{2e} \\ c_2 = p_2 - (1-v)^3 p_{1e} - v^3 p_{2e} \end{aligned} \tag{7}$$

Eq. (6) can be written as matrix form:

$$\begin{bmatrix} 3(1-u)^2 u & 3(1-u)u^2 \\ 3(1-v)^2 v & 3(1-v)v^2 \end{bmatrix} \begin{pmatrix} q_1 \\ q_2 \end{pmatrix} = \begin{pmatrix} c_1 \\ c_2 \end{pmatrix} \tag{8}$$

Assumption of a regular distance between the points, we can determine easily the value of t for each inner point on the curve, because we already knew t = 0 at the first point, and t = 1 at the last point. For example: t = 0, t = 1/3, t = 2/3, t = 1 with four points p_{1e}, p_1, p_2, p_{2e}, respectively. In order to deal with points whose spacing is highly uneven, we use the chord length method (as presented in [21]) to calculate parameter value for each t_i as follows: At first, we assign:

$$d_0 = ||p_1 - p_{1e}||; d_1 = ||p_2 - p_1||; d_2 = ||p_{2e} - p_2|| \tag{9}$$

where d_0, d_1, d_2 are represented for the distance between the points. Then, we put: $d = d_0 + d_1 + d_2$. Therefore, at the end we calculate values of u and v as follows:

$$u = \frac{d_0}{d}; v = \frac{d_0 + d_1}{d} \tag{10}$$

We have coordinates x of $p_{1e} < q_1 < q_2 < p_{2e}$, respectively. The determinant of (8) on the left most side is:

$$\begin{aligned} \boldsymbol{det} &= 3(1-u)^2 u * 3(1-v)v^2 - 3(1-u)u^2 * 3(1-v)^2 v \\ &= 9uv(1-u)(1-v)[v-u] \end{aligned} \tag{11}$$

As described above, $0 < u < v < 1$. Hence, the **det** > 0. In the next step, we multiply the inverse on the left of both sides of (8), and we get:

$$\begin{pmatrix} q_1 \\ q_2 \end{pmatrix} = \frac{1}{\boldsymbol{det}} \begin{bmatrix} 3(1-v)v^2 & -3(1-u)u^2 \\ -3(1-v)^2 v & 3(1-u)^2 u \end{bmatrix} \begin{pmatrix} c_1 \\ c_2 \end{pmatrix} \tag{12}$$

Note that the inverse will cancel the matrix on the left side of (8). The inverse of (8) is obtained by swapping the top-left and the bottom-right elements, and dividing each element by the determinant. Because the determinant is nonzero, it means existing an inverse matrix (by theorem of linear algebra). At the end, we compute q_1 and q_2 by combining between (7), (10), and (12) as follows:

$$\begin{aligned} q_1 &= \frac{1}{det}\left(3(1-v)v^2 * c_1 - 3(1-u)u^2 * c_2\right) \\ q_2 &= \frac{1}{det}\left(-3(1-v)^2 v * c_1 + 3(1-u)^2 u * c_2\right) \end{aligned} \tag{13}$$

(e) *Filling missing points*: In this section, we compute missing points and fill them in the hole. Our data points are structured in the 3D grid. Therefore, for each missing point p_i we can determine its x- and y-coordinates based on the boundary points of the hole. Figure 4 is considered as an example: The coordinates *x, z* (we are processing on the assumption of the plane *xoz*) of three missing points (p_{i1}, p_{i2}, p_{i3}) are determined as follows: $p_{i1}(3, z_{pi1})$, $p_{i2}(4, z_{pi2})$, $p_{i3}(5, z_{pi3})$. After computing q_1 and q_2 by using (13) in the previous step, we apply (4) to compute for each missing point p_i as follows:

$$p_i(t) = (1-t)^3 p_{1e} + 3(1-t)^2 t q_1 + 3(1-t)t^2 q_2 + t^3 p_{2e} \tag{14}$$

Following the above computation, we have already the coordinates (*x, z*) for each point: p_{1e}, q_1, q_2, and p_{2e}. We now apply (14) and x value of each p_i to compute its value t_i. After that, we reuse (14) to compute coordinate z_i for each p_i as follows:

$$z_{pi} = (1-t_i)^3 z_{p1e} + 3(1-t_i)^2 t_i z_{q1} + 3(1-t_i)t_i^2 z_{q2} + t_i^3 z_{p2e} \tag{15}$$

Applying this process and repeating it on each curve following the plane *xoz* and *yoz* to the end, ee obtain a set of missing points p_i (i.e., the intersection points between the curves on both the planes *xoz* and *yoz*). They will be inserted into the hole (see Fig. 4).

(f) *Refining the hole:* After filling the hole, each missing point p_i has two values of coordinate z from the two curves (one follows plane *xoz*, and one follows plane *yoz*). We will check if these two values of *z*-coordinate are the same (or approximate), and they are considered as one. In the contrary, depending on the values between them and their neighbors on the hole boundary points, we can choose one of the existing *z* values; or the new *z* value can be computed based on the average value between the two existing *z* values; or even between the two *z* values and their neighbors of the hole boundary points. At the end, each hole is filled with a set of p_i such that the local curvature of the surface is well preserved (see Fig. 4). The corresponding algorithm for filling the hole is as follows:

```
Program HoleFilling(S)
  for each hole on the surface S do
    determine boundary points p_b
     determine extended boundary points p_eb
    //fill the hole
    for each curve following the plane of xoz or yoz) do
      compute Bezier curve C based on p_b and p_eb
      compute control points q_i based on C
       compute and insert missing points p_i
      based on q_i and C end for
    end for
    //refine the hole
     for each inserted point p_i on the surface patch do
      adjust z value between the curves (following xoz
      and yoz plane)and the z value of their neighbors
    end for
  end for
End.
```

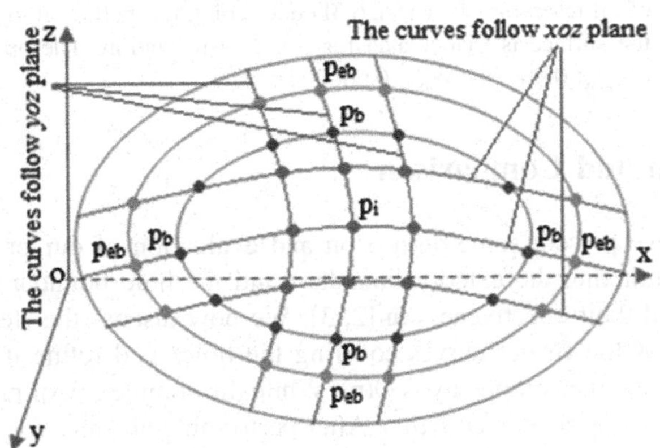

Fig. 4. Refining the hole by adjusting the z value for each p_i

4 Implementation and Results

In this section, we present the implementation of our algorithms and obtained results. We use MeshLab [22] based on VCG library [23], C++, and run on the computer (Core i7 2.6, 8 GB of RAM). We test our algorithm on various data sets. They are the elevation surfaces of oil reservoir with some holes. The local curvatures of these holes are not too high. For this reason, we also test our algorithm on some terrain surfaces created by ourselves with a higher local curvature on the holes. The total processing time is very fast, and the approximated errors are small. We use the Metro tool [24] to compute and measure the approximation errors between the input surface S1 (before filling the holes) and the output surface S2 (after filling the holes in S1). We measure both the maximum errors $\Delta_{max}(S1, S2)$, i.e., the two-sided Hausdorff distance, and the mean error $\Delta_{avg}(S1, S2)$, i.e., the area-weighted integral of the distance from point to surface (detailed in [25]). The obtained results are shown in Figs. 5, 6, 7, and Table 1. Figure 5 shows the exterior boundary, the boundaries of the holes, and the measured results proved the local curvatures of the surface patches are well preserved. Figure 7 is a surface of 3D point clouds of oil reservoir. The hole is filled and kept the local curvature of the surface.

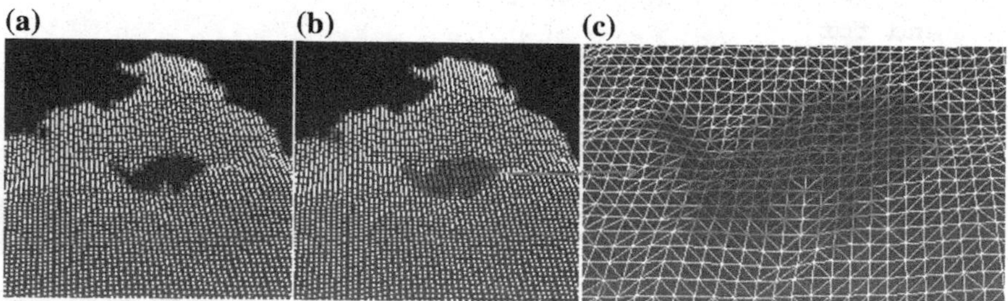

Fig. 5. A surface of oil reservoir (with 15626 3D data points): **a** before filling the hole; **b** after filling the hole; **c** the surface is triangulated for better visualization. The approximation error between **a** and **b** is Δ_{max}: 0.009 and Δ_{avg}: 0.000002

5 Discussion and Comparison

In this section, we present some discussion and evaluation of our proposed method. The steps for extracting the exterior boundary and the hole boundary of the surface have been proved their effectiveness in [2, 3]. We now discuss the step for filling the holes. We process the Bezier curves covering the holes and refine them to save the processing time, and the complexity is O(n). While the complexity of processing on the Bezier surface patch is consumed $O(n^2)$. After performing all steps for filling the holes on the surfaces, the total processing time for each surface has been shown in Table 1, included the approximation errors. We processed directly in a 3D point cloud, so the processing time is faster comparing to the methods [11, 12]. Other methods in [13, 14] are also based on the triangular mesh for filling the holes. Therefore, the computation of these methods is based on both two steps (triangulation and hole filling). The other methods [15, 16, 26] fills the holes by computing the Bezier patches and NURBS surface, directly on the triangular mesh and that leads to time consuming. The obtained results of our proposed method adapted our expectation. However, our algorithm is efficient to fill the holes of the elevation surfaces. We have tested our algorithm on the terrain surface (see Fig. 6 with many different hole locations). If the hole is on the top of a high mountain, we adjust the *z* values between the curves and choose the best one that approximates the hole boundary points of the surface. We have also performed our algorithm on the complex surfaces (with some very large holes inside), and the results in Figs. 6 and 7 illustrated that the initial shapes of these surfaces are well preserved.

Table 1. Time and comparison of approximation errors of our holes filling algorithm

Input surfaces (point number)	Number of holes	Processing time (ms)	Δ_{max}	Δ_{avg}
2294	31	127	0.015	0.002
15626	13	78	0.009	0.000002
22017	274	1697	0.0069	0.00006
23354	2	203	0.003	0.000001
37426	320	4766	0.012	0.00004
65773	4	142	0.000007	0.000001

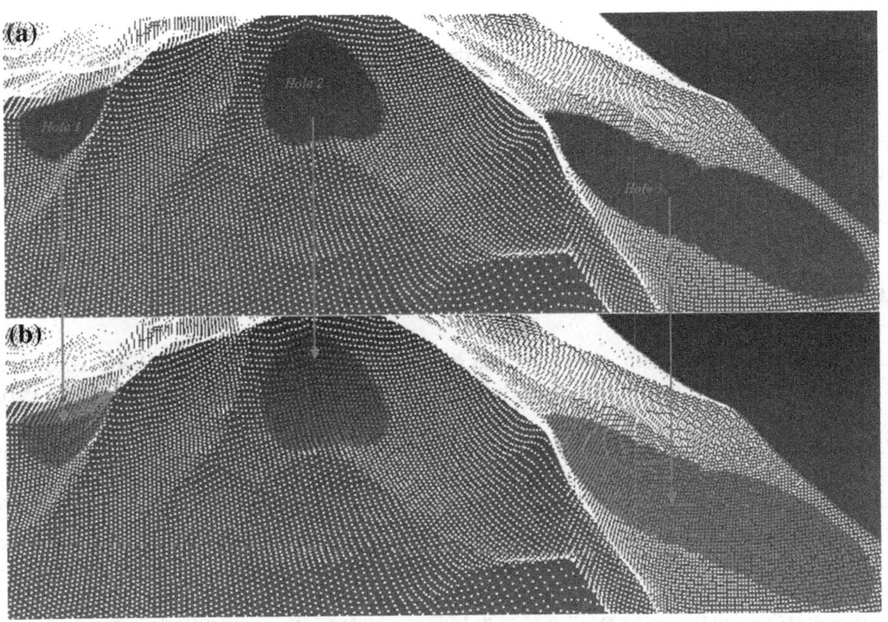

Fig. 6. Terrain surface. After filling the holes, the local curvatures are well preserved. The approximation errors: Δ_{max}: 0.000007 and Δ_{avg}: 0.000001

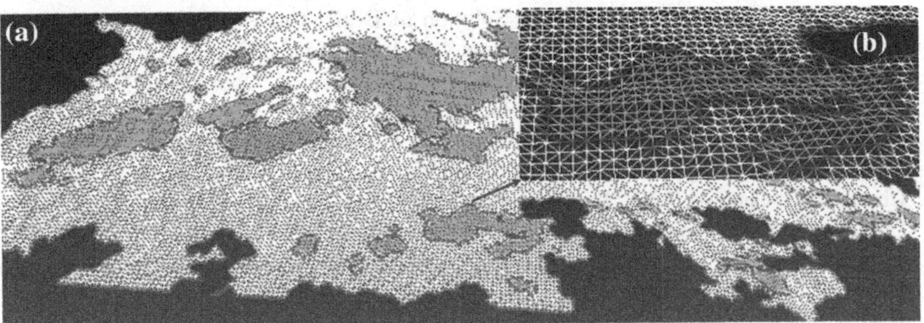

Fig. 7. A surface of 37426 points, 320 holes. **a** after filling the hole. **b** a part of the triangulated surface. The approximation error: Δ_{max}: 0.012 and Δ_{avg}: 0.00004

6 Conclusion

In this research, we have presented our method to fill the holes on a surface of 3D point clouds. The method has performed directly on a 3D point cloud. The previous works for determining the holes are very important [2, 3]. Their advantages are evaluated better comparing to the methods [27, 28]. The step for filling the holes is based on the computation of the reverse Bezier curve. This step can compute missing points of a Bezier surface patch and fill them into the hole. The obtained results have shown that the surface has been reconstructed approximating the input one. Although the idea of our proposed method in this research is not new, the processing time of the algorithm has proved its efficiency. Especially, the method is proposed to fill the holes directly on the surface of 3D point clouds without triangulating step. For the future work, we will research and improve our method to process on a varied data set and apply the method to both open and close surfaces.

Acknowledgements. This research is funded by Vietnam National University Ho Chi Minh City (VNU-HCM) under grant number C2016-28-07. We would like to thank VNU-HCM for the fund.

References

1. P.Verney, "Interprétation géologique de donneés sismiques par une méthode supervisée basée sur la vision cognitive", PhD Thesis, École Nationale Supérieure des Mines de Paris, 2009.
2. Van Sinh Nguyen, Trong Hai Trinh, Manh Ha Tran, "Hole Boundary Detection of a Surface of 3D point clouds", Proceedings of International Conference on Advanced Computing and Applications (ACOMP 2015). Pages. 124–129, IEEE ISBN-13: 978-1-4673-8234-2, 2015.
3. V.S. Nguyen, A. Bac, M. Daniel, "Boundary Extraction and Simplification of a Surface Defined by a Sparse 3D Volume", The third international symposium on information and communication technology SoICT, pp. 115–124, ACM-ISBN: 978-1-4503- 1232-5, 2012.
4. Pierre Bézier, "History of Bezier curve", https://en.wikipedia.org/wiki/Pierre Bézier.
5. David F.Rogers, "An Introduction To NURBS With Historical Perspective", ISBN-10:1-55860-669-6, Academic Press, 2001.
6. Syed Belal, Shashank.Kr.Tripathi, Shashank Kaswdhan, Md. Nadeem Akhtar, Satish Kumar Dwivedi, Rahul Singh, "Inverse Point Solution of Bezier Curve", International Journal of Scientific & Engineering Research, Volume 4, Issue 6, ISSN 2229-5518, June 2013.
7. Marta Szilvasi-Nagy, "Filling Holes with B-spline Surfaces", Journal for Geometry and Graphics Volume 6, No. 1, Pages. 83-98, 2002.
8. Licio H.B, "Efficient computation of Bezier curves from their Bernstein-Fourier representation", Journal of Applied Mathematics and Computation, Vol.220, pp. 235–238, 2013.
9. Chong P.L, Cheong K.Y "Approximation of free-form curve-airfoil shape", Journal of Engineering Science and Technology, Vol. 8, No. 6, pp. 692–702, 2013.
10. Van Sinh Nguyen, "3D Modeling of elevation surfaces from voxel structured point clouds extracted from seismic cubes", PhD Thesis, Aix-Marseille University, 2013.
11. Ji Ma, "Surface reconstruction from unorganized point cloud data via progressive local mesh matching", PhD Thesis, School of Graduate and Postdoctoral Studies, The Univerity of Western Ontario, 2011.

12. Nam-Van Tran, "Traitement de surfaces triangule´es pour la construction des mode`les geologique structuraux", PhD Thesis, Universite´ de la Me´diterrane´e, 2008.
13. Van Sinh NGUYEN, Manh Ha TRAN, Ba Cong NHAN, "A Complete Method for Reconstructing an Elevation Surface of 3D Point Clouds", REV Journal on Electronics and Communications, IEEE ISSN 1859–378X, Vol. 4, No. 34, Pages. 85–91, May 2015.
14. Yann Quinsat, Claire Lartigue, "Filling holes in digitized point cloud using a morphing-based approach to preserve volume characteristics", International Journal of Advantages Manufacturing Technology, 81(1), pp. 411–421, 2015.
15. Hongwei Lin, Wei Chen, Hujun Bao, "Adaptive patch-based mesh fitting for reverse engineering", Journal of Computer-Aided Design, Vol. 39(2007), pp. 1134–1142, 2007.
16. Amitesh.K, Alan .S, Yasushi I, Douglas. R and Bharat.S, "A Hole-filling Algorithm Using Non-uniform Rational B-splines", in Proceedings of the 16th International Meshing Roundtable, Springer Berlin Heidelberg, 2008, pp. 169–182, 2007.
17. Xiao J.Wu, Michael Y.Wang, B.Han, "An Automatic Hole-Filling Algorithm for Polygon Meshes" Computer Aided Design And Applications, Vol. 5, Issue. 6, Pages. 889–899, 2008.
18. Gen Li, Xiu-Zi Ye, San-Yuan Zhang, "An algorithm for filling complex holes in reverse engineering" Journal of Engineering with Computers archive, Vol. 24 Issue 2, pp 119–125, Springer-Verlag London, UK, 2008.
19. Pave Chalmoviansky, Bert Juttler, "Filling Holes in Point Clouds" Mathematics of Surfaces Lecture Notes in Computer Science, 2003, Vol. 2768/2003, Pages. 196–212, 2003.
20. Zhiyi.Z, Xian.Z, Huayang .Z, "A sampling method based on curvature analysis of cubic Bezier curve", Journal of Computer, Vol. 9, No. 3, Pages. 595–600, March 2014.
21. J. Kozak, M. Krajnc, "Geometric interpolation by planar cubic polynomial curves", The journal Computer Aided Geometric Design, 24 (2007), pp. 67–78, 2007.
22. "Meshlab", http://meshlab.sourceforge.net/, 2016.
23. ISTI, "The Visualization and Computer Graphics", http://vcg.isti.cnr.it/vcglib/, 2016.
24. P.Cignoni, C.Rocchini, R.Scopigno, "Metro: Measuring error on simplified surfaces", The Eurographics Association 1998, 17(2), 1998.
25. M. Pauly, M. Gross, L.P. Kobbelt, "Efficient simplification of pointsampled surfaces", Visualization VIS IEEE 2002, Page.163–170, ISBN: 0-7803-7498-3, 2002.
26. John A. Gregory and Jianwei Zhou, "Filling polygonal holes with bicubic patches", Journal of Computer Aided Geometric Design, 11(1994), Page. 391–410, 1994.
27. A.Sampath, J.Shan, "Building Boundary Tracing and Regularization from Airborne Lidar Point Clouds", Photogrammetric engineering and remote sensing, Volume. 73, Number. 7, Pages. 805–812, 2007.
28. Shen Wei, "Building Boundary Extraction Based on LiDAR Point Clouds Data", The International Archives of the Photogrammetry, Remote Sensing and Spatial Information Sciences, Vol. 37, Part. B3b, 2008.

Keyword Extraction from Hindi Documents Using Document Statistics and Fuzzy Modelling

Sifatullah Siddiqi(✉) and Aditi Sharan

School of Computer and Systems Sciences, Jawaharlal Nehru University, New Delhi, India
{sifatullah.siddiqi,aditisharan}@gmail.com

Abstract. In this paper, we put forward a novel unsupervised, domain independent and corpus independent approach for automatic keyword extraction. Our approach combines the document statistics of frequency and spatial distribution of a word in order to extract the keywords. We have extracted keywords from Hindi documents using document statistics and utilized the power of fuzzy logic to combine those document statistics effectively for better results. Further, we use this information to frame fuzzy rules for keyword extraction. Main advantages of our approach are that it uses the fuzzy membership for the variables instead of dealing with crisp thresholds and corpus independent setting of fuzzy membership boundaries. Our work is especially significant in the light that it has been implemented and tested on Hindi which is a resource poor and underrepresented language.

1 Introduction

Keywords of a document are a set of representative words that give higher-level information regarding the content of the document to the user. Keywords help in quickly analysing big collections of textual data and are useful in many areas such as text mining, information extraction and natural language processing techniques. In spite of developments in these fields, generating a focused set of keywords is still a challenge, especially in resource-poor languages such as Hindi. Broadly speaking, there can be different approaches for automatic keyword extraction, each having its own pros and cons, but there are four major approaches, viz. ***Rule-based linguistic approaches, Statistical approaches, Machine Learning approaches and other approaches*** [1].

Statistical methods for the extraction of keywords from documents have certain advantages over linguistic-based approaches such as the same approach can be applied to many different languages without the need to develop different set of rules each time for a different language. Due to the availability of large amount of unlabelled data in different languages and the subsequently arising need of finding keywords for such documents, there is a need to find approaches which are unsupervised, language independent as well as corpus independent. In this paper, we attempt to suggest a keyword extraction approach that does not require document corpus and is

V. Bhateja et al. (eds.), *Information Systems Design and Intelligent Applications*, Advances in Intelligent Systems and Computing 672,
https://doi.org/10.1007/978-981-10-7512-4_35

unsupervised and statistical in nature, so that it can be applied to any language. Our approach is based on Fuzzy rules, which exploit the information from document corpus and uses the power of Fuzzy sets for extracting keywords from the documents.

Our main motivation for using fuzzy logic comes from the difficulties of threshold setting in statistical approaches. In any statistical approach, we have to set a threshold on a statistical parameter that demarcates its belongingness to a category of interest. For example, one observation is that the words with medium frequency and high standard deviation (SD) of position vector are good candidates for being a keyword [2]. But we know that there are no strict crisp lower and upper boundaries which can be assigned to words having medium frequency or high standard deviation. Thus, setting a fuzzy threshold seems to be an obvious choice here. Thus, we have tried to capture the fuzziness (presence of words medium and high) incorporated in statistical approaches and utilized the power of Fuzzy sets to develop a fuzzy rule-based approach that uses statistical information for extracting keywords. We have applied our approach to perform keyword extraction from Hindi documents.

There are various limitations and handicaps present while working with a language like Hindi such as:

1. Hindi language is highly inflected and provides rich and challenging set of linguistic and statistical features.
2. Hindi is a resource-poor language. Annotated corpora, named dictionaries, POS taggers, etc., are still not available in required quality and quantity. In contrast to this, for English and many other European languages, we have a large set of tools and corpora available to work on.
3. Even very basic preliminary resource such as list of stopwords for Hindi is not available and whatever is available is neither complete nor authentic.
4. Another problem is that it is difficult to use stemming to reduce the size of vocabulary.

Considering the above factors, Hindi becomes a good choice for application of our approach. As far as we know, this is one of the preliminary attempts to use Fuzzy logic for keyword extraction using document statistics. Some of notable works done in keyword extraction by utilizing document statistics are mentioned in next section.

2 Related Work

Luhn [3] argued that words with medium frequencies in the document are more important than words with low or high frequencies. Rare words with low frequencies and the common words with high frequencies have little contribution to the content of the text.

Discrimination value analysis [4] was proposed by G. Salton et al. which ranked the words based on their discriminatory power to differentiate between the documents of a collection from each other. Value of a word is assigned based on the difference in average separation among various documents which occurs when a given term is assigned as keyword for content representation. Words with the greatest separation were taken to be the best words.

It was found empirically that in general the important words of a text have a tendency to attract each other and form clusters [2]. Standard deviation of the distance between successive occurrences of a word was used to quantify this self-attraction and rank the words of the document.

The problem of finding and ranking the relevant words was also attempted by using statistical information related to the spatial use of the words [5]. Shannon's entropy was used for this purpose.

Generalized level statistical analysis of quantum disordered systems was utilized to extract keywords from literary texts [6]. Frequencies of the words in the document along with their spatial distribution were used in the analysis.

2.1 Background and Motivation

A work by Luhn [3] suggested that medium-frequency terms of a text are good representatives of the content of the text while higher frequency terms are not information-bearing words (stopwords) and lower frequency words are not important enough to be considered in analysis by virtue of being not frequent enough. Although the statement is correct, there are many words in higher frequency category which are good keywords and they are eliminated through this approach. Also there is no crisp frequency threshold which can demarcate between higher frequency words and medium-frequency words and also between medium-frequency words and low-frequency words. This situation lends itself to fuzzy analysis of word frequency.

Ortuno et al. [2] found that a relevant word is repeated more often in specific sections of the text and thus a clustering pattern emerges while visualizing the word occurrence. The non-relevant words and stopwords were found to be randomly distributed, and they approximated a Poisson distribution. To find out words with good clustering and hence a good representative keyword, the standard deviation of intermediate distance vector was calculated. Higher deviation words were taken as keywords. But standard deviation approach suffers from outlier problems. The presence of very few outliers or even a single outlier can distort the result and give undue weightage to the word under consideration and can report non-keywords as keywords thereby decreasing the result quality. Another weakness of this approach is that the low-frequency words which are not informative can even beat higher frequency words which are informative but have lower standard deviation. Thus, this approach is not reliable enough to judge between all kinds of frequency words.

We present some examples from a famous Hindi novel "Godan" by Premchand. Figures 1 and 2 show the distribution pattern of a keyword and a stopword in the text, respectively. The word "**होरी**" is a central character in the text and hence a keyword and "**गया**" is a stopword. We have selected these words with similar frequencies to better appreciate the differences in distribution pattern of these words.

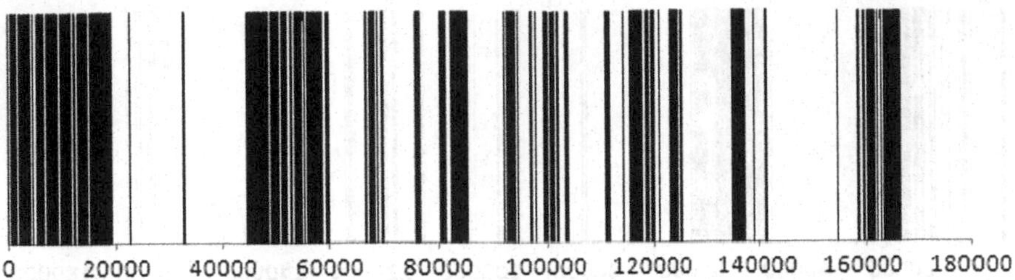

Fig. 1. Word = "होरी", frequency = 623, SD = 4.015, Category = Keyword

Fig. 2. Word = "गया", frequency = 677, SD = 1.166, Category = Stopword

However, the approach is not flawless and provides incorrect results too in some cases. Let us consider another word "बाँस" which is a non-relevant word and is given a higher standard deviation than many other keywords of the text. The distribution is shown in Fig. 3 and provides the example of a non-relevant word extracted as keyword due to presence of outliers in data and weakness of standard deviation method. Comparing this non-relevant word with another keyword "धनिया" (Fig. 4) from the text, we see that this keyword is ranked lower than it even though the clustering present in the word "धनिया" is more reliable and visual than "बाँस". Thus, there is a scope of improvement in this approach.

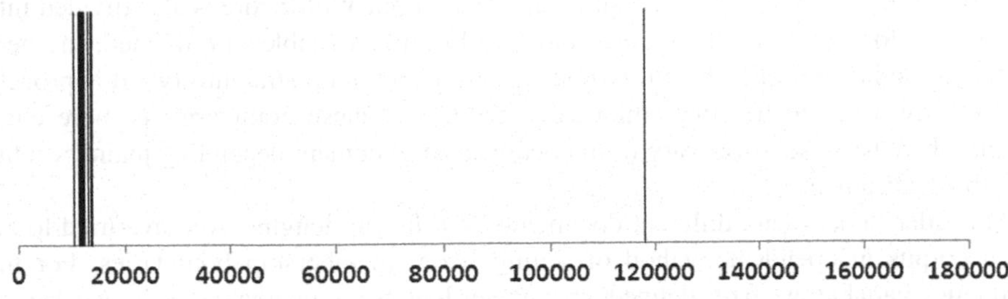

Fig. 3. Word = "बाँस", Frequency = 22, SD = 4.331, Category = Non-Keyword

Fig. 4. Word = "धनिया", Frequency = 355, SD = 4.125, Category = Keyword

In this paper, we try to provide an improved approach for keyword extraction. Our approach considers the same parameters frequency and standard deviation of the word. Our use of fuzzy rules is based upon the following observations:

1. Words with medium frequency are more important than lower frequency or highest frequency words. The use of medium frequency suggests the genuine applicability of fuzzy logic.
2. Words with high standard deviation of intermediate distances are keywords while those of low deviation are non-keywords. The use of high standard deviation suggests the genuine applicability of fuzzy logic.
3. An appropriate combination of frequency and SD scores is important to decide if a word is keyword or not.

3 Proposed Approach

The two input parameters to our fuzzy system are frequency and standard deviation, and one output parameter is WordScore, which assigns a score to each word.

The input variable frequency is divided into three categories; low frequency, medium frequency and high frequency. Similarly, the second input standard deviation is also divided into two parts; low and high. The output WordScore is also divided into three parts low, mid and high. Once input and output variables are defined, the next step is to define the range for fuzzy boundaries. There is no straightforward approach, and mostly it has to be done empirically. Setting of these boundaries is more challenging here because it will vary from document to document depending mainly on the length of document.

In order to deal with different documents of differing lengths, we have tried to set up a corpus independent method of setting fuzzy membership boundaries. For the frequency variable, we have defined a parameter K which demarcated the boundaries of membership functions low, medium and high. K is defined as the token to type ratio for a document. If there are a total of N number of words in the text and T number of unique words, then $K = N/T$. The first number of range of decreasing half triangular function represents the top, and the second number represents the end of the function's range. Similarly, the first two numbers in the range of trapezoidal function represent the starting and ending points of the upward slope of the trapezoidal function and last two numbers representing the downward slope of the trapezoid. Similarly for standard deviation, we have again set up corpus independent boundary settings.

The fuzzy system variables along with their membership functions are shown in Table 1, and their graphs are shown in Figs. 5, 6 and 7.

Table 1. Details of membership functions of the fuzzy variables for keyword extraction

Variable name	Membership functions	Nature	Range
Frequency	Low	Decreasing half triangular	[1, 3 K]
	Medium	Trapezoidal	[2, 15, 25, 40 K]
	High	Half trapezoidal	[25, 40, 40, >40 K]
Standard Deviation	Low	Decreasing half triangular	[0, 3]
	High	Half trapezoidal	[2, 3, >3]
WordScore	Low	Decreasing half triangular	[0, 0.2]
	Medium	Triangular	[0.175, 0.75]
	High	Half trapezoidal	[0.71, 1]

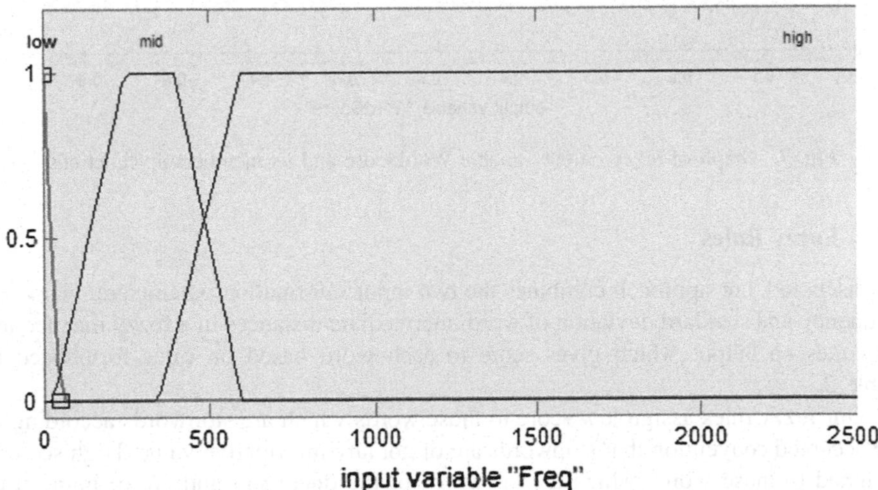

Fig. 5. Graph of fuzzy input variable frequency and its membership functions

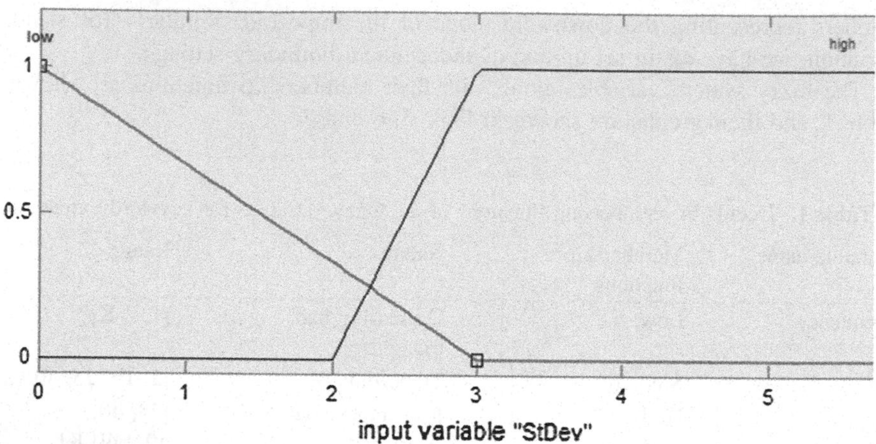

Fig. 6. Graph of fuzzy input variable standard deviation and its membership functions

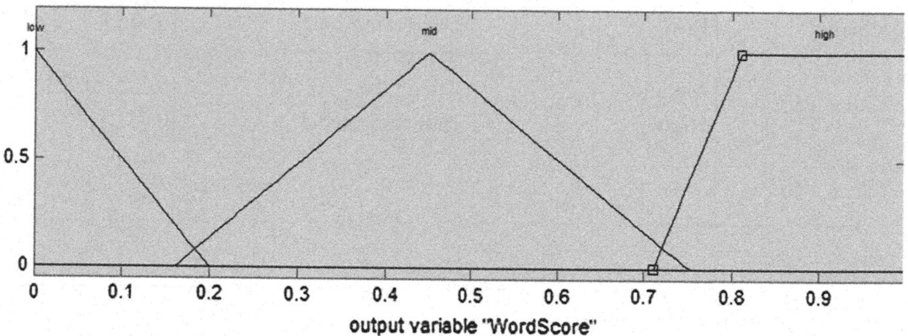

Fig. 7. Graph of fuzzy output variable WordScore and its membership functions

3.1 Fuzzy Rules

As discussed, our approach combines the two input information parameters, viz. word frequency and standard deviation of word intermediate distances in a fuzzy manner and generates an output which gives score to each word based on rules formulated in Table 2.

Our fuzzy rules assign low score to those words which are stopwords according to the accepted convention that stopwords are of not any informative value. High score is assigned to those words which are keywords of the document and are of high informative value. Medium score is assigned to rest of the words which are termed as non-relevant words. It should be noticed that number of stopwords in a text are limited though they can be much larger than number of keywords of the text while the most number of words in the text fall into the non-relevant category or medium score words because they are obviously more informative than stopwords but not good enough to be included in keyword category. It can be seen from the membership function of output variable WordScore that medium membership function (words that are non-relevant but not stop words) has the largest range which is because bigger chunk of words in the text fall into this category. It evidently includes words with low and very low frequency. Generally, removing low-frequency words from the analysis reduces the vocabulary of the text to a great extent.

Table 2. Rule set for keyword extraction

1.	If Frequency is low then WordScore is medium
2.	If Frequency is medium and standard deviation is high then WordScore is high
3.	If Frequency is medium and standard deviation is low then WordScore is medium
4.	If Frequency is high and standard deviation is low then WordScore is low
5.	If Frequency is high and standard deviation is high then WordScore is high

4 Experiments and Results

4.1 Data Set Preprocessing and Related Information

In order to perform our experiment, we have selected three Hindi books to perform the experiments with our approach. All the punctuation marks were removed from the documents to perform the standard deviation analyses on them. The books used are:

"**Godan**" by Premchand and it's a famous Hindi novel.

"**Bharat ki Nadiya**" a text on rivers of India published by National Book Trust of India

"**Bharat ka Prakritik Bhuvigyan**" a text on physical geography of India published by National Book Trust of India.

There are no standard data sets in Hindi for evaluation of keyword extraction algorithms. Due to the absence of any benchmark data set, we tried to create our own benchmark. We have used "Bharat ki Nadiya" and "Bharat ka Prakritik Bhuvigyan" since they have indexes at their back and these indexes have been treated as the benchmark for performance evaluation of both the approaches. Those words which matched with entries in index were treated as keywords. Table 3 shows the information regarding the data sets. We have compared our approach with standard deviation-based approach suggested by Ortuno et al. [2].

Table 3. Information about the data sets used in analysis

Document title	No. of words	No. of unique words	Value of K
Godan	167707	11160	15
Bharat ki Nadiya	58947	7276	8
Bharat ka Prakritik Bhuvigyan	51507	6417	8

Figures 8, 9 and 10 show the running precision values for both the approaches, and it is evident that our fuzzy approach performs better than standard deviation.

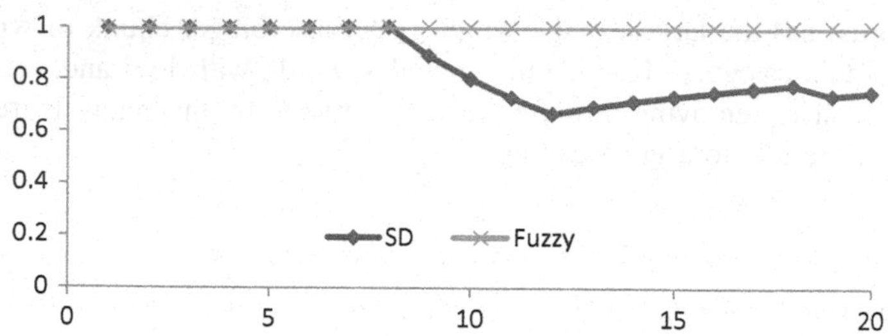

Fig. 8. Running precision curves for the keywords extracted via Fuzzy and Std-Dev approach for document "Godan". Horizontal axis represents the rankings of words

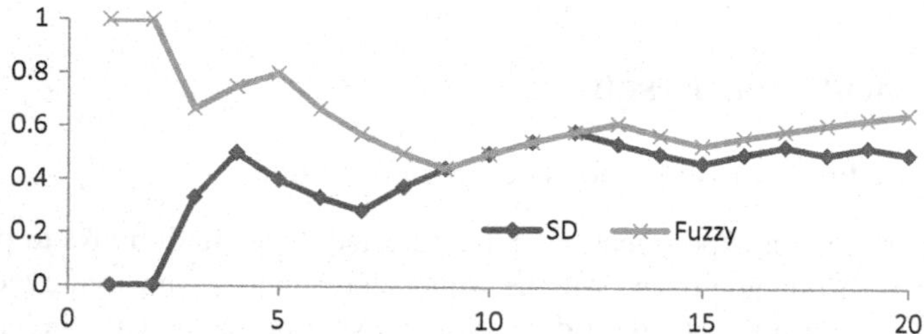

Fig. 9. Running precision curves for the keywords extracted via Fuzzy and Std-Dev approach for document "Bharat ki Nadiya". Horizontal axis represents the rankings of words

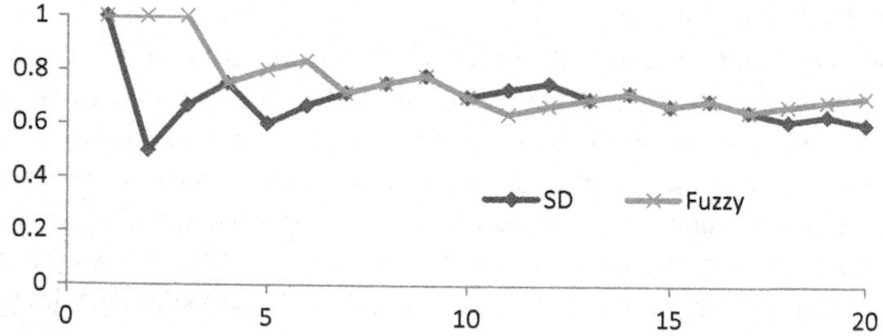

Fig. 10. Running precision curves for the keywords extracted via Fuzzy and Std-Dev approach for document "Bharat ka prakritik bhuvigyan". Horizontal axis represents the rankings of words

5 Conclusion

In this paper, we have combined two statistical parameters, viz. word frequency and standard deviation of word intermediate distance in a fuzzy manner to generate a relevance score for each word. Our approach is unsupervised, language independent as well as corpus independent. We have tested our approach on three different texts of Hindi language and found good results. Precision values of our approach are quite encouraging; our approach gives better results than simple standard deviation method; it is very much adaptable to being used in different languages and its applicability is especially significant in case of resource-poor languages.

References

1. Zahang, C., Wang, H., Liu, Y., Wu, D., Liao, Y., & Wang, B.: Automatic Keyword Extraction from Documents Using Conditional Random Fields, Journal of CIS (2008), pp. 1169–1180.
2. Ortuño, M. et al.: Keyword detection in natural languages and DNA, Europhys. Lett. (2002).
3. Luhn, H. P.: A Statistical Approach to Mechanized Encoding and Searching of Literary Information, IBM Journal of Research and Development, 1 (4). (1957) pp. 309–317.
4. G. Salton, C. S. Yang, Yu, C. T.: A Theory of Term Importance in Automatic Text Analysis, Journal of the American society for Information Science, 26(1), (1975) pp. 33–44.
5. Herrera, J.P., Pury, P.A.: Statistical keyword detection in literary corpora, The European physical journal, (2008).
6. Carpena, P. et al.: Level statistics of words-Finding keywords in literary texts and symbolic sequences, Physical Review E, (2009).

Improving Data Hiding Capacity Using Bit-Plane Slicing of Color Image Through (7, 4) Hamming Code

Ananya Banerjee(✉) and Biswapati Jana

Department of Computer Science, Vidyasagar University, Midnapore 721102, West Bengal, India
{anaanya.2011, biswapatijana}@gmail.com

Abstract. Achievement of high-capacity data hiding with good visual quality is an important research issue in the field of steganography. In this paper, we have introduced RGB color image and bit-plane slicing for data hiding through Hamming code using shared secret key. We partitioned the color image into (3×3) pixel blocks and then decomposed into three basic color blocks. Again each color blocks are sliced up to four bit-plane starting from LSB plane. Now, a segment of three bits secret data is embedded within each bit-plane depending on a syndrome calculated using hamming code. As a result, 36 bits secret data can be embedded within (3×3) pixel block and achieve a high payload capacity with good visual quality compared with existing schemes.

Keywords: Steganography · Hamming code · Least significant bit (LSB) Bit-plane · Data hiding

1 Introduction

Steganography is the art and science of hidden data communication. Many data hiding schemes [1–10] are developed in last few decades. Some of them [6, 9, 10] are better in terms of security and imperceptibility. The data hiding schemes are useful in many application areas to solve the problem of ownership identification, copyright protection, authentication, verification, and more. The main aims of data hiding schemes are to ensure extraction of secret data and recovery of original object from stego media. On the other hand, data should stay hidden in stego image even if the eavesdropper tampered the stego or degrading through natural phenomenon like transmission resampling, compression or filtering, etc. The main drawbacks of data hiding schemes are not to provide a good solution in such cases. The degree of distortion will be high due to increase of data embedding capacity that should be balanced mathematically using spread spectrum. The data embedding in color image is considered to be more unsuspicious and secured, and less exploration has been done till today in this research area using hamming code.

Data hiding through matrix coding has been introduced by Crandall [2], and Westfield [7] implemented F5 algorithm based on matrix encoding using hamming code. Kim and Shin [8] suggested a data embedding procedure for halftone image. The

V. Bhateja et al. (eds.), *Information Systems Design and Intelligent Applications*, Advances in Intelligent Systems and Computing 672,
https://doi.org/10.1007/978-981-10-7512-4_36

scheme provides good capacity but poor visual quality. Based on matrix encoding, 'Hamming +1' method has been developed by Zhang et al. [3]. The embedding capacity is increased in 'Hamming +1' scheme by $\frac{k+1}{2^k}$ bpp. Chang [5] suggested an improved data embedding procedure using Hamming code which can hide (k + 1) bits of message in 2^k pixels with at most one change. Kim and Yang [9] developed 'Hamming + 3' data hiding scheme which can embed k + 3 bits within $2^k - 1$ pixels by at most two change. In 2016, Cao et al. [10] proposed their algorithm which can preserve stego image quality under high embedding capacity. The visual quality and data embedding capacity of Cao et al. are 37.90 dB PSNR and payload 3 bpp, respectively. Also Jana et al. [11] proposed reversible data hiding scheme through dual image with 53 dB PSNR and payload 0.14 bpp. After studying this literature, we have found that there is a chance to improve data embedding capacity and visual quality through Hamming code for color images. Here, we have proposed an improved information hiding scheme using (7, 4) Hamming code for color images. We have divided R, G, and B color pixels in bit-plane [12] starting from LSB to LSB-3 (up to four bit-plane) into (3 × 3) blocks and then applied hamming code-based data hiding scheme. In this scheme, 36 bits are embedded within 9 pixels which are more higher than other existing hamming code-based scheme, and in parallel, it maintains high visual quality.

2 Motivation and Objectives

1. Till date, data embedding algorithms are implemented and tested using grayscale images. But, we have implemented hamming code-based data hiding scheme for color images using basic color channel (that is R, G, and B) and their bit-plane.
2. We have introduced the concept of bit-plane. That means each R, G, and B color pixels are divided into four bit-plane starting from LSB to LSB-3.
3. Using Hamming code, Cao et al. [10] achieved data hiding scheme up to 3 bpp payload. But we have implemented our proposed algorithm for 4 bpp where the value of PSNR is always greater than 39 dB.

3 Proposed Method

I is considered as the cover image of size (M × N), and I′ is the marked image with data $D = \{d_1, \ldots, d_X\}$ are embedded, where $d_i \in \{0, 1\}$, $1 \leq i \leq X$. Here, H is a parity check matrix of the Hamming code. Let H is

$$H = \begin{vmatrix} 0 & 0 & 0 & 1 & 1 & 1 & 1 & 0 & 0 \\ 0 & 0 & 1 & 0 & 1 & 1 & 0 & 1 & 0 \\ 0 & 0 & 1 & 1 & 0 & 1 & 0 & 0 & 1 \end{vmatrix},$$

Embedding capacity is an important metric for data embedding. It is measured by how many secret bits can be embedded into a cover image. The embedding capacity is calculated as [10] $ER = L/M \times N\,bpp$, where L is the length of the secret message.

Before embedding the secret data, we take 36-bit secret key k_1 which is known to both the sender and the receiver, to encrypt the secret data bit using symmetric key encryption. We have taken each pixel block of size (3 × 3), and four bit-plane of each pixel is used to embed the data, which result in (3 × 3) × 4 = 36 bits of data (D_1) in one iteration. As an additional security measure, instead of choosing the cover image pixel block serially, we will use pseudo random number generator (PRNG) function with a secret predefined seed k_2 (which is only known to the sender and the receiver) to determine the next available block for embedding. Since this seed will be known to the sender and receiver only, the generated unique pattern of pixel block selection can be used in embedding and extraction process securely. The data embedding procedure is enlisted in Algorithm 1, and the data extraction procedure is depicted in Algorithm 2.

Algorithm 1: Data embedding process Input: Color cover image I (M × N), secret data bits D, Hamming matrix H, secret key k_1, and seed value k_2

Output: A stego image I′ (M × N).

Step 1: Collect random sequence of pixel blocks of size 3 × 3 from $I_{M \times N}$ using PRNG (k_2). Say the pixel blocks are $X_1, X_2, \ldots, X_{MN}$.
Step 2: Convert X_i into three separate RGB color blocks X_{iR}, X_{iG}, X_{iB}.
Step 3: Convert each X_i 's into binary form.
Step 4: Perform bit-plane slicing of each X_i 's up to four bit-plane starting from LSB that is $X_{iR(LSB)}, X_{iR(LSB\text{-}1)}, X_{iR(LSB\text{-}2)}, X_{iR(LSB\text{-}3)}$.
Step 5: Take $c = X_{iR(LSB)}$ and calculate the syndrome $S_1 = (H \times (c)^{T})^{T}$.
Step 6: Perform $D_1{}' = (D_1 \oplus k_1)$; k_1 = 36 bit length and D_1 is also same length
Step 7: Take three bits secret data $d_i = \{d_1, d_2, d_3\}$ from $D_1{}'$ where $d_i \in \{0, 1\}$.
Step 8: Calculate $S_2 = (d_i \oplus S_1)$; if $S_2 = 0$, no change, otherwise flip a bit at the positional value of S_2 and generate H′.
Step 9: Compute $S_3 = (H' \oplus c)$ and store the data.
Step 10: Replace the matrix(c) with S_3 and update $X_{iR(LSB)}$.
Step 11: Repeat Step 4–10 using $X_{iR(LSB\text{-}1)}$, $X_{iR(LSB\text{-}2)}$, and $X_{iR(LSB\text{-}3)}$.
Step 12: Repeat Step 5–11 to embed secret data on X_{iG} and X_{iB} color blocks.
Step 13: Repeat Step 2–12 to embed secret data on each and every random sequence of (X_i's) of pixel blocks.
Step 14: Finally, after combining each stego block, we get stego image (I′) of size (M × N).
Step 15: End.

Algorithm 2: Data extraction process Input: Stego image I′(M × N), Hamming matrix H, secret key k_1, and seed value k_2

Output: Original secret message D.

Step 1: Use PRNG with predetermined seed k_2 to determine the stego pixel of random sequence X'_i of size [3 × 3] from stego image I′.
Step 2: Separate RGB components into $X'_{iR}, X'_{iG}, X'_{iB}$.

Step 3: Convert into binary form of each X'_{iR}, X'_{iG}, X'_{iB}.
Step 4: Perform four bit-plane slicing of each X'_i 's starting from LSB, that is, $X'_{iR(LSB)}$, $X'_{iR(LSB-1)}$, $X'_{iR(LSB-2)}$, $X'_{iR(LSB-3)}$.
Step 5: Take $c' = X'_{iR(LSB)}$ and calculate the syndrome $S' = (H \times (c')^{T})^{T}$.
Step 6: Concatenate syndrome S′ with data unit of D′, that is, $D' = D' \| (S')$.
Step 7: Repeat Step 4–6 using X'_{iG} and X'_{iB}.
Step 8: Compute $D_i = D' \oplus k_1$.
Step 9: Repeat Step 2–8 using next random sequence of X_i block.
Step 10: Concatenate D_i 's, we get original secret message D.
Step 11: End.

The original color image is divided into R, G, and B color image as shown in Fig. 1. Then every color image is divided into (3 × 3) RGB color pixels. Then the secret bits are embedded within the image pixels. The stego (3 × 3) RGB color pixels are generated. After that, the color pixels are combined and formed the stego image of Lenna which is also (512 × 512) pixels.

3.1 Numerical Illustration

Example 3.1.1: Data embedding

1. Let I be a color image block with (3 × 3) pixel. $D = \{d_1, d_2, \ldots, d_{36}\}$ = {0, 1, 0, 1, 0, 1, 0, 0, 1, 1, 0, 1, 1, 1, 0, 0, 0, 1, 1, 0, 1, 0, 0, 1, 1, 0, 1, 0, 1, 0, 0, 0, 1, 0, 1, 0}. k_1 = {0, 0, 1, 0, 1, 1, 0, 0, 1, 0, 0, 0, 1, 0, 1, 0, 0, 0, 1, 1, 1, 0, 0, 1, 0, 0, 1, 0, 1,0, 1, 0, 1, 0, 1, 0} and ER = 36/(3 × 3) = 4 bpp and the cover image pixels are as follows:

$$I_{3\times 3} = \begin{vmatrix} 9277330 & 9276816 & 9277072 \\ 9276816 & 9343124 & 9343381 \\ 9211793 & 9409173 & 9409173 \end{vmatrix}$$

and $D' = D \oplus k_1$ = {0, 1, 1, 1, 1, 0, 0, 0, 0, 1, 0, 1, 0, 1, 1, 0, 0, 1, 0, 1, 0, 0, 0, 0, 1, 0, 0, 0, 0, 0, 1, 0, 0, 0, 0, 0}

2. Divide into three RGB image pixel blocks as shown below.

$$R = \begin{vmatrix} 141 & 141 & 141 \\ 141 & 142 & 142 \\ 140 & 143 & 143 \end{vmatrix} \quad G = \begin{vmatrix} 143 & 141 & 142 \\ 141 & 144 & 145 \\ 143 & 146 & 147 \end{vmatrix} \quad B = \begin{vmatrix} 147 & 145 & 145 \\ 145 & 149 & 150 \\ 146 & 150 & 151 \end{vmatrix}$$

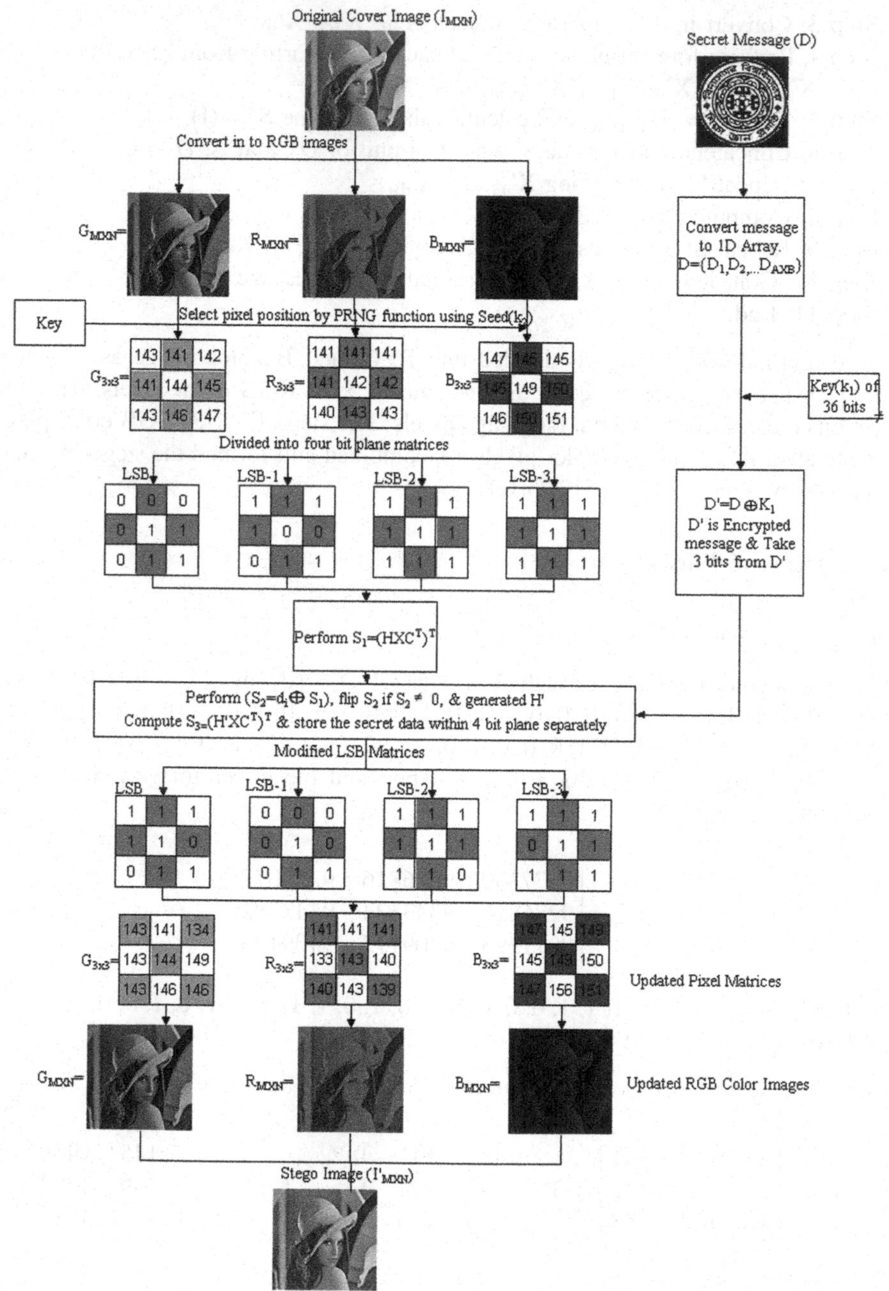

Fig. 1. Pictorial diagram of the proposed data hiding scheme

3. Take red image pixel block and transform into binary number matrix.

$$R = \begin{vmatrix} 10001101 & 10001101 & 10001101 \\ 10001101 & 10001110 & 10001110 \\ 10001100 & 10001111 & 10001111 \end{vmatrix}$$

4. Divide it into four bit-plane matrices starting from LSB.

$$R_{LSB} = \begin{vmatrix} 1 & 1 & 1 \\ 1 & 0 & 0 \\ 0 & 1 & 1 \end{vmatrix} \quad R_{LSB-1} = \begin{vmatrix} 0 & 0 & 0 \\ 0 & 1 & 1 \\ 0 & 1 & 1 \end{vmatrix}$$

$$R_{LSB-2} = \begin{vmatrix} 1 & 1 & 1 \\ 1 & 1 & 1 \\ 1 & 1 & 1 \end{vmatrix} \quad R_{LSB-3} = \begin{vmatrix} 1 & 1 & 1 \\ 1 & 1 & 1 \\ 1 & 1 & 1 \end{vmatrix}$$

5. Read the LSB matrix and form a 1D matrix. c = [1 1 1 1 0 0 0 1 1]
6. Calculate the syndrome $S_1 = H \times (c)^T = \begin{vmatrix} 0 & 0 & 0 & 1 & 1 & 1 & 1 & 0 & 0 \\ 0 & 0 & 1 & 0 & 1 & 1 & 0 & 1 & 0 \\ 0 & 0 & 1 & 1 & 0 & 1 & 0 & 0 & 1 \end{vmatrix}$

$$\times \begin{vmatrix} 1 & 1 & 1 & 1 & 0 & 0 & 0 & 1 & 1 \end{vmatrix}^T = \begin{vmatrix} 1 \\ 0 \\ 1 \end{vmatrix}$$

7. Transpose the syndrome and XOR with the secret data bit, i.e., [1 0 1] ⊕ [0 1 1] = [1 1 0] which match with the fifth column of Hamming matrix.
8. Generate the code H′ = [0 0 0 0 1 0 0 0 0] and XOR with the original code c.

   S_3 = [1 1 1 1 0 0 0 1 1] ⊕ [0 0 0 0 1 0 0 0 0] = [1 1 1 1 1 0 0 1 1].

9. Transform into a new LSB matrix.

$$R'_{LSB} = \begin{vmatrix} 1 & 1 & 1 \\ 1 & 1 & 0 \\ 0 & 1 & 1 \end{vmatrix}$$

10. Similarly, compute the LSB-1, LSB-2, and LSB-3 matrices as follows:

$$R'_{LSB-1} = \begin{vmatrix} 0 & 0 & 0 \\ 0 & 1 & 0 \\ 0 & 1 & 1 \end{vmatrix} \quad R'_{LSB-2} = \begin{vmatrix} 1 & 1 & 1 \\ 1 & 1 & 1 \\ 1 & 1 & 0 \end{vmatrix} \quad R'_{LSB-3} = \begin{vmatrix} 1 & 1 & 1 \\ 0 & 1 & 1 \\ 1 & 1 & 1 \end{vmatrix}$$

11. Update all four modified binary matrices to their corresponding position in original red pixel matrix.

$$R'_{3\times3} = \begin{vmatrix} 10001101 & 10001101 & 10001101 \\ 10000101 & 10001111 & 10001100 \\ 10001100 & 10001111 & 10001011 \end{vmatrix} = \begin{vmatrix} 141 & 141 & 141 \\ 133 & 143 & 140 \\ 140 & 143 & 139 \end{vmatrix}$$

12. Similarly, get updated green and blue pixel matrices.

$$G'_{3\times3} = \begin{vmatrix} 143 & 141 & 134 \\ 143 & 144 & 149 \\ 143 & 146 & 146 \end{vmatrix} \quad B'_{3\times3} = \begin{vmatrix} 147 & 145 & 149 \\ 145 & 149 & 150 \\ 147 & 156 & 151 \end{vmatrix}$$

13. Finally, the stego image block will be $I'_{3\times3} = \begin{vmatrix} 9277330 & 9276816 & 9275028 \\ 8753040 & 9408660 & 9213333 \\ 9211794 & 9409179 & 9409179 \end{vmatrix}$

Example 3.1.2: Data extraction

1. The marked image sized I′ of size 3 × 3 is shown below

$$I' = \begin{vmatrix} 9277330 & 9276816 & 9275028 \\ 8753040 & 9408660 & 9213333 \\ 9211794 & 9409179 & 9409179 \end{vmatrix}$$

2. Divide into three RGB image pixel blocks.

$$R = \begin{vmatrix} 141 & 141 & 141 \\ 133 & 143 & 140 \\ 140 & 143 & 139 \end{vmatrix} \quad G = \begin{vmatrix} 143 & 141 & 134 \\ 143 & 144 & 149 \\ 143 & 146 & 146 \end{vmatrix} \quad \mathrm{B} = \begin{vmatrix} 147 & 145 & 149 \\ 145 & 149 & 150 \\ 147 & 156 & 151 \end{vmatrix}$$

3. Take red image pixel block and transform into binary numbers.

$$\begin{vmatrix} 10001101 & 10001101 & 10001101 \\ 10000101 & 10001111 & 10001100 \\ 10001100 & 10001111 & 10001011 \end{vmatrix}$$

4. Divide it into four bit-plane matrices starting from LSB.

$$R_{LSB} = \begin{vmatrix} 1 & 1 & 1 \\ 1 & 1 & 0 \\ 0 & 1 & 1 \end{vmatrix} \quad R_{LSB-1} = \begin{vmatrix} 0 & 0 & 0 \\ 0 & 1 & 0 \\ 0 & 1 & 1 \end{vmatrix}$$

$$R_{LSB-2} = \begin{vmatrix} 1 & 1 & 1 \\ 1 & 1 & 1 \\ 1 & 1 & 0 \end{vmatrix} \quad R_{LSB-3} = \begin{vmatrix} 1 & 1 & 1 \\ 0 & 1 & 1 \\ 1 & 1 & 1 \end{vmatrix}$$

5. Read LSB matrix and form a 1D matrix. c = [1 1 1 1 1 0 0 1 1]
6. Calculate the syndrome $S_1 = H \times (c)^T =$

$$\begin{vmatrix} 0 & 0 & 0 & 1 & 1 & 1 & 1 & 0 & 0 \\ 0 & 0 & 1 & 0 & 1 & 1 & 0 & 1 & 0 \\ 0 & 0 & 1 & 1 & 0 & 1 & 0 & 0 & 1 \end{vmatrix} \times \begin{vmatrix} 1 & 1 & 1 & 1 & 1 & 0 & 0 & 1 & 1 \end{vmatrix}^T = \begin{bmatrix} 0 \\ 1 \\ 1 \end{bmatrix}$$

7. Transpose the syndrome to get secret data bits d = [0 1 1]
8. Repeat the above steps until we do not get the secret data bits. Concatenate all the data bits to get the data, that is, D′ = {0, 1, 1, 1, 0, 1, 0, 0, 1, 1, 0, 1, 1, 1, 0, 0, 0, 1, 1, 0, 1, 0, 0, 1, 1, 0, 1, 0, 1, 0, 0, 0, 1, 0, 1, 0}.
9. XOR the modified secret data with secret key k_1 to get the original secret data bits, that is, D = {0, 1, 0, 1, 0, 1, 0, 0, 1, 1, 0, 1, 1, 1, 0, 0, 0, 1, 1, 0, 1, 0, 0, 1, 1, 0, 1, 0, 1, 0, 0, 0, 1, 0, 1, 0}

4 Experimental Result and Comparison

The scheme is implemented using NetBeans IDE 8.0 on standard color images to measure the performance. The standard cover images are collected from image database of SIPI [13].

The quality of the stego images is measured using mean square error (MSE) and peak signal to noise ratio (PSNR) [14, 15]. Figure 2 shows the stego image when ER = 4 bpp.

$$MSE = \frac{1}{M \times N} \sum_{i=1}^{M} \sum_{j=1}^{N} [I(i,j) - I'(i,j)]^2 \tag{1}$$

$$PSNR = 10 \log_{10} \frac{255^2}{MSE} (\text{dB}) \tag{2}$$

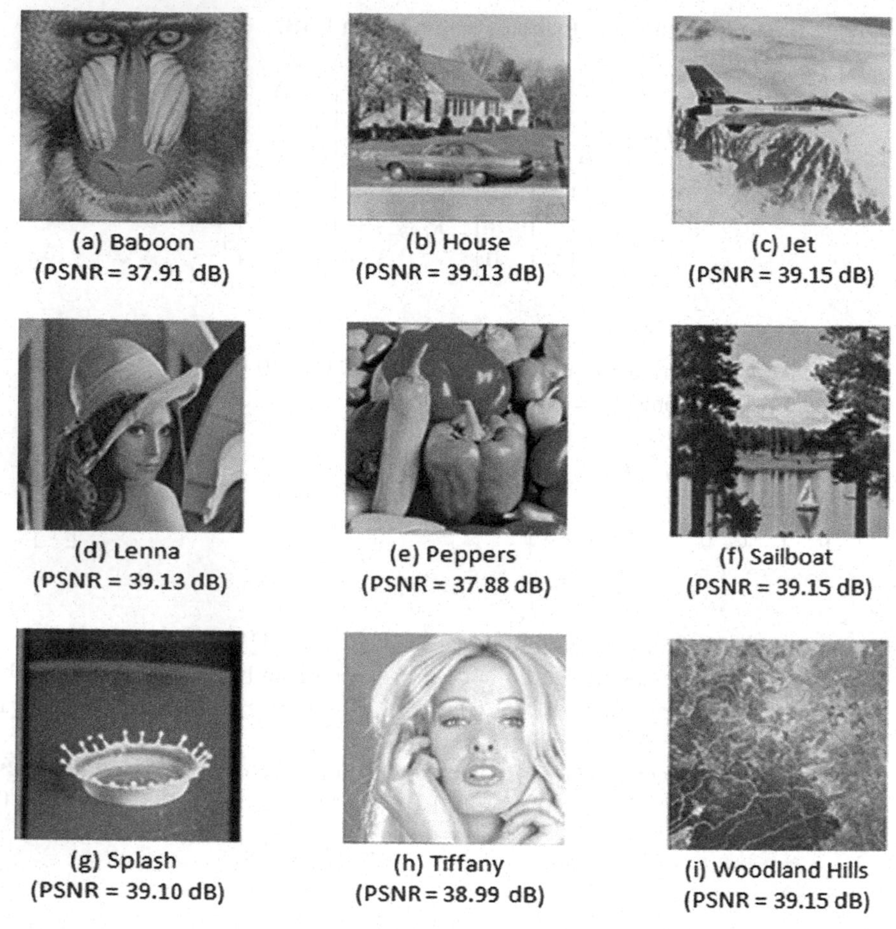

(a) Baboon (PSNR = 37.91 dB) (b) House (PSNR = 39.13 dB) (c) Jet (PSNR = 39.15 dB)

(d) Lenna (PSNR = 39.13 dB) (e) Peppers (PSNR = 37.88 dB) (f) Sailboat (PSNR = 39.15 dB)

(g) Splash (PSNR = 39.10 dB) (h) Tiffany (PSNR = 38.99 dB) (i) Woodland Hills (PSNR = 39.15 dB)

Fig. 2. Stego images of size (512 × 512)

The following tables represent the comparison of PSNR values of stego images generated by different methods with varying payload, and we achieve better PSNR value every time compared to existing methods (Tables 1, 2 and 3).

Table 1. Comparison of PSNR-ER with existing methods for ER = 1 bpp

	Lenna	Baboon	Tiffany	Peppers	Jet	Sailboat	Splash
Matrix encoding [2]	47.02	47.02	47.02	47.02	47.04	47.01	47.03
Nearest code [5]	47.02	47.02	47.03	47.02	47.04	47.01	47.03
Hamming +1 [9]	45.14	45.14	45.10	45.14	45.14	45.14	45.13
Cao et al. [10]	51.14	51.14	51.15	51.14	51.15	51.14	51.14
Proposed scheme	57.31	57.31	58.56	57.31	57.56	57.57	57.57

Table 2. Comparison of PSNR-ER with existing methods for ER = 2 bpp

	Lenna	Baboon	Tiffany	Peppers	Jet	Sailboat	Splash
Matrix encoding [2]	33.10	33.08	33.09	33.05	33.10	33.06	33.18
Nearest code [5]	33.11	33.07	33.09	33.06	33.10	33.07	33.18
Hamming +1 [9]	20.62	20.85	19.78	20.63	19.98	20.27	20.54
Cao et al. [10]	41.61	41.60	41.57	41.60	41.67	41.59	41.62
Proposed scheme	50.39	50.38	51.55	50.36	51.63	51.62	51.62

Table 3. Comparison of PSNR-ER with existing methods for ER = 3 bpp

	Lenna	Baboon	Tiffany	Peppers	Jet	Sailboat	Splash
Matrix encoding [2]	19.80	19.77	19.87	19.89	20.07	20.09	19.82
Nearest code [5]	19.80	19.77	19.87	19.89	20.08	20.09	19.81
Hamming +1 [9]	–	–	–	–	–	–	–
Cao et al. [10]	37.92	37.92	37.91	37.92	37.98	37.89	37.94
Proposed scheme	44.04	44.05	45.16	44.01	45.28	45.29	45.26

Table 4. Comparison of PSNR-ER, SSIM, and SD with existing methods for ER = 4 bpp

	Lenna	Baboon	Tiffany	Peppers	Jet	Sailboat	Splash
Proposed scheme	37.89	37.91	38.98	37.88	39.14	39.16	39.11
SSIM	0.93057	0.97676	0.93865	0.93022	0.9392	0.96248	0.9244
Standard deviation	0.08159	0.11102	0.175409	0.046307	0.0768	0.05825	0.02812

5 Security Analysis

Security analysis is an important factor of data hiding process. In this paper, we have used two levels of security to enhance our proposition from security perspective. First, we take a 36 bits secret key and encrypt the secret data bits using symmetric key encryption. As it is only known to the sender and receiver, the third party will not be able to decrypt it without knowing the secret key. In second level of security, we have taken a secret seed which is also known to the receiver and sender only. Using this seed, we generate a sequence of unique numbers with the help of PRNG function. We have taken the cover image pixel blocks according to the generated numbers. So

Table 5. RS analysis of stego image with ER = 4 bpp

	Lenna	Woodland Hills	House	Peppers	Jet	Sailboat
R_m	18137	17288	18109	17427	18294	17508
R_{m1}	19056	18152	19237	18151	19637	18123
S_m	14555	15772	14805	15396	14427	15509
S_{m1}	13868	15006	13983	14825	13362	15053
RS value	0.0491	0.0493	0.0592	0.0395	0.0736	0.0324

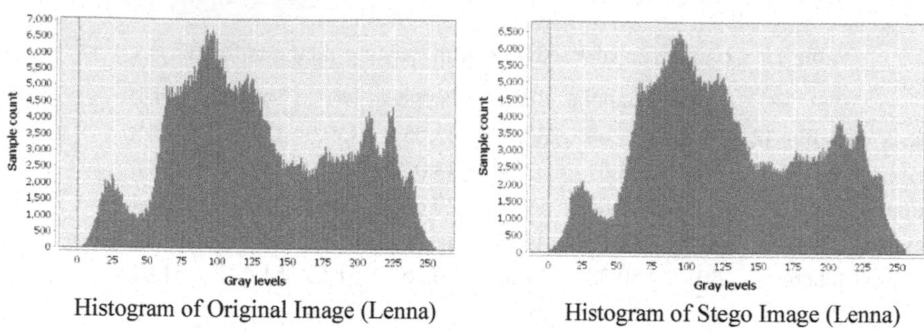

Fig. 3. Histogram of original and stego image

without knowing this seed, no one will be able to predict the number sequence.

We also verified our algorithm against some standard measurement like SSIM, standard deviation, RS analysis, histogram analysis. The structural similarity (SSIM) index is a method for measuring the similarity between two images. The SSIM index can be viewed as a quality measure of one of the images being compared, provided the other image is regarded as of perfect quality. From Table 4, it is observed that the SSIM values of all test images are nearer to 1. Standard deviation is used to measure the amount of variation between original and stego images. Here, we achieve SD neared to zero which means that the stego image and cover image are similar in nature. We also analyze our stego image through RS analysis [16]. In Table 5, it is shown that the values of R_m and R_{m1}, S_m and S_{m1} are almost equal and the ratio of R and S lies around 0.05, which is very small, so that we can conclude that our proposed scheme is secure against RS attack. Figure 3 represents the histogram of the original cover image and the stego image. It is shown that the shape of the histogram almost remains same after embedding high-capacity secret data. So we can say that our proposed method is robust against histogram attack.

6 Conclusion

In this paper, we introduced a novel secure data hiding scheme using Hamming code for RGB color image. Bit-plane slicing of the each RGB color cover image block is also introduced to increase data hiding capacity over grayscale image. So the data embedding rate is raised up to 4 bpp which is greater than other existing schemes. In our algorithm, PSNR is also high compared to existing schemes that mean we generate a better visual quality stego image. From security perspective, we introduced a shared secret key to find suitable bit pattern through XOR operation during data embedding as well as data extraction. The cover image block has been chosen in random location through PRNG of shared seed value which also enhances security in our proposed scheme. We have tested our stego image with RS analysis, histogram analysis, SSIM, SD method and observed that the proposed scheme is preferable for data embedding where visual quality and security constraint needs to be maintained for high payload. In future, the scheme has been extended to enhance security, capacity, and quality in different domain for video-based steganography.

References

1. Wang, R. Z., Lin, C. F., & Lin, J. C. (2000), Hiding data in images by optimal moderately-significant-bit replacement. Electronics Letters, 36(25), 2069–2070.
2. Crandall R (1998), Some notes on Steganography, Posted on Steganography mailing list. http://os.inf.tudresden.de/westfield/Crandall.pdf.
3. Zhang, W., Wang, S., & Zhang, X. (2007). Improving embedding efficiency of covering codes for applications in steganography. IEEE Communications Letters, 11(8).
4. Huffman, W. C., & Pless, V. (2010). Fundamentals of error-correcting codes. Cambridge university press.
5. Chang, C. C., & Chou, Y. C. (2008, January). Using nearest covering codes to embed secret information in grayscale images. In Proceedings of the 2nd international conference on Ubiquitous information management and communication (pp. 315–320). ACM.
6. Liu, Y., Chang, C. C., & Chien, T. Y. (2017). A Revisit to LSB Substitution Based Data Hiding for Embedding More Information. In Advances in Intelligent Information Hiding and Multimedia Signal Processing (pp. 11–19). Springer International Publishing.
7. Westfeld, A. (2001, April). F5—a steganographic algorithm. In International workshop on information hiding (pp. 289–302). Springer Berlin Heidelberg.
8. Kim, C., Shin, D., & Shin, D. (2011, April). Data hiding in a halftone image using hamming code (15, 11). In Asian Conference on Intelligent Information and Database Systems (pp. 372–381). Springer Berlin Heidelberg..
9. Kim, C., & Yang, C. N. (2014). Improving data hiding capacity based on hamming code. In Frontier and Innovation in Future Computing and Communications (pp. 697–706). Springer Netherlands.
10. Cao, Z., Yin, Z., Hu, H., Gao, X., & Wang, L. (2016). High capacity data hiding scheme based on (7, 4) Hamming code. Springer Plus, 5(1), 175.
11. Jana, B., Giri, D., & Mondal, S. K. (2016). Dual image based reversible data hiding scheme using (7, 4) hamming code. Multimedia Tools and Applications, 1–23.
12. Banik, B. G., & Bandyopadhyay, S. K. (2017). Image Steganography Using BitPlane Complexity Segmentation and Hessenberg QR Method. In Proceedings of the First International Conference on Intelligent Computing and Communication (pp. 623–633). Springer Singapore..
13. University of Southern California, The USC-SIPI Image Database, 2015 http://sipi.usc.edu/database/database.php.
14. P. Gupta, et al., "A Modified PSNR Metric based on HVS for Quality Assessment of Color Images," Proc. of IEEE International Conference on Communication and Industrial Application (ICCIA-2011), Kolkatta (W.B.), India, no. 23, pp. 96–99, December, 2011.
15. P. Gupta, et al.,"A New Model for Performance Evaluation of Denoising Algorithms based on Image Quality Assessment," Proc. of (ACM ICPS) CUBE International Information Technology Conference & Exhibition, Pune, India, pp. 5–10, September, 2012.
16. Fridrich J, Goljan M, Du R (2001) Invertible authentication. In: Photonics West 2001-Electronic Imaging (pp. 197–208). International Society for Optics and Photonics.

Optimization of Fuzzy C-Means Algorithm Using Feature Selection Strategies

Kanika Maheshwari(✉) and Vivek Sharma

C.S.E. Department, Samrat Ashok Technological Institute, Vidisha, India
engg.kanikamaheshwari@gmail.com

Abstract. In the era of Digital World, everything has a cost whether the cost is in terms of money, time, space, or data. Big data is a term used to define very large volume of data which possesses a lot of varieties in it. In the following paper, we are presenting feature selection strategy to optimize fuzzy-based clustering over very large data. By using selective features method, we can reduce number of features used to classify the following dataset; thus, the reduction of dimensions/features can help in optimizing iteration count, space, time as well as minimize objective function for the following dataset. In final observation, we found out the reduction in iteration count as well as time in comparison to literal fuzzy c-means algorithm that is 10.65 s and 9.84 s, respectively, for pen digits and cement dataset in comparison to PCA that is 8 s and 9 s and EFA that is 5.74 s and 8 s, respectively, for both the datasets.

Keywords: Very large data · PCA · EFA · Feature selection · Optimization

1 Introduction

In the era of Digital World, everything has a cost whether the cost is in terms of money, time, space, or data. Big data is a term used to define very large volume of data which possesses a lot of varieties in it; lots of techniques are available to handle crisp data, but it is impractical in real-life situations to do crisp partition of data. Big data includes fuzziness in it which reflects varied nature of inside data; for knowledge discovery, it is necessary to find out relation among variables of dataset and its fuzziness with the optimization of necessary parameters. Literal fuzzy c-means algorithm does the same; it clusters data and find out relationship among them, but it is not optimized for a big set of database. We are presenting a method to optimize performance of fuzzy c-means algorithm over very large data by reducing dimensions of data.

With the emerging era of internet, terabytes of data generating every minute all over the world in the form of user logs, transactions, images, videos and so on. The craze of being on social media or a techno savvy one leading people to be online, day of internet increasing day by day for home business to big Enterprises all things are getting online. The generation of such huge amount of data results in the demand of large storages for the enterprises, while finding out hidden patterns behind such huge amount of data is another challenge. It is quite impractical to go with the traditional methods to process such huge amount of data that is known as big data in today's world. Facebook, Twitter, LinkedIn, Google, Amazon are few of the examples of the companies those

V. Bhateja et al. (eds.), *Information Systems Design and Intelligent Applications*, Advances in Intelligent Systems and Computing 672,
https://doi.org/10.1007/978-981-10-7512-4_37

need to store petabytes of data on daily basis; of course, the data store is in the raw form and required lots of preprocessing to bring out hidden patterns out of that, initial step of preprocessing any raw data is clustering where data are roughly clustered on the basis of similarity among them without any initial class labels. Clustering or cluster analysis is a machine learning methodology which helps in identifying group of similar objects that is known as cluster [1]. Basically, clustering can be divided into two types: soft clustering (fuzzy based) and hard clustering (crisp based) [2]. In hard clustering, a variable can be on side whether it can belong to the cluster or it may not that results in 0 or 1 case, while in soft clustering, a variable can belong to more than one cluster with varied degree of membership. Recently, researchers advised the use of partitional algorithms in order to cluster very large database as their computational requirements are low in comparison to others [3].

In last decade, researchers introduced more than a bunch of partitional algorithms; one among them is literal fuzzy c-means (LFCM) [4]; following algorithm takes entire dataset for clustering which makes it sufferer for big data or very large data clustering purpose. Other than that, most of the algorithms proposed work on random sampling methodology; some of them among are corsets [5], CLARA [6], CURE [7]; thus, the sample set taken by the above-mentioned algorithms does not do the representation of entire data set which is again a drawback; we are addressing some of these drawbacks in our proposed methodology.

2 Dimensionality Reduction

In LFCM, data are processed in the form of full dataset which results in more complexity parameters especially time cost is the one among them; to overcome the following drawback, we worked on the following technique. Generally most of the algorithms now work on data subset as a representative of whole data for clustering which result in overlapping cluster centers, as very large data due to its size have many dimensions, so reduction of data dimensions can result in the optimization of literal fuzzy c-means algorithm. Here, we are mainly focusing on two dimension reduction or transformation techniques those are principal component analysis (PCA) as well as exploratory factor analysis (EFA); literal fuzzy c-means algorithm will run on the dataset after following dimensional reduction process which will result in time as well as space optimization that is a must in today's world.

3 Principle Component Analysis

Principal component analysis is mainly a method of reducing data dimensions; in literal fuzzy c-means algorithm, whole dataset is taken as an input which is not good in case of large dataset. Thus, the process of data dimension reduction enhances the clustering results as well reduces time of data processing. By selecting the particular principal components, we can generate a new data set as a representative of full dataset. PCA transforms the data to a very new coordinate system in the way that the variable with the greater variances is considered as a first principal component and so the others. We are working with the PCA using covariance method; following is the procedure we followed:

- Load a dataset: $X = (X_1 \ldots X_2 \ldots X_3 \ldots X_4 \ldots . X_n\}'$
- Calculate mean of the dataset: $\mu_x = M\{X\}$
- Find covariance matrix for the following dataset: $CV_x = M\{(X - \mu_x)(X - \mu_x)'\}$
- Identify eigenvectors of the following covariance matrix: $CV_x \; e_i = \lambda_i e_i$, where $i = 1, \ldots, n$
- Next, to sort the eigenvectors with the low-order eigenvalues and choose k no. of eigenvectors with the largest eigenvalues to form a new dimensional matrix W.
- Now, the final transformed matrix will be: $Y = W' \times X$

4 Exploratory Factor Analysis

Exploratory factor analysis can be more fruitful when talking in terms of fuzziness; it gives a more exploratory dataset which provides a more explanatory correlation between different variable. Thus, it helps in dimension reduction, which ultimately increases the cluster quality and decreases computing time. Its main purpose is to focus on the shared variance (Fig. 1).

- Load dataset: $X = X_1 \ldots X_2 \ldots X_3 \ldots X_4 \ldots . X_n$
- Find common factors among different variables: $F = F_1 \ldots F_2 \ldots F_3 \ldots . F_4 \ldots F_m$
- The intention of the following analysis is to find the maximum correlation among different variables
  $X_j = a_{j1}F_1 + a_{j2}F_2 + \ldots\ldots\ldots\ldots a_{jm}F_m$

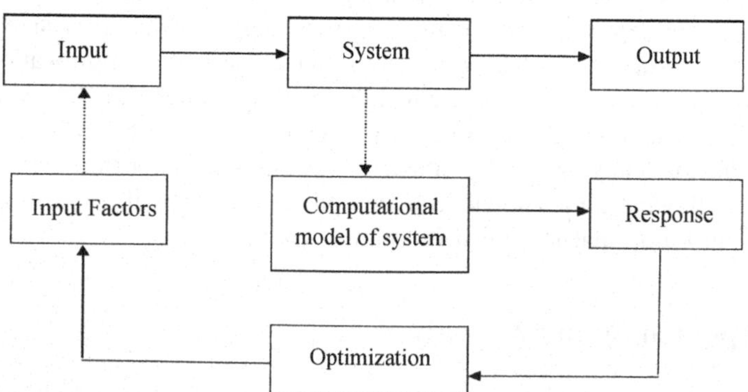

Fig. 1. Optimization procedure

5 Mathematical Comparison Between PCA and EFA

In order to compute the correlation between components, EFA takes little bit more time than PCA. While EFA and PCA do, moreover, the same kind of work, the only difference between the two is that in PCA, components look over the overall shared variance, while EFA, moreover, focuses on the shared parts of the variances. PCA looks over the variance of variables rather than a correlation among them while EFA looks over the correlation among variables. Other than these methods, a simple correlation algorithm can also provide reduced dataset, which can also help in far better results; we are using EFA and PCA in our analysis because these are basically available algorithms that are supportable in most of the tool where PCA focuses on the direction of data toward the maximum variance, while EFA focuses on some sort of smaller set of underlying factors. Mathematically, PCA and EFA have a lot more similarities than is immediately apparent. After taking k principal components in the form of eigen vectors with all the k no. of eigen values included with the highest scores in the form of a new matrix Y. Where W is the new diagonal matrix with highest eigen values. Loading matrix for PCA can be calculated as $L = KW^{1/2}$. Correlation matrix for dataset X can be calculated as $C = 1/n\, X^T X$; If the value of K is set to the maximum value of CV_x, C can be considered as $C = L\, L^T$; for smaller values of k, correlation matrix's values will approximate itself, which helps in minimizing sum of squared errors for the output, which provides a similarity scenario of PCA to EFA. Following mathematical equation proves that PCA and EFA both does not ignore squared error measures. We can determine Loadings of matrix numerically only, in both PCA and EFA thus loadings & factor scores respectively are calculated in such a way that they provide a reduced set of features for further cluster analysis.

6 Literal Fuzzy C-Means Algorithm

- The only objective of a literal fuzzy c-means algorithm is to minimize the following objective function as effective as possible:
- where $\|X_i - c_j\|^2$ is a Euclidean distance between i data points and j cluster centers.
-

$$J_m = \sum_{i=1}^{N} \sum_{j=1}^{c} u_{ij}^{m} \|x_i - c_j\|^2 \; 1 \leq m < \infty$$

(a) **Traditional Fuzzy C-Means Approach**

- Initialize $U = [u_{ij}]$ matrix, $U^{(0)}$
- Calculate the center

$$U_{ij} = 1/(\|x_i - c_j\| / \| x_i - c_k\|)^{2/m-1}$$

- Update $U^{(k)}$, $U^{(k+1)}$
- If $\left\|U^{(k+1)} - U^{(k)}\right\| < \mathcal{E}$, then STOP; value of $\mathcal{E}$ is set as default.

(b) **Proposed Methodology**

- Organize dataset X
- Perform PCA/EFA for resultant new dimension's matrix: $X_d \rightarrow X_m$
- Calculate the centers (run over the literal FCM).

$$C_j = \sum_{i=1}^{N} \frac{u_{ij}^m \cdot x_i}{u_{ij}^m}$$

- Update $U^{(k)}$, $U^{(k+1)}$

$$U_{ij} = \frac{1}{\sum_{k=1}^{c} \left(\frac{\|X_i - C_j\|}{\|X_i - C_k\|} \right)^{\frac{2}{m-1}}}$$

- If $\left\|U^{(k+1)} - U^{(k)}\right\| < \mathcal{E}$, then
- *STOP;*
- *Else go to step 4.*

7 Experiment and Output

We performed the following experimentation on Intel core machine with 8 GB memory in MATLAB R2010b environment on Windows 7, 32-bit operating system. The dataset has been taken from UCI machine learning repository; following is the details of dataset (Table 1):

Table 1. Basic information about the datasets

Datasets	Samples	Features
Pen-based recognition of handwritten digits	10992	16
Cement	9279	9

(a) **Comparison of FCM and Our Proposed Methodology**

The following graph is showing low variance of components; the maximum variance visible is 20% which is quite less, so the following data required more no. of components in order to be the representative of full data set, while in some of the datasets, variance is relatively high where the process can be more useful (Tables 2, 3, 4, 5, 6 and 7) (Figs. 2, 3, 4, 5, 6, 7, 8, 9, 10, and 11).

Table 2. Before PCA/EFA: pen-based recognition of handwritten digits

M	Objective function	Time (s)	Iteration count
1.75	81698382.196742	10.65	100

Table 3. After PCA: pen-based recognition of handwritten digits

M	Objective function	Time (s)	Iteration count
1.75	81698382.196742	08	43

Table 4. After EFA: pen-based recognition of handwritten digits

M	Objective function	Time (s)	Iteration count
1.75	21956.360278	09	40

Table 5. Before PCA/EFA: cement data

M	Objective function	Time (s)	Iteration count
1.75	21102410.247013	9.284	100

Table 6. After EFA: cement data

M	Objective function	Time (s)	Iteration count
1.75	216156.722862	8	37

Table 7. After PCA: cement data

M	Objective function	Time (s)	Iteration count
1.75	0.880069	5.744	56

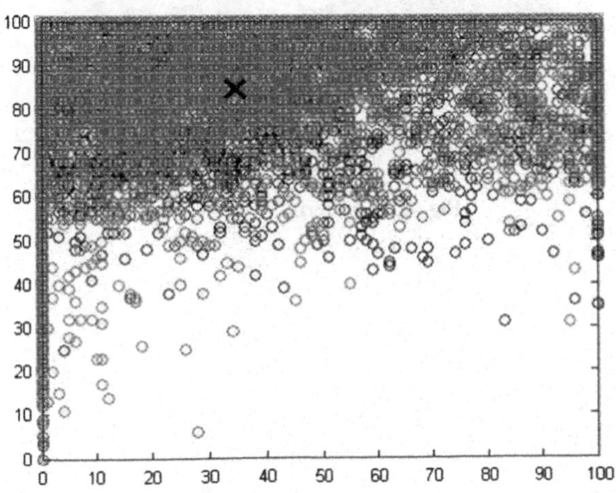

Fig. 2. Clustering before PCA/EFA

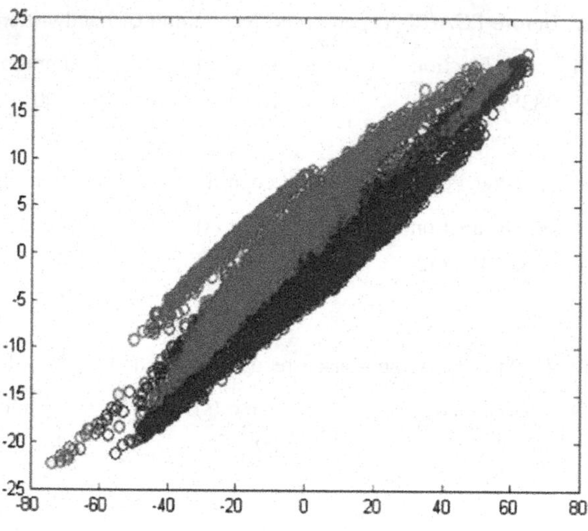

Fig. 3. Clustering after PCA

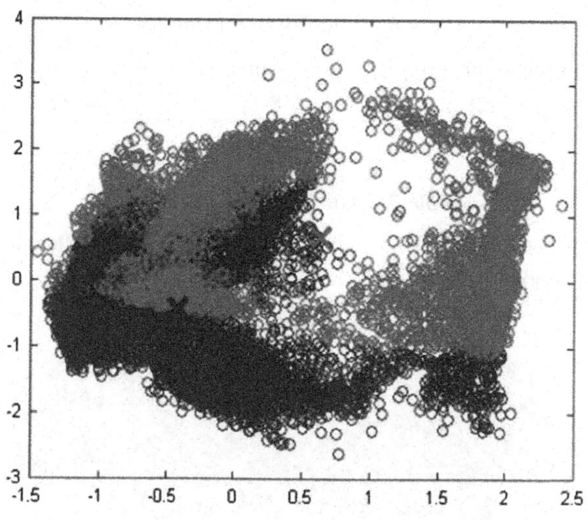

Fig. 4. Clustering after EFA

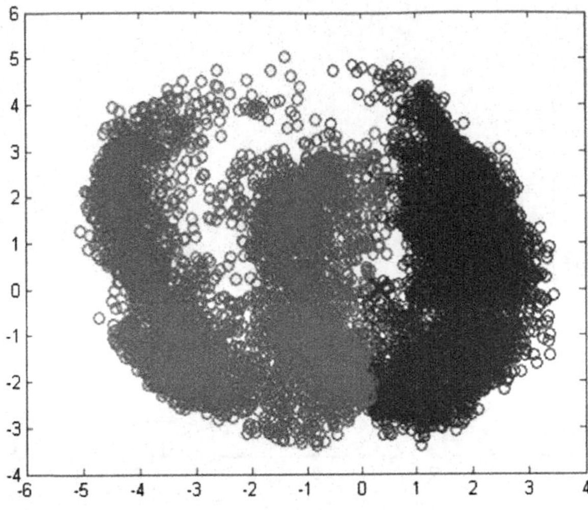

Fig. 5. Clustered z scores

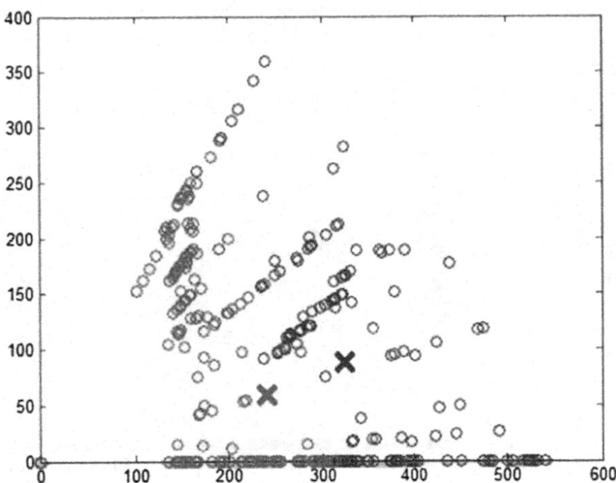

Fig. 6. Clustering before PCA

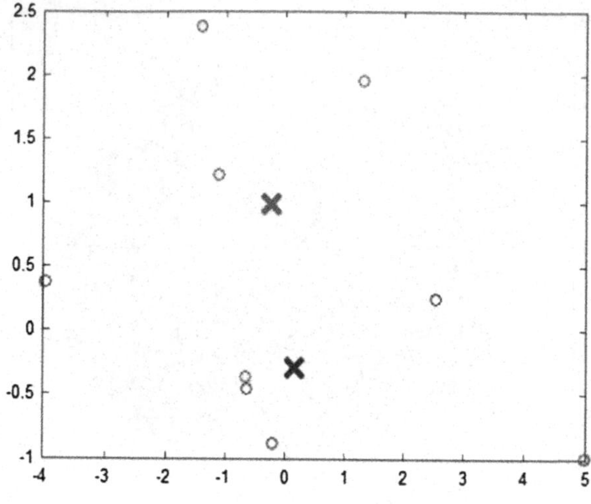

Fig. 7. Clustering after PCA

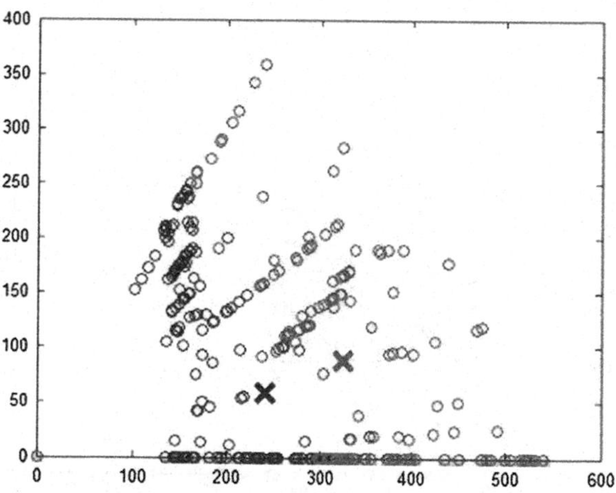

Fig. 8. EFA score plotting

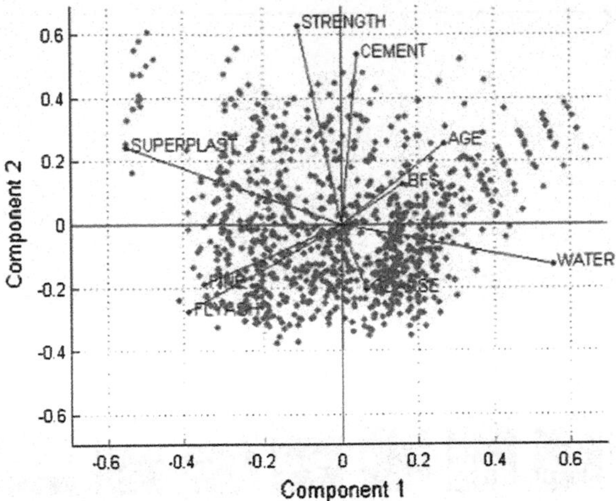

Fig. 9. PC variance

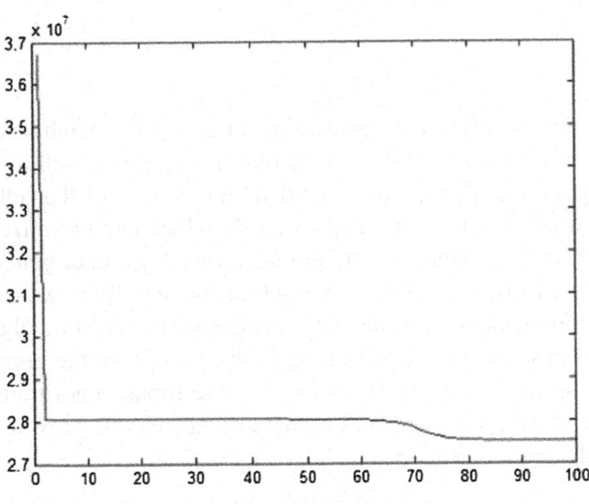

Fig. 10. FCM time graph

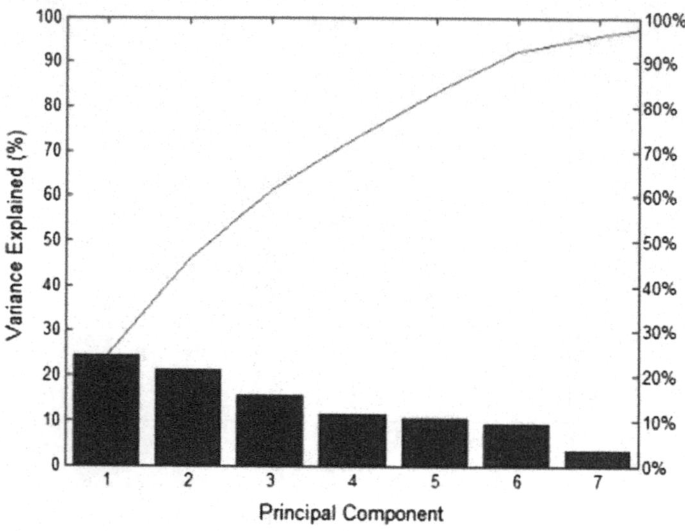

Fig. 11. PC variance percentage

8 Conclusion

In the following era, handling data generating in terabytes, petabytes, and so on is a challenging task. Such huge amount of data requires space as well as time to process, which itself is again a challenge; traditional methods are not that efficient to handle such large amount of data; thus, need of methods which can optimize time as well as space for such very large datasets is required. Nowadays, data generated contains n number of dimensions; by using feature reduction methodology, we can optimize time as well as space complexity of various data clustering/classification algorithms. We are applying the most basic feature reduction algorithms to prove the ground truth; results can be enhanced up to great level by applying more robust algorithms. In that era, to filter unnecessary variables in order to optimize results can be proved as a boon in the data analytics industry.

9 Future Work

Optimization of algorithms for handling such very large dataset is almost emerging as a challenge; we introduced here feature reduction technique; other than such methodology, one can introduce weight factor while processing such large datasets. Weight factors will ultimately help in selecting number of clusters from each dataset which will optimize result to an another level.

References

1. Neha Bharill and Aruna Tiwari.: Handling Big Data with Fuzzy Based Classification Approach, *Advance* Trends in Soft Computing WCSC 2013, Springer International Publishing Switzerland (2014).
2. KIMBERLY L. ELMORE & MICHAEL B. RICHMAN: Euclidean Distance as a Similarity Metric for Principal Component Analysis, American Meteorological Society (2001).
3. Har-Peled, S., Mazumdar, S.: On coresets for k-means and k-median clustering. In:Proc. ACM Symp. theory Comput., pp. 291–300 (2004).
4. Kaufman, L., Rousseeuw, P.: Finding Groups in Data: An Introduction to Cluster Analysis. Wiley Blackwell, New Work (2005).
5. Guha, L.S., Rastogi, R., Shim, K.: CURE: An efficient clustering algorithm for large databases. Inf. Syst. 26(1), 35–58 (2001).
6. Havens, T.C., Bezdek, J.C., Leckie, C., Hall, L.O., Palaniswami, M.: Fuzzy c-Means Algorithms for Very Large Data. IEEE Trans. Fuzzy Systems 20(6), 1130–1146 (2012).
7. A.M. Hafiz & G.M. Bhat.: Handwritten Digit Recognition using Slope Detail Features, International Journal of Computer Applications (0975–8887) Volume 93 – No 5, May 2014.

Model-Based Test Case Prioritization Using UML Activity Diagram and Design Level Attributes

Shaswati Dash(✉), Namita Panda, and Arup Abhinna Acharya

School of Computer Engineering, KIIT University, Bhubaneswar, Odisha, India
shaswatidash156@gmail.com, {npandafcs, aacharyafcs}@kiit.ac.in

Abstract. This paper presents a prioritization technique that prioritizes the test cases using different design attributes like cohesion, coupling, the number of database access, and non-functional requirements. First, the system requirements are modeled using UML activity diagram (AD). The AD is turned into activity diagram graph (ADG), and the ADG is traversed to find out the test scenarios that are identified by the linearly independent paths in the ADG. Depending upon the different design attributes, weights for each node in the graph are identified and a final priority value is assigned to each node. The nodes executed for every test are identified, the priority value (PV) of the nodes is summed up, and the test case is assigned with a final priority value. Finally, depending on the priority value, the test cases are prioritized. The efficiency of the suggested approach is evaluated using the APFD metric.

Keywords: Average Percentage of Faults Detected (APFD) · Regression testing · Testing · Test cases · UML Activity Diagram

1 Introduction

Nowadays, software testing is the vital process to test the software that acts as per user's requirements [1]. To increase the quality of software product, software testing is required. Regression testing is a type of testing when a modification occurs by adding new features or fixing bugs [1–3]. It is impractical to rerun each and every test case of a software due to exponential risk in time and efforts. So, test case prioritization (TCP) is used [4–6]. It is a method that arranges test cases in a particular order [3, 5, 6]. The earlier performance of important test cases can shrink the regression testing cost [5, 6].

TCP is of two types [6, 7]. In this paper, we have considered four different design attributes like coupling, cohesion [3], database, and non-functional requirements to prioritize the test cases.

V. Bhateja et al. (eds.), *Information Systems Design and Intelligent Applications*, Advances in Intelligent Systems and Computing 672,
https://doi.org/10.1007/978-981-10-7512-4_38

The test cases are generated from UML activity diagram [8–10]. In UML, activity diagram [8] is the vital diagram to point out dynamic aspects of the system. AD is turned into ADG, and the ADG is traversed to find out the test scenarios [9]. Those are identified by the linearly independent paths [3, 10]. Depending upon the different design attributes, the weights are assigned for each node, which summed up the priority values of nodes. Depending on the highest priority value, test cases are arranged [6, 7]. APFD metric [2, 11] is used to measure the efficiency of the proposed approach.

The other sections of the paper are discussed below. Related works are described in Sect. 2. The proposed methodology is described in Sect. 3. Lastly, in Sect. 4, comparison table with related work is discussed. Conclusion and future work are discussed in Sect. 5.

2 Related Work

Ahmed et al. [4] present the software testing suite prioritization using multi-criteria fitness function. In this paper, genetic algorithm is used to automatically prioritize the test case. Here proposed method is multi-criteria fitness function and applied multiple control flow coverage metric. Releasing faults and severity, we can measure the metrics. Zeng et al. [11] describe Historical Based dynamic test case prioritization for requirement proportion Regression Testing. The test cases are ordering depending on the historical data. In the history contained approach it is relay on prioritized requirements. The existing historical data in regression testing are evaluated based on the historical data. Galeebathullah and Indumathi [5] describe a novel approach for controlling a size of a test suite with simple technique. This paper introduces set theory that helps to reduce test suite. Janes [13] describes test case generation and prioritization: a process mining approach, which is based on the execution of the extracted parts with the help of input given by the user.

3 Proposed Methodology

In this suggested methodology, the UML AD is considered for system modeling [9, 10]. Then, the AD is converted into ADG to generate test cases [10]. The input to the system is given. We describe the control flow graph [3]. This procedure is described above. Figure 1 shows a proposed framework in which we discuss the test case generation techniques.

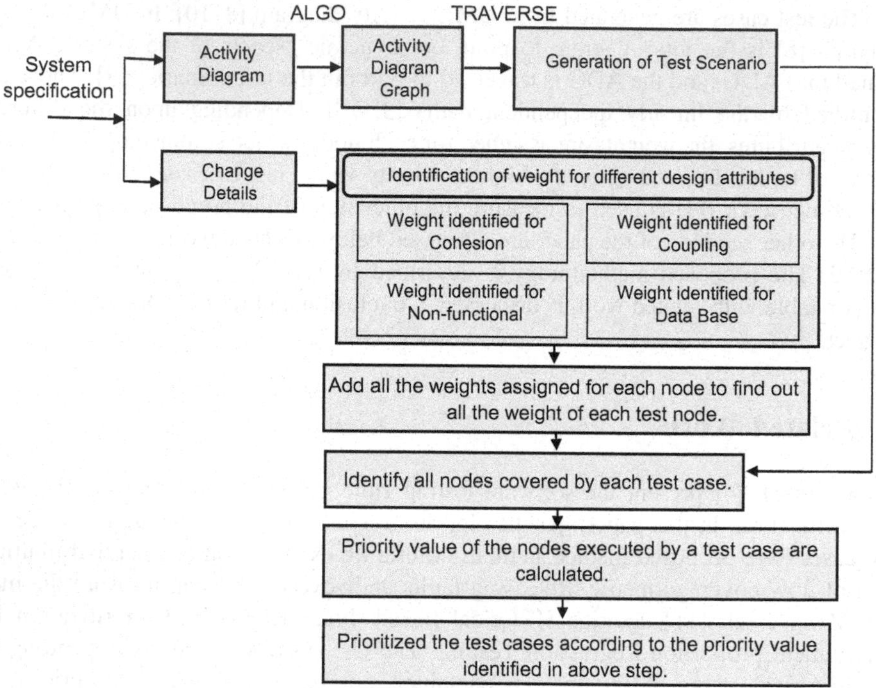

Fig. 1. Proposed framework for test case prioritization

3.1 Steps for the Proposed Approach

For our proposed method, in first step, the test cases are generated from UML AD. The generated test cases are represented by a linearly independent path in which different nodes are present and each node is connected by edges with each other. In second step, weights are assigned for generated nodes depending on different design attributes. After assigning weights, final priority value is identified by each node. In third step, all the final priority values of nodes are added and all nodes covered by each test case are identified. In fourth step, the test cases are prioritized depending on the highest priority value [6, 7].

Step 1: Test case generation from UML AD.
Different inputs are given to model a system and graph is generated.

Step 2: Identify the weights for different design attributes.
After generating the nodes from ADG, the weights for each node are identified in the graph. Final priority value is assigned to each node. For better understanding, here we design four tables and assign weights to each node depending upon the different design attributes. The mapping of weight with the corresponding type is given in Table 1, Table 2, Table 3, and Table 4, respectively.

In Table 1, we take different types of cohesions [10].

Table 1. Assigning weight for different types of cohesion

Node	Types of cohesion	Weight
N1	Coincidental	1
N2	Logical	2
N3	Temporal	3
N4	Procedural	4
N5	Communicational	5
N6	Sequential	6
N7	Functional	7

In Table 2, we describe coupling attribute [10]. We assume a value of 10 for context coupling and 2 for data coupling.

Table 2. Assigning weight for different types of coupling

Node	Types of coupling	Weight
N1	Data	2
N2	Stamp	4
N3	Control	6
N4	Common	8
N5	Context	10

In Table 3, we take five types of non-functional attributes like security, portability, reliability, stability, and usability and assign weights to different nodes.

Table 3. Detail of weight assigned for different types of non-functional requirements

Node	Non-functional	Weight
N1	Security	10
N2	Portability	7
N3	Reliability	8
N4	Stability	5
N5	Usability	2

In Table 4, we assign values according to the database accuracy. By using the metric functional point (FP) [3], we take the scale value (0–10).

Table 4. Number of table assigned by database

Node	No. of table	No. of database
N1	4	4
N2	10	10
N3	5	5
N4	6	6
N5	8	8

Step 3: Add the assigned weight and identify all nodes covered by each test case.
After the final priority value is assigned to each node, the nodes are executed for each test case. The priority value of nodes is added as a result the test case gets final priority value. Priority value of nodes is summed up, and the test is assigned with a final priority value.

Step 4: Prioritize the test cases.
Finally, depending upon the priority value, the test cases are prioritized in descending order [7].

3.2 A Case Study: Withdraw Money from ATM

For generation of test, activity diagram is used that is turned into activity diagram graph (ADG) and generates the test scenarios [10]. Here, we take an example of the structure of ATM activity diagram. In this machine, a user can insert a valid ATM card. If the ATM card is correct, then the options show as ENTER PIN or it ejects card and returns card to the user. When the option ENTER PIN is visible to the user, the user enters the authorized pin number given by the bank. If the user puts the valid pin number, then the user gets an option ENTER THE AMOUNT. If the balance is greater than the entered amount, the user takes money from slot and debits the account. If the balance is less than the entered amount, it shows the balance present in the account. If the user enters invalid pin number, then the transition is canceled. If the user wants to check balance, then the user clicks on the option CHECK BALANCE. Then, user enters the valid pin number and the balance is viewed by user.

3.3 Working Process of the Proposed Method

Here, we take an ATM activity diagram to briefly discuss the test case generation process.

Step 1: Generation of test cases from the UML activity diagram
AD helps to generate test cases to withdraw money from ATM machine use case. First, we have to draw a UML AD of ATM machine by applying several conditions. After that, the ADG is generated from ATM machine AD [9, 10]. From ADG, different nodes are found and graph is generated [10]. Then, weights are assigned for each node. Figure 2 shows an AD for removal of money from ATM machine. Following diagram is drawn with the different activities.

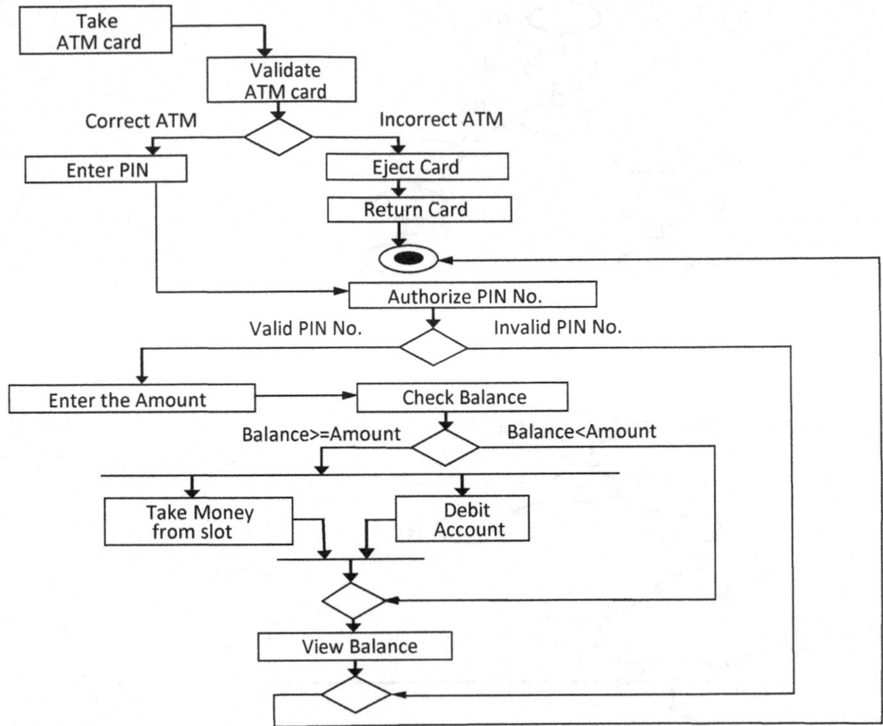

Fig. 2. Activity diagram for withdrawal of money from ATM machine use case [12]

Step 2: Identify the weights for different design attributes

Here, we named the generated 24 nodes as N1, N2… N24. ADG is traversed to generate test scenarios [10]. According to four design attributes, we design the activity diagram graph of ATM machine. Then, depending upon the different design attributes, nodes are assigned with weights and obtained its final value (Fig. 3).

Step 3: Add the assigned weights for each node and identify all nodes covered by every test case

Depending upon the four design attributes, each node is assigned with weights. These weights are found by adding the different values of four design attributes. Final priority value is assigned to each node, and values are summed up. Here, N represents nodes.

Calculation of different nodes:

$$\text{Node(N)} = \text{weight of cohesion} + \text{weight of coupling} + \text{weight of non} - \text{functional}$$

In Table 5, the weights found by adding the different values depending upon different design attributes are given.

Different paths that are found from the above ADG graph are as follows:

1. 1, 2, 3, 4, 6, 9, 10, 11, 13, 14, 15, 16, 18, 20, 21, 22, 23, 24 = TC_a
2. 1, 2, 3, 4, 6, 9, 10, 23, 24 = TC_b

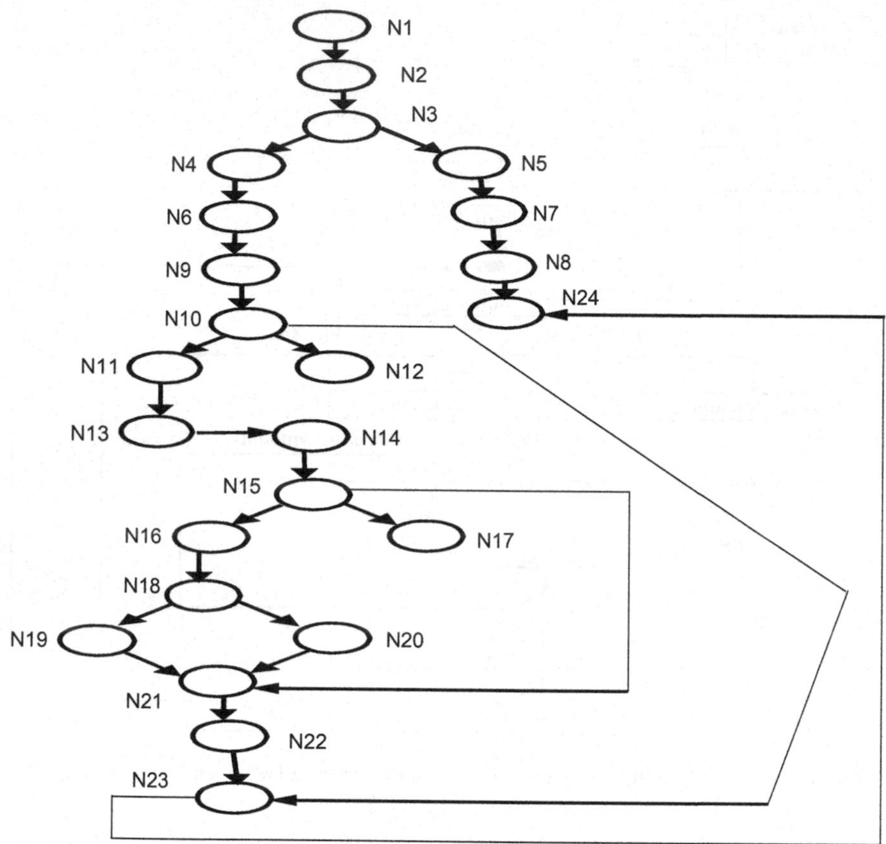

Fig. 3. Activity diagram graph from ATM machine [10]

3. 1, 2, 3, 5, 7, 8, 24 = TC_c
4. 1, 2, 3, 4, 6, 9, 10, 11, 13, 14, 15, 21, 22, 23, 24 = TC_d
5. 1, 2, 3, 4, 6, 9, 10, 11, 13, 14, 15, 16, 18, 19, 21, 22, 23, 24 = TC_e

Put the values of nodes in different paths that are found from ADG. Those paths are identified as test cases. The priority value of nodes is summed up. Final priority value is given to every test case.

Here, we put the values of nodes to calculate test cases.

1. 25 + 21 + 22 + 23 + 25 + 21 + 37 + 22 + 24 + 25 + 20 + 19 + 27 + 22 + 21 + 25 + 23 + 25 = 427
2. 25 + 21 + 22 + 23 + 25 + 21 + 37 + 23 + 25 = 222
3. 25 + 21 + 22 + 25 + 24 + 19 + 25 = 161
4. 25 + 21 + 22 + 23 + 25 + 21 + 37 + 22 + 24 + 25 + 20 + 21 + 25 + 23 + 25 = 359

Table 5. Detail of final weight of nodes

Name of the node	Weight of different attributes	Value
N1	7 + 8 + 10	25
N2	7 + 6 + 8	21
N3	3 + 14 + 8	22
N4	7 + 8 + 8	23
N5	7 + 8 + 10	25
N6	7 + 8 + 10	25
N7	6 + 8 + 10	24
N8	4 + 7 + 8	19
N9	7 + 6 + 8	21
N10	6 + 23 + 8	37
N11	7 + 7 + 8	22
N12	7 + 0 + 8	15
N13	7 + 0 + 8	24
N14	7 + 8 + 10	25
N15	3 + 19 + 8	20
N16	7 + 4 + 8	19
N17	7 + 0 + 8	15
N18	3 + 16 + 8	27
N19	7 + 7 + 10	24
N20	7 + 7 + 8	22
N21	5 + 8 + 8	21
N22	7 + 8 + 10	25
N23	7 + 8 + 8	23
N24	7 + 8 + 10	25

5. 25 + 21 + 22 + 23 + 25 + 21 + 37 + 22 + 24 + 25 + 20 + 19 + 27 + 24 + 21 + 25 + 23 + 25 = 429

Generate test are $TC_a = 427$, $TC_b = 222$, $TC_c = 161$, $TC_d = 359$, $TC_e = 429$.

Step 4: Arrange the test cases

According to the highest value, the test cases are ordered. The highest important test is written first than the lowest important test. We give high priority to the highest result and low priority to the lowest result [13].

Arrange the test in decreasing order:

$$TC_e = 429, TC_a = 427, TC_d = 359, TC_b = 222, TC_c = 161$$

Here, the TC_a, TC_b, TC_c, TC_d, and TC_e are known as test cases.

Table 6. Calculation of non-prioritized test case [2]

	TC_a	TC_b	TC_c	TC_d	TC_e
Fa		*			*
Fb			*	*	
Fc	*	*		*	*
Fd			*		
Fe				*	*

Table 7. Calculation for prioritized test case [2]

	TC_e	TC_a	TC_d	TC_b	TC_c
Fa	*			*	
Fb			*		
Fc	*	*	*	*	
Fd					*
Fe	*		*		

Then, we calculate APFD value for prioritized and non-prioritized test cases. Lastly, we prove that the prioritized value is greater than non-prioritized value. Then, we calculate APFD for non-prioritized test and prioritized test [2, 7, 13] (Table 6).

APFD metric for non-prioritized test case:

$$\text{APFD} = 1 - \frac{2+3+1+3+4}{5*5} + \frac{1}{2*5} = 0.58$$

APFD metric for prioritized test cases (Table 7)

$$\text{APFD} = 1 - \frac{1+3+1+5+1}{5*5} + \frac{1}{2*5} = 0.66$$

Hence, it is proved that calculation of APFD for non-prioritized test case is less efficient than APFD for prioritized test cases.

4 Table Comparison with Related Work

See Table 8.

Table 8. Comparison results

Author name	Model used	Design dependency used (Yes/No)	Historical data store (Yes/No)	Metric used for performed evaluation	Optimization technique used or not (Yes/No)
Ahmed et al. [4]	Model based	Yes	No	Multiple control flow coverage metric	Yes
Indumathi et al. [5]	Set theory	Yes	No	SSR metric (TS size reduction)	Yes
Zeng et al. [11]	History-based approach	No	Yes	APFD and fault detection rate (FDR) metric	Yes
Janes et al. [13]	Model based	Yes	Yes	No metric	Yes
Our approach	Cohesion, coupling, non-functional, database	Yes	No	APFD	Yes

5 Conclusion and Future Work

Here, a planned perspective is presented for generation of the test from the AD by applying several design attributes to generate and prioritize the test cases. In our approach, we generate nodes from ADG and then generate test cases and calculate those values using three several design attributes that are cohesion, coupling, and non-functional requirements. In the future, we will work on database access to generate test cases. Depending upon those attributes, weights are identified that are assigned for each node in the graph. Each node is assigned by final priority value. We add the priority values of nodes, and each test case is obtained by final priority value. With the help of APFD metric, we calculate the prioritized and non-prioritized values of test cases. Lastly, it is proved that prioritized test case is better than non-prioritized test case. In the future, we use more design attributes to calculate different nodes dependency and work on prioritization of test cases.

References

1. Budha G., Namita P., Arup A. A.: Test Case Generation for Use Case Dependency Fault Detection. Electronics Computer Technology (ICECT), 2011 3rd International Conference on Vol. 1. IEEE, 2011.
2. Naresh C., Software Testing Principles: Practices. Oxford University Press, 2010.
3. Rajib M.: Fundamentals of software engineering. PHI Learning Pvt. Ltd., 2014.

4. Ahmed A. A., Mohamed S., Essam K.: Software testing suite prioritization using Multi-Criteria Fitness Function. 2012 22nd International Conference on Computer Theory and Applications (ICCTA), Alexandria, 2012, pp. 160–166.
5. Galeebathullah B., Indumathi C. P.: A Novel Approach for controlling a Size of a Test Suite with Simple Technique. Int. J. Comput. Sci. Eng 2 (2010): 614–618.
6. Samia J., Dip N., Sharfuddin M.: Test Case Prioritization Based on Fault Dependency. International Journal of Modern Education and Computer Science (IJMECS) 8.4 (2016): 33.
7. Bogdan K., George. K., Luay. H. T.: Application of System Models in Regression Test Suite Prioritization. 2008 IEEE International Conference on Software Maintenance, Beijing, 2008, pp. 247–256.
8. http://www.uml.org/.
9. Monalisa S., Debasish K., Rajib M.: Automatic Test Case Generation from UML Models. In Information Technology, (ICIT 2007). 10th International Conference on. IEEE, 2007, pp. 196–201.
10. Debasish K., Debasis S.: A Novel Approach to generate test cases from UML Activity Diagrams. Journal of Object Technology, vol. 8, no. 3, pp. 65–83, 2009.
11. Xiaolin W., Hongwei Z.: History-Based Dynamic Test Case Prioritization for Requirement Properties in Regression Testing. Continuous Software Evolution and Delivery (CSED), IEEE/ACM International WORKSHOP on. IEEE, 2016.
12. https://www.google.co.in/search?q=uml+activity+diagram+for+ATM+withdrawal+process&tbm=isch&tbo=u&source=univ&sa=X&ved=0ahUKEwiAmd7qmcHTAhXLvI8KHcj1BHoQ7AkIRQ&biw=1366&bih=662.
13. Andrea J.: Test Case Generation and prioritization: A Process-Mining Approach. 2017 IEEE International Conference on software Testing, Verification and validation workshops (ICSTW), Tokyo, Japan, 2017, pp. 38–39.

An Ensemble-Based Approach for the Development of DSS

Mrinal Pandey(✉)

Manav Rachna University formally Manav Rachna College of Engineering, Faridabad, India
mrinalpandey14@gmail.com

Abstract. A typical classification problem pertaining to DSS can be solved by employing any classification algorithm such as Bayesian classifiers, neural network, decision tree. But, existing single classifier-based predictive modeling has limited scope to provide a generalized solution for different learning contexts. In this paper, an ensemble-based classification approach using voting methodology is proposed for the decision support system. The proposed ensemble-based system combines three heterogeneous classifiers, namely decision tree, K-nearest neighbor, and aggregating one-dependence estimator classifiers using product of probability voting rule. This paper presents a comparative study of the proposed voting algorithm with the other well-known classifiers for 15 standard benchmark datasets and proved that the proposed method achieves better accuracy for most of the datasets.

1 Introduction

A decision support system is software-based interactive information system that supports decision-making activities. A classical decision support system is used to design for the specific purpose based on the particular decision processes, set of methods, and techniques such as simulation-based DSS, forecasting-based DSS, and analysis-based DSS. The objective of a DSS is to enhance the decision-making capabilities of the decision makers. Traditional techniques of data analysis were not sufficient enough to all kinds of problems and thus result in lacking the decision-making process. Therefore, data mining was introduced as a new interdisciplinary field for data analysis and decision-making process. A data mining-based decision support system extracts the meaningful information from the database and provides the solution to various decision-making situations. Moreover, it hides the complexity of the typical KDD process and enables the decision maker to take decision based on algorithmic approach.

In modern predictive analytics, the concept of multiple classifiers became popular and considered as new emerging field in the area of decision support systems for the improvement of the performance of the classifier [1]. Therefore, we have implemented ensemble-based DSS by combining three classifiers, namely decision tree (J48) [2], AODE (aggregating one-dependence estimator) [3], and K-nearest neighbor classifiers

V. Bhateja et al. (eds.), *Information Systems Design and Intelligent Applications*, Advances in Intelligent Systems and Computing 672,
https://doi.org/10.1007/978-981-10-7512-4_39

using voting methodology. This algorithm is originally developed for the prediction of academic performance of the engineering graduate students. The proposed voting algorithm is also applied on 15 standard benchmark datasets, and the results are compared for the verification and validation of the proposed approach using tenfold cross-validation in this research.

2 Ensemble Methods

An ensemble consists of a set of individually trained classifiers such as neural networks or decision trees whose predictions are combined when classifying novel instances [4]. Ensemble techniques became popular from last two decades in the area of classification and prediction. The concept of ensemble is obtained from the real-life scenario where vital decisions can be taken on the basis of opinion of the majority among the experts instead of a judgment of solo expert [5]. Perfect ensembles consist of a set of heterogeneous classifiers with high-accuracy and diversify errors obtained by each classifier. In such case, all the classifiers in an ensemble will make different kinds of mistakes and the total inaccuracy will be minimized, while in case of homogenous ensemble where each and every classifier is same in nature will prompt to produce the same kind of errors during the modeling, and there will be no change in inaccuracy. Therefore, the result of the integration of multiple models varies upon the degree of divergence between classifiers and merely helpful if classifiers are considerably different from each other [5, 6]. In most of the studies, ensembles have been found more accurate rather than a single classifier. A significant amount of research has been done in recent years, and good results were reported in [7–10]. Ensemble methodologies were successfully applied in many fields such as finance, health care, bioinformatics, manufacturing.

The ensemble learning process is defined as "Ensemble learning is a process that uses a set of models, each of them obtained by applying a learning process to a given problem. This set of models (ensemble) is integrated in some way to obtain the final prediction" [11]. This definition covers both supervised and unsupervised learning techniques. This ensemble-based approach decomposes the input space into several subgroups and trains each model independently for each subgroup [7].

However, there is no perfect classification of ensemble learning, and authors classify ensemble according to their research. Four methods of ensembles, namely bagging, boosting, stacking, and error-correcting output codes, were presented in [12], while [13] covers five methods, namely BMA, committees, boosting, tree-based models, and conditional mixture models [14]. Similarly, research [15] covers boosting (AdaBoost and Stumping), bagging, and the mixture of expert method, while authors of [16] presented seven methods for integrating multiple learners, including voting, error-correcting output codes, bagging, boosting, mixtures of experts, stacked generalization, and cascading. In another research, ensemble method was classified into two categories as homogenous ensemble and heterogeneous ensembles [17].

3 Literature Review

Recently, the work has been carried out by the many researchers to build the decision support system for various domains of data mining. An ensemble-based decision support system for predicting heart attack was proposed in [18]. Authors combine three classifiers, namely Naïve Bayes, SVM, and decision tree using majority voting and applied it to the heart diseases data obtained from the UCI repository. A novel ensemble-based DSS was proposed by Salih and Abraham [19] for the healthcare monitoring system. In this research, the data were obtained from the wearable sensors. The authors experimented on homogenous classifiers as well as heterogeneous classifiers and finally combine three classifiers, namely decision tree, random forest, and random tree using voting methodology. Three algorithms, namely Naïve Bayes tree, decision tree, and instance-based learner, were combined and tested against on 28 UCI benchmark dataset [20]. The proposed method was also compared with the bagging and boosting. A DSS tool has been developed for the prediction of students' performance for the discipline of mathematics using voting technique [21]. This tool was proposed to support in decision-making activities for admission procedure as well as to provide right feedback and assistance to the students about their progress. A DSS for the prediction of students' academic performance was developed by employing regression technique [22]. A prominent decision support tool with the help of data mining techniques was developed [23]. In another research, a DSS based on Bayesian network for the prediction of students' performance was developed with the use of case-based mechanism [24]. An ensemble-based DSS was proposed using the stacking methods by combining three classifiers, namely decision tree, Naïve Bayes, and backpropagation version of the neural network, and the algorithm was tested against the various standard benchmark datasets [25].

4 Methodology

In current research, 15 datasets from UCI repository were selected, and then, proposed voting algorithm was applied to all 15 datasets to determine the behavior of the proposed model. These datasets cover different domains, including pattern recognition (Zoo), image recognition (ionosphere), medical (hepatitis, lung cancer, lymphography, diabetes), and commodity trading (Credit-g). Table 1 shows the detailed description of the datasets including number of instances, categorical features, number of numerical features, and number of classes for respective dataset. The tenfold cross-validation is used for validation and testing purpose. Each dataset was divided into 10 equal size subsets, each time one subset is used as a testing dataset, and nine subsets are used for the training dataset. The open-source code written by Witten and Frank [12] is used for conducting the experiments. A two-tailed t-test was also applied to determine the statistical comparisons of the proposed model with other existing models for all 15 datasets with 95% of the confidence level. Additionally, experimenter interface of open-source data mining tool weka [26] was used for statistical comparison of the 15 datasets against the proposed voting classifiers as well as other well-known classifiers.

Table 1. Detailed description of the dataset

S.N	Datasets	Instances	Categorical features	Numerical features	Classes
1	Anneal	898	32	6	5
2	Credit-g	1000	13	7	2
3	Diabetes	768	0	8	2
4	Hepatitis	155	13	6	2
5	Ionosphere	351	34	0	2
6	kr versus kp	3196	36	0	2
7	Lung cancer	32	57	0	3
8	Lymph	148	15	3	4
9	Mushroom	8124	22	0	2
10	Nursery	12960	8	0	5
11	Segment	2310	0	19	7
12	Sick	3772	22	7	2
13	Supermarket	4627	216	0	2
14	Vote	435	16	0	2
15	Zoo	101	16	1	7

In order to construct the ensemble-based model, three algorithms, namely K-nearest neighbor (IBK), AODE, and decision tree (J48), were used as base classifiers. The DT and KNN classifiers are used to enhance the globalization and flexibility for different types of datasets, while AODE was introduced to deal with the attribute-independence problem of the Naïve Bayes classifier; besides it has weaker-independence assumption, it often develops considerably more accurate classifiers than Naive Bayes at the cost of a modest increase in the amount of computation (AODE). Therefore, AODE is used to provide more flexibility for the attributes in current research. All three aforementioned classifiers were combined using product of probability voting rule as mentioned in Eq. (1), and the predicted class can be selected by the maximum vote gain by all the combined classifier as shown in Eq. (2). It has been assumed that product of probability in nonzero.

$$F_j(x) = \prod_{n=1}^{N} V_{n,J}(x). \tag{1}$$

$$\text{Predicted class}_{\text{final}} = \text{argmax}\, F_j(x). \tag{2}$$

where n: number of classifiers and j: class.

Figure 1 shows the detailed methodology and experimental setup. In this experiment, each individual classifier (IBK, AODE, and J48) generates their hypotheses h1, h2, and h3, respectively. For each output class, a posteriori probability is generated by the individual classifier which is multiplied to find the product of probabilities and the class is represented by the maximum of a posterior probability. This posterior probability is selected to be the voting hypothesis (h*) for the final decision.

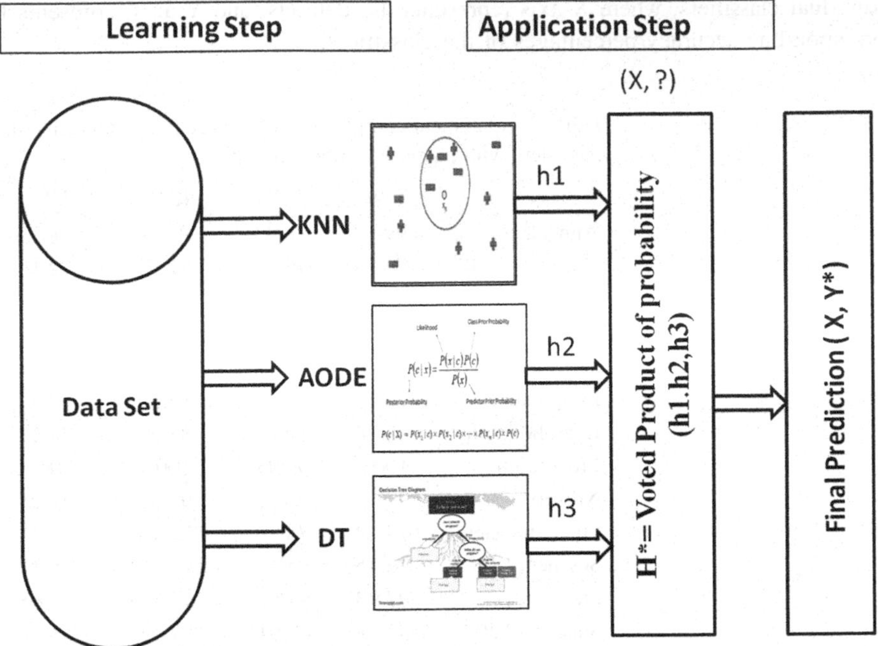

Fig. 1. Methodology and experimental setup

5 Results and Comparisons

Tables 2 and 3 show the results of the statistical t-test comparisons on 15 datasets along with the voting algorithms and other well-known classifiers during the experiments. In Tables 2 and 3, the term "v" means that the particular algorithm performs significantly better than the proposed method and * means that proposed algorithm is significantly better than the compared algorithm. The last row of these tables shows the aggregated results in the form (x/y/z), where "x" is the number of times a particular algorithm wins and "z" is the number of times that particular algorithm losses, while "y" is the total number draws for that particular algorithm. In Table 2, the proposed ensemble classifier is compared with the accuracy of the individual classifiers, namely AODE, IBK, and DT (J48). The last row of this table shows the aggregated results. The projected ensemble is considerably more precise in accuracy rather than IBK in five datasets among the 15 datasets, while it has an extensively higher error rate in one dataset. The proposed ensemble is also considerably more correct than DT (J48) in two datasets from the 15 datasets, while there is only one dataset that has an extensively higher error rate. Furthermore, the presented ensemble is appreciably more correct than AODE in three datasets, and only one dataset has higher inaccuracy among the 15 datasets. However, there is no statistically significant difference for AODE, IBK, and J48 of 11, 9, and 2 datasets as compared to individual classifiers, but the error rate is also minimized. Figure 2 depicts the graphical representation of the comparisons with individual classifiers, where X-axis represents the datasets, and Y-axis represents the corresponding accuracy percentages of the classifiers.

Table 2. Statistical comparisons of the accuracy (%) of the three classifiers with proposed voting classifier

Dataset	Vote	AODE	IBK	J48
Anneal	99.33	98.33	99.22	98.77
German_credit	74.8	77.1	72.5	72.1
Hepatitis	84.54	87	83.25	82.63
Ionosphere	93.17	92.6	93.74	89.17
kr versus kp	99.25	91.24*	96.28*	99.44
Lung cancer	74.17	74.17	67.5	77.5
Lymphography	83.76	87.05	83.67	78.33
Mushroom	100	99.95	100	100
Nursery	97.58	92.71*	98.38 v	97.05*
Pima_diabetes	77.74	78	77.74	78.26
Segment	96.75	95.58	94.68*	95.32*
Sick	97.93	97.43	97.43*	97.85
Supermarket	37.58	63.91 v	37.13*	63.71 v
Vote	95.66	94.27	92.44*	96.33
Zoo	95.18	95.09	96.18	92.18
x/y/z		1/11/3	1/9/5	1/12/2

Next, the proposed voting method is also compared with other well-known classifiers as well as AdaBoost ensemble. Table 3 shows the results of this extended part of this experiment. Similar to previous experiment, the last row of Table 3 shows that the results of the ensemble proposed in this research are considerably more correct in terms of accuracy rather than NB in six datasets among the 15 datasets, while there is radically higher inaccuracy in one dataset. The projected ensemble is also appreciably more precise than KSTAR, OneR, ZeroR, Decision Stump (DS), AdaBoost algorithms with 5, 7, 11, and 8 datasets, respectively, among the 15 datasets, while there is only one dataset that has extensively higher error rate for six classifiers, namely KSTAR, OneR, ZeroR, DS, NBT, and AdaBoost. Similarly, there is only one dataset for which NBT is less accurate than the projected algorithm, and there is no noteworthy difference for 13 datasets using NBT. Additionally, projected voting ensemble is appreciably more precise than Decision Stump version of the AdaBoost (Homogenous classifier) classifier with eight datasets.

Figure 3 depicts the graphical representation of the second experiment. Six well-known classifiers and one homogenous classifier are compared with the proposed voting method. X-axis shows the datasets, and Y-axis represents the corresponding accuracy of the classifiers.

Table 3. Statistical comparisons of the accuracy (%) of seven well-known classifiers with proposed voting classifier

Dataset	Vote	NB	KSTAR	OneR	ZeroR	DS	NBT	AdaBoost (DS)
Anneal	99.33	96.66*	99.33	83.63*	76.17*	77.16*	98.33	83.63*
German_credit	74.8	75.8	74.2	71.7	70	70	75	71.9
Hepatitis	84.54	85.13	80.54	83.13	79.38	79.38	84.46	86.38
Ionosphere	93.17	90.6	93.74	87.78	64.10*	83.19*	92.32	89.47
kr versus kp	99.25	87.89*	97.03*	66.46*	52.22*	66.05*	97.09	93.84*
Lung cancer	74.17	77.5	67.5	87.5	71.67	74.17	77.5	77.5
Lymphography	83.76	83.67	85	74.81	54.76*	75.48	82.29	74.14*
Mushroom	100	95.83*	100	98.52*	51.80*	88.68*	100	96.20*
Nursery	97.58	90.32*	96.77*	70.97*	33.33*	66.25*	97.49	66.25*
Pima_diabetes	77.74	77.87	77.74	74.74	65.11*	74.74	77.22	78.52
Segment	96.75	91.52*	94.42*	65.80*	14.29*	28.57*	94.33*	28.57*
Sick	97.93	97.3	97.24*	96.55*	93.88*	96.55*	97.45	96.79*
Supermarket	37.58	63.71 v	63.71 v	67.21 v	63.71 v	64.40 v	63.71 v	74.86 v
Vote	95.66	90.14*	93.36*	95.64	61.38*	95.64	95.64	95.41
Zoo	95.18	93.18	96.09	42.64*	40.64*	60.45*	92.18	60.45*
x/y/z		1/8/6	1/9/5	1/7/7	1/3/11	1/6/8	1/13/1	1/6/8

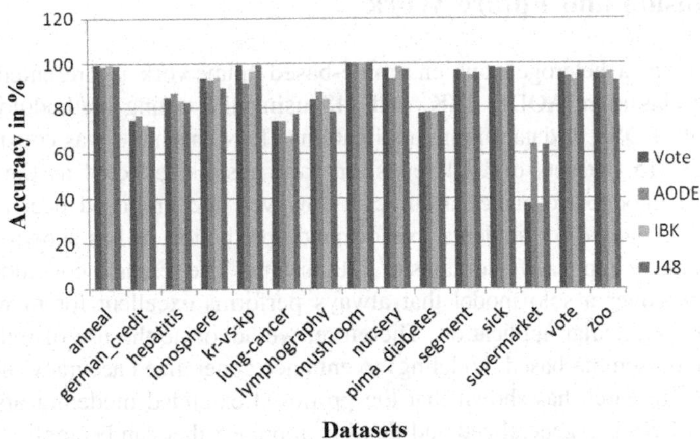

Fig. 2. T-test comparisons results for 15 datasets of three classifiers and voting

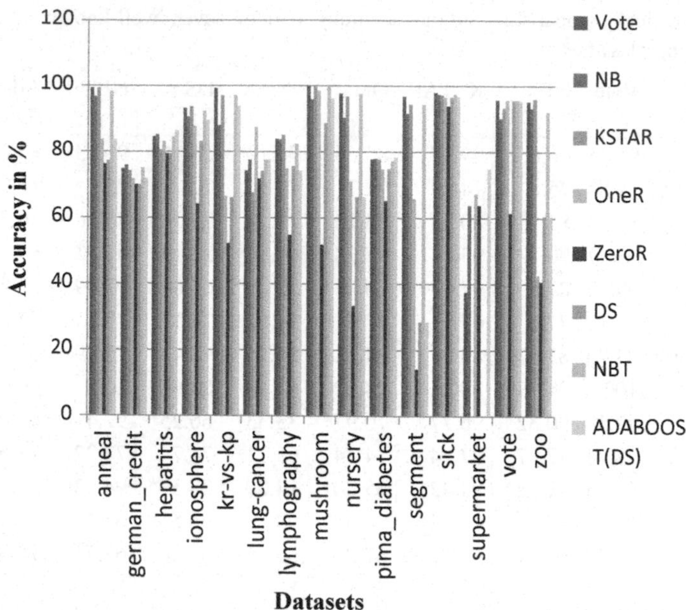

Fig. 3. T-test comparisons results for 15 datasets of seven other well-known classifiers and voting

6 Conclusion and Future Work

In this research, a heterogeneous ensemble-based framework is presented that integrates three classifiers AODE, IBK, and J48 using the voting methodology for the development of DSS in data mining applications. This ensemble was constructed and tested against 15 datasets of UCI repository and results reflected as the consistent behavior and good performance accuracy. However, the proposed decision support system was developed for students' performance prediction. It has shown better performance in many applicative domains of data mining. The research concludes that it is difficult to discover a solo model that always performs excellent for many learning contexts for a particular application. Therefore, we advocate the use of multiple classifier system (ensemble-based modeling) to enhance generalized accuracy of predictive learning. The research has shown that the proposed extended modeling approach for developing a DSS is a generalized and flexible approach that can be applied to various applications and performs better in most of the cases. In this research, only accuracy is used as a parameter for model evaluations. In future research, the proposed approach will be tested against various other parameters such as recall, precision, true positive rate, and false positive rate. Moreover, other dataset as well as other ensemble methods such as stacking and bagging will be tested against the proposed voting methods.

References

1. Turban, E. Decision support and expert systems: Management Support Systems. Englewood Cliffs, N.J., Prentice Hall (1995).
2. Quinlan J.R. C4. 5: Programs for machine learning. Morgan Kaufmann, San Francisco, CA, USA (1993).
3. Webb, G. I., Boughton, J. R., & Wang, Z. Not so naive Bayes: aggregating one-dependence estimators. Machine learning, 58(1), 5–24.
4. Kwon, O. A new ensemble method for gold mining problems: Predicting technology transfer. Electronic Commerce Research and Applications, 11(2), (2012). 117–128.
5. Wan, Shaohua, Yang., H. Comparison among Methods of Ensemble Learning. International Symposium on Biometrics and Security Technologies (2013).
6. Mendes-Moreira, J., Soares, C., Jorge, A. M., & Sousa, J. F. D. Ensemble approaches for regression: A survey. ACM Computing Surveys (CSUR), (2012), 45(1), 10.
7. Witten, I. H., Eibe Frank, and Mark A. Hall Ensemble Learning. Data Mining Practical Machine Learning Tools and Techniques (2011).
8. Liu, Y., Yao, X. & Higuchi, T. Evolutionary ensembles with negative correlation learning. Evolutionary Computation, IEEE Transactions on, 4(4), (2000), 380–387.
9. Breiman, L. Random forests. Machine. Learning. 45(1), (2001) 5–32.
10. Rodriguez, J.J., Kuncheva, L., Alonso, C.J., Rotation forest: A new classifier ensemble method. IEEE transactions on pattern analysis and machine intelligence (2006), 1619–30.
11. Dietterich, T. G. Machine-learning research. AI magazine, 18(4), (1997).
12. Witten, I. H., & Frank, E. Data Mining: Practical machine learning tools and techniques. Morgan Kaufmann, ed.2. (2005).
13. Bishop, C. M. Pattern recognition and machine learning. Springer, (2006).
14. Kumar, G., & Kumar, K.. The use of artificial-intelligence-based ensembles for intrusion detection: a review. Applied Computational Intelligence and Soft Computing, (2012) 21.
15. Marsland, S. Machine Learning: An Algorithmic Perspective, (2009), CRC Press.
16. Alpaydin, E. Introduction to machine learning. MIT press, (2014).
17. Rooney, N., Patterson, D., Anand, S., & Tsymbal, A. Dynamic integration of regression models. In Multiple Classifier Systems (2004). Lecture Notes in Computer Science, vol. 3077. 164–173. Springer Berlin Heidelberg.
18. Bashir, S., Qamar, U., & Javed, M. Y. An ensemble based decision support framework for intelligent heart disease diagnosis. In Information Society (i-Society), International Conference on (2014, November) 259–264. IEEE.
19. Salih, A. S. M., & Abraham, A. Novel Ensemble Decision Support and Health Care Monitoring System. Journal of Network and Innovative Computing, 2(2014), 041–051.
20. Gandhi, I., & Pandey, M. Hybrid Ensemble of classifiers using voting. In Green Computing and Internet of Things (ICGCIoT) (2015). 399–404. IEEE.
21. Livieris, I. E.,. Mikropoulos T. A., Panagiotis. A decision support system for predicting students' performance. Themes in Science and Technology Education. vol 9, (2016) 43–47.
22. Kotsiantis, S. B. Use of machine learning techniques for educational proposes: a decision support system for forecasting students' grades. Artificial Intelligence Review, 37(4), (2012) 331–344.
23. Luan, J. Data Mining Applications in Higher Education. A chapter in the upcoming New Directions for Institutional Research (2001), 1st Ed., Josse-Bass, San Francisco.
24. Nguyen Thi Ngoc Hien and Haddawy, Peter. A Decision Support System for Evaluating International Student Applications. In 37th ASEE/IEEE Frontiers in Education Conference, IEEE, (2007).

25. Kotsiantis, S. B., & Pintelas, P. E. A hybrid decision support tool-using ensemble of classifiers (2004).
26. M. Hall, E. Frank, G. Holmes, B. Pfahringer, P. Reutemann, Ian H. Witten. The WEKA Data Mining Software: An Update; SIGKDD Explorations, Volume 11, Issue 1, 2009.

Classification of Amazon Book Reviews Based on Sentiment Analysis

K. S. Srujan(✉), S. S. Nikhil, H. Raghav Rao, K. Karthik, B. S. Harish, and H. M. Keerthi Kumar

Department of Information Science and Engineering, Sri Jayachamarajendra College of Engineering, Mysuru 570006, Karnataka, India
{snkr.is.2017, ssnikhilkumar, raghav.rao9, jckarthikk, hmkeerthikumar}@gmail.com, bsharish@sjce.ac.in

Abstract. Since the dawn of internet, e-shopping vendors like Amazon have grown in popularity. Customers express their opinion or sentiment by giving feedbacks in the form of text. Sentiment analysis is the process of determining the opinion or feeling expressed as either positive, negative or neutral. Capturing the exact sentiment of a review is a challenging task. In this paper, the various preprocessing techniques like HTML tags and URLs removal, punctuation, whitespace, special character removal and stemming are used to eliminate noise. The preprocessed data is represented using feature selection techniques like term frequency-inverse document frequency (TF–IDF). The classifiers like K-Nearest Neighbour (KNN), Decision Tree (DT), Support Vector Machine (SVM), Random Forest (RF) and Naive Bayes (NB) are used to classify sentiment of Amazon book reviews. Finally, we present a comparison of (i) Accuracy of various classifiers, (ii) Time elapsed by each classifier and (iii) Sentiment score of various books.

1 Introduction

Amazon, one of the most popular e-commerce companies, has its presence across the globe. It allows customers to express their opinions by rating and reviewing the products they purchase. By the virtue of its popularity, it gathers huge amount of data every day. These reviews can be in the form of text or photos. Analysis of these reviews attracts researches from all over the world.

Sentiment analysis is the process of determining the opinion or feeling of a piece of text [1,2]. The opinion can be either positive, negative or neutral [3]. Sentiment analysis uses techniques like natural language processing (NLP), computational linguistic to identify subjective information. Opinion mining or sentiment analysis is employed by either lexicon based, machine learning based or hybrid/combined analysis. Lexicon-based approaches focus on dictionary for predicting sentiment, whereas machine learning-based approaches focus on testing data against trained data for prediction of opinion. Hybrid approaches combines

V. Bhateja et al. (eds.), *Information Systems Design and Intelligent Applications*, Advances in Intelligent Systems and Computing 672,
https://doi.org/10.1007/978-981-10-7512-4_40

features of both. Human beings are pretty good at determining the sentiment. We can look at the review and immediately know if it is negative or positive. Companies across the world have implemented machine learning to do sentiment analysis automatically. It is useful for gaining insight into customer opinions. After analyzing the reviews, we can identify customer's opinion about the product [4–6]. By sentiment analysis, companies can build recommendation systems or better targeted marketing campaigns. The complexity in sentiment analysis includes removing noisy data from raw dataset, selecting suitable features for representation and choosing appropriate classifier.

In this paper, we apply various preprocessing methods and use different classifiers to classify book reviews as either positive or negative class. The main objective of this work is to present a comparative study on different classifiers based on accuracy and processing time taken by each classifier. Finally, we focus on comparing sentiment score for different books. There are two basic sentiments positive and negative, and eight basic emotions namely joy, anger, sadness, trust, surprise, disgust, fear and anticipation.

The rest of the paper is organized as follows: in Sect. 2, we give a brief overview of the literature that has been done related to sentiment analysis. Preprocessing steps and classification models are presented in Sect. 3. The detailed experimental analysis is presented in Sect. 4. Finally, Sect. 5 concludes with future work.

2 Literature Survey

Plenty of research has been undertaken in the domain of sentiment analysis and text classification. Categorization of sentiment polarity is one of the fundamental problem in sentiment analysis [7–10]. Given a part of text, the task is to categorize whether the text is positive or negative. There are different levels of sentiment polarity classification namely the entity or aspect level, the sentence level and document level [11]. The entity type focuses on what exactly people like or dislike from their opinions. The document level considers the polarity, i.e. positive or negative sentiment for the entire document, while the sentence level deals with each sentence's sentiment categorization.

Many researchers [3,12–16] across the globe have conducted research based on supervised, semi-supervised and unsupervised machine learning approaches. Bhatt et al. in [14] proposed a system for sentiment analysis on iphone 5 reviews. The methodology integrates various preprocessing techniques to reduce noisy data like HTML tags, punctuations and numbers. The features are extracted using part-of-speech (POS) tagger and rule-based method are applied to classify the reviews into different polarity. Emma et al. in [15] explore the role of text preprocessing on online movie reviews. The various preprocessing techniques such as data cleaning, stemming and removal html tags are used to remove noisy data. The irrelevant features are eliminated by using chi-square feature selection technique. The Support Vector Machine (SVM) is used to classify the reviews into positive or negative classes. In [3], Tripathy et al. presented a comparison of different classifiers based on accuracy for movie review dataset. The methodology incorporated various preprocessing techniques to reduce noisy data like whitespaces, numbers, stop word removal and vague information removal. The features are extracted and represented by count vectorizer and TF–IDF. Naive Bayes (NB) and Support Vector Machine (SVM) are used to classify the data as positive or negative. By comparing accuracy of NB with SVM, SVM achieved accuracy of 94%. Turney et al. [13,17] presented an unsupervised learning algorithm for rating a review as thumbs up or thumbs down. The algorithm uses POS tagger to extract the adjectives or adverbs from epinion customer reviews. The pointwise mutual information and information retrieval (PMI-IR) algorithm is used to calculate the semantic orientation of each phases and classify the review based on the average semantic orientation phrases.

Mohammad et al. in [12] developed a system to generate emotion association lexicon of words. The system incorporates Amazon's crowd service platform called mechanical turk. The emotion lexicon consists of list of words and their associations with two sentiments (negative and positive) and eight emotions (sadness, joy, trust, surprise, disgust anger, fear and anticipation). The syushet package [18] in R Language implements emotion lexicon developed by National Research Council (NRC), Canada.

In this work, we present a comparative study on (i) Accuracy of various classifiers, (ii) Time elapsed by each classifier, (iii) Sentiment score of various books. In next section, we discuss methodology and classifiers used in our experiment.

3 Methodology

The Amazon book reviews dataset [19] is considered for our analysis. The methods include preprocessing, representation and classification. The unstructured Amazon book reviews are preprocessed using various preprocessing techniques like data cleaning, URL/HTML tag removal, punctuation and number removal, stop word removal and stemming. The preprocessed text are represented using TF–IDF representation model. Classifiers like K-Nearest Neighbours (KNN), Random Forest (RF), Naive Bayes (NB), Decision Trees and Support Vector Machine (SVM) are used to classify the dataset into positive and negative class.

3.1 Preprocessing

Data preprocessing involves transforming raw data into a coherent format. Data preprocessing is a proven method for refining data. The raw dataset looks like "4.0/gp/customer-reviews/R2BBCQKO693KA4?ASIN=1491590173 FiveStars <span class=a-size-base review-text>Great read beginning to end. Mr. Weir knows his science </span>" which contains noise and vague information and hence needs to be cleaned. Following are the preprocessing techniques incorporated for Amazon book review dataset.

Data cleaning: it is a process of finding and eliminating inaccurate, useless and corrupt records from the Amazon book review dataset. Abstract contents like "gp/customer-reviews//R2BBCQKO693KA4?ASIN" were eliminated from the dataset as they do not depict any sentiment.

Removal of HTML tags and URL's: Amazon book review dataset has many html tags like "<span>", "<br>" and URLs having prefix like "http", "ftp", "https". They do not convey any sentiment and hence it is removed.

Punctuation and special character removal: punctuation such as full stop (.), comma (,) and brackets () are used in writing to separate sentences are removed as they do not denote any sentiment. Along with that, special characters like "%", "#", "$" used to denote cost, percentage, comment are also eliminated.

Removal of numbers and white-spaces: Amazon book reviews dataset contains page numbers, date, etc. They are eliminated as they do not convey any sentiment. Whitespaces including "/t" (tab) are eliminated as they do not indelicate any sentiment.

Stemming: stemming is used to extract the root of a word. For instance, the root word or stem of words such as "satisfaction", "satisfied" and "satisfying" is "satisfy". Stemming reduces indexing size around 30–50% [20].

Stop word removal: stop words are mostly prepositions, articles, conjunctions like the, is, at, which, and on which appears repeatedly in a sentence but they do not denote any sentiment. Hence these stop words, which do not convey any meaning independently, are removed.

We convert all the letters to lower case like "STAR" to "star", which eliminates redundant words in dataset. By enforcing various preprocessing methods, we are removing noisy data. Further, preprocessed text are given suitable representation for the classification.

3.2 Representation

Representation is the important step in sentiment classification [21]. Generally, raw data contains noise and is refined by applying various preprocessing methods. The preprocessed data is converted into term document matrix (TDM) which computes frequency of each word. Feature extraction methods like bag of words and TF–IDF are implemented on the TDM. TF–IDF contains the two factors TF and IDF, by multiplying these two factors we will get the TF–IDF score of a word. TF score assigns weight to most frequently occurring words in the book review dataset. IDF is the scaling factor which assigns weight to the least frequent words in the dataset. For rare and frequent words, this score is less compare to the other words. We can eliminate them by ignoring the words with less TF–IDF scores.

3.3 Classification

Classification is the method of grouping target function into different classes. The focus of the proposed work is on sentiment analysis. We use five classifiers like K-Nearest Neighbours (KNN), Random Forest (RF), Naive Bayes (NB), Decision Tree and Support Vector Machine (SVM). The various classifiers are evaluated to give a comparative study of accuracies for grouping dataset into positive or negative. Each classifier has its significance and utility, but few perform better for a given dataset. For example, NB is a probabilistic classifier which uses the properties of Bayes theorem assuming strong independence between the features [22]. One of the advantage of this classifier is that it requires small amount of training data to calculate the parameters for prediction. SVM classifier represents each review in vectorized form as a data point in the space. SVM is used to analyse the complete vectorized data, and the intent behind training of model is to find a hyper plane [13]. KNN is a non-parametric method used for classification. It takes k closest training set as input and outputs as class membership [10]. Random Forest is an ensemble method which constructs multiple decision trees during training time and outputs label of the class [23].

4 Experimental Evaluation

In this section, we present detailed description of dataset, experimental procedures carried out and discussion on the results obtained.

4.1 Dataset Description

Our work focuses on Amazon book review dataset [19] available at UCI repository. The dataset consists of 213,335 book reviews of eight most popular books namely, "Gone Girl", "The Girl on the Train", "The Fault in our Stars", "Fifty Shades of Grey", "Unbroken", "The Hunger Games", "The Goldfinch" and "The Martian".

Table 1. Number of positive class and negative class reviews for each book

	The Martian	The Goldfinch	Fifty Shades of Grey	Gone Girl	The Fault in Our Stars	Unbroken	The Girl on The Train	The Hunger Games
Negative	523	3708	11939	9504	4410	322	7400	628
Neutral	3010	5388	2494	5011	1206	3110	4163	3185
Positive	19038	13765	18544	27459	30228	22444	25576	20214
Total	22571	22861	32977	41974	35844	25876	37139	24027

Table 1 depicts the number of positive class and negative class reviews for each book. Total number of reviews is given in the fourth row for each column. Each entry contains four attributes, review score, tail of review URL, review title and HTML of review text, and each entry is separated by a newline character. An example of a review is shown in Sect. 3.

4.2 Experimentation

In the experiment, we consider two class problem, i.e. positive and negative class. Two thousand sample reviews for each book is selected randomly such that it has about 50% positive opinions and 50% negative opinions. Each review consists of rating and text reviews by the user. These reviews are grouped into two classes-positive and negative based on their ratings. If the review has a rating of 1 or 2, it is grouped into negative class and the reviews consisting of ratings 4 and 5 are grouped as positive. The corpus is divided in the ratio of 60:40 for training and testing, respectively. The various preprocessing techniques like data cleaning, URL and HTML tag removal, whitespace removal, stop word removal, stemming are applied on the dataset . The training and testing corpuses are converted to individual term document matrix (TDM), and frequency of each word is computed. We use bag of words model to build a corpus, from the preprocessed data. TF–IDF is used for representation. The TDM is then fed as an input to various classifiers such as KNN, SVM, Naive Bayes, Decision Tree and Random Forest. The accuracy of each classifier is computed by dividing the number of correct classification with the total number of tuples in the input data set.

$$Accuracy = \frac{Number\ of\ correct\ classification}{Number\ of\ tuples\ in\ the\ test\ set} \tag{1}$$

In Table 2, we observe that Random Forest gives better result in two class classification problem for the dataset. The TDM consists of feature vector values called points. Significant improvements in classification accuracy have resulted from building an ensemble of trees and letting them to vote for the most popular class. In order to grow these ensembles, often random vectors are generated. On an average, Random Forest was able to classify for about 88.86% accurately for eight books and reached maximum accuracy of 94.72% for the book "Unbroken". "Goldfinch" and "Gone girl" got highest accuracy for KNN. This is because KNN is non-parametric, i.e. it makes no assumption about the data distribution. Those book reviews has points near to the neighbour. Hence, achieved highest accuracy.

Table 2. Accuracy achieved using various classifiers for different books

Book	KNN	Decision tree	Naive Bayes	SVM	Random forest
1. The Hunger Game	85.50	88.44	88.44	89.24	**89.64**
2. The Girl on Train	**84.51**	82.60	82.60	82.60	86.20
3. Goldfinch	**90.00**	75.40	81.20	84.20	84.00
4. Gone Girl	84.91	66.88	79.68	79.68	**82.45**
5. The Martian	86.58	53.01	85.94	85.94	**91.16**
6. Fifty Shades of Grey	86.22	50.80	52.20	65.80	**86.60**
7. Unbroken	71.90	84.64	94.62	94.60	**94.72**
8. The Fault in Our Stars	87.10	57.00	93.80	93.80	**94.40**

We also compare the time elapsed for two class classification problem. Time taken by classifier to group into positive and negative class for different books is shown in Fig. 1.

Customer can write a review in a single sentence or in a paragraph. If reviews have multiple sentences, then the time taken to classify is significantly high. Hence, some books take more time than other books. Figure 1 shows time taken in seconds for classifying either as positive or negative class for different books. Amongst all the classifiers implemented, Random Forest takes more execution time, because it needs to grow multiple trees or ensembles.

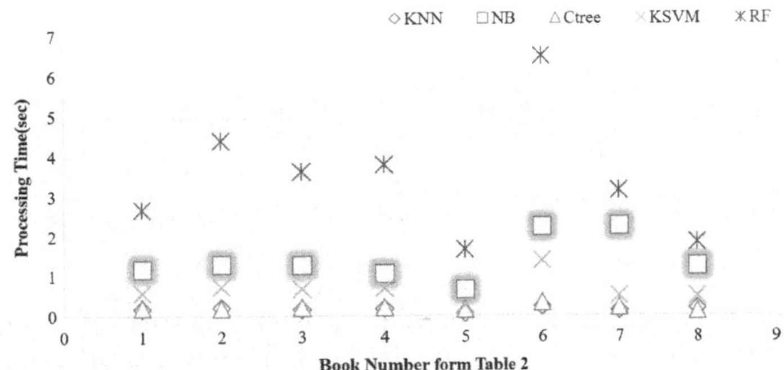

Fig. 1. Time taken by classifier

Sentiment analysis involves the study of various sentiments expressed by the customer. These sentiments can be related to anger, sadness, joy, trust, surprise, disgust, fear and anticipation [10]. Syuzhet package [18] in R language is used for our analysis. The input data is preprocessed using various data cleaning techniques. Term document matrix (TDM) is created which indicates frequency of each word involved in the text.

After constructing the TDM, we fetch sentiment word from text using National Research Council (NRC), Canada, emotion lexicon dictionary. Later we count sentiment words by category. Finally, we generate sentiment score for each book.

Figure 2 shows the sentiment score of the book "The Hunger Games", which is an adventure novel, thus evoking emotions such as positive, anticipation and emotions related to trust. There is also a sense of trust amongst the readers. Single word can be mapped to multiple emotion. For example, abandoned is mapped to anger, fear and sadness. But abandonment is mapped to anger, fear, sadness and surprise. This mapping is present in NRC emotion dictionary.

$$Accuracy = \frac{Number\ of\ Word\ count\ for\ particular\ emotiom}{Total\ Number\ of\ Word\ count\ for\ that\ book} \quad (2)$$

Table 3 illustrates the percentage of each emotions evoked amongst its readers by a book. Percentage is obtained by dividing word count of different emotions

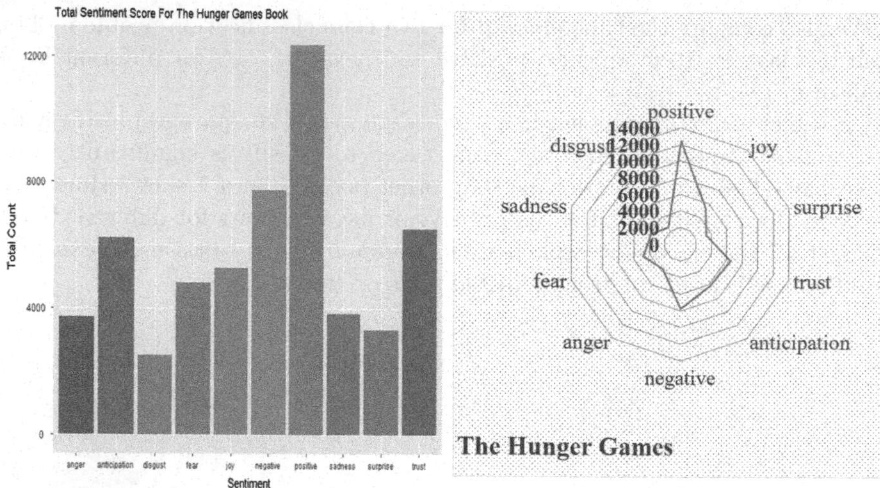

Fig. 2. Illustrates sentiment score for the book "The Hunger Games"

such as joy, anger, surprise, anticipation, sadness and trust expressed by the readers to total word count for a particular book. It shows, clearly, that thrillers such as "Gone Girl", "The Girl on Train" are high on words which are used to express surprise, fear and anticipation and books such as "Fifty shades of grey" and "Gone girl" have generally evoked negative, anger and disgust amongst its readers. Novels such as "Fault in our stars" and "Unbroken" made the readers feel sad, positive and trust. Adventure books such as "The Hunger Games", "The Martian" and "The Goldfinch" garnered more words on positive and anticipation.

4.3 Discussion

In this paper, we have evaluated various preprocessing techniques, feature extraction for classifying the text as either positive or negative class. We present a comparative study of (i) Accuracy of classifiers, (ii) Time elapsed by each classifier and (iii) Sentiment score of various books. On the basis of results obtained, we can observe from two Random Forests compared to Naive Bayes, which offers consistent and marked improvements in accuracy but requires more processing time. If we have multiple points in a low-dimensional space, then KNN is considered as a good choice. Thus, we infer that SVM with a kernel and Random Forest (RF) are the best choices for most of the problems. Random Forest is known to be a reliable and efficient algorithm. It is versatile robust and performs well when it comes to classification task. In general, they require very little feature engineering and parameter tuning. By computing sentiment score, we can get to know what kind of that emotion customers evoked after reading a book. Comparison helps for judging the books based on emotion. Thus, proposed model

Table 3. Percentage of distribution of each emotions evoked amongst readers by a book

	The Hunger Games	The Girl on Train	Goldfinch	Fault in Our Star	Gone Girl	Martian	Fifty Shades	Unbroken
Positive	21.96	21.94	27.00	22.04	21.72	26.81	19.77	24.00
Joy	9.35	8.65	11.47	11.59	8.94	10.38	9.20	9.07
Surprise	5.91	6.23	5.99	5.04	5.75	5.81	4.82	4.67
Trust	11.57	11.42	13.60	11.56	11.50	13.45	11.81	13.66
Anticipation	11.04	11.26	12.15	8.89	11.06	11.32	10.30	8.69
Negative	13.77	14.52	11.93	14.17	14.19	12.40	16.92	13.17
Anger	6.60	5.86	4.15	5.39	6.54	4.62	6.79	6.17
Fear	8.53	6.91	5.21	6.69	7.26	6.06	6.45	9.38
Sadness	6.80	7.84	5.11	8.95	7.42	5.78	7.07	7.04
Disgust	4.46	5.36	3.39	5.69	5.62	3.38	6.87	4.16

can be incorporated along with recommendation algorithms for giving better insights of a product for online customers.

5 Conclusion and Future Scope

Sentiment analysis is important for any online retail company to understand customer's response. Reviews need to be analysed for building recommendation algorithms to target customers based on their needs. Amazon, one of the e-commerce giant, generates huge amount of feedback data. In our work, we present the analyses of customer book reviews from amazon.com. This work presents a comparative study of accuracy of various classifiers. Random Forest was able to give average accuracy of 90.15% for six books. By analyzing processing time of various classifiers, Random Forest ranked the highest. Sentiment analysis by making use of NRC Emotion Lexicon has helped us to gauge the predominant sentiment amongst the readers of various bestselling books on Amazon. The sentiments were classified on the basis of different emotions. After classification, we also analysed the results graphically and arrived at the conclusion as to which book evokes what kind of emotion predominate amongst its readers.
The scope of this work can be extend by incorporating various feature selection techniques like mutual information (MI), chi-squared test and information gain for better representation. We can use hybrid classifiers like SVM with other combination to enhance accuracy. Better recommendation algorithm can be built by considering emotion evoked by customer reviews.

References

1. Pang, B., and Lee, L.: A sentimental education: Sentiment analysis using subjectivity summarization based on minimum cuts. In Proceedings of the 42nd annual meeting on Association for Computational Linguistics, pp. 271 (2004).
2. Pang, B., and Lee, L.: A sentimental education: Sentiment analysis using subjectivity summarization based on minimum cuts, in Proceedings of the 42nd annual meeting on Association for Computational Linguistics. Association for Computational Linguistics, pp. 271 (2004).
3. Tripathy, Abinash, Agarwal,A. and Santanu Kumar Rath.: Classification of Sentimental Reviews Using Machine Learning Techniques. Procedia Computer Science, Vol. 57, pp. 821–829 (2015).
4. Smithikrai, C.: Effectiveness of teaching with movies to promote positive characteristics and behaviors. Procedia-Social and Behavioral Sciences, Vol. 217, pp. 522–530 (2016).
5. Shruti, T., and Choudhary, M.: Feature Based Opinion Mining on Movie Review. International Journal of Advanced Engineering Research and Science, Vol. 3, pp. 77–81. (2016).
6. Hu, M., and Liu, B.: Mining and summarizing customer reviews. In Proceedings of the tenth ACM SIGKDD international conference on Knowledge discovery and data mining ACM, pp. 168–177 (2004).
7. Pang, B., and Lee, L.: Opinion mining and sentiment analysis. Foundations and Trends in Information Retrieval, Vol. 2, No. 1–2, pp. 1–135 (2008).

8. Chesley, P., Vincent, B., Xu, L., and Srihari, R. K.: Using verbs and adjectives to automatically classify blog sentiment Training, Vol. 580, No. 263, pp. 233 (2006).
9. Choi, Y., and Cardie, C.: Adapting a polarity lexicon using integer linear programming for domain-specific sentiment classification. In Proceedings of the 2009 Conference on Empirical Methods in Natural Language Processing, Vol. 2, pp. 590–598 (2009).
10. Tan, L. K. W., Na, J. C., Theng, Y. L., and Chang, K.: Sentence-level sentiment polarity classification using a linguistic approach. In International Conference on Asian Digital Libraries Springer Berlin Heidelberg, pp. 77–87 (2011).
11. Liu, B.: Sentiment analysis and opinion mining. Synthesis lectures on human language technologies, Vol. 5, No. 1, pp. 1–167 (2012).
12. Mohammad, S. M., and Turney, P. D.: Crowdsourcing a wordemotion association lexicon. Computational Intelligence, Vol. 29, No. 3, pp. 436–465 (2013).
13. Turney P. D.: Thumbs up or thumbs down? Semantic orientation applied to unsupervised classification of reviews. In Proceedings of the 40th annual meeting on association for computational linguistics. Association for Computational Linguistics, pp. 417–424 (2002).
14. Bhatt, Aashutosh, Chheda,H., Gawande,K.: Amazon Review Classification and Sentiment Analysis. International Journal of Computer Science and Information Technologies, Vol. 6, No. 6, pp. 5107–5110 (2015).
15. Haddi, Emma, Liu,X. and Yong Shi.: The role of text pre-processing in sentiment analysis. Procedia Computer Science, Vol. 17, pp. 26–32 (2013).
16. Anand, D., and Naorem, D.: Semi-supervised Aspect Based Sentiment Analysis for Movies Using Review Filtering. Procedia Computer Science, Vol. 84, pp. 86–93 (2016).
17. Mohammad, S. M., and Turney, P. D.: Emotions evoked by common words and phrases: Using Mechanical Turk to create an emotion lexicon. In Proceedings of the NAACL HLT 2010 workshop on computational approaches to analysis and generation of emotion in text Association for Computational Linguistics, pp. 26–34 (2010).
18. https://cran.r-project.org/web/packages/syuzhet/syuzhet.pdf
19. https://archive.ics.uci.edu/ml/datasets/Amazon+book+reviews
20. Vijayarani, S., Ilamathi, M. J., and Nithya, M.: Preprocessing Techniques for Text Mining-An Overview. International Journal of Computer Science and Communication Networks, Vol. 5, No. 1, pp. 7–16 (2015).
21. Trstenjak, B., Mikac, S., and Donko, D.: KNN with TF-IDF based Framework for Text Categorization. Procedia Engineering, Vol. 69, pp. 1356–1364 (2014).
22. McCallum A., Nigam.K .:A comparison of event models for naive bayes text classification, in AAAI-98 workshop on learning for text categorization, Citeseer, Vol. 752, pp. 41–48 (1998).
23. Ho T.K.: Random decision forests. In Document Analysis and Recognition, Proceedings of the Third International Conference on IEEE, Vol. 1, pp. 278–282 (1995).

A Secured Framework for Encrypted Messaging Service for Smart Device (Crypto-Message)

Shopan Dey[1], Afaq Ahmad[2], Anil Kr. Chandravanshi[1], and Sandip Das[1](✉)

[1] Department of Electronics and Communication Engineering, University of Engineering and Management, Jaipur, India
{shopan222,anilchandravanshi039,info.sandipec}@gmail.com

[2] Department of Computer Science and Engineering, University of Engineering and Management, Jaipur, India
amantech005@gmail.com

Abstract. The involvement of technology in our life makes it more advance and provides access on our fingertip. It provides us with the capability to get connected with people and explore the information on the topics which is very beneficial for the ease of life. Thus our lives are dependent on several mobile chatting applications which offer different security to user and chatting details but leads to increase in vulnerabilities and risk of attack on data. As in sensitive business and legal conversation data security is most important for preventing from unwanted hacking activities. To overcome this kind of situation, it is proposed an encrypted messaging protocol for secure conversation. In the present world, there is a lot of encrypted messaging applications, but all those are based on a software generated encryption key along with SQLite database which is used to store the message of respective users which are not secure and the messages of any user can be obtained by a third party. But the proposed protocol uses a user-defined password for SHA-2 hash generation which is used as the key of AES-256 encryption for encrypting the message during the transition of message; on the other hand, our software stored encrypted message which will be decrypted by user in time of accessing the conversation by the introduction of user-defined key only. Through this approach, it can create a more secure atmosphere for the transition of data across the globe.

Keywords: Crypto-Message · Encrypted messaging · Secure framework AES-256 encryption · Google cloud messaging (GCM)

1 Introduction

There are a lot of android messaging applications that are used nowadays for communication purpose not only in everyday life but also in business purpose. Therefore, security and privacy of such conversation or transfer of message through those messaging applications is becoming an important issue. Therefore, there arises a need to

V. Bhateja et al. (eds.), *Information Systems Design and Intelligent Applications*, Advances in Intelligent Systems and Computing 672,
https://doi.org/10.1007/978-981-10-7512-4_41

provide security to such application. The messaging application proposed in this paper provides such security. In this, an encrypted application named Crypto-Message is proposed, where all the messages will be encrypted and only the end user will have the authentication to view those messages by providing a unique key which will be given to the user at the time of registration (Fig. 1).

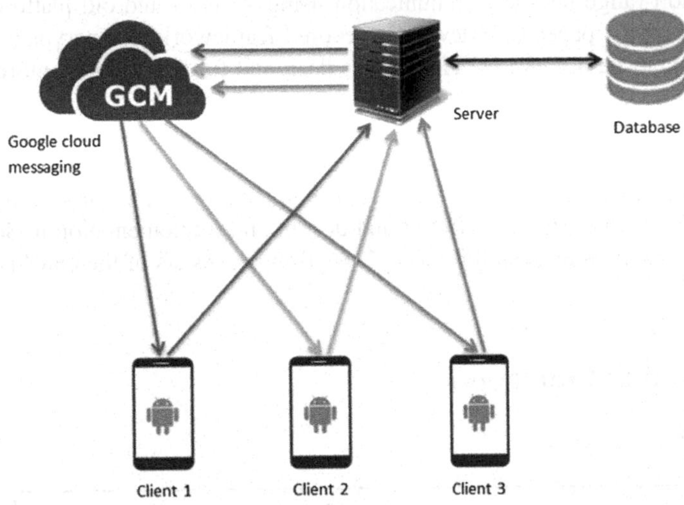

Fig. 1. Framework structure

A paper on crypto-cloak protocol by Dijana Vukovic proposed an application on crypto-cloak protocol implementing as a plug-in for the spark IM which is an open source messaging application in java programming [1]. Telegram being the most popular chat platform for its security also poses some disadvantages, and in [2], some modification of the protocol that telegram uses is shown which can ensure end-to-end security. In a framework for peer to peer networking using Wi-Fi direct in android proposed by Athanasios V. Vasilakos proposed a framework consisting of two major components namely connection establishment protocol and group management protocol. Group management protocol is mainly Wi-Fi direct topology by a conventional star network. This protocol distributes peer IP addresses for the addition and removal of peer from the group in transport layer [3]. In a paper, prototype design of PFC (Private Facebook Chat) is proposed by Chris Robison, which provides convenient browser-based secure instant messaging within Facebook Chat [4]. In the analysis [5] about the network traffic features and also proposes a rule-based classifier for detection of malwares in android. As none of the existing malware detection techniques have focused on the network traffic features for detection of malicious activity. Social networking services adopt centralized architecture by which Internet connectivity is prerequisite for each user to exploit those services, and centralized servers are used for storage and processing of all application/context data. Omniscient centralized server may cause serious privacy concern, due to the fact that it collects and stores all users.

Transmitting of large data creates pressure on the architecture, which is neutralized by using technique such as Wi-Fi direct-based P2P communication [6]. Localization based on Google Geo coding, Chat, and file-sharing component supporting intermittent transmission as prototype done on android platform. In wireless communication, a proposal of creating network among the android devices using Bluetooth is given as it would be power efficient and cost-effective. This approach leads to build a chatting room for short-range mobile communication using APIs of android platform [7].

The aim of this paper is to develop a secured framework for encrypted messaging service for smart device. An HTTP server is designed for storing user information and controlling authentication and message traffic system. An android application for client side consists of mainly multiple user interfaces for registration, authentication, and instant messaging. AES-256 encryption and decryption module is used and an Internet service routine for HTTP protocol. The networking framework is introduced in Sect. 2, followed by data encryption in Sect. 3 and design and implementation in Sect. 4. The experimental result is discussed in Sect. 5. Section 6 consists of the conclusion of this work (Fig. 2).

2 Networking Framework

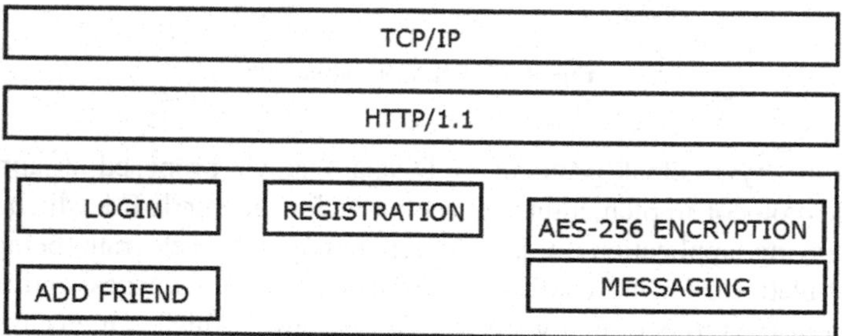

Fig. 2. System architecture

The framework proposed in this paper, HTTP server, is responsible for the operation of authorization and message traffic. To establish communication between server and client, a general hypertext transfer protocol port (port 80) is being used in World Wide Web for sending request to server. In our framework the application layer protocol is designed using HTTP which allows performing within for Internet protocol. The transport layer protocol in CryptoMsg framework is done by using transmission control protocol (TCP). The process of sending request to the server by client is carried through POST method. In terms of computing, HTTP protocol offers several supporting request methods by the World Wide Web as POST request method is one of them. As per design, the POST request method is responsible for requesting the Web server to accept and to store the requested message which is enclosed with data within its body.

2.1 Data Packaging

registration or login data

username	password	GCM registration id

message data

sender id	destination id	encrypted message	data & time

Fig. 3. Packet configuration

A packet consists of control information and user data which is also known as the payload [8]. In the above proposed framework, there are three types of data packets that are used. On the top is registration packet, followed by authorization packet, and at the bottom is the message packet as shown in Fig. 3. The registration packet consists of necessary user information which is required for user registration on server and a registration id of GCM service. Authorization packet consists of login credentials (username, password) and GCM registration id. Lastly, the message packet consists of sender id, recipient id, encrypted message, and date and time. All packets are received by an acknowledgment packet (Fig. 4).

2.2 GCM Server

To communicate with any android device there is a need of synchronous with Google cloud messaging (GCM) server arises. The GCM server is used for identifying each

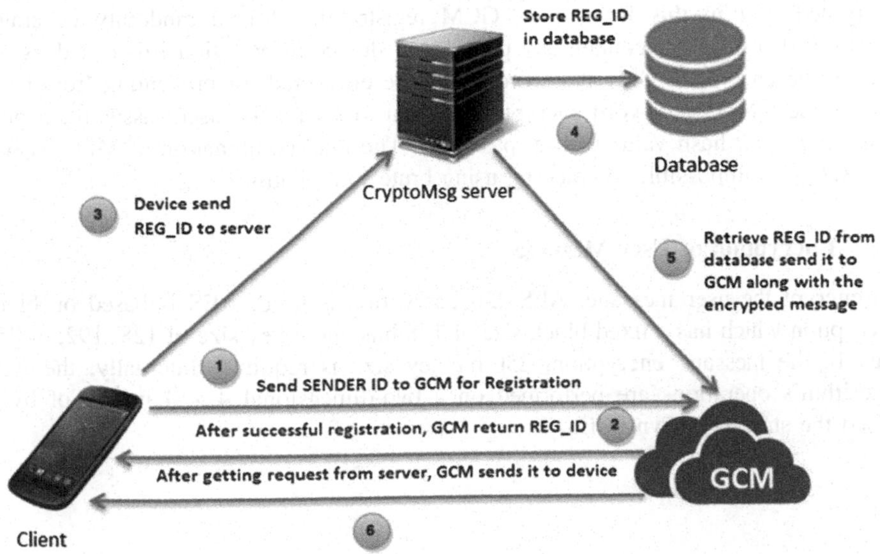

Fig. 4. Workflow of GCM

device and thus communicates with them. To connect with server client, a device needs a unique registration id from GCM. The GCM registration id is a randomly generated identifier that would allow a developer to discover personal identity of the user [9]. To get a registration id, the subscription API sends a request to the GCM. In response to that, the GCM sends the registration id to client. Now the client device includes the registration id into data packet and sends the packet to server. After receiving the data packet, the server gets the registration id from the packet and stores into server database. And this registration id is used to identify the device. In the time of communication sever API sends data packet and registration id to the GCM service. Then, GCM enquired and delivered the data packet to that device according to the registration id.

3 Data Encryption

The data encryption is an integral part of any application. In our above-mentioned framework, to make the data secure, AES-256 encryption technique is used. Till date this encryption technique is the most secure encryption technique making our chat application a secure an efficient among the other chat applications present in the market. In our chat application, encryption is done in two steps, namely Encryption of User Information and Encryption of User Message which are as follows.

3.1 Encryption of User Information

In the framework of Crypto-Message, there is no need to verify mobile number or e-mail id to create an account. But user needs to enter a unique username and a password. The client application also sends a registration id for GCM service along with user information. So there are three types of data that will be stored on the server database. To make a more secured framework, user information should also be encrypted. But in this framework, GCM registration id is a randomly generated identifier that does not contain any personal or device information [9] so it does not need to be encrypted. Only password should be encrypted for preventing from brute force attack. MD5 password encryption is used to secure the user password. It generates a *128-bit* hash value of user password. The total combination of MD5 hash is 2^{128} which is impossible to crack by using brute force login.

3.2 Encryption of User Message

To encrypt the user message, AES-256 encryption is used. AES is based on block encryption which has a fixed block size of 128 bits and a key size of 128, 192, or 256 bits. In the message encryption, 256-bit key size is required. Internally, the AES algorithm's operations are performed on a two-dimensional 4×4 matrix of bytes called the state as shown in Eq. 1.

$$S = \begin{matrix} S_{0,0} & S_{0,1} & S_{0,2} & S_{0,3} \\ S_{1,0} & S_{1,1} & S_{1,2} & S_{1,3} \\ S_{2,0} & S_{2,1} & S_{2,2} & S_{2,3} \\ S_{3,0} & S_{3,1} & S_{3,2} & S_{3,3} \end{matrix} \qquad (1)$$

In AES encryption fourteen cycles of repetition of transformation rounds for 256 bits key that convert the input, known as plaintext, into the final output, known as cipher text. Each round consists of several different stages (Fig. 5).

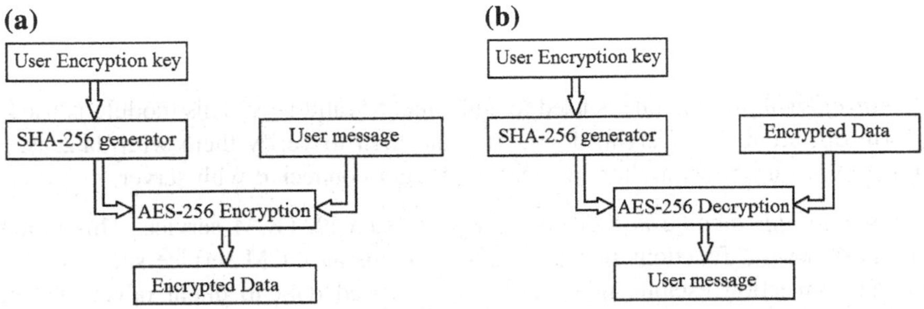

Fig. 5. Block diagram of message **a** encryption and **b** decryption

Encryption

For AES 256 bits key size encryption process [10] starts with *keyExpension*. After this *IntialRound* is performed where just have only one stage named as *AddRoundKey*. Then execute thirteen time repetitions of regular round which is consist of four different stages are named as *SubBytes*, *ShiftRows*, *MixCoulmns* and *AddRoundKey*. The 14th round is final round which is like as regular round but have no *MixColumns* stage. Have only *SubBytes*, *ShiftRows*, *AddRoundKey*.

Decryption

AES 256 bits Decryption process is same as encryption. But all processes will be inversed, and *SubBytes*, *ShiftRows*, and *MixCoulmns* function will be replaced with *InvSubBytes*, *InvShiftRows*, and *InvMixCoulmns*.

4 Designs and Implementation

The proposed Crypto-Message application consists of two parts namely server end and client end. Our android application is designed in such a way that if by any means the server end is hacked, no user information will be at risk due to the unique design and implementation of our application which is discussed below.

4.1 Server End

In implementation, cloud computing is being used to establish our CryptoMsg server through Amazon Web Services (AWS) with address of http://52.34.233.63/. Here, essential Web application which controls user registration, authorization, and message traffic is developed using PHP programming language, and SQL language is used to create link between MySql database and Web application. CryptoMsg server consists of four modules which are named as register.php, androidlogin.php, function.php, and mtcs.php.

The above-mentioned modules are defined as follows:

register.php module is used to receive user information from client device and store them on database. _POST is used to get user data from HTTP packet.

androidlogin.php module is used to authenticate valid user. This module is used to perform operation for receiving login data and then to verify them with database; if login data is correct, then client device will be get connected with server.

functions.php module is used to get connected with GCM service. This module consists of several functions that are required for using GCM service such as authorization, connection establishment, and sending used data to client device through GCM server.

mtcs.php (message traffic control system) module is used to receive message packet from client device and forward it to specific client device as per the recipient id. This module receives data of four types within the message packet, and those are sender id, recipient id, message, and message date/time. Then, this module finds GCM registration id from database for certain recipient id and forwards the message packet to that user through GCM service.

4.2 Client End

For the client end user, a development on an android application named as Crypto-Message messenger is carried out. Crypto-Message android-based messaging application is encoded with AES-256-bit encryption technique. This encryption technique developed is an open-source platform for the development of android application to make it compatible with an android device. Crypto-Message application offers highly secure chatting platform as it uses AES-256-bit encryption technique to encrypt messages/data. This messages/data can only get decrypted by user-defined key. In this application, several necessary steps are followed to make it fully functional. Firstly installation of Crypto-Message application into the user's android device. After a successful installation, the application is opened. Now to get started with our application, the user needs to turn on the Internet connection to register as a new user by providing an input of user's username and the password which are user defined to get authorized registration. Now, after a successful registration, the user needs to provide unique username and password to login into the user's Crypto-Message account. After a successful login, a page comes up that consists of a list of friends of the user with whom the user can exchange data. Along with the list of friends, it also provides an option of ADD which allows the user to add friends. Now, before the user starts to exchange data/information with user's friend in user's friend list, there is a requirement of user-defined unique AES encryption key which is declared by the user and known by the user and the friend with whom the user shares it, as it is necessary to have same AES encryption key on both sides for proper decryption process to generate desired output message/data. Only authorized friend with proper AES encryption key defined by user can access the message/data, else it is impossible to access those message/data by any third party. Make sure that Internet connection remain ON, while sending the data/message to avoid error of server not responding.

5 Result and Discussions

This paper proposes the framework named Crypto-Message which uses AES security. After designing a prototype of our framework, multiple analyses are performed on the different security and privacy-preserving issue. After the completion of all the analyses, the conclusion drawn is that our Crypto-Message framework is more secure than other currently available messaging frameworks. In case of Crypto-Message server is been hacked the user information and message will remain secure. The reason of this is that all user information in the database is encrypted by client end. Moreover, user message is encrypted with a military-grade encryption by a user-defined key. Evens this key will not be stored on server or android device. If the raw data is monitored that is received by server, then it is seen that all data is encrypted.

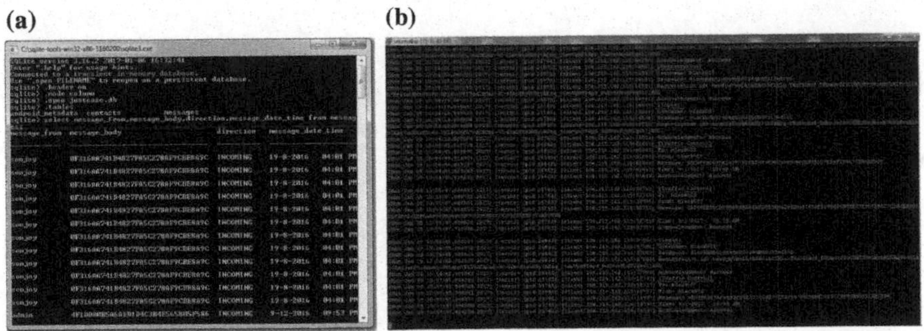

Fig. 6. **a** CryptoMsg message data on local device database and **b** traffic on server side

Figure 6b shows the received data on server end. Here, the first data packet is the registration data. In this packet, client sends username, password and registration id, where user password is encrypted with MD5 hash encryption. The username and registration id are not encrypted because there is no personal detail of the respective users. In the next step, login data packet is received. In these received packets, the configuration of login data packet is same as registration data packet. After which received packet is declared as message data where receiving of sender id, recipient id, message with time and date is executed. Here encryption with the involvement of user-defined key is performed on the packet consist of message data on client end, to generate an encrypted data packet which will only get decrypted by the introduction of user-defined key, known by the user only. This increases the possibilities that original message cannot be get decrypted by any middle man attack even if they got full authority access in server end.

6 Conclusion

In this paper, a new framework for enabling end-to-end encrypted messaging over HTTP protocol is developed. The key components of the framework are a connection establishment protocol and a traffic management protocol which make it more secure and convenient. As the connection establishment protocol allows only authorized device to connect to the server, it offers two modules. Among the two, one is registration module and the other is authorization module. The registration module is being used to add a new user through a unique username and a password. Moreover, there is no need for any e-mail id or phone number verification to create an account. The authorization module present in the above-mentioned framework performs a role for secure login into user account. The traffic management protocol controls over message encryption and traffic system of encrypted messaging and decryption of messages. Through our protocol, user message is encrypted using AES-256 encryption which makes it secure and impossible to crack. As in future work, it may consider the following upgradation. Firstly including multimedia streaming and files transfer protocol and the other one is to including in a voice calling feature.

Appendix

Snapshots of android application Fig. 7.

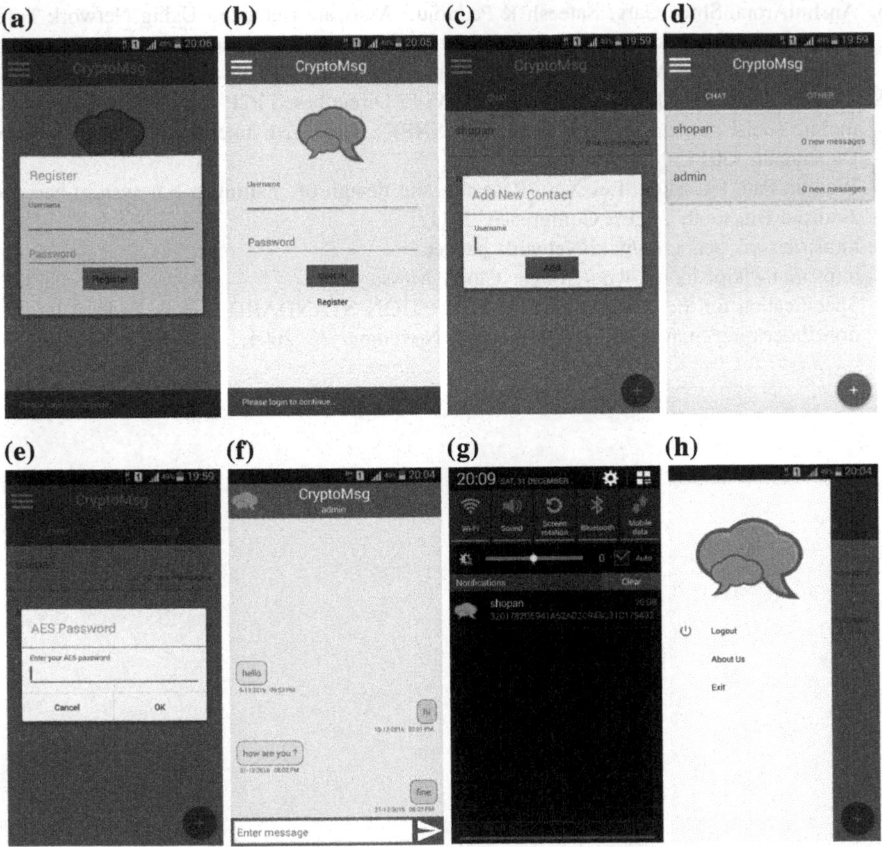

Fig. 7. Screenshot of android application: **a** register a new user, **b** user login, **c** add a new friend, **d** friend list, **e** encryption key input, **f** chat activity, **g** new message notification, and **h** side drawer activity

References

1. DijanaVukovic, DaniloGligoroski, ZoranDjuric, "CryptoCloak Protocol and the Prototype Application", 2015 IEEE Conference on Communications and Network Security (CNS), 28–30 Sept. 2015.
2. Job J, Naresh V, K Chandrasekaran, "A modified secure version of the Telegram protocol", 2015 IEEE International Conference on Electronics, Computing and Communication Technologies (CONECCT), 10–11 July 2015.
3. Ahmed A. Shahin, and Mohamed Younis, "A Framework for P2P Networking of Smart Devices Using Wi-Fi Direct", IEEE 25th International Symposium on Personal, Indoor and Mobile Radio Communications, 2014.
4. Chris Robison, Scott Ruoti, Timothy W. van der Horst, "Private Facebook Chat", ASE/IEEE International Conference on Social Computing, 2012.

5. AnshulArora, Shree Garg, Sateesh K Peddoju, "Malware Detection Using Network Traffic Analysis in Android Based Mobile Devices", 8th International Conference on Next Generation Mobile Applications, Services and Technolog, 2014.
6. Yufeng Wang, Qun Jin, jianhuawang, "A Wi-Fi Direct based P2P application prototype for mobile social networking in proximity (MSNP)", IEEE 12th International Conference on Dependable, 2014.
7. Weihua Pan, FucaiLua, Lei Xu, "Research and design of chatting room system based on Android Bluetooth", IEEE Conference, 2011.
8. https://en.wikipedia.org/wiki/Network_packet.
9. https://en.wikipedia.org/wiki/Google_Cloud_Messaging.
10. Specification for the ADVANCED ENCRYPTION STANDARD (AES), Federal Information Processing Standards Publication 197, November 26, 2001.

Robust Adaptive Backstepping in Tracking Control for Wheeled Inverted Pendulum

Nguyen Thanh Binh[1], Nguyen Anh Tung[2], Dao Phuong Nam[2(✉)], and Nguyen Thi Viet Huong[3]

[1] Thuyloi University, Hanoi, Vietnam
ntbinh@tlu.edu.vn
[2] Hanoi University of Science and Technology, Hanoi, Vietnam
nam.daophuong@hust.edu.vn
[3] Thai Nguyen Industrial College, Hanoi, Vietnam

Abstract. The previous papers (Olfati-Saber PhD. thesis, 2001) [1], (Khac Duc Do, Gerald Journal of Intelligent and Robotic Systems 60(3), 2010) [2] (Wei et al. Automatica, 841–850 1995) [3] applied feedback linearization or global change coordinates to separate system model into rotate and straight movement, leading to difficulty in control design for uncertain parameter systems. Additionally, paper (Culi et al. Nonlinear control of a swinging pendulum. Automatica 31(6), 1995, 851–862) [4] presenting robust adaptive law ignored time-varying inertia matrix to design control Lyapunov function (CLF) easily, leading to wrong theoretical proof. In this paper, we propose a new adaptive law and a new controller to control WIP system, in that the error tracking of heading angle and position is bounded and tilt angle converges to the small arbitrary ball of origin and tracking position, which is not be ensured (Culi et al. Nonlinear control of a swinging pendulum. Automatica 31(6), 1995, 851–862) [4] (Li et al. Automatica 2010, 1346–1353) [5]. Moreover, time-varying inertia matrix is considered to choose a proper CLF via backstepping technique. The simulation results are implemented to demonstrate the performances of the proposed adaptive law and controller.

1 Introduction

In [2], the wheeled inverted pendulum described includes a pair of identical wheels, a chassis, wheel actuators, an inverted pendulum, and a motion control unit, in that the pair of wheels and the inverted pendulum are supported by chassis. The wheel actuators rotate the wheels with respect to the chassis. The wheel actuators are controlled by the motion control unit to move the vehicle and to stabilize the inverted pendulum.

The dynamic model and several control methods are presented in [6]. In this content, we use dynamic model of WIP built by Newton–Euler method. This model is separated into three subsystems, in that the subsystems describe two angle and straight. Authors in [7,8] approached to control inverse pendulum

V. Bhateja et al. (eds.), *Information Systems Design and Intelligent Applications*, Advances in Intelligent Systems and Computing 672,
https://doi.org/10.1007/978-981-10-7512-4_42

based on energy function, but the disadvantage of them is that disturbances impacting on the considered system are ignored and pendulum always fluctuate around the origin. Olfati-Saber [1] proposed coordinate transformation to change Cart-Pole model to strict forward system to apply nest saturation method. This transformation would not be effective, when some parameters are uncertainties or disturbances appear. Therefore, K.D. Do [2] used the similar transformation in [1] and combines with nest saturation method, disturbance observer to steady error to converge to the origin asymptotically with the assumption of zero straight accelerator of Cart. Researchers in [8] applied properties of nonholomic system and the backstepping technique, but their drawback is an assumption of satisfying state constraints. The instantaneous switching of control input is proposed in [9], nevertheless, the position of Cart is not able to stabilize at the desired point, all these previous papers do not mention uncertain parameters. The papers in [4] researching into adaptive control have some weakness, which are time-dependent inertia matrix, adaptive law, and control law, leading to the wrong of Lyapunov function control candidate.

In this paper, we propose new adaptive law with time-dependent inertia matrix and a new controller to control WIP system with uncertain parameters via Backstepping technique in [10]. The dynamic system model is separated into three subsystems including two angle subsystems and straight movement subsystem with underactuated control input as [5]. The proposed controller is designed to control two angle subsystems, in that error of tracking heading angle and its reference and tilt angle are bounded, the remainder will be stable following this subsystem.

2 Dynamic Newton–Euler Model of WIP System

In this chapter, the dynamic model of WIP is described by Newton–Euler with nonholomic constraint as follows:

$$M\left(q\right)\ddot{q}+C\left(q,\dot{q}\right)\dot{q}+F\left(\dot{q}\right)+G\left(q\right)+D=B\left(q\right)\tau+J(q)^{T}\lambda \tag{1}$$

$$\dot{x}sin(\varphi)-\dot{y}cos(\varphi)=0 \tag{2}$$

Where $q=[\alpha,\ \varphi,\ x,\ y]^{T}\in R^{4\times1}$ is variable of WIP including α is tilt angle, φ is heading angle, and x,y are its position. $M\left(q\right)\in R^{4\times4}$ is inertia matrix, $C\left(q,\dot{q}\right)\in R^{4\times4}$ is Coriolis and Centrifugal forces, $F\left(\dot{q}\right)\in R^{4\times1}$ is a vector of viscous friction forces, $G\left(q\right)\in R^{4\times1}$ is a vector of gravitational forces, $D\in R^{4\times1}$ is a vector of external disturbances, $B\left(q\right)\in R^{4\times4}$ is a matrix of input transformation, τ is vector of torque input, $J(q)^{T}$ is Jacobian matrix, and λ is Langrange multiplier.

In order to eliminate constraint (2) in (1), we use the transformation in [10], then (1) is rewritten as follows:

$$M_{1}\left(\eta\right)\ddot{\eta}+C_{1}\left(\eta,\dot{\eta}\right)\dot{\eta}+F_{1}\left(\dot{\eta}\right)+G_{1}\left(\eta\right)=B_{1}\left(\eta\right)\tau+D_{1} \tag{3}$$

Table 1. Parameters and variables of WIP system

Parameter	Symbol	Variable	Symbol
Distance between two wheels	$D\,(\mathrm{m})$	Heading angle of pendulum	$\varphi\,(\mathrm{rad})$
Radius of wheel	$R\,(\mathrm{m})$	Tilt of pendulum	$\alpha\,(\mathrm{rad})$
Moment of inertia of the wheel about y-axis	$I_\omega\,(\mathrm{kgm}^2)$	Torque control in left and right wheel	$\tau_L, \tau_R\,(\mathrm{Nm})$
Moment of inertia of chassis and pendulum about z-axis	$I_m\,(\mathrm{kgm}^2)$	Position of chassis	$x, y\,(\mathrm{m})$
Mass of load and chassis	$m\,(\mathrm{kg})$	Disturbances impacting on two wheels	$d_L, d_R\,(\mathrm{Nm})$
Mass of wheel	$M_\omega\,(\mathrm{kg})$		
Gravity acceleration	$g\,(\mathrm{m/s}^2)$		
Distance between central point of load and chassis	$l\,(\mathrm{m})$		
Coefficient of viscous friction force	μ_1, μ_2, μ_3		

where variables $\eta = [\alpha, \varphi, s]^T$ and parameters in (Table 1) of WIP system are substituted into (3), and we have a new matrices in dynamic Newton–Euler equation as (4) and (5):

$$M_1(\eta) = \begin{bmatrix} m_{11} & 0 & m_{13}(\alpha) \\ 0 & m_{22}(\alpha) & 0 \\ m_{13}(\alpha) & 0 & m_{33} \end{bmatrix}; C_1(\eta) = \begin{bmatrix} 0 & c_{12}(\alpha, \dot{\varphi}) & 0 \\ c_{21}(\alpha, \dot{\varphi}) & c_{22}(\alpha, \dot{\alpha}) & 0 \\ c_{31}(\alpha, \dot{\alpha}) & 0 & 0 \end{bmatrix} \tag{4}$$

$$G_1(\eta) = \begin{bmatrix} g_1(\alpha) \\ 0 \\ 0 \end{bmatrix}; F_1(\dot{\eta}) = \begin{bmatrix} f_1(\dot{\alpha}) \\ f_2(\dot{\varphi}) \\ f_3(\dot{s}) \end{bmatrix}; D_1 = \begin{bmatrix} 0 \\ d_1 \\ d_2 \end{bmatrix}; B_1\tau = \begin{bmatrix} 0 \\ \tau_1 \\ \tau_2 \end{bmatrix} \tag{5}$$

where
$m_{11} = ml^2 + I_m, m_{13}(\alpha) = ml\cos(\alpha), m_{22}(\alpha) = I_m + 2D^2\left(M_\omega + \frac{I_\omega}{R^2}\right) + ml^2\sin^2(\alpha)$, $m_{33} = m + 2M_\omega + \frac{2I_\omega}{R^2}; c_{12}(\alpha, \dot{\varphi}) = -\frac{1}{2}ml^2\dot{\varphi}\sin(2\alpha), c_{21}(\alpha, \dot{\alpha}, \dot{\varphi}) = \frac{1}{2}ml^2\dot{\varphi}sin(2\alpha)$, $c_{22}(\alpha, \dot{\alpha}) = \frac{1}{2}ml^2\sin(2\alpha)\dot{\alpha}, c_{31}(\alpha, \dot{\alpha}) = -ml\sin(\alpha)\dot{\alpha}, g_1(\alpha) = -mgl\sin(\alpha)$, $d_1 = 2D(d_L - d_R), d_2 = (d_R + d_L), \tau_1 = \frac{2D}{R}(\tau_L - \tau_R), \tau_2 = \frac{1}{R}(\tau_L + \tau_R)$, $f_1 = \mu_1\dot{\alpha}, f_2 = \mu_2\dot{\varphi}, f_3 = \mu_3\dot{s}$

Remark 1. The properties of the matrix $M_1(\eta)$ and $C_1(\eta, \dot{\eta})$ are:
(a) $\dot{M}_1 - 2C_1$ is skew matrix
(b) M_1 is symmetric positive define matrix $\delta_{\min} \le \|M_1\|_2 \le \delta_{\max}$
(c) $0 < m_{ii\min} < m_{ii} < m_{ii\max}$

By substituting (4) and (5) into model (3), we have three equations describing system model as follows:

$$m_{11}\ddot{\alpha} + m_{13}\ddot{s} + c_{12}\dot{\varphi} + g_1 + f_1 = 0 \tag{6}$$
$$m_{22}\ddot{\varphi} + c_{21}\dot{\alpha} + c_{22}\dot{\varphi} + f_2 = \tau_1 + d_1 \tag{7}$$
$$m_{13}\ddot{\alpha} + m_{33}\ddot{s} + c_{31}\dot{\alpha} + f_3 = \tau_2 + d_2 \tag{8}$$

The state variable is defined as $z_1 = \varphi$, $z_2 = \dot{\varphi}$, $z_3 = \alpha, z_4 = \dot{\alpha}$ to change model (6) and (7) to become:

$$\dot{z}_1 = z_2 \tag{9}$$
$$\dot{z}_2 = -\frac{c_{21}}{m_{22}}z_4 - \frac{c_{22}}{m_{22}}z_2 - \frac{f_2}{m_{22}} + \frac{\tau_1}{m_{22}} + \frac{d_1}{m_{22}} \tag{10}$$
$$\dot{z}_3 = z_4 \tag{11}$$
$$\dot{z}_4 = -\frac{m_{33}}{m_{11}m_{33} - m_{13}^2}(c_{12}z_2 + g_1 + f) + \frac{m_{13}}{m_{11}m_{33} - m_{13}^2}(c_{31}z_4 + f_3) - \frac{m_{13}}{m_{11}m_{33} - m_{13}^2}d_2 - \frac{m_{13}}{m_{11}m_{33} - m_{13}^2}\tau_2 \tag{12}$$

The nonlinear parts in (10) and (12) is separated into product of parameter vector and vector function of state variable (13) and (14) to build adaptive law with ξ_2, ξ_4 defined in (45):

$$c_{21}z_4 + c_{22}z_2 + f_2 - \frac{\dot{m}_{22}\xi_2}{2} = \theta_1^T H_1(\eta, \dot{\eta}, \varphi_r, \dot{\varphi}_r) \tag{13}$$

$$\frac{m_{33}}{m_{13}}(c_{12}z_2 + g_1 + f) - m_{13}(c_{31}z_4 + f_3) - \frac{\dot{p}\xi_4}{2} = \theta_2^T H_2(\eta, \dot{\eta}, \alpha_r, \dot{\alpha}_r) \tag{14}$$

Where

$$\theta_1^T = \begin{bmatrix} \frac{ml^2}{2} & \mu_2 \end{bmatrix}$$

$$\theta_2^T = \begin{bmatrix} -m_{33}l & -m_{33}g & \frac{m_{33}\mu_1}{ml} & \frac{m^2l^2}{2} & -ml\mu_3 & \frac{-m_{11}m_{33}}{2ml} & \frac{-ml}{2} \end{bmatrix} \tag{15}$$

$$H_1(\eta, \dot{\eta}, \varphi_r, \dot{\varphi}_r) = \begin{bmatrix} \sin(2\alpha)\dot{\alpha}\left[\dot{\varphi} - (k_1(\varphi - \varphi_r) + \dot{\varphi}_r)\right] & \dot{\varphi} \end{bmatrix}^T$$

$$H_2(\eta, \dot{\eta}, \alpha_r, \dot{\alpha}_r) = \Big[\sin(\alpha)\dot{\varphi}^2 \;\; \tan(\alpha) \;\; \tfrac{\dot{\alpha}}{\cos(\alpha)} \;\; \sin(2\alpha)\dot{\alpha}^2 \;\; \cos(\alpha)\dot{s}$$

$$\tfrac{\sin(\alpha)}{\cos^2(\alpha)}\dot{\alpha}(\dot{\alpha} - \dot{\alpha}_r + k_3(\alpha - \alpha_r)) \;\; \sin(\alpha)\dot{\alpha}(\dot{\alpha} - \dot{\alpha}_r + k_3(\alpha - \alpha_r)) \Big]^T \tag{16}$$

$$p = \frac{m_{11}m_{33} - m_{13}^2}{m_{13}} \tag{17}$$

By setting (13) and (14), the state-space model (9), (10), (11), and (12) become:

$$\dot{z}_1 = z_2 \tag{18}$$

$$\dot{z}_2 = -\frac{\theta_1^T H_1 + 0.5\dot{m}_{22}\xi_2}{m_{22}} + \frac{\tau_1}{m_{22}} + \frac{d_1}{m_{22}} \tag{19}$$

$$\dot{z}_3 = z_4 \tag{20}$$

$$\dot{z}_4 = -\frac{\theta_2^T H_2 + 0.5\dot{p}\xi_4}{p} - \frac{d_2}{p} - \frac{\tau_2}{p} \tag{21}$$

Assumption 1. The disturbances impacting on two wheels of system and all reference signals $\varphi_r, s_r, \dot{\varphi}_r, \dot{s}_r$ are bounded

$$|d_1| \le d_{1\_\max}, |d_2| \le d_{2\_\max} \tag{22}$$

$$|\varphi_r| \le \varphi_{r\_\max}; |\dot{\varphi}_r| \le \dot{\varphi}_{r\_\max}; |s_r| \le s_{r\_\max}; |\dot{s}_r| \le \dot{s}_{r\_\max} \tag{23}$$

3 Robust Adaptive Backstepping Design for φ, α—Subsystem

The heading angle error is defined as $\xi_1 = z_1 - \varphi_r$ to substitute into (18) with virtual input α_1 in (26) and ξ_2 in (25)

$$\dot{\xi}_1 = \xi_2 + \alpha_1 \tag{24}$$

$$\xi_2 = z_2 - \dot{\varphi}_r - \alpha_1 \tag{25}$$

$$\alpha_1 = -k_1\xi_1 \tag{26}$$

$$\dot{\xi}_2 = -\frac{\theta_1^T H_1 + 0.5\dot{m}_{22}\xi_2}{m_{22}} + \frac{\tau_1}{m_{22}} + \frac{d_1}{m_{22}} - \ddot{\varphi}_r - \dot{\alpha}_1 \tag{27}$$

Then, virtual input (26) is substitute into (24)

$$\dot{\xi}_1 = -k_1\xi_1 + \xi_2 \tag{28}$$

The Lyapunov candidate function is chosen as:

$$V_1 = \frac{1}{2}\xi_1^2 \tag{29}$$

$$\dot{V}_1 = -k_1\xi_1^2 + \xi_1\xi_2 \tag{30}$$

The proposed adaptive law:

$$\dot{\hat{\theta}}_1 = -\kappa_1\hat{\theta}_1 - \Gamma_1 H_1\xi_2 \tag{31}$$

$$\dot{\hat{m}}_{22} = -\kappa_2\hat{m}_{22} - \beta_1\vartheta_1\xi_2 \tag{32}$$

The proposed controller:

$$\tau_1 = \hat{m}_{22}\vartheta_1 + \hat{\theta}_1^T H_1 \tag{33}$$

$$\vartheta_1 = -k_2\xi_2 - \xi_1 + \ddot{\varphi}_r + \dot{\alpha}_1 \tag{34}$$

Selecting the Lyapunov function candidate V_2 as follows:

$$V_2 = V_1 + \frac{1}{2}m_{22}\xi_2^2 + \frac{1}{2}\tilde{\theta}_1^T\Gamma_1^{-1}\tilde{\theta}_1 + \frac{1}{2\beta_1}\tilde{m}_{22}^2 \tag{35}$$

where: $\tilde{\theta}_1 = \theta_1 - \hat{\theta}_1$, $\tilde{m}_{22} = m_{22} - \hat{m}_{22}$

$$\dot{V}_2 = \dot{V}_1 + m_{22}\xi_2\dot{\xi}_2 + \frac{1}{2}\dot{m}_{22}\xi_2^2 + \tilde{\theta}_1^T\Gamma_1^{-1}\dot{\tilde{\theta}}_1 + \frac{1}{\beta_1}\tilde{m}_{22}\dot{\tilde{m}}_{22} \tag{36}$$

Substituting (27) and (30) into (36)

$$\begin{aligned} \dot{V}_2 = & -k_1\xi_1^2 + \xi_1\xi_2 + m_{22}\xi_2\left(-\frac{\theta_1^T H_1 + 0.5\dot{m}_2\xi_2}{m_{22}} + \frac{\tau_1}{m_{22}} + \frac{d_1}{m_{22}} - \ddot{\varphi}_r - \dot{\alpha}_1\right) \\ & + \frac{1}{2}\dot{m}_{22}\xi_2^2 + \tilde{\theta}_1^T\Gamma_1^{-1}\dot{\tilde{\theta}}_1 + \frac{1}{\beta_1}\tilde{m}_{22}\dot{\tilde{m}}_{22} \end{aligned} \tag{37}$$

The proposed control law in (33) and (34) is replaced in (37)

$$\begin{aligned} \dot{V}_2 = & -k_1\xi_1^2 - m_{22}k_2\xi_2^2 - (m_{22} - 1)\,\xi_1\xi_2 + \xi_2\left(\hat{m}_{22}\vartheta_1 - \tilde{\theta}_1^T H_1 + d_1 - m_{22}\vartheta_1\right) \\ & + \tilde{\theta}_1^T\Gamma_1^{-1}\dot{\tilde{\theta}}_1 + \frac{1}{\beta_1}\tilde{m}_{22}\dot{\tilde{m}}_{22} \end{aligned} \tag{38}$$

The proposed adaptive law in (31), (32), and $\dot{\tilde{\theta}}1 = -\dot{\hat{\theta}}_1$, $\dot{\tilde{m}}_{22} = \dot{m}_{22} - \dot{\hat{m}}_{22}$ is employed in (38):

$$\dot{V}_2 = -k_1\xi_1^2 - m_{22}k_2\xi_2^2 - (m_{22}-1)\,\xi_1\xi_2 + \xi_2\left(-\tilde{m}_{22}\vartheta_1 - \tilde{\theta}_1^T H_1 + d_1\right)$$
$$+\,\tilde{\theta}_1^T \Gamma_1^{-1}\left(\kappa_1\hat{\theta}_1 + \Gamma_1 H_1\xi_2\right) + \frac{1}{\beta_1}\tilde{m}_{22}\left(\dot{m}_{22} + \kappa_2\hat{m}_{22} + \beta_1\vartheta_1\xi_2\right) \tag{39}$$

$$\dot{V}_2 = -k_1\xi_1^2 - m_{22}k_2\xi_2^2 - (m_{22}-1)\,\xi_1\xi_2 + \xi_2 d_1 + \tilde{\theta}_1^T\Gamma_1^{-1}\kappa_1\hat{\theta}_1$$
$$+\,\frac{1}{\beta_1}\tilde{m}_{22}\left(\dot{m}_{22} + \kappa_2\hat{m}_{22}\right) \tag{40}$$

$$\dot{V}_2 \le -\left(k_1 - \frac{|m_{22}-1|}{2}\right)\xi_1^2 - \left(m_{22}k_2 - \lambda_1 - \frac{|m_{22}-1|}{2}\right)\xi_2^2 - (1-\lambda_2)\,\Gamma_1^{-1}\kappa_1\left\|\tilde{\theta}_1\right\|_2^2$$
$$-\left[\frac{(1-\lambda_3)\,\kappa_2}{\beta_1} - \frac{m^2 l^4\delta}{4\beta_1^2}\right]\tilde{m}_{22}^2 + \frac{d_{1\,\max}^2}{4\lambda_1} + \frac{\Gamma_1^{-1}\kappa_1\left\|\theta_1\right\|_2^2}{4\lambda_2} + \frac{\kappa_2}{4\beta_1\lambda_3}m_{22}^2 + \frac{z_4^2}{\delta} \tag{41}$$

The new variable ξ_3 is error between z_3 tilt angle and its virtual reference α_r—defined in next section to ensure tracking position. $\dot{\xi}_3, \xi_4$ are selected and virtual input α_2 in (44) as follows:

$$\xi_3 = z_3 - \alpha_r \tag{42}$$
$$\dot{\xi}_3 = \xi_4 + \alpha_2 \tag{43}$$
$$\alpha_2 = -k_3\xi_3 \tag{44}$$
$$\xi_4 = z_4 - \alpha_2 - \dot{\alpha}_r \tag{45}$$

The Lyapunov function candidate for ξ_3 is chosen as:

$$V_3 = \frac{1}{2}\xi_3^2 \tag{46}$$
$$\dot{V}_3 = \xi_3\left(\xi_4 - k_3\xi_3\right) = -k_3\xi_3^2 + \xi_3\xi_4 \tag{47}$$

The proposed adaptive law for α—subsystem:

$$\dot{\hat{\theta}}_2 = -\kappa_3\hat{\theta}_2 - \Gamma_2 H_2\xi_4 \tag{48}$$
$$\dot{\hat{p}} = -\kappa_4\hat{p} - \beta_2\vartheta_2\xi_4 \tag{49}$$

The proposed controller for α—subsystem:

$$\tau_2 = -\hat{p}\vartheta_2 - \hat{\theta}_2^T H_2 \tag{50}$$
$$\vartheta_2 = -k_4\xi_4 - \xi_3 + \dot{\alpha}_2 + \ddot{\alpha}_r \tag{51}$$

From (45), we have:

$$\dot{\xi}_4 = -\frac{\theta_2^T H_2 + 0.5\dot{p}\xi_4}{p} - \frac{\tau_2}{p} - \frac{d_2}{p} - \dot{\alpha}_2 \tag{52}$$

Theorem 1. *By using controller (33) and (50) and adaptive law (31), (32), (48), and (49), the closed-loop system (24), (27), (43), and (52) is globally uniformly stable.* $\xi_1, \xi_2, \xi_3, \xi_4$ *converge to ball centered at the origin, the radius of this ball can be chosen arbitrarily small by adjusting control parameter* k_1, k_2, k_3, k_4. *Moreover, the error estimation parameters* $\tilde{\theta}_1, \tilde{\theta}_2, \tilde{m}_{22}, \tilde{p}$ *are uniformly bounded.*

Proof. The Lyapunov function candidate for (24), (27), (43), and (52) subsystem is chosen to proof their stability as follows:

$$V_4 = V_2 + V_3 + \frac{1}{2}p\xi_4^2 + \frac{1}{2}\tilde{\theta}_2^T \Gamma_2^{-1}\tilde{\theta}_2 + \frac{1}{2\beta_2}\tilde{p}^2 \tag{53}$$

where $\tilde{\theta}_2 = \theta_2 - \hat{\theta}_2, \tilde{p} = p - \hat{p}$ and Γ_2, β_2 are positive scalar:

$$\begin{aligned}\dot{V}_4 =& \dot{V}_2 + \dot{V}_3 + p\xi_4\left(\frac{-\theta_2^T H_2 + 0.5\dot{p}\xi_4 - \tau_2 - d_2}{p} - \dot{\alpha}_2\right)\\ &+ \frac{1}{2}\dot{p}\xi_4^2 + \tilde{\theta}_2^T \Gamma_2^{-1}\dot{\tilde{\theta}}_2 + \frac{1}{\beta_2}\tilde{p}\dot{\tilde{p}}\end{aligned} \tag{54}$$

Substituting (47) and (50) into (54):

$$\begin{aligned}\dot{V}_4 =& \dot{V}_2 - k_3\xi_3^2 + \xi_3\xi_4 + \xi_4\left(-\left(\tilde{\theta}_2^T H_2 + \tilde{p}\vartheta_2\right) - pk_4\xi_4 - p\xi_3 - d_2\right)\\ &+ \tilde{\theta}_2^T \Gamma_2^{-1}\dot{\tilde{\theta}}_2 + \frac{1}{\beta_2}\tilde{p}\dot{\tilde{p}}\end{aligned} \tag{55}$$

Substituting (48) and (49) into (55):

$$\begin{aligned}\dot{V}_4 =& \dot{V}_2 - k_3\xi_3^2 + \xi_3\xi_4 + \xi_4\left(-\left(\tilde{\theta}_2^T H_2 + \tilde{p}\vartheta_2\right) - pk_4\xi_4 - p\xi_3 - d_2\right)\\ &+ \tilde{\theta}_2^T \Gamma_2^{-1}\left(\kappa_3\hat{\theta}_2 + \Gamma_2 H_2\xi_4\right) + \frac{1}{\beta_2}\tilde{p}\left(\kappa_4\hat{p} + \beta_2\vartheta_2\xi_4\right)\end{aligned} \tag{56}$$

$$\dot{V}_4 = \dot{V}_2 - k_3\xi_3^2 - pk_4\xi_4^2 - (p-1)\xi_3\xi_4 - \xi_4 d_2 + \tilde{\theta}_2^T \Gamma_2^{-1}\kappa_3\hat{\theta}_2 + \frac{1}{\beta_2}\tilde{p}\kappa_4\hat{p} \tag{57}$$

$$\begin{aligned}\dot{V}_4 \le& \dot{V}_2 - \left(k_3 - \frac{|p-1|}{2}\right)\xi_3^2 - \left(pk_4 - \lambda_4 - \frac{|p-1|}{2}\right)\xi_4^2 + \frac{d_{2\max}^2}{4\lambda_4}\\ &- \tilde{\theta}_2^T \Gamma_2^{-1}\kappa_3\tilde{\theta}_2 + \tilde{\theta}_2^T \Gamma_2^{-1}\kappa_3\theta_2 - \frac{1}{\beta_2}\tilde{p}\kappa_4\tilde{p} + \frac{1}{\beta_2}\tilde{p}\kappa_4 p\end{aligned} \tag{58}$$

$$\begin{aligned}\dot{V}_4 \le& \dot{V}_2 - \left(k_3 - \frac{|p-1|}{2}\right)\xi_3^2 - \left(pk_4 - \lambda_4 - \frac{|p-1|}{2}\right)\xi_4^2 + \frac{d_{2\max}^2}{4\lambda_4}\\ &- (1-\lambda_5)\Gamma_2^{-1}\kappa_3\left\|\tilde{\theta}_2\right\|_2^2 + \Gamma_2^{-1}\kappa_3\frac{\|\theta_2\|_2^2}{4\lambda_5} - \frac{(1-\lambda_6)}{\beta_2}\kappa_4\tilde{p}^2 + \frac{1}{\beta_2}\kappa_4\frac{p^2}{4\lambda_6}\end{aligned} \tag{59}$$

$$\begin{aligned}\dot{V}_4 \le& -\left(k_1 - \frac{|m_{22}-1|}{2}\right)\xi_1^2 - \left(m_{22}k_2 - \lambda_1 - \frac{|m_{22}-1|}{2}\right)\xi_2^2 - \left(k_3 - \frac{k_3^2}{2\delta} - \frac{|p-1|}{2}\right)\xi_3^2\\ &- \left(pk_4 - \lambda_4 - \frac{1}{2\delta} - \frac{|p-1|}{2}\right)\xi_4^2 - (1-\lambda_2)\Gamma_1^{-1}\kappa_1\left\|\tilde{\theta}_1\right\|_2^2 - \left[\frac{(1-\lambda_3)\kappa_2}{\beta_1} - \frac{m^2 l^4\delta}{4\beta_1^2}\right]\tilde{m}_{22}^2\\ &- (1-\lambda_5)\Gamma_2^{-1}\kappa_3\left\|\tilde{\theta}_2\right\|_2^2 - \frac{(1-\lambda_6)\kappa_4\tilde{p}^2}{\beta_2} + \frac{d_{1\max}^2}{4\lambda_1} + \frac{d_{2\max}^2}{4\lambda_4}\\ &+ \frac{\Gamma_1^{-1}\kappa_1\|\theta_1\|_2^2}{4\lambda_2} + \frac{\kappa_2 m_{22}^2}{4\beta_1\lambda_3} + \frac{\Gamma_2^{-1}\kappa_3\|\theta_2\|_2^2}{4\lambda_5} + \frac{\kappa_4 p^2}{4\lambda_6\beta_2}\end{aligned} \tag{60}$$

In order to satisfy Theorem 1, coefficients in (60) of controllers and adaptive laws are selected as:

$$\delta = k_3; 0 < \lambda_2, \lambda_3, \lambda_5, \lambda_6 < 1; \lambda_1, \lambda_4 \quad \textit{is selected arbitrarily} \tag{61}$$

$$\begin{aligned} &k_1 > \frac{|m_{22} - 1|}{2}, k_3 > |p - 1|, k_4 > \frac{\lambda_4 + \frac{1}{2k_3} + \frac{|p-1|}{2}}{p} \\ &k_2 > \frac{\lambda_1 + \frac{|m_{22}-1|}{2}}{m_{22}}, \kappa_2 > \frac{m^2 l^4 \delta}{4\beta_1 (1 - \lambda_3)} \end{aligned} \tag{62}$$

Note that the parameter and disturbance terms m_{22}^2, p^2, $\|\theta_2\|_2^2$, $\|\theta_1\|_2^2$, $d_{1\max}^2$, $d_{2\max}^2$ in (60) are bounded and because λ_1, λ_4 in (61) can be chosen arbitrarily and coefficients k_1, k_2, k_3, k_4 can be selected largely under condition (62), the state variables $\xi_1, \xi_2, \xi_3, \xi_4$ are able to converge to arbitrary small ball of origin in Theorem 1. Moreover, because of existing negative terms $-(\bullet)\tilde{m}_{22}^2$, $-(\bullet)\tilde{p}^2, -(\bullet)\left\|\tilde{\theta}_1\right\|_2^2, -(\bullet)\left\|\tilde{\theta}_2\right\|_2^2$ in (60), the error estimation parameters $\tilde{\theta}_1, \tilde{\theta}_2, \tilde{m}_{22}, \tilde{p}$ are uniformly bounded.

4 Tracking Control Design for s—Subsystem

Applying the controller (50) to s—subsystem (8) to achieve a new model (63):

$$\begin{aligned} &\frac{m_{11}m_{33} - m_{13}^2}{m_{11}} \ddot{s} = -\hat{p}\vartheta_2 - \hat{\theta}_2^T H_2 - c_{31}\dot{\alpha} - f_3 \\ &+ \frac{m_{13}}{m_{11}} (c_{12}\dot{\varphi} + g_1 + f_1) + d_2 \end{aligned} \tag{63}$$

$$\vartheta_2 = [(k_4 k_3 + 1)\alpha_r + (k_4 + k_3)\dot{\alpha}_r + \ddot{\alpha}_r] - (k_4 k_3 + 1)\alpha - (k_4 + k_3)\dot{\alpha} \tag{64}$$

By setting virtual input (65), model (63) becomes (66)

$$\hat{p}[(k_4 k_3 + 1)\alpha_r + (k_4 + k_3)\dot{\alpha}_r + \ddot{\alpha}_r] = -k_5 tanh(s_e) - k_6 tanh(\dot{s}_e) \tag{65}$$

$$\begin{aligned} &\ddot{s}_r = 0; \dot{s}_e = \dot{s} - \dot{s}_r; s_e = s - s_r \\ &p\ddot{s}_e + \mu_3 \dot{s}_e + k_6 tanh(\dot{s}_e) + k_5 tanh(s_e) = \mu_3 \dot{s}_r - \hat{\theta}_2^T H_2 - c_{31}\dot{\alpha} \\ &+ \hat{p}[(k_4 k_3 + 1)\alpha + (k_4 + k_3)\dot{\alpha}] + \frac{m_{13}}{m_{11}} (c_{12}\dot{\varphi} + g_1 + f_1) + d_2 \end{aligned} \tag{66}$$

Because α, φ–subsystem reach its stable state following Theorem 1, the right side of equation (65) bounded derives $\alpha_r, \dot{\alpha_r}$ in its left side is bounded, leading to boundary of $\alpha, \dot{\alpha}$. The right side of equation (66) is bounded leads to that the left side of (66) is global asymptotic stable with the Lyapunov function (67) and k_5, k_6 are chosen arbitrarily.

$$V_5 = \frac{1}{2} L_1 \dot{s}_e^2 + \int_0^{s_e} L_2 \frac{\tanh(s_e)}{p} d\sigma \tag{67}$$

$$\dot{V}_5 = L_1 \dot{s}_e \ddot{s}_e + \frac{L_2}{p} \dot{s}_e \tanh(s_e) \tag{68}$$

$$\dot{V}_5 = \frac{L_1}{p} \dot{s}_e \left(-\mu_3 \dot{s}_e - k_5 \tanh(\dot{s}_e) - k_6 \tanh(s_e) + f(\bullet)\right) + \frac{L_2}{p} \dot{s}_e \tanh(s_e) \tag{69}$$

$$\textit{Selecting} \quad L_1 k_5 = L_2 \tag{70}$$

$$\dot{V}_5 = -\frac{\mu_3 L_1}{p} \dot{s}_e^2 - \frac{k_5 L_1}{p} \dot{s}_e \tanh(\dot{s}_e) + \frac{L_1}{p} \dot{s}_e f(\bullet) \tag{71}$$

Because $f(\bullet)$—right side of (66) is bounded, there exist k_5 such that (71) negative with $|\dot{s}_e| \geq \varepsilon$, in that k_5 is chosen as (72) where $F \geq |f(\bullet)|$.

$$k_5 \geq \frac{F}{|\tanh(\varepsilon)|} \tag{72}$$

5 Simulation Results

Scenario simulation: The objective of this work is to guarantee $x, \dot{x}, \theta, \dot{\theta}$ to track given desired strategies $x_d, \dot{x}_d, \theta_d, \dot{\theta}_d$ with unbalanced initial state. The good performance of simulation result demonstrates the ability of the proposed control law. The parameters are chosen as [2] to simulate above-considered WIP system: $D = 0.15\,\text{m}$, $R = 0.25\,\text{m}$, $I_\omega = 1.5\,\text{kgm}^2$, $I_M = 2.5\,\text{kgm}^2$, $m = 7$ kg, $M_\omega = 1$ kg, $g = 9.8\,\text{m/s}^2$, $l = 1.2\,\text{m}$, $\mu_1 = 0.01$, $\mu_2 = 0.01$, $\mu_3 = 0.01$. The initial value is selected as: $\dot{\varphi}(0) = 0\,\text{rad/s}$, $\alpha(0) = 0\,\text{rad}$, $\dot{\alpha}(0) = 0\,\text{rad/s}$, $s(0) = 0\,\text{m}$, $\dot{s}(0) = 0\,\text{m/s}$. The coefficients of controller are set as: $\kappa_1 = 1$, $\kappa_2 = 1$, $\Gamma_1 = 1, \beta_1 = 1$, $k_1 = 15$, $k_2 = 15$, $\kappa_3 = 1$, $\kappa_4 = 1$, $\Gamma_2 = 1$, $\beta_2 = 1$, $k_3 = 80$, $k_4 = 40$, $k_5 = -600$, $k_6 = -150$, $d_1 = rand(\bullet) + 20\sin(\omega t)$, $d_2 = rand(\bullet) + 20\sin(\omega t + \pi/2)$.

This system is able to self-balance with tilt angle and straight velocity, heading angle tracking their reference signals under bounded disturbances after about twenty seconds. Control forces are continuous and bounded (Fig. 1).

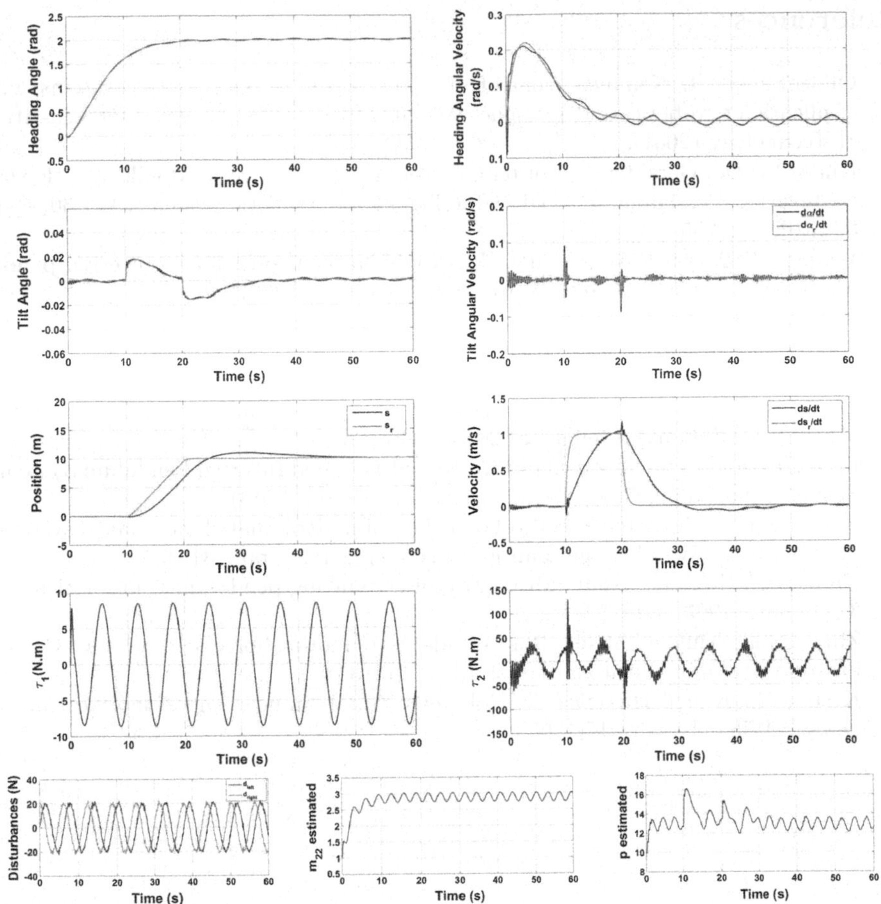

Fig. 1. Simulation results

6 Conclusion

The proposed method in this paper including two new adaptive laws and two new controllers solves a problem of time-dependent inertia matrix in CLF in previous papers as [1]. Moreover, the considered WIP system can balance when the massive and central point of load standing is unknown. Additionally, effects depending on time from the environment such as wind, human, and contact surface impact on WIP system, these situations are appropriate reality, but the considered system is able to stable automatically without manual adjust.

References

1. Olfati-Saber, R. Nonlinear control of under actuated mechanical systems with application to robotics and aerospace vehicles. PhD. thesis, Massachusetts Institute of Technology (2001).
2. Khac Duc Do, Gerald Seet Motion Control of a Two-Wheeled Mobile Vehicle with an Inverted Pendulum, Journal of Intelligent and Robotic Systems, Vol 60, Issue 3, 2010, 577–605.
3. Wei, Q., Dayawansa, W., Levine, W.: Nonlinear controller for an inverted pendulum having restricted travel. Automatica, 841–850 (1995).
4. R. Culi, Ji Guo, Zhaoyong Mao Adaptive backstepping control of wheeled inverted pendulums models Nonlinear Dynamic, 2015. Chung, C., Hauser, J. Nonlinear control of a swinging pendulum. Automatica, Vol 31, Issue 6, 1995, 851–862.
5. Zhijun Li, Yunong Zhang Robust adaptive motrion/force control for wheeled inverted pendulums Automatica, 2010, 1346–1353.
6. Li, Z., Yang, C., Fan, L. Advanced Control of Wheeled Inverted Pendulum Systems, Springer Publishing Company, Incorporated, Berlin (2013).
7. Mark Spong: Energy based control of a class of underactuated mechanical systems, in Proc. IFAC World Congr.,San Francisco, CA, 1996, pp. 431–435.
8. Chung, C., Hauser, J.Nonlinear control of a swinging pendulum. Automatica, Vol 31, Issue 6, 1995, 851–862.
9. Zhijun Lia, Yunong Zhang Robust adaptive motion/force control for wheeled inverted pendulums Automatica, 2010, 1346–1353.
10. Krstic, M., Kanellakopoulos, I., Kokotovic, P.: Nonlinear and Adaptive Control Design. Wiley, New York (1995).

Bandwidth Enhancement of Microstrip Patch Antenna Array Using Spiral Split Ring Resonator

Chirag Arora[1(✉)], Shyam S. Pattnaik[2], and R. N. Baral[3]

[1] Krishna Institute of Engineering & Technology, Ghaziabad, UP, India
c_arora2002@yahoo.co.in

[2] National Institute of Technical Teachers' Training and Research, Chandigarh, India
shyampattnaik@yahoo.com

[3] IMS Engineering College, Ghaziabad, UP, India
r.n.baral@gmail.com

Abstract. This communication presents a spiral split ring resonator (SSRR) loaded four-element microstrip patch antenna array for 5.8 GHz WiMAX applications. The unloaded antenna array resonates at 5.8 GHz with gain of 7.15 dBi and bandwidth of 480 MHz. Whereas, when the patches of this array are loaded with SSRR, the bandwidth increases to 700 MHz with almost no variation in gain. The proposed antenna array has been designed on 1.48 mm thick FR-4 substrate and simulated in FEM-based HFSS commercial electromagnetic simulator.

Keywords: Antenna array · Metamaterial · Permeability · WiMAX

1 Introduction

Microstrip patch antennas are among the most common antenna types that are used in present era, particularly in the frequency range of 1–6 GHz, as they have compact size, planar structure, and low cost. But they possess low gain and narrow bandwidth due to various losses like surface wave losses, dielectric losses, conductor losses [1]. Several approaches have been used to reduce these losses, such as use of high impedance surfaces [2], use of photonic band gap structures [3], use of hybrid substrates [4], use of superstrates [5]. But use of all these techniques makes the antenna bulky and difficult to fabricate.

After the realization of left-handed metamaterials by Smith et al. [6], researchers have tried various techniques to use their novel properties for performance enhancement of patch antennas, such as loading the traditional antennas with the single negative or double negative metamaterials in various ways such as loading the patch with split ring resonators (SRR) [7, 8], using metamaterial superstrates [9], etching metamaterial slots on ground plane of the substrates [10]. Integration of metamaterials with conventional antennas has many advantages as these materials not only enhance gain and bandwidth of patch antennas, but also produce subwavelength resonance [11],

V. Bhateja et al. (eds.), *Information Systems Design and Intelligent Applications*, Advances in Intelligent Systems and Computing 672,
https://doi.org/10.1007/978-981-10-7512-4_43

resulting in size reduction of conventional antennas while retaining the two-dimensional planar structure of the antenna. But in some applications, like radars, where the gain and bandwidth improvement obtained by these techniques with single patch element is still insufficient, the arrays of patch antennas are used, which can further be loaded with various types of metamaterials to obtain the desired results. As per the best knowledge of the authors, till now very less attention has been paid toward the metamaterial loading of arrays. Therefore, to take the initiative, in [12–14] different methods have been adopted by the authors to incorporate metamaterial with patch antenna arrays for improvement of antenna performance parameters.

Further in this paper, the authors have proposed four-element microstrip patch antenna array and then loaded it with two SSRR. This loading is done in such a way that the proposed antenna still remains planar and its overall size also remains unchanged. This communication is methodically divided into four segments. The design of the traditional and proposed antenna array is discussed in second section; the third section gives the simulated results of designed array, and finally, the article is ended with conclusions in fourth section.

2 Antenna Array Design

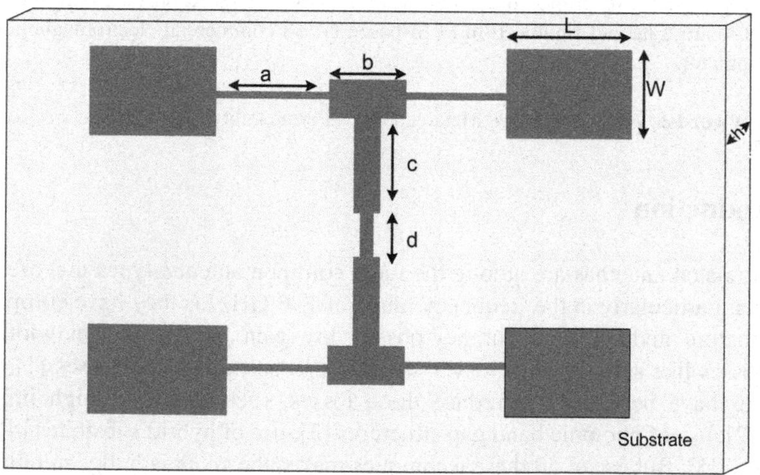

Fig. 1. Geometric outline of the traditional patch antenna array

Figure 1 depicts top view of a corporate resonator fed traditional microstrip patch antenna array. This conventional antenna array is composed of four microstrip patches, each of dimensions W = 11.9 mm × L = 15.6 mm. The two patches are connected to each other with the help of quarter wave transformers of dimensions b = 7.55 mm and a = 5.52 mm. The dimensions of patch and quarter wave transformer are calculated using transmission line equations [15–17]. FR-4 substrate of thickness h = 1.48 mm, loss tangent = 0.01, and dielectric constant of 4.3 is used to design this array, and its feeding is done by a 50 ohm connector. Figure 2 shows the proposed antenna array in which two SSRR are loaded to the conventional antenna array, in such a way that the size of proposed antenna array remains exactly same. Figure 3 presents the geometric sketch of this SSRR, used for loading the conventional array, and its dimensions are presented in Table 1. This loading of the antenna array has been done in such a way that two-dimensional planarity of the array is retained. The simulations are done using FEM-based commercial electromagnetic software Ansys HFSS.

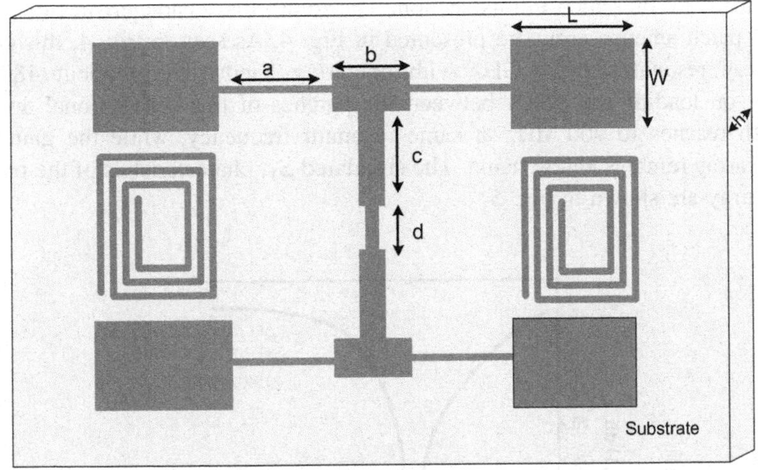

Fig. 2. Proposed metamaterial loaded microstrip patch antenna array

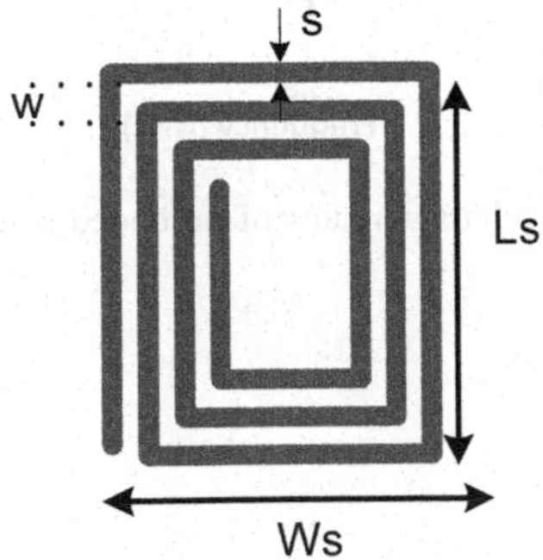

Fig. 3. Geometry of metamaterial unit cell

Table 1. Dimensions of SSRR

Parameters of SSRR	Dimensions (mm)
Length of outer ring (Ls)	15
Width of outer ring (Ws)	15
Width of ring (w)	1
Gap between two rings (s)	2

3 Simulated Results and Analysis

The simulated results of the conventional and designed metamaterial loaded patch antenna array are presented in this section. The return loss characteristics of the conventional patch antenna array are presented in Fig. 4. As seen in Fig. 4, this conventional array resonates at 5.8 GHz with operating bandwidth of about 480 MHz. However, on loading the SSRR between the patches of this conventional array, the bandwidth reaches to 700 MHz at same resonant frequency, while the gain of the proposed array remains almost same. The simulated S_{11} characteristics of the proposed antenna array are shown in Fig. 5.

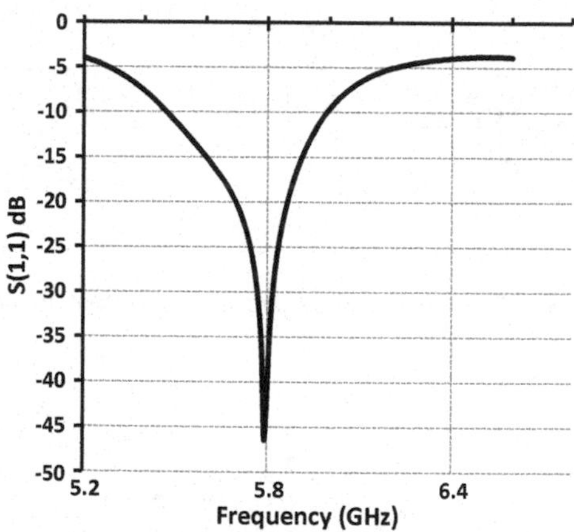

Fig. 4. S_{11} characteristics of unloaded antenna array

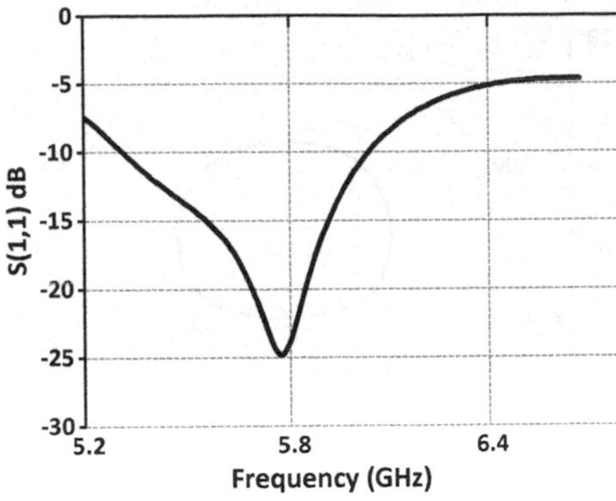

Fig. 5. S_{11} characteristics of SSRR loaded antenna array

The radiation patterns in the elevation plane of traditional and SSRR loaded antenna array are shown in Fig. 6 and Fig. 7, respectively. As seen from these figures, the traditional antenna array resonates at 5.8 GHz with gain of about 7.15 dBi, whereas the metamaterial loaded proposed antenna array also resonates at 5.8 GHz with gain of about 7.0 dBi, which is almost same as that of the unloaded antenna array. On loading the SSRR to the traditional patch antenna array, it interacts with the electric field and offers negative permeability at resonant frequency.

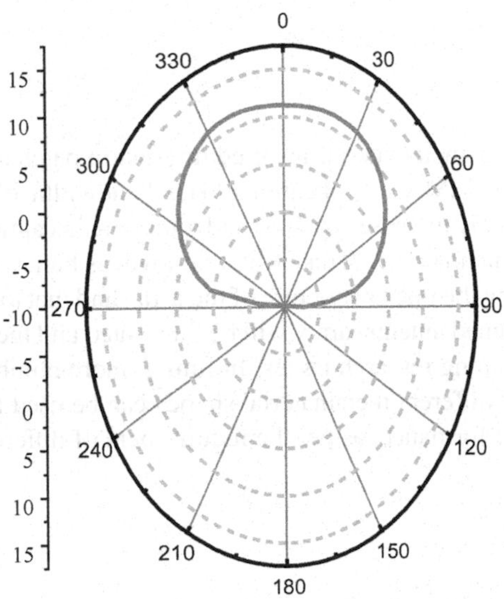

Fig. 6. Elevation plane radiation pattern of traditional antenna array at 5.8 GHz

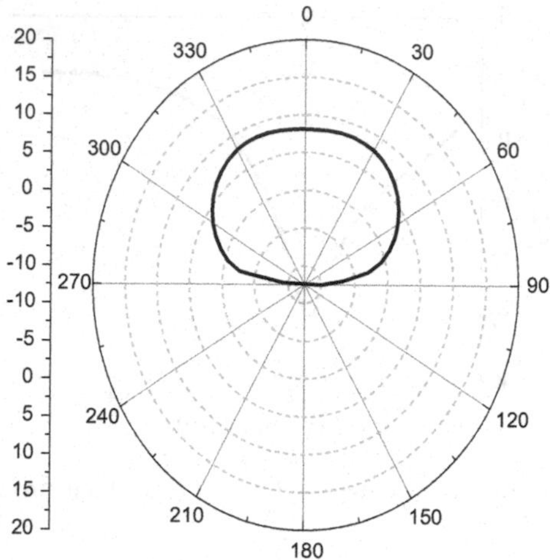

Fig. 7. Elevation plane radiation pattern of SSRR loaded antenna array at 5.8 GHz

By placing the SSRR in varying magnetic field, the time varying magnetic flux generated by the antenna induces current on SSRR. It results in large electric field across the gap capacitance at the splits and mutual capacitance between the split rings. This capacitance of SSRR is quite large to match with the inductance of the patch antenna array, which results in the reduction of mutual coupling between the patches of the array.

4 Conclusions

In this communication, a metamaterial-inspired microstrip patch antenna array has been proposed for 5.8 GHz WiMAX applications. The bandwidth of the proposed SSRR loaded patch antenna array increases by 46% and gain remains almost same, as that of the unloaded traditional antenna array. This design is novel as bandwidth improvement has occurred by maintaining the planar structure of the array and that too at no extra hardware cost and size. The designed antenna array will be fabricated and the measured results will be given in the forthcoming research issues. In future, more number of elements can be added to this array and different metamaterial shapes can be used for loading purpose to obtain the enhanced performance, as per the requirement of different applications.

References

1. Waterhouse, R.B.: Microstrip Patch Antennas: A Designer's Guide. MA: Kluwer Academic Publishers, Boston (2003).
2. Wang, C., Yan, D.B., Yuan, N.C.: Application of High Impedance Electromagnetic Surface to Archimedean Planner Spiral Antenna. Microwave Optical Technology Lett. 49(1), 129–131 (2007).

3. Li, Z., Xue, Y.L., Shen, T.: Investigation of Patch Antenna Based on Photonic Band-Gap Substrate with Heterostructures. Hindawi Publishing Corporation. 2012, 1–9 (2012).
4. Albino, A.R., Balanis, C.A.: Gain Enhancement in Microstrip Patch Antennas using Hybrid Substrates. IEEE Antennas Wireless Propag. Lett. 12, 476–479 (2013).
5. Attia, H., Yousefi, L.: High-Gain Patch Antennas Loaded With High Characteristic Impedance Superstrates. IEEE Antennas Wireless Propag. Lett. 10, 858–861(2011).
6. Smith, D.R., Padilla, W.J., Vier, D.C., Nasser, S.C.N., Schultz, S.: Composite Medium with Simultaneous Negative Permeability and Permittivity. Phy. Review Lett.; 84(18), 4184–4187 (2000).
7. Joshi, J.G., Pattnaik S.S., Devi, S.: Geo-textile Based Metamaterial Loaded Wearable Microstrip Patch Antenna. International J. Microwave Optical Technology. 8(1), 25–33 (2013).
8. Gao, X.J., Zhu, L., Cai, T.: Enhancement of Gain and Directivity for Microstrip Antenna using Negative Permeability Metamaterial. AEU International J. Electronics Communications. 70(7), 880–885 (2016).
9. Razi, Z.M., Rezaei, P., Valizade, A.: A Novel Design of Fabry- Perot Antenna using Metamaterial Superstrate for Gain and Bandwidth. AEU International J. Electronics Communications. 69(10), 1525–1532 (2015).
10. Pandeeseari, P., Raghavan, S.: Microstrip Antenna with Complementary Split Ring Resonator Loaded Ground Plane for Gain Enhancement. Microwave Optical Technology Lett. 57(2), 292–296 (2015).
11. Alu, A., Bilotti, F., Engheta, N., Vegni, L.: Subwavelength, Compact, Resonant Patch Antennas Loaded with Metamaterials. IEEE Trans. Antennas Propag. 55(1), 13–25 (2007).
12. Arora, C., Pattnaik, S.S., Baral, R.N.:SRR Inspired Microstrip Patch Antenna Array. J. of Prog. in Electromagnetic Res. C. 58 (10), 89–96 (2015).
13. Arora, C., Pattnaik, S.S., Baral, R.N.: Microstrip Patch Antenna Array with Metamaterial Ground Plane for Wi-MAX Applications. In: Proceedings of the Springer Second International Conference on Computer and Communication Technologies. AISC, vol. 381, pp. 665–671, Springer, India (2015).
14. Arora, C., Pattnaik, S.S., Baral, R.N.: Metamaterial Superstrate for Performance Enhancement of Microstrip Patch Antenna Array. In: 3rd IEEE International Conference on Signal Processing and Integrated Networks (SPIN-2016), pp. 775–779, IEEE Press, India (2016).
15. Garg, R., Bhartia, P., Bhal, I., Ittipiboon, A.: Microstrip Antenna Design Handbook, Artech House, UK (2001).
16. Balanis, C.A.: Modern Antenna Handbook John Wiley & Sons, New York, USA (2011).
17. Pozar, D.M.: Microwave Engineering, John Wiley & Sons, New York, USA (2008).

An Ensemble Classifier Approach on Different Feature Selection Methods for Intrusion Detection

H. P. Vinutha(✉) and B. Poornima

Bapuji Institute of Engineering & Technology, Davangere, Karnataka, India
{vinuprasad.hp, poornimateju}@gmail.com

Abstract. Knowing a day's monitoring and analyzing events of network for intrusion detection system is becoming a major task. Intrusion detection system (IDS) is an essential element to detect, identify, and track the attacks. Network attacks are divided into four classes like DoS, Probe, R2L, and U2R. In this paper, ensemble techniques like AdaBoost, Bagging, and Stacking are discussed which helps to build IDS. Ensemble technique is used by combining several machine learning algorithms. Selection of features is one of the important stages in intrusion detection model. Some feature selection methods like Cfs, Chi-square, SU, Gain Ratio, Info Gain, and OneR are used in this paper with suitable search technique to select the relevant features. The selected features are applied on AdaBoost, Bagging, and Stacking with J48 as a base classifier and along with that J48 and PART are used as single classifies. Finally, results are shown that the use of AdaBoost improves the classification accuracy. Experiments and evaluation of the approaches are performed in WEKA data mining tool by using benchmark dataset NSL-KDD '99.

Keywords: Intrusion detection system · Feature selection · Ensemble techniques · WEKA · Classification accuracy

1 Introduction

Intrusion detection is a process of monitoring and analyzing event of network traffic for the signs of intrusion. Big amount of data contained on the network day by day increases the intrusions. The prevention technologies like firewall and access controls are failed to protect networks and systems from the increase of complicated attacks. Intrusion detection system (IDS) is becoming an essential element to detect, identify, and track the attacks. IDS are able to scan the network activity to recognize the attacks. There are two different intrusion detection approaches called misuse detection system and anomaly detection system. In misuse detection, the attacks are determined on the basis of pattern that are based on the known intrusions. In anomaly detection, the attacks are determined on the basis of patterns that take the deviation from normal behavior of the system. To monitor the network, IDS has an alarm system; it generates an alarm to notify that the network is under attack. It can generate four different types of attacks like True positive when legitimate attack occurs, False positive when no attack occurs, False negative when actual attack occurs, and True negative when no

V. Bhateja et al. (eds.), *Information Systems Design and Intelligent Applications*, Advances in Intelligent Systems and Computing 672,
https://doi.org/10.1007/978-981-10-7512-4_44

attack occurs. So the main focus of the intrusion detection system is to increase the detection accuracy and minimize the false alarm rate.

Deploying an effective intrusion detection system is a challenging task, because dataset contains larger number of irrelevant features and redundant features. If IDS examines the entire data feature to detect intrusion, analysis becomes difficult because large number of features make it difficult to detect the suspicious behavior pattern. This reduces the learning performancc and computational efficiency. So, before applying any data mining techniques like classification, clustering, association rule, and regression on the dataset, it is necessary to reduce the dimensionality of the data. A preprocessing step called feature selection is used to reduce the dimensionality. Once features are selected the classification techniques are applied on the reduced dataset to increase the performance and efficiency. Instead of using single classifiers, an ensemble classifiers are used which combine multiple classifiers to improve the accuracy and performance.

In this paper, we have discussed the different feature selection algorithms and compared the results for number of selected features and number of features removed. Then different classifiers are applied on the reduced data for all the feature selection algorithms, and the results are compared to show the best classifier. The experiment is conducted on NSL-KDD'99 dataset which was developed by Massachusetts Institute of Technology (MIT) in 1999 which is an advanced version of KDDCUP'99. The dataset contains 125973 single connection records with no redundancy with 41 features and 5 classes; they are classified as normal and 4 category of attacks: Denial of service attack (DOS), Probing attack (Probe), Remote to Local attack (R2L), and User to Root attack (U2R) [1].

Section 2 explains the related work. Section 3 discusses about feature selection algorithms. Section 4 gives the details of ensemble classifiers. Proposed method, results, and discussions are explained in Sects. 5 and 6. Section 7 concludes the paper.

2 Related Works

K Umamaheshwari et al. in this paper [2] author has proposed classification techniques on KDD-99 dataset which is a model finding process that is used for portioning the data into different classes. They have evaluated the performance of a comprehensive set of classifier algorithms Random forest, Random tree, and j48, etc. WEKA is used to compare the performance, and finally, they have concluded that Random tree algorithms produce better accuracy.

Rajender Kaur et al. [3] propose a method to deal with large amount of features which represents the whole dataset. They have done some feature selection and machine learning approaches to design the intrusion detection systems which are going to classify the network traffic data into intrusive traffic and normal traffic. Estimation is done for seven classification algorithms like Bayes Net, Naïve Bayes, J48, Random Forest, OneR, PART, and Decision Tree for tenfold cross-validation on KDD-99 dataset. It is also recommended that rule-based J48, RandomForest, and OneR classifiers are used for the detection of various attack classes.

In papers [4–7], authors have proposed different methods to deal with ensemble techniques like AdaBoost, Bagging, and Stacking. These ensemble techniques are combined with different machine learning algorithms. The experiments showed that the better results are obtained for ensemble technique combined with other algorithms than the use of single machine learning algorithms. They have used KDD-99 dataset for an experimental purpose. The preprocessing step is done on the dataset using different feature selection algorithms.

In papers [8, 9], authors have discussed various feature selection algorithms applied on dataset to select the relevant features which is used to classify the accuracy of classifiers. There are different categories of feature selection techniques like Filter method, Wrapper method, and Embedded method. The feature selection algorithms like Cfs subset, InfoGain, Gain Ratio, Filtered Attribute, Randomized hill climbing, Genetic algorithms are discussed in the paper. They have analyzed those set of algorithms with the use of different search methods, and some of the relevant features are selected by removing the irrelevant features. The best selected features are applied on different classifiers to show the improvement of performance and accuracy.

3 Feature Selection

To build an intrusion detection model, feature selection is one of the most important steps. Network data contain large number of features, but it is not good practices of using all these features. Because in the network, it is necessary to reduce the processing time to achieve the higher detection rate and accuracy. Feature selection is one of the important data preprocessing techniques in data mining. Feature selection is also known as an attributes selection method. There are three feature selection methods like Filter method, Wrapper method, and Embedded method but Filter and Wrapper methods are the commonly used methods. In Filter approach, features are selected without depending on any classifier. In Wrapper method, features are selected with the dependent on classifier [10]. Comparing the filter method, wrapper method is more time consuming because it is strongly coupled with induction algorithm which repeatedly calls the subset of features to evaluate the performance.

3.1 Importance of Feature Selection

- To reduce the size of the problem.
- Removal of irrelevant features which improves the performance of learning algorithms.
- Reduction of features reduces the storage requirement.

In this paper, we have concentrated on six different feature selection methods like Correlation Attribute Evaluation (Cfs), Chi-squared, Symmetrical Uncertainty (SU), Gain Ratio, Information Gain, and OneR. The search techniques like Best-First search and Rankers method are used with feature selection algorithms to rank the features. The ranking denotes how useful the feature which is to be classified [10].

3.2 Correlation-Based Feature Selection (Cfs)

In this algorithm, filter method is used to select the attributes. Cfs measures the individual feature by heuristic approach. Maximum value obtained for correlated and irrelevant features is avoided. Equation 1 is used to calculate the irrelevant and redundant features [11].

$$F_s = \frac{N r_{ci}}{N + N(N-1) r_{jj}} \tag{1}$$

where N indicates the number of feature in the subset, r_{ci} says the mean feature correlation with the class, and r_{jj} means average feature inter-correlation.

3.3 Chi-Squared (X^2 Statistic)

Chi-square feature selection algorithm uses the filter method, and it calculates the need of independence between term and class for one degree of distribution freedom [12]. The expression is as Eq. 2 [12].

$$X^2_{(t,c)} = \frac{D * (PE - MQ)^2}{(P+M) * (Q+N) * (P+Q) * (M+N)}. \tag{2}$$

where D indicates the total number of documents, P says the number of documents of class C containing term t, Q means the number of documents containing t occurs without C, M indicates the number of documents class C occurs without t, N is the number of documents of others class without t [12].

3.4 Symmetrical Uncertainty (SU)

In this algorithm, set of attributes are calculated by measuring the correlation between feature and target class [13], and it is given in Eq. 3 [13].

$$SU = \frac{H(X) + H(Y) - H(X/Y)}{H(X) + H(Y)}. \tag{3}$$

where H(X) and H(Y) = entropies based on the probability associated with each feature and class value, respectively, and H(X, Y) = The joint probabilities of all combinations of values of X and Y [13].

3.5 Gain Ratio Attribute Evaluation

In this method, Gain Ratio is measured by evaluating the gain with respect to the split information. The equation is given in Eq. 4 [14].

$$\text{Gain Ratio}(A) = \text{Gain}(A)/\text{Split Info}(A) \tag{4}$$

3.6 Information Gain

In this, score is found based on how much maximum information is obtained about the classes when we use that feature. The Information Gain equation is shown in Eq. 5 [13],

$$IG(X) = H(Y) - H(Y|X) \tag{5}$$

where H(Y) and H(Y|X) say the entropy of Y and the conditional entropy of Y for given X, respectively [13].

3.7 One Rule (OneR)

This is a simple classification algorithm. Classification rule is very simple and accurate in this. One level decision tree is generated by this. The rule with the smallest error rate in the training data is selected for each attribute [15].

4 Ensemble Methods

In Ensemble method, the performance of classifier is improved by combining the multiple single classifiers. Compared to single classifier, ensemble techniques are more effective and efficient. Divide and conquer approach are used in ensemble methods [16]. In this method, complex problem is divided into small subproblems which are easy to analyze and solve. Advantage of this approach is that they can get more accuracy than single algorithm. Base model is used to classify the data. In this paper, we have evaluated three different ensemble classifier techniques called Boosting, Bagging, and Stacking are used with J48 and PART classifiers [16].

4.1 Bagging

Bagging is also known as Bootstrap Aggregation. It is the simple ensemble method used to improve unstable classification problem. Variance of a predictor is reduced by this method [17]. N number of training set are created by selecting one point of the training set without the replacement of N examples. N indicates the size of original training set. Each of these datasets is used to train a different model [18].

4.2 AdaBoost

To construct a strong classifier, AdaBoost algorithm is used which is one of the most widely used Boosting techniques. The performance of individual classifiers is constructed by AdaBoost classifier [19]. It improves the performance of weak classifier by its ensemble structure. In boosting method, a set of weights is maintained across the dataset. The objects acquire more weights to classify by forcing subsequent classifier to focus on them. These methods work well by running the learning algorithm repeatedly and then combining the classifier to produce the single classifier [19].

4.3 Stacking

Stacking is the method in which we combine various classifiers to increase the efficiency. The combination of classifier is done step by step where the output of first classifier is given as an input to the second classifier. In stacking, whole dataset is divided into n number of partitions. Out of these n numbers of partitions consider two disjoint sets to use it for the first classifier. If it is S_{ij} then, i denotes number of partitions and j denotes the two disjoint set. Stacking works in two stages. Stage1 is a base learner where dataset is used on various models. A new dataset is obtained and instances of that dataset are used for prediction purpose. In stage 2, it takes the new dataset as input and gives the final output [17].

5 Proposed Methodology

Figure 1 shows the general methodology used to get the best classifier on different feature selection methods for intrusion detection system. Firstly, classify the attack types know as Normal, DoS, Probe, U2R, and R2L and save the dataset into an ARFF file format. In this proposed work, performance is analyzed by using data mining tool called WEKA3.6. The NSL-KDD'99 dataset which contains 125973 labeled connection records is used to analyze the performance. Full dataset is applied on the attribute selection algorithms called Cfs, Chi-square, SU, Information Gain, Gain Ratio, and OneR to compute the feature selection and to evaluate the classification performance on each of these feature sets. We have selected three meta-classifiers called AdaBoost, Bagging, and Stacking, and one decision tree classifier called J48 and one rule-based classifier called PART are used with full training set and tenfold cross-validation for testing purpose.

The different parameters are used to analyze the result of classification model with True positive (TP) rate, False positive (FP) rate, Kappa statistics, ROC area, Classification Accuracy. The confusion matrix summarizes the number of instances calculated normal or abnormal by the classification model.

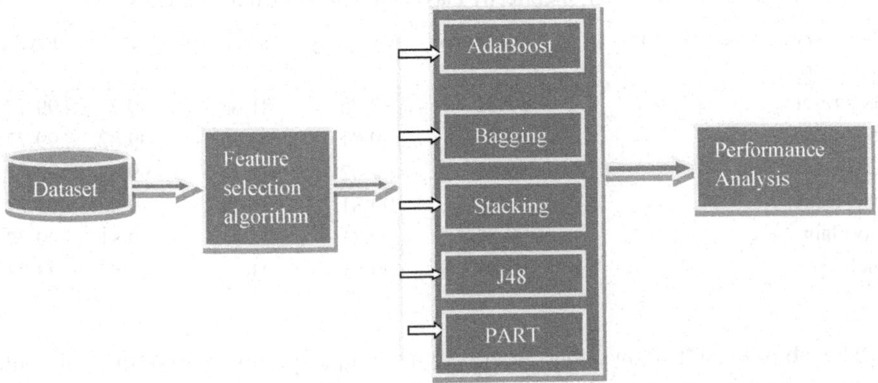

Fig. 1. Proposed methodology

6 Experimental Results and Analysis

In this experiment, we have analyzed various feature selection approaches with the help of different search methods; then the best subset of features is selected to perform on classifiers. The comparative analysis of each classifier for all the feature selection algorithms is given. The optimally selected subset of features is used on classifiers. Numbers of features removed are listed in Table 1, and they are observed to remove commonly selected features among all the approaches. They are considered separately and again to perform on classifiers.

Table 1. Number of features selected and list of features removed by different feature selection approaches

Feature selection approach	No. of features selected	List of features removed
Best-first + CfsSubsetEvl	11	1, 2, 7, 8, 9, 10, 11, 13, 15, 16, 17, 18, 19, 20, 21, 22, 23, 24, 26, 27, 28, 31, 32, 33, 34, 35, 36, 38, 40, 41, 42
Ranker + Chi-square	28	22, 14, 17, 13, 11, 18, 8, 16, 9, 19, 15, 7, 21, 20
Rankers + SU	27	22, 8, 13, 16, 17, 14, 19, 11, 15, 9, 7, 21, 20
Ranker + Gain Ratio	33	13, 40, 16, 19, 15, 24, 21, 7, 20
Ranker + Info Gain	30	13, 16, 17, 14, 11, 19, 18, 15, 9, 7, 21, 20
Ranker + OneR	30	7, 14, 20, 22, 19, 21, 9, 15, 18, 16, 17, 11

Number of features selected is used on three ensemble classifiers and two single classifiers. The ensemble classifiers are AdaBoost, Bagging, and Stacking. The single classifiers are J48 and PART. The percentage of correctly classified values is given in the Table 2.

Table 2. Comparison of correctly classified values for different classifiers

Classifiers/Feature selection approaches	AdaBoost	Bagging	Stacking	J48	PART
Cfs subset	**99.81**	99.75	81.66	99.77	99.77
Chi-square	**99.89**	99.81	93.39	99.82	99.77
SU	**99.89**	99.82	93.71	99.82	99.77
Gain Ratio	**99.89**	99.81	93.29	99.83	99.77
Info Gain	**99.89**	99.81	92.91	99.81	99.77
OneR	**99.89**	99.81	93.10	99.59	99.77

The above result shows that AdaBoost technique performs best than all other classifiers in accuracy rate. After observing the result of individual feature selection approaches, we have considered the number of features removed from the list. In those features, some of the common features are selected which have less importance in all the techniques and are removed to see the performance. Table 3 shows the commonly selected attributes to be removed. Table 4 shows the correctly classified values for commonly selected features removed.

Table 3. Number of commonly selected features to remove

List of features selected commonly to remove	22, 14, 17, 13, 11, 18, 16, 19, 15, 7, 21, 20, 9

Table 4. Comparative result of different classifiers for commonly selected features

Classifier	TP rate	FP rate	Precision	Recall	ROC	Accuracy
AdaBoost	**0.999**	**0.001**	**0.999**	**0.999**	**1**	**99.89**
Bagging	0.998	0.001	0.998	0.998	1	99.81
Stacking	0.938	0.027	0.933	0.938	0.955	93.76
J48	0.998	0.002	0.998	0.998	0.999	99.77
PART	0.998	0.001	0.998	0.998	0.999	99.82

In the above result of different classifiers for commonly selected features shows that AdaBoost classifier is best among five selected classifiers. Table 5 shows confusion matrix of Ensemble AdaBoost algorithm with total of each class type with accuracy. Accuracy is calculated as the sum of correct classification divided by the total number of classification.

Table 5. Confusion matrix of ensemble AdaBoost algorithm

	Normal	DoS	Probe	R2L	U2R	Total	Accuracy
Normal	67306	11	10	7	9	67343	99.94
DoS	11	45913	3	0	0	45927	99.96
Probe	33	2	11631	0	1	11667	99.69
R2L	20	0	0	954	3	977	97.64
U2R	22	0	0	1	36	59	61.01
Total	67392	45926	11644	962	40		
Accuracy	99.87	99.97	99.73	99.16	90.00		

7 Conclusion and Feature Work

In this research work, we have performed a set of experiment on classifiers at benchmark NSL-KDD'99 dataset contains 41 features. In order to remove irrelevant features from the larger dataset, the Cfs, Chi-square, SU, Gain Ratio, Info Gain, and

OneR feature selection algorithms are used. The performance of three Ensemble classifiers like AdaBoost, Bagging, and Stacking and two single classifiers like J48 and PART is compared using classification accuracy. By considering the removed features from all the algorithms, some of the common features are selected to remove and experiment is performed on the classifiers. Empirical result of experiment shows that AdaBoost classifier gives the better result. As a feature enhancement, Ensemble classifiers can be used with some other classifier as a base learning algorithm.

References

1. Vinutha H P and Poornima B: A Survey—Comparative Study on Intrusion Detection System, IJARCCE, Vol 4, Issue 7, July 2015, ISSN 2778-1021.
2. K. Umamaheswari and S. Janakiraman: Machine Learning in Networking Intrusion Detection System, ARPN journal of Engineering and Applied Sciences, Vol 11, No 2, January 2016, ISSN 1891-6608.
3. Rajender Kaur, Monika Sachdeva and Gulshan Kumar: An Empirical Analysis of Classification Approaches for Feature Selection in Intrusion Detection, IJARCSSE, Issue 9, Vol 6, September 2016, ISSN: 2277 128X.
4. Samah Osama M Kamel, Nadia H Hegazi, Hany M Harb, Adl Y S Tag El Dein, Hala Hala M Abd El Kader: AdaBoost Ensemble Learning Technique for Optimal Feature Subset Selection, IJCNCS, Vol 4, No 1, January 2016, 1–11.
5. Annkita Patel, Risha Tiwari: Bagging Ensemble Technique for Intrusion Detection System, International Journal for Technological Research in Engineering, Issue 4, Vol 2, December 2014, ISSN 2347-2718.
6. Riyad A M and M S Irfan Ahmed: An Ensemble Classification Approach for Intrusion Detection, Vol 80, No 2, October 2013.
7. Snehalata S Dongre and Kail K Wankhade: Intrusion Detection System Using New Ensemble Boosting Approach, International Journal of Modeling and Optimization, Vol 2, No. 4, August 2012.
8. S. Vanaja and K Ramesh Kumar: Analysis of Feature Selection Algorithms on Classification: A Survey, International Journal of Computer Application, Vol 96, No 17, June 2014.
9. Theyazn H Aldhyani, Manish R Joshi: Analysis of Dimensionality Reduction in Intrusion Detection, International Journal of Computational Intelligence and Informatics, Vol. 4, No. 3, October–December 2014.
10. Sheena, Krishan Kumar, Gulshan Kumar: Analysis of Feature selection Techniques: A Data Mining Approach, International Journal of Computer Applictions, ICAET 2016, IJCA2016 (1):17–21.
11. S Vanaja and K Ramesh Kumar: Analysis of Feature Selection Algorithms on Classification: A Survey, IJCA, Vol 96, No 17, June 2014.
12. Subhajit Dey Sarkar, Saptarsi Goswami: Empirical Study on Filter based Feature Selection Methods for Text Classification, International Journal of Computer Applications (0975 – 8887), Volume 81, No. 6, November 2013.
13. Zahra Karimi, Mohammad Mansour and Ali Harpunabadi: Feature Ranking in Intrusion Detection Dataset using Combination of Filtering Methods, IJCA, Vol 78, No 4, September 2013.
14. Megha Aggarwal, Amritha: Performance Analysis of Difference Feature Selection Method in Intrusion Detection: IJSTR, Vol 2, Issue 6, June 2013.

15. Krishan Kumar, Gulshan Kumar, Yogesh Kumar: Feature Selection Approach for Intrusion Detection System, International Journal of Advanced Trends in Computer Science and Engineering (IJATCSE), Vol. 2, No. 5, Pages: 47–53 (2013) Special Issue of ICCECT 2013.
16. Iwan Syaif, Ed Zaluska, Adam Pruge-Bennett, Gary Wills: Application of Bagging, Boosting and Stacking to Intrusion Detection, Vol 28, Issues 1–2, pp, 2012.
17. NilufarZaman D P, Gaikwad: Comparision of Stacking and SVM method for KDD dataset, International Journal of Engineering Research and General Science, Issue 3, Vol 3, part 2, May–June 2015.
18. Neeraj Bisht, Amir Ahmad and A K Pant: Analysis of Classifier Ensembles for Network Intrusion Detection Systems. CAE, Vol 6, No 7, February 2017.
19. Jasmina D. Novakovic and Alempije Velijovic: AdaBoost as classifier Ensemble in classification problems, INFOTEH-JAHORINA Vol. 13, March 2014.

Study of Image Segmentation Techniques on Microscopic Cell Images of Section of Rat Brain for Identification of Cell Body and Dendrite

Ashish Kumar, Pankaj Agham(✉), Ravi Shanker, and Mahua Bhattacharya

Department of Information Technology, ABV Indian Institute of Information Technology, Gwalior 474010, India
{ranavat1991,pankajagham1,rsmiet60,mahuabhatta}@gmail.com

Abstract. Present paper illustrates the study and comparison of different image segmentation techniques on microscopic cell images—a part of computerization for cell image analysis. The process of segmentation is highly required for analysis and to study the behavior of live cell structure. The error is less in the computerized system of cell image analysis as compared to the manual system. Region growing, region split and merging, FCM, k-mean, and hybrid clustering segmentation technique are used for comparison. Hybrid clustering gives better results than other techniques in terms of accuracy and time, while region spit and merging and FCM give poor results. For performance evaluation, some parameters are used.

1 Introduction

Analysis of fluorescent microscopic cell images is required to get information [1–3] of the characteristics of the cell and its behavior in different environments. The experiments show how the morphology and structure of cell change under varying conditions. Experiments on live cell images using manual process may be error prone and complex in an analysis. Hence, automation/semi-automation is the preferred technique to avoid the complexity and error during experiment and analysis [3, 4]. During the development of the automatic and computerized approach for microscopic cell image analysis, one of the most crucial and prior steps is cell image segmentation [4] to understand the biological process related to image application. Present paper concerns to understand the behavior of live cell characteristics/morphology under the different environmental condition where the rat brain cell images have been considered as a subject of the experiment. Segmentation [5–12] is the procedure of partitioning an image into distinct regions containing each pixel with similar color intensity. Segmentation procedure should be blocked when the region of data input has been divided. Basically, segmentation has two types: discontinuity based and similarity based. In discontinuity-based approach, image partition is carried out with some abrupt changes in intensity or gray level of image and edge-based segmentation technique belongs to

V. Bhateja et al. (eds.), *Information Systems Design and Intelligent Applications*, Advances in Intelligent Systems and Computing 672,
https://doi.org/10.1007/978-981-10-7512-4_45

this category. Similarity-based image segmentation is categorized as thresholding technique [5, 6], region-based technique, and clustering-based technique. It is very difficult to segment nontrivial image. Precise segmentation of medical images is very significant for the exploration and identification of anomalies in dissimilar parts of the body. It becomes important since it further helps in classification of the segmented regions as non-threatening or malicious. In region-growing approaches, we start from any pixel of an image and then form a group with the other pixels which are adjacent to that pixel and similar in their intensity values. The intensity value of all pixels in the same group is same. In split and merge an approach, the image is being split into a number of different components on some defined criteria. After splitting the image, the region is merged which is adjacent and similar in intensity value [13]. Hybrid Clustering is also used which is integration of two techniques that are FCM and k-mean.

The main objective of our work is to provide comparison between different segmentation techniques on microscopic cell image. Images used in this paper are cell images of rat brain for identification of cell body and dendrite. Segmentation techniques that are used in this paper are unsupervised in nature. Some parameters are used for performance analysis of different segmentation techniques. The arrangement of paper is as follows—Sect. 2 deliberates different image segmentation algorithms implemented on cell images, Sect. 3 provides experimental methodology and results of segmentation techniques on microscopic images of rat brain cell, Sect. 4 contains the performance evaluation of segmentation processes and discussion, finally Sect. 5 is the conclusion of the work.

2 Implementation of Segmentation Techniques on Microscopic Cell Images of Section of Rat Brain

In this section, different segmentation techniques are presented which are being implemented on microscopic cell images of a section of rat brain cell images.

2.1 Region-Based Segmentation Operation

The region in an image is defined as a group of pixels that are connected and having similar properties. Basically, region-based segmentation can be classified as (i) region growing (ii) region splitting and merging. A more comprehensive study on region-based segmentation can be found in [13].

2.1.1 Region Growing

Seed points are selected in region growing based on some predefined criteria. Region is growing around the seeds point. This method groups the pixels or subregion into larger region based on some criteria. This conditions or criteria can be anything. It may intensity, color, or texture. If the neighbor's pixel is similar to seeds point, then it is added in seeds point. When neighbouring's pixels are not similar to the seed points and do not satisfy the predefined criteria, then the processes of region growing stops. The explanation of algorithm is given below.

Algorithm for Region Growing:

Step1. Select pixel and seed point.
Step2. Calculate Euclidean distance d(p, p2) to measure pixel similarity (homogeneity) using Eq. (1).

$$d(x, y) = \sqrt{(x_2 - x_1)^2 + (y_2 - y_1)^2} \quad (1)$$

where (x_1, y_1) and (x_2, y_2) are pixel coordinate of p1, p2.
Step3. Choose homogeneity threshold (T) between the pixel p1 and p2.
(a) If d(p, p2) < T, then homogeneous.
(b) If d(p, p2) > T, then not homogeneous.
Step4. Merge the pixel if they are homogeneous otherwise do not merge them.
Step5. Repeat this process until there are no more changes in the threshold.

2.1.2 Region Splitting and Merging Technique

In split and merge technique, splitting and merging of images are taken place based on some criteria. These criteria may be intensity, color, texture, or anything. Splitting of image is taken place when intensity of whole image is not uniform. If the whole image is uniform, then do nothing. If the intensity of the whole region is similar, then do nothing, else further partition the region into quadratic regions. Repeat this process until partition can be uniform or smallest partition size permissible is achieved. If the partition is completed then check the adjacent partition. If they are similar, then merge them together to form bigger set. An extended explanation of both the algorithm is illustrated in details below [14].

Algorithm for Region Splitting

Step1. First step is the initialization process where the segmented image is considered.
Step2. Next step is to split the image into four parts and check predicate for all four parts. Rule for predicate is given below.

$$\mathrm{P(r)} = \begin{cases} true & if\ H(r) \in D \\ false\ otherwise \end{cases}$$

where H is evaluating function that evaluates the homogeneity of a region r and D is range of values for H.
Step3. Continue the process until all four regions satisfy the above condition.

Algorithm for Region Merging.

Step1. First step is the initialization process. In this process, every pixel is given a unique label.
Step2. Next step is to check whether predicate condition is true for the region so f interest or not. If it is true for any two neighboring regions, then both these neighbors are given same label.
Step3. The process is continued until all regions are merged.

The split and merge procedure

Step1. It splits the region into four mutually exclusive quadrants any region R_i for which $P(R_i)$ = FALSE.

Step2. Merge those regions R_j and R_k which are adjacent to each other and satisfying the condition $P(R_j \text{ U } R_k)$ = TRUE. (the quadtree structure may not be preserved).

Step3. Repeat step 1 and step 2, till convergence.

2.2 Clustering-Based Segmentation Technique

Clustering is process of partitioning a data into number of clusters such that data in same group are similar than other group.

2.2.1 K-Mean

K-mean clustering divides the data into k number cluster. It classifies data into k number of disjoint data. It contains two phase. In the first phase, it calculates the centroid and in the second phase, it takes each point of the cluster which has the nearest centroid from respected data point [15]. K-Means is an iterative approach that classifies the input data into many modules based on their expected distance from each other [16]. Mathematically, it can be represented as sum of squared error given in Eq. (2).

$$\mathrm{j} = \sum_{i=1}^{k} \sum_{x_j \in w_j}^{k} \left\| x_j - v_i \right\| \tag{2}$$

where v_i = centroid of cluster i.

Algorithm for k-mean:

Let x be the the set of data point and v the set of center.

Step1. Select no of clusters randomly.

Step2. For each pixel, compute its distance from each cluster center.

Step3. Allocate pixel to cluster whose distance is minimum to it.

Step4. Reallocate new center by averaging the pixel in the cluster using Eq. (3).

$$v_i = \frac{1}{c_i} \sum_{j=1}^{c_i} x_i \tag{3}$$

where c_i represents the number of data points in *i*th cluster.

Step5. Repeat steps 2 and 3 until divergence is attained.

2.2.2 Fuzzy C-Means Clustering (FCM)

FCM is an iterative clustering approach [17–19]. FCM works on similarity-based segmentation operation. This method gives an optimal c partition by minimizing the weighted within group sum of squared error objective function J FCM using Eq. (4).

$$j = \sum_{m=1} \sum_{i=1} u_{ij}^{q} d(x_k, v_{ij})^2 \tag{4}$$

Algorithm for FCM:-
Let x be the set of data point and v the set so f center.

Step1. Select randomly c cluster center.
Step2. Compute degree of membership where1 $\leq$ i $\leq$ n and 1 $\leq$ j $\leq$ n as in Eq. (5).

$$u_{ij} = \frac{1}{\sum_{m=1}^{c} \frac{d_{ij}}{d_{im}\left(\frac{2}{k-1}\right)}} \quad where\ d_{ij} = \left|\left|x_j - v_i\right|\right| \tag{5}$$

Step3. Compute center for j = 1, 2, 3, 4… using Eq. (6).

$$v_{ij} = \frac{\sum_{i=1}^{n} (u_{ij})^m x_i}{\sum_{i=1}^{n} (u_{ij})^m} \tag{6}$$

where j = 1, 2, 3, 4…, c.
Step4. Go to step3 until the convergence is not satisfied.

2.2.3 Hybrid Clustering

Hybrid clustering is a clustering segmentation technique which is used to partition the data into number of cluster. It is combination of FCM and k-mean clustering technique. Hybrid clustering techniques get advantage of k-mean in term of computational time and get advantage from FCM in term of accuracy. The main idea of integration of FCM and k-mean is to minimize the iteration of FCM by initializing the right cluster to the FCM clustering technique, due to this execution time of algorithm is reduced and gives better results. Due to initializing, right cluster to FCM hybrid clustering gives better result in less execution time in comparison of FCM technique. In the hybrid clustering initializing the number of clusters k, number of iteration and converges condition, so that the clustering process stops. The center of clustering is calculated by the given Eq. (7).

$$\mu = (1\text{:no of cluster}) * \text{m}/(\text{no of cluster} + 1) \tag{7}$$

where μ is the initial mean and m is defined by given Eq. (8).

$$\text{m} = \max(\text{image intensity}) + 1. \tag{8}$$

Flowchart of hybrid clustering technique is shown in the Fig. 1. This technique starts with initializing the number of clusters (k), maximum iteration, and minimum iteration. Then compute cluster centre for each cluster using Eq. (7). After that, assign the pixel to its nearest cluster based on minimum distance. Repeat the process till the condition of clustering converges.

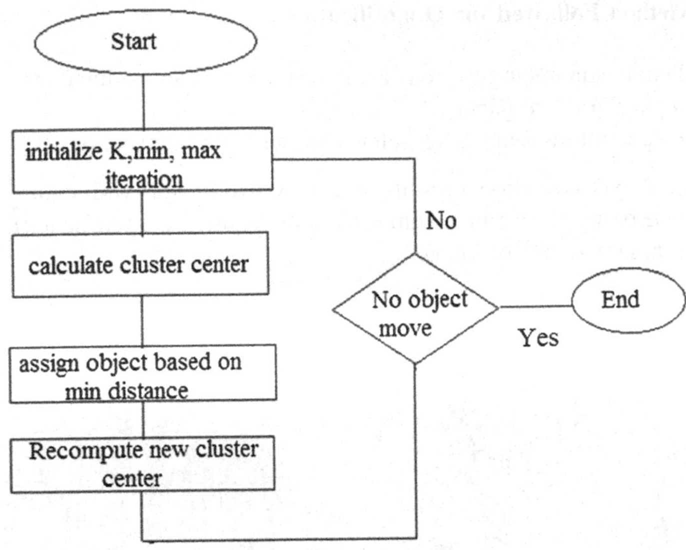

Fig. 1. Flowchart of hybrid clustering technique

3 Experimental Methodology and Results of Segmentation Techniques

In this section, we described the techniques of data preparation and image acquisition for the behavioral and morphological study of rat brain cell. The implementation of the proposed algorithm is shown in the following section.

3.1 Preparation of Data and Image Acquisition

In present work, microscopic images of rat brain cells have been used under control environment (sham control). Some protective agents are used to stimulate more dendrites. The objective is to explore any morphological and behavioral changes (in the cell body and in dendrite/hair-like branches) under the varying condition of the environment. Present work is related to the methodology development for the computerized study of change in cell structure. A typical neuron is with a lot of dendritic branches with a cell body. These dendrites move in all directions in the brain. However, sections are taken for acquisition of images in one plane. We need to measure the length of these branches and then to compare with the images under any environmental change. Moreover, the change in neuron count may also take place with environmental change. However, the most accurate method of counting would be taking images of this stained brain in 10–15 planes by Z-movement of the microscope and merging all the images.

The data preparation and the image acquisition are done by the team of School of Life Sciences, Manipal University, as collaborative partner of this study.

3.2 The Method Followed for Quantifications

Step1: Camera lucida tracing neurons are traced using a camera lucida tracing device on a plain A4 size paper.
Step2: Scoring dendritic branching points and dendritic intersections.

In the Fig. 2, we have shown results on two sets of images with control stage and after using some protective agent to stimulate more dendrites (experimental). The figure is describing these two sets of images.

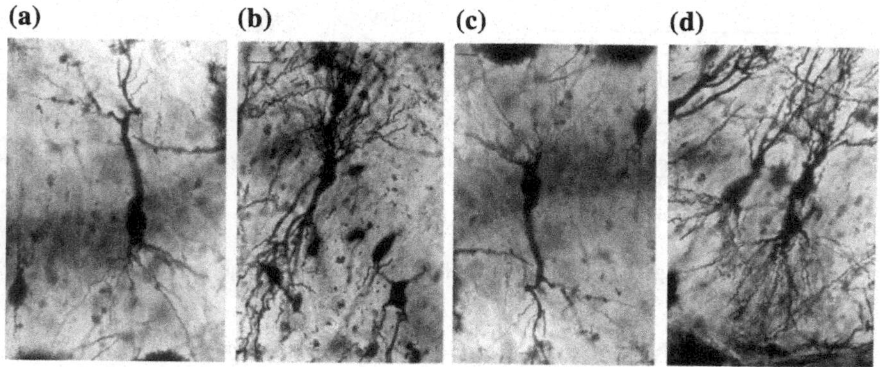

Fig. 2. **a** Control-1 **b** Experimental-1, **c** Control-2, and **d** Experimental-2

3.3 Results on Implementation of Different Image Segmentation Techniques on Microscopic Images of Rat Brain Cell

Results of different segmentation techniques are presented in Table 1 for microscopic images of rat brain cell with normal and experimental data. In this paper, several unsupervised segmentation techniques are used to segment the cell images. Four rat brain cell images are used for evaluation. These images are processed through all segmentation technique in which FCM and region split and merging give poor results to segment the images and k-mean, region growing, and hybrid clustering give a better result as compared to other techniques. But overall, hybrid clustering gives better results in terms of accuracy and time complexity, because it takes advantage of both FCM and k-mean. FCM does not give good results because initial cluster centers are chosen randomly and sometimes initial chosen cluster centers may be wrong.

In the Fig. 3, the segmentation results obtained from the algorithm: (a) Region growing, (b) k-mean, (c) FCM, (d) split and merge and (e) hybrid clustering are shown. Hybrid clustering gives good result because initial cluster centers are not chosen randomly, and due to this, the number of iterations is less. The evaluation matrices table is provided below in Table 1.

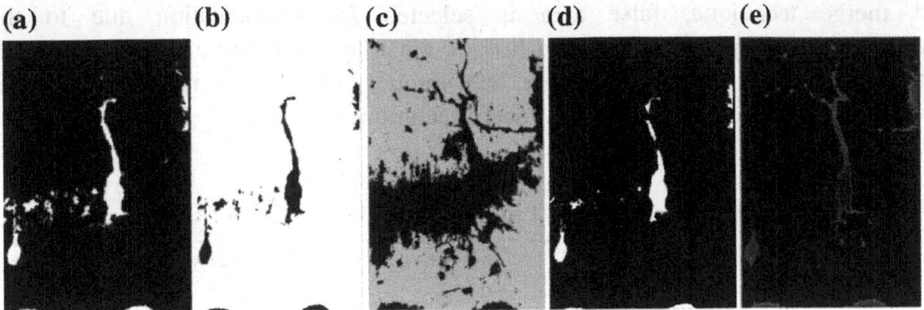

Fig. 3. **a** Region growing, **b** k-mean, **c** FCM, **d** split and merge, **e** hybrid clustering

4 Performance Evaluation of Segmentataion Techniques and Discussion

Various parameters are used to compute for performance evaluation. These are accuracy, sensitivity, specificity, peak signal to noise ratio (PSNR), border error, mean square error (MSE), and time. These parameters can be calculated based on ground true of the image.

True Positive: select those pixels which are used for segmentation.

Accuracy can be calculated by Eq. (9).

$$\text{accuracy} = \frac{\text{tp} + \text{tn}}{\text{tp} + \text{tn} + \text{fp} + \text{fn}} \tag{9}$$

Sensitivity can be calculated by Eq. (10).

$$\text{sensitivity} = \frac{\text{tp}}{\text{tp} + \text{fn}} \tag{10}$$

Specificity can be calculated by Eq. (11).

$$\text{specificity} = \frac{\text{tn}}{\text{tn} + \text{fp}} \tag{11}$$

There are five segmentation techniques used for comparison in present work. For each technique, four parameters are computed. Comparison of segmentation technique occurs based on ground truth. In this paper, we can use two sets of images which are control and experimental images. From the original image, ground truth image is prepared by selecting threshold value. For calculation of accuracy, sensitivity and specificity number of TP, TN, FP, and FN pixels are used.

Due to complex nature of rat brain cell image, segmentation technique is over-decentralized. Due to over-centralization of segmentation, some technique gives over-segmentation. This problem occurs when false area is selected for segmentation. But in hybrid clustering, these false areas are not selected. In fuzzy c-mean and split and merge technique, false area is selected for segmentation due to this over-segmentation problem occurs in fuzzy c-mean and split and merge technique. In k-mean and region growing technique, less number of false areas is selected as compared to fuzzy c-mean and split and merge technique.

Table 1. Performance evaluation

Technique/metrics	Accuracy	Sensitivity	Specificity	Time
RG	95.81	91.66	96.84	6.0
RSM	90.87	88.94	92.81	4.5
FCM	75.65	64.53	86.43	5.7
KM	93.10	92.81	95.42	3.9
HC	96.23	92.48	97.39	2.73

Fuzzy c-mean also take more execution time because in fuzzy c-mean cluster center are chosen randomly. It also takes more iteration as compared to hybrid clustering technique. In hybrid clustering technique, selection of cluster center is manual.

It is shown in Table 1 that hybrid clustering gives better result as compared to other technique. KM and hybrid clustering have good statistics in all the parameters, and they segment the dendrite part of all images with better accuracy as compared with other segmentation techniques. The FCM does not give better results if the background region has variation in intensity (Fig. 4).

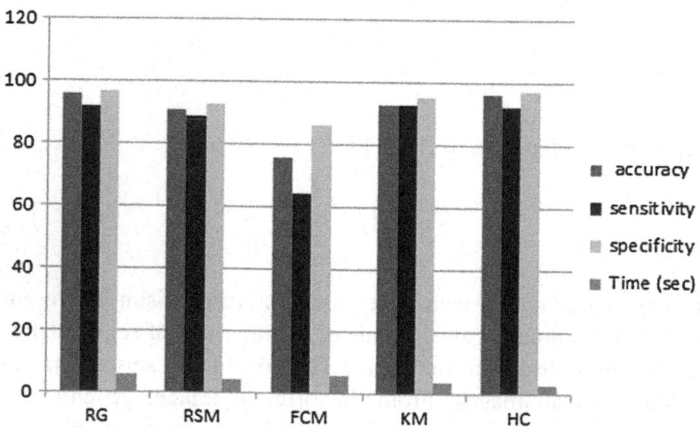

Fig. 4. Bar chart of performance evaluation

5 Conclusion

Present work evaluates comparison among different image segmentation techniques on microscopic cell images of a section of rat brain image with control environment. This is part of the computerize study of (specific for dendrite cell images) cell image analysis, quantification, identification of cell body, and dendrites. It has been observed that all algorithms do not provide same output for the same image. Therefore, the user has to choose algorithms specific to the input images. This work can be useful for those who work on segmentation of microscopic cell images.

Acknowledgements. This work is supported by Department of Science & Technology and Department of Telecommunication, Government of India, for morphological and behavioral study of cell under different environmental condition in collaboration with School of Life Sciences, Manipal University, Karnataka, India.

References

1. Peng, H.: Bioimage informatics: a new area of engineering biology. Bioinformatics, (2008), 1827–1836.
2. Meijering, E., Dzyubachyk, O., Smal, I., & van Cappellen, W. A.: Tracking in cell and developmental biology, Seminars in cell & developmental biology Vol. 20, (2009) 894–902.
3. Kaur, S., & Sahambi, J. S.: Curvelet initialized level set cell segmentation for touching cells in low contrast images, Computerized Medical Imaging and Graphics, (2016), 46–57.
4. Meijering, E.: Cell segmentation: 50 years down the road [life sciences]. IEEE Signal Processing Magazine, (2012), 140–145.
5. Senthilkumaran, N., and S. Vaithegi.: IMAGE SEGMENTATION BY USING THRESHOLDING TECHNIQUES FOR MEDICAL IMAGES. NO. 3, SEPTEMBER (2014).
6. Liu, L., & Sclaroff, S.: Region segmentation via deformable model-guided split and merge, In Computer Vision, 2001. ICCV, Proceedings. Eighth IEEE International Conference on (2001), Vol. 1, pp. 98–104.
7. John, A. J., & James, M. M. A,: New Tri Class Otsu Segmentation With K-Means Clustering In Brain Image Segmentation.
8. Pal, N.R. and Pal, S.K.: A review on image segmentation techniques. Pattern recognition, (1993), 1277–1294.
9. Sadri, A.R., Zekri, M., Sadri, S., Gheissari, N., Mokhtari, M. and Kolahdouzan, F.: Segmentation of dermoscopy images using wavelet networks, IEEE Transactions on Biomedical Engineering, 1134–1141.
10. Gajanayake, G. M. N. R., Roshan Dharshana Yapa, and B. Hewawithana: Comparison of standard image segmentation methods for segmentation of brain tumors from 2D MR images, Industrial and Information Systems (ICIIS), International Conference on. IEEE, (2009).
11. Pham, D. L., Xu, C., & Prince, J. L.: Current methods in medical image segmentation 1. Annual review of biomedical engineering, (2000), 315–337.
12. Kim, J. B., & Kim, H. J.: Multiresolution-based watersheds for efficient image segmentation. Pattern recognition letters, (2003), 473–488.

13. Yu, P., Qin, A. K., & Clausi, D. A.: Unsupervised polarimetric SAR image segmentation and classification using region growing with edge penalty, IEEE Transactions on Geoscience and Remote Sensing, (2012), 1302–1317.
14. Aneja, K., Laguzet, F., Lacassagne, L., & Merigot, A.: Video-rate image segmentation by means of region splitting and merging, In Signal and Image Processing Applications (ICSIPA), 2009 IEEE International Conference on (2009), (pp. 437–442). IEEE.
15. Andrew Moore: "K-means and Hierarchical Clustering– Tutorial Slides" http://www2.cs.cmu.edu/~awm/tutorials/kmeans.html.
16. Brian T. Luke: "K-Means Clustering", http://fconyx.ncifcrf.gov/~lukeb/kmeans.html.
17. Karmakar, G. C., Dooley, L. S., & Rahman, S.: A survey of fuzzy rule-based image segmentation techniques, (2000).
18. Y. Gefeng, O. Xu and L. Zhishenh, "Fuzzy Clustering Applications in Medical Image Segmentation", ICCSE-11, (2011), 826–829.
19. Bezdek, James C. Pattern recognition with fuzzy objective function algorithms. Springer Science & Business Media, (2013).

Performance Evaluation of Machine Learning and Deep Learning Techniques for Sentiment Analysis

Anushka Mehta, Yash Parekh(✉), and Sunil Karamchandani

Electronics and Telecommunication Department, Dwarkadas J. Sanghvi College of Engineering, Vile-Parle (W), Mumbai 400056, India
{anushka96mehta, yparekh26}@gmail.com,
skaramchandani@rediffmail.com

Abstract. Since the proliferation of opinion-based web content, sentiment analysis as an application of natural language processing has attracted the attention of researchers in the past few years. Lot of development has been brought in this domain that has facilitated in achieving optimal classification of text data. In this paper, we experimented with the widely used traditional classifiers and deep neural networks along with their hybrid combinations to optimize relevant parameters so as to obtain the best possible classification accuracy. We conducted our experiments on labeled movie review corpus and have presented relevant results and comparisons.

1 Introduction

Sentiment suggests an opinion that reflects one's feelings. Since the past decade, sentiment analysis and classification as an application of natural language processing have gained a lot of attention because of the emergence of user reviews and opinions found online. This can be attributed to the explosion of ecommerce websites, social media, blogs that has enabled people to voice their opinions through varied internet platforms. Traditionally, the sentiment classification engines were designed for applications such as topically categorizing documents, labeling the data with the underlying sentiment for obtaining a crisp summary of the text and so on [1]. Today, sentiment classification is not just limited to this domain and is being leveraged by political parties, tech giants, trading firms to better understand the ongoing trends, demands and obtain feedback from public opinion [2]. Sentiment analysis is not just limited to detection of polar words (negative and positive sentiments) in a document but also to understand its inherent contextual features.

Deep learning algorithms have shown impressive performance in natural language processing applications [3]. These models do provide a lot of advantages in terms of word representations, intuitive learning of features, and so on. Subsequent sections provide more information on the deep learning paradigm.

In general, sentiment analysis consists of tokenization of word inputs, suitable feature extraction, and classification using different models [4]. In our experiments, we have performed sentiment analysis using the traditional SVM classifier and several

V. Bhateja et al. (eds.), *Information Systems Design and Intelligent Applications*, Advances in Intelligent Systems and Computing 672,
https://doi.org/10.1007/978-981-10-7512-4_46

deep neural networks on a large set of movie reviews. For assessing the performance of deep leaning algorithms, the investigation of traditional method provides a credible baseline. We also combined the basic models with n-gram language model.

2 Resources

2.1 Labeled dataset for binary classification

We obtained our dataset from [5] which is an open dataset, now publicly available for download from the website. The labeled data consists of 50,000 movie reviews scraped from the universal review platform, *Internet Movie Database (IMDb).* The data was split into training and test sets each containing 25,000 unique reviews.

3 Methodology

3.1 Word Vectorization and Language Models

Word representations are central to almost all NLP applications. The words taken as input to the NLP systems are commonly represented as indices or vectors and mapped in a high-dimensional vector space based on efficient syntactic and semantic analysis of the given text documents. These real-valued word vectors are used as features to understand the lexicon. The words occurring in our dataset have been indexed according to their order of frequencies. For example, the index 1 represents the most commonly occurring word in the corpus. On parsing the sentence and indexing the words, we selected top 5000 words instead of the entire corpus for training all our networks.

Various neural network language models (NNLM) have been proposed over the years that employ distributed representation of features [6–8]. These models effectively outperform and also solve the problems with high dimensionality pertinent to most statistical models such as *interpolated n-grams* models [6].

We have coupled *n-gram (unigram and bigram) models* with our networks and observed subsequent changes in the accuracy levels. N-gram refers to a contiguous sequence of n words. This may also facilitate of most frequent n-grams instead of all the words in the corpora for efficient classification. It calculates probability of the nth word, given the occurrences of previous (n − 1) words.

3.2 Classical Machine Learning Techniques

Traditional classification methods that are widely used include Bayesian classifier such as Naïve Bayes [9], maximum entropy classification [10], logistic regression, and support vector machines (SVM). For these algorithms, standard bag-of-features approach is used that uses a set of ‘m’ predefined features occurring in the document d and assigns a class or category c to the document.

Support vector machines (SVM), a supervised classifier was introduced by V. Vapnik in the year 1995 [11] and has proved to be more robust and effective compared to the probabilistic models [12]. It can work efficiently on large vector spaces and gives better resistance against data overfitting [13]. We trained some SVM models on adding a linear kernel and a nonlinear RBF (Gaussian) kernel as plug-ins to the classical SVM. The results show that the accuracy relies on obtaining optimal values for class C and γ parameters.

For the given dataset, the linear kernel is seen to have outperformed both, RBF kernel and the classical SVM in terms of classification accuracy. With large number of features present, using a nonlinear kernel does not seem to improve the performance. Hence, we try obtaining the best accuracy using a linear kernel by searching for the optimum value of C. The accuracy for different values of C is mapped in the Figs. 1 and 2.

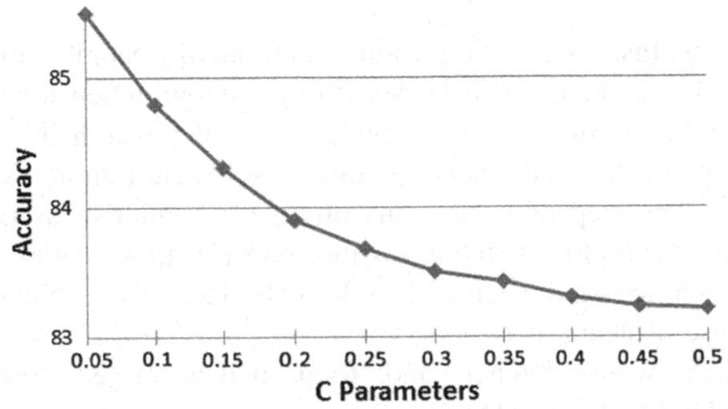

Fig. 1. Accuracies for various values of C Bigram SVM model

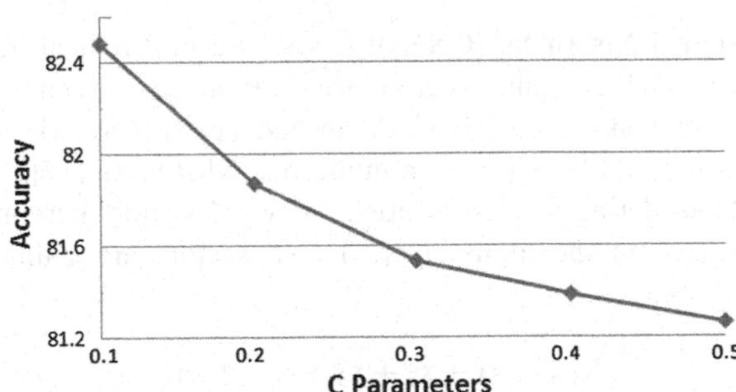

Fig. 2. Accuracies for various values of C for a unigram SVM model

3.3 Upgrading to Deep Learning Methodologies

Deep learning is a buzzword making rounds in the machine learning community due to the development and improvements brought in structure of neural networks. It has led to machine learning systems moving toward the discovery of multiple levels of representation that enables it to learn sophisticated features on its own. The heavy dependence of machine learning algorithms on choice of feature representation led to intensive, undue reliance on meticulously crafted rules as discussed above, with multilevel representation learning, neural networks provide flexible and easily learnable framework for representing the linguistic, semantic information specific to the corpus [14]. By simply supplying more input data, these models train themselves without any need for hand-crafted features.

These deep neural network (DNN) architectures basically are neural networks consisting of multiple hidden layers and a number of hidden nodes within these layers that are specified while defining the network. A suitable network hence can be designed for meeting the requirement of the task. The networks can adapt to various tasks with minimum changes made to its parameters. In our article, we evaluate performance of two such DNNs, namely convolutional neural network (CNN) and recurrent neural network (RNN) and figure if the performance is enhanced by usage of Long short-term memory (LSTM) units on our IMDb dataset.

Word Embeddings: Instead of taking features as input, the neural network takes input in the form of word embeddings, which are nothing but low-dimensional dense vectors. Word embeddings is one of the most exciting fields of research in NLP using deep learning. Raw corpus is fed to the network and on sufficient training, semantics of the lexicon are learnt. The map of words thus obtained is intuitive as the semantically similar words are placed close together on the map [7]; thus, stating that the cosine distance between these word vectors is relatively less. This achievement can be attributed to the use of neural networks.

For example, the words 'kitchen' and 'food' will be placed close to each other compared to 'kitchen' and 'cat' [15].

In our neural networks, we selected 5000 most frequently occurring words and on truncating the sequence length, the input was fed to the embedding layer. Output of the embedding layer gives a three-dimensional tensor.

Convolutional Neural Networks (CNNs): CNNs are hierarchical networks that do not rely on parsers and compute vectors for sub-phrases (even sub-phrases that grammatically do not make sense) [16]. A unified neural network architecture was proposed such that it could be applied on numerous NLP tasks [17].

Input layer—Considering k word vectors, 'n' word vectors pertaining to the sentence are concatenated in the input layer. These vectors are d-dimensional dense vectors.

$$X_{1:n} = x_1 + x_2 + x_3 + \cdots + x_n. \tag{1}$$

Convolutional layer—A filter or a window also called as *w-gram* is defined and is moved over the input sequence. The window vector and input sequence are convolved which is meant to extract features from the given sequence. The generally accepted window size is 2 (even 3 is used at times). We defined different filter counts along with varied window/kernel sizes for our experiment. The subsequent accuracies are shown in Figs. 1 and 2. On deciphering the optimum number of filters that should be used, further changing window sizes did not avail any performance change. We used *ReLU* as the activation function for the layer. A feature map is derived at this layer.

$$C_i = [C_1 + C_2 + \cdots + C_{n-w+1}] \in R\neg^{n-w+1}. \quad (2)$$

Max-pooling layer—The convolutional layer produces an output depending on the dimension of the input sequence. This layer captures the relative significance of each vector and also makes the sentence sizes uniform.

$$x_j = \max(C_{1,j}, C_{2,j}, \ldots)(j = 1, \ldots, d). \quad (3)$$

Regularization for additional efficiency: An additional layer can be used to perform regularization. It is used to avoid the overfitting problem for such architectures. Application of dropout at each layer of the network has shown good results [16].

The sentence vector is then fed to the connected neural network. We used *sigmoid* as the activation function. Satisfactory accuracy was obtained for our dataset which is shown in the Sect. 4.

Table 1. Change in performance observed on varying kernel and filter parameters

Filter size	Kernel size	Training accuracy		Test accuracy
		Epoch-1	Epoch-2	
50	3	78.62	88.96	86.54
75	3	79.47	89.77	86.98
100	3	79.35	89.95	87.25
100	4	80.64	91.07	87.78
100	5	80.65	91.65	87.68

Recurrent Neural Networks (RNNs): RNNs are sequential models that condition the network on the previous state along with the current state of the input. This results in propagation of its effects till the end of the sequence. During forward computation, gradients calculated at each time step are propagated backward for error calculations [18].

The weight matrix has an effect on the gradient value for each time step, which may explode or vanish depending on whether the weight values are small or large. Hence, vanishing gradient and gradient explosion pose as two major drawbacks encountered during training [19].

These networks are computationally complex and the cost increases with more number of epochs and layers. The accuracy obtained with RNN was relatively low. However, there seems to be a considerable drop in error and subsequent increase in training accuracy as epochs increase as shown in Fig. 3. The number of epochs is related to the number of rounds of optimization that are applied during training. Thus, the error on training data will decrease with more number of epochs. Due to high computational cost, it becomes difficult to train the model for further epochs.

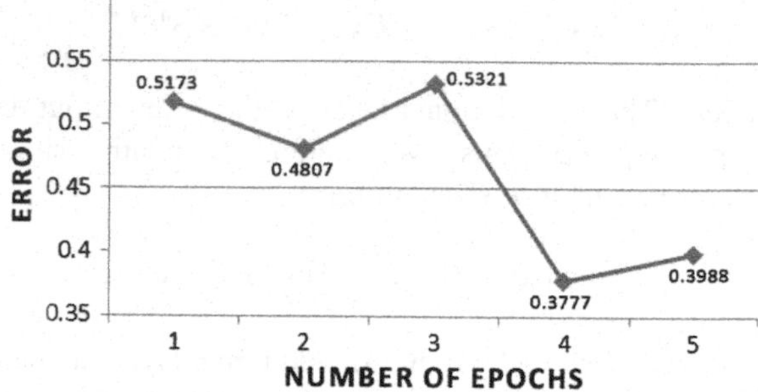

Fig. 3. Decrease in error with more epochs

Our model was trained for 5 epochs on 50 layers with *sigmoidal* activation function. RNN model was also trained on performing bigram, the results of which are shown in Sect. 4. In many experiments performed so far, CNN has been seen to perform better than GRU and LSTMs [20, 21]. However, RNN and LSTM showed comparable results, referring to [22].

Long Short-Term Memory (LSTM): LSTM is a gated memory cell and a variant of RNN as mentioned above. The three gates, namely input gate, forget gate, and output gates use sigmoidal activation function and decide whether information is to be added or removed. The memory of previous steps is retained for longer intervals as compared to RNNs, unless it is overwritten or asked to be forgotten. It uses two forms of memory retention—*Short Term memory* and *Long-term memory* [23].

The components of this memory cell with its corresponding functions are described here:

Memory cell (c_t)—stores the information
Input gate (i_t)—decides whether current input should be retained
Forget gate (f_t)—decides what to do with past hidden state
Output gate (o_t)—decides how much cell is to be exposed.

The term $h_{t-1,}$ the previous hidden state and the input given as $x_{t.}$ initially are used to generate information c_t. Thus, it evaluates the final memory cell values and hidden state values.

This network evidently has a complex structure compared to the other neural nets but the results obtained on combination with CNN and RNN structures showed enhancement in the accuracy. We experimented with a combination of 50 LSTM for all the variants of the model. On incorporating an LSTM layer in the existing CNN architecture, considerable enhancement in the performance on our corpora was observed; the results are shown in Sect. 4 (Table 2).

4 Results and Discussions

Table 2 shows the comparison of results for different techniques used in this project.

We notice that although SVM models showed appreciable accuracy, it accounts for some drawbacks compared to the neural networks, one of them being the additional step of deciphering the suitable parameters to avoid misclassification. Within SVMs, we compared the performance of a linear kernel and that of a nonlinear kernel and discovered that the performance of the linear classifier surpassed that of RBF by more than 30%. Also, the linear kernel trains faster. Although showing lower performance compared to DNNs here, SVMs can produce better results given the usage of fitting preprocessing and feature engineering techniques (SVM lin + Unigram: **85.50%**).

Among the DNN structures, the CNN structures and its variants have shown highest accuracy (vanilla CNN: **87.78%**). On coupling with bigram and unigram techniques, the result shows better performance gain for CNN. The highest accuracy obtained was with bigram technique, **87.84%**. For RNN, unigram model showed the best results (**85.88%**). We also discovered that the non-gated RNN model showed much lower performance gain (**73.87%**). As expected, the vanilla LSTM model outperformed vanilla RNN by almost 10%. The combination of LSTM and CNN gave better results as shown in Table 2.

Table 2. Sentiment analysis results of different machine learning techniques and deep neural networks in terms of accuracy (%) on movie review dataset

Machine learning methods			
Models	Parameters		Test accuracy
SVM (linear kernel)	C = 0.7		84.32
SVM (lin + Bigram)	C = 0.1		82.48
SVM (lin + Unigram)	C = 0.05		**85.50**
SVM (RBF Kernel)	C = 0.7, γ = 0.7		50.38
Deep neural networks			
Models	Training Accuracy		Test Accuracy
	Epoch-1	Epoch-2	
CNN	80.64	91.07	87.78
CNN + Unigram	81.38	89.32	87.07
CNN + Bigram	82.17	98.98	**87.84**
CNN + LSTM	80.82	89.42	86.78
CNN + LSTM + Unigram	83.24	90.62	87.12
CNN + LSTM + Bigram	81.27	98.93	85.93
LSTM	76.53	84.42	83.21
RNN	73.64	78.89	73.87
RNN + Unigram	74.10	89.45	85.88
RNN + Bigram	70.71	98.60	79.32

5 Conclusion

In this project, we investigated the performance of binary sentiment classification on various models described in Table 2. It gave an idea of applicability of these methods for different sentiment classification methods. On our dataset, we found the performance of deep learning architectures such as CNN, LSTM, and its combinations to be comparable with that of certain traditional methods.

We also observed that the performance of SVM depends heavily on the hyper-parameter, C. The optimal value of this hyper-parameter can be found with the help of grid search algorithm. Also, feature engineering becomes an integral part of machine learning and these require to be engineered specific to the application. However, with diverse fields adopting sentiment analysis and the multitude of subsequent corpora available today, manual engineering of features gets more complex. If the corpus contains unstructured data such as that found in tweets, the syntactic errors can lead to poor learning of the input text.

CNN trained faster and consumed less memory as compared to RNN and LSTM. Unlike SVM, hyper-parameters like filter size and kernel size do not affect the performance of CNN by a large margin as shown in Table 1. In case of RNN, it became difficult to train this complex model beyond certain number of epochs and also put limitations for the hidden layers that could be added. Also, increasing the number of LSTM cells was found to be impeding the training process as it increased the load on the memory. Further enhancements can be achieved in DNNs by optimizing the network parameters and usage of better language models such as Word2Vec.

The results obtained in this project pertain to the task of sentiment classification and the performance of these models may vary with the application. SVM classifier coupled with preprocessing may perform better than DNNs for certain NLP applications.

References

1. Pang B, Lee L. Opinion mining and sentiment analysis. Foundations and Trends® in Information Retrieval. 2008 Jul 7;2(1–2):1–135.
2. Bollen J, Mao H, Zeng X. Twitter mood predicts the stock market. Journal of computational science. 2011 Mar 31;2(1):1–8.
3. Du T, Shanker VK. Deep Learning for Natural Language Processing.
4. Pak A, Paroubek P. Twitter as a Corpus for Sentiment Analysis and Opinion Mining. InLREc 2010 May 19 (Vol. 10, No. 2010).
5. Maas AL, Daly RE, Pham PT, Huang D, Ng AY, Potts C.: Large Movie Review Dataset, http://ai.stanford.edu/~amaas/data/sentiment/
6. Bengio Y, Ducharme R, Vincent P, Jauvin C. A neural probabilistic language model.Journal of machine learning research. 2003;3(Feb):1137–1155.
7. Mikolov T, Chen K, Corrado G, Dean J. Efficient estimation of word representations in vector space. arXiv:1301.3781. 2013 Jan 16.
8. Morin F, Bengio Y. Hierarchical Probabilistic Neural Network Language Model. InAistats 2005 Jan 6 (Vol. 5, pp. 246–252).

9. Lewis DD. Naive (Bayes) at forty: The independence assumption in information retrieval. In European conference on machine learning 1998 Apr 21 (pp. 4–15). Springer Berlin Heidelberg.
10. Berger AL, Pietra VJ, Pietra SA. A maximum entropy approach to natural language processing.Computational linguistics. 1996 Mar 1;22(1):39–71.
11. Cortes C, Vapnik V. Support-vector networks. Machine learning. 1995 Sep 1;20(3):273–97.
12. Pang B, Lee L, Vaithyanathan S. Thumbs up?: sentiment classification using machine learning techniques. In Proceedings of the ACL-02 conference on Empirical methods in natural language processing-Volume 10 2002 Jul 6 (pp. 79–86). Association for Computational Linguistics.
13. Joachims T. Text categorization with support vector machines: Learning with many relevant features. Machine learning: ECML-98. 1998:137–42.
14. Bengio Y. Deep learning of representations: Looking forward. InInternational Conference on Statistical Language and Speech Processing 2013 Jul 29 (pp. 1–37). Springer Berlin Heidelberg.
15. Arora S, Li Y, Liang Y, Ma T, Risteski A. Rand-walk: A latent variable model approach to word embeddings. arXiv:1502.03520. 2015 Feb 12.
16. Srivastava N, Hinton GE, Krizhevsky A, Sutskever I, Salakhutdinov R. Dropout: a simple way to prevent neural networks from overfitting. Journal of Machine Learning Research. 2014 Jan 1;15(1):1929–58.
17. Collobert R, Weston J. A unified architecture for natural language processing: Deep neural networks with multitask learning. In Proceedings of the 25th international conference on Machine learning 2008 Jul 5 (pp. 160–167). ACM.
18. Mikolov T, Karafiát M, Burget L, Cernocký J, Khudanpur S. Recurrent neural network based language model. In Interspeech 2010 Sep 26 (Vol. 2, p. 3).
19. Bengio Y, Simard P, Frasconi P. Learning long-term dependencies with gradient descent is difficult. IEEE transactions on neural networks. 1994 Mar; 5(2):157–166.
20. Wen Y, Zhang W, Luo R, Wang J. Learning text representation using recurrent convolutional neural network with highway layers. arXiv:1606.06905. 2016 Jun 22.
21. Yin W, Schütze H, Xiang B, Zhou B. Abcnn: Attention-based convolutional neural network for modeling sentence pairs. arXiv:1512.05193. 2015 Dec 16.
22. Chung J, Gulcehre C, Cho K, Bengio Y. Empirical evaluation of gated recurrent neural networks on sequence modeling. arXiv:1412.3555. 2014 Dec 11.
23. Hochreiter S, Schmidhuber J. Long short-term memory. Neural computation. 1997 Nov 15;9 (8):1735–80.

Improved RAKE Models to Extract Keywords from Hindi Documents

Sifatullah Siddiqi(✉) and Aditi Sharan

School of Computer and Systems Sciences, Jawaharlal Nehru University, New Delhi, India
{sifatullah.siddiqi,aditisharan}@gmail.com

Abstract. In this paper, we have proposed several improved versions of rapid automatic keyword extraction (RAKE) algorithm for extracting keywords from Hindi documents. As RAKE requires a stopword list to generate the set of candidate keywords, which is unavailable in Hindi, we have constructed the Hindi stopword list for this purpose. We have found some weakness in keyword scoring measures of RAKE and proposed several models such as N-RAKE, SD-RAKE, NSD-RAKE, and WOS-RAKE to improve upon the effectiveness of RAKE. We have found that our modifications yield better results in general than original RAKE.

1 Introduction

RAKE is a very popular algorithm for keyword extraction [1]. It is based on the observation that keywords generally contain multiple words, but they seldom involve punctuation marks or stopwords, such as the words and, the, and of, or, etc., and other words with minimal lexical meaning.

The beauty of RAKE algorithm is that it provides us with a natural way of generating candidate keywords for the text. Candidate keywords are the set of potential single word or multiword sequences which are given scores according to some scoring criteria. Section 1.1 explains the working of RAKE algorithm while Sect. 1.2 describes its candidate keyword scoring method.

1.1 RAKE Algorithm Description

Consider the following sample paragraph in Hindi as given in Table 1.

Table 1. Sample Hindi paragraph

आजकल लोग देश में स्वास्थ्यकर पेय पदार्थ जैसे पानी और दूध के स्थान पर सॉफ्ट ड्रिंक्स लेना पसंद करते हैं । इन्हें हिंदी में शीतल पेय कहा जाता है और इन्हें कई नामों से जाना जाता है । इन्हें सोडा , पॉप , कोक , सोडा पॉप , फिजी ड्रिंक , टॉनिक , सेल्जर , मिनरल , स्पार्कलिंग वाटर , लॉली वाटर या कार्बोनेटेड पेय के नाम से भी जाना जाता है । इनमें से ज्यादातर में पानी होता है जो कि कार्बोनेटेड भी होता है , स्वीट्नर और आम तौर पर एक फ्लेवरिंग एजेंट भी शामिल होता है ।

V. Bhateja et al. (eds.), *Information Systems Design and Intelligent Applications*, Advances in Intelligent Systems and Computing 672,
https://doi.org/10.1007/978-981-10-7512-4_47

Input to RAKE algorithm requires the list of stopwords of the concerned language and a set of word delimiters. The text is partitioned into its candidate keywords list by breaking the text at stopwords and punctuation marks. This would give a set of contiguous words. Applying this approach in above paragraph would result in the text as presented in Table 2.

Table 2. Set of candidate keyphrases from RAKE

देश—स्वास्थ्यकर पेय पदार्थ—पानी—दूध—स्थान—सॉफ्ट ड्रिंक्स—पसंद—हिंदी—शीतल पेय—नामों— सोडा—पॉप—कोक—सोडा पॉप—फिजी ड्रिंक—टॉनिक—सेल्जर—मिनरल—स्पार्कलिंग वाटर—लॉली वाटर—कार्बोनेटेड पेय—नाम—पानी—कार्बोनेटेड— स्वीट्नर —एक फ्लेवरिंग एजेंट—शामिल

1.2 Candidate Keyword Weighting

As RAKE gives candidate keywords which are generally multiple word units, the weights of keywords are calculated by summation of constituent word scores. Three different weightings used in RAKE are as follows: frequency, degree, and degree/frequency.

The process of weighting in these different measures can be explained as follows:

Suppose a word ***w*** occurs in four candidate keywords (1) ***a b w***, (2) ***w a c b***, (3) ***a w***, and (4) ***w*** in the document, where **a, b, c,** and ***w*** are distinct words from the text.

Freq (***w***) is the simple count of the no. of times word w occurs in the document; therefore, its value in above example is 1 + 1+1 + 1 = 4.

Deg (***w***) measures the no of words with which a particular word w occurs in the candidate keywords, and it favors words which occur often and in longer keywords. Therefore, the degree of word w is 3 + 4+2 + 1 = 10.

Deg (***w***)/Freq (***w***) ranks those words higher which generally occur within longer keywords. Thus, Deg/Freq value of word ***w*** in the above situation is 10/4 = 2.5

For an N-word candidate keyword of the form "W_1 W_2… W_N" with corresponding constituent weights X_1 X_2… X_N, the overall score of the candidate keyword is determined by summing of constituent weights, i.e., $\sum_{i=1}^{N} X_i$.

2 Motivation

Though RAKE provides us with a natural way to generate candidate keywords for the text, there are several problems with RAKE algorithm which are as follows:

1. The biggest impediment in applying RAKE to any language is the requirement of the availability of a stopwords list of that language. The quality of list determines to a great extent the quality of keywords generated from the algorithm.

2. Important single word keywords have low score as compared to longer keywords (since because there is a simple addition of individual scores in multi-word keywords which might work well in technical kind of documents due to their nature or any such kind of document where longer keywords are more important and relatively more prevalent) which might be even less important than former. Smaller keywords would unavoidably have lower scores than longer keywords.
3. It appears that technical and literary texts have differences in average keyword length and RAKE favors longer length keywords invariably.
4. Different scoring measures give different results, and any one of them can perform better depending on the type and domain of the document.
5. For two candidate keywords AB and BA, it is evident that even if both are present in text, but they are not equally important but are ranked by RAKE of equal importance in text.

3 Modified RAKE Models

3.1 Length Normalized RAKE

Longer phrases give more information as compared to shorter phrases, and thus, former is preferred over latter in many situations, and RAKE invariably gives more weightage to longer candidates, but in many cases, some smaller candidate may be more important than longer ones such as in different contexts, the various terms such as ***sports car, विद्युत , बाढ़ नियंत्रण आयोग, फरक्का बांध*** or ***प्रदूषण नियंत्रण*** may be more important and more frequent than the terms ***hybrid sports car, विद्युत आयोग, गंगा बाढ़ नियंत्रण आयोग, फरक्का बांध परियोजना, केंद्रीय प्रदूषण नियंत्रण बोर्ड*** respectively, though RAKE would assign larger weights to latter words. It was observed that in all the three cases of scoring, candidate keywords ranking in RAKE, viz. word frequency freq (w), word degree deg (w), and word degree/word frequency deg (w)/freq (w), are not working satisfactorily for Hindi text. We therefore propose a modification in ranking schemes. Instead of just simply adding up the scores of constituent words of a candidate keyword and ranking them in decreasing order of their overall scores, we divide the score of a candidate keyword with its corresponding phrase length.

Thus, for an N-word candidate keyword of the form "$\mathbf{W_1\ W_2\ W_3 \ldots W_{N-1}\ W_N}$" with corresponding constituent weights $\mathbf{X_1\ X_2 \ldots\ X_N}$, etc., the modified score of the N-word keyword is $\sum_{i=1}^{N} \frac{X_i}{N}$. Thus, the three scoring measures of RAKE are changed from **frequency** to ***frequency/phrase length***, from **degree** to ***degree/phrase length***, and from **degree/frequency** to ***degree/(frequency * phrase length)***. This modification in scoring candidate keywords improves upon the original scoring measures is shown in Table 3. Some keywords from the document Godan are shown, and their ranks in original RAKE algorithm on all the three measures are shown, and also, their correspondingly improved rankings by normalizing the scoring measures are shown.

Table 3. Rank comparisons of keywords in RAKE and normalized RAKE for document Godan (Freq = frequency, Deg = degree, PhLen = phrase length)

Keyword	Freq	Freq/PhLen	Deg	Deg/PhLen	Deg/Freq	Deg/(Freq * PhLen)
होरी	230	1	395	6	10662	10392
मालती	554	15	726	31	10167	9459
गोबर	659	18	941	119	10668	10410
धनिया	929	39	1136	163	10642	10315
झुनिया	1475	276	1709	518	10864	10724
गाय	1864	836	2098	999	10161	9428
रायसाहब	1303	236	1562	389	10841	10664
पंडित दातादीन	1839	1711	1959	1816	5904	6960
मिस्टर तंखा	2106	1965	1797	1654	1981	1614
मेहता	695	19	913	82	10548	10020

It can be observed from Table 3 that significant improvements are made in rankings of keywords after normalizing the scoring measures. Figure 1 shows the comparison of ranks of various keywords in different weighting schemes.

3.2 RAKE Hybridization with Word Distribution (SD-RAKE and NSD-RAKE)

We propose a new measure to improve upon the quality of top-scoring candidate keywords of RAKE. The intuition behind this scoring measure is that the component words of a keyword should also be important while deciding about the importance of a candidate keyword. We hybridize RAKE algorithm with word distribution-based model approach [2]. The idea is that word occurrence pattern for a keyword is in general different from that of non-keywords. For a non-keyword, its occurrence pattern should be random in the text and no significant clustering should occur, while for a keyword, its occurrence pattern should indicate some kind of clustering since a keyword is expected to be repeated more often in specific contexts or portions of text.

We find the overall score of a candidate keyword by summing up the standard deviation score of the constituent words in the keyword. Therefore, if the standard deviation values of components of an N-word candidate keyword of the form "$\mathbf{W_1\ W_2\ W_3 \ldots W_{N-1}\ W_N}$" are $\mathbf{SD_1\ SD_2 \ldots\ SD_N}$, etc., then the new improved score would be $\sum_{i=1}^{N} SD_i$ (say model, SD-RAKE). The normalized version of this score would be $\sum_{i=1}^{N} \frac{SD_i}{N}$ (say model, NSD-RAKE).

3.3 Word Order Sensitive RAKE (WOS-RAKE)

In this section, we propose an approach to handle the word order insensitivity of RAKE algorithm. Consider a situation in which there are candidate keywords of the form A, AB, BA, ABC, CBA, BAC, ACB, BCA, CAB, where A, B, and C are different single words in the document. We have limited the size of candidate keywords to length 3 for our purpose. The following observations are in order:

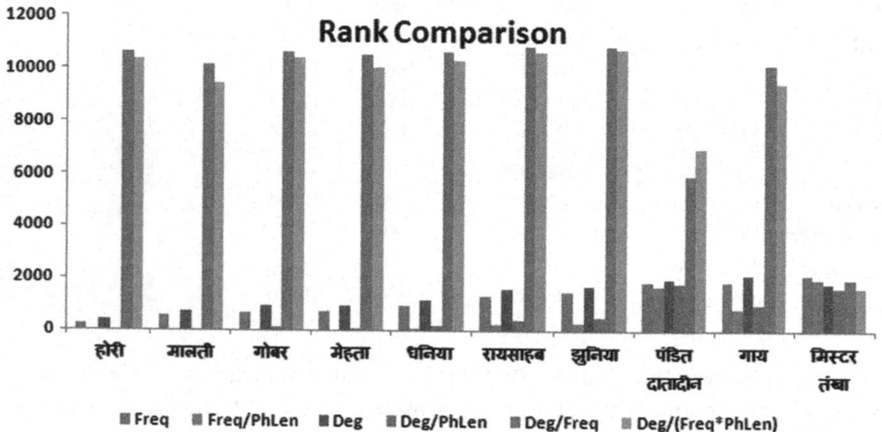

Fig. 1. Ranks of various keywords in different weighting schemes. Lower rank signifies a better ranking scheme

It is evident that AB and BA though being candidate keywords comprising of same words are not equally important but are ranked by RAKE of equal importance in text. For example, in document "Godan," there were few phrases of length 2 which were also present in reverse order such as ***होरी धनिया*** and ***धनिया होरी, रुपये हाथ*** and ***हाथ रूपये,होरी दिन*** and ***दिन होरी*** and were given same score by RAKE.

Similarly, any permutation of ABC (six permutations in all) is ranked by RAKE of equal importance due to simple summation of scores of individual words A, B, and C. But we know that all permutations of ABC cannot be equal in terms of their corresponding importance. For example, the phrases ***जल विद्युत उत्पादन, जल उत्पादन विद्युत, विद्युत उत्पादन जल*** from the document "Bharat ki nadiya" comprise of same three words ***जल,विद्युत,*** and ***उत्पादन,*** but are all different in terms of their importance and applicability and are given equal scores in RAKE.

Another concern is that though longer phrases such as ABC or AB are generally thought of being more informative and important than smaller phrases such as AB or A, but there might be very often practical situations in which smaller phrases such as AB or A are of more importance than longer phrases such as ABC or AB. For example, in some contexts, grid computing might be more important than grid computing environment and "grid" being more important than "grid computing." In the document "Godan," the word ***होरी*** is a keyword and candidate keywords from RAKE such as ***होरी रूपये*** and ***होरी अस्सी रूपये*** though are not keywords but are given higher scores than होरी.

Thus, RAKE due to its simple summation of scores of comprising words will give scores to the candidates A, AB, ABC in order **ABC > AB > A** under any condition.

Our modification of RAKE scoring by phrase length normalization suggested in Sect. 3.1 cannot deal with this scenario. The normalization would again rank all permutations equally. Also, the suggestion of ranking our candidate keywords by their standard deviation summation and corresponding normalization suffers from the same problem. Thus, we suggest another ranking scheme which is able to handle such situations effectively.

Consider again the situation of candidate keyword set **A, AB, BA, ABC, CBA, BAC, ACB, BCA, CAB**. We define a phrase scoring function **S(p)** for a phrase **p** as follows:

$$\mathbf{S(p)} = \mathbf{L(p)} * \mathbf{F(p)} \tag{1}$$

where **F(p)** is a function of phrase frequency and **L(p)** is a function of phrase length which can be constructed according to the nature of text. For our purpose we define L (p) = phrase lentgth and F(p) = frequency of phrase. Therefore the resultant score is;

$$\textbf{Phrase Score} = \textbf{Phrase Length} * \textbf{Phrase Frequency} \tag{2}$$

Thus, for example, if A occurs 20 times and AB occurs 7 times and BA once and ABC occurs 3 times, then their respective scores are 20, 14, 2, and 9.

4 Experiments and Results

In order to perform our experiment, we have selected three Hindi books to perform the experiments with our approach. In order to judge the efficiency of our approach, we have compared the results with standard deviation approach. The details of the three books are as follows:

"**Godan**" by Premchand and it is a famous Hindi novel.

"**Bharat ki Nadiya,**" a text on rivers of India, published by National Book Trust of India

"Bharat ka Prakritik Bhuvigyan," a text on physical geography of India, published by National Book Trust of India

In absence of any benchmark data set, especially in case of Hindi language, we tried to create our own benchmark. We have used "Bharat ki Nadiya" and "Bharat ka Prakritik Bhuvigyan" since they have indexes at their back, and these indexes have been treated as the benchmark for performance evaluation of both the approaches. The evaluation of results so obtained was performed in the following manner:

A candidate keyword was selected as a keyword of the text if all of the constituent words present in it were found in the index entries, e.g., if the candidate phrase is ABC where A, B, and C are separate words, then candidate ABC is termed as a keyword if all of A, B, and C are found in the index entries.

4.1 Results on Godan

Top ten ranked keywords on various ranking are as follows:

Freq: बाकी सौ रुपए होरी, होरी अस्सी रुपए, पचीस रुपए होरी, होरी रुपए, गोबर सौ रुपए, होरी गोबर, होरी दिन, दिन होरी, धनिया होरी, होरी धनिया

Deg: बाकी सौ रुपए होरी, गोबर सौ रुपए, होरी अस्सी रुपए, पचीस रुपए होरी, होरी रुपए, पाँच सौ रुपए निकलते, पाँच सौ रुपए, चार सौ रुपए, दिन रुपए, तीन सौ रुपए

Deg/Freq: हिया जरत रहत दिन, मन भाए मुड़िया हिलाए, सीटियाँ बजानी शुरू कीं, सिपाही पीली वर्दियाँ डाटे, गोबर चमाचम बूट पहने, तहमद चढ़ाए ताल ठोंक, रार मत बढ़ाओ बेटा, बेचारी अधमरी लड़कोरी औरत, नौकर उजले कुरते पहने, सुंदर गेहुँआँ रंग सँवला

Norm Freq: होरी, होरी रुपए, रुपए, होरी गोबर, होरी दिन, दिन होरी, धनिया होरी, होरी धनिया, होरी काम, रुपए हाथ

Norm Deg: रुपए, होरी रुपए, दिन रुपए, रुपए हाथ, हाथ रुपए, होरी, सौ रुपए, होरी दिन, दिन होरी, रुपए झुनिया

Norm Deg/Freq: चलेंगी, हिया जरत रहत दिन, मन भाए मुड़िया हिलाए, अमको अंग्रेज, सीटियाँ बजानी शुरू कीं, साढ़े नौ बजे, सिपाही पीली वर्दियाँ डाटे, महीन साड़ियाँ, गोबर चमाचम बूट पहने

SD-RAKE: मिस मालती मेहता, खन्ना मिस मालती, मिस्टर मेहता शिकार खेलने, मिस्टर तंखा मालती, सहसा मिस मालती कार, मालती मेहता, युवती मिस मालती, सिलिया मालती, मालती बरसों खन्ना, जहाँ मिस मालती जायँ

NSD-RAKE: चौधरी, मेहता, मालती मेहता, सिलिया, सिलिया मालती, मालती, मिर्जा मेहता, सिल्लो, नोहरी, सिलिया धनिया

WOS-RAKE: होरी, गोबर, मालती, मेहता, धनिया, काम, रायसाहब, रुपए, झुनिया, गाँव

4.2 Results on Bharat Ki Nadiya

Top ten ranked keywords on various ranking are as follows:

Freq: नदी जल प्रवाह, लाख हेक्टेयर नदी जल, नदी घाटी जल, कावेरी नदी जल, ब्रह्मपुत्र नदी जल, जल यमुना नदी, भारतीय नदी जल, अंतर्राष्ट्रीय नदी जल बोर्ड, जल कृष्णा नदी, कृष्णा नदी जल
Deg: लाख हेक्टेयर नदी जल, नदी जल प्रवाह, नदी घाटी जल, भारतीय नदी जल, कावेरी नदी जल, अंतर्राष्ट्रीय नदी जल बोर्ड, ब्रह्मपुत्र नदी जल, नदी जल विवाद, नदी जल संबंधी, जल यमुना नदी
Deg/Freq: बायां किनारा षुख्य नहर, काकरापार राइट वेक कनाल, काकरापार राइट बैक कनाल, गोविन्द वल्लभ पंत सागर, उकई लेफ्ट बैंक कनाल, अमृत पीकर असुर सर्वशक्तिमान हिमालयी अन्वेपक स्वामी प्रवानंद, कैंद्रीय जल ससाधन मंत्री, शहीद चंद्रशेखर आजाद परियोजना, अस्तरण क्षेपित कंकरीट बिछाने
Norm Freq: नदी, नदी जल, जल, नदी क्षेत्र, नदी जल प्रवाह, नदी घाटी जल, कावेरी नदी जल, ब्रह्मपुत्र नदी जल, जल यमुना नदी, भारतीय नदी जल
Norm Deg: जल, नदी जल, नदी, नदी क्षेत्र, नदी जल प्रवाह, नदी घाटी जल, भारतीय नदी जल, कावेरी नदी जल, ब्रह्मपुत्र नदी जल, नदी जल विवाद
Norm Deg/Freq: प्रभारी मंत्री, धाराएं, बायां किनारा षुख्य नहर, काकरापार राइट वेक कनाल काकरापार राइट बैक कनाल, गोविन्द वल्लभ पंत सागर, संयुक्त उपक्रम, उतेर, देर, उकई लेफ्ट बैंक कनाल
SD-RAKE: गंगा बाढ़ नियंत्रण आयोग, ब्रह्मपुत्र बोर्ड ब्रह्मपुत्र घाटी, दामोदर बाढ़ जांच समिति, टिहरी बांध परियोजना टिहरी, कावेरी जल विवाद अधिकरण, असम पश्चिम बंगाल उत्तर, मेगावाट विद्युत उत्पादन क्षमता, कार्य कैंद्रीय जल आयोग, आकलन राष्ट्रीय बाढ़ आयोग, मेगावाट जल विद्युत उत्पादन
NSD-RAKE: उद्गम, असम, दक्षिण, न्यायालय, किमी, पहुंचती, बेतवा, दामोदर, कावेरी, दक्षिण गंगा
WOS-RAKE: किमी, नदी, जल, भारत, बहती, निर्माण, मी, उत्तर प्रदेश, दिशा, नदियों

4.3 Results on Bharat Ka Prakritk Bhuvigyan

Top ten ranked keywords on various ranking are as follows:

Freq: किमी उत्तर स्थित, उत्तर स्थित विस्तृत क्षेत्र, दक्षिण स्थित क्षेत्र, किमी पश्चिम स्थित, दक्षिण दक्षिण पश्चिम दिशा, दिशा उत्तर पश्चिम दक्षिण, हिमनद हिमालय क्षेत्र, वर्ग किमी क्षेत्र, किमी उत्तर पश्चिम, किमी दक्षिण दिशा

Deg: उत्तर स्थित विस्तृत क्षेत्र, दक्षिण स्थित क्षेत्र, किमी उत्तर स्थित, किमी पश्चिम स्थित, हिमनद हिमालय क्षेत्र, दक्षिण दक्षिण पश्चिम दिशा, दिशा उत्तर पश्चिम दक्षिण, असम क्षेत्र स्थित, निकट स्थित कोटागाज हिमनद, स्थित अधिकेंद्र क्षेत्र

Deg/Freq: मुड़कर पटकाइ बम मिजो, अराकान योमा भूसीवन प्रक्षेत्र, सिंधुत्सांगपो संधि रेखा प्रक्षेत्र, सिधुत्सांगपो संधि रेखा प्रक्षेत्र, प्रमुखतया बलुआ पत्थर स्कार्प, गुरला मान्धाता पर्वत शिखर, तीन तदनुरूपी हिमनदीय अमगामिता, प्राकृतिक गैस आयोग दारा, नागपुर छिंदवाड़ा बालाघाट क्षेत्र, खातू लवण झील स्थित

Norm Freq: किमी, किमी स्थित, स्थित, मी, किमी हिमनद, हिमनद स्थित, स्थित हिमनद, मी हिमनद, क्षेत्र, किमी उत्तर स्थित

Norm Deg: स्थित, हिमनद स्थित, स्थित हिमनद, क्षेत्र, किमी स्थित, दक्षिण स्थित क्षेत्र, हिमनद हिमनद, हिमनद, किमी उत्तर स्थित, किमी हिमनद

Norm Deg/Freq: विन्यस्त, कार्ल्सबर्ग कटक, कुचामान, रूस, यथा, चलायमान, लघु, अराकान योमा, इल्मेनाइट, खनिजों

SD-RAKE: हिमनद दूसरे हिमनद, तीसरा हिमनद पुनमा हिमनद, हिमनद हिमनद, हिमनद अनुदैर्ध्य घाटी हिमनदों, नदियां हिमनद आच्छादित पर्वतमालाओं, किमी लंबा हिमनद, किमी लंबा कुकसेल हिमनद, छोटे छोटे सहायक हिमनद, हिमनद हिमालय क्षेत्र, लंबा अनुदैर्ध्य हिमनद

NSD-RAKE: हिमनद हिमनद, हिमनद, हिमनद दूसरे हिमनद, लंबा हिमनद, कंचनजंघा हिमनद, किमी हिमनद, हिमनद सागर, मी हिमनद, सहायक हिमनद, हिमनद प्रवाह

WOS-RAKE: मी, किमी, स्थित, दक्षिण, उत्तर, ऊंचाई, किमी लंबा, प्रवाहित, हिमनद, क्षेत्र

4.4 Precision-Recall Values on Datasets

Precision values of the various models are given in Tables 4 and 5.

Table 4. Precision values at 10th, 20th, 30th, 40th, and 50th ranks for the various models for Bharat ka Prakritik Bhuvigyan

Rank	Freq	Deg	Deg/freq	Norm freq	Norm deg	Norm deg/freq	SD-RAKE	NSD-RAKE	WOS0-RAKE
10th	0.1	0.1	0.1	0.1	0.3	0.2	0.2	0.3	0.4
20th	0.05	0.1	0.05	0.35	0.3	0.35	0.15	0.35	0.25
30th	0.033	0.067	0.067	0.367	0.367	0.3	0.1	0.367	0.267
40th	0.05	0.075	0.075	0.3	0.375	0.275	0.15	0.325	0.3
50th	0.04	0.08	0.06	0.3	0.36	0.24	0.12	0.32	0.24

Table 5. Precision values at 10th, 20th, 30th, 40th, and 50th ranks for the various models for Bharat ki Nadiya

Rank	Freq	Deg	Deg/freq	Norm freq	Norm deg	Norm deg/freq	SD-RAKE	NSD-RAKE	WOS0-RAKE
10th	0.7	0.6	0	0.9	0.9	0	0.5	0.5	0.4
20th	0.45	0.5	0	0.65	0.65	0.05	0.5	0.5	0.6
30th	0.4	0.367	0.033	0.533	0.5	0.033	0.567	0.567	0.533
40th	0.375	0.4	0.05	0.5	0.525	0.025	0.45	0.45	0.45
50th	0.36	0.38	0.04	0.52	0.54	0.02	0.42	0.42	0.44

5 Conclusion

We have proposed various modifications to the RAKE algorithm to improve its effectiveness and efficiency. A stopwords list of Hindi language was constructed for this purpose. Our first model N-RAKE incorporated the important aspect of effect of phrase length on the quality of extracted keywords, and it substantially improved the number and quality of keywords which were extracted in frequency and degree while having not much effect in degree upon frequency case.

Spatial distribution information of words in the text was used to identify the keywords, and this knowledge was incorporated model SD-RAKE. Normalized model of SD-RAKE which is NSD-RAKE yielded even better performance. Another model WOS-RAKE was proposed to include the importance of word order in candidate keywords. Overall, the proposed models outperformed the original RAKE algorithm and yielded better results.

References

1. Ortuño, M., Carpena, P., Bernaola-Galván, P., Muñoz, E. and Somoza, A.M., "Keyword detection in natural languages and DNA", Europhys. Lett. 57, (2002), pp. 759–764.
2. Rose, S., Engel, D., Cramer, N., & Cowley, W., "Automatic keyword extraction from individual documents", Text Mining: Applications and Theory, John Wiley & Sons Ltd., (2010).

An Information Systems Design Theory for Knowledge-Based Organizational Learning Systems

Vishnu Vinekar(✉)

Department of Information Systems and Operations Management,
Fairfield University, Fairfield, CT, USA
vvinekar@fairfield.edu

Abstract. The field of information systems (IS) is divided into "design science," which is more engineering oriented, and "natural science," which is more management oriented. This paper believes that such a conflict is detrimental to the cause of research in IS, especially because both design and natural science have a wealth of knowledge to give to and to learn from each other, and a symbiotic relationship between the two research interests could tremendously increase the contribution of IS research as a whole. As an example, this paper uses theory from natural science to guide building and evaluating an artifact using design science. Specifically, I develop an information systems design theory to provide a prescriptive conceptual design of an organizational learning system using design science that addresses the shortcomings of on knowledge management systems and organizational learning systems identified in natural science research.

1 Introduction

The field of information systems (IS) is divided into design science and natural science [16]. Design Science is more engineering-oriented and focused on building new technology, while natural science is more management-oriented and focused on people's behavior and problems. While March and Simon [16] state that such a dichotomy is "not intrinsically harmful," this paper believes that such this dichotomy is detrimental to the cause of furthering knowledge in IS, mainly because both design and natural science have a wealth of knowledge which could tremendously increase the contribution of not only each research stream, but also contribution of IS research as a whole. Natural science studies artifacts produced by design science and finds strengths and weaknesses of those artifacts. These can then be used to guide design science. Therefore, this paper takes a stance that this dichotomy between natural and design sciences should be actively mitigated by both design science and natural science research, both actively focusing on inculcating research findings from each other to guide their work. Furthermore, research in each stream should conclude with contributions for the other stream. As an example, this paper uses theory and findings from natural science, i.e., management theory, to guide building and evaluating an artifact using design science. The construction of such an artifact can contribute back to the

V. Bhateja et al. (eds.), *Information Systems Design and Intelligent Applications*, Advances in Intelligent Systems and Computing 672,
https://doi.org/10.1007/978-981-10-7512-4_48

natural science it builds on by generating new hypothesis to test: whether such an artifact addresses the issues identified by natural science, and how this may be explained. This paper strongly urges IS researchers in both design and natural camps, to actively inculcate similar reciprocal practices in their research. Through such a symbiosis, we hope to improve IS research as a whole.

1.1 Context: Knowledge Management (KM) and Organizational Learning (OL)

The recent focus on knowledge management (KM) and organizational learning (OL) has propelled the need for information technology (IT) to facilitate and support its practice. However, natural science has identified several issues with these systems. One of the central issues is that present uses of IT have a greater emphasis on incremental learning [17], which is more suited for periods in which the environment is relatively stable, linear, and predictable [7]. In this case, the utilization of IT for "exploitative learning," i.e., refining of knowledge of old certainties, results in learning benefits. However, when the environment is dynamic, and change is discontinuous, nonlinear, or radical, the organization may need to focus on "explorative learning" or "double-loop learning" [5], in which it questions its assumptions and explores new possibilities. The use of IT may hinder double-loop learning, as the organization now requires to "unlearn" what it has previously learned. Here, IT would need to focus less on incremental learning and more on explorative learning, i.e., creating and facilitating diversity, variety, and change.

Both natural science research and design science research in OL have usually taken either the exploitative or explorative approach and limit its scope to OL at either lower, middle, or higher levels of management. This paper makes a contribution in this regard by developing an information systems design theory (ISDT) for an OLS that is capable of enhancing learning at all levels of management. In this model, both exploitative learning and explorative learning occur at all levels, and the learning at these levels is complementary. OL occurs through both top-down and bottom-up processes, and conflict between upper and lower levels is resolved through exploitative and explorative learning at middle layers of management.

2 Information Systems Design Theory (ISDT)

Design is a central topic in the field of IS. In pedagogy, primary emphasis is placed on design in degree programs in IS, and in practice, design skills are highly valued among IS professionals. However, in research, developing theory for design has not received much focus in natural or design science research. This paper makes a significant contribution to IS research by developing a design theory for an OLS. Since design theories are prescriptive, this model also makes a contribution to practitioners, as development of OLS is still in early stages.

Design theory is defined as "a prescriptive theory based on theoretical underpinnings which say how a design process can be carried out in a way which is both effective and feasible" [24]. An ISDT is defined as "a prescriptive theory which integrates normative and descriptive theories into design paths intended to produce more effective information systems"

An ISDT has two aspects—one dealing with the design product and the other dealing with the design process. The first aspect the design product includes—meta-requirements, meta-design, and kernel theories. Meta-requirements describe a class of goals to which the theory applies. Meta-design describes a class of artifacts hypothesized to meet the meta-requirements. Kernel theories are theories from natural and social sciences governing design requirements. This paper applies design theory to develop an organizational learning system (OLS). The scope of the paper is limited to the design product. Design process is not part of the scope because organizational learning is an emergent knowledge process (EKP), and the design process of the ISDT developed by Markus et al. [17] should be applicable.

3 Knowledge-Based OLS: Design Product

The design product of this ISDT is a Knowledge-based OLS. The development of the design product begins with meta-requirements and meta-design. This section details these two components of the design product for a Knowledge-based OLS.

3.1 Meta-Requirements

There are five meta-requirements identified for this ISDT, all based on the findings of natural science: (1) exploitative learning, (2) explorative learning, (3) balancing exploitative learning and explorative learning, (4) environmental scanning, and (5) aligning learning at different levels.

3.1.1 Exploitative Learning

OLS developed using the structural systems approach [12] has abilities to improve capabilities for exploitative learning. Exploitative learning is important as it allows organizations to build on previous successes and failures. Organizational memory is beneficial for exploitative learning, as it helps an organization analyze what it had done in similar situations in the past, and allows it to make decisions for the future based on previous outcomes. This helps an organization to capture its best practices, document them in detail, and repeat these practices elsewhere, e.g., Infosys' KMS built on the principle of "learn once, use anywhere" [9]. The intent of these systems is to capture the experiential knowledge gained by employees, store them in repositories, and make it accessible to the entire organization via an intranet [2, 14]. Several of these systems limit their scope to ground employee and technical staff levels and not to top management. The design of these systems assumes that improving access to knowledge would facilitate use of that knowledge [11]. Some studies attempt to improve these systems by adding contextual information to internal knowledge, e.g., [1, 10, 15, 20]. These systems enable an organization to reduce process variance, as underperforming processes can be identified and replaced with best practice processes. They also enable single-loop learning [5], in which an organization gradually refines its own knowledge for greater efficiencies. Therefore, the first requirement of an OLS is acquisition and distribution of knowledge for exploitative learning.

3.1.2 Explorative Learning

Although the systems-structural perspective dominates research and practice, it has criticisms. OL that occurs through acquiring and distributing knowledge may result in organizational memories that have a negative influence on individual and organizational performance [2, 19]. At the individual level, it results in decision-making bias. Individuals attempt to replicate past successes by applying previous solutions in situations where they not applicable [4]. At the organizational level, it may reinforce single-loop learning and may lead to organizational cultures that are resistant to change. It seems ironic that successful organizations fall into "competency traps" in which they cling to "success formulas" and best practices that are no longer applicable. Heavy reliance on IT at Batterymarch Financial Management and Mrs. Fields' Cookies provided these companies great competitive advantages. However, the advantages produced by IT were short-lived as the same technologies impaired their capacity for OL [19]. When these organizations faced complex environmental changes, they found themselves deficient in their capacity to learn about the new environment. Consequently, they could not adapt to the changes, leading to their decline. Companies that focus exclusively on exploitation of their current knowledge are very likely to face failure when change is necessitated [6, 7]. Current knowledge is poor guide for future actions when conditions change [19]. Exploitative learning is therefore advantageous in the short term but self-destructive in the long term [15]. Therefore, organizations need to develop capabilities for explorative learning, to enable them to innovate and adapt to change. This requires double-loop learning [5] in which the organization questions the basis of its present knowledge, challenges its own assumptions, and looks for alternatives to its present processes. This exploration allows for innovation and radical change. Hence, we arrive at a second requirement: the OLS needs to challenge its present knowledge, its assumptions, and look for alternatives to its present knowledge to enable innovation and change.

3.1.3 Balancing Exploitative Learning and Explorative Learning

Just as exploitative learning has its negative and positive effects, explorative learning too has negative effects. An exclusive focus on explorative learning increases variance. The vast majority of innovations fail, and the organization that keeps questioning its knowledge fails to exploit its successes. Therefore, instead of focusing on systems that support either exploitative or explorative learning, it would be beneficial to have a system that supports both exploitative learning and explorative learning so that organizations can face periods of stability as well as periods of change [6, 7]. However, there may be several challenges in accomplishing this. The two processes of exploitative learning and explorative learning are opposing forces [21]; the former tends to reduce variance and the latter tries to increase variance. The two forms of OL are self-reinforcing and tend to overcome the other [15]. These are competing activities that produce opposing forces on OL that may never be fully resolved [19]. Successful organizations are those that mange to balance exploitative learning and explorative learning. Therefore, the third requirement of an OLS is that the system would need to balance different approaches to OL so that they coexist and are mutually reinforcing.

3.1.4 Environmental Scanning

In addition, OLS has tended to focus either at top management or at ground employees and technical staff. Research at these levels suggests that the requirements at these two levels are very different. At lower levels, OL may be largely dependent on internal knowledge sources e.g., [1, 10, 14, 20] while at higher levels, there is an additional need for external knowledge for strategic decision-making [3]. External influences can dramatically influence an organization's survival and success in a competitive environment. Threats and opportunities affecting an organization's future include the threat of new entrants, newer technologies, changes in legal regulations and market trends. Several organizations have been adversely affected because they failed to recognize these threats and opportunities in their external environment early enough to respond to them. The most dangerous of these are changes driven by a competitor, usually a new entrant into the market [13]. These include K-Mart missing the extent of Wal-Mart's penetration, Xerox's failure to recognize Canon's impact, and US automakers' dismissal of the threats posed by Honda and Toyota. Therefore, we see that identification of these threats is crucial to an organization's survival. Environmental Scanning is the acquisition, internal communication, and use of information about issues in an organization's external environment that may influence an organization's future [3, 25]. This process identifies the potential pitfalls in the environment that may affect the organization's future. Learning about the environment and changes in the environment helps managers identify threats and opportunities. Environmental scanning should not be reduced to merely gathering information; it is more about potential future impacts than present ones. The knowledge gained through environmental scanning is then used to shape internal processes to counter these threats and to exploit identified opportunities. There is an overwhelming consensus that environmental scanning is a basic input to business strategy [23]. Recognizing opportunities and threats that others miss is an important source of competitive advantage. Empirical evidence shows that CEOs of high-performing companies scan more frequently and broadly than those of low-performing companies [8]. The fourth requirement of the OLS is to support an organization's environmental scanning needs.

3.1.5 Aligning Learning at Different Levels

Past research has made valuable contribution in OL at separate levels of management within the organization. As these two layers have very different knowledge needs, we need a greater understanding how these different knowledge structures coexist within an organization. However, few studies have focused on how learning at each layer can support the others. Research indicates that actions at macro-economic, managerial, and technical levels are aligned in periods of stability, but their actions are misaligned in periods of instability [19]. Misalignment in actions in OL between top management and ground employees has a potential to severely hinder organizational effectiveness. Research needs to address this issue and of aligning OL. The fifth requirement for an OLS is that it needs to align the learning activities between top management and ground-level employees.

3.2 Meta-Design

Figure 1 shows the modules of the meta-design and their correspondence to the meta-requirements. A conceptual model of the system is shown in Fig. 2.

3.2.1 Balancing Exploitative Learning and Explorative Learning at the Ground Employee Level

As no single system might meet this requirement, this may need to be met through a combination of systems—a group decision support system (GDSS) for within-group explorative learning and for between-groups, a knowledge repository for exploitative learning, a knowledge forum for explorative learning, and a knowledge repository as a fallback for individual knowledge seekers.

Organizational projects are frequently done in groups, and IS can support this process through group decision support systems (GDSS). GDSSs [22] can be used to effectively create helpful task conflict and avoid "groupthink," in which the group hastily adopts a consensus without analyzing it effectively. GDSS-generated task conflict may help explorative learning.

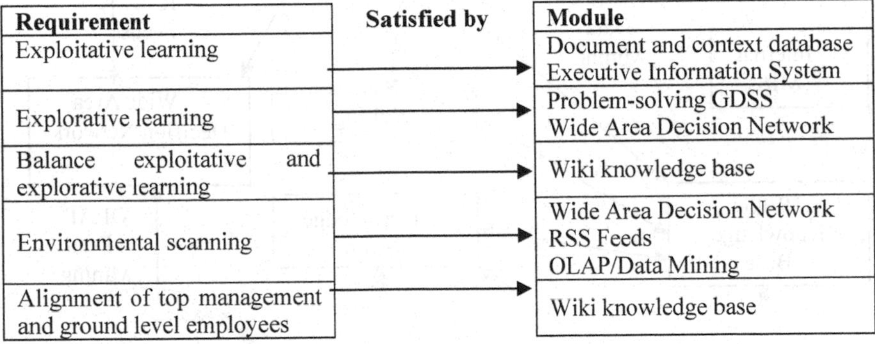

Requirement
Satisfied by
Module
Exploitative learning
Document and context database
Executive Information System
Explorative learning
Problem-solving GDSS
Wide Area Decision Network
Balance exploitative and explorative learning
Wiki knowledge base
Environmental scanning
Wide Area Decision Network
RSS Feeds
OLAP/Data Mining
Alignment of top management and ground level employees
Wiki knowledge base

Fig. 1. Organizational learning system (OLS) modules

When knowledge is needed from outside a team, a knowledge repository [1, 14] that automatically captures the documents generated through daily business activities, as well as the context they were captured in, would serve to enhance exploitative learning. Despite management interest in KMS, knowledge repositories suffer from disuse. This may be due to three key factors [14]: First, workers do not see an immediate benefit from the use of the system, and therefore, avoid using them. Second, sharing knowledge requires a common mental framework between the source and the recipient. Third, organizations do not have a knowledge-sharing culture replete with motivation and incentive to share knowledge. In addition, these systems assume that increasing access to knowledge increase its use and do not take into account the effects of intellectual demands and learning orientation [11].

Knowledge Directories [18] and Knowledge Forums (like Quora.com) can address these issues. A directory of knowledge retainers [18] contains meta-knowledge of knowledge retainers or subject matter experts (SMEs), including their education, experience, and reputation, which the seekers can use to decide which retainer to target for their information, thus addressing knowledge demand rather than the supply. This scheme provides significant advantages: Knowledge is stored tacitly instead of explicitly, it allows employees to use personal contact to obtain required knowledge, and it also takes care of the problem of changing knowledge. However, while a knowledge directory may be used by the knowledge seeker to find a suitable knowledge retainer, it has the following disadvantages—while some retainers of knowledge may be willing to help when contacted, others may feel pressurized if they are constantly contacted. Furthermore, recipients may have to answer similar questions several times, decreasing their productivity with their day-to-day routine. Some retainers may be contacted for knowledge more than others.

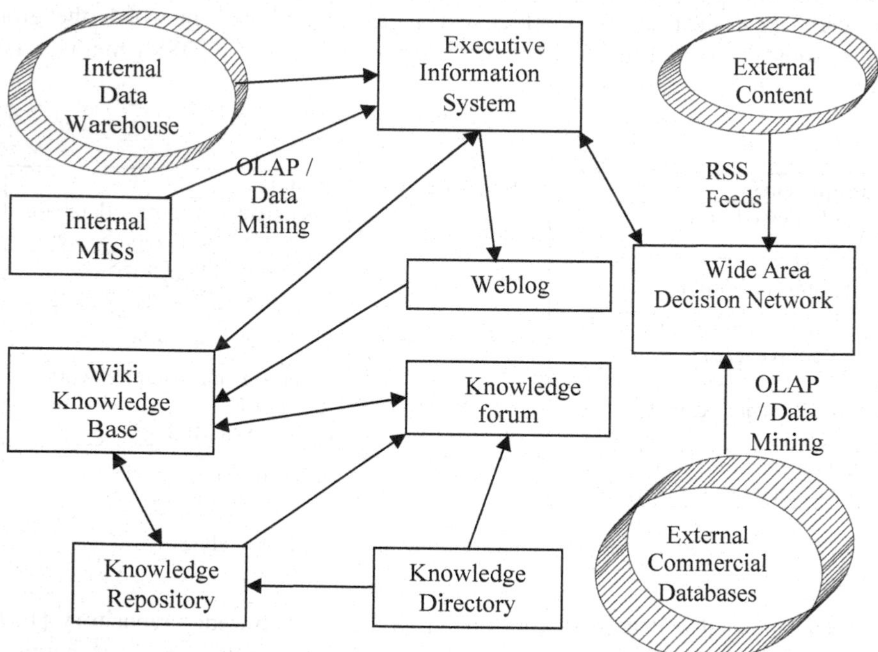

Fig. 2. Conceptual model of the organizational learning system (OLS)

Therefore, a more advantageous method may be if the knowledge seeker posts the knowledge requirements in a "Knowledge Forum," and the directory alerts the knowledge retainers that their knowledge is required, leaving it to individuals to respond. These knowledge forums allow the seeker to receive multiple answers to their query. This would enable knowledge retainers to analyze their own and each other's assumptions and learn through the experience. As tacit knowledge is made explicit through the knowledge forum, this would also help lower the number of queries that retainers may receive. Knowledge seekers may search the forum for similar problems, and the knowledge forum may alert them of similar queries posted earlier. A very important advantage of this knowledge forum is that employees see immediate benefits. They receive personalized answers for their knowledge requirements, and they also see previous knowledge seekers posts and answers. In addition, the knowledge forum may be connected to the knowledge repository so that knowledge retainers may provide hyperlinks to any suitable documents in the repository. Hence, the knowledge forum, integrated with the knowledge directory and the knowledge repository, addresses the three main causes for lack of use of KMSs: It emphasizes knowledge demand rather than knowledge supply, provides immediate benefits, shared context, and motivation, and thus may help to create a knowledge-sharing culture within the organization.

3.2.2 Balancing Exploitative Learning and Explorative Learning at the Top Management Level

The top management needs to look for internal strengths and weaknesses, as well as external threats and opportunities. For example, Honda turned GM into one of the biggest money-losing enterprises in history. Apple's garage-manufactured "toy" computers cut into IBM's market share. From research on organizational environments and sources of information [8], primary sources of knowledge of the external environment are personal contacts both inside and outside the organization. However, research in IS has focused mainly on providing knowledge from external sources of explicit knowledge, e.g., [25]. While connecting to external knowledge such as commercial databases and really simple syndication (RSS) feeds from external content providers may be beneficial for environmental scanning; the OLS will also have to support tacit knowledge sources by enabling managers to communicate with their personal network. This may be achieved through a wide area decision network (WADN), which would also allow for asynchronous communication, as personal networks may extend beyond time zones. One of the primary problems of personal internal contacts is that management may receive filtered information, as people do not want to deliver bad news in person [13]. The use of computer-mediated communication reduces this filtering, and employees may feel more comfortable delivering information on much-needed change semi-anonymously through a WADN. It would allow employees and partners to log in with aliases when they want to deliver unpleasant information. This allows employees to challenge management's assumptions and allows for double-loop learning.

3.2.3 Aligning Top Management Learning with Ground Employee Learning

Since organization learning is a top-down as well as a bottom-up approach, middle levels of management need to play a crucial role in supporting and integrating these conflicting processes. However, IS research is lacking here, with most studies analyzing either top-down learning or bottom-up learning, but not both. It is important to align top management learning efforts with ground-level employee learning, as misalignment may occur between the levels in times of environmental uncertainty [19]. Therefore, a KM layer is needed between top management and ground-level employees so that both levels of learning have a common ontology, framework, and direction. The KM layer will need to be actively monitored and changed frequently and will need input from all levels of the organization. This paper posits that a meta-knowledge wiki may meet these requirements. Wiki technology enables documents to be written collectively in a simple markup language using a Web browser, and anyone in the organization can add or edit the content. This enables a knowledge base to be created very rapidly with pages highly interconnected via hyperlinks. All aspects of the organizations knowledge may be defined in the wiki knowledge base.

Weblogs maintained by top management input may also serve as useful input into the wiki knowledge base. This KM layer may help integrate the top-down process with the bottom-up process by providing a common ontology. If there are conflicting issues about any specific knowledge instance, provisions for disambiguation and dispute help with explorative learning. The knowledge base built up this way enables knowledge to be made explicit and refined through exploitative learning.

4 Conclusion

By integrating these different perspectives into a single ISDT, this model provides enhanced support for OL. The system promotes learning at different levels of management and uses learning at different levels to support each other through both explorative OL and exploitative OL. This study has important implications for practice in the benefits of taking a more integrated approach to support both top-down and bottom-up processes. It addresses negative effects organizations face while using IT for learning, namely becoming overly reliant on exploitation and resistance to change. This study also addresses issues of motivation and incentives for individuals to contribute to OL, providing immediate benefits to motivate system use and providing long-term benefits for future organizational knowledge goals.

This study has important implications for research as well. By integrating fragmented research in learning systems, it shifts the focus from refinements in systems working at either the top management level or the ground employee level to understanding issues facing bringing these different learning systems to complement each other. Research also needs to focus on not just improvements in systems for either exploitative or explorative learning, but also on how these two conflicting processes can coexist and even support each other. Future research needs to test and address these issues of how a symbiotic relationship can be achieved between exploitative and explorative learning in an organization, as well as between top management learning and ground employee learning.

This paper has an important implication for researchers in IS in general. The field of IS is divided into design science and natural science. This paper takes a stance that this dichotomy can be more symbiotic. Both design science and natural science research need to actively focus on inculcating research findings from each other to guide their work and should conclude with reciprocal contributions. This paper uses theory from natural science to guide building and evaluating an artifact using design science. Furthermore, the construction of this artifact also contributes to natural science by generating new hypothesis to test: whether such an artifact addresses the issues identified by natural science, and how this may be explained. This paper strongly urges IS researchers, in design and natural camps alike, to actively inculcate similar reciprocal practices in their research, and through such a symbiosis, improve IS research as a whole.

References

1. Ahn, H.J., Lee, H.J., Cho, K., Park, S.J. Utilizing Knowledge Context in Virtual Collaborative Work. *Decision Support Systems* 39:4 (2005) 563–582.
2. Alavi, M., Leidner, D.E. Knowledge Management and Knowledge Management Systems: Conceptual Foundations and Research Issues. *MIS Quarterly* 25:1 (2001) 107–136.
3. Albright, Kendra S. Environmental Scanning: Radar for Success. Information Management Journal 38:3 (2004) 38–44.
4. Argyris, C. Teaching Smart People How to Learn. Harvard Business Review 69:3 (1991) 99–109.
5. Argyris, C. Double loop learning in organizations. Harvard Business Review 55:5 (1977) 115–125.
6. Benner, M.J., Tushman, M.L. Exploitation, Exploration, and Process Management: The Productivity Dilemma Revisited. Academy of Management Review 28:2 (2003) 238–256.
7. Benner, M.J., Tushman, M.L. Process Management and Technological Innovation: A Longitudinal Study of the Photography and Paint Industries. Administrative Science Quarterly 47:4 (2002) 676–706.
8. Daft, R.L., Sormunen, J., Parks, D. Chief Executive Scanning, Environmental Characteristics, and Company Performance: An Empirical Study. Strategic Management Journal 9:2 (1988) 123–139.
9. Garud, R., Kumaraswamy, A. Vicious and Virtuous Circles in the Management of Knowledge: The Case of Infosys Technologies. MIS Quarterly 29:1 (2005) 9–33.
10. Gordon, M.D., Moore, S.A. Depicting the Use and Purpose of Documents to Improve Information Retrieval. *Information Systems Research* 10:1 (1999) 23–37.
11. Gray, P.H., Meister, D.B. Knowledge Sourcing Effectiveness. Management Science 50:6 (2004) 821–834.
12. Hine, M.J., Goul, M. The Design, Development, and Validation of a Knowledge-based Organizational Learning Support System. *Journal of Management Information Systems* 15:2 (1998) 119–152.
13. Huffman, B.J. Why environmental scanning works except when you need it. Business Horizons 47:3 (2004) 39–48.
14. Kwan, M.M., Balasubramanian, P. KnowledgeScope: Managing Knowledge in Context. *Decision Support Systems* 35:4 (2003) 467–486.
15. March, J.G. Exploration and Exploitation in Organizational Learning. Organization Science: A Journal of the Institute of Management Sciences 2:1 (1991) 71–87.
16. March, S.T., Smith, G.F. Design and Natural Science Research on Information Technology. *Decision Support Systems* 15 (1995) 251–266.
17. Markus, M.L., Majchrzak, A., Gasser, L. A Design Theory for Systems That Support Emergent Knowledge Processes. *MIS Quarterly* 26:3 (2002) 179–212.
18. Nevo, D., Wand, Y. Organizational Memory Information Systems: A Transactive Memory Approach. *Decision Support Systems* 39:4 (2005) 549–562.
19. Robey, D., Boudreau, M-C. Accounting for the Contradictory Organizational Consequences of Information Technology: Theoretical Directions and Methodological Implications. Information Systems Research 10:2 (1999) 167–185.

20. Roussinov, D., Zhao, J.L. Automatic Discovery of Similarity Relationships through Web Mining. *Decision Support Systems* 35:1 (2003) 149–166.
21. Sheremata, W.A. Centrifugal and Centripetal Forces in Radical New Product Development Under Time Pressure. Academy of Management Review 25:2 (2000) 389–408.
22. Sia, C-L., Tan, B.C.Y.; Wei, K-K. Group Polarization and Computer-Mediated Communication: Effects of Communication Cues, Social Presence, and Anonymity. Information Systems Research, 13:1 (2002) 70–90.
23. Temtime, Z.T. Linking environmental scanning to total quality management through business planning. Journal of Management Development 23:3 (2004) 219–233.
24. Walls, J.G., Widmeyer, G.R., El Sawy, Omar A. Building an Information System Design Theory for Vigilant EIS. Information Systems Research 3:1 (1992) 36–59.
25. Wei, C-P., Lee, Y-H. Event detection from online news documents for supporting environmental scanning. Decision Support Systems 36:4 (2004) 385–401.

Image Saliency in Geometric Aesthetic Aspect

Dao Nam Anh(✉)

Electric Power University, Hanoi, Vietnam
anhdn@epu.edu.vn

Abstract. This paper introduces a geometric aesthetic approach for the analysis of visual attention to extract regions of interest from images. Modulation awareness, such as that perceived by visual features, can be represented by attractive proportions of visual objects. Together with supporting techniques such as similarity estimation and lighting condition manipulation, the aesthetic geometry-based analysis can be implemented to form refined attentive shifting observed from image scenes. In this paper, we propose robust kernels which comply with the golden ratio for analysis of aesthetic attractiveness which can raise visual awareness. Properties and relations of points and regions are evaluated by the corresponding kernels for images scenes. We also establish robust likelihood reasoning for the kernels with respect to human aesthetic attraction. The experimental results with a benchmark show the efficiency of the proposed method for identifying region of visual interest in images.

1 Introduction

The field of the visual attention [1] is making substantial movement as demonstrated by the reinforcing efficiency of present techniques on saliency benchmarks. Robust estimation of preferential focus on particular regions of the scene has been used for visual search [2], object detection [3], and support of object recognition [4]. Recently, robust features of visual attention employing statistics either on points or spatial regions have been applied for segmentation [5], gaze prediction [6], video attention [7], and image retrieval [8]. The approach was presented to judge whether a point or a region is focused by visual attention.

Very often, the variety in image scale and variations of viewing condition is involved in datasets differ the precision of detection. Classical saliency models would make mistakes in such cases. The correspondence of geometric query problems [9] for saliency evaluation can be outlined back to the methods of point location [10] by the estimation of fixation/interest points [11] or searching interest points [12]. The points are of great interest for attention which is discovered by some explicit mathematical property. The range search [13] has its similarity with objects of interest [14] or salient region selection [15]. For instance, specifically observed objects or regions can be devoted to object recognition and marching. In this work, we study specially the detection of regions in high visual attention by learning relevance to the visual attention with human aesthetic attraction, a part of which could be the attraction by the geometric golden ratio [16]. The attractive proportion is observed regularly in the dimensions of patterns of human face, plants, and animals. Since the aesthetic attraction with the

V. Bhateja et al. (eds.), *Information Systems Design and Intelligent Applications*, Advances in Intelligent Systems and Computing 672,
https://doi.org/10.1007/978-981-10-7512-4_49

geometric rule is watched frequently in nature [18], it can facilitate the detection of objects in scene of nature with animals. To address this, we analyze the pattern of images in the point of aesthetic geometric view about how pixels are related to each other in case they belong to an object of the golden ration and how to estimate the level of relevancy. A novel baseline algorithm is proposed from our analysis for the detection of saliency. To the best of our learning, this work is one of the first efforts to represent the human visual fixation by building a method of saliency from implementation of the geometric aesthetics to patterns of images. Our proposed algorithm is robust with variation of lighting condition, and it was carefully tested with a benchmark [17] specialized for saliency experiments.

In summary, the contributions of this paper are to: (1) analyze aesthetic geometric aspects to understand how the attractive proportion of dimensions of objects influences salient features formulated by this article; (2) introduce a new method for saliency detection by investigating the proposed image features; and (3) perform experiments to show capability and efficacy of the method. The remainder of the paper is organized as follows: Sect. 2 reports extensive study of current models related to our work. Section 3 examines the geometric aesthetics in saliency detection and describes the new algorithm. Section 4 outlines our experimental results, followed by discussion, while Sect. 5 concludes the paper.

2 Prior Works

The most evident application of geometric rules of composition is enhancement of aesthetic aspect of photography. In a multimedia tool [19], user's criteria are described by manual object segmentation and inpainting which guide on how to modify the photograph using spatial decomposition. Our approach of study of the aesthetic attraction is aimed for other objective: to allocate the region of interest for images with assumption that the arrangement of natural pattern follows the rule of golden ratio in some degrees. To gain estimation of the head pose for human faces, a set of image features is proposed by the concept of golden ratio [20]. The method requires location of eyes and nose in image as input for the evaluation. Analysis of attractiveness of faces is presented in [21] where attractiveness of a face is rated by the aspect of symmetry, neoclassical canons, and golden ratio. In our application, location of salient region is estimated for general images using the golden ratio for aesthetic attraction estimation without initial spatial constraints.

The golden ratio is studied for the prediction of contrast variability in the dynamic magnetic resonance imaging [22]. The geometric rule is employed for the reconstruction of window which is flexibly adapted according to the amplitude of cardiac motion. Golden step in Cartesian track with spiral profile ordering for heart coronary magnetic resonance angiography is addressed in [23], where the angular step between two consecutive spiral interleaves is given by the golden ratio. In pursuit of salient object detection in natural scenes, region covariance descriptors, derived from simple image features [24], express image patches. Relying on regression approach for multi-level image segmentation and learning, regional descriptors covering contrast background are designed to form the saliency map [25]. Low-level features of luminance and color are used to specify salient regions [26]. The local structures in our method are represented also by regional descriptors but in a different concept which is based on composition rule for aesthetic attraction judgment. The method presented here uses the geometric aesthetic attractiveness and was developed in response to a need for detection of salient objects where background is texture—based and complex in most cases which raise difficulty in study. The next section describes the method in detail.

3 Proposed Method

In the aesthetic view, attractiveness or beauty is a very important feature for perception. When a person is seeing a scene, he/she tries to integrate visual data into a framework of intentions. The explicit attractiveness of parts of the scene engages visual intention of the observer and plays a significant role in fixation of eyes. Related to aesthetic statement of ancient philosophers, ideal notion of attractiveness is derived from harmony, symmetry, proportions, and geometry. The effect of attractiveness and the underlying cause for its occurrences have been mentioned by the golden section or golden rule with mathematical base [27]. Thus, our approach uses the aesthetic attractiveness to feature the saliency and explicitly measures the aesthetic attractiveness by the occurrence of the golden rule in images. Given a color image $u\colon \Omega \rightarrow \Re^3, \Omega \subset \Re^2$ which is considered with two spatial dimensions $x = (x_1, x_2) \in \Omega$, the saliency map of the image is formulated briefly (1) by distribution of attractiveness denoted by a, in a relationship with the golden rule (g).

$$saliency(u) := p_u(a) = p_u(a/g)p_u(g) + p_u(a/\neg g)p_u(\neg g) \tag{1}$$

We assume that some parts of images are pursued by the composition rule, which makes $p_u(g) > 0$ for the images and consequently gives formula (2). It is important to train data to predict the level of agreement of the golden rule with the attractiveness $p_u(a/g)$ to estimate saliency by (2):

$$saliency(u) = p_u(a/g)p_u(g) + p_u(a/\neg g)(1 - p_u(g)) \tag{2}$$

Estimation of $p_u(a/g)$ and $p_u(a/\neg g)$ can be given by support vector machines (SVM) [28] in Sect. 3.3. Related to the distribution $p_u(g)$, our new contribution is the analysis of golden rule for structure of images in this section. The distinct value of golden ratio φ is determined from the equation applied for a particular rectangle illustrated in Fig. 1a. Making images in nature is usually not supported well by lighting condition, so some image parts are strongly darker of lighter than the rest, and some image region having the aesthetic proportion might loss its attractiveness due to the lighting condition. A pre-operation for balancing light is needed to implement that produce light balanced u_l. The image region followed by the golden rule can be wrapped up by a rectangle having its size satisfying the ratio φ. The rectangle is determined by four points demonstrated in Fig. 1b, where the black point has three types of relations to blue points: horizontal, vertical, and diagonal. To gain saliency estimation in geometric aesthetic aspect by the golden rectangle, we now analyze members of the rectangle, namely the horizontal, vertical, and diagonal lines and evaluate their contribution to saliency.

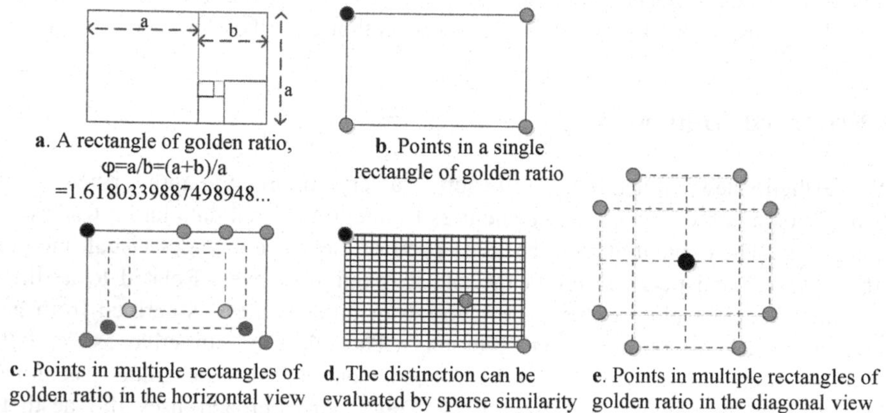

a
b
a
a. A rectangle of golden ratio,
φ=a/b=(a+b)/a
=1.6180339887498948...
b. Points in a single
rectangle of golden ratio
c. Points in multiple rectangles of
golden ratio in the horizontal view
d. The distinction can be
evaluated by sparse similarity
e. Points in multiple rectangles of
golden ratio in the diagonal view

Fig. 1. Points of a rectangle of the golden ratio

3.1 Horizontal and Vertical Lines of Golden Rectangle

If a salient object is covered by a golden rectangle drawn by continuous lines in Fig. 1c, other sub-golden rectangles drawn by dotted lines can be found inside the original one. The blue and green pixels may stay in object's border and observed with the same image features such as color or gradient. The yellow and violet pixels whose location is inside the object can have the same image features.

This similarity of local features can be seen for elements in background of image, meaning that the similarity is not unique for salient objects. Consequently, the type of similarity is not suitable to search salient objects. Inspection for vertical lines gives the same remark. However, let us see the feature in other point of view. If a sparse net of pixels (3) is arranged by vertical and horizontal lines through whole image like Fig. 1d, then

$$net(x) = \{y, \; y = x + w_1 \cdot k, k \; : \; integer\} \tag{3}$$

By taking difference of feature in pixel x to pixels of the net distributed over whole image, the distinction can be evaluated by the estimation of similarity for pixel in the sparse net (4):

$$diff(u(x)) = mean_{y \in net(x)} \|u_l(y) - u_l(x)\|^2 \tag{4}$$

The $u_d(x)$ in (5) presents evaluation of distinction for spatial locations, which are potential to be salient. Equations (4) and (5) support to save the computing time by controlling the number of points of the net:

$$u_d(x) = diff(u(x)) \tag{5}$$

Thus, checking pixels in a sparse net prepared by vertical and horizontal lines gives us the way to evaluate the distinction which enables for checking geometric aesthetic proportion in whole image.

3.2 Diagonal Lines of Golden Rectangle

An important processing stage is investigating the golden ratio φ which can be found when looking each pair of pixels. Denoting 2D dimensional distance between two pixels *x*, *y* by (d_1, d_2), a ratio *r*(*x*, *y*) can be calculated by (6). Figure 2e illustrates some possible positions of *y* by blue dots for pixel *x* marked by a black dot. We now formulate condition of spatial golden ration $r(x, y) \approx \varphi$ in the Gaussian form (7). The formula presents the degree of satisfaction of pixels *x* and *y* to the golden ratio.

$$r(x,y) = \begin{cases} \max(d_1/d_2, d_2/d_1), & d_1 \neq 0, d_2 \neq 0 \\ \infty, & otherwise \end{cases}, (d_1, d_2) = x - y \tag{6}$$

$$g_\sigma(x,y) = \exp\left(-\frac{(r(x,y)-\varphi)^2}{2\sigma^2}\right) \tag{7}$$

If two pixels have the same level of distinction u_d defined by (5), then their similarity of distinction is the base for assigning them to the same object. The measurement of distinction similarity u_d is described in Gaussian form (8). So that the probability *P*(*G*/*x*, *y*) evaluating if two pixels *x* and *y* belong to the same object and satisfy golden ratio can be presented by function *p*(*x*, *y*) in (9) with two components: the condition of the golden ratio φ (7) and the distinction similarity (8):

$$s_{\sigma_r}(x,y) = \exp\left(-\frac{(u_d(x)-u_d(y))^2}{2\sigma_r^2}\right) \tag{8}$$

$$P(G/x,y) = p(x,y)s_{\sigma_r}(x,y) \tag{9}$$

Above analysis of golden rectangle allows evaluating the probability $p_u(g)$ at *x*, meaning *x* belongs to some rectangle of golden ratio by checking distinction similarity of *x* with all other pixels *y*, which stay in diagonal lines and keep the golden ratio with *x* by (10). The accumulation of u_d in global domain Ω may provide poor performance and wrong detection as a visual object usually consists of pixels in some local region. In order to improve the performance and manage local search, we propose to embed a robust spatial condition which uses a weight inversely proportional to the distance between *x* and *y* by (11) with spatial deviation σ_s:

$$h(x) = \sum\nolimits_{y \in \Omega} u_d(y)p(x,y) \tag{10}$$

$$d_{\sigma_s}(x,y) = \exp(-\frac{(x-y)^2}{2\sigma_s^2}) \tag{11}$$

Thus, the feature of golden ration *f*(*x*) can be presented in form (12) of a convolution filter C_G (12) which accumulates the distinction similarity u_d weighted by three components: the condition of the golden ratio φ (7), the distinction similarity (8) which are grouped by *p*(*x*, *y*) in (9), and the spatial distance *d*(*x*, *y*) by (11). The parameter w_2 (13) sets the limitation of spatial distance for checking local region while the factor *e* (*x*) is to keep value of *f*(*x*) in a range [0, 1].

$$f(x) = u_d \cdot C_G = \frac{1}{e(x)} \sum\nolimits_{|x-y| \le w_2} u_d(y) p(x,y) d_{\sigma_s}(x,y) \tag{12}$$

$$e(x) = \sum\nolimits_{|x-y| \le w_2} p(x,y) d_{\sigma_s}(x,y) \tag{13}$$

3.3 Learning Saliency in Geometric Aesthetic Aspect

Hence, the golden ration feature *f*(*x*) prescribes $p_u(g)$, the level of agreement related to the golden ratio. To estimate relation of attractiveness to the golden ratio feature, we apply SVMs to train a dataset of manually selected saliency region. The training is performed to optimize the border of region of golden ratios by curvature penalty by comparing the border with saliency border provided by experts. This allows estimating the probabilities $p_u(a/g)$ and $p_u(a/\neg g)$ from (2) by learning examples with extracted golden features $p_u(g)$ and salient region $p_u(a)$. Then, learned kernels allow making prediction (14) of attractiveness from golden features feasible:

$$m(x) = saliencypredict(f(x)) \tag{14}$$

A major golden rectangle now is designed in accordance with *m*(*x*). Note that the edges of the rectangle are not always parallel or orthogonal with image frame. Figure 2a illustrates two examples for the circumstance where salient regions are shown by gray color. However, a tangential golden rectangle *t*(*x*) which has a pair of edges being paralleled with image frame border that always exists like the rectangles with black borders shown inside gray rectangles in Fig. 2a. In general, the number of golden rectangle is not limited by 1. Figure 2b shows a salient region that consists of two

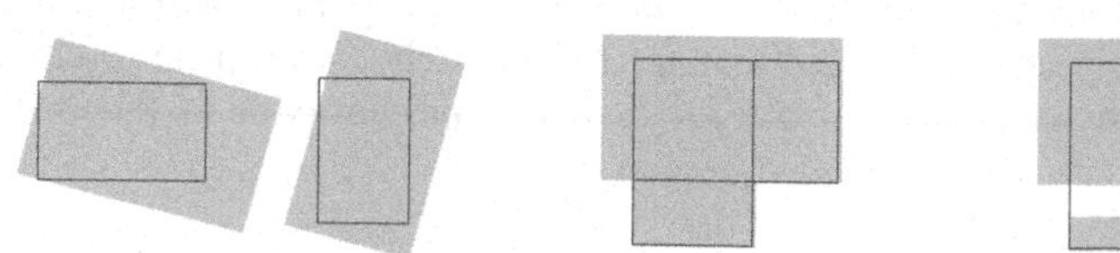

a. Objects marked in gray is observed by golden ratios partially b. Object inside central square is covered by multiple golden rectangles c. Small gray object is not covered fully by black golden rectangle

Fig. 2. Orientation and position of the golden rectangle

rectangles which go cross each other. In Fig. 2c, the rectangle in black border is not fully filled by salient region. Minor rectangles can be ignored like cases in Fig. 2b and c.

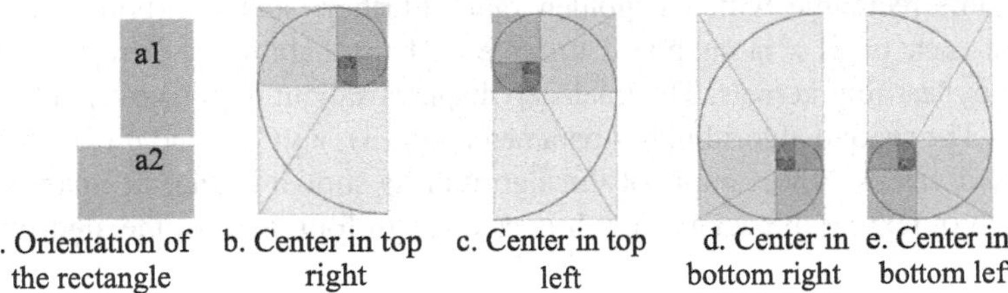

Fig. 3. Setting the golden rectangle for salient region: The orientation of the rectangle can be one of two possible options: a1—portrait and a2—landscape in (**a**). In case of orientation a1, the center of the rectangle can be case (**b**), (**c**), (**d**), or (**e**)

Once the main rectangle of the golden ratio is detected, its sub-rectangles can be explored by analysis of distribution of the intensity $f(x)$ inside the rectangle $m(x)$. There are two options of layout: portrait and landscape (Fig. 3a). By dividing the rectangle into four equal parts and comparing the total intensity of each path, the weight center of the rectangle can be allocated. Weight center for the rectangle in portrait layout can be in top right, top left, bottom right, or bottom left (Fig. 3b–e). The similar way is applied for landscape layout of the golden rectangle in Fig. 3a2. As final result, a golden rectangle $v(x)$ is defined by the geometric aesthetic aspect for the image.

3.4 The Algorithm

Here, an algorithm for detection of saliency using the aesthetic attractiveness by golden ratio is designed in Fig. 4. We exploited the composition aspect of images to search

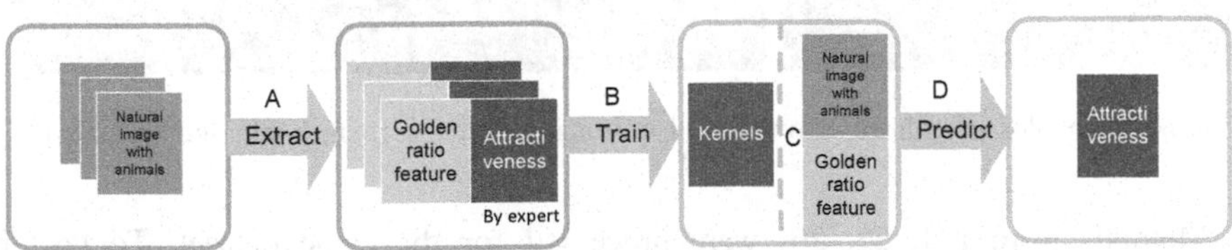

Fig. 4. Algorithm 1. Visual attention by training attractiveness with golden ratio (ATGR)

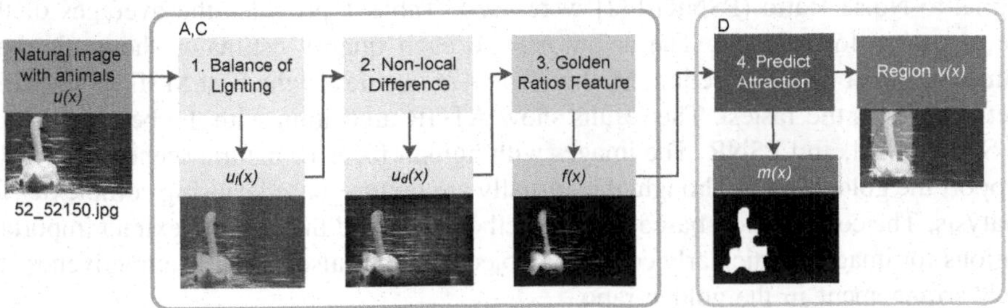

Fig. 5. Algorithm 2. Extraction of golden ratio feature (EGRF)

salient regions with aesthetic attractiveness by learning process that uses golden ratio feature detected by Algorithm 2 (Fig. 5). Specifically, after filtering image by the third step, the $f(x)$ presents region whose pixel x has high probability on staying in a diagonal line of some rectangle with the golden ratio. Further, step 4 exports the region $m(x)$ where each pixel x is supposed to have high probability of being attractiveness followed by learning kernels. The final step displays salient region covered by a golden rectangle. The second algorithm has parameters w_1 (3), w_2 (12), and σ (7) which can be set by fixed values. An example of the algorithm is shown in Fig. 5: Image $u(x)$ from [17] is rich of texture in background. It goes through four steps of the algorithm, and a rectangle in ratio φ is found based on saliency map $m(x)$.

4 Experiments

The proposed algorithm was evaluated on a dataset of 14.442 images from saliency benchmark [17] which provides three manual inputs for each test image. The dataset is divided randomly into two sets for training and prediction. Figure 6 illustrates examples which show golden ratio for scenes of human faces, man, and animals.

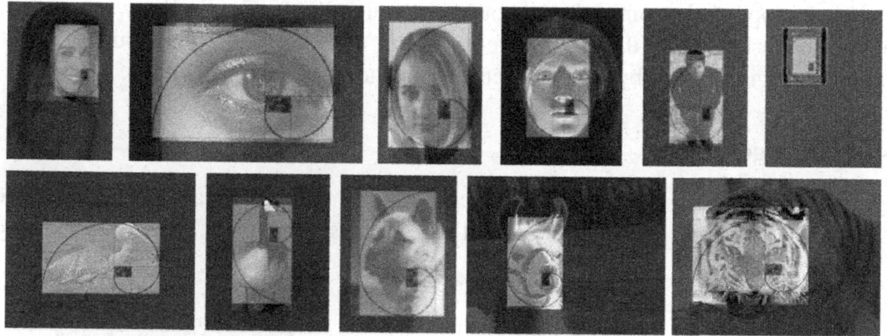

Fig. 6. Examples of golden ratio detected in images of faces, man, book, and animals [17]

Methods from [24, 25, 26] were processed for the same dataset. To test the objective performance of ATGR algorithm, Mean Squared Error (MSE) [28], Sum of Absolute Difference (SAD) [30], Structural SIMilarity (SSIM) [31], and Peak Signal-to-Noise Ratio (PSNR) [32] were used. Table 1 provides the averages of the measurement for our test. The arrow next to each quality estimator shows the best values in top or down direction. Method from [26] achieved the best MSE score, while method [25] is the fastest. The results show ATGR algorithm with the best evaluation of SAD, SSIM, and PSNR. The images with human faces or natural scenes potentially support the golden ratio. Though, they usually are texture-based causing complexity for analysis. The composition-based ATGR method enriched the ways to extract important regions for images particularly contained objects which raise aesthetic attractiveness by their arrangement in the golden ratio.

Table 1. Performance statistics [29]—the best scores are printed in bold

Method	Method [24]	Method [25]	Method [26]	ATGR
MSE ↓	0.025	0.160	**0.010**	0.338
SAD ↓	0.288	0.238	0.329	**0.219**
SSIM ↑	0.965	0.968	0.960	**0.970**
PSNR ↑	54.9	55.0	54.1	**55.5**
Time ↓	15.2	**2.6**	34.7	11.5

In our experiment, images were resized to have equal width of 256 pixels. Then, nonlocal difference is estimated with the net of nonlocal neighbors. The sparseness of the net is regulated by parameter w_1 (4). As small as w_1, the exactness may be improved, but computing time is more consumed. Figure 7 shows results of the algorithm with parameters. Figure 7b is the result for the net with more points than the net in Fig. 7a. All quality measures for Fig. 7b are better than Fig. 7a, but the case in Fig. 7b spent more time. The parameter w_2 (13) regulates the size of frame for the local convolution. As large as w_2, the number of neighbors for checking is increased; therefore, more neighbors are involved in consideration. If the image is free of noise and background objects are insignificant, then quality may be increased. Otherwise, increment of w_2 does not improve the detection quality. Figure 7c and d shows that SAD for $w_2 = 50$ in Fig. 7d is better than the case of $w_2 = 10$ in Fig. 7c, but MSE score is upgraded.

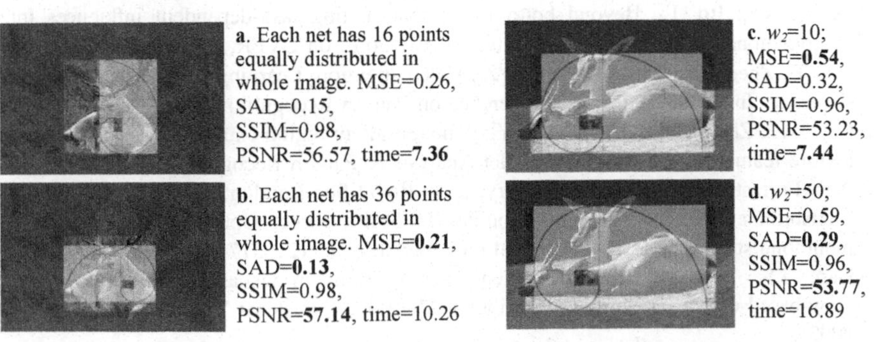

a. Each net has 16 points equally distributed in whole image. MSE=0.26, SAD=0.15, SSIM=0.98, PSNR=56.57, time=**7.36**

b. Each net has 36 points equally distributed in whole image. MSE=**0.21**, SAD=**0.13**, SSIM=0.98, PSNR=**57.14**, time=10.26

c. w_2=10; MSE=**0.54**, SAD=0.32, SSIM=0.96, PSNR=53.23, time=**7.44**

d. w_2=50; MSE=0.59, SAD=**0.29**, SSIM=0.96, PSNR=**53.77**, time=16.89

Fig. 7. Effect of using variable parameters for golden similarity detection

5 Conclusions

We provided scientific analysis for computations and novel practical technique that uses the geometric aesthetic aspect for the detection of region in visual interest. Within the expectation of compatibility of objects in images to the geometric composition rule of golden ratio, it seems possible that the objects can be detected by getting the salient map. This leads us to a novel-efficient approach solving the saliency problem: By following the effect of similarity caused by the composition rule, we can trim the search space from whole image to a specific region of visual interest. In particular, we proposed a new algorithm for saliency detection based on analysis of aesthetic attractiveness by the golden ratio of image patterns. Experimental results demonstrate that our geometric method achieved remarkable performance for human faces and natural images with wild animals. We believe, however, that architectural arrangement like temples of Ancient Greeks has a high degree of compliance with the ratio. Thus, we may be able to bring our understanding of the implementation of geometrical composition for the architectural imaging in future research.

Acknowledgements. Author thanks anonymous referees for their valuable comments.

References

1. Itti, L., Koch, C., A saliency-based search mechanism for overt and covert shifts of visual attention, Vis. Res., Vol. 40, Issues 10–12, pp. 1489–1506, 2000
2. Navalpakkam, V., Itti, L., An integrated model of top-down and bottom-up attention for optimizing detection speed. IEEE Comp. Soc. Conf. on Comp. Vis & Pattern Rec, 2006
3. Avraham T., Lindenbaum M., Esaliency (extended saliency), meaningful attention using stochastic image modeling, IEEE Trans. Pattern Anal. Mach. Intell., Vol. 32, Iss 4, 2010
4. Liu, T., Yuan, Z., Sun, J., et al., Learning to detect a salient object, IEEE Trans. Pattern Anal. Mach. Intell., Volume: 33, Issue: 2, 2010
5. Mahadevan, V., Vasconcelos, N., Spatiotemporal saliency in dynamic scenes, IEEE Trans. Pattern Anal. Mach. Intell., Volume 32, Issue 1, pp. 171–177, 2010
6. Peters, R.J., Itti, L., Beyond bottom-up, incorporating task-dependent influences into a computational model of spatial attention. IEEE Conf. on CVPR, pp. 1–8, 2007
7. Liu, T., Zheng, N., Ding, W., Yuan, Z., Video attention, Learning to detect a salient object sequence. Int. 2008 19th Inter Conference on Pattern Recognition, pp. 1–4, 2008
8. Cai, J.-Z., Zhang, M.-X., Chang, J.-Y., A novel salient region extraction based on color and texture features. Int. Conf. on Wavelet Analysis & Pattern Recognition, 8–15, 2009
9. A.R. Forrest, Computational geometry, Proc. Royal Soc London, 321, series 4, 1971
10. Edelsbrunner, Herbert, Guibas, Leonidas J., Stolfi, Jorge. Optimal point location in a monotone subdivision. SIAM Journal on Computing 15 (2), 317–340, 1986
11. Hare, J.S., Lewis, P.H., Scale saliency, applications in visual matching, tracking and view-based object recognition. Distributed Multimedia Systems Vis Inf Sys, pp. 436–440, 2003
12. Maver, J., Self-similarity and points of interest, IEEE Trans Pattern Anal Mach Intell. 2010 Jul; Vol 32, Issue 7, pp. 1211–1226, 2010
13. Matoušek, Jiří, Geometric range searching, ACM Comp Surveys 26 (4), 421–461, 1994

14. Zhang, Q., Xiao, H., Biologically motivated salient regions detection approach. Second Int. Symp. on IITA, Vol. 2, pp. 1100–1104, 2008
15. Kadir, T., Brady, M., Saliency, scale, and image description, Int. J. Comp. Vis., 45, 2000
16. Pacioli, Luca. De divina proportione, Luca Paganinem de Paganinus de Brescia, 2001
17. Liu Tie, Jian Sun, Nan-Ning Zheng, Xiaoou Tang, Heung-Yeung Shum. Learning to Detect A Salient Object. IEEE Trans on Patt Analysis and Machine Intel 33(2), 2011
18. Green, C. D. All that glitters, a review of psychological research on the aesthetics of the golden section, Perception, vol. 24, pp. 937–968, 1995
19. S Bhattacharya, R Sukthankar, M Shah. A framework for photo-quality assessment and enhancement based on visual aesthetics. Proc. of 18th ACM Int Conf on Mm, 2010
20. Gianluca Fadda, Gian Luca Marcialis, Fabio Roli, Luca Ghiani, Exploiting the Golden Ratio on Human Faces for Head-Pose Estimation. ICIAP (1) pp. 280–289, 2013
21. K Schmid, D Marx, and A Samal. Computation of a face attractiveness index based on neoclassical canons, symmetry, and golden ratios. Pattern Recogn. 41(8), 2710–17, 2008
22. Winkelmann S, Schaeffter T, Koehler T, Eggers H, Doessel O. An optimal radial profile order based on the Golden Ratio for time-resolved MRI. IEEE Trans MI 26(1), 2007
23. Prieto C, Doneva M, Usman M, Henningsson M, Greil G, Schaeffter T, Botnar RM. Highly efficient respiratory motion compensated free-breathing coronary MRA using golden-step Cartesian acquisition. J Magn Reson Imaging. 41(3), pp. 738–746, 2015
24. Erkut Erdem, Aykut Erdem. Visual saliency estimation by nonlinearly integrating features using region covariances. Journal of Vision, 13(4):11, 2013
25. Wang, J., Jiang, H., Yuan, Z. et al., Salient Object Detection: A Discriminative Regional Feature Integration Approach, Int J Comput Vis CVPR, 2016
26. R. Achanta, F. Estrada, P. Wils and S. Süsstrunk, Salient Region Detection and Segmentation, ICVS, Vol. 5008, Springer Lecture Notes in Comp Sc., pp. 66–75, 2008
27. Birch Fett, An In-depth Investigation of the Divine Ratio. Montana Mathematics Enthusiast, Vol. 3 Issue 2, p. 157, 2006
28. Press, William H., Teukolsky, Saul A., Vetterling, William T., Flannery, B. P. Section 16.5. Support Vector Machines. Numerical Recipes: The Art of Sci Computing, 2007
29. Wackerly, Dennis, Mendenhall, William, Scheaffer, Richard L. Mat Stat with App (7 ed.) USA: Thomson Higher Education. ISBN 0-495-38508-5, 2008
30. Iain E. Richardson. H. 264 and MPEG-4 Video Compression: Video Coding for Next-generation Multimedia. Chichester: John Wiley & Sons, ISBN: 978-0-470-86960-4, 2003
31. Dosselmann, Richard, Yang, X Dong. A comprehensive assessment of the structural similarity index. Signal, Image and Video Processing 5 (1): 81–91, 2009
32. Huynh-Thu, Q., Ghanbari, M. Scope of validity of PSNR in image/video quality assessment. Electronics Letters 44 (13): 800, 2008

Delay-Based Reference Free Hardware Trojan Detection Using Virtual Intelligence

S. Kamala Nandhini(✉), S. Vallinayagam, H. Harshitha, V. A. Chandra Shekhar Azad, and N. Mohankumar(✉)

Department of Electronics and Communication Engineering, Amrita School of Engineering, Amrita Vishwa Vidyapeetham, Coimbatore, India
nandhini_skn@yahoo.co.in, {vallinayagamshankar, hharshitha19, azadbsr3095}@gmail.com, n_mohankumar@cb.amrita.edu

Abstract. Virtual instrumentation is a powerful tool that has been largely left unexplored in the domain of hardware security. It facilitates creation of automated tests to detect the presence of Trojans in a circuit thereby reducing the chance of human errors and the time required for testing. The presence of a stealthy Trojan in large VLSI circuits could lead to leakage of confidential information even in high-security applications such as defense equipment. Here, we propose the usage of virtual instrumentation to detect the presence of a delay-based Trojan in a circuit. Our results confirm that VI-based systems provide a cheap, self-sufficient, easy-to-use interface, and flexible scheme which can be easily modified to accommodate any VLSI circuit. This can also be used in other detection techniques without the need for use of complex systems.

1 Introduction

A hardware Trojan (HT) is a malicious modification of any circuit with the intent of disrupting the normal functioning of the circuit and cause any range of adverse effects such as denial of service, modification of output, leakage of information, or even destruction of the chip. These hardware Trojans can be inserted in any part of a chip, especially in vulnerable nodes. This can lead to a wide number of malicious effects such as information leakage, denial of service, faulty functioning, modification of logic, excess delay. The presence of HT at such vulnerable nodes would hinder detection techniques. These Trojans, when used to leak confidential information to a third party, may lead to huge economic loss and also a serious threat to public safety. With the number of gates on an IC increasing at a very rapid rate and the advent of fast-paced technology, it is becoming increasingly difficult to detect the presence of a Trojan in a large-scale VLSI circuitry. Many methods have been explored to detect the presence of a hardware Trojan.

In this paper, we present an HT detection technique based on the use of virtual instrumentation (VI). VI is a combination of hardware and software used to simulate the functioning of different components. This is advantageous when the real components are costly and their complexity is high. We are using the VI software LabVIEW

V. Bhateja et al. (eds.), *Information Systems Design and Intelligent Applications*, Advances in Intelligent Systems and Computing 672,
https://doi.org/10.1007/978-981-10-7512-4_50

developed by National Instruments along with myRIO, a real-time embedded board. The board requires LabVIEW and contains a processor and FPGA, thus providing the edge, of not only programming the I/O ports but also being used for data acquisition.

With the size of integrated VLSI circuits increasing to such an extent as to accommodate millions of gates in a single chip, it is becoming extremely vulnerable to HT attacks. These chips are used in almost every application from household devices and mobile phones to highly confidential military and space applications. Most manufacturing processes are outsourced to different countries around the world. Thus, these chips are highly vulnerable to HT insertion. Many different methods have been developed to detect Trojans such as self-consistency-based approach, invasive methods, reverse engineering. These involve high level of complexity and take much more time to process and execute.

Various surveys have been conducted in the domain of hardware security. The hardware detection techniques can be classified as detection, diagnosis, and prevention [1]. Many methods such as side-channel analysis, logic test, obfuscation, encryption, and split manufacturing have been proposed. The main parties involved in the design of an IC are Foundries, SoC designers, IP vendors, EDA tool vendors, and IC end users [2]. The Trojan may be inserted at any of these stages causing an underlying risk for the whole IC design cycle. One of the major challenges is that most existing HT detection techniques rely on the golden chip which is difficult to obtain, even does not exist. Therefore, HT detection and diagnosis without reference model could be in an urgent demand.

Among different Trojan detection methodologies, the main methods are classified as side-channel analysis and Trojan activation [3]. Side-channel signals, mainly including timing and power, can be used for Trojan detection. However, the amount of power that a Trojan draws may be so small that it could be enveloped by noise and process variation effects. Trojan activation method, however, fails to detect Trojans that are placed at vulnerable nodes that are not easily accessible. Timing analysis presents a much more feasible solution to the detection of HT.

In order to better understand hardware Trojans, an efficient method of classification is necessary. A new classification based on prevention and detection techniques has been proposed in [4], which gives a detailed description of the various existing detection techniques. HT is a threat to the confidentiality of chips as they leak confidential information at runtime. An efficient architecture has been proposed in [5] called Runtime Trust Neural Architecture (RTNA), based on Adaptive Resonance Theory neural network (ART1). This when incorporated with the SOC architecture can inhibit it from being compromised. Four typical cases are present in time-based analysis. The four cases are No Delay-Delay (ND-D), No Delay-No Delay (ND-ND), Delay-Delay (D-D), and Delay-No Delay (D-ND). We can make use of this approach for hardware Trojan detection without the need for a golden IC. In our proposed model, we use virtual instrumentation. This method uses clock cycles directly without the need for complex coding and thereby reduces development time.

An alternate approach to HT detection is the detection of Trojan using Gate-Level Characterization (GLC). This method is based on the measurement of side-channel parameters, like leakage power [6]. This method is generally used to detect if a Trojan is present without the use of a golden IC. This method would require the use of additional software to obtain the power characteristics of any circuit and more computational effort. Hence, timing analysis-based approach is used here.

A technique involving a PIC16f877a for Trojan detection has been proposed in [7]. This technique uses a microcontroller (PIC16f877a)-based system to detect the presence of Trojan. The weights of the circuits are initialized to zero. The circuit receives inputs from PIC, and the corresponding outputs are again fed back. Now the accuracy is determined by voting, and the weights are calculated and are displayed using a display system. This system works as a robust detection system for malicious circuits. myRIO is similar to a microcontroller-based system with similar implications when applied in the field of hardware security.

Use of virtual instrumentation reduces the high cost of buying traditional instruments and provides low maintenance cost, flexibility, graphic display, and many more features. LabVIEW is a graphic programming environment used for managing and controlling industrial manufacturing stations [8]. Modeling an efficient Trojan detection mechanism can be facilitated efficiently by using virtual instrumentation concepts.

The use of myRIO by National Instruments as a platform for hands-on training in systems and control education is good due to its portability, low cost, and great functionality [9]. It enhances determinism and prevents highly negative results by being aware of the hardware's limitations. Using myRIO, which is effectively an FPGA along with a processor, we can verify the functionality of our Trojan detection scheme which uses virtual instrumentation to detect the presence of a Trojan in a VLSI circuit.

Developed using VI software platform LabVIEW, a system that focuses on a fault testing device employed in fire control system was developed [10]. The fault testing software application implements the display of diagnosis, fault causes, and troubleshooting results. A person with little understanding of the system can also carry out rapid online testing, diagnosis, and for performing maintenance activities on the system. A very similar approach is used to detect the presence of Trojan, and thus we may be able to provide an easy-to-use mechanism that can be implemented at a very low cost.

In [11], a new path delay-based HT detection technique is proposed that can be used to overcome the effect of process variations. It proposes to use a clock glitcher to sample erroneous signal values. However, it can only detect path delays greater than 8.25 ns. In [12], shadow register-based time delay analysis is studied in detail. This technique also requires a golden IC and cannot detect extremely small variations in time. Such problems are overcome in this model proposed.

2 Proposed Technique

The primary technique we will be used to detect the presence of a Trojan is timing analysis [5]. Based on the delay produced by the circuit, we can define four cases based on which we can decide whether a Trojan is present or not. Table 1 shows the four cases considered (Fig. 1).

Table 1. Criteria for deciding presence of Trojan based on delay obtained

Case	Delay criteria	Inference
CASE I	No Delay-No Delay (ND-ND)	The time required for two consecutive iterations is the same. Hence, no detection of trojan
CASE II	No Delay-Delay (ND-D)	This shows abnormality in the circuit but this could be due to other circuit parameters. Hence presence of Trojan is not absolute
CASE III	Delay-Delay (D-D)	This may occur due to circuit aging or due to the possibility of no change in input bits. Hence, no Trojan activity
CASE IV	Delay-No Delay (D-ND)	The presence of Trojan is confirmed since there is a no delay followed by delay criteria

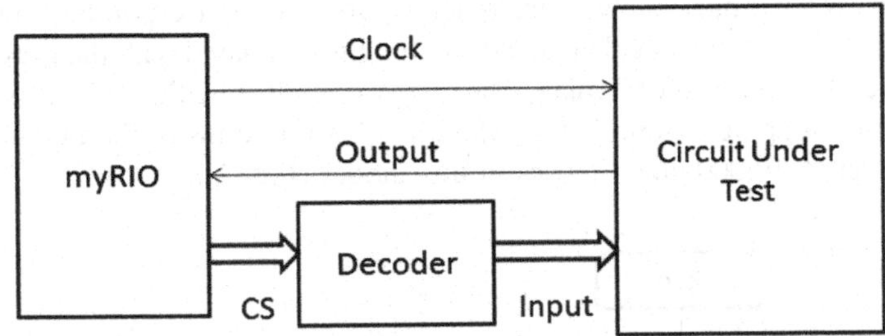

Fig. 1. Block diagram of system

In CASE I, the time taken for two consecutive toggles is the same and there is no delay. This shows that there is no Trojan activity. In CASE II, the delay is higher in the second iteration than in the first iteration. In this case, there need not necessarily be a Trojan present since the additional delay may be due to other parameters such as process variations. This does indicate some abnormality in the circuit but presence of Trojan is not absolute. In CASE III, there is a delay in obtaining output but the delay is the same for two consecutive iterations. This happens when there is no change in inputs, hence no Trojan activity. In CASE IV, there is a delay in the first iteration followed by no delay in the second iterations. As we know that a delay-based Trojan stays active for a limited period of time, specific pattern of inputs, or placed at nodes that get activated as a rare event, the Trojan will get deactivated in consecutive iterations. Thus, we observe the occurrence of a Delay followed by a No Delay condition confirming the presence of a Trojan.

VLSI circuits generally consist of a large number of gates that define the functionality and an IP core. If a Trojan is inserted, it can make modification and steal information to the same. In our model, we have considered a hardware Trojan which alters the delay of the circuit depending on the input to the Trojan. This input can either be our primary input, input at internal nodes, or inputs to the IP core of the IC. Thus, this information may be compromised and misused.

For testing our model, we have considered ISCAS 89 benchmark circuits. First, we initialize the input such that the output of the circuit under test is 1. Now the inputs are changed, and clock pulses are given. The number of clock pulses given to the circuit is counted until the output toggles to 0. The time taken for the output to be generated is calculated using the Eq. (1).

$$D = N * T \tag{1}$$

where

D represents delay,
N represents number of clock cycles,
T represents time period of the clock.

Since myRIO has only 8 I/O pins, we use a decoder to accommodate the necessary number of inputs. The clock signal is generated by myRIO. Hence, the time period can be identified. With similar analysis for other inputs, their corresponding output delay time is calculated. The newly calculated delay time is compared with the previous delay time to detect the presence of Trojan. The test is completed either when the Trojan is detected or the stopping criteria is met which is given in terms of the number of clock pulses. It is set to 10,000 clock pulses in this model (Fig. 2).

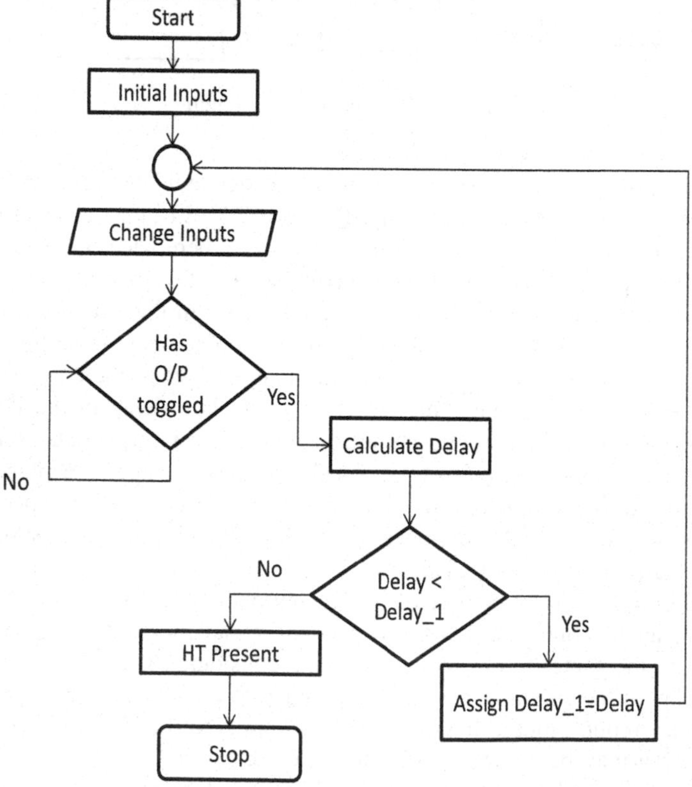

Fig. 2. Flowchart of process flow

3 Results

Figure 3 shows LabVIEW model for delay-based Trojan detection. The output of the circuit is fed to this module which calculates the delay based on when the output changes (Fig. 4).

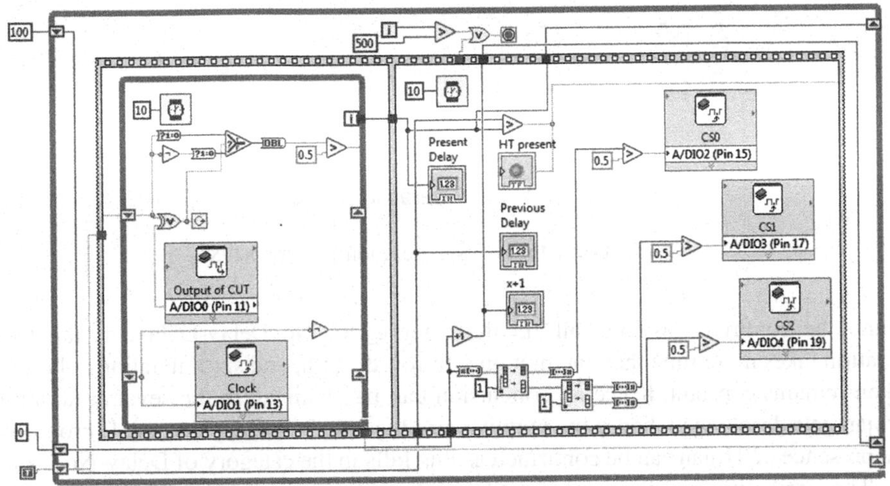

Fig. 3. LabVIEW model for delay-based Trojan detection

The plots above show the time taken for each circuit to produce the output for the corresponding inputs when a clock pulse of 10 kHz was used. In the case of S298, in Fig. 5, the output takes 3 clock triggers in first two iterations and then it takes 5 clock pulses to produce the output in third iteration. With this change, the presence of Trojan

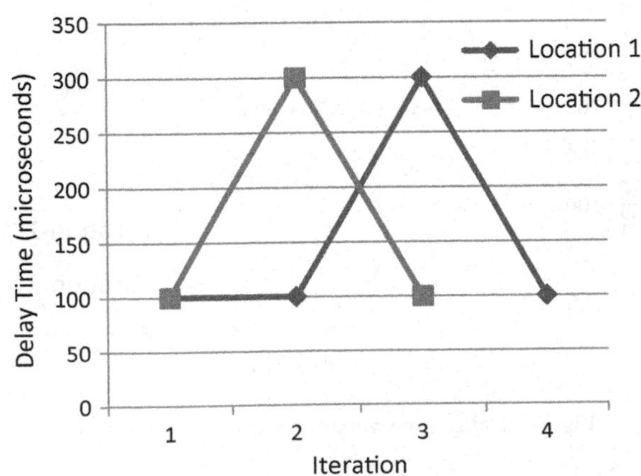

Fig. 4. Delay time versus iteration plot for S27

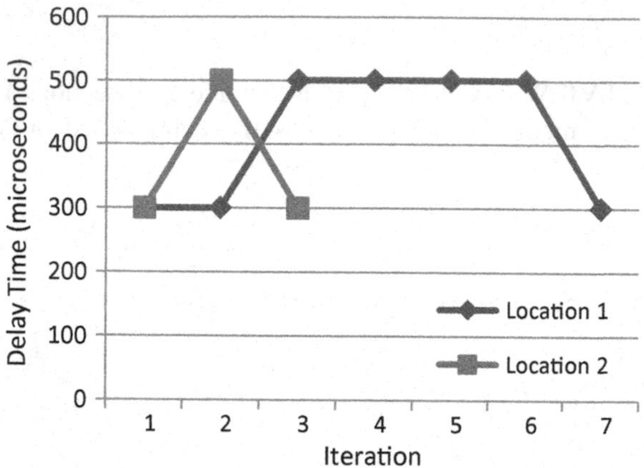

Fig. 5. Delay time versus iteration plot for S298

cannot be confirmed as this falls in the category of No Delay-Delay, as the third iteration takes more time than the first one. In fourth, fifth, and sixth iterations, the time taken remains constant; hence no conclusion can be obtained. In the seventh iteration, Trojan gets deactivated; hence the output is produced in 3 clock cycles itself. From this, the presence of Trojan can be confirmed as this falls in the category of Delay-No Delay.

The similar process was carried out by placing the Trojan in two different locations in different circuits, and results were obtained. Figure 4 shows the delay time versus iteration for S27, and Fig. 6 shows the delay time versus iteration for S344.

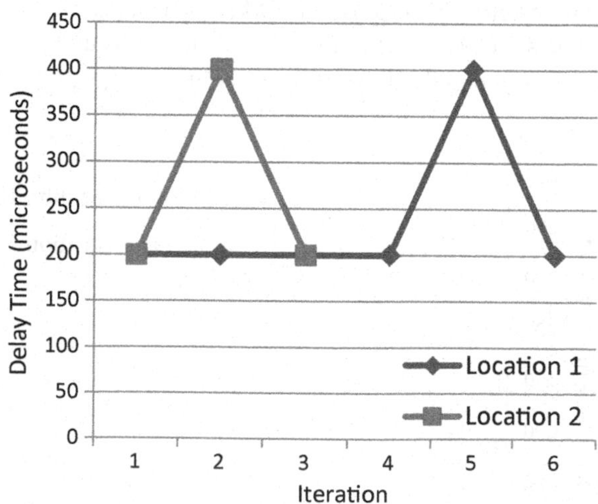

Fig. 6. Delay time versus iteration plot for S344

4 Conclusion

The proposed system uses a timing-based detection strategy to identify the presence of a Trojan in any given circuit. This is done without the use of a golden IC or reference IC thereby making the system self-sufficient. In the event of a Trojan not being identified even after 10,000 clock cycles which is our stopping criteria, the Trojan will not be detected. Also, if there is a Trojan that produces a "stuck at 0" or "stuck at 1" condition, the Trojan will not be identified by the system. This may bring down the accuracy of Trojan detection in this model. This system is a noninvasive Trojan detection methodology that can be easily implemented in a very short period of time. It can be used for run-time monitoring as we use the clock cycles taken for toggling to occur during the functioning of the circuit. It is also highly cost efficient and can also provide a user interface that is easy to understand and use. Use of virtual instrumentation reduces the high cost of buying additional hardware and provides flexibility, low maintenance cost, and advanced resources for graphic display. The use of myRIO by National Instruments provides an excellent platform due to its low cost, portability, and high functionality.

5 Future Work

The main advantage of VI-based system is that different methods of analysis can be carried out simultaneously or sequentially thereby providing a powerful and robust hardware Trojan detection system. This method of timing-based analysis for Trojan detection can be extended to other sequential circuits as well. Also with small variations made to the proposed VI model, it can be used to identify Trojans using other side-channel techniques. Methods to detect Trojan infected combinational circuits can also be developed.

References

1. M. Tehranipoor, H. Salmani, X. Zhang, X. Wang, R. Karri, J. Rajendran and K. Rosenfeld, "Trustworthy Hardware: Trojan Detection and Design-for-Trust Challenges", Computer, vol. 44, no. 7, pp. 66–74, 2011.
2. H. Li, Q. Liu and J. Zhang, "A survey of hardware Trojan threat and defense", Integration, the VLSI Journal, vol. 55, pp. 426–437, 2016.
3. M. Tehranipoor and F. Koushanfar, "A Survey of Hardware Trojan Taxonomy and Detection", IEEE Design & Test, pp. 1–1, 2013.
4. B. Amin, S. TaghiManzuri, and H. Afshin, "Trojan Counteraction in Hardware: A Survey and New Taxonomy", Indian Journal of Science.
5. K. Guha, D. Saha, and A. Chakrabarti, "RTNA: Securing SOC architectures from confidentiality attacks at runtime using ART1 neural networks," 2015 19th International Symposium on VLSI Design and Test, 2015.
6. D. K. Karunakaran and N. Mohankumar, "Malicious combinational Hardware Trojan detection by Gate-Level Characterization in 90 nm technology," Fifth International

Conference on Computing, Communications and Networking Technologies (ICCCNT), 2014.

7. R. Bharath, G. A. Sabari, D. R. Krishna, A. Prasathe, K. Harish, N. Mohankumar, and M. N. Devi, "Malicious Circuit Detection for Improved Hardware Security," Communications in Computer and Information Science Security in Computing and Communications, pp. 464–472, 2015.
8. M. A. Lemos, D. M. Brunini, G. Botura, M. A. Marques, and L. C. Rosa, "Virtual instrumentation: A practical approach to control and supervision process," Proceedings of 2011 International Conference on Computer Science and Network Technology, 2011.
9. S. D. Ruben, "Respect the implementation: Using NI myRIO in undergraduate control education," 2016 American Control Conference (ACC), 2016.
10. L. Guoliang, Y. Xiaoqiang, L. Hongwei, and H. Yu, "Fault Testing Device of Fire Control System Based on Virtual Instrumentation Technology," 2012 International Conference on Computer Science and Service System, 2012.
11. B. Robisson, "Resilient hardware Trojans detection based on path delay measurements," 2015 IEEE International Symposium on Hardware Oriented Security and Trust (HOST), 2015.
12. I. Exurville, L. Zussa, J.-B. Rigaud, and D. Rai and J. Lach, "Performance of delay-based Trojan detection techniques under parameter variations," 2009 IEEE International Workshop on Hardware-Oriented Security and Trust, 2009.

Compact RCA Based on Multilayer Quantum-dot Cellular Automata

Nuriddin Safoev and Jun-Cheol Jeon(✉)

Department of Computer Engineering, Kumoh National Institute of Technology, 61, Daehak-ro, Gumi, Gyeongbuk 39177, South Korea
nuriddinsafoyev@gmail.com, jcjeon@kumoh.ac.kr
jcjeon33@gmail.com

Abstract. The full adder is a basic logical circuit that performs an addition operation in any arithmetic circuits. In our paper, a bettered 1-bit full adder is proposed based on quantum-dot cellular automata (QCA). It confirms that the proposed circuit works properly and it can be used as a highly efficient design in arithmetic circuits. Finally, by connecting four 1-bit QCA full adders, we present a 4-bit ripple carry adder successfully. Our design has been realized with a simulation tool QCA Designer. The performance of our structure has significantly achieved improvements in terms of cell count, occupied area, and delay.

1 Introduction

Nanotechnology is extremely small dimension technology in which the construction of matter is controlled at nanometer scale. Scientific research in the field of semiconductor electronics industry has promptly and consistently reduced due to device sizes and power consumption for several decades. Therefore, scientists think that "complementary metal-oxide semiconductor (CMOS) technology is approaching to its physical limit according to the scientific prognosis" [1]. These kinds of limitations have led to significant efforts to find relevant alternatives among the proposed solutions, such as "tunneling phase logic (TPL), single-electron tunneling (SET), and quantum-dot cellular automata (QCA)." They all have received considerable attention. Especially, "QCA seems to be the most promising technology among the alternatives for CMOS technology" [2].

QCA is one of the models of quantum computation that "computes with cellular automata and it consists of quantum-dot element array." Concept of this technology was proposed by Tougaw and Lent [3]. The phenomenal characteristic of this technology is that "logic states are not stored at the voltage level like normal electronic devices, but they are represented as logic states by a quadratic-shape cell. The cell is a nanoscale mechanism that is capable of encoding data by two freelance electron configurations. The cells should be placed in a straight line at nanoscales to provide right functionality. Hence, the real performance of these devices for producing imperfection and maladjustment plays serious role in circuit quality" [4]. Data transmission of the technology is realized using QCA clocking. It is managed by a potential barrier. According to under control of potential barriers, "QCA clock raises and lowers

V. Bhateja et al. (eds.), *Information Systems Design and Intelligent Applications*, Advances in Intelligent Systems and Computing 672,
https://doi.org/10.1007/978-981-10-7512-4_51

the tunneling barriers. The clocking technique is composed of four phases: locking, locked, relaxing, and relaxed" [5].

Over CMOS technology, QCA has several preferences. Some of the benefits contain high-speed switching, high-level circuit density, and low power consumption. The expectation is that "all these benefits can bring extreme result in the progress of powerful and efficient computers."

The most frequently used component in arithmetic units of CPU is full adder. Basically, the function of full adder is addition, and also it is utilized as a core part in other arithmetic operations such as subtraction, division, and multiplication. In that case, designing efficient full adder in QCA is a great achievement to increase developing process of QCA arithmetic circuit aspects [6, 7].

This paper presents less hardware complexity full adder using three-input exclusive OR gate (TIEO) which was proposed by Ahmad et al. Furthermore, using the proposed full adder, a 4-bit ripple carry adder has been designed with significant features. It can be easy to extend serial adder as ripple carry adder. In comparison with existing approaches, the proposed design has been improved in terms of space and time complexity.

This paper is organized as follows. In Sect. 2, we provide information about basic concept of QCA technology as well as 1-bit QCA full adder. Section 3 shows the architectural designs of proposed full adder. Based on this full adder, 4-bit ripple carry adders have been proposed. Finally, we analyze the improvement of the proposed designs in Sect. 4.

2 Related Researches

2.1 QCA Concept

The quantum cell is "basic building block of this technology, and it is composed of four quantum dots." As illustration, "two polarized basic quantum cells are shown in Fig. 1. Each cell is made from four quantum dots that are placed in a square structure." "Two electrons that are free to tunnel between adjacent quantum dots charge the cell." These two arrangements indicate cell polarizations, "such as P = +1 and P = −1." Binary data can be converted into a coded form, and by using cell polarization P = +1 to represent logic "1" and P = −1 to represent logic "0", data is propagated [8, 9].

A chain of cells in a line builds a QCA standard wire. In QCA wire, due to coulomb interactions between cells, the binary information is propagated from input to output. The wire may be formed by arranging cells one by one as shown in Fig. 2a. "If the

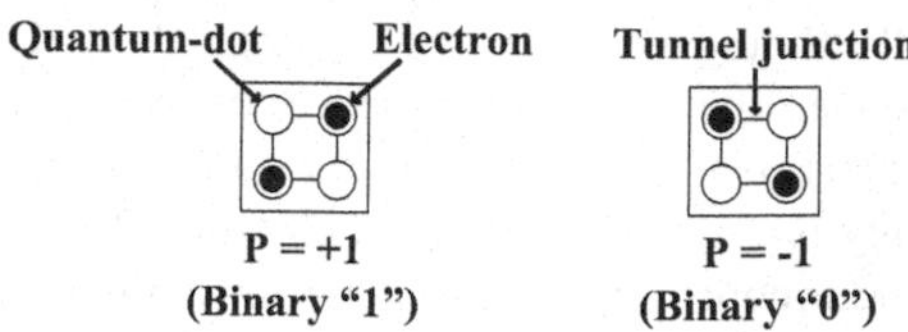

Fig. 1. Two polarizations of quantum cells

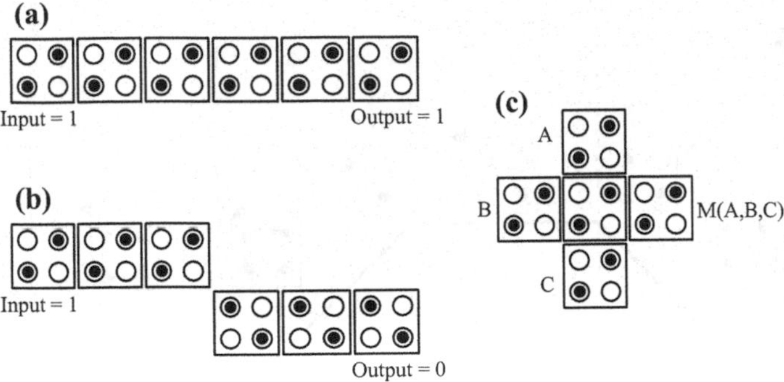

Fig. 2. QCA principal structures: **a** standard wire, **b** inverter, and **c** 3-input majority gate

input is one, then the signal is transferred from the input to the output binary one or if the input is zero, then zero is transferred to the output".

Various logic circuits can be productively designed based on QCA principal structures. "Inverter and majority gate are mostly used to design any circuit." "As shown in Fig. 2b, the input polarization is inversed when the signal transmits to the output cell. Presented inverter is simple inverter" [9]. Majority gates play the most important role in conventional circuit design. In this case, complexity of the structure has been relatively associated with the cell configuration of QCA fundamental gates. Figure 2c illustrates the building form of the majority gate, and "it is constructed by three input cells (A, B, and C), one inside cell, and one output cell." The logic equation of the majority gate is defined as:

$$M(A, B, C) = AB + BC + AC; \tag{1}$$

Basic building architecture of any digital system is made from logic operations. "General terms AND and OR gates will be obtained by polarizing one input of the majority gates." If one of the input cell, for example, input C is set to zero, the remaining two input cells A and B realize logic AND operation. Conversely, "input C is set to binary one, and input cells A and B realize logic OR operation." It is illustrated in Eq. (2) and Eq. (3), respectively [8, 10].

$$M(A, B, 0) = AB; \tag{2}$$

$$M(A, B, 1) = A + B; \tag{3}$$

As we said before, "all logical function can be implemented using combination of inverter and majority gates."

2.2 Clocking in QCA

The clock pulse mechanism in QCA technology gives permission for computing a sub-array. Then "it bricks the sub-array as well as polarization of the output cell sets as an input for the next sub-array." It means that modifying phase is as potential change.

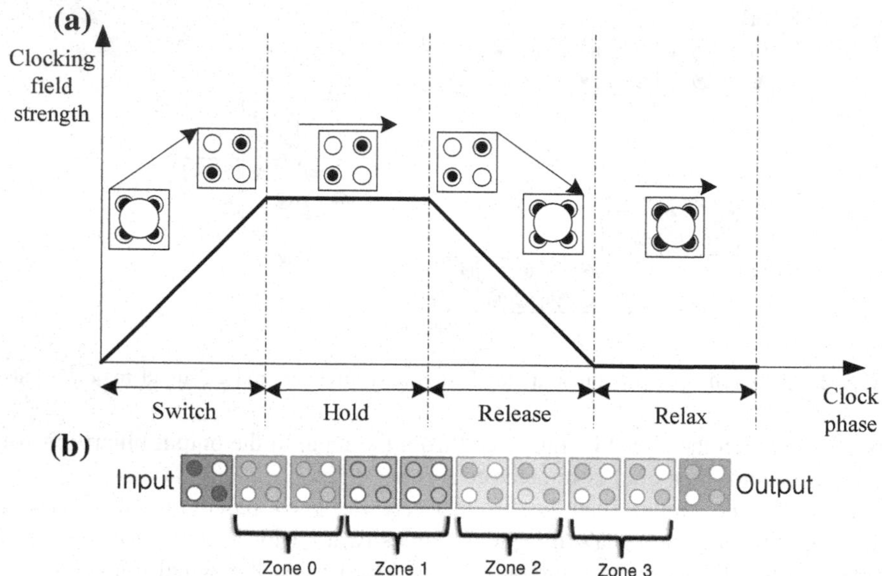

Fig. 3. QCA clocking: **a** four clock phases and **b** basic principle of four clock zones

The control of tunneling barriers between quantum dots is a function for electron transmission between the dots. Therefore, "QCA clocking controls information flow and provides with power to run the circuit." Figure 3a demonstrates clocking of a QCA cell in a zone. "Switch, hold, release, and relax are clocking phases of QCA." The phases of QCA clocking are changed "when the potential barriers are raised or lowered. For a credible signal transmission and functional gain, the clocking is crucial" [11–13].

In the switch phase, "the inter-dot barrier rises progressively and the cell condition becomes one of the two ground polarization states." High-rise inter-dot barrier is held in the hold phase. "In the release and relax phases period, the inter-dot barriers are relatively low and the excess electrons start their acting." QCA cells remain unpolarized in these two phases. There is no need of any extra additional wiring for the clocking cells because this property naturally exists inside each cell.

2.3 Wire Crossing in QCA

Some points exist with "wire crossing in QCA" circuit design. Basically, "there are two types of wire crossing technique, namely multilayer and coplanar wire-crossing techniques." Tougaw and Lent [14] proposed a coplanar crossover method. In this method, horizontal and vertical wires transmit the given value, one or zero, respectively, as shown in Fig. 4a.

Implementation of coplanar wire crossing is based on 45° rotated horizontal cells. "The horizontal wire should consist of odd number of cells because of recovering the input value."

Multilayer crossover method is extensively used way for wire crossing. The construction of this technique requires two extra cell layers, middle and upper cell layers,

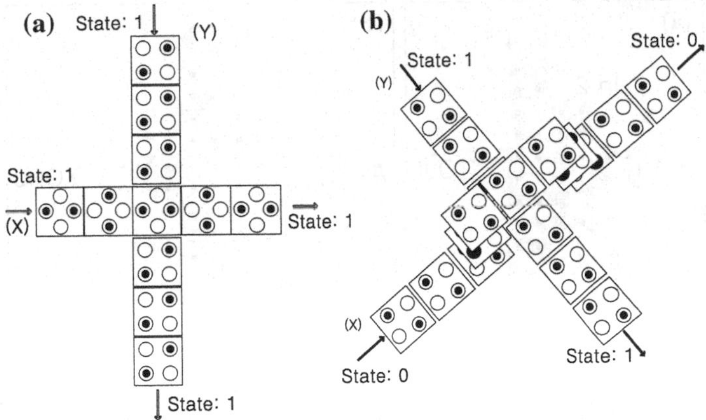

Fig. 4. Wire crossing in QCA: **a** coplanar crossover and **b** multilayer crossover

as shown in Fig. 4b. The stability of coplanar crossover is less compared to the multilayer crossover [15]. Furthermore, multilayer QCA structures mostly occupy less area than planar circuits.

2.4 QCA Full Adder

In digital circuit, the addition operation is performed by adders. Other kind of operations such as subtraction, division, and multiplication are ordinarily constructed using adders. A "full adder is a circuit that has three inputs (A, B, and Cin) and two outputs (Carry and Sum)."

We review various typical equations of "1-bit full adders" which have been presented in other papers. The first proposed formulation was presented in Eq. (4). However, its QCA clocking concept is not considered [14]. It is composed of "five conventional majority gates and three inverters." Majority-based carry formula has been achieved significantly as shown below:

$$\begin{aligned} Carry &= MG(A, B, Cin); \\ Sum &= M\left(M(A, \bar{B}, Cin),\, M\left(A, B, \overline{Cin}\right),\, M(\bar{A}, B, Cin)\right); \end{aligned} \tag{4}$$

After that, other formulation is proposed for Sum. The modification of the first sum is proposed by [16] with clocking concept. It has "four conventional majority gates and three inverters."

The Eq. (5) is "another formula for sum, and it has three regular majority gates and two inverters with five clocking phases" [17].

$$Sum = M\left(\overline{Carry}, M\left(A, B, \overline{Cin}\right),\, Cin\right); \tag{5}$$

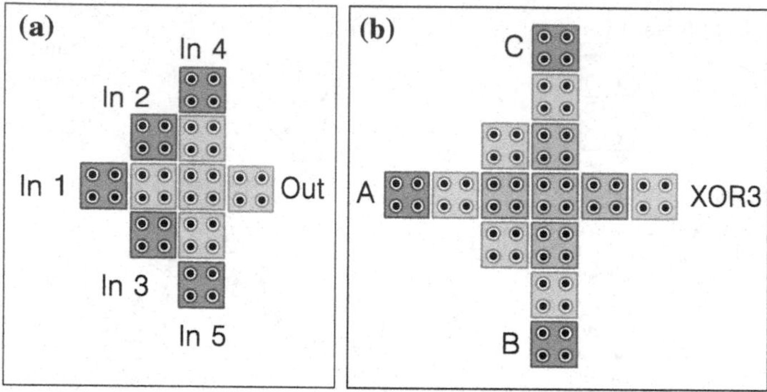

Fig. 5. **a** 5-input majority gate and **b** explicit interaction 3-input XOR gate

After the proposed various 5-input majority gates [18], a new formulation for sum is presented by researchers.

$$Sum = M\left(\overline{Carry}, \overline{Carry}, A, B, Cin\right); \tag{6}$$

It gives significant effect for hardware complexity. In this case, the most QCA full adders are implemented based on this equation.

Our proposed design for full adder depends on gate state three-input exclusive OR (TIEO) gate which was introduced in paper [19]. Its form is similar with single-layer five input majority gate [20]. Design of gate is presented in Fig. 5.

The implementation of 3-input XOR gate is constructed with explicit interaction between QCA cells. The gate consists of only 14 cells with 0.5 clock cycles. By fixing one of the inputs in Fig. 5b, namely input "*A*" to polarization "−1", simple 2-input XOR gate can be constructed.

The gate [19] can be effectively converted to a 3-input XOR gate with the same latency, cell count, and area to those of its 2-input XOR gate implementation. The presented design has "achieved significant improvements in terms of cell count," latency, and area aspects.

Less wire crossing in QCA circuit is the most important gain. In 1-bit QCA adder, the sum can be achieved with 3-input XOR gate. Hence, to use "the sum bit of the full adder," QCA gate condition can bring good solution for complex structures.

3 Proposed Structures

In this section, a low-complexity full adder is presented. Logical block diagram of the structure has been presented in Fig. 6. The implementation of full adder is constructed by two gates in three layers. In the main layer, carry bit is formed by conventional 3-input majority gate. In the upper layer, we have used 3-input XOR (TIEO) gate which is shown in Fig. 7.

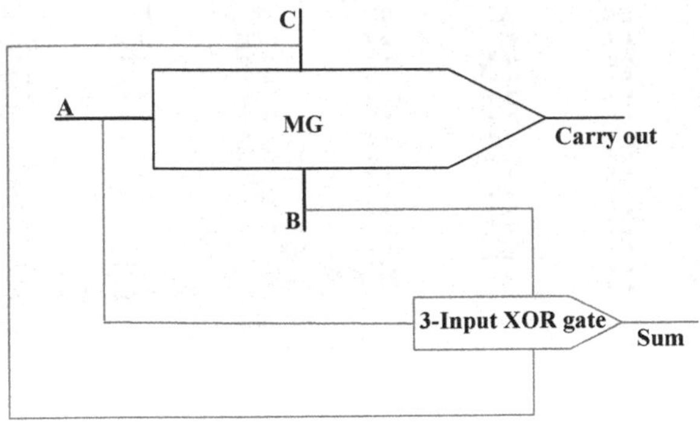

Fig. 6. Logical diagram of proposed full adder

The presented 1-bit full adder has 31 standard cells and 0.5 clock latency. Coherence vector and bistable approximation engines of "QCADesigner tool version 2.0.3" [21] are used to verify circuit functional behavior. The adder's simulation waveform is illustrated in Fig. 8.

The construction of the presented full adder in QCADesigner tool is demonstrated in Fig. 7, separately. "Layer 1" contains three input cells and carries out line as a main cell layer. In the "Layer 3," sum operation is performed by TIEO gate as upper cell layer. There are three cells in "Layer 2" that makes a link between "Layer 1" and "Layer 3." This kind of "full adder" that is based on three-input XOR gate brings more area, power, and delay efficiency. Moreover, construction method is also not complex and perplexed as well. The cost and wiring complexity are greatly reduced.

We focus on a design of "ripple carry adder (RCA)" in QCA. It is constructed by cascading 1-bit full adder blocks serially. Basically, it is simple to design 4-bit RCA. In fact, the ripple carry adder performance is not relatively fast, like carry look ahead adder, because "the carry bit" must be calculated from the former "full adder" [21]. Despite that the RCA has its own position for arithmetic circuits and it is designed with

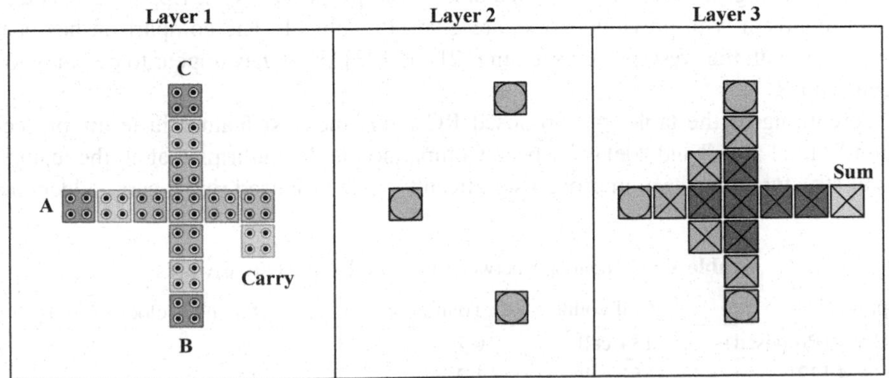

Fig. 7. Construction of proposed full adder

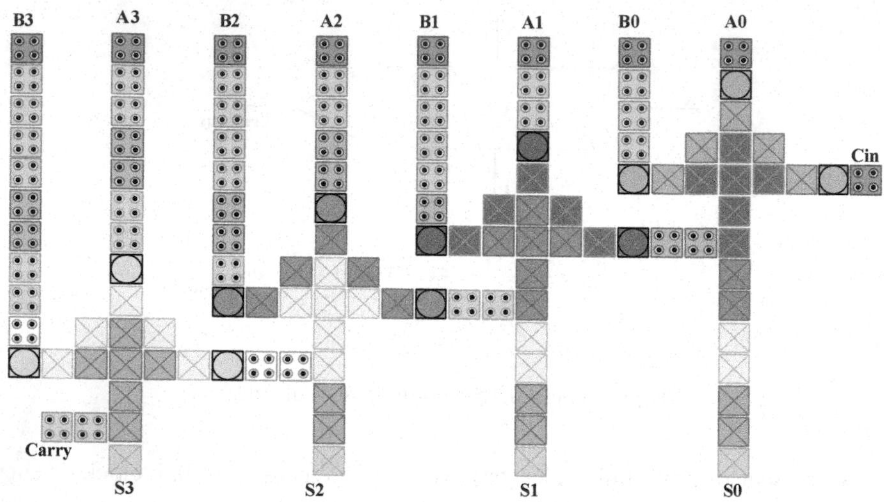

Fig. 8. QCA 4-bit ripple carry adder

only n simple full adders which are connected simply and regularly. It is also the cheapest one among the other adders.

The 4-bit size "ripple carry adder" layout is presented in Fig. 8. It uses 184 cells to design the circuit and requires 5 clock phases for getting the result. In this design, totally four conventional majority gates and four TIEO gates are used. Its occupation area is equal to approximately 0.1 μm.

4 Analysis

The functionality of the presented QCA full adder is simulated by QCADesigner tool. All parameter and condition of this tool are kept by default.

We have compared the "performance" and "achievement" of our proposed work with the existing prior designs. The features of the proposed "4-bit ripple carry adder" and comparison with previous works are given in Table 1. The comparison has been carried out with the "results" reported in [22] and [23] since they appear to be "the most recent" on RCA.

According to the table, the proposed RCA has the best feature in terms of "cell count," "total area," and "delay" aspect. Comparison table confirms that all the features of our structure have been improved significantly. The proposed ripple carry adders use

Table 1. Comparison between proposed and existing works

Approach	Cell count	Total area (μm)	Latency (clock phases)
RCA-4 (Proposed)	184 cells	0.1	5
RCA-4 [22]	651 cells	1.2	17
RCA-4 [23]	339 cells	0.25	7

72% and 46% less cells in the comparison to [22] and [23], respectively. Moreover, the proposed RCA has less delay than the other existing designs. As shown in Table 1, the number of used cells occupies less area and also it can be able to optimize some lines of the input, if required for the integration circuits.

5 Conclusion

In this paper, we have considered multilayer architecture of full adder with realization of 4-bit RCA. The presented full adder is further analyzed with the implementation of RCA that is compared with previous circuits. In our circuit, the input and output cells are located very smoothly. It means that it can be easily extended by any other circuits, and also the proposed circuit has the best complexity in area and time. Expanded adders are demanded for future works.

Acknowledgements. This work was supported by the National Research Foundation of Korea (NRF) grant funded by the Korea government (MSIP) (NO. NRF-2015R1A2A1A15055749).

References

1. C.S. Lent, P.D. Tougaw, W. Porod and G.H. Bernstein: "Quantum Cellular Automata", Nanotechnology, Vol. 4, No. 1 (1993) 49–57
2. M.B. Tahoori, J. Huang, M. Momenzadeh and F. Lombardi: "Testing of Quantum Cellular Automata", Vol. 3, No. 4 (2004) 432–442
3. J.C. Jeon: "Extendable Quantum-dot Cellular Automata Decoding Architecture Using 5-input Majority Gate", International Journal of Control and Automation, Vol. 8, No. 12 (2015) 107–118
4. H. Cho and E.E. Swartzlander: "Adder and Multiple design in Quantum-dot Cellular Automata", IEEE Trans. Comput., Vol. 58, No. 6 (2009) 721–727
5. R. Zhang, K. Walus, W. Wang and G.A. Jullien: "A Method of Majority Logic Reduction for Quantum Cellular Automata", IEEE Trans. Nanotechnology, Vol. 3, No. 4 (2004) 443–450
6. J. C. Jeon and K. Y. Yoo: "Low-power exponent architecture in finite fields", IEE Proceedings-Circuits, Devices and System. Vol. 152, No. 6 (2005) 573–578
7. S. Perri, P. Corsonello and G. Cocorullo: "Area-delay Efficient binary adders in QCA", IEEE Trans, Vol. 22, No. 5 (2014) 1174–1179
8. K. Makanda and J.C. Jeon: "Combinational Circuit Design Based on Quantum-dot Cellular Automata", International Journal of Control and Automation, Vol. 7, No. 6 (2014) 369–378
9. A.O. Orlov, I. Amlani, G.H. Berstein, C.S. Lent and G. L. Snider: "Realization of a Functional Cell for Quantum-dot Cellular Automata", Science, New Series, Vol. 277, No. 5328, (1997) 928–930
10. K. Walus, G. Schulhof, G. A. Jullien, R. Zhang, and W. Wang: "Circuit design based on majority gates for applications with quantum-dot cellular automata," in Conf. Rec. 38th Asilomar Conf. Signals, Systems and Computers, Vol. 2 (2004) 1354–1357
11. C.S. Lent and B. Isaksen: "Clocked Molecular Quantum-dot Cellular Automata", IEEE Trans, Vol. 50, No. 9 (2003) 1890–1896

12. J. S. Lee and J. C. Jeon: "Design of Low Hardware Complexity Multiplexer Using NAND Gates on Quantum-dot Cellular Automata", International Journal of Multimedia and Ubiquitous Engineering, Vol. 11, No. 12 (2016) 307–318
13. S. Hashemi, M. Tehrani, K. Navi: "An efficient Quantum-dot Cellular Automata full adder", Sci. Res. Essey, Vol. 7 (2012) 177–189
14. P.D. Tougaw and C. S. Lent: "Logical devices implemented using Quantum-dot Cellular Automata", Journal App. Phys, Vol. 75 (1994) 1818–1825
15. N. Safoev and J.C. Jeon: "Low area complexity demultiplexer based on multilayer Quantum-dot Cellular Automata", International Journal of Control and Automation, Vol. 9, No. 12 (2016) 165–178
16. K. Navi, R. Farazkish, Sayedsalehi, A.M. Rahimi: "A new Quantum-dot Cellular Automata Full Adder", Microelectronics Journal, Vol. 41, No. 12 (2010) 820–826
17. W. Wang, K. Walus, G. A. Jullien: "Quantum-dot Cellular Automata Adders", IEEE Trans. Nanotechnology (2003) 461–464
18. M. R. Azghadi, O. Kavehei, K. Navi: "A novel design for Quantum-dot Cellular Automata cells and full adders", Journal of Applied Sciences, Vol. 7 (2007) 3460–3468
19. F. Ahmad, G.M. Bhat, H. Khademolhosseini, S. Azimi, S. Angizi, K. Navi: "Toward single layer Quantum-dot Cellular Automata adders based on explicit interaction of cells", Journal of Computational Science, Vol. 16 (2016) 8–15
20. K. Navi, S. Sayedsalehi, R. Farazkish, M.R. Azghadi: "Five-input majority gate, anew device for Quantum dot Cellular Automata", Journal of Computational and Theoretical Nanoscience, Vol. 7, No. 8 (2010) 1546–1553
21. K. Walus, T.J. Dysart, G.A. Jullien and A.R. Budiman: "QCADesigner: A rapid design and simulation tool for Quantum-dot Cellular Automata", IEEE Trans. Nanotechnology, Vol. 3, No. 1 (2004) 26–31
22. H. Cho and E. Swartzlander: "Adder design and analyses for Quantum-dot Cellular Automata", IEEE Trans. Nanotechnology, Vol. 6, No. 3 (2007) 374–383
23. V. Pudi and K. Sridharan: "Low complexity design of Ripple Carry and Brent-Kung adders in QCA", IEEE Trans. Nanotechnology, Vol. 11, No. 1 (2012) 105–119

Control of the Motion Orientation and the Depth of Underwater Vehicles by Hedge Algebras

Nguyen Quang Vinh(✉)

Academy of Military Science and Technology, 17, Hoang Sam, Cau Giay, Hanoi, Vietnam
vinhquang2808@yahoo.com

Abstract. In this paper, we present an application of the hedge algebra controller in control of the orientation and the depth of underwater vehicles. The experiments simulated on computers are done to prove the effectiveness and the feasibility of the proposed algorithm of the neural controller under different actions such as the noise in the measuring devices, the influence of the flow to the motion of underwater vehicles.

1 Introduction

The inertial navigation algorithm generates navigation parameters with errors. Those errors increase over time because the drifts are integral twice. To overcome those errors, the additional information from other navigation systems is often used to reduce the mentioned errors [1, 2]. The use of a linear Kalman filter for solving the problem mentioned above has been studied by many authors [3, 4]. However, the basic drawback of this method is to make two successive stages: the inertial navigation algorithm and the Kalman filter algorithm, which cause the increment of computational time and the number of calculations affecting the response time of the system. In this paper, the authors propose an application of the extended Kalman filter (EKF) combined with measuring devices to remove the drifts and measuring errors.

Hedge algebras (HAs) solve effectively problems in uncertain environments; therefore, researchers often use them in control and automation areas. HAs have been studied in some identification and prediction problems and have given significant successes in control areas (such as in some approximation and control problems of with simple models [5, 6]). However, the use of HA in control problems is not popular. The study for the successful application of a HA controller will confirm more effectiveness of HA theory and open application possibilities in practice. We started with this purpose and designed a HA controller applied in designing the orientation and depth controller under the influence of errors of measuring devices and invisible flows to the motion of the vehicle. The simulation results are the basis of studying and learning improvement of the control and experiment quality.

V. Bhateja et al. (eds.), *Information Systems Design and Intelligent Applications*, Advances in Intelligent Systems and Computing 672,
https://doi.org/10.1007/978-981-10-7512-4_52

2 The Dynamic Model of Autonomous Underwater Vehicles

The motion of autonomous underwater vehicles (AUVs) is described in the body frame $G_bX_bY_bZ_b$ that has origin coinciding with the center of gravity G_b (Fig. 1). The linear and angular velocity vector is denoted in the body frame by symbol

$$\underline{\chi} = [u, v, w, p, q, r]^T \tag{1}$$

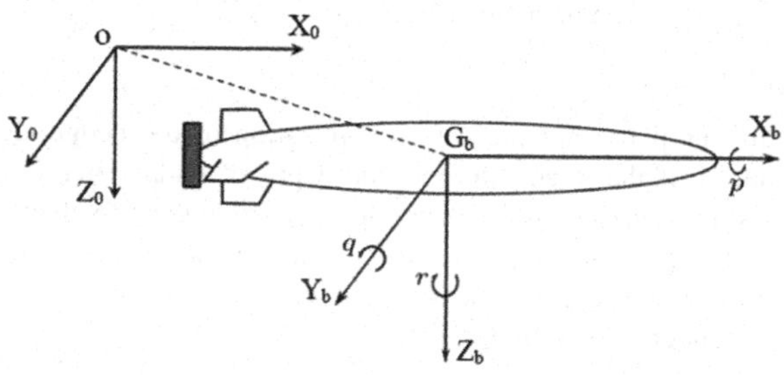

Fig. 1. Reference frames for AUVs

AUVs have a short working distance and move slowly. As a result, the local frame $OX_0Y_0Z_0$ can be considered as a navigation frame.

The position and Euler angles of AUVs in the local frame are denoted here by symbol

$$\underline{\eta} = [x, y, z, \gamma, \vartheta, \psi]^T \text{then } \dot{\underline{\eta}} = J(\underline{\eta})\underline{\chi} \tag{2}$$

where $J(\underline{\eta})$ is the direction cosine matrix that can be determined through four Rodring–Hamilton parameters $\lambda_0, \lambda_1, \lambda_2, \lambda_3$ as follows:

$$J(\underline{\eta}) = \begin{bmatrix} 2\lambda_0^2 + 2\lambda_1^2 - 1 & 2\lambda_1\lambda_2 - 2\lambda_0\lambda_3 & 2\lambda_1\lambda_3 + 2\lambda_0\lambda_2 \\ 2\lambda_1\lambda_2 + 2\lambda_0\lambda_3 & 2\lambda_0^2 + 2\lambda_2^2 - 1 & 2\lambda_2\lambda_3 - 2\lambda_0\lambda_1 \\ 2\lambda_1\lambda_3 - 2\lambda_0\lambda_2 & 2\lambda_2\lambda_3 + 2\lambda_0\lambda_1 & 2\lambda_0^2 + 2\lambda_3^2 - 1 \end{bmatrix} \tag{3}$$

Motion equations of the AUVs are described in the body frame [7]

$$M_{RB}\,\dot{\underline{\chi}} + C_{RB}(\underline{\chi})\underline{\chi} = \underline{\tau}_{RB} \tag{4}$$

where M_{RB} is the inertia matrix, $C_{RB}(\underline{\chi})$ is the Coriolis matrix, and $\underline{\tau}_{RB}$ is a generalized vector of external forces and moments

$$\underline{\tau}_{RB} = M_A\dot{\underline{\chi}} + C_A(\underline{\chi})\underline{\chi} + D(\underline{\chi})\underline{\chi} + L(\underline{\chi})\underline{\chi} + g(\underline{\eta}) + \underline{\tau} \tag{5}$$

where $g(\underline{\eta})$ is the buoyancy forces and moments, M_A is the inertia matrix of added mass, $C_A(\underline{\chi})$ is the Coriolis matrix of added mass, $D(\underline{\chi})$ is the potential damping matrix, $L(\underline{\chi})$ is the force and moment parameter matrix of rudder, and $\underline{\tau}$ is the forces and moments of rudder and propellers.

From (2), (4), and (5), the 6-DOF dynamic equation can be written as:

$$\begin{cases} M\dot{\underline{\chi}} = C(\underline{\chi})\underline{\chi} + D(\underline{\chi})\underline{\chi} + L(\underline{\chi})\underline{\chi} + g(\underline{\eta}) + \underline{\tau} \\ \dot{\underline{\eta}} = J(\underline{\eta})\underline{\chi};\ \underline{y} = \underline{\eta} \end{cases} \tag{6}$$

where $M = M_A - M_{RB}$, $C(\underline{\chi}) = C_A(\underline{\chi}) - C_{RB}(\underline{\chi})$, and $\underline{y}$ is the output system.

From (2), we have

$$\underline{\chi} = J(\underline{\eta})^{-1}\dot{\underline{\eta}} \tag{7}$$

$$\dot{\underline{\chi}} = J(\underline{\eta})^{-1}[\ddot{\underline{\eta}} - \dot{J}(\underline{\eta})J(\underline{\eta})^{-1}\dot{\underline{\eta}}] \tag{8}$$

Substituting (7) and (8) into (6) yields

$$M_\eta(\underline{\eta})\ddot{\underline{\eta}} = C_\eta(\underline{\chi},\underline{\eta})\dot{\underline{\eta}} + g_\eta(\underline{\eta}) + \underline{\tau}_\eta \tag{9}$$

where $M_\eta(\underline{\eta}) = J(\underline{\eta})^{-T}MJ(\underline{\eta})^{-1}$, $C_\eta(\underline{\chi},\underline{\eta}) = J(\underline{\eta})^{-T}[\underline{M}J(\underline{\eta})^{-1}\dot{J}(\underline{\eta}) + C(\underline{\chi}) + D(\underline{\chi}) + L(\underline{\chi})]$, $g_\eta(\underline{\eta}) = J(\underline{\eta})^{-T}g(\underline{\eta})$, and $\underline{\tau}_\eta = J(\underline{\eta})^{-T}\underline{\tau}$. From (2) and (9), the AUV dynamic equations in the body frame with multiple inputs and multiple outputs are rewritten as:

$$\begin{cases} \ddot{\underline{\eta}} = M_\eta(\underline{\eta})^{-1}C_\eta(\underline{\chi},\underline{\eta})\dot{\eta} + M_\eta(\underline{\eta})^{-1}g_\eta(\underline{\eta}) + M_\eta(\underline{\eta})^{-1}\underline{\tau}_\eta \\ \dot{\underline{\eta}} = J(\underline{\eta})\underline{\chi};\ \underline{y} = \underline{\eta} \end{cases} \tag{10}$$

2.1 Guidance

Desired output is generated from coordinates of two consecutive way points. Desired trajectory is a straight line between (i − 1)th way point and ith way point by using line of sight (LOS) techniques. LOS vector from the AUVs to ith way point in the horizontal plane is calculated according to the definition of the angle [3]

$$\Psi_r = \arctan\left(\frac{y_i - y}{x_i - x}\right) \tag{11}$$

where (x, y) is position of AUVs in the horizontal plane.

Similar LOS from AUVs to ith way point in the vertical plane determined

$$\vartheta_{ri} = \arctan\left(\frac{z_i - z}{l_i - l}\right) \tag{12}$$

where

$$l = \sqrt{x^2 + y^2};\, l_i = \sqrt{x_i^2 + y_i^2} \tag{13}$$

The next waypoint is selected when the AUVs lie within the sphere of acceptance with a radius R around the ith waypoint

$$R \geq \sqrt{(x_i - x)^2 + (y_i - y)^2 + (z_i - z)^2} \tag{14}$$

2.2 An Algorithm for Determining the Navigation Parameters of AUVs

Four-parameters Rodring–Hamilton $\lambda_0, \lambda_1, \lambda_2, \lambda_3$ are determined by solving the following equations [8]:

$$\begin{aligned} 2\dot{\lambda}_0 &= -p\lambda_1 - q\lambda_2 - r\lambda_3, 2\dot{\lambda}_1 = p\lambda_0 + r\lambda_2 - q\lambda_3, \\ 2\dot{\lambda}_2 &= q\lambda_0 - r\lambda_1 + p\lambda_3, 2\dot{\lambda}_3 = r\lambda_0 + q\lambda_1 - p\lambda_2. \end{aligned} \tag{15}$$

Navigation parameters (components of velocity V_x, V_y, V_z and position of AUVs) are determined by solving the following equations [8]:

$$\dot{\mathrm{V}}_{\mathrm{x}} = (2\lambda_0^2 + 2\lambda_1^2 - 1)\mathrm{A}_x + (2\lambda_1\lambda_2 - 2\lambda_0\lambda_3)\mathrm{A}_y + (2\lambda_1\lambda_3 + 2\lambda_0\lambda_2)\mathrm{A}_z \tag{16}$$

$$\dot{\mathrm{V}}_{\mathrm{y}} = (2\lambda_1\lambda_2 + 2\lambda_0\lambda_3)\mathrm{A}_x + (2\lambda_0^2 + 2\lambda_2^2 - 1)\mathrm{A}_y + (2\lambda_2\lambda_3 - 2\lambda_0\lambda_1)\mathrm{A}_z \tag{17}$$

$$\dot{\mathrm{V}}_{\mathrm{z}} = (2\lambda_1\lambda_3 - 2\lambda_0\lambda_2)\mathrm{A}_x + (2\lambda_2\lambda_3 + 2\lambda_0\lambda_1)\mathrm{A}_y + (2\lambda_0^2 + 2\lambda_3^2 - 1)\mathrm{A}_z + g \tag{18}$$

$$\dot{x} = V_x, \dot{y} = V_y, \dot{z} = V_z \tag{19}$$

where $\mathrm{A}_x, \mathrm{A}_y, \mathrm{A}_z$ are values of the ideal accelerometers.

Actually, angular rate sensors and accelerometers measure included slowly changed parameters and noise

$$\omega_x^n = p + b_1 + w_x, \omega_y^n = q + b_2 + w_y, \omega_z^n = r + b_3 + w_z. \tag{20}$$

$$A_x^n = A_x + c_1 + w_{ax}, A_y^n = A_y + c_2 + w_{ay}, A_z^n = A_z + c_3 + w_{az}. \tag{21}$$

where $b_1, b_2, b_3, c_1, c_2, c_3$ are slowly changed parameters that indicate gyro drift, and $w_x, w_y, w_z, w_{ax}, w_{ay}, w_{az}$ are the white noise.

In this paper, the authors propose an application of EKF to estimate navigation parameters based on the information provided by angular rate sensors, accelerometers, magnetometers, speedometers, and pressure sensors.

Assuming that the motion of AUVs is described by discrete dynamic equations:

$$X_k = F_{k-1}(X_{k-1}) + G(X_{k-1})\omega(k), F = (f_1, \ldots f_n)^T, Z_k = h(X_k) + v(k), h = (h_1, \ldots, h_m)^T \tag{22}$$

where X_k, X_{k-1} are states vector X (n-dimensional vector) at step k and k − 1, G is the noise matrix, ω is the white noise vector (one-dimensional vector), F_{k-1} is the function vector F at step k − 1, $f_1, f_2, \ldots f_n$ are nonlinear functions with the state vector X, Z_k is the observed output vector h (m-dimensional vector, m ≤ n) at step k that is measured by measuring devices, $h = (h_1, \ldots, h_m)^T$, and v is a white noise vector (m-dimensional vector) with zero mathematical expectation.

$\omega \sim N(0, Q_k)$, $M[\omega(j), \omega^T(k)] = Q(k)$, $v \sim N(0, R_k)$, $P_k = \varepsilon([X_k - \hat{X}_k][X_k - \hat{X}_k]^T)$ $M[v(k)] = 0$, $M[\omega(k)] = 0$ where ε is the mathematical expectation.

Procedures of estimating state vector algorithm X based on the observed state vector (measured) by Kalman are as follows:

$$\begin{aligned} &\hat{X}_k^{(-)} = F_{k-1}(\hat{X}_{k-1}^{(+)}), \Phi_{k-1} = \frac{\partial F_{k-1}}{\partial X}\Big|_{X=\hat{X}_{k-1}^{(+)}}, H_k = \frac{\partial h_k}{\partial X}\Big|_{X=\hat{X}_{k-1}^{(-)}}, \hat{Z}_k = h_k(\hat{X}_k^{(-)}), \\ &K_k = P_k^{(-)} H_k^T (H_k P_k^{(-)} H_k^T + R_k)^{-1}, P_k^{(-)} = \Phi_{k-1} P_{k-1}^{(+)} \Phi_{k-1}^T + G(k-1)Q(k-1)G^T(k-1), \\ &\hat{X}_k^{(+)} = \hat{X}_k^{(-)} + K_k(Z_k - \hat{Z}_k), P_k^{(+)} = (I - K_k H_k) P_k^{(-)}. \end{aligned} \tag{23}$$

where I is the unit matrix. Conditions assess if and only if rank of the matrix $\|H^T \Phi^T H^T \ldots (\Phi^T)^{n-1} H^T\|$ are equal to the order of the dynamical system, mean

$$rank\|H^T \Phi^T H^T \ldots (\Phi^T)^{n-1} H^T\| = n \tag{24}$$

To process EKF, the functions $F_{k-1}(X_{k-1})$, $h(X_k)$ in the expression (22) are established, from which we can build a state transition matrix $\Phi_{k-1} = \frac{\partial F_{k-1}}{\partial X}\Big|_{X=\hat{X}_{k-1}^{(+)}}$ and a measuring matrix $H_k = \frac{\partial h_k}{\partial X}\Big|_{X=\hat{X}_k^{(-)}}$. We define

$$\begin{aligned} &x_1 = \lambda_0, x_2 = \lambda_1, x_3 = \lambda_2, x_4 = \lambda_3, x_5 = b_1, x_6 = b_2, x_7 = b_3, x_8 = c_1, x_9 = c_2, \\ &x_{10} = c_3, x_{11} = V_x, x_{12} = V_y, x_{13} = V_z, x_{14} = z, X = (x_1, x_2, \ldots, x_{14})^T. \end{aligned} \tag{25}$$

From the discretization of Eqs. (15–19), we have functions $(f_1, f_2, f_3, f_4, f_{11}, f_{12}, f_{13}, f_{14})$ of $F_{k-1}(X_{k-1})$

In the case $b_1, b_2, b_3, c_1, c_2, c_3$ are slowly changed, we have

$$x_j(k) = f_j(X_{k-1}) = x_j(k-1),\ j = 5 \div 10 \tag{26}$$

To determine the state transition matrix Φ_{k-1} by taking partial directive $\partial f_i/\partial x_j, (i = 1,\ldots,14; j = 1,\ldots,14)$, we have:

$$\Phi = [\partial f_i/\partial x_j] = [\phi_{ij}], i = 1,2,\ldots,14; j = 1,2,\ldots,14 \tag{27}$$

If the measuring noise of angular rate sensors and accelerometers is uncorrelated with the white noise, then $\omega = [w_x, w_y, w_z, w_{ax}, w_{ay}, w_{az}]^T$ is the noise impact on the right side of dynamical Eq. (22). Note that matrix $G = [g_{ij}]$ $i = \overline{1,14}$, $j = \overline{1,6}$.

As mentioned above, AUVs have magnetometers that measure z_1, z_2, z_3 in the body frame of the earth's magnetic field vector where AUVs are operating. Suppose that the earth's magnetic field where AUVs are operating has not changed and the values following navigation frame are B_x, B_y, B_z. It is easily found that z_1, z_2, z_3 and B_x, B_y, B_z are closely related to each other only via the direction cosine matrix that mean

$$[z_1, z_2, z_3]^T = J(\underline{\eta})^T [B_x, B_y, B_z]^T \tag{28}$$

If the speedometers measure three velocity components in the body frame (using the speedometers mounted directly on the AUVs as DVL Doppler), then:

$$[z_4, z_5, z_6]^T = J(\underline{\eta})^T [V_x, V_y, V_z]^T \tag{29}$$

The pressure sensors measure the depth z_7 of AUVs.

Assuming that the measuring noise of the magnetometers, speedometers, and pressure sensors is white noise, we have:

$$z_1 = h_1(X_k) + v_{T1}, z_2 = h_2(X_k) + v_{T2}, z_3 = h_3(X_k) + v_{T3}, z_4 = h_4(X_k) + v_{Vx}, \quad z_5 = h_5(X_k) + v_{Vy},\ z_6 = h_6(X_k) + v_{Vz},\ z_7 = h_7(X_k) + v_z.$$

where v_{T1}, v_{T2}, v_{T3} are noise of the magnetometers, v_{Vx}, v_{Vy}, v_{Vz} are noise of the speedometer, and v_z is noise of pressure sensor.

To determine measured matrix H_k by taking partial directive $\partial h_i/\partial x_j$, we have $H = [h_{ij}],\ i = 1,2,\ldots,7,\ j = 1,2,\ldots,14$.

If the angular rate sensors and accelerometers are not correlated with each other, the Q matrix is a diagonal matrix, and diagonal values are the variance of the angular rate sensors and accelerometers. If the magnetometers, speedometers, and pressure sensors are not correlated with each other, then R is a diagonal matrix, and diagonal values are the variance of magnetometers, speedometers, and pressure sensors. Thus, we have fully qualified to perform Kalman filter algorithm (23).

3 The Hedge Algebra Controller

Every linguistic variable is characterized by a complex with five components (X, T(X), U, R, M), where X is the variable name, T(X) is the set of linguistic values of variable X, U is the reference space of the basic variable u, each linguistic value is considered as a fuzzy variable on U combined with basic variable u, R is the syntax rule generating language values for set T(X), M is the semantic rule assigning each linguistic value in T (X) with a fuzzy set on U. A hedge corresponding to X is a complex AX = (Dom(X), C, H, $\leq$), where C is the set of generators, H is the set of hedges, and relation "$\leq$" is the induced semantic relation on X.

Hedge algebras are a development based on logic thinking about languages. One has to define related functions discretely for the input–output relation by fuzzy logic, and meanwhile, HA has an algebraic structure of the form of the function relation, which allows us to construct a such infinitely big set of linguistic values that the obtained structure simulates well the semantic of the language and helps men in deduction processes.

The hedge algebra-based controller (HAC) consists of three blocks as in Fig. 2:

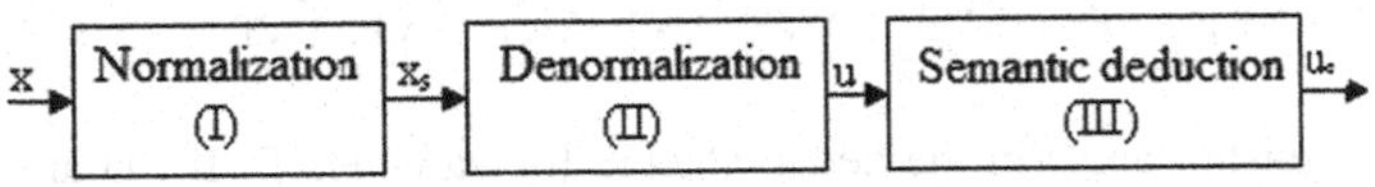

Fig. 2. Scheme of the HAC

where x is the input value; x_s is the input semantic value; u is the control value; and u is the control semantic value. Block I (normalization) transforms linearly from x to x_S; Block II (semantic deduction) (SQMs and HA-IRMd) performs the semantic interpolation from x_s to u_s based on the quantity semantic map and the rule conditions; and Block III (denormalization) transforms linearly u_s to u.

Each HAC consists of two inputs and one output. The first input is coordinate x coming in the control denoted by E, and the second input is the velocity along direction x denoted by IE. The output of the controller is denoted by.

We choose the complex of computation parameters: G = {Negative (N), Positive (P)}; H– = {Little (L)}; H+ = {Very (V)}. For a type of underwater vehicles, hedges are chosen as in Table 1 [1]:

Table 1. Chof parameters for variables E, IE, U

		Input (E) Input2 (IE) Output (U)		Output (U)	
H		μ(h)		μ(h)	
H–	Little	α	0.35	α	0.55
H+	Very	β	0.65	β	0.45

The control laws for the linguistic labels of the HAC are expressed in Table 2.

Table 2. Control laws

u		E								
		VVN	VN	N	LN	W	LP	P	VP	VVP
IE	VVN	VVN	VVN	VVN	VVN	VVN	VN	N	LN	W
	VN	VVN	VVN	VVN	VVN	VN	N	LN	W	LP
	N	VVN	VVN	VVN	VN	N	LN	W	LP	P
	LN	VVN	VVN	VN	N	LN	W	LP	P	VP
	W	VVN	VN	N	LN	W	LP	P	VP	VVP
	LP	VN	N	LN	W	LP	P	VP	VVP	VVP
	P	N	LN	W	LP	P	VP	VVP	VVP	VVP
	VP	LN	W	LP	P	VP	VVP	VVP	VVP	VVP
	VVP	W	LP	P	VP	VVP	VVP	VVP	VVP	VVP

where the linguistic labels in the hedge algebra for variables E, IE, U are as follows: Very Very Negative (VVN), Very Negative (VN), Negative (N), Little Negative (LN), W, Little Positive (LP), Posititve (P), Very Positive (VP), Very Very Positive (VVP).

The memory table of the quantitative combination is as follows:

u		E								
		0.128	0.167	0.287	0.379	0.495	0.618	0.697	0.817	0.884
IE	0.079	0.0455	0.0455	0.0455	0.0456	0.0455	0.1012	0.224	0.3762	0.5089
	0.136	0.0455	0.0455	0.0455	0.0455	0.1022	0.2249	0.3761	0.5039	0.6237
	0.298	0.0455	0.0455	0.0455	0.1015	0.2249	0.3761	0.5089	0.6237	0.7751
	0.378	0.0455	0.0455	0.1012	0.2248	0.3759	0.5089	0.6237	0.7751	0.8986
	0.5089	0.0455	0.1012	0.225	0.3761	0.509	0.5238	0.775	0.8983	0.9542
	0.61	0.1012	0.2247	0.3761	0.5087	0.6231	0.7751	0.8987	0.9542	0.9542
	0.713	0.2249	0.3761	0.5081	0.7236	0.7755	0.8986	0.9542	0.9542	0.9542
	0.8187	0.37	0.5091	0.0237	0.7751	0.8912	0.9543	0.9543	0.9542	0.9542
	0.8919	0.505	0.6237	0.7752	0.8987	0.9542	0.9542	0.9544	0.9542	0.9542

4 The Orientation Control of Underwater Vehicles

Following Zhang et al., one can use the target function for the controller in the following form [7]:

$$E_{k1} = \frac{1}{2}\left[\rho_1\left(\psi_k^d - \psi_k\right)^2 + \lambda_1\delta_{RK}^2 + \sigma_1 r_k^2\right]$$

where δ_{RK} and r_k, respectively, are the steering angle and the changing orientation rate of the underwater vehicle at the moment k, respectively; constants ρ_1, λ_1, and σ_1, respectively, are the proportion coefficient, the feedback coefficient, and the orientation differential coefficient, respectively. The HAC is described as in Fig. 3.

5 The Control of the Depth of Underwater Vehicles

Similar to (4), we choose the target function for the pitch angle controller as follows

$$E_{k2} = \frac{1}{2}\left[\rho_2\left(z_k^d - z_k\right)^2 + \lambda_1\delta_{SK}^2 + \sigma_2 w_k^2 + k_2\theta_k^2\right]$$

where ρ_2, λ_2, σ_2, and k_2 are constants; z_k^d, z_k are the wanted depth and the real depth; δ_{Sk} is the steering angle of the depth helm; w_k is the rate along the vertical orientation; and θ_k is the dive angle of the underwater vehicle (added to restrict the dive angle during change of the depth).

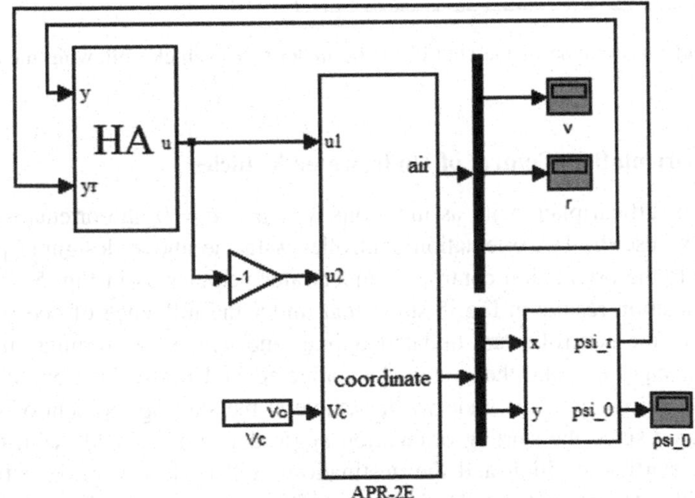

Fig. 3. Simulation scheme of the orientation control of the underwater vehicle using HA

6 Simulations, Computations, Discussion

To test the algorithms above, we use the input data of underwater vehicle APR-2E (Fig. 4).

The wanted trajectory is the line connecting point ASWs touching water (the starting position) and the pickup shot point (the destination) determined before throwing ASM. Suppose that the starting position of ASM has coordinates $(x_0, y_0, z_0) = (100, 100, 20)$ with beginning state angles $(\psi_0, \vartheta_0, \gamma_0) = (0°, -18°, 3°)$, the finishing point has coordinates $(x_2, y_2, z_2) = (1100, 1100, 120)$. External noise is the influence of ocean flows considered at the moment 20 s: $[u_c, v_c, w_c]^T = [4, 4, 0.5]^T$.

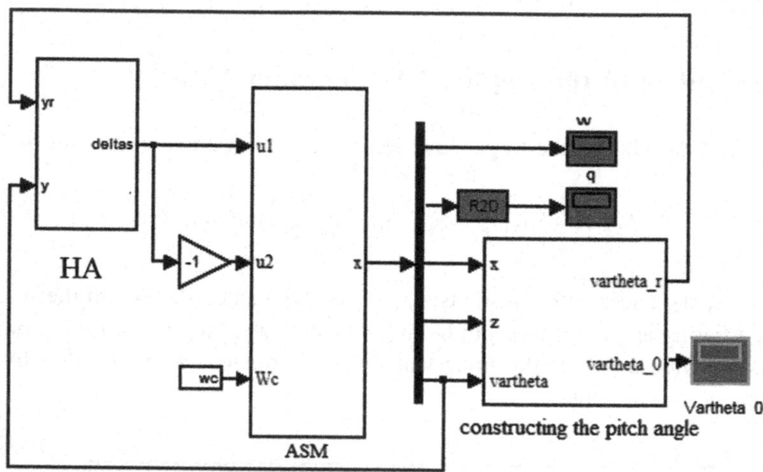

Fig. 4. Simulation scheme of the control of the underwater vehicle following the pitch angle using HA

6.1 The Orientation Control of Underwater Vehicles

Consider the vertical plane with assumptions $w = p = q = 0$, the orientation angle is small [4]. We use the HA orientation controller with the above designing parameters responsible to the orientation control of underwater vehicles as in Fig. 5.

The simulation results in Fig. 5 show that under the influence of ocean flows the parameters of the controller are updated online, and hence, the turning angle of the helm will change to make the system less affected and catch the wanted trajectory quickly. The trajectory of the underwater vehicle at the starting moment of simulation has a deviation since the starting orientation angle does not coincide with the line of sight of the starting position and the destination (LOS). However, since the system adapts quickly after short time, the trajectory of the underwater vehicle will coincide with the wanted trajectory.

6.2 The Pitch Angle Control of Underwater Vehicles

We consider the pitch angle with the control signal which is the steering angle $u = \delta_s$ of two helms δ_{s1}, δ_{s2} which are supposed to rotate together, i.e., $\delta_{s1} = -\delta_{s2}$. The simulation scheme constructed from the kinematic equation has the simulation parameters of anti-submarine missile APR-2E and the pitch angle computed from the navigation system as the system output.

The application of the HA controller to simulate the pitch angle control of underwater vehicles gives results as in Fig. 6. From the simulation results, one can see that at the starting moment of the simulation, the pitch angle of the underwater vehicle does not coincide with the wanted pitch angle, and therefore, there is a trajectory error. However, the trajectory of the system responses the wanted trajectory quickly. The system is able to adapt well under the influence of the ocean flow.

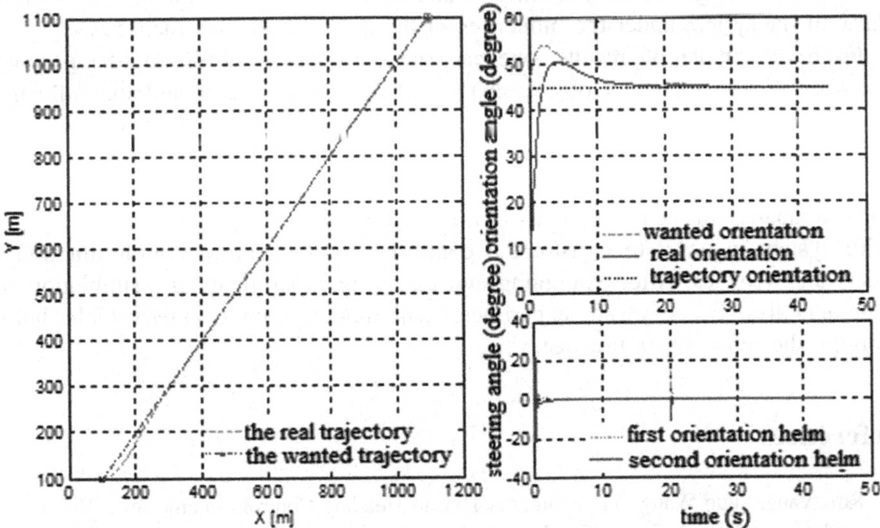

Fig. 5. Simulation results of the ASM orientation control

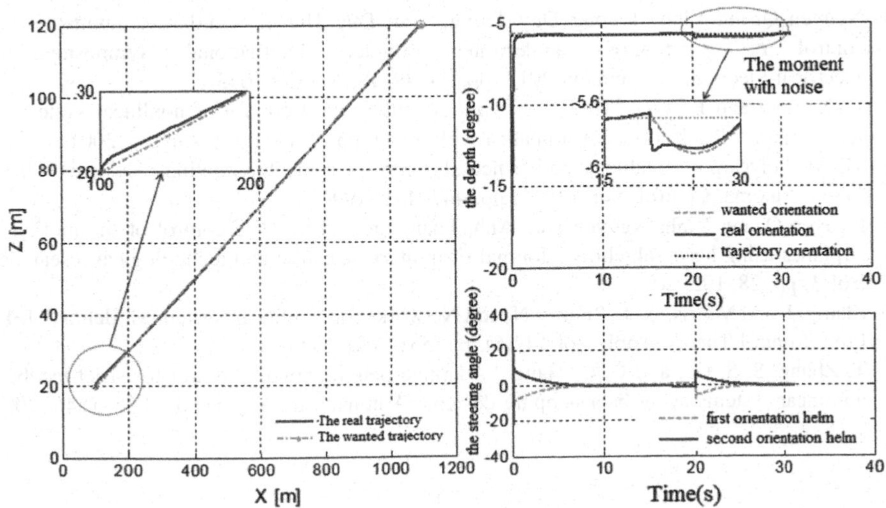

Fig. 6. Simulation results of the pitch angle control of ASM control by the pitch angle

7 Conclusions

In this paper, we have presented the system of the motion equations of an underwater vehicle controlled by two orientation helms and two depth helms of the 6th degree of freedom under the influences of the flow and measurement noise. We have applied HA to simulate the single channel orientation and depth control and analyzed the adaptive ability of the system under the influences of the flow and measurement noise.

From the simulation results, one can see the adaptive ability of the proposed controller to the controlled object, change of the environment, and the influences appearing in the system as well. The HA controller can be applied to underwater vehicles in general, torpedos in particular, and more complicated control tasks as well.

In the future, we will do researches in more various conditions and in more difficult tasks and analyze the adaptive ability of the controller and also prove the stability of the entire system in order to construct a controller for the entire motion function of underwater vehicles. After examining and assessing experiments and simulations on computers, it is necessary to do experiments with models of underwater vehicles before applying the controller in practice.

References

1. Rui Wang, Shuo Wang, Yu Wang, "A Hybrid Heading Control Scheme for a Biomimetic Underwater Vehicle," in 26th International Ocean and Polar Engineering Conference, Rhodes, Greece: ISOPE, 2016, pp. 619–625.
2. Thor I. Fossen, "Maritime Control Systems: Guidance, Navigation and Control of Ships, Rigs and Underwater Vehicles," Marine Cybernetics, Trondheim, Norway, 2002, ISBN 82-92356-00-2.
3. Nguyen Quang Vinh, Truong Duy Trung, Tran Duc Thuan, "Guidance, navigation and control of Autonomous underwater vehicle," International Symposium on Electrical-Electronics, Vietnam, 2013, pp. 44–49, ISSN 1859-1043.
4. J. Q. Gong and B. Yao, "Neural network adaptive robust control of nonlinear systems in semi-strict feedback form," Automatica, vol. 37(8), pp. 1149–1160, August, 2001.
5. M. M. Polycarpou "Stable adaptive neural control scheme for nonlinear systems," IEEE Trans. Automa. Control, vol. 41 (3), pp. 447–451, 1996.
6. Nguyen Quang Vinh, Nguyen Duc Anh, Phan Tương Lai, "The control of the motion of airplanes using hecke algebras," Journal of military scientific and technological researches, Vol 3, pp. 28–102, 2015.
7. Zhang, J. and Morris, A. J., "Fuzzy Neural Networks for Nonlinear System Modeling," IEEE Proc.-Control Theory Appl. Vol 142(6), pp. 551–556, 1995.
8. T. Zhang, S. S. Ge, and C. C. Hang, "Adaptive neural network control for strict-feedback nonlinear systems using backstepping design," Automatica, vol. 36, pp. 1835–1846, 2000.

The Relation of Curiosity and Creative use of IT

Nhu-Hang Ha[1(✉)], Yao Chin Lin[2], Nhan-Van Vo[1], Dac-Nhuong Le[3(✉)], and Kuo-Sung Lin[2]

[1] International School, Duy Tan University, Danang, Vietnam
{hanhuhangdtu, vanvonhan}@gmail.com
[2] Department of Information Management, YuanZe University, Taoyuan, Taiwan
[3] Haiphong University, Haiphong, Vietnam
Nhuongld@hus.edu.vn

Abstract. It is essential to identify factors influencing the creative use of IT that can enhance organizational performance. Based on self-determination theory, our approach investigates the intermediate role of employees' curiosity in the relationships between inherent psychological needs and employees' creative use of IT. By using the quantitative method, this study develops a model regarding the creative use of IT. To test the model, surveys are distributed to employees working for organizations located in one of the biggest city in Vietnam. The results indicate that curiosity has partial mediation in the relation between inherent psychological needs and creative use of IT. The findings of this paper are expected to shed a new light into the study of users' behavior, especially in their creative use of IT. This paper also provides valuable insight to practitioners on how to improve their employees' creative use of IT.

Keywords: Creative use of IT · Curiosity · Perceived autonomy support
Perceived IT competence · Perceived relatedness · Self-determination theory

1 Introduction

Recently, most of studies have examined system usage by using quantitative methods focusing on the use of itself [1–4], recently more attention have been given to the qualitative aspect of it with regards to how the systems are used [5, 6]. As there is greater user discretion over how an IT is used, researchers are particularly interested in the innovative use of IT [7–10]. With technologies becoming more malleable and flexible, the importance of IT use is to accrue competitive advantage that has been further elevated. However, critics have suggested that competitive advantage may only accumulate through creative ways of utilizing technologies by IT users [11]. They need to use the technology as an indispensable part of their work and take one step further to expend effort in exploring the IT in a creative way [7–10]. Thus, the issue of discovering innovative ways of exploring the IT is one of important missions for those

V. Bhateja et al. (eds.), *Information Systems Design and Intelligent Applications*, Advances in Intelligent Systems and Computing 672,
https://doi.org/10.1007/978-981-10-7512-4_53

who take responsibility for implementing technologies to leverage business value of an IT [2]. In a work setting, IT usage has been conceptualized as users employing IT effectively to attain relevant goals. Creative use of IT has been conceptualized as a higher level usage behavior that goes beyond the routine use of IT by exploiting existing features [9].

Previous researchers identify a variety of motivational factors leading to creative [1–4]. We argue that curiosity, a type over idiosyncratic motivation, is specifically applicable among the context regarding innovative use of IT. Curiosity has been perceived as a strong motive of behavior, initiate action to explore a person's environment, or searching for information to resolve uncertainty [12]. We contend that curiosity, being conceptualized as the inner desire that drives people to learn and investigate, is a salient prerequisite for exploratory behaviors such as creative use of IT. However, less attention has given to explore the relationship between curiosity and creative use of IT. Curiosity being considered as a salient contributor to creative behavior, it would be important for managers to know under what working conditions curiosity can be stimulated. For this purpose, self-determination theory (SDT) is employed in this study as a theoretical lens to examine environmental factors that induce curiosity. SDT is an approach that is used to study human motivation and personality development. SDT argues that intrinsic motivation such as curiosity is more likely to be active in contexts characterized by a sense of autonomy, competence, and relatedness [13]. The purpose of this study is twofold: to investigate the role of curiosity in stimulating creative use of IT and to identify the contextual factors for facilitating optimal functioning of the propensities for curiosity from an SDT perspective. The findings of this paper are expected to shed a new light into the study of users' behavior, especially in the use of IT in a creative manner and also to provide insights to practitioners on how to improve their employees' creative use of IT.

2 Background Theory and Hypothesis Development

2.1 Creative Use of IT

Mills and Chin developed the concept of "*Creative Use*" to describe "*the implementation of novel and useful ways of applying organizational systems to solve business problems*" [9]. Some work on IT innovation and diffusion look at the usage process as a source of creativity; however, there appears to be plenty of space for significant further study in this area. In this study, we define creative use of IT as the way users leverage IT features in non-imitable ways to support their tasks [14]. Users can use one feature or function of IT to complete multiple tasks or they also can use multiple features of the system to finish one task as well as multiple tasks. With minor changes or different thinking, the users can use simple and available functions to finish their tasks more effectively.

2.2 Perceived Autonomy Support

Perceived autonomy support is can be defined as the extent to which the user perceives the spiritual support and material support from their organization [15]. When an organization encourages users to leverage IT functions, users consider their actions as permitted and commended. Thus, they intend to get highly involved in doing things rather than worrying about being blamed by management. Therefore, given the right conditions and stimuli, users will be more inclined to generate and implement creative ideas of using IT [9]. Thus, we propose that:

- *H1: The perceived autonomy support has positive impact on creative use of IT.*

2.3 Perceived IT Competence

Perceived IT competence is the abilities to perform specific IT tasks for diverse applications [16]. Individuals with high IT competence enjoy using IT at a higher degree and experience and less anxiety. Creative use of IT begins with creative ideas from a creative person. The transformation from creative ideas of using IT into practices that support user performance depends on users' IT competence. The more IT competences users have, the more creative ways of using IT they can apply. Therefore, we propose that:

- *H2: The perceived IT competence has positive impact on creative use of IT.*

2.4 Perceived Relatedness

Perceived relatedness is defined as "one's assessment of whether or not people important to him or her feel the behavior should be performed" [17]. It is also classified into external and interpersonal influence. In this study context, we focus on interpersonal influence [18]. Deci and Ryan posit relatedness plays an important role in motivating individuals to work since they are in a secure and supportive environment [13]. So, we predict that:

- *H3: The perceived relatedness has positive impact on creative use of IT.*

2.5 Curiosity

Curiosity can be defined as the inner desire that drives people to learn and investigate new things, it is also a salient prerequisite for exploratory behaviors such as creative use of IT [19]. Moreover, users feel IT is easy to use, it helps them to arouse their curiosity. In addition, most of the users' works in organization involve interaction with other people, including with their boss or colleagues. Users can be influenced by the surrounding people's actions as they listen to what people say or observe what people do, and they may be intrigued and have questions when they want to decide whether they should follow other people to pursue the creative use of IT or not. Therefore, the hypotheses related to curiosity were posted:

- *H4: The perceived autonomy support has positive impact on curiosity.*
- *H5: The perceived IT competence has positive impact on curiosity.*
- *H6: The perceived relatedness has positive impact on curiosity.*

There is a relationship between curiosity and creative manner. If the users think about new ways to use IT which have not been applied, they are willing to find the new solution. Thus, we propose that:

- *H7: Curiosity has positive impact on creative use of IT.*

Creativity research also suggests two relevant lenses for studying individuals' creativity: personality and environment [3]. Given the right conditions and stimuli, individuals intend to explore and implement creative ideas [20]. However, in some cases, even though the company provides enough support for users, and users are surrounded by people that recommended and encouraged them to apply their ideas, or users have high IT competence, they still do not apply IT in creative ways. Even if users receive ideas or suggestions from their surrounding people, there is nothing to guarantee that they will follow their friends or colleagues' ideas to use IT in creative ways. Thus, we posit that there is an indirect relationship between creative use of IT and the perceived autonomy support, perceived IT competence, perceived relatedness. Moreover, we argue that curiosity plays a mediating role in this relation. The curious people are naturally motivated to explore new ways to solve problems [21]. Therefore, despite the availability of resources, if the users are not curious, they may or also may not use IT in creative ways. Therefore, we propose that:

- *H8: Curiosity mediates relation between perceived autonomy support and creative use of IT.*
- *H9: Curiosity mediates relation between perceived IT competence and creative use of IT.*
- *H10: Curiosity mediates relation between perceived relatedness and creative use of IT.*

Based on the above discussion, we propose the research model in Fig. 1. As the figure shows, we assume perceived autonomy support, perceived IT competence, and perceived relatedness to affect users' curiosity. Furthermore, we posit that curiosity influences the creative use of IT. In addition, in this model, we consider two control variables which are task complexity and technology flexibility.

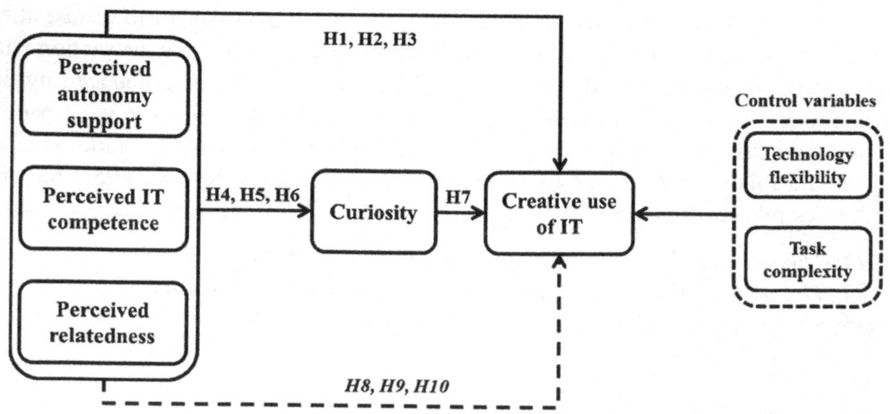

Fig. 1. Our research framework

3 Methodology

3.1 Sampling

A cover letter and the questionnaire were sent to 1000 randomly selected employees who are working for companies in Da Nang City of Vietnam. Procedures suggested by Dillman to increase response rate were used [22]. We assure that their responses would be kept confidential. In the first and second surveys, there are 27 valid responses. To increase the response rate, we conducted another survey. The total number of responses from the three rounds was 126. However, after initial checking, we found seven samples with data points that are too far-off from the mean than what is deemed reasonable. So, our final sample size is 119. Demographic features of the sample are shown in Table 1.

Table 1. Demographic information

Items	Categories	Number	Percentage
Age	< 20	3	3
	20–35	104	87
	36–50	12	10
	> 50	0	0
Gender	Male	56	48
	Female	63	52
Computer experience	< 1 year	3	3
	2–3 years	6	5
	4–5 years	6	5
	6–7 years	15	12
	> 8 years	89	75
Highest education	High school	3	3
	Bachelor	74	62
	Master	36	30
	Higher than Master	6	5
	Other	0	0
Total sample size: 119			

3.2 Measurements

Most of items were selected from previous studies in related fields. These items were measured on five-point Likert scales ranging from "strongly disagree" to "strongly agree." The items used to measure each variable are in English and listed in Table 2. We found that with a different task and IT system, the requirements of creative use of IT would be different. In this study, two variables, task complexity and system

Table 2. Research variables and measurements

Constructs	Items
Perceived autonomy support (PAS)	My company always encourage me to use existing IT (PAS1) My company are providing most of the necessary assistance and resources for us to get used to existing IT quickly (PAS2) I am always supported and encouraged by my company to make decision related to use of existing IT (PAS3)
Perceived IT competence (PIC)	I feel comfortable using existing IT on my own (PIC1) I can easily operate existing IT on my own (PIC2) I understand how to use existing IT even if there was no one around to help me (PIC3)
Perceived relatedness (PR)	My peers/colleagues/friends think that I should use the existing IT (PR1) People I know think that using existing IT was a good idea(PR2) People I know have influenced me to try out existing IT (PR3)
Curiosity (CU)	Using the IT excites me to explore more functions(CU1) Using the IT makes me want to investigate more functions(CU2) Using the IT arouses my imagination about other functions(CU3)
Creative use of IT (CUT)	I have discovered new uses of existing IT to enhance my work performance (CUT1) I have used existing IT in novel ways to support my work (CUT2) I developed new methods/ways based on existing IT to support my work (CUT3)

flexibility, are used as control variables. Campbell and Quinn found that task complexity was strongly related to creative output [23, 24]. Te'eni showed that the level of perceived complexity of the system had a relationship with the effectiveness of which the system was used [25].

Table 3. Standard loading, item–construct correlation, and reliability estimate

Constructs	Items	Factor loading	t-statistic	ITC
Perceived autonomy support Alpha: 0.933	PAS1	0.903	46.424	0.575
	PAS2	0.961	42.752	0.643
	PAS3	0.944	45.439	0.535
Perceived IT competence Alpha: 0.828	PIC1	0.892	53.263	0.637
	PIC2	0.854	55.364	0.613
	PIC3	0.830	52.435	0.535
Perceived relatedness Alpha: 0.752	PR1	0.780	45.973	0.533
	PR2	0.878	56.587	0.655
	PR3	0.757	47.383	0.443
Curiosity Alpha: 0.923	CU1	0.7911	45.896	0.679
	CU2	0.943	44.546	0.633
	CU3	0.939	41.946	0.579
Creative use of IT Alpha: 0.926	CUT1	0.926	48.708	0.702
	CUT2	0.932	50.178	0.642
	CUT3	0.942	46.454	0.660

Note: All significant at $p < 0.05$.

Table 4. Skewness, kurtosis, CR, square roots of AVE, and construct correlation

Construct	Mean	M3	M4	CR	PAS	PIC	PR	CU	CUT
PAS	3.829	−0.564	0.151	0.955	**0.936**				
PIC	4.00	−1.232	3.104	0.894	0.596	**0.909**			
PR	3.91	−0.889	1.967	0.848	0.540	0.510	**0.867**		
CU	3.722	−0.665	0.809	0.951	0.222	0.289	0.470	**0.960**	
CUT	3.952	−1.0130	1.628	0.953	0.236	0.531	0.399	0.768	**0.962**

Note: Diagonal elements (bold) are the square root of AVE between the constructs and their measures. Off-diagonal elements are correlations between constructs. M3: Skewness; M4: Kurtosis. K-S: Kolmogorov–Smirnov statistic with Lilliefors significance correction (all $p > 0.05$).

4 Data Analysis and Results

By using partial least squares (PLS) analysis, we run data to test the items' reliability, convergent validity, and discriminant validity. The results of Table 3 show that all constructs have high reliability. In addition, as shown in Table 4, most constructs have good distribution.

Three models have been tested. Model 1 consisted of control variables, technology flexibility (TF), and task complexity (TC), while Model 2 consisted of all predictors or testing the relation between independent variables and dependent variables (creative use of IT). Model 3 fully tested all variables included in Model 1 and Model 2, and the mediating variable—curiosity. Table 5 presents the standardized regression coefficients (β), R^2, change in R^2 (ΔR^2), and effect size. In addition, the magnitude of the effect size has a direct impact on the power of the statistical test and helps researchers determine

Table 5. Alternative model, R^2, and effect size

Structural paths	Model 1		Model 2		Model 3	
	β	T-statistic	*β*	T-statistic	*β*	T-statistic
TF → CUT	−0.247	3.028	−0.325	4.230	−0.126	2.071
TC → CUT	**0.140***	1.388	−0.065	0.6454	−0.096	1.390
PAS → CUT			−0.233	2.588	−0.133	1.730
PIC → CUT			**0.476*****	3.413	**0.440*****	4.594
PR → CUT			**0.405*****	3.127	−0.0003	0.041
PAS → CU					−0.094	0.751
PIC → CU					**0.108***	0.766
PR → CU					**0.467*****	3.152
CU → CUT					**0.673*****	9.294
R^2	0.082		0.444		0.738	
Differ of R^2			0.362		0.294	
Test of diff R^2			40.544		32.634	
Effect size			large		large	

Note: *$p < 0.05$; **$p < 0.01$; ***$p < 0.001$. TF: technology flexibility; TC: task complexity; PAS: perceived autonomy support; PIC: perceived IT competence; PR: perceived relatedness; CU: curiosity; CUT: creative use of IT.

whether the observed relationship is meaningful [26]. By adding three SDT factors, R^2 of Model 2 increases by 36% in variance explained. R^2 increases significantly ($f^2 = 0.65$, $F = 40.54$, $p < 0.001$), suggesting that organizations environment play an important role in explaining creative use of IT. Model 3 established for detecting mediating effects accounts for 74% of the variance of creative use of IT, increasing by 30%. R^2 increases significantly ($f^2 = 1.12$, $F = 40.54$, $p < 0.001$), indicating the presence of mediation effects.

With regard to the hypothesis testing, as shown in Table 5, while H_2, H_3, H_5, H_6, H_7, and H_9 are supported, H_1, H_4, H_8, and H_{10} are not (Model 2 and Model 3). This means that curiosity does significantly impact the creative use of IT and mediates the relationship between perceived relatedness and creative use of IT. However, curiosity has no mediating effects on the relation between perceived autonomy support, perceived IT competence, and creative use of IT. The results also show that while perceived relatedness and perceived IT competence do affect curiosity, perceived autonomy support has no influence on curiosity. In regards to control variables, while task complexity is shown to have significant positive impact on creative use of IT, IT flexibility does not impact creative use of IT (Model 1).

5 Discussion

This study investigates the role of curiosity in stimulating creative use of IT and to identify the contextual factors for facilitating optimal functioning of curiosity propensities. First, we suggest that SDT is useful in explaining how the basic inherent needs associated with intrinsic motivation and external environment can influence users' behavior. It reveals that surrounding people induce stronger intrinsic motivations in terms of curiosity, which further influences the users' creative behavior when using IT. We find that autonomy support and IT competence have no impact on users' curiosity. The probable explanation for this might be that when users are able to access what they need (IT facilitation and encouragement), exploring new ways of using the existing IT becomes unnecessary because they already have solutions to accomplish the tasks and there is no motivation to figure out novel ways. The findings also show that curiosity plays partial mediating role in the pursuit of creative use of IT. If we take curiosity out of the model, we found that IT competence has a stronger impact on the creative use of IT than relatedness, and these two factors are more important than autonomy support. However, when we put curiosity as mediating factor, the results are different. Curiosity plays a full mediating role in the relation between perceived relatedness and creative use of IT; however, it has a partial mediating role in the relation between perceived IT competence and creative use of IT. These results open a new door for the researcher to investigate other mediating variables besides curiosity in the relation between perceived IT competence and creative use of IT since we know that mediating effects exist in this relation. However, it is not necessary to dig into the mediating factor in the relation of perceived relatedness and creative use of IT. What the researchers need to do is to focus on the ways to enhance perceived relatedness. In addition, creative use of IT is different from intention to use or actual use. It does not only happen in post-adoption use but can happen at any time and at any task.

Therefore, in investigating factors that influence the creative use of IT, different approaches should be used. This study shows that curiosity has a significant impact on the creative use of IT. This is consistent with previous research on curiosity and exploratory behaviors [27]. It is necessary to find solutions to enhance users' curiosity to leverage creative behavior in using IT since curiosity acts as a force that motivates people to act and think in new ways [12]. In this study, curiosity is revealed to be influenced by perceived IT competence and perceived relatedness. This implies that curiosity can be enhanced by both internal and external factors. Therefore, curiosity cannot be viewed only as inherited in a person's nature, but it is something that can be enhanced by introducing stimulating factors in the work environment.

6 Conclusion

The results of our study offer insights into several issues that deserve further investigation. Curiosity can be separated into two types: sensory curiosity (occurs when physical factors attract the attention of users) and cognitive curiosity (is evoked when users believe that it may be useful to modify existing cognitive structures). In the future, we will expand our research framework to test the influence of different types of curiosity on creative use to build up a more comprehensive model in this topic. In addition, longitudinal studies could be performed to understand how the milieu of settings leads to the emergence of curiosity. Thirdly, this study uses SDT as the theoretical lens to investigate the environmental factors that contribute to the development of curiosity. Future studies may, based on their respective contexts and theories, further investigate the influence of other factors on curiosity like a reward system. Lastly, this study focuses solely on the impact of curiosity on the creative use of IT. Other factors based on different theory may be applied to expand our understanding of how to enhance the creative use of IT.

Declaration. Due permissions and consent are taken by authors from International School, Da Nang, Vietnam. Authors' declare that informed consent has been taken by all participants involved in this study.

References

1. Agarwal, R. (2000). Individual acceptance of information technologies. In R. W. Zmud (Eds.), *Framing the domains of IT management: Projecting the future... through the past* (pp. 85–104). Cincinnati, OH: Pinnaflex Education Resources.
2. Agarwal, R., & Karahanna, E. (2000). Time flies when you're having fun: Cognitive absorption and beliefs about information technology usage. *MIS Quarterly, 24*(4), 665–694.
3. Ahuja, M. K., & Thatcher, J. B. (2005). Moving beyond intentions and toward the theory of trying: Effects of work environment and gender on post-adoption information technology use. *MIS quarterly, 29*(3), 427–459.
4. Burton-Jones, A., & Straub, Jr. D. W. (2006). Reconceptualizing system usage: An approach and empirical test. *Information Systems Research, 17*(3), 228–246. https://doi.org/10.1287/isre.1060.0096.

5. De Guinea, A. O., & Markus, M. L. (2009). Why break the habit of a lifetime? Rethinking the roles of intention, habit, and emotion in continuing information technology use. *MIS Quarterly, 33*(3), 433–444.
6. Jasperson, J. S., Carter, P. E., & Zmud, R. W. (2005). A comprehensive conceptualization of post-adoptive behaviors associated with information technology enabled work systems. *MIS Quarterly, 29*, 525–557.
7. Al-Natour, S., & Benbasat, I. (2009). The adoption and use of IT artifacts: A new interaction-centric model for the study of user-artifact relationships. *Journal of the Association for Information Systems, 10*(9), 661–685.
8. Bardone, E., & Shmorgun, I. (2013). Ecologies of creativity: smartphones as a case in point. *Mind & Society, 12*(1), 125–135. https://doi.org/10.1007/s11299-013-0121-9.
9. Mills, A., & Chin, W. (2007). Conceptualizing creative use: An examination of the construct and its determinants. In *AMCIS 2007 Proceedings*, Paper 289. Retrieved September 8, 2016, from http://aisel.aisnet.org/amcis2007/289.
10. Salovaara, A., Helfenstein, S., & Oulasvirta, A. (2011). Everyday appropriations of information technology: A study of creative uses of digital cameras. Journal of the American Society for Information Science and Technology, 62(12), 2347–2363. https://doi.org/10.1002/asi.21643.
11. Davidson, E. J., & Chismar, W. G. (2007). The interaction of institutionally triggered and technology-triggered social structure change: An investigation of computerized physician order entry. *MIS Quarterly, 31*(4), 739–758.
12. Kashdan, T. B., & Silvia, P. J. (2009). Curiosity and interest: The benefits of thriving on novelty and challenge. In C. R. Snyder & S. J. Lopez (Eds.), *Oxford Handbook of Positive Psychology* (pp. 367–374). New York, NY: Oxford University Press.
13. Deci, E. L., Koestner, R., & Ryan, R. M. (1999). A meta-analytic review of experiments examining the effects of extrinsic rewards on intrinsic motivation. *Psychological Bulletin, 125*(6), 627–668. http://psycnet.apa.org/doi/10.1037/0033-2909.125.6.627.
14. Jasperson, J. S., Carter, P. E., & Zmud, R. W. (2005). A comprehensive conceptualization of post-adoptive behaviors associated with information technology enabled work systems. *MIS Quarterly, 29*, 525–557.
15. Zhao, L., Lu, Y., Wang, B., & Huang, W. (2011). What makes them happy and curious online? An empirical study on high school students' Internet use from a self-determination theory perspective. *Computers & Education, 56*(2), 346–356. http://dx.doi.org/10.1016/j.compedu.2010.08.006.
16. Compeau, D. R., & Higgins, C. A. (1995). Computer self-efficacy: Development of a measure and initial test. *MIS Quarterly, 19*(2), 189–211.
17. Roca, J. C., & Gagné, M. (2008). Understanding e-learning continuance intention in the workplace: A self-determination theory perspective. Computers in Human Behavior 24(4), 1585–1604. http://dx.doi.org/10.1016/j.chb.2007.06.001.
18. Bhattacherjee, A. (2000). Acceptance of e-commerce services: the case of electronic brokerages. *IEEE Transactions on Systems, Man, and Cybernetics-Part A: Systems and Humans, 30*(4), 411–420. http://dx.doi.org/10.1109/3468.852435.
19. Litman, J. (2005). Curiosity and the pleasures of learning: Wanting and liking new information. *Cognition & Emotion, 19*(6), 793–814. http://dx.doi.org/10.1080/02699930541000101.
20. Amabile, T. M. (1983). The social psychology of creativity: A componential conceptualization. *Journal of Personality and Social Psychology, 45*(2), 357–376. http://psycnet.apa.org/doi/10.1037/0022-3514.45.2.357.

21. Reio, Jr. T. G., Petrosko, J. M., Wiswell, A. K., & Thongsukmag, J. (2006). The measurement and conceptualization of curiosity. The Journal of Genetic Psychology 167(2), 117–135.
22. Dillman, D. A. (2000). *Mail and internet surveys: The tailored design method* (vol. 2). New York, NY: Wiley.
23. Campbell, D. J. (1988). Task complexity: A review and analysis. *Academy of management Review, 13*(1), 40–52. https://doi.org/10.5465/AMR.1988.4306775.
24. Quinn, E. (1980). Creativity and cognitive complexity. *Social Behavior and Personality: An International Journal, 8*(2), 213–215. http://dx.doi.org/10.2224/sbp.1980.8.2.213.
25. Te'eni, D. (1989). Determinants and consequences of perceived complexity in human-computer interaction. *Decision Sciences, 20*(1), 166–181. https://doi.org/10.1111/j.1540-5915.1989.tb01405.x.
26. Hair, J. F., Black, W. C., Babin, B. J., Anderson, R. E., & Tatham, R. L. (2006). Multivariate data analysis. Upper Saddle River, NJ: Prentice Hall.
27. Kashdan, T. B., Rose, P., & Fincham, F. D. (2004). Curiosity and exploration: Facilitating positive subjective experiences and personal growth opportunities. *Journal of Personality Assessment, 82*(3), 291–305. http://dx.doi.org/10.1207/s15327752jpa8203_05.

Advanced Interference Aware Power Control (AIAPC) Scheme Design for the Interference Mitigation of Femtocell Co-tier Downlink in LTE/LTE-A and the Future 5G Networks

Kuo-Chang Ting[1], Chih-Cheng Tseng[2], Hwang-Cheng Wang[3(✉)], and Fang-Chang Kuo[3]

[1] Department of Business Administration and Department of Information Engineering, Minghsin University of Science and Technology, Xinfeng Hsinchu 30401, Taiwan
kcting82@gmail.com

[2] Department of Electrical Engineering, National Ilan University, Yilan 26047, Taiwan
tsengcc@niu.edu.tw

[3] Department of Electronic Engineering, National Ilan University, Yilan 26047, Taiwan
{hcwang, kfc}@niu.edu.tw

Abstract. LTE femtocell base stations (FBSs) can not only expand the coverage of wireless communication systems but also increase frequency reuse. Therefore, they play an important role in the development of future 5G networks. However, the growing use of FBSs has brought a serious issue of inter-FBS interference (also referred to as co-tier interference) due to their easy and convenient installation. In this article, we propose an advanced systematic approach to reduce FBS co-tier downlink interference under the scenario that FBSs are densely deployed in an environment compared to our past proposed scheme. A femtocell user equipment (FUE) can detect interference from FBSs and send an alarm signal to each FBS which interferes with it. To minimize the negative effects of interference, power levels are reduced for FBSs once they identify themselves as excessive interference (EI) cells defined as FBSs that have received a number of alarm signals above the number of its served UEs. Our simulations show that the throughput of our proposed scheme can be boosted by at least 120% compared to that of our previous proposed IAPC scheme at cost of somewhat little-reduced power consumption, because it could effectively reduce co-tier downlink interference in shared spectrum femtocell environments, thereby improving system performance.

Keywords: Alarm signal · Co-tier downlink interference · FBS-excessive interference · Femtocell · Power control

V. Bhateja et al. (eds.), *Information Systems Design and Intelligent Applications*, Advances in Intelligent Systems and Computing 672,
https://doi.org/10.1007/978-981-10-7512-4_54

1 Introduction

Since 2010, the targets of the fifth generation (5G) wireless communication technologies are set to attain 1000 times higher mobile data volume per unit area, 10–100 times higher number of connecting devices and user data rate, 10 times longer battery life, and 5 times reduced latency [1–3] by 2020. In order to attain the goals set above, femtocells and other small cell technologies play an important role in the next generation of wireless network. We owe to the fact that at least 80% of the wireless connections are performed indoors [4]. Femtocell network cannot only expand the coverage, off-load the burden of the macrocell base station (MBS) but it can also enhance the total capacity by means of the backhaul connection with ultra-wide bandwidth and large factors of frequency reuse. However, the growing deployment of FBSs has brought a serious issue of inter-FBS interference (also referred to as co-tier interference) due to their easy and convenient installation. This problem is especially serious when the FBSs are densely deployed in an urban area. Every household deploys its own femtocell to increase the transmission rate but doing so may induce serious co-tier interference at the same time due to residence proximity. In fact, cross-tier interference arising from the FBSs and MBSs also becomes a critical issue due to the deployment of femtocells. To cope with the co-tier or cross-tier interference, various power control schemes to migrate the penalty have been proposed [5, 6, 7]. Game theory such as Stackelberg game [8, 9] has been applied in power control and resource allocation negotiation between the FBS and MBS through the pricing mechanism so that leader and followers can achieve balance in terms of throughput, outage probability, spectrum efficiency, and so on. The fractional frequency reuse (FFR) scheme proposed in [10] is used to tackle the co-channel interference problem in OFDMA network. Simulations show that FFR method can increase SINR value in all scenarios considered. However, this scheme suffers from the uneven distribution of FBSs and the frequency reuse is also restricted. In order to tackle the problem for the densely deployed femtocell networks listed above, the enhanced inter-cell interference coordination (eICIC) [11–13] has been extensively studied and adopted as a standard for LTE-A and surely will also be a standard for the future 5G networks. This scheme is centralized and complex compared to our scheme. The small cell group muting (SCGM) scheme proposed in [14] is based on the eICIC technology and is used to mitigate the interference among neighboring small cells. The key motivation of this scheme is that one subframe of LTE-A is composed of 14 OFDM symbols for each resource block (RB). Some OFDM symbols are muted for some FBSs and others are not so that the interference can be mitigated or even be avoided. Simulation results show that the scheme can boost the average throughput by 7–15%. In this article, we propose a systematic approach to reduce co-tier downlink interference for FBSs under the scenario that FBSs are densely deployed. Our proposed scheme, advanced interference aware power control (AIAPC) is used to suppress the co-tier interference among FBSs. The scheme is distributed and very easy to implement. The algorithm code developed for this scheme can be integrated into the control firmware of FBSs.

In the next parts of this article, system model of our proposed scheme will be presented and addressed in Sect. 2. Simulation results and discussions will be addressed in Sect. 3. Conclusions and the future works are given in Sect. 4.

2 System Model

The access methods of FBSs can be divided into three types,open subscriber group (OSG), closed subscriber group (CSG), and hybrid subscriber group (HSG). In OSG, all FUEs can connect to all FBSs so that the MBSs can decrease their service loading from the requests of UEs. On the other hand, the FUEs must login in the FBS by their registration accounts in CSG. HSG is with the characteristics of OSG and CSG. In this article, we assume the access method is OSG. The spectrum allocations can be divided into two types: split spectrum reuse and shared spectrum reuse. In the split spectrum reuse, the spectrum is portioned into two parts. One part is used by MBSs, and the other part is used by FBSs. On the contrary, all spectrum or bandwidth is shared by FBSs and MBSs in shared spectrum reuse. In this article, we assume the partition scheme is shared spectrum reuse. We also assume that there are totally M FBSs and N FUEs located in the considered environment with densely deployed FUEs and FBSs. We focus on the co-tier interference and the interference among the FBSs, and MBSs is assumed to be very small so that it can be ignored. Our scheme assumes that once a FUE senses interference, it will broadcast a warning message to all FBSs within its range. In our previous proposed scheme, IAPC [15], if the number of warning messages received by a FBS is greater than a given threshold, it should be aware of the fact that it is causing interference to an excessive number of FUEs and will identify itself as an FBS-Excessive Interference (FBS-EI) source. Under the circumstance, it should reduce its power level to mitigate the interference toward the FUEs of other FBSs through a distributed FBS-EI detection algorithm. However, if the number of FUEs served by this FBS is numerous, its total throughput also decreases as well. Our latter simulations show that our proposed advanced interference avoidance power control (AIAPC) scheme can increase by at least 120% compared to that of IAPC. The key idea is that if the number of warning messages, that is, the number of FUEs suffering from interference from one FBS is greater than the number of served FUEs for one FBS, this FBS can reduce their power level so that the total capacity can increase. The benefit of reducing interference toward the FUEs of other FBSs can be greater than the loss of decreasing its throughput. This scheme is also distributed and very easy to implement. On the contrary, femtocell management system (FMS)-based schemes such as those in [11] and [12] use a centralized control by a managing server. This server must collect a large amount of data in order to determine the source of interference through a special algorithm. Hence, these centralized approaches might be rather complicated to implement. In order to simplify the process of looking for the interference sources, our scheme is based on the observation that, if the coverage of an FBS overlaps with that of neighboring FBSs, the FBS will likely interfere with the FUEs served by neighboring FBSs. The interfered FUEs will send out warning messages. The number of warning messages received by an FBS can be used as an indicator of the severity of interference. The procedures of find a FBS-EI source mentioned above is illustrated in Fig. 1. Figure 2 shows that if the IAPC is applied, the service area of FBS-5 overlaps with those of FBS-1, FBS-2, FBS-3, and FBS-4. Therefore, FBS-5 will receive a lot of warning messages from the FUEs served by its neighboring FBSs and FBS-5 and other possible FBS-EI will be aware that they are FBS-EI sources if the number of the warning messages received from other FUEs is greater than or equal to the threshold. In this scenario, if the threshold number is set to 1, FBS-5 and FBS-3 indicated in Fig. 2 will reduce their power level to mitigate the interference toward the FUEs served by other FBSs. However, if the AIAPC is applied, only the FBS with the number of the served FUEs less than the number of FBS-EI warning messages it receives must reduce its power level; as a result, the total capacity can increase compared to that of IAPC because once the FBS serving many FUEs reduces their power level, the total throughput of all FUEs served by this FBS will be reduced also.

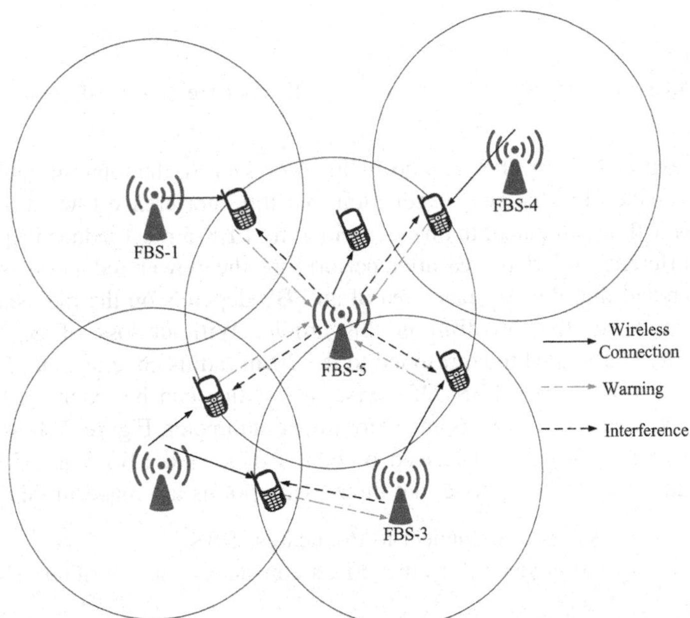

Fig. 1. Illustration of the procedures for identifying an FBS-EI source

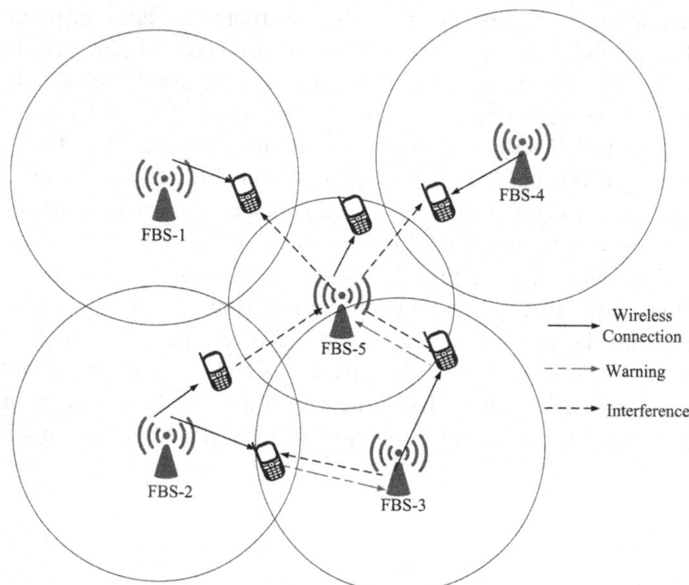

Fig. 2. Illustration of FBSs' interference on FUEs after the power reduction of FBS-5

On the contrary, if the FBS with no FUEs to be served, the total throughput of this FBS is not affected by this power reduction, but the interference toward other neighboring FBSs will be mitigated. In this scenario, only FBS-5 must reduce its power level and the interference overlap area after performing the power reduction is shown in Fig. 2. It is noted that the coverage area of an FBS depends on the radiosensitivity of FUEs, which is set to −56 dBm in this article. Without loss of generality, the radiosensitivity is assumed to correspond to the 20 m radius coverage and the transmit power of 23 dBm for each FBS. Of course, the radius can be extended to 25 m or farther. In fact, it can be treated as an environment parameter. Figure 3 shows that only FBS-5 reduces its power level instead of both FBS-3 and FBS-5 in AIAPC if the threshold number is set to 1. Note that three assumptions are made in this scheme:

(1) Every FUE is assumed to connect to the nearest FBS.
(2) Every FUE in the coverage of other FBSs can detect the interference from these FBSs.
(3) Every FBS will receive the FUEs' warning messages if they are in the coverage of these FUEs. The messages receptively received by one FBS from the same FUE are treated the same; thus, the number of warning messages does not change.

We define several notations in Table 1 and use these notations to describe our AIAPC algorithm proposed in this scheme as in Fig. 3.

The signal strength $RSS_{j,\ j\_k}$ received by FUE_{j_k} when FBS-j transmits its message with the power P_j and encounter a path loss $L_{j,\ j\_k}$ can be given by

$$RSS_{j,j\_k} = P_j - L_{j,j\_k}. \tag{1}$$

It is noted that if there is no power control performed by FBS-j, $P_j = P_{npc}$. Hence, RSS_{j,j_k} can also be given by

$$RSS_{j,j\_k} = P_{npc} - L_{j,j\_k}. \tag{2}$$

Once the power reduction control is performed by FBS-j, the signal strength RSS_{j,j_k} received by FUE_{j_k} with the largest path loss is equal to R_{min} based on the definition listed in Table 1 and (2). Hence,

$$R_{\min} = P_j - L_{j,j\_k}. \tag{3}$$

In order to be sure that all FUEs of the FBS-EI source, FBS-j can be served, the power P_j must be constraint to

$$P_j = P_{npc} + (R_{\min} - \min_{k \in F_j} RSS_{j,j\_k}). \tag{4}$$

If the power reduction controlled by the AIAPC scheme is performed, the total interference received by FUE_{i_k} from the download of other FBSs, I_{i_k} can be given by

$$I_{i\_k} = \sum_{j=1, j \neq i}^{M} I_{j,i\_k}. \tag{5}$$

Table 1. Notations used in AIAPC and IAPC schemes

Notation	Description
P_j	The transmission power level of the FBS-j in dBm
P_{npc}	The transmission power level of FBS with no power control (NPC) in dBm
RS_{i,j_k}	The receiving signal strength of FUE k of FBS-i (or FUE_{i_k}) from FBS-j in dBm
I_{j,i_k}	The interference received by the FUE_{i_k} from FBS-j in dBm
K	The ID of FUE served by FBS
η	Thermal noise
$R_{\min}$	The minimal Received Signal Strength (RSS) for each FBS to receive any message
RSS_{j,j_k}	The RSS for FUE_{j_k} received from FBS-j
L_{i,i_k}	The path loss from FBS-i to FUE_{i_k} in dB
C_{i_k}	The capacity of the kth FUE of FBS-i
|FBS-i|	The number of FUEs served by FBS-i
$LMAX_i$	The largest path loss among all FUEs served by FBS-i
NWM_i	Number of warning messages received by FBS-i from FUE_{i,j_k} and $i \neq j$ so far for each run
D_{i,i_k}	Distance of FBS i to the FUE k

```
Define: M FBSs, N FUEs, ∑_i^M |FBS-i| = N
Define:  i = 1 to M, k = 1 to |FBS-i|.
Input: |FBS-i|(the number of served FUEs for each FBS i)
Output: The new power distribution of femtocell.
  Begin a Run
  For 1=0 to M //initialize NWM to zero
  Begin
     NWM_i = 0
  End
  For i=1 to M
  Begin
    k ← 1
    LMAX_i = L_{i,i_k} = P_npc - RSS_{i,i_k}
   For k=2 to |FBS-i|
   L_{i,i_k} = P_npc - RSS_{i,i_k}
     If L_{i,i_k} > LMAX_i then
     Begin
       LMAX_i = L_{i,i_k}
     End
    Begin
        If (FUE_{i_k} senses interference from FBS-j) then
        Begin
          Send a warning message to FBS-j (i≠j)
        End
    End
  For i=1 to M
  Begin
       //The following algorithm is implemented by
individual FBS
       If FBS-i receives interference from FUE then
         Begin
           NWM_i ← NWM_i + 1
         End
       If NWM_i >= |FBS-i| then
        Begin
           P_min ← R_min+LMAX_i
           If (P_i > P_min)
        P_i ← P_i -1
        End
  End
```

Fig. 3. Our proposed AIAPC scheme

The received signal strength of FUE_{i_k}, $RSS_{i,\ i\_k}$ can be given by

$$RSS_{i,i\_k} = P_i - L_{i,i\_k} \tag{6}$$

after suffering a path loss $L_{i,\ i\_k}$. The path loss can be given by

$$L_{i,i\_k}(dB) = 37 + 30\log_{10}(D_{i,i\_k}) \tag{7}$$

based on 3GPP TR25.952 [16]. The SINR of FUE_{i_k} can be given by

$$SINR_{i\_k} = RSS_{i,i\_k}/(I_{i\_k} + \eta). \tag{8}$$

Hence, the total capacity C can be estimated by

$$C = B\sum_{i=1}^{M}\sum_{k=1}^{|FBS-i|}\log_2(1 + SINR_{i\_k}) \tag{9}$$

where B denotes the bandwidth, and the capacity is estimated based on the well-known Shannon capacity theorem for single input and single output (SISO).

3 Simulation Results and Discussion

The simulation parameters are listed in Table 2. In this study, suppose there are totally 100 FBSs and 300 or 400 FUEs randomly distributed over a 200 × 200 m^2 area. Any FUE will connect to a FBS which is the closest to itself so that it can get the best channel gain from this FBS. A total of 10 runs of simulations are performed to investigate the behavior of IAPC and AIAPC and the results are shown in Fig. 4.

Table 2. Simulation parameter values

Parameter	Value
Map range	200 m × 200 m
Number of FBSs (M)	100
Number of FUEs (N)	300, 400
FBS radius of coverage	20 m
FBS transmit power (Max)	23 dBm
Bandwidth	10 MHz
Frequency	2 GHz
Minimal sensitivity to receive (R_{min})	−65 dBm
T-value	1, 2, 3, 4

The value T in this article denotes the threshold number of warning messages received by one FBS to reduce its power level for IAPC. If T is equal to one, it implies that any FBS receiving a warning message will treat itself as a FBS-EI source and will reduce its transmission power immediately. However, in this scenario, all FBSs are inclined to reduce their power frequently so that the SINR of their served FUEs decreases on average also. On the contrary, the AIAPC does not reduce their power level unless the number of its served FUEs is less than the T-value so the total capacity will not reduce. Figure 4 shows that the total throughput can be boosted by 34.5 and 44.2% when the number of FUEs is 300 and 400, respectively, for AIAPC compared to that of NPC. On the contrary, the total throughput improvements are only 15.7 and 22.7% for IAPC when the number of FUEs is 300 and 400, respectively. Hence, the throughput can be increased by 120.0% at least depending on the total number of FUEs. We compare the power reduction ratio between AIAPC and IAPC, and the results are shown in Fig. 5. To our surprise, when the number of FUEs is 300, the power reduction ratio of AIAPC is somewhat larger than that of IAPC. We owe it to the fact that the results shown in Fig. 5 are the average of all T-value scenarios. We illustrate the impact of T-value on the power reduction as in Fig. 6. Hence, our schemes, no matter IAPC or AIAPC could reduce the total power consumption by ranging from 32.9 to 75.4%; thus, these could be considered to be green communication schemes.

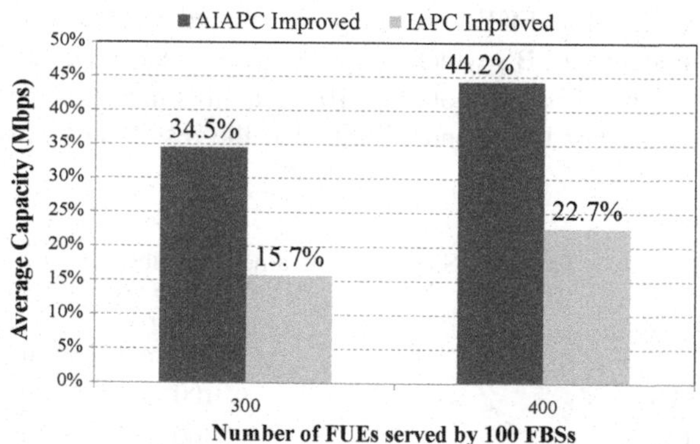

Fig. 4. Throughput improvement comparison between AIAPC and IAPC compared to that of NPC

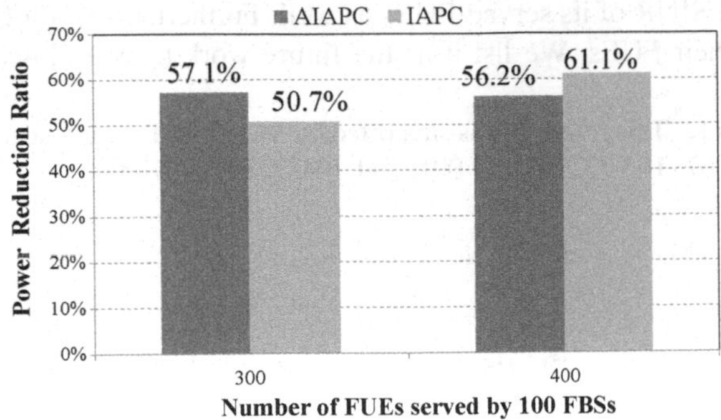

Fig. 5. Comparison of power reduction ratio between AIAPC and IAPC compared to that of NPC

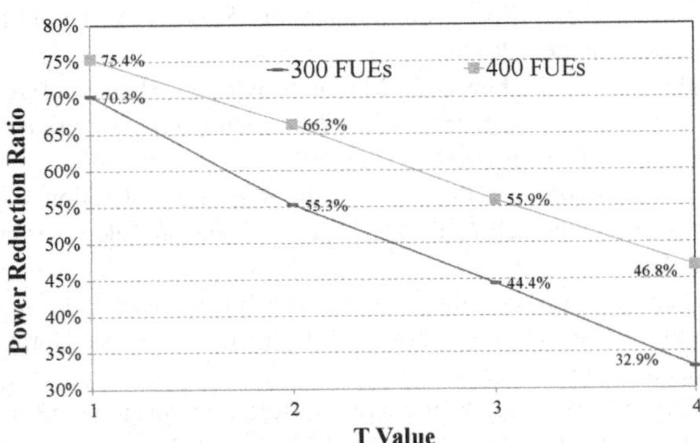

Fig. 6. Effect of *T*-value on the power reduction for IAPC

4 Conclusions

In this article, we propose a distributed AIAPC algorithm to reduce the downlink interference under the co-channel shared spectrum environment for the femtocell and the future 5G networks so that the average SINR and capacity can increase by at least 34% on average. This algorithm is distributed and easy to implement as stated before. The centralized control system such as femtocell management system (FMS) might have better performance. However, it is really very hard for FMS to control and collect the distributed data for each FBS. Furthermore, the mechanism of power control also can reduce the total power consumption to be an ideal green communication paradigm [17]. Despite the fact that there are many advantages in AIAPC, there is still a border problem for any shared spectrum femtocell system. In other words, if FUEs reside on the border between two neighboring cells under shared spectrum femtocell environments, the power control scheme has its limitation. The FBS can perform the power reduction to mitigate their interference toward the FUEs of its neighboring cell, but it might reduce the SINR of its served FUEs as well. Furthermore, the FBS might lose its service toward their FUEs. We list it as the future work to be followed up.

Acknowledgements. This research was supported by MOST of Taiwan under contract numbers 104-2221-E-197–007, 105-2221-E-159-001, and 104-2221-E-197-009.

References

1. F. L. Luo and C. Zhang, "5G standard development: technology and roadmap," in Signal Processing for 5G: Algorithms and Implementations, 1, Wiley-IEEE Press, https://doi.org/10.1002/9781119116493.ch23, 8 Aug, 2016.
2. P. T. Dat, A. Kanno, N. Yamamoto, and T. Kawanishi, "5G transport networks: the need for new technologies and standards," in IEEE Communications Magazine, vol. 54, no. 9, pp. 18–26, September 2016.
3. M. Agiwal, A. Roy, and N. Saxena, "Next generation 5G wireless networks: A comprehensive survey," in IEEE Communications Surveys & Tutorials, vol. 18, no. 3, pp. 1617–1655, third quarter 2016.
4. V. Chandrasekhar, J. G. Andrews, T. Muharemovic, Z. Shen, and A. Gatherer, Power control in two-tier femtocell networks, *IEEE Transaction on Wireless Communication.*, vo. 8, no. 8, pp. 4316–4328, Aug. 2009.
5. H. S. Jo, C. Mun, J. Moon, and J.-G. Yook, "Interference Mitigation Using Uplink Power Control for Two-Tier Femtocell NetWorks," IEEE Trans. Wireless Comm., vol. 8, no. 10, pp. 4906–4910, Oct. 2009.
6. V. Chandrasekhar, J. Andrews, and Z. Shen, T. Muharemovic, and A. Gatherer, "Power control in two-tier femtocell networks," IEEE Trans. Wireless Comm., vol. 8, no. 8, pp. 4316–4328, Aug. 2009.
7. L. Zhang, L. Yang, and T. Yang, "Cognitive interference management for LTE-A femtocells with distributed carrier selection," IEEE 72nd VTC 2010-Fall, pp. 1–5, September 2010.
8. C. C. Tseng, C. S. Peng, S. H. Lo, H. C. Wang, F. C. Kuo, and K. C. Ting, "Co-tier uplink power control in femtocell networks by Stackelberg game with pricing," Global Wireless Summit, Aalborg, Denmark, May 11–14, 2014.
9. Y. S. Liang, W. H. Chung, G. K. Ni, I. Y. Chen, H. Zhang, and S. Y. Kuo, "Resource allocation with interference avoidance in OFDMA femtocell network," IEEE Trans. Vehicular Technology, vol. 61, no. 5, pp. 2243–2255, June 2012.
10. T. I. Giovany, U. K. Usman, and B. Prasetya, "Simulation and analysis of interference avoidance using fractional frequency reuse (FFR) method in LTE femtocell," ICOICT, pp. 192–197, 2013.
11. C. Bouras, G. Kavourgias, V. Kokkinos, and A. Papazois, "Interference management in LTE femtocell systems using an adaptive frequency reuse scheme," 2012.
12. Y. P. Zhang, S. Feng, P. Zhang, L. Xia, Y. C. Wu, and X. Ren, "Inter-cell interference management in LTE-A small-cell networks," IEEE 77th VTC-Spring, pp. 1–6, June 2013.
13. H. Zhou, Y. Ji, X. Wang and S. Yamada, "eICIC Configuration Algorithm with Service Scalability in Heterogeneous Cellular Networks," in IEEE/ACM Transactions on Networking, vol. 25, no. 1, pp. 520–535, Feb. 2017.
14. N. Arulselvan, M. Chhawchharia and M. Sen, "Time-domain and frequency-domain muting schemes for interference co-ordination in LTE heterogeneous networks," IEEE International Conference on Advanced Networks and Telecommunications Systems, pp. 1–6, December 2013.

15. Hwang-Cheng Wang, Fang-Chang Kuo, Chih-Cheng Tseng, Bo-Wei Wang, Kuo-Chang Ting, "Improving LTE Femtocell Base Station Network Performance by Distributed Power Control", Universal Journal of Electrical and Electronic Engineering, vol. 4, no. 5, pp. 113–119, 2016.
16. 3rd Generation Partnership Project; Technical Specification Group Radio Access Network; TDD Base Station Classification (Release 2000), ftp://www.3gpp.org/tsg_ran/TSG_RAN/TSGR_12/Docs/PDFs/RP-010457.pdf.
17. A. De Domenico, E. C. Strinati, and M.-G. Di Benedetto, "Cognitive strategies for green two-tier cellular networks: a critical overview," in M. S. Obaidat, A. Anpalagan, and I. Woungang (eds.), Handbook of Green Information and Communication Systems, chap. 1, Academic Press, 2013.

Design and Implementation of Unauthorized Object and Living Entity Detector with PROTEUS and Arduino Uno

Samridhi Sajwan, Shabana Urooj(✉), and Manoj Kumar Singh

Electrical Engineering Department, Gautam Buddha University, Greater Noida 201310, Uttar Pradesh, India
{samridhi5869,shabanabilal,manojsingh.singh19}@gmail.com

Abstract. This paper presents and implements an unauthorized object detector using ultrasonic waves. Ultrasonic refers to the waves with frequencies over 20 kHz. It is insensitive to human ears having audible perception range of 20 to 20 kHz. Thus, ultrasonic waves are useful for distance measurement in driverless cars. The distance can be measured by using two techniques; pulse echo and phase measurement methods. In this work, pulse echo method is considered where the target is detected via 'non-contact' technology. The Ultrasonic Module keeps monitoring and checks the echo reflected back by the entity to display the exact angle and distance on the screen. The novelty of this work is the development of an alert system that not only detects the target but also measures and displays its exact distance. PROTEUS software has been used for simulation. Hardware is also developed and tested to detect any entity in the prescribed range.

Keywords: Arduino Uno · ATmega328 · Ultrasonic sensor · Servo motor PROTEUS 8

1 Introduction

In routine life, several situations occur where it is required to keep a track over an area for security purpose or to avoid trespassing. Employing human labor for this is an expensive and also unreliable way to monitor for 24 × 7. So to avoid this problem, ultrasonic radar has been developed for unauthorized human/animal or object detection. It is certainly a reliable and efficient method for instantaneous measurement of distance [1]. Ultrasonic sensors operate by emitting ultrasonic waves in a rapid succession, henceforth they are very versatile in distance measurement at very low cost [2]. This system can monitor an area of limited range and alert authorities with an alarm. For continuous monitoring, an Arduino Uno circuit has been used which is connected to an ultrasonic sensor mounted on a servo motor. The ultrasonic module keeps monitoring the surroundings and checks the echo. As soon as the sensor detects an object, the data of detection is processed and sent to the concerned authorities. Thus, ultrasonic sensor is a reliable and convenient way of continuous monitoring of a particular area. Presently, sound wave technology is the most frequently used underwater positioning

V. Bhateja et al. (eds.), *Information Systems Design and Intelligent Applications*, Advances in Intelligent Systems and Computing 672,
https://doi.org/10.1007/978-981-10-7512-4_55

technology [3]. Also ultrasonic devices have application in making an approximate 2D shape for fixed objects placed around the device based on trigonometric rules [4].

This proposed model can also be installed at a grid or a substation where huge area is to be monitored by a security guard. There an ultrasonic sensor can keep a sight over the entire section which provides an easy way of detection and distance measurement with suitable accuracy. Earlier a 'Portable Walking Distance Measurement System' was also developed by using ultrasonic wave characteristics and has proven approximately 90% of accuracy [5].

2 Theory of Operation

This chosen application is based on the reflection principle; the echo reflected back by the object is utilized for distance measurement. The ultrasonic wave propagation velocity in the air is approximately 340 m/s at 15 °C of air or atmospheric temperature. The HC-SR04 ultrasonic sensor is a 4-pin device. It is able to detect the objects like bats and dolphins by using sonar system.

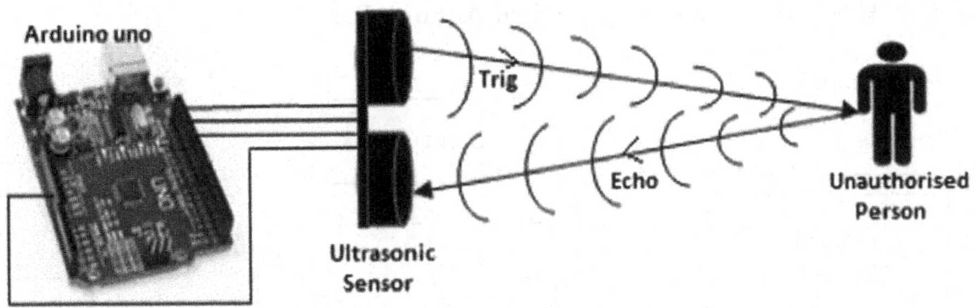

Fig. 1. Working diagram of ultrasonic module

To determine the distance, an outgoing signal called 'trig' must be high for duration of at least 10 μs. When a trig signal is applied, it produces 8 cycles of ultrasonic burst at 40 kHz. The echo pin is high from the moment of sending the signal to receiving it. Immediately after hitting the entity, the echo is received by the receiver (Fig. 1).

Further, this signal is amplified, filtered, and converted into digital form [6]. Thus, the distance can be calculated by measuring the time between the sending pulses and receiving echo. Usually, this signal is not affected by factors, such as light, electromagnetic waves, and dust [7].

Object distance can be calculated from the following formula,

$$\text{Distance} = (\text{time taken} * \text{ultrasonic velocity})/2$$

i.e.,

$$x = (v * t)/2. \tag{1}$$

The ultrasonic waves travel to and fro from the object, hence the whole distance is divided by two.

Arduino continuously sends the trig signal, and the distance of the target can be measured continuously without any delay. After one measurement, flag gets clear and the module gets ready for a new measurement [8].

3 Block Diagram of the Proposed System

Block diagram of ultrasonic sensor with Arduino is shown in Fig. 2. In the block diagram, ultrasonic sensor is used to measure the distance of the object. The sensor output is processed through Arduino, and the measured results can be seen on the computer screen. The sensor is mounted on a servo motor. Servos are essential parts to control the position of objects, i.e., to rotate the sensor. The motor controller receives a signal (from the Arduino) which decides the movement of servo motor to a desired position. This is done via programming in Arduino IDE.

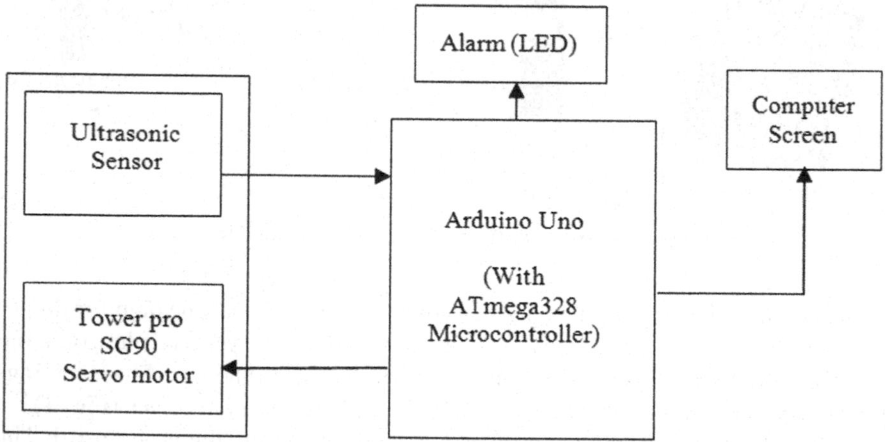

Fig. 2. Block diagram of ultrasonic sensor with Arduino Uno

Thus, it is also used to find the obstacles with the exact distance as in the case of UltraCane. UltraCane is the stick used by blinds to detect obstacles in their path. Vibrating buttons on the stick are used to sense the obstacles.

4 Software Implementation

In this work, PROTEUS 8 has been used to measure the distance of an object from the ultrasonic module. Since ultrasonic sensor cannot be mounted on the servo motor in the software, a potmeter (POT-HG) is used to vary the object distance from the ultrasonic module. The components used in the PROTEUS simulation are:

1. Arduino Uno (Simulino Uno)
2. Ultrasonic sensor (Ultrasonic V2.0B)
3. Potmeter (Pot–HG: 1 kΩ)
4. LCD (LM016L)
5. LEDs (D_1, D_2) and Buzzer
6. Resistors (R_1, R_2).

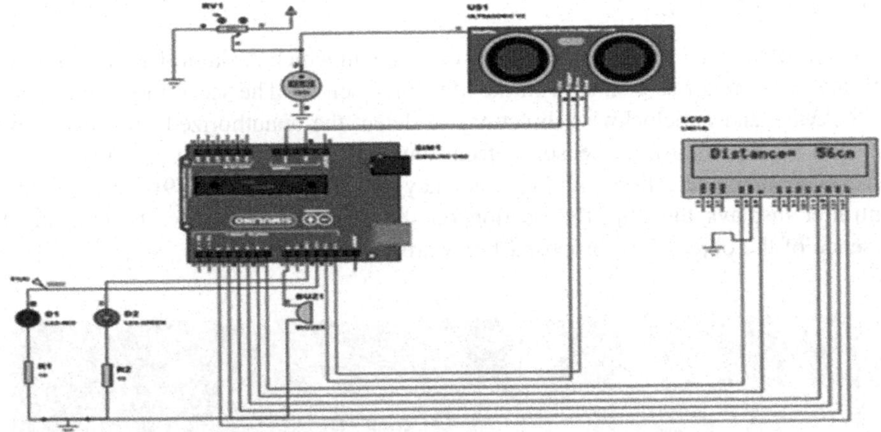

Fig. 3. Software setup after simulation with 56 cm distance

The circuit shown in Fig. 3 consists of Arduino Uno (can be called as the brain of the proposed system), an ultrasonic sensor, and an LCD display. The ultrasonic sensor is attached with the potmeter. The distance between the potmeter and the ultrasonic sensor is the actual distance of the object from the sensor.

Here, Arduino Uno has 14 digital input/output pins, out of which 11 pins are used in this work. Two pins are used for the ultrasonic sensor, six pins are used to control the LCD, and other three are used for object indication.

4.1 Arduino (ATmega328 Microcontroller)

The first and the foremost component used in any electronic circuit is the microcontroller. In this project, Arduino Uno has been used. It is a microcontroller board based on the ATmega328. The Arduino board contains everything itself that are required to support the ATmega328, and to operate, it can simply be connected to a computer with a Universal Serial Bus (USB) cable.

5 Hardware Setup

The next step includes hardware designing of the unauthorized object or any living entity detector. This setup includes the following components:

1. Arduino Uno—R3 board
2. Ultrasonic module: HCSR-04
3. TowerPro servo motor SG90
4. Breadboard and jumper wires
5. LED
6. Arduino IDE
7. Processing 2.1 IDE.

The experimental setup has simple connections in which Arduino Uno is connected with the servo motor as well as with the ultrasonic sensor. The servo motor may rotate in clockwise and anticlockwise directions to detect the unauthorized object or entity. As soon as the ultrasonic sensor detects any entity, its distance is calculated by non-contact range detection with high accuracy and stable readings [9]. Thus, Arduino controller displays the object's location on the screen. LED glows to indicate the presence of the object. The proposed hardware is shown in Fig. 4.

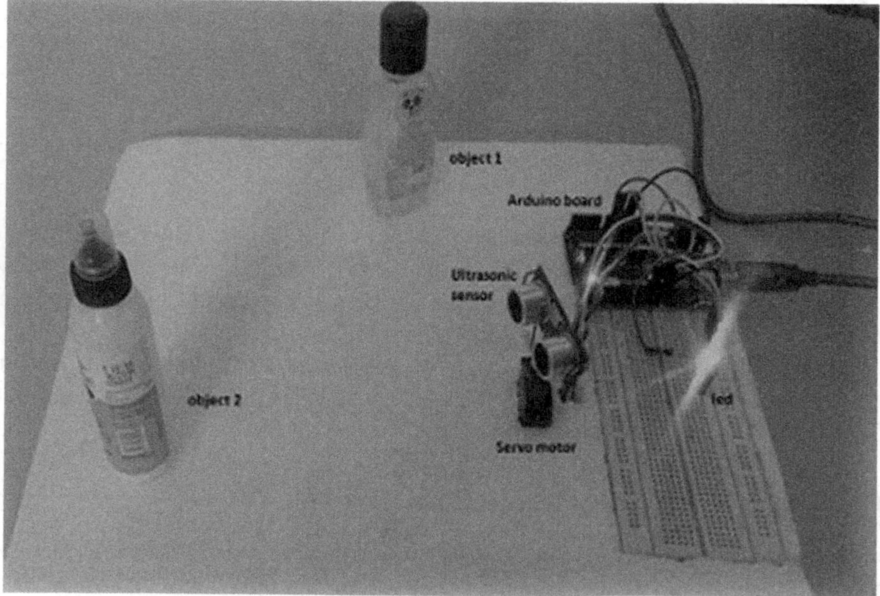

Fig. 4. Hardware setup (LED glowing as an indicator to detect object 2)

The loop code to implement the proposed system in Arduino IDE for rotation from 0° to 180° is given as follows:

```
void loop()
{
servo.write(40);
delay(1000);
servo.write(90);
delay(1000);
servo.write(120);
delay(1000);
for(servoAngle = 0; servoAngle< 180; servoAngle++)
    {
servo.write(servoAngle);
delay(30);
distance = calDistance();
sdistance = distance;
if (sdistance<=15)
    {
digitalWrite(ledpin, HIGH);
    }
else
    {
digitalWrite(ledpin, LOW);
    }
Serial.print(servoAngle);
Serial.print(" degree ");
Serial.print(distance);
Serial.print(" cm ");
}
}
```

Figure 5 shows the radar screen on detection of object 2 at 14 cm and 133°. Tower Pro SG90 servo motor is used because it is directly powered from the Arduino via USB and can also rotate from 0° to 180° as per requirement. Servo motor acts as a steering wheel of a car as it controls the position of the sensor. The proposed system is able to detect the object measure its accurate position.

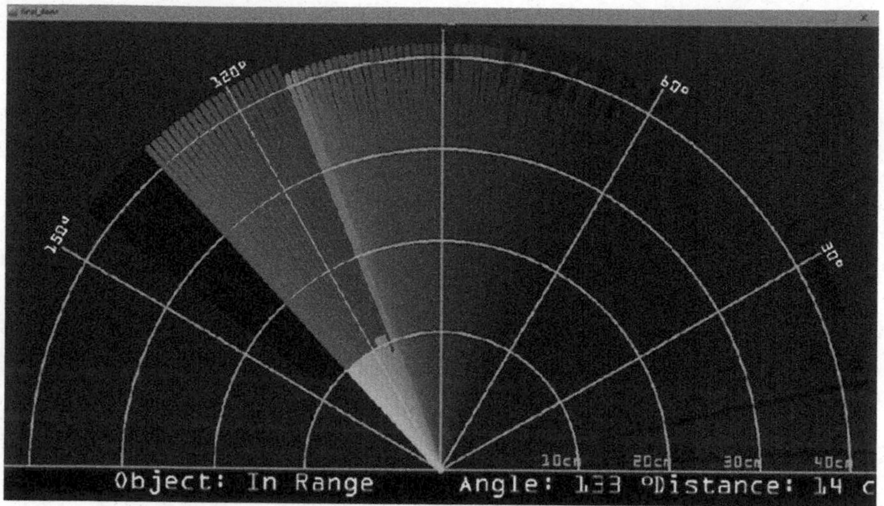

Fig. 5. Radar screen on detection of object 2 at 14 cm and 133°

6 Conclusion

Distance measurement can be done in many ways. Radar is also a commonly used distance measurement system, but it is 3 to 4 times more costly than the proposed model. The model used for distance measurement is of low cost and user-friendly as it alerts the user with the exact position of the object. This module is featured along with the computer display, and it measures distance and its angle. Simultaneously, an alert will be given regarding the unauthorized entity detection. This is a reliable and an inexpensive method for continuous monitoring of a particular region for several applications. The work can be further extended as a system in which distance measurement of an object or an obstacle is required. It can also be installed at grids or localities in terms of security. The system can be modified for driverless technology. It can measure the distance of the nearby vehicle to protect collision. Moreover, with some advancement, this system can replace CCTV cameras and can also be used for home security and related systems.

References

1. Rajan P Thomas, Jithin K K, Hareesh K S, Habeeburahman C A, Jithin Abraham, "Range Detection based on Ultrasonic Principle," International Journal of Advanced Research in Electrical, Electronics and Instrumentation Engineering; Vol. 3, Issue 2, February (2014)

2. A. K. Shrivastava, A. Verma, and S. P. Singh: Distance Measurement of an Object or Obstacle by Ultrasound Sensors using P89C51RD2. International Journal of Computer Theory and Engineering, Vol. 2, No. 1 February, 1793–8201, (2010)
3. C. C. Chang, C. Y. Chang, and Y. T. Cheng, "Distance measurement technology development at remotely tele-operated robotic manipulator system for underwater constructions," IEEE International Symposium on Underwater Technology, pp. 333–338, (2004)
4. Rechie Ranaisa Dam, Hridoy Biswas, Shuvrodeb Barman, Al-Quabid Ahmed, "Determining 2D shape of object using ultrasonic sensor," 3rd International Conference on Electrical Engineering and Information Communication Technology (ICEEICT), (2016)
5. Y. Jang, S. Shin, J. W. Lee, and S. Kim, "A preliminary study for portable walking distance measurement system using ultrasonic sensors," Proceedings of the 29th Annual IEEE International Conference of the EMBS, France, pp. 5290–5293, (2007)
6. G. Benet, J. Albaladejo, A. Rodas, P.J. Gil, An intelligent ultrasonic sensor for ranging in an industrial distributed control system, in: Proceedings of the IFAC Symposium on Intelligent Components and Instruments for Control Applications, Malaga, Spain, May, pp. 299–303 (1992)
7. H. He, and J. Liu, "The design of ultrasonic distance measurement system based onS3C2410," Proceedings of the IEEE International Conference on Intelligent Computation Technology and Automation, Oct, pp. 44–47, (2008)
8. Marius Valerian Paulet, Andrei Salceanu, Oana Maria Neacsu, "Ultrasonic Radar," International Conference and Exposition on Electrical and Power Engineering, 20–22 October, Iasi, Romania. (EPE 2016)
9. Aritra Acharyya, " Foundations and Frontiers in Computer, Communication and Electrical Engineering" Proceedings of the 3rd International Conference on Foundations and Frontiers in Computer, Communication and Electrical Engineering (C2E2-2016), Mankundu, West Bengal, India, 15–16 January (2016)

A Simulink-Based Closed Loop Current Control of Photovoltaic Inverter

Nidhi Upadhyay[1(✉)], Shabana Urooj[1], and Vibhutesh Kumar Singh[2]

[1] Gautam Buddha University, Greater Noida, Uttar Pradesh, India
nidhiupadhyayindia@gmail.com, shabanaurooj@ieee.org
[2] University College Dublin, Belfield, Dublin, Ireland
vibhutesh.k.singh@ieee.org

Abstract. In this paper, a system is proposed for maintaining alternating current with the desired characteristics of a closed loop configuration photovoltaic (PV) system. The generated output current from the PV system is highly dependent on the temperature and intensity of the solar radiation. The proposed system overcomes these critical issues by using a closed loop current control, resulting in an alternating current (AC) output of constant frequency and amplitude. The proposed system consists of a photovoltaic cell array, current controlled inverter, closed loop current control and LC filter. The closed loop strategy helps to get nearly ideal AC output. Low pass filtering is employed to further enhance the AC response. The system is developed and verified in MATLAB–Simulink.

Keywords: Inverter · Photovoltaic · Simulink · Current control · Closed loop

1 Introduction

A photovoltaic system finds its use worldwide for generating power. Numerous research is going on to find its new applications, increasing energy efficiency, improving material durability, etc. The photovoltaic system utilizes the sunlight to generate electricity. This is one of the cleanest forms of energy available, and that too never ending, i.e. a renewable source of energy. A photovoltaic system consists of the photovoltaic cell as its basic building block, which is made up of a semiconductor material like doped silicon (Si) or other composite semiconductor materials like gallium arsenide (GaAs). The scale of these photovoltaic systems could be as huge as a solar farm that generates megawatts of electric power in a clear, sunny day or as small as it could fit into various handheld devices like calculators, personal digital assistants, or even could be integrated to house roof itself, to further supplement the household's power supply.

Major components of the proposed system are the photovoltaic cell array, current controlled inverter, closed loop current control, and low pass filter. About 0.1 presents the general component design and connection block diagram of the proposed system. Here, firstly a direct current (DC) output generated by PV cell array is converted into an alternating current with the help on an inverter. To maintain the characteristics like amplitude and frequency, the output of the load is fed back to the closed loop control.

V. Bhateja et al. (eds.), *Information Systems Design and Intelligent Applications*, Advances in Intelligent Systems and Computing 672,
https://doi.org/10.1007/978-981-10-7512-4_56

In addition, low pass filtering is applied to further enhance the output (Fig. 1). The next section describes some works in the area of closed loop current controlling.

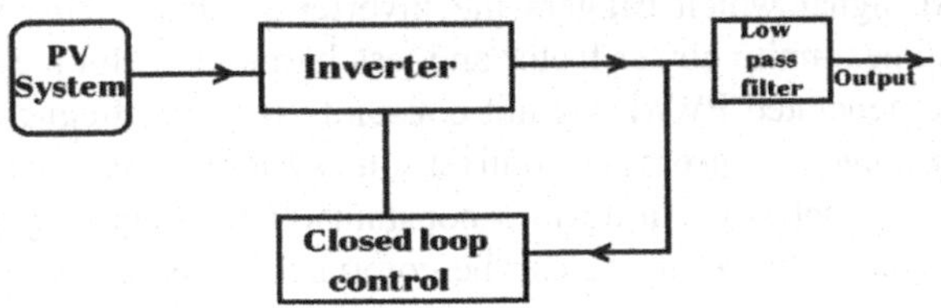

Fig. 1. A diagrammatic representation of the proposed system

2 Related Work

A variety of work has been found in literature in the field of closed loop current controlling. Some of the work includes PV parallel resonant DC link soft switching inverter using hysteresis current control by [1], which is carried out by using a hysteresis current controller, in which voltage controlling is done by proportional–integral (PI) controller, comparator, and a DC link switch controller. The focus is on soft switching which should be able to reduce the switching losses and makes the circuit fast.

In direction of the design, development, and performance of a 50 KW grid-connected PV system with three-phase current controlled inverter is reported [2]. The system ties a PV plant of 50 kW grid, which consists of the solar cell, DC/AC inverter, utility grid, and a control scheme including PWM inverter using D-Q axis transformation. More controlling strategies used along with inverters, which eventually enhance the AC output [3]. One of such controlling strategies is used in this work. Here, ENS represents the German abbreviation of mains monitoring units with allocated switching devices. This system is a digital version of a PV inverter with different control strategy and an embedded technique to measure the grid impedance. By injecting inter-harmonic current and measuring the voltage response, it is possible to estimate the grid impedance at the fundamental frequency [4]. A PI current control algorithm is implemented in digital signal processing chip to keep the current injected into the grid sinusoidal and to have a high dynamic performance with variable weather conditions.

ANFIS-based system has proposed a solution to interface and delivers maximum power from the PV power generating system in standalone operation. The stated maximum power point tracking method offers a fast dynamic response with high precision [5]. The closed loop control of quasi-Z-source regulates the shoot through duty ratio and the modulation index to effectively control the power and maintain the strict current and frequency requirement.

With the inputs of prior work done by various authors, the proposed system presents an effective technique to allow single-phase current controlling to be implemented. This achieves a substantial reduction in the variation of the magnitude and frequency of the inverter's current. In upcoming sections, the operation and control of the inverter are described, together with simulation and experimental results.

Out closest work can be seen in [4], in which two reference signals are compared to generate a PWM signal which triggers the inverter switches. This work is tending towards operating near unity power factor and less harmonics. But the major drawback of [4] is that, the generated PWM has not constant switching frequency that leads to variable output voltage or current. In contrast, our system generates a constant PWM switching frequency that is resulting in a constant output frequency and making the system more accurate. This accuracy can be compared to the output graph of [4] and the following shown in Fig. 9.

3 Methodology

As solar radiation strikes over the PV cell array, there is a momentum transfer from sunlight photon, resulting in an electromotive force (EMF). It is a simple case of energy conservation; here the light energy gets converted into electric energy. The stimulated electrons result in a flow of charge, which in turn generates current over time. In this way, the current flows when a solar cell array is connected to an electrical load. Output voltage of any silicon-based PV cell is about 0.5–0.6 V of DC under open circuit condition. The current (and power) output of a PV cell depends on size (surface area), and it is proportional to the intensity of sunlight striking the surface of the cell [6].

A PV cell in theory and simulation could be modelled as a current source, with a diode connected in parallel. The internal resistance of the cell could be thought as a resistance connected in series. In open circuit condition, there is no current flowing, but a voltage appears across the terminals. Figure 2 shows a PV cell array current generation when connected to a load.

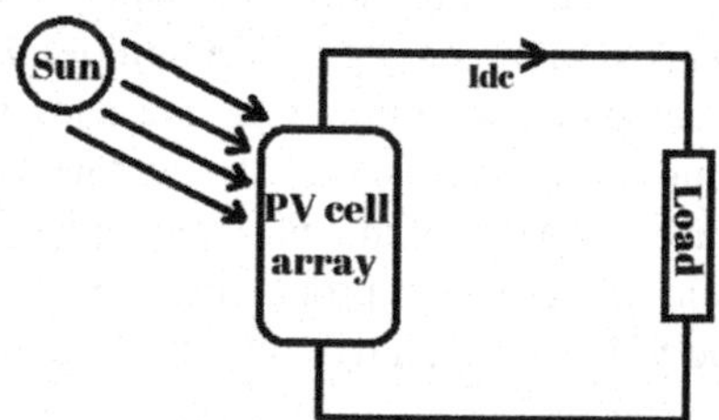

Fig. 2. A diagrammatic representation of a photovoltaic system

Figure 3 is the equivalent circuit model of a single solar cell to which load is not connected. Here, I_{ph} = photovoltaic generated current, I_o = diode, reverse saturation current, I_c = output current for PV module, V_c = PV cell voltage and R_s = series resistance due to internal resistance of PV cell.

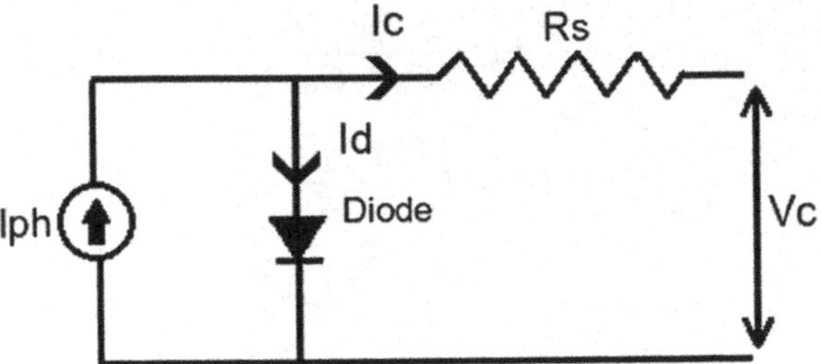

Fig. 3. Equivalent circuit of single solar cell

Consequently, by further analysis, the expression for the PV cell voltage could be elaborated to equation [7].

$$V_c = AkT_c ln \frac{(I_{ph} + I_o - I_c)}{I_o} - R_s I_c \tag{1}$$

Here from [1],

A = Curve fitting factor; 1.6
k = Boltzmann constant; 1.38×10^{-23}
T_c = Operating temperature; 293 K.
I_{ph} = Photovoltaic generated current; 0.005 A
I_o = Reverse saturation current of diode; 0.002 A
I_c = Output current from PV module; 3–5 A
R_s = Series resistance; 0.0001 Ω

Standard maximum output ratings of a single PV cell could be looked upon, as follows:

$$I_m = 3\,\text{A}$$

$$V_m = 0.5\,\text{V}$$

$$P_m = 1.5\,\text{W}$$

Our functional simulation circuit for the PV cell array consists of 36 PV cells connected in series. The generated output rating of single PV array is 18V and 3A. This output is obtained by applying the above-mentioned values of curve fitting factor (A), Boltzmann constant (k), photovoltaic cell generated current (I_{ph}) at operating temperature of 293 K. Output current and voltage can be varied by changing the independent variables of the Eq. (1).

To achieve the voltage range of 144–162 V, a total of 8–9 PV cell arrays need to be combined in series. Since the cell arrangements are in the series, the same amount of current will be generated, i.e. 3 A. The current generation of a single PV cell can be understood by the following mathematical expression (2).

$$I_{ph} = \frac{I_{phx}}{C_{ti} * C_{si}} \tag{2}$$

$$C_{si} = 1 + \frac{S_x - S_c}{S_c} \tag{3}$$

$$C_{ti} = 1 + (T_x - T_a)/S_c \tag{4}$$

where I_{phx} is the new value of the generated current using the correction factors C_{ti} (3) and C_{si} (4).

Here,

S_c = reference solar irradiation
S_x = new irradiation level operating condition.
= 0.06, a constant.
T_x = temperature at any instant, in °C.
T_a = ambient temperature, in °C.

The most important component of the system under consideration is the closed loop current control, which is represented in Fig. 4.

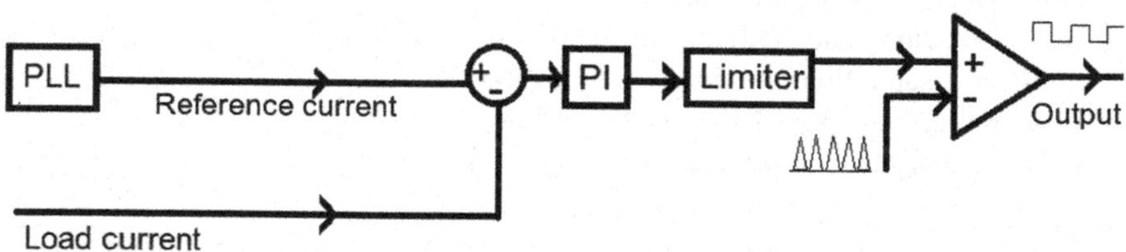

Fig. 4. Proposed closed loop current controller

This component consists of phase locked loop (PLL), PI controller, and current limiter. A PLL (Fig. 5) is used to generate the reference signal and also used to phase synchronize the load voltage and generated a reference current.

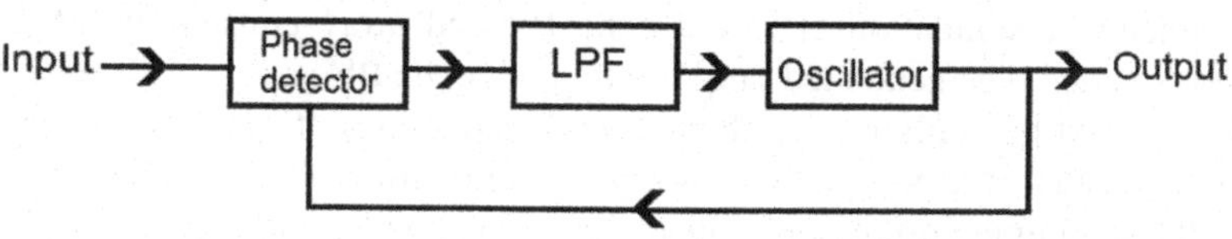

Fig. 5. A phase locked loop block diagram

PLL is a combination of a phase detector, low pass filter, and oscillator. PLL is used to generate the output reference current with same phase as load current, which system requires [8]. The phase detector compares the phase of the generated reference current with the load current. Bringing the output signal back towards the input signal for comparison, in a feedback loop, low pass filters will confine the frequency range up till the desired cut-off frequency, above which all the frequency components are attenuated. An oscillator produces a periodic signal. In this case, the alternating current is generated. The PI controller's gain G_{PI} is defined as an equation [5, 7].

$$G_{PI} = \frac{K_I}{s} + K_P \tag{5}$$

The PI controller shows the combined properties of the proportional (P) and integral (I) controllers. As K_I and K_p, both improve the accuracy and with least steady-state error together.

The PI controller enhances the output by minimizing the error between the reference and the actual current. The error is given to a current limiter. Due to the current limiter, the maximum signal and minimum signal of the PI controller resulting value are limited to a constant range.

The resulting limiter signal is fed to a comparator with a continuous running triangular wave to produce PWM signals [9] for the inverter switch. This PWM signal will lead to the PV inverter for triggering the insulated gate bipolar transistor (IGBT) switches of the inverter. After switching of inverter, a constant frequency and amplitude current will appear when LC filter [10] is used across the load. Figure 6 represents the intended system under discussion.

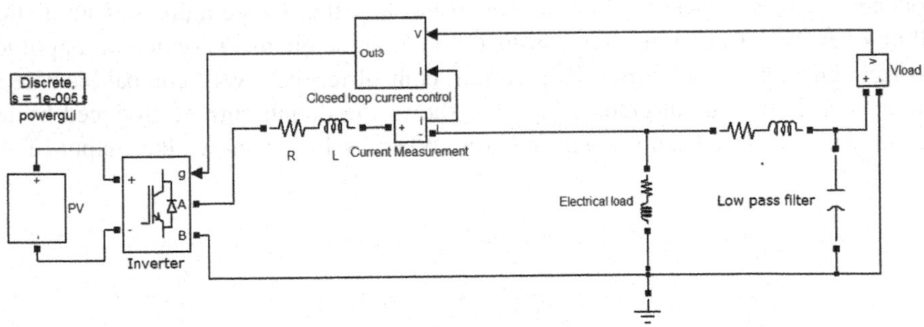

Fig. 6. A PV inverter with proposed closed loop current control

4 Results and Discussion

The PV system is the driving part of the system, which gives a DC output. This DC is converted into AC by using a current controlled inverter. Closed loop current control and LC filter will improve the output to obtain a nearly pure sinusoidal wave. Figure 7 shows the output of the PV system generating a DC output. The output current (3A), voltage (149.5 V) and thus power obtained are 448.5 W.

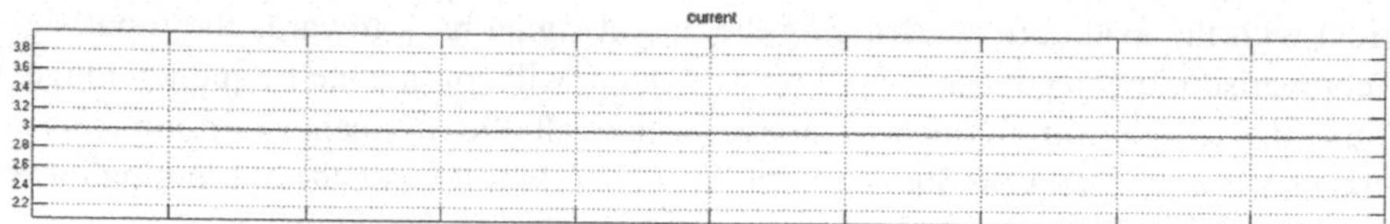

Fig. 7. Output DC current of photovoltaic system

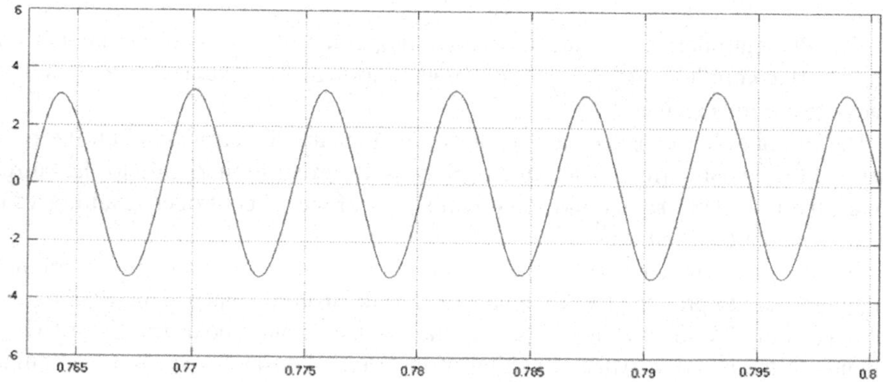

Fig. 8. Output current of PV inverter without closed loop control

The main reason for variable output after the inverter is the variable DC output produced by a PV system, which is due to the fact that temperature and irradiation falling over PV system is also not constant. So to handle out this variation of amplitude and frequency, proposed closed loop current control results a very compatible role as can be seen in the output graphs. Figure 8 shows the output current produced by the system without closed loop current control. Whereas Fig. 9 shows the output of the

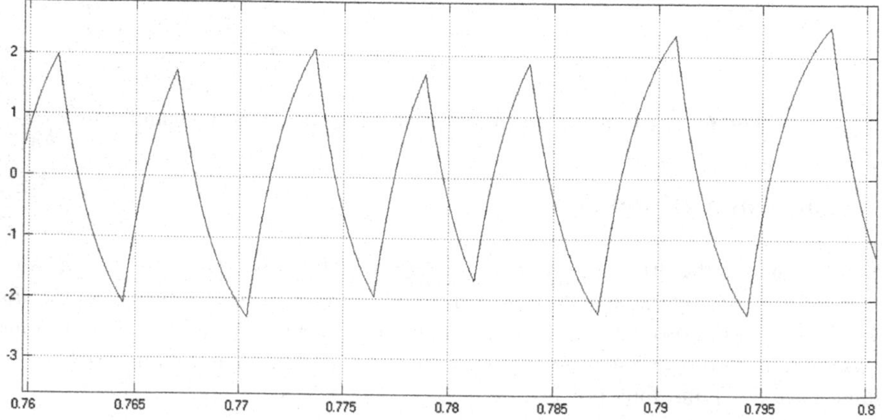

Fig. 9. Output current of PV inverter with closed loop control and filter

system when the closed loop controlling is used along with the implementation of suitable LC filter. Thus, output current will tend towards ideal alternating features.

5 Conclusions

In this proposed system, a different control strategy is implemented with the PV inverter which provides high-quality sinusoidal output current. The proposed closed loop control works well against the variation of amplitude and frequency of the current at the load. In a practical system like [11], some problems like noise and harmonics [12] could affect the quality of the output produced by the system, in addition to the variable environmental factors. When this system, with proposed closed loop control, is used, an improved output, which clearly evident from the output, is obtained as compared to the case of a system without a controller.

6 Future Work

This proposed controlled scheme works well with low ratings PV system, e.g. about till 200 V and 10 A current. But in the case of higher rating generation system, the output is affected by some undesired characteristics like variable output frequency. A direct enhancement would be to eradicate this problem.

References

1. Kim, Young-Ho, Jun-Gu Kim, Young-Hyok Ji, Chung-Yuen Won, and Yong-Chae Jung. "Photovoltaic parallel resonant DC-link soft switching inverter using hysteresis current control." In Applied Power Electronics Conference and Exposition (APEC), 2010 Twenty-Fifth Annual IEEE, pp. 2275–2280. IEEE, 2010.
2. Hwang, I. H., K. S. Ahn, H. Ct Lim, and S. S. Kim. "Design, development and performance of a 50 kW grid connected PV system with three phase current-controlled inverter." In Photovoltaic Specialists Conference, 2000. Conference Record of the Twenty-Eighth IEEE, pp. 1664–1667. IEEE, 2000.
3. Asiminoaei, L., Teodorescu, R., Blaabjerg, F., & Borup, U. (2005). A digital controlled PV-inverter with grid impedance estimation for ENS detection. IEEE Transactions on Power Electronics, 20(6), 1480–1490.
4. Selvaraj, Jeyraj, and Nasrudin A. Rahim. "Multilevel inverter for grid-connected PV system employing digital PI controller." IEEE Transactions on Industrial Electronics 56.1 (2009): 149–158.
5. Abu-Rub, H., Iqbal, A., Ahmed, S. M., Peng, F. Z., Li, Y., & Baoming, G. (2013). Quasi-Z-source inverter-based photovoltaic generation system with maximum power tracking control using ANFIS. IEEE Transactions on Sustainable Energy, 4(1), 11–20.6.
6. Kumar, L. Siva Chaitanya, and K. Padma. "Matlab/Simulink Based Modelling and Simulation of Residential Grid Connected Solar Photovoltaic System." International Journal of Engineering Research and Technology. Vol. 3. No. 3 (iMarch-2014). IJERT, 2014.

7. Gupta, Ankit, Pawan Kumar, Rupendra Pachauri, and Yogesh Kumar Chauhan. “Effect of Environmental Conditions on Single and Double Diode PV system: Comparative Study.” International Journal of Renewable Energy Research (IJRER) 4, no. 4 (2014): 849–858.
8. Ciobotaru, Mihai, Remus Teodorescu, and Frede Blaabjerg. “Control of single-stage single-phase PV inverter.” EPE Journal 16, no. 3 (2006): 20–26.
9. Aneesh Mohamed, A. S., Anish Gopinath, and M. R. Baiju. “A Simple Space Vector PWM Generation Scheme for Any General n-Level Inverter.” IEEE TRANSACTIONS ON INDUSTRIAL ELECTRONICS 56.5 (2009): 1649.
10. Kim, Jaesik, Jaeho Choi, and H. Hong. “Output LC filter design of voltage source inverter considering the performance of controller.” Power System Technology, 2000. Proceedings. PowerCon 2000. International Conference on. Vol. 3. IEEE, 2000.
11. Singh, Vibhutesh Kumar, Sanjeev Baghoriya, and Vivek Ashok Bohara. “HELPER: A Home assisted and cost Effective Living system for People with disabilities and homebound Elderly.” Personal, Indoor, and Mobile Radio Communications (PIMRC), 2015 IEEE 26th Annual International Symposium on. IEEE, 2015.
12. Castilla, Miguel, Jaume Miret, José Matas, Luis García de Vicuña, and Josep M. Guerrero. “Control design guidelines for single-phase grid-connected photovoltaic inverters with damped resonant harmonic compensators.” IEEE Transactions on industrial electronics 56, no. 11 (2009): 4492–4501.

Three-Phase PLLs for Utility Grid-Interfaced Inverters Using PSIM

Kartik Kamal[1(✉)], Kamal Singh[2], Shabana Urooj[1], and Ahteshamul Haque[3]

[1] Electrical Engineering Department, School of Engineering, Gautam Buddha University, Greater Noida, UP, India
kartik.kamal18@gmail.com, shabanaurooj@ieee.org

[2] All India Radio, Prasar Bharti, New Delhi, India
s.kamal15@rediffmail.com

[3] Department of Electrical Engineering, Jamia Millia Islamia, New Delhi, India
ahtshm@gmail.com

Abstract. This paper deals with the simulation model of synchronous rotating reference frame and trigonometric phase lock loop (PLL). For grid-connected inverters, a synchronization control technique is required for maintaining high power quality and efficiency of the system. There are many phase lock loop algorithms used for synchronization like enhanced PLL, power PLL, quadrature-based PLL. This paper presents two phase lock loops for utility grid-connected inverters. The circuits are simulated using PSIM simulation package, the generated phase angle of the PLL as its output is converted into a sine wave by adding a sine block, and results have been analyzed and discussed by the suitable input and output waveforms.

Keywords: Phase-locked loop (PLL) · Synchronous rotating reference frame (SRF) PLL · Trigonometric PLL · Coordinate transformation · Phase detection

1 Introduction

When we connect renewable energy sources to the utility grid, there exists power quality and synchronization problem. The inverter output parameters must be synchronized in phase and frequency of the utility network. For achieving this synchronism, we require a control technique called phase lock loop. PLL is a closed-loop technique, and it produces an output signal which is synchronized in phase and frequency of the input reference signal. Every PLL has three common blocks—phase detector, loop filter, and voltage-controlled oscillator [1]. Most of the PLLs differ from each other in phase detector block. Here presented two PLLs have two different phase detection techniques. SRF PLL uses coordinate transformation as phase detector block, and in trigonometric PLL, a multiplier is used as phase detector (Fig. 1).

If the synchronization of the inverter output parameters is not done with the reference to the grid signals, damage to load of consumer, poor power quality of supply, unstable grid conditions, etc., can happen.

V. Bhateja et al. (eds.), *Information Systems Design and Intelligent Applications*, Advances in Intelligent Systems and Computing 672,
https://doi.org/10.1007/978-981-10-7512-4_57

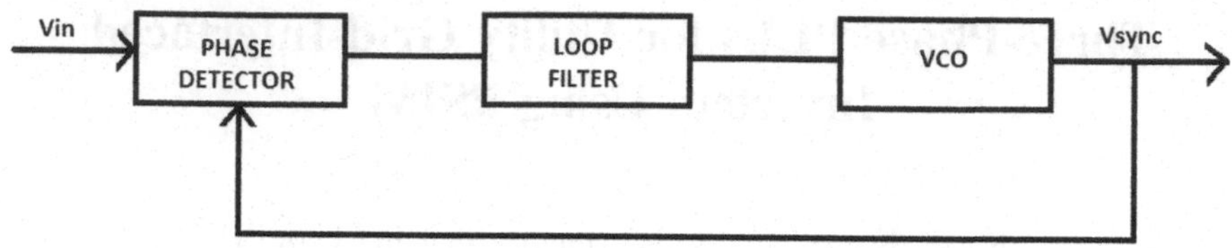

Fig. 1. Block diagram of phase-locked loop

2 System Description

This paper has two proposed PLL circuits simulated in PSIM software. There are two types of PLL circuits simulated here. Both of these PLL's basic structure and working principle remain same, as they are constantly comparing the phase of the PLL output with the reference signal and generating the signal in the phase lock with the input reference signal. The proposed PLL circuits are simulated in PSIM software, and the results are analyzed by their waveforms.

2.1 Synchronous Reference Frame PLL

The most commonly used PLL in three-phase grid synchronization is SRF PLL due to its simple structure, fast, accuracy, and easy implementation. Block diagram of SRF PLL is shown in Fig. 2. It has four blocks—Clarke transformation block, Park transformation Block, PI controller, and integrator.

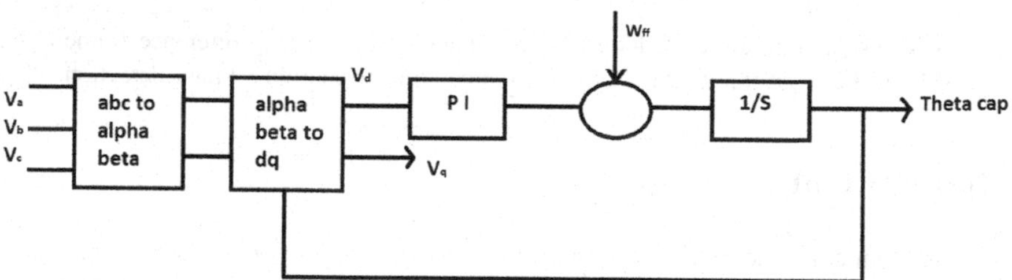

Fig. 2. Block diagram of the synchronous reference frame phase-locked loop

It uses coordinate transformation (abc to dq) as phase detector. The three-phase grid voltages V_{abc} are sensed and converted in dq reference frame by first Clarke and then Park transformation. The direct and quadrature components of the sensed voltage are given [2] by the following matrix:

$$\begin{bmatrix} V_q \\ V_d \end{bmatrix} = \begin{bmatrix} \cos\theta^\wedge & -\sin\theta^\wedge \\ \sin\theta^\wedge & \cos\theta^\wedge \end{bmatrix} \begin{bmatrix} U\cos\theta \\ U\sin\theta \end{bmatrix} \quad (1)$$

$$\begin{bmatrix} V_q \\ V_d \end{bmatrix} = \begin{bmatrix} U\cos(\theta^\wedge - \theta) \\ U\sin(\theta^\wedge - \theta) \end{bmatrix} \quad (2)$$

Hence, there are two components V_d and V_q of the transformed grid voltages in direct—quadrature—components [3], and θ is the phase of the input and $\theta^\wedge$ is the output phase of the PLL. V_d is called the error components, and V_q is the amplitude component. When the generated output phase of the PLL is exactly equal to the phase of the input reference signal, the V_d becomes zero and the V_q remains only amplitude. V_d is used as the error signal in detecting the phase lock condition, and V_q has only magnitude in the steady state and hence the quadrature component has no use. To put PLL in lock state, we need to reduce the V_d component to zero which means the output $\theta^\wedge$ must be equal to θ. Then, V_d will become zero and the V_q becomes only magnitude. The PI controller is used to regulate this error signal to zero [4]. The output of the PI controller is the angular frequency; this angular frequency is integrated into an integrator to obtain the phase angle θ of the grid voltage [5]. Then, this phase angle is given as feedback to the coordinate transformation block. This process repeats until θ is equal to $\theta^\wedge$. The phase lock signal is obtained when we add a sine block to the obtained phase angle. This output signal is a unit signal, so we have to add a proportional gain block to obtain the phase lock signal of desired amplitude with the reference grid signal. A novel renewable energy system has been proposed in [6].

2.2 Trigonometric PLL

It is a simple PLL which uses the multiplier and trigonometric functions like sine and cosine as phase detector [7]. The phase voltages of the utility grid are fed into multipliers. The $S_1(t)$ and $S_2(t)$ are two signals fed into multipliers as shown in Fig. 7. If the two signals S_1 and S_2 are in phase difference of 90°, the PLL is locked.

$$S_1(t) = A_1 \sin[\omega t + \theta_1(t)] \quad (3)$$

$$S_2(t) = A_2 \cos[\omega t + \theta_2(t)] \quad (4)$$

$$S_3(t) = S_1(t) \cdot S_2(t)$$

$$S_3 = A_1A_2 \sin[\omega t + \theta_1(t)] \cos[\omega t + \theta_2(t)]$$

$$S_3 = A_1A_2/2 \sin[\theta_1(t) - \theta_2(t)] + A_1A_2/2 \sin[2\omega t + \theta_1(t) + \theta_2(t)] \quad (5)$$

The output signal of the multiplier consists of two parts; the first one is error signal and the other signal is harmonic components which are filtered out in three-phase PLL [8]. Then, the remaining signal is the error signal which contains the phase information of the signal.

$$S_e(t) = A_1A_2/2 \sin[\theta_1(t) - \theta_2(t)] \quad (6)$$

$$\text{If } \theta_1(t) = \theta_2(t)$$

$$S_e(t) = 0$$

If the error signal $S_e(t)$ is equal to zero, it represents the lock state of the PLL. So, we have to tune the PI controller in such a way that the signal $S_e(t)$ will become zero.

3 Simulations and Results

The proposed circuit is simulated in PSIM software package. Figure 3 shows SRF PLL; the three-phase utility grid voltage V_{abc} is fed from the abc to dq block. By tuning the PI controller, we have to reduce the V_d to zero. Thus, the output of PLL is locked in phase with the reference utility grid voltage. Figure 4 shows the output of the SRF PLL which is successfully tracking the input signal. The V_d is reduced to zero as shown in the Fig. 5.

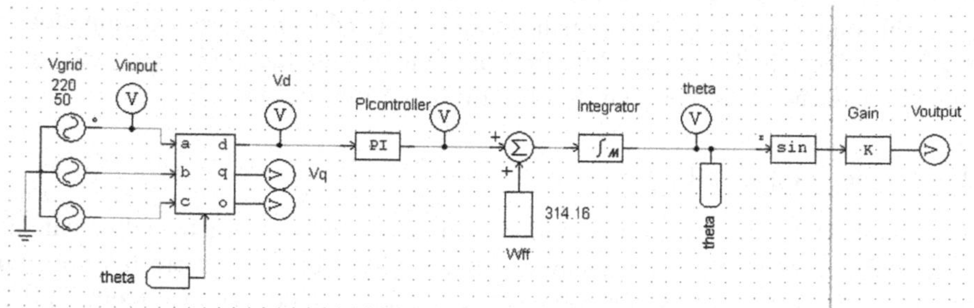

Fig. 3. Simulation model of synchronous reference frame phase lock loop

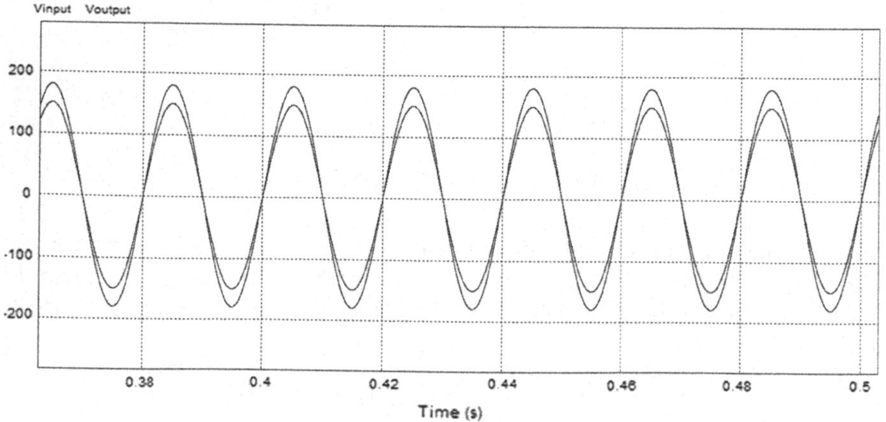

Fig. 4. Simulation results for the operation of SRF PLL with input voltage V_{in} = 220 V

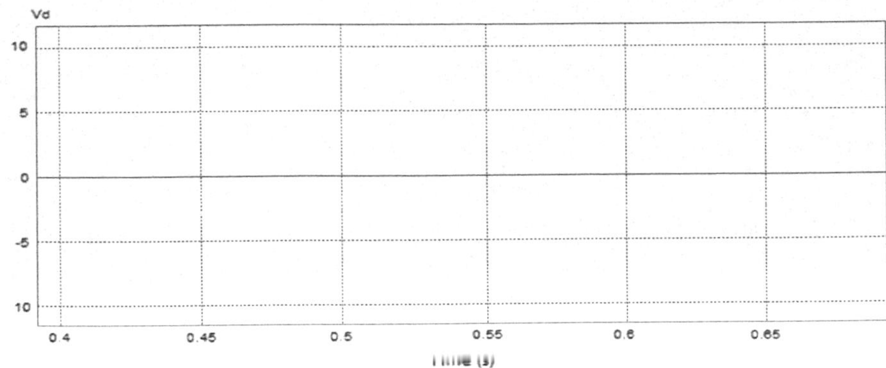

Fig. 5. Waveform of error signal $V_d = 0$

Table 1. Values of PI controller and proportional gain

Type of PLL	Voltage and frequency	K_p and time constant of PI controller	Value of gain, K
SRF PLL	220 V and 50 Hz	$K_p = 11$, $\tau = 3.3u$	150
Trigonometric PLL	440 V and 60 Hz	$K_p = 10$, $\tau = 0.01$	380

The utility grid voltage is taken to be 220 V and 50 Hz. The output of PLL is a unit signal, that's why a proportional gain is added and the value of K is 150 (Table 1).

Both the PLLs are giving good synchronization with the reference voltage but give unit amplitude voltage, that's why we have to add a gain to maintain magnitude level. Figure 6 shows the output of the PI controller which is the angular frequency, and it must be integrated into the integrator to obtain the phase angle.

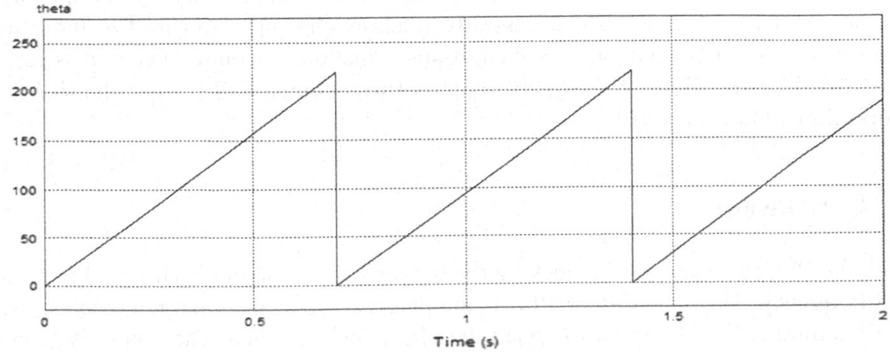

Fig. 6. Output theta of SRF PLL having grid voltage equal to 440 V and 60 Hz

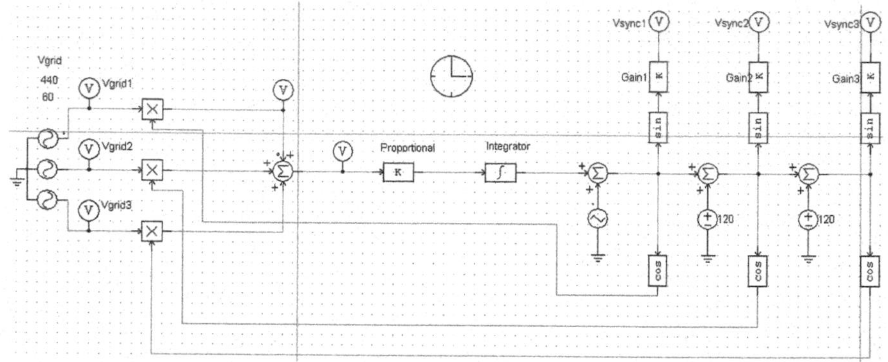

Fig. 7. Simulation model of trigonometric PLL having grid voltage equal to 440 V and 60 Hz

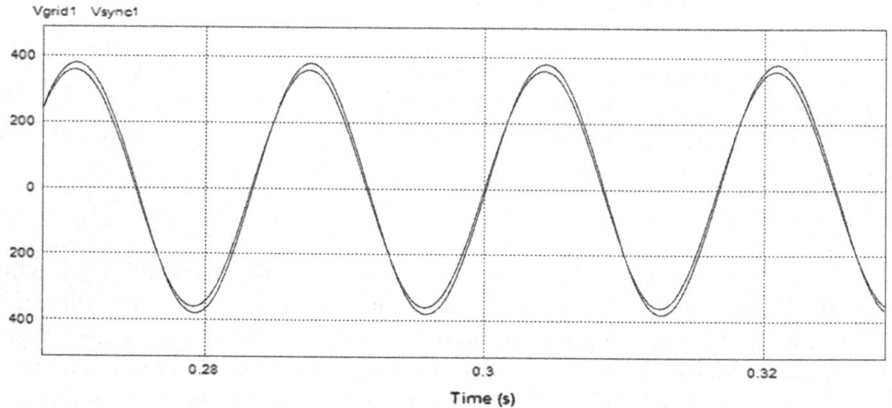

Fig. 8. Trigonometric PLL output and input

The trigonometric PLL is simulated, and the circuit is simple and easy to implement as shown in Fig. 7. It is also successfully tracking the input signal. The hardware requirement is simple, and no coordinate transformation is required hence it is faster than SRF PLL [8]. Figure 8 shows the output of trigonometric PLL when the sine and proportional block is added.

4 Conclusion

Both the PLLs are successful in locking the output signal with the input signal in phase and frequency. The output of the PLL is the phase angle of the input reference signal, and we must add a sine block to obtain the phase-locked signal. The output is of unit amplitude, so we have to add a gain block to maintain the PLL output in magnitude with the input. Both are performing well. They can be practically realized and used in synchronization of PV and wind-connected grid system. PLL is checked at different magnitude and frequency; it is able to track the input reference signal of different frequency and magnitude. Trigonometric PLL is faster as they require no abc to dq transformation which adds delay in the control loop. The practical realization of trigonometric PLL is easy and cheaper.

References

1. Angelina Tomova, Mariya Petkova, Mihai Antchev, Hristo Antchev.: Computer investigation of a sine and cosine based phase locked loop for single phase grid connected inverter, International Journal of Engineering Research and Applications (IJERA) ISSN: 2248–9622, www.ijera.com, Vol. 3, Issue 1, January–February pp. 726–728, 2013.
2. Vikram Kaura, Vladimir Blasko.: Operation of a Phase Locked Loop System Under Distorted Utility Conditions, IEEE Transactions on Industry Applications, vol. 33, no. 1, January/February 1997, S 0093–9994(97)00988-2.
3. Xiao-Qiang GUO, Wei-Yang WU, He-Rong GU.: Phase locked loop and synchronization methods for grid interfaced converters: a review, PRZEGLAD ELEKTROTECHNICZNY (Electrical Review), ISSN 0033–2097, R.87 NR 4/2011.
4. Ahmed Abdalrahman, Abdalhalim Zekry and Ahmed Alshazly.: Simulation and Implementation of Grid-connected Inverters, International Journal of Computer Applications (0975–8887), Volume 60– No. 4, December 2012.
5. Natesan, S. and Venkatesan, J.: A SRF-PLL Control Scheme for DVR to Achieve Grid Synchronization and PQ Issues Mitigation in PV Fed Grid Connected System, (2016) Circuits and Systems, **7**, 2996–3015.
6. Adel Saleh, **Shabana Urooj**.: Remedy of Chronic Darkness & Environmental effects in Yemen Electrification System using Sunny Design Web, International Journal of Renewable Energy Research Vol. 7, No. 1, pp 285–291, 2017.
7. Mihail Antchev, Ivailo Pandiev, Mariya Petkova, Eltimir Stoimenov, Angelina Tomova and Hristo Antchev.: PLL for Single Phase Grid Connected Inverters", International Journal of Engg. & Technology, (IJEET) Volume 4, Issue 5, September–October, pp. 56–77, 2013.
8. W. Phipps, M.J. Harrison and R. M. Duke.: Three-Phase Phase-Locked Loop Control of a New Generation Power Converter, ICIEA 2006, 0-7803-9514-X/06, IEEE 2006.

Occlusion Vehicle Segmentation Algorithm in Crowded Scene for Traffic Surveillance System

Hung Ngoc Phan, Long Hoang Pham, Duong Nguyen-Ngoc Tran, and Synh Viet-Uyen Ha(✉)

School of Computer Science and Engineering, International University, Vietnam National University HCMC, Block 6, Linh Trung, Thu Duc District, Ho Chi Minh City, Vietnam

hvusynh@hcmiu.edu.vn

Abstract. Traffic surveillance system (TSS) is an essential tool to extract necessary information (count, type, speed, etc.) from cameras for further analysis. In this issue, vehicle detection is considered one of the most important studies as it is a vital process from which modules such as vehicle tracking and classification can be built upon. However, detecting moving vehicles in urban areas is difficult because the inter-vehicle space is significantly reduced, which increases the occlusion among vehicles. This issue is more challenging in developing countries where the roads are crowded with 2-wheeled motorbikes in rush hours. This paper proposes a method to improve the occlusion vehicle detection from static surveillance cameras. The main contribution is an overlapping vehicle segmentation algorithm in which undefined blobs of occluded vehicles are examined to extract the vehicles individually based on the geometry and the ellipticity characteristics of objects' shapes. Experiments on real-world data have shown promising results with a detection rate of 84.10% in daytime scenes.

Keywords: Occlusion detection · Blob splitting · Vehicle segmentation · Traffic surveillance system

1 Introduction

The past decade has seen the explosion of intelligent and expert systems, especially in the area of transportation management. TSS has gained popularity among researchers and authorities. A key aspect of TSS is to derive the traffic information (count, average speed, and the density of each vehicle type) for further analysis related to traffic management and planning. In this context, many studies have been conducted in developed countries where the transportation frameworks are constructed primarily for automobiles. These systems [1,2] were

V. Bhateja et al. (eds.), *Information Systems Design and Intelligent Applications*, Advances in Intelligent Systems and Computing 672,
https://doi.org/10.1007/978-981-10-7512-4_58

developed with the advanced equipment and sensors to optimize the incoming signal including radar, infrared camera. However, in developing countries, the application of these systems has trouble with high-cost and incompatible infrastructures. On the contrary, the vision-based TSSs which are built from computer vision and image processing techniques [3–5] have shown more superior capability with lower cost and are easier to maintain. Moreover, they are extremely versatile as algorithms can be designed to cope with a broad range of operations such as detecting, identifying, counting, tracking, and classifying vehicles.

In vision-based TSS, vehicle detection is one of the most critical operations since it plays a major role in localizing moving vehicles in traffic videos. This vital tool is a well-defined mechanism that has been studied in a considerable amount of works. Some recent works by Ha and Pham et al. [5,6] have shown remarkable outcomes in detecting and classifying vehicles in urban areas. However, urban vehicle detection has its challenges that cannot be easily addressed by these methods due to the occurrence of vehicle occlusions. The problem is more challenging in rush hour when the traffic gets slower, and the inter-vehicle space is significantly reduced, which increases the occlusion among vehicles. In addition to this, the immense density of motorbikes is the main cause for chaos on urban roads in Vietnam. Furthermore, the non-rigid shapes of this kind of vehicles vary widely when they move throughout the scene. As a consequence, the traffic becomes more complicated and awkward for vehicle detection.

In this paper, we propose a robust vehicle detection algorithm that handles the occlusions of 2-wheeled motorized vehicles in crowded traffic scenes. Our work is an extension of [5,6] which presented a robust TSS comprising of three main components: background subtraction, vehicle detection, and classification. The main contribution of our method is introducing an overlapping vehicle segmentation algorithm which has been developed with a data-driven approach on real-world data. Like previous studies, we use background subtraction to model the background, from which moving vehicles can be detected. Then, blobs of overlapping vehicles are identified based on the geometric characteristics and the spatial features of objects' shapes. This assessment is done by using a decision tree constructed from a large training set of 10,000 vehicle images captured in Ho Chi Minh City, Vietnam. Once occluded vehicles are extracted, we proceed with the overlapping vehicles segmentation process. We propose a novel segmentation method that performs exhaustive checking and pairing of defect points in the object contours. The blobs resulted from each cut are validated with the vehicle model [5] consisting of vehicle size P^C, dimension ratio R^{di}, density ratio R^{de}, and ellipticity characteristics. Experiments have shown promising results with high vehicle detection rates of 84% on the considered data.

The rest of this paper is structured as follows. Firstly, Sect. 2 provides an extensive scrutiny of background subtraction model and describes vehicle detection module which we exploit in our proposed method. This will be followed by Sect. 3 where the main contribution in this research, overlapping vehicle segmentation, will be presented. Finally, experiments and discussion are stated in Sect. 4 to evaluate, to summarize our work, and to conclude the paper.

2 Vehicle Detection

In video-based approach, the aim of TSS is to process traffic videos which are captured from static pole-mounted cameras. In order to achieve this goal, the input data goes through several intermediate procedures. This section discusses background subtraction model and vehicle detection module.

At the first phase, it is typical to construct a background as well as to extract the moving objects from input videos. However, the background image is not always available as a certain frame in received data. More precisely, background subtraction is an indispensable procedure of modeling static scenery from which we separate the moving objects for further processes. The outcome of this operation which is utilized to evaluate the accuracy and the effectiveness of the model comprises two masks: background and foreground image. While the background is marked as black areas, the fields on the foreground are white blobs which imply the objects that are on the move. Particularly, in the empirical environment, the background image may be affected by external factors including illumination changes, camera vibration, and noise caused by sluggish or motionless vehicles. In order to overcome these issues, we make use of the background subtraction model which is proposed by Nguyen et al. [7] to create a stable background in many circumstances. In addition to this, by narrowing down the examining area, we eliminate unexpected objects and reduce the number of blobs that we need to investigate. Figure 1a–c illustrates the background construction and the background subtraction procedure.

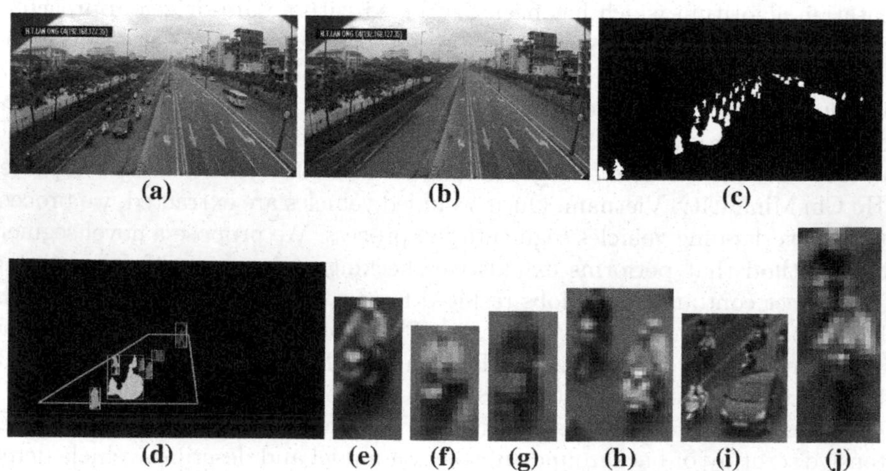

Fig. 1. **a** Original image. **b** Background image. **c** Foreground image from background subtraction process. **d** Extracted blobs of moving objects inside the examining area. **e–j** Extracted blob images

After moving objects are detached from traffic scene, in order to derive useful features from conveyances, each blob of vehicle has to be characterized individually. Regarding this issue, we adopt the construction of vehicle detection which was proposed by Ha and Pham et al. [5,6,8]. Using this approach, they take advantage of ellipse fitting model, which was discussed by Fitzgibbon and Fisher [9], to bound the isolated blobs of vehicles. As an illustration, Fig. 1d exemplifies the result of this manner. Apparently, their solution has a superiority of lower computation time when exploiting six ellipticity properties of detected blobs. The characteristics of ith candidate at tth frame are outlined and illustrated in Table 1 and Fig. 2.

Table 1. Vehicle's measurement

Symbol	Description
$E_i(t)$	Vehicle's bounding ellipse
$\theta_i^E(t)$	Bounding ellipse rotation angle
$E_i^W(t)$	Bounding ellipse width
$E_i^H(t)$	Bounding ellipse height
$P_i^E(t)$	Total pixel of bounding ellipse
$P_i^C(t)$	Vehicle size

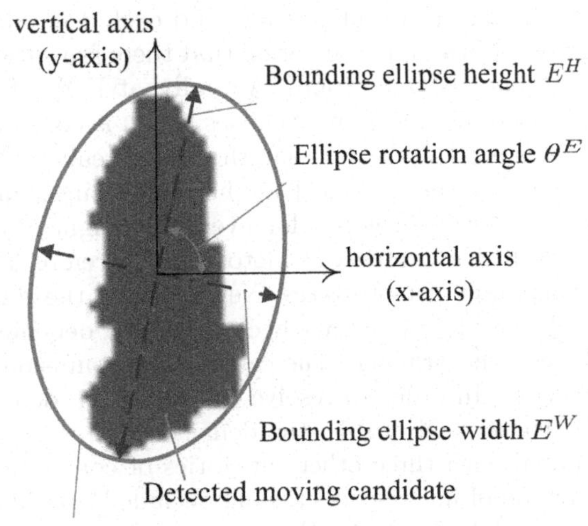

Fig. 2. Ellipticity properties

3 Overlapping Vehicle Segmentation

3.1 Occlusion Detection

In the current investigation, from the identified set of moving objects, the occlusion blobs continue to be sorted out to prepare for the further segmentation process. In this research, we categorize the detected blobs of vehicles from Sect. 2 into four classes: light (bike and motorbike), medium (car, sedan, and 12-seater bus), heavy (truck, trailer, 16-to-50-seater bus), and occluded (blob of overlapping vehicles).

Following the essence of vehicle classification, we construct an evaluation procedure generating a tuple of three informative geometric features that depict the four kinds of vehicle blobs. The first assessment is vehicle size P^C, which is the total number of pixels bounded by vehicle's contour. The second measurement is density ratio R^{de}, which is achieved by calculating the proportion of vehicle size to the bounding ellipse size P^E. The last appraisal is dimension ratio R^{di}, that is, the ratio of bounding ellipse width E^W to height E^H:

$$R_i^{di}(t) = \frac{E_i^W(t)}{E_i^H(t)}; R_i^{de}(t) = \frac{P_i^C(t)}{P_i^E(t)} \tag{1}$$

From the above literature, several empirical experiments have been executed. More precisely, the blobs of vehicles which are the outcomes from vehicle detection in Sect. 2 are extracted the necessary properties and labeled for the analysis process. Figure 3a shows a scatter plot which illustrates the distribution of vehicles' features in 3D space containing three axes that correspond to three defined attributes (P^C, R^{di}, R^{de}). In this figure, the green, yellow, blue, and red dots, respectively, describe light, medium, heavy, and occluded vehicles. The scatter diagram intelligibly confirms an observation that there is a substantial separation in dimension ratio among these groups. Actually, at FOV of surveillance cameras, the length of medium and heavy vehicles which is considered as the height of detected blobs seems to be shorter than the one in reality. On the other hand, there is a disparity between the vertical and horizontal linear measurement when considering motorized two-wheelers. Moreover, the height of candidates in this class is also affected by the presence of motorists. Therefore, R^{di} of automobiles is greater than the value of motorbikes. Nevertheless, the dimensions of blobs of obscured vehicles notably fluctuate because of the dependence on the state of occlusion, which is the primary issue causing the confusion of distinguishing with the other classes. In order to resolve this problem, density ratio R^{de} is a noteworthy investigation. Obviously, the density ratio of occlusion blob is the least when compared with three other categories of conveyances. The cause of this fact is the portion of inner spaces among occluded candidates that form the gaps interleaving in blobs detected. However, another essential aspect of handling the borderline candidates is still indecisive. By inspecting the vehicle size P^C, four groups are obviously separated by surfaces which are parallel to the plane (R^{di}, R^{de}). In other words, considering the size of blobs, there is a manifest difference between any two classes.

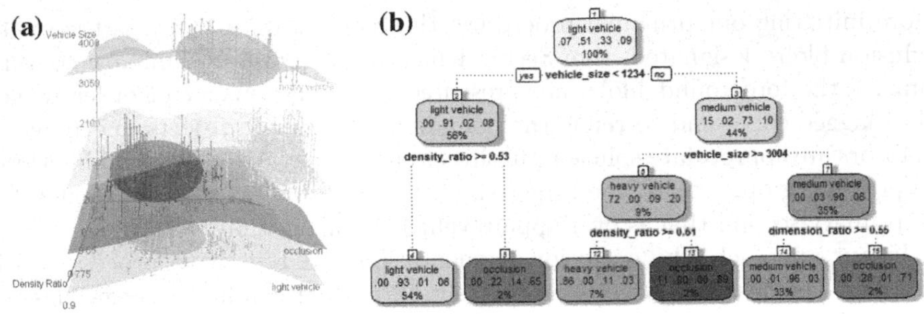

Fig. 3. a The scatter plot of vehicles from dataset VVK1. **b** The decision tree categorizing the vehicles and occlusion part in dataset VVK1

With this inspection of informative features in separating four classes of blobs detected, a set of tuples (P^C, R^{di}, R^{de}) which are extracted from an amount of data is utilized to construct a categorizing structure. In this research, the decision tree approach that is presented by Thomas [10] is adopted to build up a classification model. In the literature, the decision tree is a method classifying a batch of discrete samples. More precisely, the input data is distributed on an architectural tree with a group of nodes consisting of a root node, some internal nodes, and several leaf nodes. This structure can be obtained by using a top-down, greedy search algorithm, especially ID3, which focuses on finding out the best classifiers among the candidate's attributes to disassemble a large collection into smaller identified groups via a statistical assessment. Figure 3b shows the decision tree that corresponds to our training dataset. Starting at the root node, on the left branch, the light vehicles and the occlusion blobs share the same size criteria, but they distinguish each other with the specification of the density ratio. On the contrary, at the right branch, the remaining candidates are distributed with a particular vehicle size evaluation. Continuously, the smaller subset finally completes the compartmentalization through appropriate constraints on the density and the dimension ratio. From these observations, the decision tree can be utilized as a mechanism resolving both occlusion detection and vehicle classification.

3.2 Vehicle Segmentation

Once the occlusion blobs of vehicles are categorized as an exceptional instance of experimental subjects, at this stage, we continue to examine the bundle segmentation of detected candidates into the individual vehicles. Dealing with this issue, we present a robust solution to handle the overlapping vehicle segmentation.

As stated earlier in the introduction of this research, our proposed method is a video-based manner that processes directly on a sequence of captured images from surveillance cameras. Undeniably, pictorial signals received through the system can be affected by a variety of external factors including camera vibration, illumination change, and interference from other devices. For these reasons, before initiating our principal procedure, the received data which is the set of occlusion blobs designated from Sect. 3.1 has to undergo a preliminary convention. As the foreground blobs are presented in binary image, in order to get rid of jagged edges and to refine the bounding contours of objects, morphology operators are appropriate solutions in this circumstance. After undertaking these pre-processing operations, the target subjects are steady and stable enough for the primary section that is overlapping vehicle segmentation.

In our approach, the occlusion segmentation is a repetitive process which partitions the obscured objects alternately. In normal conditions, regardless of small defects, the appearances of individual vehicles are considered as curving outward shapes. Therefore, in case of overlapping among vehicles, there are concave spots on the edge of detected blob, which is a corollary of the overlapping phenomena among moving conveyances. These positions are effective indications for blob splitting. Hence, in the first step of this method, we construct a convex bounding polygon ρ for the border of each detected blob μ_k that comprising a set of k vertices p_i, which is mathematically indicated as:

$$\rho = \left\{ \sum_{i=1}^{k} \lambda_i p_i | p_i \in \mu_k \wedge \lambda_i \geq 0 \wedge \sum_{i=1}^{k} \lambda_i = 1 \right\} \tag{2}$$

In this matter, an optimal algorithm is thoroughly presented by Sklansky [11]. Step (a) at each iteration in Fig. 4 illustrates the convex bounding polygon of occlusion blob. From that outcome, we continue to determine potential points for later process. In particular, this set σ_m consists of m concave spots which are on the outer boundary of examining objects but do not belong to the polygon ρ. Step (b) shows two selected concave spots that are utilized to segment the occlusion blob at the initial stage. In this figure, two attributes are presented to describe the importance of a detected locality. The first characteristic, denoted by a red line segment, is defect width that is the length of the bounding convex polygon's edge of inspecting spot. The other property, indicated as a green line,

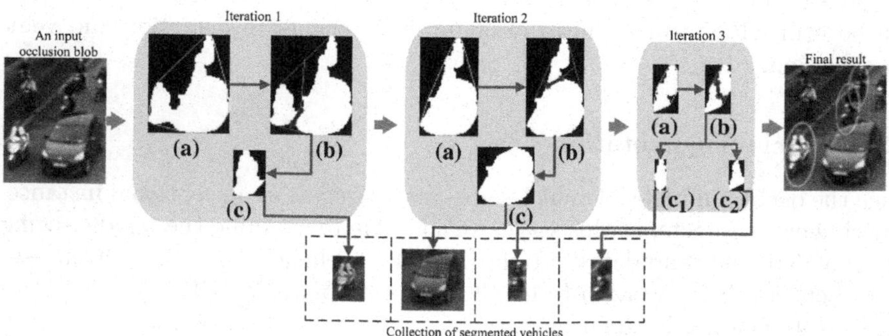

Fig. 4. The overlapping vehicle segmentation. **a** Convex bounding polygon computation. **b** Concave spot localization. **c** Segmented individual vehicles

is defect depth which is the distance from the concave spot to the midpoint of the corresponding bounding convex polygon's edge.

Furthermore, in this investigation, to select the correct segment points and to eliminate unnecessary impurities, two constraints attained from practical experiments are considered on detected set σ_m:

(1) The defect width must be greater than certain threshold.

$$\|\rho_i - \rho_j\| \geq TH_1 \text{ where } \rho_i, \rho_j \in \rho \quad (3)$$

where TH_1 is the minimum defect width and set to 7 pixels.

(2) The defect depth is limited with an interval bounded by two thresholds.

$$TH_2 \leq \left\| \sigma_t - \left(\frac{\rho_i + \rho_j}{2}\right) \right\| \leq TH_3 \text{ where } \rho_i, \rho_j \in \rho \text{ and } \sigma_t \in \sigma_m \quad (4)$$

where TH_2 and TH_3 are, respectively, the minimum and the maximum defect depths that are alternately set to 3 pixels and 30 pixels.

Afterward, for each pair of defined points, we form a cutting line which is utilized to separate obscured vehicles as step (c) at each iteration in Fig. 4. Subsequently, the segmented candidate after reconstructing the necessary vehicle's measurement is verified through the decision tree mentioned in Sect. 3.1. This manner takes 2–17 iterations of processes to attain the convergence of results. The procedure ends at the stage when all individual vehicles are identified. The rightmost image in Fig. 4 shows the final result of segmentation procedure.

4 Experiments and Discussion

Experiments have been performed on the selected data to evaluate the proposed method. In these examinations, the testing datasets are captured from static pole-mounted surveillance cameras in Ho Chi Minh City at the rate of 30 fps with a resolution of 640 × 480 to assess the accuracy objectively and to test the performance of our method during crowded scenes. Technically, these cameras are set up at the height of 8–9 m and inclined at an angle of 12°–15°. In addition to this, the testing system has a configuration consisting of Intel Core i7 2630QM and 8 GB of RAM.

In the previous studies, Ha and Pham et al. [5] demonstrated a robust classification algorithm in daytime surveillance environment with a remarkable accomplishment. In this paper, we continue to improve prior attainments by initiating a novel method to proceed the occlusion in vehicle detection. Table 2 summarizes the results of the previous studies and our proposed method. Both solutions are examined on three different datasets HMD01, NTL01, and COL01 which depict the different levels of occlusion. Dataset HMD01 was captured on the highway during rush hour in the morning. In this dataset, because there is lane separation between motorbike and automobile, two approaches mainly concentrate on detecting and classifying overlapping vehicles of the same group. Dataset NTL01, which describes the high density of vehicles in the residential area, is

Table 2. Comparison of results between Ha's method and our proposed method

			Ha's method [5]				Our method			
Dataset	Class	Actual	Detect	Deviation	Percent (%)	T.Acc (%)	Detect	Deviation	Percent (%)	T.Acc (%)
HMD01	1	498	250	248	50.20	54.06	427	71	**85.74**	**84.60**
	2	68	47	21	69.12		56	12	**82.35**	
	3	7	3	4	42.86		6	1	**85.71**	
NTL01	1	652	364	288	55.83	62.54	538	114	**82.52**	**84.97**
	2	168	115	26	68.45		144	12	**85.71**	
	3	30	19	2	63.33		26	4	**86.67**	
COL01	1	574	392	182	68.30	66.97	483	91	**84.15**	**82.70**
	2	32	21	11	65.63		26	6	**81.25**	
	3	0	0	0	–		0	0	–	
Overall Accuracy			61.19				**84.09**			

also tested to assess a complicated circumstance with mixed-flow lanes. Dataset COL01 plays a major role in scrutinizing the higher level of occlusion among the light and the medium conveyances. Figure 5 shows some examples of three experiments.

Fig. 5. Examples of occlusion vehicle detection and classification result with three experimental datasets. Green, yellow, blue ellipses, respectively, indicate light, medium, and heavy vehicles. On the 1st column: results from dataset HMD01. On the 2nd column: results from dataset NTL01. On the 3rd column: results from dataset COL01

As shown in Table 2, the results obtained from the analysis of two methods on different datasets show that our algorithm improves the overlapping vehicle detection and accurately classifies the segmented candidates into three classes with confidence up to 85%. When compared with Ha's method whose studies are most relevant to ours, in intricate circumstances where there is a considerable presence of obscured vehicle, our overlapping vehicle segmentation algorithm significantly enhances the capability of TSS by 20% and increases adaptation in complicated situations. The proposed method can detect at least 82% of moving overlapping motorbikes and over 81% of medium vehicles. In particular, this result primarily depends greatly on the verification procedure as we mentioned in Sect. 3.1. Moreover, our approach of overlapping vehicle segmentation relies much on the geometric characteristics and the ellipticity attributes of detected candidates on the whole. Hence, in some minor cases, unexpected objects are erroneously gathered as desired vehicles such as pedestrians, rudimentary means of transportation, and slabs of foreground faults caused by sudden illumination change in background subtraction model. In general, our method not only provides the result with high accuracy, but also retains the low deviation when classifying light, medium, and heavy vehicles. This result is only achieved by the contributing an effective vehicle detection, a steady verification model, and a robust segmentation algorithm.

Besides outperforming on the outcome with an overall accuracy of roughly 84.1%, the proposed method maintains a small amount of computational time on the whole solution. To be more specific, the tests were performed with a processing of approximately 9,000 frames in each execution. With the high density of occlusion which requires such more computations, the process rate is around 26 fps because the system needs to run extraordinary computation. On the contrary, with a lower degree of occlusion as in dataset HMD01, the frame rate can reach up to over 27 fps. Accordingly, our method can process traffic data in real-time and possibly integrates into the existing TSS.

5 Conclusion

This paper proposes a new method for overlapping vehicle segmentation to handle vehicle detection in circumstances of occlusion. The contribution of our research is a vision-based approach which utilizes the typical geometric features and the ellipticity characteristics to localize the conveyances inside the occlusion blob individually and to classify vehicles into 3 classes: light, medium, and heavy vehicles. In this investigation, the experiments on the suggested algorithm show some promising of results with the average accuracy over 84% and robust adaptability to the real-time performance at the overall frame rate of 27 fps.

Acknowledgements. This research is funded by International University—Vietnam National University Ho Chi Minh City (VNU-HCM) under Grant Number SV2016-IT-05.

References

1. R. Mobus and U. Kolbe: Multi-target multi-object tracking, sensor fusion of radar and infrared. IEEE Intelligent Vehicles Symposium, pp. 732–737 (2004)
2. U. Meis, W. Ritter and H. Neumann: Detection and classification of obstacles in night vision traffic scenes based on infrared imagery. Proceedings of the 2003 IEEE International Conference on Intelligent Transportation Systems, 2, 1140–1144 (2003)
3. Duong Nguyen-Ngoc Tran, Tien Phuoc Nguyen, Tai Nhu Do, and Synh Viet-Uyen Ha: Subsequent Processing of Background Modeling for Traffic Surveillance System. International Journal of Computer Theory and Engineering, vol. 8, no. 3, pp. 235–239 (2016)
4. Synh Ha, Huy Hung Nguyen, Tu Kha Huynh and Phong Ho-Thanh: Lane detection in intelligent traffic system using probabilistic model. 16th Asia Pacific Ind. Eng. Manag. Syst. Conf. (APIEMS 2015), 1856–1863 (2015)
5. Ha, S.V.-U., Pham, L.H., Phan, H.N., Ho-Thanh, P.: A robust algorithm for vehicle detection and classification in intelligent traffic system. 16th Asia Pacific Ind. Eng. Manag. Syst. Conf. (APIEMS 2015), 1832–1838 (2015)
6. Ha, S.V.-U., Pham, L.H., Tran, H.M., Ho-Thanh, P.: Improved vehicles detection & classification algorithm for traffic surveillance system. J. Inf. Assur. Secur., 9, 268–277 (2014)
7. Nguyen, T.P., Tran, D.N.-N., Huynh, T.K., Ha, S.V.-U.: Disorder detection approach to background modeling in traffic surveillance system. J. Sci. Technol. Vietnamese Acad. Sci. Technol., 52, 140–149 (2014)

8. L. H. Pham, T. T. Duong, H. M. Tran and S. V. U. Ha: Vision-based approach for urban vehicle detection & classification. 2013 Third World Congress on Information and Communication Technologies (WICT 2013), pp. 305–310, Hanoi (2013)
9. Andrew W. Fitzgibbon and Robert B. Fisher: A buyer's guide to conic fitting. In Proceedings of the 6th British conference on Machine vision (BMVC '95), David Pycock (Ed.). BMVA Press, Surrey, vol. 2, pp. 513–522, UK (1995)
10. Thomas M. Mitchell: Decision Tree Learning. In: Machine Learning (1 ed.), pp. 52–80. McGraw-Hill, Inc., New York, NY, USA (1997)
11. Jack Sklansky: Finding the convex hull of a simple polygon. Pattern Recognition Letters, vol. 1, pp. 79-83, ISSN 0167–8655 (1982)

Anomaly Detection in a Crowd Using a Cascade of Deep Learning Networks

Peng Qiu[1], Sumi Kim[2], Jeong-Hyu Lee[3], and Jaeho Choi[4(✉)]

[1] School of Computer Engineering, Nanjing Institute of Technology, Nanjing, China
[2] Seoyeong University, Gwangju, Korea
[3] Department of SW Engineering, CBNU, Jeonju, Korea
[4] Department of Electronic Engineering, CAITT, Chonbuk National University, Chonju 561-756, Korea
wave@jbnu.ac.kr

Abstract. Anomaly detection allows to detect whereabouts of aberrant objects. In this paper, we propose anomaly detection using two connected neural networks. At the front, the convolutional neural network is used to extract visual features and the recurrent neural network implemented using a long short-term memory (LSTM) is followed to track and detect anomaly. In comparison to the conventional CNN and RNN method, the proposed method is capable of faster learning and is able to effectively detect anomaly objects.

1 Introduction

Anomaly detection and localization of crowded environments such as subways, universities, stadiums, and airport terminals have been a recent interest in computer vision field. Generally, anomaly detection aims to find an abnormal pattern or behavior in the environment or data that do not conform to expected behavior. In crowded scene anomaly detection problems, anomalies have either rare shape or motion [1]. State-of-the-art approaches learn regions or patches of normal videos as reference model. In fact, these reference models include normal events or shapes of every region of training data. In testing phase, researchers consider regions that differ from the normal model as being anomaly. These nonconforming patterns are not easy to define due to distinctive applications involved and difficult to prepare large datasets of training samples that cover the different domain.

Conventional techniques for detecting anomalies include object detection, localization, segmentation, and tracking. However, crowd analysis is a challenging problem involving a higher object density, which introduces the problems of severe inter-object occlusion, small object size, and similar object appearance that makes conventional techniques inefficient [2]. In addition, another hindrance comes from the complexity of crowd behavior which involves goal-driven activities, communication with the surrounding environment as well as other people, and various emergent behaviors. These

V. Bhateja et al. (eds.), *Information Systems Design and Intelligent Applications*, Advances in Intelligent Systems and Computing 672,
https://doi.org/10.1007/978-981-10-7512-4_59

obstacles are uncommon in non-crowded scenes where the number of individual's activities identified. But, in densely crowded scene, the number of moving objects leads to serious occlusions which make individuals tracking difficult and sometimes strain even for human observers.

Our work is inspired by the extraordinary capacity of the extraction of patches, detecting, and tracking of objects using convolutional neural network and long short-term memory. Recent studies show that deep architectures are able to achieve better generalization ability compared to hand designed-based approaches, such as histogram of gradients [3] or histogram of optic flow [4]. The other importance of deep learning is the ability to learn multiple levels of representations with increasing abstraction, for instance, using auto-encoders are able to effectively characterize the data distribution and learn useful representations which are not achieved by shallow methods such as PCA or K-means [5]. However, deep architectures have not been studied in depth for abnormal detection. Before applying the model in our problem, we should answer some critical questions: (1) how to generate data that is efficient to use in deep model; (2) how to design a model which accompanies a data structure such as static data, sequential data, and spatial data; (3) how to derive statistically sound decision criteria for anomaly detection purpose.

Due to the wide application of CNN such as image classification, object detection, and activity recognition, it is more efficient than the conventional methods [6]. However, CNN lucks simplicity due to complex and time-consuming of training process and the large dataset may not have been seen in the real-world anomaly detection problem. In addition, CNN performance is slow when using block-wise method. To mitigate this problem, researcher proposed a Faster-RCNN in object detection that adopts full, convolutional image features to the detection platform, which enables high-quality region proposals. Using fully convolutional layers in CNN reduces the computation costs in feature extraction, but still its speed is in question for anomaly detection problem.

2 Related Works

In recent two decades, numerous approaches have been proposed for aberrant detection in crowded scene, and they can be categorized as follows:

(1) Trajectory-based techniques: An object is determined anomalous when it does not follow learned normal trajectories, and it has lower frequency of occurrence.
(2) Motion-based techniques: Compared to the normal crowd, the abnormal has noticeable motion pattern.

In the trajectory-based method, it first tracks objects in a scene and then trains the model from trajectories to discern abnormal activities by detecting deviations [7]. Each moving object is tracked over a sequence of inferred tracking states such as velocity, appearance, direction, shape, and position. This trajectory information is used to identify between abnormal and normal situations. In [8], the occlusion and the segmentation problems of trajectory method are mitigated by using adaptive particle sampling and Kalman filtering. The abnormal object is localized by classifying its trajectory. Wu et al. [9] used chaotic invariant features of Lagrangian particle trajectories, and the method is efficient and practical for a range of densely crowded to sparse scenes, yet it needs exhaust tracking for each representative particle. In [10], articulated ellipsoids are used to model human appearance and a Gaussian distribution to design the background for classification. Cui et al. [11] focused on tracking interest points to extract the normal/abnormal crowd pattern using the interaction energy potentials (IEP). The method does not depend on human detection and segmentation rather tracked spatiotemporal interest points to model group interaction. As it is discussed in [12], the appearance and motion patches extracted separately by handcrafted model to feed into auto-encoder. To get a better result, we used deep learning instead of handcrafted methods. Hence, our approach detects and tracks each individual using CNN and LSTM, which is fused with auto-encoder for better anomaly detection.

For the second category, motion patterns are identified using optical flow variation [13–15] or pixel/blob change [16, 17]. Recently, motion-based technique dominates in densely crowded anomaly detection. For example, Andrade et al. [18] proposed spectral clustering which is obtained by performing principle component analysis on the optical flow fields. Cong et al. proposed a multi-scale histogram of optical flow to represent the motion patterns for image sequence [14]. Unfortunately, low-level descriptors especially optical flow, pixel change histograms, and background subtraction are not reliable where occlusion is apparent, which is enough for detecting anomaly scenes from crowded places. The method for obtaining representative trajectories is effective and performs well for densely crowded scenes as well as crowded scenes. Representative trajectories also help to build time series data and to construct chaotic modeling of a scene. Using the trained sparse dictionary, the reconstruction error is calculated to identify the anomaly.

In spite of using deep learning methods like auto-encoder [19] and CNN, anomaly detection in crowded scene has many challenges: the numerous number of object movements weakens the local anomaly detector; due to the unusual appearance and temporary presence, it is difficult to model abnormal events; obtaining training dataset that covers all scenario make abnormal detection infeasible in many applications. Therefore, [20] uses both trajectory methods and motion-based techniques to detect anomaly. Although the proposed method relied on the traditional handcrafted method, the result did not show much improvement like [21] which extract low-level feature.

3 Detection and Tracking

Anomaly events in video data usually refer to irregular shapes and motions which deviate from the normal pattern. Thus, detecting and discerning shape and motion are important tasks for anomaly detection and localization. For object detection, we use convolutional neural network to extract patch. The detected object from CNN becomes an input for LSTM [2]. Thus, our design combines both trajectory technique and motion-based technique to detect abnormal events.

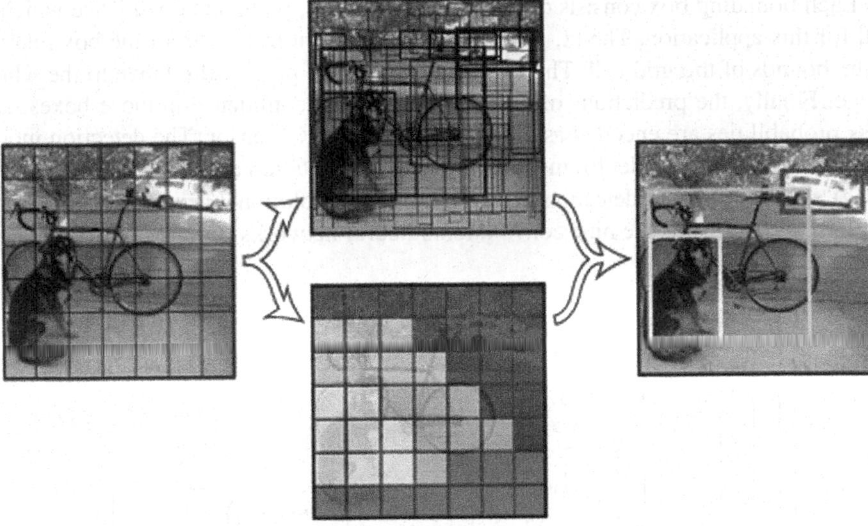

Fig. 1. Detection as a regression problem and dividing the image into even grids

The first phase of anomaly detection is extraction and detection of objects using convolutional neural networks. Among different types of CNN, we choose you-only—look-once (YOLO) detector, which associates class probabilities and spatially separated bounding boxes. In one evaluation, a single neural network predicts class probabilities and bounding boxes. As a result, the whole detection pipeline can be optimized end-to-end directly for detection purpose. But we modified YOLO to suit for our anomaly detection by ignoring the class probabilities which is void in multi-tracking application.

Starting from image pixels to bounding box coordinates, YOLO reframes object detection as a single regression problem. YOLO trains on full images and directly optimizes detection performance. The reason we chose YOLO from CNN is because it is extremely fast which is suitable for our differentiating abnormal objects. Unlike sliding window and region proposal-based techniques, YOLO sees the entire image during training and test time so it encodes contextual information about their appearance. The generalizable representations of objects in YOLO, besides being computational inexpensive, which is hundred times faster than Faster-RCNN [22, 23] is the main reason to be modified and included in our proposed system.

As shown in Fig. 1, the image data is divided into S × S sub-block grids. Each grid cell predicts B bounding boxes and confidence scores for those boxes. These confidence scores reflect how confident the model is that the box contains an object and also how accurate it thinks the box is that it predicts. For our design, anomaly detection, we reject confidence scores to be transferred to next module which is LSTM since it does not add any practical value to our output.

Each bounding box consists of five predictions: x, y, w, h, and confidence which is null for this application. The (x, y) coordinates represent the center of the box relative to the bounds of the grid cell. The width and height are predicted relative to the whole image. Finally, the predictions of the bounding boxes, confidence in those boxes, and class probabilities are encoded as a $S \times S \times (B * 5 + C)$ tensor. The detection model is based on the GoogLeNet [6] model for image segmentation and classification. Unlike GoogLeNet, the YOLO detection network has 24 convolutional neural network layers. But in this design, we use nine convolutional neural networks for fast object detection.

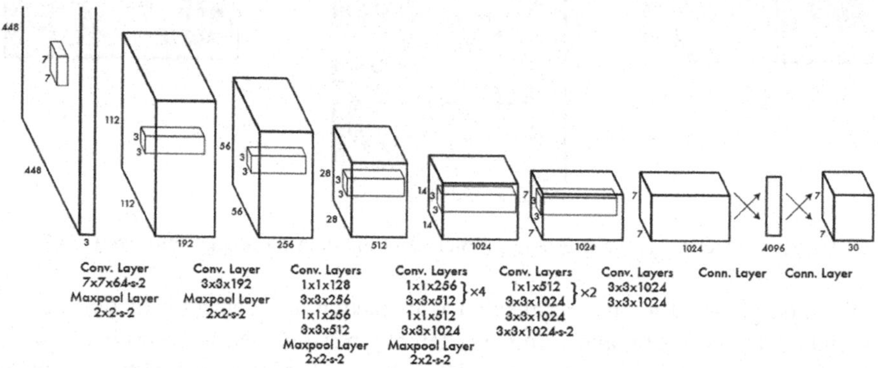

Fig. 2. Detection of CNN architecture

Therefore, we pre-train our network on small ImageNet rather than the common ImageNet dataset.

As shown in Fig. 2, the convolutional layers at the first steps of the network extract features from the input image while the last two fully connected neural network layer predicts the output probabilities and coordinates. This regression property of the network differs from many convolutional neural network models. While the main aim of our system is the detection of anomalies, this can be done based on successful detection and tracking of the objects. In this work, we focus on crowded places where many people gathered and different objects like vehicles, bicycles occasionally scene. While there exist some algorithms [19, 24] for detection and tracking of multiple objects, majority of them are not successful due to intricate motions in crowded scenes. The overview of the tracking procedures is illustrated in Fig. 3. We choose YOLO to collect rich and robust visual features, as well as preliminary location inferences, and we use LSTM in the next stage as it is spatially deep and appropriate for sequence processing.

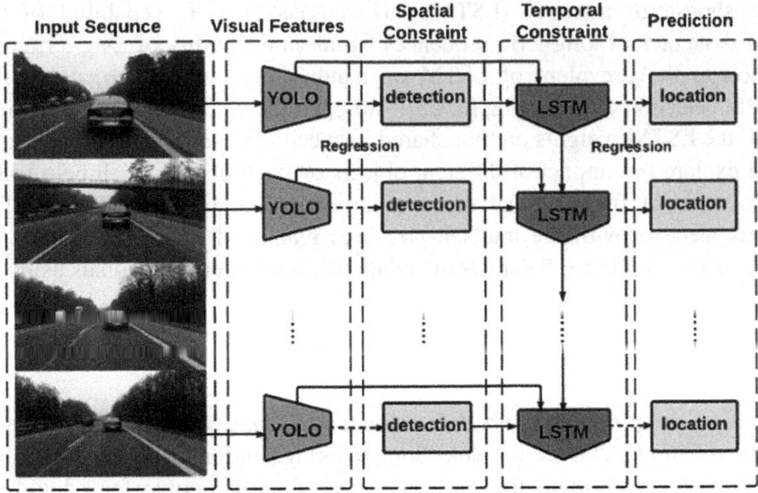

Fig. 3. Overview of detection and tracking part

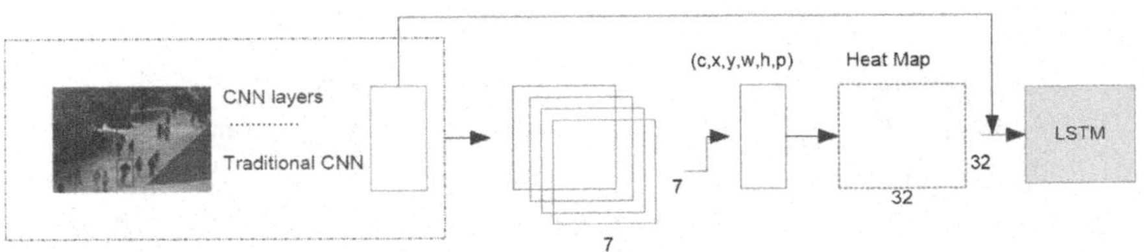

Fig. 4. Overview of the proposed anomaly detection

4 Anomaly Detection

The aforementioned challenges concerning anomaly detection are addressed in this model. Unlike the traditional long short-term memory which focuses on the individual spatial or temporal features, our model includes the relationship between an individual with neighbors property to detect abnormal objects. As shown in Fig. 4, the detected objects preprocessed with heat map are embedded with local features. Each object is defined by its coordinate (x, y) and its size (w, h). We observe the positions of all the objects from time 1 to T and forecast their positions for the next time instants. In [22], this task is deeply discussed as a sequence of generation problem, where the input sequence corresponds to the observed positions of an object and we are interested in generating an output sequence at different time instants. An object moves with different patterns in crowded scenes. Spatially, the objects may exist in different position and temporally; velocity, acceleration, and direction identify each objects from one another. We need a model that can understand the difference and evaluate the threshold in which an object identified as abnormal.

Long short-term memory (LSTM) has demonstrated its capability of learning nonlinear dynamics of long-term sequences based on the ability of long-term memory. In addition to the prevalent of LSTM on handwriting and speech, it is becoming common in detection and tracking [6, 25]. Although the object learns and predicts their positions, the LSTM weights are not shared between objects and across all sequences. Here, we explore the impact of different objects on each other, which help to localize the aberrant object from normal scene as a result. The basic coherent LSTM unit updates its memory with the tracklet, size, and feature of its own, together with its neighboring agents. Figure 5 depicts the relationship between individuals using LSTM.

5 Experiments and Results

In this section, the proposed method is evaluated by using simulated experiments. We have used different datasets for detection, tracking, and identification of abnormal objects. The dataset contains the simulated flow count of people for a building. Our system is implemented in Python using Tensorflow and runs at 20fps/60fps for

YOLO/LSTM, respectively, with 3.4 GHz Intel Core i5-4970 and NVIDIA Geforce 950 GPU. To aid in reproducing fast abnormal detection, we use small dataset for both CNN and LSTM network (Figs. 6 and 7).

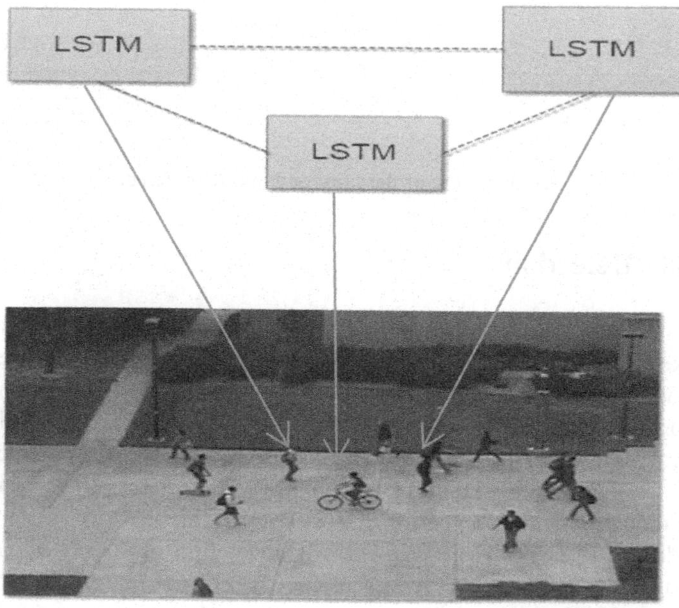

Fig. 5. Relationship between individuals using LSTM

The detection model CNN is trained on ImageNet dataset and also tuned on VOC dataset which can detect 20 different classes of objects. After the detection, LSTM will perform on that. A subset of videos picked from benchmark, where the targets belong to these classes. According to experimental results of benchmark methods in this approach, the average difficulty of OTB-30 is harder than that of the full benchmark.

Fig. 6. Tracking in street for unseen sequences using CNN

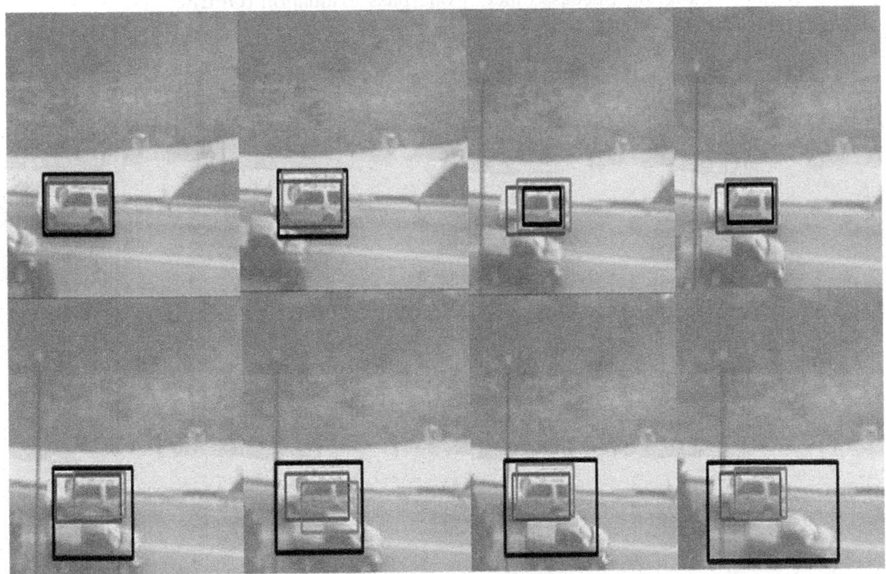

Fig. 7. Red indicates ground truth, blue indicates CNN detection, and green indicate LSTM prediction results

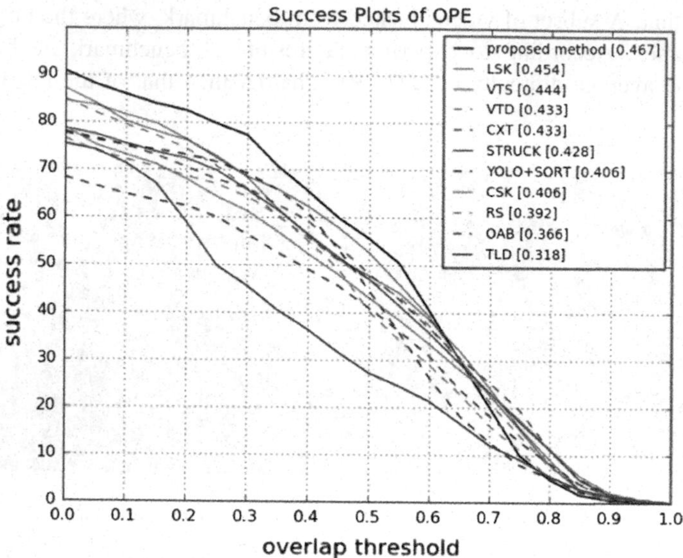

Fig. 8. Success plots of one pass evaluation (OPE)

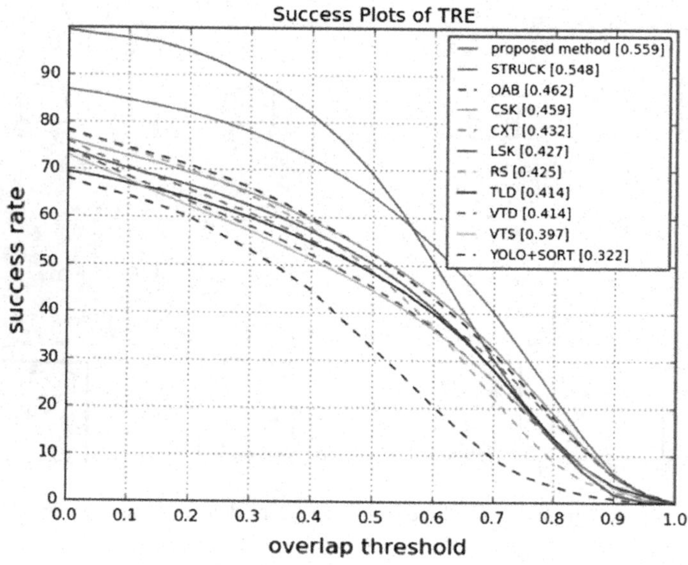

Fig. 9. Success plots for temporal robustness evaluation (TRE) of our proposed method

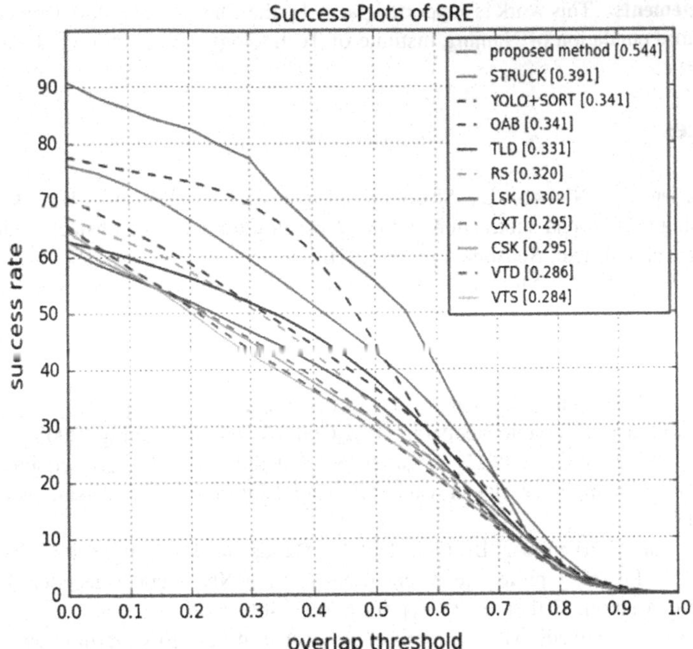

Fig. 10. Success plots for spatial robustness evaluations (SRE)

LSTM model is trained and tested using OTB-30 datasets; among them, we use 80 percent for training and the rest for testing. As shown in Fig. 8, we demonstrate the generalization ability of LSTM regression using one pass evaluation (OPE). Comparing to conventional tracking methods, the proposed method performs better as shown in Figs. 9 and 10.

From the experiments, one can conclude that the LSTM is spatially deep and capable of interpreting the visual features and detecting objects takes from CNN. Unlike CNN which is deep only spatial form, LSTM is temporally deep by exploring temporal features with locations of bounding box.

6 Conclusion

In this paper, we have detected anomaly objects using LSTM as encoder decoder and as prediction model to estimate the future tracklet of each object. Prior to abnormal detection, we have extracted features using CNN and detected bounding box of individual objects. Unlike classifier-based approaches, our detection method considered as a regression problem of class probability of each object with its bounding box. Finally, we have presented a novel approach, a coherent network in LSTM, to capture the spatiotemporal behavior of individual objects which is compared to other moving objects to discern abnormal motion or size from the normal one.

Acknowledgements. This work is supported partly by the Chonbuk National University, Korea, and the Youth Foundation of Nanjing Institute of Technology, China (No. QKJA201603).

References

1. Christiansen, P., Nielsen, L., Steen, K., Jorgensen, R., Karstoft, H.: DeepAnomaly: Combining background subtraction and deep learning for detecting obstacles and anomalies in an agricultural field. Sensors, Vol.16 (2016).
2. Graves, A., Jaitly, N., Mohamed, A.: Hybrid speech recognition with deep bidirectional LSTM. IEEE Workshop on ASRU (2013) 273–278.
3. Dalal, N., Triggs, B.: Histograms of oriented gradients for human detection. CVPR (2005) 886–893.
4. Vincent, P., Larochelle, H., Bengio, Y., Manzagol, P.: Extracting and composing robust features with denoising autoencoders. ACM Conf. Machine Learning (2008).
5. Bengio, S., Vinyals, O., Jaitly, N., Shazeer, N.: Scheduled sampling for sequence prediction with recurrent neural networks. Advances in Neural Information Processing Systems (2015) 1171–1179.
6. Cho, K., Van Merri¨enboer, B., Gulcehre, C., Bahdanau, D., Bougares, F., Schwenk, H., Bengio, Y.: Learning phrase representations using RNN encoder-decoder for statistical machine translation. (2014) arXiv preprint arXiv:1406.1078.
7. B.T. Morris, B., Trivedi, M.: A survey of vision-based trajectory learning and analysis for surveillance: IEEE Trans. on Circuits and Systems for Video Technology. Vol. 18 (2008) 1114–1127.
8. Cheng H., Hwang, J.: Integrated video object tracking with applications in trajectory-based event detection. J. Vis. Comm. Image Rep. Vol. 22 (2011) 673–685.
9. Wu, S., Moore, B., Shah, M.: Chaotic invariants of lagrangian particle trajectories for anomaly detection in crowded scenes. Proc. IEEE Conf. Comp. Vis. Patt. Rec. USA (2010) 2054–2060.
10. Zhao, T., Nevatia, R.: Tracking multiple humans in crowded environment. CVPR (2004) 406–413.
11. Cui, X., Liu, Q., M. Gao, M., Metaxas, D.: Abnormal detection using interaction energy potentials. Proc. IEEE Conf. Comp. Vis. Patt. Rec., USA (2011) 3161–3167.
12. Xu D., Ricci, E., Yan, Y., Song, J.: Deep Representations of Appearance and Motion for Anomalous Event Detection. CVPR (2015).
13. Mehran, R., Moore, B., Shah, M.: A streakline representation of flow in crowded scenes. Proc. Eur. Conf. Comp. Vis., Greece (2010) 439–452.
14. Cong, Y., Yuan, J., Liu, J.: Abnormal event detection in crowded scenes using sparse representation. Pattern Recognition, Vol. 46 (2013) 1851–1864.
15. Saligrama V., Chen, Z.: Video anomaly detection based on local statistical aggregates. Proc. of IEEE Conf. Comp. Vis. Pattern Rec., USA (2012) 2112–2119.
16. Benezeth, Y., Jodoin, P., Saligrama, V., Rosenberger, C.: Abnormal events detection based on spatio-temporal co-occurrences: Proc. IEEE Conf. Comp. Vis. Patt. Rec., USA (2009) 2458–2465.
17. Wu, S., Wong, H., Yu, Z.: A Bayesian model for crowd escape behavior detection. IEEE Trans. Circuits Syst. Video Tech., Vol. 24 (2014) 85–98.
18. Andrade, E., Blunsden, S., Fisher, R.: Modeling crowd scenes for event detection. ICPR (2006) 175–178.

19. Heilbron, F., Escorcia, V., Ghanem, B., Niebles, J.: A large-scale video benchmark for human activity understanding. CVPR (2015).
20. Khodabandeh, M., Vahdat, A., Zhou, Z., Hajimirsadeghi, H., Roshtkhari, M., Mori, G.,Se, S.: Discovering human interactions in videos with limited data labeling. (2015) arXiv preprint arXiv:1502.03851.
21. Hou, J., Wu, X., Yu, F., Jia, Y.; MultiMedia event Detection via Deep Spatial Temporal Neural Networks. Beijing Laboratory of Intelligent Information Technology, School of Computer Science (2016).
22. Ren, S., He, K., Girshick, R., SunFaster, J.: R-CNN A. Graves. Generating sequences with recurrent neural networks. (2013) arXiv preprint arXiv:1308.0850.
23. Sucheta, C., Lovekesh, V.: Anomaly detection in ecg time signals via deep long short term memory networks. IEEE International Conference on DSAA (2015).
24. Gregor, K., Danihelka, I., Graves, A., Wierstra, D.: Draw: A recurrent neural network for image generation. (2015) arXiv preprint arXiv:1502.04623.
25. Redmon, J., Divvala, S., Girshick, R., Farhadi, A.: You only look once: Unified, real-time object detection. CVPR (2016).

HAPMAD: Hardware-Based Authentication Platform for Malicious Activity Detection in Digital Circuits

Venkata Raja Ramchandar Koneru[1(✉)], Bikki Krishna Teja[1], Kallam Dinesh Babu Reddy[1], Makam V. GnanaSwaroop[1], Bharath Ramanidharan[2], and N. Mohankumar[1(✉)]

[1] Department of Electronics and Communication Engineering, Amrita School of Engineering, Amrita Vishwa Vidyapeetham, Coimbatore, India
{venkata.ramchandar1, bkteja07, dineshkallam21, swaroop.makam7}@gmail.com, n_mohankumar@cb.amrita.edu
[2] Robert Bosch Engineering and Business Solutions Ltd, Bengaluru, India
bharathr0705@gmail.com

Abstract. Hardware Trojans pose a major threat to the security of many industries and government agencies. This paper proposes a non-destructive method for detection of Hardware Trojans named HAPMAD—Hardware-Based Authentication Platform for Malicious Activity Detection in Digital Circuits using a combination of enhanced voting algorithm and hybrid voting algorithm (power and time analysis). Detecting a hardware Trojan is a cumbersome task. In most cases, the hardware Trojan can be detected only by activating it or by magnifying the Trojan activity. The detection efficiency using enhanced weighed voting is promising but in the case of Trojan circuits where there is no change in the output of the circuit; enhanced weighed voting algorithm fails to detect the Trojans effectively. Therefore, a two-phased voting algorithm is performed to increase the detection efficiency. ISCAS '85 Benchmark circuits were used to test the efficiency of the proposed technique.

Keywords: Hardware security · Hardware trojans · Voting methods Delay arcs

1 Introduction

Hardware Trojan is a malicious circuit which when inserted may alter the functionality or leak the information of the primary circuit. In the recent years, most of the VLSI chip manufacturing companies are outsourcing the manufacture of their designs; this is helping in cost reduction. At the same time, outsourcing increases the possibility of insertion of Hardware Trojans. The Hardware is not designed in the same way as it is done in the original fabrication houses, due to outsourcing the design is made by the SOC designers, the foundries fabricate the chip, and the EDA tools are used in many critical circuit designs split fabrication [1]. So there is a possibility of insertion of a Hardware Trojan by SOC designers or foundries or EDA tool vendors. Since the end

V. Bhateja et al. (eds.), *Information Systems Design and Intelligent Applications*, Advances in Intelligent Systems and Computing 672,
https://doi.org/10.1007/978-981-10-7512-4_60

user does not have the detailed information of the chip, it is difficult for end users to detect the Hardware Trojan.

There are two types of Trojans, and they are functional and parametric. In the former type, the gates are added or deleted from the original circuit, and in the later type, the original circuit is modified by reducing connecting wire's thickness, exposing the chip to radiation or weakening the flip-flops, etc. When a hardware Trojan is inserted it does not completely alter the output of the circuit but it changes the output of the circuit only at a particular input combination or after running particular amount of cycles because of this it is difficult to detect the Trojan while testing the circuits [1].

A Golden Reference is a Trojan-free circuit; the Trojan is detected by comparing this Golden Reference circuit with other circuits. In cases where the golden chip is not available, some VLSI testing methods help in formulating sophisticated detection algorithms.

2 Literature Survey

Side channel analysis [2] has been considered as one of the most used HT detection technique due the fact that hardware Trojans will have effect on the power consumption and signal delay. When a hardware Trojan is embedded in a circuit, it will add an extra leakage power to the circuit. This additional circuitry which is added will cause a change in static and dynamic power of the total circuit. The power consumed by the gates in their idle state is known as static power while consumption during the state transition is either from high to low or low to high is known as dynamic power [3].

In very large circuits, the effect of Trojans on the power consumption will be very less in magnitude and in order to carry out power analysis using very large circuits is complex; hence, partition-based approaches have been proposed. In this partition-based technique, the whole circuit is divided into small pieces of circuits which are called segments [1]. Moreover, the difference in power profile of infected and Trojan-free circuit is a negligible quantity, so in order to effectively categorize the circuits, computational algorithms like PCA are used [4]. An active-driven test pattern is generated which helps to maximize the effect of Trojan in these segments and then the power anomalies are measured in these segments.

Signal delay also plays and equal role in detecting the Hardware Trojan. When a Trojan circuitry is added to the original circuit, there will be significant change in the net delay and path delay. Using this method for detecting the Trojan sometimes can give a false positive Trojan detection because some Trojan circuits are very small in magnitude when compared to the whole circuit, the process variation delay will dominate the Trojan delay in the circuit. By foundational observation, we can notice that the impact of process variation and the effect of Trojan on path delay have lot of differences among them, since impact of Trojan on delay is unidirectional and it always increases the net total delay [5]. Hence, if we generate a specific vector for a particular path so that the generated vector increases the path's delay, then the presence of the Trojan can be detected by measuring the delay at the output of the chosen path. In this paper, a signal delay time to aid the enhanced voting technique for Trojan detection is used.

In very large circuit design, the side channel effects will mask the Trojan detection due to the noise and process variation an alternative method is proposed in S. Wei and M. Potkonjak's approach. In this approach, the signature of the supply current of a single part of the circuit is used as a reference to compare the signature of different part of the circuit to eliminate the process noise. This method uses an approach of generating a vector which is based on the region and this approach maximizes the effect of Trojan by reducing the background current [6]. Since our method is used on small circuits, we can use this approach at the later stage in order to eliminate the background current and process noise.

Even though there are many HT detection methods employed to protect the circuit against malicious Trojans, all these methods are mostly based on comparing the reckoned IC with a Trojan-free IC known as golden circuit. The golden IC may not be available when we use an IP core from a third party vendor. In this method, the need for golden IC is eliminated. This is achieved by using different voting techniques; the first technique is simple voting method [7, 8]. In this method, the output from a same functional IC with different manufactures is taken. All these ICs are given the same input data and the outputs of these ICs are connected to a voting unit. In the voting unit, by employing a bit level majority voting, we can obtain a majority consensus results. Since it is comparing the different IC's output at the bit level, it produces efficient results in terms of security. However, this technique fails when majority of the Trojan infected ICs are greater than uninfected ICs. The second technique is weighed voting method. This method always selects the ICs with higher weight. Initially, the weight of all the ICs is set to zero after each iteration; the weighted voting algorithm recalculates the weight of each IC and the weights are increased, respectively. This method is effective when there are fewer ICs for comparison. From all these methods mentioned, the prime focus of this paper is to use enhanced voting technique along with power and path delay measurements.

In Hany A.M et al. [8] and Bharath R. et al. [9] paper, only enhanced voting technique is used to detect the Trojan but when there is no change in the output then the detection of Trojan is not possible by using the method proposed in the above-mentioned papers. So in this paper, hybrid voting technique is used for detection of Trojan. Hybrid voting technique contains enhanced weighted voting method along with power and time analysis. Power and time analysis alone can be used for detecting the Trojan but the procedure is complex and it requires expensive software and computing tools. So in this paper, combination of enhanced voting along with power and time analysis is used for Trojan detection.

3 Methodology

In the HAPMAD approach, the methods used are enhanced voting technique and hybrid voting technique. Here, a PIC16f877A microcontroller is used along with Zybo FPGA board, the PIC contains the voting algorithm and the Zybo board contains the circuits that are to be tested [9].

Enhanced voting checks the output of the circuit only. In case of externally activated Trojans, the output does not change during testing. Parametric Trojans do not change the output of the circuit. There may be other Trojans that perform other malicious activities rather than changing the output of the circuit. In these cases, enhanced voting method cannot be used to identify the Trojan [9]. To address this issue, a hybrid voting technique is used. This is similar to enhanced voting but also performs time and power analysis and gives the output based on these measures this is called two-phased voting algorithm which is shown in Algorithm.1.

Two-phased voting pseudocode:

Algorithm.1. Two-phased voting algorithm

```
Begin

step1: Initialize W0=0(weights of the inputs that generate 0),
       W1=0(count the weights of the inputs that generate 1).
step2: Assign output of circuits as inputs.
step3: Count W1.
step4: Count W0.
Step5: If, W1 > W0.
Step6: Correct input = 1.
Step7: Increment W1.
Step8: Decrease W0.
Step9: If, W1 of all circuits are equal.
step10: No Trojan detected, go to step22.
step11: Else, Trojan is present, go to step22.
step12: If, W1 < W0.
step13: Calculate number of circuits producing 1 and number of circuits producing 0.
step14: If, number of circuits producing 1 > number of circuits producing 0.
step15: Go to step 6
step16: Else, correct input = 0.
step17: Increment W0.
step18: Decrease W1.
Step19: If, W0 of all circuits are equal
step20: No Trojan detected, go to step22
step21: Else, Trojan is present, go to step22
step22: Calculate Total Dynamic Power(TDP) and Cell Leakage Power(CLP).
step23: If, TDP=CLP.
step24: Assign i=0
step25: Else, assign i=1
step26: Calculate Net Delay Arcs (NDA) and Cell Delay Arcs (CDA).
step27: If, NDA=CDA.
step28: Assign j=0.
Step29: Assign j=1.
step30: If, i=1 and j=1.
step31: Trojan is present
step32: Else , Trojan is not present.
end
```

3.1 Power Analysis

In power analysis, we calculate total dynamic power (TDP) and cell leakage power (CLP).

$$Total\ Dynamic\ Power = cell\ internal\ power + net\ switching\ power \tag{1}$$

The dynamic power and static power are the main components of the power consumed by a CMOS circuit. When a transistor is not in process of switching, the power consumed during that time is called static power. It is calculated by the below given formula.

$$Power\ static = Current\ static \times V_{dd} \tag{2}$$

where V_{dd}—*supply voltage,*

current static—the overall current flowing in the device.

Usually, CMOS circuits have low static power. The power used by the device for transition of states that is from (0–1) to (1–0) is termed as transient power. The amount of power used by the load capacitance to get charged is called capacitive load power.

$$\begin{aligned} Power\ dynamic &= Capacitive\ load\ power + transient\ power \\ &= (C + C_L) \times V_{dd} \times f \times N \end{aligned} \tag{3}$$

where

load of the capacitance: C_L
internal capacitance of the chip: C
operating frequency: f
amount of switching bits: N

The amount of dynamic power consumed increases with the increment of speed and frequency of the circuit. It also increases if any new gates are added to the circuit. Hence, this variation will aid the weighed voting technique in detecting the Trojans even if the Hardware Trojan is not active.

3.2 Time Analysis

DC Shell is used to analyse the path delay of the circuit. Cell and net delay arcs are used to calculate the total path delay. Before calculating path delay, each path is broken into timing arcs by the prime time. Each timing arc gives us the information of cell delay and net delay.

$$Total\ path\ delay = net\ delay + cell\ delay \tag{4}$$

A track of edge sensitivity is kept by the prime time while analysing the path and each path is analysed with rising input and also is analysed with falling input. Logic gates are the reason for the cell delay arcs and connecting wires are responsible for the formation of net delay arcs.

So we have calculated total dynamic power, net switching power, cell delay arcs and net delay arcs for each circuit in addition to weighted voting. Now even if the votes for the infected circuit is the same as the true circuit and we can identify the Trojans by any change in the following parameters

- Total Dynamic Power (TDP)
- Cell Leakage Power (CLP)
- Cell Delay Arcs (CDA)
- Net Delay Arcs (NDA)

3.3 Hardware-Based Implementation

For the hardware-based implementation, benchmark circuits with and without Trojans are used for testing. The clock diagram of hardware based implementation is shown in Fig. 1.

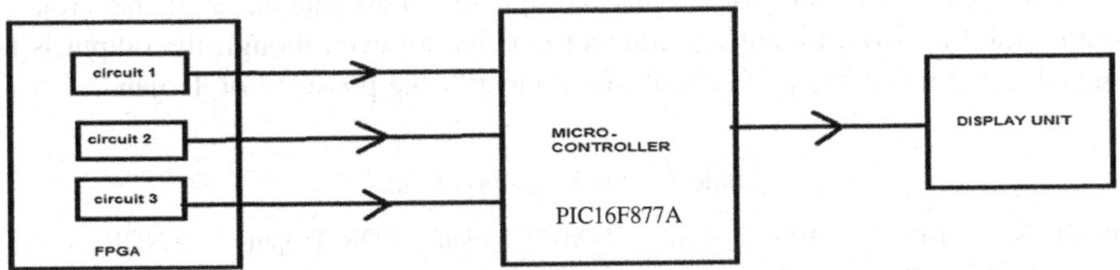

Fig. 1. FPGA-based standalone system model [9]

4 Results

4.1 RTL Diagram of Weighed Voting for Three C432 Circuits Along with Standalone System

Fig. 2 shows the Hardware Implementation Design, the weighed voting algorithm is implemented in the Microcontroller and the result is seen in the display unit, the inputs for the controller are given from the circuits that are implemented in FPGA. Figure 2 also shows that the HAPMAD algorithm can reside inside a FPGA and can be used for monitoring the CUTs or IPs in the same FPGA.

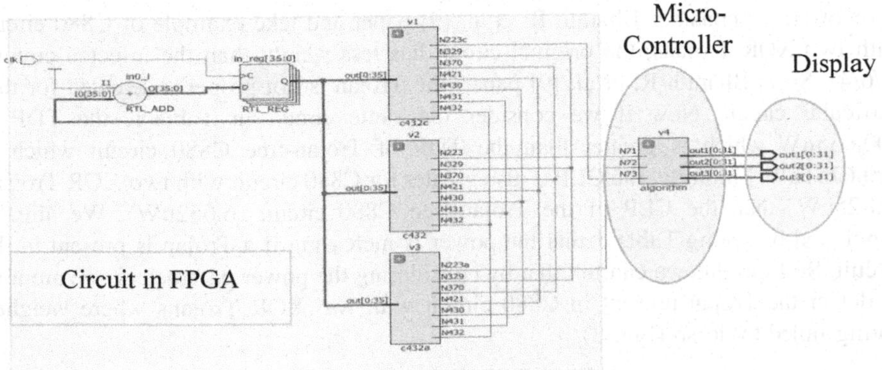

Fig. 2. RTL Diagram

4.2 Weighed Voting Algorithm Results

The weights of various circuits are calculated using weighed voting algorithm as shown in Bharath R. et al. [9] paper. The weight of Trojan-free circuits is greater than the weights of the infected circuits, a few errors have been risen in light of the fact that the quantity of contaminated circuits is more in number than the quantity of Trojan-free circuits. This problem can be solved by incrementing the number of chips that are being tested.

4.3 Power Analysis Results

The dynamic power and cell leakage power are calculated by using Synopsis Design Compiler.

From Table 1, we can observe that the dynamic power increases if the Trojan is present and it is also different for different circuits. So even though the output is not changed, we can use this power analysis to identify the presence of Trojan.

Table 1. Power analysis results

Circuit	No Trojan		AND Trojan		NAND Trojan		XOR Trojan		XNOR Trojan	
	TDP (uW)	CLP (uW)	TDP (uW)	CLP (uW)	TDP (uW)	CLP (uW)	TDP (uW)	CLP (uW)	TDP (uW)	CLP (uW)
C17	3.4566	0.106	3.350	0.12	1.775	0.05	2.97	0.129	2.881	0.140
C432	116.29	3.071	119.0	3.14	115.4	3.07	119	3.197	122.0	3.180
C499	660.89	8.157	663.9	8.24	669.4	8.25	672	8.292	670.4	8.190
C880	290.67	6.652	293.1	6.73	289.6	6.57	300	6.825	301.8	6.907
C1355	737.78	9.033	740.3	9.11	726.7	9.01	729	9.021	736.5	8.991
C1908	583.86	8.666	580.0	8.68	589.7	8.73	584	8.766	582.8	8.739
C2670	647.08	11.23	655.5	11.4	647.9	11.2	656	11.48	657.2	11.48
C3540	952.57	19.37	967.2	19.2	946.0	19.1	962	19.31	960.7	19.31
C5315	1.6257	25.05	1.623	25.1	1621	25.1	1628	25.31	1628	25.31
C6288	4129	43.98	4119	43.9	4119	43.9	4119	43.99	4119	43.99
C7552	2764	39.84	2772	39.9	2768	39.8	2771	40.11	2771	40.11

Now if we consider Bharath R. et al. [9] paper and take example of C880 circuit with two XOR Trojans, the original circuit has less weight than the infected circuits (6/6/4). So in Bharath R. et al. [9] paper, the Trojan is not properly detected for that particular circuit. Now if we consider the same circuit in Table 1, the TDP is 300.635uW which is greater than the TDP of Trojan-free C880 circuit which is 290.6751uW. Similarly, the CLP is also greater for C880 circuit with two XOR Trojans (6.825uW) than the CLP of the Trojan-free C880 circuit (6.652uW). We already concluded by seeing Table 1 that the power is increasing if a Trojan is present in the circuit. So from this we can tell that by considering the power parameters, we are able to detect the Trojan present in C880 circuit with two XOR Trojans where weighed voting failed to do so (Fig. 3).

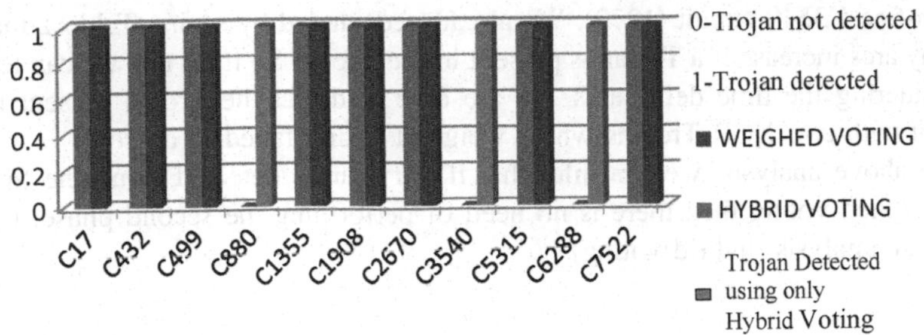

Fig. 3. Analysis of two-phase voting method

4.4 Time Analysis Results

Cell delay arcs (CDA) and net delay arcs (NDA) for each circuit are calculated using Synopsis Design Compiler.

From Table 2, we can see that as the number of gates increases due to the addition of Trojan circuit the cell delay arcs and net delay arcs increases. Discrepancies observed are due to automatic optimization performed by Synopsis Primetime.

Table 2. Time analysis results

Circuit	No Trojan		AND Trojan		NAND Trojan		XOR Trojan		XNOR Trojan	
	CDA	NDA	CDA	NDA	CDA	NDA	CDA	NDA	CDA	NDA
C17	18	9	21	11	10	5	18	10	19	11
C432	472	251	488	260	483	256	487	259	501	266
C499	1618	335	1634	343	1653	346	1663	343	1650	340
C880	1078	434	1087	436	1098	444	1138	444	1103	438
C1355	1552	408	1550	417	1537	413	1526	411	1565	410
C1908	1504	414	1504	423	1534	424	1531	425	1546	424
C2670	1916	616	1929	622	1930	623	1948	626	1972	626
C3540	2831	1272	2855	1277	2836	1273	2862	1280	2862	1280
C5315	3691	1359	3705	1366	3701	1364	3707	1371	3707	1371
C6288	7737	3804	7718	3795	7718	3795	7718	3795	7718	3795
C7552	5841	1714	5851	1718	5847	1716	5853	1731	5853	1731

The cell delay arcs and net delay arcs vary with the addition of Trojan to the circuit. This will help in the detection of Trojan especially when the weights of the circuits remain same after weighted voting.

Now if we consider Bharath R. et al. [9] paper and take example of C3540 circuit with two XOR Trojans, the original circuit has less weight than the infected circuits (6/6/4). So in Bharath R. et al. [9] paper, the Trojan is not properly detected for that particular circuit. Now if we consider the same circuit in Table 2, the CDA is 2862 which is greater than the CDA of Trojan-free C3540 circuit which is 2831. Similarly, the NDA is also greater for C3540 circuit with two XOR Trojans (1280) than the NDA of the Trojan-free C3540 circuit (1272). We already concluded by seeing Table 2 that the time delay arcs increase if a Trojan is present in the circuit. So from this we can tell that by considering the time delay arcs, we are able to detect the Trojan present in C3540 circuit with two XOR Trojans where weighed voting failed to do so.

From the above analysis, we can infer that if a Trojan is detected using the first phase, i.e. weighed voting then there is no need of performing the second phase, i.e. power and time analysis (hybrid voting).

5 Conclusion

If weighed voting algorithm is alone used for Trojan detection, then it is not possible to detect the Trojans that do not alter the output of the circuit. If power and time analysis is only used to detect the Trojan, then the process is complex and it requires the usage of tools which are highly expensive. So a two-phased voting scheme comprising of enhanced weighed voting and hybrid voting called HAPMAD is used, so that the complexity and cost of analysis is reduced and also detects the Trojan effectively. The only drawback of this technique is a minimum of three ICs are required to carry out the process. The same approach can be employed for verifying the security of IPs. Moreover, run-time detection is also possible. HAPMAD is also scalable; it can be used for combinational circuit of any complexity.

References

1. Tehranipoor M, Koushanfar F. A survey of hardware Trojan taxonomy and detection. In: IEEE Design and Test, vol. 27; pp. 10– 25, (2010).
2. Du D, Narasimhan S, Chakraborty R, Bhunia S. Selfreferencing: a scalable side-channel approach for hardware Trojan detection. In: Proceedings of the 12th International Conference on Cryptographic Hardware and Embedded Systems, CHES'10; p. 173– 87, (2010).
3. Dhineshkumar.K, Mohankumar.N, Malicious combinational hardware Trojan detection by gate level characterization in 90 nm technology. In: Proceedings of International Conference on Computing, Communication and Networking Technologies, China, pp. 1–7 (2014).
4. N. Mohankumar, M. Nirmala Devi. Improving the Classification Accuracy in Detecting Hardware Trojan in ALU Using PCA. In: Indian Journal of Science and Technology, vol. 9, Special Issue 1, (2016).
5. B. Cha and S. Gupta, "Efficient trojan detection via calibration of process variations," in Test Symposium (ATS), 2012 IEEE 21st Asian, Nov 2012, pp. 355–361.
6. S. Wei and M. Potkonjak, "Scalable hardware trojan diagnosis," Very Large Scale Integration (VLSI) Systems, IEEE Transactions on, vol. 20, no. 6, pp. 1049–1057, (2012).
7. G. Latif-Shabgahi, J.M. Bass, S. Bennett, History-Based Weighed Average Voter: A Novel Software Voting Algorithm for Fault-Tolerant Computer Systems, Proc. PDP2001: 9th Euromicro Workshop on Parallel and Distributed Processing, February 7–9, Mantova, Italy, (2001).

8. Hany A.M. Amin, YousraAlkabani, Gamal M.I. Selim. System-level protection and hardware Trojan detection using weighed voting. In:Journal of Advanced Research 5, p. 499–505, (2014).
9. R. Bharath, G. Arun Sabari, Dhinesh Ravi Krishna, Arun Prasathe, K. Harish, N. Mohankumar, and M. Nirmala Devi," Malicious Circuit Detection for Improved Hardware Security," Springer International Publishing Switzerland, CCIS 536, pp. 464–472, (2015).

Sparse Nonlocal Texture Mean for Allocation of Irregularity in Images of Brain

Dao Nam Anh(✉)

Electric Power University, Hanoi, Vietnam
anhdn@epu.edu.vn

Abstract. The medical image analysis for irregularity studies has always been a refreshing research topic for the need of efficient and precise diagnosis. A new method based on patch analysis for detection of disorder in images of brain is introduced with machine learning techniques. In the method, a sparse nonlocal texture mean filter is proposed to evaluate the similarity of each spot in the image. The spot-based similarity allows initial identification of place of abnormality which is then refined by the support vector machines to efficiently perform extraction of disorder's region. Experimental results on a benchmark's real data are assessed and compared objectively to ensure sufficient certainty of the method.

1 Introduction

In medical imagery, location of irregularities in images is a significant aspect for prognosis of disease, and the extensive analysis on the characteristics of irregularities like their size, number of lesions can be performed for specific diagnosis. Generally, the image analysis procedures work on multichannel image signals with therapeutic knowledge, and such image irregularity detection methods are usually designed in integration of appropriate clinical practice with respect to the diseases disorders [1]. It is clear that the structure of brain is sophisticated and complex with its billions nerve cells connected in a complicated network. The irregularity detection in brain images provides essential consideration of disorder of brain structure by diseases though this is a challenging task.

Current methods for the task involve the assumption that the different levels of medical image intensity are associated with variety of tissues. In images of brain, it is assumed that diversity of brain areas is related to variation of textural patterns [2]. The assumptions are implemented in region detection methods including level-set evolution with region competition [3], multispectral histogram model adaptation for region growing [4], and fuzzy connectedness [5]. In recent literature, researchers presumed that the borders of areas of brain disorder yield visualization of edges which allow implementation of gradients of intensity to uncover irregular regions. Some contour-based methods are provided with the assumption: supervised boundary identification [6], fluid vector flow [7], and content-based active contour model [8]. Within the above context, current work presents a proper method for pointing out irregularity of image by a specific filter, which is performed by a sparse nonlocal texture mean. The

V. Bhateja et al. (eds.), *Information Systems Design and Intelligent Applications*, Advances in Intelligent Systems and Computing 672,
https://doi.org/10.1007/978-981-10-7512-4_61

mentioned above assumptions on intensity, textural patterns, and the gradients are considered in the proposed method. The implementation of textural patterns has been made possible by central moments, and the sparse nonlocal texture mean filter is used for searching of spots' similarity. Intensity and gradients are scanned in building borders of irregularity regions. Our contributions are twofold: (1) The sparse nonlocal texture mean filter is introduced for probing scan of irregular appearance; (2) a computational method combining probing estimates with support vector machines (SVMs) [9] is introduced for detection of irregular regions of images of brain. The remainder of this paper is organized as follows: The survey of related scientific background is marked in Sect. 2. The proposed method with its modules is described and analyzed in Sect. 3. Experiments and discussion are presented in Sect. 4. Finally, conclusion and future directions are shown in Sect. 5.

2 Prior Works

This section will focus on reviewing the patch-based concepts at feasible level and in particular within irregularity detection for images of brains. Different approaches served for analysis of brain signals data are reviewed to highlight their capabilities and obtain a new practical method for this medical imaging question. In image analysis, patches play a major role since they provide information of basic texture, edge, object, and scene. Similarity between patches is key aspect for the patch classification and detection of image features, which can be global or local. The image features proposed by nonlocal means [10] are estimated by the search for similarity of patches placed in a larger neighborhood for blur removal or contrast enhancement. In particular case of segmentation for cerebral structures, a label fusion method [11] is introduced with implementation of the nonlocal patch-based similarity. Other method reported in [12] uses the local and nonlocal neighbors to search the appropriate neighbors of the voxels and adjusting their weights. Alternatively, a multiresolution framework is addressed in [13] for brain extraction which is based on the nonlocal segmentation.

Inspired by the nonlocal means, our method enables the observation of dissimilarity of local context in a patch-based weighting aspect and performs selection of patches distinctly based on larger neighborhood which is represented by a sparse network for reducing the number of sample patches in estimation. There are registration errors of many automatic segmentation methods which affect label fusion results. Relative distances between image structures and k-nearest neighbors are proposed to consider in [14] to solve this. Our work presents alternative solution by a sparse nonlocal texture mean filter and the SVMs to overcome this problem. The SVMs is performed on initial label fusion made by similarity analysis with the network of patches, enabling the elimination of registration errors. This is particular useful in similarity analysis of irregularity by probability distribution. A network of locally coherent image partitions is proposed in [15] for estimation of the probability density function which fuses the local estimates into a global optimal estimate for segmentation of brain pathologies. Our method uses a network of patches similarly, but the network is distributed; hence, our estimates are nonlocal in some aspect. To specify the regions of interest and consider regional elasticity in magnetic resonance imaging (MRI) analysis, a hybrid level-set model comprised of alternating global and local region competitions is discussed in [16]. The method remodels shear modulus distribution with piecewise constant level sets by referring to the corresponding magnitude image. A supervised method is reported by [17], where SVMs are performed for image of brain with initial allocation of the abnormal region. Alternatively, the regions of interest are defined in our method by an adaptive texture mean over a sparse filter, and the regional elasticity is analyzed by the SVMs. The scientific background of the method is described in details in next section.

3 Proposed Method

3.1 The Central Moments for Texture Analysis

The aim of this method is to detect irregularity in images of brain from just the observation of an image of brain *u(x)* which is presented by a set of four MRI scans layered in 2D spatial domain Ω: $u(x) : \Omega \to [0,1]^4, \Omega \in \Re^2$. It can be noted that each local patch of brain MRI has different types of textures. Thus, one should use the method that relies on texture analysis. The study of characterizing texture over recent years motivated many researchers to consider central moments [18], which are moments of a probability distribution of a random variable about the random variable's mean. In application for images, the mean takes the average level of intensity of the image $u(x)$ by (1), where $p(u(x))$ is probability density function.

$$global\,mean: \mu_\Omega = \frac{1}{|\Omega|}\int_\Omega up(u)du = \frac{1}{|\Omega|}\int_\Omega u(x)dx \tag{1}$$

Now, we concern ourselves, not with global mean, but with the texture analysis in local aspect of image of brain. For this sense, local patch with predefined size $(2 * w + 1)$ can be centered at each pixel x and considered as a textural pattern. This leads to definition of patch−based mean or local mean by (2).

$$local\,mean: \mu(x) = \frac{1}{(w*2+1)^2}\int_{|y-x|\le w} u(y)dy \tag{2}$$

The purpose of variance is to express the variation of intensity around the mean. In fact, the variance for texture analysis in our patch-based approach is calculated with the local mean (2) and local probability density function $p(u)$ by (3).

$$local\,variance: \sigma^2(x) = \frac{1}{(w*2+1)^2}\int_{|y-x|\le w} (u(y)-\mu(y))^2 p(u(y))dy \tag{3}$$

For indication of symmetry to the mean in the local consideration, the skewness of $m_3(x)$ is zero if the histogram is balanced about the local mean (4). In addition, it turns either positive or negative relying on whether it has been skewed above or below the local mean. A common way of writing the local kurtosis expresses the measure of flatness of the histogram of image patches. It is also called moment of the fourth order.

The moment is to specify flatness of the histogram for patches by (5). If we knew local probability density function, then local energy could be inferred using formula (6).

$$local\,skewness:\ m_3(x) = \frac{1}{(w*2+1)^2}\int_{|y-x|\le w}(u(y)-\mu(y))^3 p(u(y))dy \tag{4}$$

$$local\,kurtosis:\ m_4(x) = \frac{1}{(w*2+1)^2}\int_{|y-x|\le w}(u(y)-\mu(y))^4 p(u(y))dy \tag{5}$$

$$local\,energy:\ e(x) = \frac{1}{(w*2+1)^2}\int_{|y-x|\le w} p(u(y))^2 dy \tag{6}$$

$$local\,entropy:\ h(x) = \frac{1}{(w*2+1)^2}\int_{|y-x|\le w} p(u(y))\log|p(u(y))|dy \tag{7}$$

In addition, local entropy is also used frequently for measuring randomness of distribution by (7). The local measures (2–7) lead to a set of coupled features for describing texture in the patch level. Further, by combining the central moments into a feature vector by (8), we can make further assumptions on diversity and similarity of the local textures.

$$T(x) = [\mu(x)\sigma^2(x)m_3(x)m_4(x)e(x)h(x)] \tag{8}$$

3.2 The Filter for Sparse Nonlocal Texture Mean

Fundamentally, the local patch-based approaches search patches nearby the considered pixel to estimate the similarity of the patch centered at the pixel with the neighborhood patches. Irregular appearance cannot be found if the distance is too short, which makes edges and textures indifferent. The global approach performs search of similarity in whole spatial domain which requests high computational resources, and results are sensitive with noise. The nonlocal methods measure similarity for all patches in whole image to set weights for the mean. This way avoids the noise sensitivity though its time complexity is not linear. In fact, our approach is started from nonlocal mean filter which takes a mean of all pixels in the image, weighted by similarity of these pixels to the target pixel [19, 20]. Thus, the local mean value $\mu(x)$ from (2) can be used for measuring similarity of patches by $|\mu(y)-\mu(x)|^2$ that is applied for the nonlocal mean by (9, 10) where h is the standard deviation.

$$nonlocal\,mean\,filter:\ m(x) = \frac{1}{c_1(x)}\int_\Omega \exp\left(-\frac{|\mu(y)-\mu(x)|^2}{h^2}\right)u(y)dy \tag{9}$$

$$c_1(x) = \int_\Omega \exp\left(-\frac{|\mu(y)-\mu(x)|^2}{h^2}\right)dy \tag{10}$$

Note that it is possible to uncover the irregularity by checking similarity of patches by inheriting notion of nonlocality. By plugging in the texture feature vector $T(x)$ from (8) to (9, 10) and replacing the local $\mu(x)$, the nonlocal mean filter turns out to be nonlocal texture mean filter (11, 12). In fact, this is no accident, as we chose texture features like the primary measures for MRIs of brain so as to achieve this.

$$\textit{nonlocal texture mean filter}: t(x) = \frac{1}{c_2(x)} \int_D \exp\left(-\frac{|T(y) - T(x)|^2}{h^2}\right) u(y)dy \quad (11)$$

$$c_2(x) = \int_D \exp\left(-\frac{|T(y) - T(x)|^2}{h^2}\right) dy, D = \Omega \quad (12)$$

The adaptive non-filter is capable to make the value of $t(x)$ smooth for the regions having similar textures and to raise attention for the regions whose textures are not analogous to others. In what follows, we show how the domain D in (12) can be managed. The computing time for original model (9) and adaptive model (11, 12) are $\Theta(wn^2), n = |\Omega|$ due to the fact that similarity by $|\mu(y) - \mu(x)|^2$ is checked for all possible pairs of $(x, y), x, y \in \Omega$. This is time assuming if the image size is large. A simple potential solution to reduce the complexity to linear $\Theta(kwn), k < n$ is that nonlocal neighbors y are selected sparsely for a target pixel x keeping the same distance between neighborhood points by (13).

$$D(x) = \{x \pm qw\} \cap \Omega, q \in \Im \quad (13)$$

Obviously, a general sparse model of non − local texture mean that adds a condition for choosing neighbors makes the effect of smoothing and de-noising stronger than continuous model (11). However, this model is still time-consuming as k in $\Theta(kn)$ is large, since a procedure of selection for neighbors y is needed for each target pixel x in (13). From the practical point of view, a network in the 2D spatial domain Ω with predefined number of nonlocal neighbors (k) can be built independently of x to avoid mentioned above selection procedure. Based on the analysis, we thus keep the sparse net of neighbors simple like (14).

$$y \in D, D = \{qw\} \cap \Omega, q \in \Im \quad (14)$$

Now, the final formula of the sparse filter for nonlocal texture mean is presented by (15, 16). For clarify, $s(x)$ is a vector of values as $T(x)$ is a vector by (8). A regulation parameter c_3 is used to keep values of $s(x)$ inside the interval [0, 1].

$$s(x) = \frac{1}{c_3(x)} \int_{y \in \{qw\} \cap \Omega} \exp\left(-\frac{|T(y) - T(x)|^2}{h^2}\right) u(y)dy \quad (15)$$

$$c_3(x) = \int_{y \in \{qw\} \cap \Omega} \exp\left(-\frac{|T(y) - T(x)|^2}{h^2}\right) dy \quad (16)$$

A complete picture of similarity for a brain scan image is formed by applying the texture measures and the sparse nonlocal mean filter (15, 16). Hence, the similarity feature $s(x)$ can be extracted like a vector of values which is ranged from zero to one. As the value of $s(x)$ is low, the spot centered by the pixel x is more similar or regular with the spots of the sparse network.

3.3 Learning of Distinction for Detection of Abnormality

The previous section demonstrates filtering process for extraction of similarity feature based on texture and nonlocal neighborhood. Now, we shall concentrate on the distinction which is the backside of the similarity and investigates how region of abnormality can be detected. The problem is that while distinction is presented by a vector of continuous value in interval [0, 1], the classification of abnormality is Boolean. Denote abnormality by A, similarity by S and distinction by D, the probability of abnormality can be seen in form of an integral (17), knowing that $p(D) = 1 - s(x)$ and $s(x)$ is get from (15). It is well known that one can learn the correspondence of abnormality for distinction by machine learning. In this case, we apply the support vector machines to solve the question by maximizing the margin that separates classes. Figure 1 illustrates major steps of the solution. The first step uses the local measure (formula 2–7) to extract texture features (8) which allow the sparse nonlocal texture filter to produce distinction feature $d(x)$ by (15, 16, and 18).

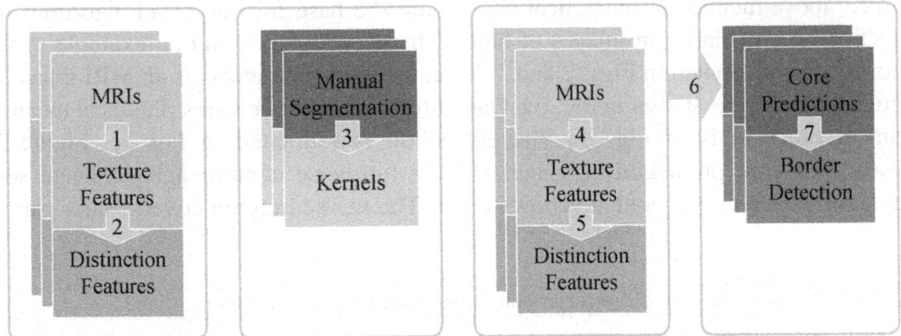

Fig. 1. Concept of the sparse nonlocal texture mean for locating abnormality

$$a(x) = p(A) = \int p(A/D)dD \tag{17}$$

$$d(x) = p(D(x)) = 1 - p(S(x)) = 1 - s(x) \tag{18}$$

To make training in the third step available, we utilize data accompanied by manual segmentation that contains classification of regions into abnormal (1) and normal (0). The trained knowledge data are saved in kernels. We then predict abnormality for

MRIs by getting texture features in step 4, distinction feature in step 5, and applying the kernels in testing phase—the sixth step in Fig. 1. Now, regions which obtain texture feature $T(x)$ by (8) and distinction $d(x)$ by (18) are classified into classes of abnormality by SVMs. Due to the nature of magnetic resonance signals, it is proper to foresee wide disparity in the values of texture features.

In addition, there are external physical factors of scanners which drive difference in a signal pattern over time. Refinement should be performed to improve classification performance. To gain the core part of abnormal region, we must ensure that the region is not incorrectly detected by mentioned external physical factors. The core part $r(x)$ is then proposed to be obtained by morphological image processing operation *imerode* (19). Then, the core part is extended by SVMs method, similarly to method in [13] which uses k-nearest neighbors to minimize registration errors of segmentation. The output $v(x)$ of the task is produced again by SVMs method. The difference of applying SVMs in this time is that we use the segmentation $r(x)$ like the test data and try to iteratively move borders of classes to maximize the margin of classes.

$$r(x) = imerode\,(d(x)) \tag{19}$$

$$v(x) = border\,regulation(r(x)) \tag{20}$$

3.4 The Algorithm

In fact, above-mentioned theoretical aspects are the base for our novel algorithm for detection of irregularity in images of brain. The pseudo-codes and an example of the algorithm are presented in Figs. 2 and 3. Here, the algorithm receives an MRI image as input; an example is shown in Fig. 3a. One of the six texture features (the local mean or moment order 1) is in Fig. 3b. The distinction is estimated in Fig. 3c. Figure 3d presents the core prediction $a(x)$ marked by a turquoise contour again manual segmentation covered by a golden yellow border. The marked region covers major part of

ALGORITHM 1. The Filter of Sparse Nonlocal Texture mean for locating irregularity (FSNT)

Input: Image of brain $u(x)$, manual segmentation $s(x)$; number of spots k, size of spot w; deviation h;
Output: The extraction of region with irregularity $v(x)$;

```
start
    for each image in train dataset u(x)          // Training
        T(x)=TextureFeature (u(x));               // by formulas (2−8)
        d(x)=DistinctionFeature (T(x));           // by (15, 16, 18)
    end
    kernels=Training(d(x), s(x));
    for each image in test dataset u(x)           // Test
        T(x)=TextureFeature (u(x));               // by (2−8)
        d(x)=DistinctionFeature (T(x));           // by (15, 16, 18)
        a(x)=Classify(d(x), kernels);
        r(x)=imerode(a(x));                       // by (19)
        v(x)=BorderRegulation(r(x));              // by (20)
    end
end
```

Fig. 2. Filter of sparse nonlocal texture mean for locating irregularity in images of brain

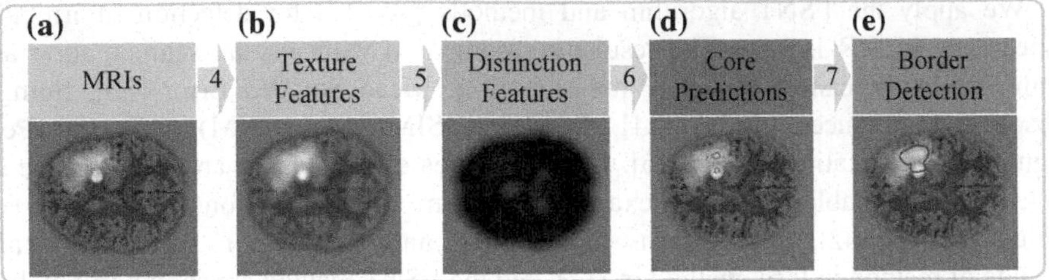

Fig. 3. Examples of detection of irregular regions in images of brain

the brain excluding the darkest part. Figure 3e demonstrates final classification $v(x)$ marked by the dark green contour. The manual segmentation inside golden yellow contour is also displayed for visual comparison.

4 Results and Discussions

To benchmark our model, we used the dataset provided for the MICCAI 2012 Challenge on Multimodal Brain Tumor Segmentation [21], which consists of 80 sets of MR images of the brain, accompanied by manual segmentation. The images sets were split randomly into two parts: One is for training, and the other is for test.

Table 1. Examples of detection of irregular regions in images of brain

	1	2	3	4	5
a.Brain image [21]					
b.Method in [16]					
c.Method in [17]					
d.Method FSNT					

We apply the FSNT algorithm and methods [16, 17] for detection of irregular appearance in MR images. The results are compared with manual segmentation and evaluated by Precision, Recall, F-measure [22], Mean Squared Error [23], Sum of Absolute Differences (SAD) [24], Structural SIMilarity (SSIM) [25] and Peak Signal-to-Noise Ratio (PSNR) [26]. Different types of irregularity are found in the set of test images. Table 1 presents examples of them: a single region (Table 1a1), two regions (Table 1a2), joint regions (Table 1a3), and split regions (Table 1a4, a5). Results of method in [16], method in [17], and the FSNT method are shown in row b, c, and d accordingly.

One can see that the irregular region inside dark green contour for case in Table 1d4 is inside the manual region marked by golden yellow contour, while the region of irregularity in case in Table 1d5 covers the manual region. Though the match is not full 100%, the method is capable to raise the attention up to irregular region expected by human experts. In brief, the sparse nonlocal texture mean filter allows catching the initial places of irregular region and saves computing time. In this test, the quality metrics and time are evaluated for each case and method and averaged for comparative overview. The statistics are reported in Table 2 with marking the best scores in bold. The arrows displayed next in each metrics show orientation to the best scores. The table demonstrates that the method of [16] obtains the highest precision; the method of [17] is the fastest and the best MSE. The best of the Recall, F-measure, SAD, SSIM, and PSNR are achieved by the FSNT method. The first point to note is that the image part presenting the brain is needed to be extracted initially to describe the region of interest by a mask $m(x)$. In a particular case where irregular appearance is usually bright, the dark image parts are not marked by the mask. This task allows analysis performed mainly on the brain part and to ignore errors that could be raised by external regions of brain. The fact that there is training data to get knowledge base on relationship of texture distinction and abnormality marks the importance of training data which cover manual segmentation by experts. The quality of prediction of abnormality depends very much on the data. There are also other texture features that were not considered in this study. Difference combination of the features is surely strong factor that needs further study. The proposed method can be applied for other medical images like lung and heart for particular medical imaging studies.

Table 2. Statistics on the quality of performance

Method	↑ Precision	↑ Recall	↑ F-measure	↓ MSE	↓ SAD	↑ SSIM	↑ PSNR	↓ Time
A. Li et al. [16]	**0.9903**	0.9645	97.6042	0.0530	0.0559	0.9938	62.8321	10.4978
B. Chandra [17]	0.9792	0.9886	98.3979	**0.0181**	0.0433	0.9934	61.1416	**3.6863**
C. FSNT	0.9825	**0.9920**	**98.7055**	0.0200	**0.0256**	**0.9980**	**65.4242**	8.1543

5 Conclusions

As mentioned in Sect. 3.1, the analysis of local, global, and nonlocal methods on the computing time and the quality of detection is the start points for setting up our new method for allocation of irregular regions in images of brain. The method proposed the sparse nonlocal texture mean filter, from which the feature of similarity is estimated by FSNT filter.

The distinction feature provides initial SVMs prediction of abnormality which is refined by second SVMs for exporting final regions of irregularity. The contribution of the paper has been to present a patch-based and contour-based implementation which allows us to retain the quality of detection while keeping linear computation cost. This motivates the further investigation into expansion of the approach to other types of medical images which may pose interesting questions to resolve.

Acknowledgements. The authors are grateful to thank the editor and the referees for their constructive comments which have led to substantial improvements in this paper.

References

1. Severino M, Schwartz E, Thurnher M, Rydland J, Nikas I, Rossi A., Congenital tumors of the central nervous system. Neuroradiology. Vol. 52, Issue 6, pp. 531–548, 2010
2. Kassner A. and Thornhill R. E., Texture analysis: a review of neurologic MR imaging applications. American Journal of Neuroradiology, Vol. 31, Issue 5, pp. 809–816, 2010
3. Ho S., Bullitt E., Gerig G., Level-set evolution with region competition: automatic 3-D segmentation of brain tumors. Pattern Recognition, Vol. 1, pp. 532–535. IEEE, 2002
4. Rexilius J., Hahn H. K., Klein J., Lentschig M. G., Peitgen H.-O,. Multispectral brain tumor segmentation based on histogram model adaptation. Proc. SPIE 6514, Medical Imaging 2007: Computer-Aided Diagnosis, 65140V, 2007
5. Harati V., Khayati R., Farzan A. Fully automated tumor segmentation based on improved fuzzy connectedness algorithm in brain MR images. Computers in Biology and Medicine, Volume 41, Issue 7, pp. 483–492, 2011
6. M Kass, A Witkin, D Terzopoulos. Snakes: Active contour models. International Journal of Computer Vision. Volume 1, Issue 4, pp. 321–331, 1988
7. Wang T., Cheng I., and Basu A. Fluid vector flow and applications in brain tumor segmentation. IEEE Tran on Biomedical Engineering, Vol 56, Iss: 3, pp. 781–789, 2009
8. Sachdeva, J., Kumar, V., Gupta, I., Khandelwal, N., and Ahuja, C. K. A novel content-based active contour model for brain tumor segmentation. MRI 30(5): 694–715, 2012
9. Cortes C, Vapnik V, Support-vector networks. Machine Learning. 20(3): 273–297, 1995
10. Buades, A., Coll, B., Morel, J.M., A non-local algorithm for image denoising. IEEE Comp Soc Conf on CVPR: Proceedings, Vol. 2, pp. 60–65, 2005
11. Coupe P., Manjon J.V., Fonov, et al. Patch-based segmentation using expert priors: application to hippocampus and ventricle segmentation. NeuroImage 54, 940–954, 2011
12. Omid Jamshidi1, Abdol Hamid Pilevar. Automatic segmentation of three dimensional brain magnetic resonance images using both local and non-local neighbors with considering inner and outer class data distances, Volume 1, Number 3, 2014
13. S F. Eskildsen, P Coupé, V Fonov, J V. Manjón, et al. BEaST: Brain extraction based on nonlocal segmentation technique. NeuroImage, Vol 59, Iss 3, pp. 2362–2373, 2012

14. Z. Wang, K. Bhatia, B. Glocker, A. de Marvao, T. Dawes, K. Misawa, K. Mori, and D. Rueckert, Geodesic Patch-based Segmentation, MICCAI, pp. 666–673, 2014
15. E I Zacharaki, A Bezerianos, Abnormality Segmentation in Brain Images Via Distributed Estimation, IEEE Trans On IT In Biomedicine, Vol. 16, No. 3, 2012
16. B.N. Li, C.K. Chui, S.H. Ong, T. Numano, T. Washio, K. Homma, et al. Modeling shear modulus distribution in magnetic resonance elastography with piecewise constant level sets. Magnetic Resonance Imaging Vol. 30, Iss 3, pp. 390–401, 2012
17. Chandra S. R. Segmentation of brain tumors, MATLAB Central, File Exchange, 2015
18. A. Papoulis, Probability, Probability, Random Variables, and Stochastic Processes, McGraw-Hill, 4th ed. International edition, 2002
19. Buades, Antoni. A non-local algorithm for image denoising Computer Vision and Pattern Recognition, 2005. 2: 60–65. https://doi.org/10.1109/CVPR, 2005
20. H. Tiwari et al., A Non-Local Means Filtering Algorithm for Restoration of Rician Distributed MRI, Emerging ICT for Bridging the Future – Proc. of the 49th Annual Convention of the Computer Society of India (CSI-2014), Vol. 2, pp. 1–8, 2014
21. MICCAI 2012 Challenge on Multimodal Brain Tumor Segmentation. http://www2.imm.dtu.dk/projects/BRATS2012/, 2015
22. Powers, David M W. Evaluation: From Precision, Recall and F-Measure to ROC, Informedness, Markedness and Correlation. J. of Machine Learn. Tech. 2 (1): 37–63, 2011
23. Lehmann, E. L., Casella, George. Theory of Point Estimation (2nd ed.). New York: Springer. ISBN 0-387-98502-6. MR 1639875, 1998
24. E. G. Richardson, Iain. H.264 and MPEG-4 Video Compression: Video Coding for Next-generation Multimedia. Chichester: John Wiley & Sons Ltd, 2003
25. Wang, Zhou, Bovik, A.C., Sheikh, H.R., Simoncelli, E.P. Image quality assessment: from error visibility to structural similarity. IEEE TIP Vol. 13 (4): pp. 600–612, 2004
26. A. Shrivastava, et al., Combination of Wavelet Transform and Morphological Filtering for Enhancement of Magnetic Resonance Images, Proc. of Int Conf on Digital Information Processing and Communications, Part-I, CCIS-188, pp. 460–474, 2011

Performance Evaluation of DC MicroGrid Using Solar PV Module

Shabana Urooj[1(✉)], Rushda Rais[1], and Ahteshamul Haque[2]

[1] Department of Electrical Engineering, Gautam Buddha University, Greater Noida, UP, India
shabanaurooj@ieee.org

[2] Department Electrical Engineering, Jamia Millia Islamia, New Delhi, India

Abstract. This paper presents the performance parameters of DC microgrid system using solar photovoltaic module. The solar power is fed through DC–DC boost converter, which is also equipped with MPPT to extract maximum solar energy. Boost converter connected along with PV module and Li-ion battery boosts up output voltage to the desired value, which is being fed to the DC bus. In order to obtain the highest efficiency and highest output power, MPPT is essential to drive the system till maximum power point. The voltage across DC bus is maintained to a certain desired value so that load can be applied to the circuit. Comparison and validation of DC output voltage are done by changing the load (resistive and inductive). Reduction of ripples in the DC output voltage is done by connecting inductors and capacitors in series and parallel, respectively. Three-phase AC source is also used and converted to DC by rectifier. Buffer circuits are used to handle the instantaneous power imbalance between source and load of the entire circuit. The system is modeled in the PSIM software, and the results obtained show that the ripples in resistive load are less as compared to the inductive load.

Keywords: DC microgrid · Solar PV energy · Boost converter · Ripples
Maximum power point tracking (MPPT) · PSIM software

1 Introduction

Due to ever increase in energy demand and concern over change in climate, there is a need to generate power from renewable energy resources. There are various renewable sources such as solar, wind, tidal, and biomass. All the sources are either DC sources or AC sources. The loads which are connected in the system are converted to DC, as a result of which there are various stages of conversion. To avoid these multiple conversions, there should be a DC source to provide power. The use of renewable resources and steady unavailability of the grid urges the use of storing electricity. DC microgrid is a combination of distributed generation sources, distributed storage devices, and distributed loads to improve reliability and quality of power supply. The components of microgrid can be connected through DC link or AC link. The concept of microgrid is about the energy storage and the utilization of distributed generation. The benefits of microgrids are high energy efficiency, quality and reliability of the delivered

V. Bhateja et al. (eds.), *Information Systems Design and Intelligent Applications*, Advances in Intelligent Systems and Computing 672,
https://doi.org/10.1007/978-981-10-7512-4_62

electric power, more flexible power network operation, and environmental and economical benefits. The distributed generation sources can be photovoltaic panels, small wind turbine, fuel cells, and distributed storage devices such as batteries, supercapacitors, and flywheel. The issue of the power quality in microgrids is an important issue due to the presence of an appreciable number of sensitive loads whose performance and life span can be adversely affected by voltage sags, harmonics, and imbalances [1]. The research has been done on several issues such as the design, control, operation, and protection of DC systems to eliminate AC–DC conversion stages. DC is used and still there is a doubt that AC is more convenient or DC. There are many benefits of DC, but there are some disadvantages also. DC is used in various fields such as spacecraft systems, data centers, telecommunication power systems, traction systems, and shipboard power systems. A model was presented to compare between the overall conversion efficiencies of AC and DC distribution topologies for residential applications by D. J. Hammerstrom. On the basis of which, each power conversion stage loses about 2.5% of the energy it converts, shows that DC systems incorporating fuel cell or other local DC generation has favorable conversion losses. There are other factors also which contradicts on using AC or DC such as sources and loads. The design of DC distribution system is very difficult, as there are many factors, which have to be considered. Many authors proposed their structures and gave different results, which led to the use of power buffer in the system to enhance transient performance of DC systems [2]. Voltage level in the DC system is an important parameter affecting performance and safety of the system [3]. DC grids are more reliable than AC grids because of multiple conversions, which add loss to the system operation. Various converters are used in order to extract maximum power from renewable power sources, to minimize power transfer between AC and DC networks, and to maintain the stable operation of both AC and DC grids under variable supply and demand conditions [4]. The bi-directional inverter operates either in grid-connection mode or rectification mode for regulating the DC-bus voltage of a DC microgrid system [5]. In order to test the performance of the DC-bus voltage regulator, voltages of varying ripple magnitude and frequency are used as test input voltages [6]. The main parameters of solar cell are short-circuit current, open-circuit voltage, efficiency, and fill factor. Maximum power point tracking is essential to operate photovoltaic system at maximum power point in order to obtain highest efficiency and output power. There are different types of photovoltaic cells on the basis of technology such as monocrystalline solar cells, amorphous solar cells, and many others. Some have long lifetime and more efficiency. There are various ambient conditions that affect the output of a PV power system like irradiance, temperature, and dust/dirt. The efficiency of a photovoltaic cell can be improved by reducing the operating temperature on the surface of cell [7].

The performance of DC microgrid is evaluated in this paper. The boost converter is connected after the solar panel and lithium-ion battery to boost the DC output voltage. The ripples in the DC output voltage are reduced by using inductors in series and capacitors in parallel with higher ratings in the DC microgrid system. Comparison and validation of DC output voltage are done by changing the load. Then, the calculation of ripple factor is done. In Sect. 2, the DC microgrid system is discussed where various parameters are evaluated. It also describes the ratings of solar module and calculation of ripples. Section 3 presents the circuit diagram, various connection schemes, and the complete PSIM models and their results.

2 Materials and Methods

DC Microgrid System

The DC microgrid system is considered in this paper. The block diagram of the proposed system is shown in Fig. 1. The figure shows only one DC line, but several DC lines can be linked depending upon the load. In the block diagram given below, the solar panel is connected to boost converter, and lithium-ion battery is also connected to boost converter. They both are connected to a DC bus. The other side of the block diagram has three-phase AC source, which is connected to rectifier (which converts AC to DC) and then connected to DC bus. Perturb and observation (P&O) maximum power point tracking (MPPT) is a technique, which is used to extract the maximum energy from the sun and to convert solar energy into electrical energy. The buffer is used to handle the instantaneous power imbalance between source and load.

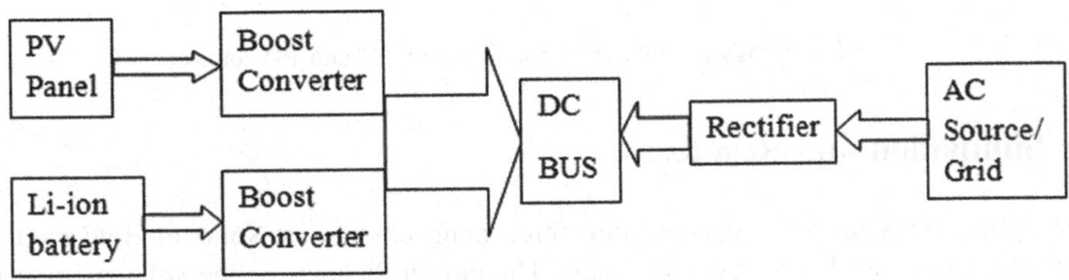

Fig. 1. Block diagram

Ratings

The ratings of the solar module and the corresponding I-V and P-V curve are shown in Fig. 2.

Maximum Power Point

P_{max} = 255.30 W

V_{max} = 29.70 V

I_{max} = 8.60 A.

Ripples

Ripples are the peak to peak to value of the output voltage. Ripple factor (RF) is the ratio of the root mean square of the ripple voltage to the absolute value of the DC component of the output voltage in the circuit.

RF = 0.2% (resistive load)

RF = 1.00% (inductive load).

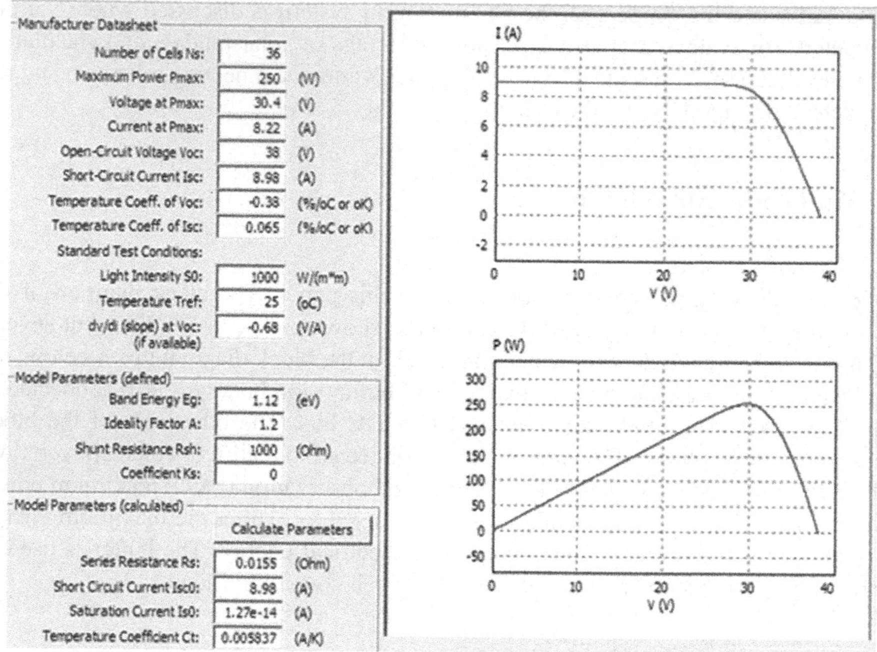

Fig. 2. Ratings of solar (PV) module and I-V and P-V curve

3 Simulation and Results

The entire PSIM model is divided into three main circuits as: solar module circuit, battery circuit, and the AC source circuit. The circuit diagram of the solar module is connected to boost converter, MPPT, and buffer and then fed to DC bus. The solar module circuit is shown in Fig. 3 (Fig. 4).

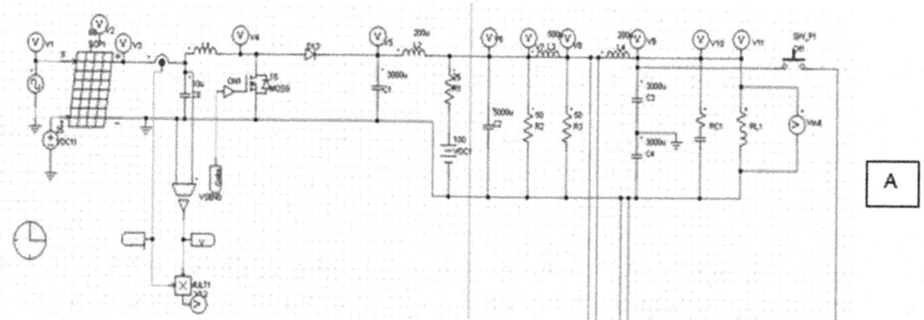

Fig. 3. Solar module is connected to boost converter

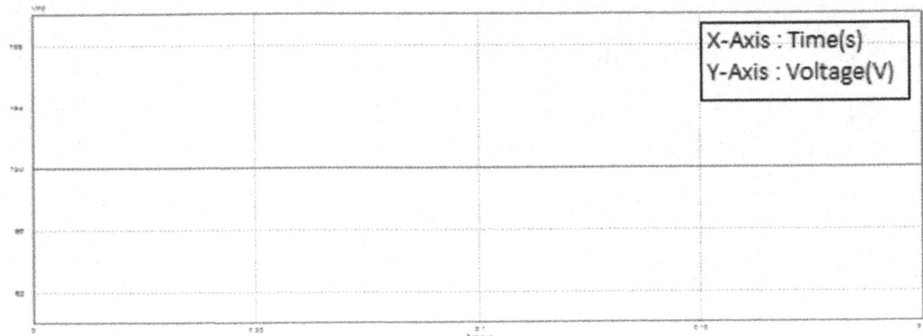

Fig. 4. Response of solar module connected to boost converter

The circuit diagram of the battery is connected to boost converter and buffer and then fed to DC bus. The battery circuit is shown in Fig. 5 (Fig. 6).

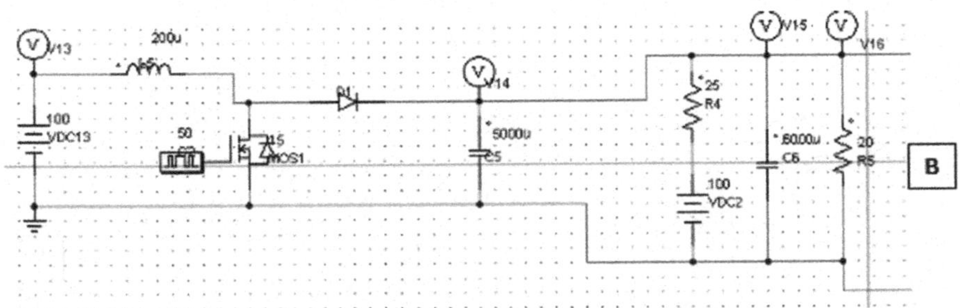

Fig. 5. Li-ion battery connected to boost converter

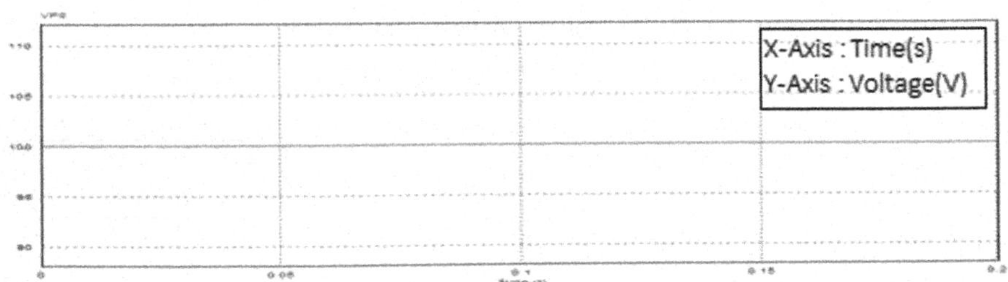

Fig. 6. Waveform of Li-ion battery connected to boost converter

The MPPT connected in the circuit is shown in Fig. 7a, b.

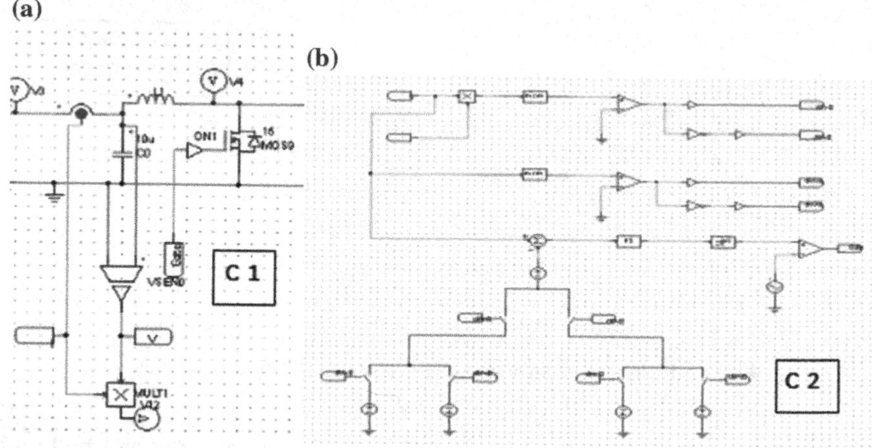

Fig. 7. **a**, **b** Maximum power point tracking (MPPT)

The circuit diagram of AC source is connected to rectifier and buffer and then fed to DC bus. The AC source circuit is shown in Fig. 8 (Fig. 9).

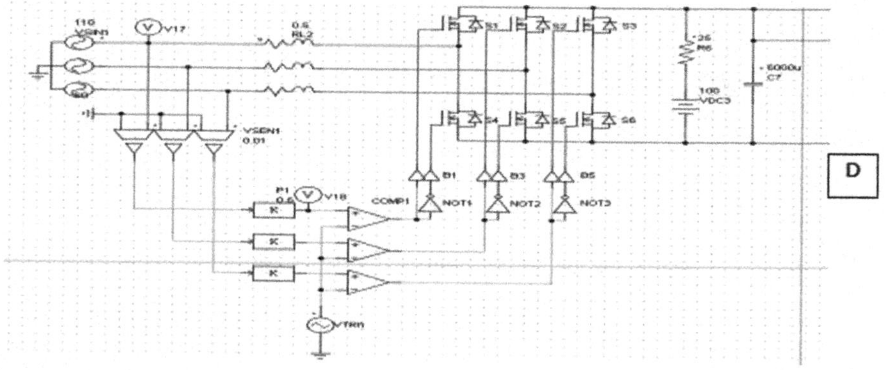

Fig. 8. AC source connected to rectifier

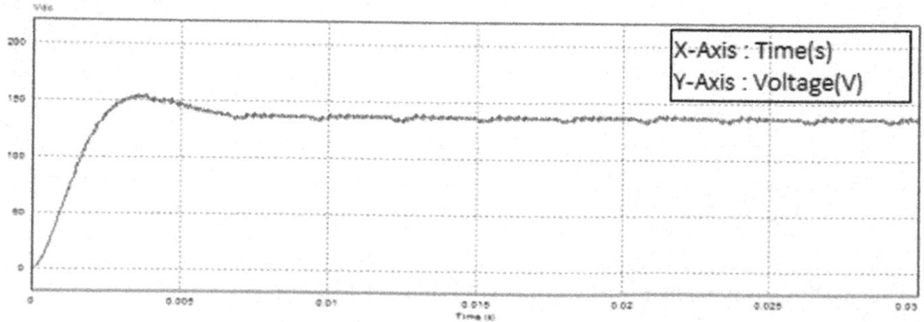

Fig. 9. Response of AC source connected to rectifier

All the circuit diagrams shown in Figs. 3, 5, 7a, b, and 8 are connected together to form complete PSIM model, which is shown in Fig. 10 (Figs. 11 and 12).

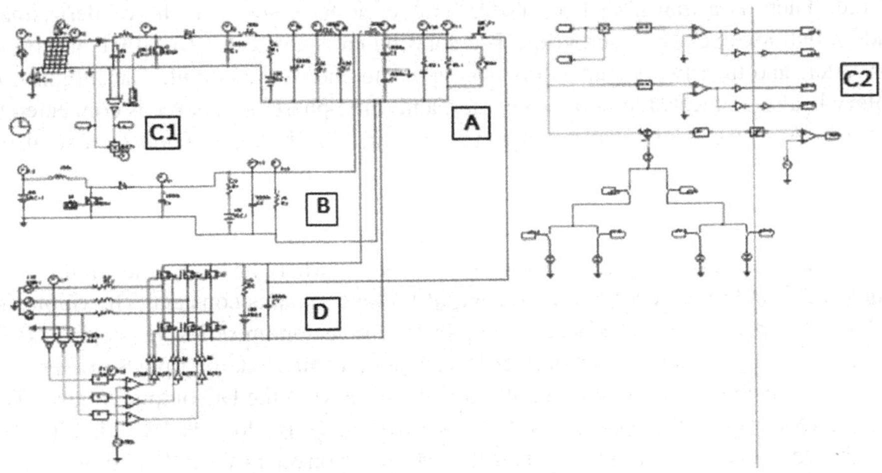

Fig. 10. Full PSIM model with MPPT

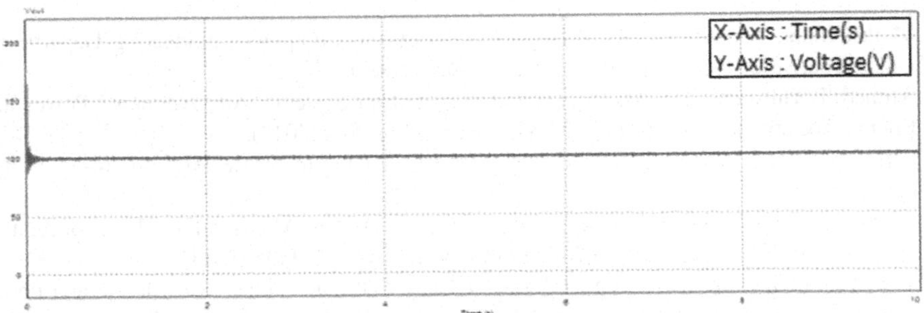

Fig. 11. Waveform of PSIM model

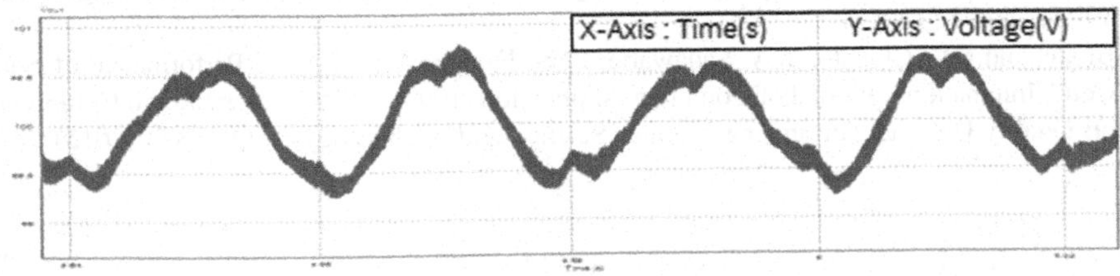

Fig. 12. Ripples in DC output voltage

4 Conclusion

Due to increase in electrical demand and excessive use of fossil fuels, there is a shift toward renewable energy sources. In this paper, sun is used as the renewable energy source. There is a transition from power conventional systems to the modern smart grid, which involves DC microgrids. The solar panel is connected to MPPT and boost converter, and then two loads are connected at the end of the circuit. The lithium-ion battery is also connected to boost converter, and three-phase AC source is connected to rectifier. At the end of all the mentioned circuits, the buffer is connected to these circuits. At last, all the circuits are connected to a DC bus. The output of solar panel and Li-ion circuit is 100 V, and the output of the circuit, which contains three-phase AC source, is 140 V, but the final output of the PSIM model is 100 V. The comparison of DC output voltage is done by changing the load, or when the load varies in the solar panel circuit at the same time, the DC output voltage remains constant. The ripples in the DC output voltage are reduced by connecting inductors in series and capacitors in parallel with higher ratings. When a resistive load is connected, the ripples are less but when an inductive load is connected the ripples increase in the DC output voltage. The ripple factor of the resistive load is 0.2%, and the inductive load is 1%. This implies that the AC component is higher than the DC component in the DC output voltage when an inductive load is connected.

References

1. Marko Gulin, "Control of DC Microgrid" Control and Computer Engineering Department, Unska Journal of University of Zagreb, Croatia. (2009).
2. Ahmed T. Ghareeb, "DC Microgrids and Distribution Systems: An Overview" Power and Energy Society General Meeting (PES), IEEE 1932–5517 (2013).
3. Sandeep Anand, "Optimal Voltage Level for DC Microgrid" IEEE 978-1-4244-5226-2/10 (2010).
4. Xiong Liu, Peng Wang, Poh Chiang Loh, "A Hybrid AC/DC Microgrid and Its Coordination Control", IEEE Transactions On Smart Grid, Vol. 2, No. 2, June (2011).
5. T.-F. Wu, C.-L. Kuo, K.-H. Sun, and Y.-C. Chang, "DC-Bus Voltage Regulation and Power Compensation with Bi-directional Inverter in DC-Microgrid Applications", IEEE 978-1-4577-0541-0/11 (2011).
6. Bakari M. M. Mwinyiwiwa," DC Bus Voltage Regulator for Renewable Energy Based Microgrid Application", World Academy of Science, Engineering and Technology International Journal of Electrical, Computer, Energetic, Electronic and Communication Engineering Vol: 7, No: 12 (2013).
7. Bhalchandra V. Chikate, Y.A. Sadawarte, "The Factors Affecting the Performance of Solar Cell", International Journal of Computer Applications (0975—8887) International Conference on Quality Up-gradation in Engineering, Science and Technology (ICQUEST) (2015).

A Customized Hardware Architecture for Multi-layer Artificial Neural Networks on FPGA

Huynh Minh Vu and Huynh Viet Thang(✉)

Danang University of Science and Technology, The University of Danang, 54 Nguyen Luong Bang street, Danang City, Vietnam
vuhm.qni@gmail.com, thanghv@dut.udn.vn

Abstract. This paper presents a novel and customized hardware architecture for the realization of artificial neural networks on reconfigurable computing platforms like FPGAs. The proposed architecture employs only one single-hardware-computing layer (namely SHL-ANN) to perform the whole computing fabric of multi-layer feed-forward neural networks. The 16-bit half-precision floating-point number format is used to represent the weights of the designed network. We investigate the scalability and hardware resource utilization of the proposed neural network architecture on the Xilinx Virtex-5 XC5VLX-110T FPGA. For performance evaluation, the handwritten digit recognition application with MNIST database is performed, which reported the best recognition rate of 97.20% when using a neural network architecture of size 784-40-40-10 with two hidden layers, occupying 91.8% FPGA hardware resource. Experimental results show that the proposed neural network architecture is a very promising design choice for high-performance embedded recognition applications.

1 Introduction

Artificial neural networks (ANNs) have attracted many applications and practical implementations in pattern recognition areas in recent years [1–7]. ANN architectures require a huge amount of parallel computing resources and memory, thereby demanding parallel computing devices like field-programmable gate arrays (FPGAs). One of the main challenges for the hardware realization of ANNs on FPGA is the data representation for weights and activation functions, for which all integer, fixed-point, and floating-point number formats can be used. While fixed-point and integer representations can bring improved execution performance in the forward computation, it is of great difficulty to train ANNs using those number formats in software to obtain the desired accuracy [5, 6]. On the other hand, floating-point number format not only can offer easier neural network training procedures in software but also can potentially bring adequate recognition accuracy and execution performance. It is shown that reduced-precision floating-point number is a suitable choice for the hardware implementation of neural networks on FPGA [3]. Preliminary results of single-FPGA

V. Bhateja et al. (eds.), *Information Systems Design and Intelligent Applications*, Advances in Intelligent Systems and Computing 672,
https://doi.org/10.1007/978-981-10-7512-4_63

implementation of ANN for handwritten digit recognition have been reported in [4], in which the 16-bit half-precision floating-point format was used.

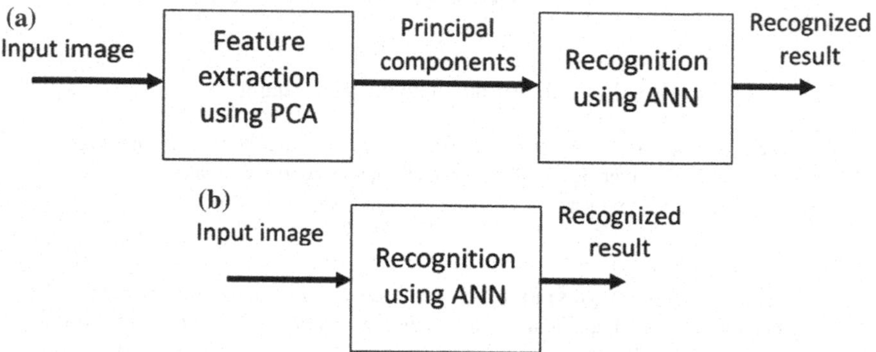

Fig. 1. Two schemes for a recognition system based on artificial neural network: **a** with feature extraction using PCA before recognition and **b** with only recognition using input image as features to artificial neural network with the removal of the feature extraction module

The functional block diagrams of a typical ANN-based image recognition hardware system are shown in Fig. 1. Two schemes for recognition system can be chosen: (a) using feature extraction (often employing principal component analysis technique) to reduce the complexity of input image data before recognition with ANN; (b) using directly input image to recognition with ANN. The first scheme is normally used when hardware-computing and storage resources of neural networks are somewhat limited; an example following the first scheme was shown in [4]. In recent years, along with the rapid development in the fields of machine learning and deep learning, the second scheme is preferable. The work presented in this paper will focus on the second scheme. The first comparison between two implementation schemes studying only *one hidden-layer* neural networks was recently carried out in [8].

In this work, we will present a novel hardware architecture for the implementation of ANNs on FPGAs with multiple hidden layers. The proposed hardware architecture uses only *one single-hardware-computing layer* to perform the whole computing fabric of multi-layer feed-forward ANNs with customizable number of hidden layers, number of neurons in each layer, and number of inputs/outputs. The MNIST database [9] is employed for performance evaluation in our work. Compared to related work in the same direction, the scientific contributions of this work are the followings: (i) This work studies performance and scalability of customized multi-layer ANNs having more than one hidden layer, normally used in deep learning; (ii) this work proposes a more efficient hardware architecture for multi-layer ANNs with respect to storage resource usage for floating-point weights; (iii) this work investigates more complicated recognition problem (using MNIST database) with larger neural network topologies.

This paper is organized as follows. Section 2 presents in detail the proposed single-hardware-layer ANN architecture. Sections 3 and 4 present experimental setup and performance results on practical FPGA device, respectively. Finally, we conclude the paper and introduce future research directions in Sect. 5.

2 Single-Hardware-Layer ANN Architecture

In a general feed-forward ANN with multiple hidden layers, each layer k receives an input vector $\mathbf{x}_k$ and is characterized by a weight matrix $\mathbf{W}_k$. For an ANN with M computing layers, the input vector is forwardly propagated to the next layer to produce an output vector $\mathbf{r}_k$, which then becomes the input vector $\mathbf{x}_{k+1}$ of the next layer $(k + 1)$, by first multiplying the weight matrix $\mathbf{W}_k$ with the input vector $\mathbf{x}_k$ and then applying the activation function f as follows:

$$\mathbf{x}_{k+1} = f(\mathbf{W}_k \mathbf{x}_k); k = 1, 2, 3\ldots, M, \tag{1}$$

where the commonly used activation function is logistic sigmoid function which is defined as:

$$f(t) = 1/(1 + e^{-t}) \tag{2}$$

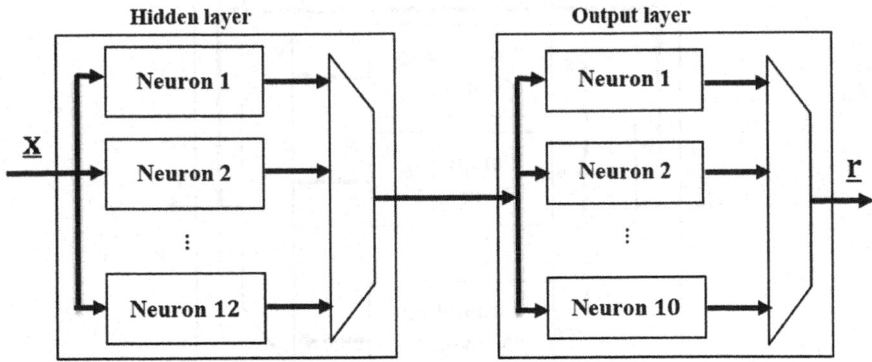

Fig. 2. An MHL-ANN architecture with 2-layer ANN: 12 neurons in the hidden layer and 10 neurons in the output layer Adapted from [4]

Since the forward propagation in ANNs is data dependent, the whole computation can be executed in a sequential manner, leading to two possible hardware architectures: (i) implementing all layers on hardware for improved performance or (ii) implementing single computing layer on hardware with shared hardware circuitry for better area utilization. The former is denoted as MHL-ANN (multiple-hardware-layer ANN), and the latter is denoted as SHL-ANN (single-hardware-layer ANN). Figure 2 shows the functional block diagram of an MHL-ANN architecture (adapted from [4]) used for handwritten digit recognition with MNIST database, in which the neural network has two computing layers implemented in parallel on hardware with 12 neurons in the hidden layer and 10 neurons in the output layer, respectively.

The hardware architecture of the SHL-ANN is presented in Fig. 3. The SHL-ANN architecture has a *hardware layer* to perform forward computation and a *control unit* to control the dataflow during forward computation. In contrast to MHL-ANN architecture, there is only one single *physical hardware layer* in the SHL-ANN architecture that consists of N *computing neurons* to perform the forward computation for all hidden layers and output layer of the desired ANN. The number of computing neuron N is identified as the maximal number of neurons of the largest layer. The functional operation of each computing neuron in the SHL-ANN architecture is as follows. Computing neuron 1 is responsible for the computation of neuron 1 in all *logical hidden layers* and neuron 1 in the *logical output layer*; those computations will be executed sequentially. Other computing neurons act the same as computing neuron 1. The outputs of one (logical) layer are fed back via an internally physical data bus (see Fig. 3) as the input to the hardware-computing layer to perform the forward computation for next (logical) layer, as described in Eqs. (1) and (2). The control unit is used to control the forward computation of logic neural layers, i.e., selecting the right weight vector and appropriate input vector—with the use of multiplexers—to the physically computing hardware layer at each computing step for each logical layer.

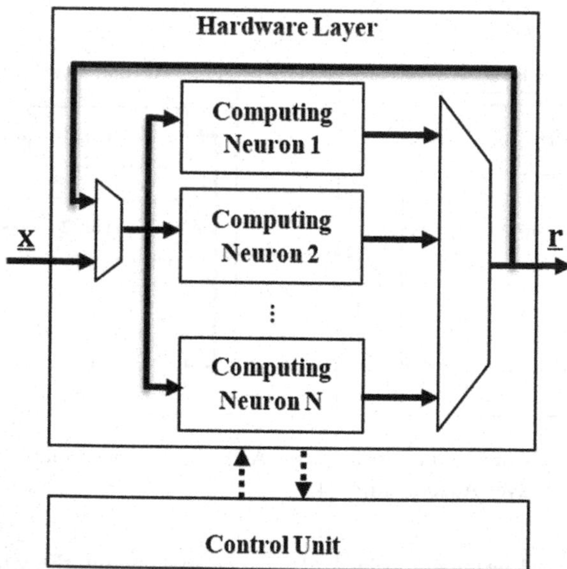

Fig. 3. Functional block diagram of the proposed SHL-ANN architecture

The hardware architecture of each computing neuron in SHL-ANN architecture is shown in Fig. 4. Each computing neuron employs (i) a MAC (multiply accumulate) unit for the multiplication of input vector and weight vector, (ii) a ROM for the storage of all weight vectors corresponding to appropriate neurons in all hidden layers and output layer, and (iii) a log-sigmoid unit for performing the activation function. For representing the weights, we use the 16-bit half-precision floating-point number format.

The IEEE-754 compatible half-precision floating-point format has 1 sign bit, 5 exponent bits, and 10 fraction bits. For implementation of MAC unit and activation function unit on FPGA, we employ the FloPoCo library [10] for the generation of VHDL code.

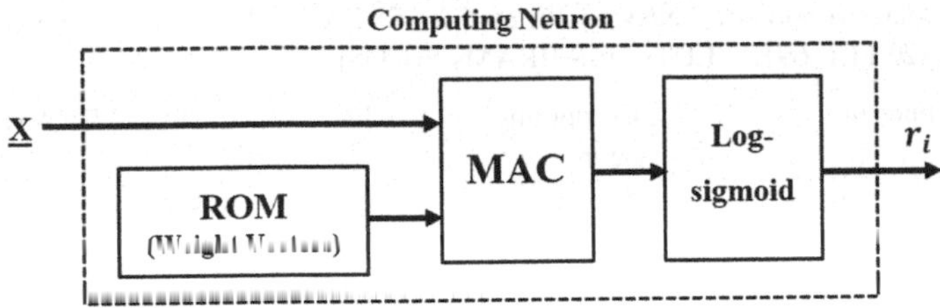

Fig. 4. Functional block diagram of each computing neuron

3 Experimental Setup

We perform handwritten digit recognition with the MNIST database [9] for performance evaluation. The MNIST database features a training set of 60,000 samples and a test set of 10,000 samples of 28 × 28 (784 pixels in total) gray-level images. There are 10 different handwritten digits from 0 to 9 in the database. As aforementioned, this work follows the recognition scheme presented in Fig. 1b. Therefore, we use 784 image pixels as inputs directly fed into the neural network, thereby allowing for removal of the feature extraction (PCA) module. As a result, the designed network has 784 inputs in the input layer and 10 neurons in the output layer. The number of hidden layers and the number of neurons in hidden layers will be varied to study the corresponding resource utilization and performance.

The neural network training is performed off-line using MATLAB on a desktop PC having 8 GB RAM and a system clock of 3.2 GHz under Windows 7. We use the back propagation with the stochastic gradient descent (SGD) algorithm for neural network training. The optimal weight matrices obtained from the training are then used for code generation of VHDL specifications for hardware implementation of SHL-ANN architectures on FPGA. We use Xilinx ISE tool 14.1 and the Virtex-5 XC5VLX-110T FPGA board (speed grade 3) for hardware synthesis and implementation.

4 Results

At first, we roughly estimate the hardware resource required to implement one computing neuron on the chosen FPGA, which reported about 1092 FFs, 1311 LUTs, 1 BRAM, and 1 DSP unit. Based on this estimation, the maximal number of computing neuron implementable on the chosen FPGA was identified as 40 computing neurons. We then vary the number of hidden layers from 1 to 3 to study the scalability of the proposed SHL-ANN architecture for the MNIST problem. For comparison with previous work, we also investigate the SHL-ANN implementations having 12 computing neurons in the hidden layers. Performance results of MHL-ANN implementation of related work [4] are also adapted. Detailed reports for FPGA resource usage are shown in Table 1. The resource utilization, performance, and MNIST recognition rates of different SHL-ANN implementations are shown in Table 2.

Table 1. Resource reports of ANNs on Virtex-5 XC5VLX-110T (speed grade 3) FPGA. Total resources: 69120 FFs, 69120 LUTs, 148 BRAMs, 64 DSPs

Network architecture	Computing neurons	FFs	LUTs	BRAMs	DSPs
MHL-ANN 20-12-10 (with PCA) [4]	22	24025	28340	22	22
SHL-ANN 784-12-10	12	13164	19055	12	12
SHL-ANN 784-12-12-10	12	13168	19549	12	12
SHL-ANN 784-12-12-12-10	12	13168	19762	12	12
SHL-ANN 784-40-10	40	44059	62579	40	40
SHL-ANN 784-40-40-10	40	44079	63454	40	40
SHL-ANN 784-40-40-40-10	40	44049	63591	40	40

Table 2. Evaluation of ANNs on Virtex-5 XC5VLX-110T (speed grade 3) FPGA

Network architecture	Resource utilization (%)	f_{max} (MHz)	Peak performance (kFPS)	Recognition rate (%)
MHL-ANN 20-12-10 (with PCA) [4]	41.1	205	404.34	90.88
SHL-ANN 784-12-10	27.6	193	18.49	93.70
SHL-ANN **784-12-12-10**	**28.3**	**193**	**18.14**	**94.14**
SHL-ANN 784-12-12-12-10	28.6	193	17.80	93.17
SHL-ANN 784-40-10	90.6	180	16.67	97.07
SHL-ANN **784-40-40-10**	**91.8**	**179**	**15.81**	**97.20**
SHL-ANN 784-40-40-40-10	92.0	178	14.98	96.76

Largest SHL-ANN architecture implementable. As shown in Tables 1 and 2, the largest SHL-ANN configuration is a 784-40-40-40-10 neural network with three hidden layers, each of which has 40 neurons, corresponding to 40 computing neurons. The 784-40-40-40-10 network architecture occupies 92% of hardware resource and is successfully implemented on the chosen Xilinx Virtex-5 XC5VLX-110T FPGA.

Scalability of SHL-ANN. Table 2 examines the scalability of the SHL-ANN architectures when increasing the number of hidden layers. Generally, the recognition rate would scale with the network size and depth. However, increasing the number of hidden layers does not necessarily bring substantially increased recognition rate for the MNIST database, as reported in Table 2 when using three hidden-layer networks. The experiments suggest that the two hidden-layer ANNs will offer the best recognition rates, corresponding to 94.14% and 97.20% when employing 12 and 40 neurons in the hidden layers, respectively.

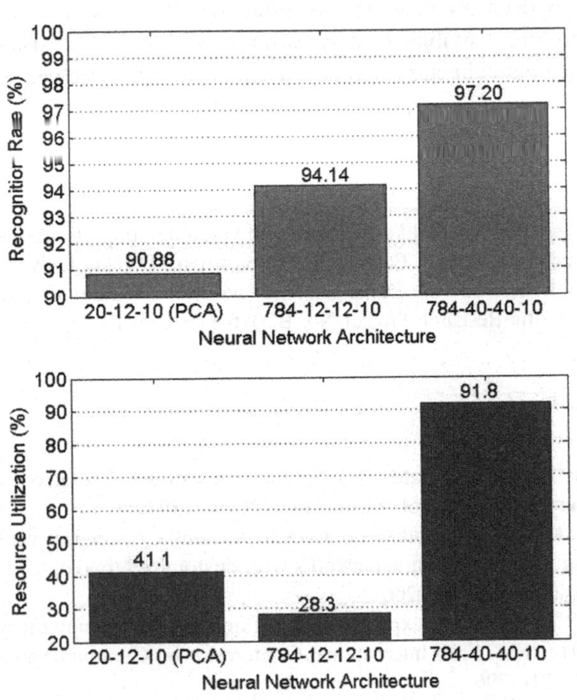

Fig. 5. Comparison among various ANN architectures implementable on Virtex-5 XC5VLX-110T FPGA with respect to MNIST recognition rate (%) and FPGA resource utilization (%)

With respect to execution performance, the peak performance of SHL-ANN architectures slightly drops down with increased network sizes, as expected. The SHL-ANN 784-40-40-10 implementation, corresponding to the best MNIST recognition rate of 97.20%, can achieve a peak execution performance of 15.81 thousand image frames per second (kFPS). Note that the MHL-ANN implementation will potentially outperform the SHL-ANN implementation as the computation of different hardware layers in MHL-ANN can be pipelined.

SHL-ANN versus MHL-ANN. We make comparison among various neural network architectures: SHL-ANN without using PCA and MHL-ANN with PCA [4]; the comparison is revealed in Fig. 5. As shown, the SHL-ANN 784-12-12-10 outperforms the MHL-ANN 20-12-10 (with PCA) in terms of both recognition rate (94.14% vs 90.88%) and FPGA resource cost (28.3% vs 41.1%). Comparison in Fig. 5 obviously shows that SHL-ANN architecture is more efficient than MHL-ANN one.

5 Conclusion

We have presented in this paper a novel and customizable hardware architecture for multi-layer neural network implementations on FPGA. With only one single computing hardware layer for performing the whole computing fabric of the neural network, the proposed architecture is an area-efficient and customized hardware design. Experiments show that the proposed architecture can offer scalable recognition rate and performance; it is, therefore, suitable for a wide range of application usage scenarios. We believe the hardware architecture will be a very promising hardware design choice for high-performance embedded recognition applications. Our future work will extend the network architecture to other recognition applications, such as face recognition and image recognition.

Acknowledgements. We would like to thank for valuable supports of our colleagues from Electronic and Telecommunication Engineering Department, Danang University of Science and Technology—The University of Danang. This work is within the framework of the Ministry-Level Scientific Research Project No. B2016-DNA-39-TT.

References

1. J. Misra and I. Saha, "Artificial neural networks in hardware: A survey of two decades of progress," Neurocomputing, vol. 74, no. 1–3 (2010) 239–255.
2. N. Nedjah, R. M. da Silva, and L. de Macedo Mourelle, "Compact yet efficient hardware implementation of artificial neural networks with customized topology," Expert Syst. Appl., vol. 39, no. 10 (2012) 9191–9206.
3. T. V. Huynh, "Design space exploration for a single-FPGA handwritten digit recognition system," in 2014 IEEE Fifth International Conference on Communications and Electronics (ICCE), (2014) 291–296.
4. T. V. Huynh, "Design of Artificial Neural Network Architecture for Handwritten Digit Recognition on FPGA," J. Sci. Technology, UDN, vol. 108 (2016) 206–210.
5. J. Park and W. Sung, "FPGA based implementation of deep neural networks using on-chip memory only," in 2016 IEEE International Conference on Acoustics, Speech and Signal Processing (ICASSP) (2016) 1011–1015.
6. Y. Umuroglu et al., "FINN," in Proceedings of the 2017 ACM/SIGDA International Symposium on Field-Programmable Gate Arrays—FPGA' (2017) 65–74.
7. "IBM Research: Neurosynaptic chips." Available: http://research.ibm.com/cognitive-computing/neurosynaptic-chips.shtml#fbid=e8AW73seOQe.
8. T. V. Huynh, "Evaluation of Artificial Neural Network Architectures for Pattern Recognition on FPGA," Int. J. Comput. Digit. Syst., vol. 6, no. 3 (2017) 133–138.
9. MNIST Database. Available: http://yann.lecun.com/exdb/mnist/.
10. FloPoCo project. Available: http://flopoco.gforge.inria.fr/.

Reconstructing B-patch Surfaces Using Inverse Loop Subdivision Scheme

Nga Le-Thi-Thu[1(✉)], Khoi Nguyen-Tan[1], and Thuy Nguyen-Thanh[2]

[1] Danang University of Science and Technology, Danang, Vietnam
lenga248@gmail.com, ntkhoi@dut.udn.vn

[2] VNU University of Engineering and Technology, Hanoi, Vietnam
nguyenthanhthuy@vnu.edu.vn

Abstract. B-patch surface is the main block to construct the triangular B-spline surfaces and has many interesting properties of the surfaces over a triangular parametric domain. This paper proposes a new method for reconstructing the low-degree B-patch surfaces using inverse Loop subdivision scheme, along with geometric approximation algorithm. The obtained surfaces are the low-degree B-patches over the triangular domain and almost cross through the data points of the original triangular meshes after several steps of the geometric approximating. Comparing with techniques use the original mesh as the surface control polyhedron, our method reconstructed B-patches with the degree reduces to 2^i times after i steps of the inverse. The accuracy of the result B-patches can be improved by adjusting the location of control points and knot vectors in each step of iterations. Some experimental results demonstrate the efficacy of the proposed approach. Because most the low-degree parametric surfaces are often employed in CAGD, mesh compression, inverse engineering, and virtual reality, this result has practical significance.

1 Introduction

In the geometric modeling, the most popular representation of the smooth surface is the subdivision surface and the parametric surface. Despite allowing to represent multi-resolution surfaces having an arbitrary topology, the subdivision surfaces are not usually used by the current modeling tools. The parametric surface is not only for representing the smooth surfaces with high continuity, stability, flexibility, local modification properties but also for providing more effective and accurate differential operator evaluations [1, 2].

Compared with the rectangular parametric surfaces, the surfaces over triangular parametric domains have most the interesting properties of the univariate splines, such as local control, affine invariance, convex hull, continuity, as well as flexibly allow joining and suit for modeling of surfaces with free shape because it is easier to separate a region into triangle domains [3, 4]. Moreover, the surfaces over a triangular domain not only possess all the important properties of univariate B-splines but also have the lower polynomial degree [2]. B-patch is a surface defined over a triangular parametric domain and is also the main block for reconstructing the triangular B-spline surfaces.

V. Bhateja et al. (eds.), *Information Systems Design and Intelligent Applications*, Advances in Intelligent Systems and Computing 672,
https://doi.org/10.1007/978-981-10-7512-4_64

Therefore, this surface has many important properties for performing a surface of 3D objects flexibly.

In general, the standard methods for reconstructing surface usually interpolate the smooth surfaces by solving linear equation systems and least-square problems [5, 6]. The reconstructed surfaces may pass through most data points, but not be very smooth. On the other hand, these methods are difficult to control locally and expensive for computing [6–8]. Recently, the iterative geometric methods have studied and improved [9–12]. They cannot only overcome limitations of the standard interpolate methods but also obtain surfaces approximated the given mesh by updating the position of the control points of these surfaces. Although the geometric methods reached interesting results, the reconstructed surfaces are both the subdivision surfaces [7, 12, 13] and the parametric surfaces over rectangular domains [5, 11, 14].

Reconstruction of the surface over triangular domains recently has studied and extended, for instance, triangular Bézier [9, 15], B-patch [8], simplex spline, and B-spline [3, 16, 17]. Some of these methods have generated global smooth surfaces, but they are difficult to control the shape locally. Moreover, because of using the given mesh as the control polygon, these techniques required the number of control points is equal to that of data points. The given meshes are almost large; thus, the degree of result surfaces is usually high.

Subdivision is a process to generate a finer mesh from a coarse mesh having free-form topology by adding new vertices and splitting old faces [18]. In contrast, from a given dense mesh, the inverse subdivision is a process to create a coarse one. Consequently, to benefit by the inverse subdivision, we employ it to simplify the given mesh and then using the result coarse mesh as the control polyhedron of B-patch.

The paper proposes a new method for reconstructing the low-degree B-patch surfaces using inverse Loop subdivision scheme, along with geometric approximation algorithm. Considering the given triangular mesh as a subdivision mesh, a sequence of the fitting B-patches successively generated after each step of the iteration. The result is that the obtained low-degree B-patches approximate to the given meshes after some steps of the geometric fitting.

Comparing with the traditional methods, the proposed approach not only avoids to solve any linear equation systems but also reconstructs the low-degree B-patch surfaces approximating to the given meshes after some of the geometric fitting steps and several of inverse subdivision times. The accuracy of the obtained B-patch surfaces can be improved by adjusting knot vectors and updating the position of control points in each of the iterations.

The rest of paper is organized as follows: Sect. 2 describes both the inverse Loop subdivision and B-patch over the triangular parametric domain. The proposed approach for reconstructing the low-degree B-patch surface is presented in Sect. 3. Section 4 shows some experimental examples. Finally, Sect. 5 gives some conclusions and possible extended works.

2 Related Works

This section describes the inverse Loop subdivision scheme and introduces B-patch surface over the triangular parametric domain. Both of them will be used for the proposed approach in next section.

2.1 Inverse Subdivision Scheme

The Loop subdivision is an approximate scheme for generating C^2-*continuous* surfaces from triangular meshes [8]. After each step of the Loop subdivision process, each triangular face of the coarse mesh is separated into four smaller ones. The vertices of obtained mesh include two types [19]: new vertices are added to the edges and old vertices are modified, namely edge-vertices and vertex-vertices correlatively (Fig. 1).

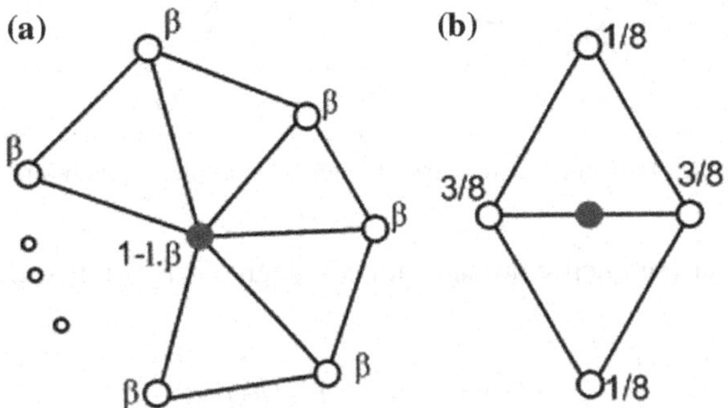

Fig. 1. Masks of the Loop subdivision for: **a** vertex-vertices and **b** edge-vertices

Letting l be a valence of vertex and β is a weight of its neighbor vertices. Then weight β is a function of valence l, and it is determined as follows [19]:

$$\beta = \frac{1}{l}\left(\frac{5}{8} - \left(\frac{3}{8} + \frac{1}{4}\cos\left(\frac{2\pi}{l}\right)\right)^2\right) \quad \text{and} \quad \alpha = 1 - l\beta \tag{1}$$

Letting i be the number of times of reversing the Loop subdivision. The position of vertices in mesh M^{i-1} can be determined from vertices of mesh M^i.

Assuming that the weights α and β correlative with the locations of edge-vertices and vertex-vertices. It is necessary to determine weights μ and η correlative with weights α and β. Based on the corresponding vertex-vertices p^i and their neighbor vertices in mesh M^i, the expression of the inverse vertex-vertices p^{i-1} in mesh M^{i-1} is determined as follows

$$p^{i-1} = \mu \cdot p^i + \eta \cdot \sum_{j=1}^{l} p_j^i \text{ with } \mu = \frac{5}{8\alpha - 3} \quad \text{and} \quad \eta = \frac{\alpha - 1}{n\left(\alpha - \frac{3}{8}\right)} \tag{2}$$

2.2 The B-Patch Surface

The B-patch surface over the triangular parametric domain is created by assigning knots to the corners of the parametric domain Δabc [3, 20]. The collection of knots corresponding to each corner is referred to as a knot vector (Fig. 2).

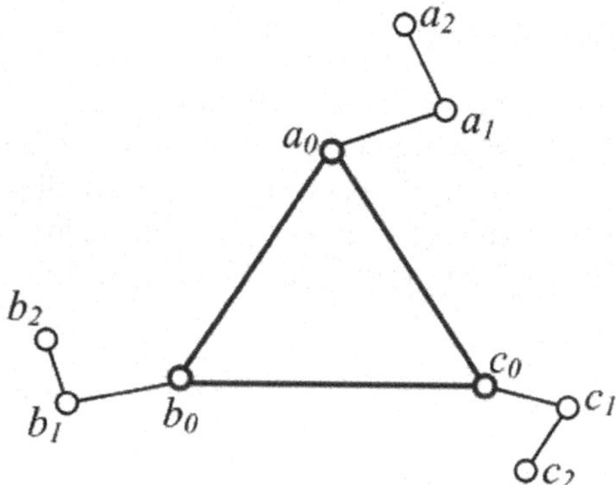

Fig. 2. Triangular parametric domain of a cubic B-patch surface

For a triangular parametric domain Δabc, a degree n B-patch surface is defined as follows [20]:

$$F(u) = \sum_{i+j+k=n} B^V_{ijk}(u)\, p_{ijk} \tag{3}$$

where

- The knot vector $V = \{a_0, a_1, \ldots, a_{n-1}, b_0, b_1, \ldots, b_{n-1}, c_0, c_1, \ldots, c_{n-1}\}$ is associated with the triangular domain $\Delta abc \equiv \Delta a_0 b_0 c_0$, and $a_0, a_1, \ldots, a_{n-1}, b_0, b_1, \ldots, b_{n-1}, c_0, c_1, \ldots, c_{n-1} \in \mathrm{R}^2$. Every triple of knots (a_i, b_j, c_k) forms a proper triangle $\Delta a_i b_j c_k$, with $0 \leq i + j + k \leq n-1$.
- The polynomials $B^V_{ijk}(u)$, $i + j + k = n$, is the normalized B-weight over knot vector V. Letting $\lambda_{ijk,d}(u)$, with $d = 0, 1, 2$, is the barycentric coordinates over the domain $\Delta a_i b_j c_k$. The polynomials $B^V_{ijk}(u)$ are defined recursively as

$$\begin{aligned} B^V_{000}(u) &= 1; \\ B^V_{ijk}(u) &= \lambda_{i-1,j,k,0} B^V_{i-1,j,k}(u) \\ &\quad + \lambda_{i,j-1,k,1} B^V_{i,j-1,k}(u) + \lambda_{i,j,k-1,2} B^V_{i,j,k-1}(u) \end{aligned} \tag{4}$$

- The coefficients $p_{ijk} \in R^3$ are called control points forming the B-patch control mesh. Letting f be the multiaffine polar form of the polynomial $F(u)$, the B-patch control points are given with blossom label as

$$p_{ijk} = f(a_0, \ldots, a_{i-1}, b_0, \ldots, b_{j-1}, c_0, \ldots, c_{k-1}) \tag{5}$$

Evaluation, subdivision and differentiation of B-patch surfaces can be computed by using the de Boor-like algorithm [20]. Figure 3a shows a possible configuration of the knot vectors. A cubic B-patch control mesh and the blending of the top three control points to generate a new point $f(a_0, a_1, u)$ is shown in Fig. 3b.

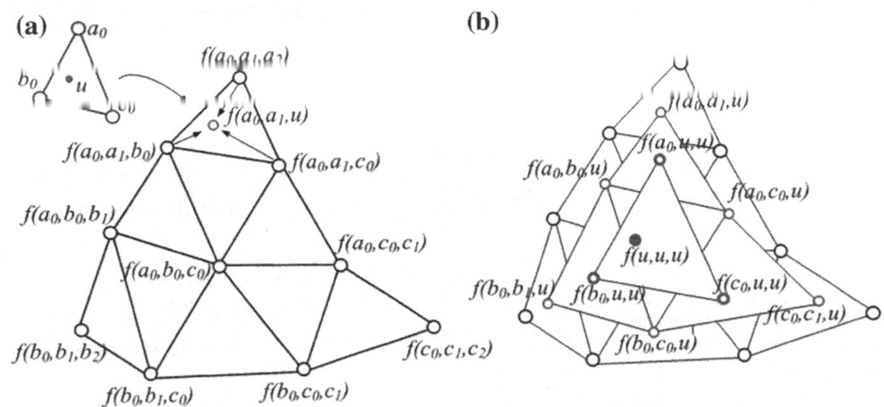

Fig. 3. Cubic B-patch: **a** blending three control points and **b** evaluating B-patch

The shape of B-patches is strongly influenced by their control meshes and knot vectors. The knot insertion algorithm is also similar to the insertion one for triangular B-spline. Consequently, the B-patch parametric surfaces inherit some interesting properties of the triangular B-spline surface such as [3, 20] convex hull property, corner vertex interpolation and tangency at multiple knots, affine invariance, and local control. These properties make B-patch attractive for the interactive smooth surface design.

3 Reconstruction of B-Patches

This section proposes a method for approximating the low-degree B-patch surfaces from the given triangular meshes.

The presented method consists of three major steps:

- Firstly, the structure of given mesh M^0 can be adapted for subdivision mesh. And then the dense mesh M^0 is simplified by using the inverse Loop subdivision scheme. After i times of the inverse subdivision process, we obtained a coarse mesh M^i.
- Next, a B-patch parametric surface S^i is generated by using the mesh M^i as a control polyhedron of B-patch. The result is that the reconstructed B-patch will have the lower degree comparing to use the original mesh M^0 as a control polyhedron.
- Finally, B-patch surface is gradually fitted for approximating the data points of the given triangle mesh. The position of control points and knot vectors of the parametric domain are also updated and adjusted to minimize the deviation between the data points of the mesh and the reconstructed B-patch in each step of the iteration.

The proposed approach can be described by the diagram in Fig. 4.

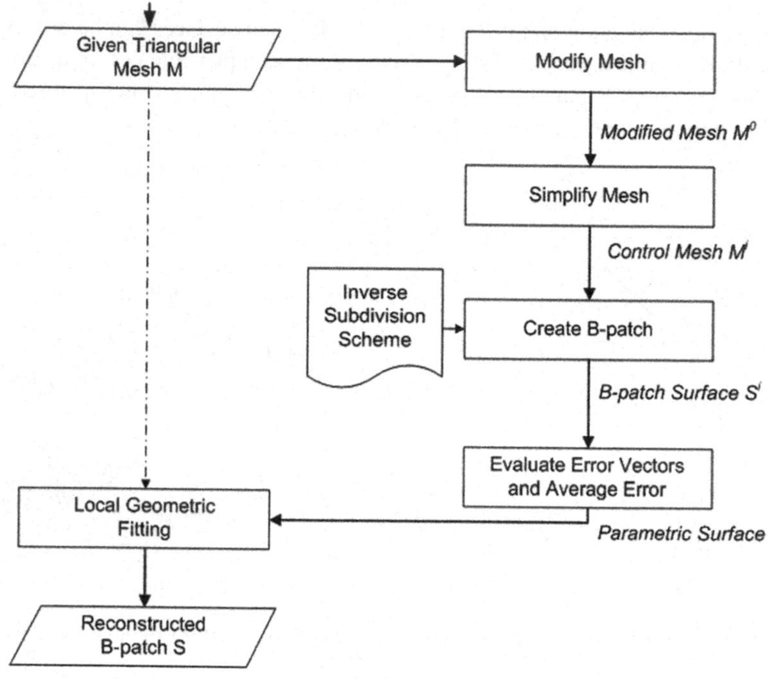

Fig. 4. Flow chart of the proposed approach

Algorithm 1. Local geometric fitting algorithm

Input: Triangular mesh M, B-patch S^i, error tolerance ε.
Output: Reconstructed B-patch surface S

$M^0 \leftarrow modiMesh(M)$ // update structure in keeping with subdivision mesh M^0
$M^* \leftarrow M^0;\ k \leftarrow 0$
Repeat
  $k \leftarrow k+1$
  $M^i \leftarrow invSub(M^*, i)$ // simplify M^* by using the inverse Loop subdivision scheme
  $D^i \leftarrow B\text{-}patchDomain(M^i)$ // update parametric domain and adjusted knotclouds
  $S^i \leftarrow B\text{-}patchSurf(M^i, D^i)$ // created B-patch surface S^i
  for each $(p_j|_{j=1..m})$ **do**
    $\varepsilon_j^k \leftarrow errVect(p_j, S^i)$ // evaluate the error vectors ε_j^i
    $p_j^* \leftarrow p_j + \varepsilon_j^k$ // compute fitted data points p_j^*
  end for
  $M^* \leftarrow triangularMesh(p_j^* |_{j=1..m})$ //create a mesh from fitted data points p_j^*
  $\varepsilon^k_{avg} \leftarrow errAvg(\varepsilon_j^k)$ //compute the average error vector ε^i_{avg}
Until $(\varepsilon^i_{avg} \le \varepsilon)$
$S \leftarrow S^i$
Return S.

Letting k be the step of the geometric fitting; ε_j^k is an error vector corresponding to each data point $p_{j|j=1...m}$ of the given mesh, these vectors are closest distances between the data points of original mesh and their corresponding points on B-patch parametric surface S^i; ε_{avg}^k is an average error vector and is computed in each step of the iteration k based on the error vectors ε_j^k.

In the geometric fitting process, the B-patches are generated and approximated to the original mesh. This process halts when the average deviation ε_{avg}^k is less than the given error ε. By this way, the result B-patch surface is the final obtained surface that passes through most the data points with the smallest average deviation. The geometric fitting algorithm of the proposed approach for reconstructing B-patch is given in Algorithm 1.

Considering the computation cost of the deviation between each data point and its corresponding point on the fitting B-patch is a constant time. There are m times of computation to be executed correlative with m data points $p_{j|j=1...m}$. Assuming that the iteration for k times, thus the value k depends on the tolerance ε. Therefore, the algorithm for reconstructing B-patch has an asymptotic complexity $\theta(m \times k(\varepsilon))$.

4 Experimental Examples

This section shows some experimental examples to make clear the effect of proposed approach. All experiment results in this paper have been obtained on a PC 2.67 GHz Intel Core i5 CPU with 4 GB RAM.

Table 1. Models of test cases

Initial mesh		Computational results					Result B-patch		
							Control mesh		Degree
#points	#faces	k	ε_{max}	ε_{avg}	$N_\varepsilon(\%)$	Time (s)	#points	#faces	
		3	0.23262	0.05413	79.215	<1			
45	64	6	0.24813	0.04052	90.006	<1	6	4	2
		9	0.20717	0.02084	90.201	1			
		3	0.22658	0.06441	79.738	4			
91	144	6	0.15767	0.03437	86.039	8	10	9	3
		9	0.10559	0.02648	89.792	14			
		3	0.38800	0.04909	76.848	8			
153	256	6	0.27048	0.03894	86.964	16	15	16	4
		9	0.20136	0.02863	90.052	23			

Letting ε_{max} be the largest deviation of the error vectors ε_j^k; ε_{avg} be an average deviation between the fitting B-patch and the data points of the given mesh; $N_\varepsilon(\%)$ is a percent of the number of data points that the obtained B-patch crosses through them. Both the given triangular meshes and reconstructed B-patches of some experimental examples are presented in Table 1.

Figure 5 shows the obtained result for reconstructing the B-patch surfaces: quadratic B-patch (*above*), cubic B-patch (*middle*), and quartic B-patch (*below*). From triangular meshes which presented in Fig. 5a and detailed in the first column (*Initial mesh*) of Table 1, we obtained the coarse meshes by applying $i = 2$ times of the Loop inverse subdivision scheme, as described in column *Control mesh* (in Table 1). Employing these coarse meshes as control meshes for creating the quadratic, cubic, and quartic B-patches. Based on each error vector ε_j^k, the location of control points is separately adjusted. B-patch surfaces can be obtained as presented in Fig. 5b after $k = 6$ steps of the geometric fitting. Figure 5b shows that the obtained B-patches cross through most data points of the initial meshes. Information on the reconstruction is listed in column *Computational results* of Table 1. After $k = 9$ steps of the local geometric fitting, we obtained the result B-patches with these zebra mapping and parametric domains along with knot vectors, are also presented in Fig. 5c and listed in the last column (*Result B-patch*) of Table 1.

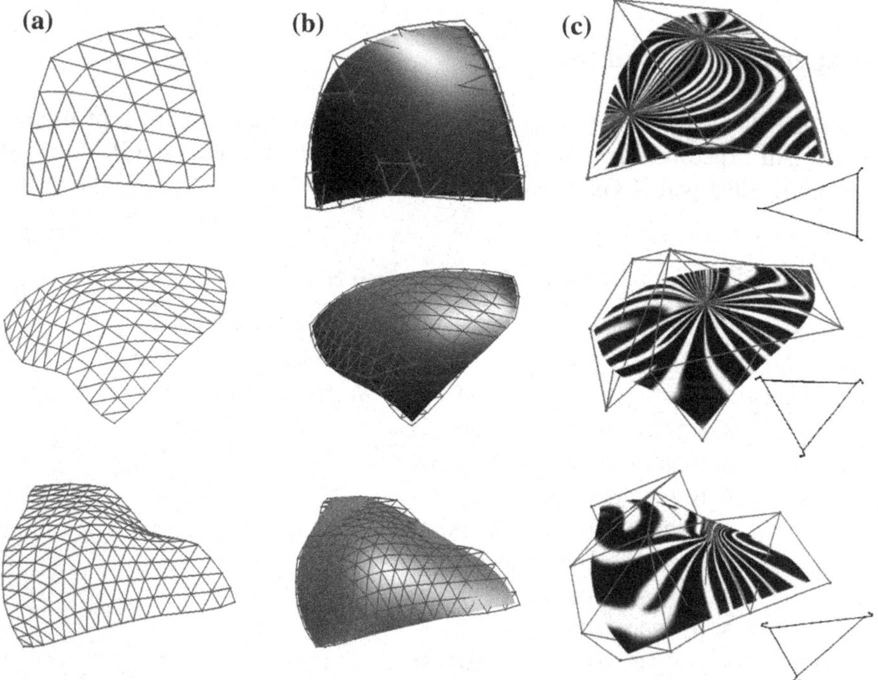

Fig. 5. Quadratic (*above*), cubic (*middle*) and quartic (*below*) B-patch surfaces: **a** Initial mesh, **b** B-patches and given meshes after k = 6 steps of geometric fitting, **c** zebra mapping of obtained B-patches, control meshes and parametric domains after $k = 9$ steps of fitting

Table 1 also presents that the low-value ε_{max} and ε_{avg}, while the rather high-value N_ε corresponding to the number of iteration k. The execution times are also proportional to the degree of the result B-patch surfaces as well as the number of iteration k.

For proving the accuracy of the reconstructed low-degree B-patches can be improved by adjusting the position of control points in each step of iteration k, we present plots of both values ε_{avg} and N_{ε} in Fig. 6. The average error values ε_{avg} drop in the first three steps and then gradually fall into the range from 0.002 to 0.003, as shown in Fig. 6a. While the values N_{ε} rapidly increase in the first three iterations, and after that, they reach to range from 89 to 90% in Fig. 6b. The plots indicate that the values ε_{avg} and N_{ε} depend on the number of iterations k. Therefore, we can obtain the result B-patches after some steps k of the geometric approximation algorithm.

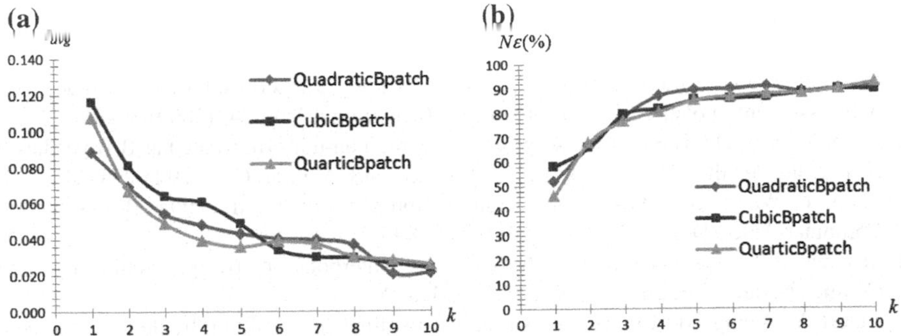

Fig. 6. **a** Average errors and **b** Convergence are proportional to the number of iteration k

5 Conclusions

Based on the inverse Loop subdivision scheme and the geometric approximation algorithm, this paper proposes a new method for reconstructing the low-degree B-patches from the triangular meshes; in particular, they are the quadratic, cubic, and quartic B-patch surfaces over the triangular parametric domain.

Our approach has the following features:

- Avoiding the disadvantages of the traditional interpolation method, for instance, linear system solution and least-square fitting problem.
- By employing the inverse Loop subdivision scheme, the control mesh of B-patch surface is simplified. Therefore, its size is only less than or equal to a quarter compared to the methods using the initial mesh as a control polyhedron of the result B-patch. Moreover, the degree of reconstructed B-patch surfaces will reduce to 2^i times after i steps of reversing subdivision.
- The accuracy of result B-patches can be improved by adjusting the knot vectors and location of control points in the reconstructing process. The reconstructed B-patches approximate to the given triangular mesh after some steps of the iterations.

The low-degree parametric surfaces often employed in geometric design and modeling tools, especially the cubic surfaces, then the reconstructed B-patches remain practical significance in some fields such as: geometric design, reverse engineering and virtual reality, especially for data compression.

References

1. Farin, G.: Curves and Surfaces for Computer Aided Geometric Design: A Practical Guide, 5th edn. Morgan Kaufmann, San Mateo (2002).
2. Greiner, G.: Geometric modeling. Lecture in Winter Term (2010).
3. Christopher, K.I.: A Geometric B-Spline Over the Triangular Domain. M.S. Mathematics thesis (2003).
4. Botsch, M., Pauly, M., Rossl, C., Bischoff, S., Kobbelt, L.: Geometric Modeling Based on Triangle Meshes. EuroGraphics (2006).
5. Deng, C., Lin, H.: Progressive and iterative approximation for least squares B-spline curve and surface fitting. Computer-Aided Design, Vol. 47, (2014) 32–44.
6. Eck, M., Hoppe, H.: Automatic reconstruction of B-spline surfaces of arbitrary topological type. In Proceedings of SIGGRAPH96, ACM Press (1996) 325–334.
7. Cheng, F., Fan, F., Lai, S., Huang, C., Wang, J., Yong, J.: Loop subdivision surface based progressive interpolation. Journal of CS and Technology, Vol. 24, (2009) 39–46.
8. Zhao, Y., Lin, H.: The PIA property of low degree non-uniform triangular B-B patches. In Proceedings of the 12th International Conference on CAD and CG (2011) 239–243.
9. Chen, J., Wang, G-J.: Progressive iterative approximation for triangular Bézier surfaces. Computer-Aided Design, Vol. 43, (2011) 889–895.
10. Maekawa, T., Matsumoto, Y., Namiki, K.: Interpolation by geometric algorithm. Computer-Aided Design, Vol. 39, (2007) 313–323.
11. Kineri, Y., Wang, M., Lin, H., Maekawa, T.: B-spline surface fitting by iterative geometric interpolation/approximation algorithms. CAD, Vol. 44(7), (2012) 697–708.
12. Nishiyama, Y., Morioka, M., Maekawa, T.: Loop subdivision surface fitting by geometric algorithms. Poster proceedings of pacific graphics (2008).
13. Deng, C., Ma, W.: Weighted progressive interpolation of Loop subdivision surfaces. Computer-Aided Design, Vol. 44, (2012) 424–31.
14. Xiong, Y., Li, G., Mao, A.: Convergence analysis for B-spline geometric interpolation. Computers & Graphics, Vol. 36, (2012) 884–891.
15. Nga, L.T.T., Khoi, N.T., Thuy, N.T.: Reconstructing low degree triangular parametric surfaces based on inverse Loop subdivision. In Proceedings of the International Conference on Nature of Computation and Communication, No. 144, (2014) 98–107.
16. Dian, P.: The Implementation of Univariate and Bivariate B-Spline Interpolation Method in Continuous. IJCSI International Journal of Computer Science Issues, Vol. 10 (2), (2013).
17. Neamtu, M.: Bivariate simplex B-splines: a new paradigm. In Proceedings of the 17th spring conference on computer graphics (2001) 71–78.
18. Zorin, D., Schroder, P.: Subdivision for Modeling and Animation. SIGGRAPH Course Notes (2000).
19. Loop, C.: Smooth Subdivision Surfaces Based on Triangles. M.S. Mathematics thesis (1987).
20. Seidel, H.P.: Symmetric recursive algorithms for surfaces: b-patches and the Boor algorithm for polynomials over triangles. Constructive Approximation, Vol. 7, (1991) 257–79.

A Rule-Based Method for Text Shortening in Vietnamese Sign Language Translation

Thi Bich Diep Nguyen[1(✉)], Trung-Nghia Phung[2], and Tat-Thang Vu[3]

[1] Graduate University of Science and Technology, Ha Noi, Vietnam
ntbdiep@ictu.edu.vn
[2] Thai Nguyen University of Information and Communication Technology, Thai Nguyen, Vietnam
ptnghia@ictu.edu.vn
[3] Institute of Information Technology, Ha Noi, Vietnam
vtthang@ioit.ac.vn

Abstract. Sign languages are natural languages with their own set of gestures and grammars. The grammar of Vietnamese sign language has significantly different features compared with those of Vietnamese spoken/written language, including the shortening, the grammatical ordering, and the emphasis. Natural language processing research on Vietnamese sign language including study on spoken/written Vietnamese text shortening into the forms of Vietnamese sign language is completely new. Therefore, we proposed a rule-based method to shorten the spoken/written Vietnamese sentences by reducing prepositions, conjunctions, and auxiliary words and replacing synonyms. The experimental results confirmed the effectiveness of the proposed method.

1 Introduction

The deaf communities in the world mostly communicate by performing gestures. The commonly used gestures were converted to sign languages since the eighteenth century. After that, the sign languages have developed gradually and recognized as the official sign languages of the deaf communities of the countries. The sign language used by the deaf community of Vietnam is Vietnamese sign language (VSL).

Although sign languages share many similarities with spoken languages, there are some significant differences between sign and spoken/written languages in grammar and linguistic properties. As a result, VSL as well as other sign languages are natural languages with their own set of gestures and grammar. For instance, American sign language (ASL) has its own grammar system (its own rules for phonology, morphology, syntax, and pragmatics), separate from that of English [1].

Sign language translation (SLT) is the system translating written text to signs or/and signs to text. The adult deaf are quite easy to show what they mean to normal hearing people after their practice of sign and body languages every day. However, people with normal hearing feel very difficult to show their ideals to the deaf since they rarely use body and sign languages. As a consequence, it seems that the translation side from text to signs is more helpful for the deaf than the side from signs to text.

V. Bhateja et al. (eds.), *Information Systems Design and Intelligent Applications*, Advances in Intelligent Systems and Computing 672,
https://doi.org/10.1007/978-981-10-7512-4_65

Since sign languages are natural languages with their own grammar and linguistic properties, SLT requires researches in natural language processing (NLP). However, there are just a few NLP researches in SLT in the world [2–6]. Especially, NLP research on VSL and Vietnamese SLT is completely new.

In [5], Humphries mentioned that shortening is one of the most important characteristics in ASL. What this means is text shortening is a critical step in NLP researches for SLT.

One recent research of Gouri Sankar Mishra and his colleagues proposed an NLP system to translate spoken English to Indian sign language (ISL) as shown in Fig. 1, including a text shortening method as shown in Fig. 2 [3]. The translation model follows a rule-based method in which a parser is used to parse the full English sentence and a dependency structure is identified from the parse tree. This structure represents the syntactic and grammatical information of a sentence. The shortened ISL sentence is generated from a bilingual ISL dictionary and a WordNet. From this shortened ISL sentence, the corresponding ISL signs are displayed.

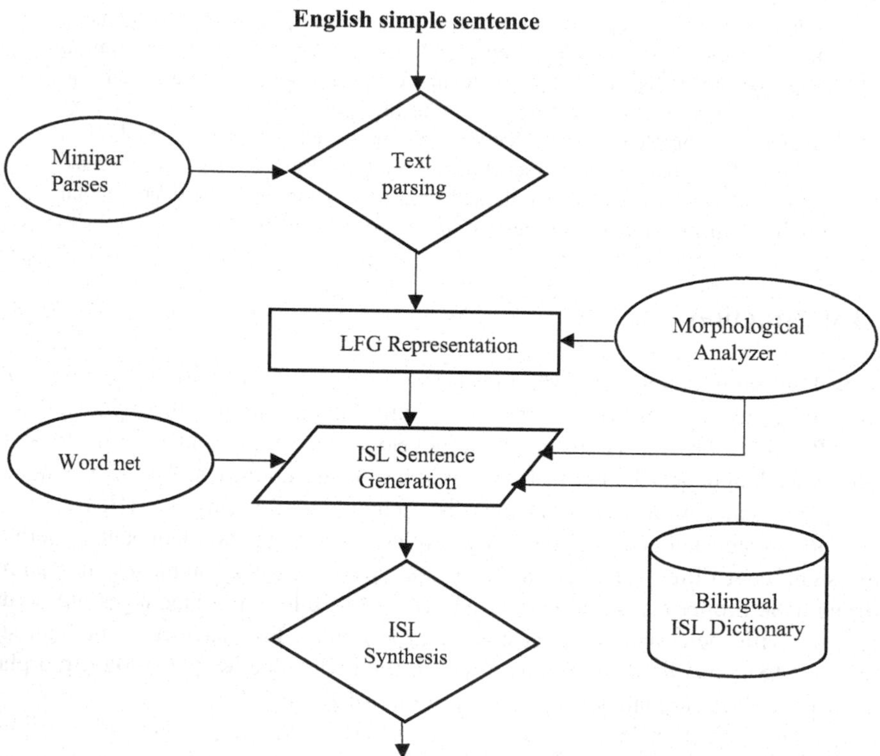

Fig. 1. Prototype machine translation system for ISL [3]

Since NLP research on VSL and Vietnamese SLT is completely new, this paper proposed a method of spoken/written Vietnamese text shortening for Vietnamese SLT.

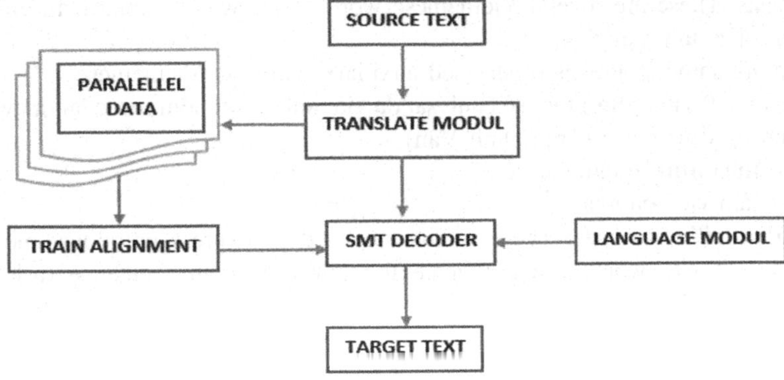

Fig. 2. Architecture of the translation model for ISL [3]

2 Linguistic Fundamentals of the Text Shortening Characteristic in VSL

VSL in the normal communication among the Vietnamese deaf has some basic characteristics. In [7], people show three most important characteristics of VSL. The first one is the shortening characteristic, in which the sentence of VSL is shorter than the sentence of the spoken/written Vietnamese language caused by the reduction of the prepositions and auxiliary words in the sentence. The second one is the grammatical ordering characteristic, in which grammatical ordering in VSL is different with that of spoken/written Vietnamese. The third one is the limited vocabulary characteristic of VSL. This characteristic is originated from the limited cognitive of the deaf. In [8], people show that there are three features in VSL, including the shortening, the grammatical ordering, and emphasis. The reduced components in VSL sentences are prepositions, conjunctions, and auxiliary words (Table 1).

Table 1. Some examples of text shortening

Full sentences in spoken Vietnamese language (in Vietnamese)	Reduced sentences in VSL (in Vietnamese)
Viết bằng bút chì	Viết ~~bằng~~ bút chì
Tôi và anh đi học	Tôi ~~và~~ anh đi học
Anh ăn cháo hay ăn cơm?	Anh ăn cháo ~~hay~~ ăn cơm?
Mặc dầu trời mưa, tôi vẫn đi học	~~Mặc dầu~~ trời mưa, tôi ~~vẫn~~ đi học
Lấy hộ chị quyển sách	Lấy ~~hộ~~ chị quyển sách

Based on the characteristics of meaning and location appearing in the sentences, auxiliary words in Vietnamese can be divided into the following categories:

- ***Stressed auxiliary words***: These words are used to emphasize words, phrases, or sentences that accompany them. They are preceded by words or phrases that need

emphasis. These are several Vietnamese words such as cả, chính, đích, đúng, chỉ, những, đến, tận, ngay,....
These are some examples of stressed auxiliary words in Vietnamese:
– Hai ngày sau, **chính** một số cảnh sát đã giải anh đi tối hôm trước lại quay về nhà thương Chợ Quán (Trần Đình Vân)
– Nó mua **những** tám cái vé
– Nó làm việc **cả** ngày lễ.

- ***Modal auxiliary words***: These words are used to express emotions, moods, or attitudes. These words often indicate the purpose of the sentence (ask, order, exclaim ...). They stand at the end of the sentence to express the doubt, urge, or exclamation. They also reveal the attitude or the feelings of the speaker, the writer. These are several Vietnamese words such as à, ư, nhỉ, nhé, chứ, vậy, đâu, chăng, ừ, ạ, hả, hử ...
These are some examples of modal auxiliary words in Vietnamese:
– Chúng ta đi xem phim **nhé**?
– Đã bảo **mà**!
– Trời có mưa **đâu**?
- ***Exclamations words***: These are words that directly express the emotions of the speaker. They cannot be used as emotional names, but as indications of emotions. They cannot be official part of a phrase or sentence, but can be separated from the sentence to form a separate sentence. They are often associated with an intonation or gesture, facial expressions or gestures of the speaker. Exclamation words can be used to call or answer (ơi, vâng, dạ, bẩm, thưa, ừ,...), and to express feelings of joy, surprise, pain, fear, anger,... (ôi! trời ơi, ô, ủa, kìa, ái, ối, than ôi, hỡi ôi, eo ôi, ôi giời ôi,...). It can be said that exclamation words are used to express sudden, strong emotions of different types.
A text shortening algorithm involves grouping the above components into a dataset that is compared to the original text in the shortening process. With the linguistic bases analyzed above, we proposed a rule-based shortening method as present below.

3 The Proposed Text Shortening Method

The core ideal of the text shortening method proposed in this paper is the use of a vocabulary of all removable words in VSL including prepositions, conjunctions, and modal auxiliary verbs.

Our current VSL dictionary contains about 3000 words where each word corresponds to a distinct sequence of sign gestures. The number of words and phrases in this VSL dictionary is much smaller than that in spoken/written Vietnamese dictionary. The words that appeared in Vietnamese dictionary but reduced in VSL dictionary are not only prepositions, conjunctions, and modal auxiliary verbs but also synonyms. Therefore, in the proposed method, if words or phrases appeared in Vietnamese dictionary but not in VSL dictionary, corresponding synonyms are searched in the VSL dictionary also.

The proposed method is described in Fig. 3 and explained step by step as follows:

Step 1. Use Vietnamese word segmentation with parsing tool Bikel, Vietnamese treebank, and VSL dictionary in the preprocessor to return the list of words and phrases.
Step 2. Find the labels of words and phrases.

- If the words and phrases are not found in the VSL dictionary, we find the corresponding synonyms and find the labels again.
- If the results are found, we move to Step 3, else we move to Step 4.

Step 3. Shorten the sentence by reducing the words or phrases with labels as prepositions, conjunctions, and modal auxiliary verbs. Then, return the shortened sentence.
Step 4. Insert unfound words and phrases to a database. In the next steps, for building a full VSL translator, each word or phrase in this database will be performed by pronouncing.

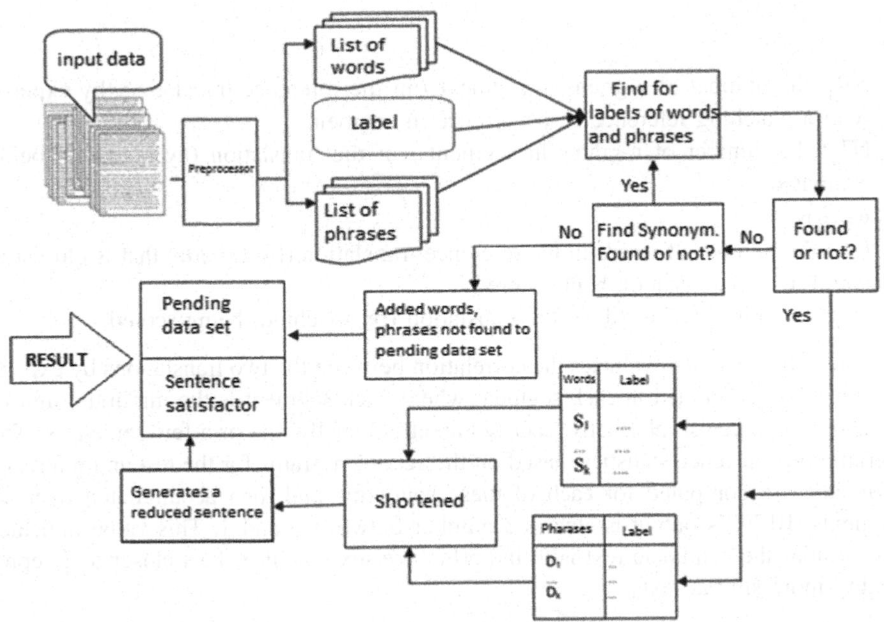

Fig. 3. Systematic diagram of the proposed method

4 Experiment Results

4.1 Evaluation Method

BLEU is a method to evaluate quality of the documents automatically translated by machine, proposed by IBM in 2002 [9] and used as the primary evaluation measure for research in machine translation in [10]. The original ideal of the method is to compare two documents automatically translated by machine and manually translated by linguistic experts. The comparison is performed by statistically analyzing the coincidence of the words in the two documents that takes into account the order of the words in the sentences using n-grams. Specifically, BLEU scores are computed by statistically analyzing the degree of coincidence between n-grams of documents automatically translated by machine and the ones manually translated by high-quality linguistic experts [11].

BLEU score can be computed as follows [11]:

$$score = \exp\left\{\sum_{i=1}^{N} w_i \log(p_i) - \max\left(\frac{L_{ref}}{L_{tra}} - 1, 0\right)\right\} \tag{1}$$

- $P_i = \frac{\sum_j NR_j}{\sum_j NT_j}$

- NR_j: the number of n-grams in segment j in the reference translation (by experts) with a matching reference co-occurrence in segment
- NT_j: the number of n-grams in segment j in the translation (by machine) being evaluated.
- $w_i = N^{-1}$
- L_{ref}: the number of words in the reference translation (by experts) that is closest in length to the translation being scored.
- L_{tra}: the number of words in the translation (by machine) being scored.

The value of *score* evaluates the correlation between the two translations by experts and machine, computed in each segment where each segment is the minimum unit of translation coherence. Normally, each segment is usually one or a few sentences. The n-gram co-occurrence statistics, based on the sets of n-grams for the test and reference segments, are computed for each of these segments and then accumulated over all segments. BLEU's output is always a number between 0 and 1. This value indicates how similar the candidate text is to the reference texts, with values closer to 1 representing more similar texts.

4.2 Evaluation Results

We built a VSL dictionary with 3000 words and phrases. For evaluation, we used 200 simple sentences extracted from on the textbooks used in the schools for deaf children. After being translated (shortened) by using the proposed method, we computed the

BLEU scores between the translated sentences and the corresponding ones conducted by one expert in VSL. The data of this study were assisted by Dr. Cao Thi My Xuan—a VLS researcher. The matching translation was evaluated by experts from the deaf and dumb children education school in Thai Nguyen province, Vietnam.

The results of computed BLEU scores are shown in Fig. 3. The ratio of sentences correctly translated by using the proposed method (corresponding with BLEU score is one) is 97.5%. A few sentences incorrectly translated and caused by semantic ambiguity will be solved in our future researches (Table 2).

Table 2. BLEU scores

ID sentence	L_{input}	NR_j	NT_j	L_{ref}	L_{tra}	BLEU score
1	3	7	7	3	3	1.000
2	5	12	12	4	4	1.000
3	8	15	15	6	6	1.000
4	9	26	20	9	7	0.7515
5	5	14	14	5	5	1.0000
…	…	…	…	…	…	….
99	7	22	16	7	6	0.8465
100	8	24	24	8	8	1.0000
…	…	…	…	…	..	…
199	7	23	23	7	7	1.000
200	6	13	18	5	6	0.9762

5 Conclusions

NLP research on VSL and Vietnamese SLT is completely new. It is known that shortening is one of the most important features of VSL. In this paper, we proposed a rule-based method to shorten the spoken/written Vietnamese text into VSL forms by reducing prepositions, conjunctions, and modal auxiliary verbs and replacing synonyms. The experimental results show that the proposed method is efficient.

In the next researches, we will continue to study other issues on VSL and Vietnamese SLT.

Acknowledgements. This study was supported by the Ministry of Education and Training of Vietnam (project B2016-TNA-27).

References

1. Fromkin, V.: Sign language: Evidence for language universals and the linguistic capacity of the human brain. Sign Language Studies. 59 (1988) 115–127.
2. Matthew P. Huenerfauth, American Sign Language Natural Language Generation and Machine Translation Systems, Technical Report Computer and Information Sciences University of Pennsylvania MS-CIS-03–32, September 2003.

3. Gouri Sankar Mishra, Ashok Kumar Sahoo and Kiran Kumar Ravulakollu, Word based statistical machine translation from english text to indian sign language, ARPN Journal of Engineering and Applied Sciences, VOL. 12, NO. 2, 2017.
4. Dasgupta T., Basu A., Prototype machine translation system from text-to-Indian sign language, Proceedings Of The 13th International Conference On Intelligent User Interfaces, Gran Canaria, Spain, pp. 313–316, 2008.
5. Humphries, T., & Padden, C. Learning American sign language, Englewood Cliffs, N.J: Prentice Hall (1992).
6. Kar P., Reddy M., Mukherjee A. and Raina A.M. INGIT: Limited Domain Formulaic Translation from Hindi Strings to Indian Sign Language, International Conference on Natural Language Processing (ICON), Hyderabad, India, 2007.
7. Pham Thi Coi, The process of language formation of the deaf children in Vietnam, PhD thesis, Institute of Linguistics, 1988 (in Vietnamese).
8. Vuong Hong Tam, Study the sign language of the deaf Vietnamese, Project report, Institute of Education Science of Vietnam, 2009 (in Vietnamese).
9. Papineni K., Roukos S., Ward T., Zhu Z-J, BLEU: A method for Automatic Evaluation of Machine Translation, Proceedings of the 20th Annual Meeting of the Association for Computational Linguistics (ACL), Philadelphia, pp 311–318, 2001.
10. Hovy E.H.: Toward finely differentiated evaluation metrics for machine translation. Proceedings of the Eagles Workshop on Standards and Evaluation, Pisa, Italy, 1999.
11. NIST report: Automatic evaluation of machine translation quality using N-gram co-occurrene statistics, 2002.

Vehicle Classification in Nighttime Using Headlights Trajectories Matching

Tuan-Anh Vu, Long Hoang Pham, Tu Kha Huynh, and Synh Viet-Uyen Ha(✉)

School of Computer Science and Engineering, International University, Vietnam National University HCMC, Block 6, Linh Trung, Thu Duc District, Ho Chi Minh City, Vietnam

tuananh05253@gmail.com, phlong@hcmiu.edu.vn, hktu@hcmiu.edu.vn, hvusynh@hcmiu.edu.vn

Abstract. Vehicle detection and classification is an essential application in traffic surveillance system (TSS). Recent studies have solely focused on vehicle detection in the daytime scenes. However, recognizing moving vehicle at nighttime is more challenging because of either poor (lack of street lights) or bright illuminations (vehicle headlight reflection on the road). These problems hinder the ability to identify vehicle's shapes, sizes, or textures which are mainly used in daytime surveillance. Hence, vehicles' headlights are the only visible features. However, the tracking and pairing of vehicle's headlights have its own challenge because of chaotic traffic of motorbikes. Adding to this is various types of vehicles travel on the same road which falsifies the pairing results. So, this research proposes an algorithm for vehicle detection and classification at nighttime surveillance scenes which consists of headlight segmentation, headlight detection, headlight tracking and pairing, and vehicle classification (two-wheeled and four-wheeled vehicles). The novelty of our work is that headlights are validated and paired using trajectory tracing technique. The evaluation results are promising for a detection rate of 81.19% in nighttime scenes.

Keywords: Traffic surveillance system · Headlights pairing
Headlights tracking · Vehicle classification

1 Introduction

The past decade has seen increasingly rapid advances in the field of computer vision which in turn has led to a renewed interest in traffic surveillance systems (TSS) [1]. Vision-based traffic monitoring systems have the capability to provide fast and reliable information that is necessary for a wide range of applications such as traffic management and congestion mitigation. The main objective is to detect interesting objects (moving vehicles, people, and so on). Other targetsinclude classifying objects based on their features and appearance

V. Bhateja et al. (eds.), *Information Systems Design and Intelligent Applications*, Advances in Intelligent Systems and Computing 672,
https://doi.org/10.1007/978-981-10-7512-4_66

(shape, color, texture, and area), counting and tracking vehicles (trajectory, motion), assessing the traffic situation (congestion, accident). While later processes are dependent on specific application requirements, the initial step of object detection must be robust and application independent [2].

Detecting moving vehicle is one of the most important applications in TSSs. Generally, there are two main types of application: daytime detection and nighttime detection. In the daytime, there are lots of features to detect vehicle such as shape, edges, bounding, or shadows of the vehicle. However, nighttime detection is more challenging because of low illuminations conditions which result in missing of vehicle feature such as shape, size, color, texture. A considerable amount of techniques has been developed for vehicle detection and classification in nighttime based on headlight and taillight of the vehicle. Chen et al. [3], Zhou et al. [4], Hajimolahoseini et al. [5] proposed many methods for tracking and detecting moving vehicle in nighttime and archived positive result with four-wheeled vehicles. These methods rely on headlight matching and pairing to detect and count vehicles. Other different methods have been proposed to improve the accuracy of taillights detection such as using support vector machine (SVM) classifier to train the shape descriptors of the taillight in [6,7]. Another aspect is O'Malley et al. [8], and Rubio et al. [9], both of them use the Kalman filtering method to track the location of the taillights and tracking the taillight spots. However, these techniques cease to work or have unreliable results in the chaotic traffic of two-wheel vehicles. Especially, in the developing countries, such as Vietnam, motorbikes usually travel in chaotic fashion such as side by side to each other, crossover, sudden lane cutting, or lane splitting. Adding to that the density of two-wheel vehicles is extremely high in the urban areas.

From the literature review, we propose an algorithm to detect and classify vehicle in nighttime based on observations on real-world data. The novelty of our work is that headlights are first validated and then paired using trajectory tracing approach. Our algorithm consists of four steps. First, bright objects are segmented using the luminance and color variation. Then, the candidate headlights are detected and validated through the characteristics of headlight such as area, centroid, rims, and shape. In the next step, we track and pair headlights by calculating the area ratio and spatial information on common vertical and horizontal of the headlight. Finally, vehicles are classified into two groups: two-wheeled and four-wheeled. Experiments have shown an effective nighttime vehicle detection and tracking system for identifying and classifying moving vehicles for traffic surveillance.

The paper is organized as follows: Sect. 2 introduces the proposed method to detect and classify vehicle via headlights trajectories matching. Section 3 presents the experimental results and discussion to illustrate the application and usefulness of the proposed algorithm. Section 4 concludes this study with a discussion.

2 The Proposed Method

In this session, we introduce our proposed method including four main steps, i.e., bright object segmentation, headlight detection, headlight tracking and pairing, and classification.

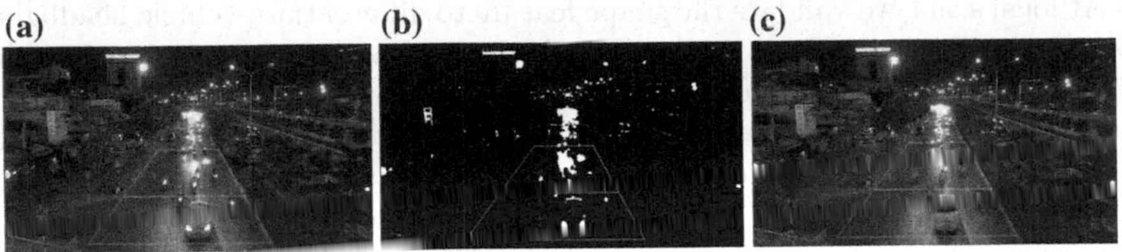

Fig. 1. Example of bright object segmentation at frame 190 on dataset DBP02. **a** Input frame. **b** Bright object segmentation. **c** Headlight detection

2.1 Bright Object Segmentation

The first step in nighttime vehicle detection and classification is to segment bright objects from the scenes. A bright object segmentation without losing bright object can reduce wrong classification of the vehicle. In nighttime traffic, the vehicles headlights and the road lights are the brightest objects and are detected at the top, left, and right side of a scene in both urban and highway environments. The vehicle lights consist of two parts: headlights of the vehicle and their reflection on the road. Therefore, we use color feature Eq. (1) to separate them from the background. However, the reflection of vehicle light on the road causes some missing count in two circumstances. The first situation is decrease counting of the vehicle when the overlapping of light reflection and other vehicle inside reflection's area. And the other is increase counting of the vehicle when the light reflection is considered as a light.

The example is shown in Fig. 1a, b. In this study, the bright object is segmented by using the following formula:

$$B = \begin{cases} 255 & if\, L > T_L\, and\, C > T_C \\ 0 & otherwise \end{cases} \quad (1)$$

where B is the segmentation result of bright objects, L is the luminance of headlight in RGB color space, C is the color variation of headlight in grayscale, $T_L = 190$, and $T_C = 20$ are the threshold of luminance and color variation which are selected when testing on the training datasets.

2.2 Headlight Detection

To detect headlights of vehicles, a validation process on bright objects is performed to specify the bright object as a headlight. We propose a global and local validation approach.

At global scale, the observation zone technique proposed by Ha et al. [10] is used to remove the other light sources at the top, left, and right side of a scene to reduce computational processing time. The observation zone is the area that the pieces of information or features of a vehicle are stable, and it does not consist of other interferential objects such as street light, traffic light, or other light sources. An example of observation zone is shown in Fig. 1a.

At local scale, we validate the shape feature to differentiate vehicle headlights from road reflections. First, bright object's centroid and rims are determined by using contour extraction and image moment [11]. Then, the bright object's centroid and rims are verified whether they are inside observation zone. Afterward, the bounding rectangle area (A) and the roundness (R_n) of each bright object can be computed as Eq. (2). The most important part of this session is the roundness of each bright object. In the past, the headlight of vehicle is almost circular but now, it has different shapes such as triangle, rounded rectangle, as shown in Fig. 2a–c. Therefore, a specific formula and parameter for the roundness are defined to solve these problems.

$$R_n = T_p * \sqrt{\frac{\sum_{i=1}^{N} (D_i - M_D)^2}{N}} \tag{2}$$

where D_i is the distance between bright object's rims to centroid, M_D is the mean distance of each bright object, N is the total rims of each bright object, and T_p is the scale parameter for the roundness of bright objects.

Then, the bright objects are determined as headlight candidates by using Eq. (3). The example of headlight detection is shown in Fig. 1c.

$$HL_c = \begin{cases} 255 & if\ T_{AL} < A < T_{AU}\ and\ R_n < T_{Rn} \\ 0 & otherwise \end{cases} \tag{3}$$

where HL_c indicates the corresponding headlight candidate result, T_{AL} and T_{AU} are the lower and upper threshold for bounding rectangle area, and T_{Rn} is the threshold for roundness of bright object.

2.3 Headlight Tracking and Pairing

The main step of the proposed method is to track and pair headlight of vehicle. Therefore, the headlight candidates are divided into four stages: entering, validating, pairing, and exiting to keep track headlight's status. The algorithm used to track and pair headlight is shown in Algorithm 1.

First, if two headlights satisfy the tracking conditions in Algorithm 1, they are segmented from other non-vehicle lights such as traffics light, reflections, and banner lights. These conditions are checked by look back up to three frames to handle missing objects. The horizontal distance between two headlights is the absolute difference between the horizontal coordinate of headlight's centroid.

$$D_x = | C_{x2} - C_{x1} | \tag{4}$$

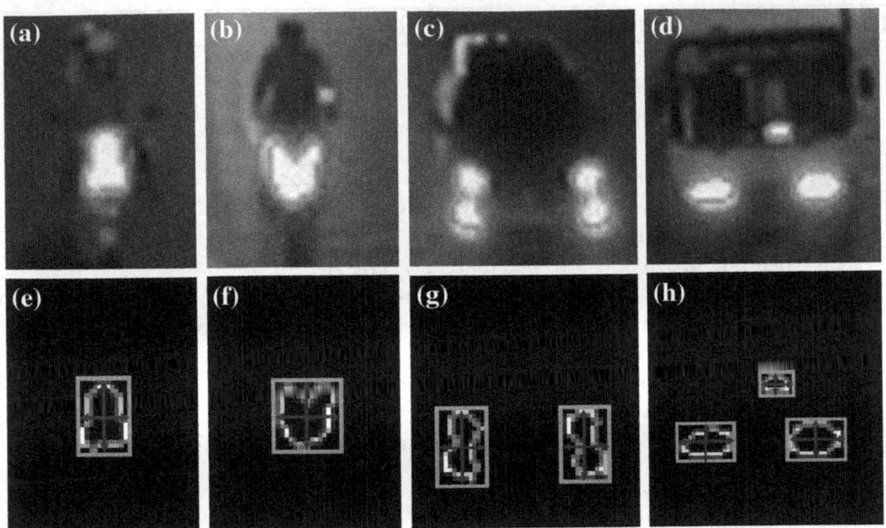

Fig. 2. Example of different headlights shapes and features. **a–d** Different headlights shapes of vehicles. **e–h** Different headlights feature of vehicles

where D_x is the horizontal distance of two headlights, and C_{x1} and C_{x2} are the horizontal coordinate of each headlight's centroid. The vertical distance of two headlights is the absolute difference between two vertical coordinates of headlight's centroid.

$$D_y = | C_{y2} - C_{y1} | \tag{5}$$

where D_y is the vertical distance of two headlights, and C_{y1} and C_{y2} are the vertical coordinate of each headlight's centroid. The area ratio is the ratio between min and max area of two headlights.

$$A_r = \frac{\min(A_{HL1} , A_{HL2})}{\max(A_{HL1} , A_{HL2})} \tag{6}$$

where A_r is the ratio of two headlight's area, and A_{HL1} and A_{HL2} are the bounding rectangle area of headlight 1 and 2.

Then, after having trajectories of each headlight, the headlight candidates are checked and determined whether they can be matched and paired. The slope of connecting line of two headlights is the ratio between horizontal distance and vertical distance.

$$S_s = \frac{D_y}{D_x} \tag{7}$$

where S_s is the slope between two headlight's distance, and D_x and D_y are the horizontal distance and vertical distance of two headlights. The speed difference is the absolute difference between speed of headlight 1 and 2 to determine they are one or two vehicles.

$$S_d = | S_{HL2} - S_{HL1} | \tag{8}$$

Algorithm 1 Tracking and pairing headlight

```
Input
HL_c: the set of headlight candidates.
HL_o: the set of headlight.
Output
M_HL: the mask consists of paired headlight's trajectories
procedure TrackingAndPairingHeadlight
    for each HL_o and HL_c do
        use Eq. 4,5,6 and 9 to calculate D_x, D_y, A_r, S_HL;
        if D_x and D_y and A_r in allowed threshold then            ▷ Track headlight
            if A_f greater than 1 then                 ▷ A_f is appearance frequency
                set speed equal S_HL;
                if A_f in range from 1 to 10 then
                    status = Validating;
                else if A_f greater than 10 and status equal Validating then
                    set status equal Pairing;
                end if
            else
                set status equal Entering;
                set speed to initial speed;
            end if
            Draw bounding rectangle and trajectories of each headlight.
            use Eq. 7,8 and 10 to calculate S_s, S_d, D_Eu;
            if S_d and D_Eu and S_s in allowed threshold then        ▷ Pair headlight
                - Determine No light zone and eliminate redundancy headlights.
                - Re-calculate speed using Eq. 9
                - Compare speed and status, then marked as a pair.
            end if
        end if
    end for
end procedure
```

$$S_{HL} = \sqrt{(C_{x1} - C_{x2})^2 + (C_{y1} - C_{y2})^2} * P_s \tag{9}$$

where S_d is the speed difference between two headlight, C^i_{x1}, C^i_{x2} and C^i_{y1}, C^i_{y2} are horizontal and vertical coordinate of ith headlight's centroid, and P_s is scale parameter for calculating speed. The Euclid distance is popular technique to perform distance analysis.

$$D_{Eu} = \sqrt{D_x^2 + D_y^2} \tag{10}$$

In the Algorithm 1, we define the "No light zone" as an area that embraces around vehicle, including the vehicle body, headlights, and region around headlights. In the "No light zone," there is only one pair of headlight (in case of a car) or one headlight (in case of a motorbike) is inside, and the other headlights are eliminated. The example is shown in the Fig. 3b, where a car consists of three headlights, but only two main headlights are paired.

2.4 Vehicle Classification

The last step in proposed method is vehicle classification. In our algorithm, vehicles are classified into two class: two-wheeled and four-wheeled vehicles. We would like to work with simple features such as measurement-based features suggested in Eqs. 4, 5, 6, 7, 8, and 10 because of their lower computational cost and storage requirements. To maintain an observation feature database for each detection instance of a vehicle.

A set of features are constructed based on the combination of features mentioned on Sect. 2.3 and the features proposed by Pham et al. [12]. Based on experiments, a set of optimal features are selected to present information of headlights. A decision tree theory proposed by [10] is constructed to classify vehicle by partitioning vehicle data into smaller part. The decision tree for dataset DBP02 is shown in Fig. 4.

In the decision tree, the green node represents four-wheeled vehicles (car), and the blue node represents two-wheeled vehicles (motorbike). In each node, the top value is the type name of the vehicle, the middle value is the proportion of vehicle type, and the bottom value is the percentage of vehicle type in the top value.

3 Experiments and Discussion

To evaluate the system, all experiments will be performed on traffic datasets captured in Ho Chi Minh City, Vietnam. The capture rate is 30 frames per second (fps) with the resolution of 640480. The system has been developed and tested on a computer comprising of Intel Core i7 6700 HQ and 8 GB of RAM. Experiments have been conducted on datasets: DBP02 and VVK01 (mixed lane for both two-wheeled and four-wheeled vehicles) to measure the effectiveness of the proposed method under different traffic conditions.

First, we will show our subjective evaluation on the datasets DBP02 and VVK01. In the Fig. 3a, our method can detect, track, and separate multiple motorbikes traveling in near same position but our method does not pair them as a four-wheeled vehicle. In the Fig. 3b, the motorbike's headlight on the left and the third headlight of bus lie nearly on the vertical axis and can become a four-wheeled vehicle. However, by using "No light zone" of our method, we can eliminate the third headlight of the bus, and we can isolate it as noises. In the Fig. 3c, our method also detect and track sudden lane changing of both two-wheeled and four-wheeled vehicles. In the Fig. 3e, our method can detect and track multiple two-wheeled vehicles running parallel. On the two Fig. 3d and f, our method also detect and track multiple vehicles running parallel.

Then, the objective evaluation of our method is done by calculating the accuracy, recall, and total accuracy that proposed in [14–16]. These datasets represent ideal traffic environments in developing countries as well as tropical regions. The table consists of the actual number of vehicles, the vehicle counted by the system, true positive (TP), true negative (TN), false positive (FP), false

Fig. 3. Experiment results on dataset DBP02 and VVK01. **a–c** Tracking of multiple vehicles on dataset DBP02. **d–f** Tracking of multiple vehicles on dataset VVK01

Table 1. Classification and counting results of our proposed method

			Our method							
Dataset	Class	Actual	Count	TP	TN	FP	FN	Acc%	Recall%	T.Acc%
DBP02	2W	834	786	731	49	55	103	**83.16**	**87.65**	**79.46**
	4W	23	27	21	4	6	2	**75.75**	**91.3**	
VVK01	2W	466	429	420	17	9	46	**88.82**	**90.137**	**82.91**
	4W	17	19	15	3	4	2	**75**	**88.24**	
Overall Accuracy%			**81.19**							

Note that: TP = true positive; TN = true negative; FP = false positive; FN = false negative; Acc = accuracy; T.Acc = total accuracy; Recall = Sensitivity.

negative (FN). Table 1 summarizes the results of our proposed method in both datasets DBP02 and VVK01.

As shown in Table 1, the accuracy, and recall of our method is not high because of the extremely high traffic density of developing countries in the urban area and the chaotic movement of the transportation such as: cross-over, sudden lane cutting, or lane splitting. Moreover, the illustration of headlight reflection on the road that embraces other vehicles causes some false detection and miss detection.

Finally, we tested the performance of proposed method regarding the processing time (in fps). Because all video sequences have been captured and stored on the testing PC so that we can neglect the delays introduced by video streaming over the network. Table 2 shows that the new system can achieve an average of 31.91 fps in processing speed. These results demonstrate the efficiency of our algorithm; it is easy to conclude that the system has real-time processing capability.

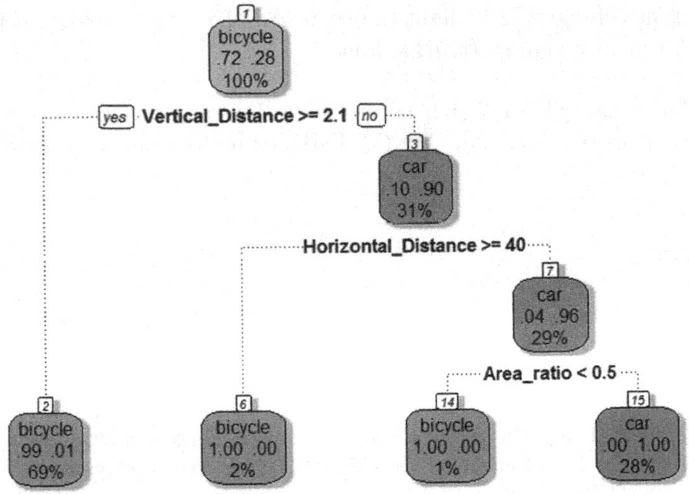

Fig. 4. The decision tree for vehicle classifying process on dataset DBP02

Table 2. Average computing time of proposed method

Dataset	Total frames	Frames s (FPS)
DBP02	1000	31.89
VVK01	1000	31.92
Average		**31.91**

In our experiment, we also figure out that illustration of headlight reflection on the road causes some false detection and miss detection. Therefore, more complicated techniques and further study are required to overcome this problem. The result is shown in Fig. 3 on sequences DBP02 and VVK01.

4 Conclusion

In this paper, we have presented the results of the current work-in-process of the nighttime vehicle detection and classification. First, the bright objects are segmented in nighttime traffic scenes, by using the luminance and color variation of bright object. Then, the candidate headlights are detected and validated through the characteristics of headlight such as area, centroid, rims, and shape. Afterward, we present a way to tracking and pairing headlight by calculating area ratio and vertical and horizontal distance of headlight. Finally, the vehicle is classified into two groups: two-wheeled and four-wheeled vehicles. The results of the experiments confirm that this algorithm can effectively solve the problems related to light illumination; chaos of traffic and parallel running motorbike is very similar to cars. Based on this framework, future studies can extend to handle different vehicle such as light (motorbikes, bikes, tricycles), medium (cars, sedans, SUV), heavy vehicle (trucks, buses).

Acknowledgement. This research is funded by International University, Vietnam National University Ho Chi Minh City (VNU-HCM) under grant number SV2016-IT-02.

References

1. S. V.-U. Ha, H.-H. Nguyen, H. M. Tran, and P. Ho-Thanh: Improved optical flow estimation in wrong way vehicle detection. Journal of Information Assurance and Security, vol. 9, no. 5, pp. 165–169 (2014).
2. B. Tian and B. T. Morris and M. Tang and Y. Liu and Y. Yao and C. Gou and D. Shen and S. Tang: Hierarchical and Networked Vehicle Surveillance in ITS: A Survey. in IEEE Transactions on Intelligent Transportation Systems, vol. 18, no. 1, pp. 25–48 (2017).
3. Y. L. Chen, B. F. Wu, H. Y. Huang and C. J. Fan: A Real-Time Vision System for Nighttime Vehicle Detection and Traffic Surveillance. IEEE Transactions on Industrial Electronics, vol. 58, no. 5, pp. 2030–2044 (2011).
4. Z. S. Shifu Zhou, Jianxiong Li and L. Ying: A night time application for a real-time vehicle detection algorithm based on computer vision. Research Journal of Applied Sciences, Engineering and Technology (2013).
5. H. S.-Z. Habib Hajimolahoseini, Rassoul Amirfattahi: Robust vehicle tracking algorithm for nighttime videos captured by fixed cameras in highly reflective environments. IET Computer Vision (2014).
6. J. Rebut, B. Bradai, J. Moizard and A. Charpentier: A monocular vision based advanced lighting automation system for driving assistance. IEEE International Symposium on Industrial Electronics, Seoul, pp. 311–316 (2009).
7. P. F. Alcantarilla, L. M. Bergasa, P. Jimnez, I. Parra, D. F. Llorca, M. A. Sotelo, and S. S. Mayoral: Automatic LightBeam Controller for driver assistance. Journal of Machine Vision and Applications, vol. 22, no. 5, pp. 819–835 (2011).
8. R. O'malley, M. Glavin and E. Jones: Vision-based detection and tracking of vehicles to the rear with perspective correction in low-light conditions. IET Intelligent Transport Systems, vol. 5, no. 1, pp. 1–10 (2011).
9. J. C. Rubio, J. Serrat, A. M. Lopez and D. Ponsa: Multiple-Target Tracking for Intelligent Headlights Control. IEEE Transactions on Intelligent Transportation Systems, vol. 13, no. 2, pp. 594–605 (2012).
10. S. V.-U. Ha, L. H. Pham, H. M. Tran, and P. Ho-Thanh: Improved vehicles detection and classification algorithm for traffic surveillance system. Journal of Information Assurance and Security, vol. 9, no. 5, pp. 268–277 (2014).
11. Wei-zhi Wang and Bing-han Liu: The vehicle edge detection based on homomorphism filtering and fuzzy enhancement in night-time environments. IEEE International Conference on Intelligent Computing and Intelligent Systems, Xiamen, pp. 714–718 (2010).
12. L. H. Pham, T. T. Duong, H. M. Tran and S. V. U. Ha: Vision-based approach for urban vehicle detection and classification. Third World Congress on Information and Communication Technologies (WICT 2013), Hanoi, pp. 305–310 (2013).
13. T. M. Michell: Machine Leaning. McGraw Hill, Ch. 3, pp. 52–80 (1997).

14. M. Sokolova and G. Lapalme: A systematic analysis of performance measures for classification tasks. Information Processing and Management, vol. 45, no. 4, pp. 427–437 (2009).
15. Tom Fawcett: An introduction to ROC analysis. Pattern Recognition Letters, vol. 27, no. 8, pp. 861–874 (2006).
16. D. M. W. Powers: Evaluation: from precision, recall and f-measure to roc., informedness, markedness and correlation. Journal of Machine Learning Technologies, vol. 2, pp.37–63 (2011).

A Scaled Conjugate Gradient Backpropagation Algorithm for Keyword Extraction

Ankit Aich(✉), Amit Dutta, and Aruna Chakraborty

St. Thomas College of Engineering and Technology, Kolkata, India
{ankitaich09, to.dutta, aruna.stcet}@gmail.com

Abstract. In modern days, it is highly important that one can get the defining content from any desirable source. When it comes to excessively large documents, it becomes an issue to effectively get the most important parts of it. Every document's main topic can be conveyed using a few defining words. This paper provides a novel approach to extract such words from a given document corpus. Domain-specific keyword extraction is the principle highlight of our work. A series of documents from a specific domain is provided to us as the working set, and identification of the top three to five words will be done to convey the documental message. Our experiments show an accuracy of 80.6%.

Keywords: tf-idf · Stemming · Scaled conjugate gradient backpropagation
Conjugate directions

1 Introduction

With the rapid advent of new websites and ever-growing data, it is imperative that one can extract the main parts of a document without the verbiage. Every document has certain words which can convey the whole message of the document. These words do not occur at a high frequency across the document corpus and maybe as less as three to five words to explain exactly the primary content of the document. These are known as *keywords*.

Keyword extraction finds use in a lot of areas from text mining to search engine optimization, natural language processing, comparison of documents, and data compression.

Fei Liu et al. in 2011 [1] have used a supervised framework to extract the keywords from meeting transcripts, where the use of a supervised learning framework to get keywords from text has been employed. They have primarily used feature forward, backward selections, and dynamic programming to get outputs. This method was improved by Anette Hulth in [2] who has classified words on the basis of n-grams and divided into exact parts of speech to get required words. This method, however, needs more linguistic knowledge. Given that the previous paper was devoid of the required linguistic knowledge, Maryam Habibie et al. in [3] have used clustering methods by making queries and using diverse reward functions. This approach mainly uses a greedy algorithm and focuses more on optimization and usually showed results on document recommendations in conversions. Work in the domain saw a slight setback when Yutaka Matsuo et al. in [4] had only single documents to use with almost no scope for extraction

V. Bhateja et al. (eds.), *Information Systems Design and Intelligent Applications*, Advances in Intelligent Systems and Computing 672,
https://doi.org/10.1007/978-981-10-7512-4_67

of keywords from multiple documents across one or multiple domains, a problem that our approach will bridge. The future scope clearly mentions that they can only hope to extend the work in future across many documents someday and relies heavily on word co-occurrence and statistical approaches. With the development of more frameworks and APPIs Claudia Marinica et al. in [5], the ARIPSO framework has been used and the paper is more inclined towards composite data mining in general. The paper also successfully points out the problems in the other papers in this field. They have used the Nantes Habitat database to conduct experiments and have used various ontologies. With frameworks still existent and the advent of tf-idf indexing and weight calculation, Menaka S. et al. in [6] use tf-idf and WordNet a process where words are classified like a dictionary. They employ Naïve Bayes and KNN methods to get results. The advantage of our process is that the scaled method does not use a line-searching technique and thus works faster. Finally, Feifan Liu et al. in [7] have used parts of speech and word clustering along with sentence salience scores to get results. They used the above-mentioned techniques to filter the text further and get results.

In this paper, we have employed the use of a scaled conjugate gradient backpropagation neural network, instead of calculating probabilities and using a Naïve Bayes as is common practice. The novelty of the paper lies in using not only a neural net to filter the results but also the fact that the scaled conjugate gradient backpropagation neural network avoids line search and thus improves search time and works faster.

For domain-specific extraction, we approached the document corpus with traditional methods to generate an output and then used a neural network with scaled conjugate backpropagation algorithm to remove more of the non-keywords from that list. We employed the use of certified documents from different domains to test, whether the keywords mentioned in those documents and our outputs yield the same results or not. We obtained accuracy, the measure of which shall be discussed further in the experimental results section.

In Sect. 2, we discussed the architecture of the neural network which is employed. In Sect. 2.1, we further extend our discussions to the mathematical model of the neural net. Section 3 describes the proposed method in intricate detail, and Sect. 4 tabulates the experimental results and mentions the accuracies. Section 5 ends the paper with conclusions and talks on further expanding and future scope of the paper.

2 Architecture of the Scaled Conjugate Gradient Backpropagation Neural Network (SCG-BPNN)

In this paper, we have employed the use of a scaled conjugate gradient backpropagation neural network. The scaled conjugate gradient backpropagation algorithm is based on conjugate directions, though it does not perform a line search at each iteration [1].

Training an SCG network stops when either of the following occurs [2].

- 100 epochs are reached.
- Performance is minimized to 0.

- Performance gradient falls below 1×10^{-6}.
- Validation performance has increased more than 5 times since it has last decreased.

In general, a backpropagation algorithm adjusts the weights in the steepest descent direction (negative of the gradient). This is the direction in which the performance function is decreasing most rapidly. It turns out that, although the function decreases fastest along the negative of the gradient, this does not necessarily produce the fastest convergence.

In the conjugate gradient algorithm, a search is performed along conjugate directions, which generally provide the user with faster convergence than steepest descent directions.

There are primarily four different conjugate gradient algorithms,

- Fletcher-Reeves
- Polak-Ribiere
- Powell-Beale Restarts
- Scaled Conjugate Gradient

Among these myriad options, we have used the last one which is scaled conjugate gradient. This was developed by Moller to avoid the time-consuming linear search of conjugate and optimal direction that occurs in the previous three cases. This algorithm may initially pose to be a bit complex but the basic idea is to combine two approaches, viz. the model-trust region approach (used in the Levenberg–Marquardt algorithm) with the conjugate gradient approach. This process trains the network as long as its weight, net input, and transfer functions are all ones with very valid derivative functions. Backpropagation is used to calculate derivatives of performance with respect to the weight and the bias variables. Figure 1 shows a block diagram representing the stopping criteria and general working of the network.

This process is faster and takes into account the results of the program that is the tf-idf scores, instead of us having to calculate the probabilities of the words and using a standard approach like a Naïve Bayes classifier or a support vector machine to implement the function of the neural network.

2.1 Mathematical Model of SCG-BPNN

The scaled conjugate gradient method uses a fast strategy and chooses the search direction and the step size more carefully by using information from the second order approximation [8].

$$\mathrm{E}\ (\mathrm{w}\ +\ \mathrm{y}) \approx \mathrm{E}(\mathrm{w})\ +\ \mathrm{E}'(\mathrm{w})\mathrm{Ty} + 1/2\mathrm{yTE}(\mathrm{w})\mathrm{y} \tag{1}$$

We try to show the quadratic approximation to E in a neighborhood of a point w by $\mathrm{Eqw}(\mathrm{y})$, so that $\mathrm{Eqw}(\mathrm{y})$ is given by

$$\mathrm{Eqw}(\mathrm{y}) = \mathrm{E}(\mathrm{w}) + \mathrm{E}'(\mathrm{w})\mathrm{Ty} + 1/2\mathrm{yTE}''(\mathrm{w})\mathrm{y} \tag{2}$$

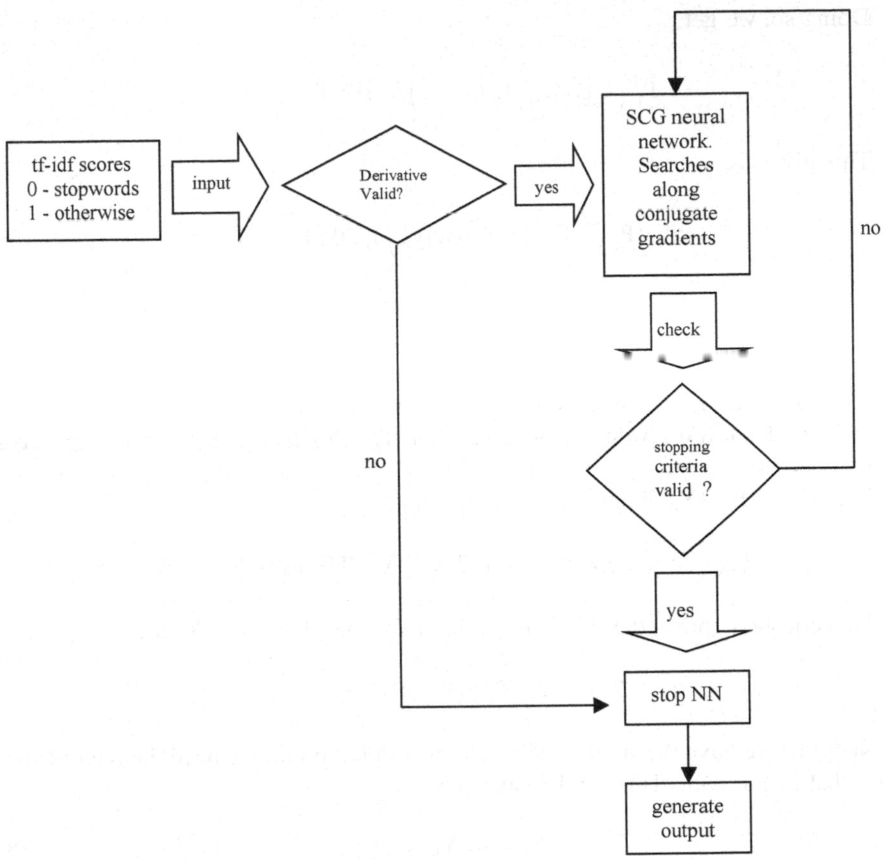

Fig. 1. Inputs, outputs, and general stopping checks of SCG-BPNN

In order to determine minima to Eqw(y), the critical points for Eqw(y) must be found, that is the points where

$$\mathrm{Eqw}(y) = E''(w)\ y + E'(w) = 0 \tag{3}$$

The critical points are the solution to the linear system defined by (3).

Let a starting point be Y_1, and $P_1 \ldots P_N$ be a conjugate system. The scaled conjugate network says that this Y_1 to Y^* can be expressed as a linear combination of the points till P_N, where Y^* is the critical point.

$$Y^* - Y1 = \sum_{i=1}^{N} \alpha ipi, \text{ where } \alpha_I \in R \tag{4}$$

Now this equation can be multiplied by $P_J^T E^N(W)$ and $E'(W)$ may be substituted for $-E^N(W)Y^*$

Doing so, we get,

$$P_j^T\left(-E'(w)-E''(w)y1\right)=\alpha jP_j^TE''(w)P_j \tag{5}$$

This gives us,

$$\alpha=(P_j^T\left(-E'(w)-E''(w)y1\right))/(jP_j^TE''(w)P_j) \tag{6}$$

Using Eqs. (4), (5), and (6), we can iteratively determine the value of the critical point which is Y*. However, we need minima to solve the equation and get the path. Y* is not always the minima of the equation, or a minimum value, it can also be more like a saddle point to the maximum one. So this value will not work.

Only if the $E^N(W)$ value is positive, then $E_{qw}(Y)$ has a distinct and universal minimum value.

This value, $E_{qw}(Y)$, is given by the equation,

$$E_{qw}(Y) = Eqw(Y^*) + 1/2(Y-Y^*)TE^N(W)(Y-Y^*) \tag{7}$$

This equation shows that if Y* is a minimum, then for every Y, the term

$$1/2(Y-Y^*)TE^N(W)(Y-Y^*) > 0$$

Now that we have the starting point and the critical point, we need the intermediate point, that is, the path. This can be calculated by

$$Y_{k+1}=Y_{k+\alpha kpk} \tag{8}$$

The values obtained from this equation have all minima for $E_{qw}(Y)$.

So, following the above process, the scaled conjugate gradient neural network successfully avoids the time-consuming line-searching techniques that are used in most networks. In the three other versions mentioned above, all three use a proper way to bypass a line search by using a search algorithm of their own. Out of the documents we had, we used 70% documents to train the network, 15% to validate the network, and rest 15% to test the network.

3 Proposed Method

The first step to working on a document to extract the required keywords is to preprocess the corpus. This helps getting rid of futile and unimportant words from single document.

To preprocess, we follow the following steps.

1. *Remove punctuations*—We first make the document devoid of punctuations and other symbols we do not need. To do this, we used Python 3.4's defined list of punctuations and removed anything and everything from the document which

belonged to this list. After doing so, we are left with the document corpus free of punctuations.

2. *Remove digits*—From the new corpus, we have to remove the digits. Again to do so, we scanned the document and removed all digits.
3. *Removing stopwords*—Every document corpus or text has a few words which are very common. For instance words like, '*this*', '*however*', '*is*'. In general, we classify auxiliary verbs, prepositions, articles, and demonstrative adjectives like "this" and "that" as stopwords. To do this, we made another list of words which are pre-defined as stopwords in Python 3.4's Natural Language Package [9]. Once this is done, we are left with the document corpus free of any verbiage. Since no text was deleted, we have acronyms and proper nouns preserved. So next we need to stem the documents.
4. *Stemming* to get unique candidate—Every word comes from a root. Stemming is the process of getting the root word from a word. According to *Turney* (2000), it has been found that an aggressive approach to stemming is more suitable for the purpose of keyword extraction compared to a non-aggressive one [10]. Out of the many stemming algorithms at our disposal, we have employed the use of the porter2 [11] algorithm for stemming, though other ones may be used too. To exemplify what we mean by stemming, we have an example in Table 1.

Table 1. Stemmed words on the right and original words on the left

Word	Stemmed word
Consignment	Consign
Consigned	Consign
Consigning	Consign
Consign	Consign

In the left-hand column, we have the words, and in the right-hand column, we show the stemmed versions. We can see that it may so happen that multiple candidates may have the same word root, and this may create ambiguity in our output, which is why we need to get rid of the duplicate roots and free the corpus to avoid ambiguous output. The process is shown in Fig. 2.

Once this is done, we have the document preprocessed and we can proceed.

The next step involves using another method followed by the application of the neural network.

As our working set of documents, we have multiple documents belonging to a single domain which will be provided by a user in any way they want. To proceed, we have to get some ideas clear.

1. *Term Frequency* (*tf*)—The number of time a word occurs in a document is called its term frequency. If it is a keyword, then the tf will be high, and if it is not a keyword

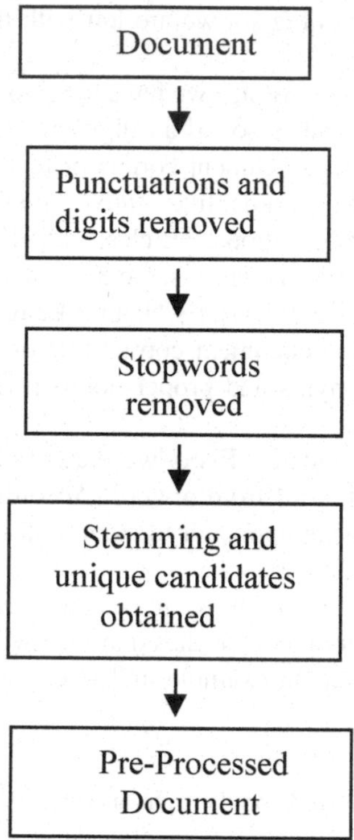

Fig. 2. Steps in preprocessing the document

then it will be low. So we find tf by considering each word and finding how many times it appears in the document.

2. *Inverse Document Frequency (idf)*—The number of times a word occurs in all documents of the domain is called the document frequency. Inverse of this is called the inverse document frequency. Now it is obvious that keywords are unique in nature, so if a word occurs many times in multiple documents then it is less likely to be a keyword. So we need the idf score. Because more the idf more the chances of it being a keyword.
3. *TF_IDF*—The product of tf-idf scores. Since both tf and idf vary directly with keywords, so higher the tf-idf more the chance of it being a keyword.

So we take all tf-idf scores of all the words in a document across documents in the domain and choose the top 3–5 (based on the user) to determine the keywords. However, it is seen that still a few words remain which are clearly not keywords.

The details will be discussed in the results section.

Using the Neural network—With the above method, it is seen that in the final output, a lot of stopwords are encountered. This cannot be filtered by simple algorithms and needs to be classified by an intelligent system. To further filter this, we use the *neural network using scaled conjugate backpropagation algorithm*.

As our input set, we choose the tf-idf scores, and for our outputs, we assigned 0 to the stopwords and 1 otherwise. The neural net was trained to classify keywords based on this data set. After training and classification, the neural net removed most stop-words and left us with what were seemingly the keywords.

In the next section, that is the experimental results section, we shall see the accuracy of our system across two domains. We shall also see what words were given as keywords and what we got. To explain what we did in the process, we have shown Fig. 3, a block diagram to exemplify.

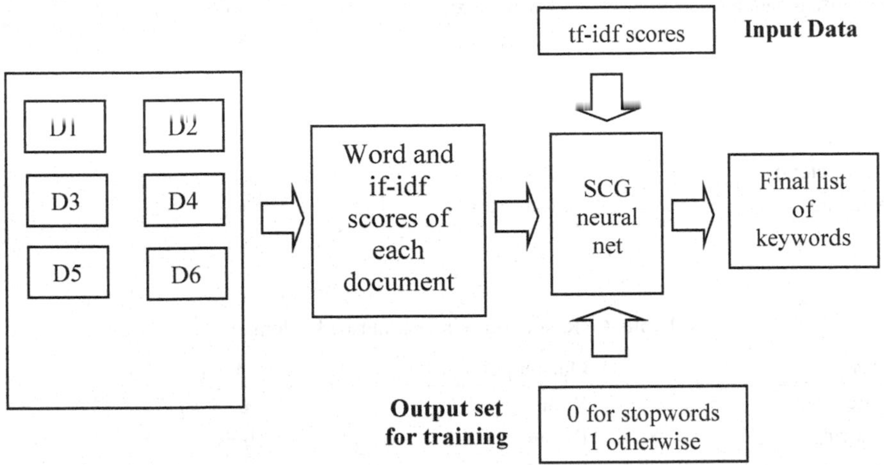

Fig. 3. Six documents D1—D6 in a common domain. We use the above method and get the words. The method is displayed pictorially here

4 Experimental Results

In the above process, we tested across a lot of documents from various sources. However, we used some published conference papers to test our work and see whether our results tally. The following are the results obtained from our experiments. The columns on the right side show the keywords as defined by the document.

The middle shows the keywords that our process gave as output and at the end of each table, we have the accuracy measure.

These documents were taken from conference papers which have been mentioned in references. The keywords on the left are the words that the paper mentioned to be keywords. We have chosen two domains to work on. They are cloud computing and wireless networks. The first table shows the cloud computing results, and the next table shows the wireless network domain (Table 2).

The accuracy was calculated to be 79.2%. This was just for the first domain. To test further, we took the help of another set of documents from another domain as shown in the table below. We have used wireless networks here, and all the words from the papers which were keywords, and words from our output has been shown in the table. In this case, the accuracy obtained was slightly higher than that of cloud computing domain. The average accuracy has been calculated and shown in Table 3.

Table 2. Shows the accuracy readings from papers in cloud computing domain

Paper	Our output	Match Y/N
Rough	Rough	Y
Set	Set	Y
Service	Service	Y
Provider	Provider	Y
Hypervisor	X	N
Full virtualization	Virtualization	50%
Cloud stack	Cloud stack	Y
Virtualization	Virtualization	Y
Balancing	Balancing	Y
Open stack	Open stack	Y
Private cloud	X	N
Hardware	Hardware	Y

Table 3. Results in wireless networks domain

Paper	Our output	Match
Route	Route	Y
Repairing	Repair	60%
Sensor	Sensor	Y
Route	Routing	62.8%

In this domain, that is wireless networks, it is seen that the accuracy of the domain is 82%. So clearly, we have a higher accuracy here. We calculate average accuracy as the mean of these two obtained accuracies. The average accuracy is seen to be 80.6% which is quite nice. These tests may be conducted over further domains as well and probably if tested over and over with more training to the neural net we may obtain a possible higher accuracy.

Final Calculated Accuracy – 80.6%

The question arises that why we used this method. The traditional tf-idf method works fine till we take into consideration the experimental accuracy. Through tf-idf out of every (N) number of words generated on an average floor (N/3) words were stop-words. This was verified by us. Many previous works have used other classifiers on tf-idf, to filter. This is where the *SCG-BPNN* comes in as a much faster and optimized solution. To compare its benefits, we list them below.

- It enables better filtering of words generated from a document corpus.
- It increases accuracy by 27% on an average.
- It works faster than most neural nets as it avoids the line search technique.
- It is more optimized.

5 Conclusions

Keyword extraction is a big issue in the modern era of Internet. This is mostly because with thousands of new companies being formed every day, it becomes difficult to make business without getting a website on top of Google's search page. And without paying the best way to accomplish that is by search engine optimization. Keyword extraction is like the stepping stones to NLP and SEO. With proper keyword extraction, both time and money can be saved. There are millions of extended approaches to what we have done. Predicting the domain from a document, to understand human speech, everything is based on this very basic principle of keyword extraction. Our idea lies in the fact that the tf-idf scores need to be further filtered to get better results. We have done this using the SCG or scaled conjugate gradient neural network. However, other approaches have been there before and will be there too in the not-so-distant future.

The future scope of our paper can be myriad. In our paper, we have provided the domain to the computer system. This may be extended later by designing a system which can automatically detect the domain and draw out different document corpuses for testing on them collectively. The main advantage of our system is its speed and accuracy. This is due to the fact that our neural network does not waste time on line-searching techniques.

References

1. Fei Liu, Feifan Liu, and Yang Liu.: A Supervised Framework for Keyword Extraction From Meeting Transcripts, IEEE Transactions on Audio, Speech, and Language Processing, vol. 19, no. 3, pp. 538, March 2011.
2. Anette Hulth.: Improved Automatic Keyword Extraction Given More Linguistic Knowledge, ACMDL, EMNLP'03, pp. 216–223, July 2003.
3. Maryam Habibie and Andrei Propescu-Belis.: Keyword Extraction and Clustering for Document Recommendation in Conversations, IEEE/acm Transactions on Audio, Speech, and Language Processing, vol. 23, no. 4, pp. 746, April 2015.
4. Yutaka Matsuo and Mitsuru Ishizuka.: Keyword Extraction from a single document using word co-occurrence statistical information, FLAIRS-2003, Florida, US pp. 392–396, 2003.
5. Claudia Marinica ad Fabrice Guillet.: In Knowledge Based Interactive Post Mining of Association Rules using Ontologies, IEEE Transactions on Knowledge and Data Engineering, vol. 22, no. 6, pp. 784, June 2010.
6. Menaka S and Radha N.: Text Classification using Keyword Extraction Technique, International Journal of Advanced Research in Computer Science and Software Engineering, Volume 3 Issue 12, pp. 734–740, December 2013.
7. Feifan Liu, Deana Pennell, Fei Liu and Yang Liu.: Unsupervised Approaches for Automatic Keyword Extraction using Meeting Transcripts Human Language Technologies: The 2009 Annual Conference of the North American Chapter of the ACL, pp. 620–628, Boulder, Colorado, June 2009.
8. M.F. Moller.: A Scaled Conjugate Gradient Algorithm for Fast Supervised Learning, Neural Networks, Vol. 6, pp. 525–533, 1993.
9. Bird, Steven, Edward Loper and Ewan Klein.: *Natural Language Processing with Python.* O'Reilly Media Inc. 2009.

10. Turney, P.D.: Learning Algorithms for Keyphrase Extraction, Information Retrieval, Vol. 2 Issue 4, pp. 303–336, 2000.
11. Willet P.: The Porter stemming algorithm then and now, Program: Electronic Library and Information Systems, 40(3). pp. 219–223 ISN 0033-0337 2006.

Probabilistic Model and Neural Network for Scene Classification in Traffic Surveillance System

Duong Nguyen-Ngoc Tran, Long Hoang Pham, Ha Manh Tran, and Synh Viet-Uyen Ha(✉)

School of Computer Science and Engineering, International University, Vietnam National University HCMC, Block 6, Linh Trung, Thu Duc District, Ho Chi Minh City, Vietnam

[illegible]hcmiu.edu.vn

Abstract. Traffic surveillance system (TSS) has seen great progress in the last several years. Many algorithms have been developed to cope with a wide range of scenarios such as overcast, sunny weather that created shadows, rainy days that result in mirror reflection on the road, or nighttime when low lighting conditions limit the visual range. However, in real-world applications, one of the most challenging problems is the scene determination in a highly dynamic outdoor environment. As also pointed out in recent survey, there have been limited studies on a mechanism for scene recognition and adapting appropriate algorithms for that scene. Therefore, this research presents a scene recognition algorithm for all-day surveillance. The proposed method detects and classifies outdoor surveillance scenes into four common types: overcast, clear sky, rain, and nighttime. The major contributions are to help diminish hand-operated adjustment and increase the speed of responding to the change of alfresco environment in the practical system. To obtain high reliable results, we combine the histogram features on RGB color space with the probabilistic model on CIE-Lab color space and input them into a feedforward neural network. Early experiments have suggested promising results on real-world video data.

Keywords: Scene recognition · Traffic surveillance system
Probabilistic model · Artificial neural network

1 Introduction

Traffic surveillance system (TSS) has become a vital tool for many intelligent traffic systems. As progress has been made in the field of computer vision and image processing [1], TSSs have been targeting to obtain an understanding of the traffic flow through extracting information (counts, speed, vehicle type, and

V. Bhateja et al. (eds.), *Information Systems Design and Intelligent Applications*, Advances in Intelligent Systems and Computing 672,
https://doi.org/10.1007/978-981-10-7512-4_68

density) [2,3]. The knowledge obtained by these systems can provide decision supports regarding traffic managements and urban planning.

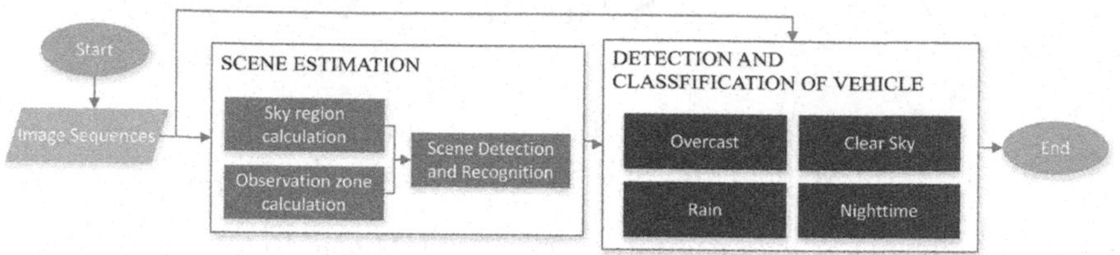

Fig. 1. Traffic surveillance system process. The scene estimation process classifies image sequences from the camera, which leads TSS to determine the appropriate method for the succeeding processes

Over the last decade, a significant amount of works has been done to make the TSSs cope with a wide range of outdoor environments. One such scenario is the vehicle detection and classification in daytime [4–6] or in the presence of shadows created by different lighting conditions [7,8]. Some works focused on detecting and pairing vehicle headlights in nighttime surveillance [9,10]. Others proposed methods to remove vehicles' reflections on the wet road in rainy conditions [11]. These algorithms have proven to work well in their specific defined scenarios. However, in real-world all-day surveillance, the outdoor scene can change drastically from overcast to sunny or rainy, or from daytime to nighttime. Moreover, in a practical system, there is a network of cameras spanning across a wide area (which are set up along the street), so the local environments of them can be different from each other. Then, the manual adjustment for each camera is not pragmatic and is time-consuming. For instance, some cameras are in system are in the overcast condition (Fig. 5-OC-2), while other cameras are in the rain condition (Fig. 5-RI-5). In addition, the small road (Fig. 5-CS-1) could be marked as the overcast, and the large one (Fig. 5-CS-3) is labeled as the clear sky while the sun is near the skyline in the early morning. There have been few studies on a mechanism to recognize the gradually changing environment, from which different algorithms can be interchanged and applied to appropriate scenes throughout all-day surveillance. This problem hinders the autonomous capability of TSSs and reduces the performance of vehicle detection and classification.

Fig. 1 illustrates the workflow of the traffic surveillance system. In addition, based on the hierarchical architecture of TSS in [1], the proposed method is in the first layer, in which the main function is to perceive traffic scenes and obtain images using visual detection. At the initial stage of this process, the proposed method receives the input image sequence from the dataset or the real-world stream. In the subsequent step, the sky region and the observation zone calculation give the several output parameters, which the determination uses to describe the current scene on the road. After that, the result of estimation would come over to the main stage, in which the TSS chooses the suitable technique for the following layers.

Fig. 2. Illustration of four typical outdoor scenes captured from one surveillance camera of our traffic surveillance system. **a** Overcast. **b** Clear sky. **c** Rain. **d** Nighttime

In this chapter, we present a scene recognition algorithm that provides TSSs with the ability to adapt different algorithms to all-day surveillance. The proposed method detects and classifies outdoor surveillance scenes into four types (Fig. 2): overcast, clear sky, rain, and nighttime. Thoroughly, the term "overcast" (OC) has come to be used to refer to the scene in which the sky is cloudy, and there is no shadow. The term "clear sky" (CS) is defined as the scene with the shadow come along with vehicles. The term "rain" (RA) is used here to refer to the wet road condition which creates the reflection of the vehicle. Moreover, the term "nighttime" (NT) indicates the evening time in which the traffic light is turned on. To accurately determine the scene, we create a scene feature model consisting of the histogram of RBG color space and the probability model of CIE-Lab color space. We then use a feedforward neural network (FFNN) to train a dataset of several all-day surveillance videos. The trained FFNN has been validated with the real-time video stream from the surveillance cameras in Ho Chi Minh City, Vietnam. Early experiments have shown promising results with the accuracy of 87.38, 90.85, 86.7, and 96.81% for overcast, clear sky, rain, and nighttime conditions, respectively.

The remainder of this paper is organized as follows. In Sect. 2, we present the detail of the proposed method, the dual sampling region we used is presented in Sect. 2.1, the probabilistic model is in Sect. 2.2, and the FFNN is in Sect. 2.3. The benchmark results to corroborate the advantages of the new method are described in Sect. 3. For more detail, we show the qualitative performance in Sect. 3.1 to prove the efficiency of the method and the computational time in

Sect. 3.2 to make sure that the TSS would run in real time including the proposed method. Finally, Sect. 4 summarizes the proposed method and concludes this paper.

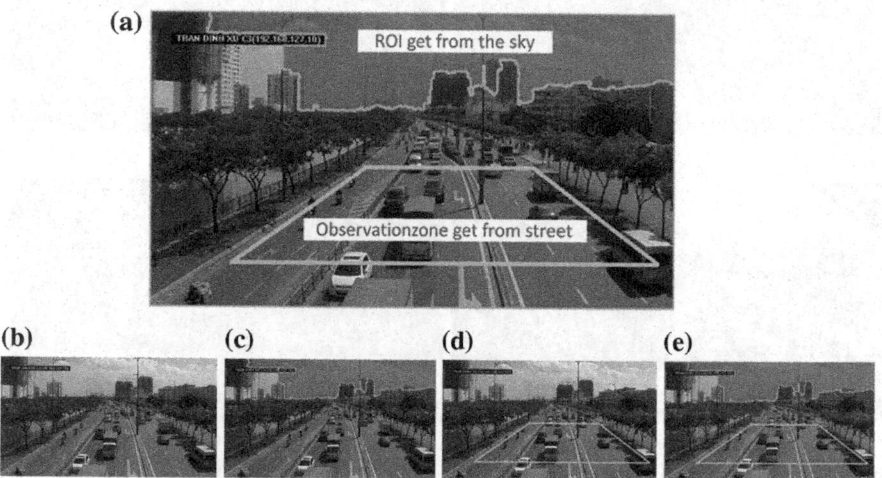

Fig. 3. Illustration of the dual sampling regions approach. **a** The detail of the dual sampling regions. The green part represents the observation zone. The pink area represents the sky region. **b** The original frame. **c** The watershed segmentation for getting the sky region. **d** The manual setting observation zone. **e** The combination of the sky region and the observation zone

2 The Proposed Method

2.1 Dual Sampling Region

In this section, we introduce our approach and its implementation. From the input frame, there are two observations in the camera field-of-view (FOV) containing the sky part and the road:

- Sky region: The color intensity of the sky is distinguishable between daytime and nighttime.
- Observation zone: The road surface has different luminosity between overcast, clear sky (cast shadows), and rainy weather (reflection).

For a dual sampling regions, as shown in Fig. 3, consisting of an observation zone and a sky region is applied. The observation zone can be obtained as suggested by [6]. It is a common practice to get the sky region using watershed segmentation in combination with the horizon line [12, 13]. The method acquires the region that covers most of the sky, because it gets whole characteristic of the sky, and avoids some small changes which would make the wrong detection. For example, the raincloud which does not cover the sun would alter the detection if the sky region is on the cloud. However, the observation zone we just covered is a part of the road, which is the area the TSS [6] used for detection. The color intensity of sky is different in the daytime and nighttime, and their histogram changes dramatically from high in the nighttime to low in the nighttime. From the daily repeat fluctuation, the TSS could accurately recognize the change in scene.

Additionally, we use the observation zone on the road to identify the condition of the road. For more detail, the shadow detection is the technique to distinguish the overcast scene from others. Furthermore, the proposed method uses the features from the observation zone and the sky region to determine the two-left scene, the rain, and the clear sky. There are many confused features between the clear sky and the rain. For example, after the rain, the sky is clear, and the road is wet, which make the reflection of the vehicle; therefore, in this case, the scene is the rain. Later, when the road is dry, the scene is the clear sky. The following Sect. 2.2 shows the technique to extract the feature from the sky region and the observation zone and the probabilistic model for the Sect. 2.3.

2.2 Scene Model

In this section, we introduce the probabilistic model (Eq. 1), from which the proposed method feeds the input for the neural network in the Sect. 2.3. Each parameter of the probabilistic model will be clearly illustrated in following paragraphs.

$$S = f\left(\sigma_L, \mu_{ab}, U_{shadow}, H_{sky}\right) \tag{1}$$

Firstly, in our approach, we extract features from the sky region obtained using the algorithm in [14] (Fig. 3c). In the sky region, we evaluate the scenes based on the histogram on the RGB color space, with H_{sky}. The histogram of the sky zone is suitable to detect the daylight and nighttime. All FOVs of the traffic surveillance cameras always have sky (Fig. 3); therefore, the features gotten from sky always are cogitated in training. In the observation zone, we apply a probabilistic model on the CIE-Lab color space. In the CIE-Lab color space, L stands for the lightness intensity, and a and b stand for the average color, from which we calculate the mean of ab value to model the change of the illumination. The shadow and the overcast have their own range value; therefore, the mean of ab is the primary value which is derived from training data as shown in Eq. 2:

$$\sigma_L = \frac{1}{K}\sum_{i=0}^{K}\left(\rho_L - \mu_L\right)^2 \tag{2}$$

where K is the number of pixels in observation zone; ρ_L is the lightness of pixel; μ_L, μ_a, and μ_b is the mean color value of all pixels; σ_L is the standard deviation of color value.

Secondly, to determine the occurrence of shadows in the current frame, the sum of μ_a and μ_b is less than or equal to the T (Eq. 3).

$$\mu_{ab} = \mu_a + \mu_b \leq T \tag{3}$$

where $T = 250$ based on practical experiments; however, in some case, the T could change, which depends on the magnitude of the two regions and the shape of two-side building.

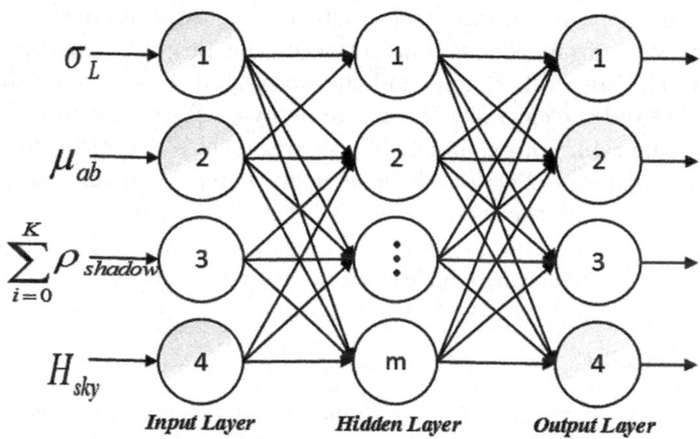

Fig. 4. A three-layer feedforward neural network

Thirdly, to get the shadow frame in binary image, the pixel which has the L color less than or equal to $\mu_L - \sigma_L/3$ (Eq. 4) is the pixel containing shadow. The sum of all pixel of the observation zone in binary image is used for feeding neural network (Eq. 5).

$$\rho_{shadow} = \begin{cases} 255, \; if \; \rho_L \leq \mu_L - \sigma_L/3 \\ 0 \end{cases} \tag{4}$$

$$U_{shadow} = \sum_{i=0}^{K} \rho_{shadow} \tag{5}$$

Lastly, for the determination of the rainy scene and the clear sky, we use combination of μ_{ab} on the observation zone and the histogram in sky region H_{sky}, and we feed both parameters for the neural network.

2.3 Feed Forward Neural Network

In the previous section, we present four parameters which affect the channel width. This section builds models to relate these parameters to detect scene S as $S = f\left(\sigma_L, \mu_{ab}, U_{shadow}, H_{sky}\right)$. The function f is actually represented by a feedforward neural network in this paper. Feedforward neural networks are preferably appropriate for modeling relationships between input variables or set of predictors and response or output variables. Otherwise stated, they are suitable for any practical mapping problem which shows how a number of input variables affect the output variable. The feed forward neural networks are the most widely studied and used neural network model in practice.

To train the network, a large amount of labeled data are required. We labeled the datasets with the scenes as mentioned above. The video data were collected from the cameras of real-world TSS in Ho Chi Minh City. Due to the time lapse video over all day, the environment transforms, which from the daylight to nighttime, and the illumination changes in the daytime in which the overcast or shadow will happen. In case of rainy day, the camera would capture the street from the dry to wet condition.

Fig. 5. Frames captured from five different cameras in the traffic surveillance system at the same time on Vo Van Kiet Avenue, Ho Chi Minh City. The first column shows the camera in the overcast condition, the second column in the clear sky, the third column in rain, and the last column in nighttime

Figure 4 shows the structure of a typical three-layer feedforward neural network. The input layer has four nodes corresponding to the inputs of four parameters. The hidden layer contains m hidden neurons, and the output layer has four nodes which are the scenes.

3 Experiments and Discussion

3.1 Quantitative Evaluation

This section illustrates the evaluation of the proposed method. We evaluate the proposed method on the datasets captured on the Vo Van Kiet Avenue in the Ho Chi Minh City, Vietnam. The videos were recorded at several locations from single pole-mounted cameras with the frame rate of 30 fps and the resolution of 640 × 360. Moreover, we also performed online testing with the video stream from the same cameras. We trained at least 10000 images per camera to get the right parameters. The scene of camera for testing illustrates in Fig. 5, the row represents for each camera in the experiment, and the column shows the scene in each case: overcast, clear sky, nighttime, and rain. The first case and the fourth case cover alley, in which the road has two lanes and takes the minor part of the frame. The building on the road edge takes the most of the image, for which the magnitude of the sky region is small. Therefore, the label of them is different from the cases, in which the sky region and the observation zone take the considerable size of the frame. In the early morning, the sun makes the long shadow of each vehicle in the large street and causes in that the shadow of the building in alley crosses all the road. Consequently, at the same time, in this condition, the label of the big road is the clear sky, but the alley is overcast. The difference in topographic makes the variation in the label.

Table 1 shows the results of quantitative evaluations, the measure approach is mention in [15]. For more detail, each column represents the scene for recognition (overcast, clear sky, rain, and nighttime). Each row of table corresponds the videos used for benchmark. The first four cases are recorded, and the fifth one is online video stream. The last row is for the average accuracy. The accuracy of recognition via three videos was more than 85%; especially, the night time was nearly 95%. For the testing in real-time dataset, the accuracy was lower than running from captured video; however, the deviation was just approximately 5%.

Table 1. Quantitative evaluation

Case	OC	CS	RA	NT
1	93.5	93.6	84.6	97.5
2	89.3	95.2	92.2	96.1
3	87.4	96	91	97.6
4	79.3	78.6	79	96.05
5	81.3	94.3	79.4	94.6
Average(%)	**86.16**	**91.55**	**85.23**	**96.35**

As can be seen, the result of the clear sky and nighttime scene got the better result than the rain and overcast scene. The overcast scene happens when the cloud passed across and nearly covers the sky. In some cases, the great cloud, almost darkened the sky, makes the method incorrectly detect the nighttime instead of the overcast, or the time of cloud covering the sky is too short, which results in the method recognizes the clear sky as an alternative. The rainy scene got the lowest results in Table 1. Because the rainy scene always comes with the thick cloud with a gloomy sky that leads to mistakenly detect the night as a substitute. Moreover, the shadow rain, which makes the reflection on the road, could be confused with the clear sky scene.

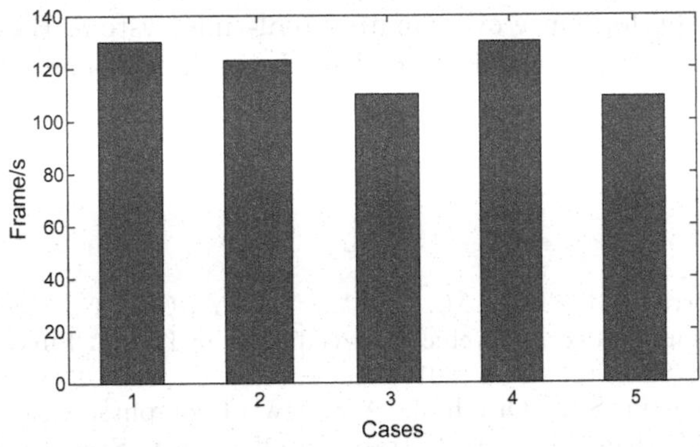

Fig. 6. Speed evaluation from the experiment. The x-axis presents each case for testing, and the y-axis describes the running time. The columns show the frame rate for each case

3.2 Speed Evaluation

The section illustrates the speed performance of the proposed method, which proves the method is fast enough for the traffic surveillance system that could run in real time. According to experiments, the numbers of frame per second we execute the method in each case are high (Fig. 6): The uppermost is 130, and the lowermost is 110 while it runs in a real-time stream. The time-consuming for the proposed method is conditional on the magnitude of the sky region and the observation zone.

The reason for different time-consuming is that the sky region and the observation zone have diverse size. The first and the fourth cases observe the small streets and the narrow part of the sky; therefore, the number of pixel of processing is fewer than other, for which the highest speed is in those cases. Moreover, with the sky regions that cover nearly one-third of the frame and the observation zones that are double in size with two cases above, the second, the third, and the fifth cases take more time for computation. However, the duration of processing is still fast enough for running in real time. To conclude, the total running time of the proposed with other processes in TSS is satisfied for performing in real time.

4 Conclusion

In the chapter, we illustrate the method to detect and recognize the scene of the environment (overcast, clear sky, rain, and nighttime). Our most significant contribution to this study is that the traffic surveillance system uses our method to handle the suitable method running in real world (motion segmentation, vehicle tracking, and classification). The novel approach using feedforward neural network with CIE-Lab color space and histogram features has shown the improvement in performance over running real-time system. Based on the proposed method, the future studies are that the technique can give more detail of the anomaly scene such as rush hour, accident,..., and the transition time between scenes.

References

1. Tian, B., Morris, B.T., Tang, M., Liu, Y., Yao, Y., Gou, C., Shen, D., Tang, S.: Hierarchical and Networked Vehicle Surveillance in ITS: A Survey. IEEE Trans. Intell. Transp. Syst. 18, 25–48 (2017).
2. Buch, N., Velastin, S.A., Orwell, J.: A review of computer vision techniques for the analysis of urban traffic. IEEE Trans. Intell. Transp. Syst. 12, 920–939 (2011).
3. Wang, X.: Intelligent multi-camera video surveillance: A review. Pattern Recognit. Lett. 34, 3–19 (2013).
4. Xiaoxu Ma, Grimson, W.E.L.: Edge-based rich representation for vehicle classification. In: 10th IEEE International Conference on Computer Vision (ICCV05) Volume 1. p. 1185–1192 Vol. 2. IEEE (2005).
5. Chen, Z., Ellis, T., Velastin, S. a: Vehicle detection, tracking and classification in urban traffic. 15th Int. IEEE Conf. Intell. Transp. Syst. 951–956 (2012).
6. Ha, S.V.-U., Pham, L.H., Phan, H.N., Ho-Thanh, P.: A robust algorithm for vehicle detection and classification in intelligent traffic system. In: 16th Asia Pacific Industrial Engineering and Management Systems Conference (APIEMS 2015). pp. 1832–1838 (2015).
7. Cucchiara, R., Grana, C., Piccardi, M., Prati, A.: Detecting objects, shadows and ghosts in video streams by exploiting color and motion information. In: Proceedings 11th International Conference on Image Analysis and Processing. pp. 360–365. IEEE Comput. Soc (2001).
8. Cucchiara, R., Grana, C., Piccardi, M., Prati, A.: Detecting moving objects, ghosts, and shadows in video streams. IEEE Trans. Pattern Anal. Mach. Intell. 25, 1337–1342 (2003).
9. Meis, U., Ritter, W., Neumann, H.: Detection and classification of obstacles in night vision traffic scenes based on infrared imagery. In: Proceedings of the 2003 IEEE International Conference on Intelligent Transportation Systems. pp. 1140–1144. IEEE (2003).
10. Chen, T.H., Chen, J.L., Chen, C.H., Chang, C.M.: Vehicle detection and counting by using headlight information in the dark environment. Proceeding of th 3rd International Conferences Intelligent Information Hiding and Multimedia Signal Processing IIHMSP 2007. 2, 519–522 (2007).
11. Ha, S.V.-U., Pham, N.T., Pham, L.H., Tran, H.M.: Robust Reflection Detection and Removal in Rainy Conditions using LAB and HSV Color Spaces. REV J. Electron. Commun. 6, 13–19 (2016).

12. A. Cohen, C. Meurie, Y. Ruichek, J. Marais, and A. Flancquart, Quantification of GNSS signals accuracy: An image segmentation method for estimating the percentage of sky, in 2009 IEEE International Conference on Vehicular Electronics and Safety (ICVES), 2009, pp. 35–40.
13. Couprie, C., Grady, L., Najman, L., Talbot, H.: Power Watershed: A Unifying Graph-Based Optimization Framework. IEEE Trans. Pattern Anal. Mach. Intell. 33, 1384–1399 (2011).
14. Wang, X.-Y., Wu, J.-F., Yang, H.-Y.: Robust image retrieval based on color histogram of local feature regions. Multimed. Tools Appl. 49, 323–345 (2010).
15. Sokolova, M., Lapalme, G.: A systematic analysis of performance measures for classification tasks. Inf. Process. Manag. 45, 427–437 (2009).

An Evaluation of Virtual Organizational Structure on Employee Performance of Selected Telecommunication Companies in Kaduna State, Nigeria

Fadele Ayotunde Alaba[1], Yunusa Salisu Tanko[1], Sani Danjuma[2], Rajab Ritonga[3], Abulwafa Muhammad[4], Tundung Subali Patma[5], and Tutut Herawan[6,7](✉)

[1] Federal College of Education, Zaria, Kaduna State, Nigeria
ayotundefadele@yahoo.com, yunusak2001@gmail.com
[2] Department of Computer Sciences, Northwest University, Kano, Nigeria
sani_danjuma@yahoo.com
[3] Universitas Prof Dr Moestopo (Beragama), Jakarta, Indonesia
rajab_ritonga@dsn.moestopo.ac.id
[4] Universitas Putra Indonesia (YPTK), Padang, Indonesia
abulwafa@upiyptk.ac.id
[5] State Polytechnic of Malang, Malang, East Java, Indonesia
subalipatma@yahoo.com
[6] Universitas Teknologi Yogyakarta, Yogyakarta, Indonesia
tutut@uty.ac.id
[7] AMCS Research Center, Yogyakarta, Indonesia

Abstract. The business environment will no doubt require firms to become even more flexible and agile to bring products and services to market at an increasing rapid pace with the advent of IT in Nigerian in the context of MTN, ETISALAT, and GLO. Traditional organizational structure is no longer capable of sustaining the needs of their relentless pace, new forms of organizing such as the virtual organization hold promise as organizational leaders experiment and learn new strategies for managing in the twenty-first century and beyond. Thus, this study seeks to evaluate the extent to which virtual organizational structure enhances employees' performance. A census survey research design was used in which data was sourced via questionnaire and the data was analyzed using descriptive statistics SPSS V.20 and regression analysis in testing the hypothesis. This study found out that virtual organizational structure has significant effect on employees' performance in some selected telecommunication in Kaduna state, and it was recommended that the organization should take advantage by improving on managerial training and study virtual organizational structure so as to enhance employees' performance.

Keywords: Virtual organizations · Employee performance · Information technology

V. Bhateja et al. (eds.), *Information Systems Design and Intelligent Applications*, Advances in Intelligent Systems and Computing 672,
https://doi.org/10.1007/978-981-10-7512-4_69

1 Introduction

Organizations in the telecommunication industry today seek to pursue certain aims and objectives in order to sustain their existence and survival and to also gain competitive advantage. However, employee performance has been quite significant and plays a vital role in achieving this organizational goal. And the means to attaining such requirements in the twenty-first century is to ensure that employees' perform at optimum level. The main goal of any organization is to enhance job performance of its employees so that it could survive in a turbulent business environment. Organizations need highly performing individuals in order to meet their goals, to deliver the products and services they are specialized in, and finally to achieve the stated goal. Therefore, performance is also important for the individual as it promotes intrinsic job satisfaction and self-actualization. An employee with greater job satisfaction and commitment has higher job performance [1, 2]. Uzonna [3] further stressed that employees' performance should be visualized on the perspective of behavior rather that result. He however stated that result-based measures are not always functional to the organization, as the employee may try to maximize results at the expense of other things. In addition, employees' performance has been conceptualized as those actions or behavior under control of the individual that contributes to the organizational goals and that can be measured according to the individual's level of [4].

The industry is currently facing unprecedented challenges in an ever dynamic and complex environment as several factors including the pace of technological innovation and the globalization of the economy have coerced businesses and industry to adapt to new challenges triggered by an ever sophisticated society characterized by an increasing demand for customized and high-quality services and products in various segments of industry. The uncertainties of business environment are still however growing, and new game plans are being drawn everyday by organizations concerned with their profitability. Creative thinking with regard to employee performance has led to various organizational structures for quality job performance; thus, virtual organization is used to describe a nation of independent firms that join together, often temporarily to produce a service or product. This is often associated with such terms as virtual office, virtual teams, and virtual leadership. The ultimate goal of the virtual organization structure is to provide innovative high-quality products or services instantaneously in response to customer demands. It actually has its roots in the computer industry when a computer appears to have more storage capacity than it really possesses, which is referred to as virtual memory. Like any other telecommunication providers, MTN Nigeria was established in 2001 to provide cellular network access and ICT solutions to millions of Nigerians both locally and internationally with over 55 million subscribers', 15 service centres, 144 connect stores, and 247 connect points located at every state of the federation. Globacom limited is a Nigeria multinational telecommunication company privately owned started on August 29, 2003, with over 25 million subscribers and network connection to 140 countries; it has its service connected to 235 network with largest roaming coverage in Africa. Similarly, Etisalat started in Nigeria on the March 13, 2008, with over 17 million subscribers with network connection to over 18 countries. The prominent theorist of traditional hierarchical organization was Frederick (1911) where hierarchical structure was designed to manage highly couple processes like automobile assembly where production could be broken down into a series of simple steps. Hierarchical corporations often control and manage raw materials to their allocation to consumers, and a centralized managerial hierarchy controls the entire production process, but with the global challenges, competition, communication, and bureaucratic barriers have made most organization into adapting to a more modern and flexible organizational structure in the twenty-first century. Studies were conducted by Brunelle [2], Askarzai et al. [5], Danzfuss [6], Moeini et al. [7] on the consistency of virtual organizations enabling capabilities and improvements in knowledge management performance in Tehran; a survey study of employee attitudes, virtual management in small and medium sized enterprises in Greater Western Sydney: virtually in work arrangement and affective organizational commitment in Canada and the impact of organizational structure of the performance of virtual teams in south Africa but no such study was conducted on virtual organizational structure on employees' performance in telecommunication in Nigeria which is gap the researcher intends to fill.

Recently in Nigeria, employees have been complaining about the stereotype form of working structure of organization with communication being from the top to bottom, lack of flexibility and creativity, poor working conditions which had affected the effective and efficiency of employees' performance [8]. Hence, the need to come up with new initiative that promotes employees performance giving comfort and effective service delivery to customers. Grenier and Metes [9] discussed the moves to new organization structure as a response to unprecedented customer expectations and alternative, global competition time compression, complexity, rapid change, and increased use of technology. Telecommunication companies, such as MTN, Glo, and Etisalat with combine 50 million subscribers making virtual organizational structure serve as an option for effective employees' performance. Virtual organization in many western countries is seen as a way of improving employee satisfaction which helps to reduce cost: increase work-life balance positively, flexible workplace with less cost than regular workplace [10, 11]. It is on this crux that the researcher seeks to evaluate virtual organizational structure on employee's performance using MTN, Glo, and Etisalat as selected from the telecommunication industry in Nigeria (Kaduna State). Therefore, this paper evaluates the extent to which virtual organizational structure has effect on employees' performance.

The rest of this paper is organized as follow. Section 2 reviews the state of art of the study. Section 3 gives an analysis of previous algorithms. Section 4 presents an alternative approach of normal parameter reduction. Section 5 shows the real-life applications of the proposed algorithm and finally, Sect. 6 presents conclusion and future work.

2 Rudimentary

The new form of organization "virtual organization" emerged in 1990 and is also known as digital organization, network organization, or modular organization. In other words, a virtual organization is a network of cooperation made possible by what is called ICT, i.e., information and communication technology which is flexible and comes to meet the dynamics of the market [12]. However, the concept of virtual organization is an emerging research topic in the context of both inter-organizational and intra-organizational relationships. Despite this fact, the existing literature on the subject provides varied and multifarious perspectives of virtual organizations and no clear delineation of its facets as shown [13]. According to Rouse in [14], she defined virtual organization or a company whose members are geographically apart, usually working by computer, e-mail, and group wircs while appearing to others to be a single organization with a real physical location. Coello in [15] sees virtual organization as any organization with non-colocated organization entities and resources, necessitating the use of re-virtual space of interaction between the people in these entities to achieve organization objectives. Virtual organization can also be defined as groups of geographically or organizationally distributed participants who collaborate toward a shared goal using combination of technologies to accomplish a task. Researchers in [9, 16] posited that the ultimate end of effective virtual organizations is the satisfactory accomplishment of the stated objective or task wherein. Similarly, the concept of virtually in work arrangement refers to executing work through the intermediary of cyberspace [17]. More specifically, it refers to a context in which work is done at a distance generally outside convention offices regardless of when and where, and in which interaction are mediated by technology [18]. Virtual organization does not need to have all of the people, or sometimes any of the people to deliver their service. The organization exists but you cannot see it. It is a network not an office. In summary, these characteristics could further buttress on the picture of what virtual organization depicts: flat organization, power flexibility, boundaryless, dynamic, ICT-inclined goal orientation, homework, information sharing, staffed by knowledge work, geographical location, etc. In addition, virtual organization can be categorized into three, i.e., telecommuters which deal with personnel working at home with computers using the network, phone lines, and modern for example MTN, Etisalat, and Glo network. Outsourcing employee/competencies this is characterized by the outsourcing of the all most core competencies, i.e., marketing, human resources, finance information system etc. Completely virtual, these are the companies metaphorically described as companies without walls that are tightly linked to a large network of suppliers, distributors, retailers, etc. Figure 1 further illustrates the virtual organization structure.

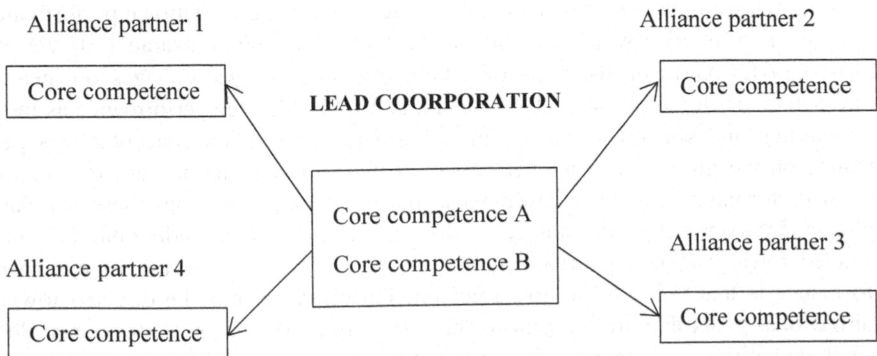

Fig. 1. Virtual organization structure [9]

Figure 1 explains how the virtual organizational structure works; the lead corporation is the core center of the organization. The core competence A and core competence B can be located in any part of the world naturally small in nature, but it outsources its jobs to other company to produce different components needed to compliment the lead cooperation. The alliance partners 1, 2, 3, and 4 can be located around the world, but they are all linked with ICT providing the Internet network communication making the corporation look like single company. Grenier and Metes in [9] described the virtual model as a lead organization that creates alliances with groups and individuals from different organizations who possess the highest competencies to build a specific product or service in a period of time.

2.1 Concept of Employee Performance

Every organization has been established with certain objectives to achieve. These objectives can be achieved by utilizing the resources like men, machines, materials, and money. All these resources are important, but out of these, the manpower is the most important as it plays an important role in performing tasks for accomplishing the goals. Organizations today require outstanding performance from employees in order to meet their goals, to deliver the products and services specialized in, and in turn achieve competitive advantage. Performance is a major, although not the only prerequisite for future career development and success in the labor market. Although there might be exceptions as high performers get promoted more easily within an organization and generally have better career opportunities than low performers [19]. Employee Performance has been vaguely defined as it does have a clear-cut delineation based on scholars' perception; it therefore bores down to how it is being viewed by researchers. Viswesvaran [20] defined work performance or employee performance as scalable actions, behavior, and outcomes that employees engage in or bring about that are linked with and contribute to "organizational goals." According to Sardana and Vrat in [21] posits that the terms "performance" and "productivity" are incorrectly used. People who claim to be discussing "productivity" are actually looking at the more general issue of performance. The "productivity" is a fairly specific concept, while performance includes many more attributes. Campbell in [22] describes employee job performance as an individual level variable, or something a single person does which is what differentiates it from more encompassing constructs such as organizational performance or national performance which are higher level variables. Viswesvaran [20] viewed employee performance on the basis of a behavior an employee engages on, geared toward achieving organization goal. Sardana and Vrat in [23] see performance as more encompassing and general. Campbell in [22] on the other hand conceptualizes performance on the ground of employees effort which differentiates it from the organizational performance. Having viewed these divergent opinions from these scholars, employee performance is the ability, skills, capabilities of an individual behavior channeled toward achieving organizational goal. Another key feature of employee performance is that it has to be goal relevant. Performance must be directed toward organizational goals that are relevant to the job or role. Therefore, performance does not include activities where effort is expended toward achieving peripheral goals. For example, the effort put toward the goal of getting to work in the shortest amount of time is not performance (except where it is concerned with avoiding lateness) and the need to be in a team group.

2.2 Review of Empirical Studies on Virtual Organization and Employee Performance

Several researchers have discussed extensively on virtual organization on employees' performance, but the researcher intends to embark on an intellectual excursion to view scholars' opinion on the construct before us. Reference [2] conducted a study on virtually in work arrangement and affective organizational commitment in Canada with a population size of 380 and sample of 139, using multiple linear regression and correlation as statistical tool for testing hypothesis. The study found out that there is significant relationship between the level of virtually of work arrangements and affective organizational commitment. However, the response from the sample size of 139 from population size of 380 is grossly inadequate for the findings to hold due to the poor level of respondents in the study under review. Another study by Moeini et al. [7] on the consistency of virtual organizations enabling capabilities and improvements in knowledge management performance in Tehran with a population size of 610 and sample response of 372, using AVES versus squared correlation, it was discovered that ICT usage significantly contribute to consistency in studied sample at a level of 0.001. Base on the study between the two variables, it was established that ICT has played a significant role in enabling capabilities and improvements in knowledge management performance. However, the sample response of 372 from the population size of 610 needs to be improved to give a fair representation. Similarly, Askarzai et al. [5] undertook a survey study of employee attitudes, virtual management in small and medium-sized enterprises in Greater Western Sydney with sample size 116 from a population of 1150, adopted explorative data analysis (EDA). The study found out that all three types of employee: non-virtual, partial virtual, and virtual employees have high job satisfaction, are highly involved in their jobs with organizational commitment. Furthermore, it was ascertained that all three employees have positive attitude toward management.

Danzfuss [6] reviewed a work on the impact of organizational structure of the performance of the virtual team, a survey research with sample size of 87 respondents in South Africa using descriptive statistics; he found out that there is no statistical significant relationship between the organizational structure and the performance of the virtual team. But in my opinion, census population should have been used, since the population is 87 and the researcher took 69 from the phone interview and a more statistical tool would have been more appropriate for more robust findings.

2.3 Theoretical Framework

So many researchers have discussed on organizational structure ranging from classical school to virtual organizational structure. The early theorists of organizational structure, Taylor, Fayol, and Weber "saw the importance of structure for effectiveness and efficiency and assumed without the slightest question that whatever structure was needed, people could function accordingly". Organizational structure was considered a matter of choice. When in the 1930s, the rebellion began that came to be known as human relations theory, there was still not a denial of the idea of structure as an artifact, but rather an advocacy of the creation of a different sort of structure, one in which the needs, knowledge, and opinions of employees might be given greater recognition. As individual productivity loss theory shall be discussed.

The individual productivity loss theory was propounded by Steiner [24]. Some inefficiencies were first defined by Steiner as motivational and coordination loss, and they were later expanded by Mueller to include relational loss. Motivational loss occurs when team members are not motivated to perform the task at hand. Another assumption of this theory is that virtual team members may suffer from motivational loss due to being isolated from the rest of the team and not having the pressures that usually accompanies face-to-face interaction. Coordination loss occurs when it becomes more difficult to coordinate the execution of the task at hand. As virtualization increases within a team, the coordination effort is required further [25]. This increase in coordination effort is attributed to the dispersion and asynchronous execution of tasks. Some of the challenges with asynchronous communication that were highlighted are: The norm is for more than one topic to be active at the same time; information overload; contributions are being made at different times; reduced linkage between responses and long-time lapses between communication cause discontinuous and disjointed discussions [26]. Relational losses defined by Mueller [27] draw from the theory of social support which says that the perception of social support is a function of the quality of interpersonal relationships experienced by the team member. This perception is built on the availability of emotional support, instrumental support, appraisal support, and informational support. This allows us to draw the following relationships between team virtualization and team performance: When team virtualization increases, individual team members find it increasingly difficult to coordinate tasks. When team virtualization increases, individual team members become less motivated to perform team-related tasks, When team virtualization increases, individual team members' perception of the availability of support decreases, and when individual productivity loss occurred, the team's overall performance decreased.

One can therefore argue that as the virtualization of a team increases the chance of individual productivity loss of a team member increase, which negatively impacts the overall performance of the work team. This theory underpins this study because it explains the relationship between virtual organization and its performance.

3 Proposed Method

3.1 Research Design

Basically, research work of any kind attempts to study in detail some aspect to a chosen area or discipline. Research methods are numerous, and each method enables the researcher to gain good basis for a thorough understanding of phenomena. Therefore, survey method as a research design shall be used in the research due to the size of the population under review and it shall be quantitative in nature.

3.2 Population of the Study

The population of the study shall consist of staff of MTN 36 Etisalat 110, and GlobaCom 60 which totaled 206. A sample is a part of the population deliberately taken to represent the population of study. Sample size of staff shall be drawn using [28] formula for adequate representation, while simple random sampling technique shall be employed in the selection of staff under the study at confidence level 95% and $P = 0.05$. Below is the formula:

$$n = \frac{N}{1 + N(e)^2}$$

where n = Sample size, N = Population size, e = Sample error Level of significance, 1 = Constant. Therefore, we have

$$n = \frac{206}{1 + 206(0.05)^2} = \frac{206}{1 + (206 * 0.0025)} = \frac{206}{1.515} = 135$$

However, to compensate for non-response, provision of 30% increased representation will be taken [29] and this gives a total of one hundred and seventy-five respondents (175) to be administered questionnaires as 30% of the 135 sample size was added to compensate for those respondents the researcher will be unable to reach in order to have an adequate representation for the study.

3.3 Sources of Data Collection

For any research to be carried out, adequate data must be collected. On this basis, primary source of data was used to source for data as questionnaire which is a research instrument mostly used in survey research to extract specific information or questions from the respondents. Therefore, a structured questionnaire with a five-point Likert's scales was used to source for data for the study under review. Secondary data was also used to compliment data obtained from the primary source such as handbook, libraries, journal, textbooks, and annual reports published by MTN, GloMobile, and Etisalat.

3.4 Method of Data Analysis

Method of data analysis consists of statistical calculations performed with the raw data to produce answers to the question or problem in the analysis of the questionnaires that were designed to find out the view of the respondents on the subject matter. Therefore, descriptive statistics and tables were used in the analysis and presentation of data with the view to enhancing accuracy. Regression analysis was used for hypothesis testing which is to either reject or accept the null hypothesis.

3.5 Validity and Reliability of Instrument

The instrument that was used for this study was subjected to the scrutiny by three panels of experts from Department of Business Administration in order to ensure the content and face validity of the instrument. The suggestions and amendments made by the experts were fully effected before the pilot study. The questionnaire for virtual organization was developed by Chumg et al. in [12], and employees performance scale was developed by Van Scotter et al. [19] both questionnaires were adopted. For the purpose of data analysis, the Cronbach's alpha measures how well a set of items (or variable) can be relied upon and to determine the reliability of the instrument, Cronbach's coefficient alpha was used to measure the instruments with both having 0.88 and 0.91 (virtual organization and employee performance). Accordingly, [30] scale reliability, posited that Cronbach alpha that are less than 0.60 are considered poor and those in the range of 0.70 are acceptable and good. Therefore, the instrument for the study under review can be relied upon.

4 Data Presentation and Analysis

4.1 Introduction

The objective of this chapter is to present and analyze the data collected in this study as it tends to explore the evaluation of virtual organizational structure on employee performance of some selected telecommunication companies.

4.2 Data Presentation and Analysis

The purpose of data presentation is to analyze the data collected to facilitate thorough understanding and further explanation of the research findings. The questionnaire comprises of three parts, Part A is the personal data of the employee, Part B deals with questions on virtual organizations, and Part C deals with employee performance. The researcher administered one hundred and seventy (175) questionnaires accordingly out of which one hundred and thirty (130) were duly completed and returned and for the purpose of this seminar paper, the completed and returned questionnaires of (96%) forms a good representation for the paper, in other words the total respondents in this particular questions shall represent hundred percent (100%).

Table 1 shows the adjusted R-square value of 0.293 indicating that 29% behavior of the dependent variable was explained by the independent variable, and the standard error estimate is 0.506. Therefore, going by the R-square, it can be stated that virtual organization structure appears a weak variable for predicting employees' performance in telecommunication in Kaduna. Furthermore, the F-statistics of the model at 54.451 is fit since its acceptable region ranges from 1.96 to above, the sig which indicated the goodness of the data really fits. The p-value of 0.00 is below the significant level of 0.05, and the F-statistics of 54.451 is large enough for the null hypothesis which stated that virtual organization structure has no significant effect on employees' performance be rejected and accept the alternative hypothesis virtual organization structure has significant effect on employees' performance. This is in tandem with findings of the findings of [2].

Table 1. Result of regression analysis

Hypothesis	Relation	Beta	Std error	*t*-value	*p*-value	Decision
HO_I	Virtual organizational	0.546	0.506	7.379	0.000	Rejected
	Structure has no significant					
	Effect on employees					
	Performance					
Cronbach's alpha						
VO	0.88					
EP	0.91					
R	0.546					
R^2	0.298					
F	54.451					
Sig	0.00					

Note N = 135 dependent variable (employee performance) $p < 0.005$**

5 Conclusion

One of the findings discovered from the study is that virtual organization structure has significant effect on employees' performance in some selected telecommunication in Kaduna state. This result is in consistent with the findings of [2]; based on the foregoing, the following recommendations are made to further strengthen the research work The organization should take advantage by improving on managerial training and study virtual organizational structure so as to enhance employees performance. Conclusively, this work is targeted toward improved and quality employee performance with specific reference to virtual structure in an organization which is a small core organizational structure which outsources major businesses functions, this allow for flexibility, aid creative thinking, encourages competition in term of quality delivery of service.

Acknowledgements. The work of Abulwafa Muhammad is supported by Universitas Putra Indonesia YPTK under research grant no 094/UPI-YPTK/RG/IX/2016. The work of Tutut Herawan is supported by Universitas Teknologi Yogyakarta Research Grant no vote O7/UTY-R/SK/0/X/2013.

References

1. Walumbwa, F.O. and C.A. Hartnell, *Understanding transformational leadership–employee performance links: The role of relational identification and self-efficacy.* Journal of Occupational and Organizational Psychology, 2011. **84**(1): pp. 153–172.
2. Brunelle, E., *Virtuality in work arrangements and affective organizational commitment.* International Journal of Business and Social Science, 2012. **3**(2).

3. Uzonna, U.R., *Impact of motivation on employees' performance: A case study of CreditWest Bank Cyprus.* Journal of Economics and International Finance, 2013. **5**(5): p. 199.
4. Ali, A., M. Abrar, and J. Haider, *Impact of Motivation on the working performance of employees-A case study of Pakistan.* Asian Journal of Research in Business Economics and Management, 2012. **2**(7): pp. 328–340.
5. Askarzai, W., Y.-C. Lan, and B. Unhelkar. *Study of Employee Attitudes towards Virtual Management in Small and Medium sized Enterprises: An Exploratory Data Analysis.* in *Proceedings of 23rd International Business Research Conference.* 2013. Citeseer.
6. Danzfuss, T.W., *The impact of organisational structure on the performance of virtual teams.* 2013.
7. Moeini, A., A.F. Farahani, and A.Z. Ravasan, *The consistency of virtual organizations enabling capabilities and improvements in knowledge management performance.* International Journal of Enterprise Information Systems (IJEIS), 2013. **9**(2): pp. 20–43.
8. Eme, O.I., A. Onyishi, and I.E.J. Emeh, *Problems of Personnel Management in Nigeria: The Nigerian Local Government System Experience.* Oman Chapter of Arabian Journal of Business and Management Review, 2012. **1**(6): pp. 36–49.
9. Grenier, R. and G. Metes, *Going virtual: Moving your organization into the 21st century.* 1995: Prentice Hall PTR.
10. Powell, A., G. Piccoli, and B. Ives, *Virtual teams: a review of current literature and directions for future research.* ACM Sigmis Database, 2004. **35**(1): pp. 6–36.
11. Sanchez, J., *Building statistical literacy assessment tools with the IASE/ISLP.* International Association for Statistical Education (IASE)/International Statistics Institute (ISI) Satellite, 2007.
12. Chumg, H.-F., et al., *Factors affecting knowledge sharing in the virtual organisation: Employees' sense of well-being as a mediating effect.* Computers in Human Behavior, 2015. **44**: pp. 70–80.
13. Humphreys, P., R. McIvor, and S. Shekhar, *Understanding the virtuality of virtual organizations.* Leadership & Organization Development Journal, 2006. **27**(6): pp. 465–483.
14. Rouse, W.B., 2005. Enterprises as systems: Essential challenges and approaches to transformation. *Systems Engineering*, *8*(2), pp. 138–150.
15. Coello, C.A.C., *Citas a Publicaciones del Dr. Carlos A. Coello Coello que aparecen en el Citation Index.* optimization, 2005. **27**(2): pp. 171–186.
16. Bjørn, P. and O. Ngwenyama, *Virtual team collaboration: building shared meaning, resolving breakdowns and creating translucence.* Information Systems Journal, 2009. **19**(3): pp. 227–253.
17. Shekhar, S. and L. Ganesh, *A morphological framework for virtual organizations.* IIMB Management Review, 2007. **19**(4): pp. 355–364.
18. Cascio, W.F. and S. Shurygailo, *E-leadership and virtual teams.* Organizational dynamics, 2003. **31**(4): pp. 362–376.
19. Van Scotter, J., S.J. Motowidlo, and T.C. Cross, *Effects of task performance and contextual performance on systemic rewards.* Journal of Applied Psychology, 2000. **85**(4): p. 526.
20. Viswesvaran, C., *Modeling job performance: Is there a general factor?* 1993, DTIC Document.
21. Sardana, G. and P. Vrat, *Performance objectives-productivity (PO-P): A conceptual framework and a mathematical model for productivity management.* Productivity, 1983. **24** (3): pp. 299–307.
22. Campbell, J.P., *The definition and measurement of performance in the new age.* Pulakos (Eds.), The changing nature of performance: Implications for staffing, motivation, and development, 1999. **399**: p. 429.

23. Sardana, G. and P. Vrat, *Productivity measurement in a large organization with multi-performance objectives: A case study*. Engineering Management International, 1987. **4**(2): pp. 105–125.
24. Steiner, I.D., *Group Process and Productivity (Social Psychological Monograph)*. 2007.
25. Montoya-Weiss, M.M., A.P. Massey, and M. Song, *Getting it together: Temporal coordination and conflict management in global virtual teams*. Academy of management Journal, 2001. **44**(6): pp. 1251–1262.
26. Ocker, R., et al., *The effects of distributed group support and process structuring on software requirements development teams: Results on creativity and quality*. Journal of Management Information Systems, 1995. **12**(3): pp. 127–153.
27. Mueller, J.S., *Why individuals in larger teams perform worse*. Organizational Behavior and Human Decision Processes, 2012. **117**(1): pp. 111–124.
28. Yamane, T., *Problems to Accompany Statistics: An Introduction Analysis*. 1967: Harper & Row.
29. Israel, G.D., *Determining sample size*. 1992: University of Florida Cooperative Extension Service, Institute of Food and Agriculture Sciences, EDIS Gainesville.
30. CDATA-Nunnally, J. and I. Bernstein, *Psychometric Theory*. 1978, New York: McGraw-Hill.

An Impact of Transformational Leadership on Employees' Performance: A Case Study in Nigeria

Yusuf Musa[1], Sani Danjuma[2], Fadele Ayotunde Alaba[1], Rajab Ritonga[3], Abulwafa Muhammad[4], Ludfi Djajanto[5], and Tutut Herawan[6,7](✉)

[1] Federal College of Education, Zaria, Kaduna State, Nigeria
yusufmusa02@gmail.com, ayotundefadele@yahoo.com

[2] Department of Computer Sciences, Northwest University, Kano, Nigeria
sani_danjuma@yahoo.com

[3] Universitas Prof Dr Moestopo (Beragama), Jakarta, Indonesia
rajab.ritonga@dsn.moestopo.ac.id

[4] Universitas Putra Indonesia (YPTK), Padang, Indonesia
abulwafa@upiyptk.ac.id

[5] State Polytechnic of Malang, Malang, East Java, Indonesia
ludfimlg@yahoo.com

[6] Universitas Teknologi Yogyakarta, Yogyakarta, Indonesia
tutut@uty.ac.id

[7] AMCS Research Center, Yogyakarta, Indonesia

Abstract. The objective of this paper is to determine the extent to which transformational leadership has impact on staff performance in Federal College of Education, Zaria, Nigeria. Statement of the problem under study was employees' affirmation to lack of direction to the organization's transformational style of leadership and an intellectual excursion was undertaken to review related literatures on transformational leadership and performance. A survey research design was used as primary data and was sourced via questionnaire and interview which was complimented with secondary data. The analysis of data was conducted using SPSS version 20, and regression was used as tool for hypothesis testing, and it was found out that there is significant relationship between transformational leadership and staff performance in the college. The study concludes that a sound and viable leadership with individual consideration at heart, encourages innovation, and creativity, and it was recommended that management should adopt fully the transformational leadership role with leadership qualities such as role modeling, perseverance, empathy, pragmatism, visionary, innovative, coaching, stimulating, and valuing employees so as to enhance staff performance.

Keywords: Transformational leadership · Employee performance · Impact Nigeria

V. Bhateja et al. (eds.), *Information Systems Design and Intelligent Applications*, Advances in Intelligent Systems and Computing 672,
https://doi.org/10.1007/978-981-10-7512-4_70

1 Introduction

Many organizations today seek to pursue certain aims and objectives in order to sustain their existence and survival and one of the greatest things that is constant in the business environment is change and change in areas such as: competitive strategies, organizational restructuring, strategic management planning, research and development, human resources development. Therefore, the need to man both human and material resources becomes pertinent which serve as catalyst to achieving organizational goals. The ability to exert influence on some group of people toward some goals is leadership, and leadership is said to be pivotal to the success or failure of any organization, though there is no consensus as to what kind of leadership is needed to manned an organization but one common assertion is that leadership is the act of influencing some individuals toward common goals. The role every manager should fill in the workplace is leadership. However, many managers often make the mistake of assuming the position of a manager for a leader and that their subordinates will automatically follow their whims and caprices. In reality, position only denotes title not leadership. Northouse [1] defines leadership as a process whereby an individual influences a group of individuals to achieve a common goal and to be an effective leader; the manager must be able to influence his subordinate in a positive way so as to reach the needed goals of the organization. This extends to the concept of transformational leadership which is related to the special ability that aims at bringing about change, innovation, renewal, and revitalization by inspiring individuals through creating and promoting achievable visions despite opposition [2]. Transformational leadership is embedded in ethics, authenticity, relationship building, personal development, trust, and subordination of self-interests to the overall interests of the organization. The complexities of change in today's business environment, organizations, institutions, etc., can be overwhelming, irrespective of their operating motives and resources. The pressures of deregulation, privatization, tax change, social renewal, and globalization have compelled most organizations to adapt to a system of leadership that will inspire employees toward the attainment of the organizational goals. The transformational leadership approach is geared toward making managers become exceptional leaders which is the ability to want people to change, to improve, and to be led as this serves as a catalyst for employees' performance. It is on this note that the researcher decides to study the impact of transformational leadership on staff performance in Federal College of Education, Zaria, Nigeria. Scholars have argued that continuous change requires employees to modify not only work routines but also social practices (e.g., relations with their managers, senior officers, and peers). To cope with the daily challenge of real-time adaptation, employees selectively retain effective elements of their performance routines and integrate them with new and more efficient ones.

Recently, In Federal College of Education, there has been issues bordering the effectiveness of leadership of the college as the management embarked on transforming the college in all ramifications while the employees asserted that they are not given fair treatment, no clear vision of change and employees development such as staff training and development and lack of employees participation in decision making. The problem under consideration in this study is to investigate the impact of the transformational leadership in the institution on staff performance.

Therefore in this paper, we determine the extent to which transformational leadership has impact on staff performance in Federal College of Education, Zaria, and find out whether the working environment is conducive for staff performance.

The rest of this paper is organized as follows: Section 2 presents transformational leadership and performance. Section 3 presents proposed method. Section 4 presents results and discussion. Finally, Sect. 5 presents conclusion and recommendation.

2 Rudimentary

2.1 Concept of Transformational Leadership

A lot has been written on leadership. Scholars, researchers have discussed extensively on leadership, leadership styles, theories of leadership, and transformational leadership is not exceptional. To use this approach in the workforce, one must first understand exactly what transformational leadership is? Sale in [3] stated that transformational leadership gives a vision and sense of mission, instills pride and gain, respect and trust which boast followers' morale to perform. According to Northouse in [1], transformational leadership is the ability to get people to want to change, to improve, and to be led. It involves assessing associates' motives, satisfying their needs, and valuing them. Therefore, a transformational leader could make the company more successful by valuing its associates. He posited that there are four factors to transformational leadership, (also known as the "four I's"): idealized influence, inspirational motivation, intellectual stimulation, and individual consideration. Each factor is discussed below.

- ***Idealized influence*** describes managers who are exemplary role models for associates. Managers with idealized influence can be trusted and respected by associates to make good decisions for the organization.
- ***Inspirational motivation*** describes managers who motivate associates to commit to the vision of the organization. Managers with inspirational motivation encourage team spirit to reach goals of increased revenue and market growth for the organization.
- ***Intellectual Stimulation*** describes managers who encourage innovation and creativity through challenging the normal beliefs or views of a group. Managers with intellectual stimulation promote critical thinking and problem solving to make the organization better.
- ***Individual consideration*** describes managers who act as coaches and advisors to the associates. Managers with individual consideration encourage associates to reach goals that help both the associates and the organization.

Effective transformational leadership results in performances that exceed organizational expectations [1]. He further stated that in 39 studies conducted in USA showed that individuals who exhibited transformational leadership were more effective leaders with better work outcomes (performance). This was true for both high- and low-level leaders in the public and private sectors. Each of the four components describes the characteristics that are valuable to the "transformation" process. When managers are strong role models, encouragers, innovators, and coaches, they are utilizing the "four I's" to help "transform" their associates into better, more productive, and successful individuals. In the same vein, Sale [3] noted that transformational leadership occurs when a leader transforms or changes his followers in three (3) important ways that together result in followers trusting the leader's performing behavior that contributes to the achievement of organizational goals and being motivated to perform at high levels, i.e., transformational leader includes subordinate awareness of the importance of their task and the importance of performing them well, he acquaints them of the needs for personal growth and development and further motivate their subordinate to work for the good of the organization rather than exclusively for their own personal gain or benefit. He went further that transformational leaders are charismatic leaders who have vision of how good things could be in an organization in contrast to how things currently are. Charismatic leaders clearly communicate the vision to their followers and the excitement and enthusiasm induce their followers to enthusiastically support the vision. Transformational leaders influence their followers by intellectually stimulating them to become aware of problem in their groups, organization, and view these problems from a new perspective and thus cause the followers to view problems differently and feel some degree of responsibility for helping to solve them.

Researches on transformational leadership ascertained that it is positively related to subordinates job satisfaction and job performance as well as organizational commitment. To further buttress home this assertion, these findings are given in different work setting considering many other factors as well. Like the work of Davenport [4], who found out that employees with internal locus of control are more committed toward organization's; similarly employees who follow transformational leaders are also more committed toward their organizations. Mert et al. [5] stated that transformational leadership style positively affect organizational commitment of followers. Another finding was given by Tseng and Kang in [6] and their study showed that there is positive and significant relationship between transformational leadership style and employees performance. Emery and Barker in [7], in their study posited that transformational leadership is positively correlated with the organizational commitment. Lo et al. in [8] stated that different angles of transformational leadership have positive relationship with organizational commitment and employees' performance. Lee in [9] conducted a research to find out effects of leadership style on organizational commitment; there results revealed that transformational style of leaders has direct bearing on commitment level of employees. Ekeland [10] also found out that transformational leadership has positive effect on follower's affective commitment as they bring a positive change in those who follow them. They are mostly vigorous, keen, ardent, and paying attention on the success of every member of the group. As past research shows that commitment is affected by employees attitude, their work behavior, motivation and performance and transformational leadership are positively linked with all these which enhance the level of commitment both at employees as well as at organizational level [11]. Lowe in [12] argued that transformational leaders transform their followers' aspirations, identities, needs, preferences, and values such that followers are able to reach their full potential. This is because followers of transformational leaders are expected to identify with their leaders and therefore are expected to have greater feelings that they can have an impact on their organization through enhancements to their psychological empowerment. This also confirmed to study of Spence Laschinger in [13] that transformational leaders get followers involved in envisioning an attractive future and inspire them to be committed to achieving that future. They build team spirit through enthusiasm, high moral standards, integrity, and optimism and provide meaning and challenge to their followers' work, enhancing followers' level of self-efficacy, confidence, meaning, and self-determination. However, Erkutlu in [14] found that transformational leadership has positive relationship with employees' commitment but when organizational culture operates between transformational leadership, it does not positively influence the organizational commitment of its employees as well as performance. Afolabi et al. in [15] did analysis on both transformational and transactional leadership style and posited that transactional leadership which is a leadership style based on carrot and stick approach is more effective when organization desires to achieve their aims and objectives as compared to transformational leadership.

This is in line with Franke and Felfe in [16] whom asserted that supervisors play a vital role in enhancing employees' performance and job satisfaction in attaining the end goals of an organizational by giving a new direction of thinking, but on other side transformational leaders can some time exploit workers using their power to achieve personal goals at the expense of the organization. In spite of these divergent views on transformational leadership, the researcher sees transformational leadership as an approach whereby leaders communicate the vision and mission of the organization by inspiring confidence and trust on the subordinate in the attainment of his full potentials and that of the organization.

2.2 An Overview of Productivity/Performance

The concept of productivity or performance is often vaguely defined and poorly understood though it is a widely discussed topic. Different meanings and interpretations had been associated with its definition. Expert from all works of life have looked at it from different perspective. Drucker in [17] stated that "productivity at organizational level (organizational productivity) relates to how well an organization is able to utilize it resources (human, material, and machine) at a minimum cost with great efficiency and effectiveness. According to Jekelle in [18] that "productivity relates to quality of work life, especially when it is defined in terms of individual internal work standard or output, it includes such behavior as low turnover, absenteeism, grievances, strikes, union activities." Drucker in [17] views productivity at the organizational level, while Jekelle in [18] assesses it from the view point of quality of work life. Productivity is best understood when both perspectives are aligned. She asserted that individual productivity of employees transcends to total organizational performance which is in tandem with view of [17]. Odu and Chukwura in [19] see productivity "as the ratio between output of wealth produced in the form of goals and services and inputs of resources used up in two process of production."

Nwagbo in [20] stated that productivity is not production nor is it performance! Nor is it result or outcome! These three variables are components of productivity's effort but they are not equivalent terms. Traditionally, most people associate productivity with production and manufacturing because that is where it is most visible and measurable.

Sardana and Vrat in [21] posit that the terms "performance" and "productivity" are incorrectly used. People who claim to be discussing "productivity" are actually looking at the more general issue of performance. The "productivity" is a fairly specific concept while performance includes many more attributes. Odu and Chukwura in [19], Nwagbo in [20], and Sardana and Vrat in [21] have similar dispositions on productivity and performance. Odu and Chukwura in [19] view it on the basis of output of wealth produced; Nwagbo in [20] and Sardana and Vrat in [21] are concerned with the way the terms "performance" and "productivity" are misunderstood. They argue that most times they are used interchangeably instead of being seen as totally different concepts. They have also asserted that economist have supported this view by defining productivity as output per labor cost but this view must be changed to incorporate all segments of work life. Government, health institutions, education service industries, and professional groups are also interested in and concerned about productivity. As a matter of fact, productivity touches all of us as customers, taxpayers, citizens, and members of organization. This broad view of productivity requires a broad definition which seems to be satisfied by Mali in [22] sees productivity as "The measure of how well resources are brought together in organization and utilized for accomplishing a set of results." According to him, productivity entails reaching the highest level of performance with the least expenditure of resources. This is simply the best outcome for efficiency and effectiveness of the employees input and output.

2.3 Relationship Between Transformational Leadership and Performance

Transformational leadership and performance has gained considerable attention by organizations due to the fact that the main theme of every organization is to enhance employee performance. Behery in [23] posited that a relation exists between transformational leadership and employee's performance as sound organizational communication act in way of fostering workforce by transmitting cultural norms from an organizational framework to an individual's way of life in the organization and by supporting style of manager plays an incredible role for increasing employee's performance. Moreover, transformational leader may be more likely to instill trust on subordinates; this can elevate member's perception of procedural justice which in turn harnesses improvement in employee's performance. Bass in [24] created a model describing how transformational leadership influences followers' performance (See Fig. 1).

Howell and Hall-Merenda in [25] suggested that transformational leadership will play an imperative role in increasing job satisfaction as well as role play to achieving organization's goal and employees' acts. Walumbwa et al. [26] expressed transformational leadership as being correlated with subordinate skills with work worth to asses' employees' performance. They train their workers, arrange meetings with their subordinates, and accept feedback from their subordinates, and in end, employee performance is added.

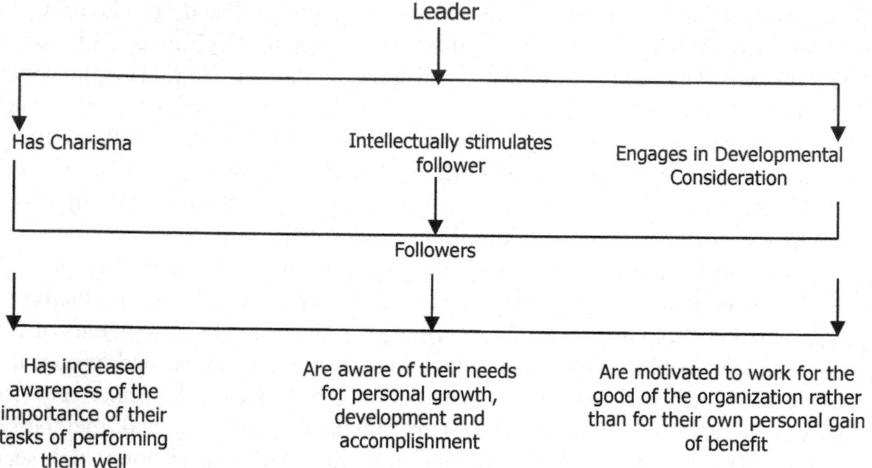

Fig. 1. Transformational leadership [24]

2.4 Theoretical Framework

Transformational leadership, however, covers a wide range of aspects within leadership; there are no specific steps for a manager to follow in becoming an effective transformational leader. Therefore, understanding the basics of transformational leadership and the four I's (idealized influence, inspirational motivation, intellectual stimulation, and individual consideration) can help a manager apply this approach. According to [1], a transformational leader has the following qualities:

- Empowers followers to do what is best for the organization;
- Is a strong role model with high values;
- Listens to all viewpoints to develop a spirit of cooperation;
- Creates a vision, using people in the organization;
- Acts as a change agent within the organization by setting an example of how to initiate and implement change;
- Helps the organization by helping others contribute to the organization.

Therefore, transformational leadership is a vital role for effective managers because leader effectiveness determines the success level of the organization. According to Hesselbein et al. in [27], organizations that take the time to teach leadership are far ahead of the competition. By becoming familiar with the transformational leadership approach and combining the four I's, managers can become effective leaders in the business world. In view of the above, transformational leadership can be applied in one-on-one or group situations. Using this approach, the manager (leader) and the associates (followers) are "transformed" to enhance job performance and help the organization to be more productive and successful. The theory of transformational postulated by Northouse in [1] will underpin this study as it explains the key components of transformational leadership as it affects employees performance.

3 Proposed Method

For any research work to be useful and meaningful it must be analyzed and interpreted in the light of the conditions under which the study was conducted. The regression as a tool was used for hypothesis testing and proper interpretation of data collected. However, the analysis of data was conducted using SPSS version 20.

3.1 Data Presentation and Analysis

The purpose of data presentation is to analyze the data collected which will facilitates comprehension and further explanation of the research findings. The questionnaire was adopted from transformational leadership questions and questions on employees' performance. The Cronbach alpha test was conducted for reliability test and the result shows that the Cronbach alpha for transformational leadership is (0.637) while for staff performance is (0.568). Therefore, the two values are greater than 0.4 which indicates good internal consistency of the items in the scale, thus the questionnaire can be relied upon. A total of three hundred and thirty-two questionnaires (332) were administered out of which two hundred and forty-eight (248) were duly completed and returned and for the purpose of this study under review, the completed questionnaire of 74.7% would be analyzed and interpreted which served as a good representation for the study.

3.2 Regression Analysis

Table 1 as follow presents the model summary results.

Table 1. Model summary results

Model summary									
Model	R	R Sq	Adjusted R square	Std. error of the estimate	Change statistics				
					R square change	F change	df1	df2	Sig. F change
1	0.825[a]	0.681	0.685	0.2143	0.088	23.823	1	246	0.000

Source SPSS output
[a]Predictors: (Constant), TL

The model summary (see Table 1) shows the adjusted *R* square value of 0.685 indicates that 68% of behavior of the dependent variable was explained by the independent variable and the standard error of the estimate is 0.214 in this case, the coefficient of determination is strong because values of *R* square is above 0.4, which is considered strong. Therefore, going by this *R* square, transformational leadership appeared a strong variable for predicting staff performance in FCE, Zaria.

Table 2 shows *p*-value of 0.00 which is far below the significant level of 0.05 and the *F*-statistics of 23.823, therefore, the *F*-statistics is large enough for this paper to reject the null hypothesis.

Table 2. Analysis of variance results

ANOVA[a]						
Model		Sum of squares	DF	Mean square	*F*	Sig.
1	Regression	6.477	1	6.477	23.823	0.000[b]
	Residual	66.885	246	0.272		
	Total	73.362	247			

Source SPSS output
[a]Dependent variable: SP
[b]Predictors: (Constant), TL

Table 3. Calculated coefficient results

Coefficients[a]						
Model		Unstandardized coefficients		Standardized coefficients	*t*	Sig.
		B	Std error	Beta		
1	(Constant)	2.793	0.226		4.386	0.000
	TL	0.305	0.063	0.297	1.881	0.000

Source SPSS output
[a]Dependent variable: SP

From Table 3, the value of *t*-statistic for transformational leadership is (1.881) and *p*-value (0.00) which is less than the critical value of (2.58) at 5% level of significant. Therefore, the hypothesis which states that there is no significant relationship between transformational leadership and staff performance is rejected and the alternate hypothesis accepted. This is in consistency with the work of [6, 10], and [12] that there is a significant relationship between transformational leadership and employees' performance.

4 Results and Discussion

In this section, we present and analyze the data collected in this study as it tends to explore the evaluation of virtual organizational structure on employee performance of some selected telecommunication companies.

4.1 Data Presentation and Analysis

The purpose of data presentation is to analyze the data collected to facilitate thorough understanding and further explanation of the research findings. The questionnaire comprise of three parts, part A is the personal data of the employee, part B deals with questions on virtual organizations, and part C deals with employee performance. The researcher administered one hundred and seventy-five (175) questionnaires accordingly out of which one hundred and thirty (130) were duly completed and returned and for the purpose of this seminar paper, the completed and returned questionnaires of (96%) forms a good representation for the paper, in other words the total respondents in this particular questions shall represent hundred percent (100%).

The regression Table (see Table 4) shows the adjusted *R* Square value of 0.293 indicating that 29% behavior of the dependent variable was explained by the independent variable and the standard error estimate is 0.506. Therefore, going by the R square, it can be stated that virtual organization structure appears a weak variable for predicting employees' performance in telecommunication in Kaduna. Furthermore, the *F*-statistics of the model at 54.451 is fit since it acceptable region ranges from 1.96 and above, the sig which indicated the goodness of the data really fits. The p-value of 0.00 is below the significant level of 0.05, and the *F*-statistics of 54.451 is large enough for the null hypothesis which stated that virtual organization structure has no significant effect on employees' performance be rejected and accept the alternative hypothesis virtual organization structure has significant effect on employees' performance.

Table 4. Result of regression analysis

Hypothesis	Relation	Beta	Std error	*t*-value	*p*-value	Decision
HO_I	Virtual organizational	0546	0.506	7.379	0.000	Rejected
	Structure has no significant					
	Effect on employees					
	Performance					
Cronbach's alpha						
VO	0.88					
EP	0.91					
R	0.546					
R^2	0.298					
F	54.451					
Sig	0.00					

Note N = 135 dependent variable (employee performance) $p < 0.005$**

5 Conclusion and Recommendation

This study concluded that transformation leadership is a leadership style that is visionary, innovative which considers, and motivates employees to commit the vision of the organization by intellectually stimulating them to achieve their goals and that of the organization. Finally, it is evident from the data analyzed and interpreted that a sound and viable leadership with individual consideration at heart encourages innovation and creativity will in turn stimulate employees in enhancing their performance toward achieving both their goals and that of the organization. Based on the findings stated above, the following recommendations should be implemented to improve employees' performance in Federal College of Education.

- Management should adopt fully the transformational leadership role with leadership qualities such as role modeling, perseverance, empathy, pragmatism, visionary, innovative, coaching, stimulating, and valuing employees' passion for growth in manning the affairs of the college. These qualities when properly harnessed will encourage employees toward improving their performances.
- Staff should be given freedom in making decisions relating to their work and employees' consultation in decisions that affects their wellbeing should be sought for so as to boast their morale and in turn enhance higher performance in the college.

For the best outcome in any organization, the relationship between the management, and employees is imperative as it leads to improved performance of employees as well as achievement of the organizational goals. Finally, it is also recommended that further studies should be carried out on transformational leadership as it relates to organizational commitment and employees job satisfaction.

Acknowledgements. The work of Abulwafa Muhammad is supported by Universitas Putra Indonesia YPTK under research grant no 094/UPI-YPTK/RG/IX/2016. The work of Tutut Herawan is supported by Universitas Teknologi Yogyakarta Research Grant no vote O7/UTY-R/SK/0/X/2013.

References

1. Northouse, P.G., *Leadership: Theory and practice*. 2012: Sage.
2. Daft, R.L. and D. Marcic, *Management: the new workplace*. 2008: Evans Publishing Group.
3. Sale, F., *Bass And Stogdills Handbook Of Leadership-Theory, Research, And Managerial Applications-BASS, BM*. 1991, Personnel Psychology Inc 745 Haskins Road, Suite A, Bowling Green, Oh 43402.
4. Davenport, J. *Leadership style and organizational commitment: the moderating effect of locus of control*. In *ASBBS Annual Conference: Las Vegas*. 2010.
5. . Mert, I.S., T. Baş, and N. Keskin, *Leadership style and organizational commitment: Test of a theory in Turkish banking sector*. Journal of Academic Research in Economics (JARE), 2010(1): pp. 1–19.
6. Tseng, H. and L. Kang. *How Does Regulatory Focus Affect Organisational Commitment? Transformational Leadership as a Mediator*. In *Proceedings of the ACME Conference*. 2009.
7. Emery, C.R. and K.J. Barker, *The effect of transactional and transformational leadership styles on the organizational commitment and job satisfaction of customer contact personnel*. Journal of Organizational Culture, Communication and Conflict, 2007. **11**(1): p. 77.
8. Lo, M.-C., T. Ramayah, and H.W. Min, *Leadership styles and organizational commitment: a test on Malaysia manufacturing industry*. African Journal of Marketing Management, 2009. **1**(6): pp. 133–139.
9. Lee, J., *Effects of leadership and leader-member exchange on commitment*. Leadership & Organization Development Journal, 2005. **26**(8): pp. 655–672.
10. Ekeland, T.P., *The relationships among affective organizational commitment, transformational leadership style, and unit organizational effectiveness within the corps of cadets at texas A&M University*. 2006, Texas A&M University.

11. Dumdum, U.R., K.B. Lowe, and B.J. Avolio, *A meta-analysis of transformational and transactional leadership correlates of effectiveness and satisfaction: An update and extension*, in *Transformational and Charismatic Leadership: The Road Ahead 10th Anniversary Edition*. 2013, Emerald Group Publishing Limited. pp. 39–70.
12. Lowe, K.B., K.G. Kroeck, and N. Sivasubramaniam, *Effectiveness correlates of transformational and transactional leadership: A meta-analytic review of the MLQ literature*. The leadership quarterly, 1996. **7**(3): pp. 385–425.
13. Spence Laschinger, H.K., J. Finegan, and J. Shamian, *The impact of workplace empowerment, organizational trust on staff nurses' work satisfaction and organizational commitment*, in *Advances in Health Care Management*. 2002, Emerald Group Publishing Limited. pp. 59–85.
14. Erkutlu, H., *The impact of transformational leadership on organizational and leadership effectiveness*. The Journal of Management Development, 2008. **27**(7): p. 708.
15. Afolabi, O., et al., *Influence of gender and leadership style on career commitment and job performance of subordinates*. Global Journal of Humanities, 2008. **7**(1/2): p. 1.
16. Franke, F. and J. Felfe, *How does transformational leadership impact employees' psychological strain? Examining differentiated effects and the moderating role of affective organizational commitment*. Leadership, 2011. **7**(3): pp. 295–316.
17. Drucker, P., *Management: Tasks, Responsibilities and Practices, Harper & Row*. New York, NY, 1974.
18. Jekelle, H. *Employment Policies and Public Service Productivity*. In *A Paper presented at the Proceedings of the First National Conference on Productivity, Published by National Productivity Centre, Abuja, Nigeria*. 1987.
19. Odu, C. and C. Chukwura. *Preliminary Investigation Into the Effects of Drilling Mud Additives, Chrome-Lignosulfocate (Spersene) and Chrome Lignite (Xp-20), on Symbiotic Nitrogen Fixation*. In *NNPC Seminar on the Petroleum Industry and the Nigerian Environment, Port-Harcourt*. 1983.
20. Nwagbo, E., *Relative performance indices of institutional agricultural credit in Funtua LGA of Nigeria*. Nigerian journal of rural development and cooperative studies, 1986. **1**(3): pp. 146–162.
21. Sardana, G. and P. Vrat, *Productivity measurement in a large organization with multi-performance objectives: A case study*. Engineering Management International, 1987. **4**(2): pp. 105–125.
22. Mali, P., *Improving Total Productivity: MBO Strategies for Business Government and Non Profit Organization*. 1978, New York: John Wiley & Sons.
23. Behery, M.H., *Retracted: Leadership behaviors that really count in an organization's performance in the middle east: The case of Dubai*. Journal of Leadership Studies, 2008. **2**(2): pp. 6–21.
24. Bass, B.M., *From transactional to transformational leadership: Learning to share the vision*. Organizational dynamics, 1990. **18**(3): pp. 19–31.
25. Howell, J.M. and K.E. Hall-Merenda, *The ties that bind: The impact of leader-member exchange, transformational and transactional leadership, and distance on predicting follower performance*. Journal of applied psychology, 1999. **84**(5): p. 680.
26. Walumbwa, F., et al. *Unlocking the mask: Understanding the multiple influence of authentic leadership*. In *University of Nebraska Gallup Leadership Institute Authentic Leadership Conference, Omaha, NE*. 2004.
27. Hesselbein, F., M. Goldsmith, and R. Beckhard, *The leader of the future*. 1996: Jossey Bass.

Future Private Cloud Architecture for Universities

Saqib Hakak[1], Gulshan Amin Gilkar[2], Guslendra[3], Rajab Ritonga[4], and Tutut Herawan[5,6](✉)

[1] Department of Computer Science and Information Technology, University Malaya, Kuala Lumpur, Malaysia
saqibhakak@gmail.com

[2] Department of Computer Engineering, Shaqra University, Shaqra, Saudi Arabia
amingulshan9@gmail.com

[3] Universitas Putra Indonesia (YPTK), Padang, Indonesia
guslendra@upiyptk.ac.id

[4] State Polytechnic of Malang, East Java, Indonesia
rajab.ritonga@dsn.moestopo.ac.id

[5] Universitas Teknologi Yogyakarta, Yogyakarta, Indonesia
tutut@uty.ac.id

[6] AMCS Research Center, Yogyakarta, Indonesia

Abstract. Cloud computing is one of the new and recently emerged research field which offers resource on demand facility. Education is must for each society, and this medium of education is imparted through schools, colleges, and universities. Universities are regarded as highest and elite learning centers. Technology is rapidly progressing day by day and so the universities have to update their technical resources too like new tools, equipment, and so on, so that students can learn new technologies and be better professionals in future. However, this is quite tedious task as it involves lot of budget considerations. Even some universities which cannot afford to cope with time due to inadequate facilities are on verge of extinction. Thus, concept of cloud computing can be implemented in Universities, and using cloud will not involve much expense as compared to owning equipment. Hence, in this paper, theoretical concept of architecture as how private cloud can be used in universities to impart better education is proposed.

Keywords: Cloud computing · Architecture · Educational institutions

1 Introduction

Cloud computing is a kind of distributed architecture that centralizes server resources on a scalable platform so as to provide on-demand computing resources and services [1, 2]. In case of cloud computing, cloud platforms are offered to clients so that they may use their own cloud as they desire by cloud computing providers. The concept is same as that of Internet service providers who provide wide quality range of services to their customers in terms of surfing speed, downloading speed, and much more other

V. Bhateja et al. (eds.), *Information Systems Design and Intelligent Applications*, Advances in Intelligent Systems and Computing 672,
https://doi.org/10.1007/978-981-10-7512-4_71

services. It is an information technology deployment model that involves entrusting data to information systems managed by external parties on remote servers or "in the cloud." [1, 3]. Currently, there are three types of services offered by cloud computing providers which are Software as a Service (SaaS), Platform as a Service (PaaS), and Infrastructure as a Service (IaaS) [1, 4]. Besides, there are four types of cloud which differ only in the type of services, those clouds offer, and these are public cloud, private cloud, community cloud, and hybrid cloud.

With the time, technology is getting more sophisticated and so as the competition. The highest learning centers, i.e., the universities need to cope with very high pace in order to equip students with the must skills so that after their University studies, they can do something for their society by their intellectual ways, skills, and thinking. And getting cloud computing concept in education system is a must step for each University [5]. With the help of cloud computing, students within universities can make use of all tools or services which a cloud will offer, and University does not have to spend much amount of money as compared to getting new tools and equipment. The present research tries to address all this critical issue by proposing a conceptual model of cloud for implementation within Universities.

The organization of paper is as follows: Sect. 2 is related to the concept of cloud computing technology along with its services and types. Section 3 is related to proposed architecture for bringing cloud computing concept in educational institutions. Section 4 is related to benefits of using cloud in educational institutions. The paper is concluded in Sect. 5.

2 Cloud Computing

The concept of cloud computing is quite simple. It is that type of computing which relies on sharing of computer resources rather than having personal devices or servers to handle applications [6]. It is completely Internet-dependent technology, and each client is assigned its own cloud with the help of which that client can use different services like services of servers, storage, manage applications, and so on as per pay service basis. Here, the word cloud is used as metaphor for the Internet [6, 7]. In Fig. 1,

Fig. 1. Cloud computing [7]

general scenario of cloud computing concept is given. Cloud is to be adapted in between Internet service providers (ISP) and end users, where end users are allowed to select any type of required service like PAAS, IAAS, SAAS, and so on.

The primary aim of cloud computing is to perform computations in most efficient way and as fast as it can, say trillions of computations per second [6], and in order to achieve this, it makes use of large number of servers running low-cost consumer PC technology with specialized connections to spread data processing chores across them, and virtualization techniques are used to maximize the power of cloud computing [6, 7]. The cloud computing has four basic deployment models, which are private cloud, public cloud, hybrid cloud, and community cloud [1]. Following is the description of these four deployment models:

2.1 Private Cloud

It is that type of cloud which is private, i.e., owned, leased, or managed by any private organization. This kind of cloud is usually dedicated to single organizations compared to public cloud, that is dedicated to multiple organizations [1]. The users or service providers who use private cloud usually have improved security as compared to public cloud although private cloud is more expensive than public cloud. However, there is no further addition in terms of legal requirements, security regulations, and bandwidth limitations as compared to public cloud. It is more secure due to optimized control of infrastructure than public cloud [1, 7]. Eucalyptus system is one of the best examples of private cloud [8].

2.2 Public Cloud

It is that type of cloud in which users have little control over the cloud, and the control is within third party. In this type of cloud, users simply use the services provided by any third party and are charged according to usage. The third party alone is responsible for all activities of cloud like installation activity, maintaining activity, or resource-providing activity to the end user. At a time, many organizations can use the same infrastructure, and users are given resources dynamically. In this kind of cloud, no access or authentication techniques can be implemented [1, 9]. One of the best examples of this kind of cloud is Google search engine.

2.3 Hybrid Cloud

It is that kind of cloud which is composed of more than one cloud deployment model. In this model, the transfer of data takes place through a hybrid model, which is composed of two or more different models in such a way that transfer of data through two models does not affect each other. The companies can outline the needed services and goals with the help of this model [8]. This model can be useful for some secure services like payroll processing or receiving customer payments [1]. However, there is a major drawback to this kind of cloud, i.e., to get different services from different sources and to implement that from a single location. Besides, even the implementation can get further difficult and complicated if there is an interaction between public and private cloud. Amazon Web services are a good example of Hybrid cloud [1, 10].

2.4 Community Cloud

This kind of cloud is just for a shared cause in which many organizations have some agreement and based on that agreement, these organizations use this kind of cloud and share the required infrastructure such as for banking purposes or any other domain [11]. This type of cloud can be managed either by group of companies or any other third party. One of the best and known to almost everyone is Facebook, which is a community cloud [12].

2.5 Services Offered by Cloud Computing

The services offered by cloud are numerous. However, the main services provided by cloud computing include: (Fig. 2).

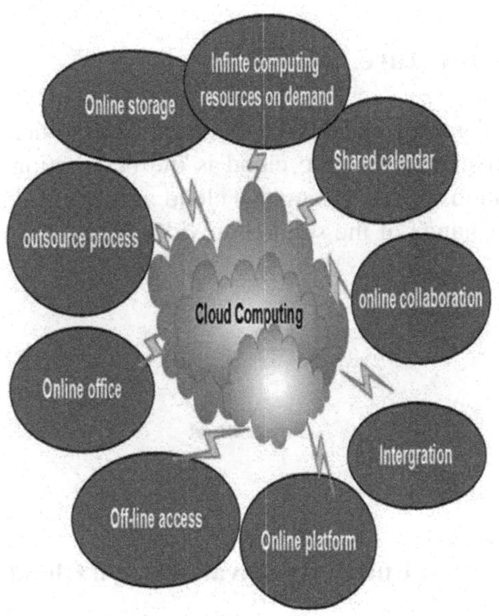

Fig. 2. Services of cloud computing [8]

2.5.1 PAAS (Platform as a Service)

This service is utilized by Web developers for the application development and deployment platform purposes over the Web. All applications with the help of programming languages and tools are developed here. And this service offered by cloud computing is very useful for the developers as they can use the equipment without requiring to have those equipment physically and can develop quality programs and then transfer to the end users via Internet, thus reducing both cost as well as complexity. This is where applications are developed using a set of programming languages and tools that are supported by the PaaS provider [8, 13].

2.5.2 IAAS (Infrastructure as a Service)

In this service, the users acquire computing resources like memory, processing power and storage from IAAS provider, and make use of resources to run and deploy their applications. It allows users to provide resources on demand without requiring any long-term commitment [8, 14].

2.5.3 SAAS (Software as a Service)

In this type of service, the users simply view the software developed by others through Web browser and offer a service to users to buy it. In other words, we can say that, in SAAS, readymade software is developed by some third party and then that particular software is made available to users for using that software. Besides, there are many advantages of this service like users do not need to bother about the version of software as the software gets updated automatically. Also, software will be globally accessible [14, 15].

3 Proposed Architecture

As mentioned above, there are many types of clouds like public, private, hybrid, and so on but in case of universities, private cloud is the best option as control is more in hands of the organization which is using this cloud as compared to public cloud, where control is more in the hands of the service provider [7].

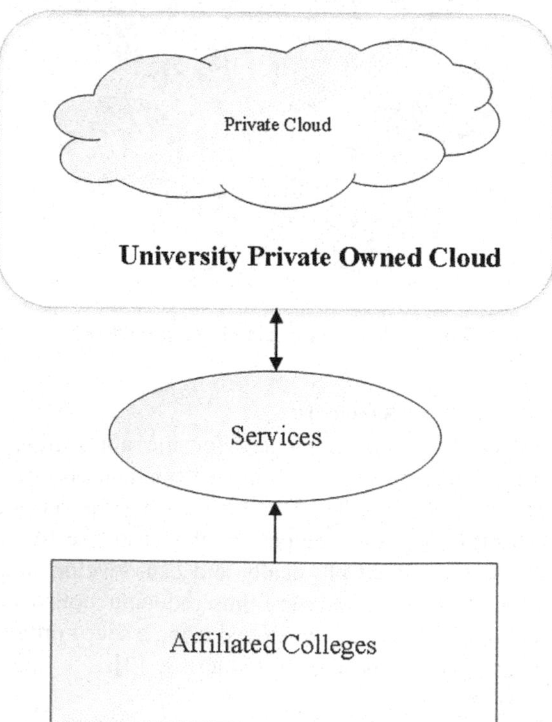

Fig. 3. General scenario of cloud computing service for University and affiliated colleges

The concept is quite simple and general as shown in Fig. 3. There will be private cloud from which University will access it all required services, and in return, other colleges affiliated with that University can also access the required services of the cloud. There can be use of proxy cloud also as shown in Fig. 4 that can communicate with affiliated educational institutions and offer them on-demand services which will considerably reduce burden of entertaining requests from main private cloud of the University.

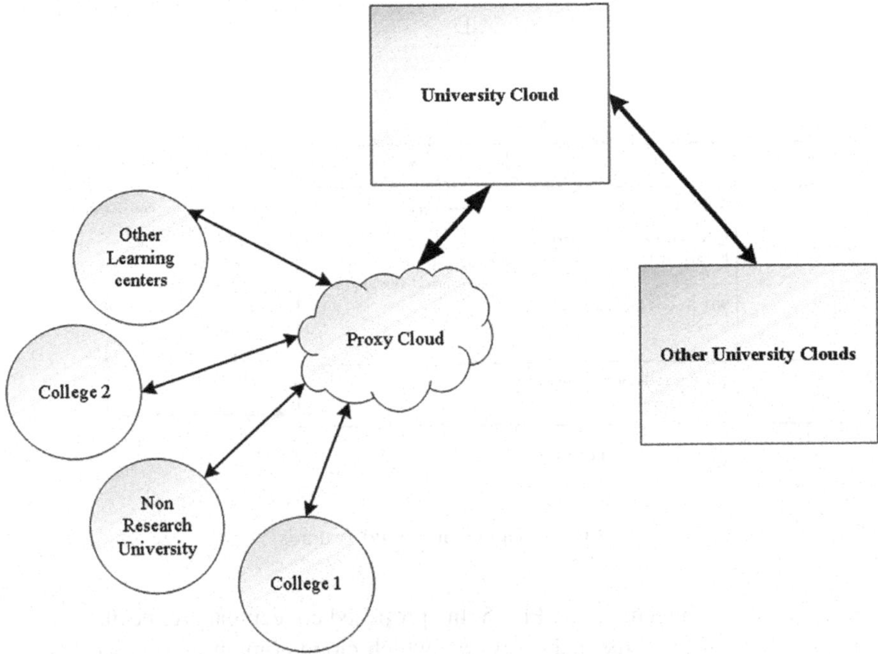

Fig. 4. Cloud computing service for University and affiliated colleges

The first step in this process will be to connect private cloud service within the main University campus itself like connecting its services with different departments and so on. And the second step will be to make use of this cloud service to the affiliated colleges of the University, and future work can be done to connect this cloud with clouds of other universities for further knowledge sharing and making education system more comprehensive and better. The main services that University cloud can provide include:

a. Access to latest software tools.
b. Access to Standard Digital Libraries like IEEE, Web of science, etc.
c. Access to E-books.
d. Access to simulators.
e. Access to networking tools.

f. Access to research grants available in affiliated universities.

These are few main services that can be implemented within the University cloud. Once implemented, the services can be added based on students and academician's feedback. The detailed overview of University Cloud is presented in Fig. 5.

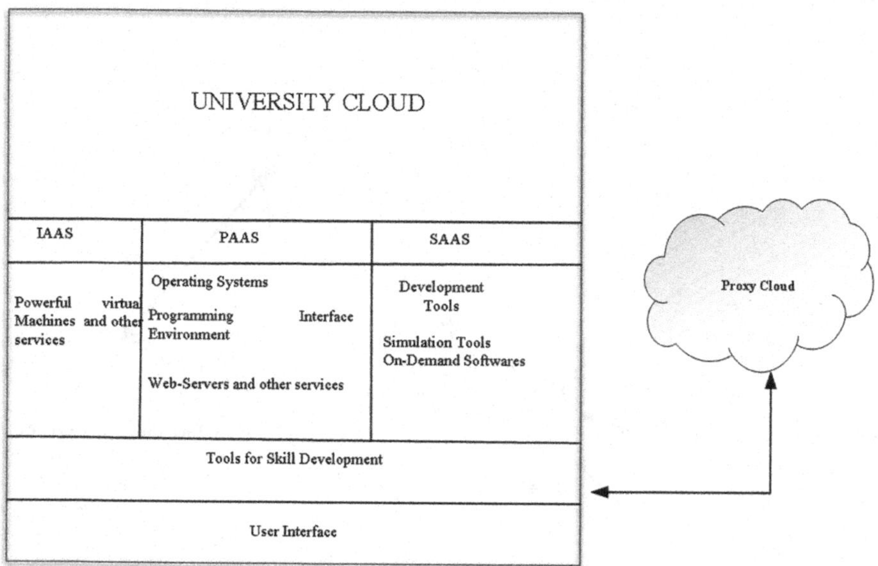

Fig. 5. University cloud features

In the above architecture, i.e., Fig. 5, the proposed conceptual architecture is given. As all are aware of the three main services, which cloud computing offers to its users, but for universities, these services must be classified further. Generally, students and faculty members do not need many services which professionals need. Usually, students need to use simple software tools for completing projects or they need some quality video tutorials for grasping any concept or they need any research material. Similarly, faculty members also do not need much heavy tools or services as compared to professionals. Thus, in order to make process simpler and without overloading the cloud servers, it will be best for cloud service providers to make their services further simpler like, in this paper, the main three services of cloud computing have been further divided into five more categories. By this classification, it will prove much useful to universities in terms of costs, functionality, speed, and so on. University does not need to attain full-fledged services of the cloud. University can only avail few major services like in *(i). Processing* area where universities can attain the benefit of using processing functionality of computers. Similarly, *(ii). reading content service,* where universities can avail the benefit having access to top tutorial, research papers, and so on. Similarly, there is another service of *(iii). innovative ideas*, which will be very useful for students. In this area, new ideas or future work-related concepts should be there so that students can get the insight of latest trends and it will illuminate them further to get involve in research and invent new things. Other two services include *(iv). Development tools* and *(v). storage services*.

All facilities will be accessed via main server of University provided by cloud. Although, within universities premises, the access to services will be faster but for colleges affiliated to that particular University which is using cloud service, the access to services can be slower depending on the traffic conditions. For that purpose, proxy cloud server can be used as already mentioned above within each college, which can hold records of most widely used services in that college, and this will also reduce burden on main server of University. For example, in one particular college, huge demand is for reading material content, and proxy cloud server of that college can store requests for that service and can fulfill this service instead of main server. These proxy servers will also have to be handled by cloud service providers but based on limited functionality.

4 Benefits of Using Cloud

There will be lot of benefits/advantages for implementing cloud computing concept at University level. Some of the major advantages are listed below.

4.1 Average Hardware Specification of Systems

Cloud computing provides many wonderful services like online storage service, which means that there is no need to buy costly hard drives. It also provides services of processing/computing, which further means universities will not need to buy costly processors which are much expensive [8]. Thus, cloud computing will save considerable amount of expenses. However, University will have to pay cloud service provider the usage charges but still renting charges are always cheaper as compared to buy new equipment. Thus, cloud computing concept for universities will provide economical as a whole.

4.2 Savings in Buying New Operating Systems

Cloud computing will also remove the overhead of buying new operating systems from the vendors and then maintenance of those systems due to the fact that cloud computing will provide itself interface to the users. Users only need open-source operating systems, which are easily and freely available. Users have to simply open their browsers using Internet, and interface will be provided by cloud provider itself.

4.3 Ease of Access to Good Quality Study Material

Access to World topmost video tutorials, quality research papers, and other useful study materials can be provided to the students as well as to the University faculty members at one place. It will save time also for the students who have to surf Internet for some basic information or sometimes trying to find some video tutorial on Web sites like YouTube takes whole day.

4.4 Benefit of Using Latest Tools/Software

One of the most important features which will be most useful for the universities will be: Universities can request to upgrade the software, hardware, and so on, which will be useful for students as well as universities. It will be useful for the students in the sense that they will be always equipped with latest software and tools. And for universities, it will be useful in the sense that University does not have to buy latest hardware tools and software as that work will be done by cloud service provider.

4.5 Ease of Access to Share Information with Clouds of Other Universities

One more important feature of implementing cloud computing concept in universities will be: Universities can share their private cloud by making it public for some duration of time so that two or more universities can share some useful information among themselves related to knowledge-related matters or so on.

4.6 End to Pirated Software

Moreover, this concept of implementing cloud computing in academic institutions can at least stop 50% of software pirating as mostly students use pirated software as they usually cannot afford buying licensed versions of software.

5 Conclusion and Future Work

The main purpose and goal of this paper is to give a notation as how cloud computing can be implemented within educational institutions. In this paper, theoretical architecture is proposed. There will be lot of advantages of implementing cloud computing concept in educational institutions. Some major advantages include access to latest and licensed tools, access to reading content, simulators, enhancement of skill development, and so on. However, for a cloud service provider, it needs huge investment in the initial stages but later on same concept will prove beneficial for the whole world. At the very initial stage, government bodies can also help in setting up this kind of infrastructure. Future work will be to make this proposed architecture more concrete by evaluating flaws on simulator and finally implementing this proposed architecture to some University for verification.

Acknowledgements. The work of Guslendra is supported by Universitas Putra Indonesia YPTK under research grant no 094/UPI-YPTK/RG/IX/2016. The work of Tutut Herawan is supported by Universitas Teknologi Yogyakarta Research Grant no vote O7/UTY-R/SK/0/X/2013.

References

1. Padhy RP, Patra MR, Satapathy SC. Cloud computing: security issues and research challenges. *International Journal of Computer Science and Information Technology & Security* (IJCSITS). 2011;1(2):136–46.
2. Armbrust M, Fox A, Griffith R, Joseph AD, Katz R, Konwinski A, et al. A view of cloud computing. *Communications of the ACM*. 2010;53(4):50–8.
3. Arpaci I, Kilicer K, Bardakci S. Effects of security and privacy concerns on educational use of cloud services. *Computers in Human Behavior*. 2015;45:93–8.
4. T Ograph B, and Morgens YR. Cloud computing. *Communications of the ACM*. 2008;51(7).
5. Tashkandi AN, Al-Jabri IM. Cloud computing adoption by higher education institutions in Saudi Arabia: an exploratory study. *Cluster Computing*. 2015;18(4):1527–37.
6. González-Martínez JA, Bote-Lorenzo ML, Gómez-Sánchez E, Cano-Parra R. Cloud computing and education: A state-of-the-art survey. *Computers & Education*. 2015;80:132–51.
7. Hakak S, Latif SA, Amin G. A review on mobile cloud computing and issues in it. *International Journal of Computer Applications*. 2013;75(11).
8. Olanrewaju RF, Hakak SI, Khalifa O. MARKCLOUD-Software for Data Confidentiality and Security in Cloud Computing Environments. *Proceedings of the 2014 International Conference on Industrial Engineering and Operations Management*, Bali, Indonesia, January 7–9, 2014.
9. Ren K, Wang C, Wang Q. Security challenges for the public cloud. *IEEE Internet Computing*. 2012;16(1):69.
10. Sotomayor B, Montero RS, Llorente IM, Foster I. Virtual infrastructure management in private and hybrid clouds. IEEE Internet computing. 2009;13(5):14–22.
11. Jensen M, J\, \#246, Schwenk r, Gruschka N, Iacono LL. On Technical Security Issues in Cloud Computing. *Proceedings of the 2009 IEEE International Conference on Cloud Computing*. IEEE Computer Society; 2009, 109–16.
12. Marinos A, Briscoe G, editors. Community cloud computing. *IEEE International Conference on Cloud Computing*; 2009: Springer.
13. Assunção MD, Calheiros RN, Bianchi S, Netto MA, Buyya R. Big Data computing and clouds: Trends and future directions. *Journal of Parallel and Distributed Computing*. 2015;79:3–15.
14. Puthal D, Sahoo B, Mishra S, Swain S, editors. Cloud computing features, issues, and challenges: a big picture. *Proceedings of the International Conference on Computational Intelligence and Networks* (CINE), 2015; IEEE Press.
15. Rittinghouse JW, Ransome JF. *Cloud computing: implementation, management, and security*: CRC press; 2016.

Towards a Privacy Mechanism for Preventing Malicious Collusion of Multiple Service Providers (SPs) on the Cloud

Maria M. Abur[1], Sahalu B. Junaidu[1], Sani Danjuma[2], Syafri Arlis[3], Rajab Ritonga[4], and Tutut Herawan[5,6(✉)]

[1] Department of Computer Science, Ahmadu Bello University, Zaria, Nigeria
{mmaburl,abuyusra}@gmail.com
[2] Department of Computer Science, Northwest University, Kano, Nigeria
sani_danjuma@yahoo.com
[3] Universitas Putra Indonesia (YPTK), Padang, Indonesia
syafri_arlis@upiyptk.ac.id
[4] Universitas Prof Dr Moestopo (Beragama), Jakarta, Indonesia
rajab.ritonga@dsn.moestopo.ac.id
[5] Universitas Teknologi Yogyakarta, Yogyakarta, Indonesia
tutut@uty.ac.id
[6] AMCS Research Center, Yogyakarta, Indonesia

Abstract. Cloud computing is cyberspace computing, where systems, packages, data and other required services (such as appliances, development platforms, servers, storage and virtual desktops) are dispensed. It has generated a very significant interest in educational, industrial and business set-ups due to its many benefits. However, cloud computing is still in its early stage of development and is faced with many difficulties. Researchers have shown that security issues are the major concerns that have prevented the wide adoption of cloud computing. One of the security issues is privacy which is about securing the personal identifiable information (PII) or attributes of users on the cloud. Although researches for addressing privacy on the cloud exist (uApprove, uApprove.JP and Template Data Dissemination (TDD)), users' PII remains susceptible as existing researches lack efficient control of user's attribute of sensitive data on the cloud. Similarly, users are endangered to malicious service providers (SPs) that may connive to expose a user's identity in a cloud scenario. This paper provides a mechanism to solve the malicious SP collusion problem and control the release of user's attribute in the cloud environment. This will require the use of policies on the SPs, where SPs are only allowed to request for attributes that are needed only to process a user's service at any point in time. This can be achieved using a combination of Kerberos ticket concept, encryption and timestamp on the attribute to be released to SPs from the identity provider (IdP), thereby helping to control attributes given to SPs for processing the release of services to users for one-time usage by the SPs and not kept for future use by them. Thus, replay attacks and blocking other SPs from accessing them are prevented. Hence, any malicious intention of assembling users' attributes by other SPs to harm them is defeated.

V. Bhateja et al. (eds.), *Information Systems Design and Intelligent Applications*, Advances in Intelligent Systems and Computing 672,
https://doi.org/10.1007/978-981-10-7512-4_72

Keywords: Cloud computing · Attributes · Privacy · Service providers and control

1 Introduction

Cloud computing is cyberspace computing, where systems, packages, data and other required services (such as appliances, development platforms, servers, storage and virtual desktops) are dispensed. It is based on pay before accessing services involved in distributing hosted facilities over the Web. Cloud computing has generated a very significant interest in educational, industrial and business set-ups due to its many gains [1–3]. However, cloud computing is in its early stage of development and is faced with many difficulties [4–8]. Researches in [9–11] have shown that security issues are the major concerns that have prevented the wide adoption of cloud computing. One of the security issues is privacy which is about securing the PII of users on the cloud [12–14]. Although researches for addressing privacy on the cloud exist: Switch (2010) added a Plugin solution called uApprove—to provide awareness of data disclosure when accessing some resource on the cloud. Orawiwattanakul et al. in [15] extended uApprove to uApprove.JP. Furthermore, Weingatner et al. in [12] added a lightweight extension on uApprove.JP called Template Data Dissemination (TDD) to tackle some privacy issues on IdP and to assist users on PII disclosure. However, users' PII remains vulnerable as existing researches require enhancements to be effective and efficient.

The general problems of cloud computing are: privacy, performance and interoperability. Privacy issues include: lack of control of user's attribute, data breaches, leaks and loss of data. uApprove, uApprove.JP and Temple Data Dissemination (TDD) were used in addressing these challenges. Despite all these solutions, the cloud is still without adequate protection. Users are endangered to malicious service providers (SPs) that may connive to expose a user's identity in a cloud atmosphere. For instance, if we have ten service providers (SPs) and each of them have partial information about a user, what measure can one put in place to prevent these SPs from colluding to profile users' attributes?

In Fig. 1, the relationship between users, SP and IdP on the cloud environment is indicated. Although researches so far have worked on securing the privacy of users' attributes on the IdPs end, the other end (i.e. from IdP to SP) does not protect privacy by itself; users are still vulnerable to malicious SPs that may collude to profile a user identity in a federated environment, Weingartner et al. in [12]. However, there are still issues to be dealt with from the SP side; this paper proposes a mechanism to control the SP, thereby preventing the collusion that may occur due to malicious activities in the cloud that cause harm to users.

Fig. 1. Relationship between user, SP and IdP

The rest of the paper is organized as follows: Sect. 2 presents a short review on existing works connected to cloud computing privacy. Section 3 presents cloud computing service models. Section 4 presents cloud computing deployment models. Section 5 presents challenges of cloud computing. Section 6 presents our proposed solution. Lastly, Sect. 7 concludes the paper.

2 Related Work

Orawiwattanakul et al. worked on user-controlled privacy protection with attribute-filter mechanism for a federated SSO environment using Shibboleth. Their proposal tackled the lack of control on PII disclosure in cloud federations in [15]. Their proposal added uApprove.JP, an extension of Shibboleth framework, that would permit users to make their choice from all optional attributes which one they wish to reveal to the SP that is being accessed. As a limitation, there is a flaw in the case of releasing user's attribute (mandatory/optional) to the SP in order to get their consent before they can access the service. An intruder can study the part through which these attributes flow and pretend to be a user, then capture the attributes and try to utilize them in order to cause harm to the stored attribute. Hence, privacy is compromised.

Sanchez et al. in [16] worked on "Enhancing Privacy and dynamic federation in IdM for consumer Cloud Computing". They proposed a new reputation protocol and implemented Enhanced Client Profile (ECP). It weighs the reputation of entities in a federation in order to support data disclosure [16]. It gives users room for checking what is being done with their data, and on that note, they could decrease or increase the reputation provided. As a limitation, their model could not demonstrate how privacy is handled in a real-life scenario [16]. Their research requires validation of the most favourable values of the parameters of the reputation model. However, their model is vulnerable to some attacks, thereby lacking some measures to fully guarantee users' privacy.

Weingartner et al. in [12] worked on "Enhancing Privacy on Identity Provider". They proposed a model for addressing some security and privacy issue called Template

Data Dissemination (TDD) with cryptography keys. Their solution is a lightweight extension on top of Shibboleth identity provider and its uApprove.JP Plugin [12]. They also attempted the problem of lack of users' awareness to their data (i.e. PII) when it is been disseminated. As a limitation, there are still issues to be dealt with at the service provider side, such as means to control attributes that were released from an IdP to a SP. Their solution is inefficient, as far as user's privacy is concern, since users' attributes are still at risk of malicious SPs that may plot to expose user's identity in a federated cloud atmosphere. Hence, the need for investigating means of enforcing user's privacy in service providers (SP).

In the light of the above, this paper provides a mechanism to solve the malicious SP collusion problem and then control the release of user's attribute in the cloud environment.

3 Cloud Computing Service Models

Cloud computing service models consist of Cloud Clients, Software as a Service (SaaS), Platform as a Service (PaaS) and Infrastructure as a Service (IaaS) [17]. As cloud computing is advancing, different vendors offer clouds that have different services associated with them. The collection of services offered put in another set of definitions is called the service model Mell et al. in [17]. Usually, cloud service model takes the following form: XaaS, where X is anything. Many cloud service models have been described here using this format. They have different strengths and are appropriate for different users and business purposes. The Service Models are presented below based on the definition of the National Institute of Standard and Technology (NIST), Mell et al. in [17].

a. Software-as-a-Service (SaaS): The consumer uses the provider's applications, which are hosted in the cloud [17].
b. Platform-as-a-Service (PaaS): Consumers deploy their own applications (home-grown or acquired) into the cloud infrastructure. Programming languages and application development tools used must be supported by the provider [17].
c. Infrastructure-as-a-Service (IaaS): Consumers are able to provide storage, network, processing and deploying resources, and controlling arbitrary software, ranging from applications to system software [17].

Following the service model, clients have different levels of control over the infrastructure management. In the SaaS model, control is normally narrowed to user-specific application configuration settings. PaaS provides control over the deployed applications and perhaps application hosting environment configurations. IaaS provides control over operating systems, storage and deployed applications [18]. Figure 2 shows the cloud computing service models.

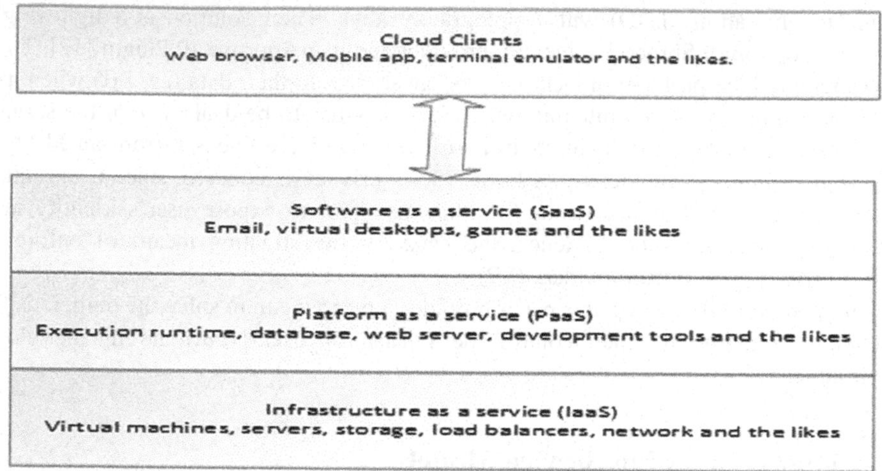

Fig. 2. Cloud computing service model [18]

4 Cloud Computing Deployment Models [17–19]

Another relevant concept of cloud computing is the cloud deployment models. The most recognized are the following four (public, private, community and hybrid), but it is important to note that other models can be developed from them.

a. Public: Resources are usually available to the general public via the Internet. In this case, "public" characterizes the scope of interface accessibility, whether or not resource usage is charged. This environment emphasizes the benefits of scalability, rationalization and operational simplicity (since the environment is hosted by a third party, i.e. the cloud provider). The main issue is security, since the environment is shared and managed by the cloud provider, and accordingly, the consumer/subscriber has little control over it.
b. Private: Resources are accessible within a private organization. This environment emphasizes the benefits of scalability, integration and optimization of hardware investments. The main issue is operational complexity, since the environment is hosted and managed by internal resources.
c. Community: Resources on this model are shared by several organizations with a common mission. It may be managed by one of the organizations or a third party [17].
d. Hybrid: This model combines the techniques of public and private clouds. A private cloud can have its local infrastructure supplemented by computer capacity from a public cloud [18, 19]. The benefits and challenges of the hybrid cloud is a combination of the items above.

In this research, the private cloud is intended to be used to actualize our solution.

5 Challenges of Cloud Computing

Cloud computing is still in its infancy and is faced with many challenges, and users are doubtful about its genuineness. Following an investigation conducted by International Data Corporation (IDC) in 2009 [20]. The most important challenges that prevented cloud computing from being widely adopted are: security challenge (which ranked highest on the survey), trust, performance issues, cloud interoperability issue, costing model, charging and service-level agreement (SLA). Figure 3 shows the cloud computing challenges based on the IDC findings:

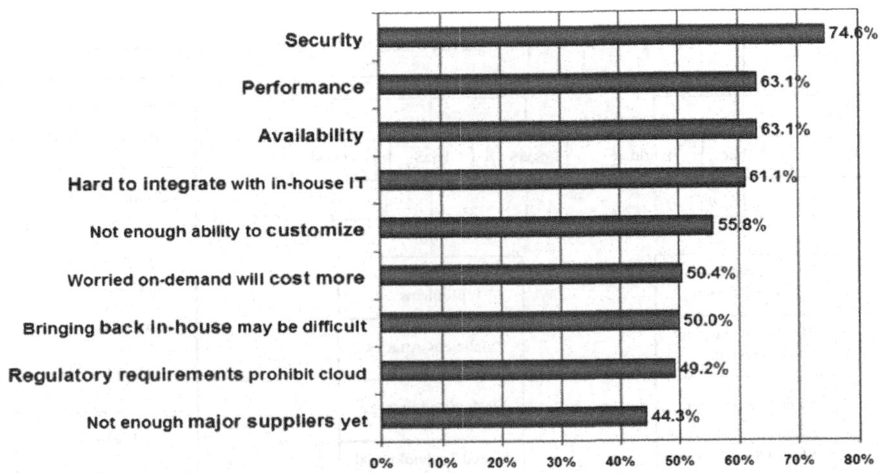

Fig. 3. Challenges to cloud computing adoption [20]

According to Rima et al. in [9], Zhou et al. in [10] and Chen and Zhao in [11], there are a lot of challenges that hamper the effectiveness and efficiency of these services such as security issues (authentication and identity management issues, privacy, trust, data confidentiality and integrity issues, non-repudiation, numerous threats, data leakages, vulnerabilities and the likes). These security issues among others are the biggest barrier to the adoption of cloud computing.

Similarly, in 2013, the Cloud Security Alliance (CSA) [21] put together a list of the nine most prevalent and serious security threats in cloud computing, known as the "*Notorious Nine: Cloud Computing Threats*". They are data breaches, data loss, account or service traffic hijacking, insecure interfaces and APIs, denial of service, malicious insiders, cloud abuse, insufficient due to diligence and shared technology [21]. Furthermore, researches have been done on the security challenges hindering the acceptance of cloud computing and these challenges directly affect the deployment models, service models and networks. They include lack of data security such as data leakage, authentication and identity management and consequent problems, malicious attacks, backup and storage, shared technological issues [22], service hijacking, virtualized machine (VM) hopping, VM mobility, VM denial of service, browser security, SQL injection attack, flooding attacks, locks and the likes. These challenges are further categorized into various groups by Parekh et al. in [22] as shown in Fig. 4. Some other security threats are phishing, password cracking and botnets.

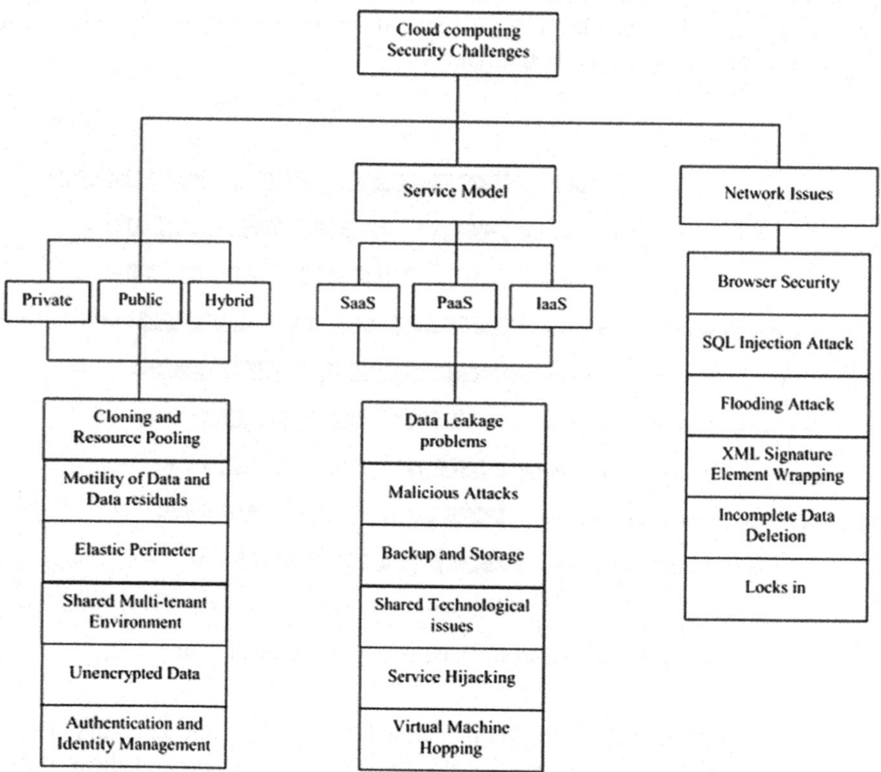

Fig. 4. Classification of security challenges [22]

6 Proposed Solution

Considering the fact that no IdP can stop or reduce the number of attributes required by an SP to process the release of resources for a user, in this paper, the following solutions were proposed below:

a. Let *n* represent number of SPs i.e. SP_1, SP_2, SP_3… SP*n* and *n* number of resources, R requested by users are represented as (R_1, R_2, R_3… R*n*) as indicated in Fig. 5.
b. We propose to use policies on all SPs, where each SP shall be allowed to request for attributes that are needed only to process a user's service at any point of time.

c. Then, we shall introduce two Kerberos tickets: T_1 for the IdP and T_2 for SP. T_1 is encrypted with IdP secret key and T_2 with the SP's secret key, the requested attributes with timestamp and a session key K_{US} for both IdP and SP. The IdP, opens T_1 extracts the IdP's secret key and sends messages to the SP; containing Ticket T_2 with the SP's secret key, the requested attributes with Timestamp and a session key, K_{US}. The SP on receiving the ticket opens the message, uses the secret key to decrypt information and releases resource to the user. At the expiration of the timestamp, the session key, K_{US}, attributes in the possession of the SPs within that timestamp becomes worthless, rendering them invalid. Even if the SP may want to play smart by decrypting the ticket and want to share user's attributes before releasing resources to the user, anything contained in the ticket is rendered invalid to anyone who receives them.

Consequently, this is to ensure that all attributes given to SPs for processing or releasing services to users are within a given timestamp and allow one-time usage, thereby preventing any malicious intention to expose users' attributes by SPs.

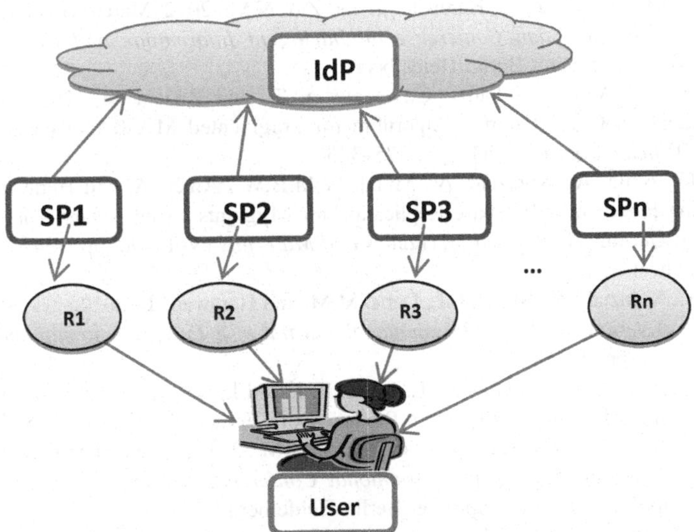

Fig. 5. Illustration the flow of resources from SPs to the users

7 Conclusion

Cloud computing is cyberspace computing, where systems, packages, data and other required services (such as appliances, development platforms, servers, storage and virtual desktops) are dispensed. It is still in its early stage of development and is faced with many difficulties. Cloud computing is characterized by security challenges, one of which is privacy concern. This includes lack of control of user's attributes, data breaches and loss of data. Researches based on Shibboleth have added uApprove,

uApprove.JP and Template Data Dissemination in tackling privacy issues; nevertheless, users' personal identifiable information remains vulnerable as existing researches lack efficient control of user's attribute of sensitive data on the cloud. Similarly, users are endangered to malicious service providers (SPs) that may connive to expose a user's identity in a cloud scenario. This paper provides a mechanism to solve the malicious SP collusion problem and control the release of user's attribute in the cloud environment. This will require the use of policies on the SPs, where SPs are only allowed to request for attributes that are needed only to process a user's resource at any point of time.

Acknowledgements. The work of Syafri Arlis is supported by Universitas Putra Indonesia YPTK under Research Grant No 094/UPI-YPTK/RG/IX/2016. The work of Tutut Herawan is supported by Universitas Teknologi Yogyakarta under Research Grant No vote O7/UTY-R/SK/0/X/2013.

References

1. Fauzi, A.A.C., Noraziah, A., Herawan, T. and Zin, N.M., 2012, March. On cloud computing security issues. In *Asian Conference on Intelligent Information and Database Systems* (pp. 560–569). Springer Berlin Heidelberg.
2. Noraziah, A., Azila, A., Fauzi, C., Herawan, T. and Zailani, A., 2015. Binary Vote Assignment on Cloud Quorum Algorithm for Fragmented MyGRANTS Database Replication. *Wulfenia Journal*, *22*(1), pp. 375–386.
3. Fauzi, C., Azila, A., Noraziah, A., Mohd, W.M.B.W., Amer, A. and Herawan, T., 2014. Managing Fragmented Database Replication for Mygrants Using Binary Vote Assignment on Cloud Quorum. In *Applied Mechanics and Materials* (Vol. 490, pp. 1342–1346). Trans Tech Publications.
4. Khan, N., Noraziah, A., Ismail, E.I., Deris, M.M. and Herawan, T., 2012a. Cloud computing: Analysis of various platforms. *International Journal of E-Entrepreneurship and Innovation (IJEEI)*, *3*(2), pp. 51–59.
5. Khan, N., Noraziah, A., Herawan, T., Ismail, E.I. and Inayat, Z., 2012b, September. Cloud Computing: Architecture for Efficient Provision of Services. In *NBiS* (pp. 18–23).
6. Khan, N., Noraziah, A., Herawan, T. and Deris, M.M., 2012c, September. Cloud computing: analysis of various services. In *International Conference on Information Computing and Applications* (pp. 397–404). Springer Berlin Heidelberg.
7. Khan, N., Noraziah, A. and Herawan, T., 2012d, September. A cloud architecture with an efficient scheduling technique. In *International Conference on Information Computing and Applications* (pp. 381–388). Springer Berlin Heidelberg.
8. Khan, N., Ahmad, N., Herawan, T. and Inayat, Z., 2012e. Cloud Computing: Locally Sub-Clouds instead of Globally One Cloud. *International Journal of Cloud Applications and Computing (IJCAC)*, *2*(3), 68–85.
9. Rima B. P., Choi E., & Lumb I. (2009): A Taxonomy and Survey of Cloud Computing Systems. In *Proc. of the 5Th International Joint Conference on INCIMS and IDC, NCM'09*, IEEE Press, 44–51.
10. Zhou M., Zhang R., Xie W., Quian W. & Zhou A., (2010) Security and Privacy in cloud: Survey. In *Proc. Of the 6Th International Conference on Semantics, Knowledge and Grids*, IEEE Press, 105–112.

11. Chen D. & Zhao H. (2012): Data Security and Privacy Protection Issues in Cloud Computing. In *Proc. of the 1st International conference on Computer Science and Electronics Engineering*, 647–651.
12. Weingartner, R. (2014) "Dissemination control of data Sensitive environment in Federated Systems". *M.Sc. Computer Science Thesis*, Department of Informatics and Statistics, Federal University of Santa Catarina, Brazil.
13. Betg′e-Brezetz S., Kamga G. B., Ghorbel M., and Dupont M.P., "Privacy control in the cloud based on multilevel policy enforcement," In *Proceedings of 2012 IEEE 1st International Conference on Cloud Networking* (CLOUDNET), IEEE Press 2012, 167–169.
14. Betge-Brezetz S., Kamga G. B., Dupont M. P. & Guesmi A. (2013). End-to-end Privacy Policy Enforcement in Cloud Infrastructure," In *Proceedings of IEEE 2nd International Conference Cloud Networking* (CloudNet 2013), 25–32.
15. Orawiwattanakul T., Yamaji K., Nakamura M., Kataoka T. & Sonehara N. (2010): "User-controlled privacy protection with attribute-filther mechanism for a federated SSO environment using Shibboleth," in *IEEE International Conference on P2P, Parallel, Grid, Cloud and Internet Computing* (3PGCIC), IEEE Press, 243–249.
16. Sanchez R., Almenares F., Arias P., D′ıaz-S′anchez D. & Marın A., (2012). "Enhancing Privacy and dynamic federation in IdM for consumer Cloud Computing," *IEEE Transactions on Consumer Electronics*, 58(1), 95–103.
17. Mell P. & Grance T. 2011. "The NIST Definition of Cloud Computing," *NIST Special Publication* 800-145 (draft), 1–7.
18. Abur, M. M., Adewale O. S., Junaidu S.B., (2015): Cloud Computing Challenges: A review on Security and Privacy issues. *Proceedings of the ACM International Conference on Computer Science Research and Innovations* (CoSRI), 89–92.
19. Pearson S. 2011 "Taking account of privacy when designing cloud computing services", HP Laboratories, Tech. Rep. HPL- 2009-54, 2009, http://www.hpl.hp.com/techreports/2009/HPL2009-54.pdf.retrieved.
20. Gens F. (2009). "New IDC IT Cloud Services Survey: Top Benefits and Challenges", IDC eXchange, Available: http://blogs.idc.com/ie/?pp730.
21. Cloud Security Alliance (2013): The Nine Notorious Threats. Top threats working group.
22. Parekh D. H. and Sridaran R., (2013): "An Analysis of Security Challenges in Cloud Computing". *International Journal of Advanced Computer Science and Applications*, Vol. 4, No.1 pp. 38–46.

Model-Based Testing for Network Security Protocol for E-Banking Application

Fadele Ayotunde Alaba[1], Saqib Hakak[2], Fawad Ali Khan[2], Sulaimon Hakeem Adewale[1], Sri Rahmawati[3], Tundung Subali Patma[4], Rajab Ritonga[5], and Tutut Herawan[6,7](✉)

[1] Department of Computer Sciences, Federal College of Education, Zaria, Nigeria
ayotundefadele@yahoo.com, sulaimonha@gmail.com
[2] Faculty of Computer Science and Information Technology, University Malaya, Kuala Lumpur, Malaysia
saqibhakak@gmail.com, fawadkn@siswa.um.edu.my
[3] Universitas Putra Indonesia YPTK, Padang, Indonesia
sri_rahmawati@upiyptk.ac.id
[4] State Polytechnic of Malang, East Java, Indonesia
subalipatma@yahoo.com
[5] Universitas Prof Dr Moestopo (Beragama), Jakarta, Indonesia
rajab.ritonga@dsn.moestopo.ac.id
[6] Universitas Teknologi Yogyakarta, Yogyakarta, Indonesia
tutut@uty.ac.id
[7] AMCS Research Center, Yogyakarta, Indonesia

Abstract. Model-based testing is one of the promising innovations to meet the difficulties required in program design testing. In model-based testing, a system under test is tried for consistency with a model that portrays the required behavior of the system. In this paper, model-based strategies are utilized for recognizing vulnerabilities as a part of system security conventions and testing for right behavior of e-banking applications in which system security conventions are actualized. The Kerberos V5 network authentication protocol is used in this research to demonstrate customers' identity to a server (and the other way around) over an uncertain connection. Password-based encryption (PBE) algorithm is used for message exchange between client and Kerberos.

Keywords: E-banking applications · Kerberos V5 · Password based encryption

1 Introduction

Model-based testing has remained effective for several kinds of application tests; however, little has been said in regard to testing for security protocols of applications and procedures [1]. Naturally, a vast number of programming applications contain bugs and errors which go through project design and implementation undetected [2, 3]. Accordingly, huge numbers of the PC applications that individuals use consistently either contain bugs or flaws [4–6]. Through the wide extent of the Internet and system

V. Bhateja et al. (eds.), *Information Systems Design and Intelligent Applications*, Advances in Intelligent Systems and Computing 672,
https://doi.org/10.1007/978-981-10-7512-4_73

administrators, the number and size of new security proprieties are outpacing the human capacity to thoroughly break down and accept them [7, 8]. To accelerate the advancement of the next generation of security proprieties, and to enhance their security [9], it is of most extreme significance to have tools that support the challenging inquiry of the security proprieties by either discovering faults or building up their accuracy [10, 11]. Ideally, these devices ought to be totally automated, robust, communicative, and easy to use, so they can bc incorporated into the security requirements during the development to enhance the security protocols.

A security protocol is a classification of operations that guarantee security of data. A security protocol is a theoretical or real protocol that implements a security-related purpose and relates cryptographic techniques [12, 13].

Network security protocols offer a concept of communications that needed to be exchanged among parties that are executing security-related process. In [14, 15], security protocols are utilized as a part of the network security to control access to limited resources that are difficult to implements accurately. Also, wrong usage of network security protocols can prompt to unwanted misfortunes as far as cash and life [16].

Security management for Global System for Mobile communications (GSM) networks using Kerberos V5 [17] through implementation, analytical models, and aid from an intermediary server creates variation of public key for Kerberos suitable for authentication protocol [18]. Both the test and model arrangements expected a portable user (mobile device) and a stationary key distribution center (KDC) and application server [19]. The arrangement includes KDCs and applications servers.

Security protocols are similarly considered by researchers to be a difficult thing that should be indicated, planned, and checked formally. Among many other researchers [20], proposed and developed novel methods such as combinatorial optimization devices to implement computerized configuration of security protocols. Analysts at the French National Institute for Research in Computer Science and Control are experimenting with new and ideally better strategies for programmed confirmation of protocols based on approximations [8, 14].

In this paper, e-banking is used as a case study to investigate the proposed model-based test and confirmation system, and it allows users to carry out businesses from their mobiles devices.

The paper is organized as follows: Sect. 2 discusses the overview of related work. Section 3 discusses the proposed e-banking and Kerberos architecture. Section 4 discusses the procedure for message exchange between client and Kerberos. Section 5 presents results and discussion. Section 6 discusses future directions. Section 6 concludes the paper.

2 Related Work

This paper reviewed related literature such as presented by [15, 19, 21–25] which explores with practical experiments the use of testing as well as authentication techniques to implement network security protocols for e-banking applications. A model-based testing system was developed to access the e-banking application

[26, 27] and also to ascertain if the application imitates to the expected behavior that is usually specified casually [28].

Standard security models proposed by [19, 21–23] all address the protocols of a protected framework reflected as an item. They appear to be less modified in modeling the security problems that emerge during the product development phase and frequently refined the relations that also emerge from the interaction among procedure and the end-product.

Security protocols in these applications stand to guarantee that data are conveyed in the way that is anticipated to interest parties. A portion of the anticipated qualities in the transmission comprises of privacy and reliability. Accuracy in developing security protocols is no longer considered as an appropriate quality, but a necessity [29]. This is because of the high stakes of accounts required in the event that vulnerabilities in the framework that are abused by an aggressor [30].

3 Proposed E-Banking and Kerberos Architecture

Kerberos utilizes a user/server architecture and offers client-to-server validation rather than host-to-host verification as shown in Fig. 1.

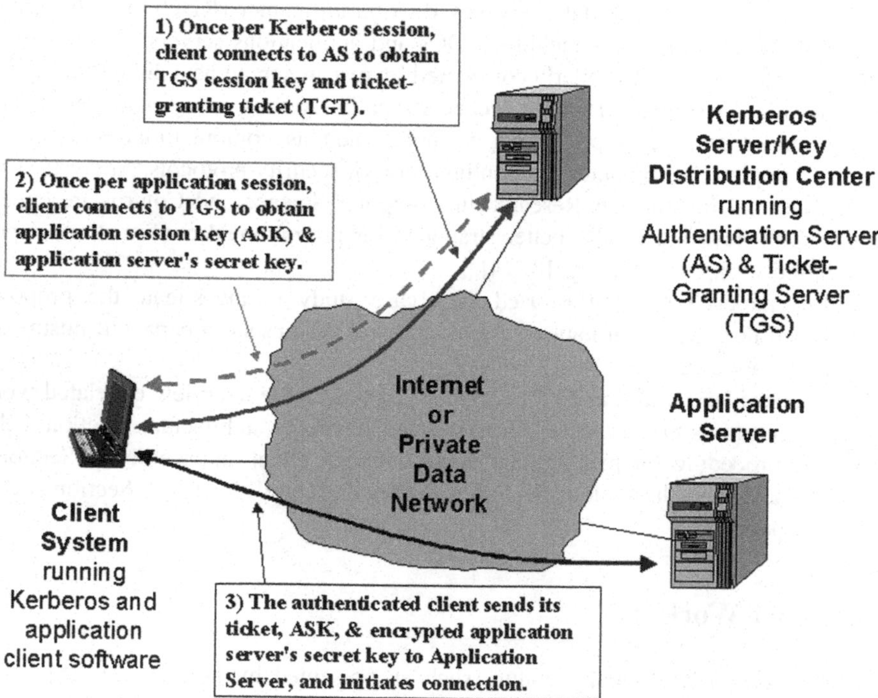

Fig. 1. Architecture overview of e-banking and Kerberos. Source Kamala, 2012

In this model, security and verification depend on secret key innovation where each host on the network is provided with a particular key. It would obviously be uncontrollable if each host needed to know the keys of every single other host, therefore, is a need to develop a safe and reliable host someplace on the network, identified as a key distribution center (KDC), the KDC identifies as well as have the keys for all the hosts in the entire network (or possibly a portion of the hosts inside a segment of the network, called a domain). Along these lines, when another hub is conveyed online, is only the KDC, and the new hub that will be configured with the hub's key. Keys can be disseminated physically or by some other secure means.

The system will provide secure communication across a distributed location by means of verification, reliability, privacy, and authorization. This was achieved by using Kerberos V5 network authentication protocol which provides authentication between communicating server using cryptography. The Kerberos V5 helps the client to demonstrate its uniqueness to the server (and vice versa) through an uncertain communication channel subsequently, since both the user and server have utilized Kerberos to demonstrate their uniqueness; now, they can also encode all of their communication to guarantee data confidentiality and reliability, and here, the server is the e-bank, a business logic server.

4 Procedure for Message Conversation Between Client and Kerberos

The data flow diagram as shown in Fig. 2 explains the steps of exchange of messages in Kerberos.

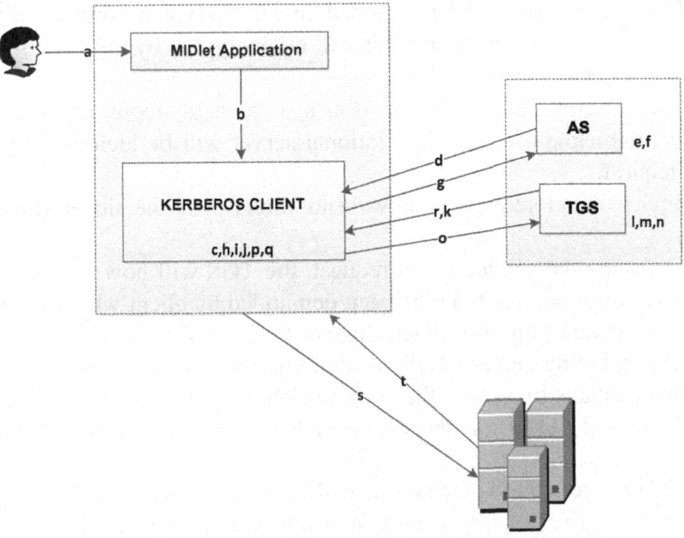

Fig. 2. Data flow diagram for Kerberos messages exchange

The conversation of Kerberos message to create a secure data transmission context is summarized in the following 1–20 phases.

(a) The client gives his or her identity (a mutual secret that is known by the client and the e-bank server only) to the MIDLet. (Note: the secret key is utilized within Java Micro Edition (J2ME) for handling client identity and it by no means traversed through any network. But only the username is sent transversely through the network).
(b) The MIDLet delivers the clients identity to Kerberos client. Kerberos client's is responsible to create a secure connecting environment for both client and the e-bank.
(c) The Kerberos client creates a demand to authentication server (AS) to provide ticket-granting ticket (TGT). A sole TGT denotes a secure session. A client has the privilege to create various sub-sessions inside a sole secure session. A TGT only demands username of the client.
(d) The Kerberos client directs the request to the AS.
(e) When the AS gets the TGT, it removes the username and gets the equivalent password inside the database. The AS then creates a TGT and wraps the TGT in a response message.
(f) The TGT is divided into two parts which include plaintext part and a ciphertext (encrypted) part, used by AS to get users password from the TGT encrypted ciphertext portion through cryptographic key.
(g) The AS then directs the response message (alongside the TGT) Kerberos client who is demanding it.
(h) After receiving the response from AS, the Kerberos client removes the TGT from the response and decodes the encrypted part of the TGT. The Kerberos client will now create a demand for a service ticket.
(i) The TGT demanded will be wrapped in an encrypted format well known as authenticator. And the authenticator will be encrypted by the clients by using the session key separated from TGT.
(j) The authenticator guarantees the user's awareness about the session key. In addition, the name of the e-bank rational server will be identified by the service ticket request.
(k) The service ticket request is now send directly to the ticket-granting service (TGS).
(l) After receiving the service ticket request, the TGS will now remove the name of the server where service ticket is been demanded by client which in turn creates a service ticket and quite not much diverse from TGT.
(m) The ticket-granting server (TGS) is also known as sub-session key. It is similar TGT that contains two parts: the cryptograph text part and the new cryptographic key and majorly used for decrypting and encrypting of service ticket from the server.
(n) Once the service ticket is created, it will be wrapped as a feedback by TGS. The feedback comprises of the cryptograph text part encrypted with session key and the new cryptograph text part which contains the sub-session key.
(o) The TGS now sends a feedback to the user.

(p) When the user has received the feedback from TGS, the user will use the session key to decrypts the cryptograph text part of feedback, remove the sub-session key as well as the service ticket.
(q) After decrypting the cryptograph text part of feedback, the user will now create a message e-banks server, demanding the serve to generate a new safe session with the user.
(r) When the new session is generated, the user now sends message directly to the e-bank server.
(s) The e-bank server will remove the requested service ticket, decrypt the cryptograph text part and also get the sub-session key (key which is well known to both the user and the server).
(t) Finally, the e-bank server will send a progressive acknowledgment to the user.

5 Results and Discussion

To start the Kerberos services, open the project name in MIDLet then selects the lunch button as shown in Fig. 3.

Fig. 3. Opening the project name

After opening the project, enter the user's login name and password, these are stored in user details in data base of the e-bank, as shown Fig. 4.

Now enter the user name and password and then press the lunch button. The verification for user and password will take place as displayed in Fig. 5.

Fig. 4. Displays user login and password text boxes

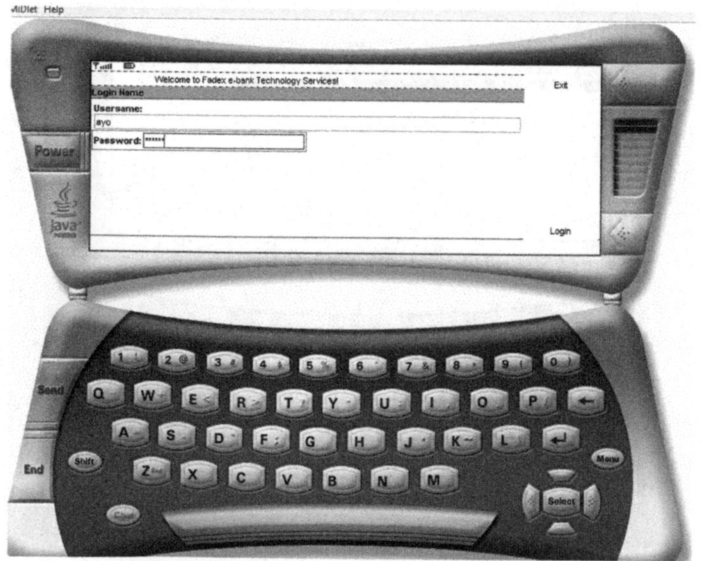

Fig. 5. Entering username and password for connection with e-bank

Since the verification has been completed as shown in Fig. 5, it displays the transaction details in the mobile phone as shown in Fig. 6. Hence, to transfer say an amount of 45000 Naira to the account number 1203431231, then choose the select button, as shown in Fig. 6.

Fig. 6. Displays transaction detail

For the "Amount Transfer," enter the account number and the amount in number and then press the lunch button (Fig. 7).

Fig. 7. Displays the amount transferred to the other account

After pressing the lunch button, it displays successful transaction on the mobile phone, and the account balance is checked (Fig. 8).

Fig. 8. Displays successful transaction

After successful "Transaction," press lunch button and press the select option button (Fig. 9).

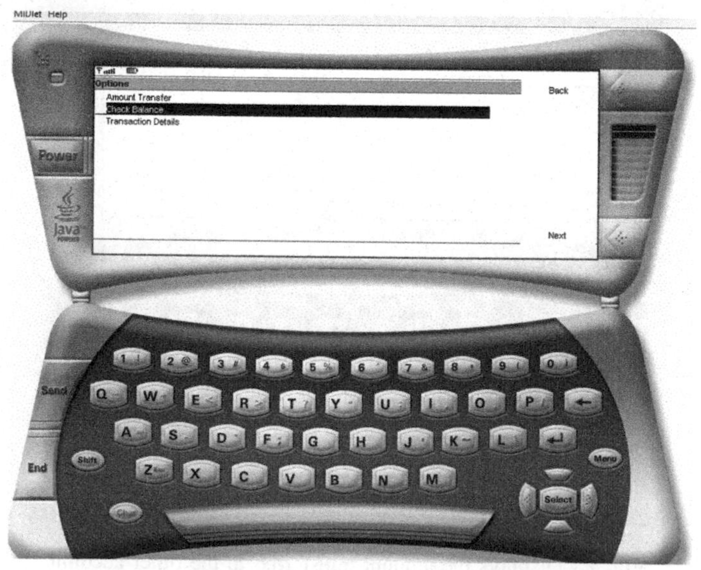

Fig. 9. Displays transaction details

The findings of this research paper established an improve algorithm and the proposed model-based testing model that will enhance network security protocols in e-banking applications by also define a way in which users can execute businesses from their mobiles devices.

6 Conclusion and Future Work

This paper meets the security requirements such as authentication, integrity, confidentiality, and authorization for e-banking. It also explains series of Kerberos messages (alongside Kerberos data layouts) that show the conversation procedures of cryptographic keys for secure communication channel between the user and the e-bank server. This system leads to a greater security for application in privacy protection and intellectual property rights protection. Wireless innovation is being created that can incredibly amplify the utilization of versatile datasets past everyday communication. As wireless operatives spread transmission capacity that users can access, it is normal that more information, by and large, and money-related exchanges specifically will happen over the systems. In any case, whatever inevitable applications advance, cryptography in some structure will be expected to shield the information from meddlers and abuse. Finally, actual implementation of e-banking application was developed and used to explore computerized confirmation.

From the basic implementation and logical models have shown that public key Kerberos is a practicable verification protocol with support from an intermediary server and can also adapt to any field of study. Therefore, there is extend to this research to other field of study that involves large number of user. In addition, both the test and model structures accepted a portable device, a static KDC, and application server. However, there is need to develop a mobile KDCs and applications servers which can be applicable in military battleground environment.

Acknowledgements. The work of Sri Rahmawati is supported by Universitas Putra Indonesia YPTK under research grant no 094/UPI-YPTK/RG/IX/2016. The work of Tutut Herawan is supported by Universitas Teknologi Yogyakarta Research Grant no vote O7/UTY-R/SK/0/X/ 2013.

References

1. Agrawal, R., Evfimievski, A. and Srikant, R. (2003). Information sharing across private databases. In *Proceedings of the ACM SIGMOD Int'l Conf. on Management of Data*, San Diego, CA.
2. Alur, R., Courcoubetis, C. and D. Dill. (1990). Model-checking for real-time systems. *Proceedings of Fifth Annual IEEE Symposium on Logic in Computer Science*, pg. 414–425.
3. Arkin, K., Thompson K., and Weinel W. (2005). *The mobile jigsaw - a collaborative learning strategy for mlearning about the environment*. Presented at mlearn; Making the connections.
4. Armando A and Compagna L. (2004). A SAT-based Model Checker for Security Protocols. *Lecture Notes in Computer Science*, pg. 13–18.

5. Armando, A., Basin, D., Boichut, Y., Chevalier, Y., Compagna, L., Cuéllar, J., Drielsma, P.H., Héam, P.C., Kouchnarenko, O., Mantovani, J. and Mödersheim, S., 2005, July. The AVISPA tool for the automated validation of internet security protocols and applications. In *Proceedings of International Conference on Computer Aided Verification* (pp. 281–285). Springer Berlin Heidelberg.
6. Basin, D., M¨odersheim, S. and Vigan`o, L.; (2005): A symbolic model checker for security protocols. *International Journal of Information Security*, 4(3):181–208.
7. Boichut, Y, Heam, and Kouchnarenko, O. (2005). *Automatic Verification of Security Protocols Using Approximations*; pg. 10, Doctoral dissertation, INRIA.
8. Viganò, L., 2006. Automated security protocol analysis with the AVISPA tool. *Electronic Notes in Theoretical Computer Science, 155*, pp. 61–86.
9. Boreale, M., Nicolas, D and Pugliese, R., (2002). Proof techniques for cryptographic processes. In *Proceedings of 14th IEEE symposium on logic in computer science*, Pg. 157–166.
10. Castellani, G.C., Quinlan, E.M., Bersani, F., Cooper, L.N. and Shouval, H.Z., 2005. A model of bidirectional synaptic plasticity: from signaling network to channel conductance. *Learning & Memory*, *12*(4), pp. 423–432.
11. El-Far, I. K and James A. Whittaker, (2001): *Model-based Software Testing*, Florida Institute of Technology, Melbourne, Florida, U.S.A.
12. Wikipedia, the free encyclopedia. http://en.wikipedia.org/wiki/formal verification. July, 2012.
13. Wikipedia2012, the free encyclopedia. http://en.wikipedia.org/wiki/security_protocol. Retrieved on 24th September, 2012.
14. Ermentrout, B., Wang, J.W., Flores, J. and Gelperin, A., 2004. Model for transition from waves to synchrony in the olfactory lobe of Limax. *Journal of computational neuroscience*, *17*(3), pp. 365–383.
15. Heckel, R. and Mariani, L., 2005, April. Automatic conformance testing of web services. In *International Conference on Fundamental Approaches to Software Engineering* (pp. 34–48). Springer Berlin Heidelberg.
16. Confora, G., Penta, M.D., Esposio, R., (2009). A brief survey of software architecture concepts and service oriented architecture. In *Proceedings of International Conference on Computer Science and Technology*, pg. 34–38, IEEE Press.
17. Martin, B., Mitrovic, A. and Suraweera, P. (2006): ITS domain modeling with ontology. *Journal of Universal Computer Science*, 14(17): pg. 2758–2776.
18. Martin, B.I. (2007): An Authoring System for Constraint-Based ITSs. Chicago, IL, USA: American Educational Research Association (AERA) Annual Meeting, pg. 102–106.
19. Kamala, K. (2012). *Security Management for GSM Networks using Kerberos* v5. SRM Institute of Science and Technology, Deemed University, 23–32.
20. Utting, M., Pretschner, A. and Legeard, B., 2012. A taxonomy of model-based testing approaches. *Software Testing, Verification and Reliability*, *22*(5), pp. 297–312.
21. Bell, G and Paulson L, (1998). Kerberos version IV: Inductive analysis of the secrecy goals. http://en.wikipedia.org/wiki/kerberos_IV.
22. Brügger, L.A., 2012. *A framework for modelling and testing of security policies* (Doctoral dissertation, ETH ZURICH).
23. Clark, E.M., Grumberg, O. and Long, D.E. (2008). Model checking and abstraction. *ACM Transactions on Programming Languages and Systems* (TOPLAS), 16(5):1512–1542.
24. Holzmann, G.J., 1997. The model checker SPIN. *IEEE Transactions on software engineering*, *23*(5), pp. 279–295.
25. Kennedy-Clark, S., and Thompson, K. (2012). Methods of analysis for identifying patterns of problem solving processes in a computer-supported collaborative environment. In

Proceedings of the 10th International Conference of the Learning Sciences (ICLS), pg. 14–18, IEEE Press.

26. Chevillat, C., Carrington, D., Strooper, P., Süß, J.G. and Wildman, L., 2008. Model-based generation of interlocking controller software from control tables. In *European Conference on Model Driven Architecture-Foundations and Applications* (pp. 349–360). Springer Berlin Heidelberg.
27. Steiner, J.G., Neuman, B.C. and Schiller, J.I., 1988, February. Kerberos: An Authentication Service for Open Network Systems. In *Usenix Winter* (pp. 191–202).
28. Thompson, K., and Kelly, N. (2012). Combining collaboration spaces: Identifying patterns of tool use for decision-making in a networked learning environment. In *Proceedings of the Eighth International Conference on Networked Learning*, pg. 5–9.
29. Terpstra, R., Pires, L.F., Heerink, L. and Tretmans. J. (2009). *Testing theory in practice: A simple experiment*. COST, 247(6):168–183.
30. Turpe, S., (2012). Point-and-shoot security design: can we build better tools for developers? *Proceedings of the 2012 workshop on new security paradigms*, ACM, 27–42.

A Framework for Authentication of Digital Quran

Saqib Hakak[1], Amirrudin Kamsin[1], Jhon Veri[2], Rajab Ritonga[3], and Tutut Herawan[4,5](✉)

[1] Faculty of Computer Science and Information Technology, University Malaya, Kuala Lumpur, Malaysia
saqibhakak@gmail.com, amir@um.edu.my
[2] Universitas Putra Indonesia YPTK, Padang, West Sumatera, Indonesia
jhonveri@upiyptk.ac.id
[3] Universitas Prof Dr Moestopo (Beragama), Jakarta, Indonesia
[4] Universitas Teknologi Yogyakarta, Yogyakarta, Indonesia
tutut@uty.ac.id
[5] AMCS Research Center, Yogyakarta, Indonesia

Abstract. Increment of Internet users accessing content related to digital Quran and Hadith has increased from past few years. This has increased the need of Quran and Hadith authentication system that can authenticate between fake and original verses. In this work, complete framework related to automatic authentication and distribution of digital Quran and Hadith verses is proposed. Authentication process is divided into two phases, i.e., verification and security. For verification part, existing and standard exact matching algorithm, i.e., Boyer-Moore algorithm, is used. In case of security phase, watermarking technique will be used to secure the verified and tested verse. Furthermore, only verification phase of proposed framework is tested as system is still under development phase. For dealing with diacritics involved in Quranic text and Hadith text, segmentation is done based on clitics using UTF-16 encoding. On testing the proposed framework with comparison to popular search engines and other related existing works, our approach is 96.8% accurate in terms of full verse detection.

Keywords: Authentication · Digital Quran · Integrity of Quran · Text authenticity · Hadith authentication

1 Introduction

Millions of Internet users read and exchange digital content available on Internet through various different means like social media, blogs, Web sites [1]. This trend of online reading and sharing information has its own threats of copyright protection, digital counterfeiting, content integrity, and authenticity issues. The area of authentication for online content has emerged as one of the most popular and challenging issues especially for sensitive content [2]. This issue is more serious related to religious scriptures like digital Quran, Hadith, and other such sensitive content [1, 3]. Digital Quran is being regarded as one of the most online sensitive and holiest scriptures for

V. Bhateja et al. (eds.), *Information Systems Design and Intelligent Applications*, Advances in Intelligent Systems and Computing 672,
https://doi.org/10.1007/978-981-10-7512-4_74

Muslims. It consists of 6236 verses divided into 114 chapters written in Classical Arabic. All these 114 chapters consist of variable number of verses or Ayats. Each surah varies with different number of verses (Ayat) [4, 5]. It is written in many ways but Uthmanic [6] style that involves more diacritics is most prominent one followed by simple Arabic style involving minimum diacritics [6]. From past few years, research work in the area of authentication of digital Quran is increasing as shown in research trend in Fig. 1 (taken from IEEE Xplore and Elsevier libraries).

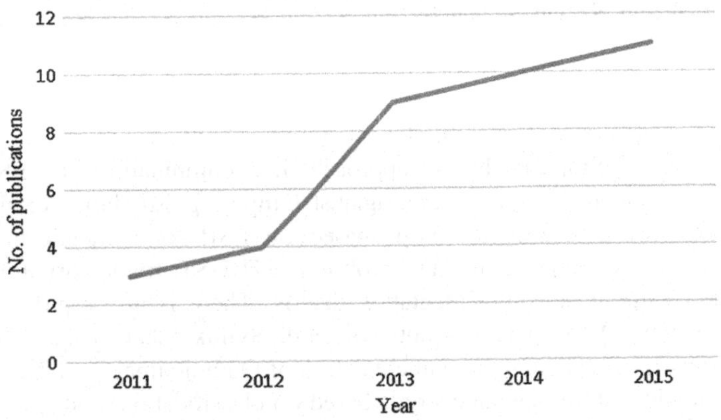

Fig. 1. Research publications in the area of Quran and Hadith authentication (IEEE Xplore, Elsevier Library)

The prime focus of these works has been the verification of Quranic or Hadith content and protection from tampering using different approaches of cryptography, string matching, and watermarking. However, there is still lot of research gaps as this field is quite new and growing. One of the main research gaps is to authenticate Quranic verses without removal of diacritics. There are lot of symbols and diacritics used in Quranic text, and modification in just one symbol changes the meaning of whole sentence/verse. The problem with diacritic text is one symbol can change complete meaning of the whole word. For example, an Arabic word كتب consisting of three consonants, i.e., ب ت ك, gives different meanings with different arrangements of diacritics [7]. Different meanings with different arrangement of diacritics like كَتَبَ means wrote, كُتُب means books, كُتِبَ means written and كَتَّبَ means Make someone to write. Thus, it makes necessary to secure and authenticate the Quranic verses available on internet.

Accordingly, to solve the disadvantages and drawbacks of the prior art, this paper presents a novel framework for authentication of Holy Quran and Hadith. More particularly, the present invention is related to an approach and a system that use exact matching technique for searching and verification of given Arabic text and use digital sign and watermarking for securing the verified content.

The organization of this paper is as follows: Related work is presented in Sect. 2. In Sect. 3, description of proposed framework is given. Results are given in Sect. 4. Finally, this work is concluded in Sect. 5.

2 Related Works

There is no proper organization with respect to works done related to authentication of digital Quran or Hadith as of now. In order to address this issue and bring order to existing works, we divided existing works into image-based format and text-based format. Existing works related to authentication of digital Quran based on image format are as follows.

2.1 Image-Based Research Works

Tayan et al. [8] proposed zero-watermarking approach for content verification and authentication. The authentication phase is based on hashing process. Document to be authenticated is embedded with a specific data sequence obtained from watermark logo. AlAhmad et al. [9] proposed hybrid approach, i.e., combination of AES and RSA algorithms for protecting digital Quran against tampering. AlAhmad et al. [10] proposed invisible watermarking technique based on LSB for authentication of Holy Quranic images. Abuhaija et al. [11] proposed ITRUST framework for Web site authentication without any implementation details. There is no proper methodology mentioned as how Web content is authenticated. Syifak Izhar et al. [12] identified fragile watermarking technique for authentication of Quranic images. Sabbah et al. [13] classified Quranic and non-Quranic words based on classification model. Classification model was developed based on support vector machine (SVM) approach. Based on SVM classification model, words were extracted from online sources. However, before extracting the words, all symbols, diacritics, non-Arabic letters were removed via filtering phase. For evaluating purposes, three parameters, i.e., accuracy, precision, and F-measure, were taken. However, it is not mentioned how the algorithm extracts the features [13]. Kurniawan et al. [14–16] proposed different approaches for protecting Quranic images from tampering. Authentication mechanism is based on hashing process. Original Quranic image is hashed to get initial authentication code. This authentication code is then encrypted using private key to get secured authenticated code. The authors evaluated the proposed approach on four images of Holy Quran. In their other related works, fragile watermarking approach to protect Quranic images based on wavelet and spatial domain is used. There is significant difference in their other two research articles. Similarly, in other works of Laouamer et al. [17] and Gutub et al., watermarking approaches are used to protect and authenticate Quranic images [18].

2.2 Text-Based Research Works

Lot of Internet users prefer to copy Quranic text from a specific source and paste it on either social media Web sites or other online blogs. It is one of the most suitable and easy-to-use approaches. Based on complexity of Quranic text, we divided its text format into two sub-categories, i.e., diacritic text and non-diacritic text. Most of the non-native speakers of Arabic rely heavily on using diacritics for reading and understanding Quran. Quranic verse with diacritics is shown in Fig. 2.

ٱلْحَمْدُ لِلَّهِ رَبِّ ٱلْعَٰلَمِينَ   ٱلحمد لله رب ٱلعلمين

Fig. 2. Diacritic Arabic Quran verse

Less work has been diverted toward text-based authentication of digital Quran compared to image-based version. This can be due to the complexity of diacritics involved in text-based format. Works based on text format of digital Quran are as follows:

Alginahi et al. [19] proposed algorithm for verification of Arabic verses along with diacritics and other symbols. Algorithm is based on SELECT query using MySQL that uses either linear search algorithm or binary search algorithm. However, in preprocessing phase, all diacritics are removed for verse identification. Alsmadi et al. [1] proposed authentication model for Quranic verses. The author claims that document control and digital signature are two mostly widely used approaches to authenticate documents. Document control is giving permission before and after publishing the document online. In digital signature, signed documents should be verified by the people who signed it. Focus has been on integrity checking. The author mentions that it is a challenge to read or parse Arabic diacritics correctly. Hashing approach is used in this research. Hash is calculated for the particular verse and that hash value is compared with hash value in database. However, there is possibility of hash collision using this approach. Sabbah et al. [20] proposed framework to detect and authenticate Quranic verses. Focus has been to increase detection accuracy of diacritic text. Accuracy on an average is 62%. However, this algorithm will not work with non-diacritical text and there is so much overhead associated while calculating weights and dividing verses into two groups. The complexity of the algorithm will increase with extremely complex diacritical texts. Alshareef et al. [21] proposed a framework for Quranic verse detection. The idea is simple where text Quranic verse is taken as input and result gets displayed whether verse is authentic or not. There are two major components in this, i.e., Quranic quote filtering and verification mechanism. In Quranic quote filtering, all Arabic diacritics and special symbols are removed. The author claims symbols, and diacritics limit the traditional search engines to provide acceptable and accurate results to the users without any valid proof or justifying the claim. Finally, after removing all symbols and diacritics, Quranic verification mechanism is used which uses regular expression SQL query to verify the text. The authors have used some single verses to evaluate the authenticity of those verses. The proposed algorithm shows 89% accuracy as compared to rest of the search engines against few words. In our assumption, this accuracy will decrease if this algorithm is used on large Arabic data set due to the fact that regular expression uses prefix–suffix approach for searching [21]. Nisha et al. [22] studied different search engines along with their limitations. A new search engine with the name of "Truth-search-now" has been proposed. Five search engines with respect to Islamic content have been taken, i.e., The Islamic search, Into Islam, Search-truth, Islami City, and Allah.pk, and evaluated based on the time taken by each search engine to find a particular query. The experimental evaluation is not conducted and there is no algorithm mentioned in the article [22].

From the above works, it can be concluded that there are still lot of issues like in terms of accuracy, precision related to detection and authenticity of Quranic verses. In order to address the issues of content integrity and security, we proposed efficient framework with the features of content verification and security.

3 Framework

As mentioned in the literature above, lot of work has been in the area of image processing as compared text, and our proposed framework is based on binary data, i.e., text as presented in Fig. 3. This framework can be used to detect fake Quranic verses from online sources like social media, online blogs, Web forms along with Hadiths also.

The framework consists of two servers, i.e., server *A* and server *B,* as shown in Fig. 3. The basic purpose of server *A* is to take input and verify whether the particular verse has been verified before by the system or not. If the verse has been verified before also, then verified copy of that verse is sent back to the user through server *A*. However, if the verse has not been verified before, then request for authentication is being passed to server *B*. The purpose of server *B* is to authenticate unverified verses and digitally sign the verified verse with a unique ID so that people can know a particular verse is authentic. Then, the verified verse is passed to server *A* with a unique ID embedded. This ID is used by the system to verify that the verse has been authenticated before if again fed to the system. Brief description of each block of our proposed framework is explained below.

3.1 Uploader

In this component, user enters Quranic verse to be authenticated along with his/her email. After getting the Quranic verse as an input from a particular user, uploader saves the email of that user into database of server *A*. The role of uploader is to segment the given Quranic verse. The segmentation is carried out using regular expressions approach. We use UNICODE-16 scheme as it is variable length encoding and suits diacritical Arabic and other complex texts. Sample Unicode representation for each Arabic character is shown in Table 1.

Finally, UTF-16-based encoded verse passes to a query component known as Decision Maker.

3.2 Decision Maker and Request Completion

Decision Maker takes the UTF-based input from uploader and looks for unique ID. If that unique ID is present, Decision Maker passes the request to Request Completion block. Request Completion block simply retrieves that ID from database and mails the verified verse back to the user through his/her email. If there is no ID present or any other unverified ID is there, Decision Maker simply passes the UTF-16-based verse to server B for authentication purposes.

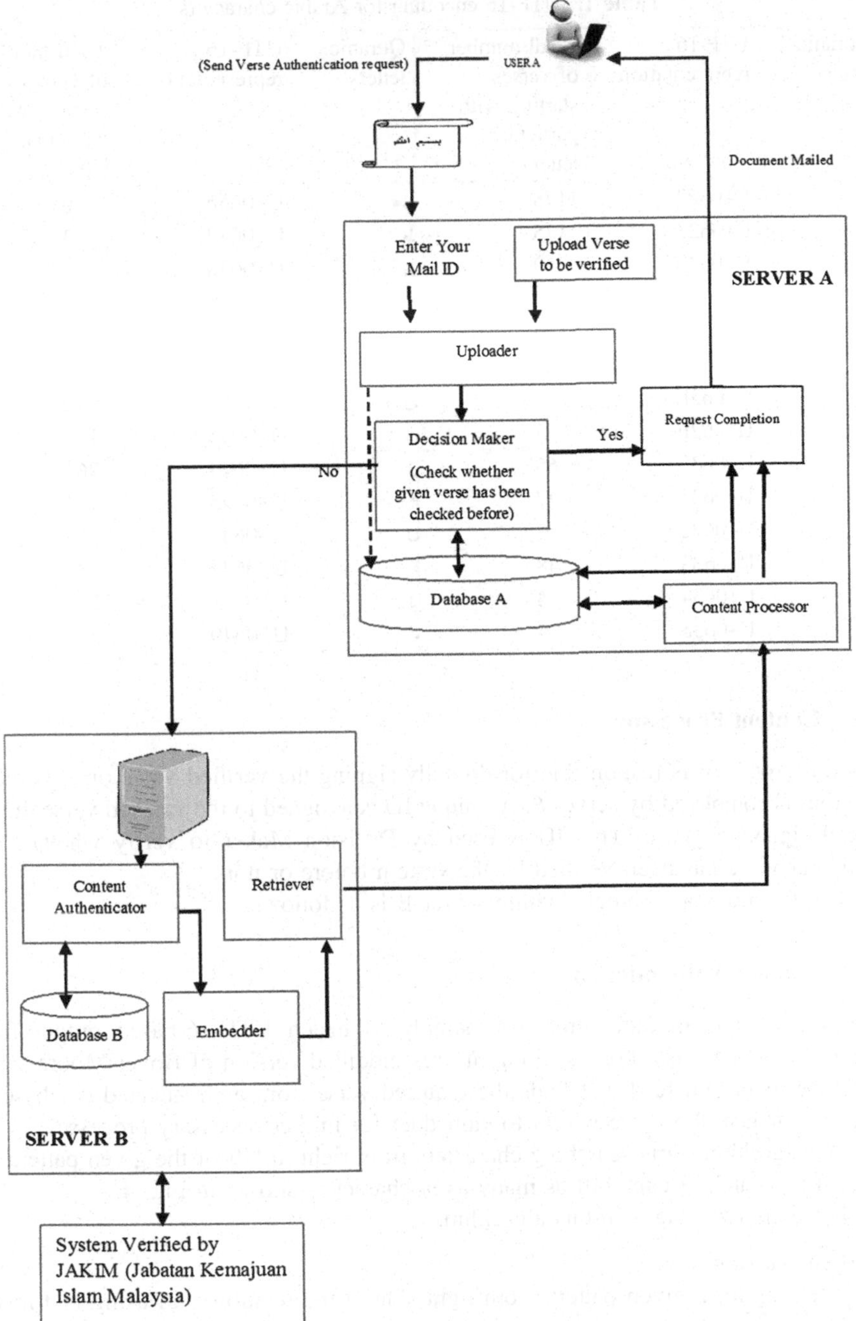

Fig. 3. Proposed framework for authentication of digital Quran

Table 1. UTF-16 encoding for Arabic characters

Quranic letters	UTF-16 representation	Total number of verses starting with particular letter	Quranic letters	UTF-16 representation	Total number of verses starting with particular letter
ا	U+0627	1178	ض	U+0636	6
ب	U+0628	175	ط	U+0637	3
ت	U+062A	59	ظ	U+0638	1
ث	U+062B	109	ع	U+0639	42
ج	U+062C	14	غ	U+063A	2
ح	U+062D	24	ف	U+0641	698
خ	U+062E	31	ق	U+0642	530
د	U+062F	3	ك	U+0643	118
ذ	U+0630	65	ل	U+0644	262
ر	U+0631	47	م	U+0645	155
ز	U+0632	3	ن	U+0646	25
س	U+0633	48	ه	U+0647	85
ش	U+0634	4	و	U+0648	2215
ص	U+0635	5	ي	U+0649	329

3.3 Content Processor

Content processor is responsible for digitally signing the verified verse once verification part is completed by server *B*. A unique ID is assigned to the verified verse during digital signature phase. This ID is used by Decision Maker to verify whether any particular verse has been verified by the system before or not.

The functioning of blocks within server B is as follows.

3.4 Content Authenticator

The role of content authenticator is simply to match UTF-16-based output from authenticated database. For matching, it uses modified version of Boyer-Moore string matching algorithm [23] that finds the required verse from authenticated database. If there is a match, the verse is sent to embedder for further necessary processes.

The algorithm starts searching characters from right to left of the given pattern. In case of mismatch, it can shift as many as *m* characters shown in Fig. 4.

There are two stages in this algorithm:

a. Preprocessing;
b. Searching for a given pattern from right side of the window and using bad match table to skip characters in case of mismatch.

Their descriptions are given below:

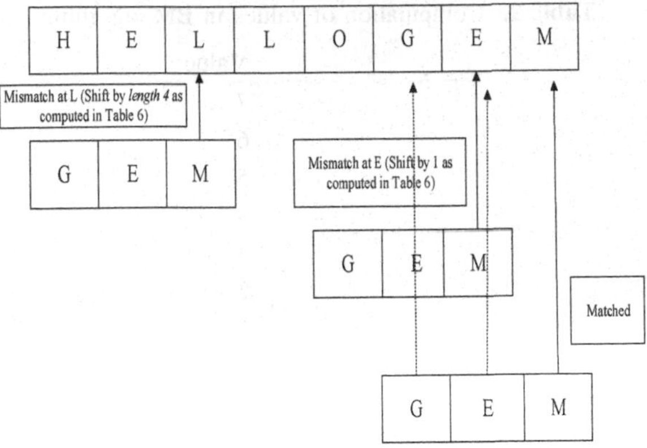

Fig. 4. Boyer-Moore algorithm

a. Preprocessing: During preprocessing stage, a table is created which gives values regarding how much shift is required in case of mismatch. This is also known as bad match table. Thus, once a character mismatch occurs, algorithm shifts to the right of the pattern according to the value given in bad match table.
b. Searching starts from the tail of the pattern, i.e., from right to left of the text as compared in naive algorithm, where search starts from left to right. The algorithm works by computing the length of search string and storing its value as default shift length. In the search string, for each character, the shift value is set as shown in Fig. 5.

Text with Index Numbering from 0-7

H	E	L	L	O	G	E	M
0	1	2	3	4	5	6	7

H	E	L	O	G	M	*
7	6(1)	5(4)	3	2	8	8

Values

Length of original text

Fig. 5. Calculated values

c. The values can be computed using *Value = Length of pattern-1-index of character* as shown in Table 2 as follows.

Table 2. Computation of values in BM algorithm

Length-1-index	Value
(1, 8-1-0)	7
(1, 8-1-1)	6
(1, 8-1-2)	5
(1, 8-1-3)	4
(1, 8-1-4)	3
(1, 8-1-5)	2
(1, 8-1-6)	1

By referring to Table 2, the values in brackets for character “E” and “L” in Fig. 4 are the current values of shifting phase. When algorithm finds another occurrence of same character twice, the previous value is replaced with new values. Therefore, 6 and 5 values of respective “E” and “L” are replaced by 1 and 4. For the last character, value is length of text. This gives time complexity $O(n + m)$ for the best case and in worst case $O(n * m)$. Here, m denotes length of pattern and n denotes length of text which is to be searched.

3.5 Embedder

Embedder is responsible for watermarking process. Once match is found, the verse is watermarked by embedding specific sequence of bits calculated using logo of Jabatan Kemajuan Islam Malaysia (JAKIM) for tampering-related issues. JAKIM is well-known religious governing body of Malaysia that will verify the full system before making it available to masses for use. Finally, the verified verse with watermark embedded passes through retriever.

3.6 Retrieval

Retriever is final phase of server B, where verified verse after watermarking phase is retrieved. Finally, complete secured verse is sent to server A for digital signature and final phase.

4 Results and Discussion

This section presents results and discussion.

4.1 Experimental Setup

In order to test our framework, we developed the initial prototype. Prototype was implemented using NetBeans IDE environment 8.02 on i-5 Intel Processor with 4 MB cache, 4 GB RAM using Windows 10 and programming language used was Java. Initial graphical user interface is shown in Fig. 6.

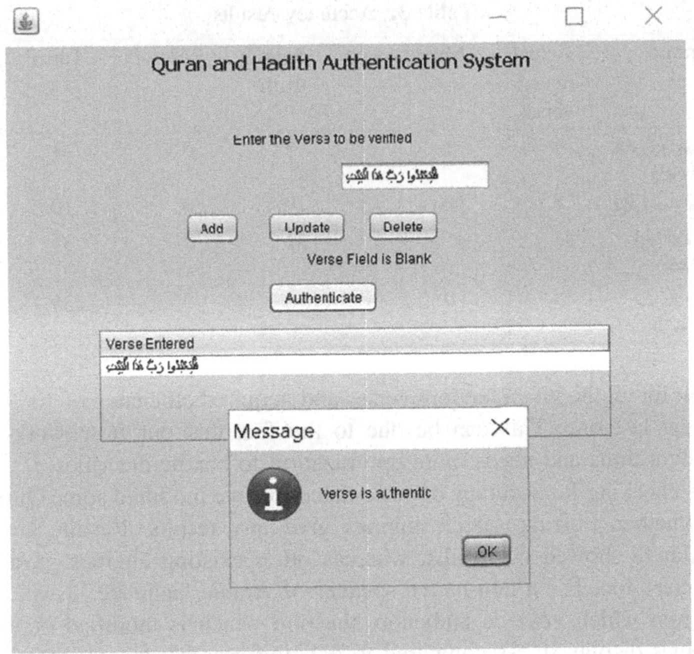

Fig. 6. Initial prototype for digital Quran authentication

We copied random Quranic verses from Internet and analyzed the output of the initial prototype. For verifying authenticity, standard Quran datasets from Tanzil.net were used [24]. The verses were tested with two parameters of actual number of verse to be verified in digital Quran and retrieved given in Table 3.

4.2 Results

For experimental part, ground truth related to tested verses was calculated manually. Each tested verse was checked manually first as how many occurrences of that verse are in digital Quran. Finally, accuracy was calculated based on given formula:

$$Accuracy = \frac{\text{Number of particular verses Found}}{\text{Total number of particular verses}}$$

Accuracy and overall results are given in Table 3 and Table 4, respectively. Results are compared from existing search engines that are popular for searching Quranic verses. Search engines include Muslim-web [25], Search-truth [26], Alawfa [27], and tanzil.net [24].

From Table 3, it can be observed that all existing search engines except Alawfa were able to retrieve verse number 1 correctly. But for verse numbers 2 and 4, existing search engines were either not able to retrieve verses completely or showed no results signifying the limitations of detecting verses accurately. Our proposed framework

Table 3. Accuracy results

V. no.	Quranic verses	Actual no. of verses	Muslim-web	Search truth	Alawfa	Tanzil	Proposed approach
1	دُحُورًا وَلَهُمْ عَذَابٌ وَاصِبٌ	1	1	1	0	1	1
2	فَاتَّقُوا اللَّهَ وَأَطِيعُونِ	8	10	10	0	10	8
3	فَبِأَيِّ آلَاءِ رَبِّكُمَا تُكَذِّبَانِ	31	31	31	0	31	31
4	بِسْمِ اللَّهِ الرَّحْمَنِ الرَّحِيمِ الم	8	0	0	0	135	7

worked well for all the given random verses and acquired efficiency of 100% for all the verses except last one. This can be due to the fact that our framework uses exact matching algorithms and needs more optimization to handle diacritics.

Besides checking for accuracy of verse detection, we modified some Quranic verses to check whether existing search engines give any results. Results are shown in Table 4. Alawfa showed no results, whereas other existing engines gave results for incorrect verses too. For a non-native speaker of Arabic language, it will create confusion for him which verse is authentic, the one which is modified or the one that search engines displayed. As compared to existing search engines, again our framework detected fake verses accurately. The fact that existing search engines give results for modified verses can be due to removal of diacritics. Most of the works in the area of Quran authentication have focused on removal of diacritics for efficient matching. Thus, if you remove any diacritic, these search engines will still show results based on character comparisons.

Table 4. Overall results

Quranic verses	Muslim-web	Search truth	Alawfa	Tanzil.net	Proposed approach
ذلك الكتاب ل ريب فيه هدى للمتقين	No results	Shows results for incorrect verse	No results	Shows results for incorrect verse	Authenticates the verse is not authentic
ذلك الكتاب ل ريب فيه هدى للمتقين	No results	Shows results for incorrect verse	No results	Shows results for incorrect verse	Authenticates the verse is not authentic
بِمَا أُنزِلَ إِلَيْكَ وَمَا أُنزِلَ مِن قَبْلِكَ وَبِالْآ	Shows results for incorrect verse	No results	No results	No results	Authenticates the verse is not authentic
الْكِتَابَ بِالْحَقِّ مُصَدِّقًا لِّمَا بَيْنَ	Matches but displays more than one result	Shows results for incorrect verse	No results	Matches but displays 11 results	Authenticates the verse is not authentic

5 Conclusion

In this paper, an efficient framework related to Quran and Hadith authentication has been proposed. Processing of whole framework is divided among two servers, i.e., server A and server *B*. Server *A* is mainly responsible for segmentation part and digital signature process, while server B is responsible for authentication and watermarking the verified content. Experiments conducted on the first initial prototype of verification phase showed promising results with accuracy up to 98.6%. No diacritics were removed in the whole process which is novelty in our framework. Previous frameworks removed the diacritics for accuracy purposes. Our future work will focus on enhancing verification phase and developing complete framework by completing second phase of security. Last phase will be to test the whole framework using different Quranic verses and Hadiths from Internet.

Acknowledgements. The work of Jhon Veri is supported by Universitas Putra Indonesia YPTK under research grant no 094/UPI-YPTK/RG/IX/2016. The work of Tutut Herawan is supported by Universitas Teknologi Yogyakarta Research Grant no vote O7/UTY-R/SK/0/X/2013.

References

1. Alsmadi, I. and M. Zarour, *Online integrity and authentication checking for Quran electronic versions*. Applied Computing and Informatics, 2015.
2. Hakak, S., Kamsin, A., Tayan, O., Idris, M. Y. I., Gani, A., & Zerdoumi, S. *Preserving Content Integrity of Digital Holy Quran: Survey and Open Challenges*. IEEE Access, 2017.
3. Khan, M.K. and Y.M. Alginahi, *The holy Quran digitization: Challenges and concerns*. Life Science Journal, 2013. 10(2): p. 156–164.
4. Ibrahim, N.J., *Automated Tajweed checking rules engine for Quranic verse recitation*. Doctoral dissertation, University Malaya).
5. Jamaliah Ibrahim, N., Yamani Idna Idris, M., Razak, Z., & Naemah Abdul Rahman, N. *Automated tajweed checking rules engine for Quranic learning*. Multicultural Education & Technology Journal, 2013. 7(4): p. 275–287.
6. Abudena, M.A. and S.A. Hameed, *Toward a Novel Module for Computerizing Quran's Full-Script Writing*. International Journal of Computer System, 2015.2(11): **p.** 469–474.
7. Kirchhoff, K. and D. Vergyri, *Cross-dialectal data sharing for acoustic modelling in Arabic speech recognition*. Speech Communication, 2005. 46(1): p. 37–51.
8. Tayan, O., Y.M. Alginahi, and M.N. Kabir, *An Adaptive Zero-Watermarking Approach for Text Documents Protection*, International Conference on Advances in Computer and Information Technology. 2013.
9. AlAhmad, M.A., I. Alshaikhli, and B. Jumaah. *Protection of the Digital Holy Quran Hash Digest by Using Cryptography Algorithms*. Advanced Computer Science Applications and Technologies (ACSAT), 2013. pp. 244–249.
10. AlAhmad, M.A., I. Alshaikhli, and A.E. Alduwaikh. *A New Fragile Digital Watermarking Technique for a PDF Digital Holy Quran*. in Advanced Computer Science Applications and Technologies (ACSAT), 2013. pp. 250–253.
11. Abuhaija, B., N. Shilbayeh, and M. Alwakeel. *Security protocol architecture for website authentications and content integrity*. in Computer and Information Technology (WCCIT), 2013. pp. 1–6.

12. Syifak Izhar, H., Jasni, M. Z., Afifah Nailah, M., & Gran, B. Syifak Izhar, H, *Localization Watermarking for Authentication of Text Images in Quran*. Advances in Information Technology for the Holy Quran and Its Sciences, 2013. pp. 24–29.
13. Sabbah, T. and A. Selamat. *Support vector machine based approach for quranic words detection in online textual content*. in Software Engineering Conference (MySEC), 2014 8th Malaysian. 2014. IEEE.
14. Kurniawan, F., Khalil, M. S., Khan, M. K., & Alginahi, Y. M. Kurniawan, F. *Authentication and Tamper Detection of Digital Holy Quran Images*. in Biometrics and Security Technologies (ISBAST),2013. pp. 291–296.
15. Kurniawan, F., Khalil, M. S., Khan, M. K., & Alginahi, Y. M. *DWT + LSB-based fragile watermarking method for digital Quran images*. in Biometrics and Security Technologies (ISBAST), 2014. pp. 290–297.
16. Kurniawan, F., Khalil, M. S., Khan, M. K., & Alginahi, Y. M. *Exploiting Digital Watermarking to Preserve Integrity of the Digital Holy Quran Images*. Advances in Information Technology for the Holy Quran and Its Science, 2013. pp. 30–36.
17. Laouamer, L. and O. Tayan, *An enhanced SVD technique for authentication and protection of text-images using a case study on digital Quran content with sensitivity constraints*. Life Science Journal, 2013. 10(2): p. 2591–2597.
18. Gutub, A. A. A., Al-Haidari, F., Al-Kahsah, K. M., & Hamodi, J. *e-Text Watermarking: Utilizing Kashida Extensions in Arabic Language Electronic Writing*. Journal of Emerging Technologies in Web Intelligence, 2010. 2(1): p. 48–55.
19. Alginahi, Y.M., O. Tayan, and M.N. Kabir, *Verification of Qur'anic Quotations Embedded in Online Arabic and Islamic Websites*. International Journal on Islamic Applications in Computer Science and Technology, 2013. 1(2): p. 41–47.
20. Sabbah, T. and A. Selamat. *A framework for Quranic verses authenticity detection in online forum*. in Taibah University International Conference on Advances in Information Technology for the Holy Quran and Its Sciences. 2013.
21. Alshareef, A. and A.E. Saddik. *A Quranic quote verification algorithm for verses authentication*. Innovations in Information Technology (IIT), 2012. pp. 339–343.
22. Nisha, S., N. Ali, and A. Shawkat Ali. *Searching quranic verses: A keyword based query solution using. net platform*. in Information and Communication Technology for The Muslim World (ICT4 M), 2014.
23. Boyer, R.S., & Moore, J. S, *A fast string searching algorithm*. Communications of the ACM, 1977: p. 762–772.
24. http://tanzil.net/#2:1. [2nd January].
25. http://quran.muslim-web.com/. [15th November].
26. http://www.searchtruth.com/. [2nd December].
27. http://www.alawfa.com/en/default.aspx. [cited 2016 2nd January].

Computing Domain Ontology Knowledge Representation and Reasoning on Graph Database

Phu Pham[1], Thuc Nguyen[2], and Phuc Do[2(✉)]

[1] University of Technology, HCM (HUTECH), Ho Chi Minh City, Vietnam
phamtheanhphu@gmail.com

[2] University of Information Technology, HCM (UIT), Ho Chi Minh City, Vietnam
{thucnt,phucdo}@uit.edu.vn

Abstract. In this paper, we solve problem of the ineffectiveness of using RDFS/OWL-stored mechanism for large-scale domain ontology. In particular, when the constructed ontology contains a huge amount of entities and semantic relations, it causes difficulties in managing and visualizing the ontological knowledge as well as low performance in data querying. We resolve these issues by using graph database as the storage mechanism for representing the constructed ontology. The approach of using graph database provides the advantages not only in better ontological data management and visualization but also in the higher performance and flexible of knowledge extracting from ontology via cypher querying language.

Keywords: Large-scale domain ontology · Ontological database-based storage mechanisms · Graph database · Cypher query language

1 Introduction

In traditional way, the ontologies are stored by using RDF/OWL (W3C) language format. This type of repository can support the reasoning of ontological knowledge through SQWRL-OWL, SPARQL-OWL (query) language [1, 2]. In fact, this storage type still has been working effectively with the small- or medium-scale (ontological data) ones. However, in case that the constructed ontologies contain a huge number of entities and interrelationships (between entities), the traditional method (also called native storage) seems unappropriated due to the difficulties in maintenance as well as low performance in data extraction [3]. The most popular way is converting the ontological stored repository to database, such as RDBMS [4, 5]. In this paper, we proposed how to use graph database in representing, storing, and reasoning the constructed ontological knowledge. Our proposal has many advantages. The paper is organized as follows: Sect. 1 introduction; Sect. 2 the structure of computing domain ontology; Sect. 3 the previous approaches and shortcoming; Sect. 4 the approach of using graph database for ontological representing and reasoning; Sect. 5 experiments and evaluations; Sect. 6 conclusions and future works.

V. Bhateja et al. (eds.), *Information Systems Design and Intelligent Applications*, Advances in Intelligent Systems and Computing 672,
https://doi.org/10.1007/978-981-10-7512-4_75

2 Structure of Computing Domain Ontology (CDO)

In past works, we select the computer science as the domain knowledge for constructed ontology (called computing domain ontology—CDO). We inspire from the work of TDC Chien [6] (see Fig. 1). The CDO's data are designed and organized as multiple-layers structure as follows:

$$CDO = \langle T, I, R \rangle$$

- **Topic Layer** $\langle T \rangle$: storing the root's concept of computer science fields which are extracted from the ACM Digital Library and Microsoft Academic Search (MAS). These concepts were merged and reorganized into hierarchical structures (taxonomy).
- **Ingredient Layer** $\langle I \rangle$: containing terms which are specified and considered typical for each topic in $\langle T \rangle$. These terms had been archived by extracting from documents which are assigned for each topic manually.
- **Relation Layer** $\langle R \rangle$: containing the relationships between terms (also called CDO's entities) in the $\langle I \rangle$. These relations were formed by analyzing the semantic relationships of each term with others in training documents. We defined five main relationship types including IS_A, PART_OF, HAS_PART, HAS_ALTLABEL, and RELATED_TO…

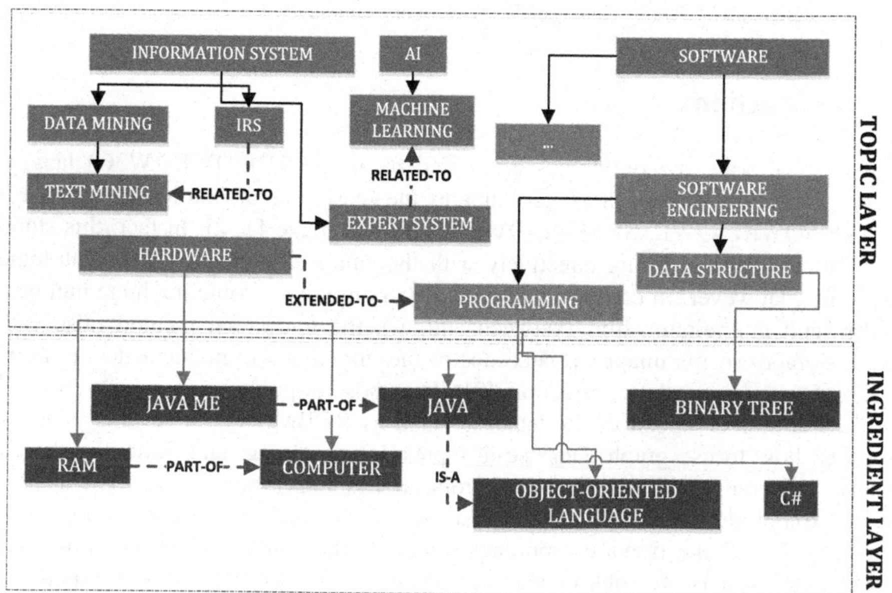

Fig. 1. Illustration of structural CDO's layers (including topic, ingredient, and relation)

In the topic layer, based on ACM computing classification system, CSS (2012), there are 12 root topics covering all related fields of computer science. One root topic has multiple subfields in each level. They have over 170 main topics. In the ingredient layer, we extracted more than 32,000 entities and the total relationships between entities. They reach over 71,000 (the highest one is HAS_ALTLABEL relation, and it occupied more than 53% for total five types of relation) (see Table 1). They cause a lot of problems for us to deal with the huge data of CDO. The key problem is that the native store method (RDF/OWL) barely working with text files to process data input and output leads to the low performance in managing as well as extracting the data from CDO through queries [3]. We intend to enrich the amount of entities up to 100,000 as well as expand the variety of relationship types [7] (MADE_OF, RESULT_OF…). Hence, we would like to find an alternative stable stored mechanism for CDO in order to ensure the effectiveness in managing as well as querying data despite the rapid increases in ontological repository size.

Table 1. Statistics of CDO extracted entities and relations data source

Descriptions	Details
Number of extracted topics	170 topics—accommodating in five levels (following the ACM CSS taxonomy)
Number of extracted entities/concepts	Over 32,000 (corresponding to 170 identified topics)
Number of extracted relations	Over 71,000, included: IS_A: 4,262 (6%) PART_OF/HAS_PART: 4,986 (7%) HAS_ALTLABEL: 37,638 (53%) RELATED_TO: 24,142 (34%)

3 The Previous Approaches and Shortcoming

3.1 Ontological RDBMS-Based Repository and Reasoning

This is the most popular approach that has storage mechanisms for constructing large-scale ontologies. The uses of RDBMS in storing ontology provide the advantages such as easier in data managing as well as maintaining and faster and more effective in querying data from ontologies by using SQL queries [8]. Several types of RDBMS can be applied to store CDO such as Microsoft SQL Server, Oracle, and MySQL… The ontological entities and all semantic relations will be represented in separated tables [5, 8] (see the ER design diagram shown in Fig. 2).

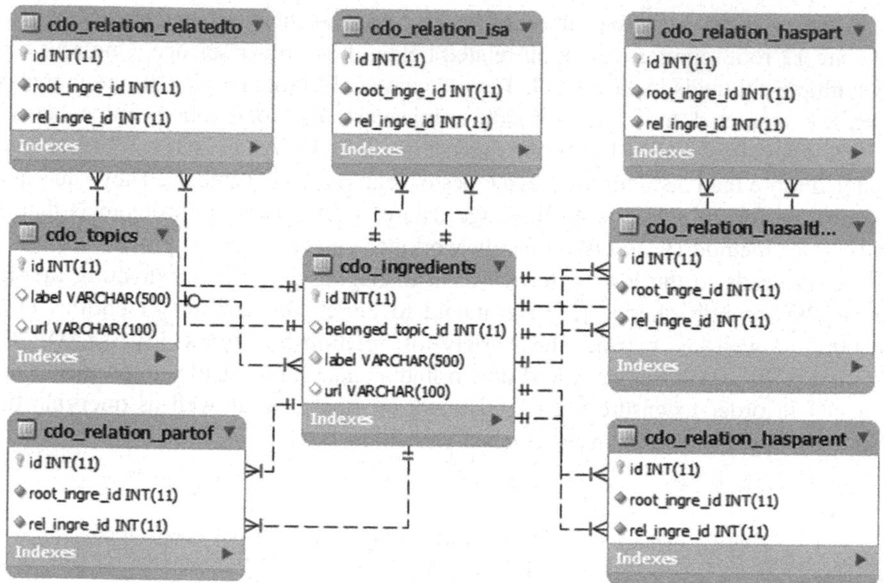

Fig. 2. Illustrate the CDO-based RDBMS repository—the ER and table diagram on MySQL database

In our case, the method for converting CDO's repository from the native store to relational databases is simple. At the beginning, we reorganize the CDO's structural layers, as follows:

- **For the** $\langle T \rangle$: Each topic class will be organized in a single table, and its properties (<id>, <label>..) are stored in fields. Each topic stored in a separate table will have its own id key for identification.
- **For the** $\langle I \rangle$: All the concepts in ingredient layer $\langle I \rangle$ were defined as the "entities" and stored on a single table with several fields to describe its properties (such as <id>, <name> <label>, <url>…). Each entity in ingredient's table will have the references to the topic it belonged (in CDO topic's table).
- **For the** $\langle R \rangle$: Each relationship type will be organized separately in each table and the relation will be represented as the key mapping between entities table and relationship table. For example, IS_A relation for *<concept_A>* to *<concept_B>* will be stored as a pair of [<id of copcent_A>—<id of copcent_B>] in table IS_A relation. Hence, with the five CDO relationship types, we have the five tables to store all the relation of concepts in $\langle I \rangle$ layer.

3.2 Evaluations and Encountered Disadvantages

- There are many disadvantages while using the RDBMS-based repository. We all know that the ontological structure is a kind of network with connections between entities; an entity can have at least one or multiple connections to others. With the relational table-based mechanism, if we conduct too much relationship (key-based references) between tables, it will slow down the database's productivity while running and performing the queries. Moreover, in case that constructed ontology had a lot of relationship types, the try of scanning through all the relations of specific entity will take a lot of effort. Because each relationship type is stored in separated table, hence the expansion in the number of relationship types is equivalent to making the queries much more complicated as well as taking more time for the query to be completed due to multiple JOIN, UNION... operations. For example, we have a sample SQL statement for extracting all the relationships of entity (or list of entities) (see Fig. 3).

```
...
UNION
    /* extracting each type of ontological relation */
    SELECT DISTINCT (REL_HP.label) AS label
    FROM cdo_ingredients AS A, cdo_ingredients AS REL_HP, cdo_relation_haspart AS HP
    WHERE A.id = HP.root_ingre_id AND REL_HP.id = HP.rel_ingre_id AND A.`label` = [INPUT_ENTITIES]
UNION
...
```

Fig. 3. Sample SQL query to extract the relation from RDBMS-based repository

We can see clearly that the operations will increase simultaneously with number of relationship types of constructed ontology—we make an experiment to evaluate and compare the execution's time of cases, when we increase the amount of relationship type for the same investigated entities. The experimental results are as follows (see Table 3, Fig. 4) (Table 2):

Table 2. Experimental environments of CDO-based RDBMS repository

Experimental environments	
Hardware	CPU: Intel(R) Core(TM) i5-4460 CPU @ 3.20 GHz (4 CPUs) RAM: 8 Gb RAM
	OS: Windows 8.1 Enterprise 64-bit (6.3, Build 9600)
RDBMS	MySQL Community Server 5.7.17, Windows 64-bit base

Table 3. Experimental results of data extraction in CDO-based RDBMS repository

Amount of input entities	Average execution time (ms) based on the amount of ontological relationship type					
	1	2	3	4	5	6
1000	59.9	75.9	107.2	164.4	240.9	258.9
2500	74.5	129.3	161.5	280.2	396.3	457
4000	106.3	164	237.4	413.2	531.1	599
8000	155	254.3	424.3	727.7	900.9	1009.7
10000	177.7	359.2	501.2	860.9	1063.6	1339.3

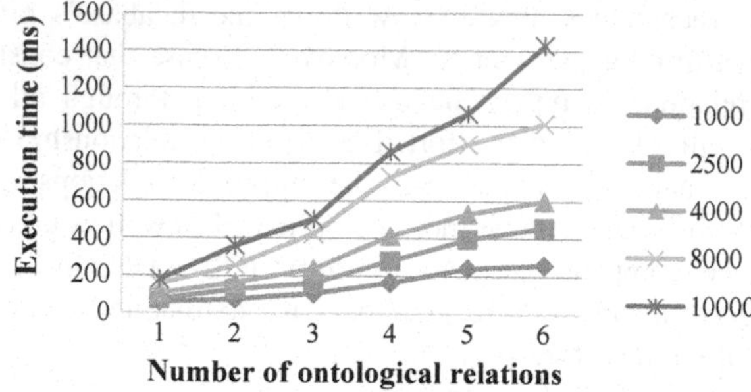

Fig. 4. Line-chart graph illustrates the changes of execution's time value for extracting data in CDO-based RDBMS repository

We see that with the same number of entities, when we increase the amount of relationship type to be scanned, the execution time will increase dramatically (nearly 0.35 times) for each increasing step. Thus, we can assume that one more scanned relationship type added to query is equal to increase the execution's time to 0.35 times —with the [n] relationship type added will raise the time for the query (to be completed) up to approximately [0.35n] (times). It is really a problem related to the performance and it needs to be concerned, especially with the ontology that has the large amount of ontological relationship type.

4 The Approach of Using Graph Database for Ontological Representing and Reasoning

4.1 Advantages of Using Graph Database

As mentioned above, the main purpose of using graph database is resolving the issues related to multiple linked data storing and querying when using RDBMS. In some cases, the use of relational table is an ineffective solution. In fact, it is designed to handle large-scale data patterns rather than the multiple linked (relations) ones, like ontological knowledge and semantic network. The reason comes from the designed architecture, and using purpose of each storage model, the RDBMS stores its data as a set of tables and columns. They cause critical wastes of resources for carrying out complex joins and self-join operations, especially when the stored data become more and more inter-related in complicated ways. We would like to find an alternative stored repository for CDO. We figured out that applying the graph database as the base storage for CDO provides us a lot of advantages such as:

- The supported features and organizations of graph database are completely appropriate with our CDO's designed structures. All the topics and concepts will be considered as entities/nodes in the graph, each entity has its own properties *<id>*, *<name>* *<label>*..., and the relationships of entities were converted to edges linking between nodes. Each kind of relation also has its own properties (IS_A, PART_OF...).

- Easier in data management and updating the content of CDO (data of node, relationship...), the processes of adding new concepts or relations can be conducted much simpler.
- Supports for visualizing the CDO ontology data via DBMS (Neo4J)'s GUI.
- The abilities to the simplifying queries' syntax and optimizing the process of data extractions via cypher query language.
- Applying the solution for finding shortest paths (between graph's nodes) in graph database to identify the potential non-direct linked concepts between terms in user's queries or documents—supporting for the processing of query's expansion or document clustering in information retrieval.

4.2 Method for Converting CDO from Native Store to Neo4J

In general, the ontological architecture is similar to the organized data model of a graph. They include entities, properties, and relationships. Hence, the structure of CDO of new proposed structure is as follows (see Fig. 5):

- **Conversions on $\langle T \rangle$ and $\langle I \rangle$**: all the topics and concepts in both $\langle T \rangle$ and $\langle I \rangle$ will be considered as data node with properties such as *<url>*, *<name>* *<label>* ...
- **Conversions on $\langle R \rangle$**: The relationships between entities will be represented as the graph's edges—each edge (or relationship between nodes). They carry the related parameters to identify which CDO's relationship type it is (IS_S, HAS_PART, RELATED_TO...).

4.3 Extracting Ontological Knowledge via Cypher Query

After converting and storing the CDO's data on Neo4J, the querying method will be switched from SPARQL to cypher query language of Neo4J. We also keep the SPARQL language to extract the information from Neo4J via supported plug-ins. However, in this research, we just focus on the uses of cypher query and its advantages in reasoning and extracting ontological data. Similar to SPARQL, we consider that all the entities and relations stored in database are formed by the triple stores with the pattern of [entity]-(relationship/predicate)-[entity]—the first element of a triple store also called as "*subject.*" Within graph database, the relationship—which connects two nodes—also has the direction (edge's direction) to indicate the "subject" of a specific triple, for example, [computer]-(HAS_PART) → [CPU], where "→" indicates that the "computer" is the subject of this triple store (see Fig. 5). Moreover, cypher query language also supports filtering, grouping, and sorting operations such as WHERE, ORDER BY, GROUP BY... which are similar to SQL language.

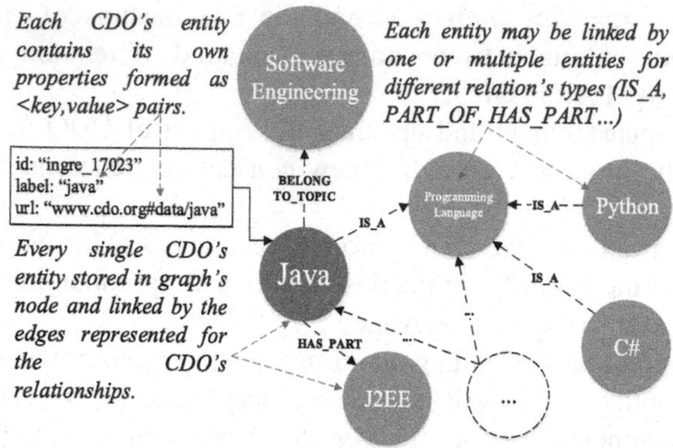

Fig. 5. Reorganizing the CDO's layers and data on graph-based structure

4.4 Applying Shortest Path Problem to Identify Non-direct Nearest Concepts

Expanding the research's scope to user's query expansion problem, the hidden or no-direct entities between two or more concepts in user's query play an important role because it can support the computer to identify exactly the knowledge's scope or topic domain that the user actually wants to point by identifying hidden concepts between existed concepts in user's queries. With the CDO's data stored in Neo4J, we can execute the cypher query language to extract all the entities between two concepts [**java**] and [**python**], as shown in Fig. 6.

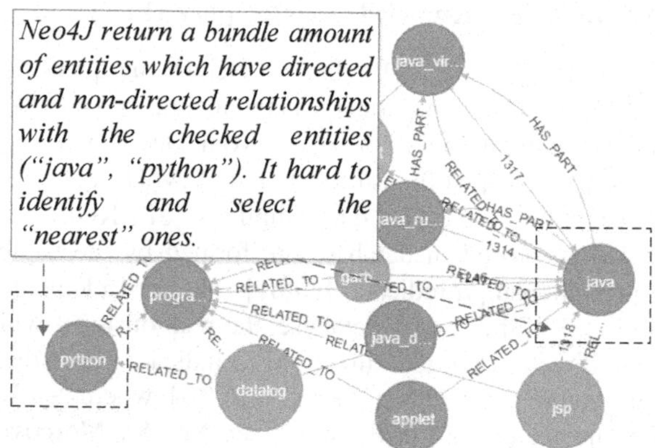

Fig. 6. Illustrate the cypher (*MATCH p = (a {label: '**java**'})-[*1..100]-(b {label: '**python**'}) RETURN p;*) query on the Neo4J

However, in case that there is a huge amount of return concepts (which have relation direct/non-direct) with two inputted concepts ("java" and "python"), it seems hard to identify which are the nearest/potential ones for selection. Hence, in order to improve the result of return concepts to be more exact and reasonable, we just need to get the concepts within the shortest path(s) in collection of paths that connect two entities "java" and "python." We can do it by executing the cypher query language, like below, and the nearest entity that we can get is "*programming_terms*" (see Fig. 7).

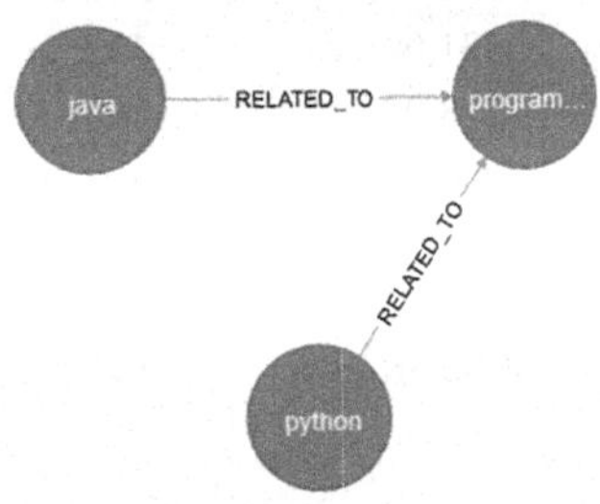

Fig. 7. Illustrate the cypher (*MATCH (a {label: '**java**'}), (b {label: '**python**'}), p = shortestPath ((a)-[*1..100]-(b)) RETURN p;*) query on Neo4J

5 Experiments and Evaluations

To evaluate the achievements of the ontological repository's conversion from the native to graph database storage, we made some experiments. These experiments were conducted by comparing the execution's time between two types of CDO's storage-based, the graph-based (Neo4J), and RDBMS-based (MySQL). (Table 4).

Table 4. Experimental environments for query performance testing of graph-based versus RDBMS-based CDO repository

Experimental environments	
Hardware	CPU: Intel(R) Core(TM) i5-4460 CPU @ 3.20 GHz (4 CPUs) RAM: 8 Gb RAM
	OS: Windows 8.1 Enterprise 64-bit (6.3, Build 9600)
RDBMS	MySQL Community Server 5.7.17, Windows 64-bit base
GDBMS (graph based)	Neo4j 3.1.1 Community Edition, Windows 64-bit base

Two types of queries (SQL and cypher) are executed to do the same purposes of getting data from CDO—and the inputted entities in each test case are increased gradually. The experimental results are given in Table 5.

Table 5. Comparisons in query execution's time between RDBMS-based (MySQL) and graph-based (Neo4J) CDO repositories

Amount of input ontological entities	GraphDb (Neo4J)	RDBMS (MySQL)	RDBMS/GraphDb (%)
100	114.09	113.45	99.44
500	221.80	148.70	67.04
1000	352.82	302.27	85.67
2000	431.09	352.36	81.74
4000	578.09	564.18	97.59
8000	986.55	1068.36	108.29
10000	942.91	1215.09	128.87
12000	1007.82	1504.64	149.30
15000	1182.36	1547.27	130.86

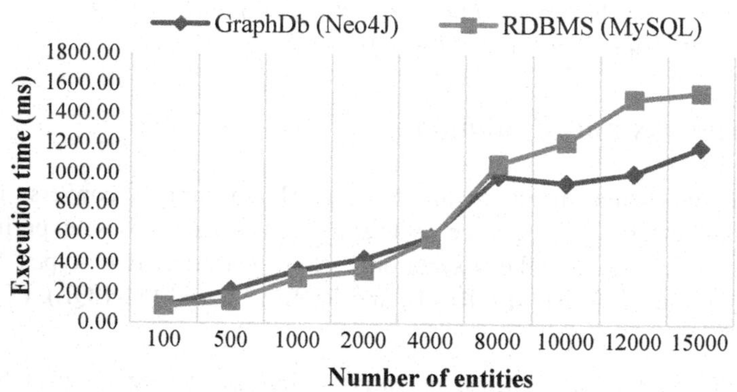

Fig. 8. Comparisons in query execution's time of different CDO repositories (Neo4J vs. MySQL)

From Table 5 and Fig. 8, we can figure out that based on the number of input entity for the most of initial tests (the range of entity amount from 100 to nearly 8,000), the gained both storage models are the same (in some cases the RDBMS model run faster a little bit than the graph database model). However, when the number of entities increases to 8,000, the graph database model gains a better performance than RDBMS. It is nearly more than 50% when the number of entities reached to over 12,000. The experimental results prove that the use of graph database model for ontological knowledge representation is potential and effective, especially with the large-scale ontologies containing a huge amount of entities and relationships.

6 Conclusions and Future Work

We propose a novel approach to use graph database for ontological knowledge representation and extraction, where the overall structure of ontology will not be changed. Our approach includes the new way for extracting ontological data via cypher query language as well as performance optimization in data querying. For future works, we focus on enriching the size of CDO including the amount of entities and relationship types much larger in order to make deeper evaluations about the stabilities and effectiveness when using graph database model for representing the ontological knowledge and semantic relations network.

Acknowledgements. This research is funded by Vietnam National University Ho Chi Minh City (VNU-HCMC) under the Grant Number B2017-26-02.

References

1. Antoniou, Grigoris, and Frank Van Harmelen: Web ontology language: OWL. In: Handbook on ontologies. Springer Berlin Heidelberg, pp. 91–110, (2009)
2. Hitzler, Pascal, et al.: OWL 2 web ontology language primer. In: W3C recommendation, 27(1), 123 (2009)
3. Stijn Heymans, Li Ma, Darko Anicic, Zhilei Ma, Nathalie Steinmetz, Yue Pan, Jing Mei, Achille Fokoue, Aditya Kalyanpur: Ontology Reasoning with Large Data Repositories. In: Springer Science + Business Media, LLC, pp. 89–128, (2008)
4. Mogotlane, Kgotatso Desmond, and Jean Vincent Fonou-Dombeu: Automatic Conversion of Relational Databases into Ontologies: A Comparative Analysis of Protege Plug-ins Performances. In: ArXiv preprint arXiv:1611.02816, (2016)
5. Gali, Anuradha, et al.: From ontology to relational databases. In: International Conference on Conceptual Modeling. Springer Berlin Heidelberg, (2004)
6. Chien. Ta Duy Cong, Tuoi. Phan Thi: Building Ontology Based-on Heterogeneous Data. In: Journal of Computer Science and Cybernetics, vol. 31, pp. 149–158 (2015)
7. Chien. Ta Duy Cong, Tuoi. Phan Thi: Identifying the semantic relations on unstructured data. In: International Journal of Information Sciences and Techniques (IJIST), vol. 4 (2014)
8. Miranker, Daniel Paul, and Juan Federico Sequeda: System for Accessing a Relational Database Using Semantic Queries. In: U.S. Patent No. 9,396,28, (2016)

An Efficient Model for Finding and Ranking Related Questions in Community Question Answering Systems

Van-Tu Nguyen[1](✉), Anh-Cuong Le[2](✉), and Dinh-Hong Vu[2]

[1] VNU University of Engineering and Technology, Hanoi, Vietnam
tuspttb@gmail.com

[2] Faculty of Information Technology, Ton Duc Thang University, Ho Chi Minh City, Vietnam
leanhcuong@tdt.edu.vn, vudinhhong@tdt.edu.vn

Abstract. The task of finding- and ranking-related questions plays the most important role for any real-world Community Question Answering (cQA) systems. This paper proposes a new method to solve this problem by considering multi-views for measuring the similarities between the input questions and the question-answering pairs in the database. Our model will investigate various aspects for understanding questions. Beside the traditional features such as bag of n-grams, we will use more efficient aspects that include word embeddings and question categories. We will use a word representation model for generating word embeddings, a question classification module for determining the category for an input question. Then all these obtained features are combined into a machine learning-based framework for getting similarity existing question-answering pairs as well as for ranking these pairs. We tested our proposed approach on the dataset SemEval 2016 and the experiment shows obtained results with the *Accuracy* and *MAP* of 80.43% and 77.43%, respectively, which are the highest accuracies in comparison with previous studies.

Keywords: Community question answering · Question classification · Question representation · Ranking question

1 Introduction

Question answering (QA) is a research topic in the field of natural language processing and also is an important application of machine learning. QA is the task of automatically obtaining the correct answers for an input question. Note that the question and the answers are presented in natural language sentences. Question answering problems can be divided into types. In the first type, given a question the task here is how to retrieve the most appropriate answering sentences from existed documents. In the second type, the question is generated

V. Bhateja et al. (eds.), *Information Systems Design and Intelligent Applications*, Advances in Intelligent Systems and Computing 672,
https://doi.org/10.1007/978-981-10-7512-4_76

from the learned model. In this paper, we focus on the first type of QA problems, in which we aim to build a model for extracting the best answers for an input question from the data of question-answering pairs from a community. Such system will help to improve the quality of any cQA sites such as Quora.[1] The cQA sites provide a common way to connect user questions with answers from the community. The benefits of cQA have been proven in [1,2] and cQA has attracted lots of attention in the fields of information retrieval and natural language processing research [3]. In cQA, when user submits an input question, if the input question is similar to the previously answered question (i.e., semantic similarity), the cQA should have a mechanism to finding and ranking the related questions from cQA sites.

This paper proposes a new method for solving the task of finding- and ranking-related questions for cQA systems. The previous studies solved this problem by measuring the similarity between the input question and the question-answering pairs in the database. The measurement is usually based on the common features such as n-grams or some other kinds of linguistic information which require deep analysis such as syntactic parsing [4–6]. To enrich the information for similarity measurement but without a complicated linguistic analysis, we will use some new aspects of information. Different from previous studies, we consider multi-views of a question so that we can obtain added knowledge for measuring the similarities between the input questions and the question-answering pairs in the database. Our model will investigate various aspects for understanding questions including traditional features such as bag of n-grams, word embeddings, and question categories.

Word embeddings are the result of word representation learning models which represent words by real-valued vectors whose relative similarities correlate with semantic similarity. Word embedding models base on statistics of word occurrences in a corpus to encode semantic information which expresses how meaning is generated from these statistics, and how the resulting word vectors might represent that meaning. In this work, we use the model [7] for generating word embeddings. For the other view of question category, we will use a question categorization module to obtain the corresponding question category for an input question. Then all these obtained features will be combined into a machine learning-based framework for getting similarity existing question-answering pairs as well as for ranking these pairs. The experimental results on the dataset provided by SemEval 2016 show that the model we proposed has significantly improved both the classification accuracy and the ranking question compare to previous studies on the same dataset.

2 Related Work

Finding- and ranking-related questions in cQA have become the important task for designing a cQA system. Many studies have explored the common ways for measuring the similarity between questions, between a question and answers

[1] https://www.quora.com/.

such as using cosine measurement between words. Moreover, some other studies proposed more advanced features and models. Cao et al. [8] classified questions into different topics and based on this feature for building a recommendation system. Duan et al. [9] extracted questions focus and use it for the similarity measuring. Some other researches such as [10,11] proposed a method of using topic modeling. Interestingly, [12,13] followed an approach of translation for question and answer pairs.

There are also some studies based on syntactic information. Wang et al. [6] used some substructures of the parsed tree as features for measuring the similarity between questions. Authors in [4,5] also used syntactic information, but with parsed tree kernel based on the KeLP platform from [14]. More recently, [15,16] used SVM rank method on distributed representation of words for the task on ranking question at the SemEval 2016 Task3 [17]. Other studies use neural network-based models for calculating the similarity between questions in cQA [18–20].

Different from previous studies, in this paper we will focus on different representations of question and answering for measuring the similarity between the input question and the database. We consider this idea as multiple views of data and exploiting them for measuring the similarities. In this paper, we will use a question categorization module to obtain the corresponding question category for an input question. Then we will extract the features based on this question categories and combine all the features into a machine learning-based framework for getting similarity existing question-answering pairs as well as for ranking these pairs.

3 Our Approach

Our proposed model uses multi-view as an approach for generating different and rich feature kinds for the tasks of question finding and ranking. To this end, our model adds one more component to the common solution, that includes question categorization and word2vec. Our model for finding- and ranking-related questions in cQA is described in Fig. 1.

3.1 Representation of Questions

Our idea is to represent a question by a rich knowledge vector based on a consideration of multi-view for questions. The following are the different kinds (i.e., different views) of representing a question.

Representation of questions using n-gram features. In this representation model, each input question-related question pair is representation as a vector of n-gram words (n = 1, 2, 3). The n-gram words for each question-question pair are extracted based on the context of the words of the question. Note that the document representation also uses this kind of features.

Representation of questions using question property. In this study, to evaluate the similarity between the two questions, we use some common properties extracted from the question such as the ratio of the number of words between

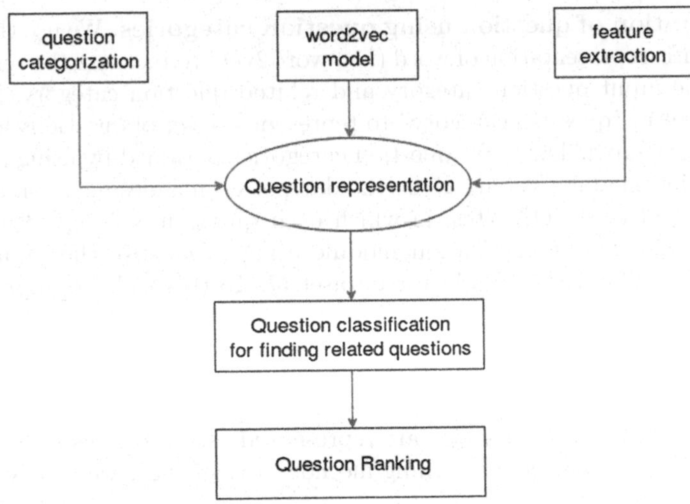

Fig. 1. General architecture of finding and ranking related questions model

the input question and the related question, the ratio of the number of sentences between the input question and the related question, bag of word, word overlap, noun overlap, name entities overlap.

Representation of question using word embeddings. In this work, we used the continuous Skip-gram model [7] of the word2vec toolkit[2] to generate vector representations of the words in dataset. First, all the sentences in input questions and related questions are tokenized and the words are then converted to vectors using the pre-trained word2vec model. Next, for each question, we averaged its word vectors in order to have a single representation of its content as this setting has shown good results in other NLP tasks (e.g., for language variety identification [21] and discriminating similar languages [22]). In this work, we use a dataset from Qatar Living (English)[3] to pre-trained word2vec model with 200-dimensional vectors, window = 5. After representing questions as corresponding vectors, we compute the similarity between the two vectors and use this similarity score as a feature. Here, the similarity between the two vectors is estimated using the cosine measurement as follows:

$$cosin\_sim(u,v) = \frac{\sum_{i=1}^{n} u_i * v_i}{\sqrt{\sum_{i=1}^{n} (u_i)^2} * \sqrt{\sum_{i=1}^{n} (v_i)^2}} \quad (1)$$

where u and v are two n-dimensional vectors, u_i is the ith element of u vector.

[2] https://code.google.com/p/word2vec.
[3] http://www.qatarliving.com/forum.

Representation of question using question categories. We use the distributed semantic representation of word (i.e., word2vec) to measure the relationship between the input question category and related question category. Note that we use the term "question category" to represent the set of questions having the same category label. The input question categories obtained by using a question categorization module. We are given the dataset Q includes question-answering pairs extracted from cQA sites, in which each question is assigned a category label. The question categorization module aims to classify the input q_o into one of the question categories in the dataset Q. To this end, we implement the following steps:

1. The first step: we prepare a training dataset including questions in dataset Q, they are assigned with category labels (the label here is question category).
2. The second step: the questions are represented as vectors (using word2vec).
3. The third step: a machine learning method is used (here we choose SVM) to learn the classifier.
4. Finally, for testing, we first represent the input question by feature vectors and use the classifier obtained at the third step to predict the label (i.e., an input question category).

Suppose that we are given the input question q_o. First, all the sentences in the related question categories and sentences in q_o are tokenized and the words are represented as vectors using a word2vec tool. Second, we align each word in related question category to the word in q_o that has the highest vector cosine similarity, as the formula below:

$$score(t_k) = \max_{1 \leq h \leq m} (cosin\_sim(t_k, b_h)) \tag{2}$$

where:

m: the number of words in input q_o. t_k: the representation vector of the word kth in the related question category. b_h: word vector representation of the hth word in input q_o. $cosin\_sim(t_k, b_h)$: is the cosine similarity of two vector representations of kth word in related question category with the hth word in input q_o.

Finally, the similarity score between related question category and input q_o is calculated as follows:

$$score(question_i) = \frac{\sum_{k=1}^{n} score(t_k)}{n} \tag{3}$$

where n is the number of words in related question category.

3.2 Question Classification

The question classification module aims to classify the input question-related question pairs. To this end, we implement the steps as follows:

1. The first step: preparing the training dataset which contains input question-related question pairs assigned with classification labels (the label here is "Relevant" or "Irrelevant").
2. The second step: representing each question in the training data as a vector of features.
3. The third step: learning the classifier using a machine learning method (here we choose SVM) based on features obtained at the second step.
4. Finally, given a test input question-related question pair, like the second step, it is converted to a vector of features and use the classifier obtained at the step 3 to get predicted label.

3.3 Ranking-Related Question

Ranking-related questions are the task of ranking questions according to their relevance to the input question q_o. This ranking is based on the similarity scores between related questions and input question q_o. These scores are generated from the question classification module in Sect. 3.2.

4 Experiments Setting

4.1 Dataset and Evaluation Metrics

In order to set up the experiment, we use the cQA datasets provided by SemEval 2016.[4] The datasets were extracted from Qatar Living (http://www.qatarliving.com/forum), which is a forum for people give their own questions and answer others' questions. Table 1 shows statistics of the dataset.

We used several measures to evaluate our models, and this measure consists of: The classification measures include $Accuracy(Acc)$, $Precision(P)$, $Recall(R)$, and F_1-$measure(F_1)$. The ranking measures include: Mean Average Precision (MAP), Average Recall ($AvgRec$) and Mean Reciprocal Rank (MRR).

Table 1. Statistics of the dataset

	Input questions	Related questions	Total
Train	267	2669	2936
Test	70	700	770

[4] http://alt.qcri.org/semeval2016/task3/index.php?id=data-and-tools.

4.2 Experiments and Results

To illustrate the effect of different feature sets, we implement the following experiments.

Experiment 1. In this experiment, we want to check the performance of system when the representation of questions just includes n-gram features. The input question-related question pairs are represent as vectors of n-gram word features, then a SVM classifier is used to classify and we use prediction scores to ranking-related questions. Finally, the results of classification and ranking-related questions are shown in Table 2.

Table 2. Finding- and ranking-related question results when representation of questions using n-gram features

Representation of question	Classification measures				Ranking measures		
	Acc	P	R	F1	MAP	AveRec	MRR
1-gram (Unigram)	62.43	42.02	33.91	37.53	55.98	76.22	61.85
2-gram (Bigram)	61.43	39.66	30.47	34.47	53.33	72.72	57.07
3-gram (Trigram)	64.29	44.97	32.62	37.81	53.69	73.31	56.49

Experiment 2. This experiment aims to check the performance of system when the representation of questions including question properties and word embeddings. The experimental results are shown in Table 3.

Table 3. Finding- and ranking-related question results when representation of questions using question properties and word embeddings

Representation of question	Classification measures				Ranking measures		
	Acc	P	R	F1	MAP	AveRec	MRR
Question properties	78.86	68.09	68.67	68.38	76.90	90.66	84.69
Question properties, word embeddings	79.43	69.43	68.24	68.83	76.97	90.13	84.09

Experiment 3. In the third experiment, we first question categorization q_o into one of the related question categories in dataset Q. We assume that questions in the same category are often more similar than the questions in the different categories. Next, we extract the features based on the input question categories and related question categories. The results of this experiment are shown in Table 4.

Table 4. Finding- and ranking-related question results when add the features based on question category

Representation of question	Classification measures				Ranking measures		
	Acc	P	R	F1	MAP	AveRec	MRR
Question properties, question categories	79.14	69.68	66.09	67.84	76.91	90.68	84.69
Question properties, word embeddings, question categories	80.43	72.22	66.95	67.84	77.43	91.08	84.45

From the results of the above experiments 1, the representation of question in vector form of the n-gram features for the classify and ranking result is not high, only the highest result with *Accuracy* and MAP measures of 64.29% and 55.98%, respectively. In experiment 2, we represent the questions as vector of features extraction from lexicon and syntax of questions such as noun, verb, adjective, entity name, word n-gram overlap between questions between two questions resulted in higher classification and ranking results than using only n-gram features. The *Accuracy* measure increased 15.14% and the MAP measure increased 20.99% compare to using only n-gram features.

In experiment 3, we performed a model using the question categorization module and experiment 2 does not use the question categorization module. The results of these two experiments (Tables 3 and 4) show that our proposed model (using the question categorization module) is more efficient, both at *Accuracy* and MAP measures. The results of Experiment 3 increase both the *Accuracy* and the MAP measure of 1% and 0.46% respectively, in comparison with the results of experiment 2. Figure 2 compares our experimental results with different models: the using the question categorization module and do not use question categorization module.

Figure 2 shows that when using the model with the question categorization module will give higher results than the model that does not use the question categorization module.

5 Comparison

It is worth to make a comparison of our proposed method with previous studies which are well known and use the same evaluation metrics as well as the datasets. Results of this comparison are presented in Table 5.

Table 5 shows that our proposal achieves the *Accuracy* and *MAP* measures of 80.43% and 77.43%, respectively, which achieves the best results in comparison with others.

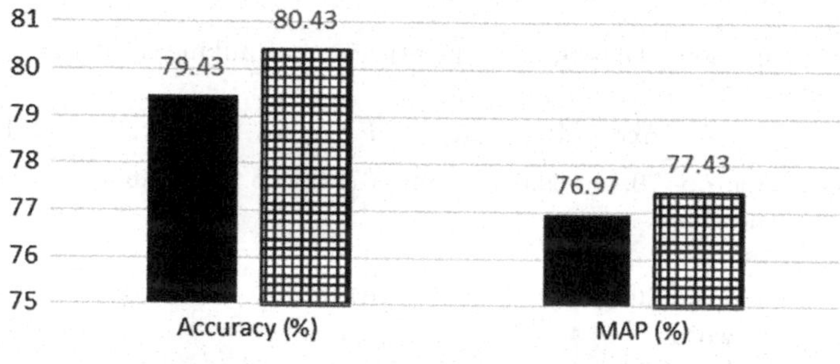

Fig. 2. Model comparison using and not using question categorization module

Table 5. Results of comparison with previous studies

Models	Classification measures				Ranking measures		
	Acc	P	R	F1	MAP	AveRec	MRR
UH-PRHLT-primary [15]	76.57	63.53	69.53	66.39	**76.70**	90.31	83.02
ConvKN-primary [4]	78.71	68.58	66.52	67.54	76.02	90.70	84.64
Kelp-primary [5]	**79.43**	66.79	75.97	71.08	75.83	91.02	82.71
SLS-primary [19]	**79.43**	76.33	55.36	64.18	75.55	90.65	84.64
ICL00-primary [23]	33.29	100	49.95	33.29	75.11	89.33	83.02
ECNU-primary [24]	72.71	100	18.03	30.55	73.92	89.07	81.48
UniMelb-primary [18]	74.57	63.96	54.08	58.60	70.20	86.21	78.58
Our	**80.43**	72.22	66.95	67.84	**77.43**	91.08	84.45

6 Conclusion

This paper has proposed a new method for finding- and ranking-related questions task of questions from cQA sites. Our proposed model uses multi-view as an approach for generating different and rich feature kinds for the tasks of question finding and ranking. The features are extracted from the different representations of the questions, such as using n-grams, using question properties, using word embeddings, question categories. The experimental results have shown that our proposed model gives higher classification and ranking results than previous studies on the same task and datasets.

Acknowledgements. This paper is supported by The Vietnam National Foundation for Science and Technology Development (NAFOSTED) under grant number 102.01-2014.22.

References

1. Jurczyk, P., Agichtein, E: Discovering authorities in question answer communities by using link analysis. ACM, pp. 919–922 (2007)
2. Li, B., Lyu, M. R., King, I: Communities of yahoo! answers and baidu zhidao: Complementing or competing?. IJCNN, pp. 1–8 (2012)
3. Bilotti, M. W., Elsas, J., Carbonell, J., Nyberg, E.: Rank learning for factoid question answering with linguistic and semantic constraints. In Proceedings of the 19th ACM international conference on CIKM, pp. 459–468 (2010)
4. Cedeno, A.B, Bonadiman, D., Martino, G.D.S: Answer and Question Selection for Question Answering on Arabic and English Fora. SemEval 2016, pp. 896–903
5. Filice, S., Croce, D.: KeLP at SemEval-2016 Task 3: Learning Semantic Relations between Questions and Answers. SemEval 2016, pp. 1116–1123 (2016)
6. Wang, K., Ming, Z., Chua, T.S.: A syntactic tree matching approach to finding similar questions in community-based qa services. ACM, pp. 187–194 (2009)
7. Mikolov, T., Chen, K., Corrado, G., Jeffrey Dean: Efficient Estimation of Word Representations in Vector Space. ICLR (2013)
8. Cao, Y., Duan, H., Lin, C.Y., Yu, Y., Hon, H.: Recommending Questions Using the Mdl-based Tree Cut Model. In Proceedings of WWW, pp. 81–90 (2008)
9. Duan, H., Cao, Y., Lin, C.Y., Yu, Y.: Searching Questions by Identifying Question Topic and Question Focus. ACL, pp. 156–164 (2008)
10. Ji, Z., Xu, F., Wang, B., He, B.: Question-answer topic model for question retrieval in community question answering. ACM, pp. 2471–2474 (2012)
11. Zhang, K., Wu, W., Wu, H., Li, Z., Zhou, M.: Question retrieval with high quality answers in community question answering. ACM, pp. 371–380 (2014)
12. Zhou, G., Cai, L., Zhao, J., Liu, K.: Phrase-based translation model for question retrieval in community question answer archives. ACL, pp. 653–662 (2011)
13. Jeon, J., Croft, W.B., Lee, J.H.: Finding Similar Questions in Large Question and Answer Archives. ACM, pp. 84–90 (2005)
14. Filice, S., Castellucci, G., Croce, D., Basili, R.: KeLP: a Kernel-based Learning Platform for Natural Language Processing. ACL-IJCNLP, pp. 19–24 (2015a)
15. Salvador, M.F, Kar, S., Solorio, T., Rosso, P.: Combining Lexical and Semantic-based Features for Community Question Answering, SemEval 2016, pp. 814–821

16. Joachims, T.: Training Linear SVMs in Linear Time. ACM SIGKDD, pp. 217–226 (2006)
17. Nakov, P., Marquez, L., Moschitti, A., Magdy, W., Mubarak, H., Freihat, A., Glass, J., Randeree, B.: SemEval-2016 Task 3: Community Question Answering.
18. Hoogeveen, D., Li, Y.: Identifying Similar Questions by Combining a CNN with String Similarity Measures. SemEval 2016, pp. 851–856 (2016)
19. Mohtarami, M., Belinkov, Y., Hsu, W.L., Zhang, Y.: Neural-based Approaches for Ranking in Community Question Answering. SemEval 2016, pp. 828–835 (2016)
20. Santos, C.D., Barbosa, L., Bogdanova, D., Zadrozny, B.: Learning Hybrid Representations to Retrieve Semantically Equivalent Questions. ACL, pp. 694–699 (2015)
21. Salvador, M.F., Rangel, F.: Language variety identification using distributed representations of words and documents. CLEF, pp. 28–40 (2015a)
22. Salvador, M.F., Rosso, P., Rangel, F: Distributed representations of words and documents for discriminating similar languages. RANLP, pp. 11–16 (2015b)
23. Wu, M.Z.Y: Translation-Based Method for CQA System. SemEval 2016, pp. 857–860 (2016)
24. Wu, G., Lan, M.: Exploring Traditional Method and Deep Learning Method for Question Retrieval and Answer Ranking in cQA. SemEval 2016, pp. 872–878 (2016)

Haralick Features-Based Classification of Mammograms Using SVM

Vikrant Bhateja[1(✉)], Aman Gautam[1], Ananya Tiwari[1], Le Nguyen Bao[2], Suresh Chandra Satapathy[3], Nguyen Gia Nhu[2], and Dac-Nhuong Le[4]

[1] Department of Electronics and Communication Engineering, Shri Ramswaroop Memorial Group of Professional Colleges (SRMGPC), Lucknow 226028, Uttar Pradesh, India
{bhateja.vikrant, gautamaman543, absoluteananya}@gmail.com

[2] Duytan University, Danang, Vietnam
{baole, nguyengianhu}@duytan.edu.vn

[3] Department of CSE, PVP Siddhartha Institute of Technology, Vijayawada, AP, India
sureshsatapathy@gmail.com

[4] Haiphong University, Haiphong, Vietnam
nhuongld@dhhp.edu.vn

Abstract. The contrast enhancement of mammograms at preprocessing stage optimizes the overall performance of a computer-aided detection (CAD) system for breast cancer. In the proposed approach, contrast enhancement is performed using a sigmoidal transformation mechanism followed by extracting a set of 14 Haralick features. For classification purposes, a support vector machine (SVM) classifier is used which sorts the input mammogram into either normal or abnormal subclasses. The performance of the classifier is estimated by calculating parameters like accuracy, specificity, and sensitivity. The performance of the proposed approach has been reported to be better in comparison to other existing approaches.

Keywords: Sigmoidal transformation · Support vector machine (SVM) Haralick features · Region of interest (ROI)

1 Introduction

Mammography is a technique using low dose of X-ray over human breasts for diagnosis and screening purposes. About 10–25% of the abnormal cases in mammograms are almost inaccurately classified by the radiologists [1]. The detection rate can be improved by using double reading; however, it has known to be very expensive and time-consuming [2]. Hence, a computer-based detection approach is designed to provide intelligent inferences over the screened mammograms and thus share the workload of the radiologist. The detection of breast cancer has improved by 8% on adoption of CAD system in the diagnosis process [3]. Although, acquisition of CAD system has

V. Bhateja et al. (eds.), *Information Systems Design and Intelligent Applications*, Advances in Intelligent Systems and Computing 672,
https://doi.org/10.1007/978-981-10-7512-4_77

attracted a lot of radiologists but further work needs to be put on improving the detection capability of the CAD system. There are various modules of a CAD system such as preprocessing, features extraction, and classification. Preprocessing module provides an improvement in visual quality of mammograms so that the ROI is clearly visible and the unwanted artifacts present in the mammograms are removed. Under features extraction process, various different characteristics regarding the texture of ROI is learned by computing some mathematical attributes for the anomalous region. Classifier is the sole brain of the CAD system which depending upon its learning experience would give the final decision. Enhancing a mammogram which comes under the preprocessing module helps the CAD system to diagnose the ROI with greater efficiency [4]. Some of the earlier works regarding the enhancement and classification of mammograms using CAD system are discussed underneath. Wu et al. [5] used a Laplacian Gaussian Pyramid algorithm followed by a contrast limited adaptive histogram equalization (CLAHE) for contrast improvement. The denoising of the mammogram was not performed due to which the SNR value was low as compared to other approaches mentioned in their work. Tzikopoulosa et al. [6] performed breast density estimation over the mammograms for its segmentation and used SVM classifier for classification. The processing time of the technique mentioned by the authors was high and the accuracy of the approach was 85.7%. cellular neural network (CNN) was used by Sampaio et al. [7] for segmentation purpose, followed by features extraction using Ripley's K function and Moran's and Geary's indexes. Finally, for classification they have used SVM classifier which had an accuracy of 90.85%. Oliveira et al. [8] introduced a methodology to classify ROI(s) as mass and no mass. Under preprocessing stage, creation of phylogenetic tree and computation of the taxonomic diversity was performed in order to extract features of the ROI. This methodology was dedicated toward detection of tumorous mass with definite shape thus, not suitable for various other irregular anomalies present in the breast. A new set of preprocessing mechanisms such as nonlinear polynomial filters [9, 10] and nonlinear unsharp masking [11, 12] have also been utilized recently. Singh and Singh [13] proposed a three-class classification of MLO and CC views of mammograms using SVM classifier. Liu and Tang [1] integrated SVM with recursive feature elimination (SVM-RFE) with a normalized mutual information feature selection (NMIFS) for classification. However, the work presented by these authors [1] has not shown any enhancement of mammograms at the preprocessing stage due to which the accuracy is low. Based on the above discussion on previous works, it can be inferred that the fall in classification performance of SVM classifier has been majorly due to lack of appropriate enrichment of mammograms during preprocessing. This is necessary as the choice of nonlinear enhancement approach [9–12] has shown to improve the contrast and edges of ROI, which may optimize the diagnostic capability of the CAD system. Also by using a proper synthesis tool on mammograms, the formulation of decision criteria for classification is simplified which further improves the performance of SVM. The proposed work concentrates over the change in the classification accuracy of a CAD system due to enhancement of mammograms. Hence, for the preprocessing module, a sigmoidal transformation algorithm is used, which improves the overall contrast of the mammogram. A set of 14 Haralick features are computed for extraction of textural data from the ROI. The final classification into two subclasses has been done by using SVM classifier. SVM formulates a decision rule during its training session; this rule is then used to classify the test images. The upcoming subsections explain detailed discussion of the methodology, analysis of results along with conclusions.

2 Proposed CAD Methodology

The CAD system in the present study consists of various modules such as preprocessing, features extraction, and classification. The input mammograms are first synthesized at preprocessing stage after which the detected ROI is obtained. These ROIs are sent to a features extraction stage which calculates some textural data regarding the detected regions. The extracted features provide information regarding the texture of the ROI which helps the classifier to form a pattern of selection and thus classification. The training and testing of mammograms are practiced through these features with the help of SVM classifier, which sorts the input mammogram into either normal or abnormal class. Block Diagram shown in Fig. 1 depicts the proposed methodology.

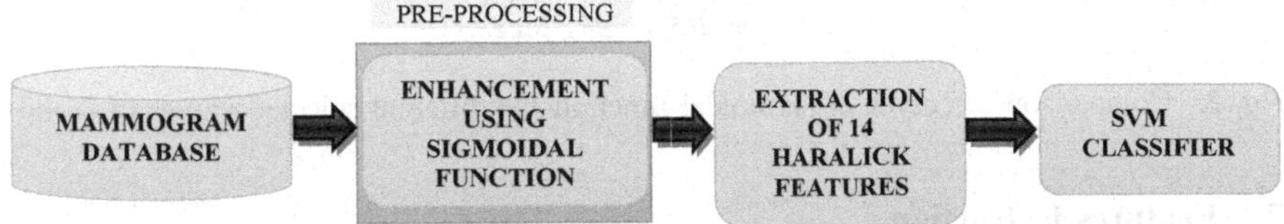

Fig. 1. Block diagram of proposed CAD methodology

2.1 Preprocessing of Mammograms

The first module as per Fig. 1 involves preprocessing stage in which the visual customization of mammograms is performed which transforms the image in such a way that the suspicious regions are clearly visible. Sigmoidal transformation function [14] has been used for enhancement whose mathematical expression is as shown in Eq. (1). It performs sigmoidal enhancement in contrast to the mammograms so that the detection procedure is less cumbersome.

$$y(x) = a[sigmoid\{k(x - b)\} - sigmoid\{-k(x + b)\}] \tag{1}$$

where the parameters a and x have their original meanings as mentioned by Gautam et al. [15]. There are two parameters b and k, where b sets the threshold for enhancement and ranges from (0, 1); and k controls the enhancement intensity and is generally seen to vary between 10 and 25. In the preexisting approaches [14–16], many researchers have selected the value of b and k for improving the nonlinear effect. The enhancement function varies greatly for different selected values of the parameters b and k, as shown in Fig. 2.

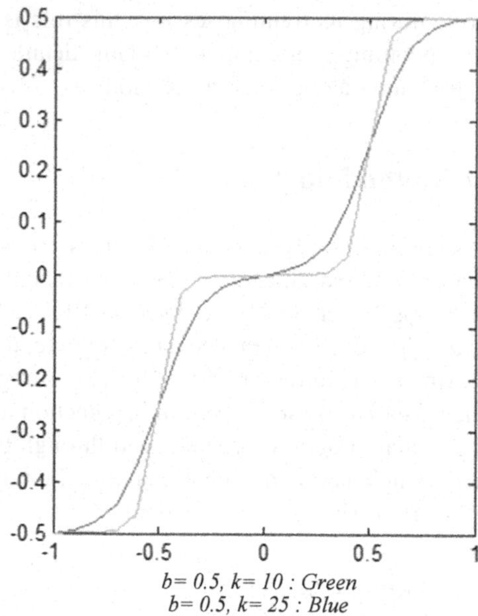

Fig. 2. Comparison of curves of sigmoidal function for different selected values of *b* and *k*

2.2 Features Extraction

Features extraction is the phase of CAD system where the mathematical attributes of an image are computed for the final classification step. The features calculated help the classifier to determine the type of anomaly present in the mammogram and provide necessary diagnostic conclusions. Different forms of anomalies show different characteristics w.r.t. shape, texture, position, orientation, and density; thus, it is necessary to compute the quantitative and qualitative properties of ROI(s) obtained from mammograms such as textural features. It is one of the most important classes of features to define the properties of an image. Haralick features define the correlation of intensity of a pixel with its neighbor. Haralick formulated a set of fourteen features of textural characteristics which were acquired from a well-known statistical method for textural features, co-occurrence matrix. These set of features describes how the image intensities in the pixel from a certain position varies with its neighboring pixels. In the proposed work, Haralick features are calculated with the help of Gray level co-occurrence matrix (GLCM) features. GLCM is a technique in which each pixel's intensity is compared with its neighboring pixel in four possible directions (0°, 45°, 90°, and 135°) of an image (Fig. 3) [17]. The 14 different features as mentioned by Haralick et al. [18] are derived from the four basic GLCM features (energy, correlation, contrast, and homogeneity). These features help in analyzing the texture and shape of the anomaly that have been segmented. The list of Haralick features that have been calculated is shown in Table 1.

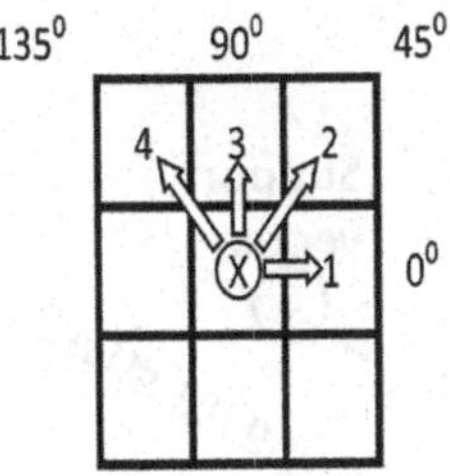

Fig. 3. GLCM pixels [17]

Table 1. Haralick features

Features	Values
Angular second moment	0.0913
Contrast	2.4613
Correlation	−0.0079
Variance	8.9146
Inverse difference moment	0.5013
Sum average	5.3648
Sum variance	31.0723
Sum entropy	3.5611
Entropy	3.6308
Difference variance	1.9031
Difference entropy	1.3142
Information measure of correlation I	−1.8157
Information measure of correlation II	0.9996
Maximal correlation coefficient	0.0033

2.3 Classification

The above-calculated 14 features are set in form of features vector and are fed to the SVM classifier for final diagnosis. SVM classifier is well known for its efficiency to classify a given dataset into two subgroups. During its training session, SVM forms a hyperplane or decision threshold plane depending upon the training dataset which sets criteria for classification as shown in Fig. 4 [19]. This figure represents the linear classification mechanism where the optimal hyperplane segregates the textural features into two groups [19]. When a test mammogram is sent as an input to the system, the textural features obtained are compared with those of the classifier's trained dataset. SVM sorts the input data into either of the two classes depending upon which side of the hyperplane does the input data lies.

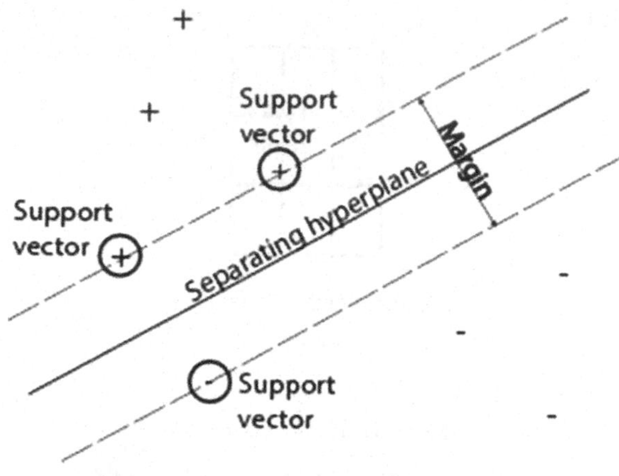

Fig. 4. Decision hyperplane structure of SVM classifier [19]

3 Results and Discussions

In proposed methodology, the digital mammograms are taken from MIAS database [20] which provides a classified set of images for cancer anomalies. During simulations, these mammograms are normalized by performing RGB to Gray conversion followed by enhancement process using sigmoidal transformation as discussed in the previous section.

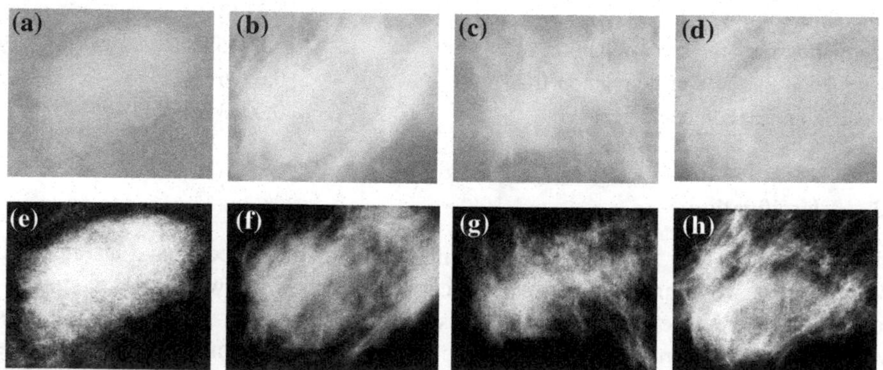

Fig. 5. **a**, **b**, **c**, **d** Original ROIs of mammograms: roi1, roi2, roi3, roi4, **e**, **f**, **g**, **h** corresponding enhanced ROIs: Eroi1, Eroi2, Eroi3, Eroi4

The results of enhanced mammograms using the proposed sigmoidal transformation function are shown in Fig. 5. Herein, Fig. 5a, b (roi1, roi2) shows original ROIs from glandular mammograms. After enhancement operation as shown in Fig. 5e, f (Eroi1, Eroi2), the background get suppressed and contrast is improved thus focused ROIs are clearly visible. Similarly, Fig. 5c (roi3) shows ROI from a dense mass and

Table 2. Validation of classifier performance

Parameters	Original ROIs	Enhanced ROIs
Training set (Abnormal = 25, Normal = 15)	40	40
Validation set (Abnormal = 44, Normal = 26)	70	70
True positive (TP)	5	42
True negative (TN)	3	23
False positive (FP)	17	3
False negative (FN)	19	2
Accuracy (%)	18.18	92.85
Specificity	0.15	0.954
Sensitivity	0.208	0.885

Fig. 5d (roi4) is from a fatty mass with corresponding contrast improved enhanced ROIs in Fig. 5g (Eroi3) and Fig. 5h (Eroi4), respectively. The performance parameters like accuracy, sensitivity, and specificity are calculated as mentioned by Bhateja et al. [21], which measures the performance of the classifier.

The accuracy of the proposed approach is reported to be 92.85% with a sensitivity score of 88.5% and specificity of 95.4%. From Table 2, it can be observed that the accuracy of the classifier for original ROI(s) is very low. This is due the fact that the image acquisition procedure for mammogram is done using low dose of X-ray which introduces noise in the image and the overall contrast of the mammogram obtained is of low quality. The obtained results of the proposed methodology are notably better in comparison to the reported works in [13, 22] as illustrated in Table 3. The results proclaim an improvement in the classification system performance and are quite promising.

Table 3. Comparison of accuracy of proposed methodology with other existing approaches

References	Classification accuracy (%)
Singh and Singh [13]	75
Li et al. [22]	86.92
Proposed methodology	92.85

4 Conclusion

In the proposed work, a sigmoidal transformation function has been incorporated for the enhancement of mammograms followed by features extraction using Haralick features which is finally used for classification using SVM classifier. From the results obtained, it can be concluded that by applying proper enhancement techniques over mammograms, classification accuracy can be optimized. The proposed approach has shown better performance than the other compared methodologies. The proposed nonlinear sigmoidal transformation technique improves the overall contrast of the ROI which further elevates the accuracy of the system. The adaptability of the mentioned enhancement technique can be improved in the future. The sigmoidal transformation approach for enhancement of mammograms can also be modeled with different classifiers for multiple-class classification.

References

1. Liu, X., Tang, J.: Mass Classification in Mammograms Using Selected Geometry and Texture Features, and a New SVM-Based Feature Selection Method, In: IEEE Systems Journal, vol. 8, no. 3, Sept. 2014.
2. Jain, A., Singh, S., Bhateja, V.: A Robust Approach for Denoising and Enhancement of Mammographic Breast Masses, In: International Journal on Convergence Computing, vol. 1, no. 1, pp. 38–49, Inderscience Publishers (2013).
3. Morton, M. J., Whaley, D. H., Brandt, K. R., Amrami, K. K.: "Screening mammograms: Interpretation with Computer-Aided Detection-Prospective Evaluation, In: Radiology, vol. 239, no. 2, pp. 375–383, Mar. 2006.
4. Bhateja, V., Urooj, S., Misra, M.: Technical Advancements to Mobile Mammography using Non-Linear Polynomial Filters and IEEE 21451-1 Information Model, In: IEEE Sensors Journal, vol. 15, no. 5, pp. 2559–2566, Advancing Standards for Smart Transducer Interfaces (2015).
5. Wu, S., Yu, S., Yang, Y., Xie, Y.: Feature and Contrast Enhancement of Mammographic Image Based on Multiscale Analysis and Morphology, In: Computational and Mathematical Methods in Medicine, vol. 2013, pp. 1–8, Hindawi Publishing Corporation (2013).
6. Tzikopoulosa, S. D., Mavroforakisb, M. E., Georgioua, H. V., Dimitropoulosc, N., Theodoridisa, S.: A Fully Automated Scheme for Mammographic Segmentation and Classification based on Breast Density and Asymmetry, In: Computer Methods and Programs in Biomedicine, vol. 102, no. 1, pp. 47–63, 2011.
7. Sampaio, W. B., Diniz, E. M., Correa Silva, A., Cardoso-de-Paiva, A., Gattass, M.: Detection of masses in mammogram images using CNN, geostatistic functions and SVM, In: Computers in Biology and Medicine, vol. 41, no. 8, pp. 653–654, Elsevier (2011).
8. Oliveira, F. S. S., Filho, A. O. C., Silva, A. C., Cardoso-de-Paiva, A., Gattass, M.: Classification of breast regions as mass and non-mass based on digital mammograms using taxonomic indexes and SVM, In: Computers in Biology and Medicine, vol. 57, pp. 42–53, Elsevier (2014).
9. Bhateja, V., Misra, M., Urooj, S., Lay-Ekuakille, A.: A Robust Polynomial Filtering Framework for Mammographic Image Enhancement from Biomedical Sensors, In: IEEE Sensors Journal, vol. 13, no. 11, pp. 4147–4156, IEEE (2013).
10. Bhateja, V., Misra M., Urooj, S.: Non-Linear Polynomial Filters for Edge Enhancement of Mammogram Lesions, In: Elsevier-Computer Methods and Programs in Bio-medicine, vol. 129C, pp. 125–134, Elsevier (2016).
11. Bhateja, V., Misra, M., Urooj, S.: Human Visual System based Unsharp Masking for Enhancement of Mammographic Images. In: Journal of Computational Science (2016).
12. Siddhartha, Gupta, R., Bhateja, V.: A Log-Ratio based Unsharp Masking (UM) Approach for Enhancement of Digital Mammograms, In: Proc. of (ACM ICPS) CUBE International Information Technology Conference & Exhibition, Pune, India, pp. 26–31, 2012.
13. Singh, D., Singh, M.: Classification of Mammograms using Support Vector Machine, In: International Journal of Signal Processing, Image Processing and Pattern Recognition, vol. 9, no. 5, pp. 259–268, IJSIP (2016).

14. Bhateja, V., Devi, S.: An Improved Non-Linear Transformation Function for Enhancement of Mammographic Breast Masses, In: Proc. of (IEEE) 3rd International Conference on Electronics & Computer Technology (ICECT-2011), Kanyakumari (India), vol. 5, pp. 341–346, IEEE (2011).
15. Gautam, A., Bhateja, V., Tiwari, A., Satapathy, S. C.: An Improved Mammogram Classification Approach Using Back Propagation Neural Network, In: Data Engineering and Intelligent Computing, Advances in Intelligent Systems and Computing 542, vol. 35, pp. 1–8, Springer (2017).
16. Alankrita, Raj, A., Shrivastava, A., Bhateja, V.: Contrast Improvement of Cerebral MRI Features using Combination of Non-Linear Enhancement Operator and Morphological Filter, In: Proc. of (IEEE) International Conference on Network and Computational Intelligence (ICNCI 2011), Zhengzhou, China, vol. 4, pp. 182–187, IEEE (2011).
17. Mohamed, H., Mabroukb, M. S., Sharawy, A.: Computer Aided Detection System for Micro Calcifications in Digital Mammograms, In: Computer Methods and Programs in Biomedicine, vol. 116, no. 3, pp. 226–35, April 2014.
18. Haralick, R. M., Shanmugam, K., Dinstein, I.: Textural Features for Image Classification, In: IEEE Transactions on Systems, Man and Cybernetics, vol. SMC-3, no. 6, pp. 610–621, IEEE (1973).
19. Hastie, T., R. Tibshirani, and J. Friedman. The Elements of Statistical Learning, second edition. Springer, New York, 2008.
20. J Suckling et al. (1994) "The Mammographic Image Analysis Society Digital Mammogram Database" Exerpta Medica. International Congress Series 1069, pp. 375–378.
21. Bhateja, V., Tiwari, A., Gautam, A.: Classification of Mammograms using Sigmoidal Transformation and SVM, In: Smart Computing and Informatics, vol. 78, Springer (2016).
22. Li, Y., Chen, H., Wei, X., Peng, Y., Cheng, L.: Mass Classification in Mammograms based on Two-Concentric Masks and Discriminating Texton, In: Pattern Recognition, vol. 60, pp. 648–656, Elsevier (2016).

A Genetic Algorithm Approach for Large-Scale Cutting Stock Problem

Nguyen Dang Tien(✉)

People's Police University of Technology and Logistics, Bac Ninh, Vietnam
dangtient36@gmail.com

Abstract. This work investigates the one-dimensional integer cutting stock problem, which has many applications in Information and Communications Technology for green objectives. The problem consists of cutting a set of available objects in stock to produce smaller items with minimum the wastage of materials. On the basis of the traditional group-based Genetic Algorithm, we solve the large-scale cutting stock problems by adding two new proposals. Firstly, we put two additional steps to the First Fit heuristic to utilize wastage stock rolls when items are added to genes. These steps are applied to initialize the first population and to perform the mutation operation between two parents. Secondly, we propose a new heuristic in the crossover operation to create new individuals. The heuristic increases good genes and decreases bad genes which appeared in the population. We use them to improve traditional Genetic Algorithm in terms of the individual's quality and the diversity of good genes in the populations. As a result, the wastage of stock rolls decreases. These heuristics are empirically analyzed by solving randomly generated instances and large instances from the literature, then results are compared to other methods. We specify an indicator to show some solutions are optimal. The numerical simulation shows that our approach is effective when it is applied to large-scale data sets, with better result in 40% of instances than the traditional cutting plane algorithm. On the other hand, we show that our approach can reach 289 optimal solutions out of 400 generated instances.

Keywords: Green algorithm · Green strategy · Combinatorial optimization · Genetic algorithm · Cutting stock problem

1 Introduction

The cutting stock problem (CSP) is an NP-complete optimization problem of green IT, which refers to the study and practice of using computing resources in an efficient, effective, and economic way. It is stated as a problem of finding the best way to cut items with a specific required quantity from a finite set of stock rolls with a minimum wastage. It has many applications in real life, especially in construction industries such as wood, windows, metal, glasses. In addition,

V. Bhateja et al. (eds.), *Information Systems Design and Intelligent Applications,* Advances in Intelligent Systems and Computing 672,
https://doi.org/10.1007/978-981-10-7512-4_78

our strategy can be further developed to apply into other optimization problems in telecommunications network, such as minimum connected dominating set problem on ad hoc wireless networks and multicast routing problem. CSP, a combinatorial optimization problem, cannot be solved by any algorithm in polynomial time unless P = NP. There are several forms of problems, which are mainly different by the optimality criteria. The problem could be defined in one, two or three-dimensional space and cuts could be implemented with or without contiguity. The more the variables, the more the complex problem appears. Thus, the search space grows exponentially as well as the time to find the optimal solution. The purpose of this study is to investigate the effects of new heuristics applied for Genetic Algorithm to solve CSP. We focus on one-dimensional form with limited stock availability, which can be stated as follows:

Given:

m: Number of rolls;
n: Number of item types;
L_j: Length of the roll jth;
l_i: Length of the item type i;
b_i: Demand for the item type i;
a_{ij}: Number of the item type i which is cut from the roll j;
y_i: 1 if the roll j is used. Otherwise, $y_i = 0$;
w_i: Wastage of the roll j.

Mathematical model:

$$minW = \sum_{j=1}^{m} w_j y_j$$

subject to:

$\sum_{j=1}^{m} a_{ij} = b_j, i = 1, \dots, n;$
$\sum_{i=1}^{n} a_{ij} \cdot l_i = L_j \cdot y_j, j = 1, \dots, m;$
$a_j \geq 0, a_j \in Z^+, j = 1, \dots, m$
$y_i \in \{0, 1\}$

Since Gilmore and Gomory [1] proposed an algorithm using column generation to solve the linear programing relaxation form of CSP, there are some others approach based on combination of this algorithm with different heuristics, or with branch and bound [2–5]. Their purpose is to solve CSP instances effectively with large number of items in many different sizes. Concerning Genetic Algorithm, there are two models proposed: order-based Genetic Algorithm and group-based Genetic Algorithm. In these two models, order-based strategy is simple to implement but it is not effective with instances with many variables. Group-based proposed by Hinterding and Khan [6] is the main approach using Genetic Algorithm to solve CSP problem recent years. However, when applying the original algorithm to instances with large number of items with many different sizes, the result is not feasible. For data sets with a large number of variables, greedy rounding heuristics [7] is proposed, but they only focused on problems of low demand and do not reach the optimal result in almost instances. Cutting plane algorithm [8], which iteratively refines a feasible set or objective function by means of linear inequalities, is considered as a best known method for large data sets in the literature. In this paper, we describe two new heuristics to improve the effect of traditional Genetic Algorithm on CSP. In traditional method, First Fit or First Fit Decrease is used to add new items to gene which is an individual of population. These two heuristics either reduce genes diversity of the population or remain large wasted material, which can be used to cut more small items. Consequently, it takes more generations for them to get feasible result, and the result is not good enough, especially in large data sets. We applied additional steps to First Fit to utilize remained stocks after First Fit, without loss of genes diversity and so that, improve quality of populations. In addition, we propose a new crossover method to help the child to get more good genes from its parents. Since traditional crossover operation is performed with several crossover points, child can get many bad genes, as well as receive limited good genes from its parents. Our approach can make child with better quality based on multiple-point crossover operation, so that the quality of child would be improved. Our numerical simulation is performed in two group of instances: randomly generated instances and instances from the literature.

2 The Proposed Genetic Algorithm

2.1 The Model

We use group-based Genetic Algorithm [6] to model the problem. A gene (group) is represented by a list of demanded items which are cut from a specific stock size. A chromosome is a set of genes in which total items of every item types satisfy the demand. An example is shown in Fig. 1.

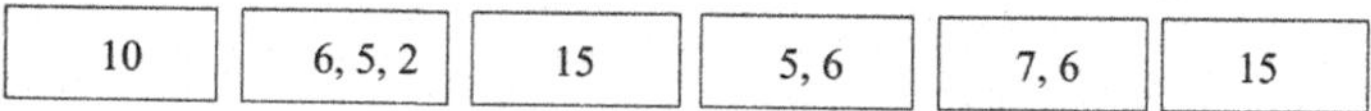

Fig. 1. A chromosome example

2.2 The Proposed Initialization

The original algorithm of Hinterding and Khan uses the First Fit heuristic to initial chromosomes. The First Fit algorithm provides a fast but often non-optimal solution, involving cutting each roll to produce the first item in which the items length is smaller than the rolls length. We note that at the end of each step in First Fit, the rolls after being cut are often discard although the length of stock rolls still can be cut to produce some small items. These stock rolls can be utilized further, since it can be used to produce other small items. In our approach, we try to save the wastage length of stock roll by finding a list of items with length smaller than rolls length, and then pick a random item from the list.

This step is only terminated if the list is empty. Our initialization method has two advantages: (1) the wastage length of rolls in First Fit algorithm is utilized and (2) distribution of items in gene remains its diversity, which is limited in First Fit heuristic. The pseudo-code of our proposed algorithm is described in Algorithm 1.

Algorithm 1: Initialization

```
Procedure MMF();
Input  : L_j: Length of the jth roll, j = 1, ... , m
         l_i: Length of the ith item type, i = 1, ... , n
         b_i: Demand for the ith item type, i = 1, ... , n
Output: a_ij: Number of the ith item type which is cut from the jth roll.
         y_i: y_i = 1 indicates that the ith roll is used. Otherwise, y_i = 0
         w_i: wastage of the jth roll.
Initialize a boolean array p[j], p[j] = false, j = 1, ... , m
Initialize a integer matrix a[i, j], a[i, j] = 0, i = 1, ... , n and j = 1, ... , m
while Total t[i] > 0 do
    Randomly pick a roll j with p[j] = false;
    p[j] = true;
    w = L[j];
    while w > 0 do
        Randomly pick t[i] with t[i] ≠ 0;
        if w > l[i] then
            w ← w − l[i];
            t[i] ← t[i − 1];
            a[i, j] ← a[i, j] + 1;
        end
        In the array l, find l[io] < w and t[io] > 0;
        if found then
            w ← w − l[io];
            t[io] ← t[io − 1];
            a[i, j] ← a[io, j];
        end
        else
            break;
        end
    end
end
```

2.3 The Crossover Operator

A state-of-the-art crossover operator. The original crossover strategy for two individuals p_1 and p_2 is proposed by Hinterding and Khan. It includes three main steps in the following:

- In p_1, choose randomly an insertion point;
- In p_2, choose randomly a segment;
- Create the child, *ch*, from p_1 and p_2:
  - The first part of *ch* is produced by copy the genes from p_1 up to the insertion point;
  - The second part of *ch* is produced by copy the segment from p_2;
  - Sequentially check each gene in the rest of p_1. A gene is only copy to *ch* if it does exceed the items demand;
  - Use First Fit algorithm to add missing items to *ch*.

This strategy has some disadvantages:

- Since the first part of p_1 is accurately copied to child, some bad genes can be to *ch*. We use the term bad genes to indicate genes with large wastage;
- The good genes from p_2s segment and the second part of p_1 may not be carried to the *ch*, when one item on the gene has been appeared on the child with enough amount. Good genes are genes with small wastage.

We propose a new crossover operator to overcome these disadvantages in the following.

The proposed crossover operator. In this approach, we first divide two parents into number of parts, each part contains only one gene. In our crossover procedure, genes in both p_1 and p_2 are first copied to a list, *g_list*, in descending order of length. We also use a list, *i_list*, which includes items which is not included in the child. Then, the smallest gene of *g_list* is copied to the child. Before copying a gene, we check whether all its items can be successfully subtracted from *i_list*. Gene is added from the parents to child only in the case: the addition does not increase the number of any item beyond the demand. At the end of crossover, if *i_list* is not empty, First Fit heuristic is then used to generate new genes (groups) from *i_list*, and the resulting gene(s) are added to the child chromosome.

The pseudo-code of our crossover operation is described in Algorithm 2.

2.4 The Mutation Operator

The purpose of mutation step is to make new genes in child, to avoid premature convergence [9]. The conventional method used in this step is to delete several genes on child and re-add missing items by First Fit algorithm [2,3]. In our work, we target a child with as many good genes as possible by the mutation. The proposed method deletes genes with large wastage and adds good genes to the child. We note that at least two genes must be deleted in order to make changes in childs chromosome.

The mutation operator for a chromosome works as follows:

- Sort genes in chromosome in descending order;
- Delete several genes with largest wastage;
- Construct an item set *i_set*, which includes items that are deleted from child;

- Use modified First Fit to add deleted items to chromosome on the basis of *i_set*.

In the first step, genes with large wastage need to be extracted from chromosome without any order, i.e., we only need to extract a number of genes with largest wastage length in any order. Quick sort is suitable algorithm to do this step, since it can divide an array into two parts, one part with larger numbers and the other with smaller numbers.

Algorithm 2: Crossover procedure

Procedure Crossover;
Input : Parent individuals $p1$ and $p2$:
$p1.size, p2.size$: Number of genes in $p1$ and $p2$;
$p1.g[i,j], p2.g[i,j]$: arrays contain values of gene lengths in $p1$ and $p2$. $g[i,j]$ is the length of the item jth in the ith gene;
$p1.w[i]$ and $p2.w[i]$: wastage length of the ith gene
Output: ch: The new chromosome.
Initialize chromosomes ch, t;
Sort the arrays $p1.w[i]$ and $p2.w[i]$ in descending order;
while $p1.size + p2.size > 0$ **do**
  Pick one item from $pk(k = 1$ or $2)$ with the smallest wasgate: $g[c]$;
  $pk.size \leftarrow pk.size - 1$;
  Insert $g[c]$ to t;
  if *t has all items with smaller than or equal to the required number* **then**
    $ch \leftarrow t$;
  end
  else
    continue;
  end
end
Initialize $gn[i], i = 1, \ldots, n, gn[i]$ is the number of the ith item in ch;
Initialize $s[i], i = 1, \ldots, n : s[i] \leftarrow b[i] - gn[i]$;
Call $MFF()$ to add missing items to ch;

2.5 Fitness Function

The fitness function for individual i, which contains the numbers of both the processed objects and the setups, is defined as:

$$cost = \frac{1}{n}\sum_{1}^{n}\left(\sqrt{\frac{wastage_i}{stock\_length_i}} + \frac{number\_wasgate}{n}\right)$$

The fitness function contains two terms. The first is to reduce the wastage, we take the square root of this term to give values near the limits of the range (0–1) extra weight. This is done as we wish to concentrate on the wastage. The second term encourages solutions where fewer stock lengths contain wastage, as this leads to better solutions.

3 The Proposed Genetic Algorithm for Large Data set

3.1 Genetic Algorithm for Small Data set

With the crossover and mutation operator described above, we can solve the problem effectively with small number of variables. The algorithm contains two steps:

- Initialize population by MFF procedure;
- Use of Mutation and Crossover operator to create a new population.

However, when applying this method for instances with large number of variable, the result is infeasible. We improved it to increase performance of Genetic Algorithms result. The improvement is described in the next section.

3.2 Genetic Algorithm for Large Data set

When applying the Genetic Algorithm for large data sets, a considerable number of genes without waste appear after several generations. This is because the diversity of items length and rolls length in large data set. Those genes can be considered as an optimal solution of a sub-CSP with 0 wastage. The main idea of our approach is to keep genes with 0 wastage in the next generation. On the other hand, items in genes with waste larger than 0 can be considered as new CSP which has smaller variables. The requirement is to produce the remaining items from the remaining stock rolls with a minimum wastage. Since this new CSP has smaller number of variables than the original problem, it can be solved by many approaches. In this work, Genetic Algorithm for small data set is used for the second sub-CSP. More specifically, in mutation operation, we only delete genes with the positive wastage, and the missing items are grouped into genes by the Genetic Algorithm for small data set. A break condition may hold when wastage length of the elism is smaller than 100, or number of generations is bigger than a particular threshold. The algorithm for large data set is described by GAF procedure as follows:

4 Simulations and Results

4.1 Simulation Description

We develop 10 classes of data sets with generated by the Random Generator CUTGEN1 [10], in order to evaluate our heuristics compared to greedy rounding heuristics version 1, 2, and 3 (RGH1, RGH2, and RGH3) [7]. These data sets have some characteristic as follows: The number of different types of stock lengths are chosen between $K = 3, 5$, and 7. Stock lengths $L_k, k = 1, \ldots, K$ is multiples of 100, and they have been randomly generated varying between 100 and 1000 length units. The values of the stock lengths's availability $k, k = 1, \ldots, K$ have been randomly generated from 1 to 50 m. The number of types of ordered items has been fixed at $m = 5, 10$ and 20. The length of item i, i has been randomly generated between $v_1 \cdot L$ and $v_2 \cdot L$, where L is the average value among the $L_k, k = 1, \ldots, K$. The parameters v_1 and v_2 have been fixed at $v_1 = 0.01$ or $v_1 = 0.1$ and $v_2 = 0.2$ or 0.8. Finally, the values for demand $d_i, i = 1, \ldots, m$ have been randomly generated between 1 and 10 units.

Moreover, we applied our algorithm for 400 instances of a second set from the literature. These instances were used by Belov and Scheihauer [8], who proposed cutting plane algorithm for large data sets. The 400 instances are generated with $K = 5, L = 10000$, and $m = 40$.

Table 1. The gap with cutting plan method for $K = 5$ and $m = 40$

Class	v_1	v_2	GAP
C11	0.001	0.2	−0.70
C12	0.001	0.3	−0.23
C13	0.001	0.4	−0.13
C14	0.001	0.5	3.15
C15	0.001	0.6	−0.45
C16	0.001	0.9	4.90
C17	0.001	0.2	2.90
C18	0.001	0.3	4.56
C19	0.001	0.4	1.89
C20	0.001	0.5	7.89

4.2 Optimization Indicator

The interesting characteristic of preceding generator method is that, there is an indicator in our result showing it is optimized. Since the stock lengths is multiple of 100, the wasted lengths of cutting methods, so that, will have the same number modulo 100. Consequently, if the wasted length of a cutting method is less than 100, we can conclude that this cutting method is optimized. However, this only indicates that the proposed method is better than existing ones if its wasted length is less than 100 and there is no conclusion whether a result is better or not if its wasted length is bigger than 100.

4.3 Simulation Results

Considering random instances, the comparison between average wastes reached by GAF, RGH1, RGH2, and RGH3 is shown in Table 1. The Table 2 compares effect on instances from the literature of GAF and cutting plane method, which is considered as a best known approach for these large-scale data sets. The gap is calculated by the formula as follows:

$$GAP = \frac{GAF's\ wastage - Cutting\ plane's\ wastage}{Cutting\ plane's\ wastage}$$

We can conclude that the GAF can reach better results in a significant portion of random instances, with seven classes, compared to two classes of RGH2 and one class of RGH3. On the other hand, the number of data sets with indicated optimal solution in GAF approach is comparable with greedy rounding heuristics, 289 and 247, respectively. Moreover, we can learn from Table 2 that the gaps are quiet small with GAF, and the negative gaps mainly appear in data sets with small items. Interestingly, the number of indicated optimal solutions in our approach with small-item data sets (C11, C12 and C13) is 186, which is better than that of cutting plan, 140. It is suggested that GAF could function well with small-item data sets rather than varied-size item ones.

Table 2. The average wastage lengths of 10 classes

Class	k	M	v_1	v_2	GAF	RGH1	RGH2	RGH3
C1	3	5	0.01	0.2	55.90	118.65	113.35	109.55
C2	3	5	0.01	0.8	168.50	709.65	689.05	765.85
C3	3	5	0.1	0.8	510.80	609.85	716.75	616.45
C4	3	20	0.01	0.2	45.75	201.85	171.50	170.25
C5	3	20	0.01	0.8	1088.35	923.75	878.45	833.40
C6	3	20	0.1	0.8	808.75	693.50	775.25	976.60
C7	3	10	0.01	0.2	21.20	341.60	161.45	465.35
C8	5	10	0.01	0.8	168.65	734.85	743.95	798.25
C9	5	10	0.1	0.8	522.20	983.95	935.15	102.00
C10	5	20	0.01	0.2	16.40	165.15	189.05	293.55

5 Conclusion

This paper dealt with the one-dimensional integer cutting and stock problem with multiple stock lengths with limited availability. In practice, the number of items can be hundreds and the number of cutting patterns can be thousands, which makes the problem difficult to be solved. In this paper, a straightforward Genetic Algorithm was proposed and produced good solutions in terms of quality for such large-scale data sets. We performed multiple-point crossover operation, which can be applied into Genetic Algorithm for many problems. We also improved First Fit heuristic to either increase genes of individuals or preserve diversity of gene in populations. Our approaches increased effect of Genetic Algorithm applied to CSP in term of waste length, with 70 in 10 classes of instances compared to conventional heuristics. Considering future research, the following problems are still open. First, the time complexity of our proposals is still high compared to traditional methods. This comes from the high complexity of performing crossover to create new individuals. Second, our method haven't yet deal with other objectives apart from minimization of the wastc of material, such as the minimization of the number of different cutting patterns or optimization problems in other fields as well as the problem of minimum connected dominating set problem on ad hoc wireless networks or multicast routing. Genetic Algorithm with some heuristics embedded is also promising to those problems. We intend to investigate on these important questions as part of our future work.

References

1. P. Glimore and R. Gomory. A linear programming approach to the cutting stock problem: Part ii. Operat. Res, 11:863–888, 1963.
2. S. A. Araujo, A. A. Constantino, and K. C. Poldi.: An evolutionary algorithm for the one-dimensional cutting stock problem. International Transactions in Operational Research, 18(1):115–127, 2011.
3. Y.-P. Cui and T.-B. Tang. Parallelized sequential value correction procedure for the one-dimensional cutting stock problem with multiple stock lengths. Engineering Optimization, 46(10):1352–1368, 2014.
4. T. Johsuke and T. Kazuhiro. A genetic algorithm using tournament crossover operation for cutting stock problem. International Journal of the Information Systems for Logistics and Management (IJISLM), 5(2), 2010.
5. J. Rietz and S. Dempe. Large gaps in one-dimensional cutting stock problems. Discrete applied mathematics, 156(10):1929–1935, 2008.
6. R. Hinterding and L. Khan. Genetic algorithms for cutting stock problems: with and without contiguity. Springer, 1995.
7. K. C. Poldi and M. N. Arenales. Heuristics for the one-dimensional cutting stock problem with limited multiple stock lengths. Computers & Operations Research, 36(6):2074–2081, 2009.
8. G. Belov and G. Scheithauer. A cutting plane algorithm for the one-dimensional cutting stock problem with multiple stock lengths. European Journal of Operational Research, 141(2):274–294, 2002.
9. C. Vanaret, J.-B. Gotteland, N. Durand, and J.-M. Alliot. Preventing premature convergence and proving the optimality in evolutionary algorithms. In Artificial Evolution, pages 29–40. Springer, 2014.
10. T. Gau and G. Wascher. Cutgen1: A problem generator for the standard one-dimensional cutting stock problem. European Journal of Operational Research, 84(3):572–579, 1995.

Fuzzy Linguistic Number and Fuzzy Linguistic Vector: New Concepts for Computational Intelligence

Vu Thi Hue[1(✉)], Hoang Thi Minh Chau[2], and Pham Hong Phong[1]

[1] Faculty of Information Technology, National University of Civil Engineering, Hanoi, Vietnam
huevt@nuce.edu.vn, phphong84@yahoo.com
[2] University of Economic and Technical Industries, Hanoi, Vietnam
htmchau@uneti.edu.vn

Abstract. This paper introduces concepts of fuzzy linguistic number (FLN) and fuzzy linguistic vector (FLV). Each FLN contains a linguistic term and a membership degree. A FLV is defined as a tuple of FLNs. Some relevant notions are also proposed, in which, similarity measures for FLNs and FLVs play central contributions of the paper. Then, based on proposed measures, a novel algorithm for classification problem is given. Finally, the rationality of our method is shown by comparative evaluations.

1 Introduction

Computing with words (CWWs) has been constituted one of the most modern trends of computational intelligence. The objects of CWWs are words and sentences appearing in natural or artificial language [1], for example: "young", "old", "good", "quite possible" or even more complex sentences as "this book is interesting but not very famous". To date, many computational models that provide theoretical foundations for CWWs have been studying [2, 3]. CWWs can be efficiently applied to circumstances in which: (1) the available information is perception based or not precise enough to use numbers; and/or (2) the information is given by words but the transformation of these words to numbers is costly and impossible [4].

In the simplest cases, the considered objects of the methodology are represented by an ordinal linguistic scale $S = \{s_0, s_1, \ldots, s_g\}$ such that [5]:

- $s_i \le s_j$ iff $i \le j$;
- $\max(s_i, s_j) = s_i$, if $s_i \ge s_j$;
- $\min(s_i, s_j) = s_j$, if $s_i \ge s_j$; and
- $neg(s_i) = s_{g-i}$, for all $s_i \in S$.

Complexity of the socio-economic issues is constantly increasing, so human cannot possess a sufficient or precise level of knowledge of the issues. In such a case, we usually have some indeterminacy and hesitation in providing our linguistic evaluations over the objects, this makes the evaluations have the characteristics of assurance,

V. Bhateja et al. (eds.), *Information Systems Design and Intelligent Applications*, Advances in Intelligent Systems and Computing 672,
https://doi.org/10.1007/978-981-10-7512-4_79

negation and uncertainty. One method to represent complex information in above situations is to combine linguistic approach with an advanced fuzzy set theory. Some of combinations are: intuitionistic linguistic number [6], intuitionistic linguistic value [7], intuitionistic linguistic label [8] and hesitant fuzzy linguistic term set [9].

In this paper, the notion of fuzzy linguistic number is given. Fuzzy linguistic vector is defined as a tuple of fuzzy linguistic numbers. It is easily shown that fuzzy linguistic number is a special case of intuitionistic linguistic number, while fuzzy linguistic vector can be seen as a generalized form of hesitant fuzzy linguistic term set. The new concepts provide a convenient way to represent the information in linguistic group decision-making (GDM) problems [10]. Moreover, fuzzy linguistic numbers and fuzzy linguistic vectors accompanied with their similarity measures give effective way to build a classification algorithm.

The paper is organized as follows. Section 2 recalls some related definitions: intuitionistic fuzzy set, hesitant fuzzy set, intuitionistic linguistic number and hesitant fuzzy linguistic term set. Section 3 introduces the concept of fuzzy linguistic number as well as its basic operator laws. In Sect. 4, based on fuzzy linguistic numbers, we propose fuzzy linguistic vector as an extension of hesitant fuzzy linguistic term set. Fuzzy linguistic vector similarity measures, are the most important concepts of the contribution, are also given in Sect. 4. Section 5 is devoted to presentation a new method for classification problem using fuzzy linguistic vectors and their similarity measures. The evaluations of proposed method are shown in Sect. 6. Finally, Sect. 7 draws a conclusion.

2 Preliminaries

In 1986, Atanassov [11] introduced the notion of intuitionistic fuzzy set (IFS) which roles one of the significant modifications of Zadeh's fuzzy set (FS) [12]. An IFS has two components: a membership function and a non-membership function, and it is different from FS which characterized by only a membership function.

Definition 1. [11] An intuitionistic fuzzy set (IFS) A on a universe X is an object of the form $A = \{\langle x, \mu_A(x), \nu_A(x)\rangle | x \in X\}$. For each $x \in X$, $\mu_A(x)$ and $\nu_A(x) \in [0, 1]$ define the degree of membership and non-membership of x in A such that $\mu_A(x) + \nu_A(x) \leq 1$.

Motivated by Atanassov's IFSs and linguistic approach, Wang et al. [6, 13] proposed intuitionistic linguistic number (ILN) as a suitable tool to deal with decision situations in which each assessment contains of not only a linguistic term but also a membership degree and a non-membership degree.

Definition 2. [6] Let S be a linguistic term set, an intuitionistic linguistic number (ILN) α is defined as $\alpha = \langle s_{\theta(\alpha)}, \mu(\alpha), \nu(\alpha)\rangle$, where $s_{\theta(\alpha)} \in S$ is a linguistic term, $\mu(\alpha) \in [0, 1]$ and $\nu(\alpha) \in [0, 1]$ are correspondingly the membership and non-membership degrees of evaluated object to $s_{\theta(\alpha)}$ such that $\mu(\alpha) + \nu(\alpha) \leq 1$.

The hesitant fuzzy sets (HFSs), first given by Torra in 2010 [14], permit the membership degrees (of an element into a set) can get some values between 0 and 1.

The HFSs are a highly relevant tool to interpreting the reluctance of experts when they make judgments.

Definition 3. [14] Let X be a fixed set, then a hesitant fuzzy set (HFS) A on X is defined as a map from X to the family of all subsets of [0,1]. That means $A = \{\langle x, h_A(x)\rangle | x \in X\}$, where $h_A(x)$, denoting the possible membership degrees of the element $x \in X$ to the set A, is a set of some values in [0, 1].

The HFS can handle quantitative circumstances when several numeric values are considered to assess the membership degree of an element to a set. However, in qualitative settings, when establishing the value of a linguistic variable, each expert may use several linguistic terms. Motivated by this idea, Rodríguez et al. [15] defined the concept of hesitant fuzzy linguistic term set (HFLTS).

Definition 4. [15] Let us consider a linguistic term set S. A hesitant fuzzy linguistic term set (HFLTS) on X is defined as an ordered finite subset of the consecutive linguistic terms of S.

3 Fuzzy Linguistic Numbers

Let us consider the discrete linguistic term set $S = \{s_0, s_1, \ldots, s_g\}$.

Definition 5. Each fuzzy linguistic number (FLN) on S has the following form

$$\gamma = \langle s_{\theta(\gamma)}, \mu(\gamma)\rangle,$$

in which, $s_{\theta(\gamma)} \in S$ is a linguistic term, $\mu(\gamma) \in [0, 1]$ is the membership degree of an evaluated object into the term $s_{\theta(\gamma)}$. We denote by $FLN(S)$ the set of all FLNs on S.

Remark 1 An ILN is reduced to a FLN when the non-membership degree is assigned to 0.

Definition 6. Consider $\gamma_1, \gamma_2 \in FLN(S)$. We define:

1. γ_1 and γ_2 are equal, $\gamma_1 = \gamma_2$, if $\theta(\gamma_1) = \theta(\gamma_2)$ and $\mu(\gamma_1) = \mu(\gamma_2)$;
2. γ_1 is smaller than γ_2, $\gamma_1 < \gamma_2$, if $\theta(\gamma_1)\mu(\gamma_1) < \theta(\gamma_2)\mu(\gamma_2)$;
3. γ_1 and γ_2 are equivalent, $\gamma_1 \sim \gamma_2$, if $\theta(\gamma_1)\mu(\gamma_1) = \theta(\gamma_2)\mu(\gamma_2)$; and
4. γ_1 is smaller than or equivalent to γ_2, $\gamma_1 \lesssim \gamma_2$, if $\gamma_1 < \gamma_2$ or $\gamma_1 \sim \gamma_2$.

Definition 7. A similarity measure for FLNs is a mapping $sim : FLN(S) \times FLN(S) \rightarrow \mathbb{R}$ satisfying:

1. Commutativity: $sim(\gamma_1, \gamma_2) = sim(\gamma_2, \gamma_1)$ for all $\gamma_1, \gamma_2 \in FLN(S)$;
2. Boundary: $0 \leq sim(\gamma_1, \gamma_2) \leq 1$, for all $\gamma_1, \gamma_2 \in FLN(S)$;
3. Reflexivity: $sim(\gamma_1, \gamma_2) = 1 \Leftrightarrow \gamma_1 \sim \gamma_2$, for all $\gamma_1, \gamma_2 \in FLN(S)$;
4. Monotonicity: if $\gamma_1 \lesssim \gamma_2 \lesssim \gamma_3$,

$$\begin{cases} sim(\gamma_1,\gamma_2) \geq sim(\gamma_1,\gamma_3) \\ sim(\gamma_2,\gamma_3) \geq sim(\gamma_1,\gamma_3) \end{cases}, \text{ for all } \gamma_1, \gamma_2, \gamma_3 \in FLN(S).$$

Definition 8. Consider $\gamma_1, \gamma_2 \in FLN(S)$,

1. $sim_1(\gamma_1,\gamma_2) = \frac{g-|\mu(\gamma_1)\theta(\gamma_1)-\mu(\gamma_2)\theta(\gamma_2)|}{g}$;
2. $sim_2(\gamma_1,\gamma_2) = \frac{\min\{\mu(\gamma_1)\theta(\gamma_1),\mu(\gamma_2)\theta(\gamma_2)\}}{\max\{\mu(\gamma_1)\theta(\gamma_1),\mu(\gamma_2)\theta(\gamma_2)\}}$,

   with the convention that $\frac{0}{0} = 1$;
3. $sim_3(\gamma_1,\gamma_2) = 1 - \frac{1-\exp\left(-\frac{|\mu(\gamma_1)\theta(\gamma_1)-\mu(\gamma_2)\theta(\gamma_2)|}{g}\right)}{1-\exp(-1)}$;
4. $sim_4(\gamma_1,\gamma_2) = 1 - \frac{1-\exp\left(-\frac{\left|\sqrt{\mu(\gamma_1)\theta(\gamma_1)}-\sqrt{\mu(\gamma_2)\theta(\gamma_2)}\right|}{\sqrt{g}}\right)}{1-\exp(-1)}$.

Theorem 1. *The mappings defined in Definition* 8 *are similarity measures for FLNs.*

4 Fuzzy Linguistic Vector

Definition 9. A fuzzy linguistic vector (FLV) V on $S = \{s_0, s_1, \ldots, s_g\}$ is defined by $V = (v_0, \ldots, v_g)$, where $v_i = (s_i, \mu_i^V)$ is a FLN on $S\,(i = 0, \ldots, g)$. The set of all FLVs on S is denoted by $FLV(S)$.

Remark 2. FLV is an extension of HFLTS. Each FLV V is degenerated to an HFLTS when μ_i^V is set to 0 or 1 $(i = 0, \ldots, g)$.
The below examples clarify the meaning of FLV.

Example 1 We consider a situation, in which experts use HFLTSs to access the feasibility of a certain project. The linguistic terms are: $s_0 = unacceptable$, $s_1 = acceptable$, $s_2 = good$, $s_3 = very\_good$ and $s_4 = perfect$. In Table 1, each number 1 (resp. 0) located at the same row with expert j and the same column with s_i means the j-th expert evaluates (resp. does not evaluate) the project by the linguistic term s_i. The membership of the project into s_i is measured as the quotient when dividing the total experts who use s_i to assess by the number of experts. Therefore, information about this project can be fully expressed by a fuzzy linguistic vector:

$$V = ((s_0, 0.4), (s_1, 0.6), (s_2, 1.0), (s_3, 0.6), (s_4, 0.2)).$$

Example 2 This example shows that a real number can be expressed by a linguistic vector through a linguistic variable [10]. Thus, the initial information is repeated in many different semantics. We consider a linguistic variable which involves a universe of discourse X, a set of term $S = \{s_0, s_1, s_2, s_3, s_4\}$ and a semantic rule M which

Table 1. Example for FLV: a FLV can be derived from a set of HFLTSs

	s_0	s_1	s_2	s_3	s_4
Expert 1	0	0	1	1	0
Expert 2	0	1	1	1	0
Expert 3	1	1	1	0	0
Expert 4	0	0	1	1	1
Expert 5	1	1	1	0	0
Total/5	0.4	0.6	1.0	0.6	0.2

associates with s_i its meaning μ_{s_i}. By this linguistic variable [16], each $x \in X$ can be interpreted as a fuzzy linguistic vector.

$$V = \left(\left(s_0, \mu_{s_0}(x)\right), \left(s_1, \mu_{s_1}(x)\right), \left(s_2, \mu_{s_2}(x)\right), \left(s_3, \mu_{s_3}(x)\right), \left(s_4, \mu_{s_4}(x)\right)\right).$$

Definition 10. For $U, V \in FLV(S)$, $U = \left(u_0, \ldots, u_g\right)$, $V = \left(v_0, \ldots, v_g\right)$,

1. U and V are equal, $U = V$, if $u_i = v_i$, for all $i = 0, \ldots, g$;
2. U is smaller than V, $U < V$, if $u_i < v_i$, for all $i = 0, \ldots, g$;
3. U and V are equivalent, $U \sim V$, if $u_i \sim v_i$, for all $i = 0, \ldots, g$;
4. U is smaller than or equivalent to V, $U \lesssim V$, if $U < V$ or $U \sim V$.

Definition 11. A mapping $SIM : LFV(S) \times LFV(S) \to \mathbb{R}$ is termed as a similarity measure between FLVs if following conditions are satisfied:

1. Commutativity: $SIM(U, V) = SIM(V, U)$, for all $U, V \in FLV(S)$;
2. Boundary: $0 \leq SIM(U, V) \leq 1$, for all $U, V \in FLV(S)$;
3. Reflexivity: $SIM(U, V) = 1 \Leftrightarrow U \sim V$, for all $U, V \in FLV(S)$;
4. Monotonicity: if $U \lesssim V \lesssim T$,

$$\begin{cases} SIM(U, V) \geq SIM(U, T) \\ SIM(V, T) \geq SIM(U, T) \end{cases}, \text{ for all } U,\ V,\ T \in FLV(S).$$

Definition 12. For $U = \left(u_0, \ldots, u_g\right)$, $V = \left(v_0, \ldots, v_g\right) \in LFV(S)$, sim is a similarity measure for FLNs on S. $w = \left(w_0, \ldots, w_g\right)$ is a weight vector such that $w_i \geq 0$, for all $i = 0, \ldots, g$ and $\sum_{i=0}^{g} w_i = 1$. We define:

1. $SIM_Q(U, V) = \left(\sum_{i=0}^{g} w_i (sim(u_i, v_i))^2\right)^{\frac{1}{2}}$;
2. $SIM_A(U, V) = \sum_{i=0}^{g} w_i\, sim(u_i, v_i)$;
3. $SIM_G(U, V) = \prod_{i=0}^{g} (sim(u_i, v_i))^{w_i}$, with the convention that $0^0 = 1$;

4. $SIM_H(U,V) = \left(\sum_{i=0}^{g} \frac{w_t}{sim(u_i,v_i)}\right)^{-1}$, with the convention that $\frac{w_t}{0} = 0$.

Theorem 2. *For all* $U, V \in FLV(S)$,

$$SIM_Q(U,V) \geq SIM_A(U,V) \geq SIM_G(U,V) \geq SIM_H(U,V).$$

Theorem 3. *If* $w_i > 0$ *for all* $i = 0, \ldots, g$, SIM_Q, SIM_A, SIM_G *and* SIM_H *are similarity measures for FLVs.*

5 Algorithm

In this section, we describe our algorithm (Fuzzy Linguistic Vector Similarity Measure-based Classification Algorithm, FLVS) in detail.

In order to deploy FLVS algorithm, each n-length vector $x = (x_1, \ldots, x_n)$ should be transformed into a fuzzy linguistic matrix (FLM) $FLM(x)$ such that

$$FLM(x) = \begin{bmatrix} (s_0, \mu_{s_0}(x_1)) & (s_1, \mu_{s_1}(x_1)) & \cdots & (s_g, \mu_{s_g}(x_1)) \\ (s_0, \mu_{s_0}(x_2)) & (s_1, \mu_{s_1}(x_2)) & \cdots & (s_g, \mu_{s_g}(x_2)) \\ \vdots & \vdots & \vdots & \vdots \\ (s_0, \mu_{s_0}(x_n)) & (s_1, \mu_{s_1}(x_n)) & \cdots & (s_g, \mu_{s_g}(x_n)) \end{bmatrix},$$

where s_i is a linguistic term and μ_{s_i} is the corresponding membership function $(i = 0, \ldots, g)$.

Procedure 1 Let us consider two vectors $x = (x_1, \ldots, x_n)$ and $x' = (x'_1, \ldots, x'_n)$. The similarity between x and x', $SIM(x, x')$ is computed as following procedure:

- Firstly, for each $j = 1, \ldots, n$, we calculate the similarity SIM_j between j-th rows of $FLM(x)$ and $FLM(x')$. In this step, the similarity measure is selected from those given in Definition 12.
- Secondly, the similar degree between x and x', $SIM(x, x')$ is obtained by taking the weighted average over SIM_t $(t = 1, \ldots, n)$. In this paper, we use the weighted arithmetic average:

$$SIM(x, x') = \sum_{t=1}^{n} w_t \, SIM_t,$$

where $w = (w_1, \ldots, w_n)$ is a weight vector, $w_t \geq 0$ and $\sum_{t=1}^{n} w_t = 1$.

Algorithm 1 (Fuzzy Linguistic Vector Similarity Measure-based Classification Algorithm, FLVS) Let $D' \subset D$ is the training set having m items. Consider an item

$I^* \in D^* = D \backslash D'$, with $x^* = \left(x_1^*, \ldots, x_n^*\right)$ is the associated attribute vector containing values of its n first attributes, where x_t^* is the value of the t-th attribute, for all $t = 1, \ldots, n$.

Step 1. For each $I_j \in D'$, $x^{(j)}$ denotes attribute vector of I_j $(j = 1, \ldots, m)$. The similarity between the items I^* and I_j is determined as the similarity between x^* and $x^{(j)}$ (see Procedure 1),

$$SIM\left(I^*, I_j\right) = SIM\left(x^*, x^{(j)}\right).$$

Step 2. Let us assume that $I_1, \ldots, I_k \in D'$ be k nearest neighbours of I^*, where k is specified by experiment $(1 \le k \le m)$. Aggregate the values of the $(n+1)$th attribute of all items I_j $(j = 1, \ldots, k)$ using the weighted arithmetic average operator:

$$\bar{y} = \sum_{j=1}^{k} \omega_j y^{(j)},$$

where $y^{(j)}$ denotes the value of $(n+1)$th attribute of the item I_j. $\omega = (\omega_1, \ldots, \omega_k)$ is the weighting vector with $\omega_j = \frac{SIM\left(I^*, I_j\right)}{\sum_{j=1}^{k} SIM\left(I^*, I_j\right)}$ $(j = 1, \ldots, k)$.

Step 3. Evaluate value y^* of the $(n+1)$th attribute of the item I^*: predicted value of the $(n+1)$th attribute of I^* is $y^* = round(\bar{y})$, where *round* is the usual round function.

6 Evaluations

In this paper, we use Iris data set [17] which contains 150 rows and 5 columns. The first 4 columns represent sepal length, sepal width, petal length and petal width (centimetres) of Iris flowers. The last column indicates 3 classes of 50 instances each (Iris setosa, Iris versicolour and Iris virginica), where each class is a type of Iris plants. We use FLVS algorithm to classify the type Iris plants based on its 4 attributes of their flowers.

Firstly, the last column is converted to 0, 1 and 2 for Iris setosa, Iris versicolour and Iris virginica, correspondingly. Secondly, each remainder column of the matrix is rescaled to [0, 1]. Then, we pick up randomly 100 rows of the matrix to form the training data and the remainder 50 rows of the matrix to form the testing data. For each row of the testing data, our objective is to predict its associated class in the last column. To do that, we calculate the similarity between that row of testing data to each of 100 rows of training data. Then, we take 10 rows which are the most similar to that row to average their 10 associated class numbering. The average number is rounded to be the class number of the testing row.

Evaluation indices: In this section, we use some statistical indices (TP-rate, FP-rate, Precision and F-measure) to compare FLVS with three other classifiers (NFS [18], RBFNN [19], ANFIS [20]). Using sim1 as similarity measure for FLNs and SIM3 as similarity measure for FLVs and setting k = 12, we conduct the FLVS for 100 iterates. The results are given in Table 2.

Table 2. Detailed accuracy by each class for four classifiers (%)

Classifier	Class	TP-rate	FP-rate	Precision	F-measure
FLVS	Iris setosa	99.7	0.0	100	99.8
	Iris versicolour	92.7	2.3	95.4	93.8
	Iris virginica	95.8	3.7	93.0	94.2
	Weighted average	96.2	2.2	96.0	96.0
NFS	Iris setosa	99.9	0.7	98.9	99.9
	Iris versicolour	99.8	1.5	90.9	95.2
	Iris virginica	88.9	2.9	89.9	94.1
	Weighted average	96.7	2.3	97.0	96.6
RBFNN	Iris setosa	99.9	0.7	99.9	99.9
	Iris versicolour	52.9	0.5	99.9	66.7
	Iris virginica	99.9	23.8	64.3	78.3
	Weighted average	83.3	7.1	89.3	82.4
ANFIS	Iris setosa	99.9	0.7	99.9	99.9
	Iris versicolour	99.8	6.9	88.9	94.1
	Iris virginica	86.7	0.5	99.9	92.9
	Weighted average	95.6	3.7	96.0	95.5

Figure 1 illustrates some statistical indices comparing FLVS with three other methods. Firstly, on TP-rate indices, FLVS is almost similar to ANFIS and NFS reaching around 96% while RBFNN is much lower (83.3%). Secondly, on FP-rates, RBFN (7.1%) is significant higher than the three remaining classifiers while FLVS is in the lowest value (2.2%). Thirdly, the results of Precision are quite similar to TP-rate, the accuracies of FLVS, NFS and ANFIS fluctuate slightly around 96%, which is much higher than that of RBFNN with 89.3%. Finally, in F-measure, FLVS is slightly lower than NFS but a little bit higher than ANFIS and much higher than RBFNN.

In conclusion, FLVS gives results which are much better than ANFIS and RBFNN yet slightly less than NFS. In the future, we will develop FLVS to foster better results

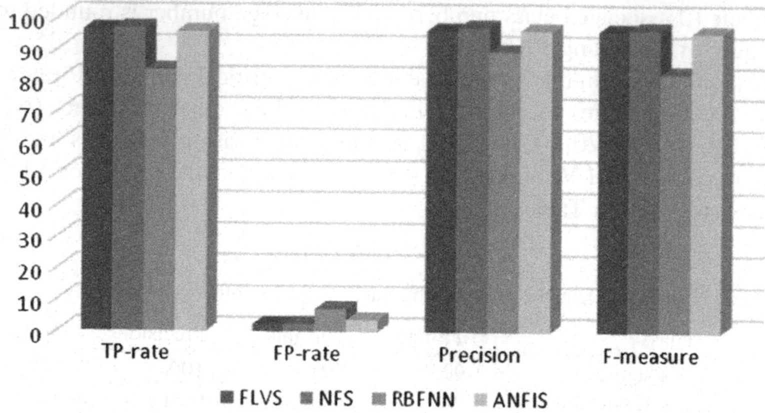

Fig. 1. Comparison of four classifiers on weighted average values

7 Conclusion

In this paper, we first propose fuzzy linguistic number (FLN) and fuzzy linguistic vector (FLV). Some elementary operator laws, similarity measures for FLNs and FLVs are also given. Based on those measures, we develop an algorithm, namely Fuzzy Linguistic Vector Similarity Measure-based Classification Algorithm (FLVS). The feasibility of the algorithm is tested on the Iris data set which is frequently used for classification. In the future, we continue to study mathematical aspect as well as applicability of FLNs and FLVs.

Acknowledgements. This research is financial supported by NAFOSTED (the Vietnam National Foundation for Science and Technology Development) under grant number 102.01-2017.02.

References

1. Zadeh, L. A.: Fuzzy logic = computing with words, IEEE Transactions on Fuzzy Systems 4 (1996) 103–111.
2. Herrera, F., Alonso, S., Chiclana, F., Herrera-Viedma, E.: Computing with words in decision making: foundations, trends and prospects, Fuzzy Optimization and Decision Making 8 (4) (2009) 337–364.
3. Martinez, L., Ruan, D., Herrera, F.: Computing with Words in Decision support Systems: An overview on Models and Applications, International Journal of Computational Intelligence Systems 3 (4) (2010) 382–395.
4. Chen, S.J., Hwang C.L.: Fuzzy multiple attribute decision making: Methods and applications, Berlin.: Springer-Verlag (1992).
5. Yager, R.: A new methodology for ordinal multiobjective decisions based on fuzzy sets, Decision Sciences 12 (1981) 589–600.

6. Wang, J. Q., Li, H. B.: Multi-criteria decision-making method based on aggregation operators for intuitionistic linguistic fuzzy numbers, Control and Decision 25 (10) (2010) 1571–1574, 1584.
7. Zhang, Y., Ma, H. X., Liu, B. H., Liu, J.: Group decision making with 2-tuple intuitionistic fuzzy linguistic preference relations, Soft Computing 16 (2012) 1439–1446.
8. Phong P. H., Cuong B. C.: Some intuitionistic linguistic aggregation operators, Journal of Computer Science and Cybernetics 30 (3) (2014) 216–226.
9. Rodríguez, R. M., Martídnez, L., Herrera, F.: Hesitant fuzzy linguistic terms sets for decision making, IEEE Transactions on Fuzzy Systems 20 (2012), 109–119.
10. Herrera, F., Herrera-Viedma, E.: Linguistic decision analysis: steps for solving decision problems under linguistic information, Fuzzy Sets and Systems 115 (2000) 67–82.
11. Atanassov, K.: Intuitionistic fuzzy sets, Fuzzy Sets and Systems 20 (1986) 87–96.
12. Zadeh, L. A.: Fuzzy Sets, Information and Control 8 (1965) 338–353.
13. Wang, X. F., Wang J. Q., Yang, W. E.: Multi-criteria group decision making method based on intuitionistic linguistic aggregation operators, Journal of Intelligent & Fuzzy Systems 26 (2014), 115–125.
14. Torra, V.: Hesitant fuzzy sets, International Journal of Intelligent Systems 25 (6) (2010), 529–539.
15. Rodríguez, R. M., Martínez, L., Herrera, F.: Hesitant fuzzy linguistic terms sets for decision making, IEEE Transactions on Fuzzy Systems 20 (2012) 109–119.
16. Zadeh, L.: The concept of a linguistic variable and its application to approximate reasoning —Part I, Information Sciences 8 (3) (1975) 199–249.
17. Fisher R. A.'s Data (1988), http://archive.ics.uci.edu/ml/machine-learning-databases/iris/iris.data.
18. Ghosh, S., Biswas, S., Sarkar, D., Sarkar, P. P.: A novel Neuro-fuzzy classification technique for data mining, Egyptian Informatics Journal 15 (2014) 129–147.
19. Hunt K. J., Haas, R., Murray-Smith, R.: Extending the functional equivalence of radial basis function networks and fuzzy inference systems, IEEE Trans Neural Networks 7 (3) (1996) 776–781.
20. Jang, J. SR., Sun CT, Mizutani, E.: Neuro-fuzzy and soft computing: a computational approach to learning and machine intelligence, USA: Prentice Hall (1997) 333–393.

An Isolated Bipolar Single-Valued Neutrosophic Graphs

Said Broumi[1(✉)], Assia Bakali[2], Mohamed Talea[1], and Florentin Smarandache[3]

[1] Laboratory of Information Processing, Faculty of Science Ben M'Sik, University Hassan II, Casablanca, Morocco
broumisaid78@gmail.com, taleamohamed@yahoo.fr
[2] Ecole Royale Navale, Boulevard Sour Jdid, B.P 16303 Casablanca, Morocco
assiabakali@yahoo.fr
[3] Department of Mathematics, University of New Mexico, 705 Gurley Avenue, Gallup NM87301, USA
fsmarandache@gmail.com

Abstract. In this research paper, we propose the graph of the bipolar single-valued neutrosophic set (BSVNS) model. This graph generalized the graphs of single-valued neutrosophic set models. Several results have been proved on complete and isolated graphs for the BSVNS model. Moreover, an essential and satisfactory condition for the graphs of the BSVNS model to become an isolated graph of the BSVNS model has been demonstrated.

Keywords: BSVNGs · Complete BSVNG · Isolated BSVNGs

1 Introduction

Smarandache [1] proposed the concept of neutrosophic sets (in short NSs) as a means of expressing the inconsistencies and indeterminacies that exist in most real-life problems. The proposed concept generalized fuzzy sets and intuitionistic fuzzy sets theory [2, 3]. The notion of NS is described with three functions: truth, an indeterminacy and a falsity, where the functions are totally independent; the three functions are inside the unit interval$]^{-}0, 1^{+}[$. To practice NSs in real-life situations efficiently, a new version of NSs. A new version of NSs named single-valued neutrosophic sets (in short SVNSs) was defined by Smarandache in [1]. Subsequently, Wang et al. [4] defined the various operations and operators for the SVNS model. In [5], Deli et al. introduced the notion of bipolar neutrosophic sets, which combine the bipolar fuzzy sets and SVNS models. Neutrosophic sets and their extensions have been paid great attention recent years [6]. The theory of graphs is the mostly used tool for resolving combinatorial problems in various fields such as computer science, algebra and topology. Smarandache [1, 7] introduced two classes of neutrosophic graphs to deal with situations in which there exist inconsistencies and indeterminacies among the vertices which cannot be dealt with by fuzzy graphs and different hybrid structures [8–10]. The first class is relied on literal indeterminacy (I) component, and the second class of neutrosophic graphs is based on numerical truth values (T, I, F). Subsequently,

V. Bhateja et al. (eds.), *Information Systems Design and Intelligent Applications*, Advances in Intelligent Systems and Computing 672,
https://doi.org/10.1007/978-981-10-7512-4_80

Broumi et al. [11–13] introduced the concept single-valued neutrosophic graphs (in short SVNGs) and discussed some interesting results. Later on, the same authors [14–17] proposed the concept of bipolar single-valued neutrosophic graphs (BSVNGs) and established some interesting results with proofs and illustrations.

The objective of our article is to demonstrate the essential and satisfactory condition of BSVNGs to be an isolated BSVNG.

2 Background of Research

Some of the important background knowledge for the materials that are presented in this paper is presented in this section. These results can be found in [1, 4, 5, 12, 13].

Definition 2.1 [1]. Let ζ be a universal set. The neutrosophic set A on the universal set ζ is categorized into three membership functions, namely the true $T_A(x)$, indeterminate $I_A(x)$ and false $F_A(x)$ contained in real standard or non-standard subset of $]^{-}0, 1^{+}[$, respectively.

$$^{-}0 \leq \sup T_A(x) + \sup I_A(x) + \sup F_A(x) \leq 3^{+} \tag{1}$$

Definition 2.2 [4]. Let ζ be a universal set. The single-valued neutrosophic sets (SVNSs) A on the universal ζ is denoted as following

$$A = \{< x : T_A(x), I_A(x), F_A(x) > x \in \zeta\} \tag{2}$$

The functions $T_A(x) \in [0, 1]$, $I_A(x) \in [0, 1]$ and $F_A(x) \in [0, 1]$ are called "degree of truth, indeterminacy and falsity membership of x in A", which satisfy *the* following condition:

$$0 \leq T_A(x) + I_A(x) + F_A(x) \leq 3 \tag{3}$$

Definition 2.3 [12]. A SVNG of $G^* = (V, E)$ is a graph G = (A, B) where

a. The following memberships: $T_A : V \to [0, 1]$, $I_A : V \to [0, 1]$ and $F_A : V \to [0, 1]$ represent the truth, indeterminate and false membership degrees of $x \in V$ respectively and

$$0 \leq T_A(w) + I_A(w) + F_A(w) \leq 3 \quad \forall w \in V \tag{4}$$

b. The following memberships: $T_B : E \to [0, 1]$, $I_B : E \to [0, 1]$ and $F_B : E \to [0, 1]$ are defined by

$$T_B(v, w) \leq \min[T_A(v), T_A(w)] \tag{5}$$

$$I_B(v, w) \geq \max[I_A(v), I_A(w)] \text{ and} \tag{6}$$

$$F_B(v, w) \geq \max[\mathrm{F_A}(v), \mathrm{F_A}(w)] \tag{7}$$

Represent the true, indeterminate and false membership degrees of the arc $(v, w) \in$ (V x V), where (Fig. 1)

$$0 \leq \mathrm{T_B}(v, w) + \mathrm{I_B}(v, w) + \mathrm{F_B}(v, w) \leq 3 \quad \forall (v, w) \in \mathrm{E} \tag{8}$$

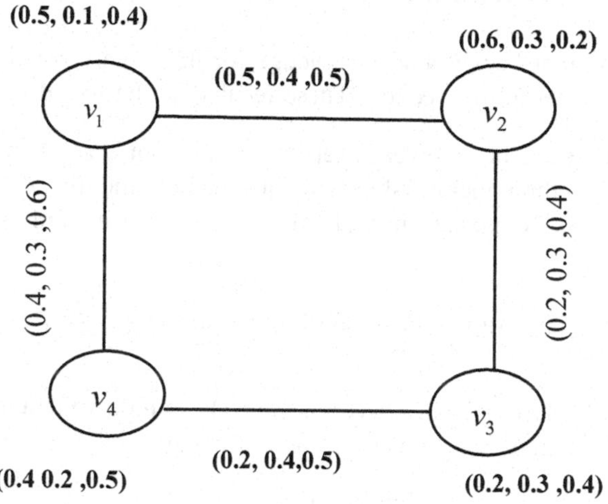

Fig. 1. SVN graph

Definition 2.4 [12]. A SVNG G = (A, B) is named complete SVNG if

$$\mathrm{T_B}(v, w) = min[\mathrm{T_A}(\mathrm{v}), \mathrm{T_A}(w)] \tag{9}$$

$$\mathrm{I_B}(v, w) = max[\mathrm{I_A}(\mathrm{v}), \mathrm{I_A}(w)] \tag{10}$$

$$\mathrm{F_B}(v, w) = max[\mathrm{F_A}(\mathrm{v}), \mathrm{F_A}(w)] \quad \forall v, w \in \mathrm{V} \tag{11}$$

Definition 2.5 [12]. Given a SVNG G = (A, B). Hence, the complement of SVNG on G^* is a SVNG $\bar{G}$ on G^* where

$$a.\bar{A} = \mathrm{A} \tag{12}$$

$$b.\bar{T}_A(w) = T_A(w), \bar{I}_A(w) = I_A(w), \bar{F}_A(w) = F_A(w) \quad \forall w \in \mathrm{V} \tag{13}$$

$$c.\bar{\mathrm{T}}_\mathrm{B}(v, w) = \min[\mathrm{T_A}(v), \mathrm{T_A}(w)] - \mathrm{T_B}(v, w) \tag{14}$$

$$\bar{\mathrm{I}}_{\mathrm{B}}(v, w) = \max[I_A(v), I_A(w)] - I_B(v, w) \text{ and} \tag{15}$$

$$\bar{\mathrm{F}}_{\mathrm{B}}(v, w) = \max[F_A(v), F_A(w)] - F_B(v, w), \forall (v, w) \in \mathrm{E} \tag{16}$$

Definition 2.6 [14]. A BSVNG of $G^* = (\mathrm{V}, \mathrm{E})$ is a partner $G = (\mathrm{A}, \mathrm{B})$ where $\mathrm{A} = (\mathrm{T}^{\mathrm{P}}_{\mathrm{A}}, \mathrm{I}^{\mathrm{P}}_{\mathrm{A}}, \mathrm{F}^{\mathrm{P}}_{\mathrm{A}}, \mathrm{T}^{\mathrm{N}}_{\mathrm{A}}, \mathrm{I}^{\mathrm{N}}_{\mathrm{A}}, \mathrm{F}^{\mathrm{N}}_{\mathrm{A}})$ is a BSVNS in V and $\mathrm{B} = (\mathrm{T}^{\mathrm{P}}_{\mathrm{B}}, \mathrm{I}^{\mathrm{P}}_{\mathrm{B}}, \mathrm{F}^{\mathrm{P}}_{\mathrm{B}}, \mathrm{T}^{\mathrm{N}}_{\mathrm{B}}, \mathrm{I}^{\mathrm{N}}_{\mathrm{B}}, \mathrm{F}^{\mathrm{N}}_{\mathrm{B}})$ is a BSVNS in $\tilde{\mathrm{V}}^2$ such that (Fig. 2)

$$\mathrm{T}^{\mathrm{P}}_{\mathrm{B}}(v, w) \leq min(\mathrm{T}^{\mathrm{P}}_{\mathrm{A}}(v), \mathrm{T}^{\mathrm{P}}_{\mathrm{A}}(w)), \quad \mathrm{T}^{\mathrm{N}}_{\mathrm{B}}(v, w) \geq max(\mathrm{T}^{\mathrm{N}}_{\mathrm{A}}(v), \mathrm{T}^{\mathrm{N}}_{\mathrm{A}}(w)) \tag{17}$$

$$\mathrm{I}^{\mathrm{P}}_{\mathrm{B}}(v, w) \geq max(\mathrm{I}^{\mathrm{P}}_{\mathrm{A}}(v), \mathrm{I}^{\mathrm{P}}_{\mathrm{A}}(w)) \quad \mathrm{I}^{\mathrm{N}}_{\mathrm{B}}(v, w) \leq \min(\mathrm{I}^{\mathrm{N}}_{\mathrm{A}}(v), \mathrm{I}^{\mathrm{N}}_{\mathrm{A}}(w)) \tag{18}$$

$$\mathrm{F}^{\mathrm{P}}_{\mathrm{B}}(v, w) \geq max(\mathrm{F}^{\mathrm{P}}_{\mathrm{A}}(v), \mathrm{F}^{\mathrm{P}}_{\mathrm{A}}(w)), \quad \mathrm{F}^{\mathrm{N}}_{\mathrm{B}}(v, w) \leq \min(\mathrm{F}^{\mathrm{N}}_{\mathrm{A}}(v), \mathrm{F}^{\mathrm{N}}_{\mathrm{A}}(w)) \quad \forall \mathrm{v}, \mathrm{w} \in \tilde{\mathrm{V}}^2 \tag{19}$$

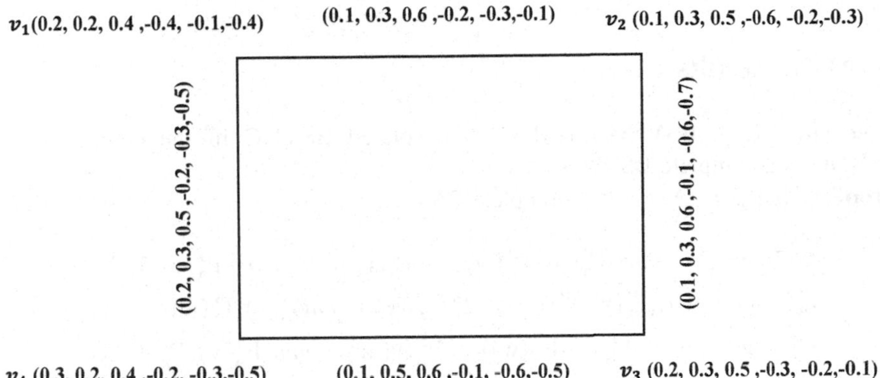

Fig. 2. BSVNG

Definition 2.7 [14]. The complement of BSVNG $\mathbf{G} = (A, B)$ of $G^* = (\mathrm{A}, \mathrm{B})$ is a BSVNG $\bar{G} = (\bar{A}, \bar{B})$ of $\bar{G}^* = (\mathrm{V}, \mathrm{V} \times \mathrm{V})$ where $\bar{A} = \mathrm{A} = (\mathrm{T}^{\mathrm{P}}_{\mathrm{A}}, \mathrm{I}^{\mathrm{P}}_{\mathrm{A}}, \mathrm{F}^{\mathrm{P}}_{\mathrm{A}}, \mathrm{T}^{\mathrm{N}}_{\mathrm{A}}, \mathrm{I}^{\mathrm{N}}_{\mathrm{A}}, \mathrm{F}^{\mathrm{N}}_{\mathrm{A}})$ and $\bar{\mathrm{B}} = (\bar{\mathrm{T}}^{\mathrm{P}}_{\mathrm{B}}, \bar{\mathrm{I}}^{\mathrm{P}}_{\mathrm{B}}, \bar{\mathrm{F}}^{\mathrm{P}}_{\mathrm{B}}, \bar{\mathrm{T}}^{\mathrm{N}}_{\mathrm{B}}, \bar{\mathrm{I}}^{\mathrm{N}}_{\mathrm{B}}, \bar{\mathrm{F}}^{\mathrm{N}}_{\mathrm{B}})$ is defined as

$$\bar{\mathrm{T}}^{\mathrm{P}}_{\mathrm{B}}(v, w) = min(\mathrm{T}^{\mathrm{P}}_{\mathrm{A}}(v), \mathrm{T}^{\mathrm{P}}_{\mathrm{A}}(w)) - \mathrm{T}^{\mathrm{P}}_{\mathrm{B}}(v, w) \tag{20}$$

$$\bar{\mathrm{I}}^{\mathrm{P}}_{\mathrm{B}}(v, w) = max(\mathrm{I}^{\mathrm{P}}_{\mathrm{A}}(v), \mathrm{I}^{\mathrm{P}}_{\mathrm{A}}(w)) - \mathrm{I}^{\mathrm{P}}_{\mathrm{B}}(v, w) \tag{21}$$

$$\bar{\mathrm{F}}^{\mathrm{P}}_{\mathrm{B}}(v, w) = max(\mathrm{F}^{\mathrm{P}}_{\mathrm{A}}(v), \mathrm{F}^{\mathrm{P}}_{\mathrm{A}}(w)) - \mathrm{F}^{\mathrm{P}}_{\mathrm{B}}(v, w) \tag{22}$$

$$\bar{T}_B^N(v,w) = max(T_A^N(v), T_A^N(w)) - T_B^N(v,w) \tag{23}$$

$$\bar{I}_B^N(v,w) = min(I_A^N(v), I_A^N(w)) - I_B^N(v,w) \tag{24}$$

$$\bar{F}_B^N(v,w) min(F_A^N(v), (w)) - F_B^N(v,w) \quad \forall v,w \in V, vw \in \tilde{V}^2 \tag{25}$$

Definition 2.8 [14]. A BSVNG $G = (A, B)$ is said to be complete BSVNG if

$$T_B^P(v,w) = \min(T_A^P(v), T_A^P(w)), \quad T_B^N(v,w) = max(T_A^N(v), T_A^N(w)), \tag{26}$$

$$I_B^P(v,w) = max(I_A^P(v), I_A^P(w)), \quad I_B^N(v,w) = min(I_A^N(v), I_A^N(w)) \tag{27}$$

$$F_B^P(v,w) = max(F_A^P(v), F_A^P(w)), \quad F_B^N(v,w) = min(F_A^N(v), F_A^N(w)) \quad \forall v,w \in V \tag{28}$$

Theorem 2.9 [13]: Let G = (A,B) be a SVNG, then the SVNG is called an isolated SVNG if and only if the complement of G is a complete SVNG.

3 Main Results

Theorem 3.1: A BSVNG = (A,B) is an isolated BSVNG iff the complement of BSVNG is a complete BSVNG.
Proof: Given $G = (A, B)$ be a complete BSVNG.

So $T_B^P(v,w) = min(T_A^P(v), T_A^P(w)), \quad T_B^n(v,w) = max(T_A^n(v), T_A^n(w)),$
$I_B^P(v,w) = max(I_A^P(v), I_A^P(w)), \quad I_B^n(v,w) = min(I_A^n(v), I_A^n(w)),$
$F_B^P(v,w) = max(F_A^P(v), F_A^P(w)), \quad F_B^n(v,w) = min(F_A^n(v), F_A^n(w)),$
$\forall v,w \in V.$

Hence in $\bar{G}$,

$$\begin{aligned} \bar{T}_B^P(v,w) &= min(T_A^P(v), T_A^P(w)) - T_B^P(v,w) \\ &= min(T_A^P(v), T_A^P(w)) - min(T_A^P(v), T_A^P(w)) \\ &= 0 \end{aligned}$$

and

$$\begin{aligned} \bar{I}_B^P(v,w) &= max(I_A^P(v), I_A^P(w)) - I_B^P(v,w) \\ &= max(I_A^P(v), I_A^P(w)) - max(I_A^P(v), I_A^P(w)) \\ &= 0 \end{aligned}$$

In addition

$$\bar{F}_B^P(v,w) = max(F_A^P(v), F_A^P(w)) - F_B^P(v,w)$$
$$= max(F_A^P(v), F_A^P(w)) - max(F_A^P(v), F_A^P(w))$$
$$= 0$$

We have for the negative membership edges

$$\bar{T}_B^N(v,w) = max(\mathrm{T}_\mathrm{A}^\mathrm{N}(v), \mathrm{T}_\mathrm{A}^\mathrm{N}(w)) - T_B^N(v,w)$$
$$= max(T_A^N(v), T_A^N(w)) - max(\mathrm{T}_\mathrm{A}^\mathrm{N}(v), \mathrm{T}_\mathrm{A}^\mathrm{N}(w))$$
$$= 0$$

and

$$\bar{I}_B^N(v,w) = min(I_A^N(v), I_A^N(w)) - I_B^N(v,w)$$
$$= min(I_A^N(v), I_A^N(w)) - min(I_A^N(\mathrm{v}), I_A^N(\mathrm{w}))$$
$$= 0$$

In addition

$$\bar{F}_B^N(v,w) = min(F_A^N(v), F_A^N(w)) - F_B^N(v,w)$$
$$= min(F_A^N(v), F_A^N(w)) - min(F_A^N(v), F_A^N(w))$$
$$= 0$$

So $(\bar{T}_B^P(v,w), \bar{\mathrm{I}}_\mathrm{B}^\mathrm{P}(v,\ w), \bar{F}_B^P(v,w), \bar{T}_B^N(v,w), \bar{I}_B^N(v,w), \bar{F}_B^N(v,\ w)) = (0,0,0,0,0,0)$

Hence $G = (A,B)$ is an isolated BSVNGs

Proposition 3.2: The notion of isolated BSVNGs generalized the notion of isolated fuzzy graphs.

Proof: If the value of $\mathrm{I}_\mathrm{A}^\mathrm{P}(w) = \mathrm{F}_\mathrm{A}^\mathrm{P}(w) = \mathrm{T}_\mathrm{A}^\mathrm{n}(w) = \mathrm{I}_\mathrm{A}^\mathrm{n}(w) = \mathrm{F}_\mathrm{A}^\mathrm{n}(w) = 0$, then the notion of isolated BSVNGs is reduced to isolated fuzzy graphs.

Proposition 3.3: The notion of isolated BSVNGs generalized the notion of isolated SVNGs.

Proof: If the value of $\mathrm{T}_\mathrm{A}^\mathrm{n}(w) = \mathrm{I}_\mathrm{A}^\mathrm{n}(w) = \mathrm{F}_\mathrm{A}^\mathrm{n}(w) = 0\mathrm{I}_\mathrm{A}^\mathrm{n}(w)$, then the concept of isolated BSVNGs is reduced to isolated SVNGs.

4 Conclusion

In this article, we have extended the notion of isolated SVNGs to the notion of isolated BSVNGs. The notion of isolated BSVNGs generalized the isolated SVNGs.

References

1. Smarandache, F: Neutrosophy, Neutrosophic Probability, Sets and Logic, Proquest Information & Learning, Ann Arbor, Michigan, USA, 105p, (1998).
2. Atanassov, K.: Intuitionistic Fuzzy Sets. Fuzzy Sets and Systems, Vol.20, (1986) 87–96.
3. Zadeh, L.: Fuzzy sets. Information and Control, 8 (1965) 338–335.
4. Wang, H., Smarandache, F., Zhang Y., and Sunderraman, R.: Single Valued Neutrosophic Sets. Multispace and Multistructure 4 (2010) 410–413.
5. Deli, I., Ali, M., Smarandache, F.: Bipolar Neutrosophic Sets and Their Application Based on Multi-criteria Decision Making Problems, in: Advanced Mechatronic Systems (ICAMechS) (2015) 249–254.
6. http://fs.gallup.unm.edu/NSS/.
7. Smarandache, F.: Refined Literal Indeterminacy and the Multiplication Law of Sub – Indeterminacies. Neutrosophic Sets and Systems, Vol.9, (2015) 58–63.
8. Gani, A., and Shajitha, B.S.: Degree: Order and size in Intuitionistic Fuzzy Graphs. International Journal of Algorithms, Computing and Mathematics, 3(3) (2010).
9. Akram, M.: Bipolar Fuzzy Graphs. Information Science, https://doi.org/10.1016/j.ins.2011.07.037 (2011).
10. Bhattacharya, P.: Some Remarks on Fuzzy Graphs. Pattern Recognition Letters 6 (1987) 297–302.
11. Broumi, S., Talea, M., Bakali, A., Smarandache, F.: Single Valued Neutrosophic Graphs. Journal of New Theory, N 10, (2016) 86–101.
12. Broumi, S., Talea, M., Smarandache, F. and Bakali, A.: Single Valued Neutrosophic Graphs: Degree, Order and Size. IEEE International Conference on Fuzzy Systems (2016) 2444–2451.
13. Broumi, S., Bakali, A., Talea, M., Smarandache, F.: Isolated Single Valued Neutrosophic Graphs. Neutrosophic Sets and Systems, Vol.11 (2016) 74–78.
14. Broumi, S., Talea, M., Bakali, A., Smarandache, F.: On Bipolar Single Valued Neutrosphic Graphs. Journal of New Theory, N11 (2016) 84–102.
15. Broumi, S., Smarandache, F., Talea, M. and Bakali, A.: An Introduction to Bipolar Single Valued Neutrosophic Graph Theory. Applied Mechanics and Materials, vol 841, (2016) 184– 191.
16. Broumi, S., Bakali, A., Talea, M., Smarandache, F.: Generalized Bipolar Neutrosophic Graphs of Type 1, 20th International Conference on Information Fusion, Xi'an, China—July 10–13, (2017) 1714–1720.
17. Hassan, Ali., Malik, M.A., Broumi, S., Bakali, A., Talea, M., Smarandache, F.: Special types of bipolar single valued neutrosophic Graphs, Annals of Fuzzy Mathematics and Informatics, Vol.14, N 1, (2017) 55–73.

Enhance Link Prediction in Online Social Networks Using Similarity Metrics, Sampling, and Classification

Pham Minh Chuan[1(✉)], Cu Nguyen Giap[2], Le Hoang Son[3], Chintan Bhatt[4], and Tran Dinh Khang[5]

[1] Hung Yen University of Technology and Education, Hung Yen, Vietnam
chuanpm@utehy.edu.vn

[2] ThuongMai University, Hanoi, Vietnam
cunguyengiap@tmu.edu.vn

[3] VNU University of Science, Vietnam National University, Hanoi, Vietnam
sonlh@vnu.edu.vn

[4] Chatotar University of Science and Technology, Nadiad, Gujarat, India
chintanbhatt.ce@charusat.ac.in

[5] Hanoi University of Science and Technology, Hanoi, Vietnam
khangtd@soict.hust.edu.vn

Abstract. Link prediction in an online social network aims to determine new interactions among its members which are probably to arise in the near future. The previous researches dealt with the prediction task after calculating similarity scores between nodes in the link graph. New links are then predicted by implementing a supervised method from the scores. However, real-world applications often contain sparse and imbalanced data from the network, which may lead to difficulty in predicting new links. The selection of an appropriate classification method is indeed an important matter. Firstly, this paper proposes several extended metrics to calculate the similarity scores between nodes. Then, we design a new sampling method to make the training and testing data based on the data created by the extended metrics. Lastly, we assess some well-known classification methods namely J48, Weighted SVM, Gboost, Naïve Bayes, Random Forest, Logistics Regressive, and Xgboost in order to choose the best method and equivalent environments for the link prediction problem. A number of open directions to the problem are suggested further.

Keywords: Link prediction · Network topology · Online social networks
Similarity metrics · Topic modeling · Classification

1 Introduction

Link prediction (LP) in an *online social network* (OSN) aims to determine new interactions among its members which are probably to arise in near future [1]. Understanding the advancement of OSNs through LP would ensure steadiness and consistency of the system while diminishing potential dangers that may threaten further [2]. Consider an example of a co-authorship network like DBLP or e-print arXiv, where nodes

V. Bhateja et al. (eds.), *Information Systems Design and Intelligent Applications*, Advances in Intelligent Systems and Computing 672,
https://doi.org/10.1007/978-981-10-7512-4_81

represent authors and links refer to the cooperation between authors having joint papers in certain time. Herein, the LP problem has to figure out whether or not two authors can have a joint publication in a given year. There are a wide variety of application areas for link prediction, such as automatic Web hyperlink creation [3], Website hyperlink prediction [4], protein–protein interaction prediction [5], or to annotate the graph in bioinformatics [6]. The link prediction was also applied in other scientific domains like Bibliography and library, in which one can use link prediction for recording link age and deduplication [7].

Most preceding studies typically in [2, 4, 8] focused on computing similarity scores between pairs of nodes in the network based on traditional metrics and then predicting the appearance of new links based on the data generated by the scores.

Regarding the first task, there were some metrics proposed for the computation of similarity scores. Adamic and Adar [9] presented some similarity metrics between nodes such as AA and Root PageRank characterized by time intervals. Munasinghe and Ichise [10] proposed a time-score index named TS to find out the influence of the relationship between the timing of the engagement and strong bonds linked to the formation of the future. A couple of metrics was used to exploit the systems in this research namely AA, JC, and CN [11], PA [12], and the TS index. To predict a new link in the network, firstly they constructed a feature vector for each node pair (link), in which the components of the feature vector are the similarity measures of a node pair. Murata and Moriyasu [13] extended the metrics called WCN, WAA, and WPA from CN, AA, and PA, respectively, by using weights and similarity. Güneş et al. [1] also extended WJC from JC.

In this paper, we propose **four new metrics (in Sect.** 2) based on the combination of the measurement of network topology and content similarity (extended metrics). They are motivated by and generalized from the existing WCN, WAA, WPA [13], and WJC [1] by considering the content similarity into the network topology for better reflection of analogousness between nodes.

After the feature vector corresponding to each pair of nodes is calculated, supervised classification methods, e.g., J48 decision tree [14], are applied to predict links which is able to occur in the future. Tylenda et al. [2] extended the local probability model with factors over time in a new predicting technique based on a graph associated with the time information that appears in the process of social development. Soares et al. [15] proposed linking forecasting methods based on the concept of temporal events. The temporal events of a networks brings useful information about how connections are formed, and therefore it should be considered in the link prediction. The network was divided into subnets corresponding to a certain time period. Then the proximity score of each node pair were proposed based on two main events that change the status of the node pair on two consecutive periods and the temporal events observed in nodes' neighborhood. Other works of link prediction can be retrieved in [16–18].

Lots of efforts have been done to explore new prediction methods that could provide better performance. However, it has been noted that the entire system can be enhanced by new similarity metrics with equivalent sampling methods and a suitable supervised classification method. That is to say, new metrics which are more adaptable to the problem and data themselves than the existing methods are designed to create the secondary data from the network. Then, a new method of sampling is given to create the training and testing sets from the data. Herein, we continue to examine several well-known classification methods besides J48 such as Weighted SVM, Gboost, Naïve Bayes, Random Forest, Logistics Regressive, and XGboost in order to choose the most suitable algorithm. The entire process would suggest the best method and equivalent environments for the problem of link prediction.

The remainder of this paper is organized as follows: Sect. 2 introduces the new similarity metrics. Section 3 presents a new sampling method from the secondary data. Section 4 examines the most suitable classification method for the link prediction problem. Finally, some conclusions and further works are drawn in Sect. 5. In order to comprehend the method better, we illustrate the new method in the application of co-authorship network throughout the paper.

2 New Similarity Metrics for Link Prediction

Assume that we have a co-authorship network as in Table 1.

Table 1. Link prediction problem (bold values are needed to be predicted)

Node	Node	Year	Paper	Node	Node	Year	Paper
1	3	97	809182	1	2	99	908043
1	2	97	812108	2	7	99	903293
2	7	98	911089	3	4	99	905132
4	5	98	911231	3	7	99	905132
4	6	98	911051	4	5	99	904044
4	7	98	911421	4	7	99	905132
5	6	98	911132	5	7	99	904144
5	7	98	911132	6	7	99	908094
6	7	98	911132				
2	**6**	**00**	**92001**	**1**	**7**	**00**	**92102**

The new metrics are designed according to the view that if two authors have many papers sharing common semantic similarity, then they are likely to write together. The component $(|t_{uz} - t_{vz}|(1 - \cos in(p_{uz}, p_{vz})) + 1)$ is attached to the new metric to verify that if the content similarity of two papers (p_{uz}, p_{vz}) is high, then the total similarity of two nodes (u, v) is also high as a result. The following defines these new extended similarity metrics.

Definition 1. The Extended Weighted Common Neighbors (EWCN) is defined as follows:

$$SIM_{EWCN}(u, v) = \frac{1}{e^{1-\cos in(p_u^{av}, p_v^{av})}} \sum_{z \in \Gamma(u) \cap \Gamma(v)} \left(\frac{\omega(u, z) + \omega(v, z)}{2(|t_{uz} - t_{vz}| \cdot (1 - \cos in(p_{uz}, p_{vz})) + 1)} \right). \tag{1}$$

$$\cos in(p_{uz}, p_{vz}) = \frac{p_{uz} * p_{vz}}{\|p_{uz}\|.\|p_{vz}\|}, p_u^{av} = \frac{1}{|\Omega(u)|}\sum_{z\in\Omega(u)} p_{uz}. \tag{2}$$

where p_{uz} is a feature vector of the paper written by authors u and z. t_{uz} and t_{vz} are the latest interactive time of common neighbor to u and v, respectively.

Definition 2. The Extended Weighted Adamic–Adar (EWAA) is defined as follows:

$$SIM_{EWAA}(u,v) = \frac{1}{e^{1-\cos in(p_u^{av}, p_v^{av})}} \sum_{z\in\Gamma(u)\cap\Gamma(v)} \left(\frac{\omega(u,z)+\omega(v,z)}{2(|t_{uz}-t_{vz}|\cdot(1-\cos in(p_{uz},p_{vz}))+1)Log(\sum_{x\in\Gamma(z)}\omega(z,x))} \right). \tag{3}$$

Definition 3. The Extended Weighted Jaccard Coefficient (EWJC) is defined as follows:

$$SIM_{EWJC}(u,v) = \frac{1}{e^{1-\cos in(p_u^{av}, p_v^{av})}} \frac{1}{\sum_{x\in\Gamma(u)}\omega(u,x)+\sum_{y\in\Gamma(v)}\omega(v,y)} \sum_{z\in\Gamma(u)\cap\Gamma(v)} \frac{\omega(u,z)+\omega(v,z)}{2(|t_{uz}-t_{vz}|(1-\cos in(p_{uz},p_{vz}))+1)}. \tag{4}$$

Definition 4. The Extended Weighted Preferential Attachment (EWPA) is defined as follows:

$$SIM_{EWPA}(u,v) = \sum_{x\in\Gamma(u)}\omega(u,x) \times \sum_{y\in\Gamma(v)}\omega(v,y) \times \frac{1}{e^{1-\cos in(p_u^{av}, p_v^{av})}}. \tag{5}$$

$$p_u^{av} = \frac{1}{|\Omega(u)|}\sum_{z\in\Omega(u)} p_{uz}. \tag{6}$$

3 Sampling Method

After generating the secondary data expressed in the format as {(A pair of nodes), EWCN, EW AA, EWJC, EW PA, class (0|1)}, we divide data into the training and testing sets by the following sampling method. Note that each pair of nodes belongs to a certain time stamp in the graph. In the context of the co-authorship network, the original data were taken from e-print collected for 12 years, from 1994 to 2005. The data include publications of astrophysics (Astro-ph), condensed matter physics (Cond-mat), and high- energy physics (theory) (Hep-th). In a co-authorship network, an author is represented by a node. Two nodes having co-authorship are indicated by an edge labeled by the time this paper published. Particularly, each edge stores the publication year of the co-authored paper. In addition, we also used the information for each article namely the paper title and abstract to construct the feature vectors for each paper. We also removed articles written by only one author since they do not have any significant meaning in link prediction.

We constructed the training data based on the periods, in which each contains four consecutive years with the candidate set containing all non-connected pairs in the first three years (see Table 2). Then, we assigned each pair of nodes that is actually linked in the fourth year to the positive class, or to the negative class otherwise. The co-authorship networks created in this manner are very sparse: 0.035, 0.0224, and 0.0207% of the

Table 2. Statistics of period data

Dataset	Periods data	Nodes	Edges	Avg. degree	Pair nodes	Pos. labels
Hep-th	**D1**(94–97)	6867	12928	3.765	1640	46
	D2(95–98)	7164	14012	3.912	2238	45
	D3(96–99)	7481	14923	3.990	2155	48
	D4(97–00)	7843	15972	4.073	2394	54
	D5(98–01)	8222	16818	4.091	2933	52
	D6(99–02)	8491	17637	4.154	3171	60
	D7(00–03)	8823	18391	4.169	3231	58
	D8(01–04)	9141	18865	4.128	3154	62
	D9(02–05)	9447	19552	4.139	2997	65
Cond-mat	**D1**(94–97)	8617	22190	5.150	1739	41
	D2(95–98)	11175	31533	5.643	2850	44
	D3(96–99)	14005	45419	6.486	5313	103
	D4(97–00)	17120	61237	7.154	12557	166
	D5(98–01)	20965	85269	8.134	21367	318
	D6(99–02)	24422	105445	8.635	40228	428
	D7(00–03)	27795	126484	9.101	47817	670
	D8(01–04)	33027	159809	9.677	71924	1157
	D9(02–05)	37653	185985	9.879	96990	1118
Astro-ph	**D1**(94–97)	8051	16035	3.983	2083	34
	D2(95–98)	10657	23086	4.333	4417	80
	D3(96–99)	13200	30683	4.649	7084	110
	D4(97–00)	15663	38402	4.904	10352	180
	D5(98–01)	17847	45105	5.055	13368	205
	D6(99–02)	19636	50306	5.124	15672	251
	D7(00–03)	21393	55184	5.159	17165	293
	D8(01–04)	23098	59731	5.172	17629	324
	D9(02–05)	24634	63735	5.175	19655	372

Table 3. Statistics of training and testing data

Prediction year	Training data	Testing data
2002	D1–D5	D6
2003	D2–D6	D7
2004	D3–D7	D8
2005	D4–D8	D9

possible links appear on average in Hep-th, Cond-mat, and Astro-ph, respectively. Since the co-authorship networks are highly sparse, it is very important to decrease a number of candidate pairs in order to make computation feasible. Besides, it is imbalance between negative and positive classes in the training set so that we used five consecutive periods to construct the training set. In the other words, to predict the set of emerged links in 2002 (using the candidate set in D6 as the testing set), the training set is generated from D1 to D5 (see Table 3). To construct the feature vectors for papers, we concatenated their titles and abstracts. We removed stop words and used *tf-idf* to choose the top distinct words as the vocabulary [19] and used the LDA source code taken from http://chasen.org/~daiti-m/dist/lda/ (MATLAB version) to get the feature vectors.

4 Assessment of Classification Methods

4.1 Classification Methods

Classification methods are used to generate the model from training data in the format of {(A pair of nodes), EWCN, EWAA, EWJC, EWPA, class (0|1)}. There are many classification algorithms for the link prediction problem. In order to experiment well-known algorithms systematically and compare with the algorithms have been applied in link prediction by other researchers, we have studied the list of algorithms including decision tree (J48), Weighted SVM, Gboost, Naïve Bayes, Random Forest, Logistics Regressive, and XGboost which are considered as common solution in the link prediction problem. We adopted the implementation of J48, Random Forest, Naive Bayes, and Logistic Regressive from WEKA environment [20]; Gboost [21]; Weighted SVM [22] for unbalanced data, and XGboost from Python. All methods were experienced using computer that has Intel Core i5-3210 M CPU @ 2.50 GHz, 4 GB RAM, and Windows 7, 64-bit operating system. The area under curve (AUC) is selected as the evaluation criteria for the problem, in which good result implies the part below the curve is approximately 1. Otherwise, the part under the curve is around 0.5 which implies that the method performance is nearly random guessing.

4.2 Discussion

(a) **The comparison between extended and existing similarity metrics**

We have made several experiments to identify the best number of topics in each dataset. In the first two datasets (Hep-th and Astro-ph), the best practical number of topics is 30 while that of Cond-mat is 60. The following Figs. 1, 2, and 3 show the comparison of AUC values of all metrics (the extended in Sect. 2 and the existing namely WCN, WAA, and WPA [13] and WJC [1]) in three datasets.

It has been shown that the extended metrics are better than the traditional ones. Among all extended metrics, EWAA and EWPA achieved the best and worst results compared with the others. The AUC values of EWAA are 0.703, 0.741, and 0.734 corresponding with the Hep-th, Astro-ph, and Cond-mat datasets. The added content similarity component in EWAA does not improve WAA as much as other extension metrics do. The significant improvement in this case is the EWJC metric with the increasing percent being 7.4% on average, while that of EWAA is only 2.9%.

There is not much difference in results between the experimental datasets. For instance, the AUC values of (EWCN, EWAA, EWPA, WWJC) on Hep-th are (0.701, 0.703, 0.607, 0.647), respectively, while those on Astro-ph and Cond-mat are (0.739, 0.741, 0.653, 0.667) and (0.730, 0.734, 0.602, 0.673), respectively. This demonstrates the stability of metrics on various types of datasets. To the end, the new extended metrics have better AUC values than the existing ones.

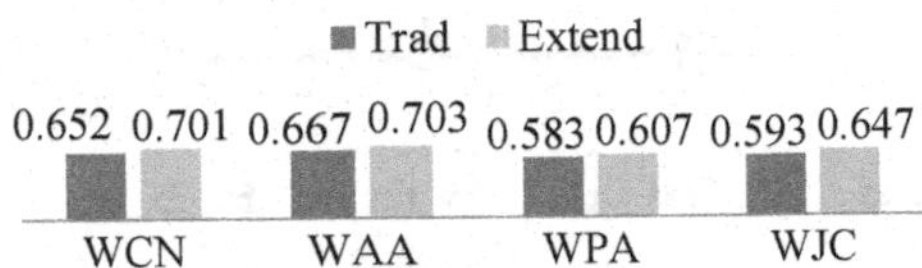

Fig. 1. Comparative results on Hep-th dataset

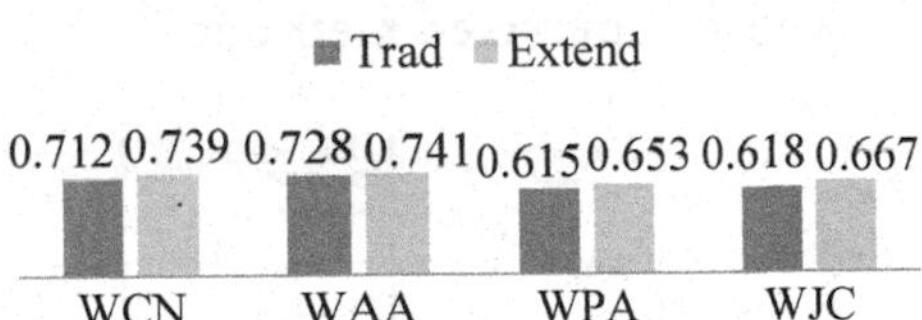

Fig. 2. Comparative results on Astro-ph dataset

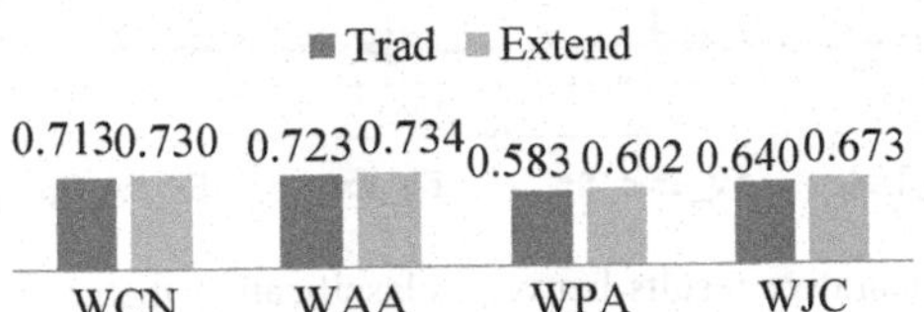

Fig. 3. Comparative results on Cond-mat dataset

(b) **The comparison between classification methods**

Again, we compare the performance of some classification algorithms on both the traditional (Fv_Trad, Fv_Trad_Sp) and the extended metrics (Fv_Ex, Fv_Ex_Sp) for non-sampled and sampled training tests (non-sampling only used one period just before that to train). The results by various datasets are shown in Figs. 4, 5, and 6. The experiments show that most algorithms have better performances when using the sampling method with 5 folds, however in few cases some are suitable with non-sampling dataset, such as Gboost and XGBoost.

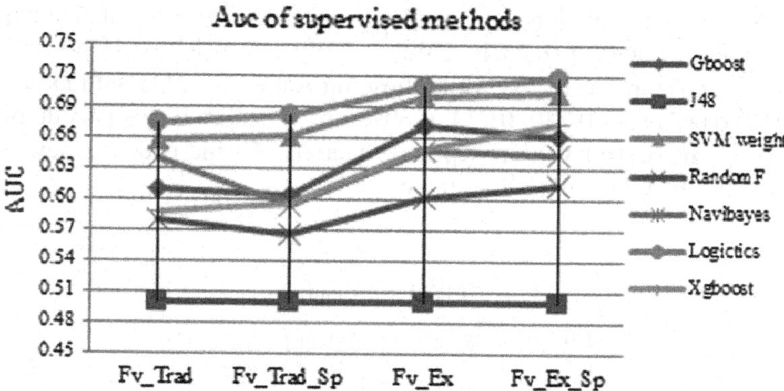

Fig. 4. Comparative results between classification methods on both the traditional (Fv_Trad, Fv_Trad_Sp) and the extended metrics (Fv_Ex, Fv_Ex_Sp) on Hep-th

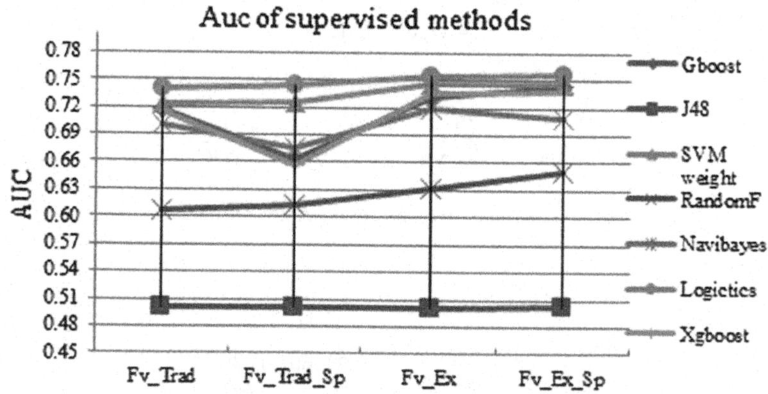

Fig. 5. Comparative results between classification methods on Astro-ph

Focusing on sampling datasets, it is clear that the performance of classification algorithms using the extended metrics is better than using the traditional ones in most cases. Figures 4, 5, and 6 demonstrate that (Fv_Trad_Sp) are always lower than (Fv_Ex_Sp) ones. From the Naïve Bayes or the J48 decision tree to Xgboost or Weighted SVM, the AUC values of the algorithms are improved when using the

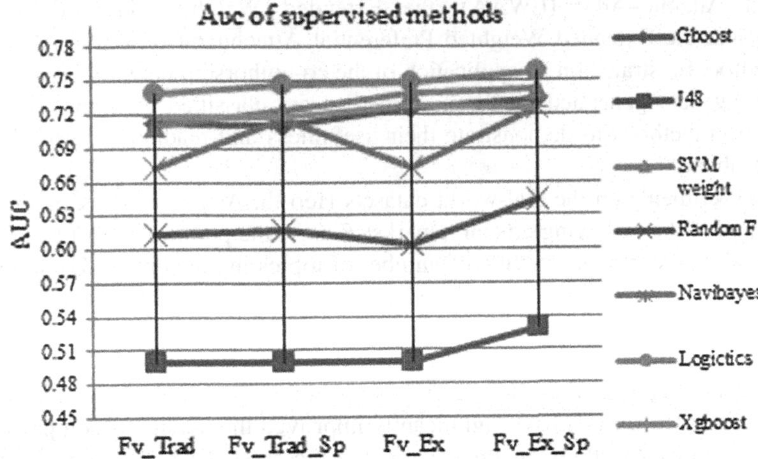

Fig. 6. Comparative results between classification methods on Cond-mat

extended metrics instead of the traditional ones. The best case of AUC witnesses the XGboost as the most efficient algorithm with the improvement percent of 13.22% on the Hep-th dataset. The improvement levels of all methods are not similar between datasets. The average AUC of all algorithms is increased about 7.31% in Hep-th dataset, 5.82% in Astro-ph dataset, and only 2.93% in Cond-mat dataset. This shows that even in a sparse dataset like Cond-mat, using the extended metrics in the classification would increase the performance.

In order to show the most suitable method for the link prediction problem, let us observe two methods: Logistics Regressive and Weighted SVM which have better performance than the other algorithms in all datasets. The AUC values of Logistics Regressive in Hep-th, Astro-ph, and Cond-mat are 0.717, 0.757, and 0.758 while those of Weighted SVM are 0.704, 0.746, and 0.743, respectively. Conversely, the J48 decision tree algorithm has the worst performance in any dataset with the average AUC value being around 0.5 in all cases. Moreover, this algorithm is not much improved by using the extended similarity metrics with the improvement percent as 0.5% excepting Cond-mat data with 5% improvement. On the other side, the algorithms which have the best improvement using the extended metrics are Gboost and XGboost. On average, their performances increased 8.31% and 8.95% respectively, while the others improved by 2.18% to 6.32%. Logistics Regressive algorithm has the best performance but its performance only enhances 2.82% due to adding content similarity.

5 Conclusions

In this paper, we investigated the link prediction in online social networks and introduced some new extended similarity metrics based on the content similarity consideration named as the Extended Weighted Common Neighbors (EWCN), the Extended

Weighted Adamic–Adar (EWAA), the Extended Weighted Jaccard Coefficient (EWJC), and the Extended Weighted Preferential Attachment (EWPA). A new sampling method, illustrated on an application of the co-authorship network, was designed. The training data generated by the new metrics are classified by some well-known classification methods to demonstrate their usefulness and readiness for the link prediction problem.

The experiments on the real-world datasets Hep-th, Astro-ph, and Cond-mat suggested some of the following crucial remarks: *Firstly*, the performance of the extended metrics is dependent on the setting of number of topics in the content. Using practical experiment with various numbers of topics in (10…100), the result showed that the optimal numbers of topics are different between datasets, particularly, 30 is the best for Hep-th and Astro-ph datasets and 60 is for Cond-mat. This suggests the number of topics that should be opted for other experiments in the same application. *Secondly*, the proposed extended metrics have significantly improved the quality as compared to the traditional metrics. Mostly often, the new metrics increases the AUC values from 2.9 to 7.4% on average compared with the traditional ones. *Lastly*, the experiments on some well-known classification algorithms namely the J48 decision tree, Weighted SVM, Gboost, Naive Bayes, Random Forest, Logistics Regressive, and XGboost specified that the extended metrics have boosted the performance of all supervised methods in any testing dataset. Among all, the Logistics Regressive is state of the art; however, Weighted SVM is a strong competitor. Logistics Regressive algorithm has highest AUC values around 0.717, 0.757, and 0.758 on Hep-th, Astro-ph, and Cond-mat datasets, respectively. XGboost is the method has the most improvement due to using new metrics with AUC value increasing to 8.92% on average.

From the conclusions found in this research, we will seek a new classification method to enhance the performance of the link prediction problem by an improvement of Logistics Regressive or Weighted SVM in further studies.

References

1. Güneş, İ., Gündüz-Öğüdücü, Ş., Çataltepe, Z.: Link prediction using time series of neighborhood-based node similarity scores. Data Mining and Knowledge Discovery 30(1) (2016) 147–180.
2. Tylenda, T., Angelova, R., Bedathur, S.: Towards time-aware link prediction in evolving social networks. Proceedings of the 3rd workshop on social network mining and analysis (2009) 1–10.
3. Adafre, S. F., Rijke, M.: Discovering missing links in Wikipedia. Proceedings of the Third ACM International Workshop on Link Discovery (2005) 90–97.
4. Zhu, J., Hong, J., Hughes G.: Using Markov models for web site link prediction. Proceedings of the Thirteenth ACM Conference on Hypertext and Hypermedia (2002) 169–170.
5. Airodi, E.M., Blei, D.M., Xing, E.P., Fienberg, S.E.: Mixed Membership stochastic block models for relational data, with applications to protein-protein interactions. Proceedings of International Biometric Society-ENAR Annual Meetings (2006) 1–34.

6. Freschi, V.: A Graph-based Semi-Supervised Algorithm for Protein Function Prediction from Interaction Maps. Learning and Intelligent Optimization. Lecture Notes in Computer Science, Vol. 5851. Springer-Verlag, Berlin Heidelberg New York (2009) 249–258.
7. Ahmed, E., Ipeirotis, P.G., Verykios, V.: Duplicate Record Detection: A Survey. IEEE Transactions on Knowledge and Data Engineering 19 (1) (2007) 1–16.
8. Soares, PRDS, Prudêncio, RBC.: Time series based link prediction. Proceedings of the 2012 International Joint Conference on Neural Networks (2012) 1–7.
9. Adamic, L.A., Adar, E.: Friends and neighbors on the web. Social networks 25(3) (2003) 211–230.
10. Munasinghe, L., Ichise, R.: Time aware index for link prediction in social networks. Data Warehousing and Knowledge Discovery. Springer Berlin Heidelberg New York (2011) 342–353.
11. Manning, C. D., Raghavan, P., Schütze, H.: Introduction to Information Retrieval. Cambridge University Press, UK. (2009).
12. Newman, M.E.: Clustering and preferential attachment in growing networks. Physical review E 64(2) (2001) 1–13.
13. Murata, T., Moriyasu, S.: Link prediction of social networks based on weighted proximity measures. Proceedings of the IEEE/WIC/ACM international conference on web intelligence (2007) 85–88.
14. Quinlan, J. R.: C4.5: programs for machine learning. Morgan Kaufmann, US (2014).
15. Soares, PR, Prudêncio, RB: Proximity measures for link prediction based on temporal events. Expert Systems with Applications 40(16) (2013) 6652–6660.
16. Papadimitriou, A., Symeonidis, P., Manolopoulos, Y.: Fast and accurate link prediction in social networking systems. Journal of Systems and Software 85(9) (2012) 2119–2132.
17. Valverde-Rebaza, J., Lopes, AA.: Exploiting behaviors of communities of twitter users for link prediction. Social Network Analysis and Mining 3(4) (2013) 1063–1074.
18. Zhu, YX., Lü, L., Zhang, QM., Zhou, T.: Uncovering missing links with cold ends. Physica A: Statistical Mechanics and its Applications 391(22) (2012) 5769–5778.
19. Blei, D., La, J.: Text mining: Theory and applications, chapter topic models. Taylor and Francis, London (2009).
20. Mark H., Eibe F., Geoffrey H., Bernhard P., Peter R., Ian H.W: The weka data mining software: an update. SIGKDD Explor. Newsl. 11 (2009) 10–18.
21. Becker, C., Rigamonti, R., Lepetit, V., Fua, P.: Supervised feature learning for curvilinear structure segmentation. Proceedings of the 16th International Conference on Medical Image Computing and Computer-Assisted Intervention (2013) 526–533.
22. Chang, C. C., Lin, C. J.: LIBSVM: a library for support vector machines. ACM Transactions on Intelligent Systems and Technology (TIST) 2(3) (2011) 27.

Fuzzy Equivalence on Standard and Rough Neutrosophic Sets and Applications to Clustering Analysis

Nguyen Xuan Thao[1(✉)], Le Hoang Son[2], Bui Cong Cuong[3], Mumtaz Ali[4], and Luong Hong Lan[5]

[1] Vietnam National University of Agriculture, Hanoi, Vietnam
nxthao@vnua.edu.vn
[2] VNU University of Science, Vietnam National University, Hanoi, Vietnam
sonlh@vnu.edu.vn
[3] Institute of Mathematics, Hanoi, Vietnam
bccuong@math.ac.vn
[4] University of Southern Queensland, Springfield Campus, 4300 Queensland, Australia
Mumtaz.Ali@usq.edu.au
[5] Thai Nguyen University of Education, Thai Nguyen, Vietnam
lanlhbk@gmail.com

Abstract. In this paper, we propose the concept of fuzzy equivalence on standard neutrosophic sets and rough standard neutrosophic sets. We also provide some formulas for fuzzy equivalence on standard neutrosophic sets and rough standard neutrosophic sets. We also apply these formulas for cluster analysis. Numerical examples are illustrated.

Keywords: Fuzzy equivalence · Neutrosophic set · Rough set · Rough neutrosophic set · Fuzzy clustering

1 Introduction

In 1998, Smarandache introduced neutrosophic set [1]; NS is the generalization of fuzzy set [2] and intuitionistic fuzzy set [3]. Over time, the subclass of the neutrosophic set [4–6] was proposed to capture more advantages in practical applications. In 2014, Bui Cong Cuong introduced the concept of the picture fuzzy set [7]. After that, Son gave the applications of the picture fuzzy set in clustering problems in [8–19]. It is to be noted that the picture fuzzy set was regarded as a standard neutrosophic set.

Rough set theory [20] is a useful mathematical tool for data mining, especially for redundant and uncertain data [21]. On the first time, rough set is established on equivalence relation. The set of equivalence classes of the universal set, obtained by an equivalence relation, is the basis for the construction of upper and lower approximation of the subset of the universal set. Recently, rough set has been developed into the fuzzy environment and obtained several interesting results [22, 23].

V. Bhateja et al. (eds.), *Information Systems Design and Intelligent Applications*, Advances in Intelligent Systems and Computing 672,
https://doi.org/10.1007/978-981-10-7512-4_82

It has been realized that the combination of the neutrosophic set and rough set achieved more uncertainty in the analysis of sophisticated events in real applications [24–26]. Bui Cong Cuong et al. [27] firstly introduced some results of the standard neutrosophic soft theory. Later, Nguyen Xuan Thao et al. [28, 29] proposed the rough picture fuzzy set and the rough standard neutrosophic set which are the results of approximation of the picture fuzzy set and standard neutrosophic set, respectively, with respect to a crisp approximation space. However, the previous researches have not defined fuzzy equivalence, a basic component in the standard neutrosophic set for the approximation and inference processes.

In this paper, we introduce the concept of fuzzy equivalence for the standard neutrosophic set and the rough standard neutrosophic set. Some examples of the fuzzy equivalence for those sets and application on clustering analysis are also given. The rest of the paper is organized as follows: The rough standard neutrosophic set and fuzzy equivalence are recalled in Sect. 3. Sections 3 and 4 propose the concept of fuzzy equivalence for two standard neutrosophic sets. In Sect. 5, we give an application of clustering and Sect. 6 draws the conclusion.

2 Preliminary

Definition 1 [27]. Let U be a universal set. A standard neutrosophic set (SNS) A on the U is $A = \{(u, \mu_A(u), \eta_A(u), \gamma_A(u))|u \in U\}$, where $\mu_A(u)$ is called the "degree of positive membership of u in A," $\eta_A(u)$ is called the "degree of indeterminate/neutral membership of u in A," and $\gamma_A(u)\gamma_A(u)$ is called the "degree of negative membership of u in A," where $\mu_A(u), \eta_A(u)\ \gamma_A(u) \in [0, 1]$ satisfy the following condition:

$$0 \le \mu_A(u) + \eta_A(u) + \gamma_A(u) \le 1, \forall u \in U.$$

The family of all standard neutrosophic sets in U is denoted by SNS(U).

Definition 2 [28, 29]. For a given $A \in SNS(U)$, the mappings $\overline{RP}, \underline{RP} : SNS(U) \to SNS(U)$, in which

$$\overline{RP}(A) = \{\left(u, \mu_{\overline{RP}(A)}(u), \eta_{\overline{RP}(A)}(u), \gamma_{\overline{RP}(A)}(u)\right)|u \in U\},$$
$$\underline{RP}(A) = \{(u, \mu_{\underline{RP}(A)}(u), \eta_{\underline{RP}(A)}(u), \gamma_{\underline{RP}(A)}(u)|u \in U\},$$

where

$$\mu_{\overline{RP}(A)}(u) = \vee_{v \in R_S(u)} \mu_A(v), \eta_{\overline{RP}(A)}(u) = \wedge_{v \in R_S(u)} \eta_A(v),$$
$$\gamma_{\overline{RP}(A)}(u) = \wedge_{v \in R_S(u)} \gamma_A(v)$$

and

$$\mu_{\underline{RP}(A)}(u) = \wedge_{v \in R_S(u)} \mu_A(u),\ \eta_{\underline{RP}(A)}(u) = \wedge_{v \in R_S(u)} \eta_A(v),$$
$$\gamma_{\underline{RP}(A)}(u) = \vee_{v \in R_S(u)} \gamma_A(v)$$

are called to the upper and lower standard neutrosophic approximation operators, respectively, and the pair $RP(A) = (\underline{RP}(A), \overline{RP}(A))$ is referred as the rough standard neutrosophic set of A w.r.t the approximation space (U, R), or A is called roughly defined on the approximation space (U, R). The collection of all rough standard neutrosophic sets defined on the approximation space (U, R) is denoted by $RSNS(U)$.

Definition 3 [21] A mapping $e : [0,1]^2 \rightarrow [0,1]$ is a fuzzy equivalence if it satisfies the following conditions:

(e1) $e(a,b) = e(b,a), \forall a, b$ in $[0,1]$,
(e2) $e(1,0) = 0$,
(e3) $e(b,b) = 1, \forall b$ in $[0,1]$,
(e4) If $a \leq a' \leq b' \leq b$, then $e(a,b) \leq e(a', b')$.

Note that (e4) satisfies iff $e(a,c) \leq \min\{e(a,b), e(b,c)\}$, for all $a \leq b \leq c$ and $a, b, c \in [0,1]$.

3 Fuzzy Equivalence on Standard Neutrosophic Set

Definition 4 A mapping $E : SNS(U) \times SNS(U) \rightarrow [0,1]$ is a fuzzy equivalence if it satisfies the following conditions:

(E1) $E(A,B) = E(B,A)$ for all $A, B \in SNS(U)$,
(E2) $E(A,B) = 0$ iff $A = 1_U$ and $B = 0_U$,
(E3) $E(A,A) = 1, \forall A \in SNS(U)$,
(E4) $E(A,C) \leq \min\{E(A,B), E(B,C)\}$ for all $\forall A, B, C \in SNS(U)$ satisfy $A \subset B \subset C$.

Example 1 Let $U = \{x_1, x_2, \ldots, x_n\}$ be a universal set and $A, B \in SNS(U)$. A mapping $E : SNS(U) \times SNS(U) \rightarrow [0,1]$, where

$$E(A,B) = \begin{cases} \frac{1}{n}\sum_{i=1}^{n} \frac{\min\{\mu_A(x_i),\mu_B(x_i)\} + \min\{\eta_A(x_i),\eta_B(x_i)\} + (1-\max\{\gamma_A(x_i),\gamma_B(x_i)\})}{2} & \textit{iff } A \neq B \\ 1 & \textit{iff } A = B \end{cases}$$

is a fuzzy equivalence of A and B.

Indeed, conditions (E1), (E2), (E3), and (E4) are obvious.

Theorem 1 Let $U = \{x_1, x_2, \ldots, x_n\}$ be a universal set and $A, B \in SNS(U)$. Let a mapping $E : PFS(U) \times PFS(U) \to [0, 1]$ is defined by

$$E(A,B) = \begin{cases} \frac{1}{n}\sum_{i=1}^{n} \frac{t_1(\mu_A(x_i),\mu_B(x_i)) + t_2(\eta_A(x_i),\eta_B(x_i)) + (1 - S(\gamma_A(x_i),\gamma_B(x_i)))}{2} & \textit{iff } A \neq B \\ 1 & \textit{iff } A = B \end{cases}$$

where t_1, t_2 are t-norm on [0, 1] and s is a t-conorm on [0, 1]; then, $E(A, B)$ is a fuzzy equivalence of A and B.

4 Fuzzy Equivalence on Rough Standard Neutrosophic Set

Here, we propose a fuzzy equivalence of the rough standard neutrosophic sets. Let $U = \{x_1, x_2, \ldots, x_n\}$, $A, B \in SNS(U)$, and $RP(A) = \left(\underline{RP}A, \overline{RP}A\right)$, $RP(B) = \left(\underline{RP}B, \overline{RP}B\right)$. For all $x_i \in U$, denote

$$\mu_{EA}(x_i) = \left|\mu_{\overline{RP}A}(x_i) - \mu_{\underline{RP}A}(x_i)\right|;$$

$$\eta_{EA}(x_i) = \left|\eta_{\overline{RP}A}(x_i) - \eta_{\underline{RP}A}(x_i)\right|;$$

$$\gamma_{EA}(x_i) = \left|\gamma_{\overline{RP}A}(x_i) - \gamma_{\underline{RP}A}(x_i)\right|;$$

$$\mu_{AE}(x_i) = \frac{\left|\mu_{\overline{RP}A}(x_i) + \mu_{\underline{RP}A}(x_i)\right|}{2};$$

$$\eta_{AE}(x_i) = \frac{\left|\eta_{\overline{RP}A}(x_i) + \eta_{\underline{RP}A}(x_i)\right|}{2};$$

$$\gamma_{AE}(x_i) = \frac{\left|\gamma_{\overline{RP}A}(x_i) + \gamma_{\underline{RP}A}(x_i)\right|}{2};$$

$$\mu_E = \frac{1}{n}\sum_{i=1}^{n}\left\{\begin{array}{l} 1 - \dfrac{\left|\mu_{\overline{RP}A}(x_i) - \mu_{\overline{RP}B}(x_i)\right| + \left|\mu_{\underline{RP}A}(x_i) - \mu_{\underline{RP}B}(x_i)\right| + \left|\mu_{EA}(x_i) - \mu_{EB}(x_i)\right|}{4} \\ - \dfrac{\left|\mu_{AE}(x_i) - \mu_{BE}(x_i)\right|}{2} \end{array}\right\};$$

$$\eta_E = \frac{1}{n}\sum_{i=1}^{n}\left\{\begin{array}{l} 1 - \dfrac{\left|\eta_{\overline{RP}A}(x_i) - \eta_{\overline{RP}B}(x_i)\right| + \left|\eta_{\underline{RP}A}(x_i) - \eta_{\underline{RP}B}(x_i)\right| + \left|\eta_{EA}(x_i) - \eta_{EB}(x_i)\right|}{4} \\ - \dfrac{\left|\eta_{AE}(x_i) - \eta_{BE}(x_i)\right|}{2} \end{array}\right\};$$

$$\gamma_E = \frac{1}{n}\sum_{i=1}^{n}\left\{\begin{array}{l} 1 - \dfrac{\left|\gamma_{\overline{RP}A}(x_i) - \gamma_{\overline{RP}B}(x_i)\right| + \left|\gamma_{\underline{RP}A}(x_i) - \gamma_{\underline{RP}B}(x_i)\right| + \left|\gamma_{EA}(x_i) - \gamma_{EB}(x_i)\right|}{4} \\ - \dfrac{\left|\gamma_{AE}(x_i) - \gamma_{BE}(x_i)\right|}{2} \end{array}\right\}.$$

Theorem 2 The mapping $E : SNS(U) \times SNS(U) \to [0, 1]$ is defined by

$$E(A,B) = \frac{\mu_E + \eta_E + \gamma_E}{3} \text{ is a fuzzy equivalence.}$$

Proof We verify the conditions for $E(A, B)$

(E1) is obvious.
(E2) $A = 1_U; B = 0_U$. Then,

$$\begin{aligned}
&\mu_{EA}(x_i) = 0, \eta_{EA}(x_i) = 0, \gamma_{EA}(x_i) = 0;\\
&\mu_{AE}(x_i) = 1, \eta_{AE}(x_i) = 0, \eta_{AE}(x_i) = 0;\\
&\mu_{EB}(x_i) = 0, \eta_{EB}(x_i) = 0, \gamma_{EB}(x_i) = 0;\\
&\mu_{BE}(x_i) = 0, \ \eta_{BE}(x_i) = 0, \ \eta_{BE}(x_i) = 1;
\end{aligned}$$

So that $\mu_E = \eta_E = \gamma_E = 0$ and $E(A,B) = \frac{\mu_E + \eta_E + \gamma_E}{3} = 0$
(E3) is obvious.
(E4) Note that, if $0 \le a \le b \le c \le 1$, then $|a-c| \ge |a-b|$ and $1 \ge |a-c| \ge |c-b| \ge 0$ so that $0 \le 1-|a-c| \le 1-|a-b| \le 1$ and $0 \le 1-|a-c| \le 1-|c-b| \le 1$. Because $A \subseteq B \subseteq C$, then $\underline{RP}A \subseteq \underline{RP}B$, $\overline{RP}A \subseteq \overline{RP}B$, and $\underline{RP}B \subseteq \underline{RP}C$, $\overline{RP}B \subseteq \overline{RP}C$. Hence, $E(A,C) \le \min\{E(A,C), E(B,C)\}$. □
Now, let $U = \{x_1, x_2, \ldots, x_n\}$, $A, B \in SNS(U)$, and $RP(A) = (\underline{RP}A, \overline{RP}A)$, $RP(B) = (\underline{RP}B, \overline{RP}B)$. For all $x_i \in U$, denote

$$\bar{E}(A,B)(x_i) = |\mu_{\overline{RP}A}(x_i) - \mu_{\overline{RP}A}(x_i)| + |\eta_{\overline{RP}A}(x_i) - |\eta_{\overline{RP}B}(x_i)| + |\gamma_{\overline{RP}A}(x_i) - \gamma_{\overline{RP}B}(x_i)|,$$
$$\underline{E}(A,B)(x_i) = \left|\mu_{\underline{RP}A}(x_i) - \mu_{\underline{RP}B}(x_i)\right| + \left|\eta_{\underline{RP}A}(x_i) - \eta_{\underline{RP}B}(x_i)\right| + \left|\gamma_{\underline{RP}A}(x_i) - \gamma_{\underline{RP}B}(x_i)\right|.$$

Theorem 3 The mapping $E : SNS(U) \times SNS(U) \to [0, 1]$ is defined by

$$E(A,B) = \frac{1}{n}\sum_{i=1}^{n}\left[1 - \frac{\bar{E}(A,B)(x_i) + \underline{E}(A,B)(x_i)}{2}\right]$$

is a fuzzy equivalence.

Proof Similar to proof of Theorem 2.

5 An Application to Clustering Analysis

Example 2 Suppose there are three types of products $D = \{D_1, D_2 D_3\}$ and ten regular customers $U = \{u_1, u_2, \ldots, u_{10}\}$. R is an equivalence and $U/R = \{X_1 = \{u_1, u_3, u_9\};$

$X_2 = \{u_2, u_7, u_{10}\}; X_3 = \{u_4\}; X_4 = \{u_5, u_8\}, X_5 = \{u_{10}\}\}$. This division can be based on age or income. We consider that each customer evaluates the product by the linguistic labels {Good, Not-Rated, Not-Good}. Thus, each customer is a neutrosophic set on the products set (Table 1). Now, we can look at similar levels of customer groups in order to strategically sell products. Therefore, one can consider the sets of equivalence classes of customers for the three product categories above. In Table 2, we obtain a rough neutrosophic information system, in which for each X_i, the upper line is $\overline{RP}X_i$ and the lower line is $\underline{RP}\, X_i (i = 1, .., 5)$.

Table 1. A neutrosophic information system

U	D_1	D_2	D_3
u_1	(0.2, 0.3, 0.5)	(0.15, 0.6, 0.2)	(0.4, 0.05, 0.5)
u_2	(0.3, 0.1, 0.5)	(0.3, 0.3, 0.3)	(0.35, 0.1, 0.4)
u_3	(0.6, 0, 0.4)	(0.3, 0.05, 0.6)	(0.1, 0.45, 0.4)
u_4	(0.15, 0.1, 0.7)	(0.1, 0.05, 0.8)	(0.2, 0.4, 0.3)
u_5	(0.05, 0,2, 0.7)	(0.2, 0.4, 0.3)	(0.05, 0.4, 0.5)
u_6	(0.1, 0.3, 0.5)	(0.2, 0.3, 0.4)	(1, 0, 0)
u_7	(0.25, 0.3, 0.4)	(1, 0, 0)	(0.3, 0.3, 0.4)
u_8	(0.1, 0.6, 0.2)	(0.25, 0.3, 0.4)	(0.4, 0, 0.6)
u_9	(0.45, 0,1, 0.45)	(0.25, 0.4, 0.3)	(0.2, 0.5, 0.3)
u_{10}	(0.05, 0.05, 0.9)	(0.4, 0.2, 0.3)	(0.05, 0.7, 0.2)

Table 2. A rough neutrosophic information system

	D_1	D_2	D_3
X_1	(0.6, 0, 0.4)	(0.3, 0.05, 0.2)	(0.4,0.05,0.3)
	(0.2, 0, 0.5)	(0.15, 0.05, 0.6)	(0.1,0.05,0.5)
X_2	(0.3, 0.05, 0.4)	(1, 0, 0)	(0.35, 0.1, 0.2)
	(0.05, 0.05, 0.9)	(0.3, 0.1, 0.3)	(0.05, 0.1, 0.4)
X_3	(0.15, 0.1, 0.7)	(0.1, 0.05, 0.8)	(0.2, 0.4, 0.3)
	(0.15, 0.1, 0.7)	(0.1, 0.05, 0.8)	(0.2, 0.4, 0.3)
X_4	(0.1, 0.2, 0.2)	(0.25, 0.3, 0.3)	(0.4, 0, 0.5)
	(0.05, 0.2, 0.7)	(0.2, 0.3, 0.4)	(0.05, 0, 0.6)
X_5	(0.1, 0.3, 0.5)	(0.2, 0.3, 0.4)	(1, 0, 0)
	(0.1, 0.3, 0.5)	(0.2, 0.3, 0.4)	(1, 0, 0)

We calculate the similarity relations on $\{X_1, X_2, X_3, X_4, X_5\}$ (based on Theorem 2) as follows:

$$R_1 = \begin{bmatrix} 1.0000 & & & & \\ 0.8111 & 1.0000 & & & \\ 0.7708 & 0.7208 & 1.0000 & & \\ 0.8278 & 0.7750 & 0.7458 & 1.0000 & \\ 0.7069 & 0.6347 & 0.7000 & 0.7819 & 1.0000 \end{bmatrix}$$

Use the maximum tree method for fuzzy clustering analysis. Firstly, the Kruskal method is used to draw the largest tree:

$$3 \underset{0.7458}{\rightarrow} 4 \underset{0.775}{\rightarrow} 2 \underset{0.7819}{\rightarrow} 1 \underset{0.8278}{\rightarrow} 5$$

The tree implies that for $\alpha \in [0, 1]$, we can classify the $U/R = \{X_1, X_2, X_3, X_4, X_5\}$ by fuzzy equivalence $E(X_i, X_j) \geq \alpha$, where $X_i, X_j \in U/R$ as follows:

+ If $0 \leq \alpha \leq 0.7458$ then it has a cluster $\{X_1, X_2, X_3, X_4, X_5\}$.
+ If $0.7458 < \alpha \leq 0.775$ then we have two clusters $\{X_3\}, \{X_1, X_2, X_4, X_5\}$.
+ If $0.775 < \alpha \leq 0.7819$ then we have three clusters $\{X_3\}, \{X_4\}, \{X_1, X_2, X_5\}$.
+ If $0.7819 < \alpha \leq 0.8278$ then we have four clusters $\{X_3\}, \{X_4\}, \{X_2\}, \{X_1, X_5\}$.
+ If $0.8278 < \alpha \leq 1$ then we have five clusters $\{X_3\}, \{X_4\}, \{X_2\}, \{X_1\}, \{X_5\}$.

Now, we calculate the similarity relations on $\{X_1, X_2, X_3, X_4, X_5\}$ (based on Theorem 3) as follows:

$$R_2 = \begin{bmatrix} 1.0000 & & & & \\ 0.8792 & 1.0000 & & & \\ 0.8542 & 0.8167 & 1.0000 & & \\ 0.8813 & 0.8563 & 0.85 & 1.0000 & \\ 0.85 & 0.7667 & 0.775 & 0.8521 & 1.0000 \end{bmatrix}$$

Use the maximum tree method for fuzzy clustering analysis. Firstly, the Kruskal method is used to draw the largest tree:

$$3 \underset{0.85}{\rightarrow} 4 \underset{0.8521}{\rightarrow} 2 \underset{0.8563}{\rightarrow} 1 \underset{0.8813}{\rightarrow} 5$$

The tree implies that for $\alpha \in [0, 1]$, we can classify the $U/R = \{X_1, X_2, X_3, X_4, X_5\}$ by fuzzy equivalence $E(X_i, X_j) \geq \alpha,$, where $X_i, X_j \in U/R$ as follows:

+ If $0 \leq \alpha \leq 0.85$ then it has a cluster $\{X_1, X_2, X_3, X_4, X_5\}$.
+ If $0.85 < \alpha \leq 0.8521$ then we have two clusters $\{X_3\}, \{X_1, X_2, X_4, X_5\}$.
+ If $0.8521 < \alpha \leq 0.8563$ then we have three clusters $\{X_3\}, \{X_4\}, \{X_1, X_2, X_5\}$.
+ If $0.8563 < \alpha \leq 0.8813$ then we have four clusters $\{X_3\}, \{X_4\}, \{X_2\}, \{X_1, X_5\}$.
+ If $0.8813 < \alpha \leq 1 < \alpha \leq 1$ then we have five clusters $\{X_3\}, \{X_4\}, \{X_2\}, \{X_1\}, \{X_5\}$.

The clustering analysis on the rough standard neutrosophic set is analogously done. We find that the clustering result by Theorems 2 and 3 is giving the same clustering results. But theoretically, computation using Theorem 3 is simpler than Theorem 2.

6 Conclusions

We have introduced the preliminary results of the fuzzy equivalences on the standard neutrosophic set and the rough standard neutrosophic set. Using these definitions, we can perform clustering analysis on the datasets of neutrosophic sets.

Further studies regarding this research can be expanded of fuzzy equivalence of topological spaces and metrics. With that, we can also build fuzzy equivalent matrix models for clustering problems for real applications.

Acknowledgements. This research is funded by Vietnam National Foundation for Science and Technology Development (NAFOSTED) under grant number 102.01-2017.02.

References

1. Smarandache, F.: Neutrosophy. Neutrosophic Probability, Set, and Logic, ProQuest Information & Learning, Ann Arbor, Michigan, USA, 105 p., 1998; http://fs.gallup.unm.edu/eBook-neutrosophics6.pdf(last edition online).
2. Zadeh, L. A.: Fuzzy Sets. Information and Control 8(3) (1965) 338–353.
3. Atanassov, K.: Intuitionistic Fuzzy Sets. Fuzzy set and systems 20 (1986) 87–96.
4. Wang, H., Smarandache, F., Zhang, Y.Q. et al: Interval NeutrosophicSets and Logic: Theory and Applications in Computing. Hexis, Phoenix, AZ (2005).
5. Wang, H.,Smarandache, F., Zhang, Y.Q.,et al., Single Valued NeutrosophicSets. Multispace and Multistructure 4 (2010) 410–413.
6. Ye, J.: A Multi criteria Decision-Making Method Using Aggregation Operators for Simplified Neutrosophic Sets. Journal of Intelligent & Fuzzy Systems 26 (2014) 2459–2466.
7. Cuong, B.C.: Picture Fuzzy Sets. Journal of Computer Science and Cybernetics 30(4) (2014) 409–420.
8. Cuong, B.C., Son, L.H., Chau, H.T.M.: Some Context Fuzzy Clustering Methods for Classification Problems. Proceedings of the 1st International Symposium on Information and Communication Technology (2010) 34–40.
9. Son, L.H., Thong, P.H.: Some Novel Hybrid Forecast Methods Based On Picture Fuzzy Clustering for Weather Nowcasting from Satellite Image Sequences. Applied Intelligence 46 (1) (2017) 1–15.
10. Son, L.H., Tuan, T.M.: A cooperative semi-supervised fuzzy clustering framework for dental X-ray image segmentation. Expert Systems With Applications 46 (2016) 380–393.
11. Son, L.H., Viet, P.V., Hai, P.V.: Picture Inference System: A New Fuzzy Inference System on Picture Fuzzy Set. Applied Intelligence (2017) https://doi.org/10.1007/s10489-016-0856-1.
12. Son, L.H.: A Novel Kernel Fuzzy Clustering Algorithm for Geo-Demographic Analysis. Information Sciences 317 (2015) 202–223.
13. Son, L.H.: Generalized Picture Distance Measure and Applications to Picture Fuzzy Clustering. Applied Soft Computing 46 (2016) 284–295.

14. Son, L.H.: Measuring Analogousness in Picture Fuzzy Sets: From Picture Distance Measures to Picture Association Measures. Fuzzy Optimization and Decision Making (2017) https://doi.org/10.1007/s10700-016-9249-5.
15. Son, L.H.: DPFCM: A novel distributed picture fuzzy clustering method on picture fuzzy sets. Expert systems with applications 42 (2015) 51–66.
16. Thong, P.H., Son, L.H., Fujita, H.: Interpolative Picture Fuzzy Rules: A Novel Forecast Method for Weather Nowcasting. Proceeding of the 2016 IEEE International Conference on Fuzzy Systems (2016) 86–93.
17. Thong, P.H., Son, L.H.: A Novel Automatic Picture Fuzzy Clustering Method Based On Particle Swarm Optimization and Picture Composite Cardinality. Knowledge-Based Systems 109 (2016) 48–60.
18. Thong, P.H., Son, L.H.: Picture Fuzzy Clustering for Complex Data. Engineering Applications of Artificial Intelligence 56 (2016) 121–130.
19. Thong, P.H., Son, L.H.: Picture Fuzzy Clustering: A New Computational Intelligence Method. Soft Computing 20(9) (2016) 3544–3562.
20. Pawlak, Z.: Rough Sets. International Journal of Computer and Information Sciences 11 (5) (1982) 341–356.
21. Fodor, J., Yager, R. R.: Fuzzy Set Theoretic Operations and Quantifers. Fundermentals of Fuzzy Sets. Klwuer (2000).
22. Dubois, D., Prade, H.: Rough Fuzzy Sets and Fuzzy Rough Sets. International Journal of General Systems 17 (1990) 191–209.
23. Yao, Y.Y: Combination of Rough and Fuzzy Sets Based on $\alpha-$ level sets. Rough sets and Data mining: analysis for imprecise data. Kluwer Academic Publisher, Boston (1997) 301–321.
24. Broumi, S. and Smarandache, F.: Rough neutrosophic sets. Italian Journal of Pure and Applied Mathematics, N.32, (2014) 493–502.
25. Broumi, S. and Smarandache, F.: Lower and upper soft interval valued neutrosophic rough approximations of an IVNSS-relation, Sisom& Acoustics, (2014) 8 pages.
26. Broumi, S. and Smarandache, F.: Interval–Valued Neutrosophic Soft Rough Set, International Journal of Computational Mathematics. Volume 2015 (2015), Article ID 232919, 13 pages http://dx.doi.org/10.1155/2015/232919.
27. Cuong, B. C., Phong, P. H. and Smarandache, F.: Standard Neutrosophic Soft Theory: Some First Results. Neutrosophic Sets and Systems 12 (2016) 80–91.
28. Thao, N. X., Dinh, N. V.: Rough Picture Fuzzy Set and Picture Fuzzy Topologies. Journal of Science computer and Cybernetics 31 (3) (2015) 245–254.
29. Thao, N.X., Cuong. B. C., Smarandache, F.: Rough Standard Neutrosophic Sets: An Application on Standard Neutrosophic Information Systems. International Conference on Communication, Management and Information Technology, in press.

One Solution for Proving Convergence of Picture Fuzzy Clustering Method

Pham Huy Thong[1(✉)], Tong Anh Tuan[2], Nguyen Thi Hong Minh[3], and Le Hoang Son[1]

[1] VNU University of Science, Vietnam National University, Hanoi, Vietnam
{thongph, sonlh}@vnu.edu.vn

[2] The People's Police University of Technology and Logistics, Bac Ninh, Vietnam
tuanqb92@gmail.com

[3] VNU - School of Interdisciplinary Studies, Vietnam National University, Hanoi, Vietnam
minhnth@vnu.edu.vn

Abstract. Fuzzy clustering on the picture fuzzy set is widely applied to practical problems. However, the assessment on its convergence has not been considered yet. In this paper, we investigate the theoretical analysis of FC-PFS especially the convergence of the algorithm. Future directions are also discussed.

Keywords: Convergence · Picture fuzzy set · Picture fuzzy clustering

1 Introduction

Fuzzy clustering on picture fuzzy set (FC-PFS) was firstly introduced in [1] to overcome the limitations of fuzzy c-means (FCM) [2]. This algorithm was directly extended from FCM based on the theory of the picture fuzzy set [3], which is a generalization of the fuzzy set [4]. Recently, there have been many researches on extensions of FC-PFS especially in [5, 6] to cope with different situations such as dealing with complex data and automatically determination of the number of clusters. Moreover, FC-PFS has been applying to many practical problems such as in stocks prediction [7], weather nowcasting via satellite sequent images using picture fuzzy rules interpolation [8, 9].

However, in order to evaluate the algorithm more comprehensively, it is necessary to investigate the convergence of this algorithm. Being mentioned in the literature that many attempts of proving the convergence for FCM such as Hall and Goldgof [10], Wu et al. [11], and Sadi-Nezhad et al. [12] were exited, there are not many papers proving the convergence of the fuzzy clustering on advance fuzzy set, especially picture fuzzy set. Proving convergence of picture fuzzy clustering algorithm plays an important role in constructing theoretical basic for applying this in many other fields.

In this paper, we investigate theoretical analysis of FC-PFS especially the convergence of the algorithm. The rest of the paper is organized into three sections. In Sect. 2, we recall the standard picture fuzzy set and picture fuzzy clustering. Section 3 proposes the proving convergence, and Sect. 4 draws the conclusions.

V. Bhateja et al. (eds.), *Information Systems Design and Intelligent Applications*, Advances in Intelligent Systems and Computing 672,
https://doi.org/10.1007/978-981-10-7512-4_83

2 Preliminary

Definition 1 [3] Let U be a universal set. A picture fuzzy set (PFS) A on the universe U is an object of the form $A = \{(x, \mu_A(x), \eta_A(x), \gamma_A(x)) | x \in U\}$ where $\mu_A(x) \in [0, 1]$ is called the "degree of positive membership of x in A," $\eta_A(x)(\in [0, 1])$ is called the "degree of neutral membership of x in A," and $\gamma_A(x) \in$ is called the "degree of negative membership of x in A" where μ_A, η_A and γ_A satisfy the following condition:

$$\mu_A(x) + \eta_A(x) + \gamma_A(x) \le 1, \forall x \in U. \tag{1}$$

Definition 2 [1] Suppose a dataset X includes N data points in d dimensions. Denotes that $\mu_{kj} = \mu_{kj}(x), \eta_{kj} = \eta_{kj}(x)$, $\xi_{kj} = \xi_{kj}(x)$, $x \in N, 1 \le j \le C, 1 \le k \le N$. C is a number of clusters. V_j is the center of cluster j, $1 \le j \le C$. m is a fuzzier number, $\propto$ is exponent coefficient used to control the refusal degree in PFS sets. The FC-PFS minimizes the objective function below:

$$J = \sum_{k=1}^{N} \sum_{j=1}^{C} (\mu_{kj}(2 - \xi_{kj}))^m \left\| X_k - V_j^2 \right\| + \sum_{k=1}^{N} \sum_{j=1}^{C} \eta_{kj}(\ln \eta_{kj} + \xi_{kj}) \to \min, \tag{2}$$

where

$$\xi_{kj} = 1 - (\mu_{kj} + \eta_{kj} + \gamma_{kj}), \tag{3}$$

with constraints:

$$\mu_{kj} + \eta_{kj} + \xi_{kj} \le 1, \tag{4}$$

$$\mu_{kj} \in [0, 1], \eta_{kj} \in [0, 1], \xi_{kj} \in [0, 1], \tag{5}$$

$$\sum_{j=1}^{C} (\mu_{kj}(2 - \xi_{kj})) = 1, \tag{6}$$

$$\sum_{j=1}^{C} \left(\eta_{kj} + \frac{\xi_{kj}}{C} \right) = 1. \tag{7}$$

Solving the objective function, we have:

$$\mu_{kj} = \frac{1}{\sum_{i=1}^{C} (2 - \xi_{ki}) \left(\frac{\|X_k - V_j\|}{\|X_k - V_i\|} \right)^{\frac{2}{m-1}}}, (1 \le j \le C, 1 \le k \le N), \tag{8}$$

$$\eta_{kj} = \frac{e^{-\xi_{kj}}}{\sum_{i=1}^{C} e^{-\xi_{ki}}} \left(1 - \frac{1}{C} \sum_{i=1}^{C} \xi_{ki} \right), (1 \le j \le C, 1 \le k \le N), \tag{9}$$

$$\xi_{kj} = 1 - (\mu_{kj} + \eta_{kj}) - (1 - (\mu_{kj} + \eta_{kj})^{\propto})^{\frac{1}{\propto}}, \ (1 \le j \le C, 1 \le k \le N), \tag{10}$$

$$V_j = \frac{\sum_{k=1}^{N} (\mu_{kj}(2 - \xi_{kj}))^m X_k}{\sum_{k=1}^{N} (\mu_{kj}(2 - \xi_{kj}))^m}, \ (1 \le j \le C, 1 \le k \le N). \tag{11}$$

The FC-PFS algorithm is presented in details below:

Picture Fuzzy Clustering algorithm	
I:	Data X with N elements in d dimensions; C is number of clusters; threshold ε; fuzzifier m; exponent $\propto$, and $maxstep \ge 0$
O:	Matrices μ,η, ξ, and centers V;
FC-PFS Methods:	
1:	$h = 0$
2:	$\mu_{kj}^{(h)} \leftarrow random$, $\eta_{kj}^{(h)} \leftarrow random$, $\xi_{kj}^{(h)} \leftarrow random$, $(1 \le j \le C)$, $(1 \le k \le N)$ satisfying constraints (4–7)
3:	Repeat
4:	$h = h + 1$
5:	Calculate $V_j^{(h)}$ $(1 \le j \le C)$ following Eq. (11)
6:	Calculate $\mu_{kj}^{(h)}$ $(1 \le j \le C)$, $(1 \le k \le N)$ following Eq. (8)
7:	Calculate $\eta_{kj}^{(h)}$ $(1 \le j \le C)$, $(1 \le k \le N)$ following Eq. (9)
8:	Calculate $\xi_{kj}^{(h)}$ $(1 \le j \le C)$, $(1 \le k \le N)$ following Eq. (10)
9:	Until $\left\| (\mu^{(h)} - \mu^{(h-1)}) \right\| + \left\| (\eta^{(h)} - \eta^{(h-1)}) \right\| + \left\| \left(\xi^{(h)} - \xi^{(h-1)} \right) \right\| < \varepsilon$ or $h > maxstep$

3 Convergence of Fuzzy Clustering on Picture Fuzzy Set

In this section, the convergence of FC-PFS is proven. Let us consider the first part of the objective function in Eq. (2).

$$J_A = \sum_{k=1}^{N} \sum_{j=1}^{C} (\mu_{kj}(2 - \xi_{kj}))^m \|X_k - V_j\|^2. \tag{12}$$

Suppose $\theta_{kj} = (\mu_{kj}(2 - \xi_{kj}))$, by (11), we have:

$$V_j = \frac{\sum_{k=1}^{N} \theta_{kj}^m X_k}{\sum_{k=1}^{N} \theta_{kj}^m} (1 \le j \le C). \tag{13}$$

By (6), we have:

$$\sum_{j=1}^{C} \theta_{kj} = 1. \tag{14}$$

The first part of J named J_A (12) becomes an objective function of FCM, and according to Wu et al. [11], J_A is convergence.

Let us consider the second one.

$$J_B = \sum_{k=1}^{N} \sum_{j=1}^{C} \eta_{kj} \left(\ln \eta_{kj} + \xi_{kj} \right). \tag{15}$$

To prove (15) converges, we have to point out that (15) is bounded and monotonic. Let us consider the boundedness of (15), we have:

$$J_B = \sum_{k=1}^{N} \sum_{j=1}^{C} \eta_{kj} \ln \eta_{kj} + \sum_{k=1}^{N} \sum_{j=1}^{C} \eta_{kj} \xi_{kj}. \tag{16}$$

Firstly, we need to prove that:

$$\sum_{j=1}^{C} \eta_{kj} \ln \eta_{kj} \leq \ln C. \tag{17}$$

Using constraint (7) with condition $0 \leq \xi_{kj} \leq 1$; $C \geq 1$, then:

$$\sum_{j=1}^{C} \eta_{kj} \leq 1. \tag{18}$$

This formula is correct if only if $\sum_{j=1}^{C} \eta_{kj} \ln \eta_{kj} - \ln C \leq 0$. We have

$$\begin{aligned} \sum_{j=1}^{C} \eta_{kj} \ln \eta_{kj} - \ln C &\leq \sum_{j=1}^{C} \eta_{kj} \ln \eta_{kj} - \sum_{j=1}^{C} \eta_{kj} \ln C \\ &= \sum_{j=1}^{C} \eta_{kj} \left[\ln \eta_{kj} - \ln C \right] = \sum_{j=1}^{C} \eta_{kj} \ln \frac{\eta_{kj}}{C}. \end{aligned} \tag{19}$$

Let us consider the inequality $\ln(x) \leq x - 1 \forall x$. We have

$$\sum_{j=1}^{C} \eta_{kj} \ln \frac{\eta_{kj}}{C} \leq \sum_{j=1}^{C} \eta_{kj} \left(\frac{\eta_{kj}}{C} - 1 \right), \tag{20}$$

with the condition $0 \le \eta_{kj} \le 1$; $C \ge 1$, then $\frac{\eta_{kj}}{C} \le 1$. Thus, $\sum_{j=1}^{C} \eta_{kj}\left(\frac{\eta_{kj}}{C} - 1\right) \le 0$ is corrected then (17) corrected. We have:

$$\sum_{j=1}^{C} \eta_{kj} \ln \eta_{kj} \le \ln C. \tag{21}$$

On the other hand,

$$\begin{cases} \xi_{kj} \le 1 \\ \eta_{kj} \le 1 \end{cases} \text{then } \eta_{kj}\xi_{kj} \le 1. \tag{22}$$

From (21) and (22), we have:

$$J_B = \sum_{k=1}^{N}\sum_{j=1}^{C} \eta_{kj} \ln \eta_{kj} + \sum_{k=1}^{N}\sum_{j=1}^{C} \eta_{kj}\xi_{kj} \le \sum_{k=1}^{N} \ln C + \sum_{k=1}^{N}\sum_{j=1}^{C} 1 = N \ln C + N\,.\,C, \tag{23}$$

$$J_B \le N \ln C + N\,.\,C. \tag{24}$$

On the other hand,

$$\eta_{kj}.\xi_{kj} \ge 0. \tag{25}$$

Let us consider the function:

$$f(x) = x\,.\,lnx \text{ where } x \in (0, 1),$$

we have

$$f'(x) = 1 + lnx,$$
$$f'(x) = 0 \Leftrightarrow x = \frac{1}{e}.$$

Let us review the changing the variable function:

$$\forall x > \frac{1}{e} \rightarrow f'(x) > 0 : \text{Covariance function in} \left(\frac{1}{e}; 1\right)$$
$$\forall x < \frac{1}{e} \rightarrow f'(x) < 0 : \text{Inverse function in} \left(0; \frac{1}{e}\right)$$

This function is minimal when $x = \frac{1}{e}$ then

$$f(x) > f\left(\frac{1}{e}\right) = \frac{-1}{e} \quad \forall x \in (0;1). \tag{26}$$

The change of variation for the function is shown in Table 1 as below.

Table 1. Change of function $f(x) = x.\ln x$

x	0	1/e	1
$f'(x)$	–	0	+
$f(x)$	↘	Minimum	↗

From (25) and (26), we have:

$$\begin{aligned} J_B = \sum_{k=1}^{N}\sum_{j=1}^{C}\eta_{kj}(\ln\eta_{kj}+\xi_{kj}) &= \sum_{k=1}^{N}\sum_{j=1}^{C}(\eta_{kj}\ln\eta_{kj}+\eta_{kj}.\xi_{kj}) \\ &\geq \sum_{k=1}^{N}\sum_{j=1}^{C}\left(\frac{-1}{e}+0\right) = \frac{-1}{e}.N.C. \end{aligned} \tag{27}$$

From (24) and (27), we have:

$$\frac{-1}{e}.N.C \leq J_B \leq N\ln C + N.C. \tag{28}$$

Let us consider the monotonic of $\{\eta_{kj}\}$. Suppose that $g(x) = \frac{1-x}{e^{-x}}$.
We have: $g'(x) = \frac{-1e^{-x}+e^{-x}(1-x)}{(e^{-x})^2} = \frac{-xe^{-x}}{(e^{-x})^2} < 0,\ \forall x \in (0,1)$.

Deducted that *g(x)* is decreased, then $\eta_{kj} = \frac{e^{-\xi_{kj}}}{\sum_{i=1}^{C} e^{-\xi_{ki}}}\left(1-\frac{1}{C}\sum_{i=1}^{C}\xi_{ki}\right)$ is decreased.

Let us consider the monotonic of $\{\xi_{kj}\}$. Suppose that $h(x) = 1-x-(1-x^{\alpha})^{1/\alpha}$
We have:

$$h'(x) = -1+\frac{1}{\alpha}(1-x^{\alpha})^{\frac{1-\alpha}{\alpha}}.\alpha.x^{\alpha-1} = -1+\left(\frac{x}{\sqrt[\alpha]{1-x^{\alpha}}}\right)^{\alpha-1}.$$

$$h'(x) = 0 \Leftrightarrow \frac{x}{\sqrt[\alpha]{1-x^{\alpha}}} = 1 \Leftrightarrow x = \sqrt[\alpha]{\frac{1}{2}}.$$

$$\text{Then}\begin{cases} f'(x) > 0 \forall x < \sqrt[\alpha]{\frac{1}{2}} \\ f'(x) < 0 \forall x > \sqrt[\alpha]{\frac{1}{2}} \end{cases}.$$

Therefore, *h(x)* is covariance function in $\left(0, \sqrt[\alpha]{\frac{1}{2}}\right)$.

Replace x by $(\mu_{kj} + \eta_{kj})$, we have $\xi_{kj} = 1 - (\mu_{kj} + \eta_{kj}) - \left(1 - (\mu_{kj} + \eta_{kj})^{\alpha}\right)^{\frac{1}{\alpha}}$ is increased function in $\left(0, \sqrt[\alpha]{\frac{1}{2}}\right)$.

Firstly, we choose $\mu_{kj}, \eta_{kj}, \alpha$ in (0, 1) satisfying $\mu_{kj} + \eta_{kj} < \sqrt[\alpha]{\frac{1}{2}}$.

Thus, the value of $\xi_{kj}^{t} > \xi_{kj}^{t-1}$. Where t is the tth loop in FC-PFS. Similarly, $\eta_{kj}^{t} < \eta_{kj}^{t-1}$.

By using inductive proof, we have come to the conclusion that $\{\xi_{kj}\}$ is increased and $\{\eta_{kj}\}$ is decreased. Consequently, J_B is proven to be bounded.

Now, we have to prove that J_B is monotonic.

$$J_B = \sum_{k=1}^{N}\sum_{j=1}^{C} \eta_{kj} \ln \eta_{kj} + \sum_{k=1}^{N}\sum_{j=1}^{C} \eta_{kj}\xi_{kj}. \tag{29}$$

Let us consider the function $J_{B1} = \eta_{kj}. \ln \eta_{kj}$. We need to prove that

$$\eta_{kj} > \frac{1}{e} \quad \text{or} \quad \eta_{kj} < \frac{1}{e}.$$

- Case 1:

$$\frac{2}{e} \le \xi_{kj} < 1.$$

Within (9), we have:

$$\eta_{kj} < \frac{e^{-\frac{2}{e}}}{\sum_{i=1}^{C} e^{-1}}\left(1 - \frac{1}{C}\sum_{i=1}^{C}\frac{2}{e}\right) = \frac{e^{-\frac{2}{e}}}{C.e^{-1}}\left(1 - \frac{2}{e}\right) = \frac{0.344}{C} < \frac{1}{e} \quad where\ C \ge 1.$$

- Case 2:

$$\frac{1}{e} < \xi_{kj} < \frac{2}{e},$$

where

$$\begin{aligned}\eta_{kj} &< \frac{e^{-\frac{1}{e}}}{\sum_{i=1}^{C} e^{-\frac{2}{e}}}\left(1-\frac{1}{C}\sum_{i=1}^{C}\frac{1}{e}\right)\\ &= \frac{e^{-\frac{1}{e}}}{C.e^{-\frac{2}{e}}}\left(1-\frac{1}{e}\right) = \frac{0.913}{C} < \frac{1}{e} \quad (\forall C \geq 3),\end{aligned} \tag{30}$$

within C = 2, we have:

$$\begin{aligned}\left|J_B^{t+1}-J_B^{t}\right| &= \left|\sum_{k=1}^{N}\sum_{j=1}^{2}\eta_{kj}^{t+1}\left(\ln \eta_{kj}^{t+1}+\xi_{kj}^{t+1}\right)-\sum_{k=1}^{N}\sum_{j=1}^{2}\eta_{kj}^{t}\left(\ln \eta_{kj}^{t}+\xi_{kj}^{t}\right)\right|\\ &= \left|\sum_{k=1}^{N}[\eta_{k1}^{t+1}(\ln \eta_{k1}^{t+2}+\xi_{k1}^{t+1})+\eta_{k2}^{t+1}(\ln \eta_{k2}^{t+1}+\xi_{k2}^{t+1})]\right.\\ &\left.-\sum_{k=1}^{N}[\eta_{k1}^{t}(\ln \eta_{k1}^{t}+\xi_{k1}^{t})+\eta_{k2}^{t}(\ln \eta_{k2}^{t}+\xi_{k2}^{t})]\right|.\end{aligned} \tag{31}$$

We have:

$$\frac{1}{e}\ln\frac{1}{e} < x\ln x < 0 \quad \forall x \in (0;1) \text{ and } 0 < \eta_{kj}.\xi_{kj} < 1. \tag{32}$$

Thus

$$\left|J_B^{t+1}-J_B^{t}\right| < \left|\sum_{k=1}^{N}\left(2-2.\frac{1}{e}\ln\frac{1}{e}\right)\right| = N\left(2-2.\frac{1}{e}\ln\frac{1}{e}\right) < \delta. \tag{33}$$

Therefore, J_B converges. The same to C = 3, we have:

$$\begin{aligned}&\left|J_B^{t+1}-J_B^{t}\right|\\ &= \left|\sum_{k=1}^{N}\sum_{j=1}^{3}\eta_{kj}^{t+1}\left(\ln \eta_{kj}^{t+1}+\xi_{kj}^{t+1}\right)-\sum_{k=1}^{N}\sum_{j=1}^{3}\eta_{kj}^{t}\left(\ln \eta_{kj}^{t}+\xi_{kj}^{t}\right)\right|\\ &< \left|\sum_{k=1}^{N}\left(3-3\frac{1}{e}\ln\frac{1}{e}\right)\right| = n\left(3-3\frac{1}{e}\ln\frac{1}{e}\right) < \delta.\end{aligned} \tag{34}$$

Consequently, J_B converges.

- Case 3:

$$0 < \xi_{kj} < \frac{1}{e}.$$

We have:

$$\eta_{kj} = \frac{e^{-\xi_{kj}}}{\sum_{i=1}^{C} e^{-\xi_{kj}}}\left(1 - \frac{1}{C}\sum_{i=1}^{C}\xi_{kj}\right) < \frac{e^{-0}}{C.e^{\frac{-1}{e}}}(1-0) = \frac{e^{\frac{1}{e}}}{C} < \frac{1}{e}\forall C \geq 4. \tag{35}$$

Thus, $\eta_{kj} \leq \frac{1}{e}$.

Combined with the monotonic of $f(x) = x.\,ln\,x$ proving above, we have come to a conclusion:

$$J_{B1} = \eta_{kj}.\ \ln\ \eta_{kj}\ \text{is monotonic.} \tag{36}$$

Let us consider $J_{B2} = \eta_{kj}.\ \xi_{kj}$. Obviously with $\eta_{kj} > \eta'_{kj};\ \ \xi_{kj} > \xi'_{kj}$ then $J_{B2} = \eta_{kj}.\xi_{kj} > \eta'_{kj}.\xi'_{kj} = J'_{B2}$, then J_{B2} is monotonic. Combined with (36), we have $J_B = J_{B1} + J_{B2}$ is monotonic. Conspicuously, J_B is not only bounded but also monotonic, then J_B is convergent. Consequently, the objective function is convergent.

4 Conclusion

We have introduced the proof for convergence of fuzzy clustering algorithm on the picture fuzzy set. It provides the theoretical basis for picture fuzzy clustering algorithm to be more comprehensive.

Further studies regarding this research can be the expansion for picture fuzzy clustering algorithm with other topological spaces and metrics.

Acknowledgements. This research is sponsored by VNU University of Science under project No. TN.17.23.

References

1. Thong, P.H., Son, L.H.: Picture Fuzzy Clustering: A New Computational Intelligence Method. Soft Computing 20(9) (2016) 3544–3562.
2. D'Urso, P.: Informational Paradigm, management of uncertainty and theoretical formalisms in the clustering framework: A review. Information Sciences, 400 (2017) 30–62.
3. Cuong, B.C.: Picture Fuzzy Sets. Journal of Computer Science and Cybernetics 30(4) (2014) 409–420.
4. Zadeh, L. A.: Fuzzy Sets. Information and Control 8(3) (1965) 338–353.
5. Thong, P.H., Son, L.H.: A Novel Automatic Picture Fuzzy Clustering Method Based On Particle Swarm Optimization and Picture Composite Cardinality. Knowledge-Based Systems 109 (2016) 48–60.
6. Thong, P.H., Son, L.H.: Picture Fuzzy Clustering for Complex Data. Engineering Applications of Artificial Intelligence 56 (2016) 121–130.
7. Thong, P. H., Son, L.H.: A new approach to multi-variable fuzzy forecasting using picture fuzzy clustering and picture fuzzy rule interpolation method. In Knowledge and systems engineering, Springer International Publishing (2015) (pp. 679–690).

8. Son, L.H., Thong, P.H.: Some Novel Hybrid Forecast Methods Based On Picture Fuzzy Clustering for Weather Nowcasting from Satellite Image Sequences. Applied Intelligence 46 (1) (2017) 1–15.
9. Thong, P.H., Son, L.H., Fujita, H.: Interpolative Picture Fuzzy Rules: A Novel Forecast Method for Weather Nowcasting. Proceeding of the 2016 IEEE International Conference on Fuzzy Systems (2016) 86–93.
10. Hall, L. O., & Goldgof, D. B.: Convergence of the single-pass and online fuzzy c-means algorithms. IEEE Transactions on Fuzzy Systems, 19(4) (2011) 792–794.
11. Wu, J., Xiong, H., Liu, C., & Chen, J.: A generalization of distance functions for fuzzy $ c $-means clustering with centroids of arithmetic means. IEEE Transactions on Fuzzy Systems, 20(3) (2012) 557–571.
12. Sadi-Nezhad, S., Khalili-Damghani, K., & Norouzi, A.: A new fuzzy clustering algorithm based on multi-objective mathematical programming. TOP, 23(1) (2015) 168–197.

Bone Segmentation from X-Ray Images: Challenges and Techniques

Rutvi Shah[1(✉)] and Priyanka Sharma[2]

[1] CPICA, Ahmedabad, Gujarat, India
rutvirshah@gmail.com
[2] Raksha Shakti University, Ahmedabad, Gujarat, India
pspriyanka@yahoo.com

Abstract. Identification of the exact location of the fracture regions in the images was done based on the active contour segmentation process. Image segmentation process identifies the regions in the images that contain the interested portions in the images. A process that identifies the fracture location in images lying on the contour extraction process is employed. The contour extraction process identifies the bone regions in the images. Extracted/fracture bone regions are identified using morphological operations and thresholding process. Morphological operations and thresholding process are used for the exact identification of the fracture location. It removes the excess regions segmented in the threading process, i.e., fracture pixels. The fracture location is identified by subtracting the two images resulting in the identification of the fracture region in the images. In this paper, a comparison of different fracture detection algorithms like region growing, level set segmentation, and active contour segmentation (proposed system) is done. Comparison analysis is done in terms of PSNR, accuracy, segmentation time, and detection ratio. Results obtained show the higher accuracy of proposed work over existing algorithm.

Keywords: X-Ray · Active contour · Morphological operations and segmentation

1 Introduction

It is necessary to classify the objects of interest from the background while analyzing medical images. One of the today's most challenging tasks in image is image segmentation. The process of segmentation is to separate the constant parts of regions in image. These parts are constant, but they also are different from one part to another. The very first discovery in the medical imaging field was X-rays. Then, different medical image types have turned up over the years. These are magnetic resonance imaging (MRI), ultrasound (US), computed tomography (CT), nuclear imaging, including single-photon emission computed tomography (SPECT), and positron emission tomography (PET). Each of these imaging methods has their own advantages and disadvantages. X-ray segmentation domain is challenging as X-ray images are

V. Bhateja et al. (eds.), *Information Systems Design and Intelligent Applications*, Advances in Intelligent Systems and Computing 672,
https://doi.org/10.1007/978-981-10-7512-4_84

complex in nature. The presence of noise or any other artifacts in X-ray images results into disconnected boundaries of region of interest. Use of different X-ray equipment creates X-rays with varied orientations, resolutions, or luminous intensities. Because of this, the quality of the segment can vary. This could influence the quality of the segmentation result. Bone regions in X-rays are often partly covered by other organs or muscle tissues. The joints between the bones can also create difficulty. Joints become significant, while we aim at segmenting the entire bone structure from the image [1]. There are various segmentation methods designed and developed. There is no such ideal method which can be globally pertinent to any kind of medical image. Here, the aim of the paper is to compare existing image segmentation methods for X-ray images.

2 Challenges in Bone X-Ray

Analysis of various issues of bone segmentation in X-ray images requires to study certain characteristics of bone images. X-ray images are different than other medical imaging methods as they contain regions which are partly covered by different organs like flesh, tissues, or muscles.

Bones are connected to each other by joints. Bone joints have to be considered while segmenting whole bone region from the X-ray image. As bones are 3-D in nature, a 2-D closed curve may give precise boundaries of bones in X-ray images.

Noise: Noise in X-ray images is a kind of noise that is known as “quantum noise.” This refers to the distinct nature of the X-ray photons that are used to produce them.

Overlapping: There are many tissues and organs that are overlapped on the bone structure in X-ray images. External problems are usually generated due to the patients.

Ambiguity: The various adjacent tissues inside of human body could have equivalent X-ray absorption rates. This results in ambiguous boundaries of the organs. In a single image, these create many non-clear sections between two neighboring organs.

Bone density variability: Diverse patients can have irregular bone densities, which comes about inconsequential unmistakable forces inside the bone districts ink their X-ray pictures. A typical patient’s picture has thick bones in their X-ray pictures, though a patient who experiences osteoporosis has low-thickness bones, which can bring about exceptionally dim bone locales. What’s more, different tissues may likewise affect the power of bone areas in the pictures.

Inter-patient form of variability: The shapes of the bones of different patients can relatively be different. For instance, the shape of the pelvis bone of a female patient is relatively different from that of a male patient. The pelvis bones of females are much wider as compared to males.

Imaging posture variability: The posture of the patient while taking X-ray images may also cause variability. These can show bones placed in different organs in different shapes due to different imaging postures.

3 Medical Image Segmentation Techniques

Common scientific image segmentation approaches may also be classified into the next classes: classical photo segmentation ways (thresholding, regions based, and edges headquartered), sample realization-headquartered, deformable models, wavelets-headquartered approaches, and atlas-situated strategies.

Traditional image segmentation methods:

Traditional methods include the following segmentation techniques:

1. Thresholding method,
2. Region-based method, and
3. Edge-based method

Thresholding is one of the less complicated segmentation methods which segment the region by thresholding the picture intensity. There are two categories of thresholding methods: international techniques and adaptive techniques. In international technique, only one threshold is chosen for obtaining segmentation in the complete photo, while in adaptive technique, different thresholds are selected separately for each and every pixel. The pixels with more prominent intensity values than the edge are named as a part of workforce A—an object of interest (with intensity value of 1)—and rest of the pixels are named as a part of workforce B—heritage (with intensity value of 0).

In adaptive technique, the given image is divided into a sequence of sub-images. Different thresholds are calculated for different sub-images. Regional adaptive thresholding analyzes intensities around each pixel. It selects a threshold value for each pixel value by measuring the intensity values of adjacent pixels. These methods are easy and quick. They are appropriate for images with bimodal intensity values. Unequal illumination in the image affects the efficiency of thresholding. Adaptive

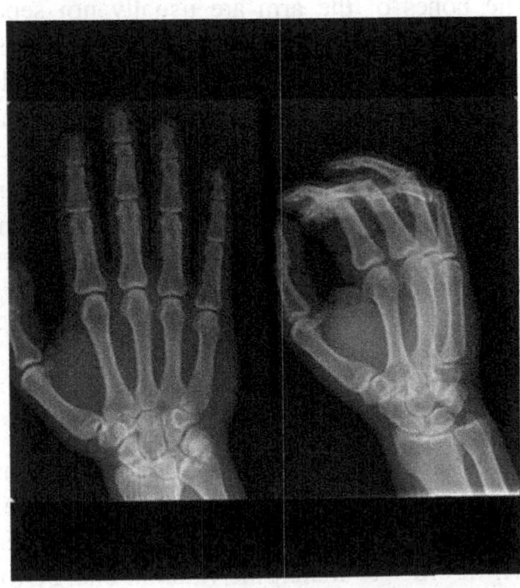

Fig. 1. An X-ray image

methods are more complicated than international approaches when compared to computations. Still, these methods can be productively used for extracting small regions or parts from a variable heritage. The adaptive threshold issued for segmentation process is used for the scan of image in Fig. 1, and the outcome is provided in Fig. 2.

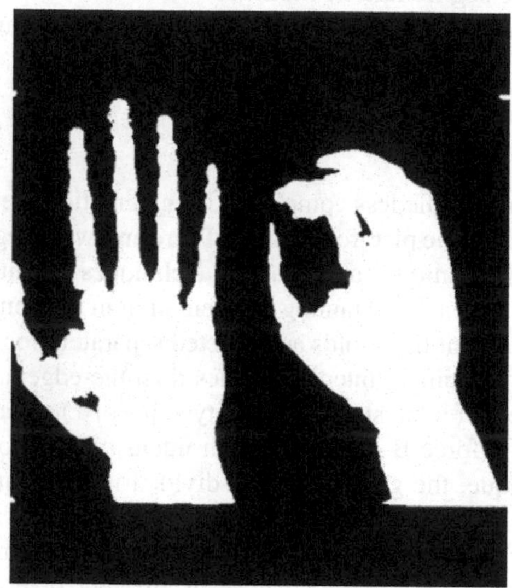

Fig. 2. An X-ray image segmentation using thresholding

In the resulting image Fig. 2, it is seen that bones are not highlighted as they are covered with flesh. The bones of the arm are usually not separated. Computerized tomography (CT) can be used to enhance this simple process. Thresholding algorithms are hardly employed in medical imaging. Region-founded methods use groups of pixels with same intensities. The fundamental area-based segmentation algorithms are region-growing segmentation and watershed algorithms. Region-growing algorithm is a simple pixel-oriented image segmentation method, which includes the identification of pixels, which are known as the seeds. The region is identified which is then developed around these seeds based on homogeneity standards. The pixels which have identical values are used to grow the region. For this selection, a statistical test is often employed.

3.1 Active Contour Segmentation

The active contour extraction process identifies the bone regions in the images based on the differences in the gray level pixels in the images. In contour extraction process, the initial seed points, internal energy, and external energy were taken as input. The initial seed points were grown in order to match the internal energy to the external energy. The initial energy term helps in the exact identification of the shape of the segmented region. Deformation of the shape and control on iteration of the process can be done with the help of the external energy term. Active contours will identify the object regions [2]. The curve will move around the boundaries of the objects and finally the exact shape around the object. The process is repetitive in iterations resulting in the identification of the exact region of input images. The number of iterations of the process depends on the image type and the grayscale concentration of the input images [3].

4 Proposed System

Proposed work includes the operations like gray conversion, active contour segmentation, segmented bone, and fracture segmentation which are explained below, Fig. 3.

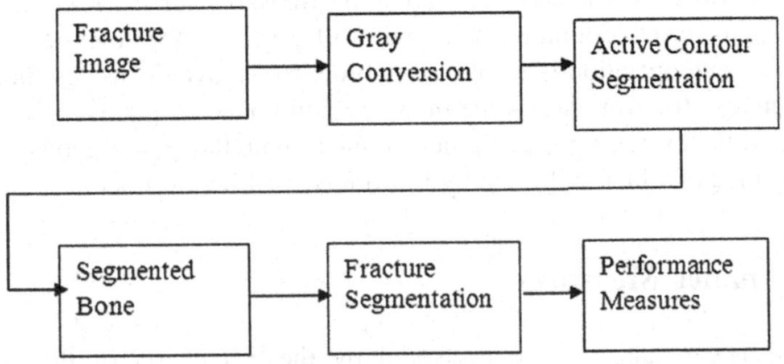

Fig. 3. Steps of proposed system

4.1 Gray Conversion

Digital images play a vital role in medical field. There is significant increase in the use of medical imaging by medical practitioners for diagnosis and decision of course of treatment. The images are produced by various up-to-date medical equipment such as MRI, CT, ultrasound, and X-ray. X-ray is the oldest and most commonly used medical equipment, as it is non-intrusive, painless, and economical [4]. The X-ray images can be used at various stages of treatment which can include fracture diagnosis and various bone cracks. The input is the bone X-ray images. The images were converted into gray scale, and the bone regions were extracted from the images based on active contour segmentation [5].

4.2 Segmented Bone

Segmentation of bone structure from X-ray image is fundamental for X-ray image analysis techniques. The main purpose of segmentation is to subdivide the various portions, so that medical practitioners can diagnose and study the bone structure, do identification of bone fracture, can measure the fracture, and decide course of treatment prior to surgery. This is a challenging task as the bone X-ray images are complex in nature. The output of segmentation algorithm can be affected by various factors like partial volume effect, intensity in homogeneity, presence of noise and artifacts, and proximity in gray level of different soft tissues [6].

4.3 Fracture Segmentation

Tissues around cracked bone are more mind boggling to perceive on the grounds that they have some extra elements to be considered. Actually, the bone pieces may have self-assertive shape and can have a place with any bone in an adjacent region, and it ends up noticeably fundamental to name every one of the parts amid the segmentation procedure. This naming procedure requires master learning. An earlier learning cannot be effectively obtained in light of the fact that it is uncommon to discover two indistinguishable cracks shapes in bone X-ray pictures. It is hence hard to figure the state of the bone parts, particularly in comminuted cracks. For our situation, cracked area in the input pictures was distinguished by discovering threshold for the specific region [7]. Intensity of gray level is found lower than 150 pixels in the fracture region. Morphological opening operation is employed in order to remove the useless portion of the segmented image. It also removes the image region boundary pixels and thus helps in the removal of the excess regions segmented along with the bone regions. The smaller blobs in the images which will thereby be formed will be the fracture regions [8].

5 Performance Measures

The performance parameters are measured by the calculation of the performance metrics like PSNR, accuracy, segmentation time, detection ratio, mean value, variance, and standard deviation. PSNR means peak signal-to-noise ratio. Its value indicates the noise ratio in the input image and the resulting image. The PSNR value must be high. Accuracy of the process denotes the percentage of exact segmentation of the input image comparing with the result. Mean value is the statistics depicting the gray level pixel range of the images. Variance is the average of the squared differences of the mean value. Standard deviation is the statistics depicting how to spread the segmented regions in the images. Segmentation time is the process of partitioning a digital image into multiple segments (sets of pixels also known as super-pixels). The objective of segmentation is to simplify and/or change the representation of an image into something that is more meaningful and easier to analyze. Detection ratio is the ratio of the number of fractured bone detected to total number of fractured bone. Its value must be high [9].

Algorithm The result of proposed system is the identification of fracture in the input image. Input image from dataset is segmented using an active contour segmentation. By using the morphological operation, we separate the fracture region from the input image. The steps for proposed system are given below.

Step 1 Random input image from dataset
Step 2 Identification of seed points

Step 3 Identification of boundaries
Step 4 Gray level conversion using Hough transform

$$r = x\cos\theta + y\sin\theta \tag{1}$$

Step 5 Segmentation of image using active contour segmentation

$$\Delta t < \frac{\Delta x}{\max\{|\mu|\}} \tag{2}$$

Step 6 Identification of fracture area from image using gray levels
Step 7 Subtraction of fracture area using morphological operations
Step 8 Identification of fracture region using red and blue circle
Step 9 Computation of PSNR, mean, standard deviation, detection ratio, and accuracy, i.e., performance measure.

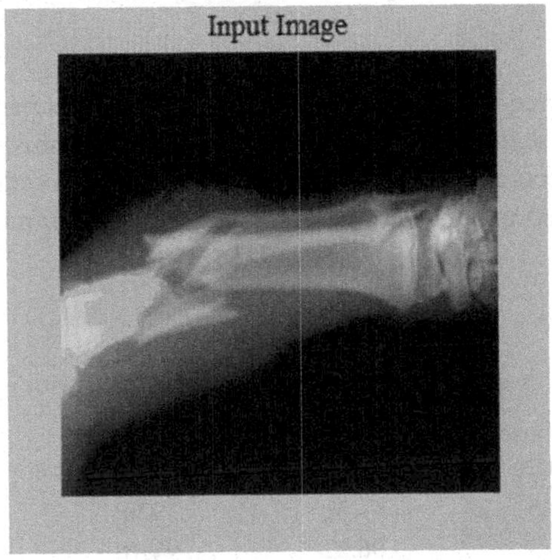

Fig. 4. Hand input image

6 Results

The proposed algorithm has been applied on dataset which consists of fractured and non-fractured bones. We compared proposed system with other algorithms like region growing, level set segmentation, and Hough transformation. We computed PSNR, mean, standard deviation, variance, accuracy, segmentation time, and detection ratio for various random images from dataset and also analyzed the result for fractured hand bone, fractured leg bone, and one without fractured hand bone for understanding the procedure.

Figure 4 is the image of hand bone which acts as an input image to the algorithm. Result denoted the fractured region by circles around them as shown in Fig. 5.

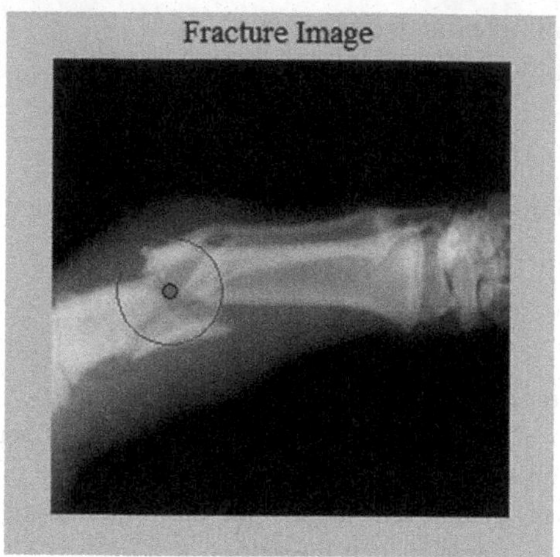

Fig. 5. Hand fracture image

Table 1 shows the comparison of different performance of parameters for sample of datasets for contour segmentation and using other algorithms. From the result, we understood that the accuracy of the code is averagely 96%. Average segmentation time is 28.38 s. We can also understand that the average detection ratio for this code is 72. 68%.

Table 1. Comparison of different fracture detection algorithm with parametric performance

Parameters	Proposed system	Region growing	Level set segmentation	Hough transform
PSNR	44.66	41.61	39.62	38.41
Accuracy	98.73	96.22	94.10	93.25
Segmentation time	28.22	30.81	32.51	35.34
Detection ratio	68.51	66.35	64.51	62.26

From the above table, we can conclude that proposed system is highly efficient and highly accurate and has high detection ratio and less segmentation time. All these parameters of different fracture detection algorithms are shown below in the form of bar chart (Chart 1).

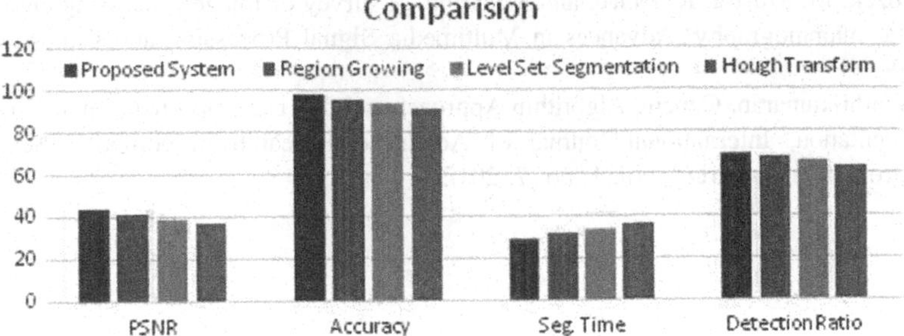

Chart 1. Different performance of parameters for sample of datasets

7 Conclusions and Future Work

The location of fracture from the image is extracted using segmentation process. The segmentation process is done using active contour segmentation process. For non-fractured images, no regions were obtained during the thresholding and the morphological operation process. The performance measures prove that the proposed method is efficient compared to the existing algorithms. Active contour extraction process can be replaced with the help of the super-pixel segmentation for minimization of the segmentation time. The super-pixel segmentation process searches around the input seed points given by the user based on the image intensity. The seed points were grown, and the exact regions around the interested portions were identified.

References

1. V. Zharkova, S. Ipson, J. Aboudarham and B. Bentley, Survey of image processing techniques, EGSO internal deliverable, Report number EGSO-5-D1_F03-20021029, October, 2002, 35p. [Online]. Available:.
2. A. Tirodkar, A Multi-Stage Algorithm for Enhanced XRay Image Segmentation, International Journal of Engineering Science and Technology (IJEST), Vol. 3 No. 9.
3. N. Komodakis, N. Paragios, and G. Tziritas, MRF energy minimization and beyond via dual decomposition, IEEE Trans. Pattern Anal. Mach. Intell. vol. 33, no. 3, Mar. 2011.
4. G. Dougherty. Medical Image Processing Techniques and Applications. Springer, 2011.
5. S. M. Ali Eslami, N. Heess and J. Winn, The shape boltzmann machine: A strong model of object shape, Proc. IEEE Comput. Soc. Conf. Computer. Vis. Pattern Recognit. Providence, RI, USA, Jun. 2012, pp. 406–414.
6. D. Kang, J. Woo, P. Slomka, D. Dey, G. Germano and C. Jay Kuo, Heart chambers and whole heart segmentation techniques: Review, J. Electron. Imag. vol. 21, no. 1, Jun. 2012.
7. S. K. Mahendran and S. S. Baboo, Enhanced automatic X-ray bone image segmentation using wavelets and morphological operators, Proc. of the International Conference on Information and Electronics Engineering, 2011.

8. J. Bozek, M. Mustra, K. Delac, and M. Grgic, A survey of image processing algorithms in digital mammography, Advances in Multimedia Signal Processing and Communications, 2009.
9. N. Senthilkumaran, Genetic Algorithm Approach to Edge Detection for Dental X-ray Image Segmentation, International Journal of Advanced Research in computer Science and Electronics Engeneering, vol. 1, no. 7, 2012.

Sparse Coded SIFT Feature-Based Classification for Crater Detection

Savita R. Gandhi(✉) and Suchit Purohit(✉)

Gujarat University, Navrangpura, Ahmedabad, Gujarat, India
drsavitagandhi@gmail.com, Suchit.s.purohit@ieee.org

Abstract. Morphological classification of impact craters upto now is done through visual interpretation which suffers from high degree of subjectivity. We are proposing a classification approach to classify a given crater into crater or non-crater class. The approach has been implemented and tested on Lunar images. We have also implemented some of the existing approaches like Hough transform, template matching and supervised classification using AdaBoost and found limitations of these. Our proposed framework uses sparse coded SIFT as local features and use SVM as classifier. We have compared this with other classification approaches employing pixels as features; using SIFT as local feature without applying sparse coding; using SIFT + sparse coding but employing KNN. Since SIFT demands huge memory requirement, we suggest a two phase process in which crater candidates are identified and SIFT features are extracted only for these identified areas. This drastically reduces memory and processing requirements. On testing, our approach is found to reduce the memory and time requirements almost to 50% yet outperforming in terms of accuracy as well as robustness against noise, occlusion, different viewing angles, and various illumination effects.

Keywords: Crater detection algorithm · Crater morphology · Scale-invariant feature transform · Sparse coding

1 Introduction

Craters are ubiquitous components found on surface of any planetary body. They are crustal depressions or areas of negative relief that result from propagation of shock waves after crash of bolides (generic term used when precise nature of impact body is not known) [1]. They are found on most of all planetary bodies but more on planets like Moon, Mars, and Mercury, and they can sustain for long due to absence of eroding entities like atmosphere. The volume as well as resolution of data acquired increases significantly with each new planetary mission. Therefore, manual extraction of all relevant information present in the images is no longer possible. Hence, there is need of automated crater detection and classification algorithms. The last decade has witnessed

V. Bhateja et al. (eds.), *Information Systems Design and Intelligent Applications*, Advances in Intelligent Systems and Computing 672,
https://doi.org/10.1007/978-981-10-7512-4_85

significant contributions in the area of crater detection. From the survey, it is found that common techniques for crater detection use supervised or unsupervised techniques. Unsupervised approaches use methods of circle or ellipse fitting, shadow matching, whereas supervised learning methods use SVM, neural networks, Adaboost, Fuzzy logic, template matching, and many more [1]. From the literature survey we identified that 3 of the techniques are widely used namely like Hough transform, template matching and supervised classification using AdaBoost. We have implemented these existing approaches and found limitations of these. The implementation has been done to as to provide common computation environment to compare them with our proposed work. Figure 1 shows crater detection using Hough transform, Fig. 2 shows detection using Haar features and Adaboost, and Fig. 3 represents crater detection using Template-matching technique. For the detection phase of our proposed approach, we have used Hough transform because it does not require any training data as in the case of Adaboost or Template matching [1].

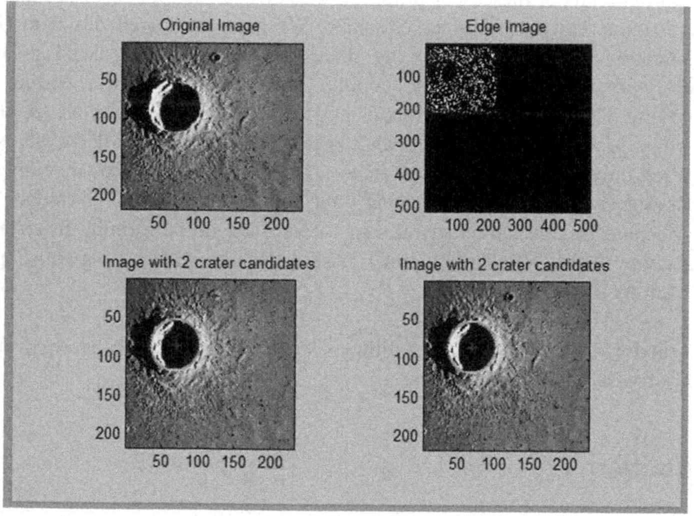

Fig. 1. Crater detection using Hough transform

The classification phase of our approach comprises of three stages namely feature extraction, coding to map local features into compact representation, and pooling to aggregate compact representation together.

The classification model used by our approach uses local features extracted using scale-invariant feature transforms (SIFT) [2]. Advantage of using SIFT is that it invariant to scaling, illumination effects, viewing angle and is robust against clutter, occlusion, and noise. SIFT descriptor calculates value of feature points in the form of 128-bit feature vector. Out of many coding methods available like histogram of SIFT popularly known as "Bag-Of-visual-words" which uses vector quantization for coding,

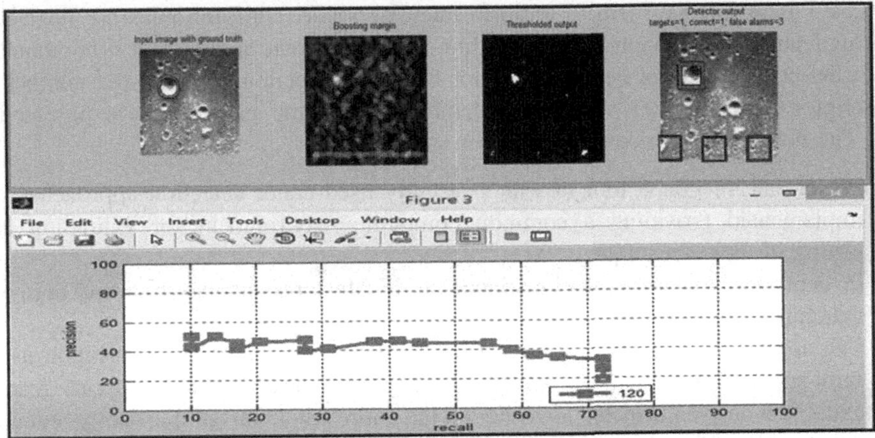

Fig. 2. Crater detection using Adaboost

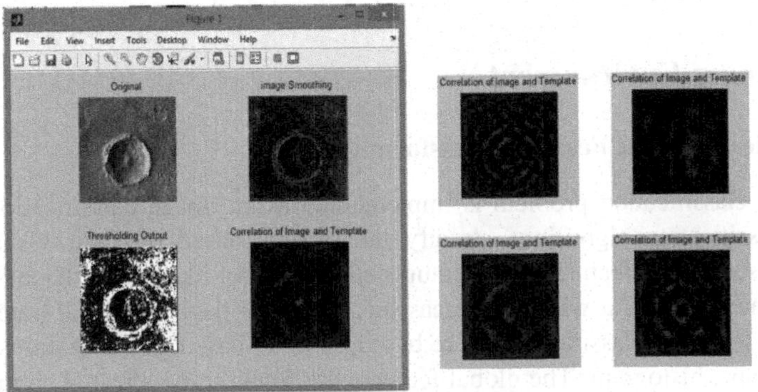

Fig. 3. Crater detection using template matching

uncertainity-based quantization, Fisher vector which uses Gaussian Mixture Model for coding and Sparse coding, we have chosen **Sparse Coding** for coding SIFT features [3]. Sparse coding converts the image feature into high-dimensional space, improves discriminative capability, and has low reconstruction error, prone to quantization error.

Since SVM classifier accepts a single feature vector as input, the sparsely coded features are pooled using spatial pyramid pooling before feeding into SVM classifier. Though many literature have shown promising results in image classification using sparse coded SIFT features but all the approaches apply SIFT extraction algorithm to the entire image. Hence, the number of descriptors is large and occupies large memory and also demands computation time during training and testing phase. Our technique first detects the craters using Hough transform approach which detects the craters and returns the center and radius of detected craters. We set all the pixels other than these detected areas and set them to uniform intensity value. Uniform areas are neglected during feature extraction in SIFT algorithm; hence, the areas of the image other than the detected craters will not be processed for keypoint detection reducing the number of descriptor almost by 50% which considerably reduce time and memory requirements.

The novel contributions of this paper are:

1. Present an integrated tool in which recently used crater detection approaches are implemented providing a common computing environment for validation and testing of those techniques.
2. Present a tool to validate the invariance of SIFT features and its superiority in image matching.
3. The approach combines detection using Hough transform and classification using semi-supervised technique. Novelty of approach lies in that it applies feature extraction only on detected areas not on the entire whole image drastically reducing the number of features extracted hence optimal in memory and time requirements. At the same time, the sparsely coded SIFT features make the approach robust to invariances, noise, and occlusion.

2 Learning Environment

2.1 Scale-Invariant Feature Transform (SIFT)

The crater classification problem is implemented as an image classification problem. Image classification algorithms classify the objects based on their visual/semantic content described by features. The features can be global like color, shape, texture and describe the image as a whole, whereas local features describes local patches. Initial researches in image classification were based on pixel-based global features like color, texture, shape, histogram. The global features are found to be affected significantly by various illumination effects, viewing angles, noise, and other distortions, and it is found that classifiers based on global features lack in accuracy [2]. Hence, recent researches are focusing on classification based on local features for quantification of visual information present in the image [2]. A good local feature should be easy to extract, distinctive, repetitive, invariant, and robust to noise occlusion and clutter. A survey on local features has proved the superiority of a local feature detector and descriptor founded by David Lowe popularly known as known as **SIFT or "Scale-Invariant Feature Transform"** [4]. "SIFT is **a technique to extract key features from images. These features do not change when image is subjected to change in size, orientation, viewpoint, illumination,** or noise levels" [2]. An inhouse tool has been developed to test the invariance of SIFT under various conditions. It allows the user to manually subject the image to different distortions and transformations and perform image matching between original and tempered image using SIFT as well as other image matching techniques like PSNR, SSIM. It is found that SIFT gives almost 99% matching after scaling and change in intensity values, whereas matching is a little less after rotation and noise, though much greater than PSNR and SSIM (Fig. 4).

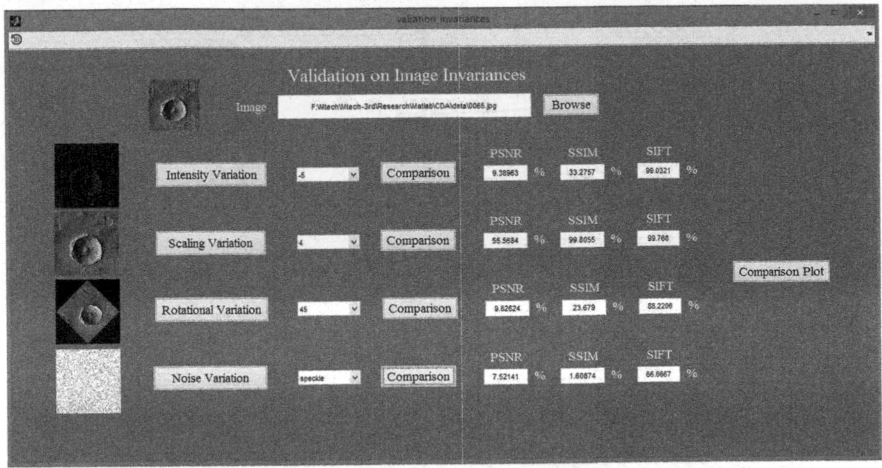

Fig. 4. Validation of SIFT invariances

Major stages of generation of SIFT feature generations as described by David Lowe are [3]:

1. Scale-space extrema generation: A scale space is representation of image at multiple resolution levels. An image in scale space is defined as L(x, y,) where σ represents the scale. A Gaussian scale space is obtained by convuling the image function with 2D Gaussian kernel given by

$$\mathrm{G(x, y)} = \frac{1}{2\pi\sigma^2} e^{\left(-\frac{x^2+y^2}{\sigma^2}\right)} \tag{1}$$

$$\text{Hence}\, \mathrm{L(x, y, \sigma) = G(x, y, \sigma) * I(x, y)} \tag{2}$$

   Mikolajczk [5] found that maximum and minimum of the Laplacian of Gaussian σ^2 ∇2G produce stable features and Linderberg [6] showed that difference-of-Gaussian can be approximated by Laplacian Gaussian; hence, Lowe's approach is to construct a difference-of-Gaussian by incrementally convuling image with Gaussian to produce an octave. An octave is collection of images which are separated by a constant k. Difference-of-Gaussian is obtained by subtracting adjacent images and is represented $\mathrm{D(x, y, \sigma)}$ (Fig. 5). After processing of octave, Gaussian image is resampled to half and initial value of σ double than the previous octave. Owing to high repeatability, initial value of σ is taken as 1.6 and number of scales as 3. To find local extrema in difference-of-Gaussian, each pixel is compared with neighboring (3 × 3) pixels in adjacent scales and it is chosen as a keypoint if it is sufficiently large or small from other pixels.
2. Keypoint localization: To remove features with low contrast and unfocalized, difference-of-Gaussian is approximated by Taylor series expansion denoted by D(x). The local extremum, $\mathrm{D}(\widehat{x})$, is used to reject low contrast point. Only keypoints with

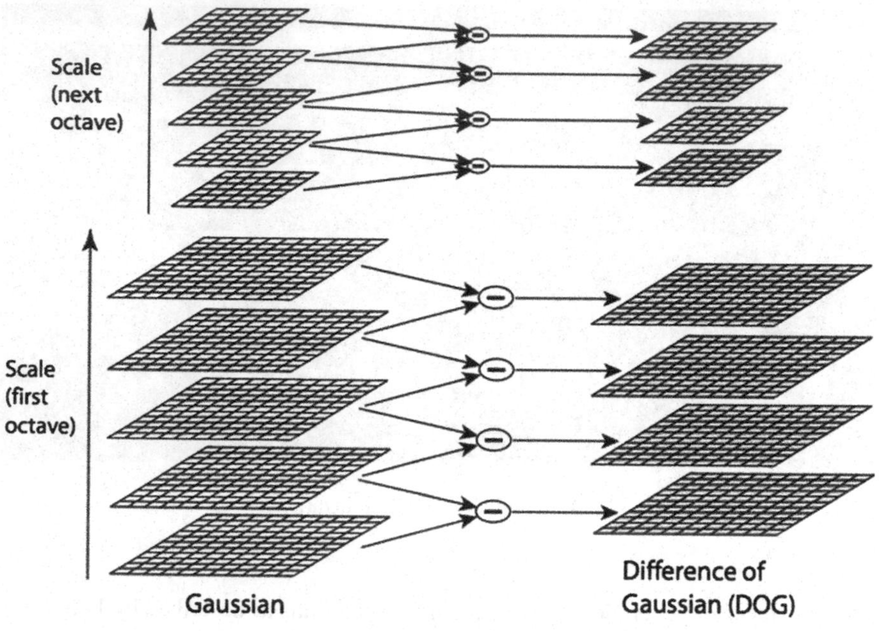

Fig. 5. Difference-of-Gaussians

$|D(\widehat{x})| > 0.3$ are kept and others are rejected. To remove edges as interest points, trace and determinant of the Hessian matrix of D are computed as

$$H = \begin{bmatrix} Dxx & Dxy \\ Dxy & Dyy \end{bmatrix} \tag{3}$$

$$\text{Trace} = \text{Tr(H)} = \text{Dxx} + \text{Dyy} = \alpha + \beta \tag{4}$$

$$\begin{gathered} \text{Determinant} = \text{Det(H)} = D_{xx}D_{yy} - \left(D_{xy}\right)^2 = \alpha\,\beta; \\ \text{where } \alpha \text{ and } \beta \text{ are eigenvalues} \end{gathered} \tag{5}$$

$$Evaluate \frac{Tr(H)^2}{Det(H)} = \frac{(r+1)^2}{r} where\ r = \frac{\alpha}{\beta}$$

All keypoints of which are >10 are likely to represent edges; hence, they are eliminated.

3. Orientation assignment of keypoints: The gradient magnitude m(x, y) and orientation θ(x, y) is computed for each sample image at a scale of keypoint detected.

$$m(x,y) = \sqrt[2]{\left((L(x+1,y\ ) - L(x-1,y\ ))^2 + \left((L(x,y+1) - L(x,y-1))^2\right)\right.}$$

$$\theta(x,y) = \tan - 1((L(x,y+1) - L(x,y-1))/(L(x+1,y) - L(x-1,y)))$$

Find the orientation of each point in the neighborhood of keypoint and assign weights according to the magnitude and build histograms of 36 bins. Peaks in the histogram indicate dominating direction of the keypoint. Any other peak which is within 80% range of highest peak is also considered as keypoint.

4. Keypoint descriptor: For every keypoint location (16×16) neighborhood, magnitude and direction of image gradients are sampled. After that the descriptor's coordinates and the gradient orientations are rotated according to orientation of keypoint. 16×16 regions are divided into 4×4 blocks. Histogram of bins is found weighted by the magnitude. Hence, a 128-dimensional vector for each keypoint is computed.

To provide invariance to illumination, feature vector is normalized to 1. Because if there is a change in image contrast, it would also produce similar change in image gradients. The contrast change will hence be neutralized by vector normalization. Similarly, if there is change in brightness, it will add same value to each image pixel. To provide invariance to illumination, feature vector is normalized to 1. Because if there is a change in image contrast, it would also produce similar change in image gradients. The contrast change will hence be neutralized by vector normalization. Similarly, if there is change in brightness, it will add same value to each image pixel and hence won't effect the gradient values. To reduce illumination effects due to camera saturation, the normalized feature vector values are thresholded to 0.2, and then renormalized to unit length.

2.2 Sparse Coding

SIFT extracts hundreds and thousands of keypoints in an image and hence represents an image by many feature vectors. The classifiers expect the image to be described as a single vector; hence, the SIFT vectors need to be quantized. One of the popular approaches is Bag-of-visual-words in which the features are quantized using flat k-means or hierarchical k-means and then computes the histogram for semantic classification. The discriminatory power of Bag-of-words suffers due to quantization errors and loss of spatial order of descriptors. An extension of Bag-of-features model was proposed in [6] called spatial pyramid matching. It divides image into $2l \times 2l$ segments scales of values $l = 0, 1, 2$ and computes histograms in each segment, and finally concatenates them to form a vector representation of the image. This algorithm is widely used in many computer vision applications, but problem of one or more vocabularies still exists; hence, we have used Sparse coding to compress and quantize SIFT vectors.

"Sparse coding is a representation of data as a linear combination of atoms (patterns) learned from the data itself. Such a collection of atoms (code words) is called Dictionary or codebooks" [7]. For an input X, Dictionary D is a set of normalized basis column vectors of size p such that there exists a vector α known as sparse coefficient vector such that

$$X = D\alpha$$

where $D = \{d1, d2, d3 \ldots dp\}$ and α should be as sparse as possible, i.e., most of the entries in α should be zero.

Sparse representation is more compact and high-level representation of the image. As compared with vector quantization, sparse coding has a low reconstruction error, more separable in high-dimensional spaces making them suitable for classification purposes. The last years have witnessed an increase in computer vision algorithms that utilize sparse coding survey of which is presented in [7]. Success of sparse coding depends on selection of Dictionary. One approach to choose Dictionary D is to choose from known transform (steerable wavelet, coverlet, contourlet, bandlets). These off-the-shelf dictionaries fail for specific images like face, digits. [8]. Current researches are focusing on Dictionary learning from a given input sample. Dictionary learning is a process to learn the space in the sample where the given signal can be best represented. Given a set of SIFT features, K random features are selected to train the Dictionary using following optimization

$$a = arg\min_{a}\|x - Da\|_2^2 + \lambda|a|_2^2$$

$$min_{D,a^{(k)}} \frac{1}{K}\sum_{k=1}^{K}\left\{\frac{1}{2}\|y - Da\|\frac{2}{2} + \lambda\Omega a^k\right\}$$

Some of the important Dictionary learning algorithms are K-SVD of Aharon [9], Olshausen and field [10], SPAMS of Mairal [11], and others [12–15]. Yang in 2011 proposed Fisher Discriminative Dictionary learning learns a structured Dictionary common to a number of classes C. The Dictionary is represented be as D = [D1, D2, D3…DC] [8] hence increasing the discriminative power. The performance was reported to be highest using this learning method in terms of high recognition rate and low error rate. We have used Mairal's online Dictionary which is available in SPAMS which contains an optimization toolbox for sparse coding. Once the Dictionary is learned from Eq. (1), it is applied to code all SIFT vectors in all images and sparse representation is generated via following optimization:

$$\forall y \in Y, min_a \frac{1}{2}\|y - Da\|\frac{2}{2} + \lambda\Omega a^k \text{ (Dictionary size} = 1024) \quad (6)$$

Therefore, for every SIFT vector, a sparse vector $a \in R^{1024\times 1}$ is learned.

2.3 Spatial Pyramid Pooling

Partition images into 4 × 4, 2 × 2, 1 × 1 segments and max-poole the sparse vectors within each of $2 \times l$ segments. Hence, a spatial pyramid representation of the image which has 1024 × 21 dimensions uniform throughout all the images suitable to be fed into the classifier [16].

3 Proposed Framework

3.1 Phase 1: Detection and Background Removal

In the detection phase, we have used elliptical Hough Transform to get the crater candidates. The detection algorithm returns center and radius of craters (Fig. 6). We set all the pixels other than this area to a uniform intensity value to reduce the number of descriptors obtained (Fig. 7).

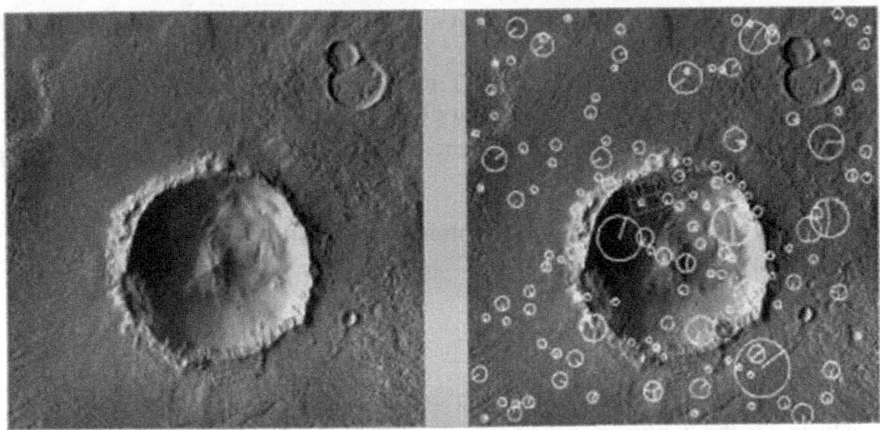

Fig. 6. Extraction of SIFT features on a crater image. Circles indicate keypoints detected

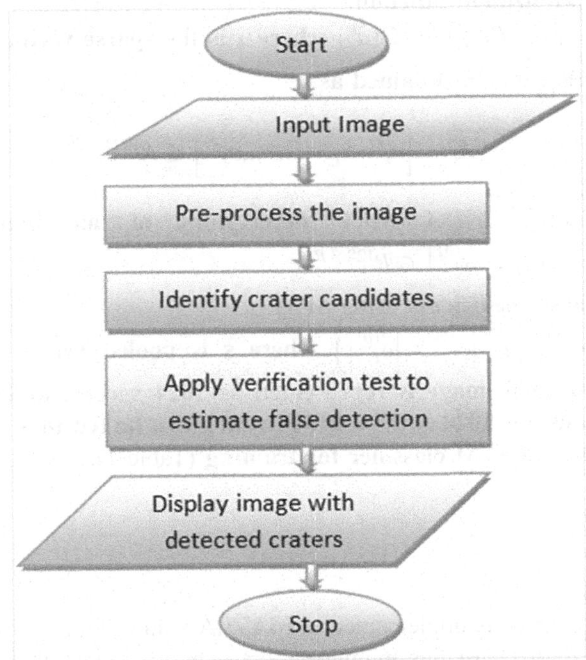

Fig. 7. Framework for crater detection

3.2 Phase 1: Classification

The images obtained from Phase 1 are classified as per following algorithm:

(a) **Input a set of images (X_i, C_i) where X_i is set of input images and C_i their respective class labels.**

(b) **For each image in the set X = {X_1, X_2..... X_N}, the SIFT descriptors are computed as**
Y = {Y_1, Y_2...Y_N} where Y_i is SIFT representation of image X_i and each Y_i contains 128-dimensional descriptors hence forming the training.
Hence, $Y_i = \left[y_i^{(1)}, y_i^{(2)} \ldots\ldots y_i^{(p)}\right] \in R^{128 \times p}$ **where p is no of descriptors in** ith **image.**

(c) **Of all y_i, choose K random descriptions and learn a Dictionary D via sparse coding by optimizing**

$$min_{D,a^{(k)}} \frac{1}{K} \sum_{k=1}^{K} \left\{ \frac{1}{2} \|y - Da\|_2^2 + \lambda \Omega\left(a^{(k)}\right) \right\}$$

where λ is the control parameter and provides trade-off between sparsity and no of atoms. $D \in R^{128 \times 1024}$ **the size of the Dictionary being 1024.**

(d) **Apply D obtained from step (c) to code all the SIFT descriptors in each image using the optimization function**
$\forall y \in Y, \min_{\propto} \frac{1}{2} \|y - Da\|_2^2 + \lambda \Omega(a^k)$ **where α is the sparse vector.**
Sparse coefficients are obtained as

$$A_i = \left[a_i^{(1)}, a_{i,}^{(2)} \ldots\ldots\ldots a_i^{(p)}\right] \in R^{1024 \times P}$$

where p are no of SIFT descriptors of an image given as $A_i = \left[y_i^{(1)}, y_i^{(2)}, \ldots\ldots\ldots y_i^{(p)}\right] \in R^{128 \times P}$

(e) **Pool the sparse coefficients using SPM**
$S_d = \max\left(\left|a_i^{(1)}\right|, \left|a_i^{(2)}\right|, \ldots\ldots, \left|a_i^{(p)}\right|\right)$ **where s is pooled vector and S_d its d_{th} element. The final image is represented as 1024 vectors in 21 scales, hence having dimensions 1024 × 21 = 21504 which can be fed to a classifier.**

(f) **Each S_d is fed to SVM classifier for training** (Table 1).

4 Results

The proposed algorithm is implemented in MATLAB, in the form of interactive software. The program first classifies an image as crater or non-crater. This extra stage is to reject those images which are some other scene images like forest, buildings. Images detected as craters are fed to intraclass classifier which classifies it into its morphological class. The intraclass classifier is implemented according to our 2-phase method. To show outperformance of our method, it is compared with other techniques like the one which uses pixel-based representation of image with SVM, one uses only SIFT descriptors and their Bag-of-words coding and SVM, one uses SIFT with Sparse Coding and knn classifier and our proposed approach which uses SIFT with Sparse coding and SVM classifier. The images for training and testing are downloaded from lpi.usra.edu.

A number of testing and training images for the stages of classification are summarized in Table 1. The dataset comprises of 455 crater images and 185 non-craters for training the 1st stage of classifier. The classifier is tested for 1st stage on 35 images of which 28 are crater and 7 are non-crater. The dataset for second stage of classification comprises of already classified craters which is used as ground truth for validation. We had 44 central dome images available and 32 multiring images available. We have used 35 central dome training images and 9 test images and 27 multiring training images and 5 as test images. The test images are totally out of training set. Experimental setting: For SIFT, extraction number of octaves is taken as 3, number of scales as 5, sigma as 1.6, k = 1.414, grid spacing as 6, patchsize as 16, maximum image size as 300. We have resized all images to 300 to maintain consistency. All the images are preprocessed into gray scale. For Bag-of-words model, Dictionary size is taken as 50. We have experimented on various Dictionary sizes for sparse coding and found accuracy to be maximum when Dictionary size is 1024. Sparse coding parameter λ is taken 10% with number of supports 10. Number of iterations for Dictionary learning is taken as 40 and random samples to train the Dictionary has been taken as 2% of total SIFT descriptors extracted. We have repeated the experiment by reshuffling the training and test images and then found the average of all the results and summarized as shown in Table 2. The classifier is tested on 35 images of which 28 are crater and 7 are non-crater for the first stage. The accuracy for the first approach which uses pixel representation is 85.71 which immediately jumps to 94.29 as the SIFT representation is used with SIFT and knn. Our proposed approach with Sparse coding and SVM shows further more accuracy of 97.14 which is highest of all methods. For the second stage, the accuracy jumps from 77.14 to 88.57 on introduction of SIFT transformation and classified using Bag-of-words approach which further escalates to 94.29 with sparse coding (Table 3). The accuracy also shows an increase when knn is replaced by svm, a valid justification for our selection of svm. We have manually induced various distortions like addition of noise, occlusion, clutter, various illumination effects to test the robustness of the approach. It is found that our proposed system is robust against salt and pepper noise, occlusion, and poor illumination (Figs. 8, 9, and 10).

Table 1. Summary of data

1st stage				
Training		Testing		
Positive	Negative	Crater	Non-crater	Total
455	185	28	7	35
2nd stage				
Training		Testing (output from first stage)		
CentralDom	Multiring	CentralDom	Multiring	Total
35	27	9	5	14

Table 2. Performance analysis of all classification methods

Method	Classifier	Morphology	TP	FP	TN	FN	Recall (%)	Accuracy (%)	Precision (%)
Classification without using SIFT but only SVM	Binary		23	0	7	5	82.17	85.71	100.00
	Intraclass	CentralDom	3	7	23	2	60.00	74.29	30.00
		Multiring	3	6	24	2	60.00	77.14	33.33
Classification using SIFT with SVM	Binary		26	0	7	2	92.86	94.29	100.00
	Intraclass	CentralDom	3	2	28	2	60.00	88.57	60.00
		Multiring	4	3	27	7	80.00	88.57	57.14
Classification using SIFT with KNN classifier	Binary		27	1	6	1	96.43	94.29	96.43
	Intraclass	CentralDom	3	2	28	2	60.00	88.57	60.00
		Multiring	3	3	27	2	60.00	85.71	50.00
Classification using sparse coding	Binary		28	1	6	0	100.00	**97.14**	96.55
	Intraclass	CentralDom	4	1	29	1	80.00	**94.29**	80.00
		Multiring	5	2	28	0	100.00	**94.29**	**71.43**

Table 3. Effect of Dictionary size on accuracy

Dictionary size	Accuracy
256	92.6
512	92.8
1024	94.29

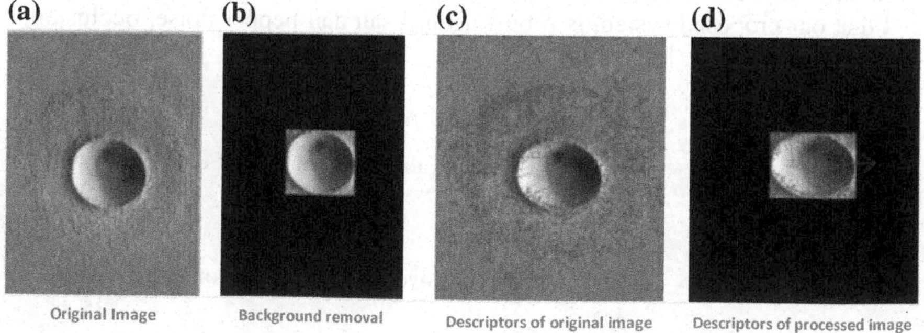

Fig. 8. (**a**) Original image. (**b**) Background removal. (**c**) Descriptors of original image. (**d**) Descriptors of processed image

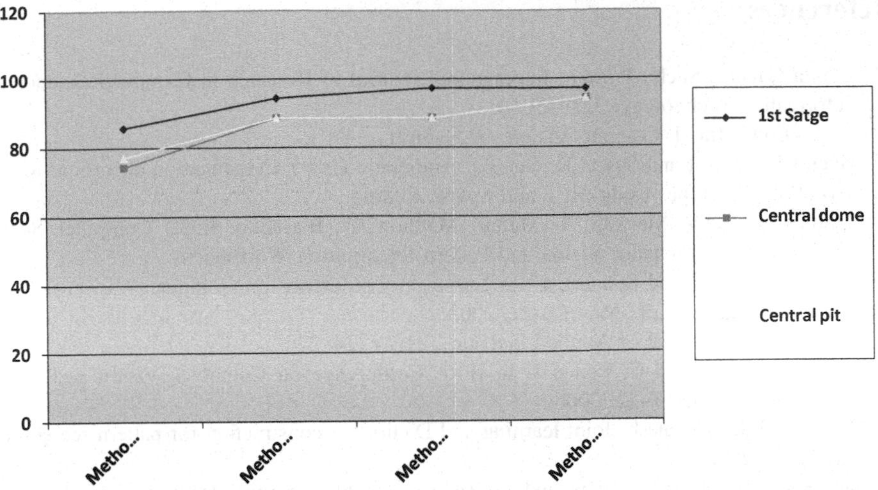

Fig. 9. Comparisons of accuracies of different stages

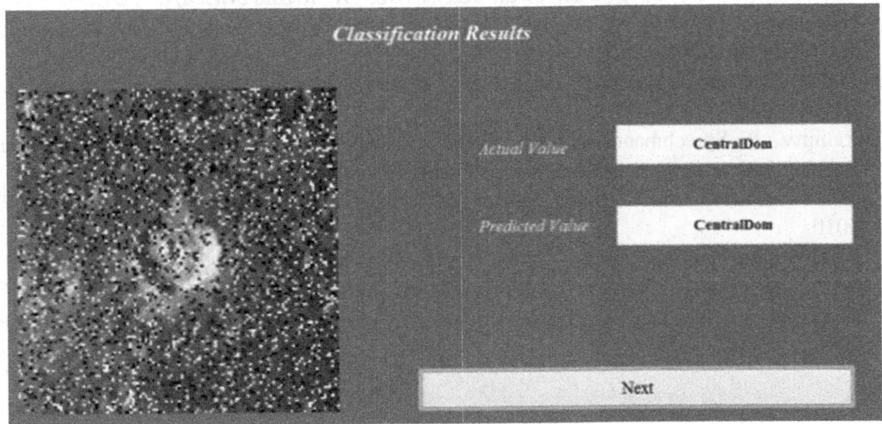

Fig. 10. Detection in presence of noise

5 Conclusion

The paper presents a comprehensive platform where different crater detection and classification techniques can be used and tested. Further, a novel crater classification approach using fusion of detection and classification approaches has been proposed, implemented, and tested. The proposed approach using SIFT and sparse coding is found to superior than other techniques and also robust in the presence of noise, occlusion, and low illumination.

References

1. Savita Gandhi, Suchit Purohit. International Journal of Research in Computer Science and Information Technology. 2319–5010.
2. D. G. Lowe. Int. J. Comput. Vision. 0920–5691.
3. Suchit S. Purohit and Savita R. Gandhi. Automatic Crater Classification according to their morphological types using SIFT and Sparse Coding.
4. Dimitri A. Lisin, Marwan A. Mattar, Matthew B. Blaschko. IEEE Computer Society Conference on Computer Vision and Pattern Recognition Workshops.
5. Mikolajczyk, K., and Schmid, C. In European Conference on Computer Vision (ECCV), Copenhagen, Denmark, May 28–31, 2002.
6. Lindeberg, T. Journal of Applied Statistics, 21(2):224–270.
7. Lazebnik S, Schmid C, Ponce J. In IEEE Conference on Computer Vision and Pattern Recognition, June 17–22, 2006.
8. Pham and S. Venkatesh. Joint learning and Dictionary construction for pattern recognition. In CVPR, 2008.
9. M. Yang, L. Zhang, J. Yang and D. Zhang. Metaface learning for sparse representation based face recognition. In ICIP, 2010.
10. J. Mairal, F. Bach, J. Ponce, G. Sapiro, and A. Zissserman Learning discriminative dictionaries for local image analysis. In CVPR, 2008.
11. Yang J-C, Yu K, Gong Y-H, et al. JOURNAL OF MULTIMEDIA, VOL. 9, NO. 1, Proceedings of IEEE Conference on Computer Vision and Pattern Recognition, pp. 1794–1801, 2009.
12. Mairal, F. Bach, J. Ponce, G. Sapiro, and A. Zisserman. Supervised Dictionary learning. In NIPS, 2009.
13. Ramirez, P. Sprechmann, and G. Sapiro. Classification and clustering via Dictionary learning with structured incoherence and shared features. In CVPR, 2010.
14. J.C. Yang, K. Yu, and T. Huang. Supervised Translation-Invariant Sparse coding. In CVPR, 2010.
15. F. Rodriguez and G. Sapiro. Sparse representation for image classification: Learning discriminative and reconstructive non-parametric dictionaries. IMA Preprint 2213, 2007.
16. Sivic and A. Zisserman. In IEEE Int. Conf. Computer Vision, Nice, France, 13–16 Oct. 2003.

eLL: Enhanced Linked List—An Approach for Handwritten Text Segmentation

M. Yashoda[1(✉)], S. K. Niranjan[2], and V. N. Manjunath Aradhya[2(✉)]

[1] Department of CSE, Bahaubali College of Engineering, Shravanabelogola, India
myashoda29@gmail.com

[2] Department of MCA, Sri Jayachamarajendra College of Engineering, Mysuru, India
aradhya.mysore@gmail.com

Abstract. In document image analysis, the interesting and challenging lie in textline segmentation of handwritten documents. This work emphasizes on developing an enhanced handwritten textline segmentation technique based on the concept of linked list. The proposed system consists of three phases namely preprocessing, enhanced linked list (eLL), and morphology processing. The experiment is evaluated on a document containing handwritten Kannada and English script, and the results are promising.

1 Introduction

In off-line handwritten character recognition (HCR) system, textline segmentation is considered to be essential stage. As it is one of the key processes because inaccurate segmented textlines lead to HCR failure. Text orientation in printed text is not variable, which often defined by shape regularity and its textlines have similar orientation, and its skewness is also similar or equal. The nature of handwritten text is fully or partially cursive and seems to tend to be multi-oriented and skewed. Primarily, textlines in handwritten are close to each other and curvilinear. Descenders and ascenders of neighbor textlines especially in south Indian documents are occasionally get mixed up. Like printed text, handwritten text's inter-word spacing is tolerable and is not formed regularly, because their distance is different. The handwritten text makes less readable due to the skewed lines with different orientation. Textline segmentation of handwritten documents is really challenging and complex due to its nature of handwritings. Challenges include (i) variation in skew between different textlines, (ii) fluctuating skew within a textline, (iii) overlapping text in pages with crowded writing where characters of adjacent textlines have overlapped, and (iv) characters in one line touches text in adjacent lines.

Textline segmentation can be roughly categorized [1] into **(i) smearing methods, (ii) horizontal projections, and (iii) bottom-up approaches.** The limitation of the smearing methods cannot handle considerably with touching and overlapping components. The main drawback of horizontal projections approaches is that it doesn't work well for skewed, curved, and fluctuating lines. Handwritten Persian/Arabic

V. Bhateja et al. (eds.), *Information Systems Design and Intelligent Applications*, Advances in Intelligent Systems and Computing 672,
https://doi.org/10.1007/978-981-10-7512-4_86

baseline tracing and straightening of textline are described in [2]. The model identifies the candidate points from white and black blocks all along the textline. By the use of candidate points, the algorithm traces the baseline, which is subsequently stretched straight horizontally. Recently, an elegant method for unconstrained handwritten textline segmentation is proposed in [3]. The approach is based on piece-wise painting algorithm (PPA). The painting technique enhances the separability between the foreground and background portions enabling for detection of textlines. In [4], the thesis on bilingual handwritten character recognizer is described. The thesis focuses on developing bilingual handwritten character recognizer for Kannada and English text. The thesis addresses the development of few novel algorithms on textline segmentation, word skew level correction, and feature extraction based on subspace algorithms and classification algorithms. Few interesting work on textline segmentation can also been seen in [5–12].

2 Proposed Methodology

In this section, we propose an improved linked list approach for text segmentation of handwritten documents.

A. ***Review on Linked List Approach*** [13]:
The method is categorized into three phases namely preprocessing, linked list, and thinning. The first part of the phase is considered to be preprocessing, where application of connected component analysis is applied to the original document image. In continuation, the components are blocked for the purpose of creating nodes. In doing so, some blocks may appear to connect two consecutive textlines or having longer heights with respect to the heights of the other blocks. These blocks indicate overlapping/touching portions. The height of these blocks will be greater or equal to twice the size of the average height of the blocks. All such longer blocks are removed temporarily, and these blocks are used in final stage of textline segmentation. In the next phase of the algorithm, the idea of linked list approach is used to group nodes together which represent a sequence. To build these sequences of textline, the method computes the center point (say "R_{mid}") of the right part of each individual node. From this point, the horizontal line is drawn with respect to the zero degree till it reaches the next nearest neighbor node or up to the end of the text page. At last, to obtain all the components, the longer height blocks (that were not considered previously) are added for subsequent process. At final phase of the algorithm, thinning operation is performed to obtain complete components distinguished as textline separators.

B. ***An Improved Linked List Approach for Text Segmentation***
The main motivations for an improved linked list approach are as follows:
(i) The standard linked list method fails to segment the textline if having more zig-zag nature.
(ii) Proposed to work for more skew (say > 40°).
(iii) Experiment to conduct on trilingual dataset comprising of Kannada, Hindi, and English.

The proposed improved approach consists of the same three phases as discussed in previous section. In the first phase, the input document image is filtered for noise using median filter. Next, by using connected component analysis algorithm, we label the components and mark those components whose height is lesser than specific threshold. Those box height less than specific threshold are debarred from further processing. Remaining components are blocked for the purpose of creating nodes for further processing. Figures 1 and 2 show the result of CCA method for its input image.

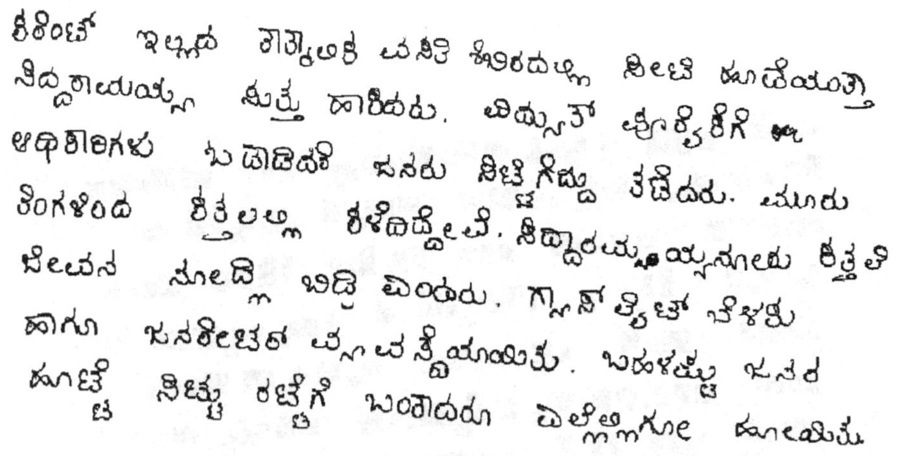

Fig. 1. Input Kannada handwritten text document

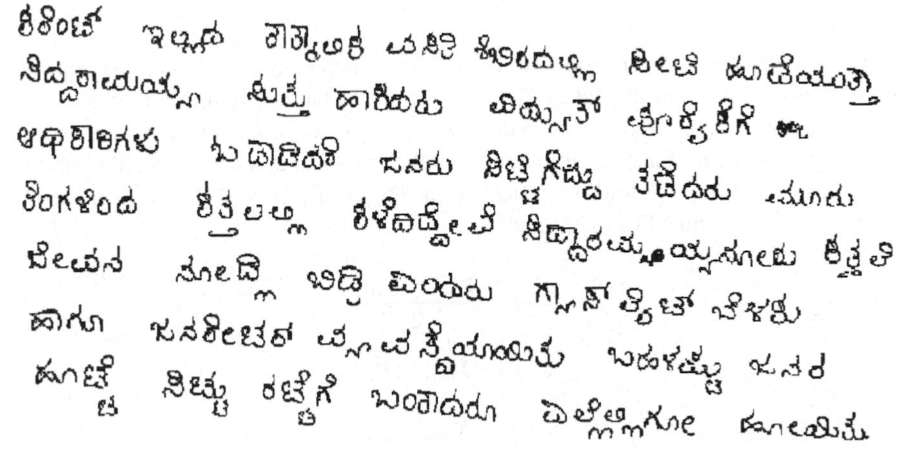

Fig. 2. Components having lesser height are debarred using CCA algorithm

In the continuation of the algorithm, some blocks may appear to connect two consecutive textlines or having longer heights with respect to the heights of the other blocks. These blocks indicate overlapping/touching portions. Generally, when two textlines touch/overlap, two or more consecutive blocks obtained from the joined lines make a longer block. The height of these blocks will be greater or equal to twice the size of the average height of the blocks. All such longer blocks are removed temporarily which is shown in Fig. 3. Figure 4 shows the final components that are considered for further processing.

Fig. 3. Components having greater heights

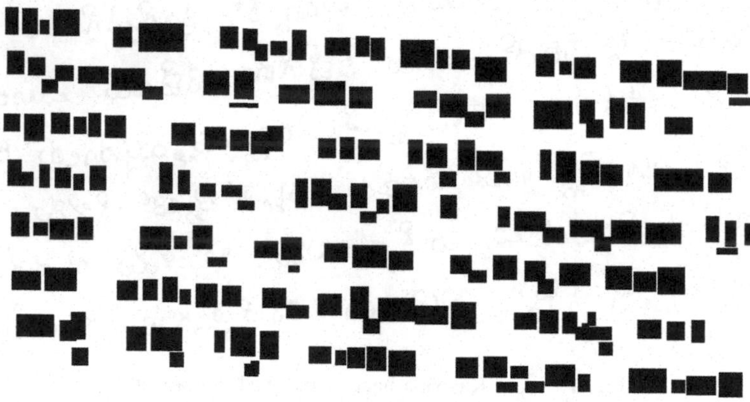

Fig. 4. Components considered for further processing

The second stage of our proposed method is the application of linked list approach. The idea is educed from the data structure concept. In data structure, linked list means group of nodes together represent a sequence; based on this concept, we call each block in the Fig. 4 as node. Each textline node together represents a sequence. To build this sequences of textline, we find the center point (R_{mid}), (R_{top}), and (R_{bottom}) of the right part of the nodes using Eqs. 2, 3, and 4.

$$Mid = \frac{Y_{max} + Y_{min}}{2} \tag{1}$$

$$R_{mid} = (X_{max}, Mid) \tag{2}$$

$$R_{top} = (X_{max}, Y_{min}) \tag{3}$$

$$R_{bottom} = (X_{max}, Y_{max}) \tag{4}$$

In standard linked list approach, from the Rmid point, horizontal line is drawn with respect to the zero degree up to the line reaches to the nearest neighbor node or up to the end of the text page is computed. This may not hold good for the textline having more skew (say 40°). In this situation, the idea of finding the components from all the three points, i.e., R_{mid}, R_{top}, and R_{bottom} or through each individual point is considered to be the major improvements incorporated in this proposed method.

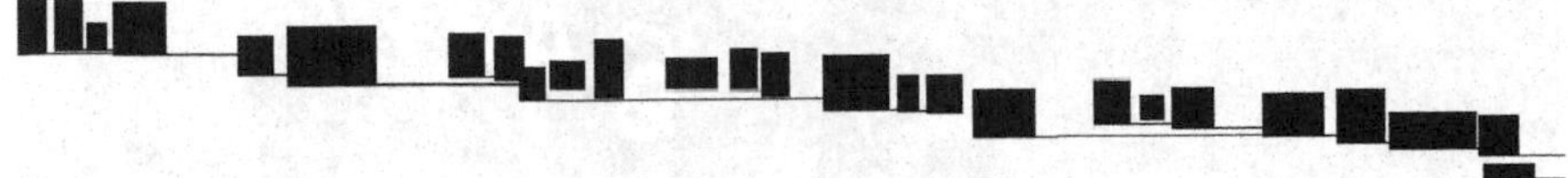

Fig. 5. Resultant of linked list for all three points

In Fig. 5, every sequence indicates the path information for final textline segmentation. Then, the sequenced image is thinned by applying the morphological thinning algorithm to obtain textline separators because morphology is a powerful tool used to find region skeleton. These line separators do not contain all the components in the textline because in preprocessing stage we have extracted longer height blocks. Hence, to obtain all the components, we have added longer height blocks into the thinned image which is obtained in our previous step. After adding a longer height block again, we applied thinning operation to obtain complete components distinguished textline separators. Finally, we add the components of the input image to the resultant complete components textline separator image. The result of final textline segmentation is shown in the Fig. 6.

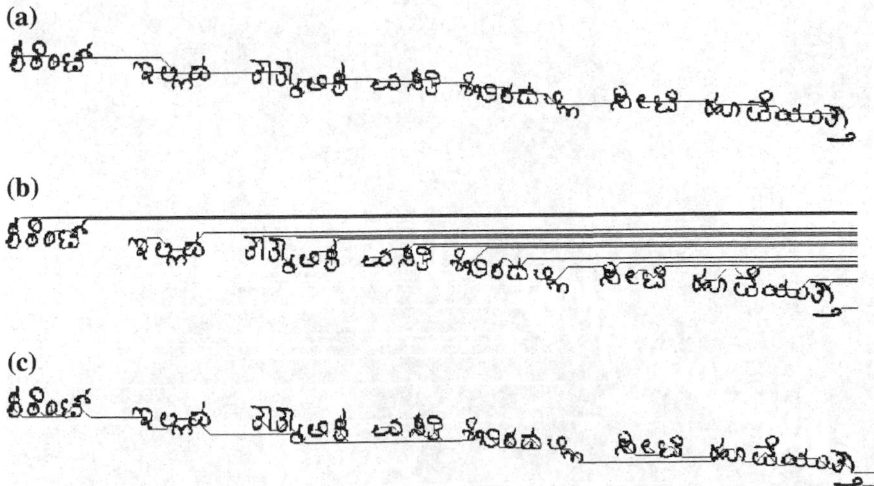

Fig. 6. Result of the proposed textline segmentation **a** Considering R_{mid} point **b** Considering R_{top} point **c** Considering R_{bottom} point

The failure case of the standard linked list method is depicted in Fig. 7. The advantage of the proposed improved linked list is shown in Fig. 8.

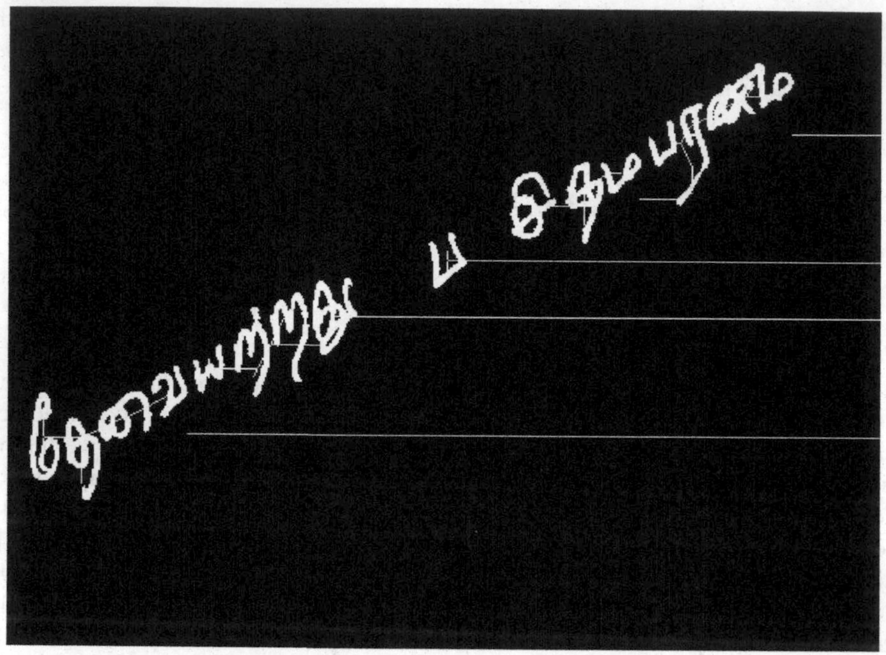

Fig. 7. Failure case of standard linked list approach

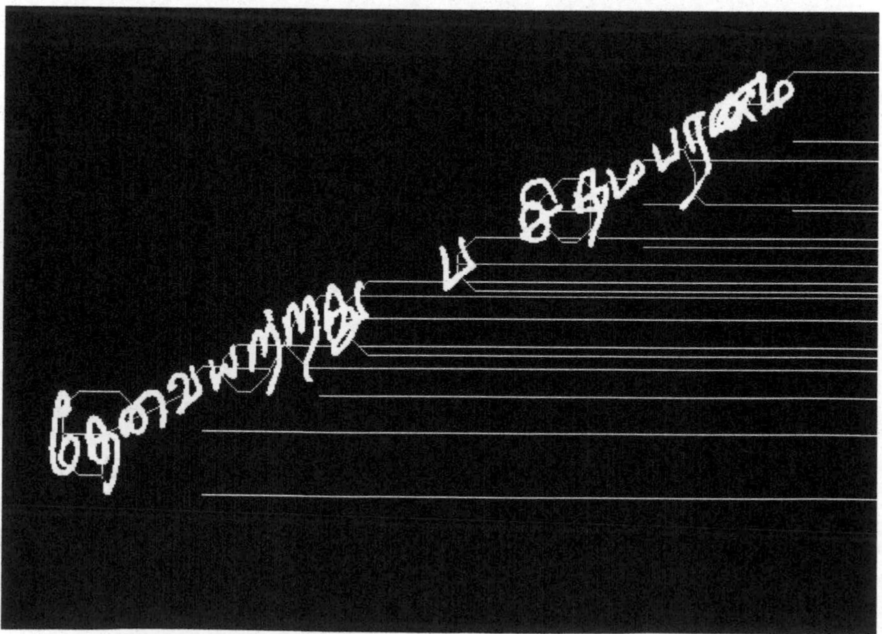

Fig. 8. Result of proposed improved linked list approach for textline having more skew

3 Experimental Results and Comparative Study

In this section, we present the result of the experiment conducted to study the effectiveness of the proposed method. For the experiment purpose, we consider the database of textline handwritten documents mentioned in [4].

Table 1 shows the recognition accuracy obtained (one2one) for proposed method. From the table, it is noticed that the proposed achieved the recognition accuracy of 92.38% for the database considered. In order to reduce the error rate of the proposed method, this is mainly due to touching components of textlines. We reduced the longer height blocks to 15% using mathematical erode operation. This procedure leads to improve in recognition accuracy of about 6%. A comparison with other existing methods is shown in Table 1. From Table, it is clear that the proposed improved linked list approach outperforms all other methods. The results of the proposed improved linked list approach show better efficiency if the textlines have more fluctuating nature and with high skew when compared to standard linked list approach.

Table 1. Recognition accuracy (%) obtained for existing methods and proposed methods on textline database considered in [4]

Method	N	One2One	Recognition (%)
PPA [3]	617	449	72.89
Improved PPA [4]	617	502	81.3
CET [4]	617	447	72.4
MCET [4]	617	507	82.17
Standard linked list [13]	617	552	89.4
Proposed enhanced linked list	617	567	92.38

4 Conclusion

In off-line HCR system, textline segmentation is considered to be essential stage. As it is one of the key processes because inaccurate segmented textlines lead to HCR failure. In this paper, we proposed an enhanced handwritten textline segmentation technique based on the concept of linked list. The proposed system consists of three phases namely preprocessing, enhanced linked list (eLL), and morphology processing. The experiment showed the efficiency of the proposed method, and the results are promising. In future, the authors plan to explore the concept of eLL to other Indian languages.

References

1. A. Nicolaou and B. Gatos. Handwritten text line segmentation by shredding text into its lines. In the proceedings of 10th Intl conference on Document Analysis and Recognition, pages 626–630, 2009.

2. P. Nagabhushan and A. Alaei. Tracing and straightening the baseline in handwritten persian/Arabic text-line: A new approach based on painting -technique. In the proceeding of Intl Journal on Computer Science and Engineering, pages 907–916, 2010.
3. Alaei, U. Pal, and P. Nagabhushan. A new scheme for unconstrained handwritten text-line segmentation. Patten Recognition, 44(4):917–928, 2011.
4. C. Naveena, Bi-lingual Handwritten Character Recognizer, Ph.D Thesis, VTU, Belagavi, Karnataka, India, 2012.
5. U. Pal ans S. Datta. Segmentation of bangla unconstrained handwritten text. In the proceedings of 7th Intl conference on Document Analysis and Recognition, number 2, page 1128, 2003.
6. S. Basu, C. Chaudhuri, M. kundu, M. Nasipuri, and D.K. Basu. Text line extraction from multi-skewed handwritten documents. Patten Recognition, 40(4):1825–1839, 2007.
7. B.B. Chaudhuri and S. Bera. Handwritten text line identification in indian scripts. In the proceedings of 10th International Conference on Document Analysis and Recognition, pages 636–640, 2009.
8. F.Yin and C.L. Liu. Handwritten chinese text line segmentation by clustering with distance metric learning. Patten Recognition, 42(4):3146–3157, 2009.
9. B. Gatos, N. Stamatopoulos, and G. Louloudis. ICDAR2009 handwriting segmentation contest. In the Proceeding of 10th International Conference on Document Analysis and Recognition, pages 1393–1397, 2009.
10. M. Liwicki and H. Bunke. Combining diverse on-line and off-line systems for handwritten text line recognition. Patten Recognition, 42(12):3254–3263, 2009.
11. G. Louloudis, B. Gatos, I. Pratikakis, and C. Halatis. Text line and word segmentation of handwritten documents. Patten Recognition, 42:3169–3183, 2009.
12. G. Louloudis, B. Gatos, I. Pratikakis, and C. Halatsis. Text line detection in handwritten documents. Patten Recognition, 41:3758–3772, 2009.
13. A Linked List Approach for Unconstrained Handwritten Text Line Segmentation", Journal of Intelligent Systems, Vol. 21, Issue 3, pages. 225–235, 2012.

Intensity Normalization—A Critical Pre-processing Step for Efficient Brain Tumor Segmentation in MR Images

S. Poornachandra[1], C. Naveena[2], and Manjunath Aradhya[3(✉)]

[1] Department of Computer Science and Applications, VTU RRC, Belagavi, India
thisispoorna9@gmail.com

[2] Department of Computer Science and Engineering, SJB Institute of Technology, Bengaluru, India
naveena.cse@gmail.com

[3] Department of MCA, Sri Jayachamarajendra College of Engineering, Mysore, India
aradhya.mysore@gmail.com

Abstract. In this paper, we present the pre-processing approaches for MRI brain scans. The magnetic bias field correction of MR images is a preliminary step and the subsequent pre-processing step of intensity normalization of MR images. Both of these pre-processing steps facilitate as promising inputs to the segmentation models, and the promising outputs from these segmentation models aid for better diagnosis and prognosis of diseases. The BRATS 2015 dataset is used in the experimental work; the intensity normalization techniques applied to this dataset yield good results in segmenting the *gliomas*, thus enhancing the further image analysis pertaining to *gliomas*.

Keywords: Gliomas · MRI · Histogram matching · N4ITK

1 Introduction

Today, the computer-aided diagnosis (CAD) has grown remarkably well in providing efficient analysis of medical images for better diagnosis and prognosis of the various diseases faced by the mankind. There are various medical imaging modalities used for detecting, diagnosing, and monitoring of various diseases. To name a few medical imaging modalities: X-ray used to detect mammogram calcifications, CT for lung cancer detection, MRI for MS lesion and brain tumor detection.

In this paper, we present the pre-processing approaches for MR images used in brain tumor analysis. Pre-processing is a critical step for MR images, which helps in subsequent steps of segmentation, classification, and analysis of brain tumor. The magnetic field bias in MR images occurs during the MR image-acquisition process, thus infiltrating the further image analysis in MRI-based diagnosis.

V. Bhateja et al. (eds.), *Information Systems Design and Intelligent Applications*, Advances in Intelligent Systems and Computing 672,
https://doi.org/10.1007/978-981-10-7512-4_87

This magnetic field bias is introduced due to the varying magnetic field strengths (ranging from 1.5–3Tesla for clinical usage) provided by different vendors of MRI scanners and also due to variation in excitation and repetition times during MR image-acquisition. Thus, for the same patient, different scanners and different acquisition protocols yield different intensities for the MRI scans, and correcting this bias is of prime importance in medical image analysis. After this, the intensity normalization is also a necessary pre-processing step in MR images; we present the intensity normalization techniques which denoise the image and enhance the quality of the MR image which aid in subsequent segmentation of *gliomas* and help the doctors in better diagnosis and treatment planning for the patients affected with brain tumor.

This paper is arranged as follows: Sect. 2 emphasizes on the various researchers working in the field of brain tumor image analysis, Sect. 3 reviews on the brain tumors and its types, Sect. 4 explains the brain tumor imaging using different MRI modalities, Sect. 5 explains about the pre-processing as a critical step for brain tumor image segmentation and analysis, and Sect. 6 describes about the techniques used for intensity normalization in MR images.

2 Literature Survey

In this section, we put forth the active researchers working in the field of brain tumor segmentation and analysis and the pre-processing steps they applied to yield better segmentation results.

Havaei Mohammad et al. [1] proposed "Brain Tumor Segmentation with Deep Neural Networks" in which as a pre-processing technique, they applied N4ITK filter for correcting the bias field and then normalized the data in each input channel by subtracting the channel's mean and dividing by the channel's standard deviation. Sergio Pereira et al. [2] proposed "Brain Tumor Segmentation using Convolutional Neural Networks in MRI Images" in which as a pre-processing step, they employed N4ITK filter for bias field correction, and to have similarity in contrast and intensity across patients, they applied the intensity normalization method proposed by Nyul et al. [3]. Laszlo et al. [4] proposed "Brain Tumor Segmentation with Optimized Random Forest," and as a pre-processing step, they employed N4 filter for inhomogeneity reduction, and intensity normalization was done by histogram linear transformation.

Jens Kleesiek et al. [5] proposed "*ilastik* for Multi-Modal Brain Tumor Segmentation" and employed Histogram Matching as a pre-processing step for MR images and also normalized each individual modality with the mean value of the CSF. Mohammad Havaei et al. [6] proposed "A Convolutional Neural Network Approach to Brain Tumor Segmentation," and as a pre-processing step, employed N4ITK bias correction and removed the highest and lowest intensities. Gregor Urban et al. [7] proposed "Multi-Modal Brain Tumor Segmentation using Deep Convolutional Neural Networks" and employed normalization technique to MR images with mean CSF values. Michael Gotz et al. [8] proposed "Extremely Randomized Trees based Brain Tumor Segmentation," N4ITK filter [9] was applied for MR images to correct the bias field, and the histograms were normalized. Joana Festa et al. [10] proposed "Automatic Brain Tumor Segmentation of Multi-sequence MR Images using Random Decision Forests" as a pre-processing step for MR images, N4ITK was employed and the intensity scale of each sequence was normalized. Hence, these are some of the active researchers working in brain tumor image analysis.

3 Brain Tumors

This section explains about the types of brain tumors, their origin of occurence, and grading of tumors by World Health Organization (WHO). Tumor is a Latin word, which means swelling. Brain tumors are the dreadful cancerous tissues which decrease the survival rate of patients affected by it, and more aggressive brain tumors result in immediate death of the patient.

Brain tumors can be classified based on their origin and infiltration. Primary brain tumors originate in the brain and do not spread to the other parts of the body, whereas metastatic brain tumors originate from other parts of the body affected with cancerous cells like kidney cancer, breast cancer, and lung cancer. Again, in the primary brain tumors, there are two types of tumors, *viz.* malignant and benignant. Malignant brain tumors are more cancerous and spread rapidly, whereas benignant tumors are less cancerous and less aggressive.

Gliomas are the primary brain tumors in adults and are malignant primary brain tumors which originate from glial cells. World Health Organization (WHO) classifies gliomas based on their aggressiveness from grades I to IV. The high-grade gliomas (HGG) are of grade III or grade IV. The grade IV gliomas are also known as Glioblastoma Multiforme (GBM) and are most common malignant primary brain tumors with a very rapid growth thus eventually leading to the subject's death [11]. Grade I and II are low-grade gliomas (LGG) which are semi-malignant tumors, less aggressive, and the survival rate of subjects affected by LGG is bit higher than HGG. In this paper, we describe pre-processing of MR images containing these cancerous *gliomas.*

4 Brain Tumor Imaging

Due to wide availability in clinics and good soft tissue contrast, MRI has become a gold standard technique for brain tumor diagnosis. By varying excitation and repetition times, MRI can produce different types of soft tissue contrast.

Brain tumors constitute only 4% of the entire brain. The pathological information of brain tumors is more complicated, hence only one MRI sequence or modality is not enough to segment brain tumors including its complex subregions like edema, necrotic cores, enhancing tumor, and other tissues in the brain. Different MRI sequences provide various pathological information for effective analysis of brain tumors. Thus, making the diagnosis, monitoring, and treatment planning easier.

The following MRI sequences are used for brain tumor detection and diagnosis.

- T1-weighted MRI
- T2-weighted MRI
- T1-weighted MRI with contrast enhancement
- Fluid-attenuated inversion recovery (FLAIR) MRI

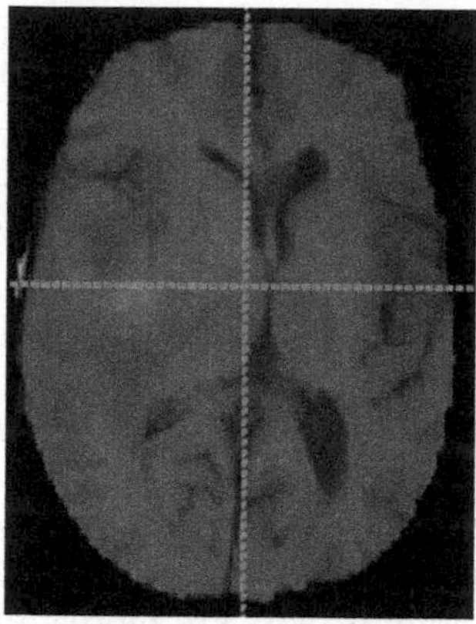

Fig. 1. T1 MRI

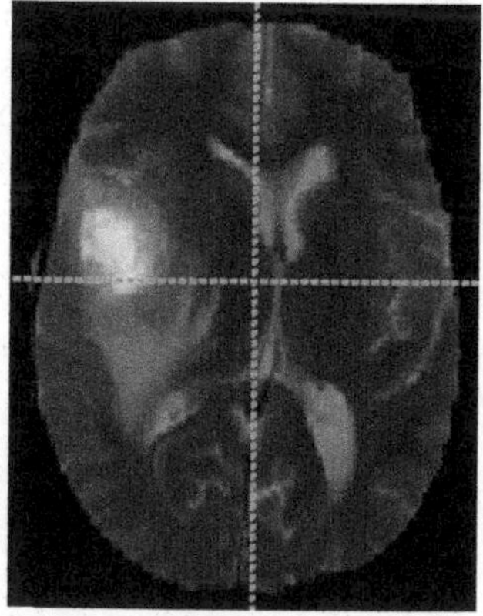

Fig. 2. T2 MRI

The T1-weighted MRI is used for labeling of healthy tissues and structural analysis. In T2-weighted MRI, the edema region surrounding the tumor can be easily differentiated.

In T1-weighted contrast-enhanced MRI, the necrotic and the active tumor regions are enhanced. The FLAIR MRI helps in seperating the edema region from CSF. Due to these capabilities, the above said four MRI modalities are widely used for brain tumor diagnosis and prognosis [12,13]. Figures 1, 2, 3, and 4 are the T1, T2, T1c and FLAIR MRI sequences from the BRATS 2015 dataset which are used to segment the brain tumor.

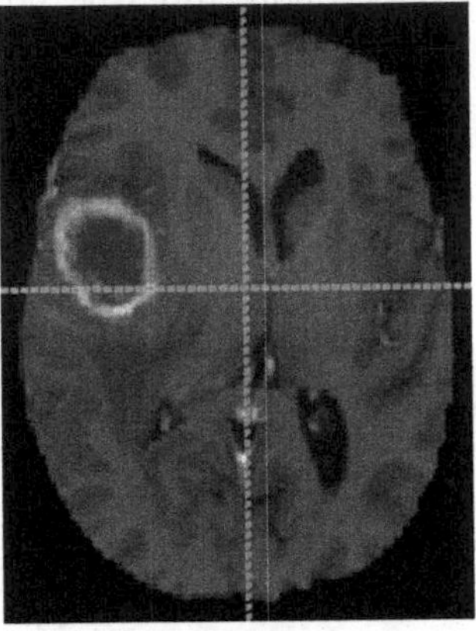

Fig. 3. T1c MRI

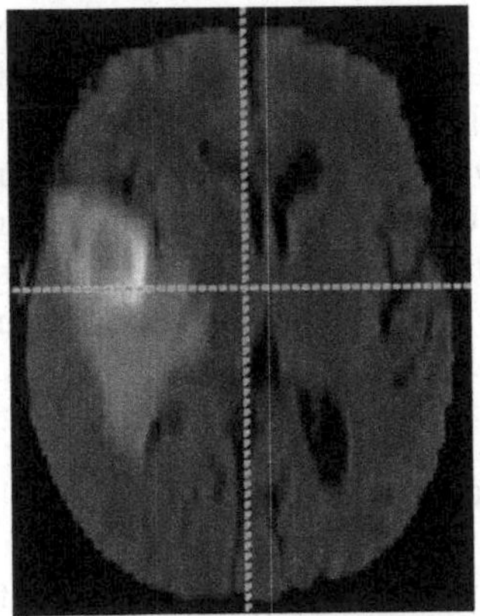

Fig. 4. FLAIR

5 Pre-processing as a Critical Step for Brain Tumor Segmentation

Segmenting the brain images bearing tumors is a cumbersome task because of the inter-scanner and intra-scanner variations in intensity of same tissue type.

First and foremost, high-grade gliomas have irregular boundaries and discontinuities for different tumor subregions. These tumor subregions can be effectively separated when different MRI modalities are combined.

Most of the segmentation approaches employ bias field correction as a primary step to eradicate the effect of magnetic field inhomogeneities during the MR image-acquisition process. Secondly, intensity normalization is also a critical step for segmentation and classification of MR images containing brain tumors. In this paper, we put forth the intensity normalization techniques for the tumor-bearing MR images.

Employing efficient pre-processing techniques yields better segmentation and classification results which in turn helps in better diagnosis of brain tumors and therefore aids in subsequent treatment planning and tumor growth monitoring. Hence, pre-processing of brain scans bearing tumors is an essential and critical step in brain tumor image analysis [12, 14, 15].

6 Methodology

In this section, we discuss about the methods applied to pre-process the MR images. The N4ITK filter was applied to correct the magnetic bias field and also normalize the MR data by subtracting the volume's mean and dividing by the volume's standard deviation within each input volume [16].

This paper describes the pre-processing approach for enhancing the image intensities in MRI. The proposed method consists of two steps. The pre-processing step of standardizing MR image intensities. The bias field corrected MR images are further pre-processed in which the reference MR image is taken and the histogram matching technique is applied to the other MR images with threshold match points. During matching, the background was excluded for all voxels whose grayscale values were less than the mean grayscale value. Further, outliers were removed by truncating the image intensities in the quantile range of 0.05–0.93. Hence, these above pre-processing techniques help in the process of segmentation and effective image analysis of brain tumors.

7 Results and Discussion

In this work, the MICCAI BRATS 2015 challenge data provided by the Virtual Skeleton Database was used. The data is acquired from MR scanners of different vendors. This dataset consists of co-registered, T1-weighted, and contrast-enhanced T1 MRIs and also T2-weighted and T2-FLAIR MRIs. As described in the previous section, the bias field-corrected images were further processed for intensity standardization.

The Histogram Matching techniques were applied to the bias field-corrected images, along with the required parameters for Histogram Matching, *viz.* histogram levels and histogram match points. The histogram levels applied were 128, and 10 histogram match points were set. The resultant output image is shown in Fig. 5. This resultant image's outliers were removed by truncating the image intensities in the quantile range of 0.05–0.93. These pre-processing techniques enable better segmentation results, yielding *Dice Score, specificity, sensitivity* of 0.81, 0.98, 0.84, respectively, for the complete recognition of brain tumor [17]. The resultant output image is shown in Fig. 6. Compared to other pre-processing techniques, these pre-processing techniques yielded good segmentation results. The pre-processing techniques were implemented using the ITK libraries [18].

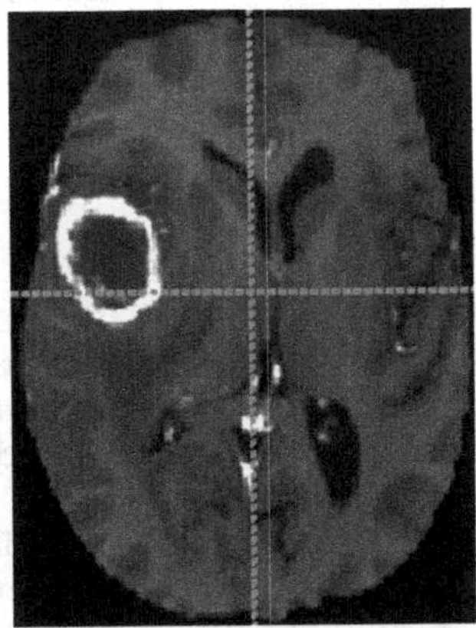

Fig. 5. Pre-processed image by Histogram Matching technique

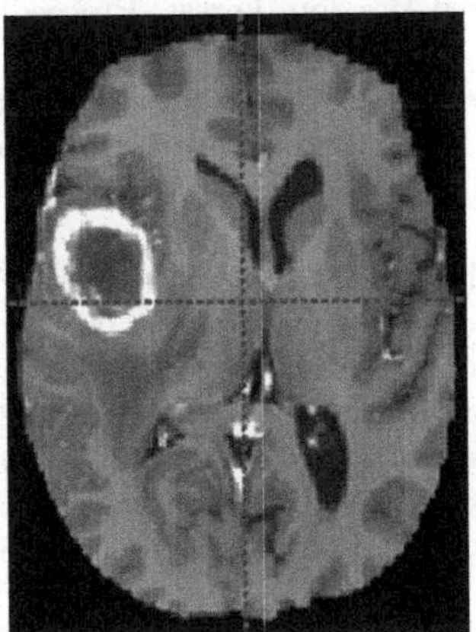

Fig. 6. Pre-processed image by truncating image intensities

8 Conclusion

In summary, we discussed the need for pre-processing the MR images before applying the subsequent image analysis techniques. Also, we discussed the brain tumors, their origin, and severity. Further, we discussed the brain imaging using MRI modality and various MR sequences used in image analysis of brain tumors. The bias field correction technique and the intensity normalization techniques as a pre-processing techniques of MR images were discussed. Thus to conclude, pre-processing of MR images is not only a necessary step but also a critical step to yield better segmentation results, and these segmentation results not only aid in assessing the tumor growth and monitoring but also help in better treatment planning, which in turn increases the survival rate of patients.

References

1. Mohammad Havaei, Axel Davy, David Warde-Farley, Antoine Biard, Aaron Courville, Yoshua Bengio, Chris Pal, Pierre-Marc Jodoin, Hugo Larochelle: Brain Tumor Segmentation with Deep Neural Networks, Medical Image Analysis, Volume 35, 18–31, 2016
2. S. Pereira and A. Pinto and V. Alves and C. A. Silva:Brain Tumor Segmentation Using Convolutional Neural Networks in MRI Images, IEEE Transactions on Medical Imaging, vol. 35, 1240–1251, 2016.
3. Nyúl, László G and Udupa, Jayaram K and Zhang, Xuan: New variants of a method of MRI scale standardization, IEEE Transactions on Medical Imaging, vol. 19, pages 143–150, 2000.
4. Lefkovits, László and Lefkovits, Szidónia and Szilágyi, László: Brain Tumor Segmentation with Optimized Random Forest, Proceedings MICCAI-BRATS Workshop, 2016.
5. Kleesiek, Jens and Biller, Armin and Urban, Gregor and Kothe, U and Bendszus, Martin and Hamprecht, F: Ilastik for multi-modal brain tumor segmentation, Proceedings of MICCAI 2013 Challenge on Multimodal Brain Tumor Segmentation (BRATS 2013).
6. Havaei, Mohammad and Dutil, Francis and Pal, Chris and Larochelle, Hugo and Jodoin, Pierre-Marc: A convolutional neural network approach to brain tumor segmentation, International Workshop on Brainlesion: Glioma, Multiple Sclerosis, Stroke and Traumatic Brain Injuries, Springer, 195–208, 2015.
7. Urban, G. and M. Bendszus and Fred A. Hamprecht and Kleesiek, J: Multi-modal Brain Tumor Segmentation using Deep Convolutional Neural Networks, MICCAI BraTS (Brain Tumor Segmentation) Challenge. Proceedings, winningcontribution, 31–35, 2014.
8. Gotz, M and Weber, Christian and Blocher, J and Stieltjes, Bram and Meinzer, Hans-Peter and Maier-Hein, Klaus: Extremely randomized trees based brain tumor segmentation, Proceedings of BRATS Challenge-MICCAI, 2014.
9. Tustison, Nicholas J and Avants, Brian B and Cook, Philip A and Zheng, Yuanjie and Egan, Alexander and Yushkevich, Paul A and Gee, James C: N4ITK: improved N3 bias correction, IEEE Transactions on Medical Imaging, vol.29, pages 1310–1320, 2010.

10. Joana Festa,Sergio Periera et al: Automatic Brain Tissue Segmentation of Multi sequence MR images using Random Decision Forests, Proceedings MICCAI-BRATS Grand Challenge, 2013.
11. Holland EC: Progenitor cells and glioma formation, Curr. Opin. Neurol., vol. 14, 2001.
12. Mohammad Havaei et al: Deep Learning Trends for Focal Brain Pathology Segmentation in MRI, Machine Learning for Health Informatics: State-of-the-Art and Future Challenges, Springer International Publishing, 125–148, 2016.
13. MICCAI-BRATS, www.braintumorsegmentation.org
14. Bauer, Stefan and Wiest, Roland and Nolte, Lutz-P and Reyes, Mauricio: A survey of MRI-based medical image analysis for brain tumor studies, Physics in medicine and biology, vol. 58, pages R97, 2013.
15. B. H. Menze et al., "The Multimodal Brain Tumor Image Segmentation Benchmark (BRATS)," IEEE Transactions on Medical Imaging, vol. 34, no. 10, pp. 1993–2024, 2015.
16. Poornachandra S,Naveena C: Pre-processing of MR Images for Efficient Quantitative Image Analysis using Deep Learning Techniques, Proceedings International Conference on Recent Advances in Electronics and Communication Technology, pages 191-195, ICRAECT 2017
17. Abdelrahman Ellawa et al: Brain Tumor Segmentation using Random Forest trained on iterative selected patients, Proceedings MICCAI-BRATS Workshop, 2016.
18. Insight Toolkit, http://www.itk.org

Indexing-Based Classification: An Approach Toward Classifying Text Documents

M. S. Maheshan(✉), B. S. Harish, and M. B. Revanasiddappa

Department of Information Science and Engineering, Sri Jayachamarajendra College of Engineering, Mysuru, Karnataka, India
{maheshan,bsharish}@sjce.ac.in, revan.cr.is@gmail.com

Abstract. This paper proposes an indexing-based classification technique to classify text documents. The most important purpose of this paper is to index the reduced feature set of text documents. To reduce the feature set, this paper uses locality preserving index (LPI) and regularized locality preserving indexing (RLPI) techniques. The reduced feature sets are indexed using B-Tree. Further, the indexed terms are matched with class indices to categorize the known text document. To reveal the efficacy of the proposed model, large experimentations are carried out on standard benchmark datasets. The outcome of the paper reveals that the presented work outperforms the existing methods.

1 Introduction

World Wide Web (WWW) is experiencing the stack of huge amount of textual information due to rapid development of Internet technologies. As a result, managing this large stack of textual information has gained lot of importance. Among the various text mining activities, text classification (categorization) has added importance in organizing and managing the textual information. The prime objective of text categorization is to categorize text content into one of the relevant categories. Many applications of text classification can be found on various domains, viz. search engines, document management system such as banking, invoices, social networking sites, blogs, and email spam filtering [1]. Unfortunately, text classification finds basic challenge in high dimensionality of feature space. Thus, to overcome this limitation, employing competent feature selection can give better document representation. Giving better representation will in turn progress the results of classification accuracy. Variety of feature selection methods and representation models are presented in the literature [2].

However, all the existing reorientation models require time complexity which is linear in nature, and thus, it is not acceptable in practical scenarios. Hence, to improvise the classification accuracy and further to support dynamicity in handling text documents, various existing data structures like KDB-Tree [3], BD-Tree [4], G-Tree [5], or Multidimensional Binary Tree could be used to index the document terms [6]. Nevertheless, the existing indexing data structures found its own challenges in handling and storing the data.

V. Bhateja et al. (eds.), *Information Systems Design and Intelligent Applications*, Advances in Intelligent Systems and Computing 672,
https://doi.org/10.1007/978-981-10-7512-4_88

Document indexing is a method of correlating documents with different search terms to its class labels. Document indexing makes the document management system easier and handles the results very effectively. In this paper, the document index terms are indexed with the help of data structure. Different indexing data structures are cited in the literature. On the other hand, each data structure has its own challenges [7] in managing the index terms. However, we used B-Tree structure to index the key terms. The reason behind using this data structure is its less complexity and simplicity. In addition, B-Tree is also used because of its ease of use and its unbiased character. Subsequent to a successful depiction for text content, the job of text categorization is to group the text files. In order to do so, diverse machine-learning classifiers are proposed in the literature: Bayes, K-NN, Centroid based, Rule based, Rocchio, NN, and SVM [8].

In this work, we presented a new representation model using dimensionality reduction techniques, indexing, and followed by classification. The proposed model contains two phases. In the first phase, stopwords are eliminated in pre-processing, and then text documents are represented using term document matrix (TDM). Unfortunately, in TDM, documents are described with very high dimension. To reduce from higher dimensionality to lower dimensionality, we used two different dimensionality reduction techniques (DRT), viz. locality preserving indexing (LPI) [9] and regularized locality preserving indexing (RLPI) [10]. Additionally, we used B-Tree to index each reduced features (terms). In the second phase, the classification of text documents is done using four different classifiers: Naïve Bayes (NB), K-Nearest Neighbor (K-NN), Centroid-based Classifier (CBC), and Support Vector Machine (SVM).

The remaining of the manuscript is well thought out as follows. Section 2 presents the literature review on various representation models and indexing. In Sect. 3, we present the proposed model. Experimental results concerning the performance study followed by the discussions are presented in Sect. 4. Section 5 concludes the paper along with the future work.

2 Literature Review

As a first stage of classification task and to give better representation for documents, stopwords are eliminated during the pre-processing stage [11]. Once the stopwords are eliminated, text documents are present with only bag of words (BOW). These BOW should be represented to machine-understandable form by using document representation model. The crucial thought of document representation model is to trim down the processing time and get better classification accuracy. Bag of word (BOW) is a well-known elementary model to characterize the text documents. BOW considers occurrence count of terms present in the text to structure a vector representation called as vector space model (VSM). VSM is an algebraic model, which symbolizes the text documents as vectors of identifiers [12]. Here, text documents are simply represented using a matrix. However, it suffers from few restrictions: high dimensionality and sparseness. The other form of representing the text documents is by using universal networking language (UNL). UNL depicts the text documents in the form of graph [13]. The graph contains a set of nodes and links. In this model, terms (words) are considered as nodes and links are the relations between the terms. The main negative aspect of UNL is building a graph for each document. For this reason, UNL's are not widely used for large datasets. Ontology representation model keeps the semantic association among the vocabulary in a text [14]. Ontology method grasps the sphere acquaintance of a word in a text document. Nevertheless, creation of regular ontology is a very complicated task due to the deficiency in the ordered knowledge base. Cavanar [15] presented another contemporary model called n-gram model. In this model, terms are used as a progression of symbols, viz: byte, character, or words. This form does not require linguistic preparation such as stopword elimination or word-stemming. However, it is extremely complex to make a decision on the quantity of grams to be considered for successful document representation.

Deerwester et al. [16] proposed latent semantic indexing (LSI), which is supported on singular value decomposition (SVD). In LSI, cosine similarity measure is used to present the precise semantic similarity. The most important purpose of LSI is to create preeminent subspace estimation to the new document space and also sink the global reconstruction error. On the other hand, LSI conserves a good number of representative features rather than discriminating features. To determine this restriction, He et al. in [9] proposed a model called LPI for text modeling. It establishes the confined geometrical organization of document space, and also it has more discriminating power than LSI. It is unsupervised learning. Cai et al. [10] claimed RLPI (RLPI) model for text modeling. On the whole, it segregates the LPI difficulty into graph implanting crisis and normalized least square crisis. Such alteration maintains missing eigen-disintegration of intense matrices, and it can really lessen both time and memory cost in computation. Harish et al. [2] proposed symbolic representation using multiple kernel fuzzy C-means. This model decides the intra-class variation problem of text classification and uses four different kernel functions to characterize the text documents.

Further, the document terms are indexed to speed up the classification process and get better accuracy. To do this, we can use any accessible hashing organization like Multidimensional Binary Trees [17], Global-Tree [5], K-Dimensional B-Tree [3], and Bounded Deformation Tree [4]. Harish et al. [6] acclaimed a text document classification method using B-Tree and also presented new data structure called status matrix. The status matrix conserves the progression of phrase occurrences in the test file. At this juncture, terms are indexed using B-Tree to keep away from sequential matching during classification task. Minnie and Srinivasan [18] discuss a mixture of indexing algorithms to classify text documents for information retrieval.

In the present work, we have used LPI and RLPI as dimensionality reduction techniques, and the same is used to give a reduced feature set representation. Additionally, LPI and RLPI-reduced features are indexed using B-Tree (knowledge base). The indexed features (stored in knowledge base) are matched with class labels. To classify the documents, various classifiers are used to match the class labels and index terms stored in knowledge base.

3 Proposed Model

The proposed approach has two phases: (i) document representation and (ii) document classification.

3.1 Document Representation

Let us consider k number of classes $C = \{c_1, c_2, \ldots, c_k\}$, where each class contains n number of documents $D = \{d_1, d_2, \ldots, d_n\}$. The set of terms (features) present in each document is represented using $T = \{t_1, t_2, \ldots, t_i\}$. Initially, stopwords are eliminated during the pre-processing stage, and then text documents are represented using term document matrix (TDM). In TDM, documents are structured with very sparse and high dimensionality. Here, every row represents a document of class, and every column represents a term and dimension of $kn \times i$. Further, we employ LPI and RLPI to reduce the dimensionality of TDM. After applying LPI and RLPI, the resultant TDM dimension is $kn \times j$ where j is the number of terms (features) selected out of i.

From these deposit of terms (features), an upturned record arrangement is created for every phrase by relating the label of the group of texts that embrace that exacting phrase. The list of group tags connected with a phrase may hold numerous group tags as it is not unusual that the text of different groups holds the identical phrase. The phrases with their associated register of group index are suggested to be accumulated in the information base to carry categorization of an unfamiliar trial text.

Thus, the proposed B-Tree contains LPI and RLPI features as index terms. The main idea of indexing is to associate documents with different search terms. The acclaimed B-Tree support scheme can be effortlessly extended toward active database, as it is extremely simple to comprise fresh documents. Adding together, introducing fresh text is as straightforward as immediately with the introduction of placing the terms into the presented set and revising the connected phrase registers. In connection to the acclaimed illustrated method, introduction of a text into the record is basically a procedure of placing the terms to the B-Tree.

In regulating to introduce a word t related to the text to be established, the balanced-tree is admitted toward discovering the position of t, with condition t is previously present in the records, the introduction difficulty is condensed toward the difficulty of receiving the register of text restructured by adding the index of terms to be introduced. Further, the record does not contain a word t, subsequently no hesitation that the terms are at node U in a B-Tree. Provisionally, U grasps fewer $(r - 1)$ words (r: order of the tree), t be plainly introduced into U in a specific pact. If not, dissimilar to usual B-Tree introduction method, the node is ultimately separated among two nodes. Within the present method, it is advised toward giving the impression on siblings node to find if any complimentary position, consequently the information preparations are able to obtain the t word contained next to the node U itself with no dividing. As an initial stage, stopwords present in every training document are removed as well as the words contained in the dictionary beside through their group information are accumulated in a Balanced-Tree intended for speedy retrieval. The complexity by means of the B-Tree is of $O(\log_r i)$, where i is the quantitative terms accumulated in the B-Tree and r is the order of the B-Tree.

3.2 Document Classification

Given a trial (query) document d_q, characterized by a term document feature vector, is estimated onto lesser dimensions using LPI and RLPI features. These features are further transformed to form feature vectors of size *m*. These *m* feature significances are used to judge against with every class representative (indices) accumulated in the knowledge base for all investigational assessment. We used standard classifiers, viz. Naïve Bayes [8, 19], K-Nearest Neighbor [20], Centroid-based Classifier [21, 22], and Support Vector Machine [8, 22]. The objective function of the above classifiers is compared with class indices accumulated in the knowledge base. The achieved identical values are used to choose to which class the agreed trial document belongs to.

4 Experimentation

4.1 Datasets

Toward validating the strength of the acclaimed scheme, we conducted experiments on the publically available three standard datasets: 20 NewsGroups, Reuters-21578, and Vehicle Wikipedia Datasets. Table 1 shows detailed information about each standard datasets.

Table 1. A summary of the standard datasets, including the number of classes/categories, number of documents, and ratio of train and test set entries

Dataset	No. of class/category	No. of documents	Train: test
20 NewsGroups	20	20000	50:50 & 60:40
Reuter-21578	135	21578	50:50 & 60:40
Vehicle Wikipedia	4	440	50:50 & 60:40

20 NewsGroups: This is one of the well-accepted regular datasets for text categorization. The dataset is a compilation of 20000 text documents, collected from UseNet posting over a period of several months in 1993 and contains the classes like electronics, politics, cricket, and religion [23].

Reuters-21578: The dataset was initially composed and labeled by Carnegie Group Inc. and Reuters Ltd, and it is an improved version of RCV1 (Reuters Corpus Volume 1). It contains 21578 Reuters news documents from 1987 [24].

Vehicle Wikipedia: This dataset contains vehicle distinctiveness mined from Wikipedia pages and contains four categories of vehicles: aircraft, trains, cars, and boats with low degree of similarity [25].

4.2 Discussion and Results

In this subdivision, we present the outcome of the testing carried out to exhibit the efficacy of the acclaimed scheme on all three benchmark datasets, viz. 20 Newsgroups,

Reuters-21578, and Vehicles Wikipedia datasets. In this method, we used two diverse dimensionality reduction methods called LPI and RLPI. Further, these LPI and RLPI features are indexed using B-Tree. We have compared classification performance using standard text classifiers. During trialing, we used 100 features (terms) represented by every representation model and performed two trials. Firstly, 50% of the texts are used for learning and the remaining 50% are used for testing. Later, the same experimentations are repeated with a split of 60:40. As an evaluation metric for goodness of the acclaimed scheme, we compute the percentage of accurateness, and the same is presented in Table 2.

Table 2 presents the performance comparison results of classifiers. In Table 2, SVM does better with RLPI compared to LPI method on all standard datasets. The SVM classifier performed better than other classifiers with outcome of 87.10% and 88.45% using both LPI and RLPI methods, respectively.

Table 2. Categorization performance using various classifiers on standard dataset

Dataset	No. of features selected	Training versus testing	Naïve Bayes	K-nearest neighbor	Centroid-based classifier	Support vector machine
20 Newsgroups	LPI—100	50 versus 50	72.80	69.05	68.40	73.45
		60 versus 40	74.95	71.35	69.25	76.10
	RLPI—100	50 versus 50	78.70	74.35	72.60	84.20
		60 versus 40	80.40	77.00	73.40	86.85
Reuters-21578	LPI—100	50 versus 50	73.60	71.20	70.25	74.60
		60 versus 40	74.10	71.65	71.50	75.35
	RLPI—100	50 versus 50	78.90	73.45	76.25	80.65
		60 versus 40	79.40	75.10	73.40	82.30
Vehicles Wikipedia	LPI—100	50 versus 50	78.50	77.35	76.40	85.60
		60 versus 40	79.90	78.40	77.50	86.20
	RLPI—100	50 versus 50	80.35	78.90	76.60	87.10
		60 versus 40	81.20	79.35	78.20	88.45

4.3 Time Complexity

As there are k classes and n documents, initially documents are preprocessed and represented in a term document matrix form of size $n \times T$ where T is the total number of features. In the next step, we used two dimensionality reduction techniques, i.e., LPI and RLPI. LPI/RLPI chooses t features from original T features, and hence, the size of TDM becomes $n \times t$. The time complexity of LPI is $n^2\left(\frac{3}{2}s + I_1 k\right) + \frac{20}{3}n^3$ [9], where s is the average number of non-zero features per document and I_1 is the number of iterations in Lanczos algorithm. The time complexity of RLPI is as follows: $nk(2I_2 s + k) + 5nkI_2$ [10] where I_2 is the number of iterations in LSQR algorithm (sparse linear equation and least square problem). The features extracted (t) from LPI/RLPI are represented using B-Tree. The time complexity to construct B-Tree is $O(\log_r t)$ [6] where r is the order of B-Tree. If a query document contains t_q terms, we require $O\left(t_q \log_r t\right)$ [6] computation to create an index at B-Tree. So the total time complexity of the proposed method using LPI becomes:

$$T_M = O\left(n^2\left(\frac{3}{2}s + I_1 k\right) + \frac{20}{3}n^3 + \log_r t + t_q \log_r t\right) \cong O\left(n^3\right)$$

Similarly, time complexity using RLPI becomes:

$$T_M = O\left(nk(2I_2 s + k) + 5nkI_2 + \log_r t + t_q \log_r t\right) \cong O(n)$$

5 Conclusion

A novel way of classifying text document is claimed in this paper. The text documents are characterized by exploiting the use of B-Tree. B-Tree is a competent index method. The major contribution of this work is to employ LPI and RLPI features to index text terms via B-Tree. Further, the class labels are compared/matched with the index terms accumulated in the knowledge base. In order to inspect the strength and performance of the acclaimed model, we performed testing on three benchmark datasets. The investigational outcome discloses that the novel representation model by means of B-Tree has achieved improved performance by SVM classifier. In future, it is also planned to index the terms by means of a variety of data structures for well-organized retrieval. Additionally, the terms in the knowledge base can also be indexed using competent data structures for efficient classification.

References

1. Sebastiani, F.: Machine learning in automated text categorization. ACM computing surveys (CSUR), Vol. 34. (2002) 1–47.
2. Harish, B. S., Revanasiddappa, M. B., & Kumar, S. A. (2015, December). Symbolic Representation of Text Documents Using Multiple Kernel FCM. In *International*

Conference on Mining Intelligence and Knowledge Exploration (pp. 93–102). Springer International Publishing.
3. Robinson J. T., The KDB tree: A search structure for large multidimensional dynamic indexes. Proceedings of ACM SIGMOD conference Ann Arbor, MI, pp. 10–18.
4. Dandamudi S. P and Sorenson P. G., 1985. An empirical performance comparison of some variations of the k-d tree and bd tree. Computer and Information Sciences. Vol. 14, no. 3, pp. 134–158.
5. Kumar A., 1994. G – tree: A new datastructure for organizing multidimensional data. IEEE transactions on Knowledge and Data Engineering, vol. 6, no. 2, pp. 341–347.
6. Harish, B. S., Manjunath, S., & Guru, D. S. (2012). Text document classification: an approach based on indexing. *International Journal of Data Mining & Knowledge Management Process (IJDKP)*, *2*(1), 43–62.
7. Punitha P., 2005. IARS: Image Archival and Retrieval Systems. Ph.D. Thesis, University of Mysore.
8. Harish, B. S., Guru, D. S., Manjunath, S.: Representation and classification of text documents: A brief review. IJCA, Special Issue on RTIPPR, (2010) 110–119.
9. He, X., Cai, D., Liu, H., Ma, W.Y.: Locality preserving indexing for document representation. In: Proceedings of the 27th Annual International ACM SIGIR Conference on Research and Development in Information Retrieval, pp. 96–103. ACM (2004).
10. Cai, D., He, X., Zhang, W.V., Han, J.: Regularized locality preserving indexing via spectral regression. In: Proceedings of the Sixteenth ACM Conference on Conference on Information and Knowledge Management, pp. 741–750. ACM (2007).
11. Uysal, A. K., & Gunal, S. (2014). The impact of preprocessing on text classification. *Information Processing & Management*, *50*(1), 104–112.
12. Salton, G., Wong, A., Yang, C.S.: A vector space model for automatic indexing. Commun. ACM 18(11), 613–620 (1975).
13. Choudhary, B., Bhattacharyya, P.: Text clustering using universal networking language representation. In: The Proceedings of Eleventh International World Wide Web Conference, pp. 1–7 (2002).
14. Hotho, A., Maedche, A., Staab, S.: Ontology-based text document clustering 16, 48–54 (2002).
15. Cavnar, W.: Using an n-gram-based document representation with a vector processing retrieval model, pp. 269–269. NIST SPECIAL PUBLICATION SP (1995).
16. Deerwester, S.C., Dumais, S.T., Landauer, T.K., Furnas, G.W., Harshman, R.A.: Indexing by latent semantic analysis. JAsIs 41(6), 391–407 (1990).
17. Bentley J. L., 1975. Multidimensional binary search trees used for associative searching. Communications of ACM, vol. 18, no. 9, pp. 509–517.
18. Minnie, D., & Srinivasan, S. (2011, December). Intelligent Search Engine algorithms on indexing and searching of text documents using text representation. In *Recent Trends in Information Systems (ReTIS), 2011 International Conference on* (pp. 121–125). IEEE.
19. Hotho Andreas, Andreas N¨urnberger, and Gerhard Paaß. A brief survey of text mining. In Ldv Forum, volume 20, pages 19–62, 2005.
20. Han Eui-Hong Sam, George Karypis, and Vipin Kumar. Text categorization using weight adjusted k-nearest neighbor classification. In Pacific-Asia conference on knowledge discovery and data mining, pages 53–65. Springer, 2001.
21. Han Eui-Hong Sam and Karypis George. Centroid-based document classification: Analysis and experimental results. In European conference on principles of data mining and knowledge discovery, pages 424–431. Springer, 2000.
22. Korde Vandana and Mahender C Namrata. Text classification and classifiers: A survey. International Journal of Artificial Intelligence & Applications, 3(2):85, 2012.

23. 20newsgroups: http://people.csail.mit.edu/jrennie/20Newsgroups/.
24. Reuters-21578: http://www.daviddlewis.com/resources/testcollections/reuters21578.
25. Isa, D., Lee, L.H., Kallimani, V.P. and Rajkumar, R., 2008. Text document preprocessing with the Bayes formula for classification using the support vector machine. *IEEE Transactions on Knowledge and Data engineering*, *20*(9), pp. 1264–1272.

Network Intrusion Detection Systems Using Neural Networks

Sireesha Rodda(✉)

Department of CSE, GITAM Institute of Technology, GITAM University, Visakhapatnam, India
sireesharodda@gmail.com

Abstract. With the growth of network activities and data sharing, there is also increased risk of threats and malicious attacks. Intrusion detection refers to the act of successfully identifying and thwarting malicious attacks. Traditionally, the help of network security experts is sought owing to their familiarity with the network technologies and broad knowledge. Recently, data mining techniques have been increasingly adopted to perform network intrusion detection. This paper presents the comparison between multi-layer perceptron and radial basis function networks for designing network intrusion detection system. Multi-layer perceptron proved to be more effective than radial basis function when applied on the benchmark NSL_KDD dataset.

Keywords: Network intrusion detection · IDS · NSL_KDD dataset

1 Introduction

The past few years have witnessed remarkable increase in the usage of networks for storing and sharing data. This led to the increase in unauthorized and malicious cyber-attacks. Network intrusion detection system (NIDS) refers to the process of detecting and thwarting abnormal or suspicious activities over a given network. Different machine learning techniques have immense use in detecting dubious activities compromising the integrity, security, and confidentiality of the data.

Intrusion detection systems may be broadly classified as belonging to either misuse detection or anomaly detection. Misuse detection formulates a model from the known attacks, whereas anomaly detection searches for behavior that deviates from the normal patterns. NSL_KDD [1] is a standard dataset used popularly for evaluating network intrusion detection systems. Techniques such as neural networks [2], k-nearest neighbor [3], C4.5 [4], and Naïve Bayes [5] have been used earlier to efficiently identify network traffic.

The advantage of using artificial neural networks (ANNs) is that it does not require priori information apart from the training data. Many applications use multi-layer perceptron (MLP) along with back propagation algorithm. Multi-layer perceptron is a feed-forward network, where a set of sensory units form the input layer, a collection of hidden layers, and an output layer. Each unit in the MLP takes the weighted sum of input values along with a sigmoid function. The output of one unit is fed as input to another connected unit in the next layer. In this paper, gradient descent algorithm is

V. Bhateja et al. (eds.), *Information Systems Design and Intelligent Applications*, Advances in Intelligent Systems and Computing 672,
https://doi.org/10.1007/978-981-10-7512-4_89

used on the network intrusion detection dataset. Radial basis function (RBF) network computes a vector prototype. The predicted value of a new instance is obtained by performing weighted sum of the RBF for each unit.

The remainder of the paper is organized as follows: Sect. 2 introduces the NSL_KDD intrusion detection dataset used for evaluating the neural network algorithms. Section 3 discusses the empirical results, followed by the conclusions in Sect. 4.

2 NSL_KDD Dataset

This paper uses the standard NSL_KDD dataset to validate the performance of two neural network-based classifiers viz. multi-layer perceptron and radial basis function. NSL_KDD is a popular network intrusion detection dataset used to validate different network intrusion detection systems. It is actually a more refined version of the KDD Cup 1999 intrusion dataset. One of the issues with the KDD Cup dataset is presence of huge amount of redundant records leading to the learning algorithm to become biased toward most frequent instances. In NSL_KDD, such redundant records are eliminated giving equal importance to minority classes as well.

The NSL_KDD dataset contains separate files for training data and test data. Both the datasets are described in terms of 42 features and a class attribute. Every instance of the dataset is labeled as normal or attack as the class attribute value. The attack belongs to one of 22 types. These attacks are broadly categorized into four attack groups namely Denial of Service (DoS), U2R (User to Root), R2L (Remote to Local), and

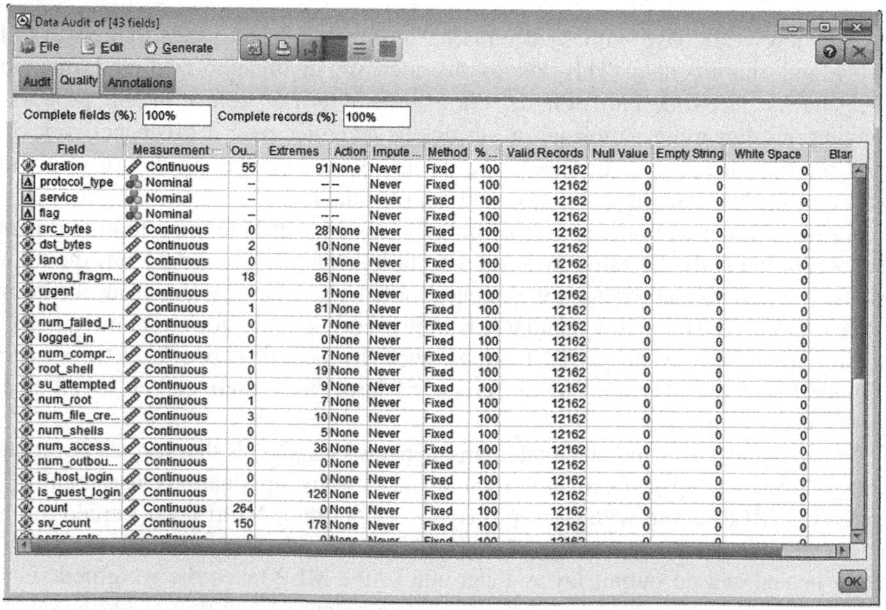

Data Audit of [43 fields]

File Edit Generate

Audit Quality Annotations

Complete fields (%): 100% Complete records (%): 100%

Field	Measurement	Ou...	Extremes	Action	Impute ...	Method	% ...	Valid Records	Null Value	Empty String	White Space	Blar
duration	Continuous	55	91	None	Never	Fixed	100	12162	0	0	0	
protocol_type	Nominal	--	--	--	Never	Fixed	100	12162	0	0	0	
service	Nominal	--	--	--	Never	Fixed	100	12162	0	0	0	
flag	Nominal	--	--	--	Never	Fixed	100	12162	0	0	0	
src_bytes	Continuous	0	28	None	Never	Fixed	100	12162	0	0	0	
dst_bytes	Continuous	2	10	None	Never	Fixed	100	12162	0	0	0	
land	Continuous	0	1	None	Never	Fixed	100	12162	0	0	0	
wrong_fragm...	Continuous	18	86	None	Never	Fixed	100	12162	0	0	0	
urgent	Continuous	0	1	None	Never	Fixed	100	12162	0	0	0	
hot	Continuous	1	81	None	Never	Fixed	100	12162	0	0	0	
num_failed_l...	Continuous	0	7	None	Never	Fixed	100	12162	0	0	0	
logged_in	Continuous	0	0	None	Never	Fixed	100	12162	0	0	0	
num_compr...	Continuous	1	7	None	Never	Fixed	100	12162	0	0	0	
root_shell	Continuous	0	19	None	Never	Fixed	100	12162	0	0	0	
su_attempted	Continuous	0	9	None	Never	Fixed	100	12162	0	0	0	
num_root	Continuous	1	7	None	Never	Fixed	100	12162	0	0	0	
num_file_cre...	Continuous	3	10	None	Never	Fixed	100	12162	0	0	0	
num_shells	Continuous	0	5	None	Never	Fixed	100	12162	0	0	0	
num_access...	Continuous	0	36	None	Never	Fixed	100	12162	0	0	0	
num_outbou...	Continuous	0	0	None	Never	Fixed	100	12162	0	0	0	
is_host_login	Continuous	0	0	None	Never	Fixed	100	12162	0	0	0	
is_guest_login	Continuous	0	126	None	Never	Fixed	100	12162	0	0	0	
count	Continuous	264	0	None	Never	Fixed	100	12162	0	0	0	
srv_count	Continuous	150	178	None	Never	Fixed	100	12162	0	0	0	

OK

Fig. 1. Description of attributes in the NSL_KDD dataset

Probe (Information Gathering). The detailed description of the attacks is presented in Rodda and Erothi [6]. The description of the attributes is shown in Fig. 1.

The training data for NSL_KDD consists of 12,162 instances out of which 5659 belongs to the attack types whereas 6503 belongs to normal category. The test dataset contains 2432 attack instances and 2569 normal instances. Figure 2 and Figure 3 depict the distribution of attacks in the NSL_KDD training and test datasets, respectively.

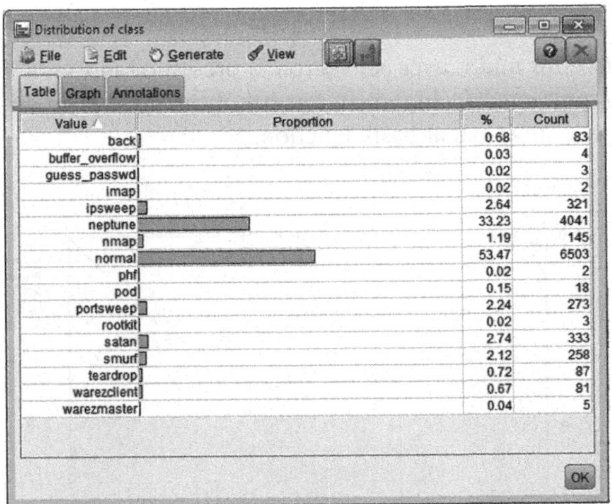

Distribution of class

Value	Proportion	%	Count
back		0.68	83
buffer_overflow		0.03	4
guess_passwd		0.02	3
imap		0.02	2
ipsweep		2.64	321
neptune		33.23	4041
nmap		1.19	145
normal		53.47	6503
phf		0.02	2
pod		0.15	18
portsweep		2.24	273
rootkit		0.02	3
satan		2.74	333
smurf		2.12	258
teardrop		0.72	87
warezclient		0.67	81
warezmaster		0.04	5

Fig. 2. Attacks in NSL_KDD training dataset

Distribution of class

Value	Proportion	%	Count
back		0.66	33
guess_passwd		0.08	4
ipsweep		3.02	151
land		0.02	1
neptune		34.31	1716
nmap		1.12	56
normal		51.37	2569
pod		0.18	9
portsweep		2.72	136
rootkit		0.02	1
satan		3.32	166
smurf		1.88	94
teardrop		0.64	32
warezclient		0.66	33

Fig. 3. Attacks in NSL_KDD test dataset

3 Experimental Results

All the experiments were executed on Intel Core i5 3.10 GHz CPU with 4 GB RAM running on 32-bit Windows operating system. The results have been obtained from IBM SPSS Statistics version 24.0. Experiments were conducted on the benchmark NSL_KDD intrusion detection dataset which is a refined and condensed version of KDD Cup 1999 dataset. The performance of two neural networks' classifiers based on multi-layer perceptron and radial basis function is presented. The efficiency of the two classifiers is evaluated in terms of accuracy and area under the curve (AUC) performance metrics.

The ROC curve for multi-layer perceptron is presented in Fig. 4 and that for radial basis function is presented in Fig. 5. It can be observed that multi-layer perceptron provides better AUC than radial basis function.

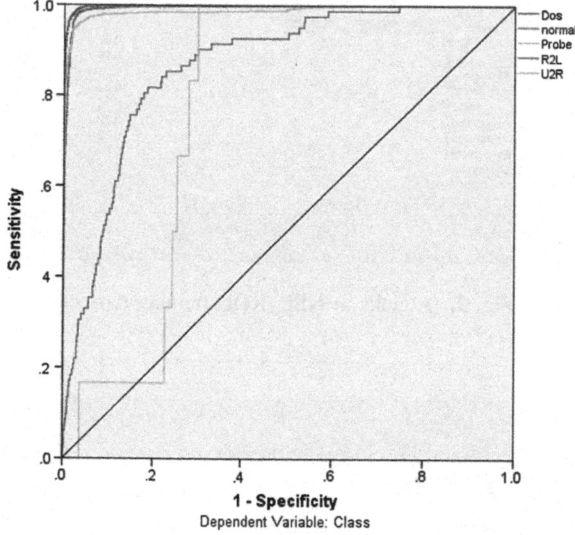

Fig. 4. ROC curve for multi-layer perceptron

It may be clearly observed from Table 1 that multi-layer perceptron returns better performance when compared with radial basis function for the NSL_KDD dataset.

The accuracy returned by the multi-layer perceptron is better than that of radial basis function networks probably because of the presence of global optimization in the multi-layer perceptron. However, both the algorithms delivered poor performance for minority classes U2R and R2L. It was observed that the instances belonging to minority classes U2R and R2L have been assigned to one of the two majority classes Normal and DoS resulting in poor accuracy for the minority classes (Table 2).

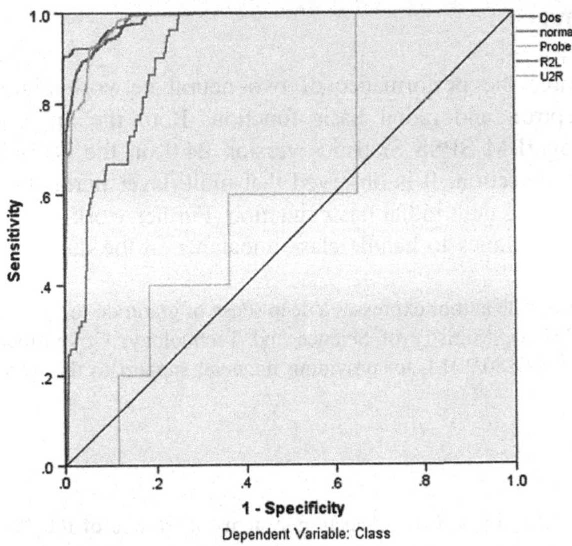

Fig. 5. ROC curve for radial basis function

Table 1. Comparison of area under the curve values

Class	Multi-layer perceptron	Radial basis function
Normal	**0.998**	0.991
DoS	**0.994**	0.989
Probe	**0.988**	0.980
R2L	**0.868**	0.812
U2R	**0.776**	0.446

Table 2. Classification performance of in terms of accuracy

Sample	Class label	Accuracy (%)	
		Multi-layer perceptron	Radial basis function
Training data	Normal	98.2	95.7
	DoS	97.9	88.4
	Probe	88.4	81.2
	U2R	0	0
	R2L	0	0
Test data	Normal	98.2	93.2
	DoS	98.0	89.5
	Probe	88.1	75.8
	U2R	0	0
	R2L	0	0
Overall accuracy (%)		98.6	91.5

4 Conclusions

This paper explores the performance of two neural network algorithms based on multi-layer perceptron and radial basis function. Both the approaches have been implemented using IBM SPSS Statistics version 24.0 on the NSL_KDD dataset for network intrusion detection. It is observed that multi-layer perceptron returned better results when compared with radial basis function. Further work must be carried out to identify suitable techniques to handle class imbalance in the dataset.

Acknowledgements. The author expresses a deep sense of gratitude to Science and Engineering Research Board (SERB), Ministry of Science and Technology, Government of India, Grant Number SB/FTP/ETA-0180/2014, for providing financial support to this work.

References

1. McHugh, John. "Testing intrusion detection systems: a critique of the 1998 and 1999 darpa intrusion detection system evaluations as performed by lincoln laboratory." ACM Transactions on Information and System Security (TISSEC) 3.4, 2000, pp. 262–294.
2. Surana, S. "Intrusion Detection using Fuzzy Clustering and Artificial Neural Network". Advances in Neural Networks, Fuzzy Systems and Artificial Intelligence, ISBN-978-960-474-379-7, 2013, pp. 209–217.
3. Dokas, P., Ertoz, L., Kumar, V., Lazarevic, A., Srivastava, J., & Tan, P. N. "Data mining for network intrusion detection. In Proc. NSF Workshop on Next Generation Data Mining", 2002, pp. 21–30.
4. Mulay, S. A., Devale, P. R., & Garje, G. V. "Intrusion detection system using support vector machine and decision tree. International Journal of Computer Applications", 3(3), 2010, pp. 40–43.
5. Panda, M., & Patra, M. R. (2007). Network intrusion detection using naive bayes. International journal of computer science and network security, 7(12), 258–263.
6. Rodda, Sireesha, Erothi, Uma Shankar Rao. "Class Imbalance Problem in the Network Intrusion Detection Systems". International Conference on Electrical, Electronics, and Optimization Techniques (ICEEOT), 2016, pp. 2685–2688.

A Study of the Optimization Techniques for Wireless Sensor Networks (WSNs)

Pritee Parwekar[1(✉)], Sireesha Rodda[2], and Neeharika Kalla[1]

[1] Anil Neerukonda Institute of Technology and Sciences, Visakhapatnam, India
pritee.cse@anits.edu.in, neeharika.kalla@gmail.com
[2] GITAM University, Visakhapatnam, India
sireesha@gitam.edu

Abstract. WSN has become one of the important technologies in the present decade. Energy consumption is the major challenge in the field of wireless sensor network. In WSN, there are some hard problems that cannot be solved in deterministic time. These hard problems can be solved by using optimization techniques. Clustering, routing, node localization, maintenance of the nodes, etc., are some of the hard problems that could be addressed. The main aim of these techniques is to provide the solution within specific time and also to minimize the consumption of the energy thus prolonging the lifetime of the network. This paper clearly describes the application of the different published optimization techniques in the field of WSN.

Keywords: Wireless sensor network · Optimization · Clustering · Routing
Base station

1 Introduction

WSN has come up with the advancement of the microelectromechanical systems (MEMs) and thus attracted many scientists to research in the field of various applications like science, military, engineering, medical [1]. Wireless sensor network (WSN) is a network that consists of huge number of sensors operating in a distributed manner autonomously. Sensors are responsible to monitor the physical environment and transmit the sensed data to the BS or to the sink either directly or in multi-hop fashion. In WSN, sensors are randomly deployed in the specific area. The sensors can either be static or mobile. If the deployment area is vast, then the sensors need to monitor for a long time and thus require more amount of energy. As sensors are tiny objects embedded with batteries, it is not possible to recharge or to replace the battery when the node dies. Thus, energy consumption of the sensor node is a major challenge in the field of WSN [1].

Sensors are the tiny devices that are equipped with the limited resources like power, energy, transmission distance. Initially, every sensor is preloaded with the limited amount of energy [2]. But due to transmission, the energy gets depleted. If the node is

V. Bhateja et al. (eds.), *Information Systems Design and Intelligent Applications*, Advances in Intelligent Systems and Computing 672,
https://doi.org/10.1007/978-981-10-7512-4_90

having minimum energy, then it cannot participate in the communication. In WSN, constant monitoring of the nodes residual energy is required.

Every node can communicate with the BS only within its communication range. If the node cannot communicate to the BS within its range, then the node transmits the data to the BS through the intermediate nodes. Thus, multi-hop communication takes place [2]. In multi-hop wireless network, the node that is near to the BS depletes more energy when compared to the nodes located far from the BS, because every node must transmit the data to the BS through this intermediate node. In order to avoid this problem, clustering technique is used in which nodes are organized in the form of clusters with one cluster head (CH) per cluster [1, 3]. The nodes transmit the data only to the CH, and the CH transmits to the BS.

Routing is another technique that consumes more energy. If the path to route the data is not best, then more energy is consumed by the node. Again same path should not be used for multiple transmissions. Another scenario is that in order to ensure proper communication and to consume less energy, location of the node information should be determined. Without knowing the nodes location, we cannot transmit the data. Thus, node localization also plays a major role in WSN. The above-mentioned problems are the hard problems that cannot be solved by using deterministic algorithms. These can be solved by the high-level procedures, and thus optimization techniques are developed. The hard problems can be solved easily by using optimization techniques which produces the best solution to the problem. Section 2 describes in brief about the optimization.

2 Optimization

Optimization is the process of either maximizing or minimizing the objective function in order to find the solution to the complex or hard problems. Hard/Complex optimization problems can be the problems which could not be solved using the deterministic methods within a specific time [4]. Optimization can also be referred to as metaheuristic optimization because it solves the optimization problems using metaheuristic algorithms. Meta means "higher level" and heuristic means "solution or procedure" [4]. That means higher-level procedure to the problems with no deterministic solution. These problems can be either single objective or multi-objective. The better is the objective function; the better will be the optimal solution. Single objective means all the particles converge at a single point and thus that point is the optimal solution. In multi-objective, the particles converge at two or more points and the best solution among them should be taken.

The metaheuristic algorithm searches for a solution in a search space. Some algorithms opt for local search while some algorithms opt for global search. The objective function is formulated by considering various parameters or metrics relating to the problem statement. Two types of metaheuristics: Single-solution-based metaheuristic and Population-based metaheuristic [4]. Single-solution-based is exploitation oriented which means confined to local search (to refine the solution). While population-based is exploration oriented which means confined to global search (to find the new good solution). Single-solution-based techniques include simulated annealing, tabu search. Population-based techniques are of two types: evolutionary algorithms and swarm intelligence. Evolutionary algorithms include genetic algorithm and differential evolution. swarm intelligence includes particle swarm optimization (PSO), ant colony optimization (ACO), artificial bee colony (ABC) optimization, bacterial foraging optimization (BFO), cuckoo search, and firefly algorithm [4].

3 Critical Analysis of the Optimization Techniques

A brief description of the published optimization techniques is provided in order to describe how the technique could be applied in the field of wireless sensor network. The techniques are as follows:

3.1 Genetic Algorithm (GA) for Clustering

In [5], Genetic Algorithm is used to find the optimum number of CHs. The algorithm operates in two stages: the set-up stage and steady-state stage. In set-up stage, BS determines the number of CHs and assigns the nodes to the CHs. In second stage, the nodes transmit the data to the CH and the CHs in turn transmit to the BS. Initially, the chromosomes are formed by the BS and then the genetic algorithm is applied to determine the best CH. Fitness/Objective function can used to calculate the fitness of an individual. The objective function used in this paper consists of transmission distance of all the nodes to the CH and from all the CHs to the BS. Here, the objective function should be maximized for minimum transmission distance. Here, an agent CH is also used to maximize the distance between CH and BS. The agent CH will be the cluster head with maximum residual energy, and it is selected from the set of CHs.

3.2 Clustering Using Bacterial Foraging Optimization (BFO)

In [1], the first step is the initialization of the bacteria and it is done randomly. The bacterium consists of some percentage of CHs. In chemotaxis, CH position is represented as node id and 2D coordinates are updated. Next swarming is performed based on the equation. Then the bacteria are ordered in ascending order based on the fitness. The top half of the bacteria is carried to the lower half of the population in order to perform reproduction. The weakest bacteria can be eliminated and again new bacterium is initialized with the new nodes. Here, the fitness function consists of remaining energy of the CH, intra-cluster distance and the distance from the CH to BS.

3.3 Routing Using Bacterial Foraging Optimization (BFO)

In [1], first the bacterium is initialized. For routing, each bacterium dimension is equal to the number of the CHs and for the BS, a position extra is added. After this, the mapping function is applied to determine the next-hop to the BS. The fitness function for routing consists of residual energy of the next-hop and the Euclidean distance between the CH to the next-hop and from the next-hop to the BS.

3.4 Artificial Bee Colony (ABC) Optimization for Clustering

In [3], ABC algorithm is used to construct the optimum CH list. Initially, the sensor node broadcasts "hello" packet. After receiving the broadcast message, each sensor node maintains a neighboring table with the id of the sensor from which broadcast message is sent and the RSSI value. Now, the nodes send their id, neighboring table information, and the residual energy to the BS. The BS now constructs the CH list by using the equation. The node with energy level higher than the threshold value is only eligible as a CH. After the construction of the CH list, clustering is done based on the RSSI values of the nodes. Then fitness is calculated. After that, ABC algorithm is run to find the best CH among the list. That means optimal number of CHs are selected from the list. Then TDMA slots are used to send the data from the nodes to the CHs. The data is routed to the BS by using CDMA MAC protocol. Here, the fitness function assumes the energy level of the CH that must be greater than the threshold. In order to consume less energy, the number of CHs should be minimized.

3.5 Routing Using PSO and V-LEACH

V-LEACH is the improvement of the LEACH protocol. In V-LEACH, a vice CH is selected from the cluster. If the cluster head in the network dies, then the vice CH becomes the new cluster head thus increasing the lifetime of the network. In [6], a cluster head is selected from each cluster whose energy is higher than the threshold. If the cluster formation takes place randomly, then one cluster may have more number of nodes while the other may have less number of nodes. To avoid this problem, particle swarm optimization (PSO) technique is used to perform uniform clustering. After the formation of clusters, CHs are selected based on the energy level of the node. As the CHs consume more energy to transmit the data to the BS, the node may die at some point. When the CH dies, the vice cluster head takes the responsibility of the CH and continues the transmission by taking the data from the different nodes.

3.6 Firefly Algorithm (FA) for Routing

In [7], firefly algorithm can be used to determine the route from the CH to BS of each cluster. First, the firefly is initialized. Here, each firefly represents the route from the CH to the BS. The dimension of the firefly is equal to the number of the CHs and for the BS, a position extra is added. In this, each position represents the next-hop of the CH in order to route the data to the BS. For every iteration, the firefly with less brightness moves toward the brighter firefly and the position of each firefly is updated. The process is repeated for the maximum number of generations. The fitness function used consists of residual energy of the next-hop, Euclidean distance from CH to the next-hop and from next-hop to the BS, node degree which is the number of the members of the CH of the next-hop.

3.7 Ant Colony Optimization Attack Detection (ACO-AD)

The algorithm in [8] is used to detect a sinkhole. Each sensor node has its ID. Four modules determine the functioning of the IDs: local packet monitoring module, local detection engine, cooperative detection engine, and local response module. Local packet monitoring is used to gather the data by listening to the neighboring nodes.

Table 1. Summary of optimization techniques in WSN

Title of the paper	Optimization technique	Challenge addressed	Problem specification
Bacterial foraging optimization algorithm for CH selection and routing in wireless sensor networks [1]	Bacterial foraging optimization (BFO)	Routing after forming the clusters	To determine the route from the CH to the BS
Bacterial foraging optimization algorithm for CH selection and routing in wireless sensor networks [1]	Bacterial foraging optimization (BFO)	Clustering	To find the optimum number of CHs
Energy efficient clustering algorithm for wireless sensor networks using ABC met heuristic [3]	Artificial bee colony (ABC)	Clustering	To increase the amount of the data transmitted to the BS and to maintain the maximum energy of each node. ABC algorithm is used to construct the CH list
A new approach based on a genetic algorithm and an agent cluster head to optimize the energy in wireless sensor networks [5]	Genetic algorithm (GA)	Clustering	To improve the remaining energy of every node by using agent cluster and to determine the location of CHs along with their number
Energy efficient routing of WSN using particle swarm optimization and V-LEACH protocol [6]	Particle swarm optimization (PSO) and V-LEACH	Clustering	PSO is used to perform uniform clustering. V-LEACH is used as fault tolerance. If the CH dies, then vice CH takes the role of CH
FARW: firefly algorithm for routing in wireless sensor networks [7]	Firefly algorithm (FA)	Routing after forming the clusters	To determine the best path from the CH to the BS
Analysis geographic routing protocol using ant colony optimization technique for WSNs [8]	Ant colony optimization (ACO)	Intrusion detection	To detect a sinkhole attack
Firefly algorithm optimization-based WSN localization algorithm [9]	Firefly algorithm (FA)	Node localization	To find the accurate position of the sensor node using DV-hop algorithm which only estimates the position of the sensor node

Based on the rules that are defined, local detection engine has a rule set for each sensor which consists of the node id of the neighboring sensor nodes and the link quality of the node that is determined by listening to the transmission of the neighboring node. Rule set do not contain the node id of the sensor itself. When the route update packet arrives at the node, the sensor compares the node id of the route update packet with the rule set. If the rule set does not contain either the node id or the link quality of the route update packet, then it is determined as a sinkhole attack and an alert is generated. ACO algorithm can be used to match the node id of the route update packet with the rule set. The node contains the list of the suspected nodes. The cooperative detection engine groups the alerted nodes together to identify the intruder. Finally, when the intruder is identified, each alerted node maintains a count to check how many times the intruder has appeared in the suspected list.

3.8 Firefly Algorithm (FA) for Node Localization

In [9], firefly algorithm is applied to find the exact position of the sensor node that is estimated by the DV-hop algorithm. First, initialize the firefly group. Then calculate the fitness of an individual and update the fluorescein value. Now, each individual adds an individual to the neighborhood set whose fluorescein value is less than its decision radius. Then the moving probability of an individual in the neighborhood set is computed. Individuals move in the direction of highest moving probability. Then fireflies position and decision radius are updated. The process is repeated till good fitness value is obtained. The fitness function consists of the computed distance between the unknown nodes and the beacon node. The summary of optimization techniques in WSN is explained in Table 1.

4 Conclusion

This paper describes about the hard problems in WSN and also provides the optimization techniques that can be applied to solve these problems. The description of how to apply the technique in WSN is also provided. The future work includes the application of the optimization techniques to clustering in order to find the optimum number of the cluster heads by minimizing the distance between the nodes. Thus, network lifetime can be increased. Each of the illustrated technique can be treated as a milestone in the domain, a unique solution or a significant improvement over an existing method. By this review, we tend to provide to the researchers a beneficial resource to explore in the field of WSN.

References

1. P. Lalwani, S. Das. "Bacterial Foraging Optimization Algorithm for CH selection and routing in wireless sensor networks." 2016 3rd International Conference on Recent Advances in Information Technology (RAIT), Dhanbad, 2016: pp. 95–100.

2. G. Gajalakshmi, G. U. Srikanth. "A survey on the utilization of Ant Colony Optimization (ACO) algorithm in WSN." 2016 International Conference on Information Communication and Embedded Systems (ICICES), Chennai, 2016: pp. 1–4.
3. A.A.Abba Ari, A. Gueroui, B. O. Yenke, N. Labraoui. "Energy efficient clustering algorithm for Wireless Sensor Networks using the ABC metaheuristic." 2016 International Conference on Computer Communication and Informatics (ICCCI), Coimbatore, 2016: pp. 1–6.
4. Ilhem Boussaïd, Julien Lepagnot, Patrick Siarry. "A survey on optimization metaheuristics." Information Sciences, Volume 237, 2013: pp. 82–117.
5. L. Aziz, S. Raghay, H. Aznaoui, A. Jamali. "A new approach based on a genetic algorithm and an agent cluster head to optimize energy in Wireless Sensor Networks." 2016 International Conference on Information Technology for Organizations Development (IT4OD), Fez, 2016: pp. 1–5.
6. A. Singh, S. Rathkanthiwar, S. Kakde. "Energy efficient routing of WSN using particle swarm optimization and V-LEACH protocol." 2016 International Conference on Communication and Signal Processing (ICCSP), Melmaruvathur, 2016: pp. 2078–2082.
7. P. Lalwani, I. Ganguli, H. Banka. "FARW: Firefly algorithm for Routing in wireless sensor networks." 2016 3rd International Conference on Recent Advances in Information Technology (RAIT), Dhanbad, 2016: pp. 248–252.
8. Rakesh Kumar Yadav, Akshita Kapila, Deepa Arora. "Analysis Geographic Routing Protocol using Ant Colony Optimization Technique for WSNs." International Journal of Innovations & Advancement in Computer Science (IJIACS), 2016: pp. 34–37.
9. Bingnan Pei, Hao Zhang, Tengda Pei, H. Wang. "Firefly algorithm optimization based WSN localization algorithm." 2015 International Conference on Information and Communications Technologies (ICT 2015), Xi'an, 2015: pp. 1–5.

Classification of Hepatitis C Virus Using Case-Based Reasoning (CBR) with Correlation Lift Metric

B. Vikas(✉), D. V. S. Yaswanth, W. Vinay, B. Sridhar Reddy, and A. V. H. Saranyu

Department of Computer Science and Engineering, GITAM Institute of Technology, GITAM Deemed to be University, Visakhapatnam, India
vikasboddu30@gmail.com

Abstract. Hepatitis is a widespread and one of the most dangerous liver diseases, affecting millions of people around the globe. Hepatitis C virus (HCV) may be present in many other body parts; however, its primary target is the liver. In this paper, a diagnosis system based on Case-Based Reasoning (CBR) with correlation lift metric for hepatitis C virus (HCV) is presented. Data mining is an extremely useful technique that can play an important role in the push towards healthcare reform. The proposed algorithm gives an efficient and precautionary way to predict the presence of hepatitis C virus which helps immensely in early diagnosis of the infection.

1 Introduction

Hepatitis is a viral infection that causes inflammation of the liver, making it one of the most dangerous diseases. Hepatitis can be caused by various sources consisting of viral infections A, B, C, and also autoimmune hepatitis, faulty liver hepatitis, spirituous hepatitis, and toxin-induced hepatitis [1–3]. Reports by World Health Organization (WHO) suggest that around 170 million people are affected by hepatitis C and 400 million people by chronic hepatitis B [1]. There are 12.2 million HCV carriers in India [4].

1.1 Hepatitis C

Hepatitis C virus was discovered by Choo and co-workers in the year 1989; since then, there has been widespread research and study on this viral epidemiology [4]. Hepatitis C virus (HCV) may be present in many other body parts; however, its primary target is the liver. Liver damage is not caused by HCV itself; it is the reaction of the immune system to the virus that causes the damage. The challenge with HCV infection is that acute hepatitis C shows negligible symptoms and develops into chronic hepatitis [5]. The problem of negligible symptoms has made the early diagnosis for the disease highly difficult. After thorough research, the primary causes for HCV infections have been classified which are blood contamination, sexual intercourse, re-usage of syringes. The irony from the research is that in many cases, transmission of HCV cannot be

V. Bhateja et al. (eds.), *Information Systems Design and Intelligent Applications*, Advances in Intelligent Systems and Computing 672,
https://doi.org/10.1007/978-981-10-7512-4_91

recognised [6]. Table 1 represents percentages of HCV-infected patients across various countries and rest of the world. The graphical representation in pie chart for Table 1 is shown in Fig. 1.

Table 1. Percentage of HCV patients in various countries and rest of the world [11]

Countries	Percentage (%)
Japan	1.2
India	1.5
Malaysia	2.3
Philippines	2.3
USA	2.5
Egypt	14.5
Rest of the world	75.7

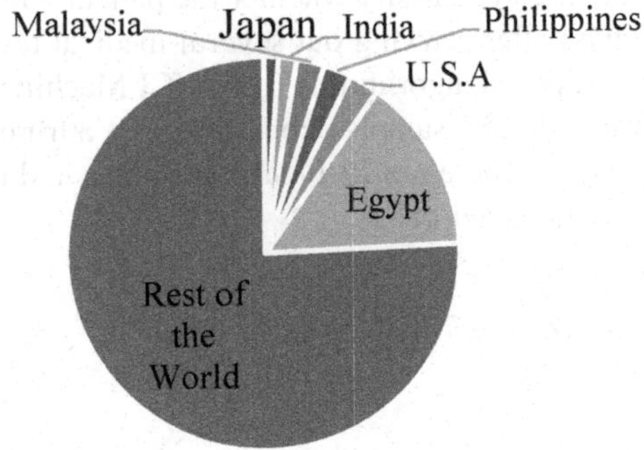

Fig. 1. Pie chart representation of percentage of HCV-infected patients in various countries and rest of the world

1.2 Data Mining

In order to identify patients affected by HCV, and to enhance the accuracy of early diagnosis, data mining techniques can be used [7]. Data mining is a method of analysing information, deriving useful hidden patterns and relationships in a huge data set. Data mining helps in development of predictive models [5, 8]. Numerous algorithms with respect to the task were developed in data mining.

In this paper, an algorithm is proposed for the classification of patients and the most effective symptom, which helps in early diagnosis and treatment. The proposed algorithm is developed with the aid of Case-Based Reasoning (CBR), successfully, with the usage of correlation 'lift' metric. First, the missing attributes in the data set are filled by applying 'statistical mode' to the remaining values of the attribute.

Further lift for every attribute with respect to the class attribute is calculated for each value. Afterwards consider a new tuple and compare it with existing tuples to check for the identical training case [9]: if none, then it searches for the most similar tuple.

The proposed algorithm was implemented using R programming language on R Studio 1.0.136. The machine on which the algorithm was developed and tested has the following configuration: Intel i5 (second generation) processor, 8 GB memory, Mac OS X Sierra(10.12.3) operating system.

This paper is organised as follows: In Sect. 2, a brief description of the data set is discussed. In Sect. 3, Case-Based Reasoning (CBR) and correlation are briefly discussed followed by description of the algorithm. In Sect. 4, experimental results are discussed and accuracy is mentioned with the help of accuracy measures. The paper is concluded in Sect. 5.

2 Data Description

The hepatitis C data set is used to classify whether the patient will be alive or dead with the attribute values derived after carrying out several medical tests. The data set taken for the development of proposed algorithm is from UCI Machine Learning Repository [7]. The database consists of 155 samples containing 19 attributes belonging to two different classes (live or die). There are 13 binary attributes and 6 numerical attributes in the data set, which are depicted below:

1. CLASS: DIE, LIVE
2. AGE: 10, 20, 30, 40, 50, 60, 70, 80
3. SEX: male, female
4. STEROID: no, yes
5. ANTIVIRALS: no, yes
6. FATIGUE: no, yes
7. MALAISE: no, yes
8. ANOREXIA: no, yes
9. LIVER BIG: no, yes
10. LIVER FIRM: no, yes
11. SPLEEN PALPABLE: no, yes
12. SPIDERS: no, yes
13. ASCITES: no, yes
14. VARICES: no, yes
15. BILIRUBIN: 0.39, 0.80, 1.20, 2.00, 3.00, 4.00
16. ALK PHOSPHATE: 33, 80, 120, 160, 200, 250
17. SGOT: 13, 100, 200, 300, 400, 500
18. ALBUMIN: 2.1, 3.0, 3.8, 4.5, 5.0, 6.0
19. PROTIME: 10, 20, 30, 40, 50, 60, 70, 80, 90
20. HISTOLOGY: no, yes.

3 Methodology

In this research, correlation lift metric is used in Case-Based Reasoning (CBR). First, this algorithm fills the missing attributes with the statistical mode of the attribute values. Further, a new tuple is considered to compare with all tuples in the database. If exactly same tuple is found, then the result of the matched tuple is the result for the new tuple. Else, lift metric is used to predict the result.

3.1 Case-Based Reasoning

In Case-Based Reasoning (CBR), the samples or tuples are stored to solve a problem as complex symbolic descriptions. In medical field, CBR is highly useful, where it considers patient case histories and treatment to diagnose new patient [9].

3.2 Correlation

A correlation measure or metric is used to expand the support-confidence framework for association rules. There are several metrics in which one can be used and 'lift' metric as one among them [9].

The lift metric is determined as follows:

$$\mathrm{Lift(A,B)}=\frac{P(AUB)}{P(A)*P(B)}. \tag{1}$$

where A and B are item sets.

$$\begin{aligned} \mathrm{P(A)} &= (\text{frequency of occurrences of the attribute values in A})/ \\ &\quad (\text{total number of attribute values}) \\ \mathrm{P(B)} &= (\text{frequency of occurrences of the attribute values in B})/ \\ &\quad (\text{total number of attribute values}) \\ \mathrm{P(AUB)} &= (\text{frequency of simultaneous occurrences of the attribute values in} \\ &\quad \text{both the attributes A and B})/(\text{total number of attribute values}) \end{aligned}$$

If Lift(A, B) < 1, item sets are negatively related.
If Lift(A, B) > 1, item sets are positively related.
If Lift(A, B) = 1, item sets are independent [9].
If Lift(A, B) > Lift(A, C), then B is more strongly related to A than C's relation with A [10]. The implication is that higher the value of the lift, stronger the relation among the attributes.

In the proposed algorithm, the lift metric is calculated for every attribute with the class attribute so that one can know the attribute which has the highest impact on to which class (live or die) the patient belongs to, which helps immensely in early diagnosis, the main problem with the hepatitis C virus infection.

3.3 Algorithm

Step 1:

The raw data given in the text format is converted to comma separated values (CSV) format.

Step 2:

The missing values in the data set are filled with the respective modes of their respective attributes.

Divide the given data set into two sets vis-à- vis training and test sets.

Step 3:

The probability of occurrence of each attribute value is calculated

Probability = (number of occurrences of a particular attribute value)/(total number of values of that attribute).

Step 4:

Lift metric for each value of the attribute with respect to the class attribute values is calculated using the Eq. (1).

Lift can be greater than 1, less than 1, or equal to 1.

Step 5:

A new tuple from the test data is taken, and the attribute values of the new tuple are compared with the attribute values of all the tuples in the training set.

Step 6:

For a tuple I in training set, if a match occurs in Step 5, the lift of that particular attribute value is considered. If lift is greater than 1, the values of the lifts are added and stored in the variable '*lift_sum*'.

If a match does not occur and the lift value of that attribute value is greater than 1, the corresponding lift values are subtracted from the variable '*lift_sum*'.

Step 7:

Repeat Step 6 for all the tuples in the training set.

The tuple in the training set with the maximum lift value is considered as the one which is similar with the tuple from the test set.

Step 8:

The corresponding class attribute value of the similar tuple in the training set is given as the predicted class value for the tuple from the test set.

Step 9:

Repeat Steps 5, 6, 7, and 8 for all the tuples in the test set to predict their values.

4 Experimental Results

The accuracy of correctly predicting that a patient will die is greater than the accuracy of correctly predicting that the patient will live. The classifier conservatively predicts that the healthy patient may be prone to the HCV infection based on the reported symptoms which will help in early diagnosis and reduce the possible risks [9].

The measure of proportion of correctly identified positives, also called sensitivity, is called **Recall**.

$$\text{Recall} = \frac{\textit{Number of True Positives}}{\textit{Number of Positives}}. \tag{2}$$

The correctly classified 'live' tuples are truc positives.
The measure of proportion of correctly identified negatives is called **Specificity.**

$$\text{Specificity} = \frac{\textit{Number of True Negatives}}{\textit{Number of Negatives}}. \tag{3}$$

The correctly classified 'die' tuples are true negatives.
The fraction of instances that are retrieved and relevant is called **Precision**.

$$\text{Precision} = \frac{\textit{number of true positives}}{\textit{number of true positives} + \textit{number of false positives}}. \tag{4}$$

The incorrectly classified 'live' tuples are false negatives.
The incorrectly classified 'die' tuples are false positives.
The harmonic mean of precision and recall is called **F-Measure.**

$$\text{F-Measure} = \frac{2}{\frac{1}{precision} + \frac{1}{recall}} = \frac{2(precision)(recall)}{precision + recall}. \tag{5}$$

The percentage of tuples of the given test set that the classifier classified correctly is called **Accuracy**.

$$\text{Accuracy} \quad = \quad \text{recall}\left(\frac{pos}{pos + neg}\right) + specificity\left(\frac{neg}{pos + neg}\right). \tag{6}$$

Here, *pos* denotes the number of positive tuples and *neg* denotes the number of negative tuples.

The proposed algorithm returned an accuracy of 73.6% with performance measures: specificity, recall, precision, and f-measure with values 0.88, 0.65, 0.92, and 0.77, respectively. The confusion matrix for the obtained results is given in Table 2, and the performance measures are shown in Table 3.

Table 2. Confusion matrix for hepatitis C virus data set

	Predicted			
Actual	Classes	Live	Die	% Correct
	Live	23	12	65.7
	Die	2	16	**88.9**
	Overall %			73.6

Table 3. Performance measures for the proposed algorithm

Accuracy measure	Value
Specificity	0.88
Recall	0.65
Precision	0.92
F-measure	0.77
Accuracy	**73.6%**

5 Conclusion

One of the most difficult jobs for a physician is diagnosing disorders. An improper diagnosis may lead the patient to death bed. In this regard, several data mining techniques are used to help the physician better diagnose the patient. In this paper, correlation metric—'*lift*' in Case-Based Reasoning (CBR)—is used to diagnose hepatitis C virus (HCV)-infected patients. First missing attributes in the training data sets are filled. Further, the correlation metric '*lift*' is calculated for every attribute value with the corresponding class attribute value. Now, a new patient (tuple) attributes are considered for classification. The new tuple will now be compared with all the tuples in training data set to find the most similar tuple; '*lift*' metric is used to classify the patient. The accuracy of correctly predicting that the patient belongs to '*die*' class is 88.9% and to '*live*' class is 65.7% which infers that *a healthy person may be prone to HCV infection and should be alert regarding the infection.* The overall accuracy returned for the proposed algorithm is **73.6%**. The proposed algorithm conservatively predicts the class (*live* or *die*) of the patient to reduce the risk with the HCV symptoms. The proposed method will be instrumental in further research in the area of hepatitis disease diagnosis.

Despite the fact that a patient is not infected with HCV, the algorithm conservatively labels the patient as an HCV-infected patient considering danger to the patient based on their condition. This led to a higher number of false positives which reduced the accuracy. However, the patients who are truly infected with HCV are correctly labelled so with an accuracy of 89%. *This will enable the patients who are not actually infected to take suitable precautionary measures.*

Acknowledgements. The authors would like to thank UCI Machine Learning Repository for providing the required data set to conduct the experiments.

References

1. WHO, Hepatitis C (Fact Sheet No. 164), World Health Organization, Geneva, 2000.
2. Hodgson S., Harrison R. F., Cross S. S., An automated pattern recognition system for the quantification of inflammatory cells in hepatitis-C-infected liver biopsies, Image and Vision Computing 24, 2006, pp. 1025–1038.
3. Moriishi K. and Y. M Atsuura, "Mechanisms of hepatitis C virus infection", Antivir. Chem. Chemother 14, 2003, pp. 285–297.

4. Dr. Sanjoy Kumar Pal, Dr. G. Choudhuri.: Hepatitis C: The Indian Scenario (2005), Dept. of Gastroenterology, Sanjay Gandhi Post Graduate Institute of Medical Sciences, Lucknow, India.
5. Huda Yasin, Tahseen A. Jilani, Madiha Danish.: Hepatitis-C Classification using Data Mining Techniques. International Journal of Computer Applications (0975–8887) Volume 24–No. 3 (2011).
6. Booth J., O Grady J. and Neuber ger J. (2001), Clinical guidelines on the management of hepatitis C, Gut 1, 11–21.
7. Mehdi Neshat, Mehdi Sargolzaei, Adel Nadjaran Toosi, and AzraMasoumi.: Hepatitis Disease Diagnosis Using Hybrid Case Based Reasoning and Particle Swarm Optimization. International Scholarly Research Network, Volume 2012, (2012) 609718.
8. AbuBakr Awad, Mahasen Mabrouk, Tahany Awad, Naglaa Zayed, Sherif Mousa, Mohamed Saeed.: Performance Evaluation of Decision Tree Classifiers for the Prediction of Response to treatment of Hepatitis C Patients. Pervasive Health 2014, Oldenburg, Germany (2014).
9. Kamber, Micheline, and Jian Pei. Data Mining, Second Edition. 1st ed. Morgan Kaufmann, 2006. Print.
10. “Lift (Data Mining)”. En.wikipedia.org. N.p., 2017. 29 Mar. Web. 2017.
11. “Hepatitis C:: The Facts: The Epidemic—Worldwide Prevalence”. Epidemic.org. N.p., 2017. Web. 29 Mar. 2017.

Software Effort Estimation Using Grey Relational Analysis with K-Means Clustering

M. Padmaja[1(✉)] and D. Haritha[2]

[1] Department of CSE, Gitam Institute of Technology, GITAM University, Visakhapatnam 530045, India
padmaja.madugula@gmail.com
[2] Department of CSE, University College of Engineering, JNTUK, Kakinada 533003, India
harithadasari9@yahoo.com

Abstract. Software effort estimation is described as a method of predicting the amount of person/months ratio to build a new system. Effort estimation is calculated in terms of persons involved per month for the completion of a project. During the launch of any new project into the market or in industry, the cost and effort of a new project is estimated. In this context, a numerous models have been developed to measure the effort and cost. This becomes a challenging task for the industries to predict the effort. In the present paper, a novel method is proposed called the Grey Relational Analysis (GRA) for estimating the effort of a particular project by considering the most influenced parameters. To achieve the same, one-way ANOVA and Pearson correlation methods are combined. Experimental results obtained with the help of clustering and without clustering by using the proposed method on the data set are presented. An attempt has been made to show the minimum error rate by using GRA for predicting the effort estimation on COCOMO 81 data set and clustered data set. The proposed method demonstrated better results compared to the traditional techniques used for estimation. The efficiency of the proposed system is illustrated through experimental results.

1 Introduction

Effort estimation plays a predominant role in the development of projects [1]. The development of a project by any company starts by making a perfect analysis about its requirements. For the same, abstraction of previous projects or analogy approach is used [2]. The estimates typically become targets when the project proceeds to the stages of detailed planning and execution. Furthermore, improper estimation can lead to a variety of problems for stakeholders. The problem with estimation is that the project manager needs to quantify the cost, effort and delivery period much before the completion of the design. The more availability of information, the better it is, as the estimates can be precise. Usually, estimation suffers due to incomplete knowledge, lack of time and various competitive pressures that are directly or indirectly put over the developer to arrive at more acceptable estimates. The same can be achieved by selecting an efficient and novel method such as GRA, as suggested in J. Deng [3].

V. Bhateja et al. (eds.), *Information Systems Design and Intelligent Applications*, Advances in Intelligent Systems and Computing 672,
https://doi.org/10.1007/978-981-10-7512-4_92

Grey System Theory (GST) helps to develop the product with completely known information and also even unknown information [4]. The generation of partially known information is practised much in software industries where incomplete information is initially provided for development. In GRA, Grey Relational Grade (GRG) specifies association with previous projects to develop the new project [5].

This paper has been structured as follows: Sect. 2 depicts related previous works, Sect. 3 presents various methods to estimate the effort, Sect. 4 illustrates assessment and experimental outcome of the introduced method and to end, Sect. 5 concludes the work and further extensions.

2 Literature Survey

Estimating the effort and cost is a prerequisite process of a software development before handling any project. The goal of a software administrator is to deliver the product efficiently with high quality. Thus, he is responsible for estimating the effort and cost efficiently in advance.

In addition, M. Padmaja and D. Haritha [6] applied GRA method on Kemerer data set and compared the results with other basic models. Then, Chao-Jung Hsu et al. [7] have proposed six weighted methods integrated with GRA on three type of data sets to estimate the effort. They prove that the weighted GRA achieves better results than non-weighted GRA. Chao-Jung Hsu et al. [8] continue this work on different data sets and compare the results with other with different distinguish coefficients and choose analogous numbers. These numbers most influenced than coefficients to estimate the effort.

Mohammad Azzeh et al. [9] have also proposed the Fuzzy Grey Relational Analysis (FGRA) model that encouraging results can be produced lesser MMRE, MdMRE and greater Pred(25) on five existing data sets. Jin-Cherng Lin et al. [10] used computing intelligence techniques, PSO on clustered data within three to four groups and estimated effort. The experimental results produced minimum error rate when compared to previous models. Jin-Cherng Lin et al. [11] and these techniques are also applied to identify key factors of projects such as ANOVA and Pearson. After recognizing the key factors, the projects are clustered through K-means clustering and finally software project effort has been estimated using particle swarm optimization (PSO). G. Chamundeswari et al. [12] have used K-means clustering on different data sets to form clusters by using MATLAB and Java. Then, the efficiency of the algorithm is proved and tabulated.

3 Methodology

In the present work, GRA assigns the grades or ranks for the projects. It tries to gaining better estimation of cost and effort for a given project. The process of system developed in this paper is explained diagrammatically in Fig. 1. GRA procedure is applied on COCOMO81 data set which consists of 63 projects and every project further having 17 parameters such as reliability (RELY), volatility (VIRT), database (DATA), complexity (CPLX), storage (STOR), execution time (TIME), turnaround time (TURN), analyst capability (ACAP), programmer capability (PCAP), analyst experience (AEXP), virtual machine experience (VEXP), language experience (LEXP), modern programming (MODP), software tools (TOOL), schedule (SCED), size (LOC) and actual effort. Here the process begins with reducing the data set by applying two methods, i.e. ANOVA and Pearson correlation method. This reduction produces most influenced parameters on which the further process takes place. Then K-means algorithm has been applied to form clusters on both the original data and the reduced data. Finally, GRA method is applied to find MMRE on both the data sets through the estimated effort of each project.

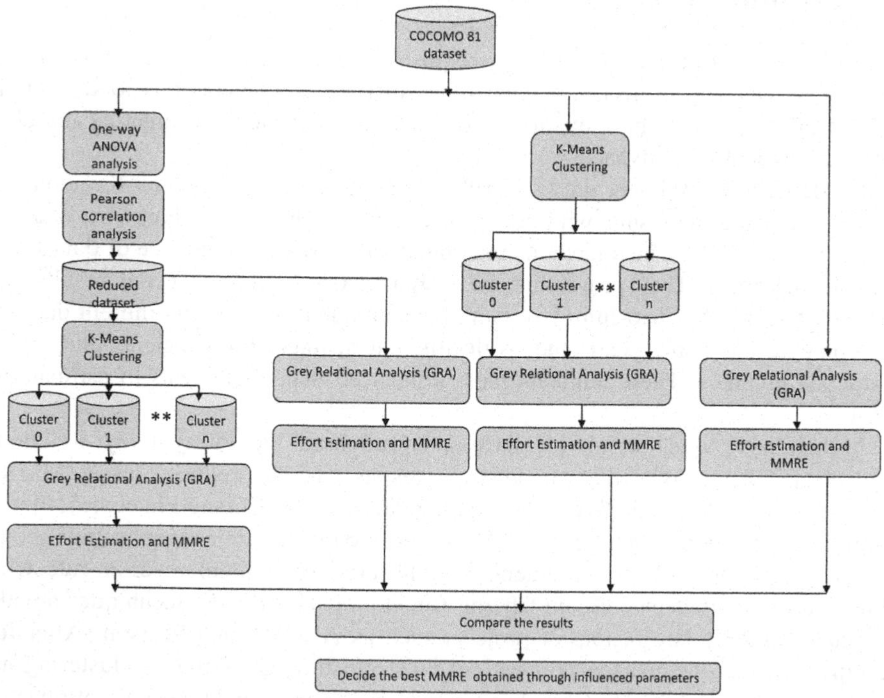

Fig. 1. Architecture of the proposed system

3.1 One-Way Analysis of Variance (ANOVA)

ANOVA is a method that analyses the factors which significantly involve providing quality. It has to be performed on each parameter with the actual effort to obtain its contribution. Contributions of the parameters are then calculated through F-value to decide which parameter has the high impact contribution presented at the significant level.

3.2 Pearson Correlation Coefficient

Correlation is described as the rate at which two variables collaborate with each other in a linear fashion. Correlation is defined in terms of coefficient which in numerical value that exhibits the direction and degree of the relation between a pair of variables, called as Pearson product-moment correlation coefficient. It is referred by *'r'* and can be computed using Eq. (1). The value of *r* ranges from −1 to +1, which affects the interpretation correspond.

$$r = \frac{\sum (X_i - \bar{X})(Y_i - \vec{Y})}{\sqrt{\sum (X_i - \bar{X})^2 \sum (Y_i - \bar{Y})^2}} \tag{1}$$

where
$\overline{X}$ is the mean of X coordinate values
$\overline{Y}$ is the mean of Y coordinate values.

3.3 K-Means Clustering

Clustering is the process of partitioning a collection of data points into groups. K-means clustering is most popularly used. It is one of the unsupervised learning algorithms. Given a data set, the K-means clustering algorithm results in a predefined *k* number of clusters. Formally, the K-means clustering prescribed [12].

3.4 Grey Relational Analysis (GRA)

GRA produces an effective value over a multiple parameters through complex relationship. In Grey Theory, with help of Grey Relational Coefficient (GRC), Grey Relational Grade (GRG) decides the selection of best parameters to implement the new project. The GRA is very effective to evaluate the effort of a project based on influenced projects.

a. Normalization

The initial step is the equivalence of the various parameters considered as duration, KSLOC and function points since they influence the effort estimation. In this paper, to normalize the parameters used, larger-the-better and smaller-the-better were used as shown in Eqs. (2) and (3). The normalized values are always between 0 and 1.

Upper-bound effectiveness:

$$x_i(k)^* = \frac{x_i(k) - \min x_i(k)}{\max x_i(k) - \min x_i(k)} \tag{2}$$

Lower-bound effectiveness:

$$x_i(k)^* = \frac{\min x_i(k) - x_i(k)}{\max x_i(k) - \min x_i(k)} \tag{3}$$

Here i = 1, 2, …, m and k = 1, 2, 3, …, n
where, $x_i(k)$ stand for the *k*th attribute in *i*th series;
$x_i(k)^*$ stand for the predicted grey relational generating of the *k*th attribute in *i*th series;
$\max x_i(k)$ signify the maximum of the *k*th attribute in *i*th series;
$\min x_i(k)$ signify the minimum of the *k*th attribute in *i*th series;

b. Grey Relational Coefficient (GRC)

GRA illustrates the relationship with historical data to implement the new project. To calculate coefficients of parameters in all projects, Eq. (4) was used.
GRC can be calculated as,

$$\gamma(x_0(k), x_i(k)) = \frac{\Delta_{\min} + \zeta\Delta_{\max}}{\Delta_{0i}(k) + \zeta\Delta_{\max}} \tag{4}$$

Here ζ is the distinguishing coefficient. Usually, it can be adjusted to implement the system.
ζ €(0, 1], i.e. 0.5

$$\Delta_{0i}(k) = |x_0(k) - x_i(k)|$$

i.e. Delta values or Deviation coefficients
$\Delta_{\min} = \min_i \min_j |x_0(k), x_i(k)|$
$\Delta_{\min} = \max_i \max_j |x_0(k), x_i(k)|$

c. Grey Relational Grade (GRG)

GRG is used to specify the relation between two sets. It shows the similarity between the project to be estimated and its historical projects. Two projects are said to maintain a closer relationship, if the GRG between them is high [6].

$$\Gamma(x_0(k), x_i(k)) = \frac{1}{n}\sum_{k=1}^{n} \gamma(x_0(k), x_i(k)) \tag{5}$$

where i € {1, 2, 3, …, n}
n is the number of response parameters

d. Grey Relational Rank (GRR)

To estimate the effort, GRR should be retrieved from the historical data with the largest GRG over all projects [6]. Irrespective of GRG, we assign ranks and decide similar projects to estimate the effort of that project.

e. Effort Prediction by Using GRA

To calculate the estimated effort of a particular project with most influenced K projects data, the following equation is applied [6].

$$\hat{\varepsilon} = \sum_{i=1}^{n} w_i * \varepsilon_i \tag{6}$$

where
$\hat{\varepsilon}$ is the predicted effort
ε_i is the actual effort of influenced project
w_i is weight is calculated by

$$w_i = \frac{\Gamma(x_0, x_i)}{\sum_{i=1}^{k} \Gamma(x_0, x_i)} \tag{7}$$

where
$\Gamma(x_0, x_i)$ is the GRG value of most influenced project
i = 1 to K-means most influenced K projects.

3.5 Evaluation Criteria

From the above proposed methods, one finds the estimated effort. Then check whether this method is efficient or not by using evaluation method, i.e. MMRE factor. MMRE means mean magnitude relative error, and it can be calculated by using Eq. (8).

$$MMRE = \frac{1}{N} \sum_{i=1}^{n} MRE \tag{8}$$

Here N means number of projects.
MRE means magnitude relative error and is calculated as follows:

$$MRE = \frac{|ActualEffort_i - EstimatedEffort_i|}{ActualEffort_i}$$

In this study, MMRE is the fitness value for the proposed method to know the error rate. The proposed model shows minimum error rate compared to other models as shown in experimental results.

4 Experiments and Results

On COCOMO81 data set, GRA method is applied to find the effort estimation and minimum error rate. To obtain the most influence parameters, both one-way ANOVA method and Pearson correlation method are combined as described in the following subsections (step 1 and step 2). Now, K-means method is implemented on both the original data set and the obtained influenced parameters from the previous step. Then, the next process in the proposed paper is applying GRA method on the original raw data and on the obtained clusters separately in order to find the effort and error rate. Hence, in conclusion it can be said that the above implementation of GRA through ANOVA and Pearson correlation method has produced efficient results, which are further discussed in the subsequent sections.

Step 1: **One-way ANOVA**

The processing starts with implementation of one-way ANOVA method to obtain contribution of the parameters with effort, and results of all parameters are shown in Table 1. From these F-values, we can select the parameters to influence the most for the estimation of effort. The tabulated values are graphically represented in Fig. 2.

Table 1. Parameter values using one-way ANOVA

Parameters	RELY	TIME	STOR	DATA	VIRT	CPLX	ACAP	TURN
F-value	8.8384	8.8363	8.8356	8.8392	8.8391	8.8369	8.8417	8.84
Parameters	AEXP	LEXP	MODP	PCAP	TOOL	VEXP	SCED	
F-value	8.8406	8.8393	8.8392	8.8409	8.8389	8.8392	8.838	

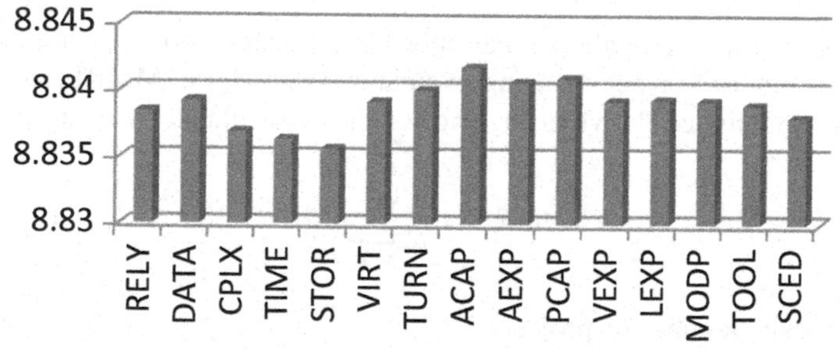

Fig. 2. COCOMO81 parameters using one-way ANOVA analysis

From the above figure, one can observe that the most influenced parameters are ACAP, PCAP, AEXP, TURN and LEXP. So these values are used to estimate the effort.

Step 2: **Pearson correlation**

To predict the effort on COCOMO81 data set, the proposed system is implemented called Pearson correlation analysis by using Eq. (1) on the 15 parameters and finally *r*-values are generated. Among the *r*-values, one can consider those parameters as the most influenced parameters that are having highest *r*-value. Hence, those parameters that hold *r*-value >0.2 are considered as the most influenced parameters in this paper. The result of Pearson correlation values is shown in Table 2 and graphically presented in Fig. 3.

Table 2. Parameter values using Pearson correlation

Parameters	RELY	TIME	STOR	DATA	VIRT	CPLX	ACAP	TURN
F-value	0.2065	0.1523	0.1045	0.4446	0.018	0.009	−0.147	0.2057
Parameters	AEXP	LEXP	MODP	PCAP	TOOL	VEXP	SCED	
F-value	−0.035	0.0875	0.2697	0.157	0.0019	0.0678	0.0211	

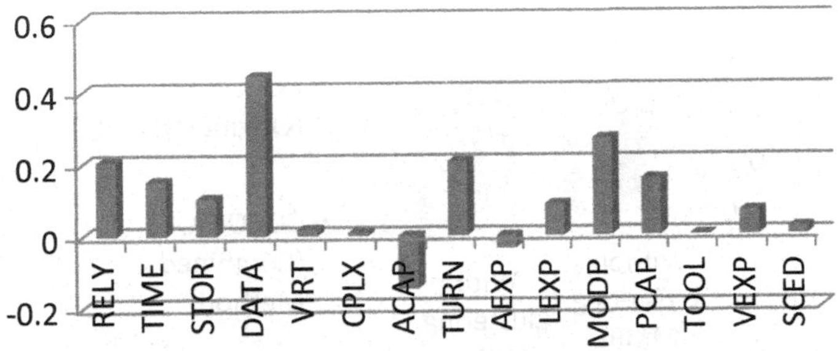

Fig. 3. COCOMO81 parameter values using Pearson correlation

This can be observed from the above figure, and the most influenced parameters are DATA, MODP, RELY and TURN. So these values are used to estimate the effort.

Step 3: **K-means clustering**

Then K-means clustering is implemented twice. K-means is applied on the entire data set for the first time which resulted in three clusters as shown in Table 3. Then K-means has been applied for the second time to cluster on the result obtained through a combined method of ANOVA and Pearson correlation method. The clustering details are shown in Table 3.

Table 3. Clustering details using K-means clustering

Cluster No	Data set	
	COCOMO81 (Original data)	COCOMO81 (Combined method)
	No. of projects	No. of projects
Cluster 0	6	26
Cluster 1	40	30
Cluster 2	17	7
Total projects	63	63

Step 4: **Grey Relational Analysis (GRA)**

GRA is used to estimate the effort on these methods by using equations mentioned in methodology Eqs. (2)–(7). Finally, the error rate is calculated for every project by using Eq. (8) and obtained 0.201752 as shown below in Table 4 and in Fig. 4.

Table 4. Evaluation results using proposed method

Data set	MMRE	
	Without clustering (MMRE)	With clustering (MMRE)
COCOMO81 (Original data set)	0.272966	0.292739
COCOMO81 (Combined method)	0.240498	0.201752

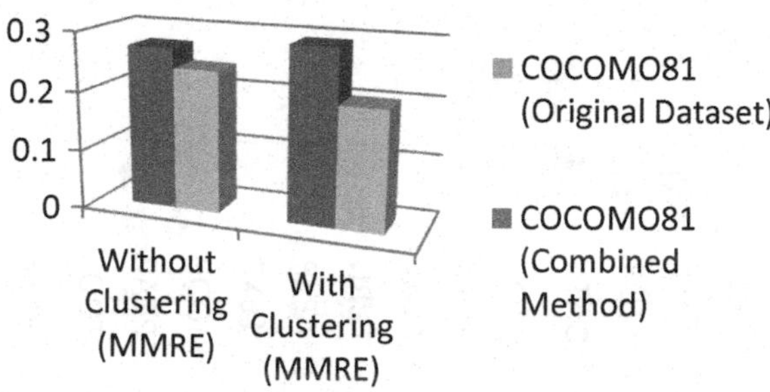

Fig. 4. Efficiency of MMRE with anticipated model

5 Conclusion and Future Work

To develop projects, effort has to be predicted before the actual software development in such a way that the newly developed project must exhibit higher quality with less cost. In order to accomplish this, the project management team must select the best model possible to achieve the above-mentioned goals and deliver the project in time. This paper states that in order to model a project, accurate and reliable effort estimation is required with respect to the project conditions. The proposed model is GRA used to estimate the effort of any project, and the experimental results prove minimum error rate. Further extension of the present work can also be done by combining the proposed model with other techniques including various effort adjustment factors of different data sets.

References

1. M. Jorgensen, “Contrasting ideal and realistic conditions as a means to improve judgment-based software development effort estimation”, Information and Software Technology, Vol. 53, Issue 12, pp. 1382–1390, Elsevier B.V, December 2011.
2. Martin Shepperd, Chris Schofield and Barbara Kitchenham “Effort Estimation using Analogy”, IEEE, 2009.
3. Deng. J “Introduction to grey system”, Journal of Grey System, Vol. 1 No. 1, pp. 1–24. 1989.
4. Qinbao Song, Martin Shepperd and Carolyn Mair, “Using Grey Relational Analysis to Predict Software Effort with Small Data Sets”, 11th IEEE International Software Metrics Symposium—METRICS 2005.

5. Sun-Jen Huang, Nan-Hsing Chiu and Li-Wei Chen, "Integration of the grey relational analysis with genetic algorithm for software effort estimation", European Journal of Operational Research, pp 898–909, 2008.
6. M. Padmaja, Dr D. Haritha, "Software Effort Estimation using Grey Relational Analysis", MECS in International Journal of Information Technology and Computer Science, 2017, Vol. 9, No. 5, May 2017.
7. Chao-Jung Hsu and Chin-Yu Huang, "Improving Effort Estimation Accuracy by Weighted Grey Relational Analysis During Software Development" 14th Asia-Pacific Software Engineering Conference, IEEE, 2007.
8. Chao-Jung Hsu and Chin-Yu Huang, "Comparison of weighted grey relational analysis for software effort estimation", Software Qual J, Springer Science + Business Media, LLC 2010.
9. Mohammad Azzeh & Daniel Neagu & Peter I. Cowling, "Fuzzy grey relational analysis for software effort estimation", Empir Software Eng, 15, pp 60–90, Springer, 2010.
10. Jin-Cherng Lin, Yueh-Ting Lin, Han-Yuan Tzeng and Yan-Chin Wang, "Using Computing Intelligence Techniques to Estimate Software Effort", International Journal of Software Engineering & Applications (IJSEA), Vol. 4, No. 1, January 2013.
11. Jin-Cherng Lin, Han-Yuan Tzeng, "Applying Particle Swarm Optimization to Estimate Software Effort by Multiple Factors Software Project Clustering", IEEE, 2010.
12. G. Chamundeswari, Prof. G. Pardasaradhi Varma, Prof. Ch. Satyanarayana, "An Experimental Analysis of K-means Using Matlab", *International Journal of Engineering Research & Technology (IJERT),* Vol. 1 Issue 5, July 2012.

Application of the Apriori Algorithm for Prediction of Polycystic Ovarian Syndrome (PCOS)

B. Vikas(✉), B. S. Anuhya, K. Santosh Bhargav, Sipra Sarangi, and Manaswini Chilla

Department of Computer Science and Engineering, GITAM Institute of Technology, GITAM Deemed to be University, Visakhapatnam, India
vikasboddu30@gmail.com

Abstract. Data mining is a powerful technology having the potential to find practical solutions to problems in diverse fields. The advent of vast information in the medical field has lead to requirement of extracting useful information through data mining techniques. Medical conditions such as the polycystic ovarian syndrome (PCOS) do not have effective diagnosis and proper treatment methods. Unfortunately, PCOS is the most common endocrinal disease which has been, till date, ignored by many. In this paper, an attempt has been made to recognize recurring patterns among symptoms of PCOS patients using frequent itemset mining. The present research also focuses on Apriori algorithm which has been used to predict those who are susceptible to the syndrome.

1 Introduction

PCOS is a severely undermined condition affecting at least one in 10 women worldwide [1]. It can cause variety of symptoms such as weight gain, ovarian cysts, difficulty ovulating, acne, facial hair, depression, anxiety [2], and heavy periods, and it may take women years to get diagnosed. In the long term, PCOS can lead to metabolic disorders, such as type 2 diabetes [3], cardiovascular disease, hormonal dysfunction, and infertility as shown in Fig. 1 [4, 5]. In fact, [1] "PCOS is the cause of more than 75% of anovulatory infertility, which is the infertility caused in a woman." Despite the seriousness of the condition, there are very little options for treatment. These treatments include the use of medicines such as metformin for controlling diabetes and treating ovulation problems; however, no cure has been discovered yet. In order to help in the diagnosis of the patient, frequency itemset mining has been used to find out the correlations between the attributes in the survey data set. The frequency itemset mining method applied Apriori algorithm which is used to find the frequency of itemsets using candidate generation.

Apriori algorithm is one of the algorithms that can be applied to a large itemset. Hence, this is proved to be one of the best tools to analyze the data obtained from survey conducted. The sole purpose of using Apriori algorithm is the fact that it generates more sets of frequent items, even though it scans the database multiple times. This process may involve significant consumption of memory, but the recurring

V. Bhateja et al. (eds.), *Information Systems Design and Intelligent Applications*, Advances in Intelligent Systems and Computing 672,
https://doi.org/10.1007/978-981-10-7512-4_93

patterns obtained from running the algorithm over the data set will give us exactly what we require, helping in diagnosis of PCOS. Also this algorithm is known for its easy to use and convenient nature and will help us find strong relations between the various attributes present in our data set.

The paper has been organized as follows: In Sect. 2, a concise description about the data set has been provided. Sect. 3 is dedicated to FIM followed by the metrics that are used for FIM in Sect. 4. In Sect. 5, Apriori algorithm and its applications have been described. In Sect. 6, experimental results are demonstrated and major symptoms observed in the survey data have been pictorially represented. Sect. 7 of this paper gives a detailed result analysis, and the limitations of the paper have been stated in Sect. 8. The conclusion has been reported in Sect. 9.

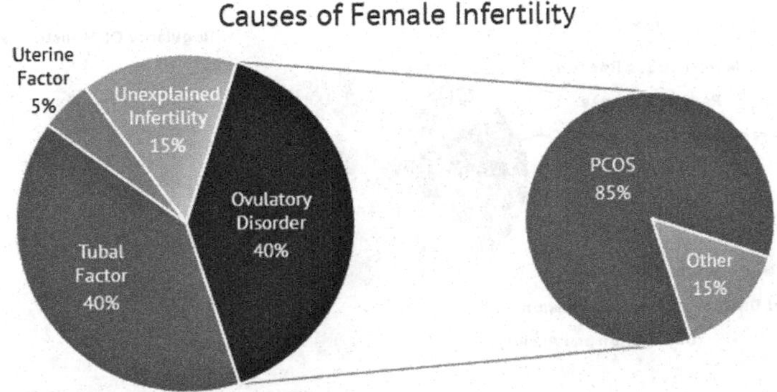

Fig. 1. Ubiquitous presence of Polycystic Ovarian Syndrome and its major symptoms

2 Data set

The data set [6] for PCOS is used for attribute interrelation which can be determined using FIM. This real-time data set has been taken from a survey conducted among 119 girls of the age range 18–22 based on their lifestyle and food habits to understand the relationship among various the traits. The database consists of 119 samples with 15 attributes belonging to two different classes (maybe or maybe not). Figure 2 depicts the survey data and their major occurrences pictorially. There are 10 binary attributes and 5 categorical attributes as shown below:

1. CLASS LABEL: MAYBE (mb), MAYBE NOT (mbn).
2. REGULARITY OF MENSTRUAL PERIODS: yes (y), infrequent menses (im), irregular bleeding (ib), heavy bleeding (hb).
3. WEIGHT GAIN: yes (y), no (n).
4. EXCESS FACIAL OR BODY HAIR: yes (y), no (n).
5. DARK AREAS ON SKIN: yes (y), no (n).
6. PIMPLES: yes (y), no (n).
7. DEPRESSION AND ANXIETY: yes (y), no (n).
8. HISTORY OF DIABETES AND HYPER TENSION: yes (y), no (n).

9. BODY WEIGHT MAINTENANCE: yes (y), no (n).
10. OILY SKIN: yes (y), no (n).
11. LOSS OF HAIR: yes (y), no (n).
12. FREQUENT EATING PLACES: hostel mess (hm), campus canteen (cc).
13. REGULAR EXERCISE: yes (y), no (n).
14. MENTAL STRESS REASON: new admission to hostel, personal problems, peer pressure, change in dietary habits.
15. FAST-FOOD INTAKE: everyday (ed), once in a week (w), once in a month (m).

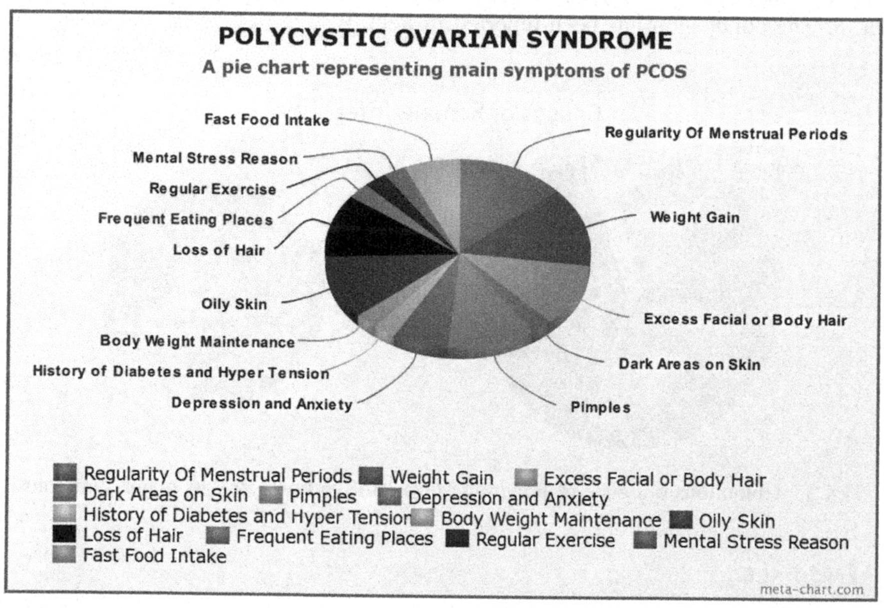

Fig. 2. Survey data attributes and major occurrences

3 Frequent Itemset Mining (FIM)

FIM [7] is a very rich topic involving frequent patterns such as item sets, subsequences, or substructures. Sets of frequently appearing items in vast data sets are called frequent itemsets. Subsequences are sequential patterns and substructures are a combination of itemsets and subsequences. FIM mainly focuses on sequences of events and patterns that can lead to the predictions of particular behaviors from the obtained data sets [8]. This interesting branch of data mining plays a vital role in the discovery of mining correlations, associations, and other such relationships in large data sets. In this paper, we have extensively used this mining task on our data sets to retrieve the required results on PCOS.

4 Metrics Used for Frequent Itemset Mining

The metrics that are used for frequent itemset mining are as follows [9]:

1. Association Rule:

   An expression of the form $X \Rightarrow Y$ *is* called an *association rule*, where X and Y are itemsets and X ∩ Y = Ø.

   X is the *body* or *antecedent* of the rule

   Y is the *head* or *consequent* of the rule.

2. Support:

   The *support* of a rule indicates how often items appear in a database. For an association rule $X \Rightarrow$ Y:

   Support(X ⇒ Y, Z) = Support(X ∪ Y, Z).

3. Confidence:

   The *confidence* of a rule X ⇒ Y is nothing but the conditional probability of having *Y* contained in a set, given that X is contained in that set:

   $$\text{Confidence } (X \Rightarrow Y, Z) = \frac{\text{Support}(X \bigcup Y, Z)}{\text{Support}(X, Z)}.$$

5 Apriori Algorithm

Apriori algorithm was proposed by R. Agrawal and R. Srikant in 1994. This seminal algorithm is a very common and well-accepted data mining approach. Its main purpose is to find frequently occurring itemsets in a transaction data set and to obtain a set of association rules from the data [7, 10]. To fulfill this purpose of finding frequent itemsets, the algorithm uses candidate generation and utilizes an iterative method called as level-wise search. In this search, n-itemsets have been used to search (n + 1)-itemsets using anti-monotonicity of itemsets, that is, "if an itemset is not frequent, any of its superset is never frequent."

In Apriori algorithm, an assumption is made that the items and data within an itemset or transaction are in lexicographical order. The set of frequent itemsets of size n have been taken as FI_n, and their candidates have been assumed as CD_n. In this algorithm, first the entire database is scanned, recurrent itemsets of size 1 are searched by gathering the count for every item, and only those satisfying the requirement of minimum support are accumulated. After which, the frequent itemsets are extracted by iterating the subsequent three steps.

1. Generate the probable candidates for frequent itemsets—C_n + 1 having n + 1 size, from the frequent itemsets of size n.
2. The support of each of the probable candidates is calculated by scanning through the database.
3. Suitable itemsets, i.e., those satisfying the minimum support requirement are then added to FI_n + 1 [10].

To generate $CD_n + 1$ from FI_n, the two steps given below must be followed:

1. Join step: Take the union of two itemsets of size n, X_n and Y_n where both the frequent itemsets have in common the first n−1 elements. Generate $Z_n + 1$ which will contain the first candidates of frequent itemsets of size n + 1.
   $Z_n + 1 = X_n \cup Y_n = \{item\_l, \ldots, item\_n{-}1, item\_n, item\_n\}$
   $X_n = \{item\_l, item\_2, \ldots, item\_n{-}1, item\_n\}$ $Y_n = \{item\_l, item\_2, \ldots, item\_n{-}1, item\_n\}$ where, $item\_l < item\_2 < \cdots < item\_k < item\_n$.
2. Prune step: In this step, all the items in $Z_n + 1$ of size n are cross-checked to ensure they are frequent. If there are any such itemsets that do not pass the minimum requirement, then $CD_n + 1$ is created by removing them from $Z_n + 1$. The reason of this step is that if there is any subset in $CD_n + 1$ that is not frequent, then it cannot be a subset of any frequent itemset of size n + 1.

It is now well established that the algorithm's main function is to calculate the frequency for candidates by examining and searching the database. It aims to attain better efficiency and hence, giving a good output by minimizing the size of candidate sets. However, there still exists a tendency for the algorithm to suffer from the cost of producing enormous candidate sets when it is faced with situations such as having recurrent or large itemsets, or very low support is given.

In the field of medicine and health care of patient, association rules have in the past offered the scope to conduct effective diagnosis. They have further enhanced the methods to acquire information and construct knowledge bases. During this research, we have observed that the Apriori algorithm has been previously applied to a repository containing the diagnosis results of diabetic patients and an attempt has been made to infer rules from the stored parameters. The results of this give an implication that the above algorithm and many such techniques applied on realistic data sets may have significant impact on the diagnostic procedure [11]. Tools have been developed based on data mining concepts for better analysis of medical information and to help in making better diagnosis [12].

6 Experimental Results

In this research paper, we are implementing Apriori algorithm using IBM SPSS Modeler v17.0 to analyze the survey data and attempt to diagnose whether the person has PCOS or not [6]. Figure 3 provides an outlook of the tool that has been used to examine the strong attributes.

The Figs. 4 through 9 are a representation of the pattern of traits that have been frequently repeated over the survey data (Figs. 4, 5, 6, 7, 8 and 9).

The rules obtained for prediction of PCOS class labels based on the metrics of Apriori algorithm—confidence and support, respectively, are given as follows:

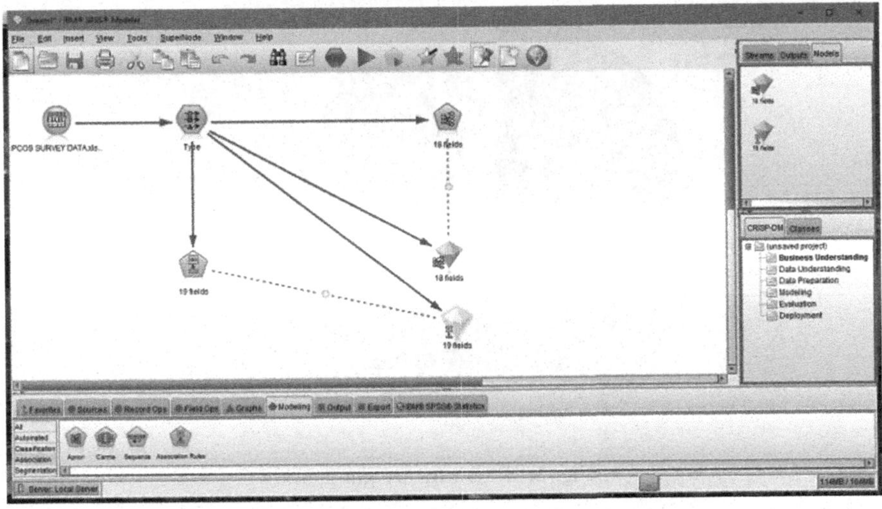

Fig. 3. Applying Apriori algorithm on survey data

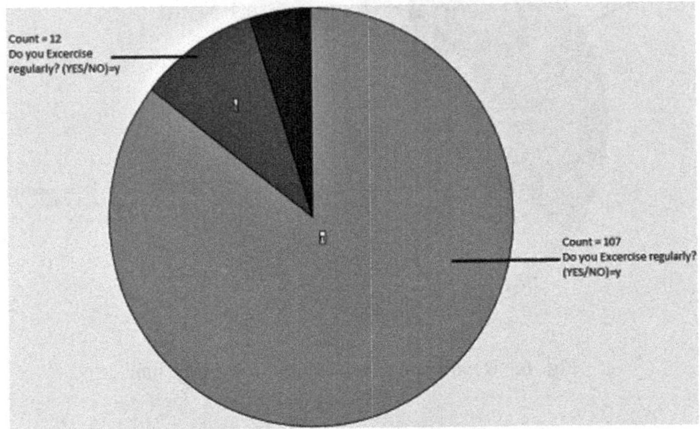

Fig. 4. Result for regular exercise

I. Prediction of Class Label Based on Confidence

- RULE 1: $A1 \wedge A2 \Rightarrow A8$ C = 100% S = 14.4% L = 5
- RULE 2: $A1 \wedge A2 \wedge A4 \Rightarrow A8$ C = 100% S = 13.6% L = 5
- RULE 3: $A1 \wedge A2 \wedge A3 \Rightarrow A8$ C = 100% S = 11.2% L = 5
- RULE 4: $A1 \wedge A2 \wedge A3 \wedge A5 \Rightarrow A8$ C = 100% S = 11.2% L = 5
- RULE 5: $A1 \wedge A2 \wedge A3 \wedge A4 \Rightarrow A8$ C = 100% S = 11.2% L = 5
- RULE 6: $A1 \wedge A2 \wedge A6 \Rightarrow A8$ C = 100% S = 10.4% L = 5
- RULE 7: $A1 \wedge A2 \wedge A4 \wedge A5 \Rightarrow A8$ C = 100% S = 10.4% L = 5
- RULE 8: $A1 \wedge A2 \wedge A4 \wedge A6 \Rightarrow A8$ C = 100% S = 10.4% L = 5

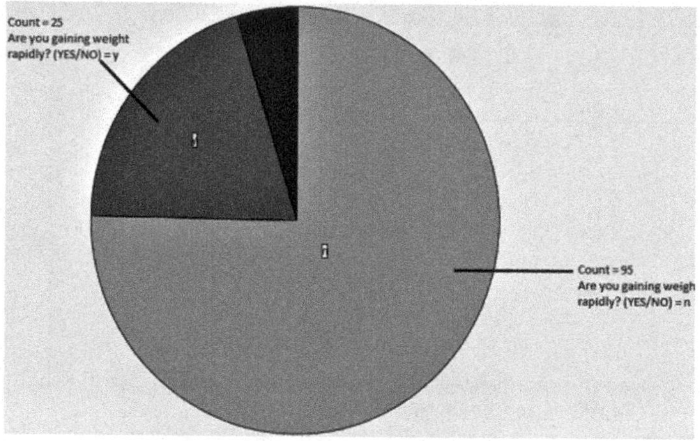

Fig. 5. Result for weight gain

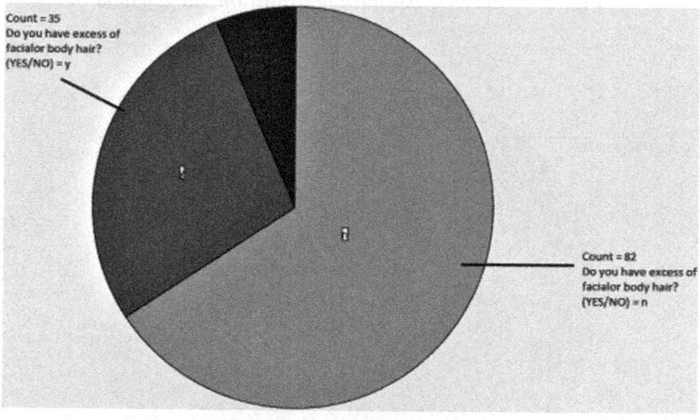

Fig. 6. Result for excess facial and body hair

- RULE 9: A1 Λ A7 ⇒ A8 C = 100% S = 9.6% L = 5
- RULE 10: A1 Λ A7 Λ A5 ⇒ A8 C = 100% S = 9.6% L = 5
- RULE 11: A1 Λ A7 Λ A4 ⇒ A8 C = 100% S = 9.6% L = 5
- RULE 12: A1 Λ A4 Λ A5 Λ A7 ⇒ A8 C = 100% S = 9.6% L = 5
- RULE 13: A1 Λ A2 Λ A7 ⇒ A8 C = 100% S = 8.8% L = 5
- RULE 14: A1 Λ A2 Λ A5 Λ A7 ⇒ A8 C = 100% S = 8.8% L = 5.

II. Prediction of Class Label Based on Support

- RULE 1: A1 ⇒ A8 S = 18.4% C = 79.31% L = 3.97
- RULE 2: A1 Λ A2 ⇒ A8 S = 14.4% C = 100% L = 5

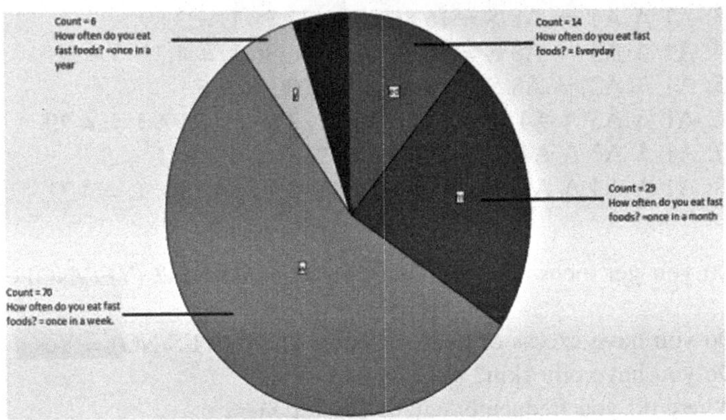

Fig. 7. Result for intake of fast foods

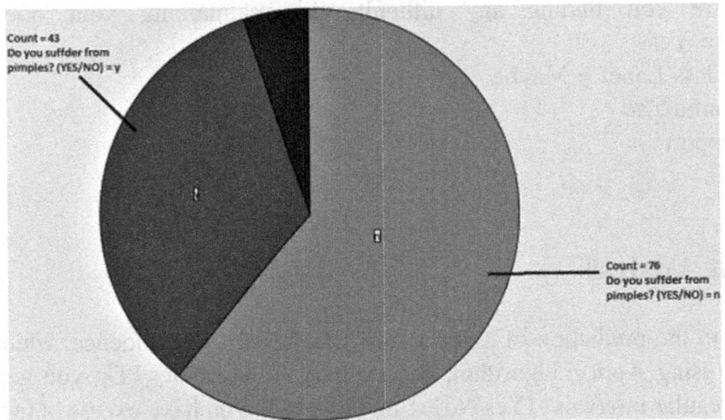

Fig. 8. Result for pimples

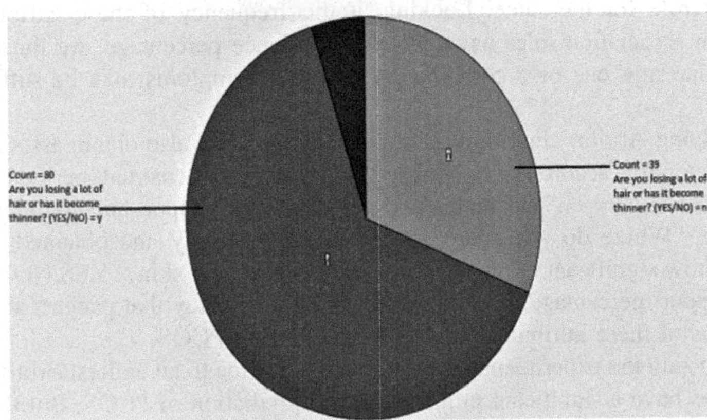

Fig. 9. Result of hair loss

- RULE 3: A1 Λ A4 ⇒ A8 S = 16.8% C = 77.78 L = 3.89
- RULE 4: A1 Λ A3 ⇒ A8 S = 15.2% C = 86.36 L = 4.32
- RULE 5: A1 Λ A2 ⇒ A8 S = 14.4% C = 100 L = 5
- RULE 6: A1 Λ A3 Λ A4 ⇒ A8 S = 14.4% C = 85.71% L = 4.29
- RULE 7: A1 Λ A2 Λ A4 ⇒ A8 S = 14.4% C = 100% L = 5
- RULE 8: A1 Λ A4 Λ A5 ⇒ A8 S = 14.4% C = 94.12% L = 4.71.

where,

A1 = Do you get menstrual periods at regular intervals? (YES/NO) = Infrequent Menses

A2 = Do you have excess of facial or body Hair? (YES/NO) = Yes

A3 = Do you have oily skin? (YES/NO) = Yes

A4 = Where do you frequently eat? = Hostel Mess

A5 = Are you losing lot of hair or has it become thinner in its strength? (YES/NO) = Yes

A6 = How often do you eat fast foods? = Once in a Week

A7 = Are you finding any difficulty in maintaining your body weight? (YES/NO) = Yes

A8 = Class Label = Maybe

C = Confidence

S = Support

L = Lift.

7 Result Analysis

According to the prediction of association rules based on confidence, which has been derived by using Apriori algorithm, we see that the attributes "Do you get menstrual periods at regular intervals? (Yes/No) = im" and "Do you have excess of facial or body hair? (YES/NO) = Y" have a major influence in forecasting PCOS. Alongside the above-mentioned two attributes, we see that "Are you finding any difficulty in maintaining your body weight? (YES/NO)" and "Where do you frequently eat?" also play an eminent role for the same. Looking at the frequency of these attributes in the above-given association rules having high confidence percentage, we thus say that a patient having any one or a combination of these symptoms may be suffering from PCOS.

By applying Apriori algorithm on our survey data, we also obtain association rules based on support, according to which "Do you get menstrual periods at regular intervals? (Yes/No) = im" attribute's and percentage of support and confidence is high followed by "Where do you frequently eat?" Interestingly, the obtained association rules also show significant frequency "Do you have an oily skin? (YES/NO) = Y" with the high support percentage. From this data, it is safe to say that patients showing any permutations of these attributes may me susceptive to PCOS.

By analyzing the experimental results, we have come to an understanding that some the attributes have a significant influence in the prediction of PCOS. But these results do not provide us with cent percent accuracy. Better diagnosis and prognosis of PCOS can be achieved by combining the results of this tool along with the clinical reports procured by the doctor.

8 Limitations

It has been proved in this paper by the authors that by applying the Apriori algorithm on the data sets, prediction of PCOS is efficient and accurate. But, the data sets taken up in this paper is insufficient for the prediction as these help only in preliminary diagnosis of the syndrome. In order to get more advanced results, clinical data along with ultra sound scan reports have to be inculcated in the existing data sets. These advanced clinical tests and scanning reports are highly expensive and hence the whole procedure of retrieval of such data is a costly affair. Apart from that, some patients might not be willing to disclose their reports. Thus, this can be done with the help of renowned organizations and hospitals. The application of Apriori algorithm on such an elaborate and extensive data set yields much better results for PCOS.

9 Conclusion and Future Works

In this paper, data mining techniques have been used to find the correlation between the attributes of the real-time survey data set and to predict the diagnosis of a given person. The application of Apriori algorithm has helped us recognize the major symptoms, such as infrequent menses, excess facial, or body hair, on the basis of which a prediction can be made. However, the results cannot give us accurate information; hence, we need to analyze clinical data and tests to precisely predict the syndrome. Thus, the patient needs to be sent for further medical examinations.

The scope of this topic for future studies and research is widespread. An automation system could be designed so as to deploy those in higher secondary schools as well as colleges to predict PCOS among the girls. This system can also be set up in the workplaces especially concentrating women working in the IT sector with the sole motive of predicting PCOS in its earlier stages. Apart from this, further studies and researches would also lead in finding a proper diagnosis of the syndrome with its prediction, thus making advancements in the field of medicine and technology.

Acknowledgements. We would like to thank Dr. Vijaya Lakshmi Chandrasekhar, Department of Obst. & Gyn., GITAM Institute of Medical Sciences and Research, GITAM Deemed to be University, Visakhapatnam, for her kind help in gathering the data and enlightening us about the relationships among various symptoms. We would also like to thank GITAM Deemed to be University Girls Hostel Kokila Sadan as well as the chief warden Prof. Dr. T. Sita Mahalakshmi for permitting us to conduct the survey and enabling us to gather information for the survey.

References

1. Fiona MacDonald. Polycystic Ovary Syndrome Might Start In The Brain, Not The Ovaries. *ScienceAlert*. N.p., 2017. Web. 28 Mar. 2017.
2. ML Wissing, MR Bjerge, AIG Olesen, T Hoest, AL Mikkelsen.: Impact of PCOS on early embryo cleavage kinetics, ELSEVIER, Reproductive BioMedicine Online (2014) 28, 508–514.
3. Antoni J. Duleba.: Medical management of metabolic dysfunction in PCOS, ELSEVIER Steroids, Steroids 77 (2012) 306–311.
4. Shady Grove Fertility. PCOS: One Size Doesn't Fit All [Web Log Post], 1 June 2016, Retrieved from "www.shadygrovefertility.com/blog/diagnosing-infertility/pcos-one-size-doesnt-fit-all/".
5. Roy Homburg.: Pregnancy complications in PCOS, ELSEVIER, Best Practice and Research Clinical Endocrinology and Metabolism, Vol. 20, No. 2, pp. 281–292, (2006).
6. PCOS Dataset source—https://github.com/PCOS-Survey/PCOSData.
7. Jiawei Han and Micheline Kamber, Data Mining Concepts and Techniques, Second Edition.
8. Anders Drachen.: Frequent Itemset and Association Rule Mining, [Web Log Post], 3 December 2012, Retrieved from "www.gameanalytics.com/blog/frequent-itemset-and-association-rule-mining-or-how-to-know-if-shirts-follows-pants-or-the-other-way-around.html".
9. Stefan Naulaerts, Pieter Meysman, WoutBittremieux, TrungNghia Vu, WimVandenBerghe, Bart Goethals and Kris Laukens.: A primer to frequent itemset mining for bioinformatics, Oxford Journals, Briefings in bioinformatics, (2015); 16(2): 216–231.
10. Xindong Wu, Vipin Kumar, J. Ross Quinlan, Joydeep Ghosh, Qiang Yang, Hiroshi Motoda, Geoffrey J. McLachlan, Angus Ng, Bing Liu, Philip S. Yu, Zhi-Hua Zhou, Michael Steinbach, David J. Hand, Dan Steinberg.: Top 10 algorithms in data mining, KnowlInfSyst, https://doi.org/10.1007/s10115-007-0114-2, (2008), 14:1–37.
11. S. Stilou, P.D. Bamidis, N. Maglaveras, C. Pappas.: Mining Association Rules from Clinical Databases: An Intelligent Diagnostic Process in Healthcare, MEDINFO (2001) V. Patel et al. (Eds).
12. M. Ilayaraja, T. Meyyappan.: Mining Medical Data to Identify Frequent Diseases using Apriori Algorithm, Proceedings of the 2013 International Conference on Pattern Recognition, Informatics and Mobile Engineering, (February 21–22).

A Novel DNA- and PI-Based Key Generating Encryption Algorithm

B. Vikas(✉), A. K. Akshay, Sai Pavana Manish Thanneeru, U. M. V. Raghuram, and K. Santosh Bhargav

Department of Computer Science and Engineering, GITAM Institute of Technology, GITAM Deemed to be University, Visakhapatnam, India
vikasboddu30@gmail.com

Abstract. Effective network security methods are essential for a private communication. In this paper, the authors achieved high security for the data to be transferred using a self-complementing algorithm. They suggested an innovative method for the key generation using a part of the DNA sequence and another part of the very large value of PI. With 3.3 billion combinations of the DNA sequence and infinitely long PI sequence, it makes it difficult for general code-breaking techniques. The values taken are converted into a key using a typical algorithm. Being a symmetric algorithm, the key is used to generate secure text/ciphertext which is shared using standard norms of the industry, ensuring secure transmission of data.

Keywords: DNA · PI · Encryption · Decryption · Cipher

1 Introduction

Communications are a major part of today's world. With increase in communications, there is an ever-growing need to do it privately. Web, the largest platform for the exchange of information, has a great deal of sensitive data. This creates the need for securing the data during transfer to protect the privacy of either the civilians or the government. Earlier, security was the primary concern only for military applications, but now the area of concern has been extended to communications occurring via the Web [1]. For data to be transferred, it must be done so through a channel. If this channel is insecure, an unauthorized third party may access or modify our data. This leads to many undesirable side effects and devoid of the user's privacy.

1.1 Cryptography

Cryptography is derived from a Greek word "krupetin" which means to hide. It is the art of protecting the information by converting the data into an incomprehensible or a meaningless format in which the message is hidden from the reader and only the intended recipient can convert it into original message [2].

DNA cryptography is based on DNA computing where message to be sent is encrypted in the form of DNA nucleotide sequence. [3] DNA computing can be

V. Bhateja et al. (eds.), *Information Systems Design and Intelligent Applications*, Advances in Intelligent Systems and Computing 672,
https://doi.org/10.1007/978-981-10-7512-4_94

effective platform for both encryption and decryption using symmetric and asymmetric keys [4].

1.2 DNA

Deoxyribonucleic acid (DNA) is a self-replicating molecule which is present in nearly all living organisms as main constituent of chromosomes. It is the carrier of genetic information [4]. The information contained in DNA helps in constructing other cells. DNA molecule forms a double-helix structure.

The four different bases of DNA are adenine (A), cytosine (C), guanine (G), and thymine (T). Adenine only bonds with thymine, whereas guanine only bonds with cytosine.

1.3 DNA Cryptography

A new field in cryptography that emerged along with the progress of DNA computing is DNA cryptography. Useful concepts like large information density and enormous parallelism that can be received from a DNA unit's sequence are taken for cryptographic purposes [5]. Having great cryptographic strength and the binding properties between the nucleotide bases A–T and C–G gives us an opportunity to create self-build structures which are a well-organized means of executing collateral molecular computations; its storing capabilities are huge. The hidden message is in the form of DNA sequence, image, audio, and video in the DNA cryptography, which is used to prevent important data from trespassers [6].

DNA computing uses various recombinant DNA techniques for computation [5]. This type of cryptography came into light and many easy and powerful algorithms are explored in order to bring DNA computing on digital level and use it on global scale.

1.4 PI and Its Significance

The ratio of circumference of a circle to its diameter is called PI. Irrespective of the size of the circle, the ratio remains constant.

Computer scientists have calculated billions of digits of PI. The value of PI starts with 3.1415926535…, but because no particular recognizable pattern emerges in successors of its digits, we could continue the calculation of the subsequent digits for millennia. Therefore, considering PI as one of the main elements for key generation is highly effective [7].

This paper is organized as follows: in Sect. 2, a brief description of symmetric key cryptography and its different techniques; in Sect. 3, public key cryptography and how it compares with symmetric key techniques; in Sect. 4, different key exchange techniques; in Sect. 5, an illustration of the different variables used in the generation of the key. In Sect. 6, the proposed algorithm is specified including a general example of the technique, with the conclusion specified in Sect. 7.

2 Symmetric Key Cryptography

The authors utilize an identical key for both encryption and decryption. The sender uses this key and employs the use of an encryption algorithm to morph plaintext into ciphertext [6]. The receiver uses a similar key and employs decryption algorithm to convert ciphertext back into plaintext [8]. Examples of a few symmetric key algorithms are as follows:

2.1 DES

Data encryption standard is one of the most widely used algorithms for symmetric encryption [2]. DES encryption has a block size of 64 bit; i.e., it takes 64-bit key as plaintext and gives us 64-bit ciphertext as output [2, 5, 8, 9]. There are many number of rounds where each round contains nonlinear substitution, bit shuffling, and exclusive OR operations.

2.2 AES

Advanced encryption standard algorithm was introduced to overcome a major defect in DES algorithm; i.e., the length of key in DES was small [10]. The block length is of 128 bits in AES algorithm. The three different key lengths allowed by AES are 128 bits, 192 bits, and 256 bits. The rounds of processing vary with the key length: 10 rounds of processing for 128-bit key, 12 rounds of processing for 192-bit key, and 14 rounds of processing for 256-bit key [2].

3 Public Key Cryptography

Public key cryptography is also called as asymmetric key cryptography. The sender encrypts the data using public key, and the receiver decrypts the data using his private key. We need not transfer the key using some secure channel, and this resolves the problem of transferring the key. In asymmetric key algorithms, two keys are generated: the public key and the private key. The public key can be used to perform encryption, whereas the private key is used to perform decryption. The public key is shared to all users, while private key is shared to their respective users.

3.1 RSA Algorithm

This is an asymmetric cryptographic algorithm proposed by Ron Rivest, Adi Shamir, and Leonard Adleman for a more secure data transmission. This algorithm uses two different keys: public key and private key. The messages are encrypted using public key. The private key decrypts the cipher data and hence should be kept private [12]. This is the standard algorithm used for asymmetric key cryptography.

4 Key Exchange

Before exchanging the cipher data between the parties, we must securely exchange the key. Key exchange is a method in cryptography in which cryptographic keys are exchanged between the sender and the receiver [13, 11]. Whitfield Diffie and Martin Hellman developed the Diffie–Hellman algorithm for key exchange. The algorithm generates the key at both ends instead of transferring. This eliminates the need for transferring the key from sender to receiver [13]. In man-in-the-middle attack, the attacker secretly alters or disrupts the communication between the sender and the receiver. Diffie–Hellman algorithm for key exchange ensures prevention of such attacks [14].

5 System Architecture

The DNA and the PI units are the prime components for generating a complex yet unique key which is used for both encryption and decryption. The type of cryptography being symmetric key cryptography, the key generated is only one (Fig. 1).

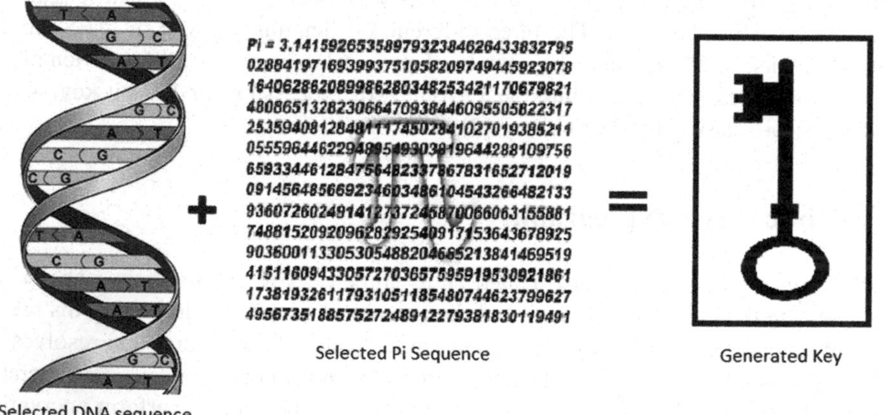

Fig. 1. Illustration of components used for key generation

6 Proposed Algorithm

An algorithm has been devised to use PI and DNA sequences to generate a key. The encryption and decryption of such an algorithm are specified below.

6.1 Encryption

Encryption is the process by which the data to be transferred is made into cipher data that is incomprehensible to others except the receiver. In our encryption algorithm, the sender follows some sequence of steps to generate the key. This generated key is used to encrypt the transfer data.

1. Start.
2. Convert the input decimal data into binary.
3. Positions are randomly chosen corresponding to the DNA and PI attributes. Positions are sent to receiver so that the key can be generated at receivers' end.
4. DNA unit is fetched from the selected position.
5. The primary units in DNA sequence are made up of A, T, G, C.
6. We consider A = 00, T = 11, G = 01, C = 10.
7. The complement of binary value of DNA sequence is calculated.
8. A sequence of numbers after the decimal point from the value of PI is picked up corresponding to the selected position.
9. The sequence of numbers is converted into binary code.
10. XOR operation is performed on the PI's binary and DNA complement binary codes generating the KEY.
11. A XOR operation is performed on the obtained KEY and the binary form of the input data, and thus, the sequence of binary digits is obtained.
12. The obtained binary output is divided into two equal parts, and then, XOR operation is performed between the two parts; the resultant is then appended to end of binary output obtained in previous step.
13. A XOR operation is performed on each of the two parts of binary output in step 9 with the resultant.
14. The output obtained in the previous step is the ciphertext.
15. End.

6.2 Decryption

Decryption is the process by which the incomprehensible cipher data received by the receiver is translated into plain understandable text. In our decryption algorithm, the receiver follows some sequence of steps to generate the key. This generated key is used to decrypt the received data.

1. Start.
2. The positions sent by the sender are used to download the respective DNA unit and PI unit.
3. The key generation algorithm is applied, and the key is generated at receiver's end.
4. The ciphertext is divided into three equal parts.
5. XOR operations are performed between third part and the remaining two parts separately.
6. The output from step 4 undergoes XOR with the key generated in step 2.
7. The obtained output is the binary code of input data.
8. End.

6.3 Example

The following section deals with the implementation of the proposed encryption algorithm specified in Sect. 6.2 and decryption algorithm specified in Sect. 6.3.

6.3.1 To Encrypt

Input decimal data: 186

1. The binary code of the input data: 10111010,
2. Range of the selected DNA and PI unit:
   Initial position:20
   Final position:23
3. Selected DNA unit is TTGA
4. The binary code for the DNA unit is 11110100.
5. The complement of the binary code of DNA is 00001011.
6. Selected PI unit is 6844.
7. The binary code for the PI unit is 1101010111100.
8. On applying XOR to the above binary codes:

```
0000000001011
1101010111100
-------------
1101010110111      ——KEY
-------------
```

9. XOR operation is applied to the obtained KEY and binary coded input data.

```
1101010110111      ——KEY
0000010111010      ——INPUT DATA
-------------
1101000001101      ——A(assumption)
-------------
```

10. The obtained binary code(A) is broken into two halves, and a XOR operation performed between the first half and the second half.

```
0110100
0001101
-------------
0111001            ——B(assumption)
-------------
```

11. The length of B is doubled by appending it with itself.

0111001 0111001

12. XOR operation is performed between the above binary code and (A).

```
  01101000001101      ——A
  01110010111001
————————————————————
  00011010110100
————————————————————
```

13. B is appended to the binary code obtained in previous step to give us the ciphertext.

"00011010110100" + "0111001"

"000110101101000111001"

6.3.2 To Decrypt

1. The receiver is provided with the initial and final positions of DNA unit and PI unit.
   Initial position: 20
   Final position: 23
2. Corresponding DNA unit and PI unit data are retrieved.
   The DNA unit: TTGA
   The PI unit: 6844
3. Data is converted into binary format.
   Binary code of DNA unit: 11110100
   Binary code of PI unit: 1101010111100
4. Complement of DNA unit binary code: 00001011. XOR operation is performed between PI unit's binary code and the above complemented code.

```
  0000000001011
  1101010111100
————————————————————
  1101010110111
————————————————————
```

5. The cipher data is broken into three equal parts.
   Ciphertext: 000110101101000111001
   Divided into → 0001101 0110100 0111001
6. XOR operation is performed between the third part and the first part, and the result is assumed to be P.

```
  0111001 – – – – – part 3
  0001101 – – – – – part 1
————————————————————
0110100 – – – – – –P(assumption)
————————————————————
```

7. XOR operation is performed between the third part and the second part, and the result is assumed to be Q.

0111001 – – – – – part 3
0110100 – – – – – part 2
———————————————
0001101 – – – – – Q(assumption)
———————————————

8. P and Q are then appended.

0110100 0001101

9. XOR operation is performed between the above acquired binary code and the KEY.

01101000001101
01101010110111– – – KEY
———————————
00000010111010
———————————

10. The obtained binary data is converted into decimal data form, and hence, the required decrypted data is found.
11. 00000010111010 is converted into decimal value as 186 (Fig. 2, Table 1).

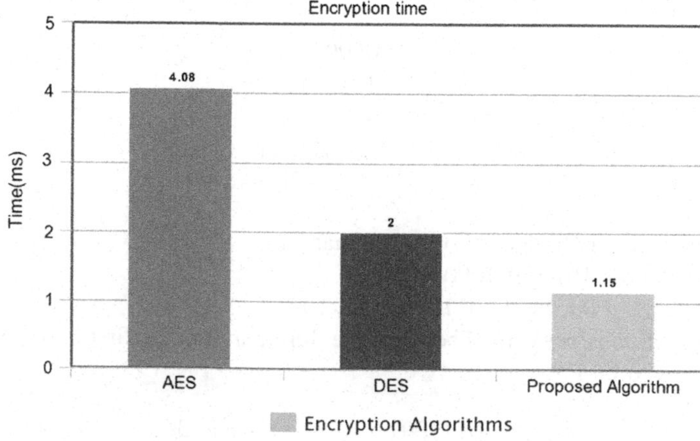

Fig. 2. Above graph shows a comparison between various algorithms and the time taken for their encryption. On the x-axis, the elements *AES, DES, and Proposed Algorithm* together comprise the various algorithms taken in comparison. On the y-axis, the time taken is mentioned

Table 1. Comparison table for AES, DES, and proposed algorithm

	AES	DES	Proposed algorithm
Encryption time (ms)	4.08	2.00	1.15
Input size (bits)	12	12	12
Block size	Fixed	Fixed	Variable

6.4 Cryptanalysis

The proposed algorithm has a drawback in the terms of MITM or man-in-the-middle attack. When sharing the key, a third party can intercept the key exchange through which he can produce a duplicate key. He intercepts the messages using the key he obtained fraudulently and re-encrypts with the duplicate key and passes on to the other end. The above-listed limitation can be closed when using a key exchange algorithm for sharing the key across public channels.

7 Conclusion

In this paper, we have reviewed various effective network security methods which are essential for a private communication. We have also introduced a new algorithm for both the encryption and decryption of data. This paper gives an insight into the different asymmetric key algorithms present and gives an ingenious way for secure data transfer using DNA- and PI-based keys that have the capability to be immune against most brute-force attacks. Implementing real-world sequences of near random data innumerable combinations of the DNA sequence and infinitely long PI sequence makes it highly complex beyond the scope of any present code-breaking algorithms. Keeping that in mind, the proposed algorithm may be used as a wrapper for other encryption algorithms.

The proposed algorithm uses subsets of the whole DNA and PI, and therefore, the chance for impersonation is next to negligible. Since the DNA sequence is randomly downloaded from the publicly available databases which have a large number of sequences, the cryptanalysis is impossible. The subsets determined by the key give us a large number of combinations that ensure the complexity of the key.

Acknowledgements. The authors are thankful to Prof. R. Sireesha, Department of Computer Science and Engineering, GITAM Institute of Technology, GITAM University, Visakhapatnam, for her continuous support and encouragement throughout the process.

References

1. PI, https://en.wikipedia.org/wiki/PI.
2. Mounika A., Pradeep M.: A Comparative Survey on Symmetric Key Encryption Techniques.: International Journal on Computer Science and Engineering Vol. 4 No. 05, pp. 877–882, IJCSE (2012).

3. Kahate, A.: Computers and Network Security. Tata McGraw-Hill (2013).
4. Ashish G., Thomas LaBean, John R.,: DNA-Based Cryptography.: Department of Computer Science and Engineering, Duke University.
5. Borda, Monica, Olga T.: DNA Secret Writing Techniques in Communication (COMM), 8th IEEE International Conference on, pp. 451–456, 2010.
6. Bibhash R.: A DNA based Symmetric Key Cryptography.: ICSSA- 24–25 (2011).
7. Tausif A, Abhishek Kumar, Sanchita P.: DNA Cryptography Based on Symmetric Key Exchange.: International Journal of Engineering and Technology (IJET) Vol 7 No 3, pp. 938–950, Jun-Jul 2015.
8. Sneha J., Rahul K.: Secure Data communication and Cryptography based on DNA based Message Encoding. International Journal on Computer Science and Engineering Vol. 98, No. 16, pp. 35–40 IJCSE (2014).
9. Harsh B., Neha S.: Cellular-genetic test data generation. In: ACM SIGSOFT Software Engineering Notes. Vol. 38 Issue 5, pp. 1–9, Sept 2013.
10. Forouzan B (2013) Data communications and networking, global edition, 1st ed. McGraw Hill, New York, N.Y.
11. Public Key Encryption and Digital Signature: How do they work? CGI (2004).
12. M., Preetha., M., Nithya.: A Study and Performance Analysis of RSA Algorithm. International Journal of Computer Science and Mobile Computing. 2, 126–139 (2013).
13. Kumar, R., C., Ravindranath.: Analysis of Diffie Hellman Key Exchange Algorithm with Proposed Key Exchange Algorithm. International Journal of Emerging Trends & Technology in Computer Science. 4, 40–43 (2015).
14. Whitfield Diffie, Martin Hellman.: New Directions in Cryptography. IEEE Transactions On Information Theory Vol IT-22, pp. 644–654, November 1976.

A Sentimental Insight into the 2016 Indian Banknote Demonetization

Rajesh Dixit Missula(✉), Shyam Nandan Reddy Uppuluru, and Sireesha Rodda

Department of Computer Science and Engineering, GITAM Institute of Technology, GITAM University, Visakhapatnam, India
rajeshdixit95@outlook.com

Abstract. On the 8 November 2016, the Government of India effectively demonetized banknotes representing the nation's two largest and most commonly used denominations: Rs. 500 and Rs. 1000. The abrupt nature of the move and the shortage of cash that followed the announcement invited a lot of polarizing opinions from the public. Social media platforms—which have now become an integral part of daily life, saw an unprecedented inflow of opinions, thereby becoming important repositories of people's views on demonetization. In this paper, an attempt has been made to understand public consensus on demonetization by utilizing data from one such social media platform—Twitter—and performing a sentimental analysis of the tweets. To this end, the R language was employed in combination with the Twitter Web API. A dictionary-based approach was taken towards classifying tweets as either positive, negative, or neutral.

Keywords: Demonetization · Sentiment analysis · Opinion mining · Twitter

1 Introduction

The Indian banknote demonetization has been one of the most polarizing events in recent history, drawing a diverse range of reactions from all sections of society. Demonetization of the Rs. 500 and Rs. 1000 notes effectively resulted in nearly 86% of currency in circulation (by value, not volume) valueless for day-to-day transactions, thereby resulting in a severe shortage of cash during the following few months. As the debate rages on regarding whether the move has positively impacted the general populace, understanding the opinions of common people is an important factor in determining the move's success. More importantly, understanding the current opinion of people is more important than ever. While initial response to the demonetization move may have been biased due to factors such as political beliefs, it is our belief that current opinions are much more reflective of the long-term views of people since ample amount of time has passed for people to have clearly understood how the move affects their daily lives.

With social media becoming such an integral part of daily life, opinions regarding demonetization were posted in large volumes onto popular social media platforms. One such platform is the microblogging service "Twitter". Tracking the right set of Twitter

V. Bhateja et al. (eds.), *Information Systems Design and Intelligent Applications*, Advances in Intelligent Systems and Computing 672,
https://doi.org/10.1007/978-981-10-7512-4_95

posts or "tweets", extracting their contents, and performing a sentimental analysis on them, can, in theory, provide us with a clear understanding of what a significant fraction of people think/thought about the demonetization move.

In this paper, the R language and the Twitter API have been used to do exactly what has been discussed above: sentiment analysis of tweets to understand the general public consensus on demonetization. Performing the analysis itself is a multistep process.

1. Accessing tweets from Twitter

The very basic component of sentiment analysis is the data itself. In this case, there is a requirement for relevant tweets to act as data. For this purpose, the Twitter API has been employed, which consists of built-in methods to extract tweets based on specified parameters.

2. Controlling source of tweets

The goal of this paper is to understand the opinions of the Indian public towards the demonetization move. Geographic constraints have thus been applied on the sources of tweets to ensure that the tweets that have been scraped originate only from users who have tweeted from locations within the Indian subcontinent.

3. Annotation extraction

"Annotation extraction" refers to the process of reading through a tweet's contents and assigning it an "annotation" or a label that indicates whether it is positive, negative, or neutral (or any other). This is the main step in sentiment analysis and is achieved through the application of dictionary-based ranking of words.

2 Related Work

Vasiliu et al. [1] have studied the social media sentiment surrounding the UK's decision to leave the European Union ("Brexit") through the use of the "Social Sentiment analysis financial indeXes" project (SSIX)—which is a collection of simple tools for social media opinion mining. Their paper discusses in length the architecture of SSIX along with platform's "X score" metrics in relevance to the Brexit issue.

Dang et al. [2] attempt to understand which groups of people are actively involved in spreading rumours on online social networks. Applying a combination of text classification, sentiment analysis, and social network analysis onto Reddit data, users' engagement with rumours is studied to classify them into one of three categories: supportive of a false rumour (or) refute a false rumour (or) joke about false rumours.

Santhiya and Bhuvaneswari [3] have analysed tweets related to crime attributes against women and children. The tweets were used to extract the different varieties of crimes being committed against women and children along with the locations from where the most number of such tweets originate from. The authors have made use of the R language and the Hadoop framework for extraction and storage of tweets.

Chang and Chen [4] aim to devise a model of "social influence" whereby sentiment analysis is used in trying to find people of real influence on social media. Retweet and reply are considered as the primary metrics in estimating the influence that a particular person possesses on social media. Celebrities, in general, were found to have higher sentiment ration than politicians. The model being proposed is expected to be particularly useful for companies looking to produce effective marketing campaigns.

Mishra et al. [5] perform opinion mining in order to understand the public consensus regarding Prime Minister Narendra Modi's "Digital India" campaign. The authors make use of the Twitter API to extract tweets related to the "Digital India" campaign and attempt to classify their polarity as either negative, positive, or neutral. A dictionary-based approach was adopted for the purpose of sentiment analysis. Half the tweets were found to be positive, 1/5th to be negative, and the rest neutral.

Sindhura and Sandeep [6] make use of Twitter data in order to perform a general analysis of various opinion mining techniques. A sufficiently detailed review of various approaches used to perform a computational analysis of opinions is performed. As part of this, feature-based opinion mining, machine learning-based opinion mining, and a ranking-based model are reviewed for their strengths and weaknesses.

Pak and Paroubek [7] explore the option of using Twitter as a corpus for opinion mining. Throughout the paper, they discuss on collection of data from Twitter and performance of linguistic analysis on the data. Following this, they build a sentiment classifier based on the multinomial Naive Bayes classifier to determine and classify sentiments of the extracted tweets as either positive, negative, or neutral.

Bollen et al. [8] try to apply behavioural economics at a large-scale/societal level to predict collective decision making and try to use this as a way to predict movements of the Dow Jones Industrial Average. As part of this, they use two tools—"OpinionFinder" and "Google-Profile of Mood States"—to analyse text content of tweets and cross-validate their results to test their ability to detect the public consensus of the presidential election and Thanksgiving Day of 2008.

Pang and Lee [9] perform a survey of various techniques and methodologies that enable opinion-based information seeking systems directly. They keep a strong sense of focus on challenges stemming from sentiment-aware systems. Along with a discussion on general sentiment analysis for opinion extraction, they also discuss on other aspects such as privacy, manipulation, and economic impact of opinion-based information access.

3 Methodology

3.1 Retrieving Tweets

The data being subject to sentiment analysis exists in the form of tweets. Retrieving tweets requires APIs and permissions that are provided by Twitter itself. It is a multistep process consisting of three sub-steps.

Steps in retrieving tweets:

1. Twitter authentication: ROAuth

ROAuth is the R language interface that allows for Open Authentication (OAuth). The sentiment analysis application developed as part of this paper utilizes ROAuth for authentication with the Twitter server before attempting to scrape any tweets. The process of authentication itself requires two keys: ConsumerKey and ConsumerSecret in addition to two tokens: AccessToken and AccessTokenSecret. These keys and tokens are provided by Twitter to every developer application.

2. Twitter Web API: twitteR

"twitteR" is an R language interface that provides access to the Twitter Web API. The actual retrieval or "scraping" of tweets is performed by this package. The most important aspect of this package with regards to sentiment analysis is the "searchTwitter()" function. This function takes up to 15 parameters such as location, search string, and language to fetch us tweets that are filtered and suitable for our specific requirements. The tweets used in this paper were all generated within the last one week (as of 7 April 2017) to ensure that the only the most contemporary opinions on demonetization are captured.

3. Location restriction

Due to the fact that demonetization discussions also occurred in Venezuela and Europe around the same time as India, demonetization tweets used in this paper were restricted to the geography of the Indian subcontinent for best results. To achieve this, Nagpur—the centre most city of India—was treated as the middle point. From here, the distance to the southernmost, easternmost, northernmost, and westernmost points of the Indian subcontinent was calculated and the average was used as the "radius" parameter in the searchTwitter() function.

3.2 Sentiment Analysis

For the actual process of determining the emotional polarity or "sentiment" behind a particular tweet, a dictionary-based approach has been used. For this purpose, two predefined lists belonging to SentiWordNet have been employed. One list contains words that indicate a "positive sentiment", and the other contains words that indicate a "negative sentiment". Each tweet is split into its constituent words, converting all letters to lowercase and removing all punctuation marks. These words are then compared with words from both the aforementioned word lists. For each positive word encountered, a positive counter variable is incremented by 1. For each negative word encountered, a negative word counter is incremented by 1. The value of the negative counter variable is then subtracted from positive counter variable, and the resultant value is subjected to a scoring model that labels tweets as one of five groups: awful, bad, neutral, good, outstanding. The positive and negative counter variables are reset for each new tweet.

The formula used to determine the sentiment of tweets is:

`Sentiment Score = Pos – Neg`

where "Pos" represents number of positive words and "Neg" represents number of negative words.

PseudoCode for Sentiment Analysis:

```
for(each tweet scraped)
do
   remove(punctuations);
  remove(control words);
  remove(numbers);
  remove(blank spaces);
  convertToLowerCase(tweet);
  word_list [n] = Split Tweet into Words;
  for(each word in word_list)
  do
    for(each word in positive_list)
    do
      if(word_list word matches positive_list word)
      positive_count++;
     done
     for(each word in negative_list)
     do
       if (word_list word matches negative_list word)
       negative_count++;
     done
    done
  sentiment_score = positive_count - negative_count;
  reset (positive_count, negative_count);
done
```

4 Results

The sentiment analysis tool built as part of this paper was run over 11,869 tweets that had the term "demonetization" contained within them while originating from within the geographical constraints of the Indian subcontinent. Upon running the application, a sample of the environment view in R-Studio is documented in Fig. 1.

Scoring schema that has been used for determining emotional polarity is shown in Table 1.

Figures 2 and 3 are histograms showing the distribution of positive and negative scores, respectively. X-axis denotes the range of scores encountered while Y-axis denotes the frequency of each score. Note that in the negative distribution graph, scores represent negative values even though the "–" symbol has not been explicitly mentioned on x-axis.

The most significant observations from Figs. 2 and 3 are that negative opinions are dominated by scores of −1 while positive opinions have a significant split between scores of +1 and scores above +1. This indicates that those who were in approval of

```
demonetization.twe… Large list (11869 elements, 7.3 Mb)
  :Reference class 'status' [package "twitteR"] with 17 fields
  ..$ text : chr "RT @_YogendraYadav: Good news: Mr RBI Gov can
  ..$ favorited : logi FALSE
  ..$ favoriteCount: num 0
  ..$ replyToSN : chr(0)
  ..$ created : POSIXct[1:1], format: "2017-04-07 06:43:49"
  ..$ truncated : logi FALSE
  ..$ replyToSID : chr(0)
  ..$ id : chr "850237661851799553"
  ..$ replyToUID : chr(0)
  ..$ statusSource : chr "<a href=\"http://twitter.com/download,
  ..$ screenName : chr "shobhesh"
  ..$ retweetCount : num 85
  ..$ isRetweet : logi TRUE
  ..$ retweeted : logi FALSE
  ..$ longitude : chr(0)
  ..$ latitude : chr(0)
```

Fig. 1. View of R-studio environment

Table 1. Scoring schema to determine emotional polarity

Score value	Assigned label
<= −2	Awful
−1	Bad
0	Neutral
1	Good
>= +2	Outstanding

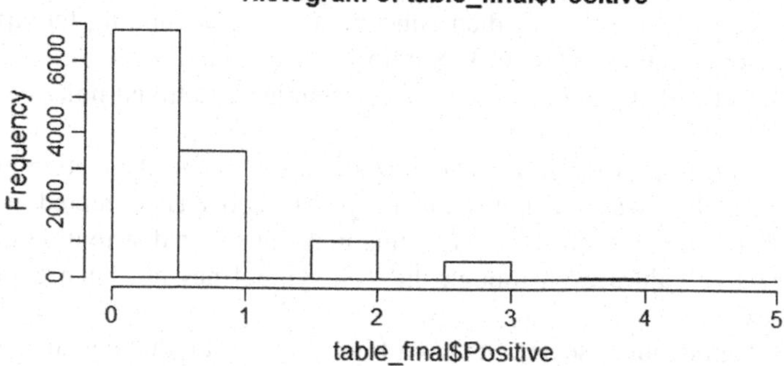

Fig. 2. Distribution of positive scores

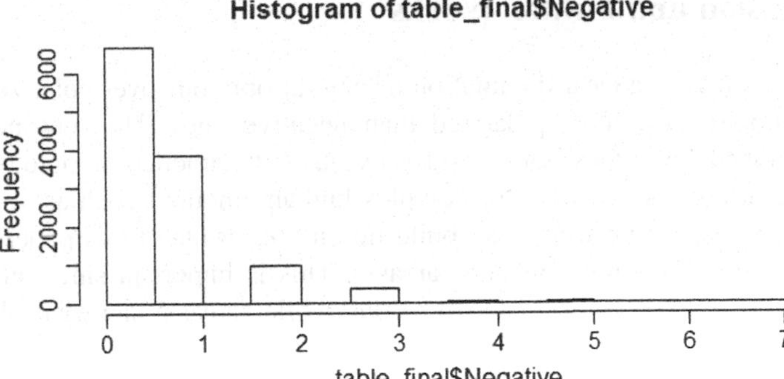

Fig. 3. Distribution of negative scores

demonetization were more likely to be strong supporters of the move while those who disapproved of it were more likely to be weakly/moderately opposed to it.

Figure 4 shows a clear domination of neutral over negative (awful + bad) and positive (good + outstanding) opinions. With negative opinions at 31% and positive opinions at 29%, the overall reception of demonetization at present can be treated as slightly more negative than positive. However, since the difference between negative and positive opinions is so minimal, they can be treated as almost equally widespread upon considering small margins of error.

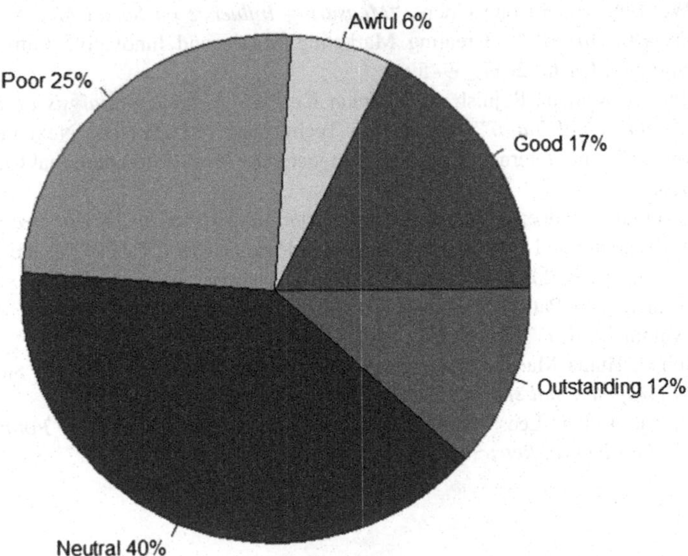

Fig. 4. Pie chart depicting share of polarities

5 Conclusion and Future Works

In conclusion, there is a clear domination of neutral opinions over polarized opinions. Positive opinions were more polarized than negative ones. The sentiment analysis model developed here does not consider tweets with emoticons included in them. Furthermore, it does not account for complex human emotions such as sarcasm being used in the tweets. Future works can build up on the results of this paper and further include provisions for emoticons and sarcasm. This is important since emoticons are common forms of expression on social media, while sarcasm can easily be misinterpreted as the opposite of what is trying to be implied.

References

1. Laurentiu Vasiliu, André Freitas, Frederico Caroli, Siegfried Handschuh, Ross McDermott, Manel Zarrouk, Manuela Hürlimann, Brian Davis, Tobias Daudert, Malek Ben Khaled, David Byrne, Sergio Fernández, Angelo Cavallini, "*In or Out? Real-Time Monitoring of BREXIT sentiment on Twitter*", SEMANTICS 2016 (2016).
2. Dang A, Smit M, Moh'd A, Minghim R, Milios E., "*Toward understanding how users respond to rumours in social media*", Advances in Social Networks Analysis and Mining (ASONAM), 2016 IEEE/ACM International Conference, 18 Aug 2016 (pp. 777–784).
3. Santhiya K, Bhuvaneswari V., "*Data Analytics Framework: R and Hadoop–Geo-location based Opinion Mining of Tweets*", International Journal of Computational Intelligence and Informatics, Vol. 6: No. 1, June 2016.
4. Chang, Wei-Lun, Guan-Rong Chen., "*Measuring Influence on Social Media: A Sentiment Perspective (An Abstract).*" Creating Marketing Magic and Innovative Future Marketing Trends. Springer, Cham, 2017. 59–59.
5. Mishra, Prerna, Ranjana Rajnish, and Pankaj Kumar. "*Sentiment analysis of Twitter data: Case study on digital India.*", Information Technology (InCITe)-The Next Generation IT Summit on the Theme-Internet of Things: Connect your Worlds, International Conference on. IEEE, 2016.
6. Sindhura Vemuri, Sandeep Y, "*Medical data Opinion retrieval on Twitter streaming data.*", Electrical, Computer and Communication Technologies (ICECCT), 2015 IEEE International Conference on. IEEE, 2015.
7. Pak, Alexander, and Patrick Paroubek, "Twitter as a Corpus for Sentiment Analysis and Opinion Mining", *LREc*. Vol. 10. No. 2010. 2010.
8. Bollen, Johan, Huina Mao, and Xiaojun Zeng. "Twitter mood predicts the stock market", *Journal of computational science* 2.1 (2011): 1–8.
9. Pang, Bo, and Lillian Lee. "Opinion mining and sentiment analysis", *Foundations and Trends® in Information Retrieval* 2.1–2 (2008): 1–135.

Encryption Model for Sensor Data in Wireless Sensor Networks

Anusha Vangala(✉) and Pritee Parwekar

Anil Neerukonda Institute of Technology and Sciences, Visakhapatnam, India
{anusha.cse, pritee.cse}@anits.edu.in

Abstract. Wireless sensor networks have become very prevalent in many industries due to its ease of implementation, high performance, and applicability in numerous areas. The widespread use of this technology brings with it the challenge of providing confidentiality to the data that wireless sensor network carries. The challenge is due to the limitation of resources of energy, memory, and computational power. This paper describes a model for encrypting the sensor data after it is collected by the sink from the sensor. This paper discusses the evolution of the model for encrypting this data from a very simplified scheme with a single key to a more sophisticated scheme which performs dual encryption over the data. The models may implement any symmetric cryptographic scheme with the encryption implemented at the sink and the decryption implemented at the base station.

1 Introduction

1.1 Architecture of WSN

A wireless sensor network consists of a number of sensors, a sink, and a base station. The sensor node has constraints of energy consumption, memory capacity, and also very limited processing capability [1]. The purpose of sensor is to capture the physical entity, such as temperature, humidity. The sensor can perform a limited amount of processing and then communicates this data to the sink. The base station collects a large amount of data obtained from the sensor nodes. It has a database to store all the collected data. It can also have components required to transform all this data into a useful and meaningful data that may be used for other applications. It also consists of a proper user interface for providing access to the database. It also implements protocols required for communication with the sink node.

The sink or the gateway reads the packets from every sensor node. It acts as an intermediary node to collect data from the sensor and then dumps the aggregated data onto the base station for further analysis (Fig. 1).

V. Bhateja et al. (eds.), *Information Systems Design and Intelligent Applications*, Advances in Intelligent Systems and Computing 672,
https://doi.org/10.1007/978-981-10-7512-4_96

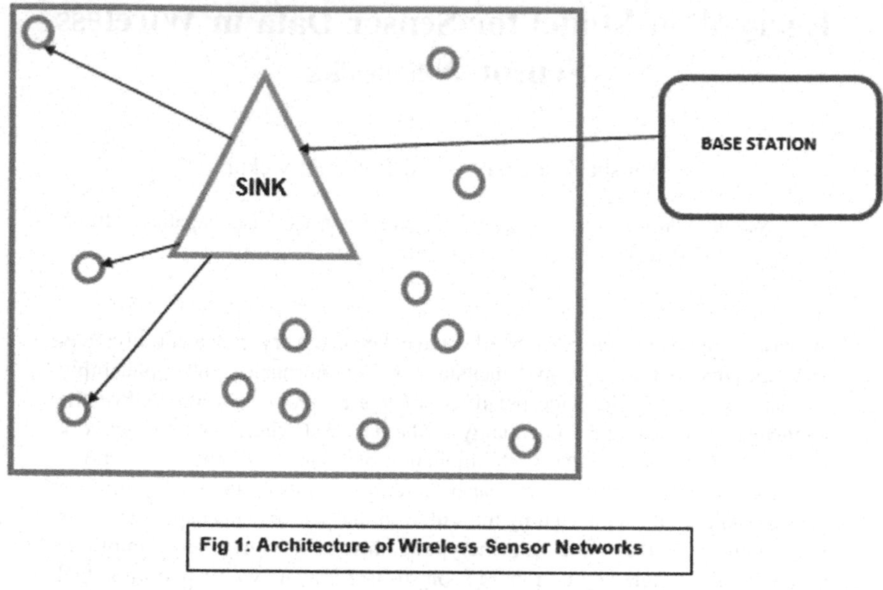

Fig. 1. Architecture of WSN

The working of the wireless sensor network is as follows [1]:

The sink moves about the network and probes each sensor to transfer the data collected to the sink. This probing may occur periodically or every time a particular event occurs. After the sink collects the data, it may transfer all the data it collected to the base station, which once again may occur periodically or when a specific event occurs. In this architecture, we can see that the data that is collected from the sensor and given to the base station has little to no security, neither in terms of confidentiality nor in terms of the authentication of the sensor.

1.2 Security Issues in Architecture

1. When the sensor data is being transferred from the sensor to the sink and also when the sink transfers the data to the base station, there are many opportunities for the adversary to compromise the confidentiality of the data. For example, the adversary could simply eavesdrop on the communication between the sensor and the sink or the communication between the sink and the base station to find what data is sensed by the sensor.
2. Another scenario is that when the sink is collecting the data from the sensors, the sink has to make sure that the sensor is authenticated to be in the sensor network from which the sink is to collect data. It is to say that if a new sensor which is not part of the sensor network tries to inject irrelevant data into the sink, it should be detected as coming from an unauthentic sensor node.

2 Related Work

Akyildiz and Mehmet [2] and Pathan et al. [3] discuss a study on how security is affected at every level of the network architecture in general, and in wireless sensor networks, in specific. It studies the different possible attacks on WSN at every layer. Kavitha and Sridharan [3] give a detailed study on the hardware and software architecture of WSN along with its protocol stack. Then it delves deep into the security aspects specific to WSN detailing the need and constraints of security in WSN providing a boundary for further work. Its study tells us that security breaches in WSN are mainly of two kinds: one, where information is released into the network, which is undesired as it facilitates a third person to overhear it; two, where data packets are inserted into the network by an adversary leading to unnecessary traffic and unclean data being collected from the sensor. Zhang et al. [4] argue that symmetric key cryptography is more suitable to encrypt sensor data due to its lesser consumption of energy compared to asymmetric cryptography. It implements various blocks and stream symmetric key ciphers and studies their energy efficiency. Toldinas et al. [5] conclude with its experiment that AES requires more energy for decryption than encryption and proposed three security profiles based on their experiments. Srivasta and Revathi [6] study the performance of AES, RC5, and blowfish using temperature, humidity, and light sensors with TinyOs on a TelosB mote for the metrics packet delivery, throughput, energy consumption, and memory consumed. Du and Chen [7] provide a holistic view of security in key distribution, time synchronization among sensors, location discovery, and routing. It explores the use of asymmetric cryptography in the context of sensor network. Panda [8] studies the limitations of sensor networks and the feasibility of using hybrid cryptography which combines both symmetric and asymmetric cryptography. Khambre et al. [9] study application of AES in various implementations. Giruka et al. [10] study the various protocols for providing authentication, key management, routing, and intrusion detection. It studies the denial of services attack and its possible remedies in detail. Kumar et al. [11] proposed an algorithm using public key cryptography to provide both confidentiality and integrity in sensor networks. Khan et al. [12] proposed an encryption model using AES and ECC in Petri net model, and the results for different modulation schemes were tested for BER and SNR.

3 Proposed Method

The proposed method consists of an encryption scheme which focuses on how to ensure the confidentiality of the data collected from every sensor by the sink.

The first algorithm uses a single key that the sink uses to encrypt the data collected from each sensor.

Scenario 1.1: Concept of a key m: The base station generates a key m and shares this key with the sink. The sink uses this key to encrypt the data collected from every sensor in the sensor network. The sink implements the encryption algorithm and the base station implements the decryption algorithm (Fig. 2).

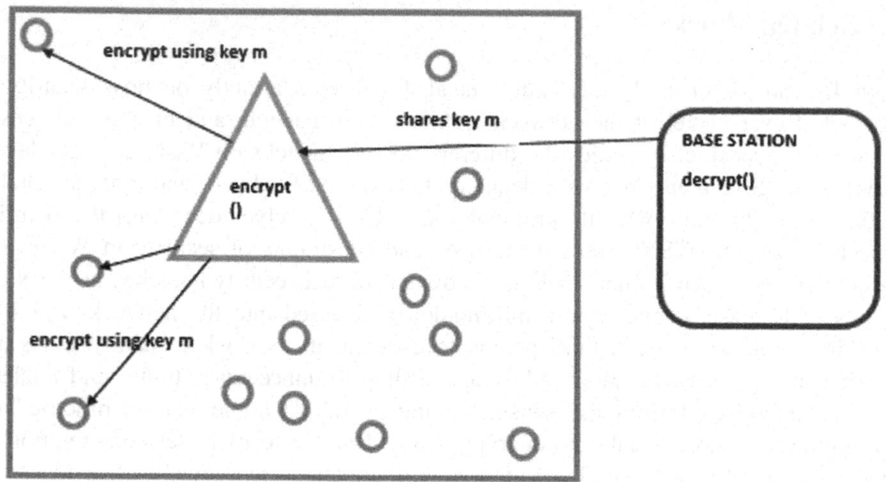

Fig. 2. Scenario 1.1: concept of a key m

Scenario 1.2: Concept of a mutable key m with mutation function: The base station generates a key m and shares this key with the sink. The sink uses this key to encrypt the data collected from the first sensor. Before applying encryption on the next sensor data, the sink applies a mutation function on it and obtains a mutated key m'. It then uses the mutated key to encrypt the next sensor data. This mutation is applied on every data collected from the sensor. The sink implements the encryption algorithm and the base station implements the decryption algorithm. The BS needs to know the mutation algorithm in order to obtain the symmetric key to be used in the decryption (Fig. 3).

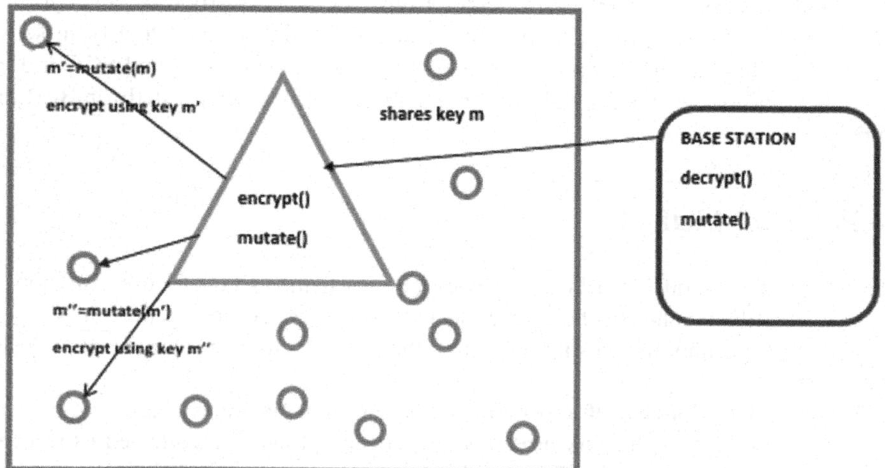

Fig. 3. Scenario 1.2: concept of a mutable key m with mutation function

Whenever a new sensor is added, it registers with the sink. And the sink later moves through the sensor field in the order of the registrations done.

Algorithm 1.1 KeyMutation

Step1 Generate a random prime integer initialSeed
Step2 Generate two random integers factor1 and factor2
Step3 l1 = intialSeed * factor1
Step4 mutatedkey2 = l1% factor2

Algorithm 1.2 encryptSensorData

Step1 call KeyMutation()
Step2 encryptWithAES()

Scenario 1.3: Concept of a stable key s along with mutable key m (dual encryption model): The base station generates a key m and shares this key with the sink. The sink uses this key to encrypt the data collected from the first sensor. Before applying encryption on the next sensor data, the sink applies a mutation function on it and obtains a mutated key m'. It then uses the mutated key to encrypt the next sensor data. This mutation is applied on every data collected from the sensor. After collecting data from a number of sensors, the sink applies encryption over the collective data from n number of sensors. This number of sensors under consideration is called the threshold factor. The threshold factor can be decided based on the number of sensors in the entire sensors filed and how many groups they can be divided into. The sink implements the encryption algorithms for both individual sensor encryption and the group sensors' encryption; the base station implements the decryption algorithms for both individual sensor decryption and the group sensors' decryption. The mutation algorithm is known to both the sink and the base station. The BS needs to know the mutation algorithm in order to obtain the symmetric key to be used in the decryption (Fig. 4).

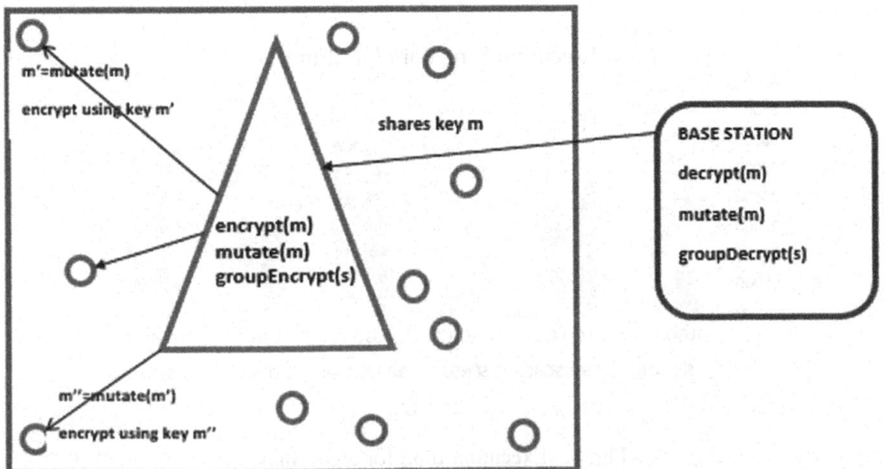

Fig. 4. Scenario 1.3: concept of a mutable key m with stable keys (dual encryption)

Algorithm 1.3 groupEncryptSensorData
Step1 count the number of sensors visited
Step2 if count == thresholdGroupSize
Step3 encryptGroupDataWithAES()

The thresholdGroupSize is fixed at 10 for now.

4 Implementation

For the implementation, the number of nodes is uniformly distributed in a region of 200 × 200. A mobile sink collects data from 10 sensors at a time and reports the data back to the sink. Every node in a region is uniquely identified by a node id.Sink reports the data of every sensor to the base station for further decision making. The data will be in an encrypted form. In the first phase, we have implemented the symmetric encryption algorithm AES. In the second phase Algorithm 1.1 and Algorithm 1.2 are run. In the third phase, Algorithm 1.3 is also run along with Algorithm 1.1 and Algorithm 1.2. The security level of each scenario has increased in order to protect the data. The observations of the algorithms are explained in the result section. [13] provides a NesC interface for AES encryption and decryption.

5 Results

The use of AES keeps the sensor data secure from statistical attacks and linear and differential attack. Brute force attack is almost impossible because of the longer size of the key used.

To generate the key, our proposed systems makes the use of random functions for initialisinginitSeed, factor 1 and factor 2 that dissipate any relation between the values (Fig. 5).

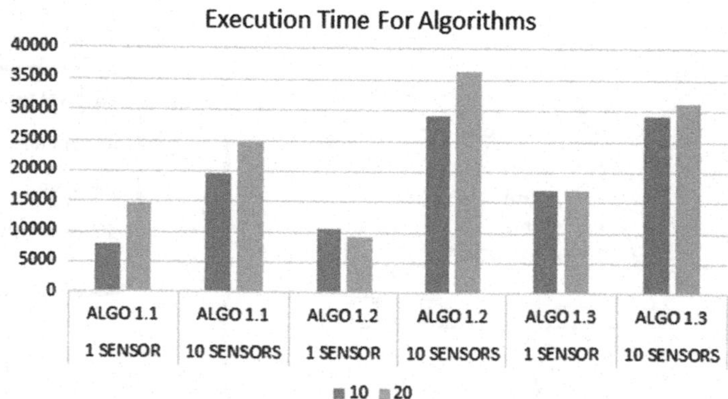

Fig. 5. Execution time for algorithms

The time taken to generate the key increases from Algorithm 1.1 to Algorithm 1.3. But this overhead is negligible over the amount of security obtained by encrypting every sensor's data with a new key. The time taken to encrypt the data using a different key for every sensor remains the same if the size of the key would remain the same for all the sensors. The overhead lies only in the mutation of the key to generate next key. This is the reason why the mutation function is kept simple in terms of the operations. Since the size of the key remains the same for every encryption, the strength of the encryption algorithm remains the same. Security is enhanced in the fact that even if an eavesdropper could obtain a key used for encryption of data of one sensor, he would not be able to obtain data from other sensors as the key itself is changed (Fig. 6).

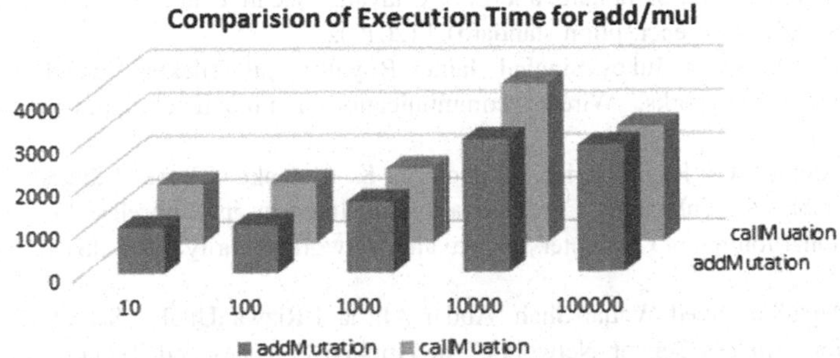

Fig. 6. Comparison of execution time for mutation algorithm

6 Conclusion and Future Work

The paper attempts to develop a model that ensures confidentiality to the sensor data. It describes three models that are built around the encryption algorithm to be used. This allows to keep the system modularised so that the algorithm for encryption itself may be changed according to the requirements of the application without affecting the model. The models increase the level of security step-by-step. This paper focuses on providing security to the transfer of data and key between the sensor and the sink. The models can be enhanced further in order to accommodate security to the key transfer between the base station and the sink.

References

1. Akyildiz, Ian F., and Mehmet Can Vuran. Wireless sensor networks. Vol. 4. John Wiley & Sons, 2010.
2. Pathan, Al-Sakib Khan, Hyung-Woo Lee, and ChoongSeon Hong. "Security in wireless sensor networks: issues and challenges." In Advanced Communication Technology, 2006. ICACT 2006. The 8th International Conference, vol. 2, pp. 6-pp. IEEE, 2006.
3. Kavitha, T., and D. Sridharan. "Security vulnerabilities in wireless sensor networks: A survey." Journal of information Assurance and Security 5, no. 1 (2010): 31–44.

4. Zhang, X., Heys, H.M. and Li, C., 2010, May. Energy efficiency of symmetric key cryptographic algorithms in wireless sensor networks. In Communications (QBSC), 2010 25th Biennial Symposium on (pp. 168–172). IEEE.
5. Toldinas, J., et al. "Energy efficiency comparison with cipher strength of AES and Rijndael cryptographic algorithms in mobile devices." Elektronik air Elektrotechnika 108.2 (2011): 11–14.
6. Srivstava, Ankit, and N. RevaththiVenkataraman. "AES-128 performance in tinoys with CBC algorithm (WSN)." International Journal of Engineering Research and Development 7 (2013): 40–49.
7. Du,Xiaojiang, and Hsiao-Hwa Chen. "Security in wireless sensor networks." IEEE Wireless Communications 15, no. 4 (2008).
8. Panda, Madhumita. "Security in wireless sensor networks using cryptographic techniques." AJER 3 (2014): 50–6.
9. Khambre, P. D., S. S. Sambhare, and P. S. Chavan. "Secure data in wireless sensor network via AES (Advanced encryption standard)." (2012).
10. Giruka, Venkata C., MukeshSinghal, James Royalty, and Srilekha Varanasi. "Security in wireless sensor networks." Wireless communications and mobile computing 8, no. 1 (2008): 1–24.
11. Kumar, Sunil, C. Rama Krishna, and A. K. Solanki. "Time Efficient Public Key Cryptography for Enhancing Confidentiality and Integrity in a Wireless Sensor Network." International Journal of Computer Science and Network Security (IJCSNS) 17, no. 1 (2017): 81.
12. Khan, Abdullah, Syed Waqar Shah, Abdar Ali, and RizwanUllah. "Secret key encryption model for Wireless Sensor Networks." In Applied Sciences and Technology (IBCAST), 2017 14th International Bhurban Conference on, pp. 809–815. IEEE, 2017.
13. http://tinyos.stanford.edu/tinyos-wiki/index.php/Security.

A Prototype Model for Resource Provisioning in Cloud Computing Using MapReduce Technique

Ananthi Sheshasaayee and R. Megala(✉)

Department of Computer Science, Quaid-e-Millath Government College for Women (Autonomous), Chennai 600 002, India
{ananthi.research,megala.research}@gmail.com

Abstract. Cloud Computing is an emerging technology in this digital world. Many organizations are starting using Cloud Computing technology for reducing their expenses. Instead of buying resources, they are renting the resources from Cloud Service Providers (CSPs) as per their need. Thus, Cloud Resource provisioning is a challenging task in the research world. Many researchers have found their own approaches for provisioning the resources in the cloud. This paper explains a new provisioning approach for large applications. It uses MapReduce technique to reduce execution delays in the job. The main aim of this model is to schedule the tasks using MapReduce technique which is a parallel programming model for distributed environment. It will maximize the customer satisfaction level (CSL) by reducing execution delays and implementing cost of Cloud.

Keywords: Cloud Computing · Cloud service provider (CSP) · Customer satisfaction level (CSL) · MapReduce · Resource provisioning

1 Introduction

With the advent of the digital era, the generation of data amount is increasing drastically. This data should be stored and shared among many clients and organizations. The client could not process the data by its own physical machine because the volume of data might be huge. Thus, the client requires some technique called Cloud Computing [1]. The Cloud offers three important services to the clients which are Infrastructure-as-a-Service (IaaS) for storage space (Amazon EC2), Platform-as-a-Service (PaaS) for application development environment (Google App Engine/Microsoft Windows Azure), and Software-as-a-Service (SaaS) for complete applications are recommended for clients (EMC Mozy) [2]. The client's can use these services by resource provisioning technique. The important aspects of Cloud Resource provision are to provide clients a bunch of computing resources for accessing the workflows and storing the data [3]. Cloud environments are handled by third persons who formulate authority to access a virtually vast quantity of resources by clients On-Demand in a pay-what-use manner, with quality of service (QoS) surety provided by Cloud Service Providers (CSPs). This facilitates the Cloud infrastructure to be

V. Bhateja et al. (eds.), *Information Systems Design and Intelligent Applications*, Advances in Intelligent Systems and Computing 672,
https://doi.org/10.1007/978-981-10-7512-4_97

balanced up and down consequently to the quantity of data to be computed. Yet, even though the hardware infrastructure is available that supports such requirements, applications performing data analytics still need better tuning in order to fully use the processing power supplied by Cloud infrastructures [4].

MapReduce [5] is one of the appropriate programming models for conveying of uses on Clouds. It is a parallel registering model intended for preparing of a lot of information in extensive computational foundations. A MapReduce application comprises of two sorts of undertakings, in particular Map and Reduce. At the point when a MapReduce application begins, Map undertakings handle input information and create the middle of the road information, which is organized as key–esteem sets. Diminish undertakings, then join every one of the qualities related to a key esteem and create the application's yield information.

1.1 Cloud Computing

A basic meaning of cloud may express that "Cloud Computing is a model for enabling convenient, universal and on-demand network access to a shared pool of configurable computing resources (network, server, storage, application, and services) that can rapidly provision and release with minimum management effort or service provider interaction" [6]. Cloud processing is that each kind of calculation can be conveyed to general society by means of the Web. It is changing the situation and furthermore

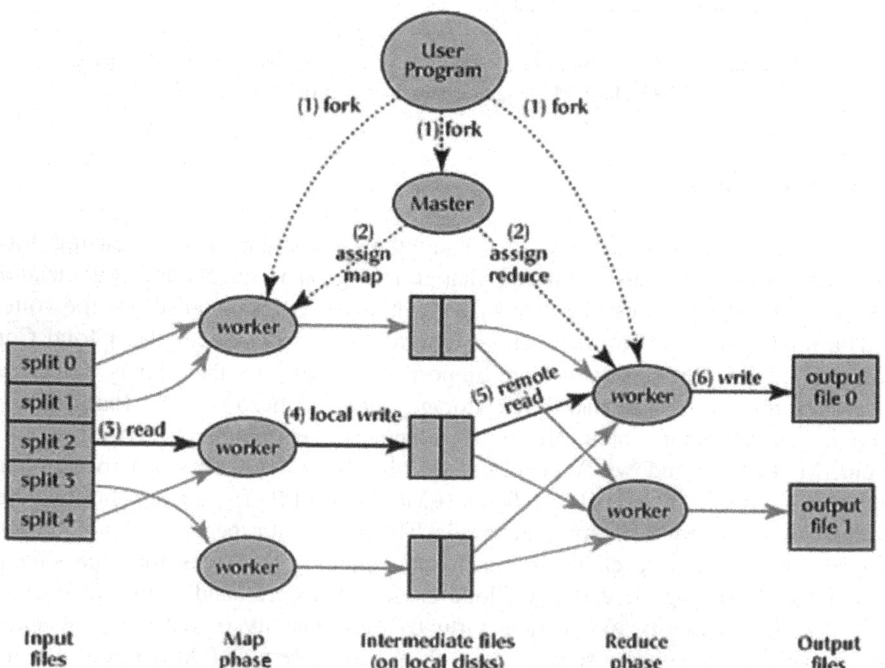

Fig. 1. MapReduce framework for Cloud Computing

influences the everyday life of a person. Any stuff can be shared over any gadget by clients by means of distributed computing with no issue. Organize transfer speed, programming, handling force, and capacity are spoken to as the figuring assets to clients as the openly available utility administrations [7].

1.2 MapReduce Model

MapReduce is a disseminated processing model. As of late, there is an assortment of engine motor receiving MapReduce. Be that as it may, with their distinction lying generally in (if not by any means) outer parts, for example, schedulers, an uncovered bone MapReduce structure can be viewed as indistinguishable. As of late, there is an assortment of MapReduce suited for embarrassingly parallel or disseminated issues, since its operations are totally free on different documents. The following diagram describes the MapReduce framework for Cloud Computing [8] (Fig. 1).

2 An Architectural Framework for Cloud Resource Provisioning

Resource provisioning is a procedure of supplying resources to the Cloud consumers for processing their jobs and storing the information. Many of the resource provisioning algorithms are being used by Cloud Service Providers (CSPs). This architectural framework describes about job scheduling using MapReduce technique for resource provisioning. MapReduce is a technique which consists Map phase and Reduce phase. The Map phase starts with input data and generates key–value pairs. Reduce phase merges all the values linked with a key and produce the application's output data and cannot start to accomplish until all Map tasks have completed.

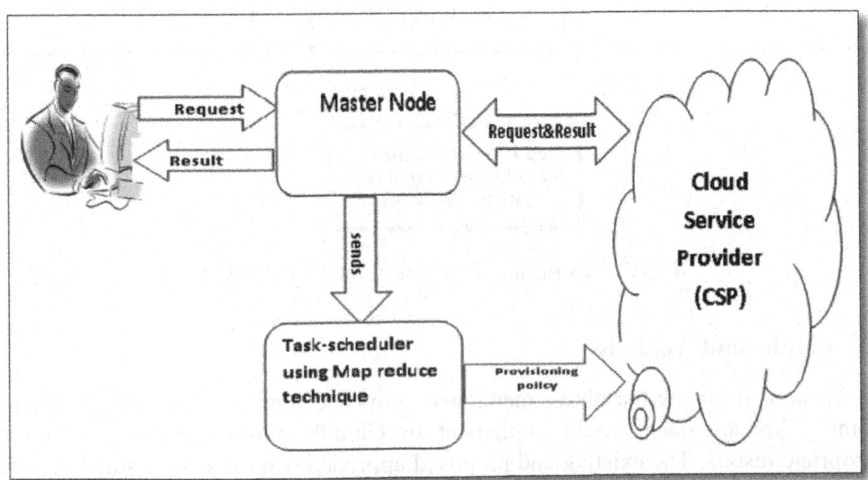

Fig. 2. An architectural framework for Cloud Resource provisioning

This MapReduce technique is widely applied for treating large quantities of data using equally large quantities of study units. The primary aim of this work is to speed up the implementation time for large amount of data as well as trimming the monetary value. Figure 2 depicts the envisioned scenario for the proposed policy. The client's request resources to Cloud Service Providers (CSPs) for their large-scale organizations, that request also saved in the Master node for Future reference. The Master node sends the request to task scheduler who schedules the tasks using MapReduce technique. The task scheduler splits large tasks into smaller nodes, and each process should have a key—value pair for processing. Finally, the output of tasks should be merged and reduced using MapReduce technique. Then Cloud Service Provider applies provisioning policy on reduced task and sends the results to clients through a Master node.

In simulation there is no user-defined function, no actual input files; though CloudSim does provide a toolkit for actual file processing, purpose of simulation, efficient in time and cost, is lost by doing so. Virtual data and workload are sent and processed, which makes the operation for key generation unnecessary. The basic structure for implementing the model is given below: (Fig. 3).

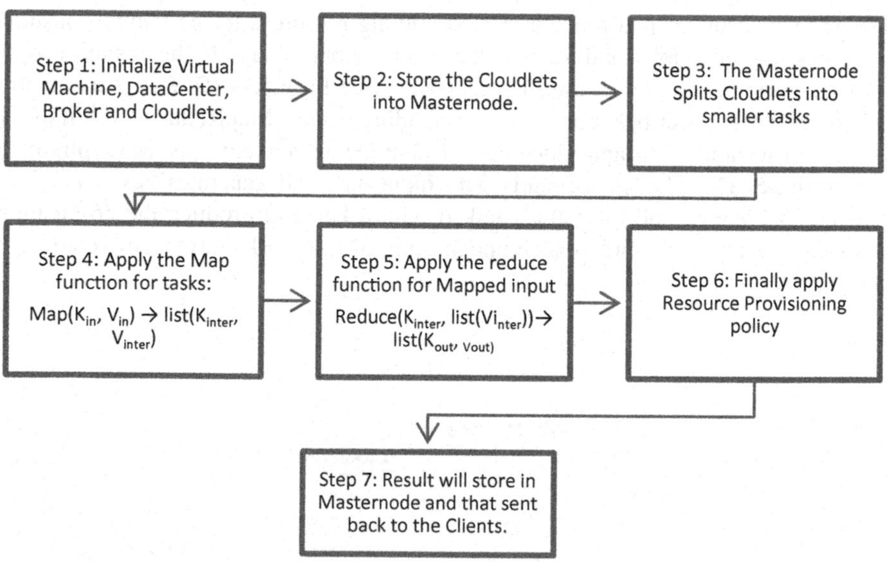

Fig. 3. A structure of proposed model (CRUMS)

2.1 Results and Analysis

This section discusses the above-mentioned proposed approach for Resource provisioning. This approach would implement in CloudSim tool to test and find the appropriate results. The existing and proposed approaches were tested using CloudSim tool. The main aim of this proposed approach is to reduce execution delays large applications. The data size is a main parameter in this approach. The requested file size of resources of each and every clients may vary. So, the proposed model (CRUMS) would decide to schedule the job by MapReduce technique and provide the appropriate resource to the client in very early with low cost. The small sample dataset has taken to test the simulation result of proposed and existing methods (Fig. 4).

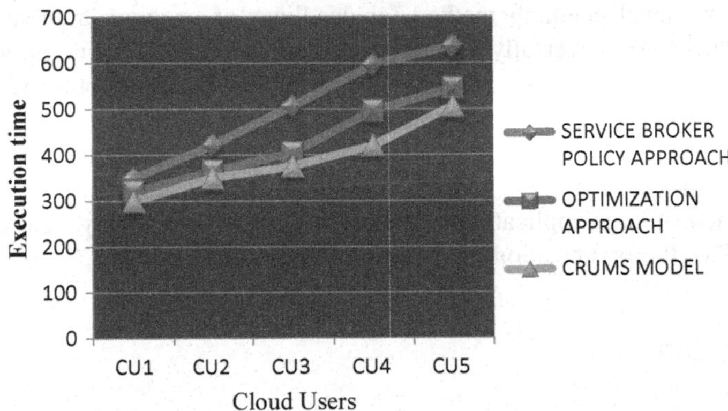

Fig. 4. Comparison of existing and proposed approach (CRUMS)

The above-mentioned graph shows that the execution delays of existing and proposed method. The proposed method reduces execution delays over existing approaches. As well as, the execution cost also reduces as per the Cloud Service Providers (CSPs) price model. It is a very useful approach for both organization and clients. This proposed model (CRUMS) also designed to follow the Service Level Agreement (SLA) which is a contract between providers and clients.

3 Related Works

The use of MapReduce Programming models within the Research community which has been increased tremendously [4]. This programming model is executed in Hadoop environment, which has been utilized as a benchmark for enhancements in the usage. There are many existing MapReduce looks into targets substantial scale; single-site situations are talked about underneath [4, 9]. According to the Tsai et al. [10], the replication model doesn't present particular procedures for resource provisioning for organizations, while our proposition oversees both resource provisioning and genuine planning (scheduling) of workloads for "MapReduce" calculation. According to Matsunaga et al. [11], Polo et al. [12], Luo et al. [13], and Fadika et al. [14], the resources from Clusters are gotten to on a best exertion premise, with the expectation of accelerating application execution. This makes the proposed foundation like grid processing frameworks. According to the Abhishek et al. [15] and Tian and Chen [16], they were the freely proposed models for ideal provisioning of the resources for large applications out in the Public Clouds, though Nikzad Babaii et al. [17] proposed a strategy for programmed setup of MapReduce design parameters keeping in mind the end goal to improve execution of uses in a Cloud. The models, in any case, were not created to address crossover Clouds components, for example, the presence of secure Cloud hubs accessible for calculation. According to the Sehgal et al. [18], building up an interoperable usage of MapReduce framework is to empower between the operations of uses that are emphatically attached to a given foundation. Dong et al. [19] proposed a two-level planning method for deadlines of ongoing large-scale applications, and it does not powerfully arrange additional asstes for meeting application due dates. KC and Anyanwu [20] proposed an approach where a confirmation control system rejects demands for executing MapReduce applications when due dates cannot be met. Rather than dismissing demands, our approach uses dynamic provisioning for assigning additional resources. To our knowledge, there is no current approach for implementation of large applications with minimum execution delays and cost, so this paper examines the organization of the proposed model (CRUMS).

4 Conclusion

Cloud Computing is a very popular technology in today's world, and many of them come across Cloud technology unknowingly if they have mail account. The growth of data is increasing drastically. Thus, most of the organization is moving their data into the Cloud for unlimited storage. For many users, resource provisioning is a crucial task in the Cloud. There are many existing approaches were used for resource provisioning. This proposed framework is too useful for them who need resources for their large-scale organizations. It uses MapReduce technique for task scheduling, and it saves execution time as well as speed up the response time.

References

1. Rajkumar, B., et al. "Cloud Computing and emerging IT platforms." Future Generation Computer Systems. Elsevier Press, Inc (2009).
2. Girase, Sagar, et al. "Review on: Resource Provisioning in Cloud Computing Environment." International Journal of Science and Research (IJSR) 2.11 (2013).
3. Nagesh, Bhavani B. "Resource Provisioning Techniques in Cloud Computing Environment-A Survey." IJRCCT 3.3 (2014): 395–401.
4. Mattess, Michael, Rodrigo N. Calheiros, and Rajkumar Buyya. "Scaling mapreduce applications across hybrid clouds to meet soft deadlines." Advanced Information Networking and Applications (AINA), 2013 IEEE 27th International Conference on. IEEE, 2013.
5. Dean, Jeffrey, and Sanjay Ghemawat. "MapReduce: simplified data processing on large clusters." Communications of the ACM 51.1 (2008): 107–113.
6. Foster, Ian, et al. "Cloud Computing and grid computing 360-degree compared." Grid Computing Environments Workshop, 2008. GCE'08. Ieee, 2008.
7. Shivhare, Hirdesh, Nishchol Mishra, and Sanjeev Sharma. "Cloud Computing and big data." Proceedings of 2013 international conference on cloud, big data and trust. 2013.
8. http://backstopmedia.booktype.pro/big-data-dictionary/mapreducehadoop/.

9. Menaga, G., and S. Subasree. "Development of Optimized Resource Provisioning On-Demand Security Architecture for Secured Storage Services in Cloud Computing." International Journal of Engineering Science and Innovative Technology (IJESIT) 2.3 (2013).
10. Tsai, Wei-Tek, et al. "Service replication strategies with mapreduce in clouds." Autonomous Decentralized Systems (ISADS), 2011 10th International Symposium on. IEEE, 2011.
11. Matsunaga, Andréa, Maurício Tsugawa, and José Fortes. "Cloudblast: Combining mapreduce and virtualization on distributed resources for bioinformatics applications." eScience, 2008. eScience'08. IEEE Fourth International Conference on. IEEE, 2008.
12. Polo, Jorda, et al. "Performance management of accelerated mapreduce workloads in heterogeneous clusters." Parallel Processing (ICPP), 2010 39th International Conference on. IEEE, 2010.
13. Luo, Yuan, et al. "A hierarchical framework for cross-domain MapReduce execution." Proceedings of the second international workshop on Emerging computational methods for the life sciences. ACM, 2011.
14. Fadika, Zacharia, et al. "MARLA: MapReduce for heterogeneous clusters." Cluster, Cloud and Grid Computing (CCGrid), 2012 12th IEEE/ACM International Symposium on. IEEE, 2012.
15. Verma, Abhishek, Ludmila Cherkasova, and Roy H. Campbell. "Resource provisioning framework for mapreduce jobs with performance goals." ACM/IFIP/USENIX International Conference on Distributed Systems Platforms and Open Distributed Processing. Springer Berlin Heidelberg, 2011.
16. Tian, Fengguang, and Keke Chen. "Towards optimal resource provisioning for running mapreduce programs in Public Clouds." Cloud Computing (CLOUD), 2011 IEEE International Conference on. IEEE, 2011.
17. Rizvandi, Nikzad Babaii, et al. "A study on using uncertain time series matching algorithms for MapReduce applications." Concurrency and Computation: Practice and Experience 25.12 (2013): 1699–1718.
18. Sehgal, Saurabh, et al. "Understanding application-level interoperability: Scaling-out MapReduce over high-performance grids and clouds." Future Generation Computer Systems 27.5 (2011): 590–599.
19. Dong, Xicheng, Ying Wang, and Huaming Liao. "Scheduling mixed real-time and non-real-time applications in mapreduce environment." Parallel and Distributed Systems (ICPADS), 2011 IEEE 17th International Conference on. IEEE, 2011.
20. Kc, Kamal, and Kemafor Anyanwu. "Scheduling hadoop jobs to meet deadlines." Cloud Computing Technology and Science (CloudCom), 2010 IEEE Second International Conference on. IEEE, 2010.

A Purview of the Impact of Supervised Learning Methodologies on Health Insurance Fraud Detection

Ananthi Sheshasaayee and Surya Susan Thomas(✉)

Department of Computer Science, Quaid-e-Millath Government College for Women, Chennai, Tamil Nadu, India
{ananthi.research, susann.research}@gmail.com

Abstract. A plethora of researches is happening in almost all sectors of insurance to improve the vitality and vibrance of its existence. As years pass, the volume of insurance policy holders increases which is directly proportional to the occurrence of frauds in these sectors. The presence of fraud is always an obstacle to the growth of an insurance organization. This paper confers the various supervised learning methodologies employed in detecting health insurance frauds.

Keywords: Health insurance · Fraud detection · Data mining · Supervised learning

1 Introduction

The health insurance sector has taken a high rise in the recent years due to the impact of the vulnerability to get hospitalized and also of the mounting hospital expense the patient had to concede. This paradigm has led nearly everyone to take health insurance policies. It gained popularity in the developing countries like India and China in the late years. This sector too faced hindrance like fraud, which retarded the profit of insurance companies, and this then paved the way for researchers to identify strategies to halt and mitigate frauds at the earliest.

Every year, millions of dollars are depleted from the insurance providers due to frauds. In order to sustain the profit, the insurance companies raise the premium amount and this in turn affects the genuine policy holders too [1]. It is estimated that frauds steal up to fifteen percent of the taxpayer amount that is used to fund the government-aided health care, making it crucial for the government agencies to find some cost-effective methods to pinpoint the fraud claims and transactions.

It is tough to eliminate these frauds and fraudsters completely, but it can be easily detectable with the incorporation of data mining methods and techniques along with artificial intelligence [2]. This mining of data and drawing patterns to recognize the odd ones gained popularity from the early 2000s.

V. Bhateja et al. (eds.), *Information Systems Design and Intelligent Applications*, Advances in Intelligent Systems and Computing 672,
https://doi.org/10.1007/978-981-10-7512-4_98

1.1 Classification of Health Insurance Frauds

Fraud in healthcare industry is just like in any other industry [3]. Fraudsters obtain full benefit of unjust profit with the help of healthcare crooks which includes patients, payers, vendors, suppliers, employers, and healthcare providers including pharmacists (Table 1).

Table 1. Illustration of various types of healthcare insurance frauds [4]

S No	Type of fraud	Area
1	Billing services not rendered	Hospital/clinic
2	Billing uncovered service as utilized service	Hospital/clinic
3	Altering dates of assistance	Hospital/clinic
4	Altering location of assistance	Hospital
5	Altering provider of assistance	Patient/customer
6	Incorrect reporting of analysis (unbundling)	Hospital/patient
7	Overutilization of assistance	Hospital/patient
8	Kickbacks/Bribery	Hospital/doctors
9	False/unnecessary issuance of drugs	Pharmacists
10	Up coding or down coding	Hospital/clinic

1.2 Statistical Insight into the Impact of Healthcare Frauds

Figure 1 shows criminal healthcare prosecutions over the last 20 years in the USA according to the TRAC report [5].

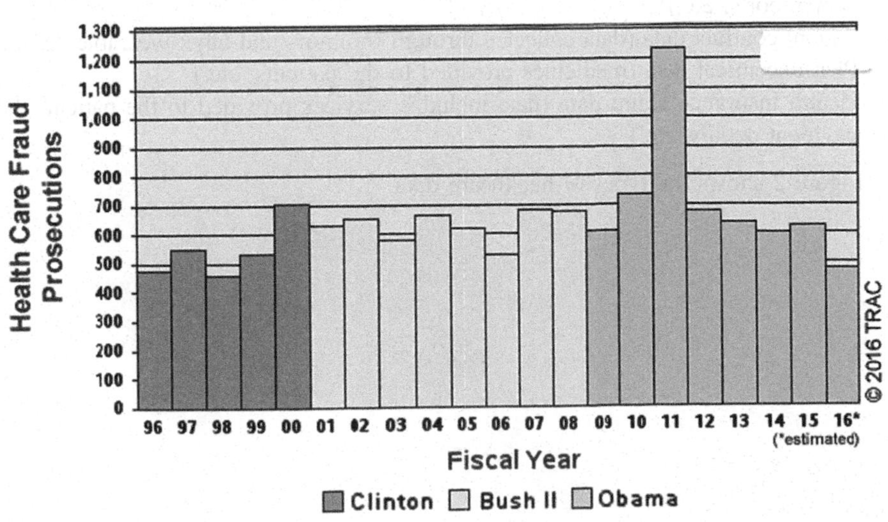

Fig. 1. Statistical representation of criminal healthcare prosecutions over the last 20 years in the USA

2 Learning Methodologies to Detect Health Insurance Frauds

Exploring data or simply "data mining" is a constant progress which involves discovery of a new fact or a hidden truth either through automated or manual procedures. It is a powerful mechanism in an exploratory research scenario where there are no fixed beliefs about the future outcomes. Data mining is the quest for new, variant, and nontrivial truths from large quantity of data. It is literally a combined effort of data analysts and the computers [6]. Solutions are obtained by harmonizing the knowledge of data analysts and the feedbacks of the search effectiveness of the computers.

2.1 Data Mining

Data mining has become extensively popular or to be precise very essential in the healthcare domain. Its application has become beneficial for all the factions in the healthcare domain; for instance, it can help health insurance providers to find hoaxs, healthcare enterprises to offer better customer relationship relations, doctors to recognize better treatment for their clients, and in course patients have more affordable services. A copious amount of data is generated here, and the traditional methods to handle this data are complicated and arduous [7]. Data mining provides efficient methods and techniques to convert these immense data into useful information for decision making.

2.2 Healthcare Data

Data used in health care can be classified mainly into four groups: [8]

(a) Hospital data (health record data of patients, medical images, laboratory and surgery reports, etc.)
(b) Patient conduct data (data collected through monitors and other wearable devices)
(c) Pharmaceutical data (medicines provided to the patients, etc.)
(d) Health insurance claim data (data includes services provided to the patient, their payment details, etc.)

Figure 2 shows the types of healthcare data.

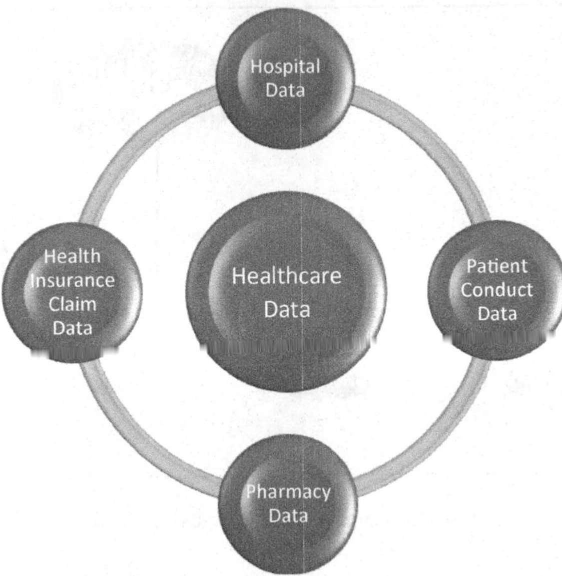

Fig. 2. Different types of healthcare data

2.3 Supervised Learning and Unsupervised Learning

The two preeminent learning methodologies are supervised learning and unsupervised learning methodologies, with the supervised learning using the trained data sets to perform mining of data while the unsupervised learning using the raw data or the real data. Fraud detection using supervised data sets is found to be more efficient and accurate [9]. But the adversity in obtaining trained data has led researchers to use raw data, i.e., using unsupervised learning. The outcome was found to be less competent. So, it is observed that supervised learning methodology has set its cardinality in detecting frauds with more effectiveness.

3 Supervised Learning in Health Insurance Fraud Detection

Supervised learning is a prominent data mining technique which has a reliable variable that is utilized either for a classification or prediction from a group of self-reliant attributes. Samples of supervised learning methodologies are naïve Bayes, linear regression, decision trees, random forest, logistic regression, and support vector machines [10]. The impediment in this type of research is that it needs some predictors or class labels to mold data for classification or prediction.

Health insurance claim studies are mostly done using supervised learning methodology (SLM). Figure 3 shows the claim process flow using SLM. Insurance claims are recorded in the database and are processed with the help of supervised learners, and claims are subjected to fraud detection models, where fraudulent claims are triggered while reports for genuine claims are generated and are subjected to payment clearing. The fraudulent claims are validating again and given to the fraud-mitigating team for further processing.

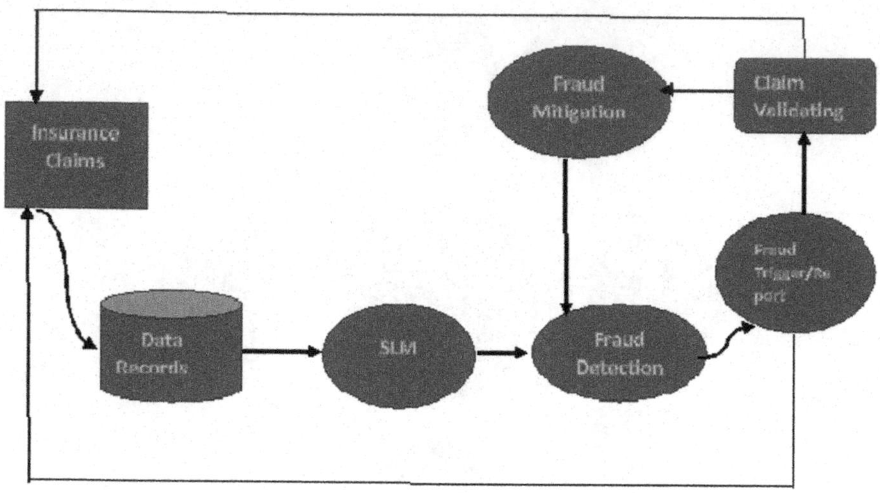

Fig. 3. Claim process flow using SLM

4 Related Works

Many studies have been carried out, implementing supervised learners to detect insurance claim frauds. This paper brushes through the works which throws an insight into the fraud detection mechanisms optimized in the insurance sector to catch hold of the fraudulent claims.

Travaille et al. [11] examine "Medicaid" fraud detection by evaluating techniques implemented in various sections from telecommunications to credit cards to healthcare that uses many machine learning techniques. They also asserted that since the divergence of health insurance data is plentiful, trained information is not easily acquired when compared to the availability in other fields.

Liu et al. [12] handle supervised learners such as multilayer perceptron (MLP) and decision trees for fraud detection in healthcare and lays a double-fold frame of anomaly revelation ideas using geolocation. In addition to that, the authors draft a supervised learning process using genetic algorithms and k-nearest neighbor (kNN) algorithm. The authors incurred that supervised learning yields more efficiency but has difficulty in procuring labeled data.

Phua et al. [13] had an in-depth study on the computerized deceit detection works collected from a span of a decade. The authors analysed that several research studies focuses mainly on complicated, nonlinear supervised algorithm whereas lighter methods such as naïve Bayes produce better and in some cases more efficient results.

Johnson et al. [14] framed an anomaly detection by implementing a multistage access to highlight hospital care fraud apparently using both private and public available data. A risk ranking is then formulated with the help of decision trees.

Joudaki et al. [15] guide an examination on the mining data learning processes fixating mainly on healthcare fraud detection. Their work discusses both supervised and unsupervised learning algorithms in the fraud detection field and mentions the need to refine supervised models to better efficiency.

Kumar et al. [16] add to the research work in fraud detection by framing a support vector machine (SVM) supervised learning for prediction of errors in insurance claims.

Ngufor et al. [17] converge to the investigation of provider fraud, concentrating in obstetrics claims, using unsupervised learners such as outlier detection with supervised learners like regression classification.

Brockett et al. [18] confined to the study considering the number of provider visits, their next visits, and the scans done without inpatient costs as the model criteria. They examined each criterion using a "ridit" scoring. Then, they utilized a double-way classification that separated claims into genuine and fraudulent.

Ortega et al. [19] suggested a many layered feed-forward neural networks to identify fraudulent claims. They used a false trigger cost as the model variable.

Liou et al. [20] proposed a model which dealt with nine types of cost-related parameters in their mining data process. They injected a classification tree, logistic regression, and neural network algorithm to spot hoax in diabetic disease claims. It was observed that the introduction of classification tree produced better results.

5 Conclusion

A wide spectrum of researches are carried out over the years to help expose frauds in the insurance field. Analysts are in search of newer and better methods to detect frauds and thus help the insurance companies to combat fraudsters in an efficient manner. Supervised and unsupervised learning algorithms are used in many cases, where both have their own advantages and disadvantages. Supervised learning has more accuracy points, but the effort to obtain labeled data is onerous. On the other hand, unsupervised learning has some negative aspect of an uncertainty since the main connection between the measured attributes of the information is unexplored. Hence, it is observed that a hybrid learning mechanism is recommended in dealing with hoax detection in the health insurance domain.

Acknowledgements. My sincere gratitude to my research guide Dr. Ananthi Sheshasayee for her expert guidance and valuable suggestions for the formulation of this paper. I thank all my well-wishers for their support and help.

References

1. Copeland, Leanndra, et al. "Applying business intelligence concepts to medicaid claim fraud detection." *Journal of Information Systems Applied Research* 5.1 (2012): 51.
2. Cortesão, Luis, et al. "Fraud management systems in telecommunications: a practical approach." *Proceeding of ICT*. 2005.
3. Berwick, Donald M., and Andrew D. Hackbarth. "Eliminating waste in US health care." *Jama* 307.14 (2012): 1513–1516.
4. http://www.fraud-magazine.com/article.aspx?id=4294976280.
5. http://trac.syr.edu/tracreports/crim/424/.

6. Gnanapriya, S., et al. "Data Mining Concepts and Techniques." *Data Mining and Knowledge Engineering* 2.9 (2010): 256–263.
7. Koh, Hian Chye, and Gerald Tan. "Data mining applications in healthcare." *Journal of healthcare information management* 19.2 (2011): 65.
8. Chandola, Varun, Sreenivas R. Sukumar, and Jack C. Schryver. "Knowledge discovery from massive healthcare claims data." *Proceedings of the 19th ACM SIGKDD international conference on Knowledge discovery and data mining*. ACM, 2013.
9. Han, Jiawei, Jian Pei, and Micheline Kamber. *Data mining: concepts and techniques*. Elsevier, 2011.
10. Bauder, Richard, Taghi M. Khoshgoftaar, and Naeem Seliya. "A survey on the state of healthcare upcoding fraud analysis and detection." *Health Services and Outcomes Research Methodology*: 1–25.
11. Travaille, Peter, et al. "Electronic fraud detection in the US medicaid healthcare program: lessons learned from other industries." (2011).
12. Liu, Qi, and Miklos Vasarhelyi. "Healthcare fraud detection: A survey and a clustering model incorporating Geo-location information." 29th World Continuous Auditing and Reporting Symposium (29WCARS), Brisbane, Australia. 2013.
13. Phua, Clifton, et al. "A comprehensive survey of data mining-based fraud detection research." *arXiv preprint* arXiv:1009.6119 (2010).
14. Johnson, Marina Evrim, and Nagen Nagarur. "Multi-stage methodology to detect health insurance claim fraud." *Health care management science* 19.3 (2016): 249–260.
15. Joudaki, Hossein, et al. "Using data mining to detect health care fraud and abuse: a review of literature." *Global journal of health science* 7.1 (2014): 194.
16. Chandola, Varun, Arindam Banerjee, and Vipin Kumar. "Anomaly detection: A survey." *ACM computing surveys (CSUR)* 41.3 (2009): 15.
17. Wojtusiak, Janusz, et al. "Rule-based prediction of medical claims' payments: A method and initial application to medicaid data." *Machine Learning and Applications and Workshops (ICMLA), 2011 10th International Conference on*. Vol. 2. IEEE, 2011.
18. Brockett, Patrick L., et al. "Fraud classification using principal component analysis of RIDITs." *Journal of Risk and Insurance* 69.3 (2002): 341–371.
19. Ortega, Pedro A., Cristián J. Figueroa, and Gonzalo A. Ruz. "A Medical Claim Fraud/Abuse Detection System based on Data Mining: A Case Study in Chile." *DMIN* 6 (2006): 26–29.
20. Liou, Fen-May, Ying-Chan Tang, and Jean-Yi Chen. "Detecting hospital fraud and claim abuse through diabetic outpatient services." *Health care management science* 11.4 (2008): 353–358.

An Improvised Technique for the Diagnosis of Asthma Disease with the Categorization of Asthma Disease Level

Ananthi Sheshasaayee[1(✉)] and L. Prathiba[2]

[1] Department of Computer Science, Quaid-e-Millath Government College for Women (Autonomous), Chennai 600002, Tamil Nadu, India
ananthisheshu@gmail.com

[2] Vels University, Chennai 600117, Tamil Nadu, India
prathiba.research@gmail.com

Abstract. The functioning of the lung tissues will be affected by asthma; the proper treatment needs to be given to the asthma patients to ensure the human's safety. In the earlier research work, vote-based ensemble classifier approach is utilized for disease diagnosis. Nevertheless, the level of the disease may differ for every patient in terms of different factors like age and environmental situations which might affect the proper treatment. This problem is resolved by presenting the asthma disease finding and level categorization technique (ADF-LCT) which is utilized to detect the various categories of asthma disease level in terms of patient's health status. In the proposed work, the Bayesian network is utilized to detect the existence of the disease by calculating the probability difference among the asthma genome profile and the input gnome details. Then, the disease level is detected by classifying patient's health details into three main categories such as low severe asthma (LSA), middle severe asthma (MSA), high severe asthma (HSA), and very high severe asthma (VHSA). The overall research of the work is executed in MATLAB simulation environment by utilizing the genome expression which proved proposed work leads to efficient prediction outcome.

Keywords: Asthma disease · Categorization of asthma disease · Level of disease · Profiling

1 Introduction

Asthma is a disease which affects airways in the lungs while inhaling and exhaling the air. The people who are affected by this disease are called as asthmatic [1]. Asthmatics generally sense these symptoms often at night and in the early morning [2]. Airways are nothing but a tube which carries air to-and-fro from the lungs [3]. It makes the airways swollen and very sensitive. Symptoms can happen every time the airways are inflamed. At times, symptoms may be mild, only minimal treatment is enough [4]. Sometimes symptoms continue to get worse. So need to get proper treatment while observing this disease at the first time [5]. Various research works have been conducted for the diagnosis of asthma disease. In [6], the authors utilized the machine learning

V. Bhateja et al. (eds.), *Information Systems Design and Intelligent Applications*, Advances in Intelligent Systems and Computing 672,
https://doi.org/10.1007/978-981-10-7512-4_99

algorithms combinations. Two kinds of methods were utilized here [7]: questionnaire and clinical diagnosis. The well-established training algorithm termed as backpropagation is utilized for multilayer neural networks [8]. The clinical decision-making tools function is enhanced by ANN technique, and it is used in medicine [9]. In [10], the authors make use of support vector machine classifier. The presented system can be utilized in asthma outcome prediction with 95.54% success. Research work [11] contains the combination of machine learning algorithms. There are random forest algorithm, AdaBoost algorithm, and artificial neural networks algorithm. In [12], the author's present approach is related to the k-nearest algorithm.

2 Finding and Categorizing Different Level of Asthma

In the earlier research work, ensemble classification approach was utilized to identify the asthma disease. But this approach fails to differentiate the various levels of this disease. In the proposed research work, the asthma disease is find out and confirmed by using the Bayesian network whose workflow illustrated in Fig. 1.

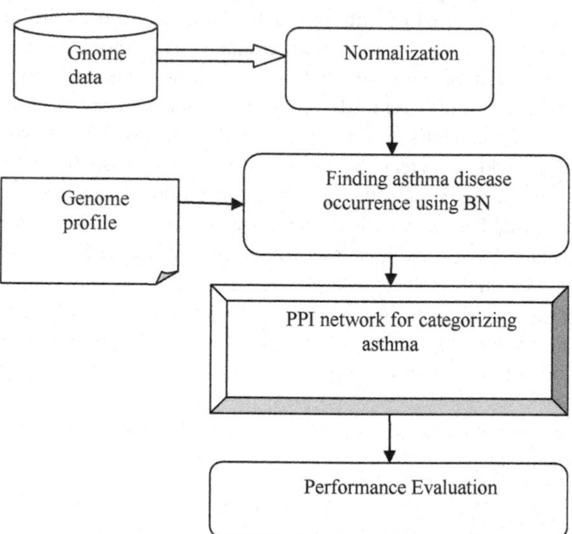

Fig. 1. Overall flow of proposed research work

2.1 Normalizing Asthma Genome Data

Normalization is an eliminating process of some sources of difference which affects the computed gene expression levels. In the proposed research work, quantile normalization is performed on the data set to eliminate the unwanted differences. Two distributions are created by using the quantile normalization in statistics. It is same as the statistical properties. The quantile-normalize is determined for both the test distribution and the reference distribution. And this is sorted for both the distribution and the reference distribution. The highest entry is considered for both the test and reference distribution, and the test distribution is affected by the reference distribution. Usually, the reference distribution is the standard statistical distribution such as the Gaussian or Poisson distribution.

3 Finding Asthma Disease Using Bayesian Network

Subsequent to normalization, asthma detection is accomplished by Bayesian network that will discover the possibility of occurence of asthma in the actual world environment. Bayesian network needs certain input to be given for contrasting and identifying the likelihood of disease happening dependent upon the input information specified. In this proposed work, genome profile is specified as input to the Bayesian network for examining the network.

3.1 Genome Profile Construction

Previous data were acquired via asthma BRIDGE. Initially, the stirred CD4+ T-cell lymphocyte RNA samples from CARE are managed by using the segment of 2,317 instanced of various tissues in the asthma disease. The Illumina human HT-12 v4 arrays are examined up to 47,036 various transcripts. (Illumina, Inc., San Diego, CA). And it is stated by manufacturer protocol. The principal component analysis of gene representation recognized two outliers. These outliers would be separated from the main downstream analysis. The entire metabolite association is examined with every 47,009 mRNA probes, and it is regulated based on the age and gender. The limma bioconductor package for the R programming language is used in this process [13]. Relationship discoveries which were because of huge leverage points were detached. Analysis was regulated for several examinations by the false discovery rate (FDR) [14].

3.2 Bayesian Network-based Asthma Disease Finding

In the Bayesian network, DAGs are combination of nodes which represent the random variable: They possibly will be noticeable quantities, suppressed variables, unfamiliar parameters or hypothesis. Edges signify conditional dependencies; the variables and the nodes are linked together and it is conditionally independent of one another. The likelihood function of the every node is considered as the input for the particular set of values in terms of their relationship among nodes. The notations of SNP and the methylation probes are ranked with significant p-value. This approach would select the increased relationship measure between the different kinds of metabolites as the parent nodes using Bayesian network analysis. These selection metabolites would then be normalized in order to make ease of data handling task. This is direct to a dataset of 64 metabolites, 19 SNPs, 19 CpG methylation sites, and 20 mRNA gene expression probes, including the age and sex, combined with the asthma control for the 20 care patients.

In MATLAB version R2013b the CGBayesNets package the data is learned by using the conditional Gaussian Bayesian network (CGBN). One of the machines learning methodologies are CGBN, and it creates the network design with the presence of statistical association's from the dataset. Likewise, the CGBN methodologies have been very successful in generating the analytical design of ICU mortality from metabolomics profiling [15]. The CGBN was studied by a less uniform Dirichlet prior to maintain the amount of edges less; a method which has been exposed to be efficient in forecast in genomic contexts [16] and persuades a sparse network with a superior complexity penalty [17]. Then, 250 bootstrap recognitions of the dataset were carried out; a method which aids right for minimum data samples by minimizing the consequence of outliers inside the population.

3.3 Diagnosing Different Stages of Asthma Using protein-protein Interaction Network

Subsequent to verifying the asthma existence, disease stage will be classified. It is necessary to classify the disease stage to give appropriate treatment for the patients those who are ill with the asthma. The patients with little severity could be dealt with less concern while a patient with more severity requires to be taken care with more care and effect. This is attained by presenting the protein-protein network communication technique that operates on gene data to discover the asthma stage on diverse phases. In the next sub part, dissimilar phases of asthma with their consequences and then the working process of protein-protein interaction network are specified.

3.3.1 Symptoms of Asthma in Diverse

Changeable indications which happen in asthma patients in their diverse disease levels are listed in the subsequent subsegments.

LSA: "Indications of a cough, wheeze, chest tension or trouble in breathing below two times a week, flare-up are concise; however, intensity possibly will differ nocturnal indications under two times a month, no indications amid flare-ups, lung function test FEV1 equivalent to or over 80% of usual values, and peak flow under 20% unpredictability AM-to-AM or AM-to-PM, day-to-day".

MSA: "Indications of a cough, gasp, chest tension or trouble in breathing three to six times a week, flare-up possibly will have an effect on movement level, nocturnal indications three to four times a month, lung function test FEV1 equivalent to or over 80% of usual values, and peak flow under 20–30% unpredictability".

HSA: "Indications of a cough, gasp, chest tension, or trouble in breathing every day, flare-up possibly will have an effect on movement level, nocturnal indications to five or more times a month, lung function test FEV1 over 60% however under 80% of usual values, and peak flow higher than 30% unpredictability".

VHSA: "Indications of a cough, gasp, chest tension or trouble in breathing constantly, Nocturnal indications regularly, Lung function test FEV1 below or equivalent to 60% of usual values, Peak flow over 30% unpredictability".

3.3.2 Finding Varying Interaction on Diverse Categories of Asthma Disease

The permanent or the transient physical contacts are indicated by using the protein–protein interactions (PPIs), and it is set up to amid two or more protein molecules subsequent of the biochemical occurrence pushed by electrostatic forces together with the hydrophobic effect. In this research work, PPINs are built to predict the varying relationship present between the different kinds of asthma severity. This is done efficiently by learning the interaction and sequence behaviour present among the different proteins that are responsible for asthma disease. The analysis of different kinds of protein would provide varying kinds of information about the asthma disease in terms of its functional behaviour. This makes ease of prediction of asthma disease, so that different growing level of asthma disease can be ensured. In this study, the gene expression data were split into four divisions in keeping with the group of samples, that is to say, LSA, MSA, HSA, and VHSA. Consequently, the protein–protein interactions which happen diverse ally among sample groupings were noticed. For instance, the protein interactions happen in LSA; however, let pass in MSA or vice versa were diverse interactions amid LSA and MSA. Like this, the diverse interactions were sensed for LSA in opposition to HSA and HSA in opposition to VHSA. To identify the diverse interactions, an interaction score was described for every PPI. At this point, the scores for diverse interactions of NC in opposition to HSA ($PPI_{cm\ (kl)}$) and MSA in opposition to SA ($PPI_{ms\ (kl)}$) were independently described in this manner.

$$PPI_{cm(k,lj)} = \frac{\text{minimum}(W_{kl(m)}, W_{kl(c)})}{\text{maximum}(W_{kl(m)}, W_{kl(c)})} \quad PPI_{ms(k,l)} = \frac{\text{minimum}(W_{kl(s)}, W_{kl(m)})}{\text{maximum}(W_{kl(s)}, W_{kl(m)})}$$

Here, c, m, and s, correspondingly, indicate LSA, MSA, and HSA, and W_{ij} is the weight described for the interaction amid protein i and protein j and was described as the Pearson correlation coefficient dependent upon their transcriptional representation profiles in this way.

$$W_{ij} = \frac{|\text{cov}(P, Q)|}{\sigma P \sigma Q}$$

Here, P and Q are representation values of genes k and l, correspondingly. In this study, in relation to their interactions scores, the diverse interactions were clustered into three dissimilar kinds: (i) category I, interactions rising for the duration of disease progress; (ii) category II, interactions which are not considerably diverse amid discrete phases; and (iii) category III, interactions fail to spot for the duration of disease progress. It can be noticed that categories I and III interactions are more possibly connected to diseases, and the proteins connected with these interactions are more probably correlated to the disease growth.

4 Experimental Outcome

In this part, the performance assessment of asthma detection and classification system is accomplished. In this part, the performance assessment of the research model is accomplished with the idea of the changeable age groups and asthma indications. The prediction accurateness of the research model called asthma disease identification and level classification technique (ADF-LCT) is matched up with our earlier work that is to say vote-based ensemble classification technique and the changing previous classification techniques in relation to their prediction values. The performance measures which are taken in this research for assessment are true positive rate (TPR), false positive rate (FPR), accuracy and F-Measure.

TPR: The true positive rate is described as the proportion of real positive that are spam subjects class properly categorized. It is the depiction of suitably categorized percentage of spam and non-spammer discovery. The TPR is described in this way:

$$\text{True Positive Rate (TPR)} = \frac{\mathrm{T_p}}{(\mathrm{T_p} + \mathrm{F_n})}$$

FPR: False positive ratio, as well called the false alarm ratio, typically denotes the likelihood of incorrectly declining the null hypothesis for a specific test.

$$\text{False Positive Rate (FPR)} = \frac{\mathrm{F_p}}{(\mathrm{F_p} + \mathrm{T_n})}$$

Accuracy: The accuracy is defined as the entire rightness of the design, and it is determined as the total of real classification parameters $(T_p + T_n)$ partitioned by the total number of classification parameters $(\mathrm{T_p} + \mathrm{T_n} + \mathrm{F_p} + \mathrm{F_n})$.

$$\text{Accuracy} = \frac{\mathrm{T_p} + \mathrm{T_n}}{\mathrm{T_p} + \mathrm{T_n} + \mathrm{F_p} + \mathrm{F_n}}$$

F-measure: F-Measure unites precision P and recall R by,

$$\mathrm{F} = 2 \cdot \frac{\mathrm{PR}}{\mathrm{P} + \mathrm{R}}$$

For assessing the classification techniques, concentrate on the F-Measure since it is a standard measure of summarizing both precision P and recall R. The comparison evaluations of these metrics are shown in the following figure.

A broad TPR assessment was carried out for diverse characteristics with the classifiers introduced above which are executed to asthma health data gathered from patients in Fig. 2. Outcomes depicted that ADF-LCT has enhanced TPR for asthma diagnosis.

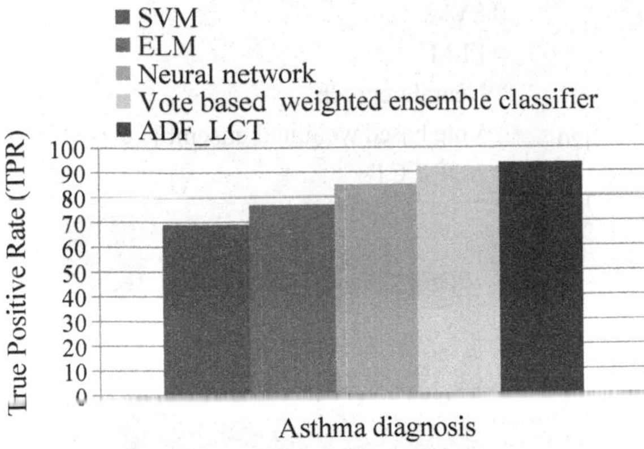

Fig. 2. True positive rate for features

An extensive FPR evaluation was carried out for diverse features with the classifiers presented above that are implemented to asthma health data collected from patients in Fig. 3. Outcomes demonstrated that ADF-LCT has improved FPR for asthma diagnosis.

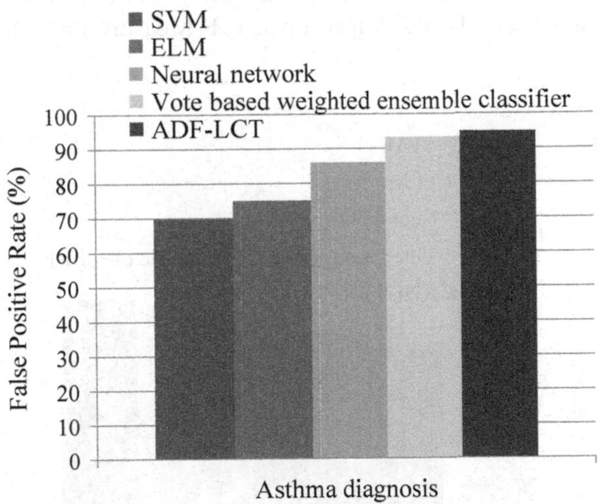

Fig. 3. False positive rate for features

The experimental results are carried out for the various features with classifiers for instance SVM and ELM, neural network and vote-based weighted ensemble classifier in Fig. 4. Outcomes expressed that ADF-LCT has progressed accurateness for asthma diagnosis.

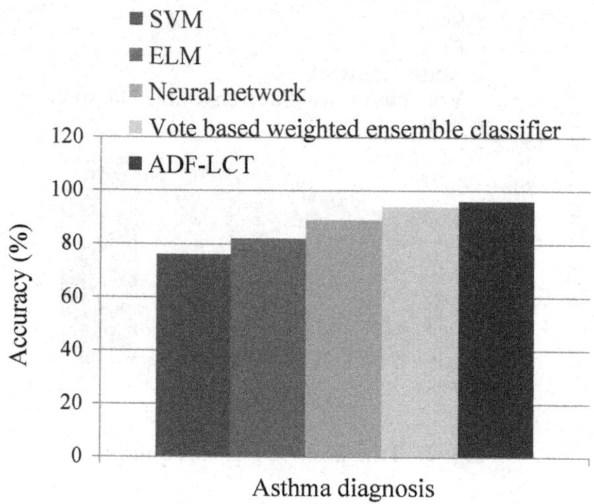

Fig. 4. Classification accuracy for features

An entire experimentation was carried out for diverse characteristics with classifiers, for example SVM and ELM, neural network and Vote-based weighted ensemble classifier in Fig. 5 for standard measure of summarizing both precision P and recall R. Outcomes expressed that ADF-LCT has superior F-Measure for asthma diagnosis.

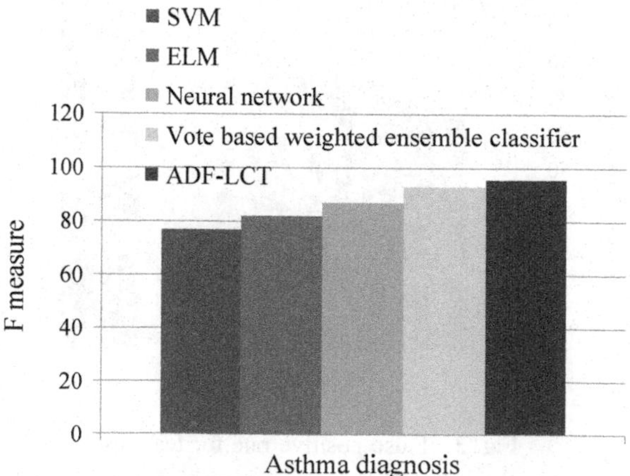

Fig. 5. F-Measure accuracy for characteristics

5 Conclusion

Asthma is very general disease that happens on more number of the people alive in the world. There is no cure identified for the asthma; therefore, the correct action requires to be given to stay away from the hazardous effects. Treatment wants to be given to the patient dependent upon their disease stage. Therefore, detecting the disease stage which is patient endured from insignificant level or higher level wants to be recognized for treating without delay. In the proposed system, Bayesian network is utilized for f the asthma existence by calculating the likelihood deviation amid the asthma genome profile and the input gnome information. Subsequent to identifying asthma existence, stage of disease is discovered for giving the superior treatment for the patients. This is accomplished by categorizing the patients' health associated detail into four types called LSA, MSA, HSA, and VHSA. This entire work is implemented in the MATLAB simulation environment for the genome representation. This performance of the proposed research methodologies is providing the enhanced results differentiated to the existing research methodologies.

Reference

1. Halm, E. A., Mora, P., & Leventhal, H. (2006). No symptoms, no asthma: the acute episodic disease belief is associated with poor self-management among inner-city adults with persistent asthma. CHEST Journal, 129(3), 573–580.
2. Holgate, S. T., Davies, D. E., Powell, R. M., Howarth, P. H., Haitchi, H. M., & Holloway, J. W. (2007). Local genetic and environmental factors in asthma disease pathogenesis: chronicity and persistence mechanisms. European Respiratory Journal, 29(4), 793–803.
3. Saini, B., LeMay, K., Emmerton, L., Krass, I., Smith, L., Bosnic-Anticevich, S.,…&Armour, C. (2011). Asthma disease management—Australian pharmacists' interventions improve patients' asthma knowledge and this is sustained. Patient education and counseling, 83(3), 295–302.
4. Sippel, J. M., Holden, W. E., Tilles, S. A., O'Hollaren, M., Cook, J., Thukkani, N.,…& Osborne, M. L. (2000). Exhaled nitric oxide levels correlate with measures of disease control in asthma. Journal of Allergy and Clinical Immunology, 106(4), 645–650.
5. Bjermer, L. (2001). History and future perspectives of treating asthma as a systemic and small airways disease. Respiratory medicine, 95(9), 703–719.
6. BDCN Prasad, P. E. S. N Krishna Prasad and Y Sagar, "An approach to develop expert systems in medical diagnosis using machine learning algorithms (Asthma) and a performance study", International Journal on Soft Computing, Vol. 2, No. 1, 26– 33, 2011.
7. Er, O., Yumusak, N., &Temurtas, F. (2010). Chest diseases diagnosis using artificial neural networks. Expert Systems with Applications, 37(12), 7648–7655.
8. Ansari, A. Q., Gupta, N. K., & Ekata, E. (2012, November). Automatic diagnosis of asthma using neurofuzzy system. In Computational Intelligence and Communication Networks (CICN), 2012 Fourth International Conference on (pp. 819–823). IEEE.
9. E. Chatzimichail, E. Parakakis and A. Rigas, "Predicting asthma outcome using partial least square regression and artificial neural networks" Advances in Artificial Intelligence, Article ID 435321, 1–7, 2013.

10. E. Chatzimichail, E. Parakakis, M. Sitzimi and A. Rigas, "An intelligent system approach for Asthma prediction in symptomatic preschool children", Computational and Mathematical Methods in Medicine, Article ID 240182, 1–5, 2013.
11. NahitEmanet, Halil R Oz, NazanBayram and DursunDelen, "A comparative analysis of machine learning methods for classification type decision problems in healthcare", Decision Analytics, vol. 1, No.: 6, 1–20, 2014.
12. TahaSamadSoltaniHeris, Mostafa Langarizadeh, Zahra Mahmoodvand, Maryam Zolnoori, "Intelligent diagnosis of Asthma using machine learning algorithms", International Research Journal of Applied and Basic Sciences, Vol. 5, No. 1, 140–145, 2013.
13. Smyth, G. K. 2005. Limma: linear models for microarray data. pp. 397–420 in R. Gentleman, V. Carey, S. Dudoit, R. Irizarry, and W. Huber, eds. Bioinformatics and computational biology solutions using R and bioconductor. Springer, New York.
14. Benjamini, Y, and Y. Hochberg. 1995. Controlling the false discovery rate: a practical and powerful approach to multiple testing. J. Royal Stat. Soc. 57:289–300.
15. Rogers, A. J., M. McGeachie, R. M. Baron, et al. 2014. Metabolomic derangements are associated with mortality in critically ill adult patients. PLoS ONE 9:e87538.
16. Jiang, X., R. E. Neapolitan, M. M. Barmada, and S. Visweswaran. 2011. Learning genetic epistasis using Bayesian network scoring criteria. BMC Bioinformatics 12:89.
17. Silander, T., P. Kontkanen, and P. Myllymaki. 2002. On sensitivity of the MAP Bayesian network structure to the equivalent sample size parameter. pp. 360–367 in R. P. Loui, and L van der Gaag, eds. Uncertainty in artificial intelligence. AUAI Press, Corvallis, Oregon.

Analyzing Online Learning Effectiveness for Knowledge Society

Ananthi Sheshasaayee and M. Nazreen Bee(✉)

Department of Computer Science, Quaid-e-Millath Government College for Women, Chennai, Tamil Nadu, India
{ananthi.research,nazreenm.research}@gmail.com

Abstract. In the recent years, there has been rapid development of integrate new technologies into educational processes. Educational system is determined to encourage the use of learning technologies that enhance and offer a successful learning outcome. An e-learning is the process of changing from instructor-interest to learner-interest. In this paper which provides an e-learning strengths, opportunities and also customizing the effectiveness of the learning system in an educational institution. Internet technologies and learning management system (LMS) are the process of communicating and interacting between instructors and students, students with students and instructors with instructors. Finally, this paper shed light on importance of online learning systems and benefits of employing e-learning system in learning perspectives and also proposed a solution to deal with the e-learning and m-learning trends.

Keywords: e-learning · m-learning · Education systems · e-learning trends

1 Introduction

Most probably, many researchers have defined that e-learning system is essential for learning strategies from different dimensions. An e-learning helps the learners to discover new knowledge and skills. Learning implies accumulate the knowledge, adding information accession, and creating sense of knowledge. In e-learning environment, most learners/students agreed that learning framework [1], learner motivation, learner's engagement, time management, and flexibility with online technologies impact the success of an online learning experience. The focus of an e-learning is the process of development of teaching/learning strategies and their resources [2]. Effective e-learning framework system involving the student experience in learning, teachers' strategies, teachers' planning and thinking, and the teaching/learning perspective. To innovate interesting learning experience in web-based learning is to integrate internet technologies into the process of learning and teaching strategies. This paper proposes ample framework for the design, implementation, and development of e-learning systems in education institution. To lead the efficiency of e-learning environment is the process to update the pedagogical principles and methods.

V. Bhateja et al. (eds.), *Information Systems Design and Intelligent Applications*, Advances in Intelligent Systems and Computing 672,
https://doi.org/10.1007/978-981-10-7512-4_100

2 Aim of the Research

- The objectives of this research work are necessarily to infer the "innovative educational technologies in learning styles."
- To generate new ideas in an e-learning environment.
- To proposed a solution to deal with the e-learning and m-learning trends.

3 Learning Design

Visual learners relate to words such as: see, look, observe, reads, and they observe and learn through pictures, diagrams, flowcharts, time lines. One of the main strengths in learning design representation is visual characteristic. The visual element enables a learning design to be summarized the pedagogical strategies. Visual learners can learn through overheads slides, visualize pictures, videos look, and graph, hands on, hands out, etc. The following pictures illustrate the elements of the visual learning strategies (Figs. 1 and 2).

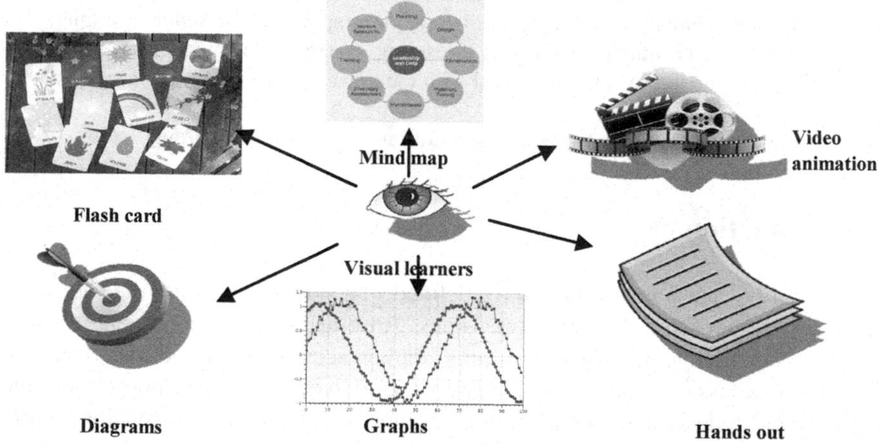

Fig. 1. Elements of visual learners

Fig. 2. Elements of verbal learners

Verbal learners are also stated that they are auditory learners and prefer a learning through in the form of a lecture or written [3].

To study and compare the visual learners and verbal learners in the learning interventions, analysis of the influence of the participants in learning styles test using Richard Felder Silverman learning style method can be employed [4].

According to this model, to analysis the index of learning style questionnaire from North Carolina State University learning style method, there are four dimensions of learning styles. There are two opposite categories (such as active and reflective) with each dimension. The information score for an each dimension with two opposite categories indicates the preference for one category or the other category. If the score for an each dimension with 2 opposite categories is 1 or 3 scoring, it was equally well balanced on the two groups of that dimension. If the score for an each dimension is 5 or 7, it specifies a adequate preference for one category of that dimension. If the score for an each dimension is 9 or 11, it denotes a strong preference for one category of that respective dimension. This learning style method is used to determine the learning style preferences among the students and also define the academic achievement levels of student's skill in learning style [4] (Fig. 3).

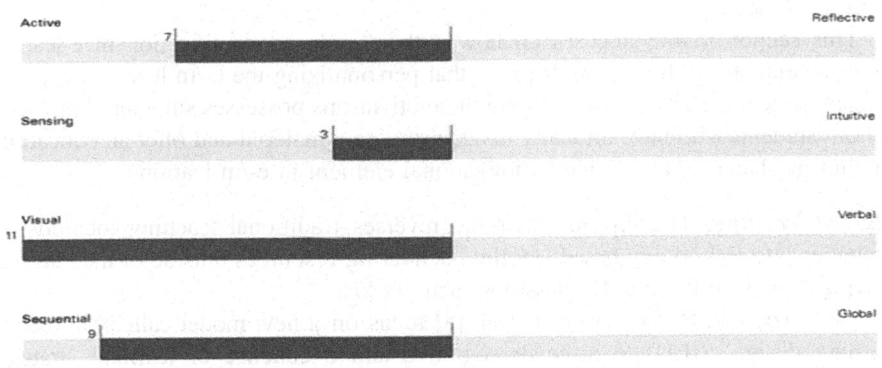

Fig. 3. Four dimension of learning styles

3.1 New Trends in E-Learning Environments

Nowadays, there are lots of new trends in e-learning system. Using a technology in learning system, it bridges the gap between a teacher and a student in different ways [5]. Technology won't replace all teachers but educators who use technology effectively will undoubtedly replace teachers/educators who do not. Modern trends in Web-based learning environment are described on the following Fig. 4.

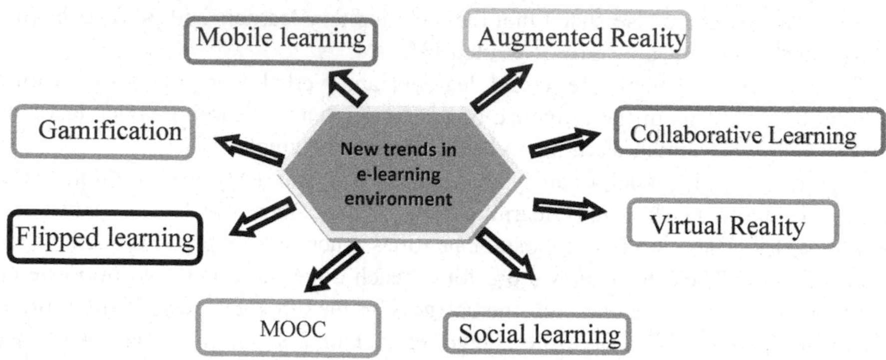

Fig. 4. New trends in e-learning system

3.2 Discussion of New Trends in E-Learning Environment on Selected Papers

Gamification: Gamification is the process of using game thinking methodology and game mechanics technologies to solve user problems and engage students in the learning strategies.

This section mainly focuses on how authors induced gamification in e-learning environments. Peter Juric et al. [6] says that personalizing the e-/m-learning approach to each student which helps to increase the motivations possesses sufficient knowledge to both students who have difficulty in acquiring new material and offer an educational gaming mechanics that provide a motivational element in e-/m-learning.

Flipped learning: The flipped classroom reverses traditional teaching methods and strategies into technology-based learning, delivering resources outside of the class and moving "homework" into the classroom activity [7].

Abdelaziz and Hamdy Ahmed et al. [8] focus on a new model called "immersive learning design" (ILD) to discover cognitive and excellence of learning strategies through flipped classrooms (FC) learning method. This model is beneficial to people who are observing the pedagogical application for virtual and blended learning environments and also helping graduate students to move forward from content learning engagement to intellectual learning engagement.

MOOC: A MOOC is a type of online course methods aimed at large-scale interactive participations and open-access online course.

Liqin and Zhang et al. [9] proposed a blend learning method to face a classroom/traditional teaching with MOOC-based e-learning on network. Using a MOOC-based blend learning method can achieve an optimization of teaching effectiveness. The growth of MOOC blend learning is described in the following graph [9] (Fig. 5).

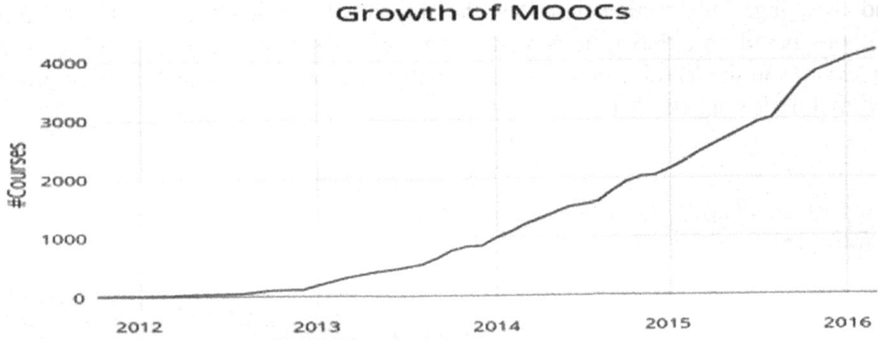

Fig. 5. Growth of MOOC-based blend learning

4 Solution to Deal with Mobile Learning Trends

M-Learning: The combination of technology and mobile mechanics is resulting in tremendous transformations of the educational world [10]. Mobile learning is a definite type of motivational and learning model [11]. Effective educational learning systems can be achieved by using mobile technology engagement system [12].

Concerning the sequence of mobile learning approaches, Zeng and Luyegu (2012) [10] states that mobile learning method organized a learning resources in both formal and informal learning contexts in higher education. In that paper, author focus on numerous dimensions of mobile learning including characterizations, hypothetical dimensions, and mobile learning engagement system in higher education.

Cheon et al. [13] states that "An investigation of mobile learning readiness in higher education based on the theory of planned behavior." In that paper, authors say that utilizing the exclusive abilities of mobile devices mobile learning is a new form of learning. This learning examined the factors affecting college student's intention to use m-learning based on the theory of planned behavior (TPB). The results provide a behavioral control into adopt mobile learning and a valuable consequences to increase college student's acceptance of mobile learning.

Huang et al. [14], in their study, researchers states that by investigate several innovative ideas to manipulative mobile learning applications. The users who using hand-held devices at hand, mobile learning application, and conceptual features will become important to the users/learners. Huang also propose a mobile learning mechanism to support collaborative learning.

Jeng et al. [15] in this study which focuses the importance of Mobile applications in teaching and learning strategies. This review study emphasized the use of essential characteristics of m-learning and also analyzed the academic learning strategies are applied in mobile learning environments.

The learning/teaching through mobile technology is the new trends in digital learning environments [16]. Basically, learning through computing devices which leads to an enormous development of learning strategies [17]. This section stated the mobile learning aids. Hence, mobile learning includes: augmented reality, microlearning, game-based learning, point in time status check, learning modules, performance aids, and blogging. This section describes the impact of mobile-based teaching/learning methods based on different tremendous dimensions as shown in Fig. 2. In Fig. 6 the dimensions in the learning environment which enhanced pedagogical learning process and mobile learner/coacher.

Fig. 6. Learning aids in m-learning environments

5 Conclusion

Along with the information and communication technology-based methods and new educational technologies are being more and more used. Nowadays Web-based learning (WBL) are used to support their own learning and also used in large variety of different learning methods. E-learning system helps to discover optimum ways of learning and optimal learning results [18]. E-/M-learning system plays a vital role in terms of measuring student's knowledge skills and disciplinary practices [12]. Using a new trend, students can emphasize the trends and make the learning processes easier and more interesting.

References

1. Miletić, Ljiljana, and Goran Lešaja. "Research and evaluation of the effectiveness of e-learning in the case of linear programming." *Croatian Operational Research Review* 7.1 (2016): 109–127.
2. Yanuschik, Olga V., Elena G. Pakhomova, and Khongorzul Batbold. "E-learning as a Way to Improve the Quality of Educational for International Students." *Procedia-Social and Behavioral Sciences* 215 (2015): 147–155.
3. Naser-Nick Manochehr. "The Influence of Learning Styles on Learners in E-learning Environments: An empirical Study." *2014 Inforamtion system- Cheer,* 2014.
4. Graf, Sabine, et al. "In-depth analysis of the Felder-Silverman learning style dimensions." *Journal of Research on Technology in Education* 40.1 (2007): 79–93.
5. Xiao-Fan Lin et al. "Trends in E-learning Research from 2002–2013: Citation Analysis."*Programming and Systems (ISPS), 2015 15th International Conferences on Advanced Learning Technologies*. IEEE, 2015.
6. Juric, Petar, Maja Matetic, and Marija Brkic. "Data mining of computer game assisted e/m-learning systems in higher education." *Information and Communication Technology, Electronics and Microelectronics (MIPRO), 2014 37th International Convention on*. IEEE, 2014.
7. Alkhabra, Saleh, and Natrah Abdullah. "Impact of Technology from Learning Environment to Organizational Practices." *International Journal of Social Sciences and Management* 3.2 (2016): 108–114.
8. Abdelaziz, Hamdy Ahmed. "Immersive Learning Design (ILD): A New Model to Assure the Quality of Learning through Flipped Classrooms." *Advanced Applied Informatics (IIAIAAI), 2014 IIAI 3rd International Conference on*. IEEE, 2014.
9. Liqin, Zhang, Wu Ning, and Wan Chunhui. "Construction of a MOOC based blend learning mode." *Computer Science & Education (ICCSE), 2015 10th International Conference on*. IEEE, 2015.
10. Zeng, Rui, and Eunice Luyegu. "Mobile learning in higher education." *Informed design of educational technologies in higher education: Enhanced learning and teaching*. IGI Global, 2012. 292–306.
11. Yau, J. Y.-K., Joy, M., & Dickert, S. (2010). A Mobile Context-aware Framework for Managing Learning Schedules – Data Analysis from a Diary Study. *Educational Technology & Society, 13* (3), 22–32. IEEE, 2016.
12. Yau, J., & Joy, M. (2010). Designing and evaluating the mobile context-aware learning schedule framework: challenges and lessons learnt. *IADIS International Conference Mobile Learning,* March 19–21, 2010, Porto, Portugal.
13. Cheon, Jongpil, et al. "An investigation of mobile learning readiness in higher education based on the theory of planned behavior." *Computers & Education* 59.3 (2012): 1054–1064.
14. Huang, Y.-M., Hwang, W.-Y., & Chang, K.-E. (2010). Guest Editorial—Innovations in Designing Mobile Learning Applications.*Educational Technology & Society, 13* (3), 1–2.
15. Jeng, Y.-L., Wu, T.-T., Huang, Y.-M., Tan, Q., & Yang, S. J. H. (2010). The Add-on Impact of Mobile Applications in Learning Strategies: A Review Study. *Educational Technology & Society, 13* (3), 3–11.
16. Esterhuyse, Maxine P., Brenda M. Scholtz, and Danie Venter. "Intention to Use and Satisfaction of e-Learning for Training in the Corporate Context." *IJIKM* 11 (2016).

17. El-Hussein, M. O. M., & Cronje, J. C. (2010). Defining Mobile Learning in the Higher Education Landscape. *Educational Technology & Society*, *13* (3), 12–21..
18. Julia Shishkovskaya et al. “EFL Teaching E-learning Environment: Updated Principles and Methods” *Procedia-Social and Behavioral Sciences* 206 (2015):199–204.

Review on Software-Defined Networking: Architectures and Threats

Sanchita Bhatia(✉), Kanak Nathani, and Vishal Sharma

Bharati Vidyapeeth College of Engineering, New Delhi, India
{sanchita.bhatia.9,kanaknathani21,vishalmtr}
@gmail.com

Abstract. Software-defined networking offers various benefits to network control and opens new ways of communication by defining powerful and simple switching elements that can use any field of a packet to determine the outgoing port to which it will be forwarded providing efficient formation, edge in performance, security and higher litheness to house inventive network proposals. This paper studies most recent developments in the exploration of software-defined networking through various architectures. Technically, all architectures are very diverse in footings of design, forwarding model, and protocol interfaces. The architectures of software-defined networking with their advantages and flaws have been discussed thoroughly followed by the beneficial provisions of software-defined networking in the field of security and as to why software-defined networking itself needs security, questioning the amount of work and research required in arena.

Keywords: Software-defined networking · Software-defined networking architecture · SDN security threats · OpenFlow · ForCES

1 Introduction

Traditional ways of managing networks using low-level and vendor's specific network components are very complicated, error-prone, and complex process. Yuefeng Wang et al. [1] so, emergence of a general paradigm for management that enables flexible management of network is needed. Through software-defined networking (SDN), the administrators are allowed to accomplish various network services through abstraction of hardware-level functionalities. It focuses on idea of providing user capability to control aspects of network by programming at an abstract level and eliminates need of inducing changes in low-level implementations. Wenfeng Xia et al. [2] As labeled according to the open networking foundation (ONF), the general architecture of SDN is composed of three layers, the control layer, application layer, and the infrastructure layer, stacking on the top of each other (Fig. 1).

V. Bhateja et al. (eds.), *Information Systems Design and Intelligent Applications*, Advances in Intelligent Systems and Computing 672,
https://doi.org/10.1007/978-981-10-7512-4_101

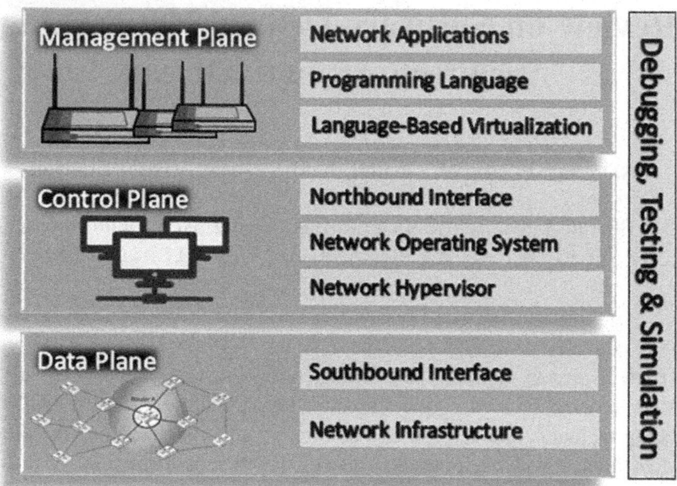

Fig. 1. Neutral architecture of SDN (ONF)

The infrastructure layer or data plane is the lowest layer made up of switching devices, which are accountable for gathering network prominence, keeping them temporarily with it, and then directing them to controllers as per directives from control plane/control layer. Data plane signifies the hardware required for forwarding or routing devices in SDN network construction. The control layer which is the central layer bonds application layer and infrastructure layer, for communication with the infrastructure layer downward. It offers utilities for controllers who ask for services to access functionalities offered by the switching devices and for the cooperation with the application layer in the upward direction; it offers various service access points in different forms. The controller represents an overview of the comprehensive network organization, giving permissions to the administrator that can be put upon protocols in the hardware system in a network. Open source-based "northbound application programming interfaces (APIs)" embody the interfaces between the SDN applications running on the top of the network platform and software segments of the controller platform. Further, certain protocols are obligatory by the controller to converse with the network infrastructure, to regulate and control the interfaces between the different sections of network apparatus, like southbound protocol. At last, top layer known as the application layer contains software-defined networking applications intended to satisfy user necessities.

Grounded on above-stated neutral architecture of software-defined networking, many reputed organizations have already adopted as in switches. Google realized OpenFlow switches for the management of WAN links between its centers of data control, OpenFlow controller, and a wide variety of SDN applications [3]. Using Switches with Network Virtualization Generic Routing Encapsulation (NVGRE) [4] for the creation of virtual networks, Microsoft Azure implemented overlay technology. ebay and Rackspace [5] developed cloud virtual networks using the Nicira's approach of the VMware.

2 Literature Review

Numerous architectures follow the elementary principle of software-defined networking for separation of the control plane and data plane and also the standardized interchange of information within the planes. However, technically they are diverse in footings of policy, forwarding model, architecture, and the protocol interfaces. Some of the architectures that modify the neutral architecture in one way or the other were analyzed. This section contains the detailed information of some of those architectures.

2.1 OpenFlow

OpenFlow was acquainted with an objective to permit easy network tryouts in a campus network. Manar Jammal et al. [6] early trials that had been done using OpenFlow intended at generating an isolated software controllable network that was dedicated on governing forwarding of packets. It is this protocol that is employed for the overall management of the southbound interface in the general architecture of SDN; it offers permits to routers and the switches too in the management layer. The first protocol that permits network administrators to adjust within the structure of switches and routers from numerous vendors in an even way so as to add and eliminate the packet flow state admissions was OpenFlow. Flow tables are used in OpenFlow which directs routers, switches, and other routing devices to manage traffic and packets. By making use of flow tables, the administrators can swiftly transform the network plan and the course of traffic. Also, some basic set of management gears are provided along with it, which can be used to control features such as packet filtering and topology modifications. Two types of OpenFlow switches are OpenFlow-hybrid and OpenFlow-only. OpenFlow-hybrid switches support a mechanism external to OpenFlow that leads traffic to any of the packet processing pipelines, along with support for consistent networking procedures. While other one being OpenFlow-only, switches support only OpenFlow operations. OpenFlow assists programming of the hardware without demanding vendors to expose the inner details of their devices.

2.1.1 OpenFlow Architecture

Plentiful OpenFlow-enabled switching devices which are governed using one or more OpenFlow controllers build the OpenFlow architecture. Transmission control protocol (TCP) is used to partition the network traffic into many flows. An OpenFlow switch is an essential element which consists of a group table and one or more flow tables. It accomplishes packet forwarding and look-ups tasks. The controller, using the OpenFlow protocol, manages the OpenFlow-enabled switch over a safe and secure OpenFlow channel which connects switches to controller. Each flow table is made up of various flow entries, which consists of match header fields to identify, counters and a set of instructions to operate on matched packets. Controller is accountable for upholding all the network policies and protocols and issuing proper directives to the devices. The switch must be able to establish communication with a controller at a user-configurable (but otherwise fixed) IP address using a user-specified port. Typically, the controller runs on a network-attached server.

2.2 ForCES

Bruno Astuto A. Nunes et al. [7] ForCES introduces two units called the control element (CE) and the forwarding element (FE) that permit ForCES procedures to converse. Providing per packet handling using the underlying hardware is the main role of FE. The job of CE is to implement control and signing functions and apply the ForCES protocol for handling packets. CEs and FEs are physically separated standard components. The protocol is built on a model similar to master–slave model, in which CEs are referred to as masters and the FEs are slaves. One of the most important components of ForCES structural design is the logical function block (LFB). The LFB is a precise serviceable unit controlled by CE through ForCES. A typical protocol permits all the CE and FE from various vendors of constituent to interconnect with one and all; hence, it becomes potential for the vendors to assimilate together the control and forwarding elements from diverse suppliers. The structure allows numerous instances of CE and FE within one network element (NE). The accumulation of many FE boundaries becomes the NE's exterior interface. With the peripheral interfaces, there should also be some sort of message so that the FE and CE can converse each other within various FEs. Below is a comparison of the two widely used SDN architectures OpenFlow and ForCES. (Table 1).

Table 1. Comparison of OpenFlow and ForCES

	OpenFlow	ForCES
Forwarding model	Flow table	Logic functional block
Building blocks	OpenFlow processing enabled through the control and forwarding elements	CE(s), FE(s), ForCES protocol, managers of FE and CE, and logical input flows
Architecture	Both network devices and architecture are changed	Network devices change, and architecture remains unchanged
Protocol	The controller-to-switch protocol runs over TLS or an unprotected TCP connection	This is a master–slave protocol in which CEs are masters and FEs are slaves
Similarity	Both protocols are used for the southbound interface and standardize information exchange between control and forwarding planes	Both protocols are used for the southbound interface and regulate information exchange between forwarding and control planes

2.3 OpenQoS

Hilmi E. Egilmez et al. [8] OpenFlow delivers resource monitoring, network perceptibility, and virtualization through management elucidations. OpenQoS is a controller architecture that permits QoS for multimedia distribution over networks of the OpenFlow. The OpenQoS approach employs dynamic routing. We do not have even a single architecture that is perfect in terms of QoS, is also most effective in providing QoS for multimedia traffic, and can be globally implemented. In general, there are two QoS architectures: IntServ and DiffServ. For routing DiffServ services use the same routing mechanism as that of the Internet. Resource reservation protocol (RSVP) is used by IntServ which communicates with any of the routing protocol to depute resources along a predefined path. In IntServ, it is difficult to apply dynamic QoS routing because it may introduce invisibility owing to relinking establishment. But with the help of OpenQoS, a totally new ranking scheme based on routing is used. To suffice the required QoS, dynamic routing is used for controlling the QoS multimedia traffic, whereas other data remain on direct path that is supposed to be the shortest. The OpenQoS is unique from the existing QoS architectures since it does not use priority queuing and resource reservation. The main advantage of this method over conventional ways is that the adverse effects of QoS such as latency and packet loss are minimized.

2.4 QosFlow

OpenFlow cannot arrange QoS constraints in an on-demand and dynamic manner. To tackle QoS problems using a manual method, a framework called QoSFlow was introduced that enables the organization of QoS in OpenFlow environment. Fei Hu et al. [9] According to QoSFlow, the management of traffic class is managed through policies. It handles the QoS resources without making any changes to the SDN architecture. QoSFlow is nothing but an enhancement of basic OpenFlow controller that gives multimedia transfer along with quality of service. The new controller, besides the standard NOX API, consists of the following constituents:

- QoSFlow agent: Creates a communication module between an administrator manager and the other two components, the manager and the monitor.
- QoSFlow manager: Manages OpenFlow domains.
- QoSFlow monitor: Monitors the OpenFlow domains.
- DB-QoSFlow client: It is accountable for making all lower-level activities onto the switch ports.

The extended versions of NOX API with QoS features are named as QoSFlow API. In QoS management tool, the activities are handled by the QoSFlow agent. As it receives any of such actions, it checks their type to select the next procedure. Then consequently, QoS actions can be deployed through packet header information.

2.5 OpenADN

Raj Jain et al. [10] Routing by means of policies of the ASP in a lively multi-cloud environment is impossible since the ISPs do not offer any services to route the information dynamically to a dissimilar server using ASP's strategies. Open Application Delivery Network (OpenADN) permits the ASPs to implement application stream of traffic management policies and delivery constraints at the level of application packets. Through OpenADN-enabled plane entities, ISPs are able to provide application delivery functionalities to ASPs. OpenADN allows the ASP controllers to effectively interact with the ISP's controller and provide ISPs with the server rules so that the ISP's controllers are able to program the control plane accordingly. With the requirement of northbound API, the OpenADN needs some advancements to the southbound API.

3 Security Analysis

The sections above provide an outlook to the network architectures of SDN with various features like managing traffic and packets dynamically, routing in a multi-cloud environment and providing application delivery functions, tackle on-demand QoS constraints using a manual approach, resource monitoring and network virtualization, providing elements that enable the procedures to converse. In this section, we discuss various good features already present in the neutral architecture of SDN that make it easy to find and take action to security-related attacks and all probable threats SDN is prone to.

3.1 Features of SDN Providing Security

- The partitioning of the data and control plane allows to establish defense experiments against large-scale attacks.
- The logically centralized controller has a worldwide view to build security policies and network-wide knowledge of the system to analyze traffic patterns for upcoming threats.
- Software-based traffic analysis encourages innovation and thus enhances the capability of switches using software-based techniques. The analysis of the traffic can be completed in real time using database and machine algorithms.
- The dynamic updation of forwarding rules aids to instantly reply to distribute denial of service attacks.
- SDN installs the rules for packet forwarding onto switching devices and blocks the attacks from the traffic from entering in network (Fig. 2).

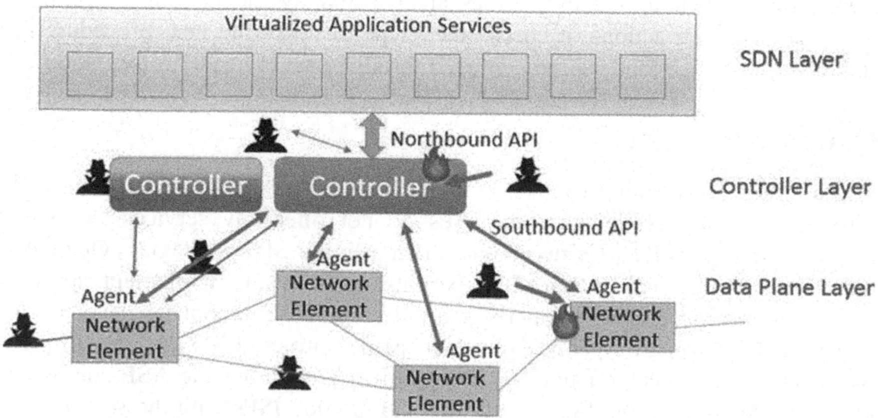

Fig. 2. Possible security threats in SDN

The above-mentioned security parameters are currently being used as defense attack system while handling denial of service attacks [11]. DDoS attacks are an attempt to make a network resource unavailable to its actually intended users.

3.2 Security Threats to SDN

Though SDN has numerous ways of protecting itself from the attacks as mentioned above, there are still some loopholes that make it vulnerable to threats which the security policy has not accounted.

- Unauthorized access: In the architecture of SDN, it is possible for many controllers and applications from multiple sources to gain access to the data plane of the concerned network by attacking control plane communications and vulnerabilities in switches and controllers. An attacker could achieve access to network resources and then transform the network operation if it impersonated a controller or application.
- Data leakage: An attacker could find the action applied to specific packet types by means of packet processing timing analysis. The attacker can discover the proactive/reactive configuration of the switches involved. An existing challenge is the storage of credentials for various logical networks in the programmable data plane in the SDN architecture. Sandra Scott-Hayward et al. [12] previously, cross-VM side channel attacks have been depicted in cloud environment.
- Data modification: If the attacker has entire system under control, it can transform flow rules present in the network devices, which would permit packets to be directed through the network to the attacker's benefit. Forged and fake traffic flow can cause immense damage to the SDN data.
- Malicious applications: An application that has been designed poorly or has bugs could unintentionally make the system vulnerable. For example, an identified bug could be misused by an attacker to initiate the application leading to an unsafe state.

4 Result

The above-mentioned content discusses the architectures following the principle of SDN implementing effective communication using various protocols. With no reference to profound study in the field of security in the architectures, related threat concerns to SDN have also been mentioned. The solution to the concerns like unauthorized access, malicious applications' intrusion is important for further growth in this field without hindrance as it would lead to secure control plane communication and no unauthorized access, filtering of fake traffic generation, reduction in the probability of failure in the system, trust initiation between the controller and administration, removal of data leakage and denial of services, tracking the user activities.

5 Conclusion

In this paper, the introductory concepts of SDN and its architectures has been explained, followed by the benefits that SDN provides in the field of security through its general architecture as demonstrated by the ONF and as to how it can be vulnerable to attacks that have not been accounted while making the security policies of the whole architecture. To overcome these threats, an evolution using existing software techniques is necessary leading to the need of more deeply understanding and doing research in this field.

References

1. Yuefeng Wang, Ibrahim Matta Computer Science Department, Boston University- SDN Management Layer: Design Requirements and Future Direction, IEEE 22nd International Conference on Network Protocols 2014, pp -555.
2. Wenfeng Xia, Yonggang Wen, Chuan Heng Foh, Dusit Niyato, and Haiyong Xie,- A Survey on Software- Defined Networking, IEEE Communications Surveys & Tutorials, 2014 pp. 4–6.
3. S. Jain et al., "B4: Experience with a globally-deployed software defined WAN," in Proc. ACM SIGCOMM Conf., New York, NY, USA, pp. 3–14. [Online], 2013. Available: http://doi.acm.org/10.1145/2486001.2486019.
4. M. Sridhavanet al., NVGRE: Network Virtualization Using Generic Routing Encapsulation. [Online] 2012. Available: http://tools.ietf.org/html/draft-sridharan-virtualization-nvgre-01.
5. P. Goransson and C. Black, Software Defined Networks: A Comprehensive Approach. 1st ed. San Francisco, CA, USA: Morgan Kaufmann Publishers Inc., 2014.
6. Manar Jammal, Taranpreet Singh, Abdallah Shami, Rasool Asal, and Yiming Li, Department of Electrical and Computer Engineering, Western University, Canada- Software-Defined Networking: State of the Art and Research Challenges, Elsevier's Journal of Computer Networks, pp – 4.
7. Bruno Astuto A. Nunes, Marc Mendonca, Xuan-Nam Nguyen, Katia Obraczka, and Thierry Turletti- A Survey of Software-Defined Networking: Past, Present, and Future of Programmable Networks, IEEE COMMUNICATIONS SURVEYS & TUTORIALS, VOL. 16, NO. 3, THIRD QUARTER 2014, pp-1619.
8. Hilmi E. Egilmez, S. Tahsin Dane, K. TolgaBagci and A. Murat Tekalp, Koc University, Istanbul, Turkey- OpenQoS: An OpenFlow Controller Design for Multimedia Delivery with End-to-End Quality of Service over Software-Defined Networks.
9. Fei Hu, Qi Hao and KeBao- A Survey on Software-Defined Network and OpenFlow: From Concept to Implementation, IEEE Communication Survey & Tutorials, VOL. 16, No. 4, Fourth Quarter 2014, pp-2194.
10. Raj Jain and Subharthi Paul, Washington University- Network Virtualization and Software Defined Networking for Cloud Computing: A Survey, IEEE Communications Magazine, November 2013, pp-30.
11. Qiao Yan, F. Richard Yu, Qingxiang Gong, and Jianqiang Li- Software-Defined Networking (SDN) and Distributed Denial of Service (DDoS) Attacks in Cloud Computing Environments: A Survey, Some Research Issues, and Challenges, IEEE Communications Surveys & Tutorials, pp-9.

12. Sandra Scott-Hayward, Sriram Natarajan, and SakirSezer- A Survey of Security in Software Defined Networks, IEEE COMMUNICATION SURVEYS & TUTORIALS, VOL. 18, NO. 1, FIRST QUARTER 2016, pp-628.

Analytical Comparison of Concurrency Control Techniques

Nabeel Zaidi[1(✉)], Himanshu Kaushik[2], Deepanshu Jain[1], Raghav Bansal[1], and Praveen Kumar[1]

[1] Amity University Uttar Pradesh, Noida, Uttar Pradesh, India
nabeelzaidi@ymail.com, {deepanshujain002, raghavbansal95}@gmail.com, pkumar3@amity.edu
[2] Shri Mata Vaishno Devi University, Jammu (Katra), India
himanshu.kaushik@smvdu.ac.in

Abstract. In database management system, there always comes a time where we need to execute various transactions simultaneously, and thus the database consistency must be maintained. The way to make sure that this consistency between the shared databases is maintained is to use concurrency control techniques. Most of the concurrency control techniques are developed using the serializability property in mind. The serializability property makes sure that the accessed data is in the manner of mutual exclusion, meaning that whenever one transaction is accessing a data, other transaction will not be able to access the same data. This paper deals with another viewpoint of various concurrency control techniques, there comparison based on the data obtained practically. Furthermore, a comparison of pessimistic, optimistic, multiversion, and two-phase locking techniques is done. We have set up an environment to analyze the performance and compare these techniques analytically.

Keywords: Transaction · Serializability · Locking · Mutual exclusion Rollback

1 Introduction

The state of the database changes as soon as a transaction takes place. For a transaction to be considered as true, it must be running in isolation. But, in shared databases, this condition may not be fulfilled as more than one transaction may take place concurrently. Thus, in order to ensure that the interaction among transactions in the system is controlled, several concurrent transactions mechanisms called concurrency control schemes are used [1].

Concurrency Control Techniques—The transaction in such techniques is either executed all at once or serially rather than executing them concurrently [2]. The schemes given below are serializable. To ensure atomicity, a special standard called serializability or isolation is used [1]. In this paper, we are going to compare the following techniques given below:

1. Lock-Based Protocols
2. Timestamp-Based Protocols

V. Bhateja et al. (eds.), *Information Systems Design and Intelligent Applications*, Advances in Intelligent Systems and Computing 672,
https://doi.org/10.1007/978-981-10-7512-4_102

3. Validation-Based Protocols
4. Multiversion Schemes.

Lock-Based Protocols—In this protocol, simultaneous access is provided to a data item using a lock mechanism. If a transaction holds a lock on a particular data item, only then it is allowed to access it. There are two locking modes in which items can be locked, either exclusive (X) mode or shared mode (S) [1]. An exclusive mode is given to a transaction that is allowed to not only read the data item but write it as well. Otherwise, the transaction only needs to read a data item that is given an S lock mode. For a transaction to be able to execute any further, its request must be granted [3]. If the lock requested by the transaction is in terms with other transactions that already hold lock on that data item, then the request is granted. A shared lock (S) on an item can be provided to more than one transaction at once. But an exclusive lock on any data item is given to only transaction at a time, and all the other transaction demanding for the same resource will be denied and they will have to wait for the current transaction to complete [1].

The Two-Phase Locking Protocol—If a transaction does not violate the property of serializability then it can always commit. But it may also lead to deadlock in a case where the requests and granting of locks are not done properly. Hence, a restriction must be imposed for transactions to be serial, in order to ensure that the operations that are conflicting are executed in a particular order. For any transaction to get new locks, it must release all the locks that it currently holds. This restriction on transaction is what is known as two-phase locking. The first phase is basically a phase of growing, the number of lock is decided by the transactions and then it acquires them. Then during the shrinking phase, it releases all the locks that it currently holds [4]. Suppose, if a transaction fails in acquiring the locks, then it must release all the locks, wait and then start over which ensures that there is no conflict in schedules of the transaction [1, 5]. Hence, it implies that all the locking protocols must be two-phased even if no information exists about the database or the transaction [3]. Further it can be divided into two broad categories:

1. Strict two-phase locking: In this, in order to ensure serializability of transactions, all the write locks are held until the transaction is complete or it aborts. The read locks maybe released previously though, i.e., as soon as the transaction terminates [5]. Hence, until abort or commit is done by the transaction it keeps on holding lock X unless a cascading rollback occurs.
2. Rigorous two-phase locking: In this technique, the transaction holds all the locks no matter S or X until it commits or aborts or a cascading rollback occurs and hence it is even stricter. In order to serialize the transaction, the order of commit is taken into consideration. There are two major drawbacks in this protocol though, which are deadlocks and starvations.

Timestamp-Based Protocols—The requests cannot be taken using any locking algorithms so that the information about the arrival and order of execution can be maintained. And therefore, algorithms which use queues and timestamps are taken into consideration [2]. This algorithm assigns a timestamp to a transaction when it starts its execution [4]. The timestamp allotted does not match the timestamp of any other transaction. This algorithm also assigns some special timestamps which are allotted to transactions that have the largest timestamps of transactions that have been read and written successfully and are known as W-timestamp and R-timestamp. And hence these timestamps become an important factor when the serializability is determined [1]. Whenever a transaction tries to access some data, its timestamps are checked with the transaction that is currently is being executed. If it is not older, then the ordering is not correct and the transaction has to abort, i.e., the transaction that started later was executed first and committed. It gives us a rule, which says that a lowered number transaction goes first and reads the data which the last committed transaction wrote [4]. This protocol gives us an order for executing conflicting read and writes operations based on their timestamps. Even though this protocol prevents deadlock but it will not be free from cascade and this might lead us to a situation where recovery is not possible [1].

Validation-Based Protocols—This algorithm depends on transaction backup which is used as a control mechanism in which the transactions are executed fully hoping that no conflicts occur between transactions [5]. This mechanism's efficiency is also dependent upon whether conflicts between transactions occur or not [6].

Validation-based protocols take help from DBMS to ensure that no conflict has taken place as they provide unlimited access of shared data during the time the transaction executes and before the time the transaction commits. And in order to resolve conflicts, the transactions are aborted [6]. The most of the transactions are read only; then, the conflict's rate becomes low. But the overhead of execution of code and delay of transactions still exist. Depending on the type of transaction, its execution takes place in different phases which are as follows:

1. Execution and Read Phase: First, all the write operations are performed on temporary local variables without updating them in a read and execution phase.
2. Validation Phase: In the next phase, some validation tests are performed in order to check that the local variables are written without serializability and hence are known as validation phase.
3. Write Phase: Validation and updation are applied to the database in the last phase and in case of any conflict the transaction is rolled back [1].

The order of phases for each transaction remains the same. No deadlocks occur in optimistic scheme as the records are not locked [2].

Multiversion Schemes—There are basically two schemes to two basic schemes to make sure of serializability. They are either an operation delay or transaction abortness altogether. For example, in case of not being written yet a read operation may be delayed or it might gets rejected since the value which was going to be read is overwritten. But if old copies of data are on the system, such situations can be avoided [7]. In a multiversion scheme, a new copy of X is created every time a write is performed on X. When there is a single operation to read the X, a single type of X is selected. This method is more flexible in controlling the order in which read and writes take place as the command of write cannot be overwritten since read may also read any version. In this type of approach, many versions of data item are kept and only the correct version is read by the read operation [6]. This scheme never rejects a read operation but it certainly requires more of storage (both primary and secondary) in order to store multiple versions. Two-phase locking and time stamping are mostly used with this scheme as it needs to perform a cleanup action in order to prevent unlimited growth of versions [7].

1. Multiversion Timestamp Ordering: When different versions of data item are made, timestamps are used to tag them. Whenever a read operation is made by the transaction, it does not have to wait; a version of data item is selected on the basis of the transaction's timestamp and is returned immediately [2]. When we write a data value X, we may want to keep both the versions instead of overwriting the old one in such case if a read request is made for X, we may supply any of the value which may support serializability the best [3, 5]. Hence, in the end, each data item X has a sequence of versions <X1, X2, …, X_m> where there are three fields in each version: the actual value of the version known as the content, the timestamp of the transaction by which the version X_k was created denoted by W-timestamp (X_k), the largest timestamp of the transaction by which the version X_k has been successfully read denoted by R-timestamp(X_k) [7] Transactions before coming in deliver the most recent version and hence when a transaction tries to write to a version that some other transaction has read, then that write operation does not succeed [2].
2. Two-Phase Locking (Multiversion): It was basically created to use the benefits of both concurrency control (multiversion) and two-phased locking protocol. These protocols work differently for both read-only transactions and updated transactions [2]. Update transaction is made to follow rigorous two-phase locking, i.e., they must hold every lock regardless of whether it is read or write till the end of the transaction [6]. In this type of locking mechanism, only two versions of the data item exist. One that was created by a transaction that has been committed and second that is created by the transaction that has got a lock over the data item. The value of data item that has been committed may be read by other transaction even while the other transaction is holding a lock over it. The transaction that has the lock over the data value can write the value to data item at any point without actually upsetting the committed value of the data item but before committing it is must to have lock which is certify for all other data item that is hold by it currently but certify locks are not yet accustomed with the lock of read, hence it is essential for a transaction to delay its commit so that all the reading operations can be completed and a certify lock can be obtained [4]. As discussed above, the multiversion two ways locking uses a rigorous two way locking protocol and hereafter they can easily be serialized on the basis of their commit level.

2 Methodologies

We have set up an environment of concurrency control using C++. For randomly storing data, any data structure can be used. Furthermore we have randomly generated several transactions. Each and every transaction generally has only either two operations, i.e., write or read which are also generated randomly. Whenever a transaction enters the system, it applies either read or write operation on some data. But there are other transactions running too simultaneously that also want to apply the same operation on the same data. We have used various protocols, and the methodology of protocols used is given below:

Two-phase Locking—Whenever an operation of transaction gets executed firstly, it is been checked that if it has a data on lock or not. If there is no lock present on the data, a lock is applied to it then. And this lock may be any of the two locks, i.e., Exclusive—Write lock or Shared—Read lock. If the operation by both of them gets lock, i.e., wither (X, X) or (S, S) or (S, X) or (X, S), execution of the transaction begins. Writing and reading of data take place while unlocking of data is taking place. Otherwise, if a lock is found by any of the two operations then the operation has gone for wait. In the process of unlocking, if timeout occurs then it goes for rollback. There are few cases in which cascading rollback too may take place. This means that the moment the transaction is being roll backed, the other transaction which has applied operation on exactly the same data shall be roll backed too otherwise commit of the transaction will be done. During the execution, we executed some transactions roughly 50 times each containing 10 operations; there were only 90 transactions that were committed while others were either in rollback or in wait as shown in Table 1. There was no serializability conflict in this protocol. However, overheads of lock were found by us.

Table 1. Two-phase locking (observations)

Transactions:	50 × 10 = 500
Rollback	185
Committed	90
Wait	225

Time Stamping—As soon as a transaction enters into the system, it is being assigned with a timestamp. In order for a transaction to read the data, its timestamp must be greater than that of the last write's timestamp only then the read operation can be executed successfully and the read timestamp is assigned to the transaction, otherwise a rollback takes place. Similarly, in order to write data, a transaction's timestamp [8] must be greater than that of the last read and write timestamp, only and only then the read is successful and the transaction's timestamp is assigned the value of the write timestamp, otherwise a rollback takes place. Sometimes a case arises where a transaction's timestamp is less than the write timestamp in such cases the transaction is ignored. When transactions with 10 operations each were executed an odd 50 times, we found only 144 were committed while a huge number of transactions rolled back as shown in Table 2. Even though there was no waiting but the number of rollbacks was quiet huge altogether. Suppose if a transaction is rolled back, then all the transactions that have used a value that was modified by the rolled back transaction (directly or indirectly) must be forced to roll back as well. This is what we call cascading rollback. Even though this protocol produces no conflict serializability but the amount of cascading rollbacks is too high. And even overheads such as read and write timestamp also come into place.

Table 2. Time stamp (observations)

Transactions:	50 × 10 = 500
Rollback	356
Committed	144
Wait	No waiting

Validation Based (Optimistic)—As discussed before, this technique is called optimistic as it works on the assumption that no interference will take place. This protocol works best for read-only transactions where no conflicts are found. First, we let all the transactions to execute locally and then we check them for conflicts using validation tests. And in case any conflict is found we use timestamps to resolve it. A transaction is allowed to execute in two cases, first if we find that the former operation is the ending timestamp of the transaction, and second if the latter operation is the starting timestamp and the former as usual the ending timestamp, otherwise the transaction is forced to roll back. When transactions with 10 operations each were executed roughly a 50 times, we found that 162 transactions out of which most were read only were committed. We also found that there were 0 conflicts and the numbers of commits were more as shown in Table 3. But, even if a single conflict occurs, it is accompanied by a lot of rollbacks. In this type of concurrency controls, all the validation is done at the end. During the phase of validation, if there is a little interference it is validated easily but if these interferences or conflicts increase then many transactions which already have completed their execution will have to be restarted and hence this concurrency technique is not good for such cases.

Table 3. Optimistic (observations)

Transactions:	50 × 10 = 500
Rollback	338
Committed	162
Wait	No waiting

Multiversion: In this technique, a new version of data item is created after every write operation that is successful if its timestamp is more than that of its last write operation, in which [9] case the value of the data item is overwritten. Read operation is always a success as some version of data is always available. When transactions with 10 operations each were executed roughly a 50 times, we found that 333 transactions out of the lot were committed while others were rolled back as shown in Table 4. An interesting finding was that the number commits exceeded the number of rollbacks. Even though the storage required was more but the conflicts which were solved by rollbacks wasted less time and hence this technique would be more suitable for large databases.

Table 4. Multiversion (observations)

Transactions:	$50 \times 10 = 500$
Rollback	167
Committed	333
Wait	No waiting

3 Comparisons

3.1 Performance Comparison

It was seen that at one point where for an update of intensive transactions, locking protocols are good but on the other hand if we are dealing with transactions with read-only operations then optimistic protocols are the best choice due to the fact that they have no useless overheads of locking data items. The performance gets somewhat hampered if there are two transactions which are incompatible because of the blocking that takes place and if a deadlock occurs the performance gets degraded even more due to thrashing [3, 9]. In such cases, timestamps are of great help as they help in determining the younger–older relationship and yield better results by improving concurrency [3]. Here, locking and optimistic approach act quiet differently, at one point where locking controls transactions by making them wait, optimistic approach controls transactions by backing them up. As discussed before, in multiversion scheme read operations are always allowed but it needs large secondary storage to save the multiple versions and hence is more suitable for systems with large database space.

3.2 Serializability

Locking is the best-suited technique in order to ensure serializability regardless of the type of transaction. It is even suitable for update-intensive transactions [1]. Timestamps can also be used to ensure serializability as they give a way to order the transactions in case of any conflict [5]. Same data items can be read by the transactions and that too without too many conflicts [5]. On the other hand, optimistic approach works on the assumption that there are no conflicts among transactions and even if there are any, they are very rare and hence it does not need any locking. Therefore, all the transactions are first validated and then executed after the serializability order has been decided. And the numbers of rollbacks reduce drastically if most of the transactions are read only.

The multiversion two-phase locking protocol serializability is given using the advantages of both multiversion and two-way locking approach, thereby providing the user with dual benefits and better serializable schedules [2].

4 Analysis and Result

We have framed our results in Table 5 which is given below:

Table 5. Transaction (average number) for various concurrency control techniques evaluated

Name	Number of times transaction is executed	Number of transaction in each execution	Transaction (rollback)	Transaction (Committed)	Transaction (Wait)
2PL	50	10	185	90	225
Timestamp	50	10	356	144	–
Pessimistic	50	10	156	344	–
Optimistic	50	10	338	162	–
Multiversion	50	10	167	333	–

The data shown above is randomly generated data as we have already discussed where even the write and read operations were also performed randomly on data. Five hundred numbers of transactions [10] were generated for the results and 50 individual run for each concurrency techniques where each execution has 10 transactions each. Based on this, we have calculated our results for no. of committed, wait, and rollback transactions for each and every technique: 2pl, Optimistic, Pessimistic, Multiversion, and Timestamp (Figs. 1, 2, 3, and 4).

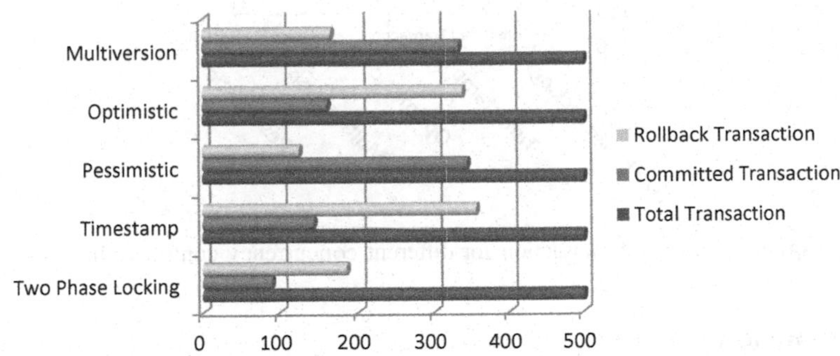

Fig. 1. Comparison of various concurrency control techniques

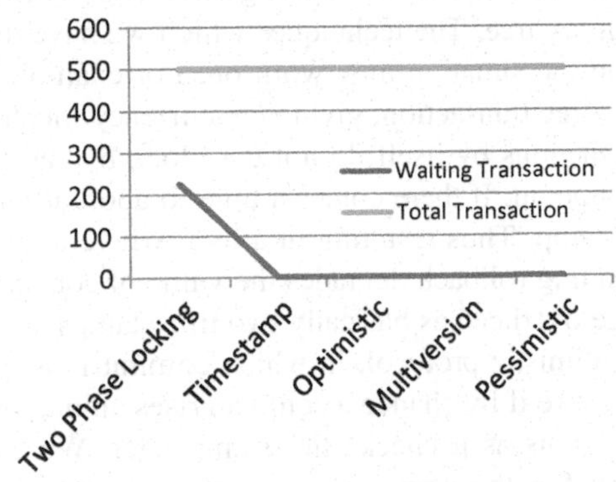

Fig. 2. Wait transaction for different concurrency control techniques

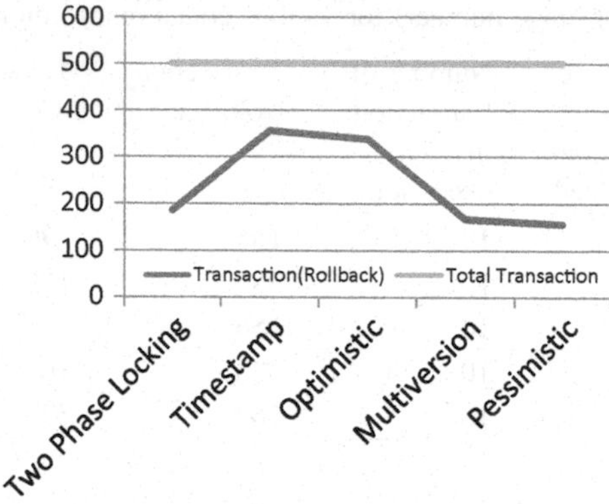

Fig. 3. Rollback transaction for different concurrency control techniques

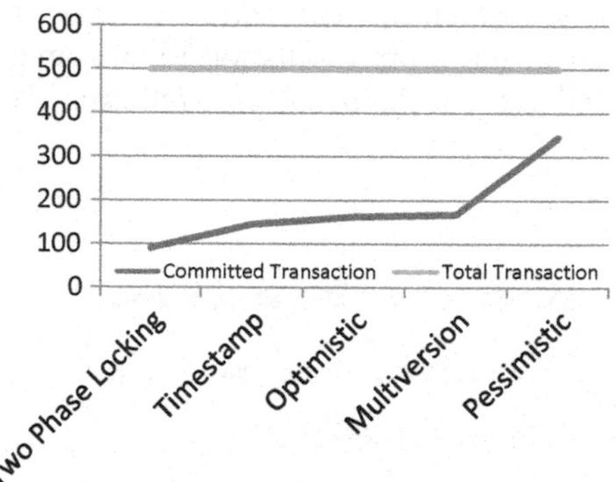

Fig. 4. Committed transaction for different concurrency control techniques

5 Conclusion

Serializability is followed by locking protocols despite the transaction type. Update tensive applications can use them despite having various overhead of locking and they are generally not deadlock free. The techniques which we have discussed in this paper are good depending on the situation they work on. For example, in timestamp protocols, there is a conflict-free transaction, giving concurrency much better than the phase locking since the transactions by itself do not get block by each other but they suffer from a huge rollback amount. If there comes a time to abort a transaction, then it must start with a new timestamp. Thus resulting in a loop where a transaction gets aborted every time. Even cascading rollback degrades the value of concurrency. There is also a disadvantage of storage overhead as basically two timestamps are maintained for every object. If we go for optimistic protocol, in which commit is only completed after the phase of validation because if by chance a conflict arises among the transactions then it can abort other transactions as it checks timestamp later. We can, in fact, use some additional information for the transaction as if we do not have the additional information then two-phase protocol can be used for serializability of conflict. In multiversion approach, an operation of read is never ever rejected and it follows serializability too. Thus, based on the various analyses, we were able to conclude the best situation where we can use the different concurrency control techniques efficiently. We also concluded that most of the times multiversion technique is the best approach possible.

References

1. JongBeom Lim, Young Sik Jeong, Doo-Soon Park, HwaMin Lee, "An efficient distributed mutual exclusion algorithm for intersection traffic control".
2. Philip A. Bernstein, Nathan Goodman, "Timestamp-based algorithms for concurrency control in distributed database systems" in Proceeding of the sixth international conference on Very Large Data Bases - Volume 6.
3. C. Mohan, Don Haderle, Yun Wang, Josephine Cheng, "Single table access using multiple indexes: Optimization, execution, and concurrency control techniques" in Volume 416 of the book series Lecture Notes in Computer Science (LNCS).
4. Michael Wei, Amy Tai, Chris Rossbach Ittai Abraham, "Silver: Ascalable, distributed, multi-versioning, Alwaysgrowing(Ag)FileSystem".
5. C. MOHAN, "ARIESIKVL: A Key-Value Locking Method for Concurrency Control of Multiaction.Transactions Operating on B-Tree Indexes" in Proceedings of the 16th VLDB Conference Brisbane, Australia, August 1990.
6. Sonal Kanungo1, Morena Rustom. D2, "Analysis and Comparison of Concurrency Control Techniques" in International Journal of Advanced Research in Computer and Communication Engineering Vol. 4, Issue 3, March 2015.
7. Praveen Kumar, Dr. Vijay S. Rathore "Improvising and Optimizing resource utilization in Big Data Processing" in the proceeding of 5th International Conference on Soft Computing for Problem Solving (SocProS 2015) organised by IIT Roorkee, INDIA (Published in Springer), Dec 18–20, 2015. PP 586–589.
8. H. T. Kung, "On optimistic methods for concurrency control" in ACM Transactions on Database Systems (TODS) Volume 6 Issue 2, June 1981.
9. Sheril Yadav "Analysis and Implementation of Business Intelligence Software for Report Bursting" International Conference in Smart Computing & Informatics (SCI-2017) held in ANITS, Visakhapatnam March 2017.
10. Rashmi Priyadarshi, Seema Rawat, Praveen Kumar An implementation of opinion mining using fuzzy inference system in Computational Intelligence on Power, Energy and Controls with their impact on Humanity (CIPECH), 2014 Innovative Applications in 2014.

Inspection of Fault Tolerance in Cloud Environment

Deepanshu Jain(✉), Nabeel Zaidi, Raghav Bansal, Praveen Kumar, and Tanupriya Choudhury

Amity University Uttar Pradesh, Noida, India
{deepanshujain002, raghavbansal95}@gmail.com, nabeelzaidi@ymail.com, {pkumar3, tchoudhury}@amity.edu

Abstract. Cloud environment is a set of various types of software and hardware that are connected with each other and works collectively to provide various services to the user as an online utility. It basically is the efficient use of hardware and software to work coherently to deliver services. Through the use of cloud computing, users are able to access files from almost anywhere or any device that have access to Internet. We are already familiar with the market of cloud computing, and how everyone is shifting to cloud because of the benefits it provides. But many of them are very less aware about the situations when there comes a failure. The task of facing the failure is not limited to the cloud providers but also to the customers. They must know what can be done when such a situation arises. Fault tolerance is basically a property that makes the system to work properly even though there is a failure. In order to be more robust and dependable, failure should be handled effectively. This paper deals with the Inspection of various fault tolerance technologies that are available. There is no existing algorithm that considers reliability and availability in fault tolerance as well. We tried to consider these things when discussing about fault tolerance. Furthermore, a brief analysis of the already proposed FTC model with some new functionally is also presented.

Keywords: Cloud · Replicating · Scheduling · Cluster

1 Introduction

In order to provide a network access to a shared pool of resources which can be enabled/disabled on demand and is universal as well as convenient, a special kind of framework is used which is known as cloud computing. Using cloud computing, we provide access to the least amount of management and very less or no interaction with the service provider [1]. One of the basic processes of using cloud as service is to be able to provide resource processing for which scheduling is necessary. Every cloud application is designed keeping business processes in mind and includes a set of abstract functions or services. And to be able to process the tasks, the system needs to allocate both the resources as well as the tasks that come to the resource. And to ensure the Quality of Service also known as QoS, there must be a Service Legal Agreement

V. Bhateja et al. (eds.), *Information Systems Design and Intelligent Applications*, Advances in Intelligent Systems and Computing 672,
https://doi.org/10.1007/978-981-10-7512-4_103

(SLAs) in place [2]. The National Institute of Standards and Technology states that the Cloud Model consists many service and deployment models along with some key essential characteristics [1].

1.1 Essential Characteristics

1. On-demand self-service—An end user can automatically get new services granted to it, may it be server or storage and that too without any interaction with the service provider [3].
2. Broad network access—The services provided can be used across many platforms ranging from thin platforms such as mobile phone to thicker platforms such as workstations and laptops.
3. Resource pooling—All the resources provided by the service provider are pooled together to serve multiple users at the same time by dynamically allocating resources according to the demand of the consumer who has no knowledge about the exact location of the end user unless it is at a higher abstraction [4] level.
4. Rapid elasticity—Services and resources provided to the users can be granted and released automatically in order to compete with the scaling may it be outward or inward. From the customer's point of view, the amount of resources may seem to be unlimited which can be granted to them at any point of time.
5. Measured service—Cloud system has a metering capability at a certain level to abstraction which can be used by them to monitor the resources, optimize the resources, and controlling them. This process is transparent to both the service provider as well as the end user.

1.2 Service Models

1. Software as a Service (SaaS)—The consumer is given the capability to use the service provider's application which is running on the existing cloud infrastructure. The client can get access to the application from various devices using Web browser as a thin interface or any other program interface. The infrastructure of the cloud is not controlled by the end user with the exception any configuration setting, that is, provided in the application itself.
2. Platform as a Service (PaaS)—The consumer is given the ability to deploy the applications that the consumer has created using tools provided by the service provider or acquired onto the cloud. The consumer has control only over the application, its configuration setting, and the environment over which the application is hosted and has no control over the cloud infrastructure.
3. Infrastructure as a Service (IaaS)—The consumer is provided with many capabilities such as storage, provisional processing, and networks among other resources using which the consumer is given the ability to run and deploy many software including operating systems. The only thing consumer can control are the resources that are provided to them and the applications they deploy, and they have no control over the infrastructure of the cloud.

1.3 Deployment Models

1. Private cloud—This type of cloud infrastructure may be owned by a particular organization and is used exclusively by the members of that organization only. It can be under the ownership of the organization or a third party or a combination of both.
2. Community cloud—This type of cloud infrastructure is used by the members of a specific community from an organization that have similar concerns. It also maybe be under the ownership of the organization to which the community belongs or a third party or maybe a combination of both. This type of infrastructure may exist either on premises or off premises.
3. Public cloud—This type of cloud infrastructure is not made for any exclusive use and is open to all. It can be under the ownership of a business, government, or any academic organization that are also responsible for its maintenance. It always exists on the premises of the organization providing the service.
4. Hybrid cloud—As the name suggests, this type of cloud infrastructure is a combination of two or more types in which the unique attributes is that each infrastructure exist, and they are bounded together by some standardized technology which allows data and application portability.

Since cloud computing allows its users to have access to resources and services for as long as they need them and that too without too many interactions, it is becoming perhaps one of the fastest growing grid technology [1, 5]. Cloud computing focuses on sharing of information among its various customers by putting up the same information on the grid of nodes [6]. For the management of resources, the following aspects must be kept in mind:

1. Infinite resources that must be made available to the consumer on their demand.
2. No commitment from the users in advance.
3. Last but certainly not the least would be the demand of high-end computing resources and that too for very short duration of time.

Implementation of fault tolerance is very important from the perspective of the service provider since it rescues the lost, improves the performance, and can also be used in failure recovery, and in order to achieve it, some mechanisms such as redundancy and replication can be used [2]. In real-time computing applications, adding a cloud infrastructure does not only mean increasing the chances of error, but it also increases the cost as the resources required to keep the replicated data increase [5]. And realizing the power of cloud, some cloud providers have even started to give real-time cloud support since the power of cloud can prove to a great added benefit to real-time applications [1]. Cloud computing infrastructure usually comprises of interconnected data centers and infinite resources which are provided to the consumer as a part of an on-demand service [7]. In order to get reliable software, fault tolerance is a necessary demand but it proves to be great difficulty to design an integrated solution since many of the user's applications are deployed on the cloud infrastructure only. The major cause of this difficulty is the complexity of the system and multiple abstraction layers due to which only limited data is available [7].

2 Resource Manager

The service provider has to keep a consistent sight of all the systems so that the resources can be assigned to each client request systematically. Hence, a database of catalogue which consists of the current state of resources is maintained by the resource manager whose main function is to observe the current state of all virtual as well as physical resources. The resource manager, therefore, keeps catalogue of each and every machine in its database along with other information of the system such as its serial number, speed of the processor, and the date the resource was issued. [7]. Given below is Fig. 1 of resource graph G(N, E) of the cloud infrastructure [8, 9] consisting of two clumps of three nodes each which are connected via network switch. In the graph, processing nodes n∈N are represented by the vertices, and to represent virtual connection between two nodes, edges of the graph are used, i.e., e∈E. Here, each node keeps the information about the virtual machine that is kept at that particular node in the vertex. Each edge and node can be further categorized into three classes which are: working(W), completely faulty(F), and partially faulty(F). Each node is marked with a particular class depending on its current state. If it is an ordinary state, it will mark W. If a failure has occurred such that the node cannot be recovered back to a normal state, it is marked as F, and the node which is currently not in use is marked as P [7].

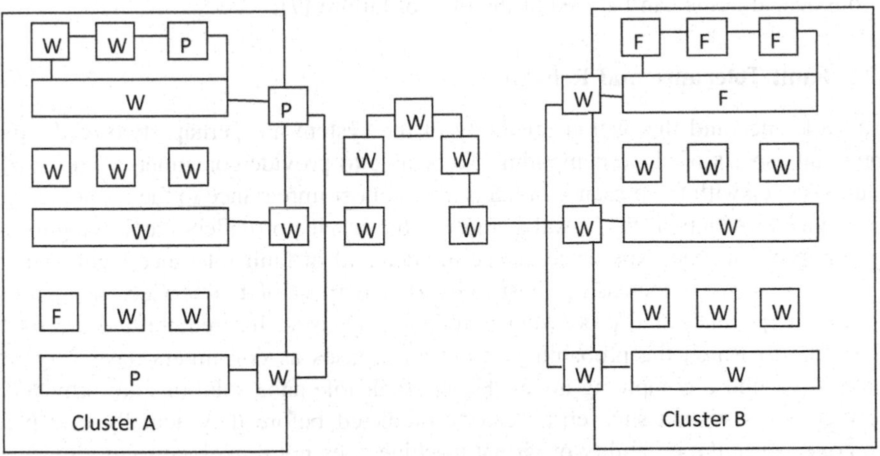

Fig. 1. Graph generated by the resource manager

Resource manager plays a very vital role in providing stability to the cost of the resource and toward the performance of the fault tolerance method that is used by the service provider [7].

3 Fault Tolerance in Cloud Computing

3.1 Basic Notation to Fault Tolerance

When a client needs fault tolerance services, he/she is enlisted by the service provider. Then, based on the requirements of the client, the service provider [10] creates a solution. While creating the solution, the balance between the following aspects must be kept in mind:

1. Fault Model—It defines the maximum capacity of the solution to handle faults and loss. In order to specify this aspect, the ability of the system to handle protocols of failure detection and the technology used is of grave importance.
2. Resource utilization—It defines the amount of resources that will be utilized in order to understand the fault. This feature is inbuilt in the system along with harsh level of failure detection and recovery.
3. Performance—The performance of any failure tolerance method is defined by the effect of the solution on the overall quality of service both at the time of failure and at times when no failure is there.
4. Redundancy—This technique is the most commonly used way of dealing with failures in a system. A failure tolerance model based on this technique replicates the components of the analytic system with the help of other resources so that these duplicated items can be used at the time of failure [7].

3.2 Fault Tolerance and Reliability

Fault tolerance and the overall reliability of the system are perhaps two of the most important aspects of cloud computing. To be able to provide consumers with the right solutions even with the presence of faults is of utmost importance to the consumers and the service providers as well. And as most of the service providers are moving toward providing a real-time experience, hence the demand of fault tolerance techniques for real-time systems is increasing drastically. But in most of the services that provide real-time experience, the processing is done on systems that remote and are on the cloud due to which the probability of error increases as consumers have very little control over those computing nodes. Hence, fault tolerance solutions are provided to the consumers so that such errors can be predicted before they actually take place. Moreover, even the reliability of virtual machine does not remain constant; it changes after each and every computing cycle [5].

Fault tolerance is done in two phases; first phase is "Effective Error Processing" in which the system is brought back to a state before the error took place, i.e., the dormant state, and the second is "Latent Error Processing." Real-time systems are characterized by two main features which separate them from any other kind of system, and these are timeliness and fault tolerance. Timeliness means that every task must complete its execution in the given time period, and fault tolerance means that the system must continue to work even if any fault arises [1]. Hence, with the growing popularity of cloud computing, service providers have come with a new design approach in which fault tolerance mechanism is provided as a module-independent service so that this service can be provided to all the users and by each and every module transparently [7].

In order to achieve fault tolerance, a set of design techniques and algorithms are applied to increase the overall dependability of the system. With new upcoming technologies, new applications arrive, and hence, new fault tolerance solutions have to be introduced as well. Earlier specific hardware and software used to be made in order to provide fault-tolerant execution of tasks but the new microprocessor chips are highly complex, and all the hardware and software are made according to pre-defined norms which are economically feasible as well. Hence, many new techniques have come up to the surface in the field of fault tolerance such as using the existing technique with RAID disks where all the information is divided among many disks, and this improves the bandwidth. Moreover, an extra disk stores the encoded information to restore the data in case of system failure. Fault tolerance techniques are also being used in parallel computers in order to detect faults and errors. Fault tolerance techniques are becoming

Table 1. Comparison of fault tolerance strategy [1]

S. No.	Strategy	Fault-tolerant technique	Programming framework	Environment	Faults detected
1.	Nicolae and Cappello (2011)	Disk-based Checkpoint	MPI	IaaS cloud	Node/network failure
2.	Hakkarinen and Chen (2013)	Diskless-based Checkpoint	NA	HPC	Process/application failure
3.	Kwak and Yang (2012)	Checkpoint	Probability analytic framework	Real-time systems	Process failures
4.	Goiri et al. (2010)	Checkpoint	Java	Virtual machine	Node failure
5.	Malik et al. (2011)	FTRT (Adaptive)	–	Real time	–
6.	Sun et al. (2013)	DAFT (Adaptive)	Java	Large-scale cloud	Works on historical failure rate
7.	Cogo et al. (2013)	FITCH (Adaptive)	Java	Large-scale cloud	–
8.	Zhang et al. (2011)	BFT Cloud (Adaptive)	Java	Voluntary resource cloud	Byzantine problems
9.	Zhao et al. (2010)	LLFT (Adaptive)	C++	Middleware	Replication faults
10.	Ko et al. (2010)	IFT(Adaptive)	Hadoop	Hadoop	Intermediate data faults
11.	Zheng (2010)	MFTLL (Adaptive)	MapReduce	MapReduce	Replication faults, stragglers detection
12.	Pannu et al. (2012)	AAD (Adaptive)	–	Local cloud	Discovers future failures

more famous day by day especially in sub-micron VLSI in order to solve major problems such as noise and improving the overall yield of the system by increasing its ability to process even with faults.

We have compared the various fault tolerance techniques which are quite famous ones as shown in Table 1.

4 Related Works

The model is already proposed and is called fault tolerance in cloud computing (FTC) [1]. We are going to analyze it more deeply and give more functioning to the model. This model tolerates the faults based on the reliability each node has. Each node to be executed is taken to be a real-time application. The model is shown in the figure given below. Here, we have "N" computing nodes or virtual machines each of which is running a different algorithm. Further, we perform an acceptance test (AT) whose result is then forwarded to the adjudication node to take a decision regarding it [2, 5]. Here, two distinct nodes are displayed, one of which contains some virtual machines on the cloud infrastructure and running different algorithms to handle real-time applications. The proposed algorithm then supplies the result again for an acceptance test to see whether it is logically valid or not. The test modules are similar to each other in every sense. If the results are valid, only then they are passed to the time checker module; otherwise, the AT modules send an exception signal reflecting the reason. The proposed scheme just not provides forward recovery but sometimes also can provide backward recovery. The other node is the adjudication node which comprises of three separate components: First, the Time Checker [TC] Module, it checks the timing of the results produced as it contains a timer that records the timestamp at which each result was produced [11]. Second is the Reliability assessor (RA) module which is used to check the reliability of each computing node. The final component is the Decision mechanism (DM) module which selects the best output based on their reliabilities.

As a cloud infrastructure consists of more than one grid, it becomes hard to tackle various security issues like confidentiality and integrity [6]. In [12], a fault-tolerant

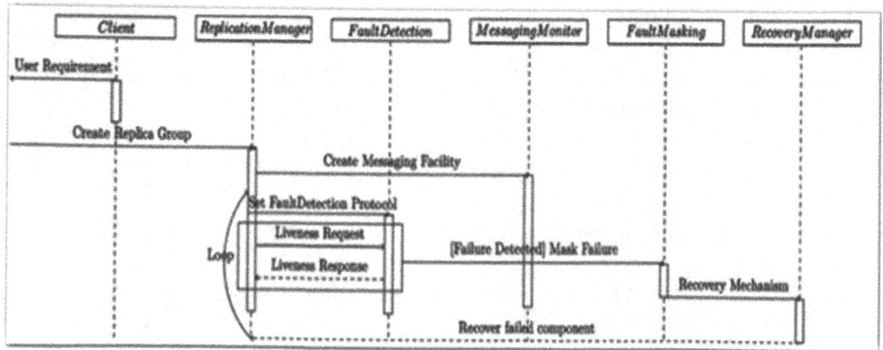

Fig. 2. FTM Kernel [1]

middleware is being proposed by the authors that can use replication to handle the faults in real-time cloud applications. In [7], in order to build a fault-tolerant protocol, we propose to use micro-protocols and use them in hierarchical order to form a system. In [6], to develop a fault-tolerant framework that is proactive as well, we propose to use a modular approach that can incorporate a requirement-specific strategy as well (Fig. 2).

The model already proposed can be used in real-time applications based on cloud infrastructure to handle the faults since the model can tolerate faults to a high extent. It has high reliability which can be dynamically configured. Moreover, we tried to introduce an approach that can be used in order to recognize generic fault-tolerant mechanisms as independent modules, the properties of each mechanism have been validated, and the user requirements have been matched with the available fault-tolerant mechanisms.

5 Conclusion and Future Work

A great analysis and design techniques are applied to create improved systems in fault tolerance. Almost every day, a new technology and applications are being developed so there is a need for new approaches to fault tolerance. Previously, it was quite easy to craft a specific hardware and software for a solution, but as technology is getting more advanced, it is getting complex too to apply a solution. Thus, there is a great deal of current research focusing on implementing fault tolerance. In this paper, we have successfully inspected the fault tolerance technology. We have discussed it with reliability and availability, and also, we have added new functionality to FTC-proposed model.

In future, we look forward to implement the proposed framework in order to measure the strength of the fault tolerance approach. Work is also proposed to create a new module called Resource Awareness Module or RAM that can help the cloud service provider's scheduler to schedule decision that is based on the characteristics of the infrastructure of the cloud system.

References

1. Dilip Kr Baruah, Lakshmi P. Saikia, "A Review on Fault Tolerance Techniques and Algorithms in Cloud Computing Environment", in International Journal of Advanced Research in Computer Science and Software Engineering Volume 5, Issue 5, May 2015.
2. Michael K. Reiter and Avishai Wool, "Probabilistic Quorum Systems", Information and Computation 170, 184–206 (2001).
3. Apolinar González-Potes, Walter A. Mata-López, Vrani Ibarra-Junquera, Alberto M. Ochoa-Brust, Diego Martínez-Castro, Alfons Crespo, "Distributed multi-agent architecture for real-time wireless control networks of multiple plants".
4. Zohaib A. Faridi, S. Rawat "Analysis and proposal of a novel Approach to collision detection and avoidance between moving objects using Artificial Intelligence" 5th Fifth International Conference on System Modelling & Advancement in Research Trends in TMU, Moradabad (UP) 25–27 Nov. 2016.

5. Alain Tchana, Laurent Broto, Daniel Hagimont, "FaultTolerant ApproachesinCloudComputingInfrastructures", in ICAS 2012: The Eighth International Conference on Autonomic and Autonomous Systems.
6. John D. Slingwine, Paul E. McKenney, "Apparatus and method for achieving reduced overhead mutual exclusion and maintaining coherency in a multiprocessor system utilizing execution history and thread monitoring".
7. Hagit Attiya[2], Alla Gorbach[3], Shlomo Moran[4], "Computing in Totally Anonymous Asynchronous Shared Memory Systems", Information and Computation Volume 173, Issue 2.
8. Praveen Kumar, Dr. Vijay S. Rathore "Improvising and Optimizing resource utilization in Big Data Processing" in the proceeding of 5th International Con-ference on Soft Computing for Problem Solving (SocProS 2015) organised by IIT Roorkee, INDIA (Published in Springer), Dec 18–20, 2015. PP 586–589.
9. Seema Rawat, Praveen Kumar, Geetika, "Implementation of the principle of jamming for Hulk Gripper remotely controlled by Raspberry Pi" in the pro-ceeding of 5th International Conference on Soft Computing for Problem Solving (SocProS 2015) organised by IIT Roorkee, INDIA, Dec 18–20, 2015. PP 199–208.
10. Sheril Yadav "Analysis and Implementation of Business Intelligence Software for Report Bursting" International Conference in Smart Computing & Informatics (SCI-2017) held in ANITS, Visakhapatnam March 2017.
11. Michael Wei, Amy Tai, Chris Rossbach Ittai Abraham, "Silver: Ascalable, distributed, multi-versioning, Alwaysgrowing(Ag) FileSystem".
12. HyoJong Lee, Shwetha Niddodi, David Bakken, "Decentralized voltage stability monitoring and control in the smart grid using distributed computing architecture" published in Industry Applications Society Annual Meeting, 2016 IEEE.

Analytical Planning and Implementation of Big Data Technology Working at Enterprise Level

Pooja Pant[1], Praveen Kumar[1(✉)], Irshad Alam[2], and Seema Rawat[1]

[1] Amity University Uttar Pradesh, Noida, Uttar Pradesh, India
ashima81k@gmail.com, {pkumar3, srawat1}@amity.edu
[2] ITM University Delhi, New Delhi, India
crytic_eye@live.com

Abstract. Today, a number of technologies are available for building of Big Data architecture deciding which technology will provide the maximum value out of the architecture depending on the extensive study of the present architecture in use, the type of data being ingested, and the desired value expected by the enterprise. In this research paper, we highlight Big Data and its characteristic, Big Data architect, various technologies involved at different levels, pipeline architecture of Big Data architecture, technologies used, planning and designing, and challenges faced while building Big Data architecture at enterprise scale.

Keywords: Big data · Apache Spark · Apache Ranger · Apache Kerberos Apache Hive · Apache Sqoop · Apache Flume

1 Introduction

Big data is a problem statement for every enterprise. Big Data is not only just large volume of transactional and non-transactional data rather processing this large volume of data. Getting the value out of this large amount of data is what makes it top the priority list of enterprises around the globe.

Building Big Data architecture requires building an entirely new data processing platform and a new architecture pipeline [1]. For optimum solution, it requires expertise at every level of the component architecture. Designing and optimizing require extensive study of what the enterprise has and what it wants. Getting the value out requires proper understanding of every technology and practice.

In this paper, we discuss the technology at every phase of the Big Data architecture pipeline. In Sect. 2, we discuss what Big Data is and why enterprise needs Big Data architecture. Section 3 provides a brief overview of the architecture of some Big Data technologies extensively used in the market. In Sect 4, we discuss planning for Big Data framework. Section 5 provides a pipeline architecture for Big Data. In Sect. 6, we discuss the challenges faced during designing and implementing Big Data architecture.

V. Bhateja et al. (eds.), *Information Systems Design and Intelligent Applications*, Advances in Intelligent Systems and Computing 672,
https://doi.org/10.1007/978-981-10-7512-4_104

2 What Is Big Data?

Doug Laney proposed the three V's for defining Big Data [2]. According to him, data with different data structure being generated at high rate is called Big Data. This concept of 3 V's was followed for over a decade, but sooner the researchers realized that the 3 V's were not sufficient for complete description of Big Data. Many new dimensions have been added to bring about the complete definition of Big Data. Big Data brought about the real value of data which was stored in enterprise databases. Organizations use Big Data for analysis and advance analytics. Big Data has a wide range of applications from banking, medical researchers to stock rate processing.

3 Big Data Technologies

Every Big Data architecture design in any organization is based on a pipeline design. This pipeline shows the connection among various technology components which describes various processes from capture of data, data processing to data visualization. The following are the various components and technologies of Big Data pipeline. There are a number of open source and paid technologies which can be installed in Big Data architecture. The most commonly used technologies in enterprises have been elaborate in this section.

3.1 Data Ingestion Technologies

Capturing the data generated in large volume and with high velocity requires technologies like Kinesis Fire hose, Apache Kafka, Apache Flume for capturing the data and storing it into database. These technologies enable ingestion of unstructured and semi-structured data into Big Data Architecture for further processing

Apache Flume

Apache Flume is the most widely used data ingestion application for enterprise. It is capable of capturing raw data from various sources and different formats. Apache Flume architecture comprises of agents, sources, sink, and channel [3]. A source can be another agent sink, client, or apache logs. The agent receives the event from the source and sends it to the sink [4] as shown in Fig. 1. There can be multiple data sources capturing or generating data at different velocity, managing the velocity, and transferring the information to the sink are responsibilities of the channel. The sink can also be in storage phase or computational phase.

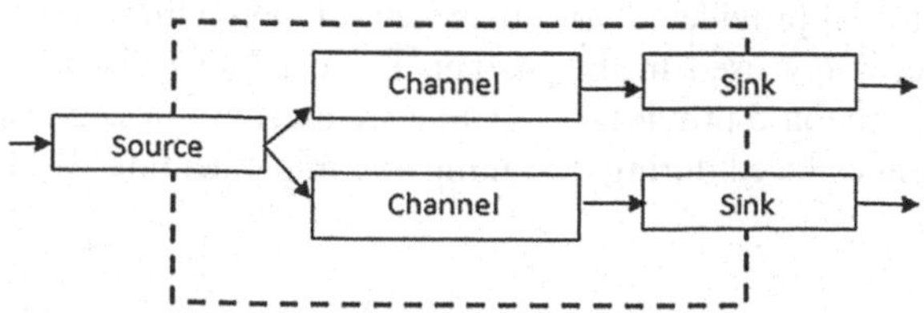

Fig. 1 Apache Flume architecture

The data being transferred to the sink passes through the channel that does not discard the data until it receives an acknowledgment from the sink, hence reducing the probability of data loss even in situation when there is failure of a sink. This feature of Apache Flume makes it highly reliable for data ingestion. Data needs to be thoroughly studied before using an agent. Different data structure different agents. A single agent can only be used by two sources if and only if both the sources have the same data structure and data format. Flume follows six protocols for the extraction of semi-structured and unstructured data.

Apache Kafka

Apache Kafka is a distributed messaging system which comprises of a producer and a consumer. The producer creates the message and sends it to the consumer as shown in Fig. 2. The message created has an offset value at its header. The data is sent through the partition, and a replica of the message is kept [5, 6]. It is the responsibility of the broker to maintain the link and the flow of the message. When the message is received by the consumer, it sends a complete status back to the broker, and after receiving this complete status, the replica of the message created is destroyed.

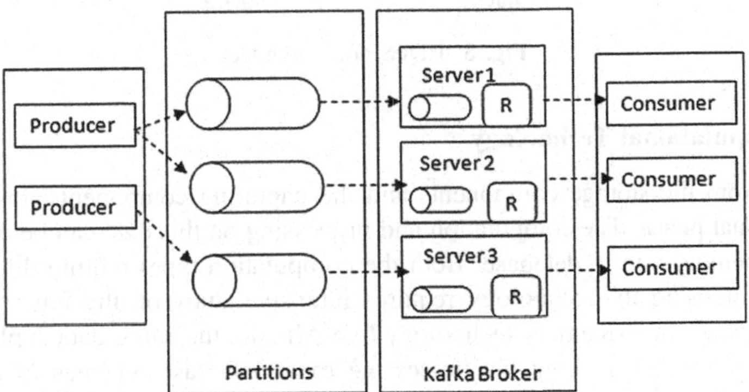

Fig. 2 Apache Kafka architecture

3.2 Storage Technology

The data stored in the legacy database and the live streaming is possible using distributed storage technologies HDFS is the most widely used distributed data storage technology present in the market which provides high availability and low hardware cost.

Hadoop Distributed File System

HDFS is a distributed file system which uses the hierarchal way of data distribution and storage. HDFS follows master slave architecture [7]. The data is distributed among various data node each of which has three replicas which will be used in case one data node fails. The data node is the slave node in HDFS architecture as shown in Fig. 3. The name node acts as the master to the name node. The name node contains the Meta data about the data nodes. There exists a replica of the name node known as the secondary name node. This secondary name node gets updated along with the name node. The name node acts not only as the master to the data node but also as a slave node to the general node; this node contains the Meta information about the name node. If in any case the name node seems to be not responding, then the general node makes the secondary name node the active name node. The non-responding name node is made the secondary name node. This enables high availability in Hadoop architecture.

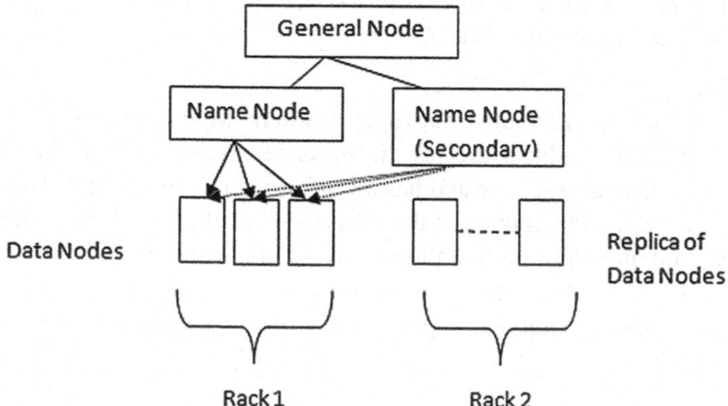

Fig. 3 HDFS node architect

3.3 Computational Technology

The data from the storage components and the capturing components flows into the computational phase. The computation and processing on this data can be done in two ways: in memory and in database. Both the computation types require different technologies. Choosing the technology requires intensive study on the future scope and budget assigned. In in-memory technology like Alluxio, the entire data is placed in the RAM and processed, making the processing extremely fast, whereas in in-database technology, the data is to be fetched using querying technologies before processing. Alluxio can work with Apache Spark as shown in Fig. 4, but doing so requires increasing the number of servers currently being used.

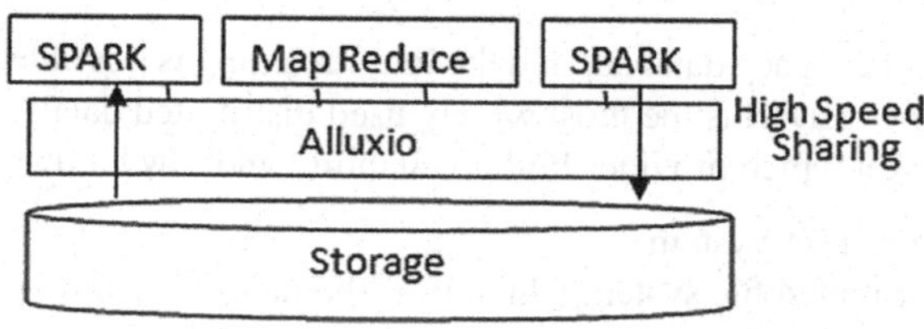

Fig. 4 Alluxio

Apache Spark

Apache Spark seems to be replacing Hadoop framework in enterprise Big Data architecture. Spark is in-memory data processing, enabling fast data processing [8–10]. Apache Spark allows scalability and high fault tolerance. Currently available spark framework consists of four libraries and a core engine as shown in the Fig. 5. The Spark SQL enables performing SQL like queries on structured data. This library also enables complicated query processing for advance analytics. Spark streaming library provides server scalability and high fault tolerance. Using of this library requires creation of a RDD or a resilient distributed datasets for any processing. MlLib library enables use of machine learning algorithms. Graph computation like GraphX supports versions of graph processing.

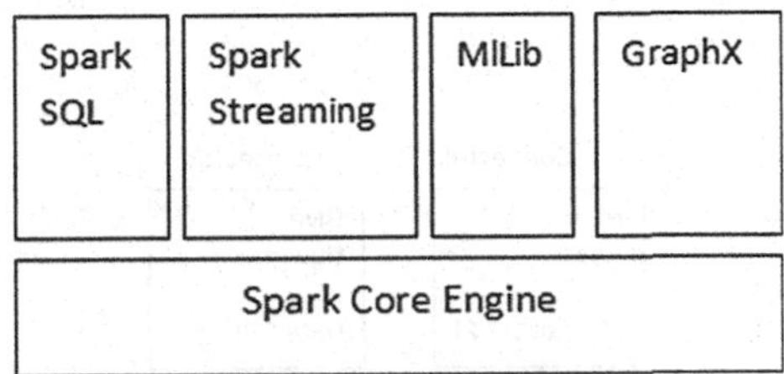

Fig. 5 Apache Spark framework

3.4 Querying Technologies

Apache Hive

Hive is a data warehouse tool placed. Apache Hive is a Horton work data platform residing on top of Hadoop for processing on structured data. It enables three data processing functions: data summarization, data analysis, and data querying [11, 12]. These functions are carried out using SQL like query on data stored in the storage

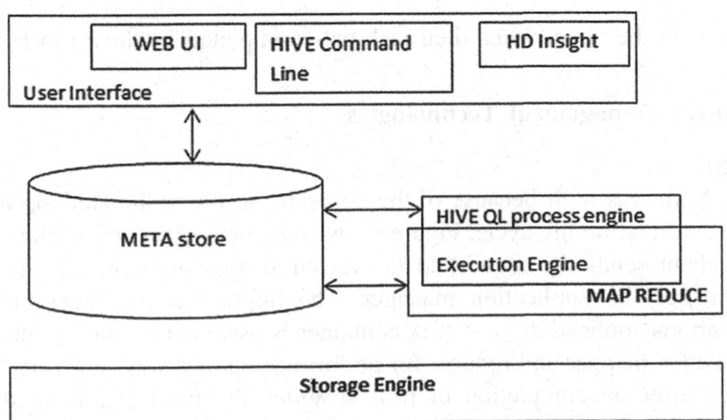

Fig. 6 Apache Hive architecture

component. Figure 6 depicts the architecture of Apache Hive. The user interface provides interaction between the client and the storage component. Client can interact with the data stored in the HDFS by using any of these methods: Web UI, hive command line, or HD insight. Hive uses Meta store for storing the schema of tables and database. HQL processing engine takes the query and matches the fields asked by the clients in the query with the table schema present. The execution engine takes the query, processes it, and generates the result similar to the result generated by a map reduce code.

Prestro

Prestro is a parallel query-processing engine also defined as a free SQL query engine. Presto and hive have similar working and schema declarations [13]. Presto enables fetching of data from multiple resources Hive, HDFS, Hbase, etc. Fetching the data requires connectors, and the task of the data node is done by workers which

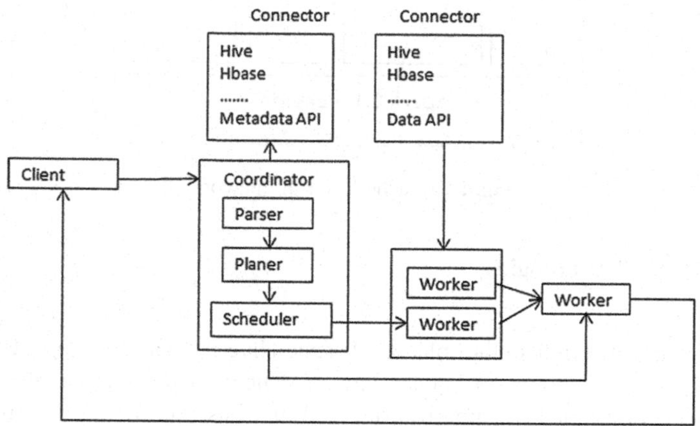

Fig. 7 Presto architecture

directly report to the client when their task gets completed as shown in Fig. 7.

3.5 Resource Management Technologies

Apache Yarn

Apache Yarn was built because of the necessity to move the Hadoop map reduce API to the next iteration life cycle. Figure shows the component architecture of Apache Yarn. The client sends the request to the resource manager [14, 15]. The resource manager initiates the application manager. The application manager manages and schedules various application master. A container is assigned by the resource manager which is used for mapper and reducer for performing its tasks as shown in Fig. 8. The application master on completion of task provides the information to the resource manager for killing of the container.

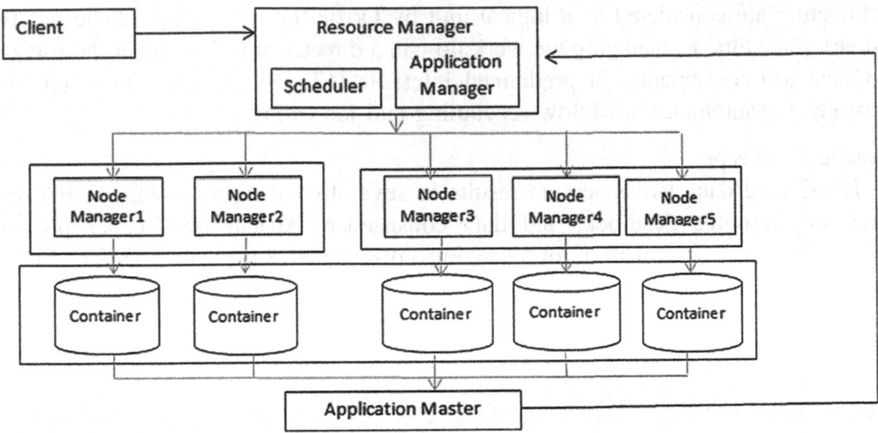

Fig. 8 Apache Yarn architecture

Apache Mesos

Apache Mesos provides API for resource management and scheduling across the entire data center. It can run on multiple frameworks and machine at the same time. Parallel processing is achieved by assigning UI to each machine. Figure 9 shows the components of Mesos architecture. The job request is sent by the client to Mesos master. Mesos master checks for the available resources and provides a list of suitable resources to the client. The clients select the resource [16]. Once the resource is selected by the client, task tracker assigns the job to the Mesos slaves. The entire coordination is done using Apache ZooKeeper.

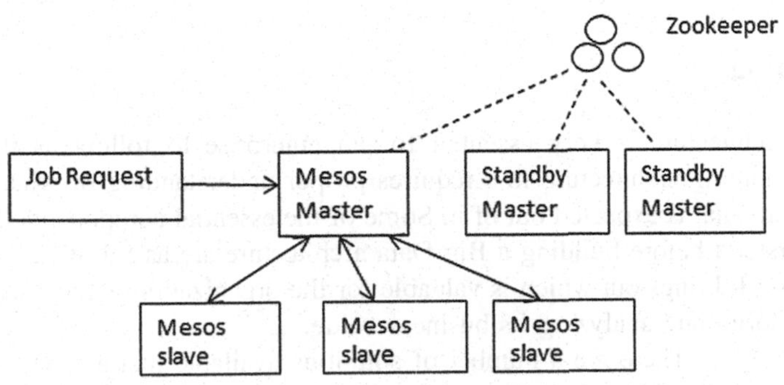

Fig. 9 Apache Mesos architecture

3.6 Management and Coordination Technologies

Apache Oozie

Apache Oozie is a work flow scheduler for various Big Data technologies and components in a Big Data pipeline. Various jobs at different levels of Big Data architecture are considered as a logical unit by Oozie for scheduling. Oozie has two primary jobs: First is managing the work flow in a directed acyclic manner, and second is Oozie job coordination at predefined intervals [17, 18]. Apache Oozie provides packages for automated workflow scheduling and job coordination.

Apache ZooKeeper

Distributed data across servers results in several challenging situations like race condition, resource deadlock, and data consistence. Apache ZooKeeper provides high-performance coordination for managing a distributed systems or a large number of hosts.

3.7 Security Technologies

Apache Shiro

In pipeline architecture, the data flows through various components. Managing the flow of the data through various components in a sequential flow is done by Apache Shiro. Apache Shiro provides port to port and connection security across the entire pipeline architecture.

Apache Kerberos

Apache Kerberos is an authentication protocol for trusted hosts on untrusted networks. Kerberos provides three level securities: server security, service security, and user security. Security is maintained by encrypting the flow of information with different keys like session key, server key, user key, and system keys. According to Kerberos, to provide proper security, the password should never be stored on the client machine. The password should never be stored in an unencrypted format even in the authentication server database. Apache Kerberos is the most widely used security protocol known to providing maximum data flow security.

4 Planning

Big Data architecture is very essential for an enterprise to follow as it acts as a roadmap. Building architecture first requires proper understanding of what the enterprise has and what is expected out of it. Some of the essential components that needed to be understood before building a Big Data architecture are as follows

Use case: It brings out which is valuable for the organization, which data needs to be given priority and analyzing its business value.

Solution Type: There are a number of solutions available that are provided in the following form: cloud based, appliance based, and service based [19, 20]. Deciding on which solution to be used requires checking the data based on privacy, governance, ingestion, and data modeling.

Dynamic nature: Big data technologies keep on changing one should design a system that should easily adapt to new technologies for future business requirement.

Security: Data in Big Data architecture is stored in database and flows through the network. Protection of data from external threats is the top most priority of the organization. Analysis of the security application needs to be done extensively. A number of security applications need to work together in order to provide security at each level.

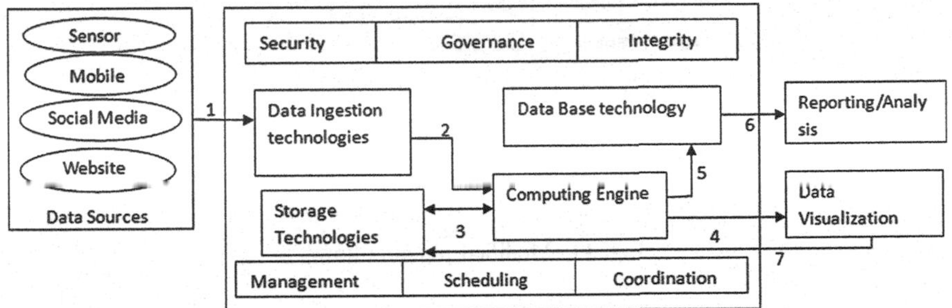

Fig. 10 Proposed frameworks

5 Proposed Framework

The proposed framework as shown in Fig. 10 provides a step-by-step view of a data-driven flow for an enterprise scale Big Data architecture.

For an enterprise, the sources of data can be many such as sensors, Web site, social media, or mobile devices. Getting the data from these sources can be done using various data ingestion technologies [18] explained in Sect. 3.1. For near real-time reporting and analysis, the data captured needs to be sent into the computing engine. The computing engine takes the data from the storage component for predictive analysis. The computing engine in our framework is comprised of Apache Spark. The machine learning programs are written using MLlib library function of Apache Spark Library. The computed data is sent to the database technologies as well as the users. The data can then be sent back to the storage technology being used. Reporting generation and analysis can be done by the organization directly from the database technologies.

The data stored in the HDFS needs to be compressed; this can be done using sequential compression technique, Snappy, Bzip2, Zip2, etc. Organizations have the need of their data from various internal and external threats. Encrypting is the most secured way of protecting the big data security applications like Apache Ranger, and Apache Kerberos provides maximum security to data. Apache Kerberos key security provides encryption of both the data in the database, the network data as well as provides encryption on the server side data. Proper sequential flow of data through various component technologies needs to be done for successful working of the pipeline architecture; this is done using Apache Oozie, whereas the linking of these components for data flow is done using Apache Shiro.

After the framework was decided, the pipeline was implemented on a two-server system as was considered beta 1. Figure 11 shows the modified architecture of the pipeline where the data gets fetched, processed, stored, and visualized.

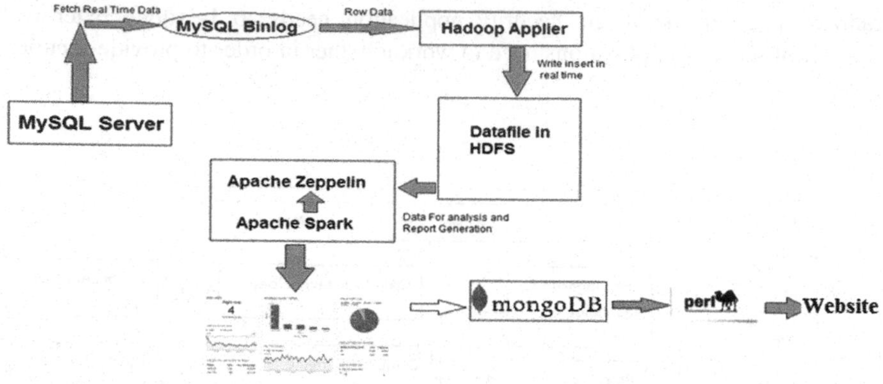

Fig. 11 Modified pipeline flow

Data in our work was stored in HDFS; this data modeled and cleaned before processing it in the compute engine as shown in Fig. 12 and later storing this processed data in MongoDB as shown in Fig. 13. Figure 14a, b depicts the loading and visualization using Zeppelin. Zeppelin enables creation of dashboards and provides clear visualization which enables direct fetching records from database technologies.

```
usersData.unpersist()

def breakkDownDate(data: String):(String) = {
        val firstHyphon = data.split(" ")
        val getDateOnly = firstHyphon(0).toString
        (getDateOnly)
}
def SHA256(text: String) : String = java.security.MessageDigest.getInstance("MD5").digest(text.g
){_ + _}

//define case class for Users table and its attributes for mapping
case class Datatest_Users(id: Int, email: String, experience: Long, regdate: String, nationality
, degree_level_highest: Int, channel_id: Int, subchannel_id: Int)

//map Split data to case class
val dataset = splitdata.map(r => Datatest_Users(Try {r(0).toInt}.getOrElse(0), SHA256(r(5)).trim
Try{r(10).toString}.getOrElse("").trim.replaceAll("""^,""","").replaceAll("""$,""",""), Try{r(11
).toInt}.getOrElse(0), Try {r(18).toInt}.getOrElse(0),Try {r(20).toInt}.getOrElse(0), Try {r(21)
```

Fig. 12 Cleaning and data modeling

```
ist split data to clear Memory
tions_split.unpersist()
s_data_register = sqlContext.sql("select resumes.id, users.email, users.experience, user
industry, users.mobile, subchannels.display_name as Channel, job_locations.name as Curre
cations, edu_levels, subchannels where users.id = resumes.uid and job_locations.id = use
bchannels.id = users.subchannel_id and subchannels.channel_id = users.channel_id ")

Config = MongodbConfigBuilder(Map(Host -> List("                    ), Database -> "baz
oncern -> "normal", SplitSize -> 8, SplitKey -> "_id"))
```

Fig. 13 Loading the data in MongoDB

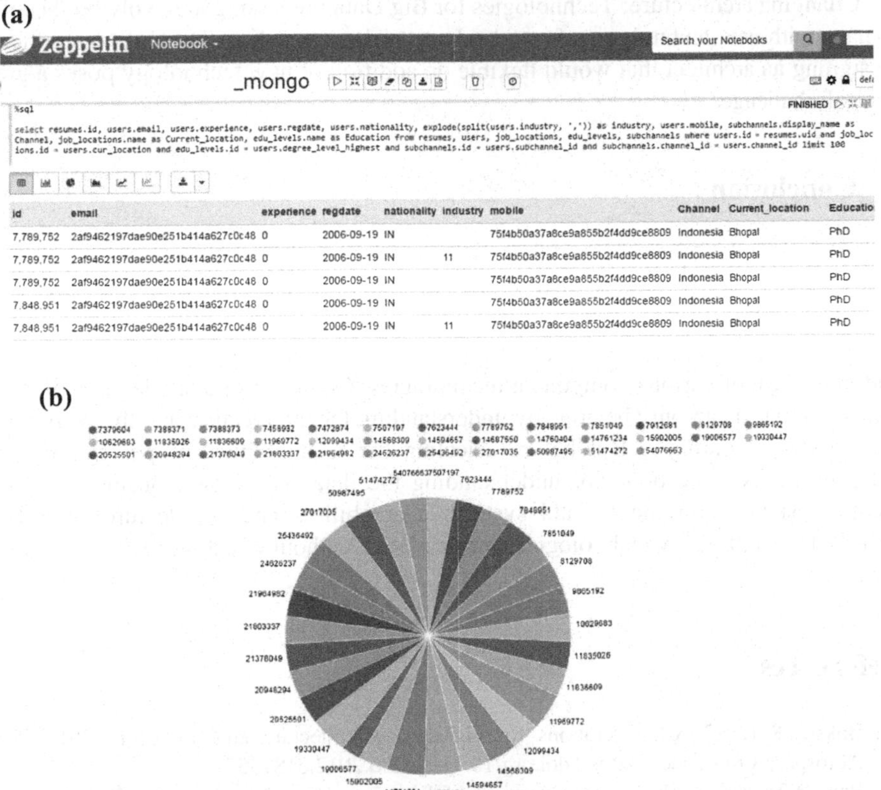

Fig. 14 **a** Fetching the data by Zeppelin from MongoDB, **b** visualization of result

6 Challenges

Data architecture once created acts as a roadmap which is followed by the entire organization. It creates significant opportunity for organization to grow. But an organization faces a number of challenges while designing and implementing the pipeline. Some of these challenges are discussed below.

Complex data: Data is collected from different platforms and sources making the system deal with great data heterogeneity. A number of times the data being collected are incomplete or incorrect affecting the quality of the result. Data modeling needs to be done in order to reduce and maintain the data quality.

Scaling: Velocity at which data is being generated is never stable. The speed at which data is being ingested is dynamically varying, posing a challenge for the processor. The speed of processor is limited, but the scaling of data is dynamic to automate the resource allocation scaling up techniques that need to be implemented at server level.

Changing architecture: Technologies for Big Data processing are evolving, keeping in pace with new technologies from database to data modeling that becomes difficult. Designing an architect that would flexible on addition of new technology poses a great deal of challenge.

7 Conclusion

Building an entire Big Data architecture requires thorough knowledge of the present technologies and the type of data being used by the present system. Designing an optimal Big Data architecture is the main idea in order to get the correct value out of the data present. Designing of Big Data architecture requires building a conceptual pipeline and placement of various component technologies. Choosing of a single technology at each component can only be done by understanding the architecture and the working of the selected technology. But before selecting the technologies, proper step-by-step planning needs to be done for understanding the data complexity, business requirements, system requirements, and systems adaptability. The architecture should be designed such that new technologies can be placed without effecting the processing of the results.

References

1. Bakshi K (2012) Considerations for big data: Architecture and approach. 2012 IEEE Aerospace Conference. https://doi.org/10.1109/aero.2012.6187357.
2. Pant P Tanwar R (2016) An Overview of Big Data Opportunity and Challenges. Smart Trends in Information Technology and Computer Communications.
3. Apache Flume Architecture. In: www.tutorialspoint.com. https://www.tutorialspoint.com/apache_flume/apache_flume_architecture.htm. 2016.
4. Apache Flume - Architecture of Flume NG: Apache Flume. In: Blogs.apache.org. https://blogs.apache.org/flume/entry/flume_ng_architecture 2016.
5. Apache Kafka Cluster Architecture. In: www.tutorialspoint.com. https://www.tutorialspoint.com/apache_kafka/apache_kafka_cluster_architecture.htm. 2016
6. Apache Kafka: Next Generation Distributed Messaging System. In: InfoQ. https://www.infoq.com/articles/apache-kafka. 2016
7. Ramesh B (2015) Big Data Architecture. Studies in Big Data 29–59.
8. Getting Started with Apache Spark. In: Mapr.com. https://www.mapr.com/ebooks/spark/03-apache-spark-architecture-overview.html. 2016
9. Spark Architecture. In: Distributed Systems Architecture. https://0x0fff.com/spark-architecture/. 2016.
10. Laskowski J. Spark Architecture · Mastering Apache Spark 2. In: Jaceklaskowski.gitbooks.io. https://jaceklaskowski.gitbooks.io/mastering-apache-spark/content/spark-architecture.html. 2016.
11. Hive Tutorial: How to Process Data with Hive. (2017). [online] Hortonworks. Available at: http://hortonworks.com/hadoop-tutorial/how-to-process-data-with-apache-hive.
12. Hive - Introduction. (2017). www.tutorialspoint.com.Available. Available at: https://www.tutorialspoint.com/hive/hive_introduction.htm.

13. Apache Presto Architecture. (2017). [online] www.tutorialspoint.com. Available at: https://www.tutorialspoint.com/apache_presto/apache_presto_architecture.htm.
14. What is Apache Hadoop YARN (Yet Another Resource Negotiator)? - Definition from WhatIs.com. (2017). [online] SearchDataManagement. Available at: http://searchdatamanagement.techtarget.com/definition/Apache-Hadoop-YARN-Yet-Another-Resource-Negotiator [Accessed 21 Feb. 2017].
15. Dawson, R. (2017). Untangling Apache Hadoop YARN, Part 1: Cluster and YARN Basics – Cloud era Engineering Blog. [online] Cloudera Engineering Blog. Available at: https://blog.cloudera.com/blog/2015/09/untangling-apache-hadoop-yarn-part-1/.
16. Apache Mesos. (2017). [online] Apache Mesos. Available at: http://mesos.apache.org/documentation/latest/architecture/.
17. Apache Oozie Tutorial. (2017). [online] www.tutorialspoint.com. Available at: https://www.tutorialspoint.com/apache_oozie/index.htm.Anon, (2017).
18. Sanchita, kalpana Jaiswal, Praveen Kumar, Seema Rawat "Using Twitter for Tapping Public Minds, Predict Trends and Generate Value "in the proceeding of Advanced Computing & Communication Technologies (ACCT), 2015 Fifth International Conference organised by ABES Ghaziabad, INDIA (Published in IEEE Explorer), 21–22 Feb. 2015. PP 586–589, DOI https://doi.org/10.1109/ACCT.2015.99.
19. Praveen Kumar, Dr. Vijay S. Rathore "Improvising and Optimizing resource utilization in Big Data Processing "in the proceeding of 5th International Conference on Soft Computing for Problem Solving (SocProS 2015) organised by IIT Roorkee, INDIA (Published in Springer), Dec 18–20, 2015. PP 586–589.
20. Sanchita, kalpana Jaiswal, Praveen Kumar, Seema Rawat "Prefetching web pages for improving user access latency using integrated Web Usage Mining "in the proceeding OF International Conference on Communication Control and Intelligent System (CCIS-2015) organised by GLA University Uttar Pradesh, INDIA (Published in IEEE Explorer), November 07–08, 2015. PP 401–405.
21. Apache Oozie - Hortonworks. (2017). [online] Hortonworks. Available at: http://hortonworks.com/apache/oozie/.

A Study of Exposure of IoT Devices in India: Using Shodan Search Engine

Nabeel Zaidi[1(✉)], Himanshu Kaushik[2], Dhairay Bablani[1], Raghav Bansal[1], and Praveen Kumar[1]

[1] Amity University Uttar Pradesh, Noida, Uttar Pradesh, India
nabeelzaidi@ymail.com, {dhairaybablani, raghavbansal95}@gmail.com, pkumar3@amity.edu

[2] Shri Mata Vaishno Devi University, Katra, Jammu and Kashmir, India
himanshu.kaushik@smvdu.ac.in

Abstract. As the world is getting digitized and Internet of Things (IoT) devices are becoming more and more popular, this has led us to an advantage of improving our life in perspective of quality. The security threat related to IP is also increasing. But unlike in other technologies, a common way to prevent these threats is not possible. And when it comes to a developing Nation such as India, it becomes more difficult to prevent from such threats. Shodan is one of the world's most acknowledged search engine. In this paper, we have given an overview of Shodan in perspective of India. Furthermore, we have evaluated various IoT devices using Shodan based on different parameter in India. We have also given our views on how these devices can easily be exploited using Shodan.

Keywords: IP address · IoT · Cyber security · Shodan search engine ICSs

1 Introduction

Being in the twenty-first century, our world is getting digitized day by day, and the technology of IT is being incorporated into devices which results in increase in the IoT devices sharply. We know that IoT devices hold a very great and promising future and economically affect about 1.9 trillion dollars. Not only its developments help nation to be prosperous but it also makes it more competitive, but because of this, it increases the threat of exposure of IP too. We can be undermined by the invasion of privacy and framework hacking of open associations. There can be other threats too like data tampering/ forgery, Denial of Service attack [1], data leakage, etc. Making countermeasures to these threats has become a priority. The Internet has been integrated into the modern society's infrastructure in the last two decades. The way the society communicates and does business is being changed by the Internet. The method of controlling the physical infrastructure of the society is also being changed by the Internet.

A large interdependent critical infrastructure of the nation is being tightly linked together for doing the critical functions of a modern, high-technology society. At the

V. Bhateja et al. (eds.), *Information Systems Design and Intelligent Applications*, Advances in Intelligent Systems and Computing 672,
https://doi.org/10.1007/978-981-10-7512-4_105

center of the society, formerly independent systems are being fused together by information infrastructure [2]. Electricity distribution is dependent on communication network as much as communication networks are dependent on it. Some elements of critical infrastructure can cause massive ripple effects to the remaining infrastructure and toward the society if it is disturbed.

In recent years, several authors (Lewis 2006, Byres 2004) have pointed out the vulnerability of critical digital infrastructure. Shodan search engine is one of the most efficient tools to visualize vulnerability of the developed economies; it can be used to search for cyber physical devices over the Internet (Matherly 2009). The detailed knowledge or inside information of the system is not needed nowadays to compromise ICSs (Weiss 2010, Byres 2004).

1.1 Background

Nature and importance of ICSs and the reasons why control systems are being exposed to the Internet are being explained in this section. Further, a little about malfunctioning issues of ICSs are also discussed. An increasing number of attacks on ICSs have been seen in recent years, including nation-state actors have been seen targeting some industrial assets (Falliere 2010). Industrial automation is being controlled digitally nowadays. These systems are also known as Industrial Control Systems (ICSs) [2]. A typical case is presented in Fig. 1. Sensors are connected to a control network through a remote terminal unit (RTU) and are monitoring a physical process. Measurement data is being received by a programmable logic controller (PLC), and

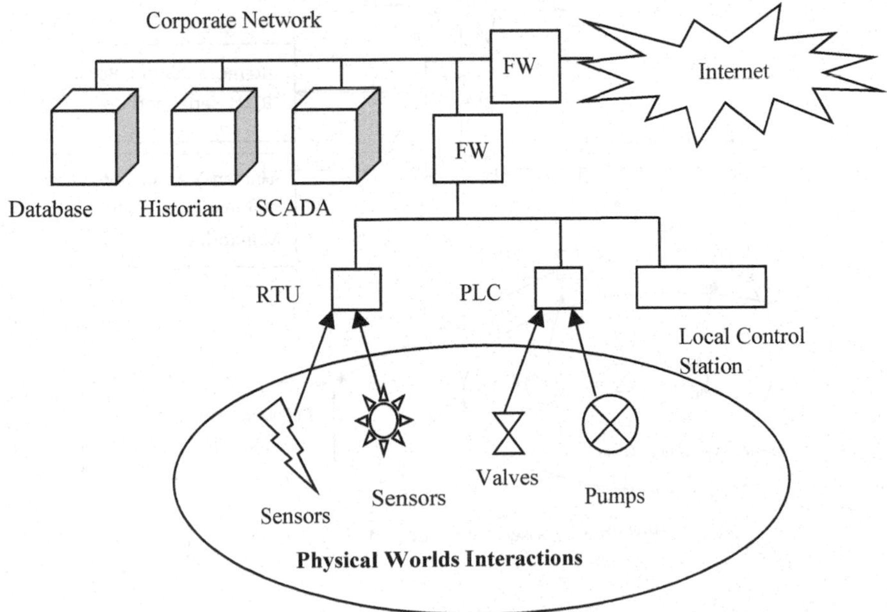

Fig. 1. Industrial control system

changes are initiated in the actuators which then further influence the process. Various types of communications are used in these types of systems, such as industrial control networks known as field buses, TCP/IP protocols used on the Web, or analog electrical connections.

High reliability requirements are needed by actual process control, as errors can be costly and hazardous. Log life spans of decades are expected by industrial systems, and unexpected results can be produced by joining old equipment with modern technologies [3, 4]. A system can be put to halt due to a malformed request to a device in the process control (Stouffer 2011). ICS asset's physical security is taken very seriously, and issues emerging from the cyber-space are also put under consideration. If an ICS is not directly connected to the Internet, a connection still can be existed through a corporate network (Weiss 2010).

Some interfaces of ICS components which might be vulnerable to outside threats are illustrated in Fig. 2. The process control network components in the figure are same as the ones in Fig. 1, except for the firewall, a gateway server, or a proxy device exists in its place to allow monitoring and remote controlling.

Usually, a field bus protocol such as Modbus is used by process control by communications, and the gateway devices work as adaptors making remote access possible. If RTU, local control HMIs, and PLC are TCP/IP capable, remote management interface is also accessible by them, for example, SSH protocols, Telnet, or HTTP, exposing the devices to the Internet.

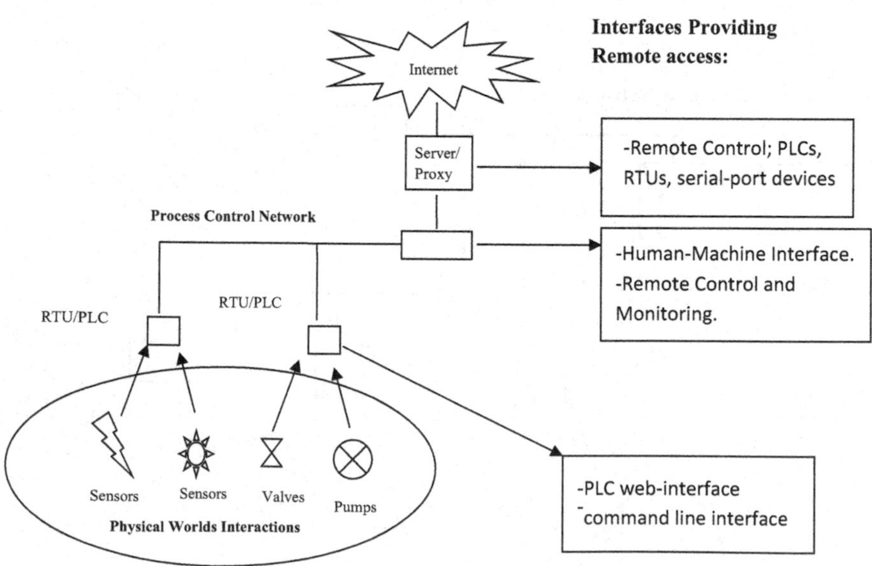

Fig. 2. Possible exposed interfaces of ICS components [2]

2 Ports Scanning

The method of searching network services on a host on the Internet is called Port scanning. Target computers receive requests on different port numbers to know whether a service is active or not on that port. It might be revealed by a port scan of target host, for example, on port number 23, a Telnet service is responsive and online, and on port 80, an HTTP service is responsive [5], and no other services are responsive on the host.

Finding open, responsive port by scanning IP addresses is easy. Having a useful scanner in the identification of devices behind the IP address, that is, required is difficult. It is not possible to make an absolute confirmation with 100% accuracy which physical device is responding, as the communication task is done by the software installed in the device: Responses and outputs can be always altered and made to be seen as something that it is not. However, a probable identification can be done about the purpose of the device by observing the response of the software. The process of comparing of software responses to the earlier known responses after gathering them for identification is known as fingerprinting.

There are multiple ways to do fingerprinting, but to identify similar devices, a unique set of fingerprints are needed. A device fingerprint database of over 2000 entries can be found in the popular open-source port scanner software Nmap (http://www.nmap.org). TCP/IP stack implementation of software in fingerprinting is used by it. It can accurately identify the devices in the fingerprint database, but when it comes to publically less-known devices, like most of the ICS devices, their TCP/IP stack is not known to Nmap, and hence, they cannot be identified. When querying services like HTTP, a high-level approach is to just looking at the headers and banners set by the devices.

For example, server software and software version are revealed in the HTTP header of the Web servers. The same is case with the ICS devices with imbedded Web servers [6]. The response message can tell the version of the device. The construction of HTTP header does not give such vast information. As a result, responses are vague, and no conclusion can be drawn out. There also exist some alternate methods to identify these devices. In Finland (Tiilikainen and Manner 2013), a research has been done to find out ICS devices which might have general HTTP and FTP responses. They can still provide with their names and other services, especially NetBIOS and SNMP messages. Such protocols (NetBIOS and SNMP message) are used as private and trusted network inside body without concerning outside queries. Due to some erroneous configuration at the local network, devices can enhance protocol advertisement outside the private network such that device can be detected by outsiders. Advertisement of unique MAC address of the device is often done by NetBIOS protocol. It further helps in resolving the issue of identifying the manufacturer of the device and also to identify a unit particularly.

3 Shodan Search Engine

In 2009, Shodan search engine was published which indexes all the protocol headers of various networks services in server (Matherly 2009). To find out specific type of devices on Internet, Shodan search engine can be used. For example ICS components.

It also helps in measuring the size of Internet-connected infrastructure problem. By providing product name and vendor information in the headers of the HTTP protocol, the vendors of ICS components made it a much easier task. Many specific protocol header queries can be sorted out by this search engine. Shodan's database can help in finding specific devices if someone knows the fingerprinting information of a particular vendor's product [6]. For example the string like "https://1.0302.Location:/Portal0000.htm."

Shodan gives seeking information to IP locations and some other data. In spite of the fact that there are a great deal of Web crawlers these days, Shodan is the most intense Web index starting at now. Shodan gives the client a chance to discover particular sorts of PCs associated with the Web utilizing an assortment of channels. Some have additionally portrayed it as an Internet searcher of administration flags [7, 8], which are meta-information that the server sends back to the customer [5]. This can be data about the server programming, what choices the administration bolsters, an appreciated message, or whatever else that the customer can discover before associating with the server. The Shodan IP Web search tool is intended to creep the Web and endeavor to distinguish and file associated gadgets. Shodan motor gathers more than 500 million IoT gadgets in one month. It likewise gathers information for the most part on Web servers right now (HTTP port 80); however, there is additionally a few information from FTP (21), SSH (22), Telnet (23), SNMP (161), and Taste (5060) administrations. Utilizing Shodan, specialists distinguished a great many Web confronting gadgets related with mechanical controls framework.

4 Internet of Things (IOT)

Internet of Things (IoT) is the system of physical items or "things" implanted with gadgets, programming, sensors, and network to empower it to accomplish more noteworthy esteem and administration by trading information with the producer, administrator, or potentially other related gadgets in view of the foundation of Universal Media transmission Union's Worldwide Models Activity. IoT permits items to be detected and controlled remotely crosswise over existing system foundation, making open doors for a more straightforward reconciliation between the physical world and PC-based frameworks and bringing about enhanced effectiveness, exactness, and monetary advantage. Everything is interestingly identifiable through its inserted processing framework which however can interoperate inside the current Web foundation [9]. Specialists evaluate that the IoT will comprise of very nearly 50 billion questions by 2020. IoT gadgets are utilized as a part of different cases. It utilized as a part of Savvy Home Framework, Wearable Gadgets, Keen Auto Gadgets, Living in Close Contact Gadgets, Brilliant Vitality Gadgets, Modern and Natural Gadgets, and so forth.

4.1 Scanning India

There were two goals of our research: to discover how vulnerable India is and to discover various devices or companies in India that are connected worldwide using Shodan.

4.2 Methodology

We started by asking Shodan to find our vulnerabilities. To match known products from different vendors, we created a list of 20 different search queries. Identifying [1] strings for industrial control systems was contained in this list. The search is significant to prove our point rather than providing an exhaustive coverage of all ICS products.

5 Results

In the first quarter of 2017 an initial Shodan search run was done and nearly 4000 ICS device from IP address were. The devices were divided by the most common use known for each particular product. SCADA/ICS systems are larger automation systems, which had open Telnet access without authentication requirements of these eight. Smaller industrial systems are referred as industrial devices, like PLCs or gateway devices used for network automation components. As especially common device type was building management system (BMS). Some of which belong to an ice rink, a hospital, and a bank office.

We have conducted a deep research, and with the help of Shodan, we were able to pinpoint several results. These may even be classified as loopholes present in technology of India.

The figures and tables below are taken exactly as the way they were extracted from Shodan. Figure 3 below shows the Top cities of India using the service banners. Here,

Top Cities	
1. Mumbai	113
2. New Delhi	111
3. Bangalore	81
4. Delhi	51
5. Chennai	33
6. Pune	29
7. Kolkata	22
8. Hyderabad	21
9. Gurgaon	16
10. Thane	15

Fig. 3. Top cities in India

we can see Mumbai and Delhi hold the first and second spot which is quite surprising in perspective of Bangalore as it is the Silicon Valley of India.

Figure 4 shows the Top services being used in India. As we can see, TELNET is the most used service in India as it is used by most of the big companies.

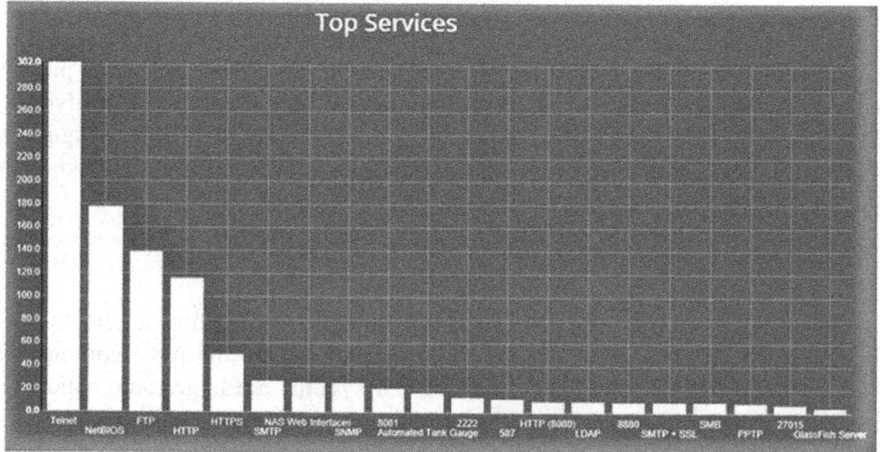

Fig. 4. Top services used in India

Figure 5 shows a comparison of the various organizations to deduce the top one. We can see Tata Communication is the top one in India, followed by BSNL but there is a huge difference between the two.

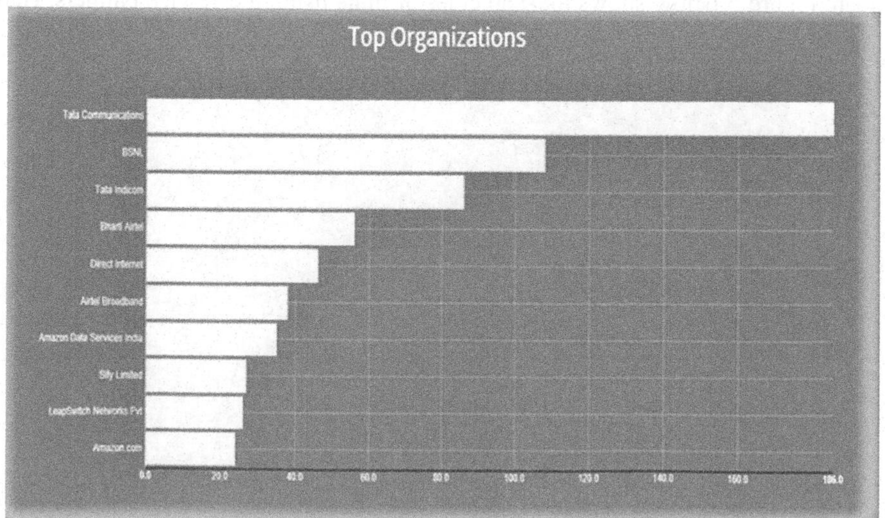

Fig. 5. Top organizations of India

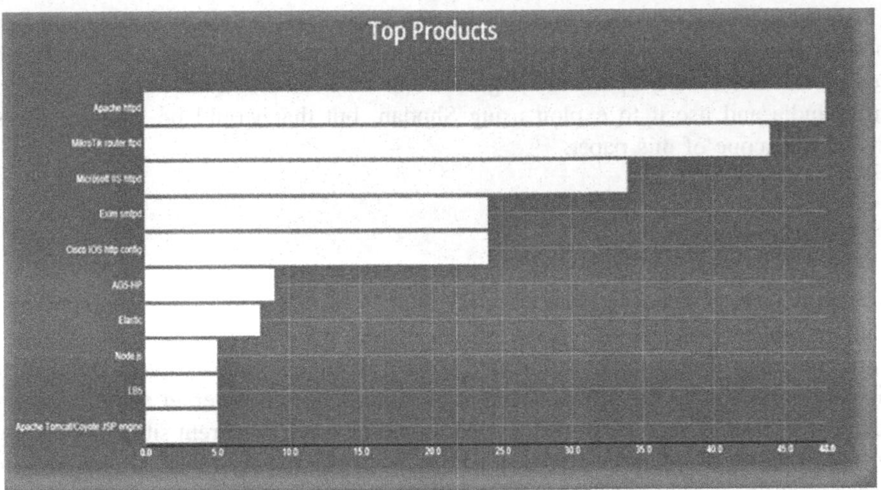

Fig. 6. Top products of organizations in India

Figure 6 shows the list of top product used by the organizations. As we can see, Apache httpd is widely used in India, while elastic, Node JS, LBS, and Apache Tomcat jsp engine too are used but are not that much used as compared to others.

In Fig. 7, we can see the top domains of India; vsnl.net.in is the mostly used domain across India. There are other domains too listed in Fig. 7.

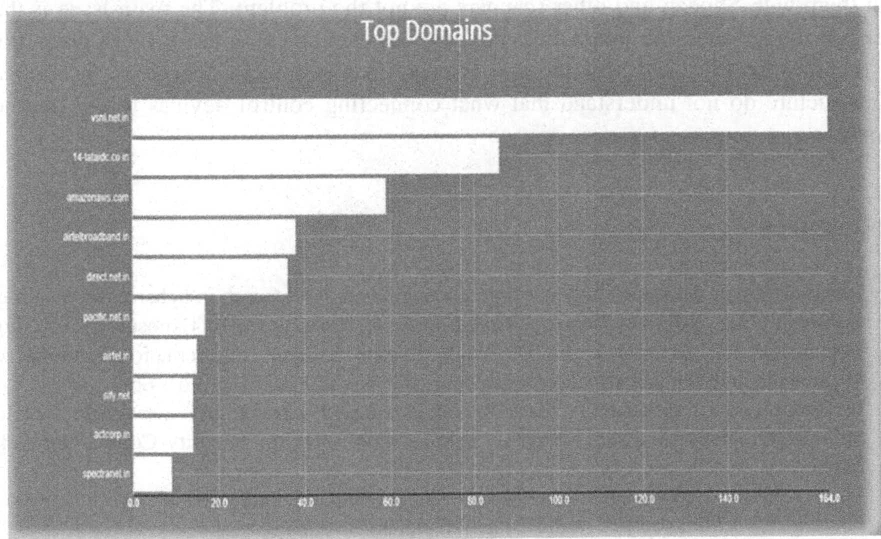

Fig. 7. Top domains of India

All the data provided in the result section is exactly the same as obtained from Shodan, and this data is collected in the month of April 2017. This data can also be used by Hackers to target particular organizations. We can also easily find an open port across India and use it to exploit using Shodan, but that would be illegal and also beyond the scope of this paper.

6 Conclusion

One of the large problems is protecting the critical infrastructure which requires many solutions, secure products, network analysis to search the weak points, and education of people in charge or regulations. A single thing is not the solution to very problem. In this paper, it is shown by us that scanning a nation's current Internet address space is technically feasible, and we argue that this should be done at current situation to gauge the vulnerability of a nation. Changes within the legislative structure are required for such proactive scanning to allow an authority to perform penetrative but controlled study of systems connected to publically accessible networks [4]. Other than analyzing the vulnerabilities, there should be a program for reporting the discoveries and acting upon them.

Vulnerability of ICS devices being connected to the Internet in so many amounts is shown by our study of the Internet. Everything from building automation to parts of critical infrastructure is included in these ICS devices. It should be assumed that it is a trend not a blip that in the future more devices will be connected, and this as a straight point should be used as security measures.

Connecting critical infrastructure control devices to the Internet and being able to find them with Shodan and other scanners are not the problem. The main issue is that they are connected to the Internet directly without even the simplest form of protection by the firewall [3]. And also the real issue is that people in charge of the critical infrastructure do not understand that what connecting control devices to the Internet means.

References

1. David Goldman, Shodan: The scariest search engine on the Internet, in cybercrime economy.
2. Yun-Seong Ko, Il-Kyeun Ra and Chang-Soo Kim, A Study on IP Exposure Notification System for IoT Devices Using IP Search Engine Shodan in International Journal of Multimedia and Ubiquitous Engineering Vol. 10, No. 12 (2015), pp. 61–66.
3. B. Genge and C. Enˇachescu, ShoVAT: Shodan-based vulnerability assessment tool for Internet-facing services, security and communication networks Security Comm. Networks 2014.
4. Timo Kiravuo, Seppo Tilikainen, Mikko Sarela, Peeking under the skirts of a Nation: Finding ICS Vulnerabilities in the critical digital Infrastructure.
5. Sajal Verma, "Searching for Fun and Profit" from, https://www.exploit-db.com/docs/33859.pdf.

6. C. MOHAN, "ARIESIKVL: A Key-Value Locking Method for Concurrency Control of Multiaction Transactions Operating on B-Tree Indexes" in Proceedings of the 16th VLDB Conference Brisbane, Australia, August 1990.
7. Praveen Kumar, Dr. Vijay S. Rathore "Improvising and Optimizing resource utilization in Big Data Processing " in the proceeding of 5th International Conference on Soft Computing for Problem Solving (SocProS 2015) organised by IIT Roorkee, INDIA (Published in Springer), Dec 18–20, 2015. PP 586–589.
8. Seema Rawat, Praveen Kumar, Geetika, Implementation of the principle of jamming for Hulk Gripper remotely controlled by Raspberry Pi in the proceeding of 5th International Conference on Soft Computing for Problem Solving (SocProS 2015) organised by IIT Roorkee, INDIA, Dec 18–20, 2015. PP 199–208.
9. Sheril Yadav, Seema Rawat Analysis and Implementation of Business Intelligence Software for Report Bursting" International Conference in Smart Computing & Informatics (SCI-2017) held in ANITS, Visakhapatnam March 2017.

Real-Time Business Analytical Model Using Big Data Strategies for Telecommunication Industry

M. Maheswaran[1(✉)] and David Asirvatham[2]

[1] Management and Science University, University Drive, Seksyen 13, Shah Alam 40100, Selangor, Malaysia
mmurugesapillai@gmail.com
[2] School of Computing & IT, Taylor's University, Subang, Malaysia

Abstract. The volume of data is growing exponentially. By 2020, about 1.7 MB of new information will be created every second for every human being on the planet. There will also be about 6 billion smartphone users and over 50 billion smart connected devices in the world. The traditional data analysis techniques will not be scalable to match the storage and processing capabilities of such high volume of data. Moreover, it becomes important to analyse these data at real-time speed. Distributed computing and platforms, such as Hadoop, will play a vital role to process these data at real-time speed to provide users with real-time reports to help users to make critical decisions. Telecommunication industry will be one of the first industries that will need to handle big data. Most of the users are connected all the time creating multiple sessions per user as well as communicating with multiple devices creating the complex high volume of data that need to be analysed. This paper will highlight the conceptual design of a real-time business intelligence model that provides insights into the telecommunication industry.

1 Introduction

Digital universe has been growing with 40% of a year into next decade, with the expansion of, not only the all-online services by people but also all the things such as smart devices communicating via the Internet. By 2020, the digital universe will be expected to reach 44 zeta bytes of data [1]. Hence, 1.7 MB of new data will be created by every human in second along with 6 billion smartphones and 50 billion smart connected devices in the world [2]. Also, social media usages and video-based services have been contributing the exponential growth in the data volume. In addition, present smartphones are being vastly improving the better Web-surfing experience resulted in increased consumption of media such as transferring live high definition videos, and

V. Bhateja et al. (eds.), *Information Systems Design and Intelligent Applications*, Advances in Intelligent Systems and Computing 672,
https://doi.org/10.1007/978-981-10-7512-4_106

content-based services are contributing the continuous increment in the data volume [3]. Traditional data analysis methods will not be sufficient to store and analyse such high volume of data. Hence, it is essential to look an alternative way to analyse the data with real-time capability.

Big data is a broad term for data sets with large, complex and unable to handle with traditional data processing applications and techniques. Main challenges are analysis, capture, data curation, search, sharing, storage, transfer, visualization, queering and information privacy [4]. Telecommunication industry is the one of most import industry that needs to use big data strategies to store and analyse the data coming with high volume, velocity, variety, etc.

1.1 Big Data Challenges for Telecom Operators

The volume of the data generation has been increasing with access technologies such as 4G/LTE, FTTH and ADSL due to the volume of operational data generated for sessions increasing dramatically. Also, most of the users are unlike making traditional voice calls, always connected with data. Hence, it is the requirements to have real-time operational capabilities to provide better services.

Varieties of data are becoming too many due to mobile devices and sensors are flooding telecom networks with data in myriad formats. Also, present call detail records (CDR) are enriched [5]. The velocity of data has been increasing day by day due to the dramatic increase in number of active devices connected to telecommunication networks, and a number of mobile service subscriptions are more than the population of a country.

1.2 Current Issues and Objectives

Due to highest competition at present, all the telecom service providers want to be innovative to maximize profit. However, most of them are suffering from making decisions in real time due to volume, variety, velocity, etc. of data coming in [3]. Hence, it is necessary to design a proper solution and make decisions in real time to achieve the goals of telecom operators as applying big data strategies are still in early phase of deployment for many of telecom operators [6].

The main objective is to design a real-time business intelligence model that provides insights into the telecommunication industry using big data strategies, and following facts need to be addressed by the proposed conceptual model

- To identify the variety of data that can be collected from the telecom industry.
- To develop a suitable platform to handle the large volume of structured and unstructured data.
- To explore real-time business insights based on the consumer data collected.
- To identify the churn in advance as the preventive basis.

- To develop an innovative business model by using information such as payment data for increased sales and match demand and offering nearby.
- To enhance the customer experiences with dynamic profiling and enhanced customer segmentation.

2 Conceptual Design of Real-Time Business Intelligence Model

Figure 1 illustrates the conceptual design of a real-time business intelligence model that provides insights into the telecommunication operators.

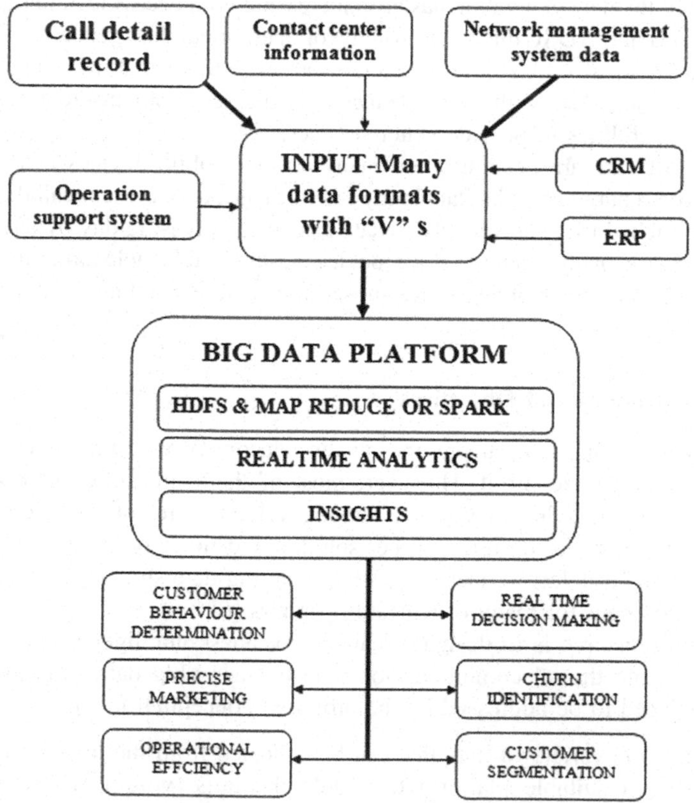

Fig. 1. Real-time business intelligence model

Since the revenue from voice traffic which was dominant before 2000 is declining and on the other hand, non-voice traffic revenue has been increasing rapidly for all the telecommunication operators globally. Therefore, it is very much appropriate to consider CDR for non-voice traffic in addition to traditional CDR, which will be dominant in the near future.

As the example of CDR, following main parameters were captured at the broadband remote access server (BRAS) for ADSL-based broadband access

- Date/Time
- Status of packet: Start or Stop
- Username for access
- Package subscribed
- Output octet for session (only for stop)
- Session-Id = "0817596553"
- PPPoE-Description = e.g.: "pppoe a4:7e:39:de:75:b8"
- Port-Id = "XXX_MSAG5200_28 atm 1/2/11/28:8.35"
- Framed-Protocol = PPP
- Event-Timestamp = "Apr 21 2016 04:55:05 IST"
- NAS-Port = 3758096459
- Unique-Session-Id = "c3dbb18e0ded8301"
- Timestamp = 1461194705.

For a single user, each session, two such logs will be created for commencing and completion of a session by the radius server. Hence, the data volumes of these log files are very high.

Block diagram of BRAS architecture for PPPoE authentication is given in Fig. 2 [7], which is mostly using by all the other broadband access technologies such as ADSL, VDSL and FTTH.

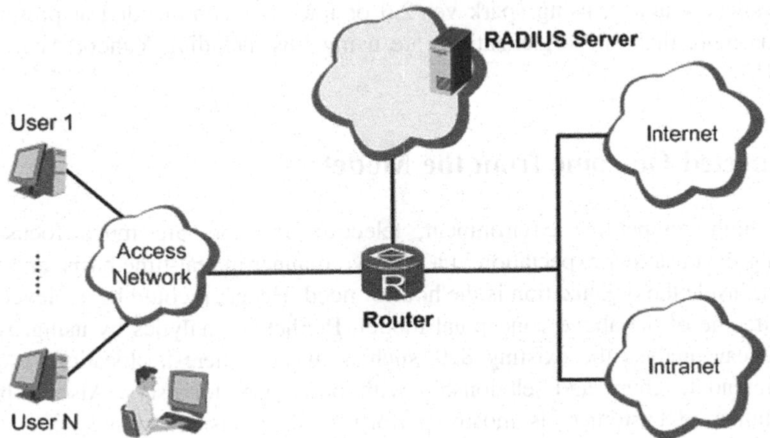

Fig. 2. Block diagram of PPPoE authentication [10]

Network management systems (NMS) are used to monitor the network nodes connected to telecommunication networks and for this business intelligence model. It is required to get the output log at the uplink port of the access node such as access gateways when required the further analysis based on requirement. Network related log outputs will be generated at NMS such as environmental, board-related exceptions and port-related issues such as abnormal conditions in uplink ports.

Contact centre will provide the customer feedback on services restoration or delivery such as customer satisfaction index, and this proposed conceptual model will explore the issues in the network in real time for the proposed solution or escalation of the issue to the relevant section of telecom operator.

Operation support system (OSS), customer relationship management (CRM) and enterprise relationship management (ERP) systems are used to keep customer related static data. OSS is having customer name and ID, services subscribed, customer contact details, service address, relevant port and circuit details for services subscribed, etc. CRM is having details related to customers and services subscribed by the customer. ERP is mainly having cost-related information such as cost related to service provision.

3 Data Processing

Distributing computing and platforms such as Hadoop and Spark will play an important role in processing data with real-time basis to take management decisions quickly [8]. Apache Spark is a most powerful data analysis engine based on Hadoop MapReduce, which assists the rapid processing of big data. It overcomes the limitations of Hadoop and emerging as the most popular framework for analysis.Considering the performance and advanced features, using spark ver 2.0 or later is recommended at present with HDFS as more than 500 organizations are using this including Yahoo, Amazon and eBay [9].

4 Expected Outcome from the Model

Present high competitive environment, telecom operators are much focusing on managing the customer expectation in a positive manner in real-time basis, and unified approach inside the organization is the hour of need. Hence, it could be achieved as the main outcome of the above conceptual model. Predictive analytics by using dynamic profiling can analyse the existing data such as usage, interests, location, economic status, intend to churn and relationship with other operator users. Also at present, segmentation of customers is mostly performing by considering the generation of revenue. Dynamic profiling and segmentation are possible with the big data strategies, which will retain with telecom operator with higher satisfaction due to consideration of all the factors beyond revenue.

Further based on usage or usage pattern, stakeholders can take initiatives to develop the particular area by considering the customer expectations and requirements. Further, from the business perspective, proper investment decision can be taken and also by increasing operational efficiency and to reduce OPEX through better decision-making. Next, it will be possible to sharpen campaign management, preemptive churn management, maximize revenue potential and reduce fraud and revenue leakages. As the important outcome, it is possible to predict unexpected and unpredicted events that might be an opportunity or threat for telecommunication operators.

5 Conclusion

The requirement of big data analytics to the telecommunication industry is well accepted. To fulfil the above, a conceptual model with real-time capabilities using big data strategies is important to achieve the goals. Validation of this model is required initially, with past data, which will help to validate the model in a real environment. Finally, the model needs to be validated with real-time data obtained from a telecommunication service provider.

References

1. IDC, The Digital Universe of Opportunities: Rich Data and the Increasing the value of Internet of Things, (2014), https://www.emc.com/leadership/digital-universe/ 2014 iview/executive-summary.htm, accessed on March 26, 2017.
2. Konstantinos Slavakis et al, Modelling and Optimization for Big Data Analytics, IEEE Signal Processing Magazine, September 2014, (2014).
3. Ari, Big data and advanced analytics in telecom: A multi-dollar billion revenue opportunity, www.heavyreading.com, (2013).
4. Lena. T. Ibrahin et al, Online Traffic Measurement and Analysis in Big Data: Comparative Research Review, American Journal of Applied Science, (2016), 13(4): 420.431.
5. Jacques Bughin, Reaping the benefits of big data in telecom, Journal of big data, Bughin *J Big Data (2016) 3:14.*
6. IDG. IDG enterprise big data research. 2014. Available at http://www.idgenterprise.com/report/big-data, accessed on March 30, 2017.
7. Guoshuai Zhao et al, Service Ratings Prediction by Exploring Social Mobile Users' Geographical Locations, IEEE Transactions on Big Data, Volume 3, Issue 1, (2017), pp 67–78.
8. Konstantin Shvachko, Hairong Kuang, Sanjay Radia and Robert Chansler, The Hadoop file distributed systems, 2010 IEEE 26th Symposium on Mass Storage Systems and Technologies (MSST), (2010).

9. Jitendra Bhatia and Preet Gandhi, Apache Spark: The Ultimate Panacea for the Big Data Era, http://opensourceforu.com/2017/01/apache-spark-the-ultimate-panacea-for-the-big-data-era/, (2017), accessed on March 29, 2017.
10. Huawei Support, BRAS Access Configuration, Huawei Support Documentation, http://support.huawei.com/enterprise/docinforeader.action?contentId=DOC1000079702&partNo=10072, access on April 15, 2017.

Dynamics of Data Mining Algorithms in Diversified Fields

Harshit Sinha[1], Jyoti Rajput[1], Achint Kaur[1], Shubham Baranwal[1], Tanupriya Choudhury[1](✉), Soniya Rajput[2], Kashish Gupta[1], and Gaurav raj[1]

[1] Department of Computer Science, Amity University, Noida, India
{harysinha, rajputjyoti4321, achintvirk9, shubhamformail, ghkashish95}@gmail.com, {tchoudhury, graj}@amity.edu

[2] Department of Computer Science, Bundelkhand University, Jhansi, India
soniya_hg@rediffmail.com

Abstract. Data mining is a concept which looks for valuable resources as examples from substantial measure of information, i.e., from a large amount of data. The theory or the information in this paper examines few of the data mining procedures, calculations, and a portion of the associations which have adjusted information mining innovation to enhance their organizations and discovered great outcomes. In this paper, it has been centered on assortment of strategies, methodologies, and diverse zones of the examination which are useful and set apart as the essential field of information mining technologies. As per prior knowledge that many MNC's and vast associations are worked in better places of the distinctive nations. This paper bestows more number of utilizations of the information mining and furthermore o centers extent of the information mining which will accommodating in the further research.

Keywords: Classification · Clustering · Analysis · Association · Application

1 Introduction

The expansion and growth of computer science and technology have created a very enormous quantity of records and enormous data in numerous zones available. The investigation databases and the current and previous information technology have expected growth in a methodology to storehouse [1] and deploy this valuable data for making or formulating of various advance decision. Data mining is a procedure for withdrawal or taking out of useful knowledge and designs (that denotes something or the other) from a very large amount of data. It is similarly known as information detection method, [1] information mining of data from data, information abstraction/withdrawal or data/design examination (Fig. 1).

Data mining is a commonsensical procedure that is cast-off exploration purpose over enormous quantity of figures, which will help us in finding the important and relevant data. The main aim of this method is to look for such patterns or designs that completely represent some state of information that is unknown [2]. When these

V. Bhateja et al. (eds.), *Information Systems Design and Intelligent Applications*, Advances in Intelligent Systems and Computing 672,
https://doi.org/10.1007/978-981-10-7512-4_107

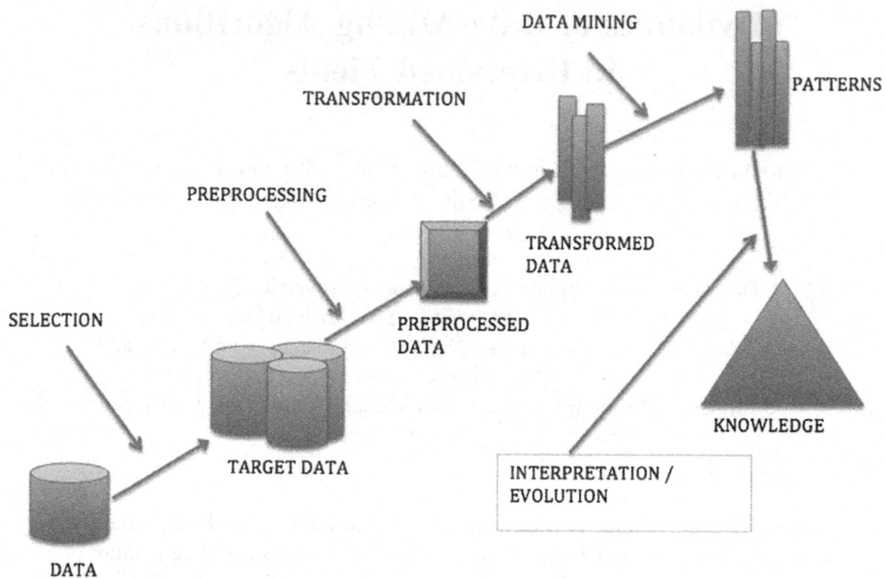

Fig. 1. Processing procedure

designs or such patterns are detected or computed then from such designs certain study or some kind of decisions on some parameters that are relevant for that particular concept or for business development is necessary can be taken.

The major steps that are considered in this are [2] as follows:

- Investigation,
- Pattern recognition,
- Implementation.

Investigation: This is the first stage of data investigation in which data is prepared and transmuted into some other alternative arrangement, and the available significant variables and also the kind of data grounded on the basis of problems are determined.

Pattern recognition: Immediately when the data is discovered, cultured and well-defined for various precise variables, the next important and available step is to formulate pattern recognition. Recognize and decide as to which pattern is to be considered which denotes the best prediction for the information that is needed [2].

Implementation: Relevant designs that give information are installed for anticipated results.

2 Procedures and Algorithms of Data Mining

In data mining, there are numerous procedures and algorithms that are available for mining of data [3].

2.1 Classification

Classification [1] is the most generally connected information mining method, which utilizes an arrangement of preordered cases to build up a model that can characterize the number of inhabitants in records on the loose. Extortion discovery and credit chance applications are especially appropriate to this kind of investigation. Godbole et al. [3] This approach oftentimes utilizes choice tree or neural system-based characterization calculations. The information arrangement handle includes learning and order. In learning, the preparation information is dissected by characterization calculation. In order, test information is utilized to appraise the precision of the grouping rules. On the off chance that the exactness is satisfactory, the tenets can be connected to the new information tuples. For an extortion recognition application, this would incorporate finish records of both false and substantial exercises decided on a record-by-record premise. The classifier-preparing calculation utilizes these prearranged cases to decide the arrangement of parameters required for legitimate segregation. The calculation gathered from the above classification is as shown in Table 1; it is then encoded and helps rearranging these factors in prototypical form known as a classifier [3].

Classification models:: *Types*

Table 1. Types of classifiers

Sno.	Features	Limitations
1. C4.5 algorithm [3]	• Models that were built by this algorithm are interpreted very easily • Implementation is very user friendly • Can use both dynamic and distinct values • Noisy data can also be considered in this algorithms	• May result in various decision trees when small data is taken • Fitting is not proper as in leads to over fitting
2. ID3 algorithm [4]	• It is more efficient then C4.5 algorithm in terms of accuracy • Provides an increase in the rate of detection and decrease in the rate of space consumption	• In this the searching time required is very large • A large amount of memory is required for storing trees
3. K-nearest neighbor algorithm [4]	• Classes do not need to be separable in linear form • Learning process in this algorithm has zero cost • Sometimes when a noisy data set is taken over then this algorithm works in a much robust manner	• In this type of algorithm, if the data set taken is large then the time taken by the nearest neighbors can be large • It is very much sensitive to the data that is noisy or irrelevant attributes • Algorithm performance credibility is dependent on the number of data sets

(*continued*)

Table 1. (*continued*)

Sno.	Features	Limitations
4. Naïve Bayes algorithm [5]	• Implementation is very easy • The rate of classification and efficiency of computation is great • Prediction of results is accurate for almost all classification and problems related to prediction [10–13]	• If the amount of data that is taken is less then the precision of the algorithm gets decreased • In order to get the best results the data set taken must be of large number
5. Support vector machine algorithm [5]	• The accuracy of this algorithm is very high • Even if the data is not linear still it will work in a better manner and will provide the results	• While training and testing, the data speed and the size are required the most • For classification in various number of cases, high complexity and high expandable memory are required
6. Neural network algorithm [5]	• It is very much easy to use as there are only few parameters that need to be adjusted • Implementation is very much easy • Wide-range applicability	• If large data set is selected then a high amount of processing speed is required • It is not easy to know how many neurons or layers are necessary to have a desired output

2.2 Clustering

Clustering can be said as ID of comparative classes of items. By utilizing bunching methods, it additionally distinguishes thick and meager districts in protest space and can find general dispersion example and relationships among information characteristics. Characterization approach can likewise be utilized for powerful methods for [6] recognizing gatherings or classes of question yet it turns out to be expensive so grouping can be utilized as preprocessing methodology for property subset choice and order. For instance, to frame gathering of clients in view of obtaining examples, to classifications qualities with comparable usefulness [6].

Clustering procedures:: *Types*

1. Partitioning methods,
2. Hierarchical agglomerative (divisive) methods,
3. Density-based methods,
4. Grid-based methods,
5. Model-based methods.

2.3 Prediction Analysis

Regression system can be adjusted for predication [6]. Regression investigation can be utilized to demonstrate the connection between at least one autonomous factors and ward factors. In information mining, autonomous factors are properties definitely known and reaction factors are what is needed to foresee. Lamentably, some genuine issues are not just expectation. For example, deals volumes, stock costs, and item disappointment rates are all exceptionally hard to anticipate on the grounds that they may rely on upon complex cooperation of different indicator factors. Subsequently, more mind-boggling procedures (e.g., calculated relapse, choice trees or neural nets) might be important to conjecture future qualities. A similar model sorts can frequently be utilized for both relapse and characterization. For instance, the classification and regression trees (CARTs) and choice tree calculation can be utilized to fabricate both order trees (to group absolute reaction factors) and relapse trees (to conjecture persistent reaction factors). Neural systems also can make both order and relapse models [6].

Regression procedures:: *Types*

1. Linear regression,
2. Multivariate linear regression,
3. Nonlinear regression,
4. Multivariate nonlinear regression.

2.4 Association Rule

Association and furthermore [5], relationship is more often than not to discover visit thing set discoveries among expansive informational indexes. This sort of discovering helps organizations to settle on specific choices, for example, list configuration, cross-showcasing, and client shopping conduct examination. Affiliation Rule calculations should have the capacity to create rules with certainty values short of what one. However, the quantity of conceivable Association Rules for a given data set is for the most part vast, and a high extent of the principles is generally of little (assuming any) esteem.

Association rules:: *Types*:

1. Multilevel association rule,
2. Multidimensional association,
3. Quantitative association rule.

2.5 Neural Networks

Neural system is an arrangement of associated information/yield units and every association has a weight give it, Amid the learning stage, arrange learns by modifying weights in order to have the capacity to foresee the right class marks of the info tuples. Neural systems have the amazing capacity to get importance or loose information and can be utilized to concentrate on designs and distinguish patterns [7] that are too intricate, either people or other PC procedures. These are appropriate for consistent esteemed sources of info and yields. For instance written by hand character revamping, for preparing a PC to articulate English content and numerous genuine business issues and have as of now been effectively connected in numerous enterprises. Neural systems are best examples or patterns in information and appropriate for expectation or estimating needs [7, 14].

3 Applications of Data Mining

Information mining is a moderately new innovation that has not completely developed. In spite of this, there are various ventures that are as of now utilizing it all the time. Some of these associations incorporate retail locations, healing centers, banks, and insurance agencies. A large number of these associations are consolidating information mining with so many things such as measurements, example acknowledgment [7], and other critical apparatuses. Information mining can be utilized to discover examples and associations that would somehow or another be hard to discover. This innovation is prominent in numerous organizations since it permits them to take in more about their clients and settle on savvy advertising choices.

Application of Data Mining:: *Future Health care*:
Information mining holds extraordinary potential to enhance well-being frameworks. It utilizes information [4] and investigation to distinguish best practices that enhance mind and diminish costs. Specialists utilize information mining approaches like multidimensional databases, machine adapting, delicate figuring, information representation, and insights. Mining can be utilized to foresee the volume of patients in each class. Procedures are produced that ensure that the patients get suitable care at the perfect place and [7] at the ideal time. Information mining can likewise assist medicinal services backup plans with detecting misrepresentation and mishandling.

Application of Data Mining:: *Education*
There is another rising field, called Educational Data Mining, worries with creating techniques that find information from information starting from instructive environments. The objectives of EDM are recognized as foreseeing understudies' future learning conduct, concentrate the impacts of instructive support, and progressing logical information about learning. Information mining can be utilized by a foundation to take exact choices and furthermore to foresee the consequences of the understudy. With the outcomes, the foundation can concentrate on what to instruct and how to educate. Learning example of the understudies can be caught and used to create strategies to show them. [2]

Application of Data Mining:: *Customer Segmentation*:
Customary [4] statistical surveying may help us to section clients, yet information mining dives in deep and expands advertise adequacy. Information mining helps in adjusting the clients into an unmistakable fragment and can tailor the necessities as indicated by the clients. Market is constantly about holding the clients. Information mining permits to discover a fragment of clients in light of defenselessness, and the business could offer them with unique offers and improve fulfillment [8].

Considering a data set of an icon aircraft considering the comments posted by the public on the social media for filtering the comments and getting out the analysis on the same:

Sentiment analysis process includes the following steps:

(i) Data file selection,
(ii) Preprocessing of data,
(iii) Calculate polarity,

(iv) Feature extraction,
(v) Predictive sentiment analysis,
(vi) Algorithm selection,
(vii) Comparative graph preparation (Fig. 2).

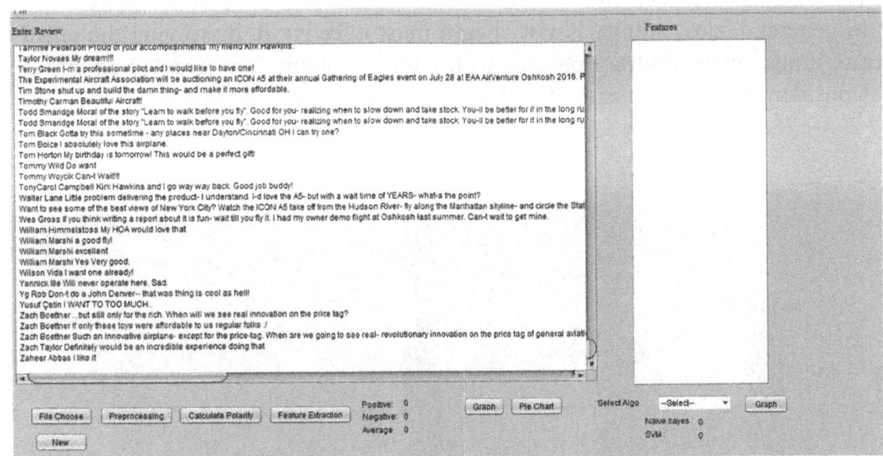

Fig. 2. ICON Aircraft data set for training and testing

The data set is taken from the company name ICON Aircraft where the data set consist of positive, negative, and neutral comments being posted by the public, and this program helps in mining the type of comments for the use of figuring out and rating their product. There were various attributes that were taken into consideration, and these are negative comments, positive comments, and neutral comments of the people that were posted on the social media for the particular company; so in order to extract the good, bad, and neutral comments, we came across with this algorithms to segregate the data.

After selecting data file, preprocessing of data has completed. Then we need to calculate polarity and extraction of features. Next step is to get predictive analysis using Naïve and SVM classifier. There is a drop box which gives us option of using Naïve Bayes classifier and support vector machine (SVM) classifier.

Considering the Naïve Bayes algorithm:

$$P(D/C)\,P(C) \tag{1}$$

$$P(C/D) = \tag{2}$$
$$P(D)$$

As D i.e. Document is common to all classes so therefore the
P(D/C) P(C) i.e. Likeliness probability matters
And Considering SVM Algorithm:

$$X2 = mX1 + C \text{ or } X2 - mX1 = C \tag{3}$$

or

$$w1X1 + w2X2 = C \text{ in general}$$

In general, lexical affinity is not always most effective in dissolving the effect on phrase; it also offers arbitrary phrases with a likely "affinity" to targeted emotions.

Considering the above example, the output that comes is (Fig. 3)

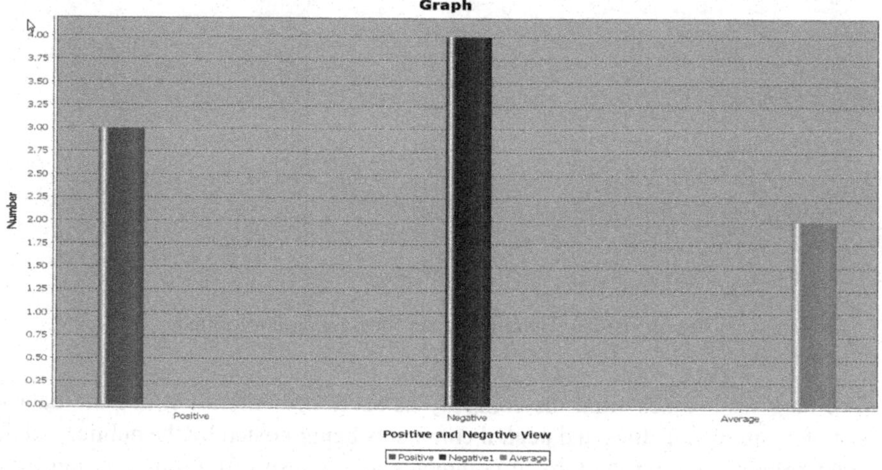

Fig. 3. Variation of negative, positive and average segregation of the data set using both algorithms

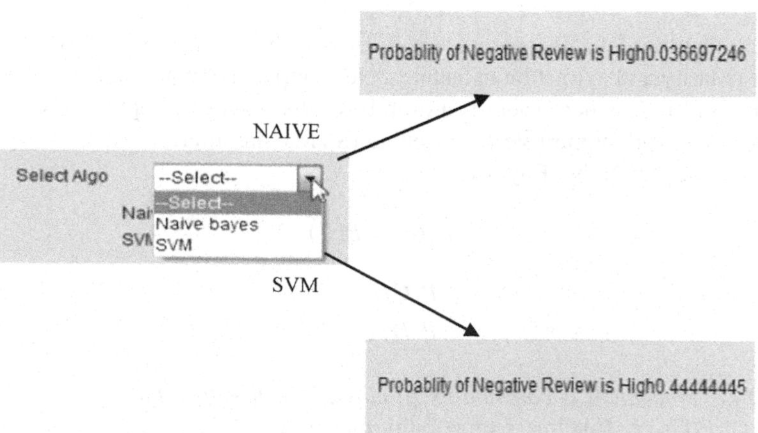

Fig. 4. Classification algorithm selection in sentiment analysis process

Figure 4 shows that through Naïve Bayes the probability of negative reviews are high, i.e., 0.036697246, and through SVM, probability of the negative probability reviews are high, i.e., 0.4444445.

Considering the results, below are the steps taken to bring out the same:

As proposed, an opining mining methodology to help new customers to make a decision for taking up of or purchasing or not purchasing a product by summarizing the reviews. The proposed approach is based on review's classification to conclude the summary of the customer's opinions [7, 8] and visualizing outcomes as well.

Our work is divided into four phases:

Phase 1: Collecting customer reviews for classification.

Phase 2: In phase 2, we process the collecting data called data processing.

It also contains the following steps:

(a) Removing stop words
(b) Stemming

Phase 3: It contains the following steps:

(a) Collect subjective sentences
(b) Extract features from subjective sentences
(c) Classifying features from feature extraction
(d) Identify sentences polarity
(e) Then finally polarity classification

Phase 4: It reviews visual summarization.

Customer reviews	Identify sentence
Remove stop words	Polarity
Stemming	Polarity classification
Feature extraction	Reviews visual
Subjective sentences	Summarization
Features classification	Customer reviews

4 Conclusion

Information mining has significance with respect to finding the examples, anticipating, disclosure of learning and so forth in various business spaces. Information mining procedures and calculations, for example, grouping, "bunching and so forth, helps in finding the examples to settle on the future patterns in organizations to develop. Information mining has wide application area nearly in each industry where the information is created that is the reason information mining is viewed as a standout amongst the most critical outskirts in database and data frameworks and a standout amongst the most encouraging interdisciplinary improvements in Information Technology, "although complex is an engaging project. Serving as the perfect market analysis tool, both organizations and individuals stand to benefit from this system. In analysis, we have collected probability of negative reviews through Naïve Bayes and SVM. As per results, SVM is used for accuracy due to high probability. A more of work has been made to check the content of the social Websites, especially the posts or the views of the users; their sentiment analysis has to be done. The main aim is to analyze the sentiments and then to compare those views with the main topics and other platforms. There is enormous amount of the posts given by the users on the social Websites. But some of the people are interested in knowing the influence and the relations among those posts. Many of the researchers find the influence of the post on these social Websites. But we cannot find the influence of these posts without knowing the sentiments of the posts. Neglecting such an abundant source of information seems like a daunting task, and hence, this sentiment analysis tool has been developed.

References

1. D. Bollegala, D. Weir, J. Carroll, —Cross-Domain Sentiment Classification Using a Sentiment Sensitive Thesaurus‖, IEEE Transactions on Knowledge and Data Engineering, 2012, pp. 1719–1731.
2. G. Tsoumakas, I. Katakis, and I. Vlahavas, "Mining Multilabel Data", Data Mining and Knowledge Discovery Handbook, 2nd ed. Springer, ISBN 978-0-387-09822-7 e-ISBN 978-0-387-09823-4, https://doi.org/10.1007/978-0-387-09823-4, 2010.
3. N. Godbole, M. Srinivasaiah, and S. Skiena, "Large-scale Sentiment analysis for news and blogs," in Proceedings of the International Conference on Weblogs and Social Media (ICWSM), 2007.
4. Nidhi R. Sharma, Prof. Vidya D. Chitre, "Opinion Mining, Analysis and its Challenges", International Journal of Innovations & Advancement in Computer Science (IJIACS), ISSN 2347 – 8616, Volume 3, Issue 1, April 2014.
5. J. Isabella, Dr. R.M.Suresh, "Analysis and evaluation of Feature selectors in opinion mining", Indian Journal of Computer Science and Engineering (IJCSE), ISSN: 0976-5166 Vol. 3 No.6 Dec 2012-Jan 2013.
6. G. Vinodhini, RM.Chandrasekaran, "Sentiment Analysis and Opinion Mining: A Survey", International Journal of Advanced Research in Computer and Communication Engineering, Volume 2, Issue 6, June 2012.
7. Arti Buche Dr.M.B. Chandak, Akshay Zadgaonkar, "Opinion Mining And Analysis: A Survey", International Journal on Natural Language Computing (IJNLC), Volume 2, No. 3, June 2013.
8. Ravi kumar V, and K. Raghuver, "Web User Opinion Analysis for Product Features Extraction and Opinion Summarization", International Journal of Web & Semantic Technology (IJWesT) Vol.3, No.4, October 2012.
9. Tanupriya Choudhury, Vivek Kumar, Darshika Nigam, Bhaskar Mandal," Intelligent Classification of Lung & Oral Cancer through Diverse Data Mining Algorithms", International Conference on Micro-Electronics and Telecommunication Engineering (ICMETE), IEEE Explore, https://doi.org/10.1109/ICMETE.2016.24, 22–23 Sept. 2016.
10. Tanupriya Choudhury, Vivek Kumar, Darshika Nigam," An Innovative and Automatic Lung and Oral Cancer Classification Using Soft Computing Techniques", International Journal of Computer Science & Mobile Computing, Vol.4 Issue.12, Dec- 2015, pg. 313–323.
11. Tanupriya Choudhury, Vivek Kumar, Darshika Nigam," An Innovative Smart Soft Computing Methodology towards Disease (Cancer, Heart Disease, Arthritis) Detection in an Earlier Stage and in a Smarter Way", International Journal of Computer Science and Mobile Computing, Vol.3 Issue.4, April- 2014, pg. 368–388.

12. Tanupriya Choudhury, Vivek Kumar, Darshika Nigam," Intelligent Classification & Clustering Of Lung & Oral Cancer through Decision Tree & Genetic Algorithm", International Journal of Advanced Research in Computer Science and Software Engineering, Volume 5, Issue 12, December 2015.
13. Gaurav Dangi, Tanupriya Choudhury, Praveen Kumar," A Smart Approach to Diagnose Heart Disease through Machine Learning and Springleaf Marketing Response", IEEE International Conference on Recent Advances and Innovations in Engineering (ICRAIE-2016), December 23–25, 2016.

Application of Genetics Using Artificial Immune System Through Computation

Harshit Sinha[1], Jyoti Rajput[1], Achint Kaur[1], Shubham Baranwal[1], Kashish Gupta[1], Tanupriya Choudhury[1(✉)], and Soniya Rajput[2]

[1] Department of Computer Science, Amity University, Noida, India
{harysinha, rajputjyoti4321, achintvirk9, shubhamformail, ghkashish95}@gmail.com, tchoudhury@amity.edu

[2] Department of Computer Science, Budelkhand University, Jhansi, India
soniya_hg@rediffmail.com

Abstract. One of the key characteristics of the human being insusceptible framework is to spot the nearness of pathogens, and all things considered there are numerous invulnerable calculation and algorithms which perform inconsistency uncovering and example acknowledgment. An extra aspect of the human safe framework is that a fitting effector comeback is created upon the identification of a pathogen "a procedure named the essential reaction. Moreover, the human invulnerable framework can recall the appropriate reaction to a specific pathogen—the auxiliary reaction. The unpredictable coordination of both the essential and optional reactions is profoundly progressive—portrayed in immunological terms as plastic. In this research work in which it will be shown an outline of the correct sections of the era of a T-helper cell essential reaction and the instruments by which it educates optional reactions and talk about how this can be computationally helpful in artificial immune framework advancement."

Keywords: Artificial immune frameworks · Essential and auxiliary reaction

1 Introduction

Artificial Immune Systems [1, 2] are a gathering of algorithms grounded on such purpose and comportment of humanoid immune system. Mass of AIS emphasizes on finding of irregularities and optimization of finding. AIS have straightway used or hired the principal and subordinate reaction tools. Till now, some have unified system into a B-cell recreation system to basically create one of the first and foremost immune system tactics through reminiscence. Safety-centered investigation done by various people made a vision of instigating a type of subordinate comeback, fabricating a computationally [3] modified or manipulated or an original faster reply to formerly confronted dangers in data configurations in various observed transportable ad hoc networks. Newly, attention in main and subordinate data replies has reduced, hypothetically with the postulation element is not interesting.

V. Bhateja et al. (eds.), *Information Systems Design and Intelligent Applications*, Advances in Intelligent Systems and Computing 672,
https://doi.org/10.1007/978-981-10-7512-4_108

The hidden immune system of main and subordinate reaction is facilitated by a class of white platelet named a T-aide cell. These cells can adjust to novel dangers, named pathogens, and to make suitable for that reaction. Contingent upon the sort of reaction which is produced, the reaction is spared in a memory framework ought to a similar pathogen be experienced later on. The sort of reaction retained by the resistant framework is reliant on the way of the essential T-partner cell reaction. The coordination of a firmly controlled and refined essential reply depends on the dynamic technique of T-assistant cell sub-sort exchanging. In this paper, it will be shown a rundown of the [3] T-assistant cell instruments, also which make this dynamic essential reaction, and how this creates and keeps up a memory of the reaction for giving future assurance. These instruments go past what is utilized beforehand in Artificial Immune System as it perspectives the way toward making an optional reaction from an essential reaction as a dynamic instrument, containing frequent administrative segments which keep up the precision of the framework [4].

2 Role of Immune System

- Shield our bodies from disease [5]
- Essential safe reaction
- Dispatch a reaction to attacking pathogens
- Auxiliary invulnerable reaction
- Keep in mind past experiences
- Quicker reaction the second time around.

3 Where Does the Immune System Get Found?

See Fig. 1.

4 For Pattern Recognition for Immune System

- The immune recognition depends on the complementarity between the coupling district of the receptor and a part of the antigen called epitope [6].
- Antibodies display a solitary sort of receptor, and antigens may introduce a few epitopes.
- This implies diverse antibodies can perceive a solitary antigen and considering Table 1, showing the difference in immune system and computational system (Fig. 2).

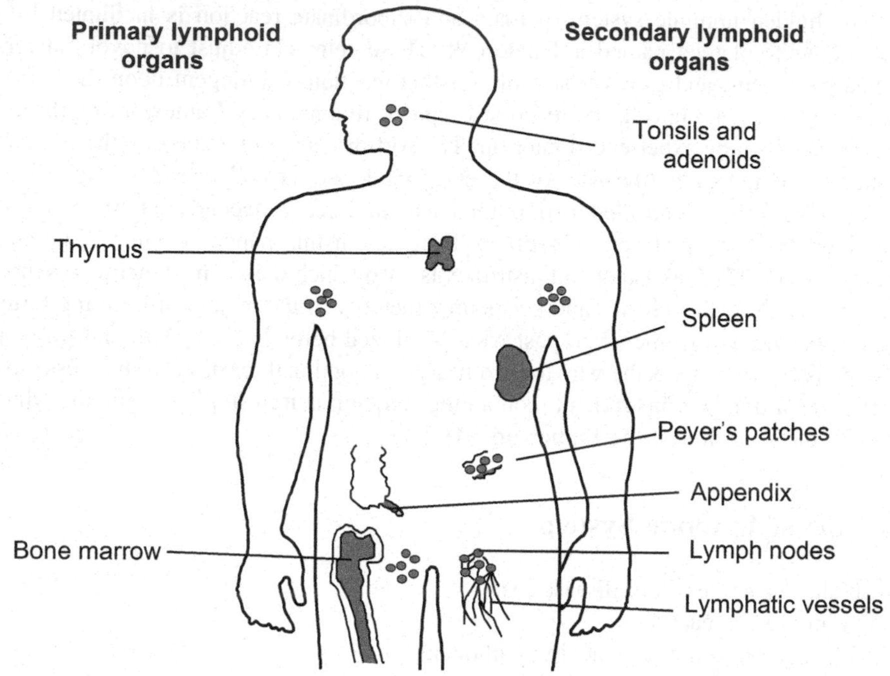

Fig. 1. Artificial Immune System

Table 1. Comparisons between GA, NN, and AIS

Parameters	GA (Optimization) genetic algorithm	NN (Classification) neural network	AIS Artificial Immune System
Components	Chromosome strings	Artificial neurons	Attribute strings
Location of components	Dynamic	Pre-defined	Dynamic
Structure	Discrete components	Networked components	Discrete components/networked components
Knowledge storage	Chromosome strings	Connection strengths	Component concentration/network connections
Dynamics	Evolution	Learning	Evolution/learning
Meta-dynamics	Recruitment of components	Construction of components	Recruitment of components
Interaction between components	Crossover	Network connections	Recognition/network function
Interaction with environment	Fitness function	External stimuli	Recognition/objective function

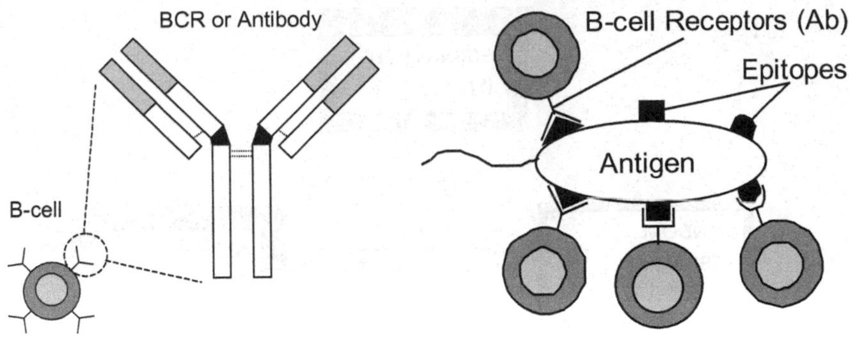

Fig. 2. B-cells and BCR antibody

5 Algorithms and Models for Basic Immune System

a. *Bone Marrow Model*:

Quality libraries are utilized to make antibodies from the bone marrow counter acting agent generation through an arbitrary link from quality libraries straightforward or complex libraries.

b. *Negative Selection Algorithms:*

Thought taken from the negative determination of T-cells in the thymus connected at first to PC security split into two sections:

Blue penciling (censoring) [7] observing.

- Each duplicate of the calculation is interesting, so that each secured area is given a one of a kind arrangement of finders.
- Location is probabilistic, as a result of utilizing diverse arrangements of indicators to secure every substance.
- A strong framework ought to distinguish any remote action instead of searching for particular known examples of interruption.
- No earlier learning of peculiarity (non-self) is required [7] (Fig. 3).

c. *Immune Network Models*:

Used immune network theory as a basis, proposed the Genetic Artificial Immune Network System algorithm

Step 1: Initialize GAINE [8]
Step 2: For each antibody and antigens
Step 3: Assign each ARB element present in the GAINE to each antigen present considering Eq. 1
Step 4: Calculating the simulation level of the present Artificial Recognition Balls (ARB) in the GAINE

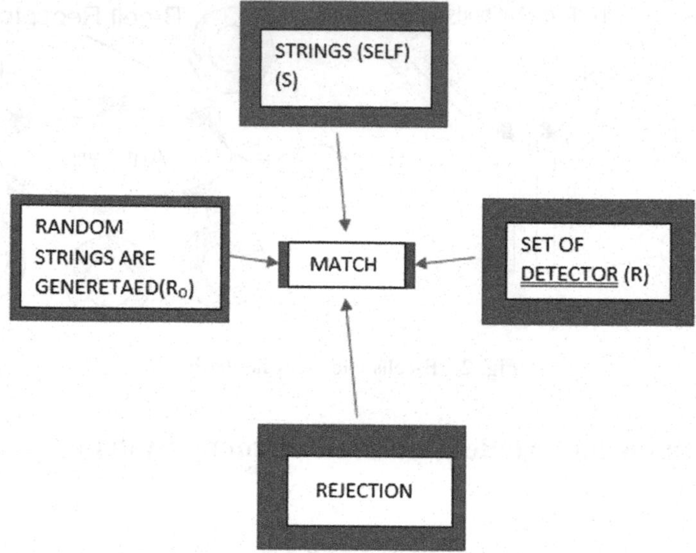

Fig. 3. Considering T-cells and considering the negative selection

Step 5: Allocate T-cells to ARBs or B-cells to ARBs depending upon the simulation model and its level
Step 6: Now, filtering further removes the weakest of the ARBs (those ones that do not hold any of the T-cells or any of the B-cells)
Step 7: Now If the ending condition or the termination meets up
Reach out or Exit
Else
Step 8: Now mutating the ARBs and cloning the same
Step 9: After the mutation and cloning, integrate the new ARBs available and out for integration into GAINE model and simulator.

Simulation model equation for the establishing the immune network model:

$$\text{Simulation} = \mathrm{C} * \left(\sum \mathrm{m}(\mathrm{b}, \mathrm{yf_j})\right) - \mathrm{P1}\left(\sum \mathrm{m}(\mathrm{b}, \mathrm{yg_j})\right) + \mathrm{P2}\left(\sum \mathrm{m}(\mathrm{b}, \mathrm{k})\right) - \mathrm{P3} \tag{1}$$

where

"N is the number of antibodies, n is the number of antigens, c is rate constant that depends on the number of comparisons per unit of time and the rate of antibody production stimulated by a comparison, a is the current B-cell object, yfj represents the jth B-cell's epitope, ygj represents the jth B-cell's paratope, and y represents the current antigen. This equation takes into account matches between neighbors (both stimulation and suppression) and antigens. The antigen-training data are being presented with a set for a number of times, and once this has completed, the antibodies are saved and can be used for classification. The authors claimed a certain amount of success for their technique, claiming a 90% success rate (on average) for classification of unseen items."

$$\text{Simulation2} = \sum 1-\text{qf} + \sum (1-\text{dis}) - \sum \text{dis} \quad (2)$$

where

"qf is defined as the distance between the ARB and the antigen in the normalized data space such that $0 \leq \text{qf} \leq 1$, and disx the distance of the xth neighbor from the ARB. The population control mechanism that replaced the 5% culling mechanism forces ARBs to compete for survival based on a finite number of resources that AINE contains; the more stimulated an ARB, the more resources it can claim. Once an ARB no longer claims any B-cells, it is removed from the AIN" (Fig. 4).

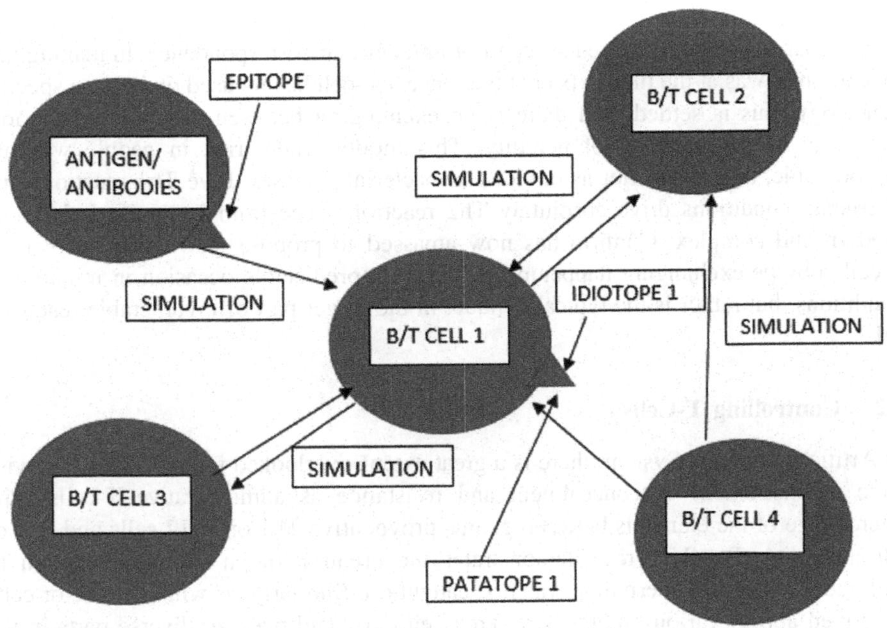

Fig. 4. Immune network model

6 Comparison of Three Algorithms

See Table 1.

7 Reactions and Plastic T-Cells

7.1 Multiplicity in T-Cells

As far as understanding T-aide cell elements, the expression "plastic" alludes to the capacity of a T-cell to receive diverse utilitarian parts and practices under various natural conditions. In immunological terms, the thought of T-cell pliancy is generally new, and there is still levelheaded discussion and guess regarding what characterizes the plastic T-cell subtypes. It might appear like insignificant semantics, yet the revelation that there is a contrast between a cell ancestry and a cell sub-sort is the thing that has driven research in the comprehension of the versatility of T-cell comebacks [9]. The pliancy is thought to have the capacity to make proper and custom fitted reactions to both experienced and un-experienced pathogens.

At first, it was expected that T-cell separation was confined to a terminally separated ancestry. In the 1980s, it was resolved in mice models that T-aide cells separate into two classes of various utilitarian subtypes to be specific T-assistant sort 1 (Th1) cells and T-aide sort 2 (Th2) cells [9]. In severe research facility conditions with mice with no earlier pathogenic introduction, this instrument is clear. The Th1 and Th2 communicated diverse cytokine star chronicles, where a cytokine is characterized as an atomic marker which is utilized as a part of intracellular correspondence. In making this revelation, it was at the time expected that once a T-cell has focused on being a specific genealogy, this is settled, and there is no exchanging between various useful parts, according to the meaning of heredity. This model when tried in people was not reproducible, and it was not as obvious as bacterial diseases drive Th1 reactions and allergenic conditions drive absolutely Th2 reactions. The truth is significantly more modern and complex. Confirm has now amassed to propose that not exclusively is T-cell subtype exchanging happening is not an unpredictable occasion in remarkable conditions, but rather is the typical conduct in the larger part of invulnerable reactions [9].

7.2 Controlling T-Cells

In Artificial Immune System, there is a great extent overlooked is that T-cells likewise have an instrument of concealment and resistance as administrative T-cells. The interchange of the elements between genius provocative Th1 or Th17 cells and that of administrative T-cells (Treg) demonstrates the creation of an essential reaction to pathogens as more modern than was first caught on. Similarly, as with the erector cells portrayed above, various subtypes of Treg cell exist and perform diverse parts in the span of the insusceptible [10] reaction. There are two primary sorts of Treg cell, characteristic Tregs (nTregs) and inducible Tregs (iTregs).

The nTregs play out a critical part in keeping up homeostasis and are capable to a limited extent for the avoidance of autoimmunity. In Artificial Immune System, it has been seen thymic choice from an absolutely negative determination viewpoint. However, positive selection additionally happens and at the crossing point of positive and negative choice discovers the arrangement of nTregs. At first, the iTreg is of another subtype, either a Th1 or Th17 cell. "Changes with regard to the area from a star incendiary to a calming domain. The nearness of IL-10 is presently accepted to be the key atom required in progressively changing T-cells to and from the iTreg sub-sort. At the point when a disease happens, the inborn safe framework deciphers both signs and antigen to the erector T-cell populaces in lymph hubs. Given adequate enactment of the T-cells, a versatile reaction is mounted, and a few instruments start, including the creation of antibodies and the enlistment and actuation of inborn" cells [10].

7.3 Active Directive Memory

The subtypes of T-cell made and kept up amid a reaction additionally impact the creation and upkeep of safe memory. Memory is characterized as the ability to store and review data from already experienced occasions and to react more rapidly and forcefully than on the main experience. Established models of T-cell memory depended on the clonal development of terminally separated Th1 or Th2 cells to prepare memory cells for quick expansion on account of a future experience. The memory T-cells get away from the apoptotic procedure of the expulsion of T-cell clones taking after the determination of infection [4]. The assurance of what T-cells progress toward becoming memory cells is dictated by the quality of the T-assistant reaction, formed by the pliancy of the erector T-cells [10]. The revelation of the energy of nTregs and iTregs however has expanded this perspective. It is "suspected that not exclusively are memory cells created to erector T-cells with administrative usefulness. This definitely changes the progression of a memory-based framework from just recalling pathogenic-related antigen yet to recollecting antigen which bounds to administrative cells." Memory administrative T-cells are a test to characterize as a "genuine" memory cell as self-antigen is communicated continually inside the body, and along these lines, the idea that memory just exists without the antigen was abused in the meaning of these memory cells. Notwithstanding, now research center procedures have propelled, and this wounds up plainly [11] conceivable prompting the characterization of a cell which delivers extensive rejoinders against self-antigen. The era and determination of an iTreg can add to this collection permitting memory to adjust and to endure "evolving self," appeared specifically in the progressions which occur amid pregnancy. Prove exists to demonstrate that as opposed to the nTregs guiding the very own T-memory cells are outfitted with extraordinary suppressive power [7]. While the correct systems are a dynamic zone of research, is intriguing that memory is not just recollecting that is contained and the awful but rather additionally having the capacity to effectively smother immune system reactions. Specifically, administrative T-memory cells are found in the outskirts and not as limited to the lymph hub zones [11]. The possibility of duality in memory of cells, driven by the pliancy of the essential reaction, to effectively keep up and elevate resistance and to give an ancient to the upkeep of homeostasis [12,13].

8 Applications

1. *Anomaly Detection*: The typical conduct of a framework is regularly portrayed by a progression of perceptions after some time. The issue of recognizing oddities, or inconsistencies, can be seen as discovering deviations of a trademark property [1] in the framework. For PC researchers, the distinguishing proof of computational infections and system interruptions is viewed as a standout among the most essential peculiarity recognition assignments [1].
2. *Virus Detection*:
   - "Okamoto and Ishida proposed a distributed approach which detected viruses by matching self-information—first few bytes of the head of a file, the file size and

path, etc., against the current host files. Viruses were neutralized by overwriting the self-information on the infected files. Recovering was attained by copying the same file from other uninfected hosts through the computer network (Fig. 5).

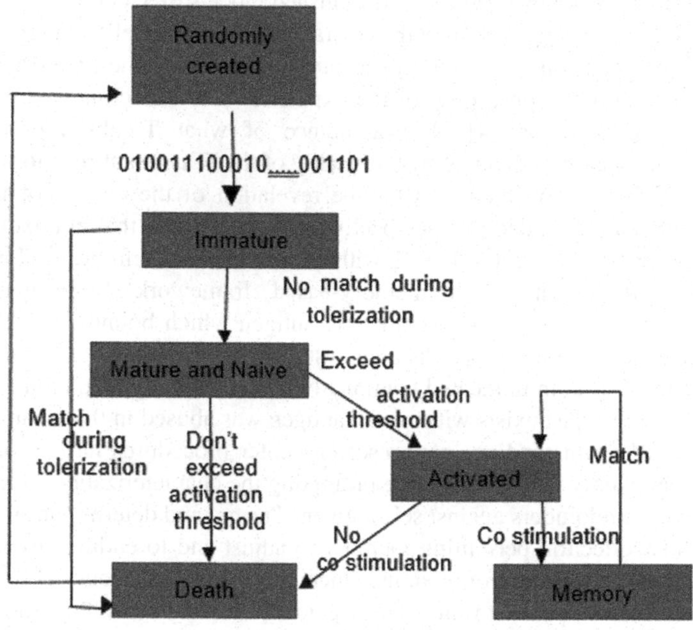

Fig. 5. Detection procedure

- Protecting the systems and personal computers from the virus that gets infected and harms the system.
- Considering the virus-free systems, coming across a concept for having a virus-free system.

Computation of Virus Detection using the Malicious Executable Detection Algorithm:

Considering the Detector Generation Algorithm and Fig. 6:

Step 1: Begin initialize Istep, Id, K = 0
Step 2: Do cutting f(ex, n) from Eg(b)
Step 3: i = 0
Step 4: while i <=n − 1d do
Step 5: Begin
Step 6: d = seq(ek, n), I, id);
Step 7: id d! = $Did then Did ← d;
Step 8: i = i+Istep

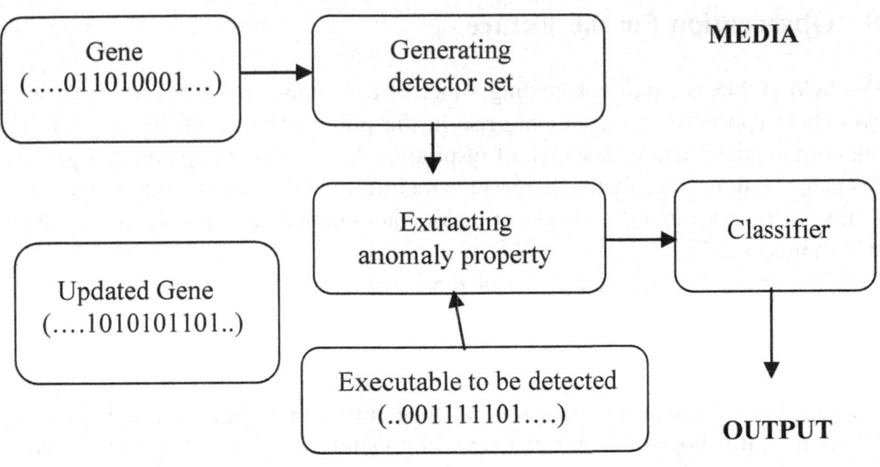

Fig. 6. Flowchart

Step 9: End
Step 10: k = k+1
Step 11: until Eg(b) is empty
Step 12: Return Did
Step 13: End

Illustration of Detector Generating Process:
File Hex Sequence: 56 32 12 0A 34 ED FF 00 2D … 00 0A 34 ED FF FA 11 00
Extracting Detector: 56 32 12
32 12 0A
12 0A 34
¦ ………………………………………¦

FF FA 11
FA 11 00

Generating Process of 24-bit Detectors with 8-bit stepsize (ld = 24, lstep = 8)

9 Conclusion

Late advances in epigenetic systems demonstrated T-assistant cell conduct to be plastic. This will give a ripe zone of motivation for making novel plastic Artificial Immune System which is dually outfitted with recognition and reaction experiences that are being studied and the capacity to powerfully create and look after memory. Essential and optional reactions have been overlooked in AIS for quite a while, yet there is a trust imposed that a new examination of these systems will prompt enhanced novel AIS which can have an enduring effect in the improvement of novel AIS.

10 Observation for the Future

"The field of AIS is rapidly expanding; in terms of Computer Science, the field is very new. There appears to be a growing rise in the popularity of investigating the mammalian immune system as a source of inspiration for solving computational problems. This is apparent in not only the increasing amount of work of the literature, but also the creation of special sessions on AIS at major international conferences and tutorials at such conferences."

"This chapter has outlined some of the major algorithms in the field of machine learning and data mining. Other work on AIS has not been covered by this chapter, and it is acknowledged by the authors that there is a large body of work emerging. The authors feel that while these algorithms are a promising start to a very exciting field of research, it is clear that these algorithms need further and more detailed testing and examination. It is hoped that this chapter will go some way into acting as a catalyst for other researchers to use these ideas and put them to the test. This field is very promising, as the algorithms created offer the flexibility of being distributed, adaptable, and in some cases self-organizing to allow for patterns in data to emerge and create a diverse representation of the data being learned. The strengths of the algorithms are clear, but as yet issues, such as scalability and areas where they are, have greater potential and have yet to be addressed."

References

1. Tanupriya Choudhury, Vivek Kumar, Darshika Nigam, Bhaskar Mandal, "Intelligent Classification of Lung & Oral Cancer through Diverse Data Mining Algorithms", International Conference on Micro-Electronics and Telecommunication Engineering (ICMETE), IEEE Explore, https://doi.org/10.1109/ICMETE.2016.24, 22–23 Sept. 2016.
2. Tanupriya Choudhury, Vivek Kumar, Darshika Nigam, "An Innovative and Automatic Lung and Oral Cancer Classification Using Soft Computing Techniques", International Journal of Computer Science & Mobile Computing, Vol. 4 Issue. 12, Dec-2015, pg. 313–323.
3. Burnet, F (1959). The clonal selection theory of acquired immunity. Cambridge University Press.
4. Burnet, F. M. (1978). Clonal Selection and After, In Theoretical Immunology, (Eds.) G. I. Bell, A. S. Perelson & G. H. Pimbley Jr., Marcel Dekker Inc., 63–85.
5. Carneiro, J and Stewart, J. (1995). Self and Nonself Revisited: Lessons from Modelling the Immune Network. Third European Conference on Artificial Life. 405–420. Springer-Verlag. Granada, Spain.
6. Carter, J.H. (2000). The Immune System as a Model for Pattern Recognition and Classification. Journal of the American Medical Informatics Association, Vol. 7, No. 1.
7. Farmer, J, Packard, N and Perelson, A. 1986. The immune system and adaptation and machine learning. Physica D. 22, pp 187–204.
8. Hunt, J and Cooke, D. 1995. An adaptive and distributed learning system based on the Immune system. 2494–2499 Proceedings of IEEE International Conference on SystemsMan and Cybernetics (SMC) IEEE.
9. Hunt, J, King, C and Cooke, D (1996). Immunising against fraud. Proc. Knowledge Discovery and Data Mining and IEE Colloquium. IEEE. 38–45.

10. Immune Networks. (2000). http://www.immunenetworks.com Ishida, Y (1996). Distributed and autonomous sensing based on immune network. 214–217 Proceedings of Artificial Life and Robotics Beppu AAAI Press.
11. Hunt, J, Timmis, J, Cooke, D, Neal, M and King, C (1998). JISYS: Development of an Artificial Immune System for real world applications. Artificial Immune Systems and their Applications. 157–186. Springer-Verlag.
12. Tanupriya Choudhury, Vivek Kumar, Darshika Nigam, "An Innovative Smart Soft Computing Methodology towards Disease (Cancer, Heart Disease, Arthritis) Detection in an Earlier Stage and in a Smarter Way", International Journal of Computer Science and Mobile Computing, Vol. 3 Issue. 4, April-2014, pg. 368–388.
13. Tanupriya Choudhury, Vivek Kumar, Darshika Nigam, "Intelligent Classification & Clustering Of Lung & Oral Cancer through Decision Tree & Genetic Algorithm", International Journal of Advanced Research in Computer Science and Software Engineering, Volume 5, Issue 12, December 2015.

Author Index

V. Bhateja et al. (eds.), *Information Systems Design and Intelligent Applications*, Advances in Intelligent Systems and Computing 672,
https://doi.org/10.1007/978-981-10-7512-4

Printed in the United States
By Bookmasters